COMPUTATIONAL NEUROSCIENCE

TRENDS IN RESEARCH 2003

Cover artist: Bonnie Callahan
Graphic design: Jim Bower

COMPUTATIONAL NEUROSCIENCE

TRENDS IN RESEARCH 2003

Edited by

Erik De Schutter

University of Antwerp
Universiteitsplein 1
B-2610 Antwerp
Belgium

2003

ELSEVIER

Amsterdam – Boston – Heidelberg – London – New York – Oxford – Paris – San Diego – San Francisco – Singapore – Sydney – Tokyo

ELSEVIER SCIENCE B.V.
Sara Burgerhartstraat 25
P.O. Box 211, 1000 AE Amsterdam, The Netherlands

First edition 2003

Library of Congress Cataloging in Publication Data
A catalog record from the Library of Congress has been applied for.

British Library Cataloguing in Publication Data
A catalogue record from the British Library has been applied for.

Proceedings from the 11th Annual Computational Neuroscience Meeting (CNS*02), held in Chicago, Illinois, July 2002
Reprinted from Neurocomputing, volumes 52-54 (2003)

ISBN: 0 444 51383 3

∞ The paper used in this publication meets the requirements of ANSI/NISO Z39.48-1992 (Permanence of Paper).
Printed in The Netherlands.

ELSEVIER
Available online at www.sciencedirect.com
SCIENCE DIRECT®
Neurocomputing 52–54 (2003) v–xv
NEUROCOMPUTING
www.elsevier.com/locate/neucom

CONTENTS

Poster presentations

Available at
www.ComputerScienceWeb.com
POWERED BY SCIENCE @DIRECT®

ELSEVIER

Neurocomputing 52–54 (2003) 13–18

NEUROCOMPUTING

www.elsevier.com/locate/neucom

Learning temporal patterns by redistribution of synaptic efficacy

Aaron P. Shon*, Rajesh P.N. Rao

Department of Computer Science and Engineering, University of Washington, Sieg Hall Box 352350, Seattle, WA 98195-2350, USA

Abstract

Recent experiments have shown that neocortical synapses exhibit both short-term plasticity and spike-timing dependent long-term plasticity. It has been suggested that changes in short-term plasticity are mediated by a redistribution of synaptic efficacy. Here we propose a simple model of the interaction between spike-timing dependent plasticity and short-term plasticity. We show that the model captures the synaptic behavior seen in experiments on redistribution of synaptic efficacy. Results from our simulations suggest that spike-timing dependent redistribution of synaptic efficacy offers neocortical neurons a potentially powerful mechanism for learning spatiotemporal patterns in the input stream.

Keywords: Synaptic plasticity; Dynamic synapses; Spike timing; Sequence learning

1. Introduction

How does the brain learn and encode spatiotemporal patterns and sequences? Data from recent in vitro experiments [2,10,11] suggest some possible mechanisms. First, neocortical synapses have been shown to be dynamic: postsynaptic responses are not simply a function of the presynaptic firing rate multiplied by a synaptic "weight" but rather, reflect the short-term history of input spike trains. Second, paired firing of pre- and postsynaptic neurons tends to redistribute total synaptic efficacy, so that a synapse responds much more strongly to the first presynaptic event than to subsequent events in a spike train [11]. Third, long-term synaptic plasticity appears to follow a

* Corresponding author.
E-mail addresses: aaron@cs.washington.edu (A.P. Shon), rao@cs.washington.edu (R.P.N. Rao).

0925-2312/03/$ - see front matter © 2003 Elsevier Science B.V. All rights reserved.
doi:10.1016/S0925-2312(02)00842-1

temporally asymmetric spike-timing dependent learning rule: a synaptic connection is strengthened if a presynaptic spike occurs slightly before a postsynaptic spike, and is weakened in the opposite case [10,3].

These experimental results raise three important questions that motivate our study: (1) What is the role of dynamic synapses in cortical information processing? (2) What is the relationship between spike-timing dependent plasticity (STDP) and the redistribution of synaptic efficacy observed in dynamic synapses? and (3) How does STDP in conjunction with dynamic synapses allow cortical neurons to learn spatiotemporal input patterns? With respect to question (1), dynamic synapses have been identified as a possible mechanism for gain control of cortical activity [2], used as memory buffers for "remembering" a history of presynaptic stimulation [9], and used to model irregular bursts of activity in the cortex [16]. With respect to questions (2) and (3), STDP has been suggested as a mechanism for learning temporal sequences and delays [1,7,14,12], although without taking dynamic synapses into account.

Here, we propose a model for STDP in dynamic synapses and show using simulations that it reproduces the basic experimental results on redistribution of synaptic efficacy. We then demonstrate how neocortical neurons can learn spatiotemporal input patterns by using STDP to adapt dynamic synapses. This relates our work to previous research on networks that allow explicit modification of synaptic delays to recognize temporal sequences [5,6], and networks of radial basis function (RBF) neurons for temporal clustering [13] (see also [8]). While short-term plasticity and STDP have both been studied in isolation, our simulation results demonstrate that these mechanisms operating together allow single neurons to extract temporal information from their input spike trains.

2. Methods

In our simulations, we used a leaky integrate-and-fire neuron with short-term synaptic depression modeled as

$$d_t^{\text{syn}} = \begin{cases} f^{\text{syn}} d_{t-1}^{\text{syn}} & \text{if presynaptic spike received,} \\ d_{t-1}^{\text{syn}} + \dfrac{\alpha}{\tau_d}(d_0 - d_{t-1}^{\text{syn}}) & \text{if no presynaptic spike received.} \end{cases} \tag{1}$$

Here d_t^{syn} represents a fractional amount of excitatory neurotransmitter available to synapse syn at time t, with d_0 a constant value of 1, α an integration rate of 0.7 ms, and τ_d a constant value of 50. Parameter f^{syn} is the fraction of neurotransmitter remaining at synapse syn after a presynaptic spike. Higher values of f correspond to weaker short-term depression. In our simulations, f is initially set to 1 for all synapses (no depression). More sophisticated models of synaptic depression could also be used (e.g. [2]), but the simpler model described above suffices for the purposes of this study.

The novel component of our model is the interaction between short-term depression and STDP. We assume that STDP modifies not only the peak synaptic conductance

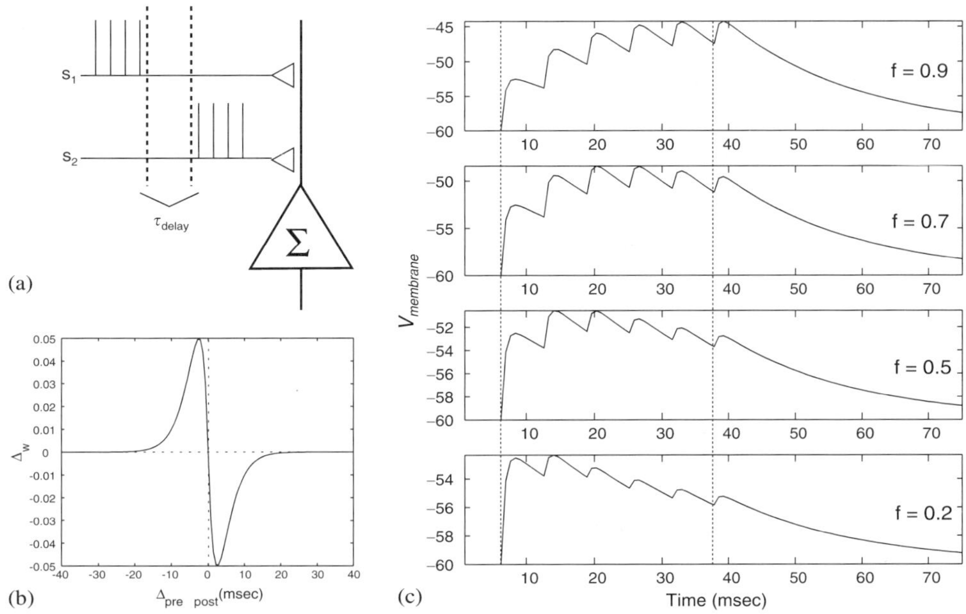

Fig. 1. Learning paradigm and redistribution of synaptic efficacy in the model: (a) A single neuron receives 2 or more spike trains spaced some time τ_{delay} apart. The objective is to learn a specific temporal pattern of spike trains. (b) A temporally asymmetric learning window updates both the peak conductance g_{peak} and the depression parameter f for each synapse according to the relative timing of pre- and postsynaptic spikes. (c) The four plots depict postsynaptic responses in the model to a fixed input spike train as a function of the depression parameter f ($=f^{syn}$ in the text). Decreasing f causes the synapse to respond much more strongly to earlier presynaptic events. Note also the gradual decrease in the overall magnitude of the responses, which is compensated for during learning by increases in g_{peak} (see Eq. (2)). This compensatory effect enhances stability and implements a redistribution of synaptic efficacy.

g^{syn}_{peak} but also the depression parameter f^{syn} as follows:

$$\Delta g^{syn}_{peak} = \beta \mathbf{w}^T \mathbf{s}^{syn}, \tag{2}$$

$$\Delta f^{syn} = -\gamma \mathbf{w}^T \mathbf{s}^{syn}, \tag{3}$$

where $\mathbf{w}$ is the vector representing the temporally asymmetric learning window (Fig. 1(b), based on [10,3]), $\mathbf{s}^{syn}$ is the vector of presynaptic spikes to synapse syn, and β and γ are gain factors (for our simulations, we used $\beta = 0.052$ and $\gamma = 0.26$). Thus, updates to g_{peak} and f are assumed to be of complementary signs, so that an increase in synaptic depression is compensated for by an increase in peak conductance. It is precisely this compensation that causes redistribution of synaptic efficacy in the model (see Fig. 1(c)).

To study the role of STDP and short-term depression in learning temporal patterns, we considered the simple case of a single integrate-and-fire neuron with two synapses. The goal was to investigate whether the neuron could learn to become selective for a specific input pattern by adapting its dynamic synapses. At the onset of each trial, one of the synapses received a train of input spikes (Fig. 1(a)) at a fixed rate r_1. After a delay τ_{delay}, the other synapse received a train of input spikes also at a rate r_2. Each spike train lasted for a specific duration (35 ms in the simulations). The model is parameterized with resting potential $V_{rest} = -60$ mV and threshold $V_{threshold} = -40$ mV. Peak conductance of each synapse before training was $g_{peak} = 0.07$ ns, with a maximum peak conductance $g_{max} = 0.2$ ns. We trained the neuron by stimulating it using input spike trains as described above for 100 trials (each lasting 210 ms), sufficient to allow synaptic parameters to converge. The specific input parameters used were: $r_1 = r_2 =$ 150 Hz with $\tau_{delay} = 15$ ms.

3. Results

Our first set of simulation results demonstrate how the combination of STDP with short-term plasticity can reproduce the experimental results on redistribution of synaptic efficacy. Recall that pairing a presynaptic spike with a postsynaptic spike that occurs a few milliseconds later causes a decrease in f^{syn} in our model (Eq. (3)). Fig. 1(c) shows the effect of decreasing f^{syn} on the postsynaptic responses to a fixed input spike train. Decreasing f^{syn} causes the synapse to respond much more strongly to earlier presynaptic events than to subsequent events in a spike train, moving the peak response to closer to the onset of the input spike train. This is accompanied by a gradual decrease in the overall magnitude of the responses, which is compensated for during learning by increase in g_{peak} (see Eq. (2)). This compensation stabilizes the learning process and captures the synaptic behavior seen in experiments on redistribution of synaptic efficacy.

Our second set of simulation results demonstrate that single neurons using short-term depression and STDP are capable of learning temporal input patterns. Fig. 2(a) shows how a neuron that is stimulated by two spatiotemporally separated input spike trains (as described in the Section 2) can redistribute its synaptic efficacies to recognize and fire in response to the temporal order of the input spike trains. Note that the neuron's output is also predictive, in the sense that it learns to fire soon after the onset of the second spike train before receiving the spike train in its entirety.

Fig. 2(c)–(f) contrast the adjustment of both f and g_{peak} with adjustment of g_{peak} alone. Again, the neuron whose synapses use short-term depression (Fig. 2(c) and (d)) learns to recognize the input sequence. More importantly, the neuron does not respond to a different ordering of the inputs. By contrast, a neuron that modifies its g_{peak} values over time, but not its f values (which remain fixed at $f = 1$) repeatedly spikes on both the training sequence and on a different ordering of the sequence (Fig. 2(e) and (f)).

This example illustrates how redistribution of synaptic efficacy allows a neuron to become selective for specific input patterns by introducing an asymmetry in synaptic excitation over time. In turn, this asymmetry leads to temporal selectivity: the neuron will fire only if the input spike trains generate a set of appropriately aligned EPSPs (according to the learned f and g_{peak} values) that push the membrane potential above spiking threshold.

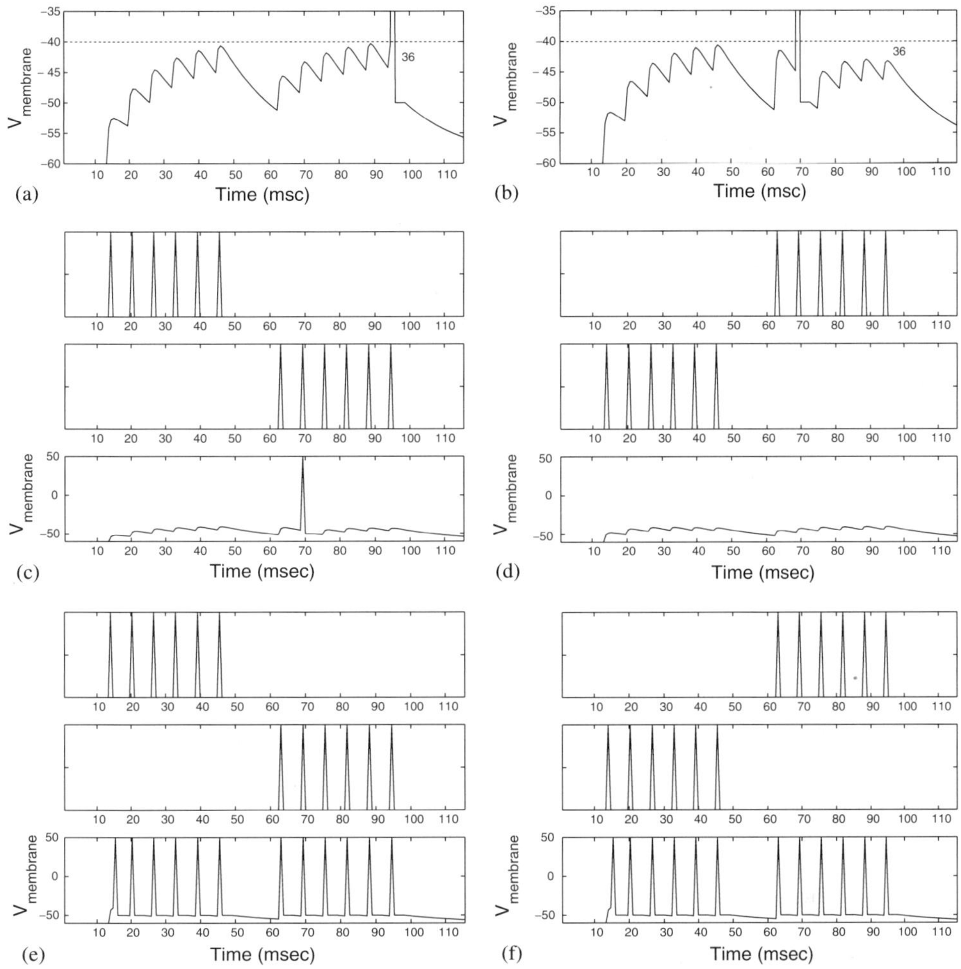

Fig. 2. Learning temporal patterns using redistribution of synaptic efficacy: (a) Before training, the neuron responds to the sequence by emitting a single spike near the end of the sequence. (b) After training, the neuron still encodes the sequence using a single spike, but the spike is now generated earlier in time, soon after onset of the second spike train. (c) A model neuron that adjusts both the depression parameter f and peak conductance g_{peak} using STDP spikes when presented with a temporal sequence used in training. (d) The same neuron does not fire when presented with a sequence that reverses the order of inputs in the training sequence. Note that the firing rate remains the same and only the temporal order is changed. (e,f) A model neuron that only adjusts peak conductance g_{peak} (as in traditional models) responds vigorously and indiscriminately to both the training sequence as well as the reverse-order sequence.

4. Conclusion

Our simulation results suggest a computational role for STDP and short-term depression in making cortical neurons selective for specific spatiotemporal stimuli. For example, in the case of moving visual stimuli, our model predicts the development of direction selectivity (Fig. 2(c) and (d)), an important property of neurons in the visual cortex (see also [4]). The question of how our neuron level model of temporal pattern learning fits into network-level models of sequence learning [1,14,15] remains a subject of ongoing investigations.

Acknowledgements

This work is being supported by the Sloan Foundation, NSF BITS grant no. 0130705, and a National Defense Science and Engineering Graduate Fellowship to APS.

References

[1] L.F. Abbott, K.I. Blum, Functional significance of long-term potentiation for sequence learning and prediction, Cereb. Cortex 6 (1996) 406–416.
[2] L. Abbott, J. Varela, K. Sen, S. Nelson, Synaptic depression and cortical gain control, Science 275 (1997) 220–224.
[3] G. Bi, M. Poo, Synaptic modifications in cultured hippocampal neurons: dependence on spike timing, synaptic strength, and postsynaptic cell type, J. Neurosci. 18 (24) (1998) 10464–10472.
[4] F. Chance, S. Nelson, L. Abbott, Synaptic depression and the temporal response characteristics of v1 cells, J. Neurosci. 18 (1998) 4785–4799.
[5] C.W. Eurich, K. Pawelzik, U. Ernst, J.D. Cowan, J.G. Milton, Dynamics of self-organized delay adaptation, Phys. Rev. Lett. 82 (1999) 1594–1597.
[6] C. Eurich, K. Pawelzik, U. Ernst, A. Thiel, J. Cowan, J. Milton, Delay adaptation in the nervous system, Neurocomputing 32 (2000) 741–748.
[7] W. Gerstner, R. Kempter, J.L. van Hemmen, H. Wagner, A neuronal learning rule for sub-millisecond temporal coding, Nature 383 (1996) 76–81.
[8] J. Hopfield, Pattern recognition computation using action potential timing for stimulus representation, Nature 376 (1995) 33–36.
[9] W. Maass, H. Markram, Synapses as dynamic memory buffers, Neural Networks 15 (2002) 155–161.
[10] H. Markram, J. Lübke, M. Frotscher, B. Sakmann, Regulation of synaptic efficacy by coindence of postsynaptic aps and epsps, Science 275 (1997) 213–215.
[11] H. Markram, M. Tsodyks, Redistribution of synaptic efficacy between neocortical pyramidal neurons, Nature 382 (1996) 807–810.
[12] M.R. Mehta, M. Wilson, From hippocampus to V1: effect of LTP on spatiotemporal dynamics of receptive fields, in: J. Bower (Ed.), Computational Neuroscience, Trends in Research 1999, Elsevier, Amsterdam, 2000.
[13] T. Natschläger, B. Ruf, Spatial and temporal pattern analysis via spiking neurons, Network: Comput. Neural Sys. 9 (3) (1998) 319–332.
[14] R.P.N. Rao, T.J. Sejnowski, Predictive sequence learning in recurrent neocortical circuits, in: S.A. Solla, T.K. Leen, K.-R. Müller (Eds.), Advances in Neural Information Processing Systems, Vol. 12, MIT Press, Cambridge, MA, 2000, pp. 164–170.
[15] D. Tank, J. Hopfield, Neural computation by concentrating information in time, Proc. Nat. Acad. Sci. USA 84 (1987) 1896–1900.
[16] M. Tsodyks, K. Pawelzik, H. Markram, Neural networks with dynamic synapses, Neural Comput. 10 (4) (1998) 821–835.

Available at
www.ComputerScienceWeb.com
POWERED BY SCIENCE @ DIRECT®

ELSEVIER

Neurocomputing 52–54 (2003) 19–24

NEUROCOMPUTING

www.elsevier.com/locate/neucom

Detection of spike patterns using pattern filtering, with applications to sleep replay in birdsong

Zhiyi Chi[a,*], Peter L. Rauske[b], Daniel Margoliash[b]

[a]*Department of Statistics, University of Chicago, 5734 University Avenue, Chicago, IL 60637-1514, USA*

[b]*Department of Organismal Biology and Anatomy, University of Chicago, 5734 University Avenue, Chicago, IL 60637, USA*

Abstract

Consolidation of memories during sleep may involve neuronal replay of daytime activity patterns. A precise replay phenomenon was observed in the song system nucleus RA. We developed statistical pattern filtering to investigate replay in HVc, where spiking activity is more variable. Long spike sequences during sleep were convolved with filters constructed from exemplar auditory spike sequences. Examining peaks in the filter responses, we found many examples of HVc spike trains with temporal patterns similar to the exemplars, confirming a replay phenomenon for HVc. Statistical pattern filtering has general utility for finding spike patterns when a referent pattern is known.

Keywords: Pattern recognition; Sleep; Birdsong; HVc

1. Introduction

In the zebra finch forebrain nucleus robustus archistriatalis (RA), neurons generate sequences of spike bursts during spontaneous sleep activity that exhibit similar temporal patterns as the pre-motor activity during singing and the auditory responses to song playback in the sleeping animal [2,3]. This sleep replay phenomenon is hypothesized to play an important role in learning and memory consolidation of the birdsong system [2,7], which gives rise to the question of whether and how other parts of the song system participate in replay during sleep. The forebrain nucleus HVc is a sensorimotor

* Corresponding author.
E-mail address: chi@galton.uchicago.edu (Z. Chi).

0925-2312/03/$ - see front matter © 2002 Elsevier Science B.V. All rights reserved.
doi:10.1016/S0925-2312(02)00746-4

center of the birdsong system. It projects to RA, and is thought to play an important role in vocal learning, auditory input integration, and higher level motor command. State dependent auditory responses have also been observed for the HVc [9,10]. We thus investigated whether the replay phenomenon also occurs in spontaneous sleep activity of HVc.

Instead of searching for individual replayed spike bursts [2], our goal was to find burst sequences that have similar temporal patterns as pre-motor activity or auditory responses [6,8]. This is challenging for two reasons. First, unlike pre-motor or auditory data, for sleep spontaneous neuronal data, there are no reliable observables to mark the occurrence of neuronal events of interest. Second, the neuronal activity in HVc has far more variability than RA activity.

2. Pattern filtering

We developed a statistical methodology using linear filtering to address the above difficulty. In general, let S be an exemplar train of spikes registered at $s_1 < s_2 < \cdots < s_p$, $s_k \in [0, \sigma]$. Given a small number $\varepsilon > 0$, fix two functions $K(x)$ and $B(x)$, with $K(x) \neq 0$ only for $x \in (-\varepsilon, \varepsilon)$ and $B(x) \leqslant 0$ for all x. The functions will be referred to as the "time window function" for a spike and the "background function", respectively, and ε the "window size". Define F on $[0, \sigma]$ by

$$F(x) = \begin{cases} \max_{1 \leqslant k \leqslant p} K(x - s_k) & \text{if } x \in J_S, \\ B(x) & \text{otherwise}, \end{cases} \tag{1}$$

where

$$J_S = \bigcup_{k=1}^{p} (s_k - \varepsilon, s_k + \varepsilon). \tag{2}$$

Let T be a data sequence of spikes registered at $t_1 < \cdots < t_N$. Regard T as a series of δ functions, i.e., $T(x) = \sum_{i=1}^{N} \delta(x - t_i)$. Define function R by

$$R(x) = \int_0^{\sigma} F(s) T(x + s)\, \mathrm{d}s. \tag{3}$$

We then search for local maximum points of R with super-threshold values and output them as plausible locations of targets, namely segments of T with temporal patterns similar to S. More specifically, given $r_0 > 0$ and $\theta > 0$, with $r_0 \approx 20$ ms in our study, each target location x satisfies $R(x) \geqslant \theta$ and $R(x) \geqslant R(t)$ for all $t \in (x - r_0, x + r_0)$ and the spike sequence in $[x, x + \sigma]$ is considered a potential target. To avoid overlapping segments to be output as targets, if $x_1, \ldots, x_n$ are plausible target locations with $I_i = [x_i, x_i + \sigma]$ intersecting with each other, we only choose one I_i, such that $R(x_i)$ is the largest among $R(x_1), \ldots, R(x_n)$.

The search proceeds rapidly because R is the linear convolution between $\Phi(s) := F(-s)$, $s \in [-\sigma, 0]$ and T, which in most cases can be computed efficiently by the fast Fourier transform. We refer to Φ as the *pattern filter* associated with S, and the convolution *pattern filtering*.

In all the cases we studied, K and B also satisfied the following conditions: (1) K is symmetric on $(-\varepsilon, \varepsilon)$, and decreases as x increases from 0 to ε; (2) $K(x) \geqslant 0$ on most part of $(-\varepsilon, \varepsilon)$, except for x close to the boundary of the interval; (3) $\min\{K(x)\colon x \in (-\varepsilon, \varepsilon)\} \geqslant \max\{B(x)\}$; and (4) $B(x) \equiv c$ for some constant $c \leqslant 0$. All the conditions can be justified by an appropriate statistical model and at the same time simplify the computation.

3. The underlying statistical model

Pattern filtering is a likelihood ratio test, as opposed to likelihood test in the unitary event analysis [4,5]. To illustrate, consider the simplest filter constructed from the time window function $K(x) \equiv \alpha > 0$, $x \in (-\varepsilon, \varepsilon)$, and the background function $B(x) \equiv -\beta \leqslant 0$. Then (3) leads to $R(x) = \#\{t_i \in T \cap [x, x+\sigma]\colon t_i - x \in J_S\} \times \alpha - \#\{t_i \in T \cap [x, x+\sigma]\colon t_i - x \notin J_S\} \times \beta$. Because $J_S \subset [0, \sigma]$, the formula is simplified into

$$R(x) = \#\{t_i \in T\colon t_i - x \in J_S\} \times \alpha - \#\{t_i \in T\colon t_i - x \notin J_S\} \times \beta.$$

Detection using this filter is derived from three statistical assumptions. First, each spike sequence is generated randomly either by the exemplar sequence $S = \{s_1, \ldots, s_p\}$ ("template") or by the background. Second, spikes in the sequence are *conditionally* independent, such that

(1) given the sequence being generated by the template, at any location within distance ε from a template spike, the probability of a spiking event is p_0, while at any other location, the probability is q_0;
(2) given the sequence being generated by the background, the probability of a spiking event is q everywhere; and
(3) conditioning on either case, the spiking events at different temporal locations are independent from each other.

Finally, $p_0 > q \geqslant q_0$. In other words, compared to background activity, in activity generated by the template, spikes are more likely to occur around template spikes, and less likely to occur elsewhere.

Note that the assumption of conditional independence is *not* one of independent firing. Also, it can be shown that both the "sliding sweeps" [1,8] and the cross-correlation approaches [2,6] can be implemented by pattern filtering. Furthermore, it is straightforward to generalize pattern filtering to multiple units, incorporating spatial as well as temporal structure of neural activity.

Returning to the filtering procedure, let T be a data spike sequence. Given x, let T_x consist of the spikes of T in $[x, x+\sigma]$. Shift the spikes in T_x by x, so that the sequence is registered onto $[0, \sigma]$, the same time frame as the template S. If n spikes of T_x fall inside J_S, and m outside J_S, and the total duration of J_S is τ, then it is not hard to get the following likelihoods

$$p_o(T_x) := P(T_x \mid T_x \text{ generated by } S) = \mathrm{e}^{-p_0\tau - q_0(\sigma-\tau)} p_0^n q_0^m$$

$$p_a(T_x) := P(T_x \mid T_x \text{ generated by background}) = \mathrm{e}^{-q\sigma} q^{n+m}.$$

Hence the log-likelihood ratio

$$L(T_x) := \log \frac{p_o(T_x)}{p_a(T_x)} = \log \frac{p_0^n q_0^m}{q^{n+m}} + C_S = n\alpha - m\beta + C_S, \tag{4}$$

where

$$\alpha = \log \frac{p_0}{q} > 0, \quad \beta = \log \frac{q}{q_0} \geqslant 0$$

and $C_S = -p_0\tau - q_0(\sigma - \tau) + q\sigma$ is a constant which only depends on the template S.

Therefore, if we define F by (1), with $K(x) = \alpha$ for $x \in (-\varepsilon, \varepsilon)$ and 0 for $x \notin (-\varepsilon, \varepsilon)$ and $B(x) \equiv -\beta$, then $L(T_x) = R(x) + C_S$. Since x runs across the entire time interval of the data spike sequence T, R consists of the log-likelihood ratios plus a constant of all the segments of T with the same duration as the template. Thus, by locating peaks in R with values larger than a given threshold, one can find all the segments that are significantly more likely to be a target than be part of the background.

4. Results

We implemented pattern filtering for HVc single unit data recorded during sleep. For each unit, the spike sequences in response to the bird's own song ('BOS') broadcast to the sleeping bird in multiple trials were selected as templates. Then, in the subsequent spontaneous activity of the unit when no acoustical stimulus was presented, segments with similar temporal patterns as the templates were searched. For the neuronal activity in HVc, several types of variability have to be taken into account: (1) the across-trial variability of the auditory response, which leads to differences across the templates; (2) the possible modulations in the time scale of the neuronal activity [8,2,6], which lead to time compression or expansion of potential targets; and (3) the transient change in firing rates, which may lead to false detections. To address the variability, we adopted a procedure involving multiple filters, time scales, and tests on the statistical significance of the targets. In most cases, it took less than 5 s to conduct detection for a spike train of 10 min duration using a single filter, thus the multi-level detection procedure was still computationally efficient.

In the example illustrated in Fig. 1, the first spike train in each group is a template, consisting of *all* the spikes during a rendition of the same "motif" (sequence of syllables) which lasted about 650 ms. The templates are aligned at the onsets of the renditions. The subsequent spike trains in each group are targets detected from the spontaneous activity, using filters constructed for the template. Since targets were detected at several time scales to account for possible time compression or expansion, they might have different durations than the templates. To make comparisons between the two, the targets are scaled in the plot, according to the scaling factors with which they were detected. For all the targets, the scaling factors are between 0.8 and 1.25. This example demonstrates the variability of the auditory response in HVc during sleep and that such variability may also be exhibited in the presumed "replay" during the spontaneous sleep activity.

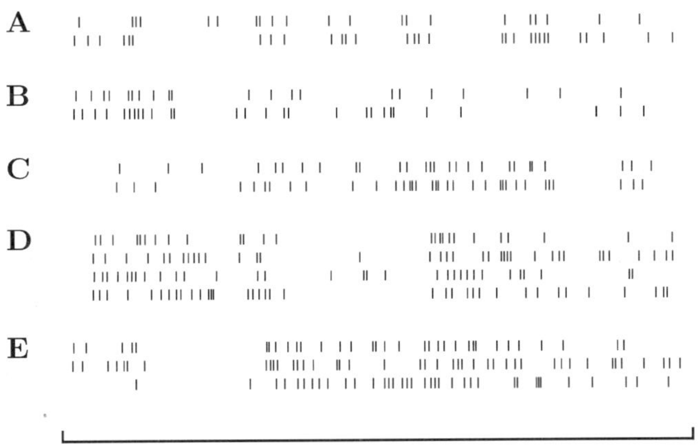

Fig. 1. Detection for a unit. In each group, the first spike train is a template and the others are targets detected using the filters for the template. The duration of a time interval (≈650 ms) during which a template was collected is illustrated at the bottom.

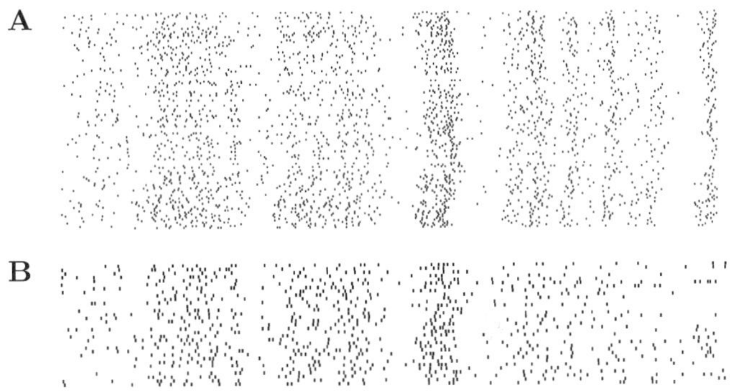

Fig. 2. Raster plots of templates (A) and detected targets (B). The targets in this case are not scaled, i.e., they have about the same duration as the templates (≈580 ms).

The second example is illustrated in Fig. 2. There were 126 responses collected while the BOS was broadcast to the sleeping bird (Fig. 2A). Again, the templates were associated with multiple renditions of a motif. While the responses were less variable than those in the previous example, they exhibited significant systematic change across the trials. For this unit, the detection was conducted without scaling the filters. From a total of approximately 25 min of recordings of spontaneous activity, we detected 33 exemplars of spontaneous activity. These are displayed in Fig. 2B. Note the similarity of matching between the responses to BOS and the spontaneous spike sequences. This example as well as the previous one is the first quantitative evidence that replay during sleep occurs in a song system nucleus other than RA.

References

[1] M. Abeles, G.M. Gerstein, Detecting spatio-temporal firing patterns among simultaneously recorded single neurons, J. Neurophysiol. 60 (1998) 909–924.
[2] A.S. Dave, D. Margoliash, Song replay during sleep and computational rules for sensorimotor vocal learning, Science 290 (2000) 812–816.
[3] A.S. Dave, A.C. Yu, D. Margoliash, Behavioral state modulation of auditory activity in a vocal motor system, Science 282 (1998) 2250–2253.
[4] S. Grün, M. Diesmann, A. Aertsen, Unitary events in multiple single-neuron spiking activity: I. detection and significance stimulus properties, Neural Comput. 14 (2002) 43–80.
[5] S. Grün, M. Diesmann, A. Aertsen, Unitary events in multiple single-neuron spiking activity: II. nonstationary data, Neural Comput. 14 (2002) 81–120.
[6] K. Louie, M.A. Wilson, Temporally structured replay of awake hippocampal ensemble activity during rapid eye movement sleep, Neuron 29 (2001) 145–156.
[7] D. Margoliash, Do sleeping birds sing? Population coding and learning in the bird song system, Prog. Brain Res. 130 (2001) 319–331.
[8] Z. Nádasdy, H. Hirase, A. Czurkó, J. Csicsvari, G. Buzsáki, Replay and time compression of recurring spike sequences in the hippocampus, J. Neurosci. 19 (1999) 9497–9507.
[9] T.A. Nick, M. Konishi, Dynamic control of auditory activity during sleep: correlation between song response and EEG, Proc. Natl. Acad. Sci. USA 98 (2001) 14012–14016.
[10] M.F. Schmidt, M. Konishi, Gating of auditory responses in the vocal control system of awake songbirds, Nat. Neurosci. 1 (1998) 513–518.

Zhiyi Chi entered Peking University, Beijing, China in 1989. He entered Brown University, Providence, Rhode Island in 1993. He received an Sc.M.from the Department of Mathematics the next year. In 1998, he graduated from the Division of Applied Mathematics with a Ph.D. for work on 'Probability Models for Complex Systems' under the supervision of Stuart Geman. Since then he has been Assistant Professor in the Department of Statistics at the University of Chicago. He is currently also a member of the Committee on Computational Neuroscience at the University of Chicago.

Peter Rauske studied physics at the University of Chicago, graduating in 1994. He is currently studying the neurobiology of birdsong production and learning using chronic recording techniques with Daniel Margoliash's group at the University of Chicago, where he is pursuing a Ph.D. in the Department of Organismal Biology and Anatomy.

Daniel Margoliash studied biology at the California Institute of Technology. After graduating in 1975 he entered the Bioinformation Systems program in Engineering Science at Caltech. Eventually joining the laboratory of Masakazu Konishi, he studied auditory responses in the birdsong system. He received an M.Sc. in 1981, and in 1983 he received his Ph.D. for work on 'Songbirds, Grandmothers, and Templates: A Neuroethological Approach'. After a brief postdoctoral fellowship with Nobuo Suga at Washington University, St. Louis, he joined the faculty of the then Department of Anatomy at the University of Chicago in 1986. He is currently Professor in the Departments of Organismal Biology and Anatomy & Psychology, and the Committees on Neurobiology & Computational Neuroscience at the University of Chicago.

Available at www.ComputerScienceWeb.com
POWERED BY SCIENCE @ DIRECT®

ELSEVIER

Neurocomputing 52–54 (2003) 25–29

NEUROCOMPUTING

www.elsevier.com/locate/neucom

At what time scale does the nervous system operate?

N. Hatsopoulos[a], S. Geman[b], A. Amarasingham[b], E. Bienenstock[b]

[a] *University of Chicago, 1027 East 57th Street, Chicago, IL 60637, USA*
[b] *Division of Applied Mathematics, Brown University, Providence, RI 02912, USA*

Abstract

A novel statistical strategy, the spike jitter method, was developed to assess temporal structure in spike trains from neuronal ensembles. Its key feature is the introduction of a null hypothesis that assumes a uniform relative likelihood of observing a spike at one temporal location versus another within a small temporal window. We applied the method to simultaneously recorded motor cortical neurons in behaving monkeys and examined the occurrence of finely timed synchrony between neuron pairs. Evidence was found for millisecond synchrony that could only be accounted for by assuming fine temporal structure in the constituent neurons' spike trains. The method was also applied to higher-order patterns.

Keywords: Time scale; Nervous system

1. Introduction

The temporal coding hypothesis states that fine spike timing in the nervous system occurs and forms an important part of the representational substrate for thought and action. It remains a controversial issue, partly because there are few mathematical methods for directly and quantitatively assessing the temporal structure in spike trains. Many existing methods are either qualitative, sensitive to artifactual sources of fine temporal structure, or limited to examining lower-order structure. For example, cross-correlation methods can be used to assess fine temporal structure between cells but are somewhat descriptive and restricted to second-order structure between pairs

E-mail address: nicho@uchicago.edu (N. Hatsopoulos).

0925-2312/03/$ - see front matter © 2002 Elsevier Science B.V. All rights reserved.
doi:10.1016/S0925-2312(02)00773-7

of cells. Moreover, quantitative techniques used in conjunction with cross-correlations such as the shift predictor can overestimate the number of expected synchronous spikes due to slow rate covariations which are known to exist [1,2,4]. To address these problems, we developed a statistical method that allows us to quantitatively test specific spike timing hypotheses without the pitfalls that plague other methods. The key feature of this novel method is to test an explicit null hypothesis regarding the operating time scale of single neurons or groups of neurons.

2. Method

The *spike jitter* method is a general statistical method that can be used to assess any statistic based on an ensemble of neuronal spike trains. In this work, we have restricted ourselves to examining the number of spikes in one spike train (the target train) that occur within a certain time interval centered on each spike occurring in a second simultaneously recorded spike train (the reference train). This time interval defines the temporal resolution of synchrony and is called the synch width, $+/-\mathbf{w}$ (from $+/-0.5$ to $+/-2$ ms).

The method tests a null hypothesis which states that given the reference train, the target train is generated from an inhomogeneous Poisson process whose rate function is constant in a collection of small windows, $+/-\mathbf{J}$ (from $+/-1$ to $+/-4$ ms) centered on each spike in the reference train. The occurrence of significant synchrony is assessed by randomly and uniformly jittering each spike in the target train within the $+/-\mathbf{J}$ window and counting the number of synchronous spikes (Fig. 1a). This is repeated multiple times (1000 times) to create a distribution of synchronous spikes under the null hypothesis. The significance is then determined by the location of the unjittered synchronous count in the distribution (Fig. 1b).

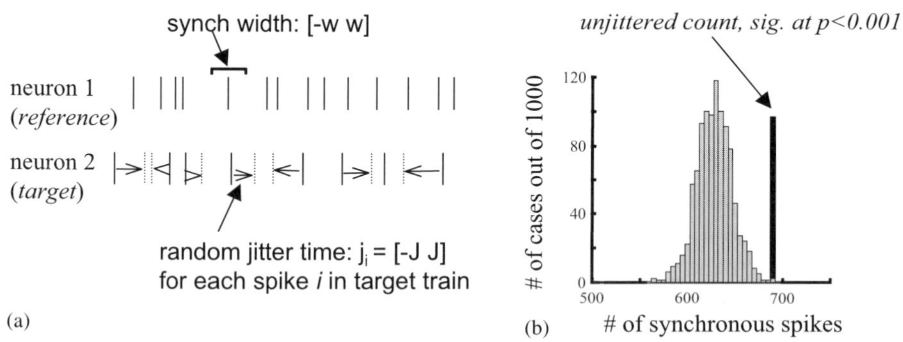

Fig. 1. (a) The spike jitter procedure involves shifting the timing of each spike in the target train by a random amount within the $+/-\mathbf{J}$ interval. The parameter, $\mathbf{w}$, defines the synchrony resolution. (b) A Monte Carlo procedure for estimating the distribution of synchronous spikes under the null hypothesis. Each realization in this distribution is the number of synchronous spikes observed after randomly jittering all the spikes in the target train once. The unjittered synchrony count is shown as a vertical line and falls to the right of the distribution in this example.

2.1. Experimental recordings and procedures

We recorded simultaneously from up to 11 neurons in primary motor cortex (MI) using silicon-based electrode arrays [3]. Four macaque monkeys were trained to perform two different reaching tasks with the hand. The first task involved moving a cursor, whose position was controlled by a two-link manipulandum in the horizontal plane, from an initial target in the center of the screen to different targets positioned around the initial target in different directions (center-out task). In the second task, the monkey was required to perform a movement sequence by moving the cursor from an initial target at the bottom of the screen, though an intermediate target, to a final target either to the left or to the right (sequence task). Information regarding the type of sequence to generate (i.e. leftward or rightward) was provided to the animal either before movement began (pre-planned) or as the intermediate target was approached (unplanned). This task also contained a control condition in which the monkey was required to stop at the intermediate target. A total of 224 cell pairs were analyzed from single units in MI based on data collected from 60 neurons recorded on 8 separate recording sessions. For both reaching tasks, data were taken from a 400 ms window centered on the start of movement and pooled over all behavioral conditions (i.e. direction in the center-out task and sequence-type, pre-planned or unplanned, and control in the sequence task).

3. Results

The spike jitter method was applied using three sets of parameter values: (1) $\mathbf{w} = 0.5$ ms, $\mathbf{J} = 1$ ms; (2) $\mathbf{w} = 1.0$ ms, $\mathbf{J} = 2$ ms; and (3) $\mathbf{w} = 2.0$ ms, $\mathbf{J} = 4$ ms. 6.7%, 7.6%, and 3.6% of all cell pairs in MI exhibited significant synchrony ($p < 0.01$) using the three parameter settings, respectively. To account for multiple comparisons, a binomial test verified that these percentages were significantly larger than the expected 1%: $p < 1.2 \times 10^{-8}$, $p < 1.92 \times 10^{-10}$, and $p < 0.002$, for each parameter set, respectively. These results provide evidence for the first part of the temporal coding hypothesis by demonstrating that precise timing of spike patterns occurs in motor cortex.

To explore whether spike patterns such as synchrony might form a representational substrate for action, we applied the spike-jitter method to data that were parsed into different behavioral conditions. Fig. 2 shows an example of different patterns of synchrony among a group of eight neurons for leftward versus rightward movements in the center-out task. Each line connects pairs of neurons that exhibit significant synchrony. These data are only suggestive of differences in synchrony because the inability to reject the null hypothesis does not imply its acceptance.

One of the attractive features of this statistical method is that it can be applied to higher-order patterns among neuronal ensembles besides pair-wise synchrony. We examined the occurrence of significant triplet synchrony among groups of 3 neurons. Fig. 3 presents the same eight neurons as shown in Fig. 2 for leftward and rightward movements in which lines connect groups of three neurons that are engaged in significant triplet synchrony.

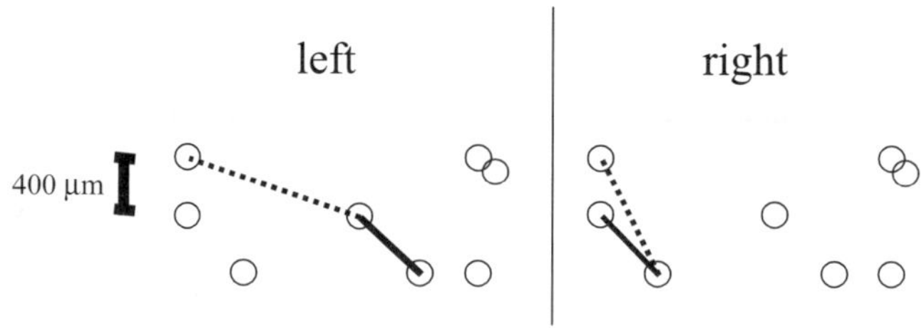

Fig. 2. The occurrence of pair-wise synchrony among an ensemble of eight simultaneously recorded motor cortical neurons arranged spatially according to the relative location of the electrode that recorded that neuron. Lines (solid and dashed) between neuron pairs indicate significant synchrony ($p < 0.01$) for leftward movements (left panel) and rightward movements (right panel). The parameters used were $\mathbf{w}$=1.0 and $\mathbf{J}$=2.0.

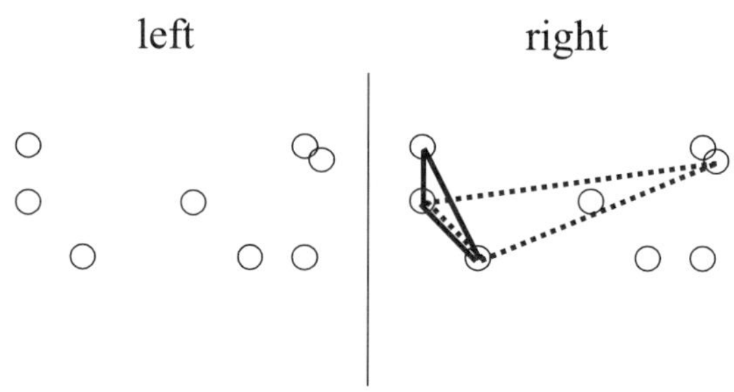

Fig. 3. The occurrence of triplet synchrony among the same ensemble of eight neurons as in Fig. 2 for leftward and rightward movements. The parameters used were $\mathbf{w} = 1.0$ and $\mathbf{J} = 2.0$.

4. Discussion

The spike-jitter method demonstrates that fine temporal synchrony as precise as +/− 0.5 ms cannot be accounted for by assuming that the exact temporal placement of the spikes (in the target train) is irrelevant. A random jitter of each spike as small as +/− 1 ms can disrupt the occurrence of the observed synchrony. This suggests that the motor cortex, at least, operates at a very fine temporal scale. Despite the well-formulated null hypothesis, the spike-jitter method imposes two restrictions on the null that may be viewed as unreasonable. First, there are intrinsic properties of neurons related to channel dynamics that can impose fine temporal structure in their spike trains and hence prevent these trains from being characterized as realizations of Poisson processes [4]. For example, neurons have an absolute refractory period and bursting properties that can impose millisecond precision on spike trains. This intrinsic temporal structure, albeit interesting, is a form of precise temporal structure that could result in excess synchrony or other higher-order patterns which we may not want to reject. Second, the null hypothesis assumes that the rate function that characterizes the generation of the target train is exactly constant within the +/− $\mathbf{J}$ jitter window. Although we have used very small jitter window lengths (as small as +/− 1 ms), the null would be rejected even for very slow varying processes characterizing the target train if enough data were collected. In future work, we will develop a modified method that will attempt to remove these restrictions.

References

[1] C. Brody, Slow covariations in neuronal resting potentials can lead to artefactually fast cross-correlations in the spike trains, J. Neurophysiol. 80 (1998) 3345–3351.

[2] C. Brody, Correlations without synchrony, Neural Comput. 11 (1999) 1553–1577.

[3] E.M. Maynard, N.G. Hatsopoulos, C.L. Ojakangas, B.D. Acuna, J.N. Sanes, R.A. Normann, J.P. Donoghue, Neuronal interactions improve cortical population coding of movement direction, J. Neurosci. 19 (1999) 8083–8093.

[4] M.W. Oram, N.G. Hatsopoulos, B.J. Richmond, J.P. Donoghue, Excess synchrony in motor cortical neurons provides direction information that is redundant with the information from coarse temporal response measures, J. Neurophysiol. 86 (2001) 1700–1716.

Available at
www.ComputerScienceWeb.com
POWERED BY SCIENCE @ DIRECT®

ELSEVIER

Neurocomputing 52–54 (2003) 31–37

NEUROCOMPUTING

www.elsevier.com/locate/neucom

Non-parametric significance estimation of joint-spike events by shuffling and resampling

Gordon Pipa*, Sonja Grün[1]

Department of Neurophysiology, Max-Planck-Institute for Brain Research, Deutschordenstr 46, Frankfurt/Main 60528 Germany

Abstract

The 'unitary event' analysis method was designed to analyze multiple parallel spike trains for correlated activity. The null-hypothesis assumes Poissonian spike train statistics, however experimental data may fail to be consistent with this assumption. Here we present a non-parametrical significance test that considers the original spike train structure of experimental data. The significance of coincident events observed in simultaneously recorded spike trains is estimated on the basis of the same spike trains, however recombined to sets of non-corresponding trials ('trial shuffling'). Resampling from this set provides the distribution for coincidences reflecting the null-hypothesis of independence.

Keywords: Spike synchronization; Trial shuffling; Significance; Bootstrapping

1. Introduction

The assembly hypothesis postulates [5] that information processing in the cortex is mediated by groups of neurons by their coordinated spiking activity and is supported by a number of experimental studies (e.g. Ref. [11]). In order to test this hypothesis, the unitary event (UE) analysis was developed [3,4] and enabled to study the relation of spike synchronization to behavioral events [2,4,9,10]. In the UE-analysis,

* Corresponding author.
E-mail address: pipa@mpih-frankfurt.mpg.de (G. Pipa).
[1] Present address: Institute for Biology, Freie Universität Berlin, Königin-Luise Str. 28/30, 14195 Berlin, Germany.

0925-2312/03/$ - see front matter © 2003 Elsevier Science B.V. All rights reserved.
doi:10.1016/S0925-2312(02)00823-8

empirically observed coincidence counts are evaluated for their significance based on expectation by chance. The significance is estimated using a Poisson distribution parameterized by the expected number of coincidences, which is a function of the product of the firing rates of the contributing neurons. The implicit assumption is that the observed spike trains follow Poissonian statistics. However, experimental data often fail to be compatible with this assumption (e.g. Ref. [8]). In the work presented here we derive a non-parametric significance test, that is based on the experimental data, while respecting the original spike train structures. This is realized by combining shuffling of trials, and subsequent resampling from the set of coincidence counts of non-corresponding trials ('CSR'-method). Thus, first we introduce the procedure of trial shuffling and the generation of the set of data containing coincidence counts from individual trial combinations. Since we aim to compare the original found in M trials, we derive in a second step the probability distribution of the sum of coincidence counts from M trials by bootstrapping [1]. This probability distribution is representing the null-hypothesis H_0 of independence which allows us to estimate the significance of coincidences detected in corresponding trials. Requirements on the number of bootstrap samples with respect to the precision of the significance estimate are discussed.

2. Destruction of intrinsic joint-events by trial shuffling

To incorporate the null-hypothesis of independence, trials of the original data are shuffled to combine non-simultaneously recorded spike trains, and yield counts of chance coincident events. The underlying assumption is that the spike trains recorded in repetitive trials, i.e. under the same stimulus or behavioral condition, are realizations of the same neuronal process [6]. The latter is not required to be stationary in time.

Number of coincidence counts resulting from all different trial combinations compose the set Ω. In case of $N = 2$ neurons, this set contains two classes of distinct subsets, i.e. the set of coincidence counts resulting from simultaneously observed trials $\Omega_s = \{\omega_s\}$, and the set of counts from non-simultaneous trial combinations $\Omega_0 = \{\omega_0\}$ (with $\Omega_0 = \Omega \setminus \Omega_s$). This can be illustrated in matrix form (Fig. 1B), where elements of Ω_s can be found on the diagonal, and elements of Ω_0 as off-diagonal entries. In case of more than two neurons ($N > 2$), we define the set of counts from trial combinations with *all* trial indices being different ('completely shuffled trials') as

$$\Omega_0 = \{\omega_{j(l_1,\ldots,l_N)} | l_h \neq l_k, \text{ with } h \neq k; h,k \in 1\ldots N \wedge l_h, l_k \in 1\ldots M\} \tag{1}$$

and contains

$$S = \binom{M}{N} \cdot N! = \frac{M!}{(M-N)!} \tag{2}$$

(M the number of trials) elements. A non-empty set Ω_0 respects the condition $M \geqslant N$, i.e. a minimum of N trials needs to be observed to allow shuffling of trials. The number

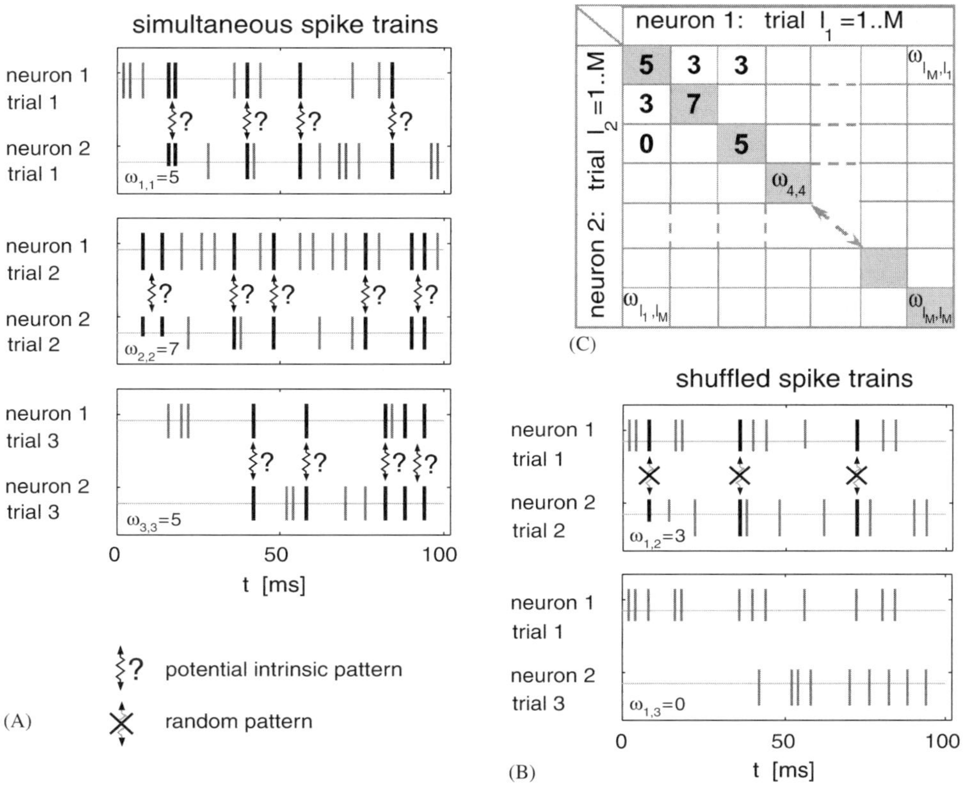

Fig. 1. Destruction of intrinsic patterns by trial shuffling illustrated for two neurons. (A) Sketch of a set of three trials (out of M) of two simultaneously recorded neurons. Coincident spike events are marked as potential intrinsic joint-spike patterns; (B) Examples of two pairs of shuffled spike trains from (A). Coincident events (marked) occur by chance only; (C) Matrix containing the number of coincident events per trial ω_{l_1,l_2} detected in all possible trial combinations of M trials. $l_1, l_2 \in 1,2,\ldots,M$ indicate trial indices of neuron 1 (horizontal) and 2 (vertical), respectively. Coincidence counts detected in simultaneous trials enter the matrix on the diagonal ($l_1 = l_2$), off-diagonal ($l_1 \neq l_2$) elements result from shuffled trials.

of elements in Ω_0 increases rapidly with increasing number of trials and neurons (Fig. 2A).

3. Generating the distribution of coincidence counts

To estimate the significance of the number of coincidences $\bar{\omega}_s$ observed in the M experimental trials, the probability distribution of the number of coincident events expected in M trials has to be estimated. For doing that we resample from the set of non-simultaneous (single) trial combinations Ω_0 all possible combinations of M elements and compute their sum $\bar{\omega}_0^*$. The total set Γ of sums of all possible combinations contains B elements (Eq. 3). The ideal estimate of the probability distribution for coincidence counts $p(\bar{\omega}_0^*)$ (see Ref. [7] for a full mathematical derivation) can be

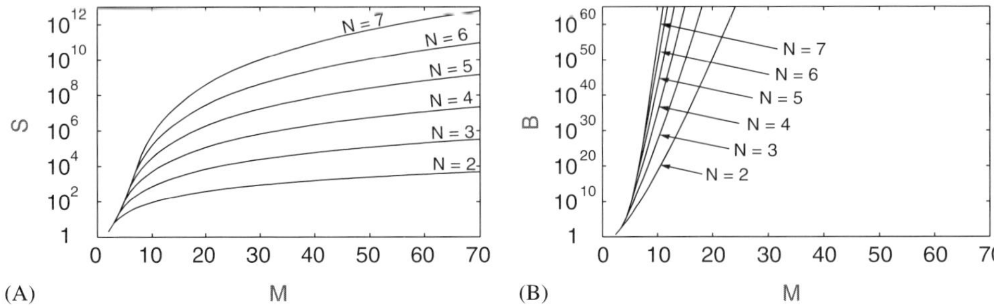

Fig. 2. (A) Number of shuffle elements (S) as a function of the number of simultaneously observed neurons (N) and the number of trials (M). (B) Number of possible bootstrap samples (B) of M non-corresponding trial combinations as a function of the number of neurons and trials (note: logarithmic scaling of the y axes).

constructed by forming the histogram (normalized) of all elements of Γ (see Fig. 3A, top):

$$p(\bar{\omega}_0^*) = \frac{(\#v \in \Gamma | v = \bar{\omega}_0^*)}{B} \quad \text{with} \quad B = |\Gamma| = S^M. \tag{3}$$

Fig. 2B illustrates that the size B of the set Γ is extremely large even for a small number of trials (M), and increases even further for a larger number of trials and neurons. Obviously, the construction of the probability distribution $p(\bar{\omega}_0)$ based on all elements of Γ is not practical in general (but see also Ref. [7]). Thus, as an alternative, we derive an estimate of (Eq. 3) on a random subset Γ_ξ of Γ consisting of $\xi < B$ elements:

$$\hat{p}(\bar{\omega}_0^*) = \frac{(\#v \in \Gamma_\xi | v = \bar{\omega}_0^*)}{\xi}. \tag{4}$$

As expected, the accuracy of (Eq. 4) depends on the sample size ξ of the chosen subset Γ_ξ (compare Fig. 3A, top to Fig. 3A, middle and bottom). The influence of the size of the subsample Γ_ξ on the significance estimation is discussed in the following.

3.1. *Significance estimation*

The significance of the empirical coincidences $\bar{\omega}_s$ detected in the simultaneously observed trials may be expressed by the probability of finding $\bar{\omega}_s$ or even more coincidence counts in case of independent spike trains ('joint-p-value', [3]). Thus, instead of using a Poisson distribution parameterized by the number of expected events [3], we now estimate the significance on the basis of the probability distribution derived by shuffling and resampling (Eq. 4):

$$\alpha^*(\bar{\omega}_s, \Gamma_\xi) = \int_{\bar{\omega}_s}^{\infty} \hat{p}(\bar{\omega}_0^*)\, d\bar{\omega}_0^* \tag{5}$$

in the following abbreviated as α^* (indicated by the area of gray bars of $\hat{p}(\bar{\omega}_0^*)$ in Fig. 3A, bottom). Since the set Γ_ξ underlying the significance estimate is a random

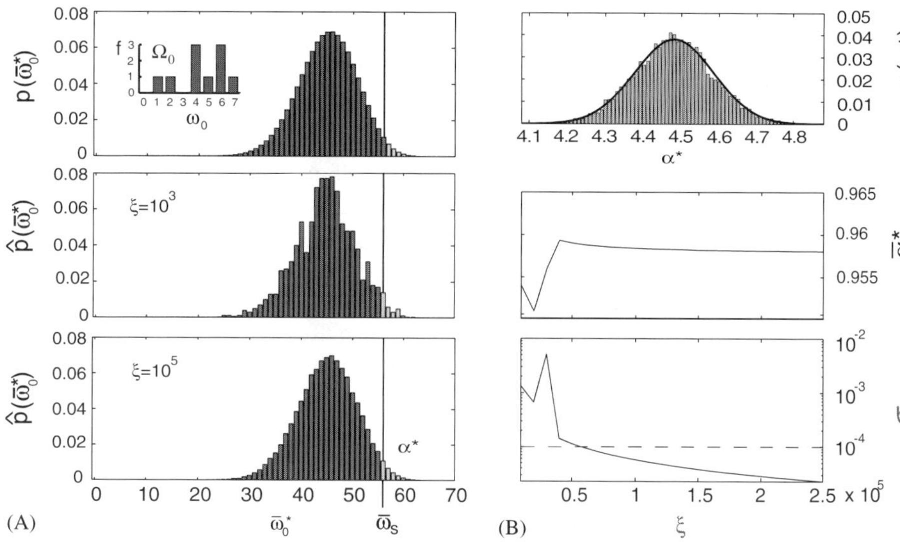

Fig. 3. (A) Probability distribution of coincidence counts from non-simultaneously recorded trials ($N = 2$, $M = 10$). Top: ideal estimate of the probability distribution $p(\bar{\omega}_0^*)$ for coincidences computed from all elements of Γ. Middle, bottom: estimates $\hat{p}(\bar{\omega}_0^*)$ of $p(\bar{\omega}_0^*)$ each based on a subset Γ_ξ of Γ consisting of ξ elements (middle: $\xi = 10^3$; bottom: $\xi = 10^5$). (B) Top: distribution of significance estimates α^* of $\bar{\omega}_S$ (area of gray bars of $\hat{p}(\bar{\omega}_0^*)$ for $\bar{\omega}_0^* \geqslant \bar{\omega}_S$ in (A)) based on 10^4 Monte–Carlo simulations. Each simulation is based on independently drawn subsets Γ_ξ, each consisting of $\xi = 10^4$ elements. Normal approximation of $p(\alpha^*)$ (solid line) with parameters $\bar{\alpha}^* = 0.0448$ and $\sigma_{\bar{\alpha}^*} = 0.0010$. Mean ((B), middle) and standard deviation (B, bottom) of five estimates of α^* as a function of ξ. The standard deviation decreases with ξ, while the mean converges to the expectation value of α^*. In this example the required precision for $\bar{\alpha}^* \pm 0.01\%$ (dashed line) is reached at $\xi = 0.6 \times 10^5$.

sample of Γ, the significance α^* itself is a random variable. Fig. 3B (top) illustrates the distribution of significance estimates α^* of $\bar{\omega}_S$, derived on the basis of random sets Γ_ξ from Γ for fix ξ. The distribution can be approximated by a normal distribution (solid line in Fig. 3B, top). Thus, the best estimate of the significance is given by its mean $\bar{\alpha}^*$, and its standard deviation $\sigma_{\bar{\alpha}^*}$ yields a measure for the confidence interval. It can be shown that $\sigma_{\bar{\alpha}^*}$ decreases with the sample size ξ of Γ_ξ (Fig. 3B, bottom). By setting a threshold on $\sigma_{\bar{\alpha}^*}$ at the required precision for the significance estimate (e.g. 0.01% as in Fig. 3B, bottom), we derive a truncation criteria for the minimal necessary size ξ of the set Γ_ξ.

4. Discussion

We have shown, that the approach of combined trial shuffling and resampling (CSR-method) allows us to estimate the significance of joint-spike events. Coincidence counts from simultaneously recorded spike trains are compared to coincidence counts resulting from shuffled trial combinations of the same data set. Since the number of possible combinations increases considerably with the number of trials (and neurons), the full set Γ of combinations can in general not be considered due to computation time and memory requirements. In Ref. [7] we discussed an approach which uses efficient combinatorial methods to estimate the significance based on Γ, which, however, is restricted to limited number of trials and low number of coincidence counts.

Here we presented a reliable estimate of the distribution of coincident events for significance testing based on random samples of subsets of trial recombinations. The precision of the significance estimate can be adjusted to a requested level by successively increasing the size of the random subset. The method integrates well in the scheme of the unitary event analysis [3,4] and provides an improved method for the significance estimation. In contrast to the approach in Ref. [3,4], (1) the assumption of a specific underlying spike train model is not required, since the null-hypothesis of independence is incorporated on the basis of the original experimental spike trains including their temporal structure. (2) Assumptions regarding stationarity are relaxed. A prerequisite of the method is that stationarity across trials is fulfilled, which actually is the definition of a 'trial' [6]. However, here the assumption of stationarity in time, implicit to most correlation analysis techniques, is not required.

Acknowledgements

We thank Wolf Singer for stimulating discussions and his support. This work was in part funded by the Volkswagen Foundation.

References

[1] B. Efron, R.J. Tibshirani, An Introduction to the Bootstrap, Chapmann Hall, London, CRC Press, Boca Raton, 1998.

[2] F. Grammont, A. Riehle, Precise spike synchronization in monkey motor cortex involved in preparation for movement, Exp. Brain Res. 128 (1999) 118–122.

[3] S. Grün, M. Diesmann, A. Aertsen, 'Unitary Events' in multiple single-neuron activity. I. Detection and significance, Neural Comput. 14 (2002) 43–80.

[4] S. Grün, M. Diesmann, A. Aertsen, 'Unitary Events' in multiple single-neuron activity. II. Non-Stationary data, Neural Comput. 14 (2002) 81–119.

[5] D.O. Hebb, Organization of Behavior. A Neurophysiological Theory, Wiley, New York, 1949.

[6] D.H. Perkel, G.L. Gerstein, G.P. Moore, Neuronal spike trains and stochastic point processes. II. Simultaneous spike trains, Biophys. J. 7 (1967) 419–440.

[7] G. Pipa, M. Diesmann, S. Grün, Significance of joint-spike events based on trial-shuffling by efficient combinatorial methods, submitted for publication.

[8] D.S. Reich, F. Mechler, K.P. Purpura, J.D. Victor, Interspike intervals, receptive fields, and information encoding in primary visual cortex, J. Neurosci. 20 (2000) 1964–1974.

[9] A. Riehle, F. Grammont, M. Diesmann, S. Grün, Dynamical changes and temporal precision of synchronized spiking activity in motor cortex during movement preparation, J. Physiol. (Paris) 94 (2000) 569–582.

[10] A. Riehle, S. Grün, M. Diesmann, A. Aertsen, Spike synchronization and rate modulation differentially involved in motor cortical function, Science 278 (1997) 1950–1953.

[11] W. Singer, A.K. Engel, A.K. Kreiter, M.H.J. Munk, S. Neuenschwander, P.R. Roelfsema, Neuronal assemblies: necessity, signature and detectability, Trends Cog. Sci. 1 (1997) 252–261.

Gordon Pipa was born in 1974. With his development of a high speed image processor he finished 3rd in the German federal competition for young scientists in 1996 (Patents: US6231185, WO99/03066, PCT01454, EP0993657, DE9701454). He is holder of the honorary medal of the Friedrich-Bläse-Foundation and of a federal scholarship. He studied Physics and Electrical Engineering at the Technical University of North Rhine–Westphalia in Aachen and received his M.Sc. in Physics in 2002 from the J.W. Goethe University (Frankfurt a.M., Germany). His Masters thesis was carried out in the Department of Neurophysiology at the Max-Planck-Institute for Brain Research (Frankfurt a.M., Germany).

Sonja Grün was born in 1960 in Germany. She received her M.Sc. in Physics from the Eberhard-Karls University in Tübingen in 1991. After associations with the Ruhr-University in Bochum and the Weizmann Institute of Science (Rehovot, Israel), she received her Ph.D. in Physics in 1996 (Ruhr-University, Bochum). From 1996 to 1997 she was a post-doctoral fellow at the Hebrew University in Jerusalem, Israel. From 1998 to 2002 she led the group for data analysis and modeling the Max-Planck-Institute for Brain Research in Frankfurt/Main, Germany. Since Dec. 2002 she is Assistant Professor for Neuroinformatics at the Freie Universität in Berlin, Germany. Her research interests are in the field of theoretical neuroscience with focus on analyis of data from electrophysiological recordings and identification of cooperative processes in biological neural networks.

Available at
www.ComputerScienceWeb.com
POWERED BY SCIENCE DIRECT

ELSEVIER

Neurocomputing 52–54 (2003) 39–44

NEUROCOMPUTING

www.elsevier.com/locate/neucom

Energy efficiency in a channel model for the spiking axon

David H. Goldberg*, Arun P. Sripati, Andreas G. Andreou

Department of Electrical and Computer Engineering, Johns Hopkins University, 3400 N. Charles St., 105 Barton Hall, Baltimore, MD 21218, USA

Abstract

We examine the spiking axon as a communication channel. We develop a first principles channel model that encompasses the noise in the axon, which manifests itself as spike jitter, and the power consumption, which arises from the activity of the Na^{+}–K^{+} pump. This model enables us to examine the trade-off between the information rate and power consumption. Using parameters from the frog myelinated axon, we determine the spike rate that corresponds to the maximum energy efficiency. This spike rate is consistent with experimental observations, which suggests that neural communication may have developed to maximize energy efficiency rather than information rate alone.

Keywords: Myelinated axon; Spike jitter; Energy efficiency; Ion channel stochasticity

1. Introduction

Recently, several researchers have suggested that energy efficiency is a guiding constraint in the evolution of neural systems [3,9,2,7]. In this paper, we examine the energy efficiency of spike communication over a myelinated axon. We determine the most energy efficient way to communicate information and compare this result to experimental observations. We also explore how fundamental biophysical parameters contribute to the energy efficiency by means of a scaling analysis.

Energy efficiency can be expressed as a function of *signal properties*, such as the interspike interval (ISI) distribution, and *channel properties*, such as the behavior of ion channels. In Section 2, we set up the optimization problem that yields the signal

* Corresponding author. Tel.: +1-410-516-8361; fax: +1-410-516-8313.
E-mail address: goldberg@jhu.edu (D.H. Goldberg).

0925-2312/03/$ - see front matter
doi:10.1016/S0925-2312(02)00770-1

properties corresponding to optimal usage of the axon. In Section 3, we determine the channel properties by developing biophysical models of the spike jitter and power consumption. In Section 4, we describe the optimization results and the scaling analysis. In Section 5, we discuss the results and conclude the paper.

2. Optimizing signal properties for energy efficiency

In this work, we examine the myelinated axon as a communication channel. The input X is the spike train that is generated at the soma. The output Y is the spike train that emerges from the axon into the presynaptic region. Our energy efficiency metric $\mathscr{E}$ (units of bits/ATP) is given by the ratio of the information rate through the axon $\mathscr{R}$ (units of bits/s) and the power consumption of the axon P (units of ATP/s) [9]:

$$\mathscr{E} = \mathscr{R}/P. \tag{1}$$

We do not know a priori what aspect of the spike trains X and Y carry information. An upper bound on the amount of information a spike train can carry can be determined by considering solely the ISIs, as a list of successive ISIs is sufficient to reconstruct the spike train in its entirety [1]. Our work makes use of this upper bound, so the signal is the ISI, and the noise is anything that alters the ISI, which we term ISI jitter.

In this context, the mutual information (in bits/symbol) is given by

$$I(X;Y) = h(Y) - h(Y|X) = h(Y) - h(N) \tag{2}$$

where $h\cdot$ indicates differential entropy. To obtain the upper bound on the information rate, we make the assumption that the ISIs are independent. In this case, the information rate $\mathscr{R}$ is simply the product of the mutual information and the spike rate λ (the number of symbols per second):

$$\mathscr{R} = \lambda[h(Y) - h(N)]. \tag{3}$$

The energy efficiency $\mathscr{E}$ becomes

$$\mathscr{E} = \lambda[h(Y) - h(N)]/P(\lambda), \tag{4}$$

where we write $P(\lambda)$ to show that the power consumption is a function of spike rate.

The maximum energy efficiency $\mathscr{E}^*$ is given by

$$\mathscr{E}^* = \max_{f_Y(\lambda)} \lambda[h(f_Y(\lambda)) - h(f_N)]/P(\lambda) = \lambda^*[h(f_Y^*(\lambda^*)) - h(f_N)]/P(\lambda^*), \tag{5}$$

where $f_Y^*(\cdot)$ and λ^* are respectively the ISI distribution and mean spike rate that realize the maximum energy efficiency. In order to maximize $\mathscr{E}$, $h(f_Y)$ must be as large as possible. Therefore, we choose the distribution that maximizes $h(f_Y)$ for a fixed spike rate λ. In the presence of a refractory period t_{ref}, f_Y^* is given by a shifted exponential interspike interval distribution with corresponding entropy [1]

$$h(f_Y^*) = \log_2[e(1 - \lambda^* t_{\mathrm{ref}})/\lambda^*]. \tag{6}$$

The maximum energy efficiency becomes

$$\mathscr{E}^* = \lambda^*\{\log_2[e(1 - \lambda^* t_{\mathrm{ref}})/\lambda^*] - h(f_N)\}/P(\lambda^*). \tag{7}$$

Now, λ^* is the only remaining free parameter. In order to find its value, we need to fill in the (unknown) fixed parameters of Eq. (7). They are f_N, the distribution of the ISI jitter, and $P(\cdot)$, the relationship between the power consumption and the spike rate. To do this, we must examine a biophysical model of the axon.

3. A channel model for the myelinated axon

We formulate a model of the ISI jitter in the myelinated axon based on ion channel stochasticity at the node of Ranvier [12]. The stochastic behavior of the ion channel population can be described by binomial statistics: For N channels, each with a single channel current $\bar{\imath}$ and probability of being open p, the coefficient of variation (CV) of the channel current is given by

$$\sigma_I/\mu_I = \sqrt{N\bar{\imath}^2 p(1-p)}/N\bar{\imath}p \approx 1/\sqrt{Np} \quad (p \ll 1). \tag{8}$$

Variability decreases as the number of ion channels increases, as the stochastic behavior of individual ion channels is averaged out.

Analysis of the mechanism of spiking at the node of Ranvier reveals that the spike time is vulnerable to the stochastic Na^+ current only during a very brief "stochastic interval" during the rising phase of the action potential (Fig. 1(a) and (b)). The duration of this interval is given by $T = C\theta/I$, where C is the node capacitance, θ is the range of voltages that delimit the interval, and I is the stochastic Na^+ current during the interval. Because the time constant of the channel noise is large compared to the threshold crossing time (Fig. 1(c)), we model I as constant during a particular interval but variable *between* intervals.

Because I is a binomial r.v. and N is large, we can approximate I as Gaussian. Then the distribution of T is given by the inverse of a Gaussian r.v. [8,13]. This yields a spike time distribution that is approximately Gaussian with standard deviation $\sigma_T = C\theta\sigma_I/\mu_I^2$ (derivation omitted) where μ_I and σ_I are the mean and standard deviation of the current during the rising phase of the action potential. The CV of the spike time is given by

$$\sigma_T/\mu_T = (C\theta\sigma_I/\mu_I^2)/(C\theta/\mu_I) = \sigma_I/\mu_I \approx 1/\sqrt{Np}. \tag{9}$$

This means that the CV of the spike time is equal to the CV of the ion channel noise, implying that the scaling behavior of ion channel noise extends to spike time variability.

We can relate the spike jitter at a single node to the ISI jitter for the entire fiber by $\sigma_{\mathrm{ISI}} = (2M)^{1/2}\sigma_T$ where M is the number of nodes in the fiber. The factor of 2 arises because the interspike interval is made up of two spikes. The entropy of the ISI jitter distribution, $h(f_N)$, is given by the entropy of a Gaussian distribution with variance σ_{ISI}^2.

We can test the validity of the spike jitter model by comparing it to measured values of ISI jitter in the frog myelinated axon. Using values of θ, μ_I, and σ_I^2 from Hodgkin–Huxley simulations of the node of Ranvier, we obtain $\sigma_{\mathrm{ISI}} = 35$ µs. The only measurement of ISI jitter in this system was made by Lass and Abeles [6], who

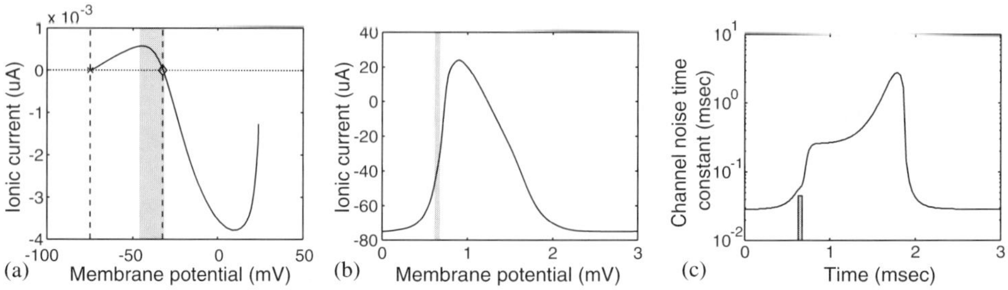

Fig. 1. The stochastic behavior of Na^+ channels is only relevant during a brief interval. Data generated by a Hodgkin–Huxley simulation. (a) I–V curve of the rising phase of the action potential. The $\times$ denotes a stable equilibrium point, corresponding to the resting state, and the $\diamondsuit$ denotes an unstable equilibrium point, corresponding to the spiking threshold [10]. The region of negative conductance is indicative of the positive feedback action of the Na^+ channels. We model the spike time as vulnerable to Na^+ channel noise only during the "stochastic interval" defined from the start of the positive feedback to the spiking threshold (shaded area). (b) The action potential with the stochastic interval (shaded region) superimposed. (c) The Na^+ channel noise time constant during the action potential (derived from a kinetic model [5]). The gray box represents the duration of the stochastic interval.

obtained $\sigma_{ISI} = 6$ μs. There are many reasons for the discrepancy, including uncertainty in the number of ion channels and variability across different frog species.

In the myelinated axon, there are two sources of power consumption: the energy expended by the Na^+–K^+ to counteract the passive leak of ions across the membrane (resting cost) and the energy used by the pumps to restore the concentration gradients following an action potential (signaling cost). Power consumption can be expressed as

$$P = a\lambda + b, \tag{10}$$

where a is the ATP expended per spike, λ is the spike rate, and b is the ATP expended per second due to resting costs. The values of these constants can be determined by considering the stoichiometry of the Na^+–K^+ pump [14].

4. Results

The expression for the maximum energy efficiency becomes

$$\mathscr{E}^* = \lambda^*\{\log_2[e(1-\lambda^* t_{ref})/\lambda^*] - (1/2)\log_2(2\pi e\sigma_{ISI}^2)\}/M(a\lambda^* + b). \tag{11}$$

The models of Section 3 enable us to derive the values for all of the channel parameters in Eq. (11) (given in the caption of Fig. 2). The only remaining free parameter is the optimal spike rate, which is found numerically. The information rate and energy efficiency are plotted as functions of spike rate in Fig. 2(a) and (b), respectively. The maximum information rate, i.e. the information capacity, is given by the peak of the curve in Fig. 2(a), whereas the maximum energy efficiency is given by the peak of the curve in Fig. 2(b). It is clear that spike rates corresponding to the two maxima are distinct.

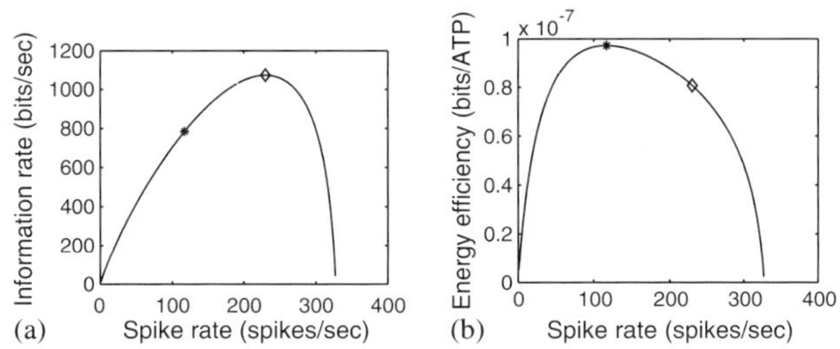

Fig. 2. Information rate and energy efficiency as a function of spike rate. Parameters as follows: $\sigma_{ISI} = 35$ μs, $t_{ref} = 3$ ms, $a = 6.42 \times 10^5$ ATP/spike, $b = 3.69 \times 10^7$ ATP/s, $M = 72$ nodes. The point corresponding to the maximum energy efficiency is denoted by the $*$ (117 Hz), and the point corresponding the information capacity is denoted by the $\diamondsuit$ (230 Hz). (a) Information rate as a function of spike rate. (b) Energy efficiency as a function of spike rate.

We are also interested in understanding of how fundamental biophysical parameters affect the energy efficiency. We must perform this analysis with caution; we cannot vary parameters at will because the kinetics of the Hodgkin–Huxley equations are voltage-dependent. Therefore, the membrane potential of the axon must remain constant while the parameters are varied. Rushton's *principle of corresponding states* provides us with way to do this: we scale the fiber diameter, nodal area, and inter-node distance by a factor S, while keeping all of the *specific* properties of the fiber constant [11].

The absolute number of ion channels increases with S, so ISI jitter goes as $S^{-1/2}$ (Eq. (9)). This gives rise to a logarithmic increase in the information rate. The power consumption is proportional to the number of ions displaced, which increases proportionally with the nodal area and in turn S. These trends alone are not a priori sufficient to establish the scaling behavior of $\mathscr{E}^*$ because the spike rate λ is a free parameter. Therefore, for each value of S, we numerically determined $\mathscr{E}^*$ and λ^*. Over the scales we examined ($S = [0.1, 10]$), λ^* had the relatively narrow range of 95 and 135 Hz. If we approximate λ as constant, then it becomes clear that as S increases, the denominator P^* grows faster (linearly) than the numerator $\mathscr{R}^*$ (logarithmically), causing $\mathscr{E}^*$ to decrease. These trends were verified numerically.

5. Discussion and conclusions

We have found that energy efficient use of the spiking axon corresponds to a shifted exponential ISI distribution with a mean spike rate of 117 Hz. In order to compare the form of the distribution with neurophysiological data, we need to consider the spiking statistics of the frog myelinated axon over the ensemble of natural behaviors. While this may prove difficult, the mean spike rate is much easier to measure. Indeed, spike rates of 100 Hz are typical of the frog myelinated axon [4], suggesting that this system may have developed to maximum energy efficiency rather than information rate alone.

We examined the scaling behavior of this system and found that despite the fact that jitter is reduced by increasing the number of ion channels participating in spike generation, the benefits of having more ion channels does not outweigh the accompanying power consumption cost. Our results can be interpreted to mean that for transmitting information over proportional distances, smaller axons are more energy efficient.

In this work, we examined axonal communication only. Spike communication entails the transformation of analog information to and from spikes, and the noise and power consumption associated with these processes can considerably affect the energy efficiency. In order to gain a more complete understanding of spike communication, we are currently performing similar analyses of spike encoding and decoding.

Acknowledgements

This work was supported by DARPA/ONR contract N00014-00-C-0315 and NSF grant EIA-0130812.

References

[1] M. Abeles, Y. Lass, Transmission of information by the axon: II. The channel capacity, Biol. Cyber. 19 (1975) 121–125.
[2] Pamela Abshire, Andreas G. Andreou, Capacity and energy cost of information in biological and silicon photoreceptors, Proc. IEEE 89 (2001) 1052–1064.
[3] J.M. Allman, Evolving Brains, Scientific American Library/W.H. Freeman, New York, 1998.
[4] W.T. Catton, Cutaneous mechanoreceptors, in: R. Llinás, W. Precht (Eds.), Frog Neurobiology: A Handbook, Springer, Berlin, 1961, pp. 629–642.
[5] B. Hille, Voltage clamp studies on myelinated nerve fibers, in: William J. Adelman Jr. (Ed.), Biophysics and Physiology of Excitable Membrane, Van Nostrand Reinhold, New York, 1971 (Chapter 12).
[6] Y. Lass, M. Abeles, Transmission of information by the axon: I. Noise and memory in the myelinated nerve fiber of the frog, Biol. Cybernet. 19 (1975) 61–67.
[7] Simon B. Laughlin, Energy as a constraint on the coding and processing of sensory information, Curr. Opinion Neurobiol. 11 (2001) 475–480.
[8] Michael W. Levine, Jeremy M. Shefner, A model for the variability of interspike intervals during sustained firing of a retinal neuron, Biophys. J. 19 (1977) 241–252.
[9] William B. Levy, Robert A. Baxter, Energy efficient neural codes, Neural Comput. 8 (1996) 531–543.
[10] Denis Noble, Applications of Hodgkin–Huxley equations to excitable tissues, Physiol. Rev. 46 (1–2) (1996) 50.
[11] W.A.H. Rusthon, A theory of the effects of fibre size in medullated nerve, J. Physiol. 151 (1951) 101–122.
[12] Elad Schneidman, Barry Freedman, Idan Segev, Ion channel stochasticity may be critical in determining the reliability and precision of spike timing, Neural Comput. 10 (1998) 1679–1703.
[13] Charles E. Smith, A comment on a retinal neuron model, Biophys. J. 25 (1979) 385–386.
[14] D. Attwell, S.B. Laughlin, An energy budget for signaling the grey matter of the brain, J. Cerebral Blood Flow Metabolism 21 (2001) 1134–1145.

Available at
www.ComputerScienceWeb.com
POWERED BY SCIENCE @ DIRECT®

ELSEVIER

Neurocomputing 52–54 (2003) 45–52

NEUROCOMPUTING

www.elsevier.com/locate/neucom

Structural and biophysical mechanisms underlying dynamic sensitivity of primary sensory interneurons in the cricket cercal sensory system

G.I. Cummins[a], S.M. Crook[b], A.G. Dimitrov[a], T. Ganje[a], G.A. Jacobs[a,*], J.P. Miller[a]

[a]*Center for Computational Biology, Montana State University, 1 Lewis Hall, Bozeman, MT 89717, USA*

[b]*Department of Mathematics, University of Maine, Orono, ME, USA*

Abstract

We constructed probabilistic models of afferent inputs and compartmental models of interneurons in the cricket cercal system to examine the effects of dendritic morphology, distribution of synaptic inputs, and membrane properties on interneuron directional tuning properties. The mean directional tuning of afferent inputs to an interneuron was an excellent predictor of its directional tuning. Location of the synapses on the interneurons' dendrites was not essential to determining tuning characteristics, but had a substantial effect on sensitivity. Thus, we conclude that both sampling of the afferent population, and anatomical distribution of synaptic inputs are important determinants of an interneuron's directional response.

Keywords: Sensory systems; Structure/function; Neural maps; Compartmental models; Gaussian mixtures

1. Introduction

The response properties of a neuron to complex patterns of synaptic inputs are controlled by a combination of factors including: (1) the electroanatomy of the cell, (2) its biophysical properties and (3) the distribution and activation pattern of synaptic inputs. We used electrophysiological experiments, statistical and compartmental modeling techniques to (1) calculate detailed, quantitative predictions of a variety of ensemble

* Corresponding author.

E-mail address: gwen@cns.montana.edu (G.A. Jacobs).

0925-2312/03/$ - see front matter © 2002 Elsevier Science B.V. All rights reserved.
doi:10.1016/S0925-2312(02)00786-5

afferent activity patterns elicited by air current stimuli, (2) define simulated synaptic input patterns onto the dendrites of compartmental interneuron models, and (3) examine the roles of dendritic morphology, distribution of synaptic inputs and membrane properties on the directional tuning properties of the neurons.

Primary sensory interneurons in the cercal system are sensitive to the direction and dynamics of air currents [6,11,18]. The interneurons receive excitatory input from an ensemble of sensory receptor afferents which form a neural map of air current direction in the central nervous system [8]. Interneurons extract and encode information about stimulus direction based on the shape and position of their dendrites within the afferent map [9]. Three independent factors could contribute to the directional sensitivity of each interneuron IN: (a) the position of its dendrites within the afferent map, (b) its selective connectivity with subclasses of afferents having specific directional sensitivities, and (c) the membrane properties of the cell.

2. Methods

Anatomical reconstructions and physiological data were used to create biophysically based compartmental models of three identified sensory interneurons. Values for R_i and C_m were set to experimentally established values. R_m was initially assumed to be uniform throughout the cell, and was set to yield the measured steady-state input resistance and complex input impedance recorded in real neurons [17]. Next, parameter domains for voltage-dependent conductances and synaptic inputs were determined using a variety of validation criteria.

The only active membrane components included in the models were Hodgkin–Huxley-type sodium and potassium conductances in the spike initiating zone (SIZ) and axon. The density and kinetic parameters of each channel (a total of 26 parameters) were determined by a genetic optimization algorithm [3]. This algorithm's fitness objective was to minimize the difference between the action potential waveform seen in the models and a physiological spike waveform, determined by averaging the waveforms of 1000 action potentials recorded intracellularly from the axons of interneurons at the point at which they exit the cercal ganglion. The resulting kinetic parameters differed substantially (by as much as a factor of 3) from the standard Hodgkin–Huxley parameters (and those included in the Neuron simulator), which are derived from squid channels.

A Gaussian mixture model was used to generate a probabilistic representation of the 3D location and density distribution of membrane surface area of the axon terminals of primary afferents. The anatomical data for this representation was derived from over 200 individually stained and reconstructed primary afferents [7,19]. This representation was used to estimate the distribution of synaptic inputs to each of the interneuron models.

The relatively large number of afferents allowed us to make a continuum approximation and formulate the problem as a density estimation of the probability density of connection between interneurons and afferents. We modeled the probability density that an afferent with certain directionality connects at a particular point along the interneuron's dendritic tree. There are several assumptions inherent in this estimation. First, we assumed that the probability density of presynaptic boutons was independent of the postsynaptic cell. Second, we assumed that presynaptic contacts were located on the varicosities of the axon terminal arbors of the afferent and that the probability of contact between an afferent and IN around spatial position $\mathbf{x}$ was proportional to the surface areas of presynaptic varicosity and postsynaptic dendritic branch in this region. In principle, the postsynaptic cell can be described in a similar probabilistic form. As a further simplification, we assumed that interneurons have a fixed structure and described the dendritic surface area in a volume around $\mathbf{x}$ with the constant S_I. Formally, the problem was expressed as describing the probability density

$$p(S_A, \mathbf{x}|A) \tag{1}$$

of jointly finding a varicosity of afferent A with surface area S_A at spatial location $\mathbf{x}$. The second assumption translates to the expression

$$p(\text{contact at } d_I|A, I) = \alpha S_I E_p S_A, \tag{2}$$

that is, the probability of a contact at dendritic segment d_I was proportional to the product of the expected surface area S_A (afferent varicosity) with S_I (the area of dendritic segment d_I). The expectation E_p was estimated in a region of interest (e.g., around a compartment of the postsynaptic cell's dendritic tree).

Estimating the above quantity required a good estimate of the underlying probability density function (1). A previous approach [8] attempted to directly estimate $E_p S_A$ using a particular density estimator (Parzen estimator). However, details of the estimation were mixed with details of the interaction model, which obscured the power of this approach. We attempted to circumvent this here by decoupling the density estimation problem from the modeling issues. In particular, we used a different density estimator. We modeled Eq. (1) as the Gaussian mixture

$$p(S_A, x|A) = \sum_{i=1}^{M} p_i N(\{S_A, x\}|m_i, \mathbf{C}_i), \tag{3}$$

where $N(\{S_A, x\}|\mathbf{m}, \mathbf{C})$ is a normal model with mean $\mathbf{m} = \{\text{mean}(S_A), \text{mean}(\mathbf{x})\}$ and covariance $\mathbf{C} = \text{cov}(\{S_A, \mathbf{x}\})$. The model order M is a free parameter of the model. For each model order, the parameters $(\mathbf{m}_i, \mathbf{C}_i)$ can be found using standard statistical estimation procedures (e.g., EM).

The density estimators for each sensory neuron class were used as the anatomical substrate for the prediction of afferent activity patterns. We constructed density estimators for 16 classes of sensory neurons, representing inputs from both cerci, with four different directional sensitivities, and two different sensory hair lengths: long hairs 900 –2000 μm, and medium hairs, 500–900 μm.

We modeled each of three identified interneurons (cells 10-3, 10-2, and 9-3) using anatomical data collected from intracellular cobalt stains of these cells [6,11]. These spatial data consisted of about 9000 x, y, z, diameter points per cell (10-3: 8114 points, 10-2: 13171 points, 9-3: 7415 points). In the default models this was also the number of compartments (nodes). These coordinates were scaled and aligned to the coordinate system of a standard cercal ganglion, as was done in previous studies [7,19]. Since the cobalt procedure results in over 20% shrinkage of the tissue [1], the anatomical data were corrected for shrinkage by comparison to measurements of dendritic branches from the same cells stained with biocytin [2].

For each of these points, we generated a frustum by connecting a circular cross-section associated with each point, to another corresponding to its parent in the dendritic tree. We used Eq. (3) to calculate the expected overlap between these frusta and each of the 16 input classes. This provided us with an estimate of joint probability distribution of afferent class and synapse location for the cell. We generated 1000 synaptic contacts to each interneuron model by drawing type/location pairs from this distribution. Spatial overlap between the IN dendritic tree and the afferent map therefore had two effects. (1) Interneurons with many dendritic processes in a region of the map dominated by afferents from a particular class received a greater number of synapses from that class. (2) Afferent classes occupying regions of the map nearer to the interneuron's spike initiating zone provided more proximal (and therefore more effective) inputs.

Synaptic events at each contact generated a localized instantaneous conductance change with a reversal potential of 0 mV and initial value of 0.02 microsiemens, which decayed exponentially with a time constant of 0.1 ms. These values were chosen so as to fit the resulting post synaptic potential to physiologically observed potentials [4,15].

Previous observations of the response of sensory afferents to air current stimuli indicate that, at low frequencies, the spike rate is roughly proportional to the deflection of the associated sensory hair along a particular preferred direction [10,14]. We assume that this deflection is proportional to the vector projection of air velocity onto the preferred direction [10,16].

We calculated the projections of air velocity onto the preferred directions of each afferent class for air current stimuli originating from 8 different directions. We generated a probability distribution by multiplying this (time varying) projection by an intensity parameter, which was constant within a stimulus, but could vary between stimuli. We then generated a list of event times for each afferent by drawing from this distribution. Afferents were assigned a 1 ms refractory period, so events occurring less than 1 ms after the last event were discarded. Previous work suggests that each afferent forms, at most, one contact with a given interneuron [15]. Consequently we generated independent event lists for each of the 1000 synapses.

Given the event times, and synaptic locations, we were able to simulate an air current stimulus by activating each synapse at the appropriate times. We were thus able to observe the output of the model interneurons in response to simulated air currents from the 8 different directions. We determined directional tuning curves for the models by presenting the first half cycle (50 ms) of a 10 Hz sine wave air current originating from each direction, and counting the resulting action potentials. The three models differed in absolute sensitivity, with 10-3 generating the most output spikes for a given input intensity. For each cell we scaled the intensity parameter of the input to a value that produced about 10 action potentials at the most sensitive direction before calculating a tuning curve. This spike rate ($\sim$ 200 Hz) is similar to the spike rates seen in actual cells responding to wind stimuli [6,11].

3. Results

The directional tuning of model interneurons in response to air currents from different directions was calculated using the methods described above. In all cases, the directional tuning curves measured from the model cells were very similar to those recorded previously in physiological experiments on these same identified cells [6,11]. Fig. 1 shows a comparison of a directional tuning curve recorded from identified IN 10-3 and a simulation of the same experiment. The histogram of the number of synaptic contacts from the classes of afferents showed a clear bias towards those afferents tuned to directions close to 315°. Also, the total number of inputs received in the first 50 ms of the stimulus for each direction correspond to those same directions. Simulations using two other identified interneurons yielded similar results. Therefore, compartmental models with passive dendrites and active membrane on the SIZ and axon appear adequate to account for the directional tuning properties of these interneurons.

In order to separate the effect of differential sampling of afferent classes by the interneuron, and differential location of the resulting synapses on the interneuron dendrite, we performed the following model experiment. Inputs were assigned to an interneuron by the procedure described above. Subsequently, the location of each contact

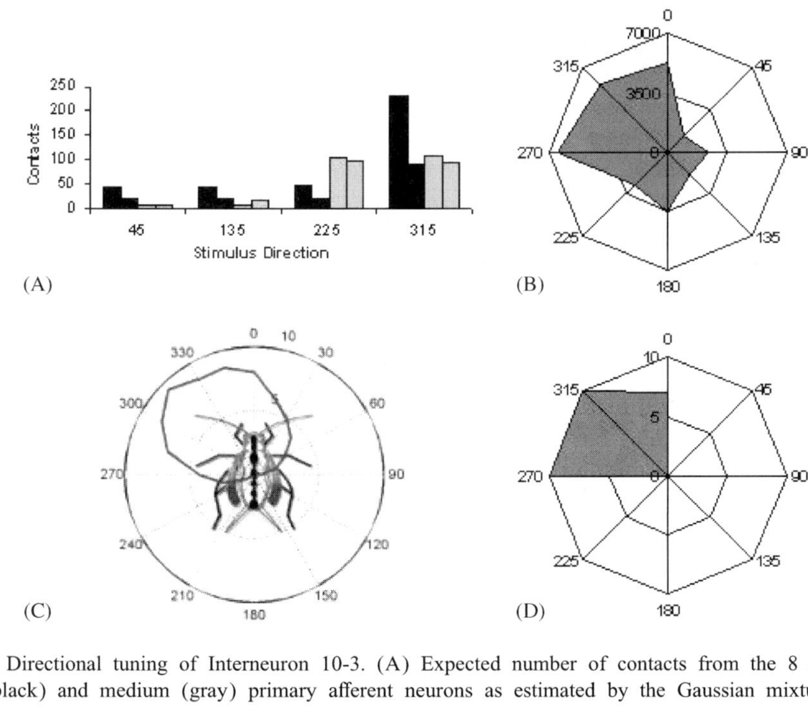

Fig. 1. Directional tuning of Interneuron 10-3. (A) Expected number of contacts from the 8 classes of long (black) and medium (gray) primary afferent neurons as estimated by the Gaussian mixture model (3). (B) The number of total synaptic events in the first 50 ms of a 10 Hz sine wave stimulus as a function of stimulus direction. (C) Directional tuning curve measured physiologically from interneuron 10-3. (D) Directional tuning curve calculated for a compartmental model of interneurons 10-3.

was reassigned to a random dendritic compartment, while the identity of the presynaptic afferent was maintained. This resulted in a model that received the same set of synaptic events in response to a given input, but in which the spatial distribution of these events on the dendrite had a random structure, rather than that specified by spatial overlap between the interneurons and specific classes of afferents previously described [8].

The models with randomized input location showed greatly decreased sensitivity to inputs. For example, our randomized 10-3 model generated no spikes to a stimulus that evoked 10 spikes from the standard model. However, if the stimulus intensity was increased sufficiently (more than a factor of 2) to evoke responses equivalent to those in the standard models at a given direction, the shape of the tuning curve (number of spikes evoked by stimuli from all other directions) differed from that in the standard models only slightly (the random models generated a few extra spikes at off-peak directions).

4. Discussion

The results of simulations reported here suggest that the directional tuning properties of primary sensory INs in the cricket cercal system are determined largely by the anatomical structure of the INs, the directional tuning properties of those afferents that provide excitatory input and the spatial distribution of those inputs to different dendritic branches. Interneurons with different dendritic structures receive different sets of excitatory inputs and thus are tuned to specific air current directions.

Relative directional tuning in the models seemed to be determined primarily by the differential sampling of afferent inputs with different preferred directions. The left 10-3 cell, for example, overlaps with a greater number of afferents sensitive to wind coming from the animal's front left, and therefore shows a directional tuning preference in this quadrant. In addition these inputs occur relatively near the cell's spike initiation zone [6], and might therefore be expected to have a stronger effect per synapse than inputs from some other quadrant, which occur on more distal regions of the dendrite.

Randomizing input locations removes the positional advantage of the preferred stimuli, but maintains their numerical advantage. Predictably, this decreases the sensitivity of the cell, by moving a large concentration of proximal inputs to more distal locations, and only moving a few distal inputs proximally, but it does not alter directional tuning. The number of preferred inputs is larger than the number of off-peak inputs such that even without more proximal placement, these inputs still dominate the cell's response.

One apparent conclusion of this input sampling is that, for low frequency unidirectional stimuli, many of the excitatory inputs to the interneurons have little effect on output. Relatively small proximal regions of the cells, with strong input representation from a small subset of the afferent types, dominate the generation of action potentials. None-the-less, the interneurons have widespread, complex dendrites, receiving many types of input.

One possible function for these "extra" dendrites is to provide resilience to damage of the sensory array. Dendrites from the three modeled cells sample inputs from both cerci.

In our simulations, the inputs from the cercus ipsilateral to the SIZ were dominant. In the event of the loss of this cercus, however, the interneurons might maintain similar directional responses (but at reduced sensitivity) by relying on smaller populations of more distal input from the contralateral cercus. Experiments in which directional tuning curves were measured from an interneuron when inputs ipsilateral to the SIZ were blocked corroborate this hypothesis [6].

A second possible function of the distal inputs is in detection of more complex, multidimensional stimuli. Although the small, distal, populations of off-peak inputs may not be sufficient to fire the cell at steady state, they could have a substantial effect in modulating the response to a more dynamic input pattern, which might activate larger regions of the dendrite at lower intensities than do the steady state stimuli.

Although the model neurons displayed robust directional tuning, in a set of preliminary simulations not reported here, our models failed to predict many of the response properties of these cells to dynamic stimuli including white noise air currents or sine wave stimuli of frequencies above 60 Hz. Physiological recordings from real cells show substantial stimulus–response coherence to stimulus frequencies less than 80 Hz, and very little coherence at frequencies above 100 Hz [18]. Our models show substantially lower coherence at low frequency, and a less sharp high frequency fall off. Since the morphology of our models, passive membrane properties, and axonal channels were fit well to large physiological data sets it is likely that these behaviors depend on the presence of active channels in the dendrites which were not included in our present models. Interneurons in the cercal system are known to have voltage activated calcium channels [12] and two types of voltage activated potassium channels distributed throughout their dendritic arbors [5]. Our future studies will examine how these additional biophysical parameters may affect the response properties of the cells to dynamic stimuli.

Acknowledgements

This work was supported by National Science Foundation grants IBN 0090966 to GAJ and SMC and MRI 9871191 to JPM. Special thanks to Zane Aldworth for providing physiology data.

References

[1] J.B. Bacon, J.S. Altman, A silver intensification method for cobalt filled neurons in whole mount preparations, Brain Res. 138 (1977) 359–363.

[2] B.J. Claiborne, D.G. Amaral, W.M. Cowan, Quantitative, three-dimensional analysis of granule cell dendrites in the rat dentate gyrus, J. Comput. Neurol. 302 (1990) 206–219.

[3] G.I. Cummins, Compartmental models and genetic algorithms provide insight into structure-function relationships in the lateral dendrite of the Mauthner Neuron, Ph.D. Thesis, University of Colorado, Boulder, 2001.

[4] G.W. Davis, R.K. Murphey, A role for postsynaptic neurons in determining presynaptic release properties in the cricket CNS: evidence for retrograde control of facilitation, J. Neurosci. 13 (1993) 3827–3838.

[5] M. Horner, P. Kloppenburg, Voltage-activated currents in identified giant INs isolated from adult crickets gryllus bimaculatus, J. Exp. Biol. 201 (1998) 2529–2541.

[6] G.A. Jacobs, J.P. Miller, R.K. Murphey, Cellular mechanisms underlying directional sensitivity of an identified sensory IN, J. Neurosci. 6 (1986) 2298–2311.

[7] G.A. Jacobs, C.S. Pittendrigh, Predicting emergent properties of neuronal ensembles using a database of individual neurons, in: G. Ascoli (Ed.), Computational Neuroanatomy: Principles and Methods, Humana Press Inc., Clifton, UK, 2002, pp. 151–170.

[8] G.A. Jacobs, F. Theunissen, Functional organization of a neural map in the cricket cercal sensory system, J. Neurosci. 16 (1996) 769–784.

[9] G.A. Jacobs, F. Theunissen, Extraction of sensory parameters from a neural map by primary sensory INs, J. Neurosci. 20 (2000) 2934–2943.

[10] M.A. Landolfa, J.P. Miller, Stimulus/response properties of cricket cercal filiform hair receptors, J. Comput. Physiol. A 177 (1995) 749–757.

[11] J.P. Miller, G.A. Jacobs, F.E. Theunissen, Representation of sensory information in the cricket cercal sensory system. I. Response properties of the primary INs, J. Neurophysiol. 66 (1991) 1680–1689.

[12] H. Ogawa, Y. Baba, K. Oka, Spike-dependent calcium influx in dendrites of the cricket giant IN, J. Neurobiol. 44 (2000) 45–56.

[14] J.C. Roddey, G.A. Jacobs, Information theoretic analysis of dynamical encoding by filiform mechanoreceptors in the cricket cercal system, J. Neurophys. 75 (1996) 1365–1376.

[15] D. Shepherd, R.K. Murphey, Competition regulates the efficacy of an identified synapse in crickets, J. Neurosci. 6 (1996) 3152–3160.

[16] T. Shimozawa, M. Kanou, Varieties of filiform hairs: range fractionation by sensory afferents and cercal INs of a cricket, J. Comput. Physiol. A 155 (1984a) 485–493.

[17] F.E. Theunissen, F.H. Eeckman, J.P. Miller, A modified Hodgkin–Huxley spiking model with continuous spiking output, in: F. Eeckman, J. Bower (Eds.), Computation and Neural Systems, Kluwer Academic Publishers, Boston, MA, 1993, pp. 9–17.

[18] F.E. Theunissen, J.C. Roddey, S. Stufflebeam, H. Clague, J.P. Miller, Information theoretical analysis of dynamical encoding by four identified primary sensory INs in the cricket cercal system, J. Neurophys. 75 (1996) 1345–1364.

[19] T.W. Troyer, J.E. Levin, G.A. Jacobs, Construction and analysis of a data base representing a neural map, Microsc. Res. Techn. 29 (1994) 329–343.

Available at
www.ComputerScienceWeb.com
POWERED BY SCIENCE @ DIRECT®

ELSEVIER

Neurocomputing 52–54 (2003) 53–58

NEUROCOMPUTING

www.elsevier.com/locate/neucom

Information processing during transient responses in the crayfish visual system☆

Christopher J. Rozell[a,*], Don. H. Johnson[a], Raymon M. Glantz[b]

[a] *Department of Electrical & Computer Engineering, Rice University, Houston, TX 77251-1892, USA*

[b] *Department of Biochemistry & Cell Biology, Rice University, Houston, TX 77251-1892, USA*

Abstract

We analyzed sustaining fiber responses in the crayfish visual system to light pulses using information processing techniques. The light pulse stimuli elicited a transient and a steady-state component in the EPSP input and in the firing rate of the spike train output. The overall information transfer of the system was very low ($\sim 10^{-4}$), with a sharp increase during the transient portion of the response followed by a steady decrease. The information transfer dynamics are consistent with a simple spike generator model that depends explicitly on stimulus changes. The present analysis also corroborates the observed light reflex behavior.

Keywords: Neural information processing; Spike generator efficiency; Crayfish visual system

1. Introduction

Sensory neurons generally encode some aspects of the information present in a stimulus and suppress others. Understanding and characterizing a sensorineural system's information processing capabilities in the presence of different stimuli is critical to progress in determining the coding structures underlying neural communication. Recently, a new theory of information processing (reviewed in Section 2) has been presented that draws from classical information theory and statistical signal processing to provide a framework for quantifying the information processing characteristics of a

☆ This work was supported by Grant MH60861 from the National Institute of Mental Health and by Grant CCR-0105558 from the National Science Foundation.

* Corresponding author. Rice University, 6100 Main Street MS 366, Houston 77025, USA.

E-mail address: crozell@rice.edu (C.J. Rozell).

0925-2312/03/$ - see front matter
doi:10.1016/S0925-2312(02)00829-9

general system [2]. We present an information processing analysis of crayfish sustaining fiber (SF) transient responses.

The retina of the crayfish compound eye has groups of photoreceptors projecting down the visual pathway in a 1:1 fashion to a column of non-spiking interneurons in the medulla. These retinotopically organized interneurons synapse on SF dendrites [3]. SF input EPSPs closely resemble graded and delayed versions of summed photoreceptor activity over a segment of the visual field. The SFs are critical from an information processing perspective because they are the first stage where an analog signal representing an outside stimulus is converted to a discrete spike train. From a signal processing point of view, the SFs represent a "front-end" for later information processing in higher centers of the brain. The fundamental problem that we address here is quantifying the efficiency (and dynamic properties) of the spike generation process.

2. Information processing analysis

The analysis framework presented in Ref. [2] is based on the concept of presenting a system with two different stimulus conditions (denoted α_0 and α_1). The input EPSP signals and output spike trains are stochastic processes (denoted X_α and Y_α, respectively, and depicted in Fig. 1) that have probability laws depending explicitly on the value of the stimulus. By examining how the input and output signals change when the stimulus values are changed, we can quantify how much information about that stimulus change is present in the signals. Because the signals are stochastic processes, we measure the distance between the probability laws that govern the input signals ($d_X(\alpha_0, \alpha_1)$) and the spike train signals ($d_Y(\alpha_0, \alpha_1)$) under both stimulus conditions. There are several information theoretic distances that measure changes in probability distributions. We use the Kullback–Leibler (KL) distance [1], which is relevant because it is the error decay rate for an optimal detector trying to distinguish the two stimuli conditions. The ratio of the output to input distances is termed the "information transfer ratio", $\gamma_{X,Y}(\alpha_0, \alpha_1) = d_Y(\alpha_0, \alpha_1)/d_X(\alpha_0, \alpha_1)$. The information transfer ratio is a single quantitative measure of the information processing efficiency of the system for the stimulus change under consideration. A fundamental result from information theory states that (for a large class of distances) the information transfer ratio will always be between zero and one ($0 \leqslant \gamma_{X,Y}(\alpha_0, \alpha_1) \leqslant 1$). A value of one represents perfect information transfer of the stimulus change and a value of zero represents total information loss.

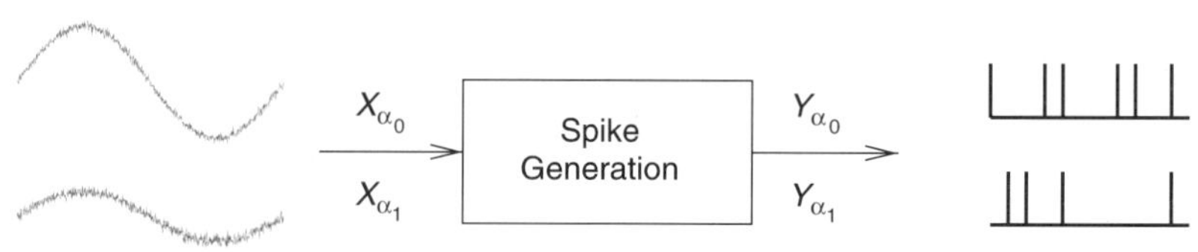

Fig. 1. The SF spike generator has an analog EPSP input (X_α) and a spike train output (Y_α) that both depend on the value of the input stimulus (α).

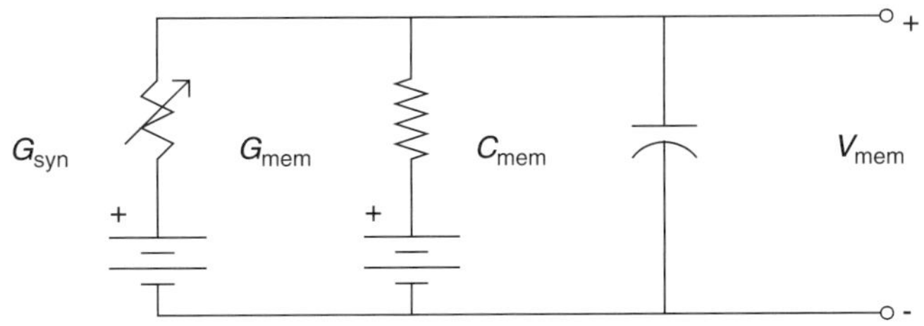

Fig. 2. The cell membrane acts as a time-varying, first-order lowpass filter.

In practice, the information transfer ratio is straightforward to estimate for the SFs. Two stimulus conditions (α_0, α_1) are presented repeatedly and the EPSP input and spike train output responses are captured through intracellular recording. The repeated trials are used to estimate the probability law governing the responses under each stimulus condition. Simple histogram estimators are formed from the binned spike responses to both stimuli and KL distances are easily calculated using these estimates. Distance estimation using spike responses has been discussed previously in Ref. [2]. The EPSPs are inherently analog signals, and a non-parametric estimate of the probability law is impossible. The compound EPSP in the SF spike generation area is the sum of many discrete events, suggesting that a Gaussian model may be a good match (which is true in our SF data). If the sampled EPSP responses under each stimulus condition are given as Gaussian random vectors with different means and variances ($X_{\alpha_0}\sim N(m_{\alpha_0}, K_{\alpha_0})$, $X_{\alpha_1}\sim N(m_{\alpha_1}, K_{\alpha_1})$), the KL distance between them will be approximately given by

$$d_X(\alpha_0, \alpha_1) \approx \frac{(m_{\alpha_0} - m_{\alpha_1})^t K_{\alpha_1}^{-1}(m_{\alpha_0} - m_{\alpha_1})}{2}. \qquad (1)$$

The mean vectors m_{α_0} and m_{α_1} are easily estimated by simply averaging the repeated EPSP responses under each stimulus condition. However, Eq. (1) indicates that we also need the term $K_{\alpha_1}^{-1}$. In short, we need to know not only the power but also the correlation structure of the membrane noise.

At rest, the biophysical properties of the impermeable cell membrane together with the embedded ion channels act electrically capacitor (C_{mem}) in parallel with a resting membrane conductance (G_{mem}). It is also known that EPSP variations are a result of a variable synaptic conductance (G_{syn}), and the relationship between the amount of the depolarization in the SF and the increase in membrane conductance has been measured by Waldrop and Glantz [5]. During the generation of an EPSP the membrane acts as the time-varying first-order lowpass filter shown in Fig. 2. Using the mean EPSPs (m_{α_0} and m_{α_1}) and the results given in Ref. [5], the time-varying increase in total conductance for the SF can be calculated. Techniques from time-varying linear system theory given an analytic expression for the correlation structure of the membrane noise and the KL distance between EPSP responses under the two stimuli conditions follows. More details about the membrane model and the correlation calculations can be found in Ref. [4].

3. Results

Several recordings were made from a crayfish SF when the stimulus was a square wave pulse of light with intensity varying from trial to trial. At the stimulus onset, the input EPSP exhibits a transient pulse of high potential for approximately 100 ms followed by a decrease to a plateau which is maintained for the duration of the stimulus. The output spike trains also exhibit a transient burst of high firing rate, followed by an adaptive response that reduces the firing rate to a steady-state value maintained for the stimulus duration. The procedures outlined in Section 2 were used to calculate KL distances between EPSP and spike responses for pairs of stimuli conditions. The time-varying information transfer ratios are shown for this dataset in Fig. 3, where the outer axis gives the log intensity of the stimuli values, and the inner plots show the time-varying information transfer ratio for a specific pair of stimuli. Note that the stimuli pair shown in the lower left corner represents the largest net stimulus change and the stimuli pairs along the diagonal (toward the upper right) have the smallest net stimulus change.

The first feature of Fig. 3 to notice is the scale of the plots. The information transfer ratio estimates have steady-state values of the order of 10^{-4}. The low value of the

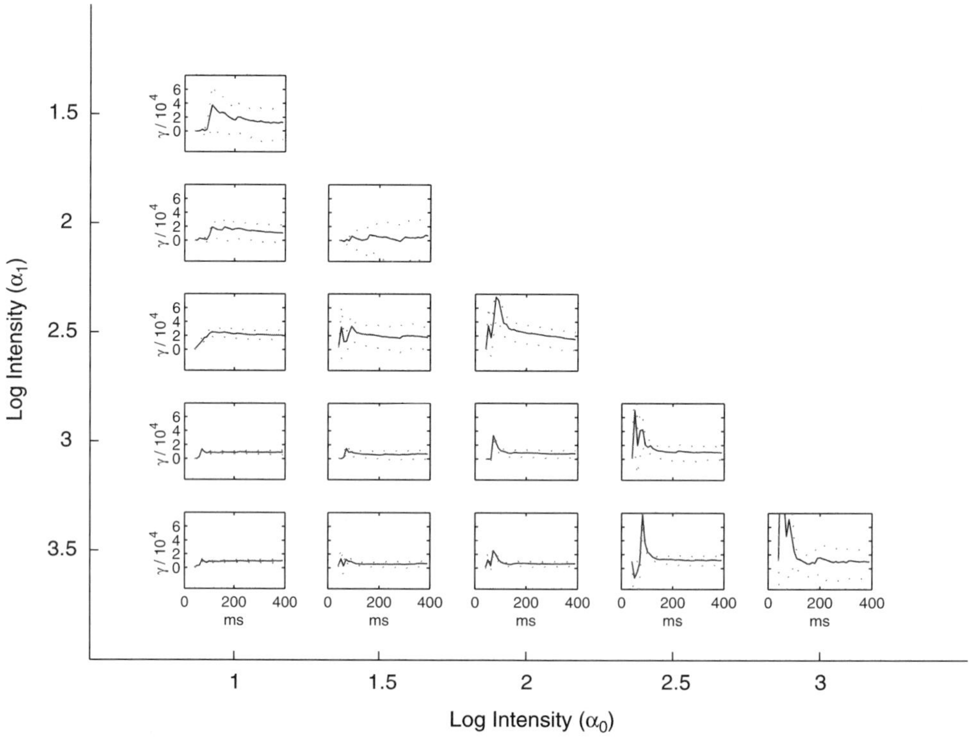

Fig. 3. Information transfer ratios for an SF. Outer axis gives the log intensity stimuli values, and inner plots show time-varying information transfer ratios for a specific pair of stimuli.

information transfer indicates that a large portion of the information about the stimulus change that is present in the analog EPSP inputs is not present in the spike train outputs. Very low information transfer values of the order seen here are also observed in other SF units and when changing other stimulus features (such as spatial frequency). A single spike train can only communicate information with each action potential, and the low information transfer ratios indicate that the low spike rates seen in crayfish SFs (often < 20 Hz) are inefficient compared to the high SNR analog EPSP inputs. The other feature noticeable in Fig. 3 is the general trend for information transfer ratios to often increase sharply after the stimulus onset followed by a slow decrease to a steady-state value. The sharp jump indicates that the system is transferring stimulus change information more efficiently when the spike rates burst to higher values, but the maximum communication rate is not sustainable when the rates fall back to steady-state levels.

We also used a simple spike generator model to investigate how information processing characteristics of a generic spike generator would depend on the properties of the encoding process. In our spike generator model, the input is a Gaussian random process ($X(t) \sim N(m(t), K)$) and the output is a Poisson random process ($Y(t) \sim Poiss(\lambda(t))$) where the rate depends on the input values $X(t)$. The output firing rate is given by an expression that depends explicitly on *changes* in the input:

$$\lambda(t) = G_s X(t) + G_t \frac{dX(t)}{dt}. \quad (2)$$

Using recorded SF inputs as $X(t)$ and estimating the steady-state and transient gains (G_s, G_t) to roughly match the firing rates seen in the SF, we are able to analytically calculate the information transfer ratio of the model system (detail of the calculations is given in Ref. [4]). When using a firing rate that explicitly depends on changes in the input, the information transfer ratio for the model (not shown) has a similar transient behavior to that observed in the SF data. The transient behavior is not a dominant feature when the output rates depend only on the value of the input (and not explicitly on input changes).

4. Conclusions

An information processing analysis of a crayfish SF responding to sudden-onset light stimuli is presented. The information transfer ratios are of the order of 10^{-4} regardless of the actual stimulus change, and reflect the inefficiency of a low-rate spike train in representing the analog EPSP signal. The information transfer ratio appears to have dynamic properties, with the SF transferring information more efficiently during the transient response. When also observing the elicited light reflex of the crayfish to the stimulus, all of the information from the stimulus producing the reflex behavior must be communicated through the SF during the first 100 ms after the stimulus onset. The reflex behavior is therefore also consistent with the dynamics of the information processing analysis presented here.

Finally, a simple spike generation model is presented where we can analytically calculate the information transfer ratio. When the output firing rates depend only on the value of the input, the information transfer ratio is not dominated by a transient and steady-state component. When the output firing rate also depends explicitly on the dynamic properties of the input, we observe the same time-varying behavior seen in the SF data. Therefore, the information transfer dynamic properties of the SF are consistent with a system that is tuned for detecting changes in a stimulus rather than absolute values.

References

[1] T.M. Cover, J.A. Thomas, Elements of Information Theory, Wiley, New York, NY, 1991.

[2] D.H. Johnson, C.M. Gruner, K. Baggerly, C. Seshagiri, Information-theoretic analysis of neural coding, J. Comput. Neurosci. 10 (2001) 47–69.

[3] M.D. Kirk, B. Waldrop, R.M. Glantz, The crayfish sustaining fibers, I. Morphological representation of visual receptive fields in the second optic neuropil, J. Comp. Physiol. 146 (1982) 175–179.

[4] C.J. Rozell, Analyzing dynamics and stimulus feature dependence in the information processing of crayfish sustaining fibers, Master's thesis, Rice University, Houston, TX, April 2002.

[5] B. Waldrop, R.M. Glantz, Synaptic mechanisms of a tonic EPSP in crustacean visual interneurons: analysis and simulation, J. Neurophysiol. 54 (3) (1985) 636–650.

Christopher J. Rozell received the M.S. in electrical engineering from Rice University in May 2002 and is currently in the Ph.D. program at Rice, with research interests in statistical signal processing applied to sensorineural systems.

Don H. Johnson received the S.B. and S.M. degrees in 1970, the E.E. degree in 1971 and Ph.D. in 1974, all in electrical engineering from the Massachusetts Institute of Technology (MIT). He is currently Professor and Chair of the Department of Electrical and Computer Engineering at Rice University. His current research areas are statistical signal processing and representation of information by neural signals.

Raymon M. Glantz received his doctorate from Syracuse University in 1967. After postdoctoral work at Caltech, he joined the faculty at Rice University in 1969, where he is currently Professor in the Department of Biochemistry and Cell Biology. His current research interests focus on information processing in the crayfish visual system.

Available at
www.ComputerScienceWeb.com
POWERED BY SCIENCE DIRECT®

ELSEVIER

Neurocomputing 52–54 (2003) 59–63

NEUROCOMPUTING

www.elsevier.com/locate/neucom

Stability analysis of entrainment by two periodic inputs with a fixed delay

Sorinel A. Oprisan, Carmen C. Canavier*

Department of Psychology, University of New Orleans, 2000 Lakeshore Dr., New Orleans, LA 70148, USA

Abstract

We were able to predict stable entrainments for an open-loop neural circuit in which an endogenously rhythmic neuron is driven by two periodic inputs separated by a fixed delay, using linear stability analysis about an assumed entrainment as well as a local linearization of the measured phase resetting curve. An example using Morris Lecar oscillators found multistable entrainments for a fixed delay. A fixed delay could be applicable to interneurons that exhibit post-inhibitory rebound bursting. This research is part of an effort to extend the applicability of our previously developed methods to analyze circuits that underlie central pattern generation.

Keywords: Phase resetting curve; Multistability; Delay

1. Introduction

Central pattern generating circuits often contain endogenous oscillators such as bursting neurons. One technique that is helpful in analyzing such circuits utilizes phase resetting curves, which can readily be generated for biological neurons. The existence of entrainment patterns can be predicted using periodicity constraints, and their stability can be determined by linearizing the phase resetting about an assumed entrainment and applying a perturbation analysis. Previous work by others dealt with entrainment by a single periodic input, or by self-entrainment of an oscillator with fixed delay [4,5,8].

Previous work from our group focused on circuits composed entirely of endogenously bursting neurons [1–3], and was restricted to cases in which a single input per cycle was received by each neuron, which limited the circuits analyzed to a ring architecture.

* Corresponding author. Tel.: +1-504-280-6775; fax: +1-504-280-6049.
E-mail address: ccanavie@uno.edu (C.C. Canavier).

0925-2312/03/$ - see front matter
doi:10.1016/S0925-2312(02)00744-0

We would like to extend these analyses to circuits in which the component neurons are not restricted to a single input per cycle, and are not restricted to endogenous bursters. As a first step, we examined the simplest possible circuit with two inputs, beginning with an open-loop circuit in which one bursting neuron drives another, but adding a nonoscillatory neuron that also synapses on the driven neuron, and produces a burst with a fixed latency after receiving a burst from the driver neuron.

2. Methods

As in our previous analyses, certain assumptions must be made in order to make the problems tractable. First, we constrained the delay plus the total duration of the two input bursts to be less than a cycle period. We assumed that the only effect of a perturbation was to move the limit cycle trajectory backwards or forwards tangentially along its path, producing either an advance or a delay. Specifically, we assumed that by the time the driven oscillator receives the second burst, it has returned sufficiently close to its unperturbed limit cycle so that the phase resetting curve still predicts the effect that a burst will produce. We also assumed that a perturbation only affected the cycle in which it received, and therefore did not affect subsequent bursts.

3. Results

The circuit contains an intrinsic oscillator (see Fig. 1) and a forced oscillator that receives two inputs: first via synaptic input from entraining neuron and the second via a delay loop with δ time constant.

The single-pulse PRC, $F^{(1)}(\varphi)$, which is a measure of the relative temporal advance/delay of the next crossing in the Poincare section when an oscillator is

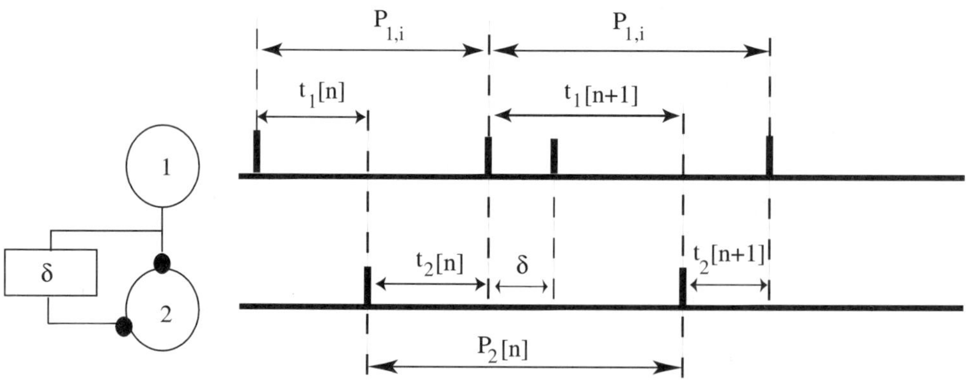

Fig. 1. The two neuron circuit (left panel) and the dynamics of the inhibitory perturbations received by the second neuron.

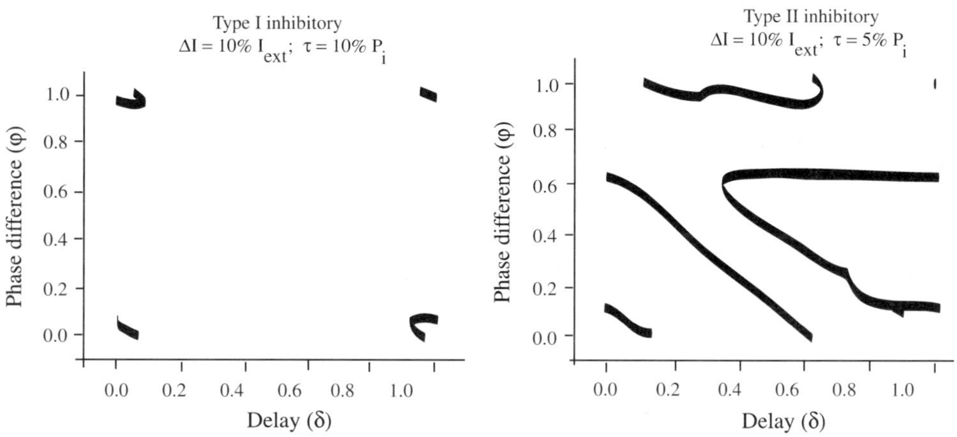

Fig. 2. The steady phase difference between the two neurons displays a stable synchronous oscillatory state for the case of two Type I Morris–Lecar model neurons coupled as seen in Fig. 1 (left panel). For the case of Type II model neurons the system shows multiple steady phase differences (right panel).

perturbed, determines the amount of phase resetting induced by two (or more) current pulses acting during the same cycle:

$$F^{(2)}(\varphi,\delta) = F^{(1)}(\varphi) + F^{(1)}(\varphi + F^{(1)}(\varphi) + \delta), \quad (1)$$

where P_i is the intrinsic period of the oscillation and δ is the normalized elapsed time between the two applied pulses. The condition $\delta < 1$ ensures that both current pulses act during the same cycle [6].

Based on multiple-inputs per cycle PRC (1) when the delay loop is present (see Fig. 1) then the entrainment timing is given by

$$t_1[n] + t_2[n] = P_{1,i},$$

$$t_2[n] + t_1[n+1] = P_2[n], \quad (2)$$

where $P_{1,i}$ is the intrinsic period of the first oscillator and $P_2[n] = P_{2,i}(1 + F^{(2)})$ is the two-inputs per cycle PRC for the second oscillator in the network. By eliminating t_2 it is possible to determine a first-order nonlinear map for the entrainment induced by the first oscillator. The steady phase difference between the two oscillators φ^* is given by

$$P_{1,i} = P_{2,i}(1 + F^{(1)}(\varphi^*) + F^{(1)}(\varphi^* + F^{(1)}(\varphi^*) + \delta)).$$

The Lyapunov exponent for the map derived from (2) is

$$\lambda = 1 - F'^{(1)}(\varphi^*) - F'^{(1)}(\varphi^* + F^{(1)}(\varphi^*) + \delta)(1 + F'^{(1)}(\varphi^*)).$$

Two identical mutually inhibitory Morris Lecar oscillators were used to test the theoretical results described above. Numerical simulations were run for a range of values of fixed delay. Fig. 2 illustrates the results of these simulations. The dark curves show stable entrainments.

For parameters in the Type I regime [6], only delays that allowed the two neurons to burst nearly simultaneously were stable. Multistability is evident when there is more than one stable pattern at a fixed delay. The multistability in this case results in part from symmetry, in that the two presynaptic bursts are identical and interchangeable. For Type II [6], more complex multistability results at a greater range of delay values. We found by numeric integration of the evolution equations that the predicted stable modes agreed with the observed stable modes.

4. Discussion

This work is a continuation and extension of a line of research in which phase resetting curves are used to analyze the stability of patterns of entrainment. This circuit is simpler than others we have studied in that there is no feedback. However, it is more complex in that a single neuron receives more than one input in a cycle, and in fact the inputs were allowed to overlap. For perturbations that cause a significant normal departure from the limit cycle, perturbations may not add in this simple fashion, for example if they cause switching to the other side of the limit cycle via a normal path [7].

The next step in our research progression is to incorporate delays into closed circuits with feedback, such as those we have previously analyzed without delay, in hopes of extending our methods to real CPGs that contain both endogenously bursting neurons and others that might be simulated as producing a burst after a delay.

Acknowledgements

This research was supported by NSF IBN-0118117 (CCC) grant.

References

[1] C.C. Canavier, D.A. Baxter, J.W. Clark, J.H. Byrne, Control of multistability in ring circuits of oscillators, Biol. Cybern. 80 (1999) 87–102.

[2] C.C. Canavier, R.J. Butera, R.O. Dror, D.A. Baxter, J.W. Clark, J.H. Byrne, Phase response characteristics of model neurons determine which patterns are expressed in a ring circuit model of gait generator, Biol. Cybern. 77 (1997) 367–380.

[3] R.O. Dror, C.C. Canavier, R.J. Butera, J.W. Clark, J.H. Byrne, A mathematical criterion based on phase response curves for the stability of a ring network of oscillators, Biol. Cybern. 80 (1999) 11–23.

[4] J. Foss, A. Longtin, B. Mensour, J. Milton, Multistability and delayed recurrent ls, Phys. Rev. Lett. 76 (1996) 708–711.

[5] J. Foss, F. Moss, J. Milton, Noise, multistability and delayed recurrent loops, Phys. Rev. E 55 (1997) 4536–4543.

[6] S.A. Oprisan, C.C. Canavier, The influence of limit cycle topology on the phase resetting curve, Neural Comput. 14 (2002) 1027–1057.

[7] S.A. Oprisan, V. Thirumalai, E. Marder, C.C. Canavier, Phase resetting in neural oscillators as a component of circuit analysis. Proceedings of the Second Joint EMBS/BMES Conference, Houston, TX USA, October 23–26, 2002, pp. 1960–1961.

[8] D.H. Perkel, J.H. Schulman, T.H. Bullock, G.P. Moore, J.P. Segundo, Pacemaker neurons: effect of regularly spaced synaptic input, Science 145 (1964) 61–63.

Oprisan received his B.Sc. in Physics from Alexandru Ioan Cuza University of Iasi, Romania, in 1987, and his Ph.D. in Theoretical Physics from the same University in 1998. He was a postdoctoral fellow and is currently a Research Specialist in the Department of Psychology at the University of New Orleans. He is also an Associate Professor at Alexandru Ioan Cuza University of Iasi, Romania. His interest include the applied nonlinear dynamics and chaos theory, statistical physics for complex systems, self-organizing systems and computational neuroscience.

Canavier received her B.E. in Chemical Engineering from Vanderbilt University in 1979, and her Ph.D. in Electrical Engineering and Computer Engineering from Rice University in 1991. She was a postdoctoral fellow and later a Research Assistant Professor in the Department of Neurobiology and Anatomy at the University of Texas Medical School in Houston. She is currently an Associate Professor in the Department of Psychology at the University of New Orleans. Her interests include the nonlinear dynamics of bursting neurons, particularly midbrain dopamine neurons, and of networks of neurons, particularly central pattern generators for locomotion located in the spinal cord, and how these dynamics ultimately contribute to behavior.

Available at www.ComputerScienceWeb.com
POWERED BY SCIENCE DIRECT®

ELSEVIER

Neurocomputing 52–54 (2003) 65–71

NEUROCOMPUTING

www.elsevier.com/locate/neucom

Burst dynamics under mixed NMDA and AMPA drive in the models of the lamprey spinal CPG

Alexander Kozlov[a,*], Anders Lansner[a], Sten Grillner[b]

[a] *Studies of Artificial Neural Systems, Department of Numerical Analysis and Computer Science, Royal Institute of Technology, Stockholm SE-10044, Sweden*

[b] *Nobel Institute for Neurophysiology, Department of Neuroscience, Karolinska Institute, Stockholm SE-17177, Sweden*

Abstract

The spinal CPG of the lamprey is modeled using a chain of nonlinear oscillators. Each oscillator represents a small neuron population capable of bursting under mixed NMDA and AMPA drive. Parameters of the oscillator are derived from detailed conductance-based neuron models. Analysis and simulations of dynamics of a single oscillator, a chain of locally coupled excitatory oscillators and a chain of two pairs of excitatory and inhibitory oscillators in each segment are done. The roles of asymmetric couplings and additional rostral drive for generation of a traveling wave with one cycle per chain length in a realistic frequency range are studied.

Keywords: Locomotion; Spinal cord; Lamprey; Modeling

1. Introduction

The lamprey swims by means of body undulations thus forming a quasisinusoidal wave of a single wavelength that travels along the body from head to tail [3]. This traveling wave is generated by local oscillators in the range of 0.1–10 Hz with intersegmental phase lag that is nearly independent of cycle period. Thus for 100 segments in the lamprey, the intersegmental phase lag is stabilized around 1% of the cycle. In experiments, the brainstem and spinal cord can be maintained in vitro while activity in the locomotor CPG is induced by electric or pharmacological stimulation. Patterns

* Corresponding author. Tel.: +46-8-7906903; fax: +46-8-7900930.
E-mail address: akozlov@nada.kth.se (A. Kozlov).

0925-2312/03/$ - see front matter
doi:10.1016/S0925-2312(02)00795-6

emerging during this "fictive swimming" are similar in their characteristics to the ones observed in vivo.

Our goal in the current study is to suggest through modeling a plausible implementation of phase and frequency control in the spinal network of lamprey. To this end, we model calcium dynamics in interneurons under mixed NMDA and AMPA drive and explore its consequences for swimming rhythm generation.

2. Description of the cell model

The oscillator equations comprise leakage, NMDA and AMPA inward currents activated synaptically and from the bath, and K_{Ca} outward current. Calcium enters the unit through NMDA channels. The model is developed by a gradual, step-by-step reduction and simplification of detailed conductance-based models of cells and synapses which have been developed earlier [1,2,4,9]. Thus the oscillator model is sufficiently simple for limited mathematical analysis and fast simulations, and still keeps direct correspondences of its parameters and state variables to characteristics measured experimentally. This makes this model different from previous simplified models [5,7,10].

Here, the dynamics of the membrane potential v and the intracellullar calcium concentration u is modeled in dimensionless form as

$$\mathrm{d}v/\mathrm{d}t = i - v + i_a + i_n + i_k + i_{sa} + i_{sn} + i_{sg}, \qquad \mathrm{d}u/\mathrm{d}t = \varepsilon(-u + i_{cn}),$$

$$i_a = a(e_a - v), \quad i_n = np(v)(e_n - v), \quad i_k = ku(e_k - v),$$

$$i_{cn} = \beta np(v)(e_{cn} - v) + \beta i_{scn}, \quad p(v) = 1/(1 + 0.014\mathrm{e}^{(1-v)/0.12}), \tag{1}$$

where i is the injected current, a and n are the conductances of the bath-activated AMPA and NMDA channels, $k = 4.5$ is the conductance of the calcium-dependent potassium channel, $e_a = 1, e_n = 1, e_k = -0.14, e_{cn} = 1.29$ are the reversal potentials, β is the calcium influx rate, $\varepsilon = 0.016$ is the reciprocal of the calcium time constant, and $p(v)$ is the opening function of the voltage-dependent magnesium block. Synaptic input has two excitatory components, i_{sa} and i_{sn} (AMPA and NMDA, respectively), and one inhibitory component, i_{sg} (glycin),

$$i_{sa} = s_a h(v_{\mathrm{pre}})(e_a - v),$$

$$i_{sn} = s_n h(w) p(v)(e_n - v), \quad \mathrm{d}w/\mathrm{d}t = (v_{\mathrm{pre}} - w)/\tau_n,$$

$$i_{sg} = s_g h(v_{\mathrm{pre}})(e_g - v), \quad h(v) = 1/(1 + \mathrm{e}^{(0.3-v)/0.1}), \tag{2}$$

where v_{pre} and v are the pre- and post-synaptic membrane potentials, s_a, s_n, s_g are the synaptic conductances, $e_a = 1, e_n = 1, e_g = -0.06$ are the reversal potentials, τ_n is the NMDA receptor activation time constant, and $h(v)$ is the synaptic transfer function. Synapse contribution to calcium influx is modeled by the current

$$i_{scn} = s_n h(w) p(v)(e_{cn} - v). \tag{3}$$

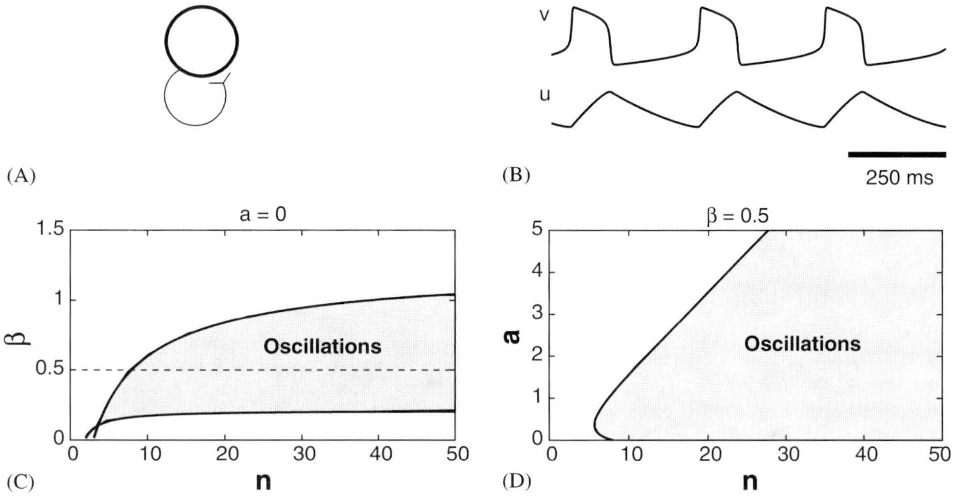

Fig. 1. Single cell dynamics. (A) A cell with recurrent excitation. (B) Typical waveforms of the membrane potential, v, and the intracellular calcium concentration, u. (C, D) Bifurcation diagrams for model (1) without recurrent excitation. Regions of periodic oscillations are shaded.

The cell model (1)–(3) is used for network simulations. All oscillators are identical, only bath drives and synaptic couplings are varied.

3. Results

3.1. Single cell dynamics

The single cell without recurrent excitation is a potential burster under bath NMDA drive. AMPA application controls the frequency of NMDA-induced oscillations so the burst frequency gradually increases with the increase of conductance a or the proportional increase of both n and a. The oscillations occur in a limited range of β, as shown in Fig. 1, so the calcium influx can be neither too high nor too low.

Recurrent excitation increases cell excitability. It shifts the region of oscillations towards smaller values of n so for the strong enough recurrent excitation the oscillations can be observed even without bath NMDA drive (not shown).

3.2. Phase locking

Excitatory synapses are able to synchronize nonidentical oscillators for sufficiently strong reciprocal coupling as shown in Fig. 2A. The oscillator with the higher intrinsic

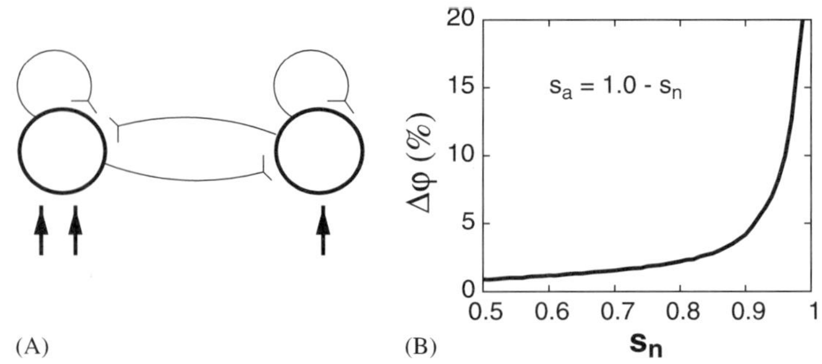

Fig. 2. Synchronization of reciprocally coupled nonidentical oscillators. (A) The left oscillator has 10% more bath AMPA drive. (B) Dependence of the phase difference between membrane potentials of the two oscillators, in percents of the cycle duration, on NMDA and AMPA synaptic conductances, s_n and s_a.

frequency due to the extra bath AMPA drive entrains the slower one so their phases get locked with the phase of the former leading. The phase difference is smaller for stronger couplings and it tends to be larger if the NMDA component dominates as shown in Fig. 2B.

3.3. E-network

A linear chain of oscillators with excitatory couplings to nearest neighbours and recurrent excitation, the E-network, models the dynamics of a piece of spinal cord split along the mid-line (L. Cangiano and S. Grillner, pers. comm.).

As shown in Fig. 3 an E-network with 100 symmetrically coupled oscillators generates forward swimming patterns in the range of 2–6 Hz with intersegmental phase lag between 0.8% and 1.3% if an additional AMPA drive is applied to the most rostral segment of the chain. This implements the 'trailing oscillator' hypothesis suggested earlier [8].

3.4. EI-network

This network consists of two E-networks reciprocally coupled via inhibitory cells as shown in Fig. 4. The EI-network features prominent descending rostro-caudal asymmetry of contralateral inhibitory connections. It is found that this asymmetry is sufficient for generation of traveling waves even without extra rostral drive as shown in Fig. 5A. Results are verified in large-scale simulations [6] as shown in Fig. 5B. This network maintains a one cycle per chain ("body") length phase lag for higher frequencies most efficiently. In the low frequency range, additional rostral drive helps to correct the intersegmental phase lag.

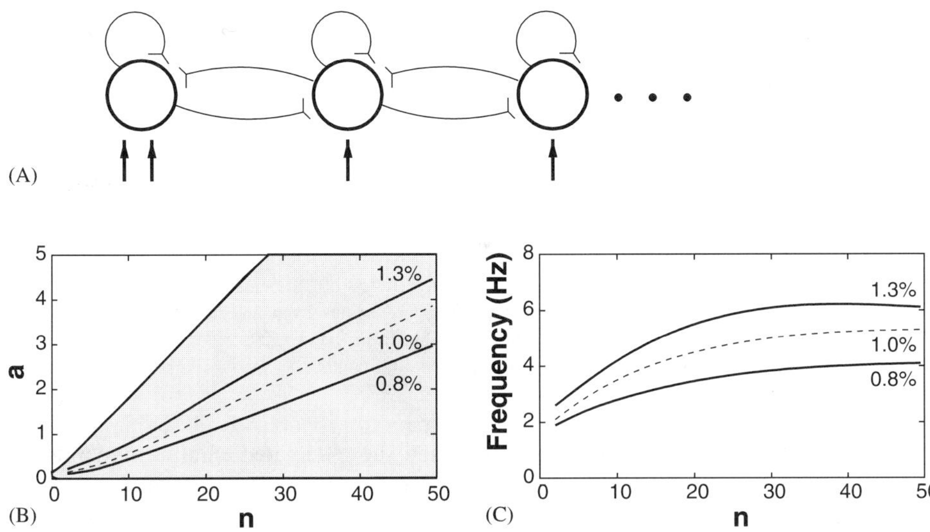

Fig. 3. Coordination of intersegmental phase lag in the chain of excitatory oscillators. (A) The most rostral oscillator has 2% more bath AMPA drive. (B) Oscillations occur within the shaded area. Lines in the region of oscillations show values of n and a corresponding to travelling waves of excitation with intersegmental phase lags of 0.8%, 1.0%, and 1.3%. Frequency of the traveling waves changes along the lines and varies between 2 and 6 Hz (C).

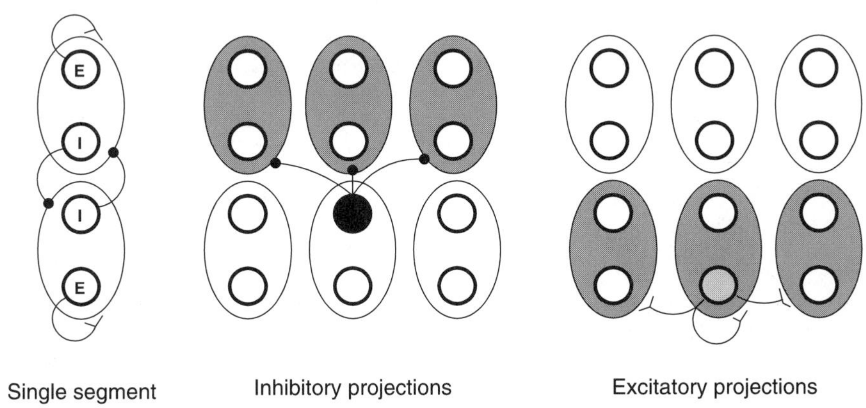

Fig. 4. Configuration of EI-network. One excitatory, E, and one inhibitory, I, oscillators build a hemisegment; two symmetric hemisegments comprise a segment. There are 100 segments in the spinal cord. Excitatory projections are symmetric, inhibitory projections are 50% weaker in the rostral direction.

4. Conclusion

Chains of coupled nonlinear network oscillators are shown to reproduce important aspects of the swimming pattern generation in the lamprey spinal CPG. Currents

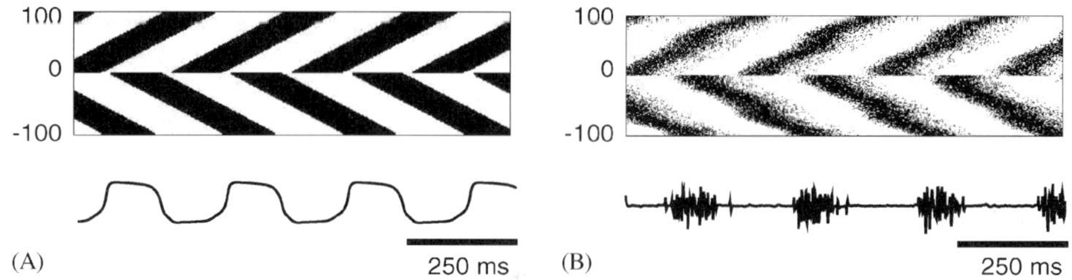

Fig. 5. Comparison of oscillations in simplified (A) and detailed (B) models of the spinal CPG of the lamprey. Time varies along the horizontal line. Excitation propagates from anterior to posterior along the left (from segment 0 to hemisegment −100) and the right (from 0 to +100) sides of the spinal cord in out-of-phase manner. Both models generate one cycle per body length.

depending on activation of NMDA receptors and the associated calcium dynamics are key determinants of the characteristics of these oscillations. Both E- and EI-networks are capable of producing an adequate swimming coordination, with the EI-configuration being most efficient.

References

[1] L. Brodin, H. Tråvén, A. Lansner, P. Wallén, Ö. Ekeberg, S. Grillner, Computer simulations of N-methyl-D-aspartate (NMDA) receptor induced membrane properties in a neuron model, J. Neurophysiol. 66 (1991) 473–484.

[2] O. Ekeberg, P. Wallén, A. Lansner, H. Tråvén, L. Brodin, S. Grillner, A computer based model for realistic simulations of neural networks I: The single neuron and synaptic interaction, Biol. Cybern. 65 (1991) 81–90.

[3] S. Grillner, T. Deliagina, Ö. Ekeberg, A. El Manira, R.H. Hill, A. Lansner, G.N. Orlovsky, P. Wallén, Neural networks that co-ordinate locomotion and body orientation in lamprey, TINS 18 (1995) 270–279.

[4] J. Hellgren Kotaleski, S. Grillner, A. Lansner, Neural mechanisms potentially contributing to the intersegmental phase lag in lamprey, I. Segmental oscillations dependent on reciprocal inhibition, Biol. Cybern. 81 (1990) 317–330;
J. Hellgren Kotaleski, S. Grillner, A. Lansner, Neural mechanisms potentially contributing to the intersegmental phase lag in lamprey, II. Hemisegmental oscillations produced by mutually coupled excitatory neurons, Biol. Cybern. 81 (1990) 299–315.

[5] N. Kopell, G.B. Ermentrout, Coupled oscillators and the design of central pattern generators, Math. Biosci. 90 (1988) 87–109.

[6] A. Kozlov, A. Lansner, S. Grillner, Large-scale models of the locomotor spinal network of lamprey, in: Abstr. Computational Neuroscience (CNS*2001), June 30–July 5 2001, San Francisco and Pacific Grove, CA, p. 69.

[7] A. Lansner, Ö. Ekeberg, S. Grillner, Realistic modeling of burst generation and swimming in lamprey, in: P.S.G. Stein, S. Grillner, A.I. Selverston, D.G. Stuart (Eds.), Neurons, Networks, and Motor Behavior, MIT Press, Tucson, USA, 1997, pp. 165–171.

[8] T. Matsushima, S. Grillner, Intersegmental coordination of undulatory movements—a 'trailing oscillator' hypothesis, Neuroreport 1 (1990) 97–100.

[9] H.G. Tråvén, L. Brodin, A. Lansner, Ö. Ekeberg, P. Wallén, S. Grillner, Computer simulations of NMDA and non-NMDA receptor-mediated synaptic drive: sensory and supraspinal modulation of neurons and small networks, J. Neurophysiol. 70 (1993) 695–709.

[10] M. Ullström, J. Hellgren Kotaleski, J. Tegnér, E. Aurell, S. Grillner, A. Lansner, Activity dependent modulation of adaptation produces a constant burst proportion in a lamprey neural network model, Biol. Cybernet. 79 (1998) 1–14.

Alexander Kozlov received his Diploma in Radiophysics and Ph.D. in Physics and Mathematics from the University of Nizhny Novgorod, Russia, in 1989 and 1995, respectively. His research in Kungliga Tekniska Högskolan, Stockholm, includes dynamics of control and pattern generating nervous systems, realistic neuron modeling and large-scale simulations.

Anders Lansner received his Ph.D in Computer Science from the Royal Institute of Technology in 1986. He was appointed Professor in Computer Science in 1999 and is manager of the SANS group (Studies of Artificial Neural Systems) founded in 1987. His research interests range from abstract neural network computation to biophysically detailed computational models of specific biological systems and processes.

Sten Grillner is a Professor of the Department of Neuroscience, head of the Division of Neurophysiology and Behavior (The Nobel Institute for Neurophysiology) of Karolinska Institute, Stockholm. His research interest is directed towards an understanding of certain aspects of the cellular and network level and how the different types of neurons combine to form networks which underlie basic modes of coordination such as control of body orientation and locomotion.

Available at
www.ComputerScienceWeb.com
POWERED BY SCIENCE @ DIRECT®

ELSEVIER

Neurocomputing 52–54 (2003) 73–77

NEUROCOMPUTING

www.elsevier.com/locate/neucom

Correlated neuronal activity can represent multiple binding solutions

Alexander Maye*

TU Berlin, Department of Electrical Engineering and Computer Science, Computer Science in Sciences and Engineering Group, Berlin, Germany

Abstract

The correlated activity of neurons is being regarded as candidate mechanism for binding the responses of different feature selective cells. Binding, however, is rarely unique, especially at levels beyond feature detection. In this article the dynamical properties of a network of coupled oscillators as a model for a mental representation of visual stimuli are investigated. Oscillator networks have been analyzed previously, but here a new coupling scheme and a biologically more relevant transfer function are used. A principal mode analysis of the activity shows that the network sensibly handles ambiguous segmentations of a visual stimulus.

Keywords: Oscillator network; Binding; Principal modes; Segmentation

1. Introduction

Multi-electrode recordings have revealed stimulus induced synchronization of cortical neurons. In addition to rate codes, the temporal correlation in the activity of neurons constitutes a coding dimension, in which the responses of cells, selective for features like edges, disparity, etc., can be dynamically integrated [5].

Oscillator networks are a popular model to explain these observations. They have been used to model, e.g., contour integration and enhancement in the visual cortex [1], odor recognition [2] and binding across different feature dimensions [4]. Most of the investigations focus on the suppression of noise, spurious edges and other artefacts.

* Corresponding author. Konrad-Zuse-Center for Scientific Computing, Takustr 7, Berlin 14195, Germany.
E-mail addresses: maye@zib.de, maye@cs.tu-berlin.de (A. Maye).

0925-2312/03/$ - see front matter
doi:10.1016/S0925-2312(02)00806-8

This implies that the suppressed information could not be perceived. However, in a noisy image one can recognize the scene, but also single spots. In general, inputs permit different interpretations, depending on the context. For example, two parallel bars of the same color can be considered as two objects or as part of a grating. Other examples are flip figures like Rubin's vase or the Neckar cube.

Here it is shown that the dynamics of a network of coupled oscillators simultaneously represents alternatives.

2. Model

On the macroscopic level, the dynamics of the mean activations of two mutually connected groups of excitatory (x) and inhibitory (y) neurons can be described by

$$\dot{x} = -\tau_x x - g_y(y) + L_0^{xx} g_x(x) + I_x + \eta_x, \tag{1a}$$

$$\dot{y} = -\tau_y y + g_x(x) - I_y + \eta_y. \tag{1b}$$

Here, τ_α $(\alpha \in \{x, y\})$ are constants that can be chosen to match refractory times of biological neurons, L_0^{xx} describes self-excitation of the excitatory population and η_α white noise, that models fluctuations within the populations. With external input I_α above threshold, the solutions of (1) are limit-cycle oscillations.

The transfer function is typically modeled by sigmoid-shaped functions. From a metabolistic viewpoint it is not desirable for real neurons to reach saturating activity in every cycle. In this model a semi-linear transfer function,

$$g_\alpha(x) = \begin{cases} m_\alpha(x - \theta_\alpha) & \text{if } x > \theta_\alpha, \\ 0 & \text{else} \end{cases} \tag{2}$$

was used to prevent neurons from reaching saturation. Conditions, under which the dynamics of (1) is stable, analytically have been derived in [7].

The external input $I_\alpha = I_0 + I_{\text{lat}}^\alpha$ consists of static input from the feature detectors I_0 and input from neighboring oscillators via lateral connections I_{lat}^α. A preliminary study of the phase plane of (1) revealed, that the excitatory coupling $I_{\text{lat}}^x = \sum_{s \in S} L_s^{xx} x_s$ synchronizes oscillators, whereas the inhibitory coupling $I_{\text{lat}}^y = \sum_{s \in S} L_s^{yy} y_s$ desynchronizes them. S is the local neighborhood of the oscillator under consideration and $L^{\alpha\beta}$ $(\alpha, \beta \in \{x, y\})$ positive weights. Generally it was found, that connections originating from excitatory neurons have a synchronizing and those from inhibitory neurons a desynchronizing effect, so all connection types could have been used. Connections between excitatory and inhibitory neurons as well as the recurrent coupling of the inhibitory neuron were left out to keep the parameter space small. The static input was applied to the excitatory neurons exclusively. Applying it to the inhibitory neurons also does not change the behavior qualitatively [1].

To demonstrate the dynamic binding of visual stimuli within the color domain, a two-dimensional lattice was used (see Fig. 1), that represents a topographic mapping of the image. At every location, a hyper-column of four oscillators encodes the spectral

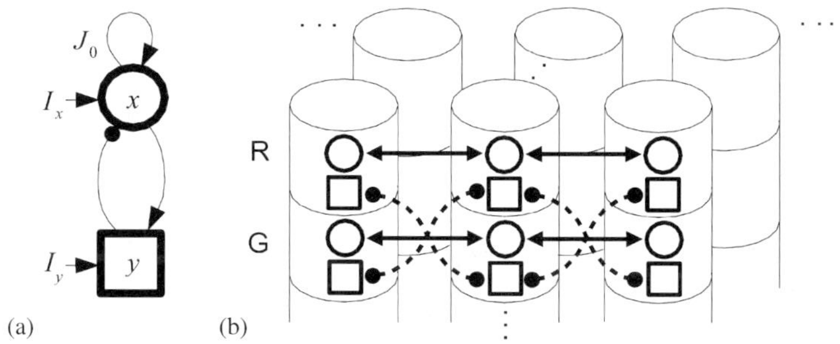

Fig. 1. (a) A single oscillator. (b) Synchronizing (solid) and desynchronizing (dashed) connections between neighboring oscillators. R and G denote the red and green channel; the scheme has to be applied to other neighbors and the remaining channels as well.

properties. The properties of the color processing system are still discussed controversially, especially the actual number of color channels [3]. Here the channels red, green, blue and yellow were chosen.

The weight structure was isotropic and local. Oscillators receiving input from the same channel were coupled with synchronizing connections, those from different channels with desynchronizing connections. The connection scheme reflects the Gestalt principles of proximity and similarity.

The simplicity of the input representation allows a straightforward application to other feature dimensions like luminance or disparity.

3. Results

The stimulus in Fig. 2(a) was applied to the network, and the system of equations (1) solved numerically. In order to show the relation to biological measurements, a cross-correlation analysis for the neurons with their receptive fields at the numbered positions in Fig. 2(a) was carried out. The results are shown in Figs. 2(b) and (d). Cross-correlation, however, displays the dynamic properties of the network only locally.

From synergetics it is well known that the dynamics of complex systems is often governed by a few dominating states and can therefore be described by a small set of the corresponding order parameters. These states are the eigen- or principal modes of the system, the corresponding eigenvalues designate, how much of the variance is accounted for by that mode.

The principal modes have been computed as the eigenvectors of the covariance matrix $\mathbf{C} = \mathbf{x}^{\mathrm{T}}\mathbf{x}$, where $\mathbf{x}$ is a matrix with the activities of the excitatory neurons after a transient phase. The modes were ordered by descending eigenvalues (Fig. 2(d)).

To assess the temporal evolution of the principal modes, the network state at any instant is considered as a weighted superposition of its principal modes $\mathbf{v}_i$, i.e. $\mathbf{x}(t) = \sum a_i(t)\mathbf{v}_i$. The time course of the order parameters $a_i(t)$ is shown in Fig. 2(f).

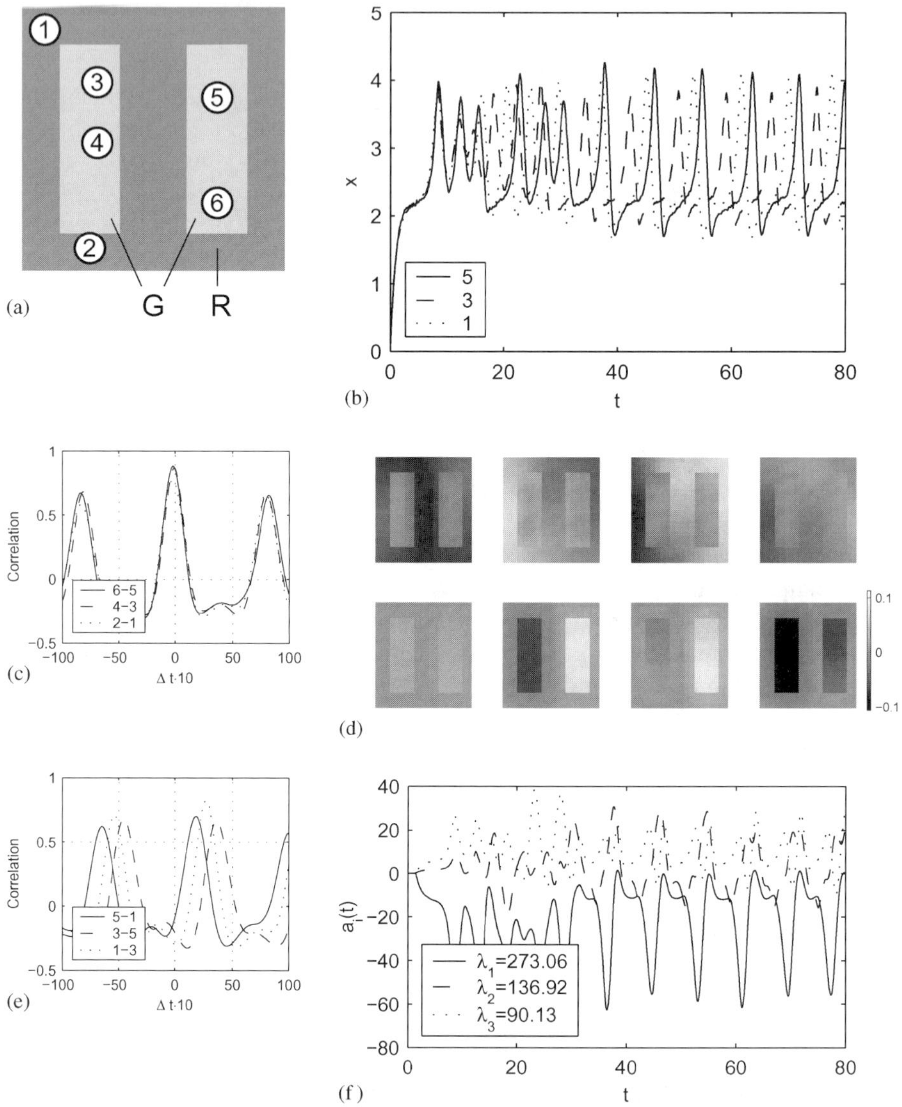

Fig. 2. (a) Stimulus. The bar displays the diameter of the neighborhood of an oscillator (S). (b) Activity $x(t)$ of three oscillators. (c) Cross-correlograms of oscillators in the same segment and in different segments (e). (d) Principal modes with the four largest eigenvalues from left to right. Only the R (top) and G (bottom) channel are shown, since the other channels receive no input from the stimulus. (f) Temporal evolution of the order parameters for the first (solid) and second (dashed) mode. Parameters: $\tau_x = \tau_y = 1$, $m_x = m_y = 2$, $\theta_x = 2$, $\theta_y = 1$, $I_0 = 2$, $L_0^{xx} = 0.1$, $\sum_{s \in S} L_s^{xx} = 1.1$, $\sum_{s \in S} L_s^{yy} = 0.3$.

4. Discussion

After a transient phase the network exhibits stable oscillatory behavior. Oscillators that represent the same segment synchronize with approximately zero phase lag, whereas those for different segments have a phase lag different from zero. This represents the two bars being different objects.

The first principal mode exhibits the difference between the foreground (the two bars) and the background. This mode considers the bars to belong to the same segment. The second mode differentiates between the bars.

The influence of these two modes, expressed by the evolution of their order parameters a, alternates (Fig. 2f). The amplitudes of the order parameters for principal modes with lower eigenvalues are smaller. Therefore their influence on the network dynamics is smaller as well. Experiments with other stimuli yielded similar results.

According to [6], clusters of synchronously activated cells can be interpreted as a logically structured conceptual representation of a visual scene. The attractivity of this theory lies in the fact that no "higher centers" and no computational facilities other than synchronized and desynchronized cells are required to achieve complex syntactic structure. Since the model developed here meets the assumptions of this theory, it gives an idea of how visual perceptions can be represented. The ability to account for ambiguous segmentations supports this view.

References

[1] Z. Li, A neural model of contour integration in the primary visual cortex, Neural Comput. 10(4) (1998) 903–940.
[2] Z. Li, J. Hertz, Odour recognition and segmentation by a model olfactory bulb and cortex, Comput. Neural Systems 11 (2000) 83–102.
[3] B. Saunders, J. van Brakel, Are there nontrivial constraints on colour categorization? Behav. Brain Sci. 20 (1997) 167–228.
[4] T.B. Schillen, P. König, Binding by temporal structure in multiple feature domains of an oscillatory neural network, Biol. Cybernet. 70 (1994) 397–405.
[5] W. Singer, Neuronal synchrony: a versatile code for the definition of relations? Neuron 24 (1999) 49–65.
[6] M. Werning, Synchrony and composition: toward a cognitive architecture between classicism and connectionism, in: B. Loewe, W. Malzkorn, T. Raesch (Eds.), Foundations of the Formal Sciences II: Applications of Mathematical Logic in Philosophy and Linguistics, Kluwer Academic Publishers, Dordrecht, 2002.
[7] H. Wersing, W.-J. Beyn, H. Ritter, Dynamical stability conditions for recurrent neural networks with unsaturating piecewise linear transfer functions, Neural Comput. 13 (2001) 1811–1825.

Available at
www.ComputerScienceWeb.com
POWERED BY SCIENCE DIRECT

ELSEVIER

Neurocomputing 52–54 (2003) 79–87

NEUROCOMPUTING

www.elsevier.com/locate/neucom

Short-term synaptic dynamics promote phase maintenance in multi-phasic rhythms

Farzan Nadim[a,b,*], Victoria Booth[a], Amitabha Bose[a], Yair Manor[c]

[a]*Department of Mathematical Sciences, NJIT, Newark, NJ 07102, USA*
[b]*Department of Biological Sciences, Rutgers University, 101 Warren St., Newark, NJ 071021, USA*
[c]*Life Science Department, Zlotowski Center for Neurosciences, Ben-Gurion University of the Negev, Beer-Sheva 84105, Israel*

Abstract

We show that in an inhibitory rhythmic network synaptic depression promotes phase constancy. As cycle period increases, the synapse recovers from depression and becomes more effective in delaying the postsynaptic cell. As a result, the delay between the pre- and postsynaptic bursts increases as cycle period increases. We discuss the dependence of the bursting phase of the postsynaptic cell on the strength and kinetics of the depressing synapse.

Keywords: Synaptic dynamics; Phase maintenance; Multi-phasic rhythms

1. Introduction

Rhythmic activities are commonly produced over a wide range of cycle frequencies. Often, these rhythmic patterns involve activities of multiple units at different times within each cycle. When cycle frequency is altered, the time delay between two events may remain constant or it may change.

Systems of the former type are latency-locked systems. Examples include terrestrial locomotion in bipeds and tetrapods [5] and insects [12], swimmeret beating in crayfish [2], pyloric–gastric interaction in crabs [10] and uropod movements in sand crabs [11]. In many systems, however, the time intervals between different events change in proportion with the cycle period. Such systems are phase-locked systems, because the

* Corresponding author.
E-mail address: farzan@newark.rutgers.edu (F. Nadim).

0925-2312/03/$ - see front matter
doi:10.1016/S0925-2312(02)00811-1

ratio of time latency and cycle period (i.e., phase) is maintained. Phase locking occurs in a variety of systems such as the Tritonia swim network [8], the pyloric rhythm of the spiny lobster [6,7] gill ventilation in the shore crab [3,14], lamprey swimming [1,13] and leech swimming [4].

Despite its importance in many motor pattern generators, the mechanism underlying phase maintenance is still unclear. Changes in the relative timings of events in a motor pattern may be caused by changes in intrinsic properties of the neurons, the strength of chemical synapses or electrical synapses. In this work we propose that short-term synaptic depression in inhibitory synapses plays a pivotal role in promoting phase constancy. We illustrate this idea using a computational model that consists of an oscillator that makes an inhibitory synapse onto a follower cell. If the synapse is non-depressing, the postsynaptic cell always starts its burst at a fixed latency after the beginning of the presynaptic burst. Hence the phase between pre- and postsynaptic cell decreases as the cycle period increases. However, when the synapse shows depression, in some frequency range the interval between post- and presynaptic bursts increases as the cycle period increases. The increase in this interval is nearly proportional to the change in period and the phase is approximately constant. We discuss how the strength and kinetics of the depressing synapse affect the phase of the postsynaptic cell.

2. The model

The model consists of an oscillator O and a follower cell F. The two cells $i = \{O, F\}$ are modeled with standard current balance equations based on the Morris–Lecar model [9]

$$V_i' = g_{L,i}(E_{L,i} - V_i) + \bar{g}_{K,i}W_i(E_{K,i} - V_i) + \bar{g}_{Ca,i}m_{\infty,i}(V_i)(E_{Ca,i} - V_i), \quad (1)$$

$$\tau_i(V_i)W_i' = w_{\infty,i}(V_i) - W_i, \quad (2)$$

where V_i is the membrane potential and W_i is a recovery variable. The cycle period of O can be modified by changing the time constant of the recovery process τ_O. When $V_O > V_{\text{thresh}}, O$ is considered to be in the active state (or bursting) and $\tau_O = T_R$. Otherwise we consider that O is silent and $\tau_O = T_L$. Similar equations hold for F, with the same V_{thresh} but distinct time constants. We change the cycle period by changing T_L. This ensures that the duration of the active state of O does not change with period.

The parameters are chosen such that, without any synaptic input, F is at a high-voltage active state. The synaptic current $I_{\text{syn}} = g_{\text{syn}}s(V_F - E_{\text{syn}})$ from O adds an additional term to the left-hand side of (1), where E_{syn} is the synaptic reversal potential, g_{syn} is the maximal synaptic conductance and s is a state variable that represents synaptic activation. The dynamics of s depend on a variable d, which measures the level of depression of the synapse. When O is in the active state s and d obey the following set of equations:

$$d' = -d/\tau_\beta, \; d(0) = d_0; \quad s' = -s/\tau_\gamma, \; s(0) = d_0 \quad (3)$$

for some initial value d_0. Note that the initial value $s(0)$ is set to d_0. Let T_A be the burst duration of O and T_I be the interburst time of O. In the silent state of

O, $d' = (1-d)/\tau_\alpha$, $d(T_A) = d_0 \exp(-T_A/\tau_\beta)$; $s' = -s/\tau_\kappa$, $s(T_A) = d_0 \exp(-T_A/\tau_\gamma)$. When O becomes active again, Eq. (3) again holds except that $s(T_I + T_A)$ is reset to be $d(T_I + T_A)$. Thus the synaptic variable s is set to the value of d exactly at the moment in time when O jumps to the active state. At all other moments in time, the two variables are decoupled. If O fires periodically, then $d_0 = d(T_I + T_A)$ which can be rewritten as $d_0 = (1-\exp(-T_I/\tau_\alpha))/(1-\exp(-T_I/\tau_\alpha)\exp(-T_A/\tau_\beta))$. This expression shows that the level of depression changes as a function of the interburst interval T_I, the burst duration T_A and of the depression (τ_β) and recovery (τ_α) time constants.

If g_{syn} is sufficiently large and the synapse is strong (d_0 is large), then the inhibitory synaptic current will prohibit F from firing until the synapse from O to F has decayed enough. Alternatively, if the synapse is weak (d_0 is small) it will have a minimal effect on the activity of F and therefore the firing of F is governed by its intrinsic dynamics.

3. Results

We will start by defining phase and will then show what parameters affect it. Let us assume that O begins its burst at $t = 0$ and has period P. Just prior to the burst of O, F is at a high-voltage active state. The effect of O's burst is to cause F to fall to its low-voltage silent state. At some later time, which we denote Δt, F returns to the active state. Therefore the phase at which F fires with respect to the onset of O's burst is $\phi = \Delta t/P$. To gain insight into what mechanisms promote phase constancy, it is crucial to understand what determines the firing time Δt of F.

Fig. 2 shows a plot of ϕ as a function of P. When the synapse was non-depressing (dotted curve), ϕ decayed as P was increased. At $P = 250$, O and F started to burst at the same time ($\phi = 1$). For P values lower than 250, the biphasic rhythm was disrupted and F remained continuously inhibited by O. In contrast, when the synapse was depressing (solid curve) Δt initially increased as P increased. In the case shown in Fig. 2, for $175 \leqslant P \leqslant 275$, a change in P produced a larger change in Δt and hence ϕ increased as P increased. We refer to the region where ϕ increased as the hump. For other P values, a change in P produced a smaller change in Δt and hence ϕ decreased as P increased. With very large values of P, the synapse fully recovered during the interburst of O. Hence, a change in P did not produce any change in Δt and ϕ decreased proportionally to $1/P$. When P was smaller than 150, the rhythm was disrupted. Fig. 2 illustrates two important points. (1) A depressing synapse, compared to a non-depressing synapse, minimizes the change in phase as period is modified. For example, in Fig. 2 a 2-fold increase of period, from 250 to 500, yields a change of phase of 50% (from 0.98 to 0.49) for a non-depressing synapse and of 16% (from 0.86 to 0.72) for a depressing synapse. (2) A depressing synapse increases the period range for which F bursts. In Fig. 2, the rhythm was disrupted (F did not burst) with P less than 150 for a depressing synapse but with a non-depressing synapse the disruption happened for P less than 250.

The main result of this work that synaptic depression promotes phase constancy is dependent on the crucial, yet simple, fact that the firing time of F is a function of the

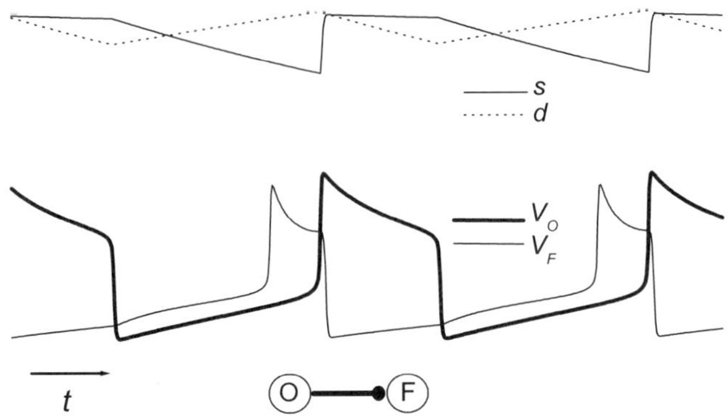

Fig. 1. The model: An oscillator O and a follower cell F are connected via an inhibitory synapse from O to F. When the synapse is modeled as depressing, the synaptic conductance is the product of the maximal conductance and s. When O starts its burst, s is set to the value of d, which represents depression and recovery.

strength of the synapse. The firing time is effectively determined by the value d_0, which gives a measure of the strength of the depressing synapse at the moment O begins its burst. The quantity Δt depends on two key factors: (1) the intrinsic dynamics of F as it evolves in its silent state, and (2) the decay time constant of inhibition τ_K. However, as we shall discuss below, each of these factors may only be relevant in determining Δt for certain intervals of P. If the synapse is weak, the firing of F is governed by its intrinsic dynamics, i.e., Δt is mainly determined by the intrinsic time constant τ_w of F during its silent state. In particular, Δt is largely independent of τ_K. Moreover, as the period P increases, by increasing T_I, Δt does not change much, since the firing time is intrinsically determined. Thus in the interval of periods between A and B in Fig. 2, ϕ behaves like $1/P$. If the synapse is strong, the main parameter affecting Δt is τ_K, the time constant for the decay of inhibition when O is silent. Notice that this parameter was irrelevant for determining Δt in the prior case. Thus if τ_K is large, Δt can increase very rapidly with small changes in P that make the synapse increase its strength. This observation accounts for the increase in phase observed between points B and C in Fig. 2. As the period P becomes very large, the synapse can no longer be considered a depressing synapse, since it would recover fully during each interburst of O. Therefore the firing time Δt would become constant so that the phase behaves like $1/P$. This accounts for the behavior of the phase curve for P values to the right of C in Fig. 2.

The key factor, thus, is the value of s and d at the beginning of the burst in O (Fig. 1). If the synapse is not depressing, this value is independent of P and hence Δt is constant. However, if the synapse is depressing Δt becomes dependent on the cycle period: as the cycle period increases, there is a longer time during which d can recover and increase towards 1. Since s assumes the value of d at the beginning of the burst of O, as the cycle period increases s becomes larger, and as a result Δt increases. In addition, the intrinsic dynamics in F (the recovery variable W_F) also play an important role, since these dynamics together with the decay rate of the inhibition determine the next firing time of F (Fig. 2).

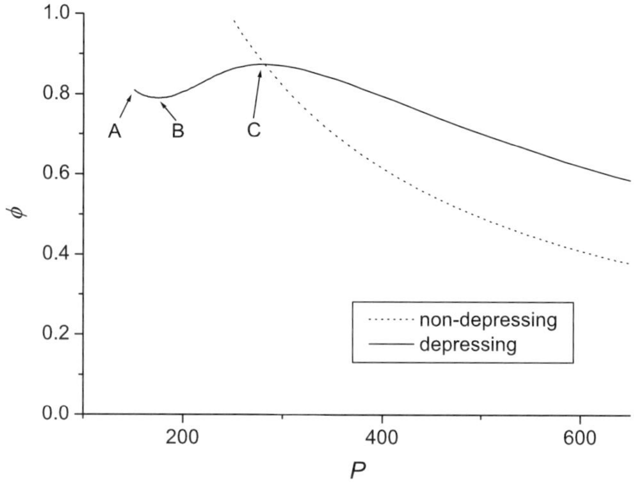

Fig. 2. Phase of F burst onset as function of cycling period: When the synapse is non-depressing (dotted line), Δt is fixed and the phase ϕ decays as period P increases. For $P < 250$, the rhythm is disrupted. When the synapse is depressing, Δt increases as P increases. In the case shown, between $P > 175$ (B) and $P < 275$ (C) Δt increases more than P, and in that range ϕ increases as P increases. For $P < 150$ (A), the rhythm is disrupted.

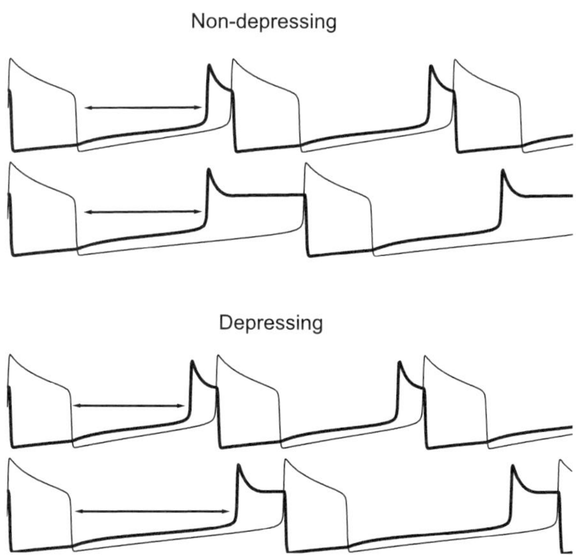

Fig. 3. A depressing, but not a non-depressing, synapse increases the burst delay as period increases: Voltage traces of O (thin lines) and F (thick lines) for different P values (from top to bottom): $P = 275$ and 365. Arrows show the time interval between onset of burst in O and onset of burst in F (Δt). *Non-depressing synapse*: Δt is constant (compare lengths of arrows). *Depressing synapse*: Δt increases as P increases.

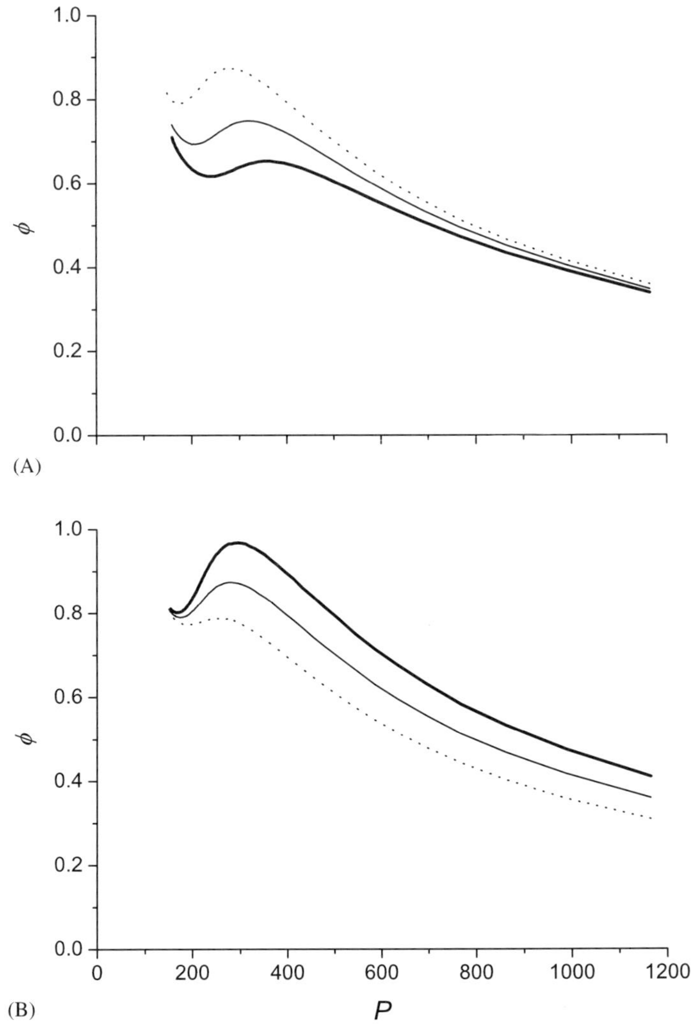

Fig. 4. Effects of dynamics on phasing: (A) The effect of the time constant of recovery τ_α : 800 (dotted line, canonical), 1000 (thin line) and 1200 (thick line). (B) The effect of the time constant of synaptic decay τ_K : 400 (dotted line), 500 (thin line, canonical) and 600 (thick line).

3.1. The sensitivity of a depressing synapse to changes in cycle period contributes to maintain phase

To illustrate the importance of a depressing synapse in maintaining phase, in Fig. 3 we compare the time courses of V_O (thick trace) and V_F (thin trace) for different cycle periods in the two cases where the synapse is depressing or non-depressing. Unless stated otherwise, we used one set of parameters that we referred to as the canonical model. The non-depressing synapse was modeled with the same equations, except that when O makes its burst s was set to a fixed value s_1, rather than the value of d (substitute $s(0) = s_1$ in Eq. (3)). The values of s_1 and g_{syn} were chosen to match the maximal value of s when the synapse was depressing and $P = 275$.

When the synapse was non-depressing, Δt remained fixed as P was increased (Fig. 3, compare the length of arrows). For the parameters used in Fig. 3, Δt was 250. At values of P smaller than Δt, the rhythm was disrupted (F did not burst) because O inhibited F before F could start a burst (not shown). When the synapse was depressing, Δt increased as P was increased. At low values of P, the rhythm was disrupted because the synapse was too weak (d did not sufficiently recover; not shown).

3.2. The effects of synaptic dynamics on the phase

We examined the effects of different synaptic parameters on the phase of F, at different cycle periods (Fig. 4). We changed the canonical model and studied the effect of the time constants of synaptic recovery (τ_α, **A**) and synaptic decay (τ_K, **B**). In Fig. 4A we increased τ_α from the canonical value (Fig. 4A, dotted line). A larger τ_α value caused less recovery of the d variable for the same interburst duration in O. Increasing τ_α caused Δt to be smaller, and therefore ϕ versus P became more proportional to $1/P$ and the size of the hump was reduced. The hump also shifted to larger P values, because the synapse needed more time to recover to the strengths at which synaptic dynamics became important. At larger τ_α values, the hump was completely eliminated (not shown). When τ_K was increased from its canonical value (Fig. 4B, thin line), the size of the hump was increased. A larger τ_K value caused slower decay of the synapse during the interburst of F. This caused Δt, and hence ϕ, to be larger. Opposite effects were seen when τ_K was decreased from its canonical value.

4. Discussion

How phase relations among components of a rhythmic neuronal network are maintained is still not understood. We propose a mechanism for promoting phase constancy that makes use of short-term synaptic depression. Using a simple biophysical model we showed that when an inhibitory synapse between an oscillator O and a follower neuron F is depressing, the phase relations depend on the cycling period P. If P is short relative to the time constant of synaptic recovery τ_α, the synapse is too weak to delay the burst of F after the burst in O is terminated. In this case, the burst time of F is determined by its intrinsic dynamics. If the intrinsic dynamics in F are relatively fast, Δt (the time between O burst and F burst) is almost constant (independent of P) and therefore the phase $\phi(=\Delta t/P)$ is inversely related to P. If P is long relative to τ_α, the synapse is strong and it maximally delays the burst of F. Again, Δt is independent of P and therefore ϕ is inversely related to P. In an intermediate range of P values, however, an increase in P results in more recovery of the synapse. This leads to a larger increase Δt. As a result, ϕ becomes less sensitive to P. We found that in some parameter ranges, ϕ may become independent of P or even increase as P is increased.

Hence the interaction between intrinsic and synaptic dynamics can contribute to the maintenance of phase, or at least to reduce the sensitivity of phase on period.

Acknowledgements

This work was supported by BSF 2001-039 (YM), NSF DMS-9973230 (AB, VB) and NIMH 60605-01 (FN).

References

[1] A. Cohen, G. Ermentrout, T. Kiemel, N. Kopell, K. Sigvardt, T. Williams, Modelling of intersegmental coordination in the lamprey central pattern generator for locomotion, Trends Neurosci. 15 (1992) 434–438.
[2] W. Davis, The neural control of swimmeret beating in the lobster, J. Exp. Biol. 50 (1969) 99–117.
[3] R. Dicaprio, G. Jordan, T. Hampton, Maintenance of motor pattern phase relationships in the ventilatory system of the crab, J. Exp. Biol. 200 (1997) 963–974.
[4] W. Friesen, R. Pearce, Mechanisms of intersegmental coordination in leech locomotion, Semin. Neurosci. 5 (1993) 41–47.
[5] S. Grillner, Control of locomotion in bipeds, tetrapods and fish, in: U. Brooks (Ed.), Handbook of Physiology: The Nervous System (Section 1)—Motor Control, American Physiological Society, Bethesda, 1981, pp. 1179–1236.
[6] S. Hooper, The pyloric pattern of the lobster *Panulirus interruptus* stomatogastric ganglion comprises two phase maintaining subsets, J. Comput. Neurosci. 4 (1997) 207–219.
[7] S. Hooper, Phase maintenance in the pyloric pattern of the lobster *Panulirus interruptus* stomatogastric ganglion, J. Comput. Neurosci. 4 (1997) 191–205.
[8] R. Hume, P. Getting, M. Del Beccaro, Motor organization of *Tritonia* Swimming. I. Quantitative analysis of swim behavior and flexion neuron firing patterns, J. Neurophysiol. 47 (1982) 60–74.
[9] C. Morris, H. Lecar, Voltage oscillations in the barnacle giant muscle fiber, Biophys. J. 35 (1981) 193–213.
[10] F. Nadim, Y. Manor, M.P. Nusbaum, E. Marder, Frequency regulation of a slow rhythm by a fast periodic input, J. Neurosci. 18 (1998) 5053–5067.
[11] D. Paul, Swimming behavior of the sand crab, *Emerita analoga* (Crustacea, Anomura). I. Analysis of the uropod stoke, Z. Vergl. Physiol. 75 (1971) 233–258.
[12] K. Pearson, J. Iles, Discharge patterns of coxal levator and depressor motoneurone of the cockroach, *Periplaneta americana*, J. Exp. Biol. 52 (1970) 139–165.
[13] T. Williams, Phase coupling by synaptic spread in chains of coupled neuronal oscillators, Science 258 (1992) 662–665.
[14] R. Young, Neuromuscular control of ventilation in the crab *Carcinus maenas*, J. Comput. Physiol. A 101 (1975) 1–37.

Victoria Booth received her Ph.D. in applied mathematics from Northwestern University. She has held positions at the National Institutes of Health and in the Mathematical Sciences Department at New Jersey Institute of Technology. She is currently an associate research professor in the Center for Applied Mathematics and Statistics at NJIT.

Amitabha Bose received his Ph.D.in applied mathematics from Brown University. He is currently an associate professor in Mathematical Sciences at New Jersey Institute of Technology.

Yair Manor received his Ph.D.in neurobiology from the Hebrew University in Jerusalem. He was a post-doctoral fellow in neurobiology at Brandeis University. He is currently an assistant professor at the Department of Life Sciences of Ben-Gurion University.

Farzan Nadim received his Ph.D.in mathematics from Boston University. He was a post-doctoral fellow in neurobiology at Emory University and Brandeis University. He is currently an associate professor in Mathematical Sciences at New Jersey Institute of Technology and Biological Sciences at Rutgers University.

Available at
www.ComputerScienceWeb.com
POWERED BY SCIENCE DIRECT®

ELSEVIER

Neurocomputing 52–54 (2003) 89–95

NEUROCOMPUTING

www.elsevier.com/locate/neucom

Modeling of spontaneous synchronized periodic activity observed in in vitro networks

Maria Marinaro, Silvia Scarpetta*

Dipartimento di Fisica "E. R. Caianiello", Salerno University, Baronissi (SA) IT INFM Sezione di Salerno (SA) IT, and IIASS Vietri sul mare (SA) Italy

Abstract

Recently Segev et al. (Phys. Rev. Lett. 88 (2002) 118102; Phys. Rev. E 64 (2001) 011920) have done long-term measurements of spontaneous activity of cortical cell neural networks placed on multi-electrode arrays. Their observations differ from predictions of current neural network models in many features. The aim of this paper is to show that the same EI network model introduced in a previous paper (Scarpetta et al., NIPS, Vol. 13, 2001) by one of us Z. Li and J. Hertz, to model driven activity and spike-timing-dependent-plasticity in cortical areas, is able to reproduce the experimental results of spontaneous activity of Segev et al. (Phys. Rev. Lett. 88 (2002) 118102; Phys. Rev. E 64 (2001) 011920) (and the observed power spectrum density (PSD) features), when we consider the model in isolation with intrinsic noise terms. Using preliminary analytical results as a guide line, we perform numerical simulations of the stochastic equations for the instantaneous firing rates. In one regime of parameters the network shows spontaneous synchronous periodic activity, and the PSD shows two peaks at the first and second harmonics, and a broad band at low frequency (indicating positive long range time correlations), in agreement with experiments. The two high peak in the PSD fades away when we increase the level of noise.

Keywords: Modeling; Periodic activity; In vitro networks

1. Introduction and motivations

To reveal the rules which govern network spontaneous activity, i.e. without any stimulation from the external world, is a preliminary requirement for the understanding of the network activity inside animal brain, when subject to continuous stimula-

* Corresponding author.
E-mail address: silvia@sa.infn.it (S. Scarpetta).

0925-2312/03/$ - see front matter
doi:10.1016/S0925-2312(02)00836-6

tion from other part of the brain. Spontaneous (and stimulation-driven) synchronized periodic activity has been observed in many in vivo and in vitro experiments. The understanding of the specific mechanisms underlying these spontaneous spatio-temporal pattern of activity is important for the comprehension of many aspects of brain activity, like central pattern generator activity, epilepsy, etc. Much efforts has been devoted recently to study in vitro neural networks, in which various numbers of neurons are grown and recorded on a multi electrode arrays, which enables the study of isolated network. Recently Segev et al. [3,4] have done accurate long-term measurements of the spontaneous activity of two-dimensional cortical cell neural networks placed on multi-electrode arrays. They observed, for a critical range of Ca concentration, periodic synchronized bursting activity, that fades away after about ~ 20 min. Periodic synchronized bursting is observed at 1 mM Ca concentration and not at higher (2 mM) and lower (0.5 mM) concentrations. Their observation differ from prediction of current neural network models in many features. They try to model the phenomena performing numerical simulations of an integrate-and-fire network model with random connections. Adding (1) dynamic threshold and (2) activity-dependent synaptic connections, they reproduce much of the observed network activity, but do not obtain a complete explanation of the experimental results. For example, their numerical simulations capture the two different activity patterns observed experimentally and the transition between aperiodic synchronized bursting versus periodic synchronized bursting when Ca concentration (and therefore model connection strengths) was increased, but they do not reproduce the transition from periodic synchronized bursting versus aperiodic activity observed experimentally when Ca concentration was increased over the critical interval. The Power Spectral Density (PSD) at the beginning shows two high peaks at first and second harmonic of the periodic network activity, which become lower and lower with time, and after ~ 20 min the PSD is almost flat. Moreover they observe that energy distribution over low frequencies has a broad band with power low decay that indicates the existence of positive long range time correlations in the sequences. These PSD features, and in particular the changes that happen in time scale of several minutes and the behavior at low frequency, have not been explained by the model of Segev et al. [3,4], and as far as we know, until now the explanation is lacking. We show that some insights and a good agreement with experiments of Segev et al. can be obtained with a simple spiking-rate excitatory–inhibitory (EI) network model with noise, as that one introduced in a previous paper [2] by one of us Z. Li and J. Hertz to study non-spontaneous driven activity and spike-timing-dependent-plasticity in cortical areas. The spatio-temporal patterns of spontaneous activity in our model is consequence of the simple dynamics of the interacting excitatory and inhibitory units, and depends critically from the synaptic organization of the EI network. In Section 2 the model is described and in Section 3 the numerical results are reported together with comments and conclusions.

2. The model

Our starting point are the instantaneous-firing-rate equations [1,2] governing the instantaneous-firing-rates $g_u(\mathbf{u})$ and $g_v(\mathbf{v})$. The state variables $\mathbf{u}, \mathbf{v}$ describe respectively the excitatory and inhibitory units; g_u, g_v are sigmoidal activation functions. The state variables u_i, v_i of the interacting excitatory and inhibitory neurons evolve throw time, following the equations of motion:

$$\dot{u}_i = -\alpha u_i - \sum_j H_{ij} g_v(v_j) + \sum_j J_{ij} g_u(u_j) + \bar{F}_i(t), \tag{1}$$

$$\dot{v}_i = -\alpha v_i + \sum_j W_{ij} g_u(u_j) + F_i(t) \tag{2}$$

where α^{-1} is a time constant (about few milliseconds, for simplicity assumed the same for excitatory and inhibitory units), and J_{ij}, W_{ij} and H_{ij} are the synaptic strengths All these parameters are non-negative; the inhibitory character of the inhibitory units v_i is expressed by the minus sign preceding H_{ij} on the right-hand side of Eq. (1). $\bar{F}_i(t)$ and $F_i(t)$ are stochastic variables that describe intrinsic noise, respectively on the excitatory and inhibitory units, not included in the statistical definition of u, v. In (cultured) interacting neurons systems noise can be due to several reasons, like thermal fluctuation, ion channel stochastic activities, and many others. We take the intrinsic noise $\bar{F}_i(t), F_i(t)$ to be uncorrelated white noise, such that $\langle F_i(t)\rangle = \langle \bar{F}_i(t)\rangle = 0$ and $\langle F_i(t)F_j(t')\rangle = \langle \bar{F}_i(t)\bar{F}_j(t')\rangle = \Gamma\delta_{ij}\delta_{t-t'}$. Apart from small fluctuations, that we neglect, the matrices **J**, **H**, **W** can be taken symmetric, indeed there is no reasons for the isolated spontaneously grown system to break this symmetry, nevertheless the connections between excitatory and inhibitory neurons are still asymmetric in this scenario, and the complete connection matrix

$$\begin{pmatrix} \mathbf{J} & -\mathbf{H} \\ \mathbf{W} & 0 \end{pmatrix}$$

is asymmetric too. A periodic synchronous bursting event with frequency ω_0, like that observed by Segev et al. [3,4], will correspond in our model to a firing rate $g(u_i)$ (and state variable u_i) such that the local Power Spectrum Density of unit i, PSD_i, and the global power spectrum $PSD = \sum_i PSD_i/N$ has a peak in ω_0. A state vector $\mathbf{u}(t) = \boldsymbol{\xi}\cos(\omega_0 t) = \boldsymbol{\xi}e^{i\omega_0 t} + \text{c.c.}$, where the ξ_i are real positive-or-null numbers, represent a synchronous periodic activity of all excitatory units, at frequency ω_0, ξ_i describes the amplitudes of the bursting on the different excitatory units. The spiking-rate of the neurons $g(\mathbf{u}(t))$ is a periodic function with period $T = 2\pi/\omega$, it reaches the highest value at time t where $\cos(\omega_0 t) = 1$. When the bursting is periodic but not synchronous, i.e. there is a phase lock between the neurons involved in the bursting, it corresponds to $u_i(t) \propto |\xi_i|\cos(\omega_0 t - \phi_i)$. We can describe both synchronous and phase-locked periodic activities by writing $u_i(t) = \xi_i e^{-i\omega_0 t} + \text{c.c.}$ and taking the ξ_i real in the first case and complex ($\xi_i = |\xi_i|e^{i\phi_i}$) in the second. A preliminary analytical solution of the noisy model in isolation (that will be reported in a paper currently in preparation) has put in evidence that the model exhibit different spontaneous regimes, which correspond to different values of the real and imaginary parts of eigenvalues of the connection matrix. In particular, when we use in the model long-range $J_{ij} = j_0/N + \varepsilon\eta_{ij}$ and $W_{ij} = W_0/N + \varepsilon\eta_{ij}$, and short-range $H_{ij} = h_0\delta_{ij} + \varepsilon\eta_{ij}$, (where $\varepsilon \ll 1$, and η_{ij} are quenched random values with zero mean), three different regimes can be distinguished, depending from the values of j_0, W_0, h_0, α:

(A) the damped fixed-point regime, where there's no spontaneous oscillatory activity
(B) the damped sub-oscillatory regime, where there's spontaneous synchronous oscillatory activity only if driven by noise (or by external periodic input)
(C) the spontaneously periodic oscillatory regime, where spontaneous activity is oscillatory, periodic and synchronous, even without noise; in this regime the activity exhibit diverging oscillations in the linearized system which become stable when non-linear terms are considered.

Model activity in regime C mimic the spontaneous synchronous periodic activity observed in cultured experiments at 1 mM Ca concentration [4].

3. Numerical results

We use sigmoidal non-linearity of excitatory neurons activation function $g_u(\mathbf{u})$, while we take for simplicity g_v as linear near the fixed point (this choice is supported by physiology). We focus on the case when g_u deviates from linearity near the fixed point $\bar{\mathbf{u}}$ due to saturation. Fig. 1A shows the function that we have used in our numerical non-linear simulations for excitatory units. The non-linear stochastic differential equations of the isolated network, with $N = 10$ excitatory and $N = 10$ inhibitory units, was numerically integrated using the Runge–Kutta method, and we calculate the PSD of excitatory activity given by $\langle \sum_i |\tilde{u}_i(\omega)|^2 \rangle / N$, using Bartlett windowing.

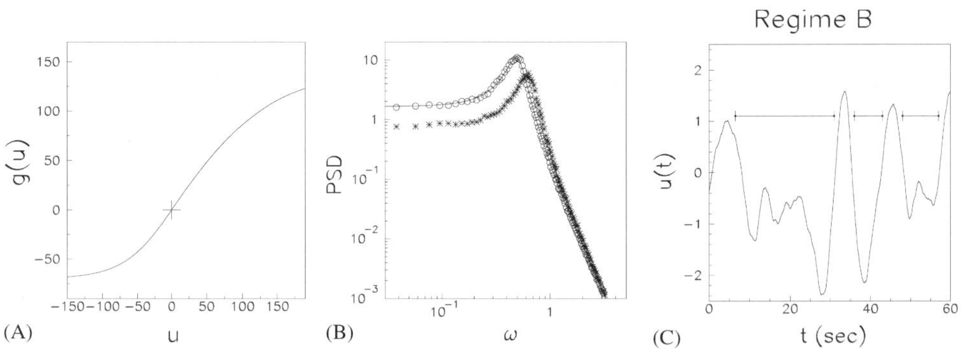

Fig. 1. (A) Activation function used in simulations for excitatory units u_i. B. Power Spectrum Density of excitatory units, in the regime B, with $\Gamma = 0.0004$. Stars shows results of numerical simulations using the activation function shown in A, circles shows simulation results for the linearized system, while solid line is theoretical prediction in linear approximation. A broad peak at $\omega \neq 0$ in the PSD is induced by noise. (B) The time behavior of the state variable $u_i(t)$, $i = 1, \ldots, N$ in the linear numerical simulation in the regime B, with $\Gamma = 0.004$. All units $u_i(t)$ shows synchronous (aperiodic) oscillatory activity. Differences in the time behavior between variables $u_i(t)$ are so small that all the lines $u_i(t)$ overlaps each other.

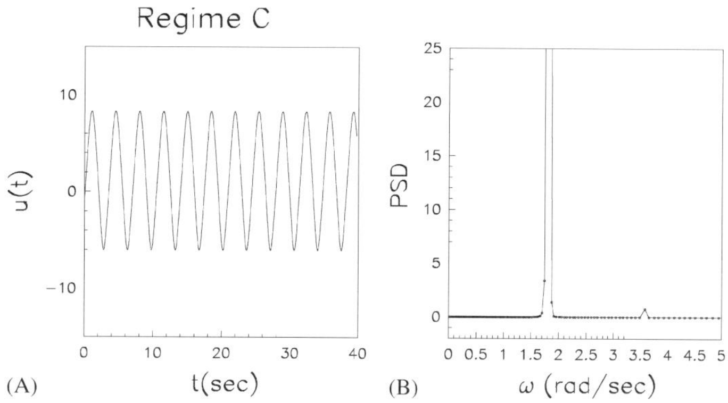

Fig. 2. (A) Time behavior of the state variables $u_i(t)$ in the noiseless nonlinear numerical simulation of the isolate EI network in the regime C, using the activation function of Fig. 1A. All the state variables $u_i(t)$ follow the same time evolution, showing clearly stable periodic synchronous activity. (B) Power spectrum density of the excitatory unit activity shown in A. The peaks in the first and second harmonic mark clearly the collective periodic activity.

In order to investigate the regime B, we use the following parameters $\alpha = 50\ \mathrm{s}^{-1}$, $J_{ij} = j_0/N = 2(\alpha - 0.1)/N$, $W_{ij} = W_0/N, H_{ij} = h_0\delta(i-j)$, and $W_0 = h_0 = \sqrt{0.25 j_0^2 + 0.25}$. Parameters are such that in the linear approximation the characteristic frequency (of the damped solution) is $\omega_0 = \sqrt{-j_0^2/4 + h_0 W_0} = 0.5$ rad/s. Without noise, the numerical simulations shows that, after a short transient, the activity of the network is stacked on the fixed point, and $u = 0, v = 0$. Interestingly, numerical simulations of the system with noise $\Gamma \neq 0$, shows synchronous collective oscillatory behavior in the neurons activity, both in the nonlinear system and in the linearized one (see Fig. 1B and C). In the regime B, noise induces synchronous oscillatory activity, where amplitude and the interval between subsequent maxima (IEI) are not constant (see Fig. 1C). This synchronous collective oscillatory behavior correspond to a broad peak in the PSD of the neurons activity at the characteristic frequency ω_0 of the system. This noise-induced collective activity (which is highly spatially correlated but decorrelates in time quite fast) mimics the aperiodic synchronous activity observed at 0.5 and 2.0 mM Ca concentration in [4].

In order to investigate the regime C, we use the following parameters $\alpha = 50\ \mathrm{s}^{-1}$, $J_{ij} = j_0/N = 2(\alpha + 0.07)/N$, $W_{ij} = W_0/N, H_{ij} = h_0\delta(i-j)$, and $W_0 = h_0 = \sqrt{0.25 j_0^2 + 0.25}$. Parameters are such that in the linear approximation the characteristic frequency (of diverging solution) is $\omega_0 = 0.5$ rad/s. Introducing the non-linearity through the function $g_u(u_i) \sim u_i - a u_i^3$, $a > 0$, make the system reach the saturation and shift the frequency to the value $\omega_0 = \sqrt{(2h_0 W_0 \alpha)/j_0 - \alpha^2} = 1.9$ rad/s. As shown in Fig. 2 the noiseless nonlinear numerical simulations shows stable synchronous bursting periodic activity **u**, with frequency about 1.9 rad/s. This behavior holds only in regime C, i.e. for a particular range of parameters; we check numerically for example that we are out of the

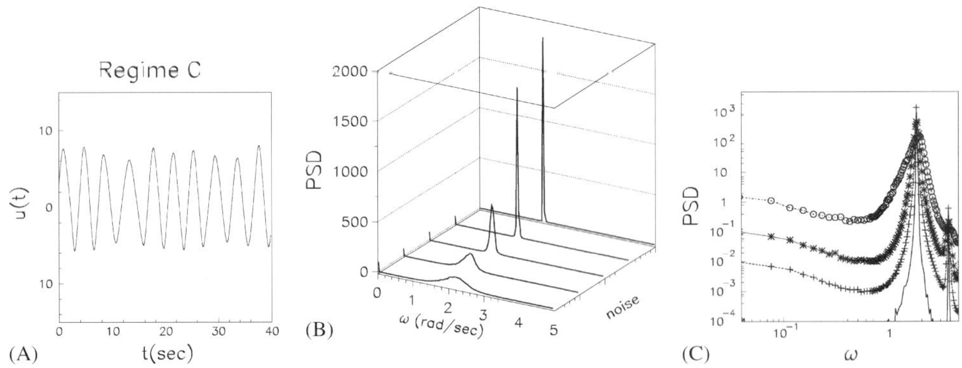

Fig. 3. (A) The time behavior of the state variable $u_i(t)$, $i = 1,\dots,N$ in the nonlinear numerical simulation of the noisy network consisting of $N = 10$ excitatory and $N = 10$ inhibitory cells, in the regime C, with $\Gamma = 0.01$. Differences in the time behavior of the state variables $u_i(t)$ are so little that all the ten lines $u_i(t)$, $i = 1,\dots,10$ overlaps each other. The activity is almost periodic, but noise decorrelates the signal over long time scales. (B) Power spectrum density of the excitatory units activity of the noisy nonlinear networks in regime C, for different noise level Γ. From the right-end to the left-front line: no-noise $\Gamma = 0.0$, 0.0001, 0.001, 0.01, 0.1. Noise make the high periodic activity fades away. (C) Log–log plot of the power spectrum density shown in (B). solid line: no-noise $\Gamma = 0.0$, dotted line with crosses: $\Gamma = 0.0001$, dashed line with stars: $\Gamma = 0.001$, Blue line: $\Gamma = 0.01$, dot–dashed line with circles: $\Gamma = 0.1$. The two high peaks in the PSD correspond to the first and second harmonics of the activity period. The low frequency power spectrum distribution shows a broad peak in $\omega = 0$.

regime using both lower ($j_0 = 99.86$) or higher ($j_0 = 103$) values of excitatory-excitatory connections (instead of $j_0 = 100.14$), leaving all the other parameters invariate. As shown in Fig. 2, the power spectral density has two high peaks at first and second harmonic of the periodic network activity, resembling the experimental results of [4]. Fig. 3B and C shows the PSD calculated numerically for different level of the noise. The two peaks at first and second harmonic in the PSD become lower and lower with noise level Γ. Fig. 3C shows clearly that the power spectrum has a broad band distribution at low frequency, very similar to the experimental results of Fig. 3 in [3]. It means that both the experimental time sequence [3] and our simulated model has positive long-range time-correlations.

Numerical results also show that noise changes the frequency of the periodic activity, albeit very slightly. This change in frequency cannot be seen from the experimental PSD, perhaps because it is covered by statistical errors of the measurement. A possible explanation of the fading of periodicity observed in time scale of several minutes in the experimental PSD may be found in the noise effects shown here. Other explanation can be found in this framework for this phenomena, for example, in addition to the noise effect, if there is plasticity of synapses driven by the spontaneous periodic activity itself, then the changes in connections parameters can change the frequency value and also make the network migrates from regime C to B and A regimes. This hypothesis is currently being investigated analytically.

References

[1] P. Dayan, L.F. Abbott, Theoretical Neuroscience, MIT Press, Cambridge, MA, 2001.
[2] S. Scarpetta, Z. Li, J. Hertz, in: T. Leen, T. Dietterich, V. Tresp (Eds.), NIPS 2000, Vol. 13, MIT Press, Cambridge, MA, 2001.
[3] R. Segev, M. Benveniste, E. Hulata, N. Cihen, A. Palevski, E. Kapon, Y. Shapira, Phys. Rev. Lett. 88 (11) (2002) 118102.
[4] R. Segev, Y. Shapira, M. Benveniste, E. Ben-Jacob, Phys. Rev. E 64 (2001) 011920.

ELSEVIER

Available at
www.ComputerScienceWeb.com
POWERED BY SCIENCE @ DIRECT®

NEUROCOMPUTING

Neurocomputing 52–54 (2003) 97–102

www.elsevier.com/locate/neucom

Synchronization of Purkinje cell pairs along the parallel fiber axis: a model

Sergio Solinas*, Reinoud Maex, Erik De Schutter

Laboratory of Theoretical Neurobiology, Born-Bunge Foundation, University of Antwerp, Wilrijk, Antwerp B-2610, Belgium

Abstract

We build a detailed model of the cerebellar molecular layer taking into account the anatomical, morphological and physiological data available to date. Based on how the percentage of input shared by nearby Purkinje cells (PCs) depends on their relative position in the PC layer we conclude that shared inhibition must play a crucial role in PC synchronization.

Keywords: Cerebellum; Purkinje cells; Synchrony; Interneuron; Inhibition

1. Introduction

Spontaneous synchronization of Purkinje cell (PC) pairs in the cerebellar cortex is implicitly predicted by many theories of cerebellar function (e.g. the beam theory of Eccles [7]). Although the percentage of excitatory and inhibitory input shared by PC pairs can reach 50% and 20%, respectively, short duration positive correlation of in vivo PC simple spike firing has been only rarely observed [6,23]. Here we build a detailed model of the cerebellar molecular layer to investigate the effect of common input on the synchronization of PC pairs.

* Corresponding author.
E-mail address: sergio@bbf.uia.ac.be (S. Solinas).

0925-2312/03/$ - see front matter © 2003 Elsevier Science B.V. All rights reserved.
doi:10.1016/S0925-2312(02)00846-9

2. Neuron models

2.1. The model Purkinje cell

Our simulations were based on the detailed PC model described previously by De Schutter and Bower [4]. This model represents the morphology of a guinea pig PC [22], and has both passive and active membrane properties. Ten voltage-gated or Ca^{2+}-dependent conductances were modeled: inactivating and persistent Na^+ channels; P-type and T-type Ca^{2+} channels; anomalous inward rectifier, delayed rectifier, persistent, and A-type K^+ channels; Ca^{2+}-activated channels of big-K^+ and K2 types. The model replicates realistic PC firing patterns recorded in vitro during current injection to the soma or dendrites (i.e. simple Na^+ spikes and dendritic Ca^{2+} spikes). Excitatory synaptic input from climbing fibers and parallel fibers (PF) was mediated via AMPA receptor channels. Inhibitory input from molecular layer interneurons (MLI) was mediated via $GABA_A$ receptors channels. In this configuration the model PC fired both dendritic Ca^{2+} spikes, elicited by the synchronous activation of many AMPA receptor channels, and realistic patterns of Na^+ spikes, elicited by a weaker excitatory input. Here, we modified the synaptic parameters to match more recent experimental data [11,21]. The average EPSC recorded in PC somata during stimulation of single granule cells has been fit by a dual exponential function with a peak current of 8.4 pA, a rise time constant of 1 ms and a decay time constant of 1.2 ms [11]. For IPSCs the average values measured 20 pA, 1.6 ms, and 9.3 ms, respectively [21]. These experiments were performed at 32°C [11] and 23°C [21]. We tuned the model PC to 37°C [4] by dividing the AMPA and $GABA_A$ time constants by 1.2 and 1.6, respectively, assuming a Q_{10} factor of 1.5 [1]. Taking into account the temperature dependency of diffusion, the peak conductances were multiplied by the same factors [10]. The synaptic peak conductances needed to generate the experimental somatic current amplitudes were recalculated in the model PC during simulated voltage clamp of the soma (holding potential −70 mV [2] and −60 mV [21], respectively), correcting this way for the incomplete space clamp of the dendritic tree and for the low input resistance of the model PC [5]. To achieve the average somatic currents measured in vitro we had to multiply the AMPA and $GABA_A$ peak conductances by a factor of 9 and 6, respectively. The final parameters were: AMPA channels peak conductance 1.3 nS, rise and decay time constants 0.6 and 1 ms, peak somatic EPSC 10 pA; $GABA_A$ channels peak conductance 3 nS, rise and decay time constants 1 and 5 ms, peak somatic IPSC 32 pA. After these corrections the model PC still replicates in vivo firing patterns. Synaptic plasticity or probability of release were not modeled.

2.2. The model molecular layer interneuron

The molecular layer interneuron (MLI: stellate or basket cell) was modeled as an isopotential spherical soma with a diameter of 10 μm [14]. A specific membrane resistance of 12.5 kΩ cm^2 [14] and a specific membrane capacitance of 1 μF/cm^2 [14] gave a passive membrane time constant of 13.5 ms [9]. The model had six types of active membrane channels: inactivating Na^+ channels, high-voltage-activated Ca^{2+} channels, delayed rectifier and A-type K^+ channels, KC Ca^{2+}-activated K^+ channels, and anomalous inward rectifier Na^+/K^+ channels. These channels were borrowed from a Golgi cell model described previously [15], since MLIs and Golgi cells have indistinguishable electrophysiological properties in vitro [18]. We tuned the model MLI to achieve a spontaneous firing rate of 18 spikes/s [9] by increasing the reversal potential of the leak current to −46 mV. The synaptic input was provided by AMPA [3] and $GABA_A$ [12] receptor channels. The peak EPSC is a free parameter of our model, the only constraint being that a single EPSC should be able to elicit a post-synaptic spike (M. Häusser, personal communication). The EPSC time course can be fit by an α-function with a time constant of 1.5 ms at 23°C [3]. The peak IPSC measures in vitro 60 pA at 21°C [12] (holding potential −60 mV, pipette filled with high Cl^- concentration solution) and the channel opening and closing time constants are 0.3 and 10.5 ms [12], respectively. We tuned the model to 37°C by dividing the AMPA and $GABA_A$ time constant by 1.8 and 1.9, respectively, and multiplying the $GABA_A$ channel peak conductance by 1.9 [10]. The final parameters were: AMPA channels peak conductance 200–400 pS, time constant 0.8 ms; $GABA_A$ channels peak conductance 1.9 nS, rise and decay time constants 0.16 and 5.5 ms, peak somatic IPSC 110 pA. Synaptic plasticity or probability of release were not modeled.

3. A local network for each Purkinje cell

In the cerebellar cortex the MLIs surround the PC dendritic tree (density 32,000 cells/mm^3 [8,13]) making GABAergic synapses onto them. The stellate cells contact the PC spiny thick dendrite. The basket cell axons make synapses onto the PC main dendrite or surround their somata with a characteristic basket like structure [19]. Both MLI dendrites and axons are mainly confined to the cerebellar sagittal plain, making only short extensions along the PF axis. The volume spanned by the dendrite (PF axis 16 μm, sagittal axis 174 μm, radial axis 140 μm) is centered on the soma while the center of the axonal volume (PF axis 50 μm, sagittal axis 266 μm, radial axis 100 μm) is dislocated by 150 μm along the sagittal axis [24]. We calculated the upper limit for the volume of the presynaptic inhibitory basin of a PC, i.e. the volume containing all the MLIs whose axons reach the PC dendritic tree. The extension of the basin along the sagittal axis equals the depth of the PC dendritic tree, 250 μm [19], plus twice the length of an MLI axon, 250 μm. Along the PF axis, the extension measures the width of the PC dendritic tree (5 μm) plus twice 50 μm. The height of our approximative PC presynaptic basin is the average thickness of the molecular layer (300 μm). From this, the basin measured 750 μm by 105 μm by 300 μm and contains approximately 750 MLIs. Conversely, an MLI axon contacts on average 15 PCs [24], and given that in the cerebellar cortex there are appropriately 10 MLIs per PC [13], each PC should receive input from 150 MLIs. Thus, each MLI inside the basin connected to the PC with a probability of 0.2. In our network model each MLI within the PC presynaptic basin had the same probability of making a synapse and the number of MLIs presynaptic to a PC is fixed to 150. Out of these, 50 MLIs from the inner third of the molecular layer connected like basket cells to single $GABA_A$ synapses on the

PC main dendrites or soma. The other 100 MLIs made on average nine synapses on the PC spiny and thick dendritic compartments like stellate cells do [24]. For these synapses, the synaptic strength was divided by the average number of contacts. From slice recordings it is known that MLIs are interconnected with each other through an average of four inhibitory synapses [12]. Since in slices some MLI axons and dendritic trees might be injured, we took this figure as a lower bound. Therefore, in our model the MLIs were coupled by an average of 10 synapses, with synaptic strength normalized by 10.

3.1. The excitatory input

In the cerebellar cortex, both PCs and MLIs receive excitatory input from granule cell axons, which originate in the granular layer and rise to the molecular layer. There they bifurcate and run parallel to each other and orthogonal to the sagittal plane giving origin to the parallel fiber system. The cerebellar granule cell population is the largest of the central nervous system. Because it is currently impossible to simulate billions of cells, we assume that only 1% of the granule cells is active at rest, which is in agreement with recent in vivo recordings [16]. Since a real PC has 175,000 PF synapses [8] the 1474 spines in our model sufficed to receive this reduced granule cell input. The same holds for MLIs, we calculated that in real animals they receive 1500 PF synapses [8] and in our simulations only 15. At the present stage of implementation the excitatory input to the PCs and MLIs is provided by random spike train generators. These elements were placed at the PF bifurcation points and connected only to AMPA channels within 2.5 mm along the PF axis and 20 μm in the other two directions [20]. The PF conduction speed was set to 0.5 m/s [17].

3.2. What inputs do Purkinje cells share?

Synchronization of in vivo PC simple spikes has only rarely been reported. In particular for wild type animals short duration correlation, within 100 ms, has not been reported for PCs separated by more than 100 μm [6,23]. Based on anatomical data [8,24] we calculated the percentage input shared by two PCs as a function of their relative position in the PC layer. Fig. 1 shows the linear drop in shared input for increasing distance between the PC somata. Adjacent PCs should share about 50% of their afferent PFs and 20% of their presynaptic MLIs, since the respective connection probabilities are 0.5 [8] and 0.2 (see above). For a PC pair aligned along the PF axis the percentage of shared excitation vanishes at the length of the PFs (i.e. 5 mm) and the percentage of shared inhibition at the depth of the inhibitory basin (i.e. 100 μm). With respect to the sagittal axis common inhibitory input is available over a broader range than shared excitation. Two PCs separated by a distance between 250 and 375 μm along the sagittal axis would still share part of their inhibition but none of their excitatory input.

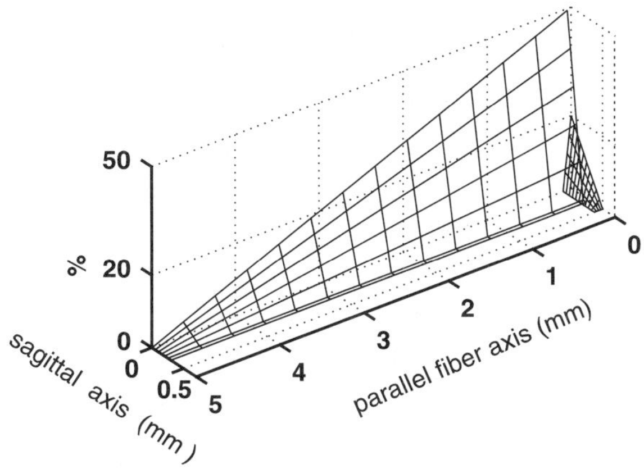

Fig. 1. Two PCs share a variable percentage of their input depending on their relative location. The plot shows the percentage of excitatory (full lines) and inhibitory (dot–dashed lines) input shared by two PCs, one positioned in the origin and another with somatic coordinates reported on the two horizontal axes.

4. Conclusions

We conclude that the common PF input to pairs of PC might not be strong enough to induce significant synchronous firing through a monosynaptic pathway. On the other hand, the common inhibitory input to PCs separated by at most 100 μm should induce synchronous firing. Nevertheless, synchronization of PCs separated by a longer distance along the PF axis may still be achieved through a disynaptic pathway.

Acknowledgements

Supported by EC QLRT-2000-02256 and HFSP RG0091/1999-B.

References

[1] P.P. Atluri, W.G. Regehr, Determinants of the time course of facilitation at the granule cell to Purkinje cell synapse, J. Neurosci. 16 (1996) 5661–5671.

[2] B. Barbour, Synaptic currents evoked in Purkinje cells by stimulating individual granule cells, Neuron 11 (1993) 759–769.

[3] B. Barbour, B.U. Keller, I. Llano, A. Marty, Prolonged presence of glutamate during excitatory synaptic transmission to cerebellar Purkinje cells, Neuron 12 (1994) 1331–1343.

[4] E. De Schutter, J. Bower, An active membrane model of the cerebellar Purkinje cell. I. Simulation of current clamps in slice, J. Neurophysiol. 71 (1994) 375–400.

[5] E. De Schutter, J. Bower, An active membrane model of the cerebellar Purkinje cell. II. Simulation of synaptic responses, J. Neurophysiol. 71 (1994) 401–419.

[6] T.J. Ebner, J.R. Bloedel, Correlation between activity of Purkinje cells and its modification by natural peripheral stimuli, J. Neurophysiol. 45 (1981) 948–961.

[7] J.C. Eccles, M. Ito, J. Szentágothai, The Cerebellum as a Neuronal Machine, Springer, Berlin, 1967.

[8] R.J. Harvey, R.M. Napper, Quantitative studies on the mammalian cerebellum, Prog. Neurobiol. 36 (1991) 437–463.
[9] M. Häusser, B.A. Clark, Tonic synaptic inhibition modulates neuronal output pattern and spatiotemporal synaptic integration, Neuron 19 (1997) 665–678.
[10] B. Hille, Ionic Channels of Excitable Membranes, Sinauer, Sunderland, Massachusetts, 1991.
[11] P. Isope, B. Barbour, Properties of unitary granule cell-Purkinje cell synapses in adult rat cerebellar slices, J. Neurosci. 15 (2002) 9668–9678.
[12] S. Kondo, A. Marty, Synaptic currents at individual connections among stellate cells in rat cerebellar slices, J. Physiol. 509 (1998) 221–232.
[13] L. Korbo, B.B. Andersen, O. Ladefoged, A. Møller, Total numbers of cell types in rat cerebellar cortex estimated using an unbiased stereological method, Brain Res. 609 (1993) 262–268.
[14] I. Llano, H.M. Gerschenfeld, Inhibitory synaptic currents in the stellate cells of rat cerebellar slices, J. Physiol. 468 (1993) 177–200.
[15] R. Maex, E. De Schutter, Synchronization of golgi and granule cell firing in a detailed network model of the cerebellar granule cell layer, J. Neurophysiol. 80 (1998) 2521–2537.
[16] T. Margrie, M. Häusser, Patch clamp recordings from cerebellar granule cells in vivo, 3rd Forum of European Neuroscience—Abstract Book, 2002, p. 126.12.
[17] E.G . Merril, P.D. Wall, T.L. Yaksh, Properties of two unmyelinated fibre tracts of the central nervous system: lateral Lissauer tract, and parallel fibres of the cerebellum, J. Physiol. 284 (1978) 127–145.
[18] J. Midtgaard, Membrane properties and synaptic responses of Golgi cells and stellate cells in the turtle cerebellum, J. Physiol. 457 (1992) 329–354.
[19] S.L. Palay, V. Chan-Palay, Cerebellar Cortex, Springer, Berlin, 1974.
[20] C. Pichitpornchai, J.A. Rawson, S. Rees, Morphology of parallel fibres in the cerebellar cortex of the rat: an experimental light and electron microscopic study with biocytin, J. Comp. Neurol. 342 (1994) 206–220.
[21] C. Pouzat, S. Hestrin, Developmental regulation of basket/stellate cell → Purkinje cell synapses in the cerebellum, J. Neurosci. 17 (1997) 9104–9112.
[22] M. Rapp, I. Segev, Y. Yarom, Physiology, morphology and detailed passive models of guinea-pig cerebellar Purkinje cells, J. Physiol. 474 (1994) 101–118.
[23] S.-L. Shin, V. Steuber, P. Fagerstedt, E. De Schutter, Correlation of simple spike activity in cerebellar Purkinje cells, Soc. Neurosci. Abstr. 2001, p. 294.2.
[24] F. Sultan, J.M. Bower, Quantitative Golgi study of the rat cerebellar molecular layer interneurones using principal component analysis, J. Comp. Neurol. 393 (1998) 353–373.

Available at
www.ComputerScienceWeb.com
POWERED BY SCIENCE DIRECT®

ELSEVIER

Neurocomputing 52–54 (2003) 103–108

NEUROCOMPUTING

www.elsevier.com/locate/neucom

Burst firing improves the detection of weak signals in spike trains

Jozien B.M. Goense[a,c], Rama Ratnam[a], Mark E. Nelson[a,b,c,*]

[a] *Beckman Institute for Advanced Science and Technology, University of Illinois, 405 N. Mathews, Urbana, IL 61801, USA*

[b] *Department of Molecular and Integrative Physiology, University of Illinois, Urbana, IL, USA*

[c] *Center for Biophysics and Computational Biology, University of Illinois, Urbana-Champaign, Urbana, IL 61801, USA*

Abstract

The detection of weak sensory signals is an important aspect of neuronal information processing. Behaviorally relevant signals are often encoded as perturbations of on-going spiking activity in primary afferents. Here, we show that a biologically plausible model, the leaky integrate-and-fire (LIAF) neuron, is capable of efficient and reliable detection of a single spike added to baseline activity. Detection performance is dependent on the statistical properties of the spike train. For the type of statistics considered here, an LIAF neuron can distinguish between a correct detection by means of burst firing, whereas false alarms tend to result in isolated spikes. The methods are illustrated by an application to electrosensory afferents of weakly electric fish.

Keywords: Signal detection; Integrate-and-fire neuron; Electrosensory system; Weakly electric fish

1. Introduction

The sensory environment of an animal is in a state of constant flux. Behaviorally relevant signals often appear at unknown times and may be so weak that they are obscured by irrelevant background activity. The task of higher-order neurons is to filter out the irrelevant information and extract what can be a weak but relevant signal. This has to be done in real time, i.e., on the basis of a continuous online examination of the

* Corresponding author. Beckman Institute for Advanced Science and Technology, University of Illinois, 405 N. Mathews, Urbana, IL 61801, USA.

E-mail address: m-nelson@uiuc.edu (M.-E. Nelson).

0925-2312/03/$ - see front matter
doi:10.1016/S0925-2312(02)00814-7

incoming spiking activity. The traditional framework for signal detection is trial-based testing employing spike count windows and a binary hypothesis test [4]. This requires repeated sampling in order to make a decision and is not suitable for online detection.

In a companion study [8], we show that the performance of a model neuron, the leaky integrate-and-fire neuron (LIAF), can approach that of a theoretically optimal online signal processing algorithm (the CUSUM algorithm) for detecting an abrupt, sustained shift in the interspike interval (ISI) distribution. Here we explore the performance of the LIAF neuron for detecting weak, transient changes in spike activity. We show that the statistical structure of the spike train can enhance the detection performance by indicating the presence of a signal with a burst of activity, whereas a false alarm (false positive) results in isolated spiking events spaced relatively far apart in time. We apply these methods to the detection of weak signals in noisy electrosensory afferents of weakly electric fish [2,7].

2. Methods

Extracellular recordings of baseline (spontaneous) P-type afferent activity were made from afferents of *Apteronotus albifrons* (black ghost knife fish). The experimental procedures are described in Xu et al. [9]. *A. albifrons* has a stable and precise EOD with a fundamental frequency that ranges from 1000 to 1400 Hz. Neurons fire at most one spike per EOD period, on an average once every three periods, making the EOD frequency a convenient sampling rate. Spike trains were thus discretized and represented as a string of 1's and 0's according to whether a spike was observed in an EOD period (see [7]). Recorded spike trains ranged from 200 to 450 s in duration. The baseline spike trains formed the "noise" data set for this study.

A weak signal was modeled by the random addition of a spike to the baseline data. The perturbed spike train was considered the "signal + noise" data set. To detect the added spike, the spike train was filtered using a leaky integrator, which simulates the LIAF neuron (Fig. 1). The integrator output was continuously compared to a fixed threshold. If a threshold crossing (hit) occurred, it was scored as a detection if the hit occurred within a time window (the signal window) following the added spike. The duration of the signal window was set to the time constant T of the leaky integrator. Detection probability (P_d) was computed by dividing the number of detections by the number of added spikes. If the hit did not occur within the signal window it was considered a false alarm. False alarm rates were computed from the baseline data set by dividing the number of false alarms by the duration of the recorded spike train. By testing over a range of thresholds an operating characteristic (OC) curve can be constructed, showing the false alarm rate versus the detection probability P_d. After the threshold crossing, the integrator input can be reset to zero, which will result in an absolute and relative refractory period. If the integrator is reset to zero, this has the effect that for a time of the order of the time constant, no new hits are likely. Reset to other values is also possible [1,5]. The extreme case occurs when the integrator is not reset after a hit. Both cases were tested in this study to determine the effect on detection performance.

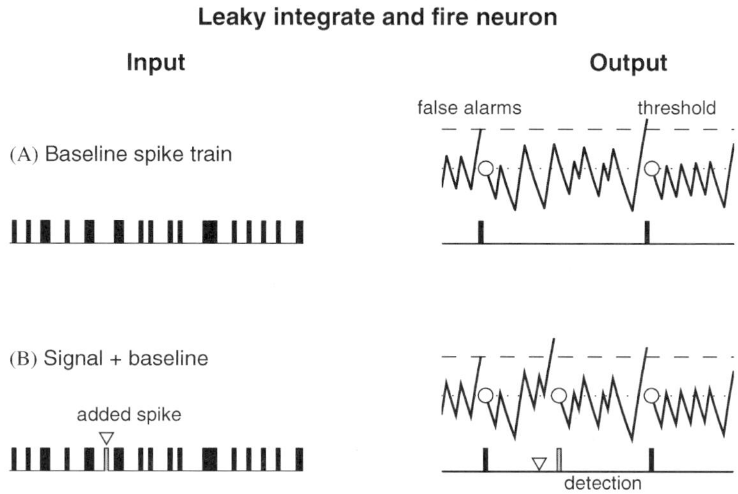

Fig. 1. An LIAF neuron can function as a signal detector. (A) Baseline spike train (left panel) is filtered by a leaky integrator (right). A threshold crossing (dashed line) results in an output spike, and the LIAF neuron is reset to the mean value of the input (dotted line). In this case, output spikes are false alarms and represent spontaneous firing of the LIAF detector. (B) A signal spike is added at an arbitrary location (triangle) to the baseline spike train (left panel); the task for the LIAF neuron is to detect the presence of this added spike. The LIAF neuron (right) generates an output spike within a few milliseconds of the added input spike (compare right panels in A and B). This particular output spike represents a correct detection.

Detections with partial resets generate burst-like activity in the LIAF neuron. We defined a burst as the presence of more than one hit within a time constant T of the leaky integrator. We evaluated the performance by considering only bursts of hits, i.e., single hits were ignored, and calculated the detection probabilities as a function of false alarm rate.

3. Results

In a simulated detection task, a P-type afferent spike train was perturbed by the addition of one spike at an arbitrary location. The goal of the detector was to determine *online* the presence of this added spike. Since such detection is presumably carried out at the next higher level of processing in the brainstem, the electrosensory lateral line lobe (ELL), we consider the IAF model to represent a neuron (pyramidal cell) in the ELL. Fig. 2 shows the signal detection performance of an IAF neuron with input a P-type afferent spike train (mean baseline firing rate for the unit shown is 449 ± 3.4 Hz, EOD $= 1061$ Hz). Spike trains from the P-type unit were filtered using a leaky integrator with time constant $T = 10$ EOD cycles (about 10 ms) and passed through a threshold element to generate spikes. The mean firing rate of the output (ELL) unit was adjusted by varying its threshold. In the absence of a signal in the input spike train, we call this the false alarm rate (see Fig. 2A). Detection

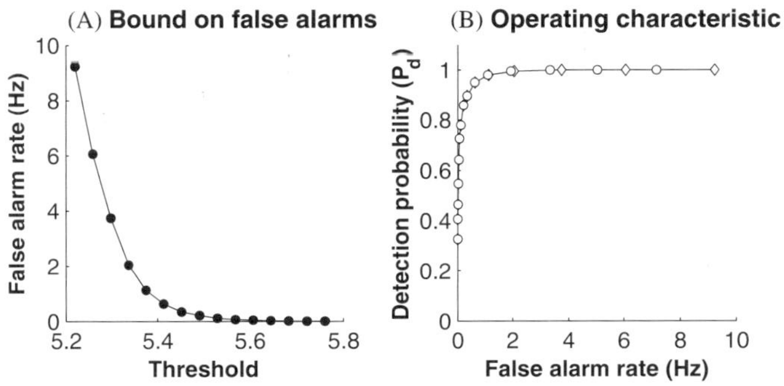

Fig. 2. Performance characterization of an LIAF as a detector. (A) False alarm rate versus threshold level of the LIAF model. (B) OC curve shows the probability of correct detection as a function of false alarm rate, when a single spike is added to an afferent spike train of *A. albifrons*. The detection probability for small false alarm rates (<1 Hz) is nearly 1, indicating that these neurons are extremely reliable detectors. The reset value of the integrator makes little difference (circles: reset to 0; lozenge: no reset).

performance of the IAF neuron was quantified using an operating characteristic (OC) curve (Fig. 2B), which measures detection probability as a function of the false alarm rate. It can be seen that the hypothetical ELL unit has near-ideal detection performance ($P_d = 1$) even for low false alarm rates (about 1 Hz).

To determine the effect of resetting the integrator, we measured detection performance by changing the reset value. This is shown in Fig. 2B for two extreme values: complete reset to 0 after firing (circles) and no reset (lozenge). The reset value affects signal detection ability only at high false alarm rates, but for the rates shown here (<10 Hz) it does not make much difference on the performance of the IAF unit. It should be noted that a good detector aims for a high probability of detection ($=1$) while keeping its false alarm rate as small as possible. For the unit shown in Fig. 2, the added spike can be detected easily with very high reliability and very low false alarm rate. Almost all units performed at comparable levels.

Raising the threshold level to reduce the false alarm rate has the undesirable effect of simultaneously lowering the detection probability (Fig. 2). One way to overcome this problem is to use bursting activity to distinguish between true detections and false alarms. This is illustrated in Fig. 3 by employing partial integrator resets (to illustrate, we consider the case of no reset). In Fig. 3A, it can be seen that for reasonably low thresholds, the number of multiple hits (i.e., bursts, lozenge) forms a significant proportion of the total number of detections (circles). However, at such low thresholds, the false alarm rate is high (see Fig. 2). On the other hand, the number of bursts during a false alarm are uniformly low over a wide range of false alarm rates (Fig. 3B, filled symbols). If we were to calculate performance based on all hits (single spikes or bursts), then at a threshold of 5.22 (false alarm rate of 9.3 Hz, Fig. 2A) the detection probability is almost 1 (Fig. 3A, circles). If only bursts are accepted, then the detection probability is only marginally lowered to 0.98 (Fig. 3A, lozenge), whereas the false alarm rate drops by a factor of 6–1.5 Hz (Fig. 3B). Thus, bursts provide a convenient

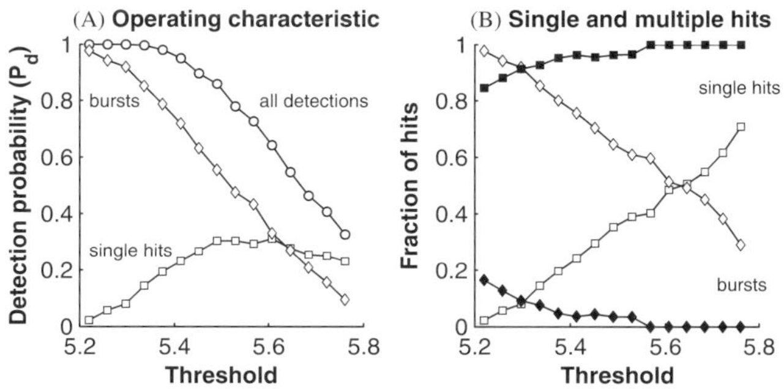

Fig. 3. Bursting activity is a reliable indicator of the presence of a weak signal. (A) Detection probability versus threshold, showing the total number of detections (circles) when there is no integrator reset. This is further partitioned into probabilities of single hits (squares) and bursts of hits (lozenge). (B) Fraction of single hits (squares) and bursts (lozenge) for detections (open symbols) and false alarms (closed symbols). The proportion of bursts during false alarms is significantly lower than single hit false alarms, whereas for detections, the opposite is true for a wide range of thresholds. This indicates that bursts are reliable indicators of a true hit, whereas false alarms are more likely to be single hits.

way to distinguish between correct detections and false alarms with a large increase in performance.

4. Discussion

This main findings of this work are that: (1) a simple leaky integrator with a threshold mechanism can function as an efficient online detector and (2) high false alarm rates are acceptable if correct detections occur in small bursts of spikes whereas false alarms occur as isolated spikes. In this case, isolated hits can be ignored with little penalty on detection performance. Gabbiani et al. [3] and Metzner et al. [6] showed that bursts of ELL neurons encode more features than isolated spikes and carry more useful information. Our finding from the viewpoint of signal detection also supports this.

The principal mechanism by which detections trigger bursts, whereas false alarms generate only isolated spikes, is based on the strong anti-correlations between adjacent ISIs in the baseline spike train [7]. Briefly, anti-correlations provide a firing rate stabilizing mechanism whereby fluctuations away from the mean rate are strongly resisted. Thus, a threshold crossing is brief with a quick return to mean firing rate. This tends to generate isolated spikes. On the other hand, adding a signal elevates the overall firing rate and so threshold crossings persist for longer duration, thus generating bursts.

Weak signals at the limit of sensory threshold may cause barely noticeable changes in the baseline firing activity of a neuron. To counter this, selection pressure may have evolved mechanisms to enhance detection performance. Among such mechanisms may be rate stabilization by anti-correlated ISIs, and consequently, the presence of bursting activity to emphasize presence of a signal. Since the nervous system processes information in stages, a hierarchy of neurons can successively filter information and refine the decision making scheme, and thereby achieve very high detection performance.

Acknowledgements

This work was supported by NSF Grant IBN-0078206 and NIH Grant R01-MH49242. We would like to thank Noura Sharabash for assistance with the data collection.

References

[1] G. Bugmann, C. Christodoulou, J.G. Taylor, Role of temporal integration and fluctuation detection in the highly irregular firing of a leaky integrator neuron model with partial reset, Neural Comput. 9 (1997) 985–1000.
[2] M.J. Chacron, A. Longtin, M. St-Hilaire, L. Maler, Negative interspike interval correlations increase the neuronal capacity for encoding time dependent stimuli, J. Neurosci. 21 (2001) 5328–5343.
[3] F. Gabbiani, W. Metzner, R. Wessel, C. Koch, From stimulus encoding to feature extraction in weakly electric fish, Nature 384 (1996) 564–567.
[4] D. Green, J. Swets, Signal Detection Theory and Psychophysics, Wiley, New York, 1966.
[5] P. Lansky, M. Musila, Variable initial depolarization in Stein's neuronal model with synaptic reversal potentials, Biol. Cybernet. 64 (1991) 285–291.
[6] W. Metzner, C. Koch, R. Wessel, F. Gabbiani, Feature extraction by burst-like spike patterns in multiple sensory maps, J. Neurosci. 18 (1998) 2283–2300.
[7] R. Ratnam, M.E. Nelson, Non-renewal statistics of electrosensory afferent spike trains: implications for the detection of weak sensory signals, J. Neurosci. 20 (2000) 6672–6683.
[8] R. Ratnam, J.B.M. Goense, M.E. Nelson, Change-point detection in neuronal spike train activity, Neurocomputing, this issue.
[9] Z. Xu, J.R. Payne, M.E. Nelson, Logarithmic time course of sensory adaptation in electrosensory afferent nerve fibers in a weakly electric fish, J. Neurophysiol. 76 (1996) 2020–2032.

Available at
www.ComputerScienceWeb.com
POWERED BY SCIENCE DIRECT

NEUROCOMPUTING

ELSEVIER

Neurocomputing 52–54 (2003) 109–115

www.elsevier.com/locate/neucom

Biases in white noise analysis due to non-Poisson spike generation

Jonathan W. Pillow[a,*], Eero P. Simoncelli[b]

[a] *Center for Neural Science, New York University, 4 Washington Pl 809, New York 10003, USA*

[b] *Howard Hughes Medical Center, and Center for Neural Science, New York University, New York, USA*

Abstract

White noise analysis methods for characterizing neurons typically ignore the dynamics of neural spike generation, assuming that spikes arise from an inhomogeneous Poisson process. We show that when spikes arise from a leaky integrate-and-fire mechanism, a classical white noise estimate of a neuron's temporal receptive field is significantly biased. We develop a modified estimator for linear characterization of such neurons, and demonstrate its effectiveness in simulation. Finally, we apply it to physiological data and show that spiking dynamics may account for changes observed in the receptive fields measured at different contrasts.

Keywords: White noise analysis; Spike generation

White noise analysis has become a widely used technique for characterizing response properties of spiking neurons in sensory systems. A sequence of stimuli are drawn randomly from an ensemble and presented in rapid succession, and one examines the stimuli that elicit action potentials. In the most widely used form of this analysis, one estimates a linear approximation to the receptive field (i.e. first-order Wiener kernel) by computing the spike-triggered average (STA); that is, the average stimulus preceding a spike (e.g. Refs. [6,8]). Under the assumption that spikes are generated by a Poisson process with instantaneous rate determined by linear projection onto a kernel followed by a static nonlinearity, the STA provides an unbiased estimate of the underlying kernel [5].

* Corresponding author.

E-mail addresses: pillow@cns.nyu.edu (J.W. Pillow), eero.simoncelli@nyu.edu (E.P. Simoncelli).

0925-2312/03/$ - see front matter
doi:10.1016/S0925-2312(02)00822-6

The white noise approach is considered to have several advantages over traditional characterization approaches, including the ability to explore a large portion of the input space, and receptive field estimation that is robust to drift or fluctuation in the responsiveness of a neuron. Despite these advantages, it has also become clear that there are drawbacks to the standard characterizations obtained with white noise methods. One such shortcoming is the well-known phenomenon that the shape of the STA varies with the amplitude (e.g. contrast) of the white noise stimuli. (e.g. Refs. [4,9,13]). This type of change cannot be explained by a linear model followed by a static nonlinearity and Poisson spike generation (the L–N-P model), since it implies a change in the linear front end. We have previously shown that nonlinear suppressive interactions such as those found in cortical neurons can explain biases in the STA, that a spike-triggered covariance analysis can be used to characterize these suppressive interactions, and that the resulting corrected model can account for the changes of STA with contrast [10].

Here, we explore another potential source of failure in white noise characterization: the assumption of Poisson spike generation. The significance of temporal dynamic (i.e. non-Poisson) properties of biological spike generation for white noise characterization of neurons has not been thoroughly analyzed [1]. However, we show that in simulated white noise experiments, a linear model which drives an integrate-and-fire spiking mechanism is inaccurately characterized by the STA. Furthermore, we show that the integrative behavior of this model can account for some of the changes in the STA estimated at different stimulus amplitudes in real neurons. Finally, we propose a new method for recovering the linear temporal filter governing neural response. We demonstrate through simulation that this approach can correctly estimate the linear kernel of a model neuron, and we also apply our method to real neural data, demonstrating that the recovered linear kernel is fairly stable with changes in stimulus contrast. We thus conclude that the recovered linear kernel may provide a more fundamental functional description of neural behavior, and might well be more directly related to the mechanisms underlying neural response.

1. Leaky integrate-and-fire model

Our analysis is based on a leaky integrate-and-fire (LIF) model. The input is convolved with a linear filter K, and this response drives a leaky integrator. When the level of this integrator reaches a threshold value, the neuron fires a spike and the integrator is reset to zero. The time evolution of the model membrane potential $V(t)$ is characterized by a single differential equation

$$\frac{\mathrm{d}V}{\mathrm{d}t} = -\frac{1}{\tau} V(t) + I(t), \tag{1}$$

[1] However, Arcas et al. have recently examined the behavior of a Hodgkin–Huxley model under white noise stimulation, and have made a number of interesting observations regarding the spike-triggered stimulus ensemble [2].

where τ is the time constant governing decay of the membrane potential, and $I(t)$ is the input current, generated by convolving the input signal $S(t)$ with the fixed kernel K:

$$I(t) = K * S(t) = \int_{-\infty}^{0} K(u)S(t-u)\,\mathrm{d}u. \tag{2}$$

This model has an analytical solution relative to the time of the most recent spike:

$$V(t) = \int_{t^-}^{t} I(u)\mathrm{e}^{(u-t)/\tau}\,\mathrm{d}u, \tag{3}$$

where t^- is the time of occurrence of the last spike before t. This dependence on the time of the previous spike (and past input to the integrator) represents a fundamental departure from L–N-P model described earlier, where the probability of firing a spike is an instantaneous function of the projection of the stimulus onto K.

2. Simulation results and comparison

We simulated a white noise analysis experiment with the model described above. In our simulations, the kernel K was chosen to be a 32-sample function whose shape loosely resembles temporal kernels measured in retinal ganglion cells. As in classical white noise experiments, we generated a random discrete stimulus $S(t)$ that was temporally white, drawing the stimulus intensity as an independent Gaussian random variable in each time step. We computed the STA as the average stimulus in the 32 time bins preceding each spike.

Fig. 1 (left) shows a plot of the actual kernel K superimposed on the STA for three different values of the membrane time constant τ. First, note that in all three cases, the

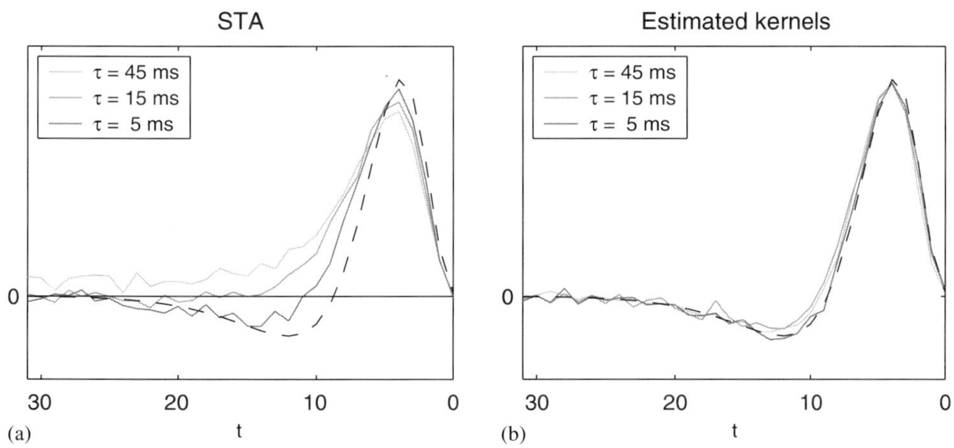

Fig. 1. Simulation of integrate-and-fire neuron. Left: STA kernels retrieved for three different contrast levels (solid lines), plotted along with the true model kernel (dashed curve). Right: Kernels recovered using our algorithm.

STA differs significantly from K. This bias reflects the integrative spiking mechanism of the LIF model, as the STA is quite close to K if the same input were given to an L–N-P model [5]. Furthermore, the discrepancy between K and the STA depends on τ. For small τ (i.e. rapid decay of V), the STA more closely resembles K, whereas larger τ (slower decay) gives rise to an STA which is smoother and more biased away from the true K. Note that although this basic effect is unsurprising, it is *not* the case that the STA shape arises simply from a low-pass filtering of K with an exponential filter. Specifically, the STA measured for a stand-alone LIF spike generator is decidedly non-exponential.

Physiological evidence indicates that at higher firing rates, the membrane conductance of neurons increases, which corresponds to a decrease in membrane time constant τ [1,3,7]. Moreover, STAs measured in real neurons at high contrast tend to be narrower than those measured at low contrasts. This suggests that an integrative spiking mechanism with time constant that depends on firing rate is at least consistent with contrast-dependent changes in the STA of real neurons.

3. Recovering the linear kernel

Assuming that the input to an integrate-and-fire spiking model is determined by projection onto a linear kernel, how can the kernel be recovered from the response to white noise stimuli? Eq. (3) provides a deterministic expression for the voltage at any time since the most recent spike. The voltage at any spike time is therefore given by

$$V(t^+) = V_{\text{th}} = \int_{t^-}^{t^+} [K * S(t)] \mathrm{e}^{(t-t^+)/\tau}\, dt, \tag{4}$$

where V_{th} is threshold, t^- is the time index of the previous spike and t^+ that of the current spike. Using Eq. (2), we can rewrite this (by switching the order of integration):

$$V_{\text{th}} = \int_{-\infty}^{0} K(u) \left[\int_{t^-}^{t^+} S(t-u) \mathrm{e}^{(t-t^+)/\tau}\, dt \right] du. \tag{5}$$

Note that, for fixed τ, this equation provides a linear constraint on K, since it expresses V_{th} as the inner product of K with the exponentially weighted S (back to the time of the previous spike). Every spike in the spike train provides one such constraint, so a discretized K can be overconstrained so long as its dimensionality is smaller than the number of spikes collected. K can easily be estimated by finding the least-squares solution to this overconstrained linear system.

In practice, one would like to estimate both τ and K simultaneously, since both are unknown for data collected in real neurons. This can be achieved simply using a nested optimization (a line search algorithm) to find the τ which minimizes the squared error in the least-squares solution for K. This algorithm is guaranteed to converge, and although the solution may be only a local minimum, in simulations it was well-behaved for a wide variety of kernel shapes and a large range of τ values. Fig. 1 (right) shows the kernels estimated for simulations conducted with three different values of τ. Close

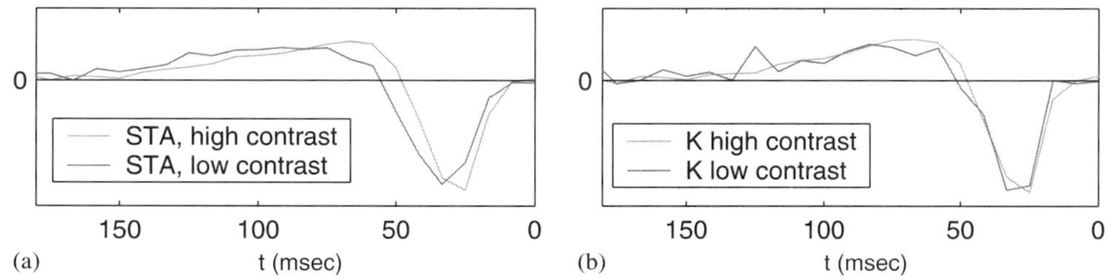

Fig. 2. Analysis of in vitro ganglion cell data in monkey retina. Left: STA estimates based on responses recorded at two different input contrast levels. Right: Kernels recovered using our procedure. The associated time constant estimates are 19.1 and 6.5 ms.

estimates of the true values of τ were also obtained. For both graphs in this figure, the stimulus contained 40,000 time samples and approximately 2000 spikes were collected for each τ.

It should be noted, finally, that this estimator for K and τ ignores a huge set of additional constraints—namely, that $V(t)$ be less than threshold at all times within the interspike interval. However, because the problem is already overconstrained by the constraint on $V(t)$ at spike times, and because the additional constraints are much harder to implement, they can be ignored. A significant improvement to the estimator may nevertheless be obtained by considering additional constraints only on the time steps immediately preceding a spike. (This can be implemented by allowing a contribution to the squared error for any pre-spike time bin where V exceeds threshold). Monte Carlo simulations exhibit rapid convergence to the true values of K and τ for this revised estimator.

4. Recovering a kernel from neural data

Our procedure for linear kernel estimation is based on an overly simplistic integrate-and-fire model for neural spike generation. We thus cannot be sure it will be applicable to real neural data. But we note that STA techniques have been used for decades to estimate linear kernels under the assumption of a Poisson spike generator. The integrate-and-fire model incorporates a dependence on the time of the previous spike and is likely to provide a more accurate description of spiking in real neurons.

We have applied our procedure directly to data drawn from a monkey retinal ganglion cell [4]. The data were recorded in vitro, using a stimulus consisting of 80,000 time samples of full-field 120 Hz flickering binary white noise. The stimulus vectors $\vec{s}$ of this sequence are defined over a 25-segment (0.21 s) time window. Two data sets were recorded, at contrasts of 32% and 64%.

Fig. 2 (left) shows STA estimates for both contrast levels. The kernels are quite different; the low-contrast STA is smoother and its peak is shifted earlier in time than the high-contrast STA. Fig. 2 (right) shows the kernels resulting from our estimation procedure. Note that the estimated kernel is now quite stable across different contrasts, a desirable property for a functional description of neural behavior. The recovered time constants of 19.1 and 6.5 ms are within ranges considered biologically plausible, although their ratio indicates a greater change with amplitude than is commonly reported for cortical neurons (e.g. Refs. [1,3,7]).

5. Discussion

Our results show that spike generation mechanisms can affect the interpretation of results obtained with white noise analysis. In particular, we have shown that even for a simple integrate-and-fire model, the temporal STA does not accurately recover the temporal linear input kernel. For this model, the magnitude of bias in the STA is influenced by the membrane conductance, which is believed to vary with stimulus strength. This amplitude-dependence of the STA mirrors changes in the STA of real neurons measured at different contrasts, and cannot be captured by an L–N-P model.

Based on this simple LIF model, we have developed a new method for the recovery of the linear kernel and integration time constant from responses to white noise stimuli. To our surprise, this kernel estimation procedure recovers a stable linear kernel when applied to data recorded from monkey retinal ganglion cells, and the associated estimates of membrane conductance are within a biologically plausible range. Finally, while not discussed here, our technique also appears to be quite robust to the presence of noise in the membrane potential.

We are currently exploring the generalization of these results to more realistic models. In particular, we have found that the incorporation of a voltage floor in the model (corresponding to an ionic reversal potential) produces an STA which is sharper and closer to the true input kernel at high contrast, independent of any changes in membrane conductance. The significance of this phenomenon, along with that of other nonlinearities associated with spike generation, remains to be analyzed.

Our results suggest a mechanistic explanation of the behaviors captured by current functional models of retinal ganglion cells (e.g. Ref. [12]), in which a nonlinear feedback signal is used to adjust the gain of the neuron. We have also previously shown that nonlinear gain control operations might account for a variety of apparent changes in receptive field properties at different contrast levels [11]. The results presented in this paper suggest that some such changes might be due to intracellular mechanisms of spike generation. It would be interesting to test such hypotheses against intracellular measurements.

Acknowledgements

We thank E.J. Chichilnisky and Divya Chander for providing electrophysiological data, and Liam Paninski and Michael Shadlen for helpful discussions. This research was supported by the Sloan–Swartz Theoretical Visual Neuroscience Center at NYU, and by the Howard Hughes Medical Institute.

References

[1] J.S. Anderson, M. Carandini, D. Ferster, Orientation tuning of input conductance, excitation, and inhibition in cat primary visual cortex, J. Neurophysiol. 84 (2000) 909–926.

[2] B.A.Y. Arcas, A.L. Fairhall, W. Bialek, What can a single neuron compute? in: T. Leen, T. Dietterich, V. Tresp (Eds.), Advances in Neural Information Processing Systems, Vol. 13, 2000, MIT Press, Cambridge, MA, pp. 75–81.

[3] L. Borg-Graham, C. Monier, Y. Fregnac, Visual input evokes transient and strong shunting inhibition in visual cortical neurons, Nature 393 (1998) 369–373.

[4] D. Chander, E.J. Chichilnisky, Adaptation to temporal contrast in primate and salamander retina, J. Neurosci. 21 (24) (2001) 9904–9916.

[5] E.J. Chichilnisky, A simple white noise analysis of neuronal light responses, Network: Comput. Neural Syst. 12 (2) (2001) 199–213.

[6] E. deBoer, P. Kuyper, Triggered correlation, IEEE Trans. Biomed. Eng. 15 (1968) 169–179.

[7] J. Hirsch, J.M. Alonso, C.R. Reid, L. Martinez, Synaptic integration in striate cortical simple cells, J. Neurosci. 15 (1998) 9517–9528.

[8] J.P. Jones, L.A. Palmer, The two-dimensional spatial structure of simple receptive fields in the cat striate cortex, J. Neurophysiol. 58 (1987) 1187–11211.

[9] K.J. Kim, F. Rieke, Temporal contrast adaptation in the input and output signals of salamander retinal ganglion cells, J. Neurosci. 21 (1) (2001) 287–299.

[10] O. Schwartz, E.J. Chichilnisky, E.P. Simoncelli, Characterizing neural gain control using spike-triggered covariance, in: T.G. Dietterich, S. Becker, Z. Ghahramani (Eds.), Adv. Neural Information Processing Systems, Vol. 14, MIT Press, Cambridge, MA, 2002, pp. 269–276.

[11] O. Schwartz, E.P. Simoncelli, Natural signal statistics and sensory gain control, Nat. Neurosci. 4 (8) (2001) 819–825.

[12] R.M. Shapley, J.D. Victor, How the contrast gain control modifies the frequency response of cat retinal ganglion cells, J. Physiol. (London) 318 (1981) 161–171.

[13] S.M. Smirnakis, M.J. Berry, D.K. Warland, W. Bialek, M. Meister, Adaptation of retinal processing to image contrast and spatial scale, Nature 386 (1997) 69–73.

Available at
www.ComputerScienceWeb.com
POWERED BY SCIENCE @ DIRECT®

ELSEVIER

Neurocomputing 52–54 (2003) 117–123

NEUROCOMPUTING

www.elsevier.com/locate/neucom

Temporal correlations of orientations in natural scenes

Christoph Kayser*, Wolfgang Einhäuser, Peter König

Institute of Neuroinformatics, ETH/UNI Zürich, Winterthurerstrasse 190, 8057 Zürich, Switzerland

Abstract

The visual system performs complicated operations such as visual grouping efficiently on its natural input. To study this adaptation to natural stimuli we measure spatio-temporal interactions of orientations in scenes with natural temporal structure recorded using a camera mounted to a cat's head. We find long range spatial and long lasting temporal correlations of orientations with collinear interactions being most prevalent and preserved over time. The spatial extent of correlations corresponds to the length of horizontal cortical connections and the temporal duration of the interactions allows co-activation of lateral and bottom up input by the same visual event.

Keywords: Natural scenes; Orientations; Gestalt principles; Image statistics; Temporal coherence

1. Introduction

In recent years processing of natural stimuli by the visual system received increased attention (cf. Ref. [12]). Indeed it was found that early stages of visual processing are specifically adapted to the structure of natural scenes [1,4]. Furthermore, laws for object perception and visual grouping, the Gestalt rules [8,14], can be linked to the statistics of natural scenes. As an example the law of good continuation, favouring collinear arrangements of orientations over parallel, was shown to have a counterpart in the interaction of orientations in still images [7,9,11]. Similar interactions of orientations are also found in contextual effects in psychophysical experiments [6,10], in surround interactions in V1 receptive fields [6] and in lateral connections in V1 [2,5]. Therefore it is of particular interest to link them to properties of natural scenes.

* Corresponding author. Tel.: +41-1-6353044; fax: +41-1-6353053.
E-mail address: kayser@ini.phys.ethz.ch (C. Kayser).

0925-2312/03/$ - see front matter © 2003 Elsevier Science B.V. All rights reserved.
doi:10.1016/S0925-2312(02)00789-0

To our knowledge, however, up to now correlations in natural scenes have only been investigated in still images. This neglects the temporal structure and it remains unclear whether these correlations persist on time scales relevant for lateral interactions in the cortex. Given the possibly long delays for tangential connections, correlations must extend over substantial temporal periods in order to fully cover the spatial extent of long-range connections. Furthermore, some of the previous studies did not report filter or correlation scales in units of degrees of visual angle leaving possible links to anatomical scales uncertain. Finally, some of the previous studies used still images captured by humans, therefore introducing a possible artistic or anthropocentric bias.

Here we address these issues and study spatio-temporal interactions of orientations in a large database of natural movies captured by a camera mounted to a cat's head.

2. Methods

We recorded movie sequences using a removable lightweight CCD-camera (Conrad electronics, Hirschau, Germany) mounted to the head of cats while taken for walks in different local environments like grassland, forest and the university campus. These videos contain a large variety of different speeds and accelerations as a result of the natural movements of the cat. Fig. 1 shows four sample images of our database. For this study a total of three animals was used and all procedures are in agreement with national and institutional guidelines for animal care.

Videos were recorded via a cable connected to the leash onto a standard VHS-VCR (Pal) carried by the human experimenter and digitised offline at a temporal resolution of 25 Hz, 320×240 pixels (1 pixels $\approx$ 12 min of arc) and 16 bit color depth. For this study videos were converted to 8-bit gray scale and 12 sequences (about 40 000 frames total) were used. Before further processing the images were normalized to zero mean.

The image statistics was investigated using oriented wavelets. Single frames were convolved with pairs of circular Gabor wavelets of 90° relative phase shift. Filters had a envelope of 20 pixel width and a spatial frequency of 7 (1/pixels). The amplitude of the orientation was computed by summing the squared amplitudes of two phase

Fig. 1. Four sample frames of our database are shown on the left. The amplitudes of the oriented energy detectors for the same frames is shown on the right. The bards indicate the orientation of the respective filters used.

shifted filters and subjecting the result to a square root, resembling a two subunit energy model. At each point the amplitudes of eight equally spaced orientations from 0° (horizontal) to 157.5° were computed. We define the 'prominent' orientation of each point by averaging the amplitude vectors (length = amplitude of filter response, orientation = orientation of the filter) of the eight filters. The resulting vector average has an orientation Θ, defining the prominent orientation of the point, and a length $A(\Theta,x,t)$, specifying the magnitude of the local orientation strength. For computational convenience these orientations Θ were binned into 16 bin between 0° and 180°. The second order statistics of these orientations was calculated assuming translation invariance of natural images. Thus correlations of two prominent orientations Θ_1 and Θ_2 were computed over all pairs of points with the same spatial separation Δx and temporal separation Δt (the mean $\langle\rangle$ runs over all points (x,t) with prominent orientation Θ_1).

$$C(\Theta_1,\Theta_2,\Delta x,\Delta t) = \frac{\langle (A(\Theta_1,x,t)-\langle A(\Theta_1,x,t)\rangle) * (A(\Theta_2,x+\Delta x,t+\Delta t)-\langle A(\Theta_2,x+\Delta x,t+\Delta t)\rangle)\rangle}{\sqrt{\langle (A(\Theta_1,x,t)-\langle A(\Theta_1,x,t)\rangle)^2\rangle} * \sqrt{\langle (A(\Theta_2,x+\Delta x,t+\Delta t)-\langle A(\Theta_2,x+\Delta x,t+\Delta t)\rangle)^2\rangle}}.$$

Correlations were computed for temporal lags from $\Delta t = 0$ to 30 frames (1.2 s) and on a spatial grid of points spaced about 2° apart. Therefore the kernels overlapped only for the smallest spatial distance used. As a control we also computed correlations using the maximally active orientation at each point instead of the 'prominent' orientation yielding similar results as reported below.

3. Results

First we investigate temporal correlations at the same point in space. Fig. 2A demonstrates that if an orientation is present at one point in time then the amplitude of this orientation in the next frames at the same point is also likely to be high. Temporal correlations are strongest for the cardinal orientations, i.e. horizontal and vertical. For the other orientations correlations decay faster but are still significant over several hundreds of milliseconds (decay time constants for 0°: > 1 s, 45°: 490 ms, 90°: 900 ms, 135°: 360 ms). Thus the presence of an oriented segment gives a strong prediction for the orientation at the same point later in time.

Next we look at the two dimensional spatial distribution of correlations as well as correlations of different orientations. Fig. 2B shows the correlations between segments of 4 different orientations (0°, 45°, 90°, 135°) situated at different relative locations in the same frame. Iso-orientation correlations (panels on the diagonal) are stronger than cross-orientation correlations. Furthermore the contour lines of the iso-orientation correlations are elongated along the direction of the particular orientation. This shows that collinear structures are more prevalent than parallel shifted contours. Also parallel contours occur more likely than T-junctions since the iso-orientation correlations are at all points stronger than the correlations of this orientation with the orthogonal. An example of how the spatial correlations decay independently of the spatial direction is

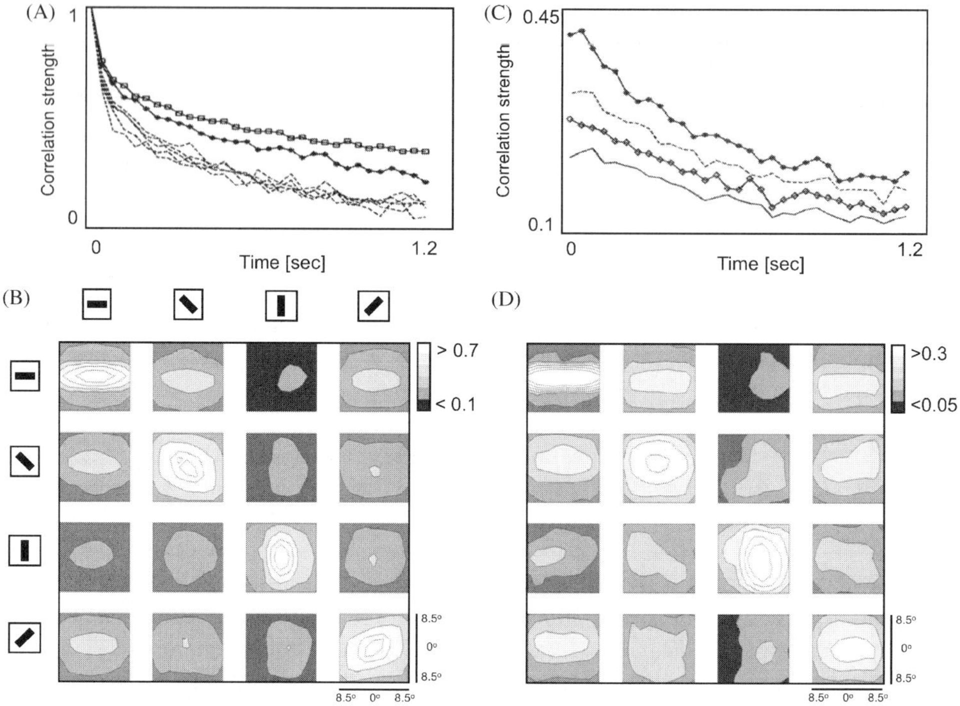

Fig. 2. (A) The correlation of orientation amplitude over time at the same pixel. Squares: 0°, stars: 90°, dashed: all other orientations (spaced 22.5°). (B) Correlations of different combinations of orientations and different spatial arrangements of the two points in the same frame. The orientations are (from top to bottom and left to right): 0°, 135°, 90°, 45°. (C) Correlations over time of points with prominent horizontal orientation but which are spatially separated by different distances independent of the relative orientation. Squares: 2.1 deg spatial distance, stars: 4.2°, dashed: 6.4°, diamonds: 8.4°. (D) Same as in B but here the two points are also separated by 400 ms in time.

shown in Fig. 2C for the horizontal (90°) orientation. Correlations decay fastest during the first 2° of spatial distance but extend well up to 8°.

Our data set allows analyzing how these spatial correlations evolve over time. Fig. 2D shows the same data as in Fig. 2B but for segments 400 ms apart in time. The spatial arrangement of correlations is the same as for zero time lag but the amplitudes decayed by a factor higher than 2. For the cardinal orientations again collinear interactions are prevalent. This is in agreement with Fig. 2A which shows that these orientations are very stable over time. Since the oblique orientations are less well correlated over time we would expect that collinearity will here be less prominent for larger time lags. Indeed the contour lines of the correlations for the oblique orientations are more circular symmetric. To quantify these changes over time we measure the aspect ratio (length/width) of the contour lines for the different time lags. Collinearity means a high aspect ratio and a loss of collinearity therefore is accompanied with a decrease

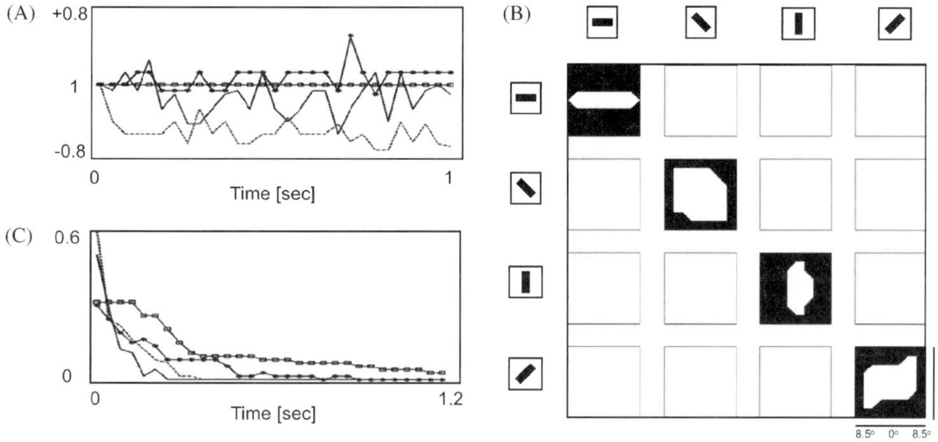

Fig. 3. (A) Relative change of the aspect ratio of the correlation contours in Fig. 2C as a function of time. Shows is the aspect ratio at each point in time divided by the aspect ratio at $t = 0$. Squares: 0°, stars: 90°, solid: 45°, dashed: 135°. (B) Areas of strong correlations. We defined spatio temporal separations with a correlation over 0.4 as strong. The figure shows these areas for the correlation diagram of Fig. 2C. (C) Shows the size of these areas relative to the total patch size over time. Lines are labeled as in A.

in aspect ratio. Using this measure, Fig. 3A shows that collinearity is preserved over long temporal lags and is strongest for the cardinal orientations.

To quantify the change in amplitude of the spatial correlations in a different way, we define areas of strong interactions by thresholding correlations. We chose a threshold of 0.4 to ensure that even for zero time-lag only iso-orientation correlations exceed this threshold (Fig. 3B). As expected from Fig. 2 the decay times are slowest for the cardinal orientations but independent of the orientation there exist points with strong correlations for at least 280 ms (Fig. 3C).

We performed controls to see how these results depend on the amount of data used. The above data were averaged over our whole database. Since one feature of our video sequences is their variety in terms of landscapes, etc. we look at the differences between different sequences. In Fig. 4 we show the correlations for one oblique orientation (135°). The mean and standard deviation over 12 video sequences is shown in Fig. 4A. The error is rather small compared to the correlation values. More importantly, the correlation surface plus minus the error (Fig. 4B) shows the same spatial structure as the mean. Also, the distinct pattern of correlations is visible in averages over shorter sequences (data not shown). Thus the distinct patterns of spatial correlations are not introduced by averaging over a large data set.

As a further control, we use filters of a different spatial scale and frequency to measure the orientation content. The filters used for Fig. 4C are twice as large as the ones used for the other experiments. The results are basically the same as with the lower frequency filters. Again collinearity is most prevalent. Therefore our results generalize over a wide range of filter parameters.

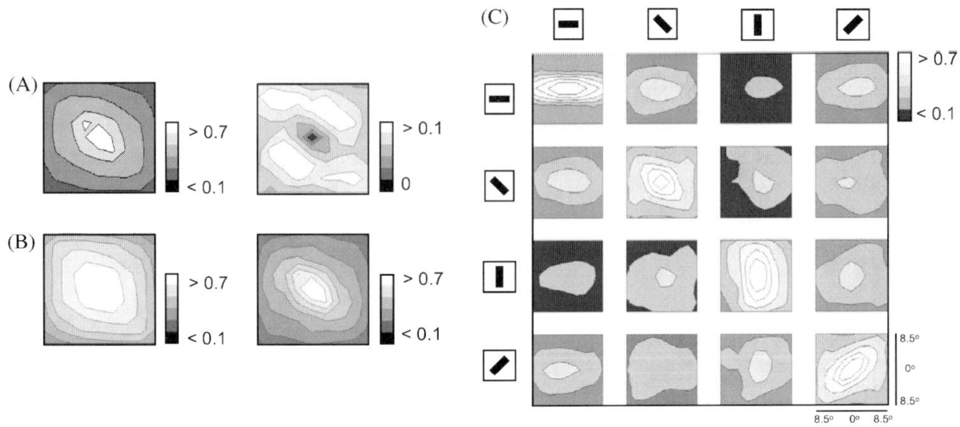

Fig. 4. (A) For an example orientation (45°) we show the mean (left) over 12 video sequences together with the standard deviation (right). (B) The mean plus/minus the standard deviation. (C) Cross orientation correlations over space for filters of a higher spatial frequency.

4. Discussion

We recorded natural image sequences from a camera mounted to a cat's head closely matching the animal's visual input. Thereby our database circumvents possible artistic or anthropocentric biases introduced in pictures and movies taken by humans. The database contains a large set of different environments, ranging from forest to grasslands and university campus. Furthermore the used sequences were recorded in different seasons and times of day providing a huge variety of lighting conditions. In respect to the temporal analysis it is worth noting that our video sequences contain natural movements of an animal, which might differ considerably from e.g. commercial movies filmed by humans.

In qualitative agreement with previous studies [7,9,11] we find spatial correlations corresponding to the Gestalt laws. For all orientations collinear contours are more prevalent than parallel contours and correlations between orthogonal orientations are weakest. However we find correlations over distances of up to 8 degrees of visual angle (Fig. 2). This is considerably larger than distances reported in previous studies. For example Kaschube et al. [7] find that already for small distances correlations are relatively weak (< 0.15 in a range from 1° to 4°). However, they do not indicate the size of their kernels in the same units. Sigman et al. [11] report similar correlations using filters of size smaller than 10 min of arc. Our higher correlation could be due to methodological differences to other studies besides the use of different and possibly larger kernels. We computed the 'prominent' orientation of a point by vector averaging the outputs of 8 oriented energy detectors. But correlations computed on these prominent orientations are very similar to correlations computed on the maximally active orientation (data not shown) a method used in Ref. [11].

The spatial distances of the correlations reported here fit well with anatomical data on long-range horizontal connections in primary visual cortex. In cat V1 8° of visual angle correspond roughly to 8 mm [13]. This is also the extent of long-range connections which preferentially connect iso-orientation domains [5] and in some mammals preferentially mediate collinear interactions [2].

In the temporal domain we find long lasting correlations of orientations to extend several hundreds of milliseconds preserving their spatial structure i.e. collinearity. These persist sufficiently long to allow bottom up and long range lateral input to be coactive and driven by the same orientated structure even given the slow speeds of lateral connections reported in Ref. [3]. Therefore the spatio-temporal interactions of orientations seem to fully cover the range of tangential connections and provide a substrate that could also guide the development of orientation maps and long-range connections in primary visual cortex.

Acknowledgements

This work was financially supported by the Centre of Neuroscience Zurich, (ZNZ), Honda R& D Europe (Germany) and the Swiss national fund (SNF grant No. 31-65415.01).

References

[1] J.J. Atick, A.N. Redlich, What does the retina know about natural scenes, Neural Comput. 4 (1992) 196–210.
[2] W.H. Bosking, Y. Zhang, B. Schofield, D. Fitzpatrick, Orientation selectivity and the arrangement of horizontal connections in tree shrew striate cortex, J. Neurosci. 17 (6) (1997) 2112–2127.
[3] V. Bringuier, F. Chavane, L. Glaeser, Y. Fregnac, Horizontal propagation of visual activity in the synaptic integration field of area 17 neurons, Sci. 283 (1999) 695–699.
[4] Y. Dan, J.J. Atick, R.C. Reid, Efficient coding of natural scenes in the lateral geniculate nucleus: experimental test of a computational theory, J. Neurosci. 16 (10) (1996) 3351–3362.
[5] C.D. Gilbert, T.N. Wiesel, Columnar specificity of intrinsic horizontal and corticocortical connections in cat visual cortex, J. Neurosci. 9 (7) (1989) 2432–2442.
[6] M.K. Kapadia, M. Ito, C.D. Gilbert, G. Westheimer, Improvement in visual sensitivity by changes in local context: Parallel studies in human observers and in V1 of alert monkeys, Neuron 15 (1995) 843–856.
[7] M. Kaschube, F. Wolf, T. Geisel, S. Löwel, The prevalence of collinear contours in the real world, Neurocomputing 38–40 (2001) 1335–1339.
[8] K. Kofka, Principles of Gestalt Psychology, Harcourt & Brace, New York, 1935.
[9] N. Krueger, Collinearity and parallelism are statistically significant second order relations of complex cell responses, Neural Process. Lett. 8 (1998) 117–129.
[10] U. Polat, D. Sagi, Spatial interactions in human vision: from near to far via experience dependent cascades of connections, Proc. Natl. Acad. Sci. USA 91 (1994) 1206–1209.
[11] M. Sigman, G.A. Cecchi, C.D. Gilbert, M.O. Magnasco, On a common circle: natural scenes and Gestalt rules, Proc. Natl. Acad. Sci. 98 (4) (2001) 1935–1940.
[12] E.P. Simoncelli, B.A. Olshausen, Natural image statistics and neural representation, Ann. Rev. Neurosci. 24 (2001) 1193–1215.
[13] R.J. Tusa, A.C. Rosenquist, L.A. Palmer, Retinotopic organization of areas 18 and 19 in the cat, J. Comp. Neurol. 185 (1979) 657–678.
[14] M. Wertheimer, Laws of Organization in the Perceptual Form, Harcourt & Brace, USA, 1938.

Available at
www.ComputerScienceWeb.com
POWERED BY SCIENCE @ DIRECT®

ELSEVIER

Neurocomputing 52–54 (2003) 125–133

NEUROCOMPUTING

www.elsevier.com/locate/neucom

Role of synaptic feedback and intrinsic voltage-gated currents in shaping cone light responses

Garrett T. Kenyon[a,b,*], Bryan J. Travis[b,c], David W. Marshak[d]

[a] *P-21 Biophysics, MS D454, USA*

[b] *Los Alamos National Laboratory, Los Alamos, NM 87545, USA*

[c] *EES-6 Geological Processes, Simulation, and Modeling, MS T003, USA*

[d] *Department of Neurobiology and Anatomy, University of Texas Health Science Center, 6431 Fannin, Houston, TX 77030, USA*

Abstract

A model of the outer primate retina was used to investigate how cone light responses could be shaped by both synaptic feedback from horizontal cells and intrinsic voltage-gated conductances. Published data was used to estimate the photoconductance, convergence/divergence ratios, synaptic transfer functions, passive membrane properties and voltage-gated conductances, the latter using a Levenberg–Marquardt search algorithm. When the cone resting potential was sufficiently depolarized, synaptic feedback from horizontal cells could account for the flash response measured experimentally. Voltage-gated conductances were not significantly activated during the flash response but rather maintained the cone membrane potential inside a narrow physiological range.

Keywords: Retina; Computer simulation; Outer plexiform layer; Horizontal cell

1. Introduction

Mammalian photoreceptors represent a potentially useful system for examining how the input/output properties of biological neurons are influenced by two factors: (1) synaptic feedback and (2) intrinsic voltage-gated conductances. Here, we use the

* Corresponding author. P-21 Biophysics MS D454, Los Alamos National Laboratory, Los Alamos, NM 87545, USA.

E-mail address: gkenyon@lanl.gov (G.T. Kenyon).

0925-2312/03/$ - see front matter
doi:10.1016/S0925-2312(02)00790-7

light-induced photocurrent as a measure of the input to the cone photoreceptor and the resulting membrane voltage as a measure of photoreceptor output. Voltage responses recorded from the inner segments of primate cones following brief light flashes are more transient than the underlying photocurrents. Photocurrents elicited by a 10 ms light flash show an average time-to-peak in dark adapted cones of 51 ms [12] for non-saturating flash intensities. In contrast, the time-to-peak of the voltage response generated under the same conditions decreases from 35 ms for dim flashes to 10 ms for brighter but still non-saturating intensities [13]. The pronounced difference in the time course of the voltage response compared to the photocurrent demonstrates that there is a non-linear input/output transformation at the very first synapse of the visual system.

In order to better understand the physiological factors that determine the input/output characteristics of cone photoreceptors when embedded within the circuitry of the outer retina, a computer model was used to investigate the relative contributions of synaptic interactions and voltage-gated conductances. To isolate the contribution due to synaptic interactions, an anatomically realistic model of the outer retina was developed in which cones and horizontal cells were initially implemented as passive elements that contained no explicit voltage-gated conductances. With physiologically reasonable assumptions, the passive model was able to reproduce the measured responses of primate cones to brief light flashes. Voltage-gated conductances that have been described in both cones and horizontal cells were then added to the model to investigate how intrinsic membrane properties might also contribute to signal processing in the outer retina. Our results suggest that for primate cones, synaptic feedback is the primary factor in shaping light responses, whereas voltage-gated conductances act primarily to keep the cone membrane potential within a relatively narrow operating range.

2. Methods

Simulations were conducted using the LCM3D simulator [16]. The model consisted of an 8×8 array of cones and a 4×4 array of H1 horizontal cells connected with convergence/divergence ratios inferred from anatomical studies [18]. Each cone made excitatory synapses ($V_{eq} = -10$ mV) onto 4 H1 cells, and each H1 cell made inhibitory synapses ($V_{eq} = -55$ mV) onto all 16 cones within its dendritic field. The assumption that horizontal cell inhibition is mediated by conventional GABA receptors is supported by evidence that mammalian cones respond to GABA [8], although alternative inhibitory mechanisms have been described in goldfish [6]. Synaptic release was modeled using Fermi functions. For cones, the transfer function had a gain of 1.75 mV ($g_{max} = 2$ nS, $V_{1/2} = -30$ mV), consistent with the kinetics of a non-inactivating calcium current in the primate cone [19] and a fourth power dependence on calcium influx [1]. The transfer function governing horizontal cell output had a gain of 15 mV ($g_{max} = 20$ pS, $V_{1/2} = -10$ mV), based on a calcium-independent release mechanism [14]. The synaptic transfer function of the model cones was thus much more non-linear than that of the horizontal cells (Fig. 1). The photoconductance of the cone ($V_{eq} = -10$ mV, $g_{max} = 58$ pS) was adapted from a realistic photocurrent

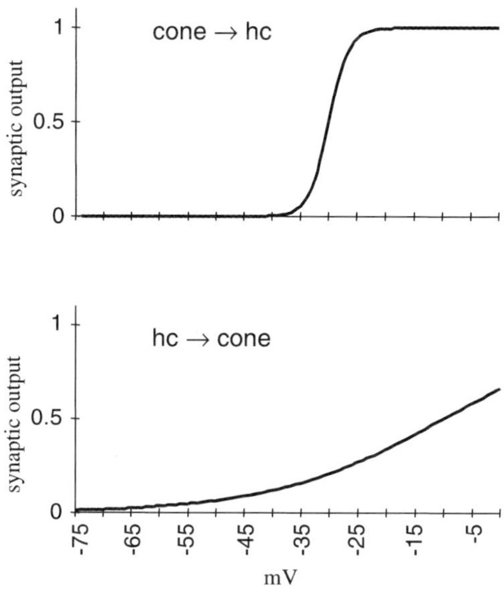

Fig. 1. Synaptic transfer functions in retinal model. Cone output to horizontal cells (top) depends non-linearly on membrane potential [1] while horizontal cell feedback to cones (bottom) is approximately linear [14].

model [12]. Passive membrane properties of the model cone ($R_{in} = 1$ G$\Omega, \tau = 10$ ms) were based on measured values in lower vertebrate photoreceptors after taking into account damage to the cell membranes resulting from electrode penetration [2,4]. The resting potential of the model cone (in the absence of a photocurrent) was set to -40 mV. This relatively depolarized resting potential, which was necessary to account for the plateau phase of the cone's flash response, might result from cGMP-gated cation channels [10,11]. The resting potentials of the H1 cells were set to -60 mV, based on cat type-B horizontal cells [17]. H1 input impedance ($R_{in} = 1$ GΩ) was based on data from cultured postnatal rabbit horizontal cells [7]. The passive H1 membrane time constant was set to 20 ms. The H1 time constant was not well constrained by available experimental data but its precise value did not affect our main conclusions.

For the active membrane model, voltage-gated conductances were implemented as previously described [5]. Hodgkin–Huxley activation and inactivation parameters were fit to voltage-clamp data recorded from primate cones [19] and to similar data from cat [17] and cultured rabbit [7] horizontal cells using a Levenberg–Marquardt search algorithm [9]. The calcium-dependent chloride current present in the primate cone was based on a previously developed model of a mammalian calcium-dependent potassium current [5], except that the dependence of the chloride current on intracellular Ca^{2+} was strongly non-linear. Space does not permit a full description of each of the voltage-gated currents included in the active model. However, our conclusions regarding the contribution of these currents to light responses are based on their general order of magnitude and should not depend critically on the details of their implementation.

3. Results

3.1. Passive model

A passive model, containing no explicit voltage-gated conductances, was able to qualitatively account for the main features of cone responses to brief light flashes (Fig. 2). For dim flashes, voltage responses are hyperpolarizations similar in duration to the underlying photocurrents. For brighter flashes, however, voltage responses develop an initial sharp transient that decays to a plateau potential several milli-volt below the dark resting potential. Such effects have been observed in primate cones in response to similar stimuli [12,13]. In the model cone, the attenuation of the voltage response at bright flash intensities is due to negative feedback from horizontal cells. During the initial response to a bright light flash, the light sensitive conductances within the cone outer segments approach zero, causing both cones and horizontal cells to become hyperpolarized. The resulting reduction in synaptic inhibition from horizontal cells causes the cone membrane potential to repolarize towards the resting potential in the absence of a dark current, assumed to be −40 mV. Repolarization is not complete, however, because of residual inhibition from horizontal cells.

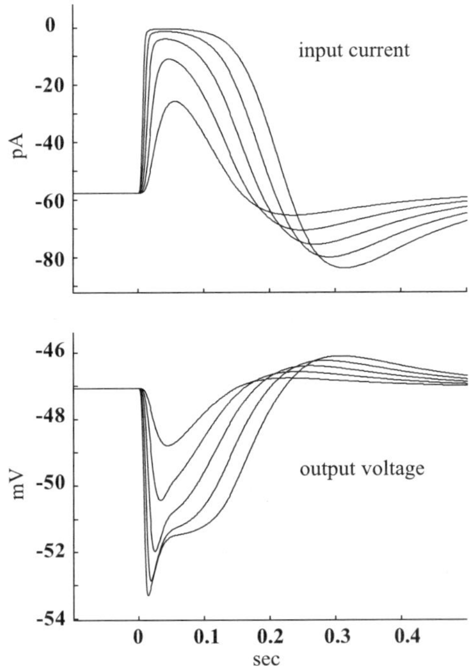

Fig. 2. A passive membrane model of the outer retina reproduces the non-linear input/output properties of primate cones [12,13]. Photocurrent (top) and membrane voltage (bottom) in response to 10 ms full-field light flashes of variable intensity.

In a previous computational study [15], it was shown that in response to a sustained hyperpolarizing current, intended to represent a light stimulus, negative feedback from horizontal cells repolarizes the membrane potential of the cone back towards its resting potential in the dark. Our results suggest, however, that when current injection is replaced by a light-sensitive decrease in a photoconductance, feedback from horizontal cells will not contribute to repolarizing the cone unless the resting potential of the isolated cone inner segment is sufficiently depolarized. Another important distinction is that in the present model the transfer function characterizing negative feedback from horizontal cells onto cones is nearly linear (Fig. 1), whereas in the previous study this transfer function was very non-linear, producing small oscillations in the responses to large current injections. Such oscillations are not evident in the light responses generated by the present model because at the peak of the light response the gain of the cone-horizontal cell circuit is small.

Examination of the photocurrent shows that there is a rebound phase of opposite polarity beginning approximately 200 ms after the onset of the flash stimulus. The depolarizing portion of the photocurrent response does not affect the membrane potential of the cone as strongly as the initial hyperpolarizing response, however, because negative feedback from the horizontal cells opposes depolarization of the cone above the dark potential. Since the dark potential of the cone is very close to the steeply rising portion of its synaptic output function, negative feedback from the horizontal cells is recruited very strongly by a depolarization of the cone above the dark potential. In contrast, feedback from the horizontal cell is recruited only weakly during the primary phase of the flash response when the cone is hyperpolarized and the synaptic output function of the cone is relatively insensitive to changes in membrane potential. This simple model therefore suggests that the feedback from horizontal cells plays a major role in shaping the voltage responses of primate cones to brief light flashes, and that the non-linearity in the synaptic transfer function from cones to horizontal cells is necessary to prevent oscillations that would otherwise result from the delayed negative feedback.

3.2. Active model

We next investigated the contribution of voltage-gated conductances to the light responses of cones and horizontal cells. Voltage-clamp studies of the isolated primate cone inner segment identified five voltage-gated currents: an A-current, a delayed rectifier, a sustained calcium current, a calcium-dependent chloride current, and an anomalous rectifier or h-current [19]. Similarly, three voltage-gated conductances have been identified in adult cat type-B horizontal cells: a large A-current, a delayed rectifier and an anomalous rectifier [17]. Using the built-in capabilities of the simulator to optimize an arbitrary number of parameters simultaneously, it was possible to construct a model cone that was able to reproduce many of the voltage-dependent effects seen in the biological data. The currents recorded from the isolated inner segment of a primate cone generated in response to a series of 10 mV voltage steps from a holding potential of −75 mV were reasonably well fit by the currents recorded from an isolated model cone in response to an analogous stimulation protocol (Fig. 3). For each of the

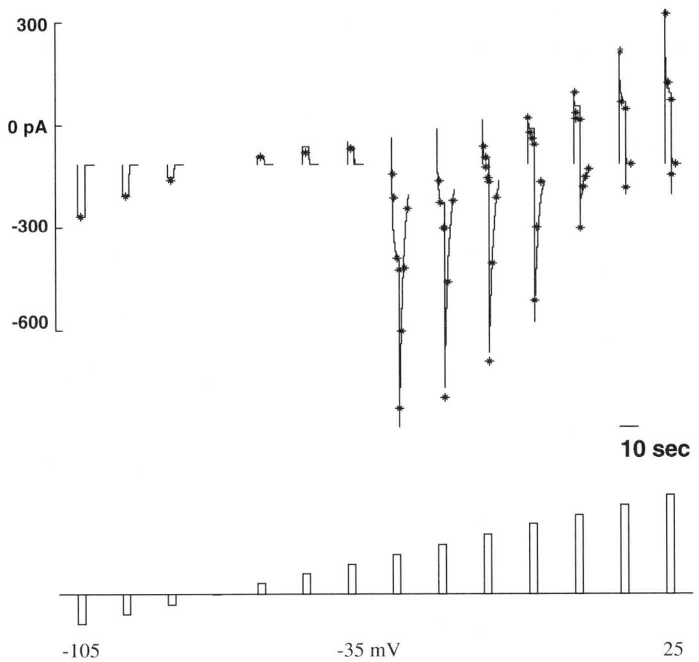

Fig. 3. Voltage-clamp protocol used to fit model cone responses (—) to experimental data (*). Bottom: Voltage-clamp protocol.

voltage-gated currents in the model, the parameters characterizing the activation and (if present) inactivation functions and associated time constants were determined by first fitting voltage-clamp studies designed to isolate the current in question. The maximum conductances for the different currents were then determined simultaneously by fitting the whole cone voltage-clamp series with the activation parameters of the individual currents held constant. A similar procedure was used to implement voltage-gated currents in horizontal cells, the details of which are less important for the conclusions derived here.

Voltage-gated currents contributed little to the output of the model cone in response to a 10 ms impulse representing a brief light flash (Fig. 4). The largest contribution to the flash response was from the photocurrent (photo), which dropped from its dark level of approximately −30 pA to nearly zero. The next largest contributor to the flash response was due to synaptic feedback from horizontal cells (hc). In the passive model, synaptic feedback from horizontal cells mediated a rapid repolarization of the cone, and this was true in the active model as well. In contrast, none of the voltage-gated currents (h, L, Cl[Ca], DR) changed by more than a few milli-volt during the flash response, less than the light-induced modulation of the leak current (leak). (The A-current was negligible in these experiments and was therefore omitted.) Furthermore, the voltage-gated currents that were present tended to cancel each other out. In particular, the outward

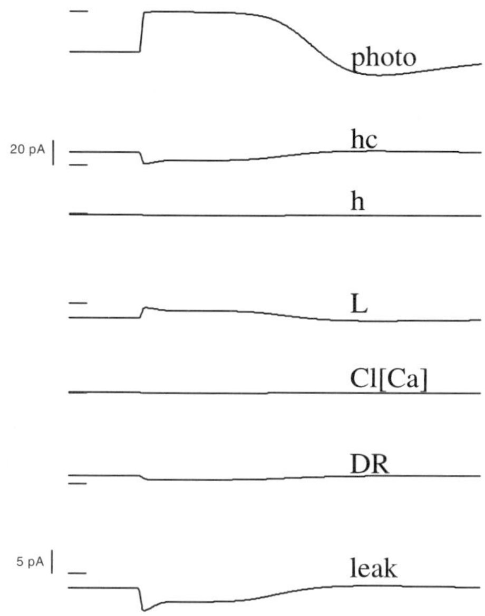

Fig. 4. Currents in model cone during flash response. Upper scale bar (20 pA): photo: photocurrent, hc: synaptic current. Lower scale bar (5 pA): h: anomalous rectifier, L: non-inactivating calcium current, Cl[Ca]: calcium-dependent chloride current, DR: delayed rectifier, leak: leak current. Horizontal scale bars (25 ms) at zero current.

current through the delayed rectifier was reduced by approximately the same amount as the inward current through the non-inactivating calcium channel, resulting in very little net change. Thus, our results indicate that voltage-gated conductances do not contribute significantly to the input/output properties of primate cones during the response to brief light flashes.

4. Discussion

By making the following three biologically reasonable assumptions, we could account for the voltage responses recorded from primate cones to brief light flashes: (1) the transfer function describing synaptic input from cones to horizontal cells is strongly non-linear [1], (2) horizontal cells make feedback inhibitory synapses onto cones governed by a transfer function that is nearly linear [14], and (3) the resting potential of the isolated cone inner segment is relatively depolarized, presumably reflecting the action of cGMP-gated channels that have been identified in the cone inner segments of lower vertebrates [10,11].

Although it was possible to account for the input/output properties of primate cones using a purely passive membrane model, we sought to understand how known voltage-gated conductances might also contribute to this transformation. Voltage-clamp studies of the primate cone inner segment have identified five voltage-gated currents [19], all of which were included in the present model. Similarly, the model also included the three voltage gated conductances that have been identified in adult cat type-B horizontal cells [17]. Including these voltage-gated channels into the model did not significantly affect the input/output properties of the cone as measured by the responses to brief light flashes. Space constraints prohibit a detailed description of how each of the voltage-gated currents was implemented, but given the limited amount of published experimental data for each current, it is unlikely that our implementation correctly accounted for the full dynamic range of primate cones. However, given that we achieved a relatively good fit to published data, the fact that none of the voltage-gated currents in the model approached even the same order of magnitude as the photo and synaptic currents makes it unlikely that our conclusions would be substantially modified by changing the details of our implementation.

In lower vertebrates, it has been reported that the repolarizing phase of the cone's light response is due in part to the action of voltage-gated currents [3]. A selective cationic conductance activated at hyperpolarized potentials, identified as an h-current, was shown to account for the repolarization following the onset of a bright light, while a potassium current activated by membrane depolarizations, termed Kx, contributed to the repolarization following onset of a dim light. The time course of these effects are much slower, however, than the very rapid repolarization seen in primate cones. It is thus unclear to what extent mechanisms identified in the lower vertebrate can be applied to the mammalian outer retina.

If voltage-gated currents indeed do not contribute to the flash response in primate cones, then one might ask what physiological function they do perform? Examination of how the different currents are activated at various potentials reveals that their major contribution occurs outside the normal physiological range, and would act in such a manner so as to drive the membrane potential back into the physiological range. Thus, we suggest that the major contribution of voltage-gated currents in the primate cone may be to ensure that the membrane potential always remains near an optimal operating point.

References

[1] D. Attwell, The photoreceptor output synapse, in: N.N. Osborne, G. Chader (Eds.), Progress in Retinal Research, Pergamon Press, New York, 1990, pp. 337–362.

[2] D. Attwell, F.S. Werblin, M. Wilson, The properties of single cones isolated from the tiger salamander retina, J. Physiol. 328 (1982) 259–283.

[3] S. Barnes, After transduction: response shaping and control of transmission by ion channels of the photoreceptor inner segments, Neurosci. 58 (1994) 447–459.

[4] D.A. Baylor, A.L. Hodgkin, T.D. Lamb, The electrical response of turtle cones to flashes and steps of light, J. Physiol. (London) 242 (1974) 685–727.

[5] J.R. Huguenard, D.A. McCormick, Simulation of the currents involved in rhythmic oscillations in thalamic relay neurons, J. Neurophysiol. 68 (1992) 1373–1383.

[6] M. Kamermans, I. Fahrenfort, K. Schultz, U. Janssen-Bienhold, T. Sjoerdsma, R. Weiler, Hemichannel-mediated inhibition in the outer retina, Science 292 (2001) 1178–1180.

[7] S. Lohrke, H.D. Hofmann, Voltage-gated currents of rabbit A- and B-type horizontal cells in retinal monolayer cultures, Vis. Neurosci. 11 (1994) 369–378.

[8] S. Picaud, B. Pattnaik, D. Hicks, V. Forster, V. Fontaine, J. Sahel, H. Dreyfus, GABAA and GABAC receptors in adult porcine cones: evidence from a photoreceptor-glia co-culture model, J. Physiol. (London) 513 (1998) 33–42.

[9] W.H. Press, B.P. Flannery, S.A. Teukolsky, W.T. Vetterling, Numerical Recipes, Cambridge University Press, Cambridge, 1986.

[10] F. Rieke, E.A. Schwartz, A cGMP-gated current can control exocytosis at cone synapses, Neuron 13 (1994) 863–873.

[11] A. Savchenko, S. Barnes, R.H. Kramer, Cyclic-nucleotide-gated channels mediate synaptic feedback by nitric oxide, Nature 390 (1997) 694–698.

[12] J.L. Schnapf, B.J. Nunn, M. Meister, D.A. Baylor, Visual transduction in cones of the monkey *Macaca fascicularis*, J. Physiol. 427 (1990) 681–713.

[13] D.M. Schneeweis, J.L. Schnapf, Photovoltage of rods and cones in the macaque retina, Science 268 (1995) 1053–1056.

[14] E.A. Schwartz, Depolarization without calcium can release gamma-aminobutyric acid from a retinal neuron, Science 238 (1987) 350–355.

[15] R.G. Smith, Simulation of an anatomically defined local circuit: the cone-horizontal cell network in cat retina, Vis. Neurosci. 12 (1995) 545–561.

[16] B.J. Travis, A computational model of one pathway in the cat sub-cortical auditory system, Math. Comput. Simulation 40 (1996) 81–99.

[17] Y. Ueda, A. Kaneko, M. Kaneda, Voltage-dependent ionic currents in solitary horizontal cells isolated from cat retina, J. Neurophysiol. 68 (1992) 1143–1150.

[18] H. Wässle, B.B. Boycott, J. Röhrenbeck, Horizontal cells in the monkey retina: cone connections and dedritic network, Euro. J. Neurosci. 1 (1989) 412–435.

[19] T. Yagi, P.R. Macleish, Ionic conductances of monkey solitary cone inner segments, J. Neurophysiol. 71 (1994) 656–665.

Available at
www.ComputerScienceWeb.com
POWERED BY SCIENCE DIRECT®

ELSEVIER

Neurocomputing 52–54 (2003) 135–140

NEUROCOMPUTING

www.elsevier.com/locate/neucom

Adaptation of the temporal receptive fields of macaque V1 neurons☆

Richard Romero[a], Yuguo Yu[a], Pedram Afshar[b], Tai Sing Lee[a,*]

[a] *Center for the Neural Basis of Cognition & Computer Science Department, Carnegie Mellon University, Pittsburgh PA 15213, USA*

[b] *Medical School, University of Pittsburgh, Pittsburgh, PA 15213, USA*

Abstract

Traditional view of macaque V1 neurons is that they are static spatiotemporal filters. This study shows that V1 neurons' temporal receptive fields (kernels) are dynamic and adaptive. In order to study the adaptation of the kernels to a variety of stimuli, we developed a regression approach to recover kernels using naturalistic stimuli. Using this technique, we studied how the kernels of the V1 neurons in macaque monkeys adapt to the frequency bandwidth of naturalistic signals. We found that the temporal frequency tunings and the gains of the receptive fields of the neurons change according to the statistical context of the stimulus environment.

Keywords: V1; Receptive field; Wiener kernel; Volterra series

1. Introduction

We refined and applied a system identification technique based on the approach of Korenberg [3] to study the adaptation of the temporal receptive cell of macaque V1 neurons to naturalistic stimuli. Earlier work on the adaptive behavior of neurons in salamander retina [7] and blowfly visual neurons [6,1] typically used Gaussian white noise as input. However, the fact that neurons are adaptive to the variance of the Gaussian white noise in those studies also suggests that their input–output relationship between

☆ This research was supported by NSF CAREER 9984706 and NIH Vision Research core grant EY08098, and a NIH grant to the Pittsburgh Supercomputing Center. Romero is supported by an NSF Graduate Research Fellowship. Afshar is supported by a NIH MD/Ph.D. Medical Scientist Training Fellowship.

* Corresponding author.

E-mail address: tai@cnbc.cmu.edu (T.S. Lee).

0925-2312/03/$ - see front matter © 2003 Elsevier Science B.V. All rights reserved.
doi:10.1016/S0925-2312(02)00799-3

white noise and natural stimulus might in fact be different. Wiener kernel (white noise) method [10,4,5] has been used to deduce the first order kernel of anesthetized cats [8] and the first and second order spatial temporal kernels of anesthetized monkeys [2]. But this method requires white noise input and a large amount of data, hence inappropriate for awake monkey experiment. Here, we would like to study the kernels of the neurons in response to naturalistic stimuli with different stimulus statistics in awake behaving monkeys.

2. Method

Korenberg's [3] fast orthogonalization algorithm for kernel identification is useful for characterizing systems using non-white and non-Gaussian stimuli with a limited amount of data. We refined this method using Singular Value Decomposition to stabilize the algorithm. The basic idea of this method is to assume the transfer function can be expressed in terms of the Volterra series, and apply the Least Means Square (LMS) regression method to obtain the best coefficients of the series relating the input and the output, taking into account the non-white nature of the stimuli, as described below.

Let $x(t)$ be the input signal and the cell's response be $y(t)$. A linear system in terms of the first and the second order kernels with memory length L is then given by

$$y(t) = h_0 + \sum_{\tau=1}^{L} h_\tau x(t-\tau) + \sum_{\tau_1=1}^{L} \sum_{\tau_2=\tau_1}^{L} h_{\tau_1,\tau_2} x(t-\tau_1) x(t-\tau_2), \qquad (1)$$

where h_0 corresponds to the mean firing rate, h_τ is the first order kernel, and h_{τ_1,τ_2} the second order kernel. We restrict all τ's to be positive, so we only consider causal filters.

This equation is easily expressed in matrix form as $Y = XH$, where time is now indexed by matrix row in Y and X. H contains the concatenation of the terms

$$[h_0, h_1, \cdots h_L, h_{1,1}, h_{1,2}, \cdots, h_{L,L}]'.$$

And row t of X is similarly

$$[1\, x(t-1) \cdots x(t-L)\, (x(t-1)x(t-1))\, (x(t-1)x(t-2)) \cdots$$
$$(x(t-L)x(t-L))].$$

The standard solution for this regression problem is $H = (X'X)^{-1}X'Y$. Because of the temporal correlations in the input $x(t)$, typical of naturalistic stimulus, the matrix $(X'X)$ is ill conditioned. Instead of directly inverting this matrix, we use the singular value decomposition $USU' = (X'X)$ where $US^{-1}U' = (X'X)^{-1}$ and S is a diagonal matrix. We include the first n largest dimensions as ranked by their eigenvalues, where n is chosen so that we account for 99% of the variance in X. In essence, the kernel is obtained by the normalizing the cross-correlation of the input and output with the covariance matrix of the input. The use of SVD and selecting the n largest eigenvectors is a type of principle component regression.

3. Experiment

In order to focus our study on the temporal receptive field, we used a moving visual stimulus with its movement limited to one dimension. The stimulus was a sinusoidal grating (Fig. 1). During the experiment, the receptive field of the recorded neurons were roughly placed at the center of the grating stimulus. The sinusoidal grating underwent a random walk in a direction perpendicular to the grating. The step sequence was drawn from a distribution such that the temporal correlations of the cos(phase) variation (which is also the intensity variation at a point in the receptive field) exhibited the $1/f$ spectrum, i.e. the power spectrum of the natural stimulus. The movement of this stimulus thus has the same statistics on temporal correlations as natural stimuli [9] and is called 'naturalistic'.

To investigate the potential adaptation of the neurons in response to the statistical structures of the stimulus environment, in different blocks of the experiment, we used the signals that have been filtered to have different frequency bandwidths. The lower end of the frequency band tested was fixed at 0.1 Hz, and the higher end can be one of the following frequencies: 6, 9, 12, 15, 18, 21, 24, and 27 Hz. Generally, we attempted to study a neuron's response to as many frequency bandwidth signals as possible in so far we could maintain the neuron and that the monkey was willing to work.

For each neuron, we selected a particular sinusoidal grating with the direction and spatial frequency that the cell was optimally tuned to. Each experimental trial began by turning on both a fixation dot and the sinusoidal grating moving back and forth along one-dimension with steps according to the specified noise distribution. The monkey needed to maintain fixation for 2200 ms at the fixation dot before it received a juice reward. Both the fixation dot and stimulus were removed from the screen for a fixed period of time before the next trial began. There were two types of stimulus presentation, random and repeated. The random trials were ones where the stimulus

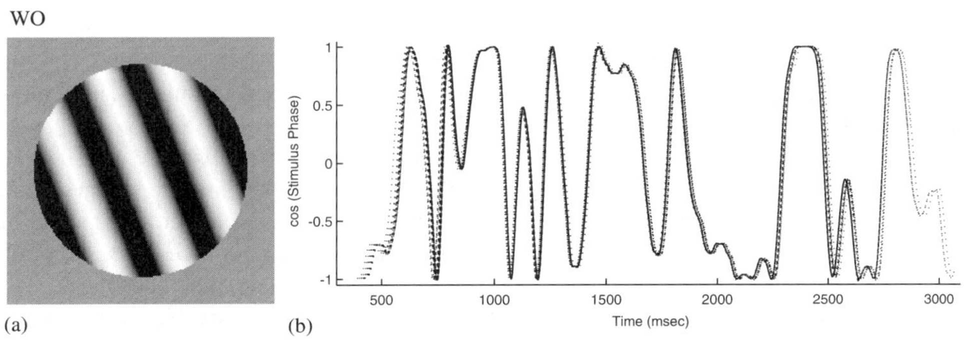

Fig. 1. (a) Example sine wave grating. The orientation and spatial frequency were chosen according to the optimal tuning of the cell. The diameter of the stimulus was 5°. (b) The series of repeated input signals presented to the monkey are shown as overlaid plots. The input signal used by the kernels is the cosine of the phase, as opposed to using the phase directly. A phase of zero (cosine of 1) corresponds to the neuron's maximum response. This phase alignment is determined empirically from the data, and computed once for each cell.

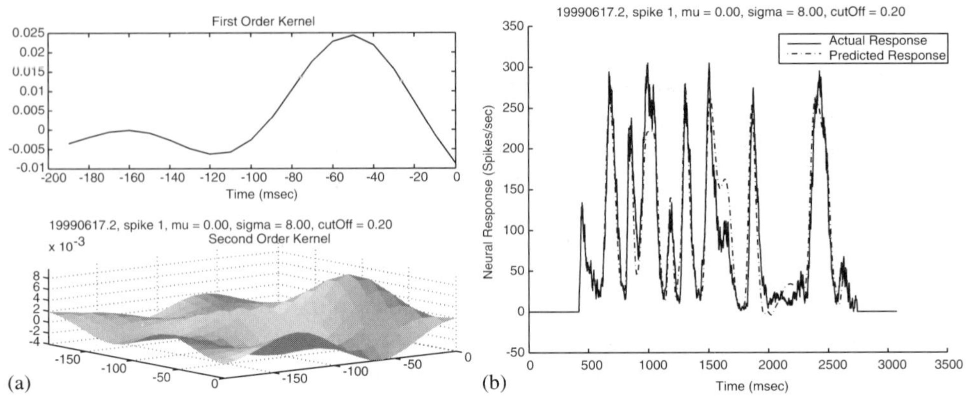

Fig. 2. (a) The first and second order kernel computed for the neuron. The second order kernel has been shown as a symmetric surface, when in reality only the upper triangular coefficients are estimated. (b) The average response over the 80 repeated trials is shown, smoothed by a 10 ms moving average filter. The estimate for this response, using the first and second kernels presented above, is also shown, showing that the prediction is fairly accurate.

motion was different each trial. The repeated trials used over and over again a particular 2.2 s movement sequence drawn from the same distribution as the random trials. The monkey was presented with 10 random sequences followed by 2 repetitions of one sequence, for a total of 40 times, providing us with neural responses to 400 trials of unique random sequences, and 80 trials of a particular stimulus sequence. Kernels estimated using only the random trials were used to predict the average post-stimulus time histograms of the repeated stimulus sequence.

4. Results

We recorded from 30 neurons from one monkey using standard single-unit chronic recording technique and 10 neurons from a second monkey using chronically implanted micro-electrode array (Bionics, Inc, Utah). Fig. 2a shows the first and second order kernels of a typical cell recovered from the random trials. The kernels have an enforced latency of 50 ms and a memory length L of 200 ms. Coefficients are estimated at 10 ms intervals. The kernels are applied to the input to yield a predicted response of the neuron for the repeated trials. Fig. 2b shows the average neural response (PSTH) to the repeated trials and the predicted response based on both the first and second order kernels. Note that the repeated trials were not in the training set. Nevertheless, the prediction was generally accurate. We found the second order kernels carry a significant amount of information that yield an improvement of at least 20 percent in predicting the neuronal responses.

When the frequency bandwidth was varied, we found that the temporal kernels of the cells changed (Figs. 3a and b). The gain of the kernel, obtained by integrating its power spectral density, reaches a maximum at an intermediate frequency bandwidth.

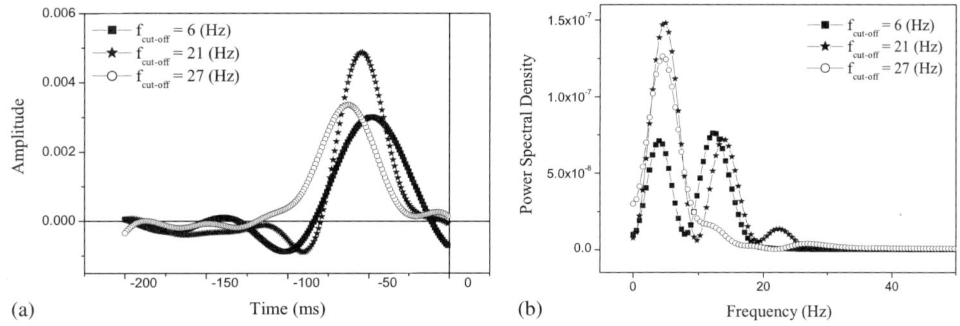

Fig. 3. (a) The temporal receptive fields of a cell recovered in the three noise stimulus environment show adaptation in the scale and amplitude of the cell's receptive field. (b) The power spectral density plots of the temporal receptive fields reveal adaptation in the frequency tuning.

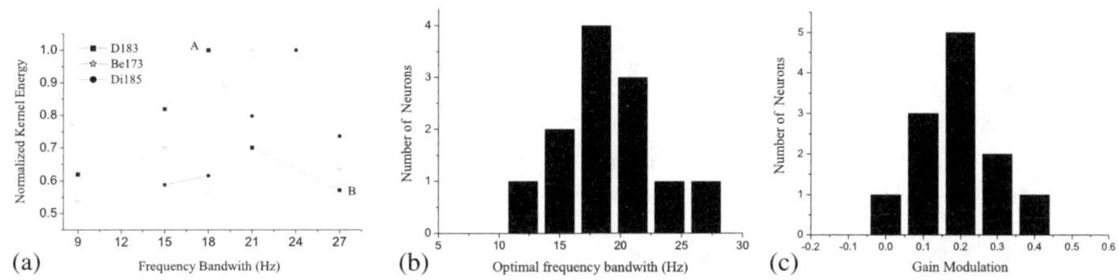

Fig. 4. (a) Three different cells' kernel gains for signals of different frequency bandwidths. Each cell's gain is normalized by its maximum gain across different bandwidths. (b) Distribution of cells in terms of their optimal frequency bandwidths. (c) Distribution of the strength of gain modulations.

The optimal frequency bandwidth varies from cell to cell. Fig. 4a shows the kernel gains of three different cells as a function of bandwidths. The kernel gain of each cell was normalized by the maximum gain of the cell. It clearly shows a tuning to the frequency bandwidth: some preferred a lower frequency bandwidth (e.g. cell d183) while others preferred a higher one (e.g. cell Di185). Fig. 4b shows the distribution of the optimal frequency bandwidth for the 10 neurons we managed to study for a long time. We evaluate the strength of the gain modulation by computing the ratio $(A - B)/(A + B)$ where A is the peak kernel gain, and B is the kernel gain for the 27 Hz frequency bandwidth signal (see Fig. 4a). The distribution of the gain modulation relative to the highest frequency bandwidth signal was shown in Fig. 4c.

The observed neuronal adaptation to frequency bandwidth may arise from the underlying temporal frequency tuning of the cells and the spectral statistics of the signals. The high frequency component of a wider band signal may exert a destructive interference on the cell's nonlinear spiking dynamics. While the causes and the purpose of frequency bandwidth adaptation remains to be elucidated, we have demonstrated in this work the feasibility and utility of our method for studying kernel adaptation in V1 neurons of awake macaque monkeys to non-white noise stimulus. This opens up a lot of opportunities for studying the adaptation of the neuronal kernels in awake monkeys under different behavioral manipulations and training.

References

[1] N. Brenner, W. Bialek, R. de Ruyter van Steveninck, Adaptive rescaling maximizes information transmission, Neuron 26 (2000) 695–702.

[2] J.P. Gaska, L.D. Jacobson, H.W. Chen, D.A. Pollen, Space–time spectra of complex cell filters in the macaque monkeys: a comparison of results obtained with pseudowhite noise and grating stimuli, Visual Neurosci. 11 (1994) 805–821.

[3] M.J. Korenberg, Identifying nonlinear difference equation and functional expansion representations: the fast orthogonal algorithm, Ann. Biomed. Eng. 16 (1988) 123–142.

[4] Y.W. Lee, M. Schetzen, Measurement of the Wiener kernels of a nonlinear system by cross-correlation, Int. J. Control 2 (1965) 237–254.

[5] V.Z. Marmarelis, Identification of nonlinear biological systems using Laguerre expansion of kernels, Ann. Biomed. Eng. 21 (1993) 573–589.

[6] F. Rieke, D. Warland, R.D. Ruyter van Steveninck, W. Bialek, Spikes: Exploring the Neural Code, MIT Press, Cambridge, MA, 1997.

[7] S.M. Smirnakis, M.J. Berry, D.K. Warland, W. Bialek, M. Meister, Adaptation of retinal processing to image contrast and spatial scale, Nature 386 (1997) 69–73.

[8] G.B. Stanley, F.F. Li, Y. Dan, Reconstruction of natural scenes from ensembles responses in the lateral geniculate nucleus, J. Neurosci. 19 (18) (1999) 8036–8042.

[9] J.H. van Hateren, Theoretical predictions of spatio-temporal receptive fields, J. Comput. Physiol. A 171 (1992) 157–170.

[10] N. Wiener, Nonlinear Problems in Random Theory, Wiley, New York, 1958.

Richard Romero was a Ph.D. student in Computer Science and the Center for the Neural Basis of Cognition, Carnegie Mellon University. He is currently the Chief Technical Officer of Eizel Technology.

Yuguo Yu is a Ph.D. in Physics from Nanjing University (2001), PRC and is currently a postdoctoral fellow in the Center for the Neural Basis of Cognition, Carnegie Mellon University.

Pedram Afshar is a MD-Ph.D. student in University of Pittsburgh.

Tai Sing Lee is a Ph.D. in Engineering and Medical Physics from Harvard University and the Harvard-MIT Division of Health Science and Technology (1993). He is currently an associate professor in Computer Science and Neural Basis of Cognition at Carnegie Mellon University.

Available at www.ComputerScienceWeb.com
POWERED BY SCIENCE DIRECT

ELSEVIER

Neurocomputing 52–54 (2003) 141–149

NEUROCOMPUTING

www.elsevier.com/locate/neucom

Functional roles of receptive field structures in the perception of orientation

Ko Sakai

Institute of Information Sciences and Electronics, University of Tsukuba 1-1-1 Tennodai, Tsukuba 305-8573, Japan

Abstract

To investigate the cortical mechanisms underlying the determination of apparent tilt, including the tilt illusion, we developed and examined a V1 network model consisting of cortical neurons and phase-coupled oscillators for neural grouping. Grouping local orientations based on V1 colinear connections, the model showed quantitative agreement with psychophysically measured apparent tilt. Computational investigations suggest that even-phase receptive fields are crucial but not exclusive for the apparent orientation of a bar. Our results further suggest importance of the feedback of grouping information from complex cells, which are synchronized through colinear connections, to simple cells with specific phases.

Keywords: Early vision; Orientation; Psychophysics; Synchronization; Spatial phase

1. Introduction

To determine the global orientation of a bar, the neural system has to group the local orientation signals extracted by neurons in the primary visual cortex (V1). If a bar junction exists, Gabor filters signal erroneous information around the intersection. Therefore, the global orientation of a bar will be misjudged, which is expected to be an origin of the tilt illusion. We investigate the cortical mechanisms underlying the neural grouping for the perception of apparent tilt, specifically in the tilt illusion.

It has been suggested that the horizontal connections exist among neurons in V1 superficial layers, most likely complex cells, as they form colinear connections [1], and that synchronization among them is responsible for contour saliency [9]. A similar

E-mail address: sakai@is.tsukuba.ac.jp (K. Sakai).

0925-2312/03/$ - see front matter © 2002 Elsevier Science B.V. All rights reserved.
doi:10.1016/S0925-2312(02)00833-0

colinear synchronization is expected to account for the neural grouping for the determination of global orientation. Given the colinear connections among complex cells, we examined recently what neural groups tend to be formed when a bar junction is shown [6]. We used a phase model as a simple and mathematically well-established mechanism to assess a tendency of grouping automatically. The results of simulations showed that the colinear synchronization realizes the exclusion of the intersection region, which agrees with grouping determined psychophysically. However, there was quantitative disagreement between the global orientations computed from the phase model and the apparent tilt measured psychophysically.

This disagreement between the model and apparent tilt might originate from that simple cells play somewhat more role than an equally weighted input to the grouping mechanism as is in that model. For instance, simple cells with even- and odd-phases might have different connections and roles for the determination of apparent tilt. It has been thought generally that even-phase simple-cells are responsible for the detection of bars, and odd-phase for the detection of edges [2]. Simple cells could be important for these detections, and complex cells could be crucial for texture processing [5] and context dependency [3]. We carried out simulations of a network model based on major properties of V1 including the colinear connections, in order to examine the functional roles of receptive field structures for the perception of global orientation of bars, specifically a role of simple and complex cells, and even- and odd-phase receptive fields of simple cells.

2. The model

A schematic illustration of the model is shown in Fig. 1, which consists of three major stages; a set of Gabor filters followed by half-wave rectification, estimation of local orientations by a winner-take-all mechanism, and coupled neural-oscillators to assess tendency of grouping. We designed the model so that the entire input field covers $3.88^\circ \times 3.88^\circ$ in visual angle by 97×97 units. The first stage of the model consists of simple-cell-like units whose receptive field is given by a Gabor filter with a variety of orientation, spatial frequency and phase preferences. For the sake of computational simplicity, no temporal property of V1 neurons is taken into account, which maintains our focus on the spatial distribution of cellular activities. For each spatial position, there are 576 types of model cells: 36 types of optimal orientation with an increment of 5°, four types of optimal spatial frequency, 17, 10, 5, and 2.5 cpd, corresponding to multiples of 0.25, 0.5, 1.0 and 2.0 times the width of stimulus bars, and four types of spatial phase, 0, 90, 180, and 270° phases of the Gabor filters. The output of Gabor filters passes through half-wave rectification with thresholding to generate the output of simple-cell units.

The second stage of the model determines local orientation. We represent the local orientation by the *local peak orientation* as a simple and biologically plausible measure [5]. The local peak orientation is the optimal orientation of the unit that responds most strongly among those units located at the same spatial position, with various orientation and spatial frequency preferences. It has been reported that the bar orientation estimated

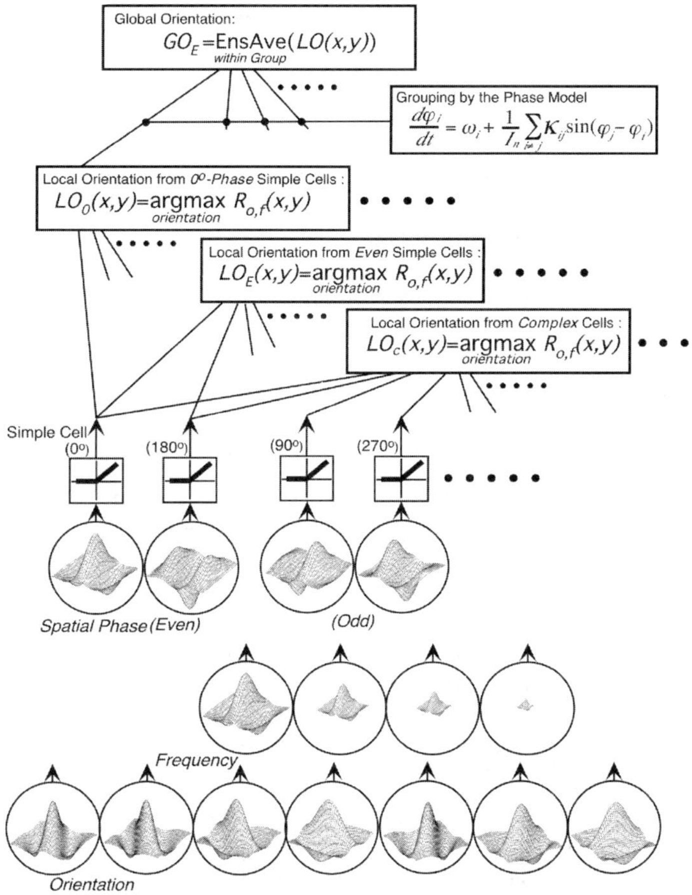

Fig. 1. An illustration of the model. The first stage consists of a set of Gabor filters followed by half-wave rectification. The second stage estimates local orientations by determining the strongest unit among others. The last stage computes the global orientation of a bar by taking the ensemble average of local orientations within the group determined by the phase model.

from the local peak orientations is similar to that computed by taking vector summation [7]. The spatial-phase invariance of model complex-cells is established at this stage. It has been shown that this pooling by a winner-take-all mechanism reproduces the spatial nonlinearity of the receptive field [8].

The third stage of the model determines the groups of cells that correspond to each of two bars. A phase model governed by Kuramoto's equation [4] is employed for the grouping process. We designed the model so that units fire periodically and the temporal phase of each unit indicates the timing of cellular firing. It is assumed that cells are grouped together if the timing of firing is synchronized. Although it is doubtful that neurons in vivo are actually governed by this equation, the phase model is a simple and suitable model to employ here because our specific aim is to examine what neural groups are formed given the colinear connections and the local orientations deduced from X-shaped stimuli that produce illusory tilt.

Synchronization will be established if coupling between two units is strong. The coupling coefficient is determined depending on three measures: (1) co-circularity, (2) relative difference in local orientation, and (3) spatial distance. The product of the three measures gives the overall coupling coefficient, which realizes the preference to colinearity [9,6]. In most cases, starting from randomized phases, the synchronization reached a stable state after 10 to 20 cycles of iterations. The last part of the stage computes the global orientation of bars by taking the ensemble average of the local peak orientations among the synchronized cells.

3. Spatial distribution of effective neuronal responses—simulation results

We examined the spatial distribution of neuronal responses effective for the perception of the global orientation of bars, with specific interests in differences among simple and complex cells, and even- and odd-phase receptive fields.

3.1. Spatial distribution of local orientations

We carried out simulations of the model to estimate the local orientations when an X-shaped stimulus is presented. The retinotopic distribution of local peak orientations computed by the complex-cell units are shown in Fig. 2(A). The responding complex-cell units are distributed retinotopically for a wide range, almost twice the width of actual bars, because complex-cells lack the spatial localization of orientation information. It is uncertain whether all complex cells hold necessary and meaningful information for the determination of bar orientations. To clarify the effects from the complex cells outside the bar region, we extracted the responses within the extent of the vertical bar from Fig. 2(A), and showed in (B). Fig. 2(C) shows the local orientations of the synchronized units that correspond to a vertical bar. The next stage of the model computes the global orientation by taking the ensemble average of these synchronized units. Complex-cell-unit responses are determined by pooling four types of simple-cell-units with different spatial phases. We computed the local orientations that is determined by pooling only even-phase units (Fig. 2(D)), and 0°-phase units (E). We can observe that the responses outside the bar extent originate mostly from odd-phase units, and that strong tilt effects originate from even-phase units, particularly 0°-phase units. For further quantitative analysis, we computed the global orientation of a vertical bar by selectively pooling units with distinct phase type.

3.2. Complex-cell responses inside the bar region

We computed the global orientation of a vertical bar by taking the ensemble average of the local peak orientations among synchronized units corresponding to the vertical bar. Disregarding the responses of complex-cell-units whose receptive field center is

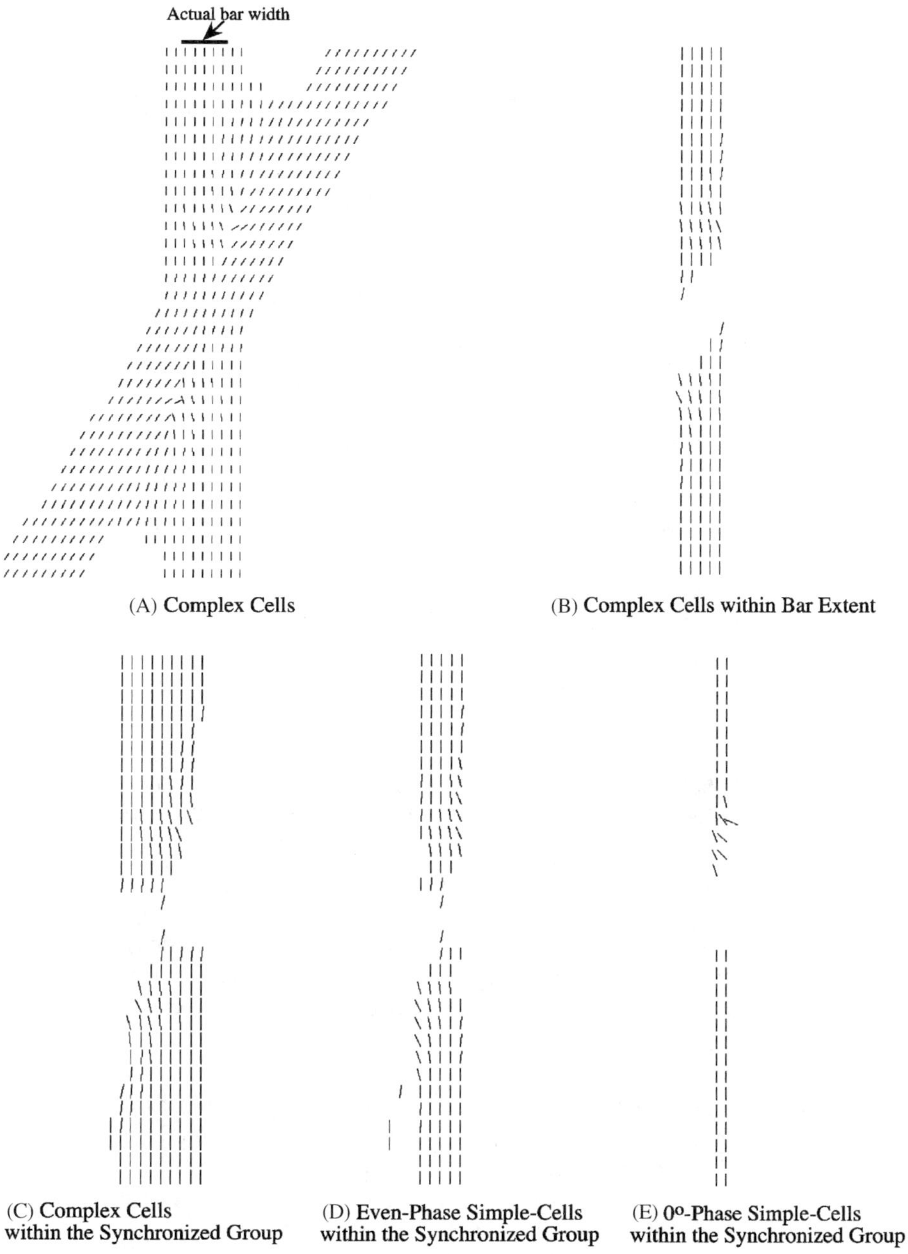

Fig. 2. The retinotopic distributions of computed local peak orientations. The tilt of a needle indicates the local peak orientation at the position. The responses less than a threshold are not shown here. Each panel shows the result of specific pooling of units as indicated in the panel.

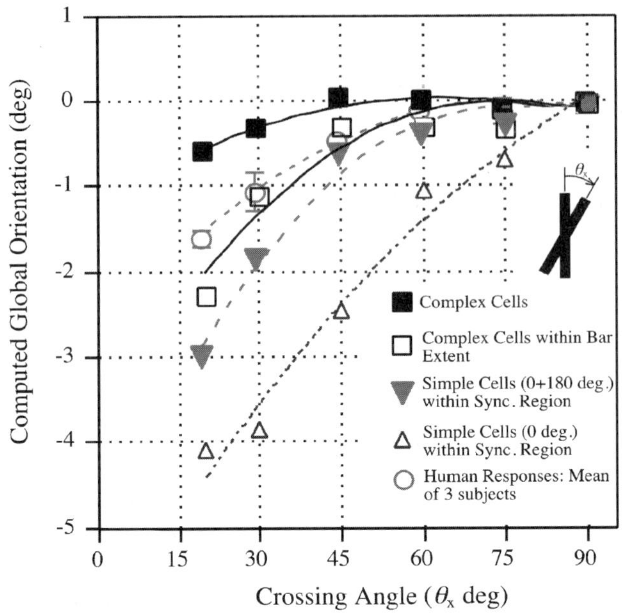

Fig. 3. The computed global orientations for stimuli consisting of two bars with identical contrast. Open circles indicate human responses psychophysically measured [6].

located outside the bar region, we also evaluated the global orientation from the synchronized units located inside the bar region. The difference between these two would clarify the significance of the responses of complex cells located inside or outside the bar region. Note that the responses outside the bar extent originate mostly from odd-phase units. The computed global orientations for various X-shaped stimuli are shown in Fig. 3. For comparison, we re-plotted the apparent tilt measured in corresponding psychophysical experiments [6]. The amount of global orientation computed from inside the bar extent is greater than that incorporates outside the bar extent, and closer to the apparent tilt psychophysically measured.

Psychophysical experiments have shown interesting characteristics in multiple-gray X-shaped stimuli; apparent tilt is independent of segmentation but dependent on contrast as shown in Fig. 4(A). It is important to confirm that the model reproduces the major human response characteristics, segmentation independence and contrast dependence, as well as the amount of apparent tilt. We computed global orientation of the vertical bar in multiple-gray stimuli from the responses of synchronized complex cells as shown in Fig. 4(B). The model reproduces segmentation independence and contrast dependence. Stimuli consisting of a variety of contrasts evoke various activities in V1-cell-units, which further deduce various local peak orientations. Therefore, even if the grouping is similar, the resultant apparent tilt is different. On the other hand, the segmentation-dependent difference in the activities is not large enough to alter the apparent tilt.

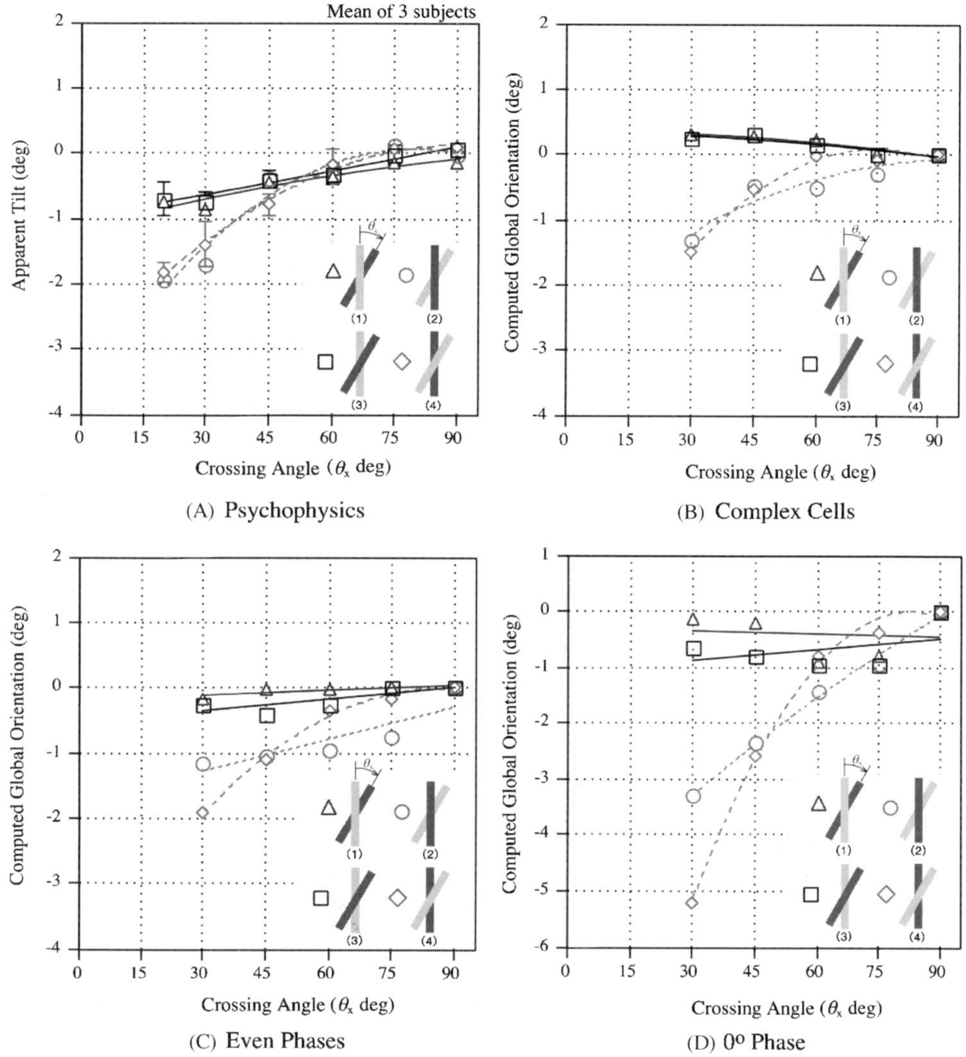

Fig. 4. The estimated global orientations for stimuli consisting of two bars with different contrasts. The configurations of the stimuli are indicated in inset. For example, insets (A) and (B) have identical segmentation but different contrast.

The above results indicate that the orientation information, carried by the complex cells whose receptive-field center is located within the bar extent, is crucial for the tilt illusion. This is natural in the sense that local orientation apart from a bar is less significant than that close to or placed on a bar. However, the complex cells whose receptive-field centers are apart from but closed to a bar are likely to respond to the bar, since complex cells hold the invariance to spatial phase. Given that a complex cell consists of a pooling of simple cells with various spatial phases [8], the responses of complex cells positioned outside a bar originate from simple cells whose receptive field has an odd-phase or anti-phase of a bar. Therefore, the results suggest that even-phase simple cells play a crucial role in the tilt illusion and the perception of global orientation of a bar. In the next section, we examine further a role of even-phase receptive fields, and a plausible cortical mechanism for a phase-specific operation in grouping.

3.3. Even-phase simple-cell responses within the region determined by the synchronization

In this section, for the ensemble average of local peak orientations to compute global orientation, we take the responses of the even-phases (0° and 180°) or 0°-phase simple-cell units located within a spatial extent of the group determined by the synchronization of complex-cell units. This mechanism is equivalent to considering feedback from the complex cells, synchronized through colinear connections, to simple cells with specific phases. Since the inclusion of both odd- and even-phase simple-cell responses is equivalent to the utilization of complex-cell responses, the contribution of odd-phase simple cells can also be easily induced from this simulation. The results will give a clue as to the roles of simple and complex cells, as well as even- and odd-phase receptive fields, in the perception of global orientation.

The computed global orientation of the vertical bar in X-shaped stimuli is overlaid in Fig. 3. The amount of the global orientation computed from even-phases is similar to the apparent tilt psychophysically measured, and that from the previous model that takes into account the complex cell responses within the bar extent. This result also suggests that the simple cells with an even-phase receptive-field account largely for the tilt illusion compared with those with odd-phase. The amount computed from 0°-phase simple-cell units is almost twice that of the models with multiple phases. This overestimation of apparent tilt suggests that the responses of the cells with phases other than the optimal one for the bar are not disregarded for the perception of global orientation. We also computed global orientation of the vertical bar in multiple-gray X-shaped stimuli, and confirmed that the models with even-phases and 0°-phase reproduce the major characteristics of human responses, namely contrast dependence and segmentation independence, as shown in Fig. 4(C) and (D), respectively.

4. Conclusions

We propose that receptive-field structures, specifically spatial phase preference, have different functional roles in the perception of global orientation including the tilt illusion. Our computational investigations indicate that even-phase receptive fields are crucial but not exclusive for the apparent orientation of a bar. This further suggests the importance of the feedback of grouping information from the complex cells, which are synchronized through colinear connections, to simple cells with specific phases.

References

[1] W.H. Bosking, Y. Zhang, B. Schofield, D. Fitzpatrick, Orientation selectivity and the arrangement of horizontal connections in tree shrew striate cortex, J. Neurosci. 17 (1997) 2112–2127.

[2] D.H. Hubel, T.N. Wiesel, Receptive fields, binocular interaction and functional architecture in the cat's visual cortex, J. Physiol. 160 (1962) 106–154.

[3] M.K. Kapadia, M. Ito, C.D. Gilbert, G. Westheimer, Improvement in visual sensitivity by changes in local context: parallel studies in human observers and in V1 of alert monkeys, Neuron 15 (1995) 843–856.

[4] Y. Kuramoto, Lecture Notes in Physics, Vol. 39, Springer, Berlin, 1975.

[5] K. Sakai, L.H. Finkel, Characterization of spatial-frequency spectrum in the perception of shape from texture, J. Opt. Soc. Am. A 12 (1995) 1208–1224.

[6] K. Sakai, Y. Hirai, Neural grouping and geometric effect in the determination of apparent orientation, J. Opt. Soc. Am. A 19 (2002) 1049–1062.

[7] K. Sakai, S. Tanaka, Retinotopic coding and neural grouping in tilt illusion, Neurocomputing 26–27 (1999) 837–843.

[8] K. Sakai, S. Tanaka, Spatial pooling in the second-order spatial structure of cortical complex cells, Vision Res. 40 (2000) 855–871.

[9] S.-C. Yen, L.H. Finkel, Extraction of perceptually salient contours by striate cortical networks, Vision Res. 38 (1998) 719–741.

Available at
www.ComputerScienceWeb.com
POWERED BY SCIENCE DIRECT

ELSEVIER

Neurocomputing 52–54 (2003) 151 – 158

NEUROCOMPUTING

www.elsevier.com/locate/neucom

Modulation of excitatory synaptic coupling facilitates synchronization and complex dynamics in a nonlinear model of neuronal dynamics

Michael Breakspear[a,*], John R. Terry[b], Karl J. Friston[c]

[a] *School of Physics and Brain Dynamics Centre, University of Sydney, 11A Wilson Street, Maroubra, NSW 2035, Australia*

[b] *Department of Mathematics, University of Loughborough, Loughborough, UK*

[c] *Wellcome Department of Imaging Neuroscience, University College London, London, UK*

Abstract

We study dynamical synchronization in a model of a neural system constituted by local networks of densely interconnected excitatory and inhibitory neurons. Neural dynamics are determined by voltage- and ligand-gated ion channels. Coupling between the local networks is introduced via sparse excitatory connectivity. With modulation of this long-range synaptic coupling the system undergoes a transition from independent oscillations to chaotic synchronization. Between these states exists a 'weakly' stable state with epochs of synchronization and complex intermittent desynchronization. This may facilitate adaptive brain function by engendering a diverse repertoire of dynamics and contribute to the genesis of complexity in the EEG.

Keywords: Synchronization; Intermittency; Neuromodulation

1. Introduction

The mechanisms and function of cooperative behavior in large-scale neuronal systems is a very active area of research. Synchronous oscillations between neurons have been proposed as a mechanism for perceptual 'binding' through the coupling of distinct neuronal populations to form dynamic cell assemblies, each encoding various aspects

* Corresponding author. Fax: +61-2-9351-7726.
E-mail address: mbreak@physics.usyd.edu.au (M. Breakspear).

0925-2312/03/$ - see front matter
doi:10.1016/S0925-2312(02)00740-3

of a perceived object [11]. Evidence of coherent oscillations has been demonstrated in neurophysiological data [4] and computational models of the cortex [7]. However, in much of this research, only linear measures of synergistic activity are employed. The nonlinear properties of excitable membranes motivates the study of nonlinear aspects of synchronicity in neural systems. In this paper, we study nonlinear interdependence in a model neural system consisting of an array of coupled small-scale neural subsystems. The types and stability of nonlinear synchronization are investigated and the possible role of these behaviors in adaptive brain function is briefly discussed.

2. Evolution equations

The approach adopted in this study is to consider the behavior of local ensembles of neurons, with dynamical variables taken as ensemble averages. The scale of each ensemble is taken as the extension of pyramidal cell dendritic arbors, approximately 300 μm (the size of cortical columns). The dynamical variables studied are the mean membrane potential of pyramidal cells, V, and inhibitory interneurons, Z, and the average number of 'open' potassium ion channels, W. The evolution equations are adapted from a study of epileptic seizures in hippocampal slices [6] consisting of coupled differential equations, which in turn are derived from the model of Morris and Lecar [8]. The mean cell membrane potential of the pyramidal cells is governed by the conductance of sodium, potassium and calcium ion through voltage- and ligand-gated membrane channels,

$$\frac{dV}{dt} = -(g_{Ca} + r_{NMDA} a_{ee} Q_V) m_{Ca} (V - V_{Ca}) - (g_{na} m_{na} + a_{ee} Q_V)(V - V_{na})$$

$$-g_K W (V - V_K) - g_L (V - V_L) + a_{ie} Z Q_Z + a_{ne} I_\delta, \tag{1}$$

$$\frac{dZ}{dt} = b(a_{ni} I_\delta + a_{ei} V Q_V), \tag{2}$$

where g_{ion} is the maximum conductance of each population of ion species if all channels are open, m_{ion} is the fraction of channels open, V_{ion} is the Nernst potential for that ion species and $Q_{V(Z)}$ is the firing rate of the excitatory (inhibitory) neurons. The fraction of open channels is determined by the sigmoid-shaped 'neural activation function',

$$m_{ion} = 0.5 \left(1 + \tanh\left(\frac{V - V_T}{\delta_{ion}}\right)\right), \tag{3}$$

where δ_{ion} incorporates the variance of this distribution. The fraction of open potassium channels is slightly more complicated, being governed by W, with

$$\frac{dW}{dt} = \frac{\phi(m_k - W)}{\tau}, \tag{4}$$

where ϕ is a temperature scaling factor and τ is a 'relaxation' time constant. Cell firing rate is also determined by sigmoid activation functions,

$$Q_V = 0.5 Q_{V\,\max}\left(1 + \tanh\left(\frac{V - V_T}{\delta_V}\right)\right), \tag{5}$$

where the $Q_{\max}$ are the maximum firing rate. An analogous term is also introduced for the inhibitory cells. The firing of these cell populations feeds back onto the ensemble through synaptic coupling to open ligand-gated channels and raise or lower the membrane potential accordingly. In the case of excitatory-to-inhibitory and inhibitory-to-excitatory connections, this is modeled as additional inputs to the flow of ions across the membrane channel, weighted by functional synaptic factors, a_{ei} and a_{ie}. In the case of excitatory to excitatory connections, the rate of firing Q_V, is assumed to lead to a proportional release of glutamate neurotransmitter across the synapse, onto two classes of ligand-gated ion channels: (1) AMPA channels, which open an additional population of sodium channels, and (2) NMDA receptors, which open an additional population of *voltage-gated* sodium and calcium channels. r_{NMDA} incorporates the ratio of NMDA to AMPA receptors.

Each of the set of Eqs. (1)–(5) govern the dynamics within each local cell assembly. Coupling between N nodes is introduced as competitive agonist excitatory action at the same populations of NMDA and AMPA receptors. Representing each node with a superscript, this is incorporated by modifying Eq. (1) to

$$\begin{aligned}\frac{\mathrm{d}V^i}{\mathrm{d}t} &= -(g_{\mathrm{Ca}} + (1-C) r_{\mathrm{NMDA}} a_{\mathrm{ee}} Q_V^i + C r_{\mathrm{NMDA}} a_{\mathrm{ee}} \langle Q_V \rangle) m_{\mathrm{Ca}} (V^i - V_{\mathrm{Ca}}) \\ &\quad - g_K W (V^i - V_K) - g_{\mathrm{L}} (V^i - V_{\mathrm{L}}) - (g_{\mathrm{na}} m_{\mathrm{na}} + (1-C) a_{\mathrm{ee}} Q_V^i \\ &\quad + C a_{\mathrm{ee}} \langle Q_V \rangle)(V^i - V_{\mathrm{na}}) + a_{\mathrm{ie}} Z Q_Z^i + a_{\mathrm{ne}} I_\delta \end{aligned} \tag{6}$$

for $i, j = 1, \ldots, N$. C parameterizes the strength of excitatory coupling between cortical columns. If $C = 0$ the systems evolve independently. $C > 0$ introduces interdependences between consecutive columns. $C = 1$ corresponds to maximum coupling, with excitatory input from outside each column surpassing excitatory input from within each column.

All physiologically measurable parameters (conductances, threshold potentials and Nernst potentials) are set to their accepted values [6], These are given in Table 1.

Table 1
Parameter values for the simulations presented in this paper

I	0.3
a_{ee}	0.4
a_{ei}	0.1
a_{ie}	1
a_{ne}	1
a_{ni}	0.4
r_{NMDA}	0.2
δ	0.001

3. Transition to chaotic synchronization

With no intercolumn coupling ($C = 0$), each of the systems exhibit aperiodic asynchronous oscillations. Using a Gram–Schmidt orthonormalization procedure, we find that the largest Lyapunov exponent for each system is approximately 0.018, indicative of chaotic motion. As expected, further exponents were identically zero and negative.

With the introduction of coupling, $C > 0$, the systems show epochs of synchronized behavior. At $C \cong 0.1$, there is a transition to fully synchronized motion, as evident in the time series, Fig. 1a where $C = 0.13$. Fig. 1b shows the concurrent values of V^1 and V^2, for a system that have become embedded in the low-dimensional 'symmetry manifold' $V^1 = V^2$. In addition to calculating the Lyapunov exponents for each system it is also possible to calculate the *transverse Lyapunov exponents*, which describe the rate of separation of the coupled systems transverse to this manifold–in other words, the time-average tendency for the two systems to synchronize or separate. For this strength of coupling, the largest transverse exponent is negative (-0.005), indicating that perturbations away from this state of identical synchronization decay and thus that the state of identical chaotic synchronization is stable to noise. In Fig. 1(c) we present the value of the 4 largest Lyapunov exponents. It can be seen that the largest transverse Lyapunov exponent (λ_1) crosses zero at $C \cong 0.1$. This represents the value at which a 'blowout' bifurcation occurs—so called because the dynamics will 'blow out' from the three-dimensional symmetry manifold into the full six dimensions [9].

We note two other observations. Increasing r_{NMDA}, which corresponds to increasing the population of NMDA receptors, decreases the coupling strength required to achieve stable synchronization. On the other hand, increasing the overall excitatory-to-excitatory synaptic connectivity (a_{ee}) has the opposite effect. This is because the effect of local excitatory feedback, which tends to increase the local Lyapunov exponents, has the effect of 'splitting' the synchronization, over-riding the opposing effect of increasing inter-columnar coupling.

4. Chaotic phase synchronization

In the above simulations, all the parameters of both systems were equal. In real physiological systems, such symmetry between systems is clearly impossible. To incorporate this physiological variability, we introduced a random mismatch of between 0% and 10% for each of the physiological parameters. When the symmetry is broken, the symmetry manifold is no longer an invariant of the governing equations and hence *identical* synchronization between the coupled systems cannot be achieved. However, other types of synchronization are possible. In this section, we briefly illustrate an example of phase synchronization between subsystems in the neural model.

In Fig. 2(a) the time series for an example simulation with $C = 0.2$ is shown. It can be seen that although the systems are not identical, the systems are not oscillating independently. Close inspection reveals that the systems are phase-locked, half a cycle out of phase. In panel (b), the concurrent values of V^1 and V^2 are plotted. It can be seen that although the systems are not identical, their evolution is contained on

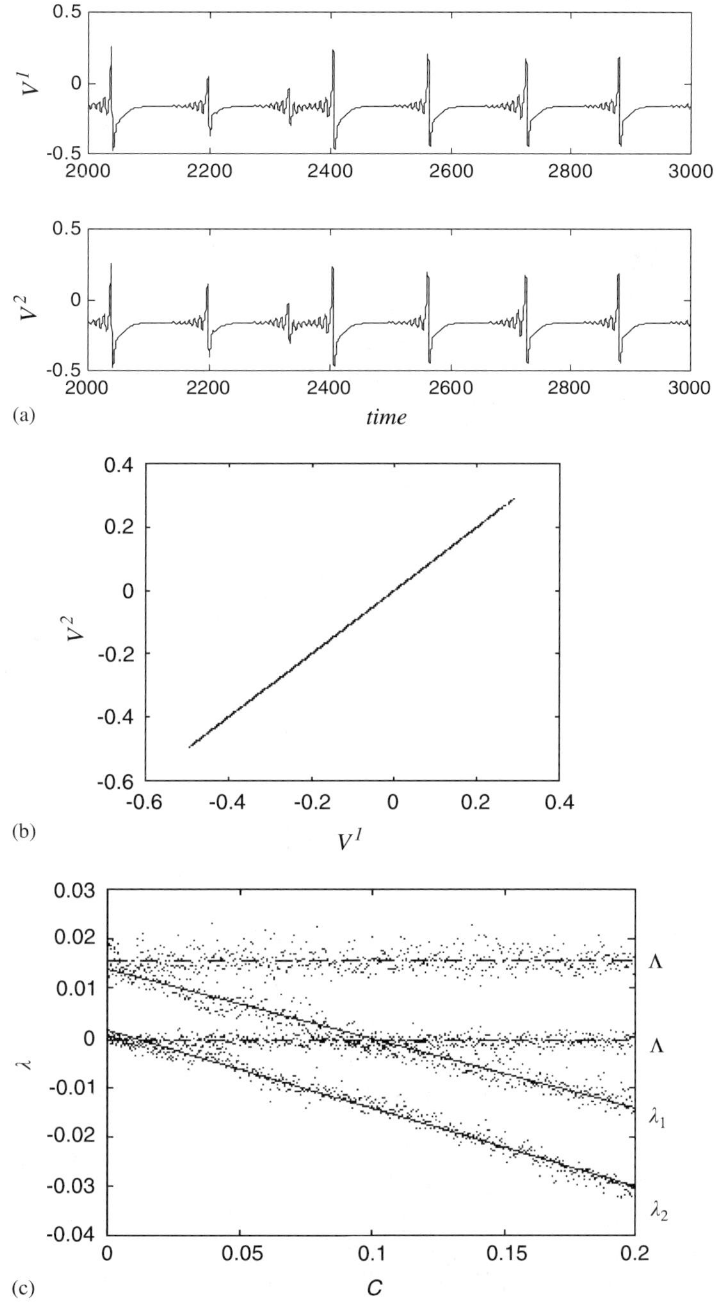

Fig. 1. (a) Time series of identical chaotic synchronization. (b) Concurrent values of V^1 and V^2. (c) Largest four Lyapunov exponents (λ_1 and λ_2 are the transverse exponents). Note the zero crossing at $C = 0.1$.

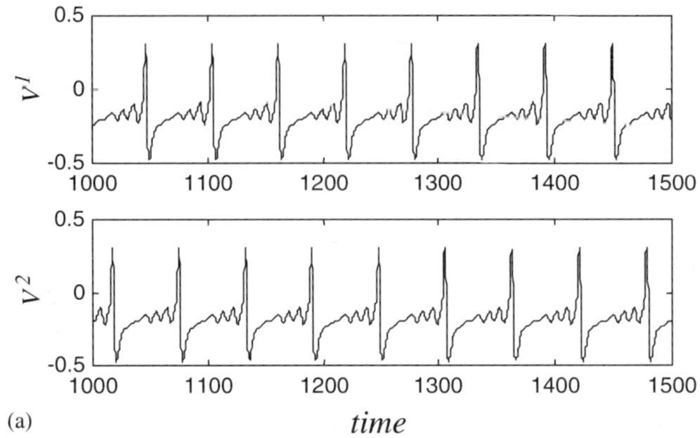

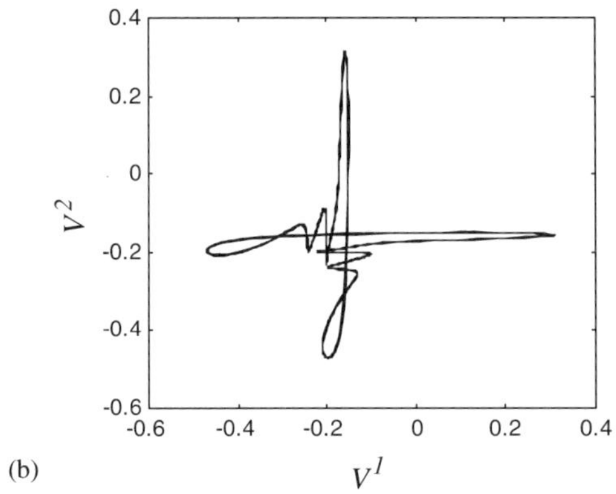

Fig. 2. (a) Time series of chaotic phase synchronization. (b) Concurrent values of V^1 and V^2.

a highly structured low-dimensional 'phase synchronization' manifold. It is possible to better quantify this aperiodic phase synchronization with the use of the Hilbert transform [10].

5. Marginal stability and chaotic transients

When the coupling strength is set close to the threshold for stable chaotic synchronization (i.e. $\lambda_1 \sim 0$) one observes epochs of varying length interrupted by bursts of desynchronization. An example of such a phenomenon is given in Fig. 3. At *time* $\sim$ 1000, it can be seen that the two time series briefly diverge from phase synchronization. Whilst space constraints prohibit further illustration, use of the Hilbert transform revealed two complete phase desynchronizations at this instance.

Such desynchronous bursts are known to be caused by *transversely unstable periodic orbits* embedded within the synchronized chaotic attractor. Desycnhronization occurs whenever the system passes close to such an unstable orbit [5]. We have previously

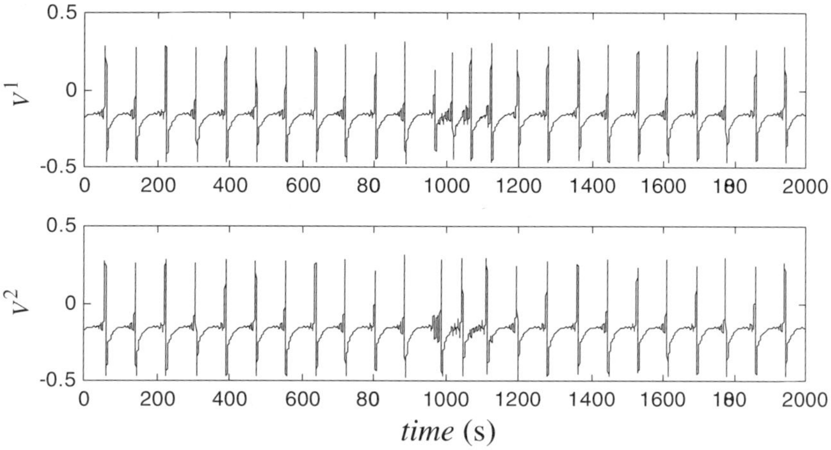

Fig. 3. Time series of weakly coupled system, showing brief desynchronization at $t \sim 1000$ s.

argued [1] that such 'marginal stability' may play a crucial role in normal brain function, permitting flexible and adaptive jumps between regional coherent states.

Analysis of larger arrays ($N > 2$) reveals that at judicious coupling strengths, the array breaks up into separate synchronized clusters of various sizes. In addition, it is possible to 'induce' synchronization between selected subsystems by increasing the coupling strength between them. This allows an extra level of complexity to be introduced into the model. For example, it may allow targeted information transfer between a partial network of the entire array.

6. Discussion

In this paper, we present a dynamical model for activity in an array of sparsely coupled local neural systems. The model incorporates the fundamental principles governing cell membrane potential, as determined by voltage and ligand-gated ion channels, and basic principles of cortical connectivity. In this way, it allows a reasonable amount of physiological complexity to be incorporated, including synaptic activation of receptor subtypes. Numerical analysis of this model reveals a transition from independent oscillations to stable chaotic synchronization, depending on the balance of local versus long-range connectivity. These functional excitatory couplings are subject to subcortical neuromodulation. Chaotic phase synchronization occurs and is robust to system noise and random parameter variations.

Behaviors observed in the regime of 'marginal stability'—i.e. when the transverse Lyapunov exponent is close to zero—are of particular interest, as they are characterized by a dynamic balance between synchronous and desynchronous dynamics in both spatial and temporal domains. Such phenomenon may facilitate the formation and dissolution of separate clusters of dynamic cell assemblies, permitting optimal and flexible adaptation to a changing external environment [3]. Whilst speculative, this conjecture is supported by the finding of occasional brief instances of nonlinear interdependence in human scalp EEG data [1,2]. Further empirical investigations are required in order to determine the cognitive correlates of these phenomena.

References

[1] M. Breakspear, Hum. Brain Mapping 5 (2002) 175–198.
[2] M. Breakspear, J. Terry, Clin. Neurophysiol. 113 (2002) 735–753.
[3] K. Friston, Philos. Trans. Roy. Soc. London B 355 (2000) 237–252.
[4] C.Gray, P. Konig, A. Engel, W. Singer, Science 338 (1989) 334–337.
[5] J. Heagy, T. Carroll, L. Pecora, Phys. Rev. E 52 (1998) R1253–R1256.
[6] R. Larter, B. Speelman, Chaos 9 (1999) 795–804.
[7] E. Lumer, G. Edelman, G. Tononi, Cereb. Cortex 7 (1997) 207–227.
[8] C. Morris, H. Lecar, Biophys. J. 35 (1981) 193–213.
[9] E. Ott, J. Sommerer, Phys. Lett. A 188 (1994) 39–47.
[10] M. Rosenblum, A. Pikovsky, J. Kurths, Phys. Rev. Lett. 76 (1996) 1804–1807.
[11] W. Singer, Putative functions of temporal correlations in neocortical processing, in: C. Koch, J. Davis (Eds.), Large-Scale Neuronal Theories of the Brain, MIT Press, London, 1995.

Available at
www.ComputerScienceWeb.com
POWERED BY SCIENCE @DIRECT®

ELSEVIER

Neurocomputing 52–54 (2003) 159–164

NEUROCOMPUTING

www.elsevier.com/locate/neucom

Evidence for slow (2–10 Hz) and gamma frequency coherence between spike trains and local field potentials in the cerebellum

Yanqing Chen*, Douglas A. Nitz

The Neurosciences Institute, 10640 John Jay Hopkins Drive, San Diego, CA 92121, USA

Abstract

A method for detecting relationships between single-unit spike trains and local field potentials (LFPs) was developed and applied to recordings from rat cerebellum. LFPs were repeatedly filtered with a shifting frequency window. The resulting traces were transformed into peak-time point processes for comparison with spike trains using "relative-phase" analysis (Biol. Cybernet., in press). Discharge of some Purkinje cells was phase-related to LFP oscillations in the 2–10 Hz and 30–50 Hz frequency ranges. This analysis method revealed hidden coherency between spike-trains and LFPs. The findings suggest that cerebellar activity is, to some extent, temporally organized according to both slow and fast rhythms.

Keywords: Cerebellum; Spike trains; Local field potential (LFP); Coherence

1. Introduction

The correlation of single-unit neuronal recordings to their associated local field potentials (LFPs) can be difficult to study because spike trains are discrete point processes and LFPs are continuous. Direct cross-correlation (spike-triggered averaging) of these mixed time series is not always appropriate because it requires averaging of multiple trials or time windows and can be dominated by the frequency having greatest power in the LFP. Thus, this method cannot determine the time series of interactions between spike-trains and LFPs and is incapable of measuring transient periods of

* Corresponding author. Tel.: +1-858-626-2069; fax: +1-858-626-2099.
E-mail address: chen@nsi.edu (Y. Chen).

0925-2312/03/$ - see front matter © 2002 Elsevier Science B.V. All rights reserved.
doi:10.1016/S0925-2312(02)00745-2

coherence. Furthermore, fluctuations of LFP frequencies exhibiting relatively high spectral power can mask significant correlations at other higher frequencies.

For these reasons, we developed a novel method by which to study correlations between mixed time series more effectively. The method, "relative-phase" analysis, was then used to address an outstanding question concerning cerebellar function. That is, is simple-spike discharge of Purkinje cells coherent, and at what frequencies, with population activity within the cerebellum as measured by the LFP?

2. Materials and method

Cerebellar, single-unit spike trains of putative Purkinje cells and their associated LFPs were recorded using stereotrodes. Recordings were made as rats performed a simple shuttling task for food reward [4]. LFPs were filtered with a shifting (0.5 Hz-step), narrow-range (2 Hz) frequency window. The filtered traces were then converted to point processes by determining peak times. The temporal relationship between these LFP "events" and single-unit action potentials were studied for each narrow frequency range. Relative-phase analysis was used to determine the instantaneous phase relationship between the converted LFPs and single-unit spike trains. Relative phase here is defined as the relative timing difference between two point processes normalized by the associated interval of one process. The time of each single unit action potential was assigned a normalized phase (from 0 to 1) based on its temporal position between two peak times of the LFP for each frequency range tested (1–50 Hz). Thus, for all frequencies between 0 and 50 Hz, an averaged plot of phase versus spike discharge can be generated. The phase associated with each spike can also be analyzed as a time series in which transient periods of phase-locking to one or more LFP frequencies may be found. Note that this method also bypasses potential artifactual cross-correlations that occur between two point processes when one or both exhibit a high level of rhythmicity [5].

Fig. 1 demonstrates the idea of the method. A trace of LFP was filtered at a specified frequency window while preserving the phase information using Matlab (Mathworks, MA). The peak (or valley) timing of the filtered LFP was identified across the whole recording. The relative phase of a spike discharge in relation to the LFP was calculated based upon the timing differences between the spike and the peak (or valley) of the LFP normalized by the adjacent inter-peak distance. The correlation between the spike and the associated LFP can be revealed by the probability distribution of the relative phase between the two. If there is no correlation, the relative phase should distribute evenly between 0 and 1. Otherwise, peaks or valleys in the phase distribution imply correlation and preferred phase relations between the discharge of the spike and the LFP at a certain frequency range.

3. Results

Strong phase relationships between spike discharge and particular LFP frequencies were found for some cells. In cells exhibiting activity coherent with the LFP, there

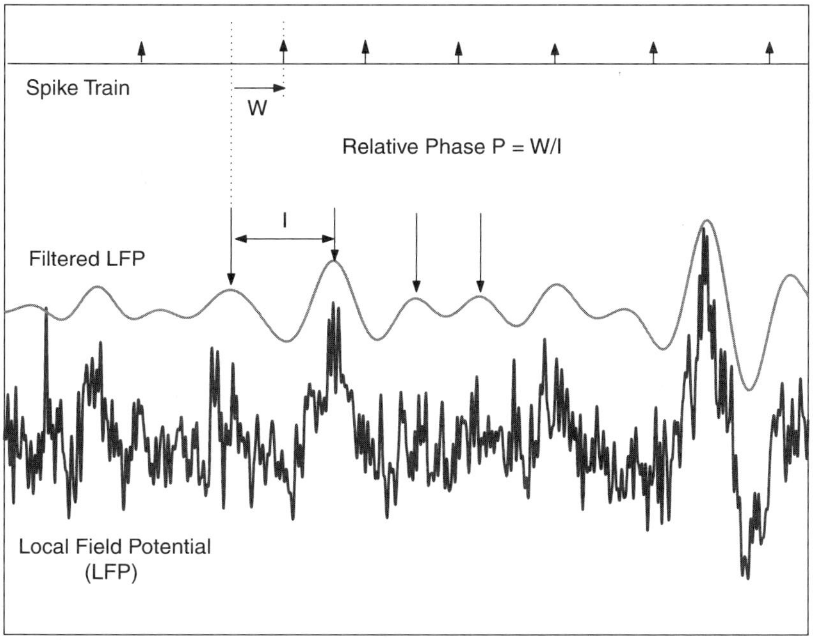

Fig. 1. An illustration of the method used in this paper to study the relation between a neuronal spike train and its associated LFP. The original LFP is filtered at a specified frequency window (e.g., 1–3 Hz in this example). The timing of the peak (or valley) of the filtered LFP is identified. The relative phase measure (P) is defined as the timing difference between a spike and the nearest peak of the filtered LFP normalized by the timing interval between two adjacent peaks. The correlation between a spike train and a LFP recording can be studied by analyzing the distribution of their relative phases calculated cross the whole spike train. Detailed discussions about the property of relative phase can be found in [2].

was a strong tendency for phase relationships to be confined to the 2–10 Hz and/or 30–50 Hz frequency ranges. Overall, of 32 cells examined (15 recordings from 6 rats), 9 cells exhibited phase relation to the LFP in the 2–10 Hz range, 4 cells exhibited phase relation to the LFP in the 20–50 Hz range. Notably, phase relationships were found for frequencies that did not stand out in power spectral analysis of the LFP or in autocorrelations of the spike train.

Fig. 2 shows an example of a LFP recording and two simultaneously recorded cerebellar spike trains. Fig. 3 shows the relative phase distributions (histograms) of these same two cells in relation to the LFP at frequency ranges from 1 to 50 Hz. Plotted are number of spikes occurring between phase 0 and 1 (equivalent to the interval between two peaks of LFP filtered at the corresponding frequency windows) from 0 to 50 Hz. The top plot shows that cell 1 has a slightly higher probability to fire close to the peaks of LFP in the 20–50 Hz frequency range. The second cell, however, did not show any phase relationship with LFP at 20–50 Hz range, but did exhibit two different preferred phase relationships, one below 3 Hz, the other between 5 and 10 Hz. A second example in Fig. 4 shows the relative phase distributions of two different

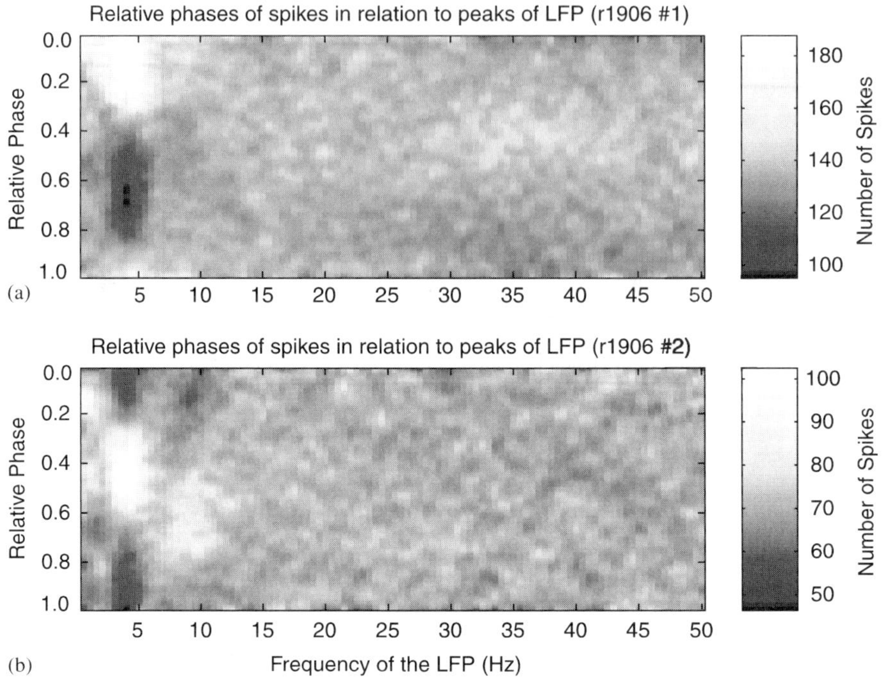

Fig. 4. Phase relations between two cerebellar spike trains recorded simultaneously and their associated LFP.

References

[1] G. Buzsaki, Theta oscillations in the hippocampus, Neuron 33 (2002) 325–340.

[2] Y. Chen, D.A. Nitz, Use of relative-phase analysis to assess correlation between neuronal spike trains, Biol. Cybernet (2003), in press.

[3] C.M. Gray, W. Singer, Stimulus-specific neuronal oscillations in orientation columns of cat visual cortex, Proc. Natl. Acad. Sci. USA 86 (5) (1989) 1698–1702.

[4] D.A. Nitz, G. Tononi, Tonic rhythmic activity of rat cerebellar neurons, Exp. Brain Res. 146 (2002) 365–270.

[5] D.H. Perkel, G.L. Gerstein, G.P. Moore, Neuronal spike trains and stochastic point processes. II. Simultaneous spike trains, Biophys. J. 7 (1967) 419–440.

[6] E.C. Walcott, Y. Chen, D.A. Nitz, Detection of rhythmic discharge of single units: comparison of cerebellar, hippocampal and cortical spike trains in behaving rats, Soc. Neurosci. Abs. 27 (2001) 828.3.

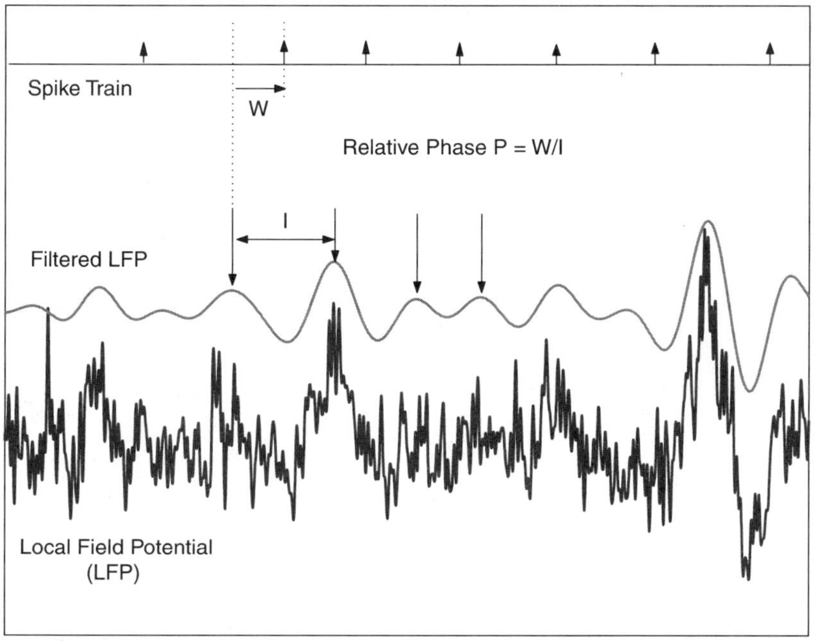

Fig. 1. An illustration of the method used in this paper to study the relation between a neuronal spike train and its associated LFP. The original LFP is filtered at a specified frequency window (e.g., 1–3 Hz in this example). The timing of the peak (or valley) of the filtered LFP is identified. The relative phase measure (P) is defined as the timing difference between a spike and the nearest peak of the filtered LFP normalized by the timing interval between two adjacent peaks. The correlation between a spike train and a LFP recording can be studied by analyzing the distribution of their relative phases calculated cross the whole spike train. Detailed discussions about the property of relative phase can be found in [2].

was a strong tendency for phase relationships to be confined to the 2–10 Hz and/or 30–50 Hz frequency ranges. Overall, of 32 cells examined (15 recordings from 6 rats), 9 cells exhibited phase relation to the LFP in the 2–10 Hz range, 4 cells exhibited phase relation to the LFP in the 20–50 Hz range. Notably, phase relationships were found for frequencies that did not stand out in power spectral analysis of the LFP or in autocorrelations of the spike train.

Fig. 2 shows an example of a LFP recording and two simultaneously recorded cerebellar spike trains. Fig. 3 shows the relative phase distributions (histograms) of these same two cells in relation to the LFP at frequency ranges from 1 to 50 Hz. Plotted are number of spikes occurring between phase 0 and 1 (equivalent to the interval between two peaks of LFP filtered at the corresponding frequency windows) from 0 to 50 Hz. The top plot shows that cell 1 has a slightly higher probability to fire close to the peaks of LFP in the 20–50 Hz frequency range. The second cell, however, did not show any phase relationship with LFP at 20–50 Hz range, but did exhibit two different preferred phase relationships, one below 3 Hz, the other between 5 and 10 Hz. A second example in Fig. 4 shows the relative phase distributions of two different

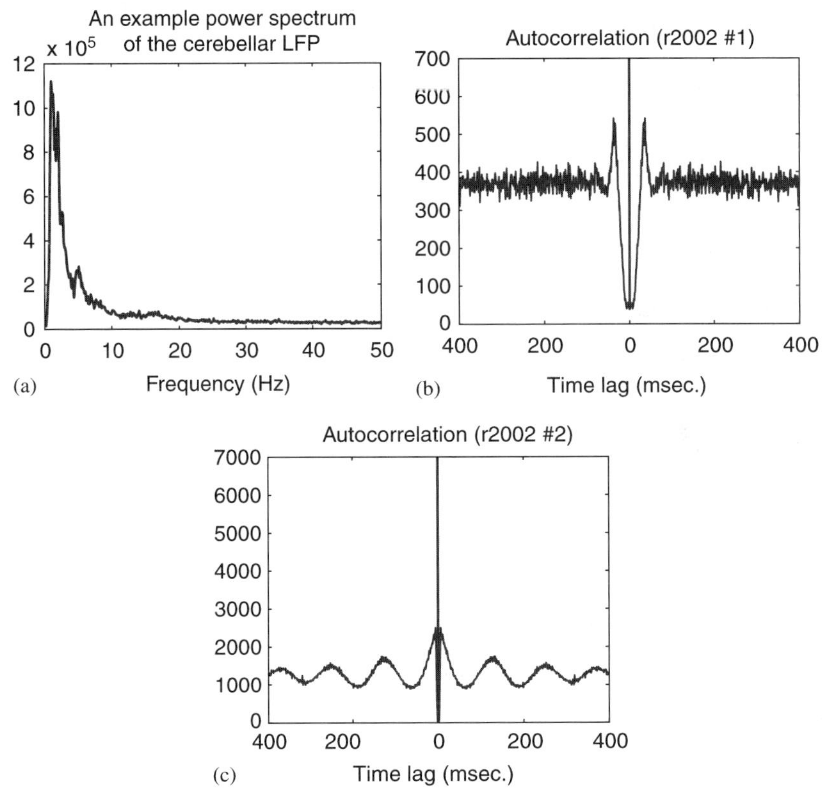

Fig. 2. Power spectrum of an example LFP and autocorrelations of two cerebellar spike trains recorded from the same wire.

cerebellar cells in relation to their associated LFP. It can be seen that they have two different phase relationships with the LFP around 5 Hz, and one cell has a tendency to fire between two peaks of LFP in the 20–50 Hz range. In both examples (Figs. 3 and 4), neither LFPs, nor the single neuron spike trains exhibited strong activity in the gamma (20–50 Hz) range. Nevertheless, the relative phase method was able to reveal clear relationships between spikes and LFPs at particular frequency ranges.

4. Summary and conclusions

The findings indicate that relative-phase analysis as applied to combined spike train/ LFP recordings is a useful tool to uncover hidden temporal relationships between single-unit activity and activity of the larger cerebellar network of which they are a small part. Similar to what has previously been shown for neocortical [3] and hippocampal [1] spike-train/LFP recordings, we have found that the simple-spike discharge of putative Purkinje cells is organized, to some extent, relative to the timing of cerebellar

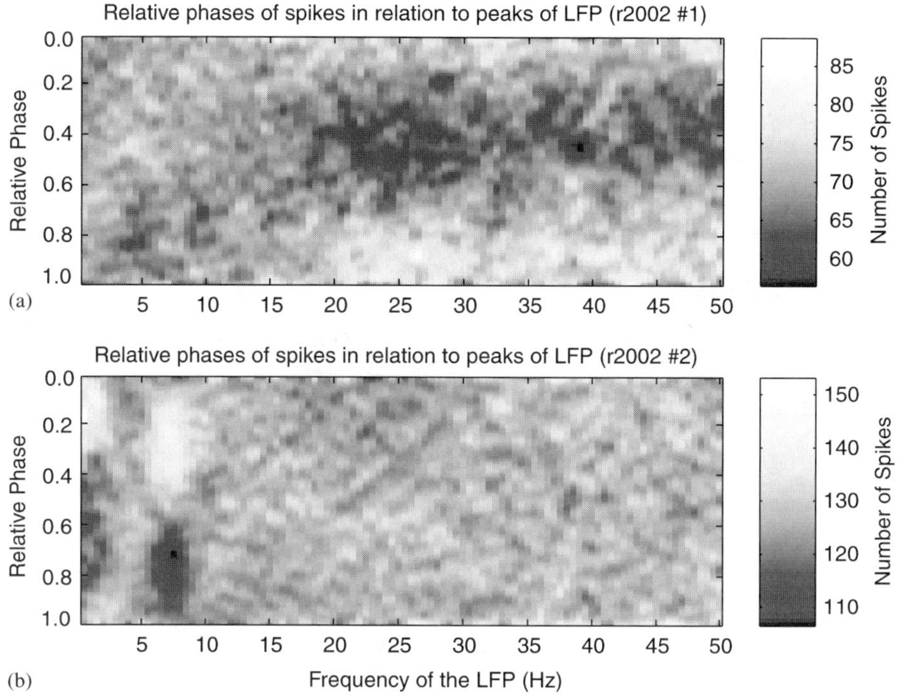

Fig. 3. Relative phase distributions of two cerebellar spike trains in relation to their associated LFP. Power spectrum of the LFP and autocorrelations of the two spike trains are shown in Fig. 1. In each plot, the horizontal axis is the frequency (0–50 Hz) of the filtered LFP, and the vertical axis is the relative phase (0–1). Gray level coded colorbars represent the number of spikes distributed between phase 0 and 1 at each frequency window of the filtered LFP.

population activity as measured by the local field potential. Notably, this finding is consistent with recent work demonstrating: (1) the existence of cerebellar neurons in lobule II which exhibit tonically rhythmic discharge patterns at either slow (approximately 4–6 Hz) or high (approximately 20–50 Hz) frequencies [4]; (2) the presence of transient periods (5–8 consecutive spike intervals) during which seemingly arrhythmic cerebellar neurons discharge spikes at regular intervals [6]. The functional role of such coherence and the extent to which it pervades the entire cerebellar network remains to be determined.

Acknowledgements

Research supported by the Neurosciences Research Foundation and the Alafi Family Foundation.

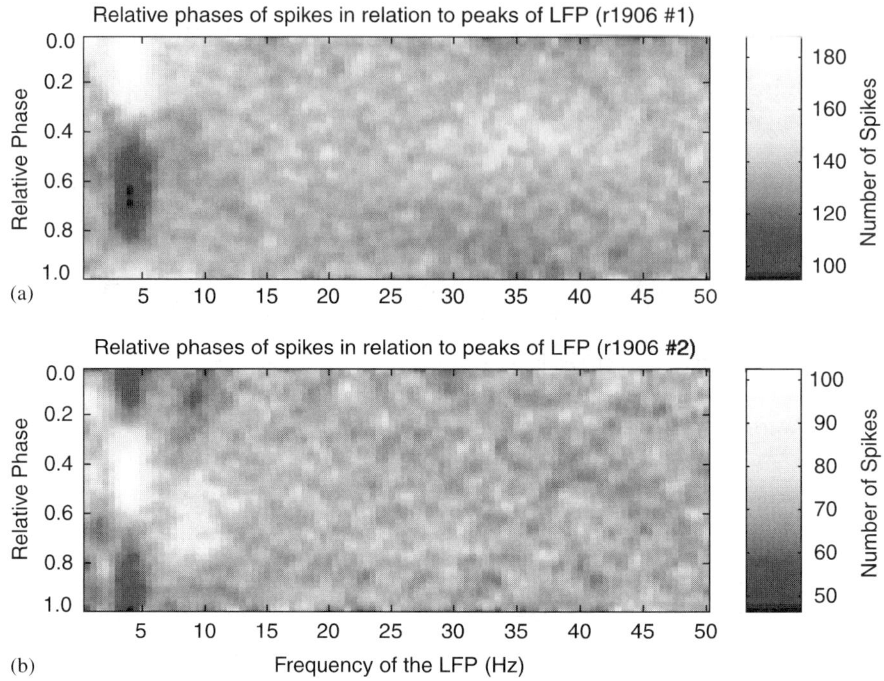

Fig. 4. Phase relations between two cerebellar spike trains recorded simultaneously and their associated LFP.

References

[1] G. Buzsaki, Theta oscillations in the hippocampus, Neuron 33 (2002) 325–340.

[2] Y. Chen, D.A. Nitz, Use of relative-phase analysis to assess correlation between neuronal spike trains, Biol. Cybernet (2003), in press.

[3] C.M. Gray, W. Singer, Stimulus-specific neuronal oscillations in orientation columns of cat visual cortex, Proc. Natl. Acad. Sci. USA 86 (5) (1989) 1698–1702.

[4] D.A. Nitz, G. Tononi, Tonic rhythmic activity of rat cerebellar neurons, Exp. Brain Res. 146 (2002) 365–270.

[5] D.H. Perkel, G.L. Gerstein, G.P. Moore, Neuronal spike trains and stochastic point processes. II. Simultaneous spike trains, Biophys. J. 7 (1967) 419–440.

[6] E.C. Walcott, Y. Chen, D.A. Nitz, Detection of rhythmic discharge of single units: comparison of cerebellar, hippocampal and cortical spike trains in behaving rats, Soc. Neurosci. Abs. 27 (2001) 828.3.

Available at
www.ComputerScienceWeb.com
POWERED BY SCIENCE @ DIRECT®

ELSEVIER

Neurocomputing 52–54 (2003) 165–168

NEUROCOMPUTING

www.elsevier.com/locate/neucom

A source of individual variation

William B Levy[a,b,*], Xiangbao Wu[a], Anthony J. Greene[c], Barbara A. Spellman[b]

[a]*Neurosurgery Department, University of Virginia, P.O. Box 800420, Charlottesville, VA 22908-0420, USA*

[b]*Psychology Department, University of Virginia, P.O. Box 800420, Charlottesville, VA 22908-0420, USA*

[c]*University of Wisconsin, Milwaukee, WI, USA*

Abstract

Even though behavioral experiments are usually conducted in a manner that minimizes variance between subjects, such variance is unavoidable—even when working with a homogenous population. A first step in producing a quantitative understanding of what causes this variation is a biologically based neural network model that reproduces the observed individual differences. Here we show that a simple model of the hippocampus can reproduce the histogram of learned performance in a homogenous population. The task and data being fit is transitive inference that is learned by college students. Individual differences mainly arise from initial connectivity differences not neuronal state initializations.

Keywords: Hippocampus; CA3; Transitive inference; Transverse patterning; Sequence learning

1. Introduction

Variation is a property of all biological processes, and it seems an unavoidable observation in cognitive psychology. Because our simple model of the hippocampus reproduces the learning curves and variability of several 'hippocampal' tasks, we decided to fit human learned performance for one of these tasks, specifically transitive inference.

* Corresponding author. Neurosurgery Department, University of Virginia, P.O. Box 800420, Charlottesville, VA 22908-0420, USA.

E-mail address: wbl@virginia.edu (W. B Levy).

0925-2312/03/$ - see front matter © 2002 Elsevier Science B.V. All rights reserved.
doi:10.1016/S0925-2312(02)00802-0

2. The model

The model used here is a hippocampal model of region CA3 (e.g. [3] for more details). The input layer corresponds to a combination of the entorhinal cortex and dentate gyrus. To make the system's operation as transparent as possible, decoding is performed by similarity comparisons rather than a CA1-subiculum-entorhinal decoding system. The CA3 model is a sparsely (10%) interconnected feedback network of 1024 neurons where all direct connections are excitatory and the network elements are McCulloch–Pitts neurons. There is an interneuron mediating feedforward inhibition and one mediating feedback inhibition. Inhibition is of the divisive form, but the system is not purely competitive because of a slight delay. Synaptic modification develops over training. The process controlling synaptic modification is a local, self-adaptive postsynaptic rule that includes both potentiation and depression aspects [4]. The network computations are all local and are contained in three equations: spatial summation adjusted by inhibition; threshold to fire or not; and local Hebbian synaptic modification (see e.g. Refs. [3,8] for details). Our computational modeling of the transverse patterning problem has been described many times (e.g., Refs. [3,5,9,10,6]. Network parameters were: $K_r = 0.055$; $K_i = 0.016$; $K_0 = 0.13$; initial weights were 0.4; threshold was 0.5; and synaptic modification rate was 0.02.

3. The transitive inference problem

In the transitive inference problem, five atomic stimuli (e.g. A, B, C, D, and E) are used. When A and B are together, A is the right answer ($A > B$). When B and C are together, B is the right answer ($B > C$) and so on. Thus, subjects and networks are taught $A > B$, $B > C$, $C > D$, and $D > E$. The critical test of TI after learning these four pairs is the response to $B?D$. Of course, the correct answer is $B > D$, and rats require a hippocampus to solve this problem (see e.g. Ref. [1]).

4. Variability

By design, only two randomizations occur for the simulations here: (1) the connectivity of each simulation, which produces the variability across simulations, and (2) the initial firing pattern, $Z(0)$, at the beginning of each training trial (see, e.g., Ref. [7]).

Training and testing are always initialized in the same way. That is, when we systematically vary the probability of this random neuronal activity at time step zero, testing uses the same random process (i.e., valued Bernoulli process independent across neurons) as was used for generating $Z(0)$ in the corresponding training trials.

5. Results

Fig. 1 shows the distribution comparing the number of correct transitivity responses (b for the BD test) that occurred in the transitive inference problem; here we are

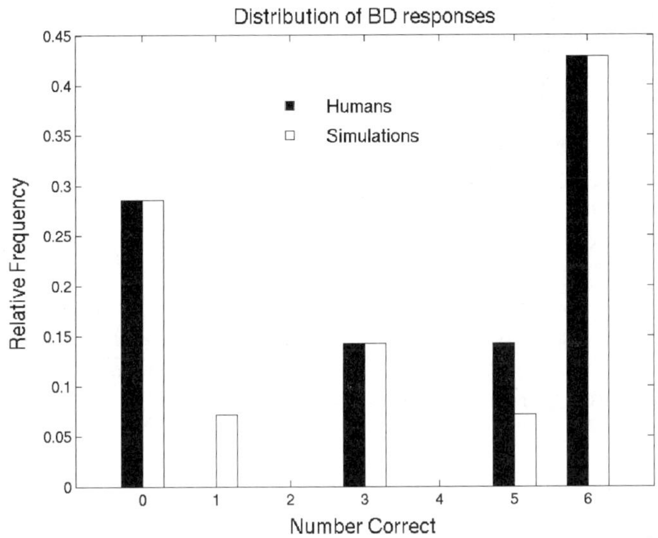

Fig. 1. Frequency distribution comparing of the number of correct transitivity (BD) responses in the inference test for seven human subjects and 14 simulations of the computer model. Despite small sample sizes, the human experiment and the computer model show remarkably similar response distributions. For both the human experiment and the computer simulations, there were four premise pairs of five atomic stimuli learned via the staged training paradigm. For the human experiment, the stimuli were five distinct Kanji characters described elsewhere (experiment 1 of Ref. [2]). For the computer model, orthogonal blocks of neurons coded each stimulus of a pair.

comparing seven human subjects and fourteen simulations of the network model. Despite small sample sizes, the human experiment and the computer model show remarkably similar response distributions. To match the experimental data, the activity level and synaptic modification constant were adjusted but all simulations are parameterized identically. The small chi square value (0.75; d.f. = 6) corresponds to very similar histograms.

6. Discussion

We have shown that a simple neural network model can produce individual variation similar to human subjects. Because the neuronal updating is deterministic, there are only two possible sources that can produce this variation; it either comes from the randomization of cell firing at the beginning of each learning trial or it comes from the initially specified random connectivity of the network.

Interestingly, the best performance is achieved at full initial state randomness. This is consistent with, and extends, our previous findings that a full initial state of randomness [7] helps learning. It is this trial-to-trial initial state activity randomization that causes variations when testing a single simulation many times: that is, initial state randomization is what produces the simulations that neither make all correct, nor all incorrect, decisions (see Fig. 1). However, because initial state randomization is negatively correlated with variation once average performance is above 50%, we conclude that connectivity randomization is the main cause of variation between simulations. By way of confirmation, when the initial state randomizations are made constant across different simulations (but still random from trial to trial), then the same variation across individual networks results. Thus, the connectivity randomization is causing the variation across simulations.

No problem is more central to the goals of psychological science than to understand the cause of individual variation, why one person is cognitively different than another. Although when asked in terms of the biological basis of cognition such a question seems hopelessly complex by virtue of possible answers, a computational model leads us to a simple, specific hypothesis—at least for the variation in learning, in a homogenous population when learning depends on the hippocampus.

Acknowledgements

This work was supported by NIH MH48161 to WBL and MH57358 to N. Goddard (with subcontract #163194-54321 to WBL).

References

[1] J.A. Dusek, H. Eichenbaum, The hippocampus and memory for orderly stimulus relations, Proc. Natl. Acad. Sci. USA 94 (1997) 7109–7114.

[2] A.J. Greene, B.A. Spellman, J.A. Dusek, H.B. Eichenbaum, W. B Levy, Relational learning with and without awareness: transitive inference using non-verbal stimuli in humans, Mem. Cog. 29 (2001) 893–902.

[3] W. B Levy, A sequence predicting CA3 is a flexible associator that learns and uses context to solve hippocampal-like tasks, Hippocampus 6 (1996) 579–590.

[4] W. B Levy, O. Steward, Synapses as associative memory elements in the hippocampal formation, Brain Res. 175 (1979) 233–245.

[5] W. B Levy, X.B. Wu, J.M. Tyrcha, Solving the transverse patterning problem by learning context present: a special role for input codes, INNS World Congress on Neural Networks, 1996, pp. 1305–1309.

[6] A.P. Shon, X.B. Wu, W. B Levy, Using computational simulations to discover optimal training paradigms, Neurocomputing 32–33 (2000) 995–1002.

[7] A.P. Shon, X.B. Wu, D.W. Sullivan, W. B Levy, Initial state randomness improves sequence learning in a model hippocampal network, Phys. Rev. E 65 (2002) (031914) 1–15

[8] A.C. Smith, X.B. Wu, W. B Levy, Controlling activity fluctuations in large, sparsely connected random networks, Network 11 (2000) 63–81.

[9] X.B. Wu, J.M. Tyrcha, W. B Levy, A special role for input codes in solving the transverse patterning problem, in: J.M. Bower (Ed.), Computational Neuroscience: Trends in Research, Plenum Press, New York, 1997, pp. 885–889.

[10] X.B. Wu, J. Tyrcha, W. B Levy, A neural network solution to the transverse patterning problem depends on repetition in the input code, Biol. Cybernet. 79 (1998) 203–213.

Available at
www.ComputerScienceWeb.com
POWERED BY SCIENCE @ DIRECT®

ELSEVIER

Neurocomputing 52–54 (2003) 169–175

NEUROCOMPUTING

www.elsevier.com/locate/neucom

Electric field modulation of synchronization in neuronal networks

E.-H. Park[a,c,*], P. So[a,c], E. Barreto[a,c], B.J. Gluckman[a,c], S.J. Schiff[b,c]

[a] *Department of Physics & Astronomy, Mail Stop 2A1, George Mason University, Fairfax, VA 22030, USA*
[b] *Department of Psychology, Mail Stop 2A1, George Mason University, Fairfax, VA 22030, USA*
[c] *The Krasnow Institute for Advanced Study, Mail Stop 2A1, George Mason University, Fairfax, VA 22030, USA*

Abstract

Motivated by the observation that applied electric fields modulate hippocampal seizures, and that seizures may be asynchronous, we modeled synaptically coupled two-compartment hippocampal pyramidal neurons embedded within an electrically resistive lattice in order to examine network synchronization properties under the influence of externally applied electric fields. Excitatory electric fields were shown to synchronize or desynchronize the network depending on the natural frequency mismatch between the neurons. Such frequency mismatch was found to decrease as a function of increasing electric field amplitude. These findings provide testable hypotheses for future seizure control experiments.

Keywords: Synchronization; Electric field; Seizures; Hippocampus; CA1

Since the 1950s, it has been generally assumed that seizures are manifestations of excessive synchrony. However, recent experiments using dual intracellular impalements in the CA1 region of rat hippocampal slices suggest that neurons may *desynchronize* as seizures initiate, and resynchronize as seizures terminate. In addition, theoretical work has recently established that sustained activity can be supported by collectively *asynchronous* states [7]. These results, in juxtaposition with recent demonstrations that

* Corresponding author.
E-mail address: epark3@gmu.edu (E.-H. Park).

0925-2312/03/$ - see front matter © 2002 Elsevier Science B.V. All rights reserved.
doi:10.1016/S0925-2312(02)00820-2

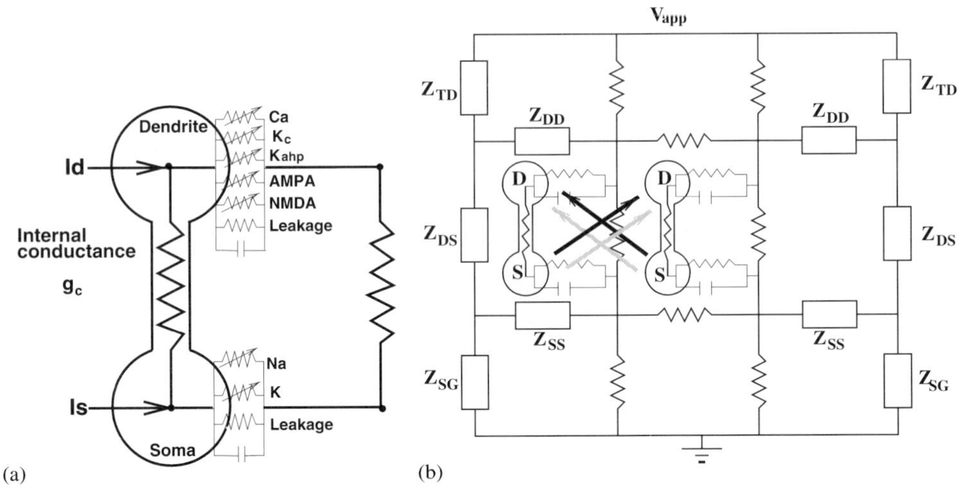

Fig. 1. (a) Currents and conductances for a single PR neuron. (b) Synaptically coupled neurons embedded within a resistive array modeling the electric properties of CA1. Black and gray arrows indicate NMDA and AMPA excitatory synapses, respectively. The terminal resistances are chosen such that the lattice is equivalent to a horizontally infinite lattice.

electric fields can modulate neuronal dynamics and suppress seizure-like activity [4,5], have motivated the present work. We seek to gain a better understanding of the synchronization properties of populations of neurons, and in particular, how such synchrony may be modulated with electric fields.

We report results based on a computational model that we have explicitly designed to include electric field interactions. Since the presence of an electric field induces spatial polarization in neurons [1,2,12], the minimum individual neuronal unit must have at least two spatially separated compartments. We therefore chose the two-compartment lumped Pinsky–Rinzel (PR) model neuron, which consists of a dendritic and a somatic compartment separated by a finite conductance [11]. For the present work, two such neurons were embedded within a lattice of resistors which models the extracellular medium, and a potential difference was imposed along the somatic–dendritic axis. In addition to interacting electrically via the resistive network, the neurons were also coupled via AMPA and NMDA synapses.

A schematic representation of the PR model neuron is shown in Fig. 1(a). Specific details can be found in Refs. [5,11]. [1] Two neurons were arranged in the resistive array as shown in Fig. 1(b), and an externally applied electric field was introduced by imposing a potential difference between V_{app} and ground. In addition to varying V_{app}, we also vary the difference Δg_{c} between the neurons' internal soma-dendrite conductances ($g_{\mathrm{c}}^{(1)} = 2.1$ mS/cm^2, and $g_{\mathrm{c}}^{(2)} = (1-\delta)g_{\mathrm{c}}^{(1)}$, where $\delta = 0$–30%).

[1] The parameters were chosen as in Ref. [11], with Id $= 0.7$ μA/cm^2, Is $= 0$, $g_{\mathrm{NMDA}} = 0.03$ mS/cm^2, $g_{\mathrm{AMPA}} = 0.0045$ mS/cm^2 and $[\mathrm{K}^+]_{\mathrm{o}} = 3.5$ mM. The potassium reversal potential relative to -60 mV is $V_{\mathrm{K}} = -38.56$ mV.

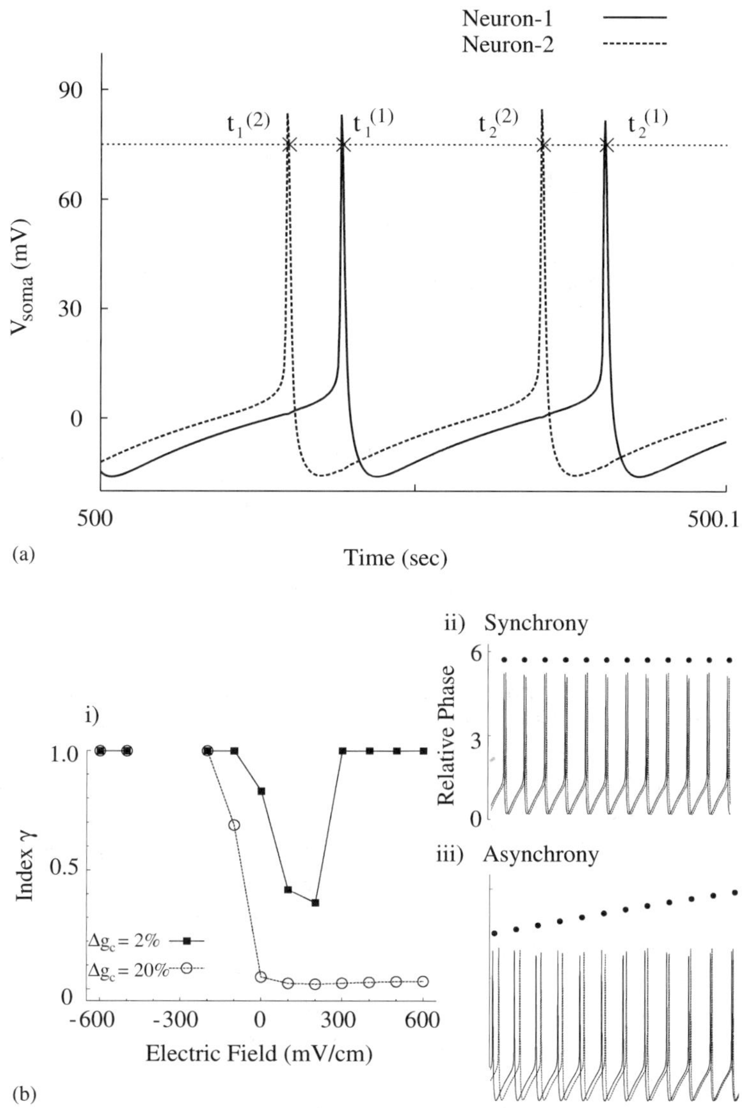

Fig. 2. (a) Tracings of the somatic voltage from two neurons within the network. (b) (i) γ vs. electric field strength for two different levels of parameter mismatch. γ and $V_{\mathrm{soma}}^{(1,2)}$ as a function of time for an electric field of 600 mV/cm in: (ii) the synchronous case, $g_c = 2\%$; (iii) asynchronous case, $g_c = 20\%$.

To characterize synchronization, we employed a phase locking index γ defined by $\gamma^2 = \langle \cos \Psi(t) \rangle^2 + \langle \sin \Psi(t) \rangle^2$ [8,10], where $\langle\, \rangle$ denotes time-average. $\Psi(t)$ is the relative phase, defined as $\Psi(t_k^{(1)}) = 2\pi(t_k^{(1)} - t_m^{(2)})/(t_{m+1}^{(2)} - t_m^{(2)})$, where k and m are event numbers for neurons 1 and 2, respectively, and the times t are defined by threshold crossings. Fig. 2(a) shows sample tracings of the somatic voltages and illustrates the procedure for

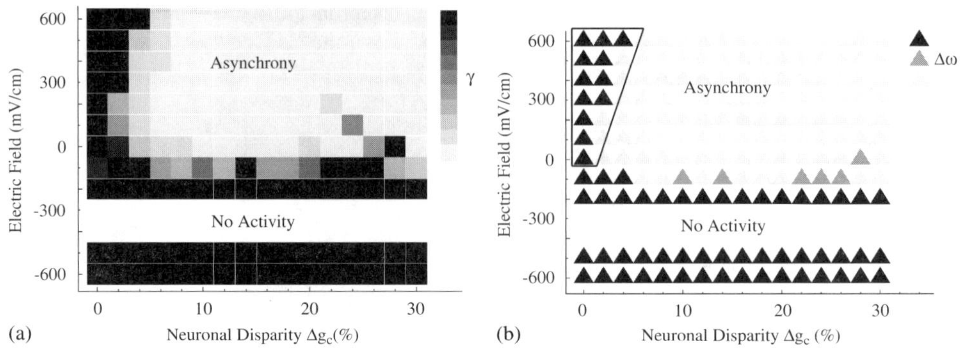

Fig. 3. (a) Synchrony phase diagram. γ is plotted in gray-scale (black = 1, white = 0) as a function of the electric field strength and the neuronal disparity. Network activity is fully suppressed by moderate inhibitory applied fields of approximately −300 to −400 mV/cm. (b) Natural frequency mismatch $\Delta\omega$ between the neurons as a function of the applied electric field and neuronal disparity (black triangle indicates $\Delta\omega < 0.1$; medium gray triangle indicates $\Delta\omega \sim 0.1$; light gray triangle indicates $\Delta\omega > 0.1$).

calculating γ. Note that γ varies between 0 and 1 for unsynchronized and synchronized activity, respectively.

Fig. 2(b) shows that by varying the strength of the applied electric field, the neurons either synchronize or desynchronize depending on the parameter mismatch Δg_c. Neurons with large parameter mismatch are observed to phase-lock for moderate inhibitory (negative) electric fields, and to desynchronize as the field becomes excitatory (positive). This is consistent with results reported by Golomb and Hansel [6], who found that in heterogeneous ensembles, increasing coupling tends to desynchronize network dynamics. However, for small parameter mismatch, we observed that sufficiently large electric fields can resynchronize the neurons.

A more complete summary of our network's dynamics is presented in Fig. 3(a). Here, we show the γ index as a function of both the applied electric field strength and the parameter mismatch Δg_c. Several features are prominent. An extensive asynchronous region is present for large excitatory fields and large neuronal disparity, again, in agreement with Golomb and Hansel's observation in Ref. [6]. Second, large inhibitory fields tend to induce phase locking even when neuronal disparity is large. Third, moderate inhibitory fields suppress collective network activity (consistent with the in vitro seizure suppression experiments of Gluckman et al. [4,5]). Perhaps the most interesting feature of this phase diagram is the boundary between the phase-locked and asynchronous states. As observed in Fig. 2(b), for small neuronal disparity there is a regime where moderately strong excitatory fields desynchronize the network, but stronger fields leads to resynchronization. This threshold is seen in Fig. 3(a) to increase with increasing Δg_c.

We now argue that this nontrivial boundary can be explained using a simple phase oscillator model [3,8,13]. In our network, the dynamics of the individual units are predominately periodic spikes or bursts, and the network is well approximated by a system of coupled phase oscillators for a significant range of parameters. Details

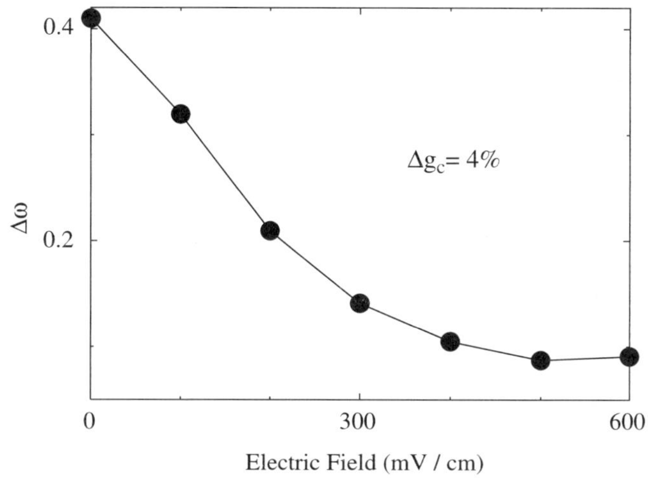

Fig. 4. Natural frequency mismatch $\Delta\omega$ as a function of the applied electric field for $\Delta g_c = 4\%$.

pertaining to the reduction of our system will be presented elsewhere [9]. Here, we show that this simple approach explains the synchrony/asynchrony boundary described above.

Fig. 3(b) shows the degree of natural frequency mismatch $\Delta\omega$ between our two individual units as a function of applied electric field and neuronal disparity. To measure $\Delta\omega$, each neuron was observed in isolation, subject only to the applied electric field. The boundary defined by $\Delta\omega = \alpha \simeq 0.1$ Hz is such that for $\Delta\omega > \alpha$, the network is in the asynchronous state, and for $\Delta\omega < \alpha$, the neurons phase lock to each other. It is remarkable that this simple criterion reproduces the observed transition boundary rather well. In particular, it correctly predicts that for small neuronal disparity ($\Delta g_c < 6\%$), increasing the (excitatory) electric field first desynchronizes and then resynchronizes the network (see the circled region in Fig. 3(b)). To clarify this point, Fig. 4 shows that $\Delta\omega$ decreases for increasing (excitatory) electric field at $\Delta g_c = 4\%$. One can understand this by noting that each individual neuron's spiking rate is a monotonically increasing concave-down function with respect to the applied electric field. Thus, as the spiking rate increases, the corresponding *incremental* increase is smaller. In the small disparity regime, the initial frequency mismatch is small enough such that as the excitatory field increases, the difference between the spiking rates of the two neurons decreases below the critical level $\Delta\omega^* = \alpha$, and phase locking is observed.

We have demonstrated that a computational model of two-compartment neurons embedded within a resistive extracellular medium exhibits both complex phase locking and asynchronous behavior as a function of the applied electric field and the degree of neuronal disparity. The observed dynamics are consistent with recent theoretical and experimental results. Most importantly, there exists a critical value of natural frequency mismatch which determines the synchronization properties of the network. We speculate that the observed phase-locking threshold may be explained in terms of a system of coupled phase oscillators.

This work was supported by NIH K02MH01493, 2R01MH50006, K25 MH01963; NSF IBN 97-27739.

References

[1] C.Y. Chan, J. Houndsgaard, C. Nicholson, Effects of electric fields on transmembrane potential and excitability of turtle cerebellar Purkinje cells in vitro, J. Physiol. (London) 402 (1988) 751–771.
[2] C.Y. Chan, C. Nicholson, Modulation by applied electric fields of Purkinje and stellate cell activity in the isolated turtle cerebellum, J. Physiol. (London) 371 (1986) 89–114.
[3] A.H. Cohen, P.J. Holmes, R.H. Rand, The nature of the coupling between segmental oscillators of the lamprey spinal generator for locomotion: a mathematical model, J. Math. Biol. 13 (1982) 345–369.
[4] B.J. Gluckman, H. Nguyen, S.L. Weinstein, S.J. Schiff, Adaptive electric field suppression of epileptic seizures, J. Neurosci. 21 (2001) 590–600.
[5] B.J. Gluckman, P. So, T.I. Netoff, M. Spano, S.J. Schiff, Stochastic resonance in mammalian neuronal networks, Chaos 8 (1998) 588–598.
[6] D. Golomb, D. Hansel, The number of synaptic inputs and the synchrony of large, sparse neuronal networks, Neural Comput. 12 (2000) 1095–1139.
[7] B.S. Gutkin, C.R. Lang, C.L. Colby, C.C. Chow, G.B. Ermentrout, Turning on and off with excitation: the role of spike-timing asynchrony and synchrony in sustained neural activity, J. Comput. Neurosci. 11 (2001) 121–134.
[8] Y. Kuramoto, Chemical Oscillations, Waves, and Turbulence, Springer, Berlin, 1984.
[9] E.-H. Park, P. So, E. Barreto, S.J. Schiff, in preparation.
[10] A. Pikovsky, M. Rosenblum, J. Kurths, Synchronization A Universal Concept in Nonlinear Sciences, Cambridge University Press, Cambridge, 2001.
[11] P.F. Pinsky, J. Rinzel, Intrinsic and network rhythmogenesis in a reduced Traub model for CA3 neurons, J. Comput. Neurosci. 1 (1994) 39–60.
[12] D. Tranchina, C. Nicholson, A model for the polarization of neurons by extrinsically applied electric fields, Biophys. J. 50 (1986) 1139–1159.
[13] A.T. Winfree, The Geometry of Biological Time, Springer, New York, 1980.

Eun-Hyoung Park received Ph.D. degree in nonlinear dynamics, physics in 2000 Dec. at Potsdam University in Germany. Since March 2001, she is now a post-doc at the Krasnow Institute at George Mason University, Virginia, working on computational modeling, specifically pyramidal cell in hippocampus.

Paul So

Ernest Barreto

Bruce J. Gluckman

Steven J. Schiff

Available at
www.ComputerScienceWeb.com
POWERED BY SCIENCE @DIRECT®

ELSEVIER

Neurocomputing 52–54 (2003) 177–183

NEUROCOMPUTING

www.elsevier.com/locate/neucom

Single-trial detection in EEG and MEG: Keeping it linear

Lucas Parra[a], Chris Alvino[a], Akaysha Tang[b], Barak Pearlmutter[b], Nick Yeung[c], Allen Osman[d], Paul Sajda[e,*]

[a] *Vision Technologies Laboratory, Sarnoff Corporation, Princeton, NJ 08540, USA*
[b] *Department of Psychology and Department of Computer Science, University of New Mexico, Albuquerque, NM 87131, USA*
[c] *Department of Psychology, Princeton University, Princeton, NJ 08544, USA*
[d] *Department of Psychology, University of Pennsylvania, Philadelphia, PA 19104, USA*
[e] *Department of Biomedical Engineering, 351 Engineering Terrace, MC 8904, Columbia University, New York, NY 10027, USA*

Abstract

Conventional electroencephalography (EEG) and magnetoencephalography (MEG) analysis often rely on averaging over multiple trials to extract statistically relevant differences between two or more experimental conditions. We demonstrate that by linearly integrating information over multiple spatially distributed sensors within a predefined time window, one can discriminate conditions on a trial-by-trial basis with high accuracy. We restrict ourselves to a linear integration as it allows the computation of a spatial distribution of the discriminating source activity. In the present set of experiments the resulting source activity distributions correspond to functional neuroanatomy consistent with the task (e.g. contralateral sensory-motor cortex and anterior cingulate).

Keywords: Linear integration; High-density electroencephalography (EEG); Magnetoencephalography (MEG); Single-trial analysis; Brain–computer interface (BCI)

1. Introduction

Trial averaging is often used to increase the signal-to-interference (SIR) ratio, for example in analysis of event-related potentials (ERPs) [3]. With the large number of

* Corresponding author.
E-mail address: ps629@columbia.edu (P. Sajda).

0925-2312/03/$ - see front matter © 2003 Elsevier Science B.V. All rights reserved.
doi:10.1016/S0925-2312(02)00821-4

sensors in high density EEG and MEG an alternative approach is to integrate information over space rather than across trials. For example in blind source separation independent signals are extracted so that noise components and artifacts can be removed [8,13,14]. However, blind methods do not exploit the timing information of external events that is often available.

In the context of a BCI system, many methods have applied linear and non-linear classification to a set of features extracted from the EEG. For example autoregressive models have been used to extract features across a limited number of electrodes, with these features combined using either linear or non-linear classifiers to identify the activity from the time course of individual sensors [9]. Others have proposed to combine sensors in space by computing maximum and minimum eigenvalues of the sensor covariance matrices to obtain a non-linear binary classification [10]. Though many of these methods show promising performance in terms of classifying covert (purely mental) processes, their neurological interpretation remains obscured.

We use conventional linear discrimination to compute the optimal spatial integration of a large array of sensors. We exploit timing information by discriminating and averaging within a short time window relative to a given external event. We demonstrate the method on three different electromagnetic brain imaging data sets: (1) predicting overt motor action (left/right button push) from 122 MEG sensors, (2) classifying covert or imagined motor activity (left/right taps) using 59 EEG sensors, (3) detecting decision errors in a binary discrimination task from 64 EEG sensors.

2. Materials and methods

2.1. Linear discrimination

Denoting $\mathbf{x}(t)$ as the M sensor values sampled at time instance t, we compute the spatial weighting coefficients $\mathbf{v}$ such that

$$y(t) = \mathbf{v}^{\mathrm{T}}\mathbf{x}(t) \tag{1}$$

is maximally discriminating between the times t, corresponding to two different experimental conditions. For example, in the prediction of explicit motor response experiments (described below) the times correspond to a number of samples prior to an overt button push. The samples corresponding to a left button push are to be discriminated from samples of a right button push. For each of N trials we may have T samples totaling NT training examples. We use conventional logistic regression [5] to find $\mathbf{v}$. After finding the optimal $\mathbf{v}$ we average over the T dependent samples of the kth trial to obtain a more robust result, $\bar{y}_k = T^{-1}\sum_{t\in T_k} y(t)$, where T_k denotes the set of sample times corresponding to trial k. Receiver operating characteristic (ROC) analysis [12] is done using these single-trial short-time averaged discrimination activities ($\bar{y}_k$).

2.2. Localization of discriminating sources

In order to provide a functional neuroanatomical interpretation of the resultant spatial weighting, we treat $y(t)$ as a source, and visualize the coupling coefficients of the source with the sensors. The strength of the coupling roughly indicates the closeness of the source to the sensor. The coupling $\mathbf{a}$ is defined as the coefficients that multiply the putative source $y(t)$ to give its additive contribution $\mathbf{x}_y(t)$ to the sensor readings, $\mathbf{x}_y(t) = \mathbf{a}\,y(t)$. Unfortunately $\mathbf{x}_y(t)$ is not observable in isolation, instead we observe, $\mathbf{x}(t) = \mathbf{x}_y(t) + \mathbf{x}_{y'}(t)$, where $\mathbf{x}_{y'}(t)$ represents the activity that is not due to the discriminating source. If the contributions, $\mathbf{x}_{y'}(t)$, of other sources are uncorrelated with $y(t)$ we obtain the coupling coefficients by the least-squares solution [7].

$$\mathbf{a} = \|\mathbf{y}\|^{-2}\mathbf{X}\mathbf{y}, \tag{2}$$

where the samples $\mathbf{x}(t)$ for different t have been arranged as columns in the a matrix $\mathbf{X}$, and $y(t)$ as a column vector $\mathbf{y}$. In general other sources are not guaranteed to be uncorrelated with the discriminating source. Therefore, the "sensor projection" $\mathbf{a}$ represents the coupling of all source activity that is correlated to the discriminating source $y(t)$. Our approach relies on the linearity of $y(t)$ and the fact that different sources in EEG and MEG add linearly [1].

2.3. Datasets for analysis

2.3.1. Prediction of motor response from MEG

This data set was provided by AT and BP. Four subjects performed a visual-motor integration task. Subjects were simultaneously presented with two visual stimuli on a CRT. Subjects were instructed to push a left hand or right hand button, depending on which side a target stimulus was present. The subject was to discover the target by trial and error using auditory feedback. Each trial began with visual stimulus onset, followed by button push, followed by auditory feedback, indicating if the subject responded correctly. The interval between the motor-response and the next stimulus presentation was 3.0 ± 0.5 s. Each subject performed 90 trials, which took approximately 10 minutes. MEG data was recorded using 122 sensor at a sampling rate of 300 Hz and high-pass filtered to remove DC drifts. Dipole fits were done using the Neuromag xfit tools, which assume a spherical head model to find a single equivalent current dipole.

2.3.2. Classification of explicit and imagined motor response from EEG

This data set was provided by AO. Nine subjects performed a visual stimulus driven finger (L/R) tapping task. Subjects were asked to synchronize an explicit or imagined tap to the presentation of a brief temporally predictable signal. Subjects were trained until their explicit taps occurred consistently within 100 ms of the synchronization signal. After training, each subject received 10 blocks of trials. Each 72-trial block consisted of nine replications of the eight trial types (Explicit vs. Imagined $\times$ Left vs. Right vs. Both vs. No Tap) presented in a random order. Trials with noise due to eye blinks were not considered in the EEG analysis. The electromyogram (EMG) was recorded to detect muscle activity during imagined movements. The 59 EEG channels were sampled at 100 Hz and high-pass filtered to remove DC drift.

2.3.3. Detection of decision errors from EEG

This data set was provided by NY. Seven subjects performed a visual target detection amongst distractors task. On each trial, subjects were presented with a stimulus for 100 ms. There were four possible stimuli, each consisting of a row of five arrows. Subjects were told to respond by pressing a key on the side indicated by the center arrow. They were to ignore the four flanking arrows. On half of the trials, the flanking arrows pointed in the same direction as the target (e.g. $<<<<<$), on the other half the flankers pointed in the opposite direction (e.g. $<<><<$). The interval between the motor-response and the next stimulus presentation was 1.5 s. Subjects performed 12 blocks of 68 trials each. The activity during an 100 ms interval prior to the response was used as the baseline. The sampling rate was 250 Hz. Following baselining, trials were manually edited to remove those with blinks, large eye movements, instrument artifacts and amplifier saturation.

3. Results and discussion

Single trial discrimination results are shown for the three different data sets and include sensor projections **a**, and detection/prediction performance using single-trial, short-time averaged $\bar{y}_k$. Performance is reported using an ROC curve computed with a leave-one-out training and testing procedure [5]. Overall performance is quantified with A_z (the area under the ROC curve).

Fig. 1 shows results for the dataset used to predict whether a subject will press a button with their left or right hand by analyzing the MEG signals in a window prior to the button push. We use an analysis window 100 ms wide centered at 83 ms prior to the button event, which at 300 Hz corresponds to $T = 30$. Fig. 1 shows the results for one subject (AT). Single-trial discrimination is shown in the ROC curve, which for this subject shows good discriminability ($A_z = 0.93$). Fig. 1 also shows the sensor projection **a** and the location of a dipole-fit for this projection. When considered with respect to the motor-sensory homunculus, these results indicate that the discrimination source activity originates in the sensory-motor cortex corresponding to the left hand.

Fig. 2 shows results for the second data set, where the goal is to detect activity associated with purely imagined motor response, a situation more realistic for a BCI system. Subjects are trained to imagine a tap with the left or right index finger synchronized to a brief, temporally predictable signal. We selected a 0.8 s time window around the time where the task is to be performed. 90 left and 90 right trials were available to train the coefficients of the 59 EEG sensors. For the nine subjects we obtain a leave-one-out performance of $A_z = 0.77 \pm 0.1$. The results for the best performing subject is $A_z = 0.90$, shown in Fig. 2. The sensor projection of the 59 EEG sensors shows a clear left-right polarization over the motor area. The results, across the nine subjects, for predicting explicit finger taps from a window 300 ms to 100 ms prior to the taps is $A_z = 0.87 \pm 0.1$. As shown in Fig. 2, sensor projections of the discrimination

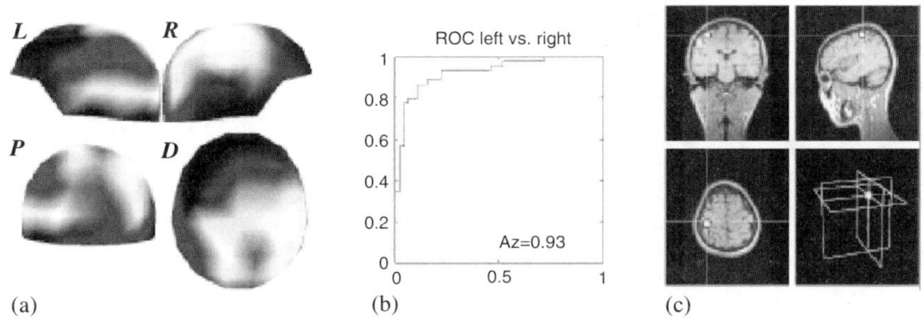

Fig. 1. MEG left/right button push prediction. (A) Sensor projections **a** for discrimination vector. (B) ROC curve for left vs. right discrimination. Area under the curve $A_z = 0.93$. (C) Dipole-fit of **a** overlaid on MRI image.

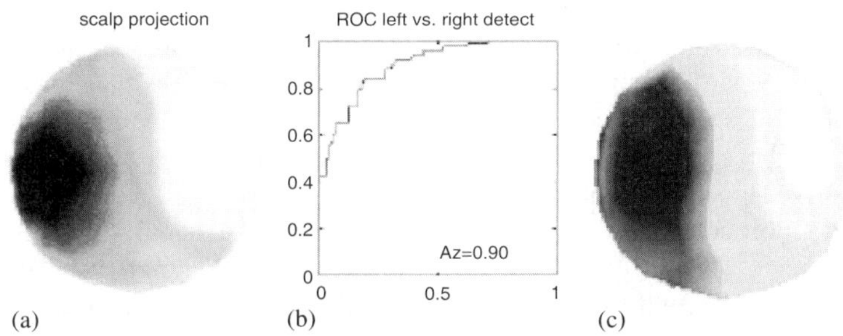

Fig. 2. Discrimination of imagined left/right finger taps. (A) Dorsal view of sensor projections **a**. (B) ROC curve for left vs. right discrimination. (C) Sensor projection of discriminating source for explicit finger tap.

vector for explicit motor response coincide with those for the imagined motor response. This is an important experimental finding supporting the development of an intuitive BCI system—signals arising from the cortical areas that encode an explicit movement can be also used for predicting the imagined movement.

Fig. 3 shows the results for the experiments where the goal is to detect the Error Related Negativity (ERN) on a single trial basis. The ERN is a negative deflection in the EEG signals following perceived incorrect responses in a discrimination task [6]. The detection of error potentials has been proposed as a means of correcting communication errors in a BCI system [11]. The ERN has a fronto-central distribution, suggesting a source in the anterior cingulate [4]. It begins around the time of the incorrect response and lasts roughly 100 ms thereafter. We use this time window for detection. Forty to eighty error trials and 300 correct trials were used for training 64 coefficients. The sensor projection, shown in Fig. 3 for one subject, is representative of the results obtained for other subjects and is consistent with the scalp topography of the ERN. The detection performance for this subject was $A_z = 0.84$ and is to be compared to $A_z = 0.63$ when detecting ERN from the center electrode where maximal activity is expected. ROC performance across all 7 subjects was $A_z = 0.79 \pm 0.05$.

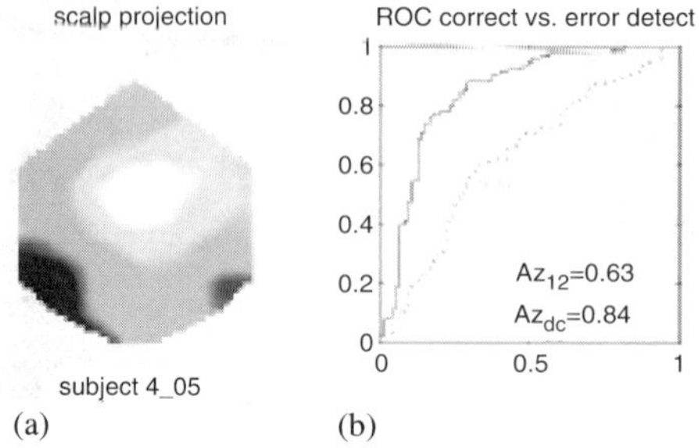

Fig. 3. Detection of decision errors with EEG. (A) Dorsal view of sensor projections. (B) ROC curve for error vs. correct trials. Solid curve corresponds to discrimination using Eq. (1) and dotted line to discrimination with center electrode (#12).

Our results demonstrate the utility of linear analysis methods for discriminating between different events in single-trial, stimulus driven experimental paradigms using EEG and MEG. A particularly important aspect of our approach is that linearity enables the computation of sensor projections for the optimally discriminating weighting. This localization can be compared to the functional neuroanatomy, serving as a validation of the data driven linear methods. In all three cases presented, we find that indeed the activity distribution correlated with the source that optimizes single-trial discrimination localizes to a region that is consistent with the functional neuranatomy. This is important, for instance in order to determine whether the discrimination model is capturing information directly related to the underlying task-dependent cortical activity, or is instead exploiting an indirect cortical response or other physiological signals correlated with the task (e.g. correlations with the stimulus, eye movements, etc.). Localization of the discriminating source activity also enables one to determine the neuroanatomical correlations between different discrimination tasks, as was demonstrated for explicit and imagined motor responses in EEG. Finally, this method is applicable to other encephalographic modalities with linear superposition of activity, such as near infrared imaging [2].

Acknowledgements

This work was supported in part by grants from the Defense Advanced Research Project Agency, the National Foundation for Functional Brain Imaging, and the National Institutes of Health (P50 MH62196 and R02 NS37528). We thank Clay Spence, Adam Gerson, and Zuohua Zhang for fruitful discussions and assistance in the data analysis, and Mimi Duvall for assistance with the figures.

References

[1] S. Baillet, J.C. Mosher, R.M. Leahy, Electromagnetic brain mapping, IEEE Signal Process. Mag. 18 (6) (2001) 14–30.

[2] D.A. Boas, et al., Imaging the body with diffuse optical tomography, IEEE Signal Process. Mag. 18 (6) (2001) 57–75.
[3] M.G.H. Coles, M.D. Rugg, Event-related brain potentials: an introduction, in: M.D. Rugg, M.G.H. Coles (Eds.), Electrophysiology of Mind, Oxford University Press, Oxford, 1995.
[4] S. Dehaene, M. Posner, D. Tucker, Localization of a neural system for error detection and compensation, Psychol. Sci. 5 (1994) 303–305.
[5] R. Duda, P. Hart, D. Stork, Pattern Classification, Wiley, New York, 2001.
[6] M. Falkenstein, J. Hoorman, S. Christ, J. Hohnsbein, ERP components on reaction errors and their functional significance: a tutorial, Biol. Psychol. 51 (2000) 87–107.
[7] S. Haykin, Adaptive Filter Theory, Prentice-Hall, Englewood Cliffs, NJ, 1996.
[8] S. Makeig, A. Bell, T. Jung, T. Sejnowski, Independent component analysis of electroencephalographic data, in: Advances in Neural Information Processing Systems, Vol. 8, MIT Press, Cambridge, MA, 1996, pp. 145–151.
[9] G. Pfurtscheller, C. Neuper, Motor imagery and direct brain-computer communication, Proc. IEEE 89 (7) (2001) 1123–1134.
[10] H. Ramoser, J. Mueller-Gerking, G. Pfurtscheller, Optimal spatial filtering of single trial EEG during imagined hand movement, IEEE Trans. Rehab. Eng. 8 (4) (2000) 441–446.
[11] G. Schalk, J. Wolpaw, D. McFarland, G. Pfurtscheller, EEG-based communication: presence of an error potential, Clin. Neurophysiol. 111 (2000) 2138–2144.
[12] J.A. Swets, ROC analysis applied to the evaluation of medical imaging techniques, Invest. Radiol. 14 (1979) 109–121.
[13] A. Tang, B. Pearlmutter, D. Phung, S. Carter, Independent components of magnetoencephalography, Part I: Localization, Neural Comput. 14 (8) (2002) 1827–1858.
[14] R. Vigario, J. Sarela, V. Jousmaki, M. Hamalainen, E. Oja, Independent component approach to the analysis of EEG and MEG recordings, IEEE Trans. Biomed. Eng. 47 (5) (2000) 589–593.

Available at
www.ComputerScienceWeb.com
POWERED BY SCIENCE DIRECT®

ELSEVIER

Neurocomputing 52–54 (2003) 185–189

NEUROCOMPUTING

www.elsevier.com/locate/neucom

Spike initiation in a hippocampal interneuron model☆

Fernanda Saraga[a,*], Frances K. Skinner[b]

[a] *Toronto Western Research Institute, University Health Network, Department of Physiology, University of Toronto, Toronto, Ont., Canada*

[b] *Department of Medicine (Neurology) and Institute of Biomaterials and Biomedical Engineering, University of Toronto, Toronto, Ont., Canada*

Abstract

Interneurons are heterogeneous in their morphologies, biophysical properties and pharmacological sensitivities. Recently, the dendrites of the oriens/alveus interneuron have been shown to contain a high density of sodium and potassium channels. These active dendrites contribute to the various signals produced by this cell. The initiation of spikes and their back-propagation into the dendritic tree has not yet been explored in interneurons. We use a computational model of an oriens/alveus interneuron with appropriate passive properties, channels, and distributions to explore where spikes initiate. We show that spike initiation depends on both the location and amount of input in our model.

Keywords: Signal propagation; Compartmental models; Ion channel kinetics; Interneurons

1. Introduction

GABAergic inhibitory interneurons consist of only ~10% of the neuronal hippocampal population but critically regulate principal cell output. Unlike the more homogeneous population of principal pyramidal neurons, interneurons are diverse in many ways including their electrophysiological responses and their ion channel content. These differences imply that interneurons should not be viewed as simple inhibitors but have functionally distinct roles depending on the interneuron type [8].

☆ This work was supported by OGSST, NSERC (F.S.) and CIHR, (F.K.S.).
* Corresponding author.
E-mail addresses: fernanda.saraga@utoronto.ca (F. Saraga), fskinner@uhnres.utoronto.ca (F.K. Skinner).

0925-2312/03/$ - see front matter © 2002 Elsevier Science B.V. All rights reserved.
doi:10.1016/S0925-2312(02)00835-4

Given this diversity in interneurons we focus on the oriens/alveus interneuron for which there is experimental data describing the channel kinetics and distributions particular to this cell [7,6]. By looking at a specific interneuron type we can begin to understand its distinctive role in the network situation.

We specifically explore the site of action potential initiation and its dependence on the strength and location of tonic input. We find that a larger input into the soma can result in a shift of the initiation site from the soma to the dendrite.

2. Model

We use the multi compartment model developed in [10,11]. Briefly, the 84-compartment neuron model is based on the morphology and characteristics of an oriens/alveus hippocampal interneuron (Fig. 1A). This model was created using the software package NEURON [4] and has a surface area of 9.884×10^{-5} cm^2. Passive properties were chosen so as to match experimental values of input resistance and membrane time constant for this interneuron type: membrane capacitance (1.3 μF/cm^2), axial resistivity (150 Ωcm), leak conductance (50 μS/cm^2), input resistance (215 MΩ), and membrane time constant (23.3 ms). Passive properties were distributed uniformly to all compartments of the model interneuron.

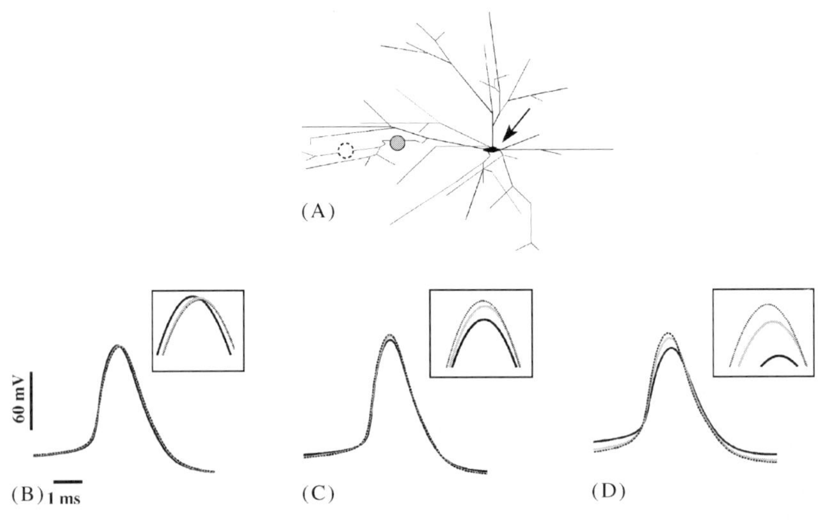

Fig. 1. Location of spike initiation depends on amount of tonic current injected into the soma: (A) Circles on compartmental model indicate the location from which voltage recordings were taken. Dotted circle corresponds to dotted line traces, grey circle to grey lines, and soma to black lines in (B)–(D). Arrow indicates location of tonic current injection; (B) $I_{ext} = 0.01$ nA, soma (black trace) spikes first; (C) $I_{ext} = 0.1$ nA, soma and dendrites spike simultaneously; (D) $I_{ext} = 0.3$ nA, distal dendrite (dotted trace) spikes first. Insets show order of spikes.

The ion channels included in the model are the Hodgkin–Huxley (HH) sodium, I_{Na}, and potassium, I_{K}, current, a transient potassium channel, I_{A}, and a hyperpolarization-activated channel, I_{h}. The equations governing these currents are described in [10]. All four channel types were included in the soma. The dendrites contained both the HH channels but with modified kinetics and conductances as in [7] ($g_{\mathrm{Na,soma/dendrites}}$= 107/117 pS/μm^2, $g_{\mathrm{K,soma/dendrites}}$=319/230 pS/μm^2, g_{A}=165 pS/μm^2, g_{h}=13.85 pS/μm^2). The density of the dendritic channels was kept constant along the dendritic tree.

3. Result

The location of spike initiation depends on both the amount and the location of the injected tonic current. Fig. 1A shows the input site as the soma (arrow), with voltage recordings taken from the soma (black trace), and two different sites on a dendrite (dotted and grey traces) in Figs. 1B–D. A small amount of tonic current (Fig. 1B; 0.01 nA) results in the soma spiking ahead of the dendrites. Increasing the amount of current injected into the soma (Fig. 1D; 0.3 nA) allows the distal dendrite to spike first. The signal then propagates towards the soma. At a current injection value of 0.1 nA (Fig. 1C), all three voltage recording locations are seen to spike at the same time. This ordering of the spikes can be explained by considering the surface areas of the compartments and the relative current that each compartment experiences. The distal dendrite has a smaller diameter and therefore a smaller surface area than the soma, or even the more proximal dendritic segment. The amount of absolute or total current needed to allow the dendritic membrane to spike is much less than that needed for the larger somatic surface area. When a large amount of current is injected into the soma (>0.1 nA), the depolarization felt by the dendrite allows it to spike before the soma. A smaller amount of current (<0.1 nA) does not have the strength to excite the dendrite and therefore the soma spikes first.

When current is injected into a distal dendrite (shown by arrow in Fig. 2A), the initiation of the spike is *always* in the soma. The signal then back-propagates into the dendrites (Figs. 2B–D). Although the absolute amount of current injected into the dendrite is the same as that in Fig. 1, the relative amount of current felt by the small dendrite is much larger than that felt by the soma previously. The dendrite therefore saturates to a depolarized value and this depolarization is propagated along the dendrite towards the soma. The spike is initiated in the soma and then traverses back along the dendrite but cannot produce the same amplitude of spike in the smaller compartments due to their persistent level of depolarization. This is most clearly seen in Fig. 2D where the distal dendritic recording (dotted trace) is depolarized relative to the somatic recording (black trace).

The time delay between somatic and dendritic spikes changes with the amount of injected current. Injected currents into the dendrite of 0.01, 0.1, and 0.3 nA produce time delays of 0.23, 0.63, and 0.53 ms, respectively in the distal dendrite (dotted traces in Figs. 2B–D). This model neuron can spontaneously fire without any current injection [10], therefore small current injections into the dendrite will not have a large effect on this firing. The entire membrane is oscillating almost at the same time without a

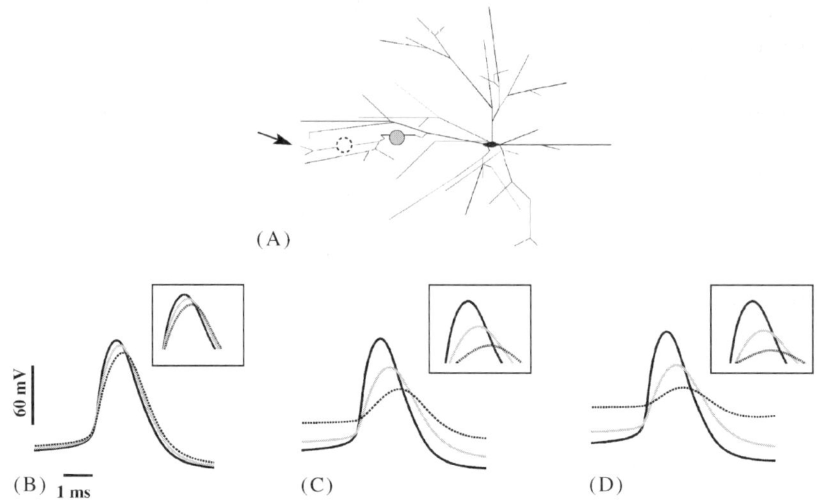

Fig. 2. Some spike precedes dendritic spike when injection site is placed at a distal dendrite. Time delay between somatic and dendritic spikes increases and then decreases with increasing current injection value. (A) Circles on compartmental model indicate the location from which voltage recordings were taken. Dotted circle corresponds to dotted line traces, grey circle to grey lines, and soma to black lines in (B)–(D). Arrow indicates location of tonic current injection; (B) $I_{ext} = 0.01$ nA, time delay between somatic spike (black trace) and dendritic spike (dotted trace) is 23 ms; (C) $I_{ext} = 0.1$ nA, time delay is 63 ms; (D) $I_{ext} = 0.3$ nA, time delay is 0.53 ms. Insets show order of spikes.

significant delay between the somatic and dendritic spikes (Fig. 2B). As the level of current injected into the dendrite is increased, the depolarization felt by the soma is passed on to the soma where a spike is initiated and then actively back-propagates into the dendrites. This is seen in the increased time delay between the somatic and dendritic spikes (Fig. 2C). With larger current injection, this time delay shortens (Fig. 2D).

4. Discussion

Recent experiments have shown that dendrites are rich in voltage-gated ion channels. The types and distributions of ion channels in hippocampal pyramidal cells has been extensively examined. The dendrites of interneurons are just beginning to be examined, but already experiments [7] have found that the density of sodium and potassium channels in the dendrites of oriens/alveus interneurons are comparable to those found in the soma with some slight differences in kinetics. This density is much larger than that found in the dendrites of cortical principal neurons implying a possible difference in action potential initiation and propagation in interneurons. Martina et al. [7] showed that the action potential initiation site could shift along the cell body depending on the injected current properties. Our model shows similar findings (Fig. 1).

The role of back-propagating signals in neurons has been examined both experimentally and in simulations [5,9,2]. These signals are highly dependent on dendritic voltage-gated ion channels. Recently, a dichotomy of the amplitude of back-propagated action potentials was seen in CA1 pyramidal neurons [2]. This dichotomy could be explained by exploring small changes in the distribution of dendritic ion channels in compartmental neuronal models. Back-propagating signals could provide a link between neuronal output and the synapse by acting as a retrograde signal to the synapse that the soma has fired. This could provide a means for the synapse to 'self-regulate' its strength in response to the back-propagated signal [3].

The ability of O/A interneurons to initiate dendritic spikes and to back-propagate somatic spikes allows them a more diverse control of the CA1 pyramidal neurons that they synapse onto. This control could result in changes in the efficacy and plasticity of entorhinal inputs onto the pyramidal cells [1] thereby having a large effect on hippocampal output.

References

[1] J.M. Blasco-Ibáñez, T.F. Freund, Eur. J. Neurosci. 7 (1995) 2170–2180.
[2] N.L. Golding, et al., J. Neurophysiol. 86 (2001) 2998–3010.
[3] M. Hausser, Current Biol. 11 (2001) R11–R12.
[4] M. Hines, N.T. Carnevale, Neural Comput. 9 (1997) 1179–1209.
[5] N. Lemon, R.W. Turner, J. Neurophysiol. 84 (2000) 1519–1530.
[6] G. Maccaferri, C.J. McBain, J. Physiol. 497 (1996) 119–130.
[7] M. Martina, et al., Science 287 (2000) 295–300.
[8] C.J. McBain, A. Fisahn, Nature Rev. Neurosci. 2 (2001) 11–23.
[9] M. Rapp, et al., Proc. Nat. Acad. Sci. 93 (1996) 11985–11990.
[10] F. Saraga, Master's Thesis, University of Toronto, 2001.
[11] F. Saraga, J. Neurosci. 2003, submitted for publication.

Fernanda Saraga is currently working on her Ph.D. at the University of Toronto. She graduated from the University of Toronto (B.Sc. (Physics) and M.Sc. (Physiology)). She is interested in creating computational models of inhibitory interneurons to understand their complex role in signal generation.

Dr. Frances Skinner is a Scientist in the Toronto Western Research Institute, University Health Network, with appointments in Medicine (Neurology), Physiology and Biomedical Engineering at the University of Toronto. She graduated from the University of Waterloo (B.Math.) and Toronto (M.A.Sc., Ph.D.) and did 4 years of postdoctoral work in Boston and California. In general, she enjoys collaborative work and is interested in cellular-based mechanisms underlying neuronal network dynamics.

Available at
www.ComputerScienceWeb.com
POWERED BY SCIENCE @ DIRECT®

ELSEVIER

Neurocomputing 52–54 (2003) 191–197

NEUROCOMPUTING

www.elsevier.com/locate/neucom

Neuron PRM: a framework for constructing cortical networks☆

Jyh-Ming Lien, Marco Morales, Nancy M. Amato*

PARASOL Lab, Department of Computer Science, Texas A&M University, College Station 77843-3112, USA

Abstract

The brain's extraordinary computational power to represent and interpret complex natural environments is essentially determined by the topology and geometry of the brain's architectures. We present a framework to construct cortical networks which borrows from probabilistic roadmap methods developed for robotic motion planning. We abstract the network as a large-scale directed graph, and use L-systems and statistical data to 'grow' neurons that are morphologically indistinguishable from real neurons. We detect connections (synapses) between neurons using geometric proximity tests.

Keywords: Cortical networks; PRM; BTS; L-system; Rectangle tree

1. Introduction

The brain has extraordinary computational power which is determined in large part by the topology and geometry of its structures. A unique instrument developed at Texas A&M University, the brain tissue scanner (BTS) [5], will enable an entire mouse brain to be imaged and reconstructed at the neuronal level of detail. Moreover, by enabling studies of the topology of cortical networks, it could provide insight into one of the

☆ This research supported in part by NSF Grants ACI-9872126, EIA-9975018, EIA-0103742, EIA-9805823, ACR-0081510, ACR-0113971, CCR-0113974, EIA-9810937, and EIA-0079874. and by the Texas Higher Education Coordinating Board grant ATP-000512-0261-2001. Morales is supported in part by a Fulbright/Garcia Robles (CONACYT) fellowship.

* Corresponding author.

E-mail addresses: neilien@cs.tamu.edu (Jyh-Ming Lien), marcom@cs.tamu.edu (M. Morales), amato@cs.tamu.edu (N.M. Amato).

doi:10.1016/S0925-2312(02)00728-2

least understood biological processes—neural computation. Using destructive sectioning and cross-sectional imaging, the BTS can scan an entire transgenic GFP/XFP-stained mouse brain in approximately 1 month. The data produced will be used to reconstruct the three-dimensional neuronal structure of the scanned tissue. While scientists have mapped small areas of cortex, we are not aware of any work mapping the entire cortical network of any mammalian species at the neuronal level of detail. Since only a small percentage (less than 10%) of neurons will be stained, the neurons reconstructed from BTS data will be augmented with synthetic neurons that are grown based on the measured biological neurons. Next, their interconnections (synapses) will be generated. The geometry and connectivity of the resulting cortical network will be studied theoretically and, in collaboration with the IBM Blue Gene project, through simulation on massively parallel machines.

In this paper, we start with geometric models of reconstructed neurons, and concentrate on the generation and connection of the anatomically realistic synthetic neurons needed to complete the cortical network. Our general strategy borrows from the *probabilistic roadmap methods* (PRMS) [4] first developed for robotic motion planning, but which have been applied to applications ranging from robotics to computational biology [7]. PRMS construct a network (roadmap) in which nodes represent feasible configurations (neurons) and edges represent valid connections (synapses). In this paper, we use L-systems [6,1] to grow synthetic neurons and describe techniques for identifying geometrically and anatomically realistic synapses. A simple prototype system, called NEURON PRM (N-PRM), for growing a synthetic cortical network has been implemented.

2. Neuron PRM framework

Our ultimate goal is to map and understand the connectivity and geometry of the cortical network. The construction process borrows from the PRM framework by using simple, local techniques to construct a representative model of a much larger unknown space. We begin our modeling by partitioning the cerebral cortex into a set of finite elements (FEs). The general system architecture is described in Fig. 1(b). During *neuron generation*, neurons are "grown" in the *cortex* using information describing the spatial distribution of neurons and their geometry contained in the cortex FE model. The neurons generated are stored in the FE's neuron database. During *synapse generation*, neurons are connected using the geometric models constructed during the *neuron generation*, and their known affinities for synapse creation (see Fig. 1(a)). Finally, a roadmap representing cortical connectivity is ready for analysis and simulation.

We now briefly describe the basic building blocks of our method: neurons, FE model, statistical databases, and the roadmap graph. A *neuron* is represented as a tree of neural components. Since neurons may be composed of tens of thousands of segments and may have thousands of synapses, the storage required to record the pre- and post-synaptic endings in the cortical network can be massive. The *finite element model of the cortex* partitions the cerebral cortex into FEs which contain local information about types, population distribution, and spatial distributions of neurons in that FE (upper right corner in Fig. 1(b).) FE models also help in visualization and speed up the

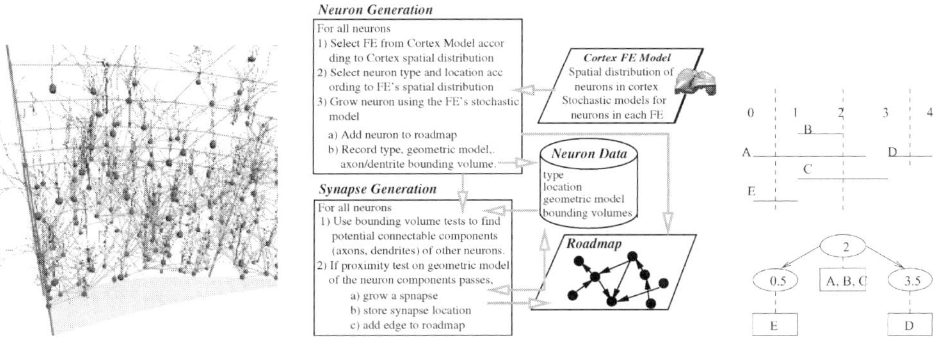

(a) (b) (c)

Fig. 1. (a) Synthetic neurons generated and connected in the cortex, (b) neuron PRM framework: finite elements, neuron generation/connection, roadmap graph, and database and (c) a 1-fold rectangle tree.

process of finding connections between neurons. Several *databases* contain information for growing and distributing neurons (see center part of Fig. 1(b)). There are three types of databases: (i) types of neurons and probability for each type, (ii) spatial distribution of each type of neuron, and (iii) statistical data for the neuron model. The information in the first two databases is unique to each FE, and so is stored with the FEs. The third database is associated with each neuron type. A *roadmap graph* (lower-right in Fig. 1(b)) is a directed graph representing an abstract cortical network. Each vertex in the graph corresponds to a neuron and each directed edge represents one or more connections (or synapses) between two neurons. Thus, the roadmap provides a hierarchical representation facilitating analysis, simulation and visualization of the cortical network.

2.1. Generation of neurons

Many different models of neurons have been proposed. Here we are interested in the morphological features of neurons and in their connectivity which will be needed to simulate their function. In [6] the use of L-systems which had been used to model plants was proposed to represent neuron morphology. L-Neuron [1], developed by Ascoli and Krichmar, uses an L-system to generate neurons. It stochastically samples parameters from experimental distributions stored in a neuroanatomical database to be used in the generation of a virtual neuron.

Neuron PRM uses a similar approach to grow neurons. An abstract neuron contains information regarding position, type, associated FE, and a unique ID. Position and type are generated randomly based on some given statistical distribution stored in the FEs database. Next, an L-system is used to create a morphologically correct synthetic neuron based on the information contained in the abstract neuron and on statistics provided by the database in the host FE for that neuron type. Currently, the statistics are calculated from a sample of SWC neurons publically available [2]. In the future, we will use reconstructed neurons provided by the BTS to extract statistics. An

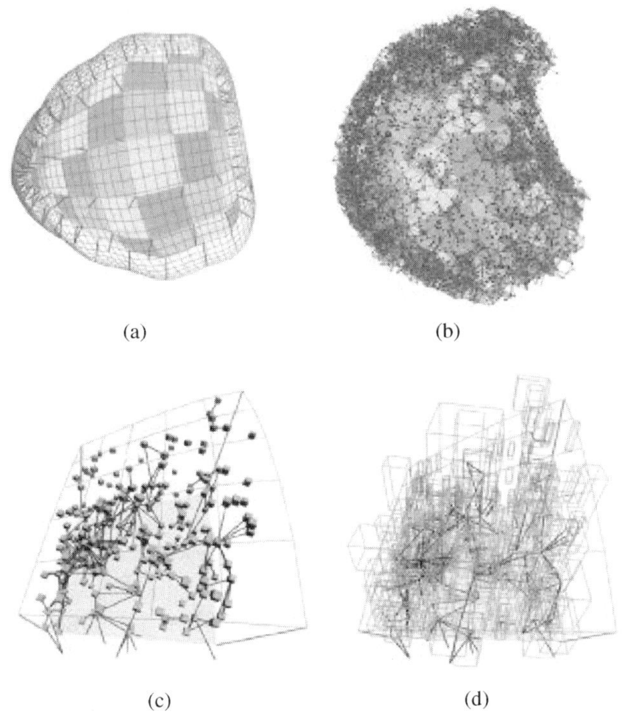

Fig. 2. (a) Empty cortical model of mouse, (b) cortical model filled with abstract neuron nodes and connections, (c) a closer view of cortical network and (d) neurons in one FE surrounded by bounding volumes.

abstract neuron and synthetic neuron pair are stored as a vertex of the cortical network (see Fig. 2(a–c)).

2.2. Generation of synapses

After the FE model has been populated with neurons as described above, the cortical network is assembled by identifying potential synapses and connecting the neurons. Statistically each neuron has thousands of synapses and they are in the order of 16 million neurons in the cortex of the mouse. Thus, a brute force strategy which tests all pairs of neurons is not feasible. To address this issue, we define simple metrics to reject neuron pairs and identify potential synapses very quickly. Most neuron pairs are quickly rejected by a filtering test which checks for intersection of their bounding volumes. Thus, detailed distance computations between the neuronal segments will only be performed for those neuron pairs that pass the bounding volume test. When the intersection of a pair of neurons is discovered, we connect their associated vertices in the roadmap (Fig. 2(d)). Three different bounding volumes were tested: spheres, boxes, and convex hulls. The bounding sphere and bounding box provide fast construction and intersection tests, but convex hull is more accurate in terms of approximating the shape of the target neuron. The strategy sketched above performs $O(n^2)$ bounding volume intersection tests, where n is the number of neurons. For the bounding box representation, we reduced this to $O(n \log n + k)$, where k is the actual number of intersections, using a method based on Edelsbrunner's *rectangle tree* [3] (see Fig. 1(c)). We implemented two versions of the rectangle tree query, one described by Edelsbrunner [3] that initiates each query from the root of the tree, and a variation we designed to solve the *all-intersection query* in a more efficient way.

After a neuron pair's bounding volumes are found to intersect, a more detailed computation will more precisely compute synapse locations. We implemented a simple algorithm based on tendencies known for (real) synapse generation. In particular, we use the fact that synapses are usually found between segments and spines. That is, for each pair of connected neurons, we compute distances from spines of one neuron to the segments of the other neuron. If the distance is less than a user specified value for synapse gap, this spine, segment pair is marked as synaptic site.

3. Experiments

To test our framework we generated a model of the mouse cortex (Fig. 2). A summary of the results is shown in Table 1. The running time for building these volumes, as expected, was much faster for the bounding sphere and the bounding box than for the convex hull. Since the bounding sphere has larger volume than the bounding box and the convex hull, there were many more potential connections identified using the spheres than were found using the other two methods. Also, due to the overly conservative bounding sphere, almost all the neurons were contained in one connected component while almost a third of the neurons are isolated using the convex hull method. For the bounding box tests, we found that the versions using the rectangle tree take 16–25% of the time required for the brute force method, depending on whether the standard query or the optimized query for computing all intersections was used.

For the connection phase, we tested 10,000 neurons with the bounding sphere, bounding box, and the convex hull. The bounding box was tested using the brute force

Table 1
Bounding volume tests for identifying (potential) connections for 10,000 neurons

Method		Bounding box			
	Bounding sphere	Brute force	RT naive	RT improved	Convex hull
Bounding volume tests (node connection)					
Time for creating bounding volume (s)	47.37	47.53	47.88	47.93	56.89
Time for connection (s)	344.46	360.1	93.17	64.1	604.23
Number of edges found	59,158	19,994	19,994	19,994	6094
Average node degree	11.8	3.9	3.9	3.9	1.2
Connected components (CC)	16	709	709	709	5089
Isolated neurons	14	540	540	540	3440

Table 2
Synapse discovery for 40,000 neurons

Input roadmap from	Bounding sphere	Bounding box	Convex hull
Cortical network refinement (synapse discovery)			
Time for finding synapse (s)	36,896.6	9,838.6	3,481.2
Number of synapse found	417,277	416,895	400,065
Avg # of synapses between two neurons	0.3	5.5	16.0
Error rate (%)	97.4%	88.0%	61.5%

approach, the original version of the rectangle tree, and our optimization to compute all intersections. To identify real synapses, we used 0.015 as the synaptic gap size (neuron height is between 0.3 and 0.9). Some results for 40,000 neurons are shown in Table 2. The roadmaps from the bounding sphere, the bounding box, and the convex hull were tested separately. The running time for discovery is linear in the number of edges of the input roadmap, so the convex hull is the fastest. Thus, since synapse identification requires both abstract connection and synapse discovery, the convex hull method is the most efficient—the first phase is more expensive, but it pays off by filtering out more neuron pairs for the second phase, in which each test is significantly more expensive. The error rate in Table 2 is defined as: $N_{\text{empty-abs-syn}}/N_{\text{total-abs-syn}}$, where $N_{\text{empty-abs-syn}}$ is the number of abstract synapses which did not contain any real synapse, and $N_{\text{total-abs-syn}}$ is total number of abstract synapses in the given roadmap. This measure gives us an indication of the accuracy of the bounding volume methods. Although all methods have error rates higher than 50%, we believe this will decrease when we are able to more densely pack the cortex with neurons. Here, we generated only 40,000 and in reality there are some 16,000,000 neurons in the mouse cortex.

4. Conclusion

We have developed a prototype system that will eventually be used to (re)construct an entire mouse cortical network containing over 15 million neurons. Our PRM-based framework, N-PRM, for constructing a hierarchical model of the cortical network uses L-System neuron generators. In this paper, we study various bounding volumes for reducing the cost of synapse identification. We found that data structures, such as the rectangle tree, which use spatial information to restrict the number of intersection tests, can lead to significant improvements in execution time. In future work, we will further investigate our proposed optimization of the rectangle tree, including its applicability to additional bounding volumes.

References

[1] G. Ascoli, J.L. Krichmar, L-Neuron: a modeling tool for the efficient generation and parsimonious description of dendritic morphology, Neurocomputing 32–33 (1–4) (2000) 1003–1011.

[2] Duke/Southampton, Archive of neuronal morphology, http://www.cns.soton.ac.uk/~jchad/cellArchive/cellArchive.html.
[3] H. Edelsbrunner, A new approach to rectangle intersections, Int. J. Comput. Math. 13 (1983) 209–229.
[4] L. Kavraki, P. Svestka, J.C. Latombe, M. Overmars, Probabilistic roadmaps for path planning in high-dimensional configuration spaces, IEEE Trans. Robot. Automat. 12 (4) (1996) 566–580.
[5] B.H. McCormick, Brain tissue scanner enables brain microstructure surveys, Neurocomputing, this issue.
[6] B.H. McCormick, K. Mulchandani, L-system modeling of neurons, in: Proceedings of the Visualization in Biomedical Computing, Rochester, MN, SPIE, Vol. 2359, 1994, pp. 693–705.
[7] G. Song, N.M. Amato, A motion planning approach to folding: from paper craft to protein folding, in: Proceedings of the IEEE International Conference on Robotics Automation (ICRA), Seoul, Korea, 2001, pp. 948–953.

Jyh-Ming Lien is a Ph.D. student working with Dr. Nancy Amato in the PARASOL Lab in the department of Computer Science at Texas A&M University. He received his B.S. in Computer Science from National ChengChi University in 1999.

Marco Morales is a Ph.D. student working with Dr. Nancy Amato in the PARASOL Lab in the Department of Computer Science at Texas A&M University. He received his B.E. in Computer Engineering and his M.E. in Electrical Engineering from the Universidad Nacional Autónoma de México in 1996 and 1998, respectively. His main interests include motion planning, robotics, computational neuroscience, and computational geometry.

Nancy M. Amato is an Associate Professor of Computer Science at Texas A&M. She received her M.S. and Ph.D. in Computer Science from UC Berkeley and the University of Illinois at Urbana-Champaign, respectively. She is an Associate Editor of the IEEE Transactions on Robotics and Automation and of the IEEE Transactions on Parallel and Distributed Computing. Her research interests include motion planning, parallel computing, and computational biology, geometry, neuroscience, and physics.

Available at
www.ComputerScienceWeb.com
POWERED BY SCIENCE @ DIRECT®

ELSEVIER

Neurocomputing 52–54 (2003) 199–207

NEUROCOMPUTING

www.elsevier.com/locate/neucom

Is there a support vector machine hiding in the dentate gyrus?

John L. Baker

School of Computational Sciences, George Mason University, 4400 University Drive, Fairfax, VA 22030, USA

Abstract

The dentate gyrus has physiological and related behavioral properties suggesting that it implements functions within the hippocampus partially associated with sensory pattern recognition. A top-down dentate gyrus model is defined in terms of an idealized support vector machine pattern recognizer constructed from spiking neurons. The resulting construction offers parallels with dentate gyrus morphology and offers explanation of some of its unique properties, in particular, the mossy fiber pathway and its connection with CA3 pyramidal cells. Derived learning rules suggest properties of the mossy fibers that might be tested experimentally.

Keywords: Hippocampus; Dentate gyrus; Mossy fiber; Support vector machine; Pattern recognition

1. Introduction

Theories of the hippocampus have assigned a variety of roles to the dentate gyrus. Associative memory models of the hippocampus typically assign the dentate gyrus a role in creating orthogonal input patterns for CA3 so that associative memory saturation is avoided [6,10]. Lesion studies suggest that the dentate gyrus has a role in pattern recognition for spatial learning [4].

Explanations as to how the dentate gyrus might create orthogonal patterns for input to CA3 vary, but a common observation is that the number of granule cells in the dentate gyrus is much larger than the comparable layer of entorhinal cortex cells, allowing the dentate gyrus to generate a sparse representation of its input. Put in other terms, the output of the dentate gyrus has higher dimensionality than its input. This increase in

E-mail address: jbaker@scs.gmu.edu (J.L. Baker).

0925-2312/03/$ - see front matter
doi:10.1016/S0925-2312(02)00733-6

dimensionality is a clue that the formalism of support vector machines (SVMs) may offer some explanation as to the underlying transformation performed in the dentate gyrus. The relative sparseness of connectivity between the dentate gyrus and CA3 is an additional clue suggesting applicability of a SVM model.

SVMs have been more completely described elsewhere [7] but are summarized here for completeness. Let $\langle w,x \rangle$ be the normal vector dot product of vectors w and x in Euclidean space. Consider the following class of linear decision functions:

$$f(x) = \mathrm{sgn}(\langle w,x \rangle + b) \in \{-1,1\}, \quad w \in \mathfrak{R}^N, \; b \in \mathfrak{R}. \tag{1}$$

Let the training vectors $x_1, x_2, \ldots,$ be vector-valued inputs and let y_i be the corresponding desired decision function value for the x_i. An optimal decision function is found by solving the optimization problem

$$\text{maximize } W(\alpha) = \sum_i \alpha_i - \frac{1}{2} \sum_{i,j} y_i \alpha_i y_j \alpha_j \langle x_i, x_j \rangle,$$

$$\text{subject to } \sum_i y_i \alpha_i = 0, \; 0 \leqslant \alpha_i \leqslant C \text{ for some constant } C. \tag{2}$$

The value C allows tolerance for when the patterns are not linearly separable. When $\alpha_i = 0$, the associated x_i does not enter into the decision rule. Those x_i for which $\alpha_i > 0$ are termed the support vectors.

SVMs generalize to non-linear decision functions by embedding the input data in a higher-dimensional space termed "feature space". Let F represent feature space and let $\Phi : \mathfrak{R}^N \to F$ be the map of the input data into feature space. Mercer's theorem states that if a function k is a positive definite kernel, there is an embedding of F in a higher-dimensional space with dot products defined as

$$\langle \Phi(x_i), \Phi(x_j) \rangle = k(x_i, x_j). \tag{3}$$

The function k is termed a Mercer kernel. Consider the case where k is a function of the dot product $\langle x_1, x_2 \rangle$, which with some abuse of notation we write as $k(x_1, x_2) = k(\langle x_1, x_2 \rangle)$. In this case, a sufficient condition for k to form a Mercer kernel is that the Taylor series expansion of k as a function of $\langle x_1, x_2 \rangle$ should have no negative coefficients [8]. This condition is also a necessary condition if the dot product is defined in an infinite-dimensional space.

An important consideration for SVMs is that most of the α_i become identically zero and thus do not need to be included in the final decision rule. Support vectors lie on the boundary of decision surface and are sparse with respect to the set of training vectors.

2. Constructing the SVM

2.1. *Scope of the model*

In describing an existence proof construction of an SVM from spiking neurons, it is convenient to use terms such as "perforant path", "granule cell" and "CA3 pyramidal" metaphorically. The process is to first derive a model from mathematical constraints and then compare the result with known physiological properties of the hippocampus.

2.2. Coding of the perforant path input

Perforant path encoding is assumed to be in the form of a population burst. Let $x_1, x_2, \ldots,$ be the various perforant path input spiking patterns as vectors where $x_i \in \{0,1\}^N$. For our purposes, each pattern is unique, that is, $i \neq j \Rightarrow x_i \neq x_j$. The index i in this case does not imply temporal sequence.

2.3. Granule cell responses

To provide a response that can be solved analytically, the granule cell is assumed to have a linear I–V response [2]. The synaptic response of cell j to input pattern x_i is determined by the dot product $\langle w_j, x_i \rangle$ where w_j is the vector of synaptic weights. To meet the requirements of the SVM formulation, for a given granule cell, w_j is fixed equal to a unique input pattern selected from among the training vectors.

Cell membrane potential follows the dynamic equation

$$C_m \frac{dV}{dt} = g_{GC}(V_R - V) + Xh_{GC}(t), \tag{4}$$

where V is the membrane potential, C_m is the membrane capacitance, V_R is the channel reversal potential, g_{GC} is the effective conductance of channels in the granule cell, $X = \langle w_j, x_i \rangle$ is the effective stimulation of the granule cell, and $h_{GC}(t)$ is the time course of the current from a single synapse.

Let $\lambda = g_{GC}/C_m$ and let V_0 be the initial membrane voltage. We assume without loss of generality that the inputs are presented at time $t = 0$. Eq. (4) can then be solved explicitly:

$$V(t) = V_R - e^{-\lambda t}(V_R - V_0) + \frac{e^{-\lambda t}}{C_m} X \int_0^t e^{\lambda s} h_{GC}(s)\, ds. \tag{5}$$

Let t_f be the time to firing as determined by the time at which $V(t_f) = V_{th}$ where V_{th} is the firing threshold. Because $h(t)$ is assumed to be short, we further assume that $h_{GC}(t)$ is negligible for all t greater than some value less than t_f, or more formally, that there is a $T_H < t_f$ such that $h_{GC}(t) = 0$ for all $t > T_H$.

Under this assumption, we can rewrite Eq. (5) as

$$V(t) = e^{-\lambda t}(X\beta + V_0 - V_R) + V_R, \quad \beta = \frac{1}{C_m} \int_0^\infty e^{\lambda s} h_{GC}(s)\, ds, \quad t > T_H. \tag{6}$$

From this we then solve for the time of firing as a change in firing time relative to some fixed time T_F where $t_f = T_F$ when $X = 0$.

$$\delta(X) = T_F - t_f = -\frac{1}{\lambda} \ln\left(1 - \frac{X\beta}{V_R - V_0}\right), \tag{7}$$

where $\delta(X)$ is the change in firing time and under the condition that $V_R > V_{th} > V_0$.

To check for the dot-product Mercer kernel property, we examine the Taylor series expansion of $\delta(X)$ noting that

$$-\ln(1-x) = x + \frac{x^2}{2} + \frac{x^3}{3} + \frac{x^4}{4} + \cdots, \qquad -1 < x < 1. \tag{8}$$

The Taylor series of $\delta(X)$ has all positive coefficients and $\delta(\langle w, x\rangle)$ is a Mercer kernel of the input data vector x and the synaptic weight vector w.

2.4. CA3 pyramidal cell responses

For analytical purposes, a simple model of pyramidal cell response is also used. As before, we assume the existence of one combined ion channel conductance and reversal potential. The membrane potential is modeled by the following dynamic behavior:

$$C_{\mathrm{m}} \frac{\mathrm{d}V}{\mathrm{d}t} = g_{\mathrm{PC}}(V_{\mathrm{R}} - V) + \sum_i w_i h_{\mathrm{MF}}(t - t_i), \tag{11}$$

where V is the pyramidal soma membrane potential, C_{m} is the membrane capacitance, g_{PC} is the net conductance of channels in the pyramidal cell, V_{R} is the reversal potential, w_1, w_2, w_3, etc. are synaptic weights for the different mossy fiber connections, h_{MF} is the time response of a mossy fiber connection in terms of the current delivered to the soma, and t_1, t_2, t_3, etc. are the times at which action potentials arrive at the various mossy fiber terminals.

We model the mossy fiber terminal as an exponential decay after the initial rise interval

$$h_{\mathrm{MF}}(t) = h_{\mathrm{N}} \exp(-\gamma(t - T_{\mathrm{N}})), \qquad t \geqslant T_{\mathrm{N}}, \tag{12}$$

where T_{N} is a time after which the decay is exponential, h_{N} is a constant chosen to scale the current appropriately, and γ the decay rate. We also have $h_{\mathrm{MF}}(t) = 0$ for $t < 0$.

Unlike the granule cell, we cannot solve explicitly for the firing time. However, we can assess whether the pyramidal cell would fire within a time interval by examining the voltage at a fixed time sufficiently long after all mossy fiber inputs would have arrived and compare that with a fixed threshold. Let T_{B} be the time at which we make our measurement, that is, we compare $V(T_{\mathrm{B}})$ with the threshold voltage V_{th}, and let

$$J_i = \frac{\exp(-\rho T_{\mathrm{B}})}{C_{\mathrm{m}}} \int_0^{T_{\mathrm{B}}} \exp(\rho s) h_{\mathrm{MF}}(s - t_i)\,\mathrm{d}s,$$

$$\delta_i(x) = T_{\mathrm{F}} - t_i, \qquad T_{\mathrm{F}} < T_{\mathrm{B}} - T_{\mathrm{N}}, \qquad \rho = g_{\mathrm{PC}}/C_{\mathrm{m}}, \tag{13}$$

where T_{F} is the nominal time at which granule cells fire. An explicit solution for $V(t)$ can be found giving a decision function of the form

$$f(x) = \mathrm{sgn}\left(\sum_i w_i J_i - b\right), \qquad b = V_{\mathrm{th}} - \exp(-\rho T_{\mathrm{B}})(V_0 - V_{\mathrm{R}}) - V_{\mathrm{R}}. \tag{14}$$

If the function J_i is a Mercer kernel with respect to inputs afferent to the associated granule cell, then Eq. (14) is in the form of a decision function of a SVM with $w_i = y_i\alpha_i$.

Using the assumed form of h_{MF} and the granule cell response model, we have

$$J_i = \frac{1}{\rho - \gamma}(P\exp(-\gamma\delta_i) - Q\exp(-\rho\delta_i)) + R\exp(-\rho\delta_i), \tag{15}$$

where P, Q, and R are constants:

$$P = C_{\mathrm{M}}^{-1}h_{\mathrm{N}}\exp(-\gamma T_{\mathrm{B}} + \gamma T_{\mathrm{F}} + \gamma T_{\mathrm{N}}),$$

$$Q = C_{\mathrm{M}}^{-1}h_{\mathrm{N}}\exp(-\rho T_{\mathrm{B}} + \rho T_{\mathrm{F}} + \rho T_{\mathrm{N}}),$$

$$R = C_{\mathrm{M}}^{-1}\exp(-\rho T_{\mathrm{B}} + \rho T_{\mathrm{F}})\int_0^{T_{\mathrm{N}}}\exp(\rho s)h(s)\,\mathrm{d}s. \tag{16}$$

We can now substitute the granule cell-derived values from Eq. (7). To determine under what conditions the result is a Mercer kernel, we examine the following individual exponential forms:

$$\exp(-\rho\delta_i) = (1 - aX_i)^{\rho/\lambda},$$

$$\exp(-\gamma\delta_i) = (1 - aX_i)^{\gamma/\lambda}, \tag{17}$$

where $a = \beta/(V_{\mathrm{R}} - V_0)$ and V_{R} and V_0 are taken from the granule cell model in Eq. (7).

Verification that $1-(1-aX)^r$ forms a dot-product Mercer kernel in X for $0 < r < 1$ and $a > 0$ follows from expansion of the Taylor series. Note that because of our assumptions regarding β, we must have $aX < 1$. A dot-product Mercer kernel does not result when $a > 0$ and $r > 1$.

There are various conditions under which the J_i could be Mercer kernels of the corresponding X_i:

(a) If $\lambda > \gamma > \rho$ and ρ is sufficiently small that the series associated with γ dominates in each Taylor series coefficient. An obvious way to achieve this is to set $\rho = 0$ suppressing leak currents. More realistically, if $\rho > 0$, possible values of T_{N} are bounded and close to zero.
(b) If $\lambda > \rho > \gamma$ and the series associated with ρ dominates with non-negative coefficients.
(c) If $\gamma = \rho$ and ρ is small, then J_i is approximately a linear function of δ_i.

Of these possibilities, $\gamma > \rho$ appears most consistent with pyramidal cell physiology and mossy fiber responses. Approximate values appear to be $\gamma = 0.1$/ms [3] and $\rho = 0.03$/ms [9]. Note that the value of λ is increased by synaptic conductances, including inhibition, and is thus may be difficult to estimate from existing electrophysiological data. If $\gamma/\lambda = 1/2$ and the granule cells have normalized synaptic weights, the resulting kernel is equivalent to a Euclidean distance metric.

The signs of mossy fiber synaptic weights have not been considered thus far. While the Lagrangian multipliers for the optimization problem posed in Eq. (2) cannot be negative, the quantity $y_i\alpha_i$ must necessarily be negative for some i. Synaptic weights for support vectors can be made positive, that is, excitatory, through the following algebraic transformation of the decision function.

$$w_i = y_i\alpha_i + C \geqslant 0, \quad k_I(x) = C\sum_i k(x,x_i),$$

$$f(x) = \operatorname{sgn}\left(\sum_i w_i k(x,x_i) - k_I(x) - b\right). \tag{18}$$

A physiological correlate to Eq. (18) would be the requirement for feedforward inhibition of the CA3 pyramidal cell stimulated by granule cell firing. In addition, the time course of the inhibition should approximate that of h_{MF}.

2.5. Learning rules

Learning corresponds with solving the optimization problem posed in Eq. (2). Platt [5] has formulated a sequential minimization algorithm (SMO), which makes optimal pairwise adjustments to the Lagrangian multipliers, or in our terms, mossy fiber synaptic weights. Given the two multiplier values α_1 and α_2, SMO makes adjustments of the form

$$\alpha_2^{new} = \max\left(L, \min\left(H, \alpha_2^{old} + \frac{y_2(E_2 - E_1)}{\eta}\right)\right),$$

$$\alpha_1^{new} = \alpha_1^{old} + y_1 y_2(\alpha_2^{old} - \alpha_2^{new}), \tag{19}$$

where

$$y_1 = y_2 \Rightarrow L = \max(0, \alpha_2^{old} + \alpha_1^{old} - C), \quad H = \min(C, \alpha_2^{old} + \alpha_1^{old})$$

$$y_1 \neq y_2 \Rightarrow L = \max(0, \alpha_2^{old} - \alpha_1^{old}), \quad H = \min(C, C + \alpha_2^{old} - \alpha_1^{old})$$

$$\eta = 2k(x_1,x_2) - k(x_1,x_1) - k(x_2,x_2), \quad E_i = f^{old}(x_i) - y_i.$$

SMO also provides for adjusting the value of the bias term b in the decision rule whenever one of the α_i is not at an extreme value.

It is doubtful that SMO can be exactly implemented within the constraints of neurological plausibility, even in an artificially defined network, but some degree of approximation may be possible at the risk of arriving at a suboptimal decision rule. In any case, the adjustment of weights is a form of competitive learning in that increasing one weight involves decreasing others to satisfy the constraint in Eq. (2).

If the maximal possible response by a granule cell, $k(x_i,x_i)$, is normalized to a constant k_{max}, the value of η can be determined from $k(x_1,x_2)$ alone and if individual component values of the vector encoded by the perforant path have known statistical properties, for example independence, then an estimate of $k(x_1,x_2)$ can be made based on the difference in granule cell firing times. In almost all cases, $\eta < 0$ and at least the direction of change in synaptic weights can be determined even if the value of η cannot, suggesting a learning rule based on relative order of granule cell firing in which MF terminals of earlier firing granule cells are facilitated at the expense of those of later firing granule cells.

Estimating a value for E_i involves knowing the desired outcome of the decision function. This could be accomplished in a number of ways, but in particular through the self-organizing characteristics of CA3 itself. This stands in contrast to the alternative assumption that the dentate gyrus, in essence, establishes the initial organization of CA3 via "detonator" cells [6]. Any supervisory stimulus for MF terminal learning rules would need to be reflected in synaptic inputs separate from the mossy fiber connections. Modulation of MF terminal plasticity by perforant path activity [12] and CA3 associative connections [1] is consistent with the type of learning proposed here.

3. What the SVM construction implies in terms of hippocampus physiology

3.1. Dentate gyrus

Implementing the SVM construction requires that the dentate gyrus have a set of precisely timed mechanisms. One plausible cycle of events is: (1) a brief inhibition is used to set the granule cell to a known membrane potential, (2) the perforant path synaptic response is integrated in the granule cell, (3) fixed excitatory stimulation is provided to the granule cell such that small input could potentially result in firing in a time-dependent manner, and (4) at time T_F in the cycle, squelching inhibition is applied to prevent late firing. A refinement of step 4 is to use lateral inhibition to suppress further firing once other granule cells have already fired and thus already determined the dominant support vectors. Such a refinement would also reduce the need for precise timing in the mossy fiber pathway. If the inhibition in step 1 is sufficiently punctate, then step 3 can occur earlier or be ongoing, reducing the requirements for precise timing of step 3 stimulation.

Because each granule cell implements a single support vector, learning patterns that require new support vectors would require either the use of previously dormant granule cells or else granule cell neurogenesis. Adding new support vectors via neurogenesis suggests that new granule cells should exhibit relatively greater synaptic plasticity than older cells, which is consistent with experimental findings [11]. Having a consistent k_{max} would imply a consistent number of synaptic connections between a granule cell and the perforant pathway.

3.2. Mossy fiber pathway

A key requirement of the mossy fiber pathway is that it accurately preserves the relative timing of firing events in the granule cells. This implies that routing of mossy fibers afferent to a given CA3 pyramidal cell follows parallel paths. Outside this parallel routing, granule cells would ideally be arranged in a pattern to ensure consistency of axon lengths. To avoid negative synaptic weight values for mossy fiber terminals, mossy fibers must also stimulate inhibitory interneurons that ultimately connect with CA3 pyramidal cells following a path parallel to the mossy fibers themselves.

3.3. Mossy fiber terminals

Mossy fiber terminals at CA3 pyramidal cells would necessarily be complex structures. The efficiency of neurotransmitter release must be very high since relatively few terminals would be active at any one time. Rise time in the synaptic current should initially be rapid (T_N near 0) with a slower decay time. In addition, the mossy fiber terminals would have to implement complex learning rules. There is nothing in the construction to differentiate presynaptic and postsynaptic effects, but the existence of a complex presynaptic organelle is consistent with the assumptions made here.

A requirement in implementing a learning rule of the form described is that the mossy fiber terminals exchange information among themselves. This argues for a physically compact arrangement and the use of a variety of extra-cellular messengers. Neurotransmitter spillover could offer one source of exchanged information, but this alone may be insufficient to implement the learning rule needed. Use of perforant path and associational connections as input to the learning rule implies a relationship between change in postsynaptic membrane potentials (or some other retrograde messenger) and MF terminal plasticity. A possibility suggested by physiology but not considered in the construction here is that neurotransmitter release time constants may be adapted under the influence of a learning rule.

3.4. CA3 pyramidal cells

In general, it is hard to predict which functions would be associated with pyramidal cells and which with the mossy fiber terminals. Retrograde messengers generated in the pyramidal cell would be a reasonable assumption, but no definite messengers are identified. The pyramidal cell is a plausible site for maintaining a consistent number of glutamate receptor sites associated with mossy fiber inputs and thus maintaining a constant sum of total excitatory weight values.

Suppression of pyramidal cell leak currents, at least during the interval during which mossy fiber signals are presented, would enhance the quality of the generated kernel function. Otherwise, the accuracy of the decision function is compromised by leakage during the period over which inputs are integrated. The assumption of a fixed time window for pyramidal cell spiking would imply the existence of periodic inhibition to reset the state of the pyramidal cell as might be expected if gamma cycle inhibition is present.

4. Conclusion

An outline for creating a support vector machine pattern recognizer using spiking neurons has been defined, providing an existence proof that support vector machines are at least not inconsistent with the constraints of neural physiology. Correspondences between the structures needed for a support vector recognizer and the novel properties of dentate gyrus and mossy fiber systems provide a rationale for these properties and suggest that there may be some merit to a functional role of the dentate gyrus as a component of a non-linear pattern recognition system. The complex MF terminal learning rules suggested here stand in contrast with an assumption of relative non-plasticity. Experimental tests should allow these possibilities to be discriminated empirically.

Acknowledgements

I would like to thank Dr. Joel Davis at the Office of Naval Research. This work was supported in part by Office of Naval Research Grant 525424.

References

[1] B.E. Derrick, J.L. Martinez Jr., Frequency-dependent associative long-term potentiation at the hippocampal mossy fiber-CA3 synapse, Proc. Natl. Acad. Sci. USA 91 (1994) 10290–10294.
[2] R.A. Fricke, D.A. Prince, Electrophysiology of dentate gyrus granule cells, J. Neurosci. 51 (1984) 195–209.
[3] D.A. Henze, N.N. Urban, G. Barrionuevo, The multifarious mossy fiber pathway: a review, Neuroscience 98 (2000) 407–427.
[4] J.-M. Lassalle, T. Bataille, H. Falley, Reversible inactivation of the hippocampal mossy fiber synapses in mice impairs spatial learning, but neither consolidation nor memory retrieval, in the Morris navigation task, Neurobiol. Learning Memory 73 (2000) 243–257.
[5] J.D. Platt, Fast training of support vector machines using sequential minimal optimization, in: B. Schölkopf, C.J.C. Burges, A.J. Smola (Eds.), Advances in Kernel Methods: Support Vector Learning, MIT Press, Cambridge, MA, 1999, pp. 185–208.
[6] R.C. O'Reilly, J.L. McClelland, Hippocampal conunctive encoding, storage, and recall: avoiding a trade-off, Hippocampus 4 (1994) 661–682.
[7] B. Schölkopf, C.J.C. Burges, A.J. Smola (Eds.), Advances in Kernel Methods: Support Vector Learning, MIT Press, Cambridge, MA, 1999.
[8] A.J. Smola, Z.L. Óvarí, R.C. Williamson, Regularization with dot-product kernels, in: T.K. Leen, T.G. Dietterich, V. Tresp (Eds.), Advanced in Neural Information Processing Systems 13: Proceeding of the 2000 Conference, MIT Press, Cambridge, MA, 2001, pp. 308–314.
[9] R.D. Traub, R.K.S. Wong, R. Miles, H. Michelson, A model of a CA3 hippocampal pyramidal neuron incorporating voltage-clamp data on intrinsic conductances, J. Neurophysiol. 66 (1991) 635–650.
[10] A. Treves, E.T. Rolls, Computational constraints suggest the need for two distinct input systems to the hippocampal CA3 network, Hippocampus 2 (1992) 189–199.
[11] S. Wang, B.W. Scott, J.M. Wojtowicz, Heterogenous properties of dentate granule neurons in the adult rat, J. Neurobiol. 42 (2000) 248–257.
[12] M.F. Yeckel, T.W. Berger, Spatial distribution of potentiated synapses in hippocampus: dependence on cellular mechanisms and network properties, J. Neurosci. 18 (1998) 438–450.

John L. Baker received his S.B. in Mathematics from MIT in 1970 and has worked professionally in the field of commercial software development. He is currently a graduate student in the interdisciplinary School of Computational Sciences at George Mason University pursuing a Ph.D. in Computational Neurobiology. His current research interests focus on computational modeling of the hippocampus.

Available at
www.ComputerScienceWeb.com
POWERED BY SCIENCE DIRECT

ELSEVIER

Neurocomputing 52–54 (2003) 209–214

NEUROCOMPUTING

www.elsevier.com/locate/neucom

Computational counting for a quantitative analysis of cells in histologically prepared brain sections

Alia Benali[a,b,*], Iris Leefken[a], David Kastrup[a], Elke Weiler[b]

[a]*Institute for Neuroinformatics, Theoretical Biology, Ruhr-University Bochum, D-44780 Bochum, Germany*

[b]*Institute for Physiology, Department of Neurophysiology, Ruhr-University Bochum, 44780 Bochum, Germany*

Abstract

We propose a reliable method for automatic counting of cells in brain sections for different antibodies (NeuN, Parvalbumin, GABA and c-Fos) and Nissl-staining. The images of stained sections are binarized by thresholding. Then regions are clustered using a common clustering algorithm. After choosing only proper sized clusters the detected cell-bodies are counted. The parameters of the algorithm are adjusted manually and remain constant for different probes. The computational cell counting method provides correct counting results, shown by a comparison of computational results and counts gained by human experimenters.

Keywords: Cell counting; Clustering; Neuronal tissue; Quantitative analysis; Threshold images

1. Introduction

The central nervous system of animals and humans consists of billions of glial and nerve cells. Depending on their developmental and physiological stage, they express different proteins and transmitters, which can be detected by antibodies. A change in the number of labelled cells corresponds to a change in the expression of proteins and transmitters and therefore a change in the function of the neuronal network. To characterize the changes it is essential to know the number of considered cells in each experiment.

The counting of stained cells can be performed directly at the microscope or based on images taken by a camera on the microscope. To gain reliable counting results

* Corresponding author.

E-mail address: alia.benali@ruhr-uni-bochum.de (A. Benali).

0925-2312/03/$ - see front matter
doi:10.1016/S0925-2312(02)00736-1

it is necessary to take many images from areas of several sections for counting. In any case, the experimenter counting the cells has to evaluate noisy data. This noise is induced by staining or by the image acquisition process. Additional errors occur with each count performed by humans. The manual cell counts include high variances due to different levels of experience in counting, due to variable staining intensity, and due to individual perception. To avoid these problems we applied a straight forward computational method for cell counting, which is calibrated manually including expert knowledge and proven to be reliable for different staining methods.

2. Materials and methods

2.1. Preparation of tissue

Sprague Dawley rats ($N = 24$) were sacrificed with an overdose of urethane and their brains were fixed by intracardial perfusion with 0.9% saline followed by 4% paraformaldehyde in 0.1 M phosphate buffer (PBS, pH 7.4). After a 4-h post-fixation at 4°C in the same fixative the brains were embedded in paraffin. Coronal sections of 5 μm were cut serially and every tenth section was placed on gelatine-coated slides, dewaxed in xylene and transferred through a descending ethanol series into PBS and stained with cresyl violet (a Nissl stain). The adjacent sections were treated immunohistochemically with antibodies against NeuN (neuronal nuclear antigen; 1:1000, Chemicon), GABA (1:4000, Sigma), Parvalbumin (1:1000, clone, PA-235, Swant) and c-Fos (1:3000, Dianova). For details see [1].

2.2. Image acquisition

Images of stained sections were acquired using a colour video camera (CCD KH 604, Heimann; Metamorph, Universal Imaging) on an Axioskop (Zeiss, Germany) at magnifications of 50–200×.

2.3. Counting method

The counting was performed on grey value images (value range = $[0, 255]$). The process of counting is separated into two modes: calibration and counting. In the calibration mode the parameter of the algorithm can be adapted according to the staining method and cell size. In the counting mode the probe is then analysed using the calibration settings and the detected cells are automatically counted.

2.3.1. Calibration

In the calibration mode thresholds for the considered grey values in the images and for a proper size for acception of clusters are adapted. The calibration is achieved by manually counting three images of each animal and marker. With the counted areas the thresholds of the program are set. Using this method expert knowledge can be directly

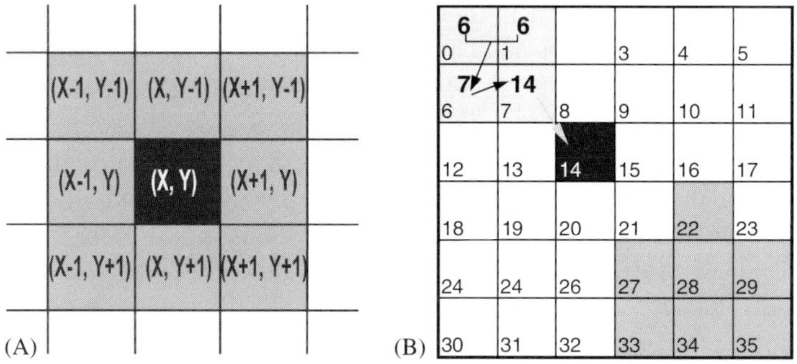

Fig. 1. Considered adjacent pixels in the applied clustering algorithm. (A) The central pixel at position (x, y)(black) is compared with its eight neighbouring pixels. (B) If adjacent pixels are set in the neighbourhood (here shown in bright grey for the analysed cluster), the current pixel and neighbouring pixels get organized into a tree with the index of its root serving as an identification during the first pass. This is done by letting each set pixel point to its parent in the tree. The small numbers show the pixel number. The dark grey region characterizes a not yet labelled, disconnected cluster.

induced into the evaluation. In the calibration mode four parameters of the algorithm are set:

- setting thresholds for binarization ($T_{\min}, T_{\max}$),
- setting cluster size ($s_{\min}, s_{\max}$).

Binarization: In order to detect the relevant cell bodies the grey value image $g(x, y)$ (x is the image row and y the image column) is transformed into a binary image $b(x, y)$ by thresholding.

$$b(x,y) = \begin{cases} 1, & T_{\min} \leqslant g(x,y) \leqslant T_{\max}, \\ 0, & \text{otherwise.} \end{cases}$$

By applying a lower threshold $T_{\min}$ very dark regions are eliminated (low grey values), like borders of the tissue. An upper threshold $T_{\max}$ serves for elimination of brighter background. The binarization is applied to separate stained regions ($b(x, y) = 1$, cells) from unstained regions ($b(x, y) = 0$) in the image.

Clustering: After binarization, the regions with a value of 1 in the binary image ($b(x, y)$) are clustered by labelling connected regions [4], resulting in a cluster image $c(x, y)$. In this applied clustering method, each set pixel in the image is joined into one region with all adjacent set pixels in the eight neighbouring pixel positions (Fig. 1). Two source image pixels $b(x_1, y_1)$ and $b(x_2, y_2)$ are considered adjacent if

$$|x_1 - x_2| \leqslant 1 \wedge |y_1 - y_2| \leqslant 1.$$

If two source image pixels are set and adjacent, the corresponding cluster image pixels will get assigned the same non-zero region label. The clustering algorithm uses two passes on the binary image: in the first from left to right, forward pass, the connectivity of the regions is established by organizing all set pixels in a connected region into a tree where the index of the root is identifying the region. The second, backward pass then assigns consecutive region numbers to elements of separate clusters.

2.3.2. Counting

To detect only clusters of proper size, the histogram $h(c(x, y))$ of the cluster image is determined. In the histogram the number of pixels of each cluster index is stored. This number represents the size of a cluster. The more pixel of one value exist the greater is the according cluster. Two parameters determine the minimal (s_{min}) and maximal (s_{max}) cluster size for acception. Those size considerations are necessary due to small disturbances or effects of tissue-borders. Improper clusters are ignored and the remaining clusters are counted.

3. Results

The reliability of the algorithm is proven by a comparison of computational results and counts gained by 18 humans observers. For this comparison 15 images from different probes and different magnifications were counted automatically and by histologists.

The examples in Fig. 2 show that the program counts for different markers just as good as the histologists.

The results of the comparison for different magnifications are shown in Fig. 3. A linear relationship between manual and automatic counts exists with a correlation coefficient of 0.93 (Fig. 3). The results of the automatic counting fall within the standard deviation of the manual counting (Fig. 3). This holds for low and high densities of cells in the images, independently of the used marker and magnification. The analysis points out that the results of the automatic counting are close to the mean value of the manual counts for each image. A further advantage of the automatic counting is the

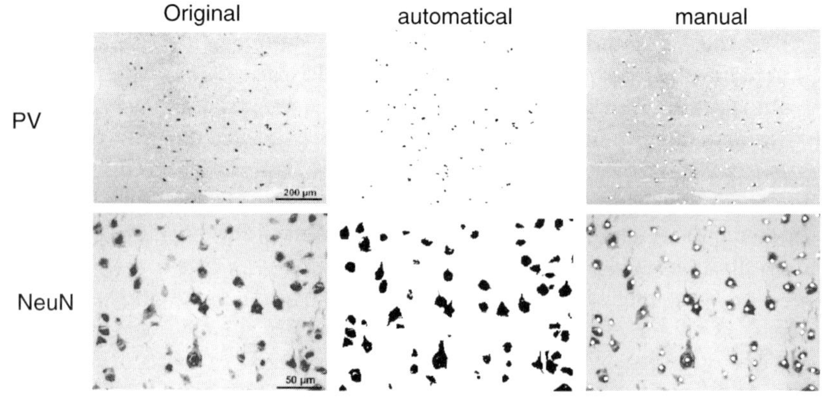

Fig. 2. Images of sections (first column) stained with parvalbumin (PV) and NeuN. In the second column are the binarized images with the accepted and automatically counted cells presented. The third column shows the manual counted cells marked by dots. Magnification see scale bars in original images.

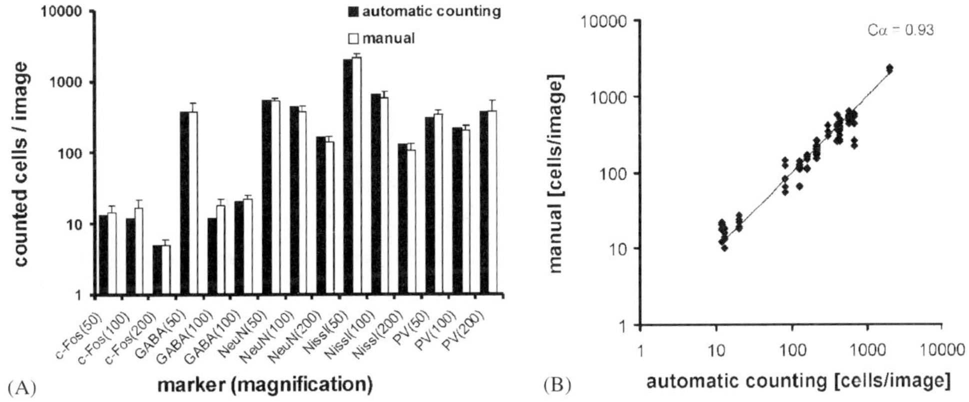

Fig. 3. Comparison of counting results of automatic vs. manual: (A) summary of counting results for all 15 evaluated images of automatic counting and mean values and standard deviations of manual counts and (B) linear correlation between automatic and manual counts. Each symbol represents the counting result of an image. A value on the x-axis corresponds to several values on the Y-axis due to the counting of the same image by several histologists.

fact that in the cell counting process the cell size is simultaneously determined by the histogram. By manual counting the cell size would have to be measured for each cell separately.

4. Conclusions

Counting cells in histologically stained sections is time-consuming and manual counting is often a tedious work with highly error-prone results. By the proposed computational algorithm the counting method produces reliable counting results, which show a high consistency, invariably the same counts are obtained for a given image, which is not the case if counts are performed by humans. Furthermore the algorithm can be used for different markers and different magnifications. Besides this program there are many other programs for analysis of biological data developed by researchers [2,3,5–7] or companies (products like MetaMorph, KS 300 or ANALY*SIS*). Most of those programs are rather specific for one application and require a good insight to adjust them for other purposes. Commercial programs normally have a wider range of applications, but most of them are very expensive and need a long training period. Additionally the user is not informed about details of the algorithms which are used for analysis. In contrast to these the used algorithm is easy to understand and can be easily adjusted to further purposes.

Acknowledgements

We are grateful to Prof. Dr. U.T. Eysel, Dr. M.A. Giese and C. Curio for review of the manuscript and helpful discussion. For assistance in recording of the images we thank Y. Benali. This work was supported by the Institute of Physiology, Department of Neurophysiology, the Deutsche Forschungsgemeinschaft (DFG) SFB-509-C4, FORUMF208/00 M122/13 (2000) and by the Institute for Neuroinformatics, Theoretical Biology, Ruhr-University Bochum, Germany.

References

[1] A. Benali, E. Weiler, U.T. Eysel, H.R. Dinse, We see light in the TUNEL of intracortical plasticity induced by microstimulation in the rat somatosensory cortex, in: Proceedings of the Fourth Meeting of German Neuroscience Society, Vol. 1, Thieme, Stuttgart, New York, 2001, p. 197.

[2] G. Henderson, B.E. Tomlinson, D. Weightman, Cell counts in the human cerebral cortex using a traditional and an automatic method, J. Neurosci. Meth. 25 (2) (1975) 129–144.

[3] B. Hutcheon, L.A. Brown, M.O. Poulter, Digital analysis of light microscope immunofluorescence: high-resolution and co-localization of synaptic proteins in cultured neurons, J. Neurosci. Meth. 96 (1) (2000) 1–9.

[4] D. Kastrup, Grouping bits to objects revisited, URL:www.neuroinformatik.ruhr-uni-bochum.de/ini/PEOPLE/dak/cluster.html , 1997.

[5] T.B. Leergaard, J.G. Bjaalie, Semi-automatic data acquisition for quantitative neuroanatomy, MicroTrace-computer programme for recording of the spatial distribution of neuronal populations, Neurosci. Res. 22 (2) (1995) 231–243.

[6] P.J. Sjöström, B.R. Frydel, L.U. Wahlberg, Artificial neural network aided image analysis system for cell counting, Cytometry 36 (1) (1999) 18–26.

[7] K. Zilles, B. Zilles, A. Schleicher, A quantitative approach to cytoarchitectonics, VI. The areal pattern of the cortex of the albino rat, Anat. Embryol. 159 (3) (1980) 335–360.

Available at
www.ComputerScienceWeb.com
POWERED BY SCIENCE @ DIRECT®

ELSEVIER

Neurocomputing 52–54 (2003) 215–220

NEUROCOMPUTING

www.elsevier.com/locate/neucom

Managing models of signaling networks

Upinder S. Bhalla*

National Centre for Biological Sciences, GKVK Campus, Bangalore 560065, India

Abstract

Signaling pathways participate in complex information processing networks. These networks handle housekeeping functions of the cell as well as specialized functions such as synaptic plasticity. I report two developments in managing such networks: a compilation of mass-action kinetic models of signaling pathways, and shared motifs in the chemistry of interactions between signaling pathways. These motifs may prove useful in abstracting signaling networks, without compromising chemical reaction details. The combination of a library of signaling pathway models, and high-level rules to connect these pathways, may simplify development of complex signaling network models.

Keywords: Mass-action chemistry; Signaling pathways; Simulation; Database

1. Introduction

Cellular signaling chemistry involves a set of networks that rival neuronal networks in complexity. These networks participate in nearly all cellular functions, from metabolism to structure to development and differentiation, but information handling is clearly one of their key attributes [5]. Some of the best studied signaling networks are those involved in synaptic plasticity [4]. Model construction is a valuable way of understanding the functioning and emergent properties of signaling networks [3]. Data management issues become severe in large networks of this kind, which can involve hundreds of molecular species and reactions. I address two related aspects of data management: a library of models, and an abstraction of connectivity between signaling pathways that facilitates modularity.

* Tel.: +91-80-363-6420; fax: +91-80-363-6662.
E-mail address: bhalla@ncbs.res.in (U.S. Bhalla).

0925-2312/03/$ - see front matter © 2003 Elsevier Science B.V. All rights reserved.
doi:10.1016/S0925-2312(02)00737-3

A first-order description of such networks using mass-action chemical kinetics is mathematically simple. Most such models are based on simple chemical rate equations such as

$$A + B \underset{k_b}{\overset{k_f}{\rightleftharpoons}} C.$$

This can be described using stereotyped systems of first-order differential equations of the form

$$d[A]/dt = k_b[C] - k_f[A][B].$$

This uniformity of representation makes it easier to devise a repository of biochemical signaling models than, for example, neuronal models. Only a limited number of parameter types need to be stored. Typical models are completely specified by a reaction scheme, the initial concentrations of molecules, rate constants, and Michaelis–Menten parameters for enzymes in the model. Although the numerical methods are very different, the same representation is also applicable to point stochastic models, where the rate constants are interpreted as reaction probabilities. Spatial and cytoskeletal factors in signaling are increasingly recognized [1,7], but even these much more complex models require the basic reaction definitions as a common denominator.

In contrast to this mathematical simplicity, signaling networks are very complex in the number of components [8]. This makes it essential to use abstracted representations of signaling pathways for humans to be able to manage models and understand their function. Block diagrams are a common form of abstraction, where signaling pathways are represented as blocks connected by arrows to form a network. Although individual signaling pathways appear to be natural candidates for encapsulating into modules consisting of sets of reactions, the interactions between pathways still require specification at the level of individual reactions. A formulation of signaling pathways and the chemistry of their interactions in an object-oriented manner would be a valuable step towards handling complex chemical circuits at the block-diagram level.

This paper discusses management of complex signaling networks such as those found at the synapse. First, a library of signaling models has been compiled into a database to provide a ready source for building blocks for such networks. Second, I describe a shared motif in the chemistry of signaling interactions that emerges from an analysis of many examples in this library [2]. Together these developments may facilitate modeling of signaling phenomena using the library of pathway modules, and interconnecting these at a block-diagram level using rules about the chemical motifs to fill in the chemical details.

2. Results

2.1. Library and database of models

A library of models of signaling pathways was developed as part of a network model of synaptic signaling pathways [3]. This library has since been extended and is now

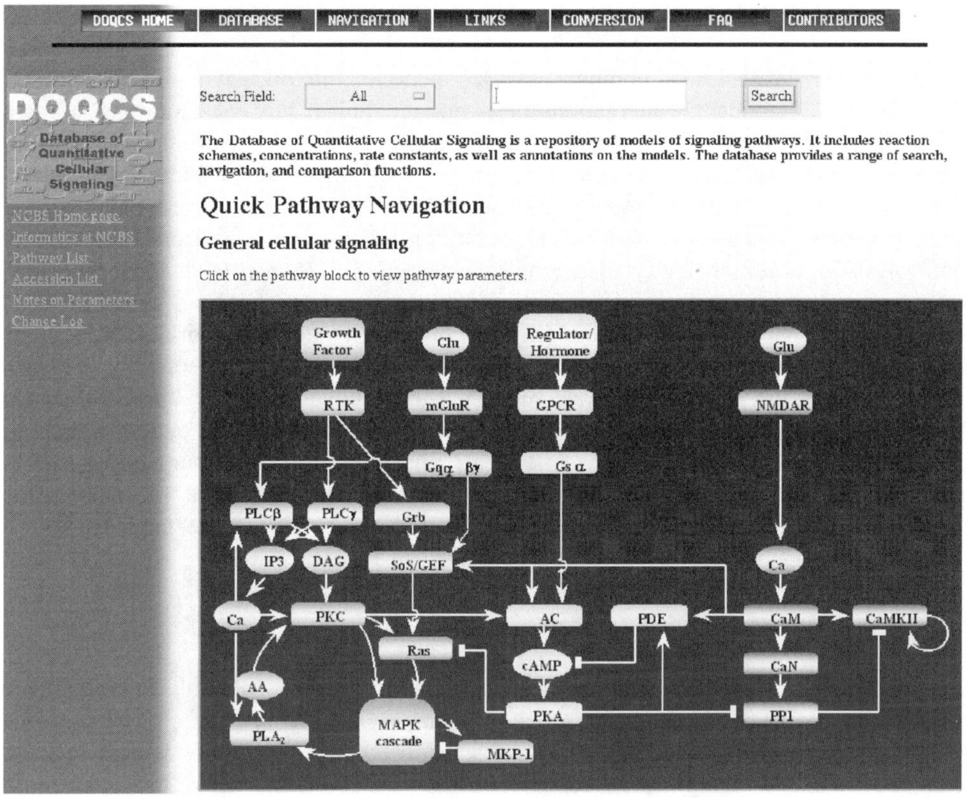

Fig. 1. Home page of DOQCS database.

on-line as the Database of Quantitative Cellular Signaling [6] (DOQCS, http://doqcs.ncbs.res.in; Fig. 1). A major goal of the database is to act as a repository of signaling models. The models in the database are heavily annotated, with the objective of encouraging reuse of models and their evolution towards greater biological accuracy and tissue specificity. All models in the database have been implemented using the simulator Kinetikit/GENESIS. Models in Kinetikit and the library consist of three chemical entities: molecules, reactions and enzymes. Molecules are characterized by an initial concentration and a flag to indicate whether they are buffered, that is, held fixed during the simulation. Reactions are characterized by a list of substrates and products, and forward and backward rate constants. Enzymes are characterized in terms of the protein molecule that exhibits the enzyme activity, a list of substrates and products, and the Michaelis–Menten constants. These chemical entities are grouped into pathways. Based on a literature search, the database now contains a significant fraction (approximately 25%) of the published models of signaling pathways.

2.2. The chemical architecture of signaling interactions

A recurring theme in developing network models of multiple signaling pathways is the difficulty in setting up interactions between signaling pathways. Examples of such interactions include the binding of an activated G-protein alpha subunit to an effector such as adenylyl cyclase, or the phosphorylation of a downstream substrate by a kinase. One the one hand, it is desirable to encapsulate all reactions in a given pathway. On the other, these chemical interactions by definition involve reactions from two or more pathways. The traditional way of setting up interactions between upstream and downstream signaling pathways is to explicitly define the reactions between constituents of the interacting pathways. The availability of complex signaling network models in the dataset has made it possible to analyze a large number of such interactions between signaling pathways [2]. This analysis suggested the following chemical motif common to all interactions analyzed: All signaling interactions consist of one or more communicating molecules from one pathway, and a replicating set of reactions and molecules from another pathway. Replicating reactions are a set of reactions at the intersection of the two pathways, which have to be repeated for every unique combination of interacting pathways. This principle is illustrated in Fig. 2.

2.3. Modularity of signaling pathways

Given the above result for signaling interactions, it becomes possible to encapsulate all signaling pathways in an object-oriented manner. The chemistry for each signaling pathway is completely internal to the pathway object. Interactions between pathways can now be set up using a well-defined interface based on the above chemical motif for interactions. One side of the interaction consists of one or more communicating molecules, and the other side consists of a reaction scheme designed to be replicated for each combination of pathways. The only additional data needed to set up any specific interaction are the parameters for the replicated reaction scheme. It is no longer necessary for one pathway to know about the internal design of any other in order to communicate. For example, it is possible for the reaction scheme of one of the pathways to be completely redesigned. Provided the interaction interface remains consistent, the second pathway can be connected to the redesigned pathway in the same manner as it was to the original.

3. Conclusion

This paper discusses two aspects of data management for complex signaling network models: the compilation of a database of such models, and an abstraction of the chemical architecture of signaling that facilitates modularity in signaling pathways and their interactions. The first aspect provides a library of models, and the second a way of easily connecting them. Together these are intended to make it possible to build network models of signaling pathways at the level of pathway modules rather than individual reactions. This may be valuable for sharing and reusing models, and

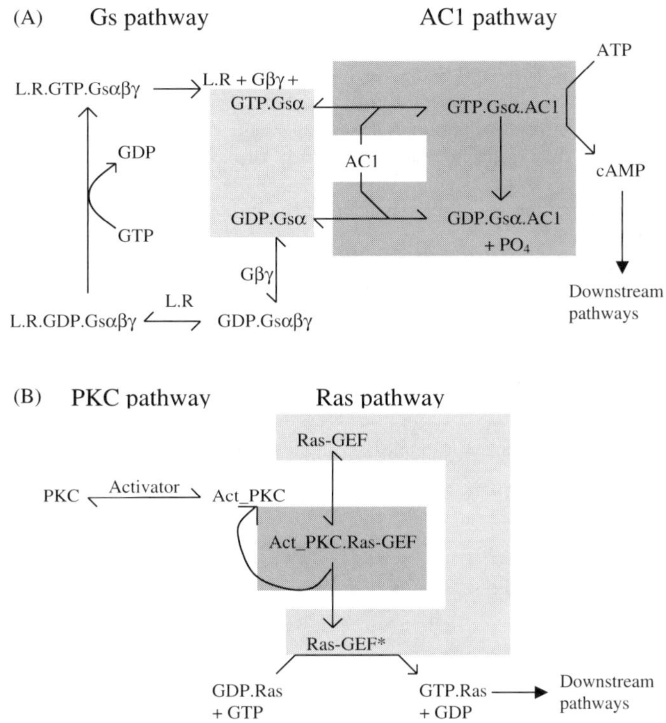

Fig. 2. Interactions between pathways expressed in terms of replicating reactions. In both examples the communicating molecules are shaded in light gray, and the replicating reactions are shaded in dark gray. (A) The molecules GTP.Gsα and GDP.Gsα (shaded in light gray) are communicating molecules from the Gs pathway. The reactions shaded in dark gray are the replicating reactions involving molecules from both the Gs and AC1 pathways. If Gs were to interact with AC2, then the same two molecules from the Gs pathway would be involved. If AC1 were to interact with a different G protein, the replicating reaction scheme would be duplicated but with different rate constants. (B) Activated PKC catalyzes the phosphorylation of Ras-GEF. The kinase catalytic reaction steps in dark gray are replicating reactions because the equivalent set of reactions would be involved for another substrate. The communicating molecules in this case are Ras-GEF and its phosphorylated state. Abbreviations: L—ligand, R—receptor, Gs—G-protein type s, GTP—guanosine triphosphate, GDP—guanosine diphosphate, ATP—adenosine triphosphate, AC—adenylyl cyclase, cAMP—cyclic adenosine monophosphate, GEF—guanine exchange factor, Act_PKC–activated protein kinase C.

focusing on the emergent computational properties of signaling networks rather than on the chemical details.

Acknowledgements

Jyoti Mishra, Sudhir Sivakumaran and H. Sridhar contributed models to the database. USB is a Senior Research Fellow of the Wellcome Trust.

References

[1] K.U. Bayer, H. Schulman, Regulation of signal transduction by protein targeting: the case for CaMKII, Biochem. Biophys. Res. Commun. 289 (2001) 917–923.
[2] U.S. Bhalla, The chemical organization of signaling interactions, Bioinformatics 18 (2002) 855–863.
[3] U.S. Bhalla, R. Iyengar, Emergent properties of networks of biological signaling pathways, Science 283 (1999) 381–387.
[4] T.V.P. Bliss, G.L. Collingridge, A synaptic model of memory: long-term potentiation in the hippocampus, Nature 361 (1993) 31–39.
[5] D. Bray, Protein molecules as computational elements in living cells, Nature 376 (1995) 307–312.
[6] S. Sivakumaran, S. Hariharaputran, J. Mishra, U.S. Bhalla, The database of Quantitative Cellular Signaling: management and analysis of chemical kinetic models of signaling networks, Bioinformatics, in press.
[7] J.R. Stiles, T.M. Bartol, Monte Carlo methods for simulating realistic synaptic microphysiology using MCell, in: E. De Schutter (Ed.), Computational Neuroscience: Realistic Modeling for Experimentalists, CRC Press, Boca Raton, 2000, pp. 87–128.
[8] G. Weng, U.S. Bhalla, R. Iyengar, Complexity in biological signaling systems, Science 284 (1999) 92–96.

ELSEVIER

Available at www.ComputerScienceWeb.com
POWERED BY SCIENCE DIRECT®

Neurocomputing 52–54 (2003) 221–226

NEUROCOMPUTING

www.elsevier.com/locate/neucom

Gain modulation and balanced synaptic input in a conductance-based neural model

A.N. Burkitt*, H. Meffin, D.B. Grayden

The Bionic Ear Institute, 384-388 Albert Street, East Melbourne, Vic. 3002, Australia

Abstract

Gain modulation of neural responses by the balanced component of the synaptic input is analyzed in the Gaussian approximation using a single compartment conductance-based neural model. The model is analyzed in the "normal operating regime," in which the output spiking rate of the neuron is equal to the spontaneous spiking rate in the absence of any stimulus. The gain in response to both additional excitatory synaptic input and injected current is found to be modulated in a non-linear way by the level of balanced synaptic input.

Keywords: Gain modulation; Integrate-and-fire neurons; Gaussian approximation; Synaptic input; Conductance-based neural model

1. Introduction

Neural gain modulation is a measure of how a modulatory input (that is independent of the stimulus) affects the amplitude of the response that a neuron generates due to the stimulus (the unmodulated input). It is a non-linear mechanism by which information is combined (or integrated) from within a single pathway or between different pathways of neural processing, which may be of sensory, motor or cognitive origin. Gain modulation provides a computational means by which neural systems may transform, combine or compare the representations that they carry of the physical world and by which they more accurately control (i.e. modulate) their output. Gain modulation plays an important role in sensory-motor integration, such as eye and reaching movements,

* Corresponding author.
E-mail address: a.burkitt@medoto.unimelb.edu.au (A.N. Burkitt).

0925-2312/03/$ - see front matter
doi:10.1016/S0925-2312(02)00742-7

and in spatial perception, as well as in auditory masking, attentional processing, object recognition and navigation. Gain modulation is one of the few neural computational principles that pervades neural processing, since it plays an important role in a number of different modalities (sensory, motor and cognitive), different brain areas (cortical and sub-cortical), and different neural functions [13].

Gain modulation has been observed experimentally to play an important role in neural processing. However, we lack an understanding of the underlying biophysical mechanisms by which gain modulation is implemented in neural systems. The central question is: "How do neurons achieve the non-linear, multiplicative behavior characteristic of gain modulation when their input–output relationship is basically integrative?" A number of possible mechanisms have been proposed [13], although they are not necessarily exclusive. One suggested mechanism involves the effect of non-linear interactions in the dendritic processing of neurons [9]. Another approach involves the effect of recurrent connections between neurons, which may give rise to non-linear interactions [11,6]. A third approach is through correlations in the synaptic input [12]. A fourth approach is that addressed in this paper, namely the effect of the balanced component of the synaptic input in providing the non-linear gain modulation of the neural response [5]. The various mechanisms are, of course, by no means mutually exclusive, e.g., the balanced component of the input could partly arise via feedback from recurrent connections.

This paper examines how balanced synaptic input can modulate neuronal gain. This modulation is a result of the effect of the balanced synaptic input upon the variance of the synaptic current, which can be varied independently of the mean synaptic current (which is determined by the difference of the excitatory and inhibitory inputs). The possible role of balanced input in neuronal gain modulation was highlighted by a recent in vitro study in which a variable current (with zero mean) was injected into a rat cortical pyramidal neuron and the gain associated with the injection of an additional constant current was measured [5]. The results indicated that the variability of the injected current affected the neuronal gain multiplicatively. In this paper, we examine gain modulation as a result of balanced synaptic inputs in a conductance-based model. The analysis will be carried out in the Gaussian approximation [3,4,1], which is most accurate for neurons with a large number of small-amplitude synaptic inputs. The effect of synaptic input over a range of input conditions (input rates, numbers of synapses and EPSP/IPSP amplitudes) is analyzed. The gain is measured as the rate of increase of the output spiking rate of the neuron in response to increased input. Two different forms of input are examined: excitatory synaptic input (i.e., in excess of the balanced component of the synaptic input) and injected current.

2. Methods

A conductance-based leaky integrate-and-fire neuron (with reversal potentials) is used in which the membrane potential $V(t)$ receives from its presynaptic inputs both excitatory and inhibitory contributions and decays in time with a characteristic time constant (the membrane time constant τ) [16]:

$$\mathrm{d}V(t) = -\frac{(V(t) - v_0)}{\tau}\,\mathrm{d}t + g_{\mathrm{E}}(V_{\mathrm{E}} - V(t))\,\mathrm{d}P_{\mathrm{E}}(t) + g_{\mathrm{I}}(V_{\mathrm{I}} - V(t))\,\mathrm{d}P_{\mathrm{I}}, \tag{1}$$

where v_0 is the reset potential, V_{E} and V_{I} are the (constant) reversal potentials ($V_{\mathrm{I}} \leqslant v_0 \leqslant V(t) \leqslant V_{\mathrm{th}} < V_{\mathrm{E}}$). $\mathrm{d}P_{\mathrm{E}}(t)$ and $\mathrm{d}P_{\mathrm{I}}(t)$ are independent temporally homogeneous Poisson processes with constant intensities $N_{\mathrm{E}}\lambda_{\mathrm{E}}$ and $N_{\mathrm{I}}\lambda_{\mathrm{I}}$, respectively, i.e., each of the N_{E} excitatory input fibers (and N_{I} inhibitory input fibers) has a rate λ_{E} (resp. λ_{I}). In the balanced neuron considered here we choose $N = N_{\mathrm{E}} = N_{\mathrm{I}}$. The postsynaptic current amplitudes g_{E} and g_{I} are chosen to be non-negative (and are identical for all excitatory and inhibitory inputs, respectively). The values of g_{E} and g_{I} are chosen so that the average excitatory and inhibitory inputs are balanced at the potential V_{B} ($V_{\mathrm{I}} < V_{\mathrm{B}} < V_{\mathrm{E}}$), denoted the *balance potential*

$$g_{\mathrm{E}}\,\frac{(V_{\mathrm{E}} - V_{\mathrm{B}})}{\theta} = g_{\mathrm{I}}\,\frac{(V_{\mathrm{B}} - V_{\mathrm{I}})}{\theta} \equiv b = b_0/\sqrt{N}, \qquad \theta = V_{\mathrm{th}} - v_0. \tag{2}$$

When the membrane potential reaches a threshold, an output spike is generated and the membrane potential is reset to its resting value v_0. In the absence of spike generation, the membrane potential reaches an equilibrium value, V_Q, about which it fluctuates with variance σ_Q. The membrane potential approaches V_Q with a time constant given by τ_Q. The values of V_Q, σ_Q, τ_Q are [7,1]

$$V_Q = \frac{v_0/\tau + r_{11}}{1/\tau_Q}, \qquad \sigma_Q^2 = \frac{\mu^2 r_{20} - 2\mu r_{21} + r_{22}}{2/\tau_Q - r_{20}},$$

$$\frac{1}{\tau_Q} = \frac{1}{\tau} + r_{10}, \qquad r_{mn} = N_{\mathrm{E}}\lambda_{\mathrm{E}} g_{\mathrm{E}}^m V_{\mathrm{E}}^n + N_{\mathrm{I}}\lambda_{\mathrm{I}} g_{\mathrm{I}}^m V_{\mathrm{I}}^n. \tag{3}$$

The analysis is carried out in the Gaussian approximation [3], in which the probability density of the membrane potential $p(v,t\,|\,v',0)$ is parameterized as

$$p(v,t\,|\,v',0) = \frac{1}{\sqrt{2\pi\Gamma(t;v')}}\exp\left\{-\frac{(v - \Upsilon(t;v'))^2}{2\Gamma(t;v')}\right\}, \tag{4}$$

where $\Upsilon(t;v)$ and $\Gamma(t;v)$ are the (time-dependent) mean and variance of the membrane potential. The Gaussian approximation is accurate in the limit of a large number of input synapses, N, which allows the probability density of the membrane potential to be evaluated using a self-consistent analysis [1]. The output spike distribution $f_\theta(t)$ obeys the renewal equation [10,2]

$$p(V_{\mathrm{th}},t\,|\,v_0,0) = \int_0^t \mathrm{d}t'\, f_\theta(t')\,p(V_{\mathrm{th}},t\,|\,V_{\mathrm{th}},t'), \tag{5}$$

where $p(v,t\,|\,v',t')$ is the conditional probability density of the membrane potential having the value v at time t, given that it had the value v' at an earlier time t'. The effect of reversal potentials is to impose a lower bound to the membrane potential, and consequently it is necessary to consider their effect when inhibition plays an important role. In the model where reversal potentials are neglected, the amplitudes of the excitatory and inhibitory post-synaptic potentials (EPSPs and IPSPs, resp.) are constant (i.e., independent of the existing membrane potential). In this case the Gaussian approximation gives results that are identical with the diffusion method [8], in which the membrane potential is modeled as a stochastic process that is characterized by its first two moments. However, the Gaussian approximation differs from the diffusion method when reversal potentials are included since it accounts for non-linear summation and allows different time courses for the mean, $\Upsilon(t;v)$, and variance, $\Gamma(t;v)$, of the membrane potential (these are given explicitly in [1]).

3. Results

An essential part of the analysis is defining the "normal operating regime" of a neuron, to ensure that the chosen parameter values correspond to biologically relevant neural behavior. This is achieved in our model by the condition that the output spiking rate in the absence of a stimulus corresponds to the spontaneously active spiking rate. This requires that there be a slight excess of excitation in the spontaneously active neuron (for the values used here, this is satisfied by $\lambda_E = \lambda_I + \lambda_P = \lambda_{spon}$ with $\lambda_P = 1$ spike/s).

The parameter values chosen are: $V_I = -75$ mV, $v_0 = -65$ mV, $V_{th} = -55$ mV, $V_E = 0$ mV, so that the ratios of the potential differences are $(v_0 - V_I) : (V_{th} - v_0) : (V_E - V_{th}) = 1 : 1 : 5.5$. The passive membrane time constant is $\tau = 10$ ms, the amplitude of the post-synaptic potentials is given by $b_0 = 1$, the spontaneous spiking rate is $\lambda_{spon} = 5$ spikes/s, the balance potential is $V_B = V_Q = -58.675$ mV, and the number of excitatory and inhibitory synapses is $N = N_E = N_I = 4000$. The results are not very sensitive to the choice of $V_B = V_Q$ and $b_0 = 1$ so long as λ_P is chosen appropriately to fulfill the above condition on the spontaneous spiking rate.

Fig. 1 shows the output spiking rate (plot a) for the parameter values given above: The solid line is for the "1X" case in which the balanced input is just the spontaneous activity, and the dashed line is for the "2X" case in which the balanced input has twice the spiking rate of the spontaneous activity. Plot b shows the gains for the "1X" and "2X" conditions. Plots c,d show the equivalent results for injected current.

4. Discussion and conclusions

There has been considerable interest in the role of balanced synaptic inputs recently, since neurons receiving balanced input produce a variability of spike times (usually measured by the coefficient of variation of the interspike interval distribution) that agrees with values observed in cortical neurons [14]. Neurons that receive predominately excitatory input, on the other hand, would be expected to have a membrane potential that is relatively smooth, resulting in regular firing and a small value of the coefficient of variation [15].

The results of this study indicate that, in the normal operating regime for the neuron, the effect upon the neuronal gain of balanced synaptic input is not simply multiplicative.

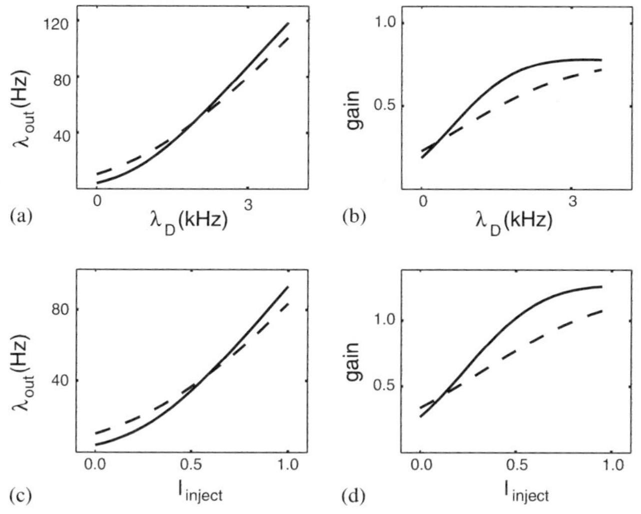

Fig. 1. Plot of results showing the output spiking rate λ_{out} and neuronal gain: (a),(b) for excitatory synaptic input with total driving input spiking rate of λ_D (the spiking rate in excess of the balanced input spiking rate, summed over all N inputs), for the "1X" (solid line) and "2X" (dashed line) cases, and (c),(d) for injected current input showing the results for the "1X" (solid line) and "2X" (dashed line) conditions.

However, the balanced inputs do give rise to a non-linear gain modulation effect, when the gain is measured both with respect to synaptic input and to injected current. Although the results presented here represent only one set of parameters, they are representative of the results over a large part of the normal operating regime of a typical cortical neuron. We are currently studying gain modulation over the entire range of biologically plausible neural parameters. We are examining particularly the dependence of gain modulation upon the parameters V_Q, τ_Q, σ_Q, and details of these studies will be published elsewhere.

Acknowledgements

This work was funded by the Australian Research Council (ARC Discovery Project #DP0211972) and The Bionic Ear Institute.

References

[1] A.N. Burkitt, Balanced neurons: analysis of leaky integrate-and-fire neurons with reversal potentials, Biol. Cybern. 85 (2001) 247–255.
[2] A.N. Burkitt, G.M. Clark, Analysis of integrate-and-fire neurons: synchronization of synaptic input and spike output in neural systems, Neural Comput. 11 (1999) 871–901.
[3] A.N. Burkitt, G.M. Clark, Calculation of interspike intervals for integrate-and-fire neurons with Poisson distribution of synaptic inputs, Neural Comput. 12 (2000) 1789–1820.
[4] A.N. Burkitt, G.M. Clark, Synchronization of the neural response to noisy periodic synaptic input, Neural Comput. 13 (2001) 2639–2672.

[5] F.S. Chance, L.F. Abbott, A.D. Reyes, Gain modulation from background synaptic input, Neuron 35 (2002) 773–782.
[6] R. Hahnloser, R.J. Douglas, M. Mahowald, K. Hepp, Feedback interactions between neuronal pointers and maps for attentional processing, Nature Neurosci. 2 (1999) 746–752.
[7] F.B. Hanson, H.C. Tuckwell, Diffusion approximations for neuronal approximations for neuronal activity including synaptic reversal potentials, J. Theor. Neurobiol. 2 (1983) 127–153.
[8] N. Hohn, A.N. Burkitt, Shot noise in the leaky integrate-and-fire neuron, Phys. Rev. E 63 (2001) 031902.
[9] B.W. Mel, Synaptic integration in an excitable dendritic tree, J. Neurophysiol. 70 (1993) 1086–1101.
[10] H.E. Plesser, S. Tanaka, Stochastic resonance in a model neuron with reset, Phys. Lett. A 225 (1997) 228–234.
[11] E. Salinas, L.F. Abbott, A model of multiplicative neural response is parietal cortex, Proc. Natl. Acad. Sci. USA 93 (1996) 6461–6474.
[12] E. Salinas, T.J. Sejnowski, Impact of correlated synaptic input on output firing rate and variability in simple neuronal models, J. Neurosci. 20 (2000) 6193–6209.
[13] E. Salinas, P. Thier, Gain modulation: a major computational principle of the central nervous system, Neuron 27 (2000) 15–21.
[14] M.N. Shadlen, W.T. Newsome, Noise, neural codes and cortical organization, Curr. Opin. Neurobiol. 4 (1994) 569–579.
[15] W.R. Softky, C. Koch, The highly irregular firing of cortical cells is inconsistent with temporal integration of random EPSPs, J. Neurosci. 13 (1993) 334–350.
[16] H.C. Tuckwell, Synaptic transmission in a model for stochastic neural activity, J. Theor. Biol. 77 (1979) 65–81.

Available at
www.ComputerScienceWeb.com
POWERED BY SCIENCE DIRECT®

ELSEVIER

Neurocomputing 52–54 (2003) 227–231

NEUROCOMPUTING

www.elsevier.com/locate/neucom

NeuroSpaces: separating modeling and simulation

Hugo Cornelis*, Erik De Schutter

University of Antwerp, Universiteitsplein 1, Antwerp 2601, Belgium

Abstract

Modeling starts with sets of mathematical equations that have been put together by people who have a good intuition of the concepts behind the biological reality. A software system for neuronal modeling thus must manage sets of equations as well as the biological concepts.

A rigorous theoretical analysis on the software requirements for such a system has led us to a new fundamental data model. Based on this data model we have implemented an extensible software system that combines mathematical and neurobiological modeling and that still reflects concepts like heterogenous components and hierarchical networks. When connected to a computation engine, the system turns into a simulator that is efficient in memory requirements, easy to use and allows complete inspection of the mathematical model.

Keywords: Modeling; Analysis; Simulator

1. Introduction

During the last twenty years traditional monolithic software design has gradually been abandoned in favor of architectures with a small extensible core or kernel. The kernel manages a set of modules and components that engage in shared activities via semantically defined interfaces [6,10]. Such systems are called collaborative systems. The current neuronal simulation packages—developed during the last decade—don't share this philosophy. It makes them hard to extend and difficult to use for large and complex neuronal models.

Recent developments try to overcome this situation in two ways: first there have been attempts to integrate the existing packages into a collaborative system [7]. Although these attempts could hold a promise, they have been unsuccessful so far because the

* Corresponding author.
E-mail address: hugo@bbf.uia.ac.be (H. Cornelis).

0925-2312/03/$ - see front matter
doi:10.1016/S0925-2312(02)00750-6

simulation package itself still suffers from its monolithic design and is difficult to interface with other software components. Second, attempts have been made to make calculation libraries accessible as object-oriented class hierarchies [14]. These attempts have no direct relation with neuronal modeling with the consequence that few people in the neuroscience community are aware of the project. The classes are sometimes ill defined or do not address a number of important issues related to neuronal modeling.

2. Problem statement

After analyzing the structural software requirements for a neural simulation package, we discovered that there is currently no data model available that covers all the modeler's needs. An example is the need to compose hierarchical networks. In Genesis such a network is composed with scripts that construct all the populations and then add links between individual cells. However, after the setup phase of the model, the topology of the network as well as the projections between the individual populations are lost i.e. the model has become a flat network instead of being a hierarchical network—the modeler's concept of the model. Put in other words, projections reside at different levels of the network hierarchy, yet the requirement to inspect *all* the connections on one cell must not be violated, even if the cell is involved in many projections. Currently no single data model is able to fulfill both of these two requirements. Other examples of problematic needs are scaling of conductances and compound models of heterogenous intracellular mechanisms. Reducing these problematic needs we ended with a set of ultimate abstract requirements.

3. Description of the work

Building further on known theoretical data structures that have been studied in informatics [12,13], we designed a new data model that is able to address all the abstract requirements (see Fig. 1 for an overview). The theoretical basis of this new data model is given by an ordered tree with relative paths to tree nodes encoded as integers and the possibility to have links between the ancestors of a node [1]. By assigning mathematical or biological semantics to the nodes of the tree, it becomes a natural representation of a detailed neuronal model of a cell or a network. The integer encoding scheme allows grouping of components at any biological level and memory efficiency via prototyping of repetitive sub-models. Trivial manipulations of the integers associated with such a relative path allow a flat representation of components of the same type such that e.g. projections in a hierarchical network can be inspected via a single connection matrix and connections converging on the same compartment can be queried.

The implementation of this data model for neuronal modeling is called NeuroSpaces. It consists of a number of independent software components that cover different needs of extensibility (Fig. 2). An example is the loadable module manager that supports the implementation of extensions covering specialized modeling semantics, such as setting

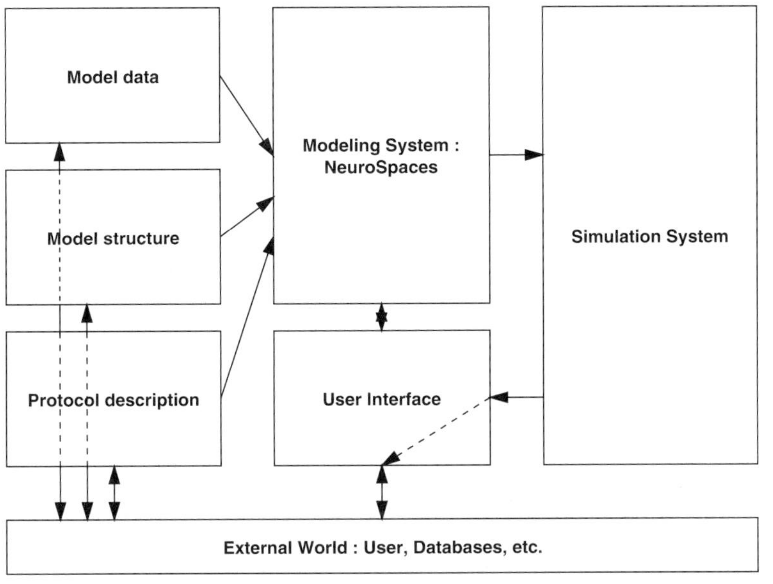

Fig. 1. Context of the design of NeuroSpaces: At the left are descriptions of the system to be simulated: the model data and structure and the experimental protocol if applicable. All these things are specified in a declarative way such that they can go into or come from a database. At the right is the simulation engine, which could be any of the currently available packages. For reasons of memory efficiency we selected the numerical solver of Genesis to do our testing. In the center, a software component—called NeuroSpaces—that only supports the setup and validation of modeling. In use it is quite close to a hierarchical database. The split between modeling and simulation allows separate specialization of software components. During model construction, the user only communicates with NeuroSpaces, during the simulation, all results are handed to the user via the user interface, such that the enormous complexity of the simulation engine remains hidden.

up three-dimensional projection schemes or randomizing parameters. Another part of the system is a file parser that gives explicit support for model reuse and grouping into libraries.

Except for computer science related research NeuroSpaces also includes a number of benefits for neuronal modelers such as memory efficiency and complete model validation. A major design principle has been the separation of the specification of the mathematical model and the way of simulating this model as done by computation engines. Computation engines can now be implemented in the most efficient way while their complexity is hidden from the user by NeuroSpaces [3].

NeuroSpaces has been linked with Genesis and its highly optimized numerical computation engine (hsolve [4,3]) and can be used to build models and simulate their behavior. Using subcellular components like dendritic segments and synaptic channels as building blocks, we constructed libraries of detailed models of the different cell types found in the cerebellar cortex. These libraries have in turn been used to build networks with the same topology as the ones currently used to investigate the function of the cerebellar cortex [5,2,11]. Since these libraries cover the complete spectrum from the subcellular up to the network level, the memory requirements of complex models and

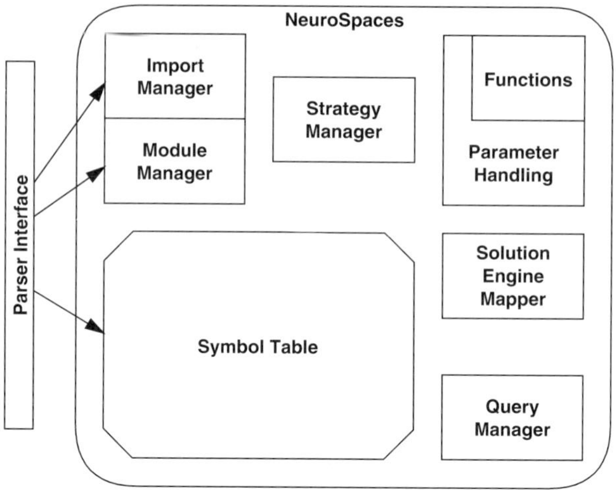

Fig. 2. The internal design of NeuroSpaces: descriptive files go through a parser that interoperates with a loadable module manager and an import manager. Then the model is stored in a compact symbol table, but using a specialized data structure it gives the illusion of a hierarchical database that contains all mathematical and algorithmic details of the model. A special module for parameter handling implements scaling of conductances and relative coordinates. The solution engine mapper allows to instantiate specialized computation engines, specific for certain model structures.

the setup time of computation engines can easily be evaluated. The structure of the network libraries can be used as a basis to build other network types.

4. Future work

After having tested this software system further our future plans are twofold: first we will be working on database integration. The libraries we are currently using are self-contained and declarative. This makes interaction possible with XML, the general agreed upon database exchange format [8]. We are planning to implement additional software to setup simulations directly from databases.

Second we are investigating how to distribute simulations on parallel machines [9]. Since the mathematical model is completely available for inspection, it can be used to calculate a workload for transparent parallelization of large simulations.

References

[1] H. Cornelis, E. DeSchutter, Treespaces, Tech. Rep., UIA, University of Antwerp, Universiteitsplein 1, Wilrijk, Antwerp, December 2001.

[2] H. Cornelis, E. DeSchutter, The purkinje cell tutorial, Available from the Genesis source code distribution. http://www.genesis-sim.org, May 2002.

[3] H. Cornelis, E. DeSchutter, Tutorial: Simulations with genesis using hsolve. Internal Document, November 2002 (This will probably be included soon in the Genesis source code distribution. http://www.genesis-sim.org.).

[4] E. DeSchutter, D. Beeman, in: the Book of Genesis, 1st Edition, Springer, Berlin, 1988 (Speeding up large simulations, Chapter 22).

[5] E. DeSchutter, J. Bower, Simulated responses of cerebellar purkinje cells are independent of the dentritic location of granule cell synaptic inputs, Proc. Natl. Acad. Sci. USA 91 (1994) 4736–4740.

[6] E. Gamma, R. Helm, R. Johnson, J. Vlissides, Design Patterns, Elements of Reusable Object–Oriented Software, Professional computing series. Addison Wesley, Reading, MA, 1995.

[7] N. Goddard, G. Hood, F. Howell, M. Hines, E. DeSchutter, Neosim: portable plug and play neuronal modelling, Neurocomputing 38–40 (2001) 1657–1661.

[8] N.H. Goddard, M. Hucka, F. Howell, H. Cornelis, K. Shankar, D. Beeman, Towards neuroml: Model description methods for collaborative modelling in neuroscience, Philos. Trans. Roy. Soc. Ser. B: Biological Sciences 356 (2001) 1–20 (Theme Issue organized and edited by Rolf Kötter on "Neuroscience databases—tools for exploring brain structure-function relationship").

[9] F. Howell, J. Dyhrfjeld-Johnsen, R. Maex, N. Goddard, E. DeSchutter, A large-scale network model of the cerebellar cortex using pgenesis, Neurocomputing 32 (2000) 1041–1046.

[10] M. Hucka, A. Finney, H. Sauro, H. Bolouri, J. Doyle, K. Kitano, Foundations of Systems Biology, MIT Press, Cambridge, MA, 2001 (The ERATO Systems Biology Workbench: an Integrated Environment for Multiscale and Multitheoretic Simulations in Systems Biology, Chapter).

[11] R. Maex, E. DeSchutter, Synchronization of golgi and granule cell firing in a detailed network model of the cerebellar granule cell layer, J. Neurophys. 80 (1998) 2521–2537.

[12] R. Sedgewick, Algorithms in C, Parts 1-4, Fundamentals, Data Structures, Sorting, Searching, 3rd Edition, Addison-Wesley, Reading, MA, 1999.

[13] R. Sedgewick, Algorithms in C, Part 5, Graph Algorithms, 3rd Edition, Addison-Wesley, Reading, MA, 2002.

[14] J.-F. Vilbert, Xnbc v9, a user friendly simulation and analysis tool for neurobiologists, Neurocomputing 38–40 (2001) 1715–1723.

Available at
www.ComputerScienceWeb.com
POWERED BY SCIENCE @ DIRECT®

ELSEVIER

Neurocomputing 52–54 (2003) 233–238

NEUROCOMPUTING

www.elsevier.com/locate/neucom

Location independence and fast conduction of synaptic inputs in neocortical neurons in vivo

Alain Destexhe*, Michael Rudolph

Unité de Neurosciences Intégratives et Computationnelles, CNRS UPR-2191, Bat. 33, Avenue de la Terrasse 1, 91198 Gif-sur-Yvette, France

Abstract

Neocortical neurons in vivo operate in a high-conductance state characterized by an intense intracellular fluctuating activity. Here we investigate how these conditions impact on the dendritic integration of synaptic inputs by using biophysical models of morphologically reconstructed neocortical pyramidal neurons. We find that the combined effect of high conductance and fluctuating activity due to the synaptic background activity may set pyramidal neurons into an integrative mode which is determined by the intensity of network activity, which is fast-conducting and in which the impact of inputs is roughly location-independent.

Keywords: Cerebral cortex; Dendritic integration; Synaptic background activity; Dendritic spikes; Computational models

1. Introduction

Active cortical states in vivo are characterized by a marked tonic increase of conductance compared to quiescent states, as well as sustained and irregular voltage fluctuations due to the presence of synaptic background activity [4,7,9]. However, how these conditions shape the integration of synaptic inputs remains unknown, mainly due to the technical difficulties of performing controlled experiments during activated states of the network in the intact brain.

Here we address this problem by using computational models of morphologically reconstructed neocortical pyramidal neurons in which in vivo conditions were simulated by random excitatory and inhibitory synaptic inputs in soma and dendrites. We show

* Corresponding author. Tel.: +33-1-6982-3435; fax: +33-1-6982-3427.
E-mail address: destexhe@iaf.cnrs-gif.fr (A. Destexhe).

0925-2312/03/$ - see front matter
doi:10.1016/S0925-2312(02)00754-3

that the presence of background activity defines a dynamic state in which mechanisms of dendritic integration are qualitatively different compared to quiescent states.

2. Methods

Simulations were performed based on several morphologically reconstructed pyramidal cells of cat cortex (layer II–III, V and VI) [1]. Different combinations of passive parameters were used, including different values of the axial resistivity (80–250 Ωcm), and different distributions of leak conductances in soma and dendrites [10].

Voltage-dependent currents included fast Na^+ and delayed-rectifier K^+ currents with different kinetics, a voltage-dependent K^+ currents, a Ca^{2+}-dependent K^+ current, a high-threshold Ca^{2+}-current, a persistent Na^+ current, as well as A-type K^+ currents. All currents were simulated by Hodgkin–Huxley type models. Various combinations and densities (e.g. 3–12 mS/cm^2 for fast Na^+, and 5–10 mS/cm^2 for delayed-rectifier K^+ currents) were tested. For details we refer to [6,8].

Synaptic currents corresponding to AMPA and $GABA_A$ receptors were simulated according to two-state kinetic models [2] with quantal conductances of 1.2 nS for AMPA and 0.6 nS for $GABA_A$ [3]. Synaptic densities were estimated from morphological studies of neocortical pyramidal cells.

Synaptic background activity was simulated by random (Poisson-distributed) release events at all synapses with average rates between 0.1 and 1 Hz at glutamatergic, and 0.55–5.5 Hz at GABAergic synapses. The release statistics was altered by including a correlation between release events [3].

For stimulation, a supplementary set of AMPA synapses was placed at different locations in the dendritic tree. Stimulation intensity was adjusted by varying the number of colocalized synchronously activated AMPA-mediated synapses.

3. Results

We first compared the effect of synaptic background activity on action potential (AP) initiation and propagation in dendrites. In quiescent conditions, superthreshold stimulation either by somatic current injection or synaptic stimulation evoked action potentials first in the axon, followed by a somatic spike which propagated back into the dendrites with declining amplitude, in accordance with experimental studies [5]. Although in-vivo-like conditions did not impact on the temporal order of axonal/somatic spike initiation, the distance of robust backpropagation of APs was smaller and often failed to reach more distal parts of the dendrites.

Under quiescent conditions, the threshold for AP initiation in the dendrites was high (Fig. 1A) compared to states with network activity (Fig. 1B). Forward-propagation of dendritic APs initiated by strong synaptic stimulation occurred in the quiescent state only over limited distances (Fig. 1C), and was sensitive to morphological or electrophysiological peculiarities, such as branch points, passive properties or distribution of active conductances. In contrast, under in-vivo-like conditions, at equivalent

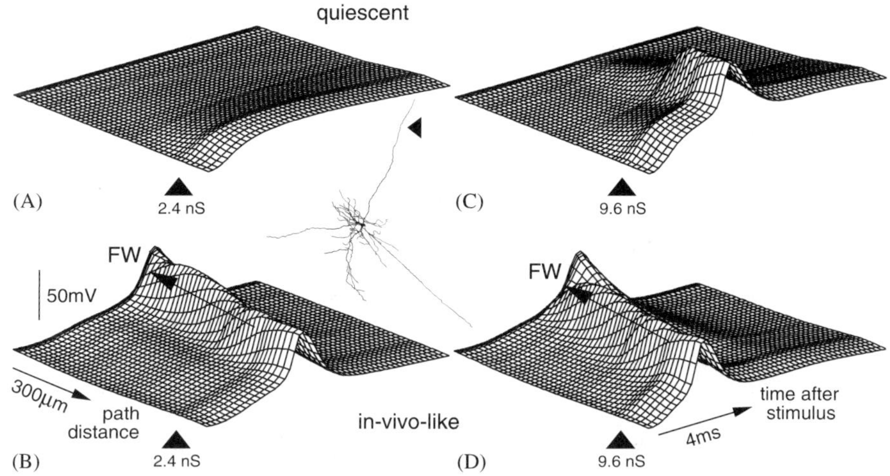

Fig. 1. Somatodendritic membrane voltage profiles for synaptic stimulation (A, B: amplitude 6 nS; C, D: amplitude 12 nS) in the distal apical dendrite of a layer V neuron (path distance 600 μm, see inset) under quiescent (A,C) and in vivo-like (B,D) conditions. Initiation and active forward-propagation of dendritic spikes evoked in the distal region is favored under in-vivo-like conditions (FW: forward propagating dendritic spikes).

voltages, dendritic APs could forward-propagate over large distances and reach the soma (Fig. 1D).

We next investigated the consequences of forward-propagating dendritic APs on the cellular response by computing the post-stimulus time histogram (PSTH) over long periods of time with repeated stimulation of single or groups of colocalized excitatory synapses. Under in vivo conditions, PSTHs obtained for stimuli occurring at different distances from the soma revealed a roughly location-independent impact of synaptic stimulation (Fig. 2A). The "efficacy" of synaptic stimulation, quantified by the probability that a somatic spike is specifically evoked by a synaptic stimulus, was found to be roughly independent of synapse location for a broad range of stimulus amplitudes subthreshold under quiescent conditions (Fig. 2B), and was robust against changes in the background intensity (Fig. 2D).

To show how this location-independent mode depends on forward-propagating dendritic APs, we selected all stimulation trials which evoked dendritic spike and found that the probability for stimulus evoked dendritic spikes increases with path distance from the soma for fixed stimulation amplitude (Fig. 2C, dashed). On the other hand, the probability that a stimulus evoked dendritic spike propagates all the way down to the soma and triggers a somatic spike was found to decrease with path distance due to the higher chance of interference with background-induced fluctuations along the path (Fig. 2C, dotted). Remarkably, these two opposing effects nearly compensated such that the probability of evoking a soma/axon AP was approximately independent on the distance to soma (Fig. 2C, solid).

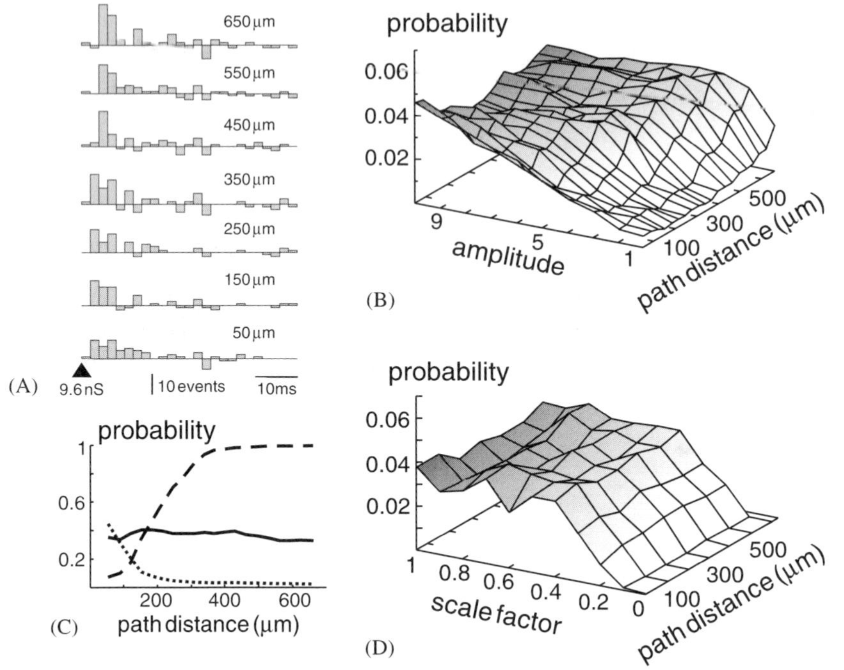

Fig. 2. Independence of the somatic response to the location of synaptic stimulation under in-vivo-like conditions. (A) Post-stimulus time histograms (PSTHs) of responses to identical AMPA-mediated synaptic stimuli (9.6 nS) at different dendritic locations. (B) Integrated PSTH (probability that a somatic spike was specifically evoked by the stimulus) as a function of stimulus amplitude (number of synchronously activated colocalized synapses with quantal conductance 1.2 nS) and distance to soma. (C) Comparison of the probability of evoking a dendritic spike (dashed) and the probability that an evoked spike translated into a somatic/axonal spike (dotted), represented as a function of the location of the stimulus (stimulus amplitude 9.6 nS). The solid line represents the multiplication of this two probabilities ($\times 10$), leading to the probability of somatic spike specifically evoked by the stimulus. (D) Integrated PSTH for various intensities of background activity obtained by scaling the release frequencies at all terminals (stimulus amplitude 12 nS).

Active network conditions also impacted on the somatic voltage deflections of dendritic EPSPs (Fig. 3A), yielding an increase of the peak height for distal stimulations under in-vivo-like conditions due to the presence of stimulation triggered forward-propagating dendritic spikes. In contrast, in quiescent conditions, the amplitude monotonically declined with increasing path distance of the stimulus. Surprisingly, the presence of dendritic spikes also showed an impact on the timing of synaptic events, yielding a marked reduction of the location dependence of the time-to-peak compared to quiescent conditions (Fig. 3B, compare black solid and black dotted). Both the synaptic background-induced high-conductance state as well as the intense fluctuating activity were found to be the determining factors of this fast-conducting state in vivo (Fig. 3B).

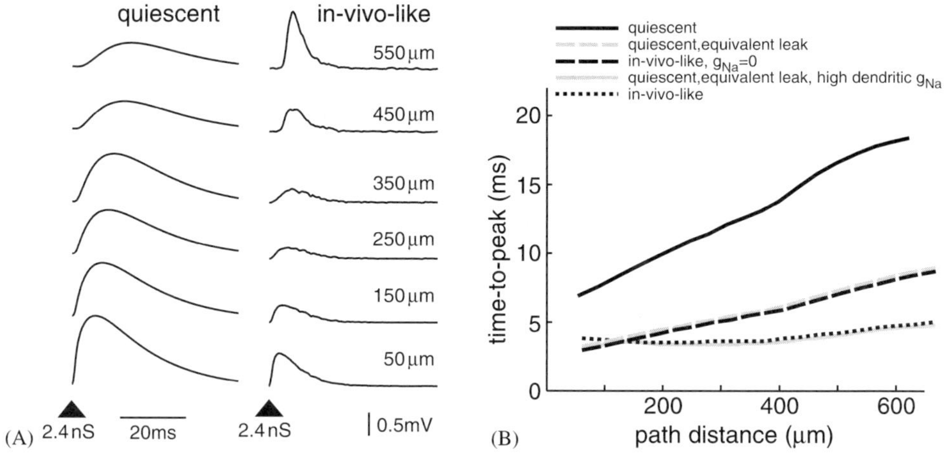

Fig. 3. (A) Somatic voltage deflections of EPSPs following synaptic stimulation (subthreshold at the soma) at different dendritic sites in the quiescent and under in-vivo-like conditions. (B) Time-to-peak for somatic voltage deflections for both conditions (dashed), and mechanism underlying fast dendritic conduction. Replacing background activity by an equivalent (light grey), or suppressing dendritic Na^+ channels (light grey dashed) led to an intermediate location-dependence of EPSP time-to-peak. High dendritic excitability together with strong synaptic stimuli evoked reliable dendritic APs and yielded a reduced location dependence of the time-to-peak in quiescent conditions (black solid), comparable to in-vivo-like conditions (black dotted).

4. Conclusions

We have investigated the impact of synaptic background activity at a level present in vivo on the dendritic integration of synaptic inputs using computational models of neocortical pyramidal neurons. The results suggest that the synaptic background activity present in vivo sets pyramidal neurons into a qualitatively different integrative mode, determined by the intensity of network activity, which is fast-conducting and in which the impact of inputs is roughly location-independent. This mode was found for different morphologies without fine-tuning the model, and was also robust to changes in active channels kinetics and distribution, passive settings as well as to the intensity of background activity (see details in Ref. [8]).

The proposed mechanism suggests that in active states in the intact brain the integration of synaptic inputs is not constrained by cellular morphology, but instead must be viewed as a state and activity-dependent dynamic property. The impact of the proposed mechanism on the spatial and temporal integration of more complex synaptic inputs, such as paired events, will be the subject of future investigations (research supported by CNRS and NIH).

References

[1] D. Contreras, A. Destexhe, M. Steriade, Intracellular and computational characterization of the intracortical inhibitory control of synchronized thalamic inputs in vivo, J. Neurophysiol. 78 (1997) 335–350.

[2] A. Destexhe, Z.F. Mainen, T.J. Sejnowski, in: C. Koch, I. Segev (Eds.), Methods in Neuronal Modeling, 2nd Edition, MIT Press, Cambridge, MA, 1998, pp. 1–26.
[3] A. Destexhe, D. Paré, Impact of network activity on the integrative properties of neocortical pyramidal neurons in vivo, J. Neurophysiol. 81 (1999) 1531–1547.
[4] E.V. Evarts, Temporal patterns of discharge of pyramidal tract neurons during sleep and waking in the monkey, J. Neurophysiol. 27 (1964) 152–171.
[5] M. Häusser, N. Spruston, G.J. Stuart, Diversity and dynamics of dendritic signaling, Science 290 (2000) 739–744.
[6] N. Hô, A. Destexhe, Synaptic background activity enhances the responsiveness of neocortical pyramidal neurons, J. Neurophysiol. 84 (2000) 1488–1496.
[7] D. Paré D, E. Shink, H. Gaudreau, A. Destexhe, E.J. Lang, Impact of spontaneous synaptic activity on the resting properties of cat neocortical neurons in vivo, J. Neurophysiol. 79 (1998) 1450–1460.
[8] M. Rudolph, A. Destexhe, A fast-conducting, stochastic integrative mode for neocortical neurons in vivo. J. Neurosci. (2003), in press.
[9] M. Steriade, Corticothalamic resonance, states of vigilance and mentation, Neuroscience 101 (2000) 243–276.
[10] G. Stuart, N. Spruston, Determinants of voltage attenuation in neocortical pyramidal neuron dendrites, J. Neurosci. 18 (1998) 3501–3510.

Available at www.ComputerScienceWeb.com
POWERED BY SCIENCE @DIRECT®

ELSEVIER

Neurocomputing 52–54 (2003) 239–245

NEUROCOMPUTING

www.elsevier.com/locate/neucom

Guiding the modeller: organizing and selecting experimental data for single cell models using the CoCoDat database

Jonas-D. Johnsen[a,b,*], Klaas Stephan[c], Jürgen Maier[d], Heiko Luhmann[d], Rolf Kötter[a,e]

[a] *C&O. Vogt Brain Research Institute, Heinrich Heine University, Düsseldorf, Germany*
[b] *Institute of Neurophysiology, Heinrich Heine University, Düsseldorf, Germany*
[c] *IME, Research Centre Jülich, Germany*
[d] *Institute of Physiology and Pathophysiology, Johannes Gutenberg University, Mainz, Germany*
[e] *Institute of Anatomy II, Heinrich Heine University, Düsseldorf, Germany*

Abstract

Collating, organizing and selecting quantitative experimental data are time-consuming tasks necessary for building and constraining biophysically realistic neuronal models. The CoCoDat (*Co*llation of *Co*rtical *Dat*a) database has been designed as an advanced environment for storing, organizing and retrieving detailed, uninterpreted quantitative data on morphology, electrophysiology and connectivity from the published literature according to neurophysiological concepts. All experimental data are linked to exact bibliographical references and detailed records of procedures used in the experiments that produced the data. We demonstrate the usefulness of CoCoDat for implementation of an example model of a layer 5 pyramidal neuron from the rat barrel cortex.

Keywords: Databasing; Microcircuitry; Quantitative experimental data; Realistic modelling

1. Data collation and selection for realistical modelling

When modelling biophysically realistic cortical microcircuits a large amount of experimental data describing cellular components and their interconnections are necessary

* Corresponding author. C&O. Vogt Brain Research Institute, Heinrich Heine University, Universitatsstr. 1 Building 22.03, level 5, room 43, Dusseldorf 40225, Germany.
E-mail address: dyhrfjel@uni-duesseldorf.de (J.-D. Johnsen).

0925-2312/03/$ - see front matter
doi:10.1016/S0925-2312(02)00756-7

to construct and constrain the models. Single neuron morphologies determining the passive membrane properties are populated with subdomain-specific distributions of ionic and synaptic conductances to constitute a model of the firing properties and functional response of the neuron. The CoCoDat database has been designed to facilitate the time-consuming tasks of collating, organizing and selecting the optimal experimental data for such modelling projects [4]. CoCoDat is implemented using the relational database approach with four major data groups under the headings *Literature*, *Experimental Methodology*, *Mapping of Recording Site* and *Experimental Findings*. All data are initially anchored in detailed bibliographical data for the publication in question. Further information is divided into records of methods and mapping, respectively, before it reunites at the level of the quantitative data records in the categories needed for modelling: *Morphology*, *FiringProperties*, *IonicCurrents*, *IonicConductances*, *SynapticCurrents* and *Connectivity*. All data in CoCoDat are entered as given in the original publication. The basic database structure is derived from the CoCoMac database of cortical connectivity in the macaque monkey [11]. CoCoDat currently contains a total of 208 quantitative experimental records in the six above-mentioned categories. These are linked to 65 recording sites from 37 publications.

2. Using CoCoDat for data selection

To illustrate the usefulness of CoCoDat in selecting data for detailed neuronal modelling, we consider an example model based on a detailed morphological reconstruction of a layer 5 regular spiking pyramidal cell from the rat barrel cortex [9]. All experimental data in CoCoDat are linked to a *RecordingSite* consisting of hierarchically organized *BrainSites* describing the location of the recording on the four levels of description: *BrainRegion*, *Layer*, *NeuronType* and *NeuronCompartment*. To extract information on the example cell these criteria are selected on the CoCoDat *SearchBoard*. An initial query of the current database contents on ionic currents from all compartments (*GM-Comp_gen*) of layer 5 (*GM-IsoCtx_L5*) pyramidal neurons (*GM-C_Pyr*) in the barrel cortex (*GM-Ctx_B*) of the rat produces no results, but when allowing for an automatic relaxation of the search-criteria by the search-routine we receive results from the closest possible match (Fig. 1).

To produce results, the original criterion for *BrainRegion* (barrel cortex) has been relaxed to its *parent* in the anatomical hierarchy, the somatosensory cortex. The four panes on the *SearchBoard* illustrate the distribution of results on the four levels of description by visualizing the hierarchies (expanded using the "+" signs on each branch-point) of *BrainSites* constituting the *RecordingSites* to which returned results are linked. Using this kind of visualization gives the user an immediate idea of the types of available results as well as the possibility to appraise the logical distance between the originally desired *RecordingSite* and the ones obtained after relaxation, thus helping the user to select the best match.

Both the automatic relaxation and general search routine utilize the hierarchical organization of *BrainSites* on the four levels of description. These hierarchies detail the anatomical parent and child(ren) of each *BrainSite* in the database, ensuring that data

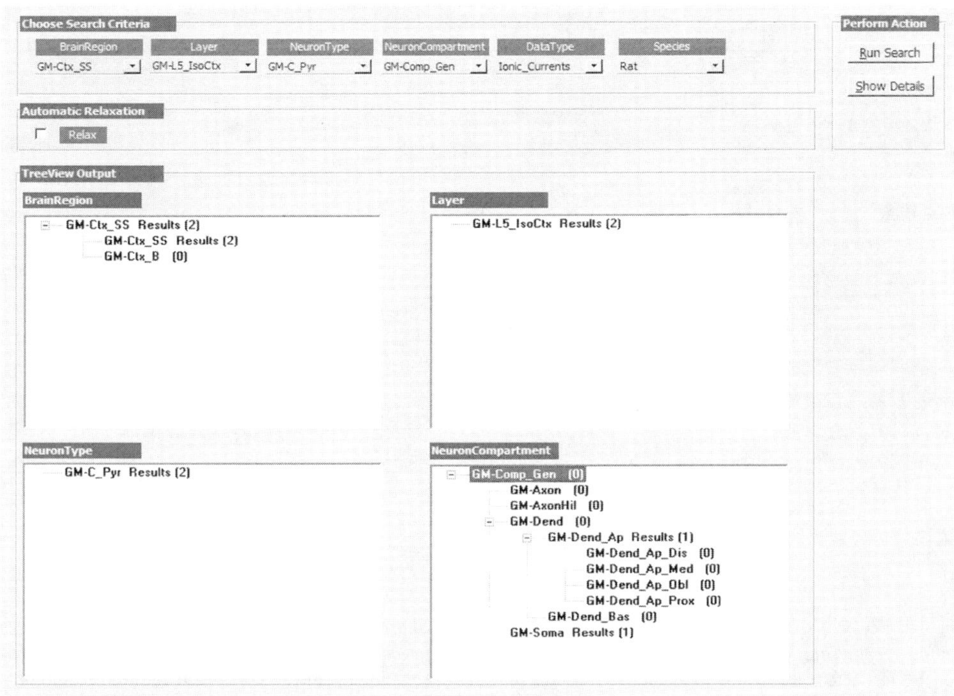

Fig. 1. The CoCoDat *SearchBoard*. The upper frame contains the search criteria (after relaxation). The four panels show the hierarchical relationships between the *BrainSites* constituting the returned *RecordingSites*. Numbers in parenthesis indicate the number of returned *RecordingSites* a *BrainSite* is part of.

linked to substructures of the initially user-specified *RecordingSite* will also be returned by the query: Specifying, e.g. dendrite as *NeuronCompartment* will also return results from basal dendrite as well as apical dendrite and subdomains thereof. When selecting automatic relaxation of one of the four levels of the description, the search routine will use the hierarchical relations to select the parental *BrainSite* in case no results are returned for the initial criteria (in the example above relaxing from barrel to somatosensory cortex). This relaxation will continue until the query produces results, or the top of the hierarchy is reached.

The detailed data can be inspected in several ways beginning with a summary detailing the full *RecordingSite* and the qualitative characteristics of the data (Fig. 2). For each *RecordingSite* the user can expand these into a detailed view containing all quantitative data, references and quotes from the published literature.

In addition to the detailed data the user can call up the detailed bibliographic and methodological information registered in the database by clicking the designated buttons on the graphical form. After inspecting the quantitative information the user can select which datasets to export to printable format by returning to the *SummaryForm* and marking these in the designated tickboxes. Similar reports can be generated for the associated literature and experimental methodology (Fig. 3).

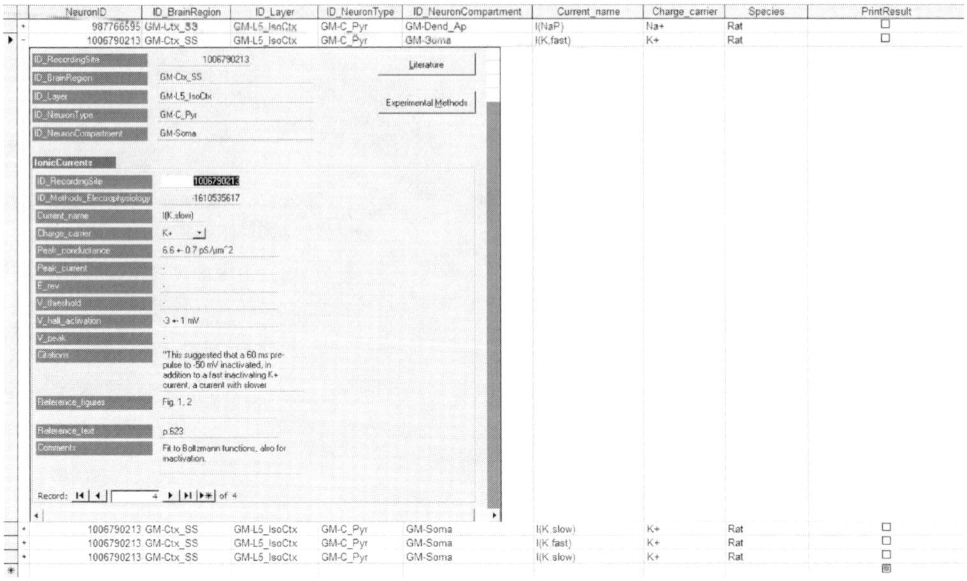

Fig. 2. The CoCoDat *SummaryForm* is opened from the *SearchBoard*. An expandable table summarizes search results and *RecordingSites*. By clicking the "+" signs a detailed view of the experimental records can be inspected. Results are selected for printable *Reports* by ticking the boxes in the rightmost column.

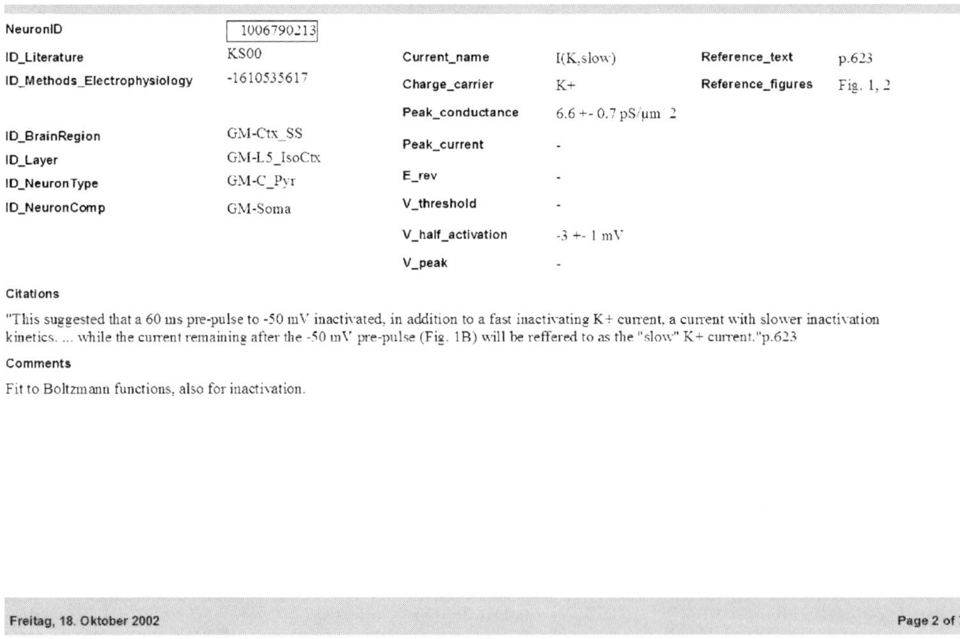
NeuronID 1006790213
ID_Literature KS00
ID_Methods_Electrophysiology -1610535617
ID_BrainRegion GM-Ctx_SS
ID_Layer GM-L5_IsoCtx
ID_NeuronType GM-C_Pyr
ID_NeuronComp GM-Soma
Current_name I(K,slow)
Charge_carrier K+
Peak_conductance 6.6 +- 0.7 pS/µm 2
Peak_current -
E_rev -
V_threshold -
V_half_activation -3 +- 1 mV
V_peak -
Reference_text p.623
Reference_figures Fig. 1, 2

Citations

"This suggested that a 60 ms pre-pulse to -50 mV inactivated, in addition to a fast inactivating K+ current, a current with slower inactivation kinetics. ... while the current remaining after the -50 mV pre-pulse (Fig. 1B) will be reffered to as the "slow" K+ current."p.623

Comments

Fit to Boltzmann functions, also for inactivation.

Freitag, 18. Oktober 2002 Page 2 of 7

Fig. 3. Reports of selected data can be automatically generated and are marked with the date of the search.

3. Implementing a first approximation model

The reconstructed morphology of a layer 5 regular spiking pyramidal neuron from the rat barrel cortex [9] was converted from NeuroLucida to GENESIS format using the software tool CVAPP [2]. To enable comparison between simulation results and experimental I–V characteristics, a general set of voltage-gated currents were sought in the CoCoDat database using records of *Experimental Methodology* describing modelling studies as an entry point. The extracted current equations were used for layer 5 pyramidal cells from visual cortex but also suited a diversity of other brain regions and cell types indicating general usefulness as a first approximation [6]. Inspection of the experimental I–V curves revealed a "sag" following hyperpolarizing current injections characteristic of the activation of an H-current. Accordingly the GENESIS implementation of a hippocampal H-current available as prototype [12] was added to the model.

The I–V characteristics of a first-step model implemented in GENESIS using the reconstructed morphology and described current equations shows a relatively good likeness to the experimental results after adjusting only somatic conductance densities in the model (Fig. 4).

Further improvement of the model necessitates the implementation of data more specific to the particular neuron modelled. Table 1 contains a summary of the results

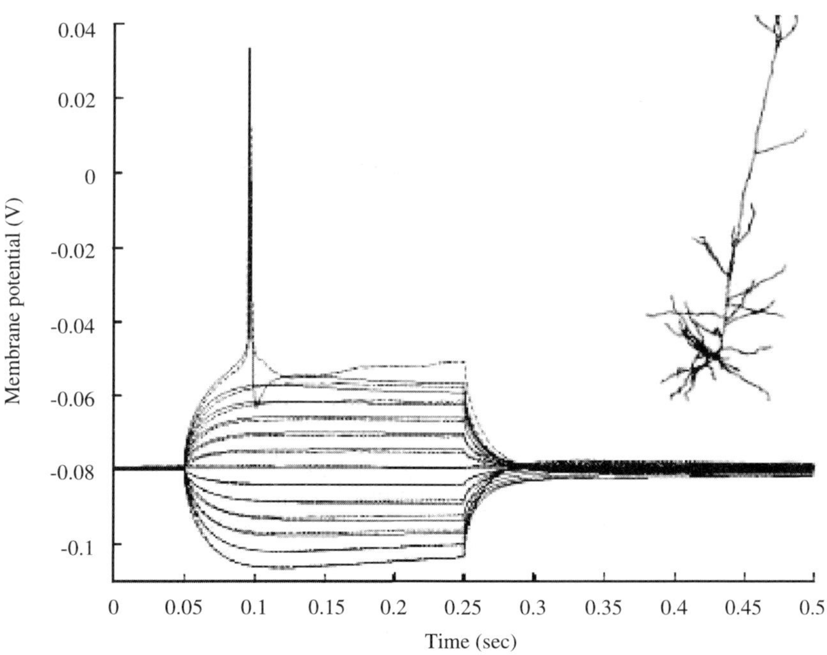

Fig. 4. Comparing the response of the model (solid line) and recorded cell (dashed line) to a series of 200 ms current injections (from −360 to 360 pA). The inset shows the reconstructed neuronal morphology.

Table 1
The experimental information extracted from CoCoDat by searching for ionic currents and conductances from layer 5 pyramidal neurons in the somatosensory cortex of rats

Recording site	Ionic currents	Ionic conductances
Somatosensory Cortex Layer 5 Pyramidal cell Soma	$I_{K,slow}$ and $I_{K,fast}$ [5]	$K_{very\ slow,\ inact.}$, $K_{slow,\ inact.}$, $K_{fast,\ inact.}$ and BK [5]
Somatosensory Cortex Layer 5 Pyramidal cell Apical dendrite	I_{NaP} [8]	K and BK [5]

returned by the earlier described query to the CoCoDat database as well as results of a similar query in the *IonicConductances* data category.

As can be seen from the tabulated results information is available on potassium currents and conductances in the neuron soma and apical dendrite [5] as well as the presence of a persistent sodium current in the apical dendrite [8]. It is likely that the implementation of the detailed information from the CoCoDat search in the neuronal model would rectify the discrepancies between experimentally obtained and simulated I–V curves shown above, as these are most evident following the action potential elicited at the strongest depolarizing current injection. This indicates that the repolarization, which is typically governed by potassium conductances, is insufficiently described in this first approximation modelling step.

4. Discussion

Several databases and data-collections with single neuron and microcircuitry data already exist, but these are either limited to a, e.g. reconstructed morphologies [1–3], tailored to a particular species [10] or provide only qualitative information and literature references [7]. CoCoDat provides a general database of microcircuitry and single cell information with direct access to quantitative data with detailed records of experimental methodology and specific references to relevant pages and figures in the original publication. The flexible tools for dynamically retrieving and visualizing datasets for neuronal structures specified on four anatomical levels of description facilitates the selection of optimal data for a given modelling project. The possibility for exporting selected datasets in printable format after inspection of the quantitative details allows for easy tracking of data available at the time of model construction.

Using the option of automatic search criteria relaxation facilitates the fast retrieval of data from neuronal structures matching the modelled object as closely as possibly, letting the user determine the least important initially selected criteria. A thorough investigation of the data space is possible by submitting one or more queries while manually selecting the parental *BrainSite* on one or more of the four levels of description after results are returned, thus giving the user an overview of closely related experimental results. By submitting a query with all four *BrainSites* set at the top-most *general* level the entire content of the database in a given *DataCategory* can be extracted for inspection.

The CoCoDat database is available for download at the URL http://www.cocomac.org as a zip-archive of an mdb file for Microsoft Access 2000. A planned future version implemented in mySQL would allow for a direct online interface as well as data submission to a central database.

Acknowledgements

This work constitutes part of a Ph.D. thesis and was supported by a Danish Research Council Fellowship to JDJ.

References

[1] G.A. Ascoli, J.L. Krichmar, S.J. Nasuto, S.L. Senft, Generation, description and storage of dendritic morphology data, Philos. Trans. R. Soc. London B 356 (2001) 1131–1145.

[2] R.C. Cannon, D.A. Turner, G. Papyali, H.V. Wheal, An on-line archive of reconstructed hippocampal neurons, J. Neurosci. Methods 84 (1998) 49–54.

[3] R.B. Gonzales, C.J. DeLeon Galvan, Y.M. Rangel, B.J. Claiborne, Distribution of thorny excrescences on CA3 pyramidal neurons in the rat hippocampus, J. Comput. Neurol. 430 (3) (2001) 357–368.

[4] J.-D. Johnsen, J. Maier, K.E. Stephan, R. Kötter, CoCoDat: collation of cortical data on neurons and microcircuitry. Systematic storage and retrieval of experimental data for biophysically realistic modeling, in: R. Kötter (Ed.), Neuroscience Databases—A Practical Guide, Kluwer Academic Publishers, Dordrecht, 2003, pp. 109–120.

[5] A. Korngreen, B. Sakmann, Voltage-gated K+ channels in layer 5 neocortical pyramidal neurons from young rats: subtypes and gradients, J. Physiol. 525 (2000) 621–639.

[6] Z.F. Mainen, T.J. Sejnowski, Influence of dendritic structure on firing pattern in model neocortical neurons, Nature 382 (1996) 363–366.

[7] J.S. Mirsky, P.M. Nadkarni, M.D. Healy, P.L. Miller, G.M. Shepherd, Database tools for integrating and searching membrane property data correlated with neuronal morphology, J. Neurosci. Methods 82 (1) (1998) 105–121.

[8] T. Mittmann, S.M. Linton, P. Schwindt, W. Crill, Evidence for persistent Na+ current in apical dendrites of rat neocortical neurons from imaging of Na+-sensitive dye, J. Neurophysiol. 78 (2) (1997) 1188–1192.

[9] D. Schubert, J.F. Staiger, N. Cho, R. Kötter, K. Zilles, H.J. Luhmann, Layer-specific intracolumnar and transcolumnar functional connectivity of layer V pyramidal cells in rat barrel cortex, J. Neurosci. 21 (10) (2001) 3580–3592.

[10] L. Stein, P. Sternberg, R. Durbin, J. Thierry-Mieg, J. Spieth, WormBase: network access to the genome and biology of Caenorhabditis elegans, Nucleic Acids Res. 29 (2001) 82–86.

[11] K.E. Stephan, L. Kamper, A. Bozkurt, G.A.P.C. Burns, M.P. Young, R. Kötter, Advanced database methodology for the collation of connectivity data on the macaque brain (CoCoMac), Philos. Trans. R. Soc. London B 356 (2001) 1159–1186.

[12] R.D. Traub, R.K.S. Wong, R. Miles, H. Michelson, A model of a CA3 hippocampal pyramidal neuron incorporating voltage-clamp data on intrinsic conductances, J. Neurophysiol. 66 (1991) 635–650.

Available at
www.ComputerScienceWeb.com
POWERED BY SCIENCE @DIRECT®

ELSEVIER

Neurocomputing 52–54 (2003) 247–254

NEUROCOMPUTING

www.elsevier.com/locate/neucom

Realtime bioelectrical data acquisition and processing from 128 channels utilizing the wavelet-transformation☆

Andre Folkers* , Florian Mösch, Thomas Malina, Ulrich G. Hofmann

Institute for Signal Processing, University of Lübeck, Ratzeburger Allee 160, 23538 Lübeck, Germany

Abstract

We propose a versatile signal processing and analysis framework for bioelectrical data, and in particular for neural recordings and EEG. Within this framework the signal is decomposed into subbands using fast wavelet transform algorithms, executed in real-time on a current digital signal processor hardware platform. The decomposition is used to perform various processing and analysis tasks. Besides fast implementation of high, band, and low pass filters, the decomposition is used for denoising and lossy, as well as lossless compression. Furthermore specific electrophysiologic analysis tasks like spike detection and sorting are performed within this decomposition scheme.

Keywords: Digital signal processor; Data acquisition; Spike detection; Wavelet transformation

1. Overview

Recording neural activity from a high number of neurons is a key issue in understanding how the brain works. Within the project VSAMUEL we developed successfully a versatile data acquisition system based on DSP boards [3]. The system is used for continuous neural data acquisiton in vivo or in vitro with a high channel count (up to 128 channels) at sampling rate $F = 50$ kHz with a precision of 16 bits per sample.

☆ This work has been supported by EU grant IST-1999-10073.

* Corresponding author.

E-mail address: folkers@isip.uni-luebeck.de (A. Folkers).

0925-2312/03/$ - see front matter © 2002 Elsevier Science B.V. All rights reserved.
doi:10.1016/S0925-2312(02)00763-4

Important online data processing tasks include filtering and spike detection and classification, but also compression, transmission and storage. We propose a signal processing framework within which these tasks can be performed in an elegant way.

2. Wavelet transform and lifting scheme

The signal is decomposed into $N+1$ subbands by a N-level wavelet transform (WT). The subbands $d_j, j=1,\ldots,N$ represent the frequency band $[F/2^{j+1}, F/2^{j+2}]$ and the subband a_N represents $[0, F/2^{N+1}]$. Fig. 1 shows the filter bank for $N=3$. In each step the signal is decomposed by applying complementary filters to a_j, i.e. a high pass $\tilde{g}$ and a low pass $\tilde{h}$, which are determined by the selected wavelet. The results of both filter operations are subsampled by a factor of two, leading to subbands d_{j+1} and a_{j+1}. Note, that the number of coefficients in d_{j+1} and a_{j+1} is equal to the number of coefficients in a_j. The wavelet transform is implemented using the lifting scheme which is faster than the standard implementation (Fig. 1). It is done in-place, and with a small modification it implements a WT that maps integers onto integers [1] while preserving the possibility of perfect reconstruction. Therefore, this implementation of the WT is well suited for realtime processing using digital signal processors (DSP).

The lifting scheme provides another point of view to the wavelet transform. Basically it consists of three stages, which are a split, a *predict*, and an *update* stage as illustrated in Fig. 2 [6]. First the signal is split such that we obtain two sequences d_j and a_j which in our case consist of sample points with odd and with even indices, respectively. Now,

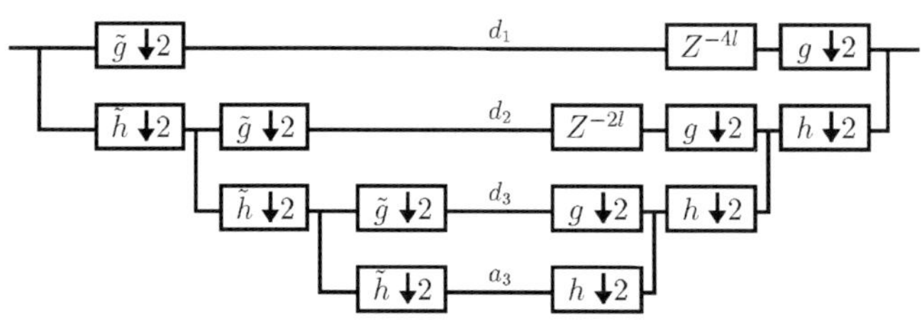

Fig. 1. Filter bank with three levels.

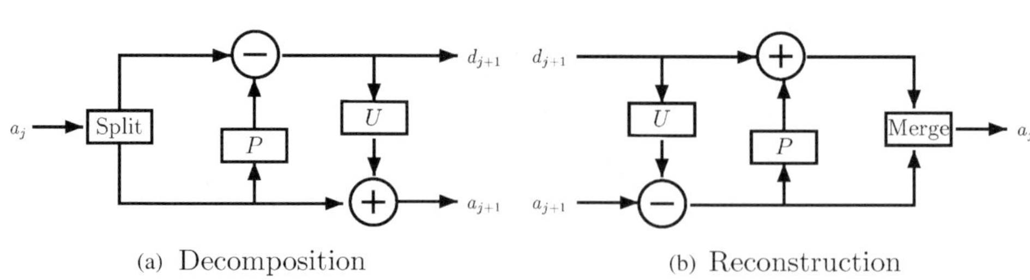

Fig. 2. Basic structure of the lifting scheme: (a) decomposition, and (b) reconstruction.

we predict the values in d_j based on a_j as $P(a_j)$. Under the assumption that the signal is continuous we have a good chance that our prediction is rather close to the actual values. We compute the difference between prediction and actual values, and keep these values which are likely to be small: $d_{j+1} = d_{j+1} - P(a_{j+1})$. In order to preserve certain properties of the original signal in the coefficients a_j, e.g. the mean value, we need the third stage, which is the update stage. Hereby, the values in a_j are modified by using an appropriate update operator on values in d_j and $U(d_j)$: $a_{j+1} = a_{j+1} + U(d_{j+1})$.

The perfect reconstruction property of the lifting scheme is obvious, because we can obtain a_j from a_{j+1} and d_{j+1} by inverting the data flow and the signs as shown in Fig. 2(b). Note, that this holds for arbitrary predict and update operators. Therefore, if these operators include a rounding to the next integer we obtain a wavelet transform that maps integer onto integers. It is possible to implement arbitrary wavelet transformations as shown in [1] by using multiple prediction and update operators successively. The respective operators are computed according to the given mother wavelet.

Using the lifting scheme and a routine optimized for our DSP, we can apply a six level Daubechies 2 decomposition filter bank on 32 channels sampled at 50 kHz in real time.

3. Filter

The decomposition allows a simple implementation of filters with different high pass, band pass, or low pass characteristics. Consider, e.g. a neural recording sampled at $F = 50$ kHz which contains both field potentials and action potentials. If it is decomposed by a 6-level WT into 7 subbands, then the field potentials are found in subband a_6. Setting the coefficients of a_6 to zero, eliminates the field potentials and corresponds to a high pass filter (Fig. 3). The respective low pass filter which eliminates the action potentials is obtained by setting the coefficients d_j for $j = 1, \ldots, 6$ to zero. A comparison of the results of both methods can be found in Fig. 3. Fig. 4 shows a comparison of the respective frequency responses. The decomposition also allows the implementation of band pass filters. Possible cut-off frequencies for band pass filter based on the WT are determined by the sampling rate and the number of levels. Arbitrary filters can be implemented if a Wavelet Packet Transform (WPT, see [8]) is utilized. However, the computation of the WPT transform involves more operations than the WT. Because the number of operations has to be as low as possible to allow real-time computation on our DSP, we currently restrict ourselves to the WT.

4. Compression and denoising

One important property of the WT is that it decorrelates the signal, i.e. the main information about the signal is collected in a few large coefficients, while the details are collected in many small coefficients. The average information content, also called

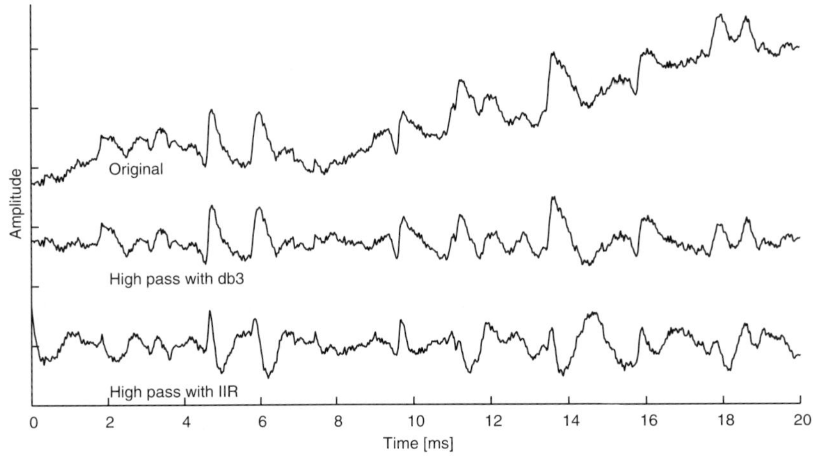

Fig. 3. High pass filtering using IIR versus elimination of wavelet approximation coefficients. The original neural recording has been decomposed with a 6 level WT using the Daubechies 4 wavelet. The coefficients of a_6 have been set to zero, corresponding to a high pass filtering with cut-off at 390.62 Hz. In a second approach the signal has been filtered by a 4-pole IIR high pass filter with cut-off at frequency 400 Hz, which was designed using the Butterworth method. Spike shapes in the IIR filtered result show significant distortions, while the spike shapes are apparently not distorted by the wavelet based high pass. Field potentials are eliminated well by both filters.

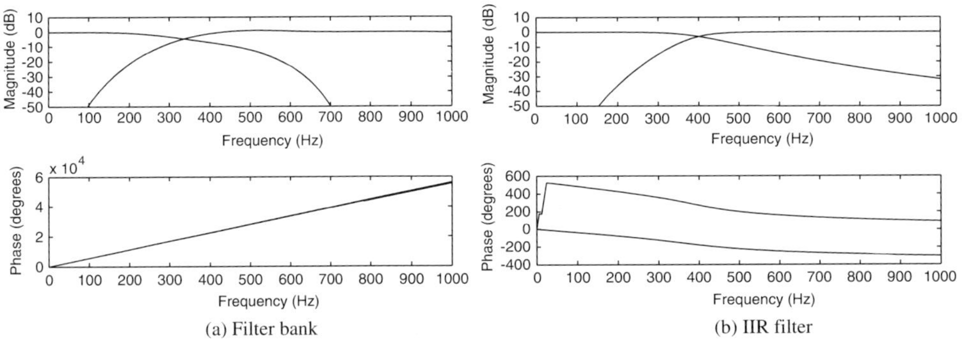

Fig. 4. Comparison of filter bank based (a) and IIR (b) high and low pass filter. For the filter bank Daubechies wavelet with four vanishing moments has been used. The IIR filter have order 4 and cut off frequency at 400 Hz. The magnitudes of the frequency responses are comparable, but the phase of the filter bank is linear while the phase of the IIR filter is nonlinear, which is the reason for the distortions found in Fig. 3.

entropy, can be computed as

$$E(s) = -\sum_{v \in \text{Values of } s} p(v)\log_2(p(v)), \tag{1}$$

where $p(v)$ is the probability of occurrence of value v within signal s. The entropy is measured in bits per sample and since our signals are sampled in 16 bit resolution,

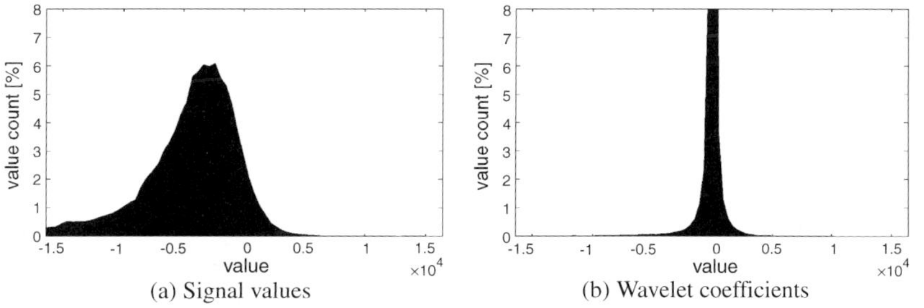

Fig. 5. Entropy reduction by wavelet transformation is illustrated by a comparison of the signal value and the wavelet coefficient histograms: (a) signal values, and (b) wavelet coefficients.

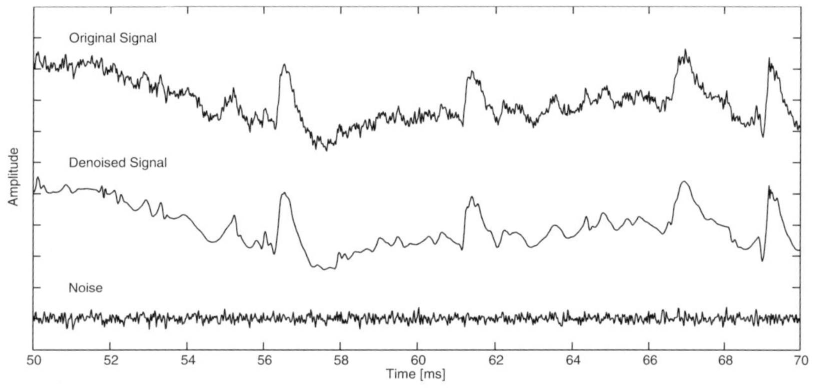

Fig. 6. Compression and denoising. The comparison of the original neural signal at the top and the denoised signal in the middle reveals no apparent distortion of the signal. This is confirmed by the difference of signal and denoised signal, i.e. the removed noise, which is shown at the bottom. The entropy of the decomposition drops from 8.34 bits per sample down to 1.04 bits per coefficient. In other words the compression rate can be improved by a factor of 8.

we can define the optimal achievable compression rate, e.g. with Huffman coding, as $E(s)/16$. The entropy of the decomposition is smaller than the entropy of the raw signal (Fig. 5). The entropy of a neural recording from a rat was quantified in [7] to be about 13.9 bits per sample and it is about 8.5 bits per coefficient for the decomposition. Therefore, the compression rate can be improved from 0.86 for the raw data down to 0.53 for the decomposition.

Another important task is the denoising of neural recordings. The typical background noise of neural recordings is mainly found in the first few levels $d_1, \ldots, d_3$ of the decomposition. Under the reasonable assumption that the background noise has a Gaussian distribution [5], a universal threshold can be found as $\delta = \sqrt{2\sigma^2 \log(n)}$ where σ^2 is the variance of the Gaussian noise and n is the length of the sequence [2]. Since the true variance σ^2 is usually unknown, it is estimated from the coefficients in d_1 which are dominated by the noise. We use the standard deviation estimator *median absolute deviation* (MAD): $\sigma^2 = \text{median}(|d_1|)/0.6745$. Using the median absolute value instead of

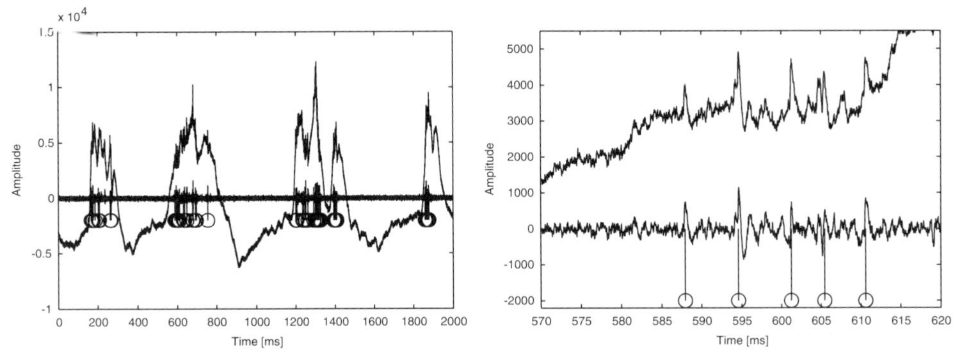

Fig. 7. Spike detection base on wavelet coefficients. Detected spikes are marked by stems. On the right an enlarged section of the signal is shown, which reveals that the signal-to-noise ratio is quite small, but still the spikes are detected well.

the mean absolute value, this estimation is robust against large coefficients representing the signal that might occur in d_1.

Compression and denoising are closely related. The thresholded decomposition which represents the denoised signal has a much lower entropy than the original decomposition and thus can be compressed with a better rate. Depending on the chosen thresholds, the compression rate can reach values below 0.1 without losing a significant part of the signal. With the universal threshold for instance we obtain a compression rate of about 0.06 for neural recordings from a rat (Fig. 6). Lossy compression which does not distort the signal significantly is particularly useful for longterm recordings.

5. Spike detection

The decomposition is used to analyse neural recordings. Spikes, for instance, are represented by a few large coefficients in subbands $d_2, \ldots, d_6$. Therefore, spike detection can be implemented by a threshold based method which uses the wavelet coefficients. In Fig. 7 such a method has been used to detect spike in a neural recording from a rat, which contains field potentials and action potentials. In [4] another method to detect spikes based on wavelet coefficients is proposed.

6. Conclusions

Altogether we can state, that our DAQ system is able to record from a high number of channels, and furthermore it can perform sophisticated processing of the incoming electrophysiological data in realtime, which in our case is a wavelet decomposition. The data obtained from the wavelet decomposition represents the original data without loss, and it provides an elegant way to compress and to denoise the signals, and also to do further processing, like, e.g. spike detection.

References

[1] I. Daubechies, W. Sweldens, Factoring wavelet transforms into lifting steps, J. Fourier Anal. Appl. 4 (3) (1998) 245–267.

[2] D.L. Donoho, I.M. Johnstone, Ideal spatial adaptation by wavelet shrinkage, Biometrika 81 (3) (1994) 425–455.

[3] A. Folkers, U.G. Hofmann, A multichannel data acquisition and analysis system based on off-the-shelf dsp boards, in: Proceedings of the EURASIP Conference on Digital Signal Processing for Multimedia Communications and Services, Budapest, September 2001.

[4] H. Nakatani, T. Watanabe, N. Hoshimiya, Detection of nerve action potentials under low signal-to-noise ratio condition, IEEE Trans. Biomed. Eng. 48 (8) (2001) 845–849.

[5] M. Sahani, Latent variable models for neural data analysis, Ph.D. Thesis, California Institute of Technology, Pasadena, California, 1999.

[6] W. Sweldens, The lifting scheme: a new philosophy in biorthogonal wavelet constructions, in: A.F. Laine, M. Unser (Eds.), Wavelet Applications in Signal and Image Processing III, Proc. SPIE 2569 (1995) 68–79.

[7] B. Weber, M. Pakura, T. Malina, K. Menne, A. Folkers, V. Metzler, U.G. Hofmann, Handling large files of multisite microelectrode recordings for the European VSAMUEL consortium. Neurocomputing 38–40(1–4) (2001) 1725–1734.

[8] M.V. Wickerhauser, in: Adapted Wavelet Analysis from Theory to Software, A K Peters, Wellesley, MA, 1994.

Andre Folkers has studied Computer Science with emphasis on Medical Computer Science at the University of Lübeck from 1994 until 2000, where he graduated with the Dipl.-Inform. degree, having performed his diploma thesis at the University of Maryland at College Park. Since 2000 he is research assistant in the Institute for Signal Processing at the Medical University of Lübeck working in the project VSAMUEL. His main research interests are advanced biosignal processing, image processing and databases, and multichannel analysis.

Florian Mösch joined the project VSAMUEL in 2001 as a student of Computer Science of the University of Lübeck. He graduated with the Dipl.-Inform. degree in 2002, having performed his diploma thesis on the development of a multichannel data acquisition system. Currently he is working in the Institute of Signal Processing at the University of Lübeck. His research fields include multichannel data acquisition and biosignal and image processing.

Thomas Malina studies Computer Science at the University of Lübeck since 1995. He was born 1974 in Teschen/Poland and joined the VSAMUEL project in 1999. His research interests are image and signal processing, in particular spike sorting, 3D-modelling and cluster analysis. He was an exchange student at the Aalborg Universitet in Denmark where he developed a data acquisition system. Currently he is working on his diploma thesis at the University of Lübeck.

Ulrich G. Hofmann received his Diploma in Physics 1993 and his Ph.D. in Biophysics 1996 from the Technical University of Munich, Germany. He stayed at the Abo Akademi, Finland, as an EU-fellow (HCM) and as Feodor-Lynen-Fellow at the California Institute of Technology. Since 1999 he is research associate ("Habilitand") at the Institute for Signal Processing of the University of Lübeck and project coordinator and initiator of VSAMUEL. His long term research aims to interface brains with computers.

Available at
www.ComputerScienceWeb.com
POWERED BY SCIENCE @DIRECT®

ELSEVIER

Neurocomputing 52–54 (2003) 255–263

NEUROCOMPUTING

www.elsevier.com/locate/neucom

A theory of neural computation

Robert L. Fry

Applied Physics Laboratory, The Johns Hopkins University, 11100 Johns Hopkins Road, Laurel, MD 20723-6099, USA

Abstract

A theory of systems is proposed that is complementary to information theory and which provides a means of understanding biological neurons. This theory, its cybernetic nature, and an engineering formalism are summarized and used to derive the architectural elements of a single-neuron system, including its operational characteristics, spatiotemporal adaptation rules, and most importantly, its computational objective. Design assumptions are tailored to obtain an architecture closely resembling that of biological neurons. The underlying theory represents a "meta-theory" in the sense that it captures information theory, systems theory, and hybrid systems such as those described by systems of single-neuron systems that internally both process and transmit information between constituent elemental "systems".

Keywords: Systems theory; Information theory; Cybernetics; Boolean Algebra; Logical questions

1. Introduction and overview

Boolean Algebra is almost universally accepted as providing the unique and internally consistent mathematical framework for the manipulation of logical expressions involving Boolean variables and literals as combined using any minimal covering set of logical operators such as {"$\sim$", "$\vee$", "$\bullet$"} or logical complementation, disjunction, and conjunction, respectively. However, one can ask the question "Why do the identities of Boolean Algebra hold and what do they mean?" Boolean Algebra cannot explain itself. At least there is no evidence of this since George Boole's original introduction of his algebraic theory [1]. The work of Cox [2–4] provides one explanation and further supports a common theory of systems that communicate information and systems that acquire and exploit it to guide action with regards to some essential purpose.

E-mail address: robert_fry@jhuapl.edu (R.L. Fry).

0925-2312/03/$ - see front matter © 2003 Elsevier Science B.V. All rights reserved.
doi:10.1016/S0925-2312(02)00765-8

The essential new idea introduced by Cox [4] is that one can postulate the existence of a logical domain of questions that lies in mutual relation to conventional Boolean Algebra and wherein the rules of Boolean Algebra also apply. Each logical domain is defined by and exists only in mutual relation to the other. Boolean logic has traditionally been viewed as static in nature. The proposition "It is raining outside!" merely represents a logical statement that by the laws of Aristotelian logic can only be true or false. However, viewed dynamically, this is an active *assertion* or proclamation by a physical *assertor* which answers some question. The physical flow of information and the conducting of actions require that every assertion has a corresponding and unique defining question, and vice versa. In this case, the assertion $r \equiv$ "It is raining outside!" implies the existence of a complementary question R that defines r. The usage of the term *complementary* in this context is different from that of simple logical complementation that for example ensures identities like $\sim\sim r = r$. We use the symbol "$\neg$" to denote this new kind of complementation operator and call it *reflection*. In this case, the defining question for the assertion r is given by the reflected question $R = \neg r$. Similarly, $r = \neg R$ defines and makes existent the assertion r. In this case, r and R are an *elementary* assertion and question, respectively. Note that r and R are not variables, but rather subjectively defined literal constants.

In general, questions and assertions are not elementary [14]. A question is more generally defined by any *system of assertions* that answer it. Thus $D \equiv$ "What side of the die will show?" is defined by the system $D \equiv \{d_1, d_2, \ldots, d_6\}$ where the assertions $d_i \equiv$ "The ith face is showing!" are elementary and comprise a system of assertions that define D. Now define the elementary questions $D_i \equiv$ "Is the ith face showing?" Then $D_i = \neg d_i$ follows. Now define the assertion d by $d = \neg D$. Then, if a system has as its *issue* the resolution of the issue D, then the same system holds as its *premise* the assertion d regarding what are possible resolutions to the issue D where $\neg D = d \equiv \{D_1, D_2, \ldots, D_6\}$. The premise d is the disjunction of all d_i while D is the conjunction of all D_i. The use of logical questions enables the manipulation of quantities corresponding to uncertainties and this is of profound theoretic and practical importance.

In a communication system, uncertainties are associated with the questions $S \equiv$ "What code will be selected and sent by the source?" and the question $R \equiv$ "What code will be decoded by the receiver given that the various source codes may have been sent?" The latter question is inclusive of the consideration of the intervening physical communication channel. Alternatively, within a systems context, the questions become $X \equiv$ "What information has been acquired?" and $Y \equiv$ "What action should be selected and conducted given the acquired information X?" The duality between channels and systems implies that the many theorems of information theory become relevant to this systems theory and vice versa [13]. In particular, it holds by way of the analogy between a communication channel that maximizes its source rate and matches it to the channel capacity and neuron system that maximizes its output decision rate and maximizes and matches its acquisition rate to its decision rate.

2. Single-neuron systems

The present theory provides a simple information-theoretic derivation and system-theoretic explanation of neural computation thereby extending previous work [5–12,15]. Before providing a mathematical overview, we first provide a qualitative and geometric perspective of the proposed computational paradigm.

One can consider the totality of binary codes received by a single-neuron "system" observed over its lifetime and which drive its operation and adaptation. The totality of presented codes or any conceivable subset can be viewed from the subjective perspective of the system as clouds of points tenuously populating an N-dimensional hypercube where N is the number of binary system inputs. The curse of dimensionality guarantees a sparsely populated space for values of N typical of biological neurons—often $>10^3$ in many cases—over any conceivable operational lifetime even including that of the universe. The adaptation process of the system [13] can be described by the construction the parameters defining an N-dimensional hyperplane having orientation λ and offset μ within the high-dimensional input space, where λ correspond to the system synaptic efficacies and μ the system decision threshold.

During adaptation, the system navigates along an information manifold defined by its $N+1$-dimensional parameter space. It does so in such a way as to arrive at local or global adaptation equilibria. At these equilibria, the system anticipates that in the future that one-half of the codes it will observe will fall on one or another of the constructed hyperplane. Those points observed by the system as falling on some specific side of the prevailing hyperplane will *probably* induce an action potential while those falling on the other side *probably* will not. We will see that it is important to consider the presence of a modulating influence that can regulate these probabilities and the rate at which the system makes errors in its output decisions—firing when it should not and not firing when it should. Once the adaptation process reaches an equilibrium, the system then realizes a condition whereby it has a maximal expected information throughput. This condition is in turn brought about through the maximization of its output decision rate and then matching this rate to its information acquisition rate.

The adaptation process serves to maximize the neuron's ability to acquire and match information derived from its input $N+1$-dimensional code space and to exploit this information to make decisions via its output code space as accurately and at as high a rate as possible. The adaptation of the parameter μ serves to maximize the decision rate while the adaptation of the parameter vector λ modifies what information is extracted by the system from the environment. The vector λ is ostensibly the measuring rod of the neuron against which it measures everything it believes can exist "outside". Alternatively, the scalar μ modulates the articulation of decisions made with regards to observed information. The system effectively learns and decides where to place a "stick in the sand" thereby deciding how it can distinguish and how it can act upon the world it subjectively postulates to exist about it [13].

We now provide a brief mathematical overview of a system-theoretic "design process" for obtaining the architecture of single-neuron system having the design objective A of maximizing its information throughput. One can graphically capture the

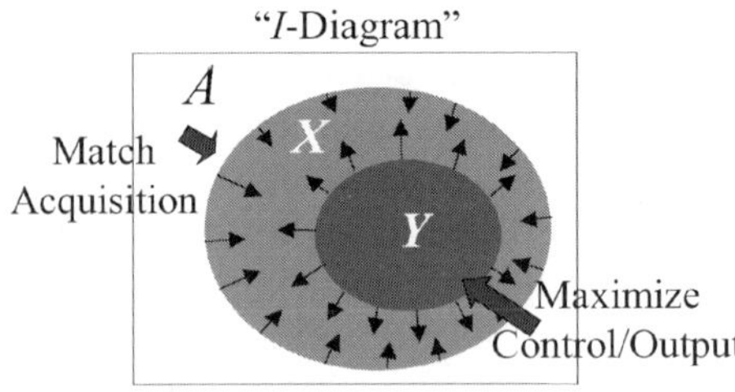

Fig. 1. I-diagram for a system attempting to match acquired information X to its decision rate Y while maximizing the rate at which it generates accurate decisions Y. Here, the logical condition $X \to Y$ holds since the region Y is contained within the region X.

computational objective of such a system through the use of an information diagram or I-diagram [20] such as shown in Fig. 1.

Suppose that the objective of the system is to reconstruct the input code X given the asserted code Y. Having done so, the information content of Y will have maximal bearing on the answering of the question X and vice versa. That is, answering the question $Y \equiv$ "What code should be asserted?" should ideally answer the questions $X \equiv$ "What code was observed?" Logically, we desire that $Y \to X$ or that the information content of the output Y contains the uncertainty of what code X was observed. That is, one would like Y to contain the uncertainty of X just as in a communication system one would like to have the question of what code was received R include the uncertainty of the question S regarding what source code was sent, or $R \to S$. Since this is architecturally impossible for the subject many-to-one single-neuron system, we then desire that the logical condition $Y \to X$ hold to the *degree* possible where *degree* means that there must exist a quantitative measure of the degree to which the logical condition holds.

Regarding Fig. 1, the objective A is achievable through two synergistic adaptation processes. The first process attempts to make the area of Y as large as possible in regards to covering A. This in effect corresponds to the maximizing of the decision rate of the system. The second process attempts to make the area X maximally conformal to Y. The arrows shown in the I-diagram in Fig. 1 depict these two adaptation processes that together serve to maximize the *degree* to which $Y \to X$. In maximizing the area Y, the system is maximizing its decision rate Y relative to the issue A which is the system decision entropy $H(Y)$. Through a conformal mapping of X to Y, the system maximizes the relevancy of the information extracted from X to Y relative to the computational issue A. This is the common information $X \vee Y$ between input X and output Y and corresponds to the mutual information $I(X;Y)$ [20]. Maximizing $I(X;Y)$ serves to guarantee that the acquired information X is matched to the need for information induced by Y.

Suppose that the single-neuron system possesses N Boolean inputs $X_1, X_2, \ldots, X_N$ with each input X_i representing a question posed by the system to its environment and answered through the existence or nonexistence of externally furnished pre-synaptic potentials. Each X_i is a "binary" question which is a question defined by a system of two complementary assertions. In this case let $X_i \equiv \{x_i, \sim x_i\}$. The defining assertions literally correspond to $x_i \equiv$ "A potential is asserted on the ith synaptic input!" and $\sim x_i \equiv$ "No potential is asserted on the ith synaptic input!" while $X_i \equiv$ "Is there or is there not a potential present?" We have excluded the additional qualifier "at a specific time" on each literal definition to keep the example simple for now. "Time" is explicitly required and is discussed briefly later where this qualifier is needed. The question of what code was observed X is determined by all questions X_i. Specifically, it is the information derived by posing in parallel to the environment all questions X_i, $i = 1, 2, \ldots, N$. This mathematically [3,4] means that the system poses the question X corresponding to the conjunction of all X_i or $X = X_1 \bullet X_2 \bullet \cdots \bullet X_N$. The question X is then defined by the system of 2^N assertions describing every possible code that can in principle be observed by the system. That is, $X \equiv \{\sim x_1 \bullet \sim x_2 \bullet \cdots \bullet \sim x_N, \ldots, x_1 \bullet x_2 \bullet \cdots \bullet \sim x_N\}$. Thus there are 2^N possible defining assertions for the question X. Since neurons can have in excess of 5000 inputs, there can be a unimaginable number of assertions ($> 2^{5000}$) in the system that defines X. Any neuron can expect to observe only an infinitesimal fraction of these possible codes over its lifetime. The question X defines the *inquiry space* of the system [13] and represents what it can ask. The inquiry space is the analog of the receiver element in a communication system.

As one might suspect, the system also has an *action space* that is analogous to the source element of a communication system. The action space corresponds to those questions that the system can answer through the selection of specific actions from a suite of possible actions. For the subject single-neuron system, the action space is extremely simple and defined by $Y \equiv \{y, \sim y\}$ where $y \equiv$ "Generate an action potential!" and $\sim y \equiv$ "Do not generate an action potential!" The question Y represents the system's query regarding what to do. The *cybernetic space* of the system is defined by the conjunction of its inquiry and action spaces as elaborated upon in [13] and so $A = X \bullet Y$ which for this system is defined by a system of assertions having exactly twice as many assertions as does the defining system for X since each assertion defining X is conjoined with both y and $\sim y$ in accordance with the definition of the conjunction of questions [3,4].

The final "design step" consists of the specification of the computational resources possessed by the system relative to its ability to extract and record information useful in guiding its two adaptation processes. It is assumed [15] that the single-neuron system cannot record or directly use previously observed input or output codes and that is a discrete memoryless system analogous to a discrete memoryless channel in information theory. However, it is assumed that the system can measure and extract certain things from its input and output codes and that it can retain certain memories of what it has seen in the past and what it did based on what it saw. In particular, we postulate that the system can measure the numerical averages $\langle x_i \bullet y \rangle$ of observed coincidences (conjunctions) of input potentials x_i with its self-generated potential decision events y. We further postulate that the system can measure the fraction of the times that it generated action potentials $\langle y \rangle$ in the past taken over the opportunities that it had to do so. Thus, it is postulated that the system is capable of measuring and recording $N + 1$ quantities that are somehow related to its $N + 1$ adjustable parameters. It is important to note that as the single-neuron system designer, one is free to postulate that the system can measure any function(s) of its inputs and its outputs. Assumptions are tailored here to yield a biologically plausible architecture.

The step of mathematically deriving the single-neuron system architecture requires the application of inductive principles that are both universal and which abide by logical consistency [3]. By universal, it is meant that the principles must apply to all systems. By logical consistency, it is meant that the system must attain a common final state of knowledge regardless of the orders of introduction of information to it as long as the information is provided in its totality. The resulting information processing rules are unique and include Bayes' Theorem and the principal of Maximized Entropy [18,19]. Given that the system measures $\langle x_i \bullet y \rangle$ and $\langle y \rangle$, the application of the principal of maximized entropy leads to a precise parametric form for the joint distribution $p(\boldsymbol{x}, y|a)$ on input codes $\boldsymbol{x}$ and output codes y where a is the premise capturing all system assumptions made thus far. In particular, the coefficients λ_i are parameters of $p(\boldsymbol{x}, y|a)$ and correspond to the synaptic efficacies λ and the Lagrange multipliers for the moment constraints $\langle x_i \bullet y \rangle$. Likewise, μ is another parameter of $p(\boldsymbol{x}, y|a)$ which also serves as the system decision threshold and the Lagrange multiplier for the moment constraint $\langle y \rangle$. The distribution $p(\boldsymbol{x}, y|a) = 1/Z \exp[-\lambda^{\mathrm{T}}\boldsymbol{x}y + \mu y]$ represents the model constructed by the system of its external world regarding what it will see and what it should do given what it has seen.

Again, the computational issue A of the system is to maximize its information throughput as realized through the dual adaptation processes prescribed by the I-diagram in Fig. 1. We thus desire that the decision entropy rate $H(Y)$ be maximized and that the acquisition rate $I(X;Y)$ (mutual information or the information shared between X and Y) be matched to this decision rate. Since the maximal amount of information that X can have in common with Y is the decision entropy rate $H(Y)$ (since $H(Y) = I(X;Y) + H(Y|X)$, it follows that $H(Y) \geqslant I(X;Y)$ since all measures are positive), then mutual information between the acquired information and possible controls can be independently maximized subject to any additionally imposed system constraints. The mutual information $I(X;Y)$ specifies the system information acquisition rate of the system, viz. the transmission rate of a channel.

Maximizing the output entropy [15] leads to the equilibrium condition for the decision threshold μ that $\mu = \langle \lambda^{\mathrm{T}}\boldsymbol{x} | y = 1 \rangle$. The quantity $\zeta(\boldsymbol{x}) = \lambda^{\mathrm{T}}\boldsymbol{x} - \mu$ is important to system operation and adaptation and represents the effective-induced somatic potential useful to the system used as a basis for the decision of whether or not to fire given observed codes $\boldsymbol{x}$. $\zeta(\boldsymbol{x})$ is formally the statistical evidence for making the decision to fire or not. In particular, it directly parameterizes the conditional probability of firing $p(y = 1|a \bullet \boldsymbol{x})$ given in Eq. (1) and which is directly obtainable from $p(\boldsymbol{x}, y|a)$:

$$p(y = 1|a \bullet \boldsymbol{x}) = \frac{1}{1 + \exp[-\zeta(\boldsymbol{x})]}. \tag{1}$$

Therefore, the evidence $\zeta(\boldsymbol{x})$ is derived solely as a measurement statistic on its input code space X and $p(y = 1|a \bullet \boldsymbol{x})$ is a sigmoidal function of the rendered evidence. Eq. (1) describes the operation of the system while $\mu = \langle \lambda^{\mathrm{T}}\boldsymbol{x} | y = 1 \rangle$ describes how the system should adapt its decision threshold indicating that μ should evolve towards the average value of the induced potential $\lambda^{\mathrm{T}}\boldsymbol{x}$ induced by input codes and as conditioned on the

[11] R.L. Fry, Cybernetic aspects of neural computation, Poster Presentation at the Sixth International Conference on Cognitive and Neural Systems, Center for Adaptive Systems and Department of Cognitive and Neural Systems, Boston University, 2001 (available from the author).
[12] R.L. Fry, Cybernetic systems based on inductive logic, in: A. Mohammad-Djafari (Ed.), Proceedings of the 20th Workshop on Maximum Entropy and Bayesian Methods, AIP, New York, 2001.
[13] R.L. Fry, The engineering of cybernetic systems, in: Proceedings of the 21st Workshop on Maximum Entropy and Bayesian Methods, AIP, New York, 2002, pp. 497–528.
[14] R.L. Fry, Course Notes for "Maximum Entropy and Bayesian Methods", Johns Hopkins University, Baltimore, MD, 1997 (available from the author).
[15] R.L. Fry, R.M. Sova, A logical basis for neural network design, in: Techniques and Applications of Artificial Neural Networks, Vol. 3, Academic Press, New York, 1998.
[16] D.O. Hebb, Organization of Behavior, Wiley, New York, 1949.
[17] E. Oja, A simplified neuron model as a principal component analyzer, J. Math. Biol. 15 (1982) 267–273.
[18] J.E. Shore, R.W. Johnson, Axiomatic derivation of the principle of maximum entropy and the principle of minimum-cross entropy, IEEE Trans. Inform. Theory 26 (1980) 26–37.
[19] J. Skilling, The axioms of maximum entropy, in: G.J. Erickson, C.R. Smith (Eds.), Maximum Entropy and Bayesian Methods in Science and Engineering, Kluwer, Dordrecht, 1988.
[20] R.W. Yeung, A new outlook on Shannon's information measures, IEEE Trans. Inform. Theory 37 (1991) 466–475.

Available at
www.ComputerScienceWeb.com
POWERED BY SCIENCE @ DIRECT®

ELSEVIER

Neurocomputing 52–54 (2003) 265–270

NEUROCOMPUTING

www.elsevier.com/locate/neucom

A novel approach to training neurons with biological plausibility

Bernadette Garner

CSSE, Monash University, Wellington Road, Clayton, Vic. 3168, Australia

Abstract

The dominant approach to training artificial neurons is to search iteratively for single numeric values that satisfy the training conditions based on error. A novel method of fully training neurons that will generalize has been developed that does not find single numeric values for weights, but instead finds regions in the weight-space that satisfy the training conditions. This method provides an alternative to the biologically implausible backpropagation for training feedforward neural networks, as it allows the McCulloch-Pitt neuron model to be used as a basis for a fast training algorithm that dynamically allocates neurons into the network.

Keywords: Biological neuron training; Constraints; SOM

1. Introduction

Experiments have shown that neurons use the step function as the activation function [1]. In other words, neurons are either activated or not. Despite recent speculation that the biological neuron is not a step function, no experimentation has demonstrated that this is the case [3,11].

When Rosenblatt [8] started training artificial neurons, he trained the neuron's weight values as single numeric values that attempt to satisfy the training conditions by iteratively modifying the weight values to eliminate error between the desired and the actual output of the neuron. This has remained the dominant method of training neurons.

Because there was no method of training neurons to learn more complex data sets than those that are linearly separable, the perceptron had to be modified to use the

E-mail address: bmg@csse.monash.edu.au (B. Garner).

0925-2312/03/$ - see front matter © 2002 Elsevier Science B.V. All rights reserved.
doi:10.1016/S0925-2312(02)00766-X

sigmoid function, hence data values that are close to boundary conditions may not accurately cause the network to fire as required.

This method of training neurons learns by finding relationships between the neuron's incoming connection weights and the neuron's threshold to produce the required output for the input the neuron is to classify. The algorithm learns by modifying the connection weights by ranges of response, which can be likened to quantization.

The algorithm being discussed in this paper uses the step function, and adds neurons into the network as required thus building its own architecture. When a single neuron cannot learn a particular input, one or more additional neurons are added to the ANN to enable the network to learn the input. This is similar to the biological brain, which can add connections between neurons in a number of ways [9]. Also this training algorithm learns in a single pass.

In this paper, Section 2 defines the neuron to be used and describes how neurons are trained with the method proposed in this paper. Section 3 demonstrates that the neuron is fully trained, in other words that the data the neuron learnt can be recovered and generalize; and Section 5 shows that the neuron can generalize. Finally, the conclusions are in Section 5.

2. New method of training neurons

This neuron being used by this algorithm is defined as the McCulloch-Pitt neuron [1], where $\mathbf{x}$ is the input vector i, n the number of inputs into the neuron, and $\mathbf{w}$ is the weight vector associated with the incoming weights of the neuron. T is the threshold of the neuron and O is the neuron's output.

$$\text{net}_n = \mathbf{x}_i \cdot \mathbf{w},$$

where $\mathbf{x} \in \{0,1\}^n$ and $\mathbf{w} \in \mathbf{R}^n$.

$$O = \begin{cases} 1, & \text{if } \text{net}_n \geqslant T, \\ 0, & \text{otherwise.} \end{cases} \tag{1}$$

There are a number of input connections into a neuron and each weight connection can be considered a dimension in the weight-space.

It has been established, through the use of sensitivity analyses, that there are many values the weights can take for each input connection into the neuron that solve the learning problem that any neural network may learn [8]. Traditional neuron training methods only find one of these weight values.

So then the region in the weight space that will allow the neuron to fire can be considered a volume. This is called the activation volume.

Relationships can be constructed based on Eq. (1) and the input vector as constraints. The constraints define relationships between the weights associated with each neuron's input connection weights and the neuron's threshold. If the neuron is to be 0 then the weighted summed input is $<$ than the neurons threshold, then the neuron will not fire, and if the output is to be 1 the weighted sum will be $\geqslant T$ and the neuron fires. This

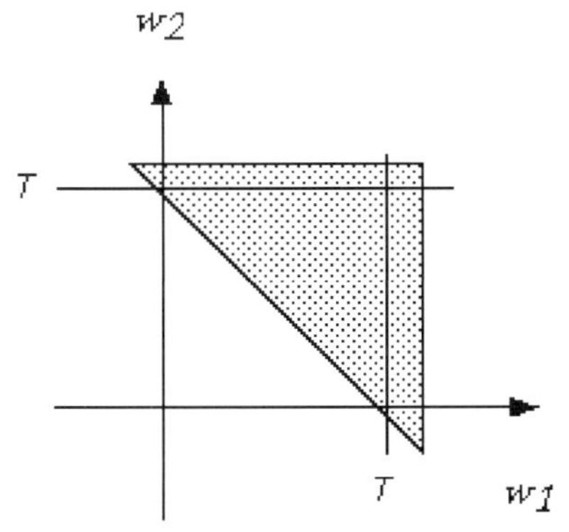

Fig. 1. The weight-space for a 2-input neuron that has learnt $\{W_1 + W_2 \geqslant T, 0 < T\}$.

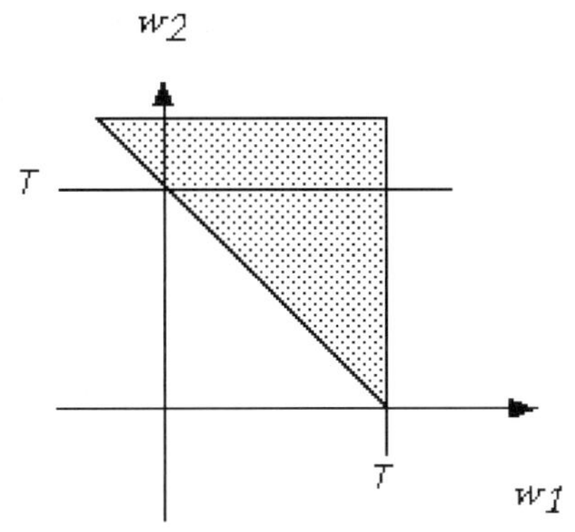

Fig. 2. The weight-space for a 2-input neuron that has learnt $\{W_1 + W_2 \geqslant T, 0 < T, W_1 < T\}$.

summarized in the following relationships:

$$\mathbf{x}_i \cdot \mathbf{w} \geqslant T \rightarrow 1$$

or

$$\mathbf{x}_i \cdot \mathbf{w} < T \rightarrow 0,$$

where i is the pattern that the neuron is currently learning.

What the neuron can learn is based on whether adding the constraint satisfies all the previously learnt constraints. Each time a constraint is applied to the weight space of a neuron the volume that causes the neuron to be activate is reduced. Hence, if a neuron learns $\mathbf{x}_i \cdot \mathbf{w} < T$, it eliminates the possibility the neuron activates when $\mathbf{x}_i \cdot \mathbf{w} \geqslant T$ and vice versa.

In an example of a 2 input neuron, if the input vector is [1 1] and this causes the neuron to be activated, then the constraint produced is $w_1 + w_2 \geqslant T$. If the neuron is to learn logical AND, then the constraints produced are $\{w_1 + w_2 \geqslant T, 0 < T\}$.

A diagram of the weight space of this is shown in Fig. 1 of what the neuron has learnt.

The neuron may then learn to classify more patterns. If the neuron has to learn [1 0] produces an output of 0, then the weight space in Fig. 2 is produced.

The neuron can still learn that either the input vector [0 1] causes the neuron to be activated or not. If it fires then the weight-space will remain unchanged but if it learns that the neuron does not fire then the weight space will be reduced to $\{w_1 + w_2 \geqslant T, 0 < T, w_1 < T, w_2 < T\}$. In this case the neuron has learnt to only be activated if both input 1 AND input 2 are present. If the neuron learns input vector [0 1] causes the neuron to be activated, then the neuron has learnt that only input 2 is required to activate it.

This algorithm learns data in a single pass, as it does not use iteration. The algorithm forms prepositional logic functions between the neuron's input connections, which is comparable to biological neurons which exhibit the property of forming different types of inhibitory and excitatory neurons [5,6,9,12].

3. Full training and generalization

Using this training method for neurons, it is possible to recover what the neurons have learnt. For instance if a 2-input neuron has learnt the following constraints $\{w_1 + w_2 < T, 0 < T, w_1 \geqslant T\}$, and it receives input vector [1 1], using Eq. (1), the output can be deduced $1 \cdot w_1 + 1 \cdot w_2 = w_1 + w_2 < T$ therefore the neuron will output 1. Hence from this example, it is seen that the network is fully trained and the network can output what it has been previously taught.

Generalization is one of the most interesting properties of neurons as it allows neurons to output responses to data is has not been trained with. Beale et al. [1] define generalization as the property that allows neurons to handle noisy input.

If a neuron is trained with input vectors with associated output

$$[1\ \ 1] \to 0,$$

$$[0\ \ 0] \to 0,$$

$$[1\ \ 0] \to 1.$$

Then the following constraint will be produced $\{w_1 + w_2 < T, 0 < T, w_1 \geqslant T\}$. During training the input vector [0 1] was not seen, but it is meant to output 0. For proper generalization, we need to be able to determine what the output of the neuron will be when [0 1] is applied to the input. When the input [0 1] is applied to the neuron, it is found we want to find what relationship $0 \cdot w_1 + 1 \cdot w_2 = w_2$ has to the threshold T.

It can be deduced from the constraints that since $0 < T$ and $w_1 \geqslant T$ that $0 < T \leqslant w_1$ however $w_1 + w_2 < T$, therefore $w_2 < T - w_1$, so $w_2 < T$, which causes the neuron not to be activated.

Hence the neuron is fully trained and can be generalized when trained with this method.

4. Training a network of neurons

Whether a neuron can learn a data pattern forms a natural criterion for adding a neuron to the network. Hence the algorithm will dynamically allocate neurons into the network only as required to learn features in the data set [7].

Because the network can learn input vectors immediately, the network can learn in a single pass. However, since it is a single pass training algorithm it requires input vectors to be sorted for it, as the rules learnt by the network depends on the order that data is presented.

The brain exhibits such a sorting in the form of brain map features on the surface of the cortex [10]. These features can be sorted artificially using SOMs [1].

Using this method of training neurons does not experience the local minima problem that artificial neural networks (ANN) exhibit since it does not feed error back or rely on gradient descent. It also produces relationships between the weights and thresholds that can be easily interpreted and adds neurons as required.

This training algorithm uses binary data, as the biological brain does. For ANN floating point data is converted to binary. The training algorithm has been implemented in prolog and provided exciting results for feedforward problems considered difficult for networks trained with backpropagation. Data sets used include the twin spiral data set [4], data sets from the UCI Repository of Machine learning databases [2] and a number of binary data set problems.

5. Conclusions

This method of training neurons explains how networks of binary data can be learnt by the simple McCulloch-Pitt neuron model and leads to a simple method of training networks. The method entails cutting planes through the neuron's weight space and then removing the regions that from the volume which cause the neuron to be activated. To learn the input vectors it converts them to constraints which allows this method to be implemented on computers, but it is assumed that planes are constructed by modifying the relationships between the neuron's weights in biological weights. This is beyond the scope of the paper. This algorithm could possibly explain why brain maps exist on the surface of the cortex.

Acknowledgements

Thanks are due to Prof. Bala Srinivasan, Monash University, for many useful comments on this text.

References

[1] R. Beale, T. Jackson, Neural Computing, an Introduction, IOP, Amsterdam, 1992.

[2] C.L. Blake, C.J. Merz, UCI Repository of Machine Learning Databases, http://www.ics.uci.edu/~mlearn/MLRepository.html, Department of Information and Computer Science, University of California, Irvine, CA, 1998.

[3] Carandini, Ferster, Membrane potential and firing rate in Cat Vi, J. Neurosci. 20(1) (2000) 470–484.

[4] S. Fahlman, C. Lebiere, The cascade-correlation architecture, Technical Report CMU-CS-90-100, Carnegie Mellon University, 1991.

[5] M.M. Francis, K.I. Choi, B.A. Horenstein, R.L. Papke, Sensitivity to voltage-independent inhibition determined by pore-lining region of the acetylcholine receptor, Biophys. J. 74 (5) (1998) 2306–2317.

[6] J.A. Freeman, D.M. Skapura, Neural Networks Algorithms, Applications, and Programming Techniques, Addison-Wesley, Clear Lake, 1992.

[7] B.M. Garner, A New Training Algorithm for Feedforward Neural Networks, in: accepted IC-AI'02, 2002.

[8] J. Hertz, A. Krough, R.G. Palmer, Introduction to the Theory of Neural Computation, Addison-Wesley, Redwood City, 1991.

[9] K. McDonald, Pictures reveal how nerve cells form connections to store short- and long-term memories in brain http://ucsdnews.ucsd.edu/newsrel/science/mccell.htm, 2001.
[10] K. Obermayer, T. Sejnowski, G.G. Blasdel, Neural pattern formation via a competitive Hebbian mechanism, Behav. Brain Res. 66 (1995) 161–167.
[11] B. Perks, Synaptic "kiss-and-run" could explain speed of neurotransmission http://news.bmn.com/news/story?day=011012story=1, 2001.
[12] J. Szentagothai, On general theories of brain organization and brain function, Naturwissenschaften 72 (6) (1985) 303–309.

Bernadette Garner is a Ph.D. candidate at Monash University, Australia, in the field of neural networks. She worked in the telecommunications industry and ran a small research group in a company she worked for.

Available at
www.ComputerScienceWeb.com
POWERED BY SCIENCE DIRECT
ELSEVIER
Neurocomputing 52–54 (2003) 271–276
NEUROCOMPUTING
www.elsevier.com/locate/neucom

A model of inositol 1,4,5-trisphosphate and calcium dynamics in single cells following metabotropic receptor activation

Greg Lemon, William G. Gibson*, Max R. Bennett

The Neurobiology Laboratory, Department of Physiology, School of Mathematics and Statistics, The University of Sydney, New South Wales 2006, Australia

Abstract

The application of external agonist to purinergic metabotropic receptors on single cells leads to transient inositol 1,4,5-trisphosphate (IP_3) and calcium (Ca^{2+}) responses. A mathematical model is formulated for the processes governing these responses, including the regulation of the receptor activity, the G-protein cascade and the cytosolic Ca^{2+} and IP_3 dynamics. This model is used to reproduce experimental observations on the extent of desensitization and sequestration of the P_2Y_2 receptor following its activation by uridine triphosphate and also the IP_3 and Ca^{2+} responses following this activation.

Keywords: Calcium; Inositol 1,4; 5-trisphosphate; Metabotropic receptor; Endoplasmic reticulum; G-protein cascade

1. Introduction

Ca^{2+} is a fundamental and universal signalling agent in living organisms, transmitting information both inside cells and between them. It is literally present at conception, when a wave of Ca^{2+} sweeps across the fertilized egg and initiates cell division; at the other extreme, excess Ca^{2+} released inside a cell can lead to its death (apoptosis). In between, it regulates many cellular processes, including secretion, contraction, proliferation, neural signalling and learning and even gene transcription [1]. Cells have internal

* Corresponding author. Tel.: +61-2-93513376; fax: +61-2-93514534.
E-mail address: billg@maths.usyd.edu.au (W.G. Gibson).

0925-2312/03/$ - see front matter © 2002 Elsevier Science B.V. All rights reserved.
doi:10.1016/S0925-2312(02)00768-3

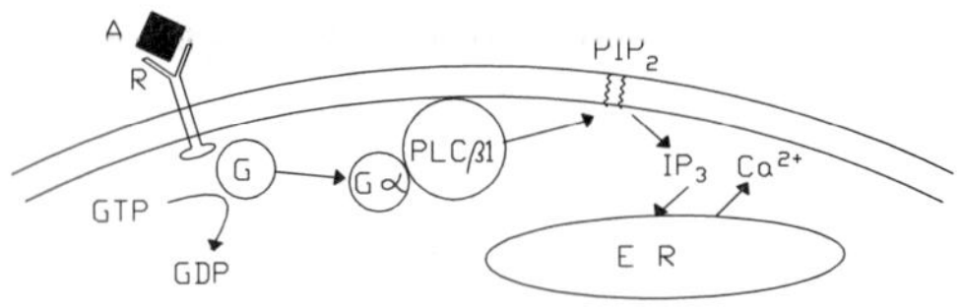

Fig. 1. Schematic diagram of the principal steps leading to Ca^{2+} release from internal stores in a cell. A transmitter molecule (A) binds to a receptor (R) in the cell membrane and thereby activates a G-protein (G) inside the cell. This sets off a cascade of events that includes the activation of the phospholipase PLCβ1 and the hydrolysis of the membrane phospholipid PIP_2, giving the second-messenger molecule IP_3. This latter molecule diffuses into the cytosol and opens Ca^{2+} channels in the endoplasmic reticulum (ER) leading to the release of Ca^{2+} into the cytosol.

stores containing a high concentration of Ca^{2+} and an important signalling mechanism involves the activated release of Ca^{2+} from these stores. This release can be initiated by the binding of an agonist to a G-protein coupled receptor, leading to a chain of events that results in the release of Ca^{2+} from the endoplasmic reticulum (Fig. 1).

We present a unified model for these processes, starting from the binding of agonist to metabotropic receptors and leading, via a G-protein cascade, to the production of IP_3 and the release of Ca^{2+} from the ER. The modelling falls into three linked modules: the first concerns the agonist–receptor binding and involves consideration of desensitization through phosphorylation and internalization (sequestration) of the receptors; the second concerns the G-protein cascade, leading to the production of IP_3 and the final section treats the IP_3-induced Ca^{2+} release from the ER.

2. Methods

The first stage involves modelling the interaction of agonist A with the receptors, R. This reversible reaction leads to the production of bound receptor AR which can then be phosphorylated to AR_P and this reversibly dissociates into A and R_P:

$$A + R \underset{k_1^-}{\overset{k_1^+}{\rightleftarrows}} AR \overset{k_p}{\rightarrow} AR_P \underset{k_2^-}{\overset{k_2^+}{\rightleftarrows}} A + R_P. \tag{1}$$

The other process involves the internalization (at rate k_e) of the phosphorylated bound receptor AR_P and its subsequent recycling (at rate k_r) to the surface as unbound receptor R. This scheme is a subset of the full cubic ternary model [6] and can be formulated as a set of differential equations involving the appropriate rate constants. The steps are described by differential equations; for example, the equation for the rate of production of bound receptor is

$$d[AR]/dt = k_1^+[A][R] - k_1^-[AR] - k_p[AR]. \tag{2}$$

Let $[R^S] = [R] + [AR]$ be the total number of unphosphorylated receptors and $[R_P^S] = [R_P] + [LR_P]$ be the total number of phosphorylated receptors. Then the above scheme reduces to

$$\frac{d[R^S]}{dt} = k_r[R_T] - \left(k_r + \frac{k_p[A]}{K_1 + [A]}\right)[R^S] - k_r[R_P^S], \tag{3}$$

$$\frac{d[R_P^S]}{dt} = [A]\left(\frac{k_p[R^S]}{K_1 + [A]} - \frac{k_e[R_P^S]}{K_2 + [A]}\right), \tag{4}$$

where $[R_T]$ is the total number of receptors, $K_1 = k_1^-/k_1^+$ and $K_2 = k_2^-/k_2^+$. In making this reduction, it is assumed that agonist binding is fast relative to the other processes and use has been made of a rapid binding approximation [5].

The next stage is the interaction of the bound receptor AR with the G-protein G·GDP leading to the replacement of GDP with GTP and its subsequent dissociation into the subunits $G_\alpha \cdot$GTP and $G_{\beta\gamma} \cdot G_\alpha \cdot$GTP then binds to a site on PLC and this activated unit binds to PIP_2 to form the complex $G_\alpha \cdot GTP \cdot PLC \cdot PIP_2$. The final step in this stage is the binding of Ca^{2+} to a catalytic site on this complex leading to the hydrolysis of PIP_2 and the production of IP_3. Schematically,

$$\begin{array}{l} AR \cdot G \rightleftarrows G \cdot PLC \cdot PIP_2 \rightarrow IP_3. \\ \qquad\qquad \uparrow\downarrow \\ \qquad\qquad Ca^{2+} \end{array} \tag{5}$$

A simplified model for the G-protein cascade is used, where the activation rate of G-protein is proportional to both the concentration of active receptor and inactive G-protein (G · GDP) and the deactivation rate is proportional to the concentration of active G-protein ($G\alpha \cdot$ GTP). Whereas the agonist bound receptor (AR) most strongly activates PLC, there is the possibility that the unbound receptor (R) may also contribute to IP_3 production, albeit at a lower rate, and this is taken to account for the basal concentration of IP_3. The resulting equation for the concentration of $G\alpha \cdot$ GTP, denoted by [G], is

$$\frac{d[G]}{dt} = k_a(\delta + \rho_r)([G_T] - [G]) - k_d[G], \tag{6}$$

where k_a and k_d are the G-protein activation and deactivation rate parameters, $[G_T]$ is the total number of G-protein molecules, δ is the ratio of the activities of the agonist unbound and bound receptor species and ρ_r is the ratio of the number of agonist bound receptors to the total number of receptors, $\rho_r = [AR]/([R_T])$

The rate of hydrolysis of PIP_2, assuming rapid kinetics for the binding of Ca^{2+}, is $r_h[PIP_2]$ where $[PIP_2]$ is the number of PIP_2 molecules and the rate coefficient is

$$r_h = \alpha\left(\frac{[Ca^{2+}]}{K_c + [Ca^{2+}]}\right)[G], \tag{7}$$

where α is an effective signal gain parameter, $[Ca^{2+}]$ is the cytosolic Ca^{2+} concentration and K_c is the dissociation constant for the Ca^{2+} binding site on the PLC molecule.

Replenishment of PIP_2 is required for IP_3 production to be maintained over extended-periods of agonist stimulation, so we have assumed that there is a recycling of IP_3 via an intracellular pool of phospholipid.

The final step to be modelled is the release of Ca^{2+} from the endoplasmic reticulum via IP_3 acting on IP_3-sensitive channels in the ER wall, and here we have used a modified version of the theory of Li and Rinzel [4].

The overall scheme is a set of equations relating agonist application to the production of IP_3 and Ca^{2+}. A number of simplifications are incorporated (in particular, the rapid binding approximation [5]) in order to concentrate on the most significant steps and reduce the number of parameters required. Those parameter whose values were not available directly from experiment were chosen in order to fit the experimental data of Garrad et al. [2] on receptor desensitization and sequestration.

3. Results

The first set of results concern surface receptor density following the application of agonist. Calculations were performed for both the transient and equilibrium surface receptor densities following the step application of uridine triphosphate (UTP) and good agreement was obtained with the experimental results [2] for both the time-dependence of the response and for the equilibrium receptor density as a function of UTP concentration (Fig. 2).

Next, IP_3 and Ca^{2+} concentrations resulting from the step application of 1 mM UTP were calculated and their time courses are shown in Fig. 3A. Also the Ca^{2+} response as a function of agonist concentration was calculated and shown to reproduce the results of [2] (Fig. 3B).

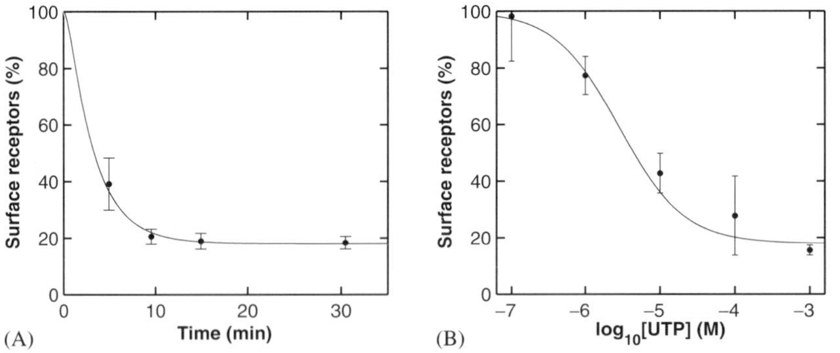

Fig. 2. Experimental and theoretical number of surface receptors as a function of time following the step application of 1 mM of UTP at time $t=0$ (panel A) and as a function of UTP concentration under equilibrium conditions (panel B). Points with error bars give the experimental data taken from Figs. 4A and 5A of [2]. In the calculations, it is assumed that 15% of the receptors are immobilized in the surface and do not recycle.

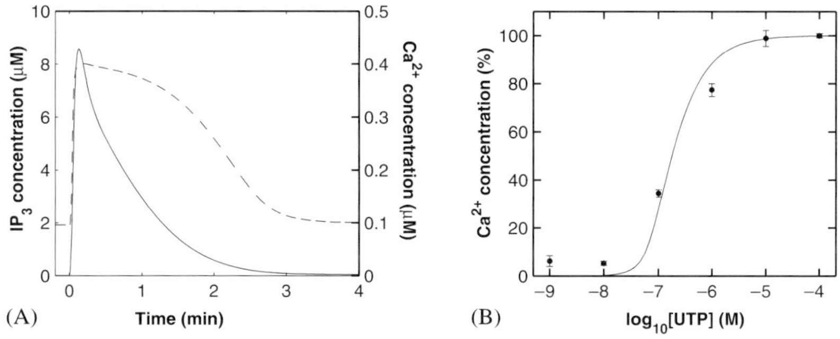

Fig. 3. Panel A shown the theoretical concentrations of IP_3 (solid line, scale on left axis) and Ca^{2+} (broken line, scale on right axis) as a function of time following the step application of 1 mM UTP at $t = 0$. Panel B shown the Ca^{2+} response, as a percentage of maximum response, as a function of the applied concentration of UTP. The points with error bars show the experimental data from [2], Fig. 3A.

4. Conclusion

This work provides a unified mathematical model of the steps leading from the activation of metabotropic receptors to the production of IP_3 and Ca^{2+}. The model has been specifically applied to systems involving the application of UTP to astrocytoma cells [2] and the application of ATP to MDCK cells [3] (results not given here) and successfully reproduces the main experimental results. However, the model is generally applicable to other systems that involve the production of IP_3 following metabotropic receptor activation, such as norandrenergic transmission in the autonomic nervous system.

References

[1] M.J. Berridge, The AM and FM of calcium signalling, Nature 386 (1997) 759–760.

[2] R.C. Garrad, M.A. Otero, L. Erb, P.M. Theiss, L.L. Clarke, F.A. Gonzalez, J.T. Turner, G.A. Weisman, Structural basis of the agonist-induced desensitization and sequestration of the $P2Y_2$ nucleotide receptor, J. Biol. Chem. 273 (1998) 29437–29444.

[3] K. Hirose, S. Kadowaki, M. Tanabe, H. Takeshima, M. Lino, Spatiotemporal dynamics of inositol 1,4,5-trisphosphate that underlies complex Ca^{2+} mobilization patterns, Science 284 (1999) 1527–1530.

[4] Y.-X Li, J. Rinzel, Equations for $InsP_3$ receptor-mediated $[Ca^{2+}]$ oscillations derived from a detailed kinetic model: a Hodgkin–Huxley like formalism, J. Theor. Biol. 166 (1994) 461–473.

[5] J. Wagner, J. Keizer, Effects of rapid buffers on Ca^{2+} diffusion and Ca^{2+} oscillations, Biophys. J. 67 (1994) 447–456.

[6] J.M. Weiss, P.H. Morgan, M.W. Lutz, T.P. Kenakin, The cubic ternary complex receptor-occupancy model I. Model description, J. Theor. Biol. 178 (1996) 151–167.

Greg Lemon is in the final stages of his Ph.D. candidature at the University of Sydney. He has degrees in Science and Engineering, with majors in Applied Mathematics and Electrical Engineering.

Bill Gibson is an Associate Professor in the School of Mathematics and Statistics at the University of Sydney. He received his Ph.D. from the University of New South Wales in 1969 and after working for many years in quantum statistical mechanics is now concerned with theoretical and computational aspect of synaptic function.

Max Bennett is Professor of Physiology in the University of Sydney and Head of the Neurobiology Laboratory. He received his D.Sc. from the University of Sydney in 1977. His research interests include the formation and function of synaptic connections as well as the history and philosophy of neuroscience.

that following three constraints are fulfilled: (1) Good approximation of the neural data by the stationary solutions of the neural model. (2) Smoothness (minimal complexity) of the fitted functions $w(x)$ and $b(x)$. (3) The neural data is approximated by *stable* solutions of the fitted neural dynamics.

4. Algorithm

The algorithm was derived by combining support vector regression [11] with methods from robust control theory (*linear matrix inequalities*) [3]. To simplify the notation in the following we will use the abbreviation $v^{(q)}(x)=f(u^{(q)}(x))$. We assume that the full functional form of this thresholded activation distribution is known. In practice it can be reconstructed from the neural data, e.g. by interpolation or nonlinear regression.

The functions $w(x)$ and $b(x)$ are represented as convolutions of two parameter functions with the Gaussian kernel $g(x)$ (cf. [11]):

$$w(x)=\int \omega(x-x')g(x')\,\mathrm{d}x', \tag{2}$$

$$b(x)=\int \beta(x-x')g(x')\,\mathrm{d}x'. \tag{3}$$

To ensure that the functional form of the estimated feed-forward and lateral connections has minimal complexity we apply the *structural risk minimization principle* by minimizing a bound for the VC dimension of these functions. Following derivations in [11] this is possible, the minimization of the following error functional:

$$E[\omega,\beta]=\frac{1}{2}\int \omega(x)^2\,\mathrm{d}x+\frac{1}{2}\int \beta(x)^2\,\mathrm{d}x+C\sum_{l,q}(\xi_l^{(q)}+\xi_l^{(q)*}) \tag{4}$$

with the positive constant C subject to the constraints (with $1\leqslant l\leqslant L$ and $1\leqslant q\leqslant Q$):

$$u_l^{(q)}-\iint \omega(x_l-x'-x'')v^{(q)}(x')g(x'')\,\mathrm{d}x'\,\mathrm{d}x''$$
$$-\iint \beta(x_l-x'-x'')s^{(q)}(x')g(x'')\,\mathrm{d}x'\,\mathrm{d}x''+\xi_l^{(q)}+\varepsilon\geqslant 0, \tag{5}$$

$$-u_l^{(q)}+\iint \omega(x_l-x'-x'')v^{(q)}(x')g(x'')\,\mathrm{d}x'\,\mathrm{d}x''$$
$$+\iint \beta(x_l-x'-x'')s^{(q)}(x')g(x'')\,\mathrm{d}x'\,\mathrm{d}x''+\xi_l^{(q)*}+\varepsilon\geqslant 0, \tag{6}$$

$$\xi_l^{(q)}\geqslant 0, \tag{7}$$

$$\xi_l^{(q)*}\geqslant 0. \tag{8}$$

Using methods from variational calculus this optimization problem can be transformed in *dual form* [11] with solutions of the form

$$\omega(x) = \sum_{l,q} (\alpha_l^{(q)*} - \alpha_l^{(q)}) \int g(x_l - x' - x) v^{(q)}(x')\,\mathrm{d}x', \tag{9}$$

$$\beta(x) = \sum_{l,q} (\alpha_l^{(q)*} - \alpha_l^{(q)}) \int g(x_l - x' - x) s^{(q)}(x')\,\mathrm{d}x', \tag{10}$$

where the coefficients $\alpha_l^{(q)}$ and $\alpha_l^{(q)*}$ are Lagrange multipliers. The dual optimization problem is a quadratic programming problem that can be written compactly by defining the vectors $\alpha^{(*)} = [\alpha_1^{(1)(*)}, \ldots, \alpha_L^{(Q)(*)}]^{\mathrm{T}}$ and $\mathbf{u} = [u_1^{(1)}, \ldots, u_L^{(Q)}]^{\mathrm{T}}$. The following function has to be maximized over the vectors $\alpha^{(*)}$

$$L(\alpha, \alpha^*) = -\tfrac{1}{2}(\alpha^* - \alpha)^{\mathrm{T}}(\mathbf{K}_v + \mathbf{K}_s)(\alpha^* - \alpha) + \mathbf{u}^{\mathrm{T}}(\alpha^* - \alpha) - \varepsilon \mathbf{1}^{\mathrm{T}}(\alpha^* + \alpha) \tag{11}$$

with the box constraints:

$$0 \leqslant \alpha_l^{(q)}, \alpha_l^{(q)*} < C. \tag{12}$$

The elements of the $(LQ) \times (LQ)$ matrices $\mathbf{K}_v$ and $\mathbf{K}_s$ are given by

$$(\mathbf{K}_v)_{ll'}^{qq'} = \iiint g(x_l - x' - x) g(x_{l'} - x'' - x) v^{(q)}(x') v^{(q')}(x'')\,\mathrm{d}x\,\mathrm{d}x'\,\mathrm{d}x'',$$

$$(\mathbf{K}_s)_{ll'}^{qq'} = \iiint g(x_l - x' - x) g(x_{l'} - x'' - x) s^{(q)}(x') s^{(q')}(x'')\,\mathrm{d}x\,\mathrm{d}x'\,\mathrm{d}x'',$$

where rows and columns of the matrices run through all combinations of the index pairs (l, q) and (l', q'). A number of additional constraints follows from the requirement that the data should correspond to stable states of the network dynamics. For each stimulus $s^{(q)}(x)$ the stability of the solution can be verified by linearizing dynamics (1) around the measured neural activity distribution $u^{(q)}(x)$. An approximation for the spatially continuous activity dynamics can be obtained by sampling the field discretely in the points x_r. From this the stability matrix of the linearized dynamics is derived as

$$\mathbf{A}^{(q)} = -\mathbf{I} + \mathbf{W}\mathbf{D}^{(q)} \tag{13}$$

with $W_{mn} = w(x_m - x_n)$ and $D_{nn}^{(q)} = f'(u^{(q)}(x_r))\Delta x_r$, where Δx_r is the discretization interval along the x dimension. Exploiting Eqs. (2), (3), and (9) it can be shown that this stability matrix is an affine function of the parameters $\alpha_l^{(q)}$ and $\alpha_l^{(q)*}$. For the stability of the dynamics it must be required that the matrix $\mathbf{A}$ has no eigenvalues with nonnegative real part. An equivalent stability condition that is more suitable for an elegant algorithmic implementation is given by the Lyapunov inequality

$$\mathbf{A}^{(q)\mathrm{T}}\mathbf{P} + \mathbf{P}\mathbf{A}^{(q)} \leqslant \mathbf{0}, \tag{14}$$

where $\mathbf{P}$ is a positive definite symmetric matrix that determines the speed of the relaxation of perturbations of the stationary solution. This matrix is preassigned to be a multiple of the unit matrix. $\mathbf{X} \leqslant \mathbf{0}$ means that the matrix $\mathbf{X}$ is negative semi-definite.

Obviously, the last equation is also affine in the parameters $\alpha_l^{(q)}$ and $\alpha_l^{(q)*}$. This allows to reformulate the stability conditions as set of linear matrix inequalities of the form:

$$\mathbf{A}_0^{(q)} + \sum_{l,q} (\alpha_l^{(q)*} - \alpha_l^{(q)})\mathbf{A}_n^{(q)} \geqslant \mathbf{0}. \tag{15}$$

Finally, by application of *Schur complements* [3], the quadratic programming problem (11) and (12) can be converted into a semi-definite programming problem

$$\begin{aligned} &\text{minimize} \quad E \\ &\text{subject to} \quad \mathbf{B}_0 + E\mathbf{B}_E + \sum_{l,q} (\alpha_l^{(q)*}\mathbf{B}_l^{(q)*} + \alpha_l^{(q)}\mathbf{B}_l^{(q)}) \geqslant \mathbf{0} \end{aligned} \tag{16}$$

with the new additional scalar parameter E. This together with Eqs. (15) and (16) define a common semi-definite programming problem that, if feasible, defines a solution that fulfills all three constraints formulated in the last section. Semi-definite programming problems are convex, i.e. have a unique solution. In addition, they can be solved efficiently with interior point methods [3,10].

5. Results

The method was evaluated with simulated neural data. A neural field with gaussian feed-forward kernel $b(x)$ and a Mexican hat-shaped lateral interaction function $w(x)$ was simulated. The stable stationary solutions for multiple input signals $s(x)$ (composed from one or multiple delta pulses along the x axes) were used as training data. Fig. 1 shows an example for the model fits obtained with three different methods: (1) *standard regularization* by modeling the functions $b(x)$ and $w(x)$ with radial basis function networks (dotted line), (2) *support vector regression* (dashed-dotted line),

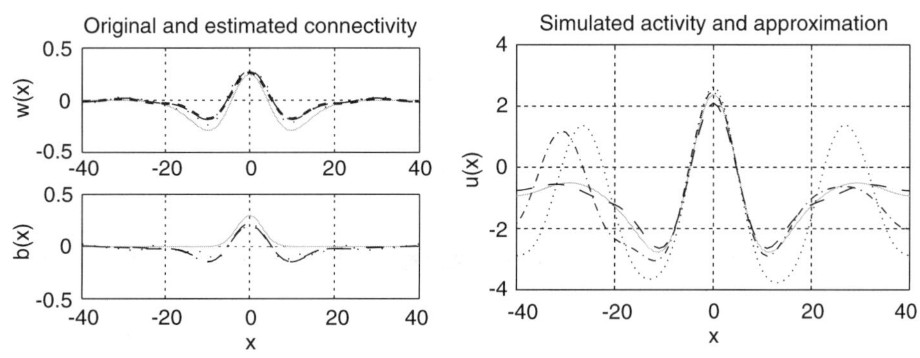

Fig. 1. Estimation results obtained with different methods for simulated data. The left panel shows the estimated kernels describing the feed-forward and feedback connectivity. Right panel shows the stable activity distributions $u(x)$ obtained by simulation of the network using the data as initial condition. The original data is plotted in grey. The dashed curves indicate the estimation results obtained with the proposed method. The results obtained using a standard radial basis function network (dotted lines) and with support vector regression [11] (dashed-dotted lines) are shown for comparison.

and (3) the proposed algorithm with *dynamic stability control* (dashed line). Only the proposed method fits a model with stable solutions that are close to the data. All three methods identify models that lead to very small equation errors for the approximation of Eq. (1). However, the solutions of the methods without dynamic stability control drift away from the data during the iteration of the fitted dynamics. This occurs in particular when the original neural dynamics is close to instability.

6. Discussion

A new learning algorithm was proposed that estimates automatically the structure of a nonlinear recurrent neural network model (neural field) from a sparse set of neural data. The stable solutions of the fitted model approximate the neural data. Only very general a priori *assumptions* about the structure of the model have to be made, including the shape of the nonlinearity and smoothness and translation invariance of the connections. Many existing dynamic neural models in neuroscience and technical applications fulfill these assumptions. Neural field models have been automatically fitted before (e.g. [4,5]). The specific property of the proposed method is that it permits an automatic estimation of such models without the knowledge of transients of the neural signals. For a variety of applications in neuroscience such transients are not easily available.

The theoretically interesting aspect of the proposed method is that it combines *structural risk minimization* with constraints resulting from a stability analysis of learned dynamical model. Constraints expressed in terms of *linear matrix inequalities* have been used for the training of recurrent neural networks without exploiting regularization (e.g. [8]). In addition, standard regularization networks have been combined with stability constraints in the context control applications [9]. The proposed approach seems very general and might be extended for other applications that require learning of regularized dynamical models with capacity control.

References

[1] S. Amari, Dynamics of pattern formation in lateral-inhibition type neural fields, Biol. Cybernet. 27 (1977) 77–87.

[2] R. Ben-Yishai, R.L. Bar-Or, H. Sompolinsky, Theory of orientation tuning in visual cortex, Proc. Nat. Acad. Sci. USA 92 (9) (1995) 3844–3848.

[3] S. Boyd, L.El. Ghaoui, E. Feron, V. Balakrishnan, Linear Matrix Inequalities in System and Control Theory, SIAM Studies in Applied Mathematics, Philadelphia, 1994.

[4] M.A. Giese, Dynamic Neural Field Theory for Motion Perception, Kluwer, Boston, 1999.

[5] C. Igel, W. Erlhagen, D. Jancke, Optimization of dynamic neural fields, Neurocomputing 36 (2001) 225–233.

[6] D.K. Lee, L. Itti, C. Koch, J. Braun, Attention activates winner-takes-all competition among visual filters, Nature Neurosci. 2 (1999) 375–381.

[7] V. Marmarelis, Wiener analysis of nonlinear feedback in sensory systems, Ann. Biomed. Eng. 19 (1991) 345–382.

[8] J. Steil, H. Ritter, Recurrent learning of input–output stable behaviour in function space: a case study with the Roessler attractor, in: Proceedings of the ICANN '99, IEE, London, 1999, pp. 761–766.

[9] J.A.K. Suykens, J. Vandewalle, B. DeMoor, Optimal control by least-squares support vector machines, Neural Networks 14 (2001) 23–35.
[10] K.C. Toh, M.J. Todd, R.H. Tütüncü, SDPT3—a MATLAB software package for semidefinite programming, Optim. Meth. Software 11 (1999) 545–581.
[11] V.N. Vapnik, Statistical Learning Theory, Wiley, New York, 1998.

Martin A Giese (born 1966) received his M.Sc. and Ph.D. in Electrical Engineering from the University in Bochum, Germany. From 1998 to 2001 he was postdoctoral researcher at the Center for Biological and Computational Learning, M.I.T. He was guiding the Boston Laboratory of HONDA R&D Americas in 2000 and 2001. Since 2001 he is head of the Laboratory for Action Representation and Learning at the University Clinic in Tübingen, Germany. His research focuses on the role of learning in the representation of complex movements and actions in brains and machines.

Available at
www.ComputerScienceWeb.com
POWERED BY SCIENCE @ DIRECT®

ELSEVIER

Neurocomputing 52–54 (2003) 285–288

NEUROCOMPUTING

www.elsevier.com/locate/neucom

Neural bases of accumulator models

Hiroshi Okamoto[a,b,*], Tomoki Fukai[b,c]

[a] *Corporate Research Laboratory, Fuji Xerox Co., Ltd., 430 Sakai, Nakai-machi, Ashigarakami-gun, Kanagawa 259-0157, Japan*

[b] *CREST of JST (Japan Science and Technology Corporation), 4-1-8 Honcho, Kawaguchi-shi, Saitama 332-0012, Japan*

[c] *Department of Information-Communication Engineering, Tamagawa University, Tamagawagakuen 6-1-1, Machida-shi, Tokyo 194-8610, Japan*

Abstract

One class of cognitive models, accumulator models (AMs), is a framework to explain cognitive mechanisms of reaction timing. In AMs, an internal signal, which grows linearly with time upon the onset of an external stimulus, is postulated. Reaction time is defined by the time at which the signal reaches a predefined threshold level. Indeed, cortical neurons with activity that grows seemingly linearly with time are largely found, which is consistent with AMs. Here we show that stochastic dynamics of a recurrent network of bistable spiking neurons can produce linear growth of neuronal activity. This suggests possible neural bases of AMs.

Keywords: Accumulator model; Reaction time; Recurrent network; Stochastic dynamics; Neuronal bistability

1. Introduction

The length of reaction time, i.e. the interval between the onset of an external stimulus (e.g. target presentation) and reaction (e.g. start of saccadic eye movement), is surprisingly long (> 100 ms) [1,11]. Obviously, this long latency cannot be accounted for only by neuronal conduction and transmission delays. This rather suggests existence of some higher-order internal processes underlying reaction timing.

* Corresponding author. Corporate Research Laboratory, Fuji Xerox Co. Ltd., 430 Sakai, Nakai-machi, Ashigarakami-gun, Kanagawa 259-0157, Japan. Tel.: +81-465-80-2015; fax: +81-465-81-8961.

E-mail address: hiroshi.okamoto@fujixerox.co.jp (H. Okamoto).

0925-2312/03/$ - see front matter © 2003 Elsevier Science B.V. All rights reserved.
doi:10.1016/S0925-2312(02)00778-6

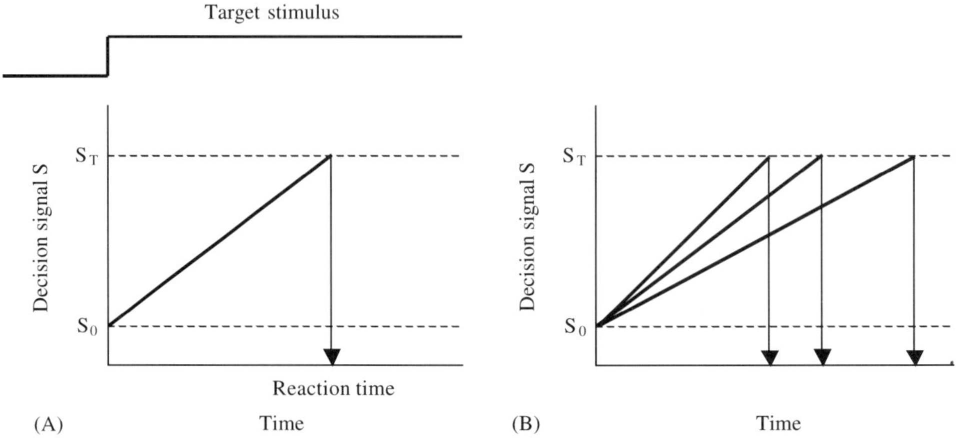

Fig. 1. (A) Outline of AMs. (B) Variable-threshold version of AMs.

Reaction timing can also be characterized by its variability. The length of reaction time largely fluctuates across trials even if the stimulation is the same in every trial [1,11].

One class of cognitive models, accumulator models (AMs) [1,7,10,11,14], well describes the long latency and the variability of reaction timing. AMs postulate a decision signal S, which represents some internal activation level (Fig. 1A). Upon the onset of an external stimulus, S starts to grow linearly from an initial level S_0 at a rate r. When S reaches a predefined threshold level S_T, reaction (e.g. saccadic eye movement) is triggered.

AMs predict two possible sources of the variability of reaction timing [1,7,11]. One is that the threshold level S_T fluctuates across trials (variable-threshold model), and the other is that so does the growth rate r (variable-rate model, Fig. 1B). These models give different experimental predictions for stochastic properties of reaction timing. It has been shown that psychological experimental results are consistent with the variable-rate models [1,7,11].

In neurophysiology, a population of neurons with the activity resembling the decision signal postulated in AMs have been found. Upon the onset of a target stimulus, the firing rates of a saccade-related neurons in the oculomotor area, principally comprising the frontal eye field (FEF) and the supplementary eye field (SEF), grow seemingly linearly with time and culminate at a start of saccade [3,4,7,8,12,13]. The growth rate appears to fluctuate across trials [3,7,12,13].

In the present study, seeking for possible neural bases of AMs, we examine neural mechanisms that can produce linear growth of the neuronal firing rate. A possible origin of the fluctuation of the growth rate is also discussed.

2. Model

We consider a model for a recurrent network of excitatory leaky-integrate-and-fire neurons. The recurrent input to each neuron is described by AMPA current originating

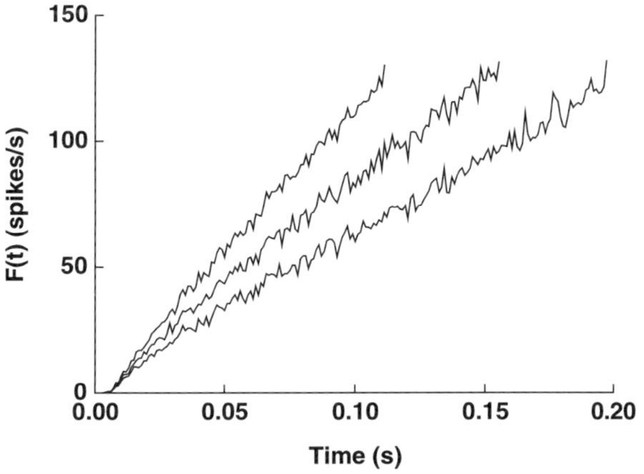

Fig. 2. Time course of $F(t)$, the output from the recurrent network. Depolarizing current I_{ex} is set on at time 0. Different lines indicate the results for different values of I_{ex}. Exact time course of $F(t)$ is roughened with 1 ms bin.

from other neurons in the network. The same gating kinetics for AMPA conductance as those used in Ref. [2] is adopted. The membrane potential of each neuron is also subject to Gaussian white noise that is uncorrelated between cells.

Specially, we assume that a neuron generates brief depolarization after it sufficiently discharges [5,6]. This after-depolarization (ADP) can make the membrane potential of a neuron bistable: In the off state, a neuron rarely discharges and ADP is inactive; in the on state, ADP activation and discharge of a neuron continue in a regenerative manner owing to the positive feedback between them [9].

The onset of a target stimulus is represented by the onset of a depolarizing current I_{ex}. Before the onset of I_{ex}, all the neurons are set in the off state. After the onset of I_{ex}, its amplitude is held constant.

3. Results and discussion

Let $F(t)$ be the output from the network at time t, defined by the firing rate averaged over all cells at time t. The time course of $F(t)$ was examined by computer simulation. The results obtained show that $F(t)$ grows seemingly linearly with time (Fig. 2). It is also shown that changes in I_{ex} across trials result in fluctuations of the growth rate without affecting the linearity (Fig. 2).

The above results suggest that the output from the recurrent network is the neural substrate of the decision signal postulated in AMs. The model is therefore expected to briefly describe a functional structure of the FEF or of the SEF.

The I_{ex} represents a target stimulus and its fluctuations across trials are a possible origin of those of the growth rate in the variable-rate version of AMs. It is highly probable that, although an external target appeared is the same in every trial, the visual cortex that processes a target stimulus and sends it to the FEF, and the SEF is modified by ‘attention’ that can easily be varied across trials. It is therefore reasonable to assume that I_{ex} fluctuates across trials.

We have demonstrated that the stochastic dynamics of bistable spiking neurons provides possible neural bases of AMs. The present study thus presents a possible link between neural mechanisms and behavioural functions.

References

[1] R.H.S. Carpenter, M.L.L. Williams, Neural computation of log likelihood in control of saccadic eye movements, Nature 377 (1995) 59–62.

[2] A. Destexhe, Z.F. Mainen, T.J. Sejnowski, Neural Comput. 6 (1993) 14–18.

[3] S. Everling, D.P. Munoz, Neuronal correlates for preparatory set associated with pro-saccades and anti-saccades in the primate frontal eye field, J. Neurosci. 20 (2000) 387–400.

[4] N. Fujii, H. Mushiake, J. Tanji, Distribution of eye- and arm-movement-related neuronal activity in the SEF and in the SMA and pre-SMA of monkeys, J. Neurophysiol. 87 (2002) 2158–2166.

[5] S. Haj-Dahmane, R. Andrade, Calcium-activated cation nonselective current contributes to the fast afterdepolarization in rat prefrontal cortex neurons, J. Neurophysiol. 78 (1997) 1983–1989.

[6] S. Haj-Dahmane, R. Andrade, Ionic mechanism of the slow afterdepolarization induced by muscariniv receptor activation in rat prefrontal cortex, J. Neurophysiol. 80 (1998) 1197–1210.

[7] D.P. Hanes, J.D. Schall, Neural control of voluntary movement initiation, Science 274 (1996) 427–430.

[8] H. Mushiake, N. Fujii, J. Tanji, Visually guided saccade versus eye-hand reach: contrasting neuronal activity in the cortical supplementary and frontal eye fields, J. Neurophysiol. 75 (1996) 2187–2191.

[9] R.C. O’Reilly, T.S. Braver, J.D. Cohen, Mechanisms of active maintenance and executive control, in: A. Miyake, P. Shah (Eds.), Models of Working Memory, Cambridge University Press, Cambridge, 1999, pp. 375–411.

[10] R. Ratcliff, T. Van Zandt, G. McKoon, Connectionist and diffusion models of reaction time, Psychol. Rev. 106 (1999) 261–300.

[11] B.A.J. Reddi, R.H.S. Carpenter, The influence of urgency on decision time, Nat. Neurosci. 3 (2000) 827–830.

[12] J.D. Schall, Neural bases of deciding, choosing and acting, Nature Rev. Neurosci. 2 (2000) 33–42.

[13] J.D. Schall, N.P. Bichot, Neural correlates of visual and motor decision processes, Curr. Opin. Neurobiol. 7 (1998) 211–217.

[14] M. Usher, J.L. McClelland, The time course of perceptual choice: the leaky, competing accumulator model, Psychol. Rev. 108 (2001) 550–592.

Available at
www.ComputerScienceWeb.com
POWERED BY SCIENCE DIRECT®

ELSEVIER

Neurocomputing 52–54 (2003) 289–294

NEUROCOMPUTING

www.elsevier.com/locate/neucom

Linking computational neuroscience simulation tools—a pragmatic approach to component-based development☆

F. Howell[a,*], R. Cannon[a,b], N. Goddard[a], H. Bringmann[a], P. Rogister[a], H. Cornelis[b]

[a] *Institute for Adaptive and Neural Computation, Division of Informatics, University of Edinburgh, 5 Forrest Hill, Edinburgh EH1 2QL, Scotland 1HN, UK*

[b] *Theoretical Neurobiology, Born-Bunge Foundation, University of Antwerp, Universteitsplein 1, Wilrijk 2610, Belgium*

Abstract

Many problems in computational neuroscience require sophisticated software systems that are beyond the development scope of a single individual or research group. Realizing such systems with minimal redundant effort requires cooperation among software developers and the adoption of design strategies and technology from the software industry.

We are working on neuroscience-specific frameworks for modelling tools aimed specifically at maximising the benefits from of investment in software development by encouraging the reuse of software components and at facilitating model development by establishing shared formats for model description.

The techniques employed include component technology for coupling parts of applications, XML file standards based on NeuroML for model and data exchange between applications and databases, and peer-peer web based indexing of models and modules.

This paper describe progress to date in the modularization of simulation and analysis functions from NEURON, Catacomb and NEOSIM.

Keywords: Neural simulation tools; Modelling; Simulation; Software

☆ Supported by: NIH (MH-57358), FWO Belgium.

* Corresponding author. Tel.: +44-131-650-3088; fax: +44-131-650-6899.

E-mail addresses: fwh@anc.ed.ac.uk (F. Howell), cannon@anc.ed.ac.uk (R. Cannon), nigel.goddard@ed.ac.uk (N. Goddard), hbringma@uos.de (H. Bringmann), paulro@anc.ed.ac.uk (P. Rogister), hugo@bbf.uia.ac.be (H. Cornelis).

0925-2312/03/$ - see front matter
doi:10.1016/S0925-2312(02)00781-6

1. Introduction

Besides its scientific goals, computational neuroscience also presents significant software engineering challenges. Modelling biological systems at the range of levels now demanded will eventually require tools ranking amongst the most advanced software systems in use.

The development of such systems is clearly beyond the resources of individual research groups, and therefore requires technology which allows each group to make maximum use of development work elsewhere. Likewise, model development would benefit substantially from the sharing and reuse of models between research groups.

The software industry is facing closely related problems in the development of software for commercial and industrial applications, and is generating a continual stream of new technology to help tackle them. We are harnessing this work in the neuroscience context to allow software developers and modelers to carry on working in their own area of expertise, but to ensure that the results of this work are more widely accessible.

2. Current state of multi-user neuroscience software

The choices facing a researcher interested in modelling at some level of computational neuroscience are to use a simulation system such as NEURON [6] or GENESIS [1] aimed at a particular type of modelling, or write their own software from scratch, frequently implemented in C, C++, Java or Matlab.

A critical issue is what researchers do when a software tool *nearly* does what is required. For example, if one would like to run a network model in NEURON but to use a feature not yet available, such as the display of a clickable connectivity graph showing when particular links are active, then the choices at present are either to ask the author to implement it or to write one's own simulation code from scratch which includes the particular display feature.

It would be more efficient if the researcher were able to write a module for the new feature in a simulator-independent way, couple it to an existing simulation environment, and make the module available for others. At present there are numerous practical barriers to this kind of component-based development which must be addressed before modular software development and reuse becomes the norm in neuroscience.

3. Working modules

There are already a number of areas where self-contained modules have been shown to work; NEURON `.mod` files for mechanisms and templates for defining reusable cells; GENESIS packages and messages, which allow a way to append user code, and the NEOSIM event interface which allows linking modules via a restricted send/handle spike interface.

Our aim is to extend the range of situations in which module-based technology can be used, to include interfaces for modules to perform:

- Synaptic dynamics.
- Ion channels.
- Artificial/ball and stick/reconstructed compartmental neurons.
- Projection patterns.
- Structured populations.

The systems biology community is faced by similar modelling problems, and has come up with the Systems Biology Workbench [7] (SBW) as a partial solution. They have a rather different starting point, with several highly developed modelling systems in widespread use and covering broadly the same problem domain. Their solution has been to implement a loose coupling between these applications based on message passing. This works well from the point of view of the user, who can run the same model under different systems, but does not reduce the software engineering requirements because each system still must be independently maintained.

4. Module technologies

Modularisation and reusable software components are widely used to develop business software, with a number of technologies available, e.g. CORBA [11], RMI [12], JavaBeans [13], .NET [10]. In the scientific community, as well as the SBW mentioned above, there is a recent effort at LLNL [8] focussed on components for parallel computations.

The approach we have adopted is based on Java archive files (.jar) to package modules into self contained units, JNI for C++/Java coupling, and the NeuroML dialect of XML for model/parameter specification.

The precise choice of component technology is less significant than making the step from developing monolithic applications to developing components. Once modules are available in some format, software bridges can be built to link the different component technologies.

5. Example interfaces

We have developed a number of interfaces for specific types of module, including `RunnableCell`, `PreSynapse`, `PostSynapse`, `Projection`. For example, the RunnableCell interface includes a few methods which a module which can receive spikes

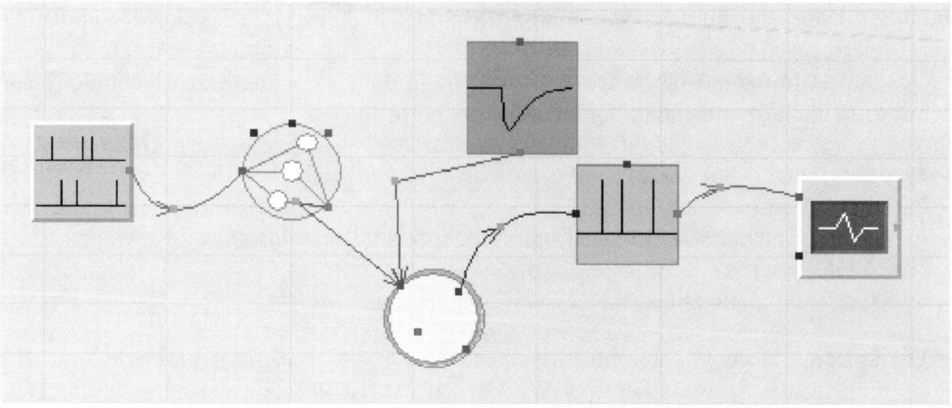

Fig. 1. A Catacomb cell model connecting components for spike generation, integration and plotting. Moving to distributed development means that modellers can develop new component types and link them to the system.

can implement:-

```
package neuroml.sim.run;
interface RunnableCell {
                    void handleSpike(double t, int inputConnection);
                    void advance(double t);
                    void reset();
}
```

Modules which implement such an interface can be developed by different people, and loaded on the fly by a module aware simulation package (e.g. NEOSIM [4]). For more documentation see the websites `www.compneuro.org/javadoc` and `www.neuroml.org`. Fig. 1 illustrates a number of modules linked using the Catacomb workbench [2].

6. Model descriptions

For simulation model descriptions, we use NeuroML [5], an XML declarative language. This has the advantages of being extensible, easy for software and people to read, and it leverages the industry investment into XML technology. Describing models in XML rather than a programming language also separates out the key parts of the model from the details of its implementation.

The key technology which is part of the NeuroML development kit is *data binding*, to allow data in an object tree (within a simulator) to be transparently mapped to/from a NeuroML file without having to implement a parser. The schema for the NeuroML sublanguage is defined using a set of class definitions, or in XML. We have also used the NeuroML development kit within a viewer for the CoCoMac connectivity database [9].

7. Peer–peer exchange

One consequence of moving from a few monolithic simulation packages to a scenario of a large number of component developers is that it becomes essential to provide tools that allow people to search for and publish modules. To address this issue we have developed peer–peer web based exchange and indexing tools (`www.neuroml.org/napster`) based on the Napster model of distributed content (people keep modules/models on their own website which they can customise and control), and centralised indexes (so it becomes simple to perform a global search for modules or models, either from a website or an application) (see Cannon et al. [3]).

We specifically avoided a centralised database because of the perceived transfer of ownership and control which takes place when items are submitted to such databases. Despite efforts to the contrary, some degree of control inevitably shifts away from the model or module creator towards the database host, a situation which is very unappealing for those investing in software development.

8. Conclusions

We have begun the process of moving from monolithic software packages to collections of modules. The parts of the puzzle we have implemented include a stable model description language (NeuroML), working Java and C++ based mechanisms for module interaction (.jar files, interfaces, JNI), and a peer-peer system for publishing and finding out about modules and models.

The process has to be evolutionary rather than revolutionary because of the large amount of time and effort invested in existing general simulation packages and the models which depend upon them. We are currently working on incrementally adding interfaces to more simulation packages, with the ultimate aim of making it easier to develop, test, distribute and explore sophisticated models of circuits and cells.

References

[1] J.M. Bower, D. Beeman (Eds.), The Book of GENESIS: Exploring Realistic Neural Models with the GEneral NEural SImulation System, 2nd Edition, Springer, New York, 1998.

[2] R. Cannon, Catacomb, `www.compneuro.org/catacomb`.

[3] R.C. Cannon, F.W. Howell, N.E. Goddard, E. de Schutter, Non-Curated distributed databases for experimental data and models in neuroscience, Network: Comput. Neural Systems 13 (3) (2002) 415–428.

[4] N. Goddard, G. Hood, F. Howell, M. Hines, E. De Schutter, NEOSIM: Portable large-scale plug and play modelling, Neurocomputing 38 (2001) 1657–1661.

[5] N. Goddard, et al., Towards NeuroML: Model description methods for collaborative modeling in neuroscience, Philos. Trans. R. Soc, Ser. B 356 (1412) (2001) 1209–1228.

[6] M.L. Hines, N.T. Carnevale, The NEURON simulation environment, Neural Comput. 9 (1997) 1179–1209.

[7] M. Hucka, et al., SBML/SBW, www.cds.caltech.edu/erato.
[8] S. Kohn, Scientific component technology at LLNL, www.llnl.gov/CASC/components.
[9] R. Kotter, et al., CoCoMac, www.cocomac.org.
[10] Microsoft, .NET, www.microsoft.com/net.
[11] OMG, CORBA, www.corba.org.
[12] Sun Microsystems, RMI, java.sun.com/products/jdk/rmi.
[13] Sun Microsystems, JavaBeans, java.sun.com/products/javabeans.

Available at
www.ComputerScienceWeb.com
POWERED BY SCIENCE @ DIRECT®

ELSEVIER

Neurocomputing 52–54 (2003) 295–300

NEUROCOMPUTING

www.elsevier.com/locate/neucom

Role of A-current in lamprey locomotor network neurons

Mikael Huss[a,*], Dietmar Hess[b], Boris Lamotte d'Incamps[b], Abdeljabbar El Manira[b], Anders Lansner[a], Jeanette Hellgren Kotaleski[a]

[a] *Department of Numerical Analysis and Computing Science, Royal Institute of Technology, Lindstedtsv 5, S-100 44, Stockholm, Sweden*

[b] *Department of Neuroscience, Nobel Institute for Neurophysiology, Karolinska Institutet, S-171 77 Stockholm, Sweden*

Abstract

A compartmental model of lamprey central pattern generator neurons was built in order to examine the effects of a fast, transient, high-voltage-activated potassium current (A-current) found experimentally. The model consisted of a soma, a compartment corresponding to the axon initial segment, and a dendritic tree. The simulation showed that the A-current was necessary for repetitive spiking in the single neuron following current injection. The functional role of adding an A-current was also examined in a network model. In this model, the A-current stabilizes the swimming rhythm by making the burst cycle duration and the number of spikes per burst less variable. All these effects are also seen experimentally.

Keywords: A-current; Lamprey; Potassium; CPG; Hodgkin–Huxley

1. Introduction

Transient A-type K^+ channels can act as a control mechanism for neuronal excitability. A transient A-type potassium current has recently been characterized in lamprey locomotor network neurons [4]. This current is high-voltage-activated and promotes repetitive firing in lamprey neurons by facilitating Na^+ channel recovery from

* Corresponding author. Tel.: +46-8-7907784.
E-mail address: hussm@nada.kth.se (M. Huss).

0925-2312/03/$ - see front matter © 2002 Elsevier Science B.V. All rights reserved.
doi:10.1016/S0925-2312(02)00784-1

inactivation. On a higher anatomical level, the A-current seems to stabilize the lamprey swimming motor pattern by keeping the cycle duration fairly constant [4]. A compartmental model of a single lamprey neuron was implemented in order to investigate the role of A-current in cell excitability. This model was based on recent experimental data. In addition, a previously developed model of a small network of lamprey locomotor neurons was modified so that the "spiking machinery" (sodium and delayed rectifier-like channels) conformed better to the new experimental data. The role of the A-current in rhythm generation by this network was then investigated.

2. Methods

A compartmental model of a lamprey neuron was developed to investigate the effect of transient A-type K^+ channels on the neurons' firing characteristics. The model was implemented in the GENESIS simulation environment [1] and included Na^+, K^+ delayed rectifier, transient A-type K^+, calcium-dependent K^+, Ca^{2+}, and NMDA channels. The intracellular Ca^{2+} that activates calcium-dependent K^+ (K_{Ca}) channels was modeled using two different intracellular pools, Ca_{AP} and Ca_{NMDA}. They activate separate K_{Ca} channels, K_{CaAP} and K_{CaNMDA} channels, respectively [2]. The simulated neuron's morphology consisted of a 32 μm diameter soma and two primary dendrites, each of which had two branches. The primary dendrites were of length 90 μm and diameter 2.5 μm, the secondary dendrites of length 148 μm and diameter 1.5 μm and the tertiary dendrites were of length 240 μm and diameter 0.75 μm. The model also included a special compartment corresponding to the initial segment of the axon. The morphology was chosen in such a way that the resulting input resistance and input capacitance would be reasonably close to the experimentally measured values (177 ± 109 MΩ for the resistance and 400–500 pF for the capacitance) [3]. There was another additional constraint: the total area of the dendritic segments was to be approximately an order of magnitude larger than the area of the soma.

Only the soma and initial segment were equipped with active ion channels. The kinetics of most ion channels were based on previous lamprey models [2], although some parameters were changed as a result of new experimental data. Gating properties of the sodium current were modified to allow the firing of a single spike when the A-current was blocked and the cell was depolarized by a pulse of current. More precisely, the inactivation curve was shifted −7 mV (so that $V_{1/2} = -51.5$ mV). Recording of the Na^+ current on isolated lamprey spinal neurons with the patch clamp technique supported this modification but also showed that the lamprey sodium current is faster than the one implemented previously [2]. The A-type potassium current was modeled as a Hodgkin–Huxley type current with three activation gates and one inactivation gate. The half-activation potential was set to −1.0 mV and the half-inactivation potential to −9.3 mV, with slope factors of 10.6 mV for the activation and −11.7 mV for the inactivation [4]. The K^+ channel was modeled in the conventional way with four activation gates. Since experimental data suggest that this current's time constant of activation is voltage independent and fluctuates around 10 ms, it was set to a constant value of 10 ms.

3. Results

Experiments have shown that the A-current is necessary for repetitive firing in single lamprey neurons [4]. It was hypothesized that the strong hyperpolarization at the end of an action potential might help the sodium channels recover from their inactivation and thus let them initiate another spike. The A-current is almost entirely responsible for the strong after hyperpolarization. Normally, the delayed rectifier is often seen in that role. In the lamprey locomotor neuron, the standard delayed rectifier is hardly activated at all during an action potential.

Fig. 1 contains two plots showing the simulated cell's response to increased current steps. In the lower plot, we see how neurons where A-current has been blocked by catechol only fire one full action potential. Fig. 2 shows results from simulations. In the upper part, the A-current is intact, and we see repetitive firing beginning with the second current step. In the lower part, blockade of A-current channels by catechol has been simulated by decreasing the A-current density by 60%—catechol removes on average 59% of the A-current [4]. The cell fires only one full action potential for each current step. It can thus be seen that A-current is indeed responsible for repetitive firing. The cell where catechol application has been simulated can occasionally fire another small spike in addition to the first one, but no more than that.

Next, the role of the A-current in a small cellular network was examined. We used a previously developed model [2] consisting of four cells with mutual inhibitory and excitatory connections. The spiking machinery in this model had to be changed to accommodate the A-current and to give the Na+ and delayed rectifier currents more realistic properties (see Section 2). The four-cell network can generate bursting patterns when simulated stimulation in the form of NMDA and/or kainate is applied. These patterns correspond to the fictive locomotion patterns observed in a lamprey spinal cord preparation when NMDA and/or kainate is present in the bath. Experiments investigating the A-current's role in NMDA swimming in [4] had shown that blocking the A-current with catechol decreases both cycle duration and the number of spikes per

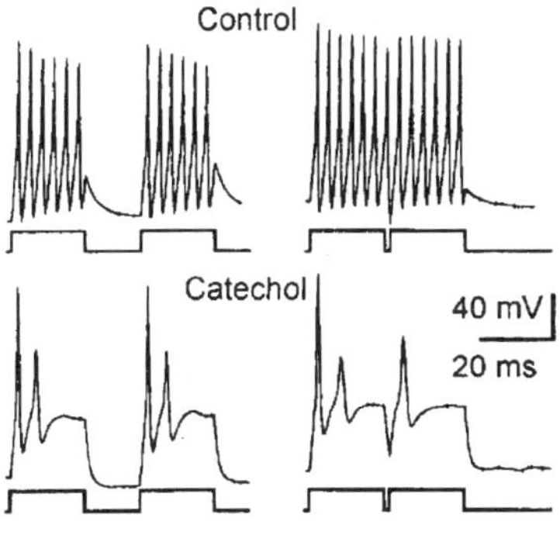

Fig. 1. Experimentally measured responses of a single cell to pairs of current pulses with a longer (left) or shorter (right) interval between the pulses.

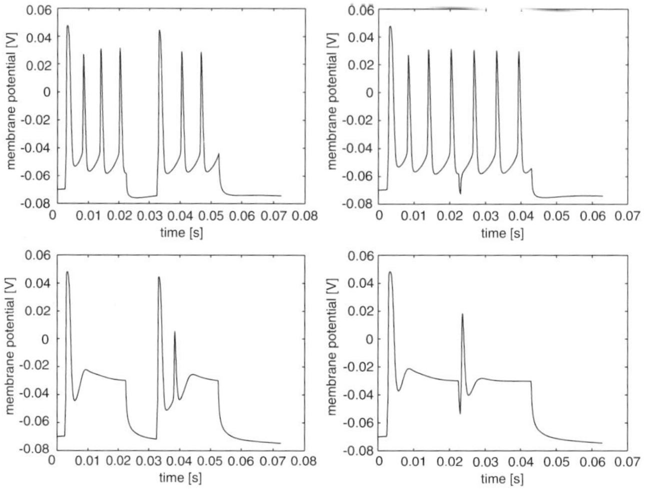

Fig. 2. Response of a simulated single cell to pairs of current pulses with a longer (left) or shorter (right) interval between the pulses. Blockage of the A-current channels has been simulated in the lower picture.

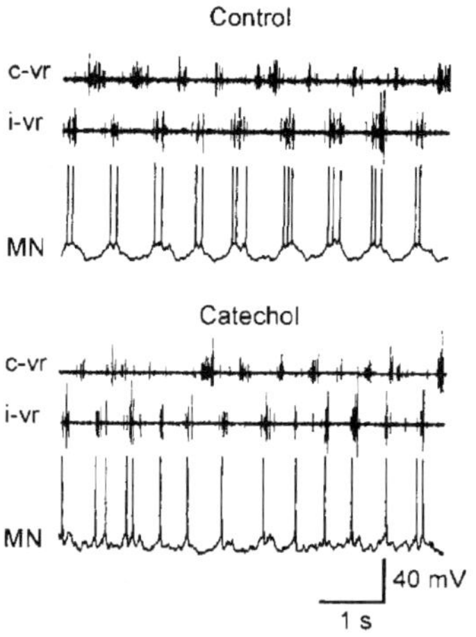

Fig. 3. Intracellular recordings from a lamprey motoneuron and from ventral roots during NMDA swimming. Above, control. Below, A-current blocked by catechol.

burst. When A-current is present, the cycles are longer and there is an increased—and arguably more stable—number of spikes per burst (see Fig. 3). The plots in Fig. 4 show what happens in the simulated four-cell network during NMDA swimming. It is apparent that A-current blockage makes the average number of spikes per burst decrease. The number of spikes per burst also varies somewhat erratically; most often

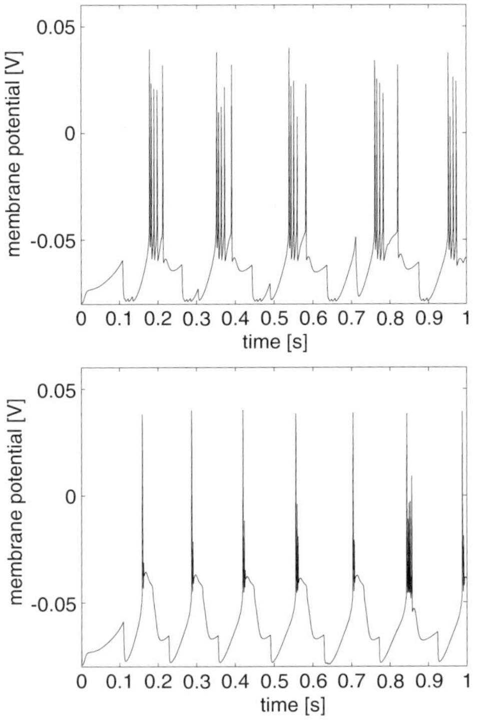

Fig. 4. Membrane potential during NMDA swimming of one of four cells connected into a network. In the lower plot, catechol blockage of A-current channels has been simulated by decreasing the A-current density by 60%.

there is only one, sometimes more. With intact A-current channels, we find a fairly consistent number of spikes per burst. As in the experimental study, the burst cycle is longer with the A-current intact.

4. Conclusions

Experiments have shown that the lamprey A-current is necessary for repetitive firing in a single cell, and that it increases both the cycle duration and the number of spikes per burst in a locomotor network. We have confirmed these effects in compartmental model simulations. During the action potential, the A-current seems to assume the role normally performed by the delayed rectifier: to rapidly bring the cell potential back to or beyond the resting potential, thus lifting the inactivation of sodium channels, enabling them to open and initiate a new action potential. The delayed rectifier in lamprey is markedly slower than the A-current and is primarily activated during bursting.

References

[1] J.M. Bower, D. Beeman, The Book of Genesis, TELOS, New York, 1998.
[2] Ö. Ekeberg, P. Wallén, A. Lansner, H. Tråvén, L. Brodin, S. Grillner, A computer based model for realistic simulations of neural networks, Biol. Cybern. 65 (1991) 81.
[3] D. Hess, unpublished observations.
[4] D. Hess, A.E. Manira, Characterization of a high-voltage-activated I_A current with a role in spike timing and locomotor pattern generation, PNAS 98 (2001) 5276.

Available at
www.ComputerScienceWeb.com
POWERED BY SCIENCE @ DIRECT®

ELSEVIER

Neurocomputing 52–54 (2003) 301–306

NEUROCOMPUTING

www.elsevier.com/locate/neucom

Synaptic plasticity, conduction delays, and inter-areal phase relations of spike activity in a model of reciprocally connected areas

Andreas Knoblauch*, Friedrich T. Sommer

Department of Neural Information Processing, University of Ulm, Oberer Eselsberg, D-89069 Ulm, Germany

Abstract

Multi-electrode recordings revealed that fast oscillatory (30–60 Hz) spike activity is often synchronized with zero phase lag, even for recording sites in distant cortical areas. We show in this simulation study that the dominance of zero-phase correlations is inconsistent with the long conduction delays measured between distant areas: For realistic delays, reciprocally connected neuron populations exhibit anti-phase rather than zero-phase correlations. We show that this inconsistency can be removed by taking into account spike-timing-dependent synaptic plasticity (STDP), as found in experiments (Science 275 (1997) 213). We demonstrate that STDP can weaken fast excitatory feedback and strengthen slower feedback with delays in the range of one oscillation period. This yields stable zero-lag oscillations, even for realistic delays.

Keywords: Synaptic plasticity; STDP; Axonal conduction delays; Synchronization; Gamma oscillations; Zero-phase-lag

1. Introduction

Neurophysiological experiments revealed evidence for long-range synchronization of fast (30–60 Hz) oscillatory activity. For example, oscillations at 50 Hz were synchronized between the two hemispheres of cat primary visual cortex apparently due to the inter-hemispheric connection via the corpus callosum [4,8]. Simulation studies of reciprocally connected neuron populations have shown that zero-lag oscillations occur only

* Corresponding author. Tel.: +49-731-50-24255; fax: +49-731-50-24156.

E-mail addresses: knoblauch@neuro.informatik.uni-ulm.de (A. Knoblauch), fritz@neuro.informatik.uni-ulm.de (F.T. Sommer).

0925-2312/03/$ - see front matter
doi:10.1016/S0925-2312(02)00792-0

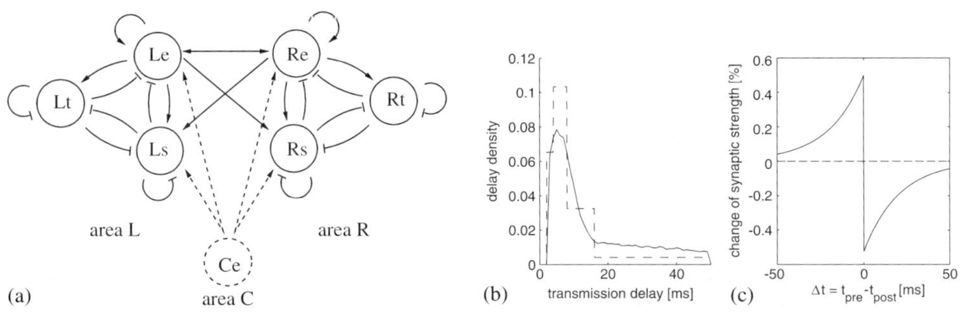

Fig. 1. (a) Network model of two reciprocally connected areas, each consisting of three neuron populations (cf. [5], → exc., ⊣ inh.). In some simulations a third area C delivered common input to areas L and R. (b) Distribution of the inter-areal conduction delays in the model (solid) and in rabbit inter-hemispheric connection of primary visual cortex (dashed) as measured from antidromic latencies (modified from [12]) (c) Modification function of STDP (parameters as in [11]).

if the mean conduction and synaptic delays are not larger than a third or a quarter of the oscillation period [10,9,6]. This implies for the example above (oscillation period 20 ms) that the mean delays for the interhemispheric connection of the primary visual cortex of cat must not exceed values of about 5–7 ms. However, conduction delay measurements have shown that the delay distribution reaches up to delay values of many tens of milliseconds ([12], Fig. 1b).

In this paper we investigate how spike-timing-dependent synaptic plasticity (STDP) can stabilize zero-lag synchrony of gamma-oscillations even at distant cortical sites. Recent experiments have shown that synaptic learning depends asymmetrically on the timing of spike pairs ([7], Fig. 1c): If a postsynaptic spike follows the presynaptic one, the synapse is strengthened, while the synapse is weakened by the inverse order. In an oscillatory regime as in the example above this has the following consequences: If two neurons from distant areas fire synchronously and the delays between the neurons are 5 ms then the postsynaptic spike precedes for about 5 ms the presynaptic spike. Consequently, this should weaken all the synaptic connections with small delays (e.g. < 10 ms) and strengthen only those with long delays (e.g. 10–20 ms).

In a simulation study we assess the dependency of inter-areal phase relations on an asymmetric learning rule and large conduction delays in more detail. We find that the specific strengthening of synapses with long conduction delays in the range of the oscillation period may contribute to zero-phase synchronization of neural activity from distant areas.

2. Methods

We examine the interaction of two reciprocally connected areas (Fig. 1a). Each area consists of three populations of 15×15 neurons (one population excitatory, the other two inhibitory, only one of them receiving extra-areal synaptic input, cf. [5]). The neuron model in this study is a standard spiking neuron similar to [5].

All connections are topographically organized (25 × 25 kernels). The probability of a synapse between two neurons is $p = 0.5$. We chose random synaptic delays $s + \gamma d + \sigma N_{0,1}$ with base delay s, distance d between the neurons and σ^2 the variance of a Gaussian $N_{0,1}$. For local connections we used $s = 0.8$ ms, $\gamma = 0.2$ ms and $\sigma = 0$. For inter-areal connections we used a bimodal delay distribution ($s_{1/2} = 5$ ms/8 ms, $\gamma_{1/2} = 0$, $\sigma_{1/2} = 4$ ms/40 ms) restricted to 2–50 ms to approximate the values found by Swadlow ([12], see Fig. 1b).

For the ratio between maximal local and inter-areal excitation we assumed a value of ten. This assumption agrees with the ratio between local and inter-areal connection densities estimated by neuroanatomical methods [3]. STDP was implemented for the inter-areal connection of excitatory neurons with the same parameters as described in [11] (see Fig. 1c).

3. Results

We examined three initial conditions for the strength of inter-areal connections on excitatory and inhibitory neurons in our model: In the so-called *balanced* regime both connections have strengths corresponding to about a tenth of local excitatory connections, while in the '*dominant-excitation*' and '*dominant-inhibition*' regimes the connections on inhibitory and excitatory neurons are weakened to a hundredth, respectively.

In the first experiment we examined the model without any synaptic plasticity. Activity from the two areas oscillates (about 50 Hz) with anti-phase for the balanced and dominant-excitation regimes (Fig. 2 top row) as expected from the large delays [10,9]. Only in the dominant-inhibition regime the two areas oscillate in phase as in the experiments. However, dominating inhibition seems to be rather unrealistic. Most previous simulation studies even ignored the presence of synapses on inhibitory neurons.

In a second experiment we repeated the simulations with activated STDP (Fig. 2 mid row). The results concerning in-phase vs. anti-phase are the same as before. The behavior of the neural dynamics is additionally embossed by the synaptic dynamics. The STDP changes synaptic efficacy as a function of transmission delay. In the case of anti-phase oscillations (balanced and dominant-excitation regimes) fast synapses with delays $<$ 10 ms and also slower synapses with delays of about 20–30 ms or 40–50 ms are strengthened while other synapses are weakened. For in-phase oscillations (dominant-inhibition regime) the situation is complementary: the fast synapses ($<$ 10 ms) are weakened, and synapses with delays corresponding to one or two oscillation periods (10–20 ms or 30–40 ms) are strengthened.

To examine whether in-phase oscillations can be stabilized by STDP with the more realistic balanced and dominant-excitation regimes, we conducted a third experiment where we synchronized the two areas during activated STDP using common input from area C (Fig. 2 bottom row). After inactivating common input, in-phase oscillations occur even in the balanced and dominant-excitation regime.

Fig. 3 shows the temporal evolution of the synaptic strength distribution during activated STDP (and common input) for the balanced regime. After about 5 s the

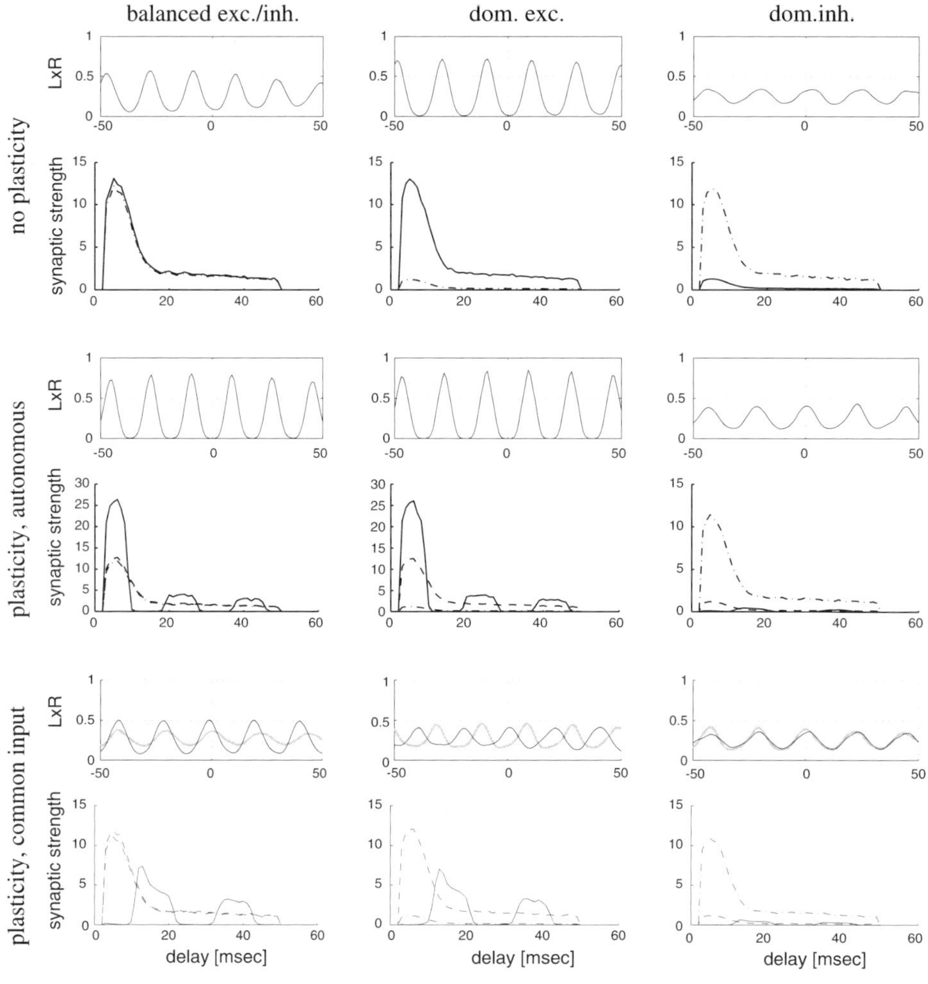

Fig. 2. Simulation results for the three regimes (balanced, dominating excitation, dominating inhibition). Upper row: no synaptic plasticity. Middle Row: Plasticity, but no common input from area C. Bottom row: Plasticity and common input from area C. The correlograms (thin black line) are computed from 5 s recordings (populations Le and Re) after 20 s of activated plasticity (for middle and bottom rows). Thick gray lines (bottom row) correspond to results after only 2 s of activated plasticity (cf. Fig. 3). Distribution of synaptic strength over delays are shown for inter-areal connections between excitatory neurons before (dashed) and after plasticity (solid), and for connections from excitatory to inhibitory neurons (dash-dotted).

distribution approaches its final shape and further changes are minor. For the balanced regime already 2 s of activated STDP are sufficient to obtain in-phase correlations (cf. Fig. 2 bottom row). In the dominant-excitation regime 3–4 s of STDP were necessary (data not shown).

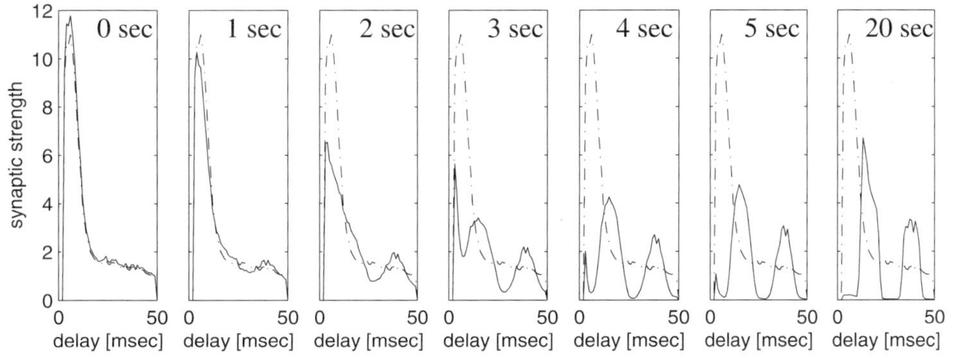

Fig. 3. Impact of STDP on synapses of different delays during oscillations. The plots show the temporal evolution of the distribution of synaptic strength (vs. conduction delay) for a simulation (balanced exc./inh.) with oscillations synchronized with zero-phase lag forced by common input. The histograms show distributions corresponding to inter-areal connections between excitatory neurons (solid), and from excitatory to inhibitory neurons (dash-dotted), and are computed (from left to right) 0, 1, 2, 3, 4, 5, and 20 s after beginning of STDP.

4. Conclusion

The standard explanation for experimentally found zero-lag synchronization of fast activity over long distances is to assume excitatory-to-excitatory connections with sufficiently small conduction delays [10,9]. However, the simulation experiments in this paper show that the distribution of delays that has been experimentally measured between distant areas [12] supports anti-phase rather than zero-lag synchronization.

Previous simulations showed that also long axonal delays can support inter-areal zero-lag synchronization if the delays are close to multiples of the oscillation period [6]. Here we showed how common input in connection with spike-timing-dependent synaptic plasticity [7] can shape the experimentally determined broad distribution of conduction delays [12] in order to support long-range synchronization. In a network model implementing the learning rule of [11] the medium-latency synapses (e.g. < 10 ms) that would produce anti-phase oscillations are weakened, while synapses corresponding to slower axons with delays in the range of one or even two oscillation periods (e.g. 20 or 40 ms) can be strengthened. This effect is complementary to other proposed mechanisms of long-range synchronization (e.g. [2,13,1]).

References

[1] A. Bibbig, H.J. Faulkner, M.A. Whittington, R.D. Traub, Self-organized synaptic plasticity contributes to the shaping of γ and β oscillations in vitro, J. Neurosci. 21 (22) (2001) 9053–9067.

[2] G.N. Borisyuk, R.M. Borisyuk, Y.B. Kazanovich, Temporal structure of neural activity and models of information processing in the brain, in: R. Miller (Ed.), Time and the Brain, Conceptual Advances in Brain Research, Harwood Academic Publishers, Amsterdam, 2000, pp. 331–349 (Chapter 13).

[3] V. Braitenberg, A. Schüz, Anatomy of the Cortex, Statistics and Geometry, Springer, Berlin, 1991.

[4] A.K. Engel, P. König, A.K. Kreiter, W. Singer, Interhemispheric synchronization of oscillatory neuronal responses in cat visual cortex, Science 252 (1991) 1177–1179.
[5] A. Knoblauch, G. Palm, Pattern separation and synchronization in spiking associative memories and visual areas, Neural Networks 14 (2001) 763–780.
[6] A. Knoblauch, G. Palm, Scene segmentation by spike synchronization in reciprocally connected visual areas. II. Global assemblies and synchronization on larger space and time scales, Biol. Cybernet. 87 (3) (2002) 168–184.
[7] H. Markram, J. Lübke, M. Frotscher, B. Sakmann, Regulation of synaptic efficacy by coincidence of postsynaptic APs and EPSPs, Science 275 (1997) 213–215.
[8] M.H.J. Munk, L.G. Nowak, J.I. Nelson, J. Bullier, Structural basis of cortical synchronization. II. Effects of cortical lesions, J. Neurophysiol. 74 (6) (1995) 2401–2414.
[9] R. Ritz, W. Gerstner, U. Fuentes, J.L. van Hemmen, A biologically motivated and analytically soluble model of collective oscillations in the cortex. II. Applications to binding and pattern segmentation, Biol. Cybernet. 71 (1994) 349–358.
[10] H. Sompolinsky, D. Golomb, D. Kleinfeld, Global processing of visual stimuli in a neural network of coupled oscillators, PNAS 87 (1990) 7200–7204.
[11] S. Song, K.D. Miller, L.F. Abbott, Competitive Hebbian learning through spike-timing-dependent synaptic plasticity, Natl. Neurosci. 3 (9) (2000) 919–926.
[12] H. Swadlow, Information flow along neocortical axons, in: R. Miller (Ed.), Time and the Brain, Conceptual Advances in Brain Research, Harwood Academic Publishers, Amsterdam, 2000, pp. 131–155 (Chapter 4).
[13] R.D. Traub, M.A. Whittington, I. Stanford, J.G.R. Jeffreys, A mechanism for generation of long-range synchronous fast oscillations in the cortex, Nature 383 (1996) 621–624.

Available at
www.ComputerScienceWeb.com
POWERED BY SCIENCE @ DIRECT®

ELSEVIER

Neurocomputing 52–54 (2003) 307–312

NEUROCOMPUTING

www.elsevier.com/locate/neucom

Registration of a 3D mouse brain atlas with brain microstructure data

Wonryull Koh*, Bruce H. McCormick

Department of Computer Science, Texas A&M University, College Station, TX 77843-3112, USA

Abstract

The brain tissue scanner http://research.cs.tamu.edu/bnl/ is being used to explore the topology and geometry of mouse brain architecture at two levels: its gross anatomy and its microstructure. Aligned volume data set acquisition using this instrument is described. A 3D atlas of mouse brain partitions the gross anatomical structures of brain by a solid model. When registered with the microstructure data, the model serves to organize and index the microstructure data. Volume element partitioning of the mouse brain data can provide a common coordinate framework for visualization and morphological modeling of neurons and can serve as containers to grow synthetic neurons.

Keywords: 3D brain atlas; Brain microstructure; Brain data registration

1. Objectives

Our objectives are to

- build an annotated solid model (3D atlas) of a mouse brain gross anatomy,
- register this model with the brain microstructure data, and thereby
- organize and index the microstructure data.

* Corresponding author.

E-mail addresses: wkoh@cs.tamu.edu (W. Koh), mccormick@cs.tamu.edu (B.H. McCormick).

0925-2312/03/$ - see front matter © 2002 Elsevier Science B.V. All rights reserved.
doi:10.1016/S0925-2312(02)00793-2

2. Introduction

We use the *brain tissue scanner* [7] to explore the topology and geometry of brain's architecture at two levels: its microstructure and its gross anatomy. Aligned volume data set acquisition using this instrument is described, and the methodology extends to volume data acquisition by confocal and two-photon microscopes as well. At the microstructure level, we have designed an exoskeleton microstructure database [4,5] that supports 3D reconstruction and modeling of neurons and cortical networks. At the gross anatomy level, we are building a solid model of the mouse brain, which exhibits cortical areas and brain nuclei and can also accommodate the gross anatomical description of encapsulated fibers in nerve tracts and blood vasculature.

The solid model of a mouse brain organizes and indexes microstructure data. We partition the solid model following the gross anatomical structures of brain. We contend that such a volume element partitioning of the 3D mouse brain model can provide a common coordinate framework for visualization and morphological modeling of neurons [1,8], and can also serve as containers to grow the synthetic neurons, using stochastic models derived from observed biological neurons. The global indexing and organization of the microstructure data at the gross anatomy level are possible because we assign each neuron to the *big voxel* (see Section 3.1) that houses its soma, and position each big voxel within a solid model of brain.

For rat and mouse brains, there are 2D brain atlases (e.g., [3,13,14,17,9]), some with more than 1000 individual structures listed. We are constructing 3D atlases for the rat and mouse brains, modeled from both the 2D structures and additional partitioning by hexahedral finite elements [1]. This solid modeling, annotated by neuroanatomical nomenclature, will facilitate later integration of our system with other anatomical, electrophysiological, and neurochemical database systems (e.g., [11,12,19]).

3. Methods

3.1. Collection of the volume data

The mouse brain is embedded in a specimen block that is 15 mm *A–P* × 12 mm *M–L* × 9 mm *D–V* (Fig. 1). The brain tissue scanner [7] concurrently sections and scans the brain tissue at a nominal 0.5 μm thickness with a nominal 0.625 mm effective knife width. For horizontal scanning, the volumetric data is obtained from scanning a total of 360,000 aligned sections.

The brain tissue scanner [7] uses knife-edge scanning to generate an aligned image stack from consecutive serial sections. Each stroke of knife is the full length of the specimen block, and therefore the length of each section is 15 mm. The width of each section is nominally 0.625 mm for the 40× objective, 2.5 mm for the 10× objective. The effective knife width of 0.625 mm represents the digitized portion of the field of view (FoV) of the 40× objective, i.e., the diameter of the largest circular area that can be simultaneously viewed by the linear sensor array of a line-scan camera. For horizontal color and monochrome scanning using the 40× objective, each section

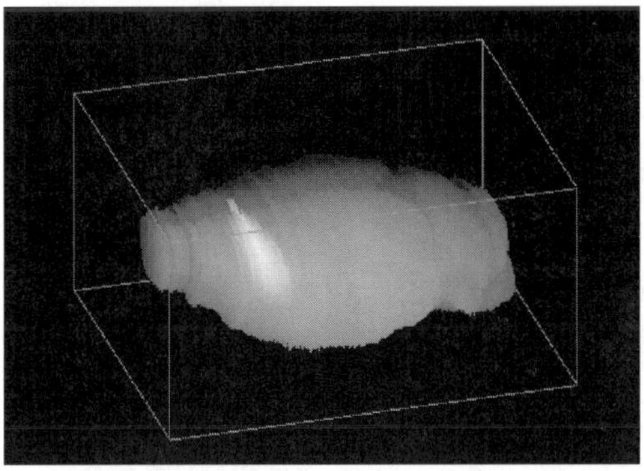

Fig. 1. Schematic of a mouse brain embedded in a specimen block.

(15 mm × 0.625 mm × 0.5 μm) line sampled at 305 nm intervals yields an image that is 49,152 line samples of 2048 pixels. Assuming 3 bytes/pixel, each color image then contains 0.302 GB of data in the worst case. This data size is less than 2 GB (the memory size on our servers), and therefore the data can be kept concurrently in memory for filtering and extraction of the regions of interest (ROIs). Moreover, when the strips are cut parallel to *M–L* or *D–V* axis, the data size contained in each image section decreases, and the data acquisition can be easily accommodated. For storage and data organization, each section is divided into square images (0.625 mm × 0.625 mm) by uniformly dividing the length of each section by the width of the knife. For example, for a specimen block length of 15 mm, each section is partitioned into 24 images. Each square image is stored in an image stack that constitutes a *big voxel* whose dimensions are 0.625 mm × 0.625 mm × 0.625 mm.

3.2. Organization of the volume data

The volumetric data is organized as a collection of $24 \times 20 \times 15 = 7200$ big voxels where each big voxel is cubic and 0.625 mm × 0.625 mm × 0.625 mm in size (Fig. 2). This collection of big voxels forms a block (15 mm × 12.5 mm × 9.375 mm) whose dimensions are nearly identical to those of the specimen block in which the mouse brain was embedded for scanning. Following the conventions employed by the National Library of Medicine's Visible Human Project [6,10], we can then visualize our volume data as serial coronal slices (perpendicular to the *A–P* axis). We can also assemble sagittal slices (perpendicular to the *M–L* axis) or horizontal slices (perpendicular to the *D–V* axis). The big voxels located at the outer boundaries of the block are treated as "dark" big voxels, i.e., as big voxels that contain no data but that facilitate storage and organization of the big voxels.

The FoV of the objective determines the voxel dimensions of 0.625 mm. Hence, when the volumetric data is obtained from confocal or two-photon microscopy, or from the BTS with a different effective knife width, the physical dimensions of the big voxel and the total number of big voxels would change depending on the objective.

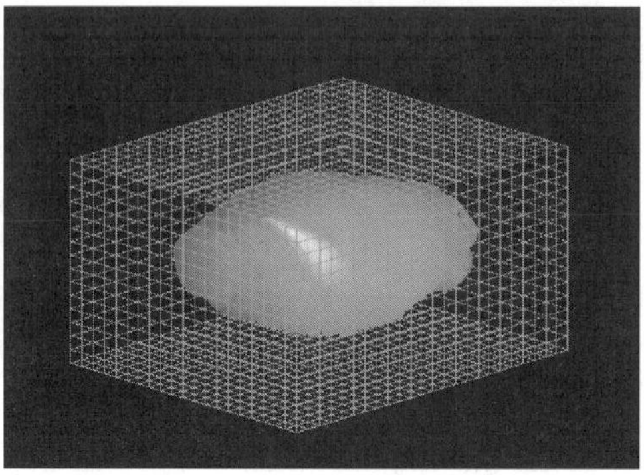

Fig. 2. Volumetric data organization as a collection of 7200 *big voxels*.

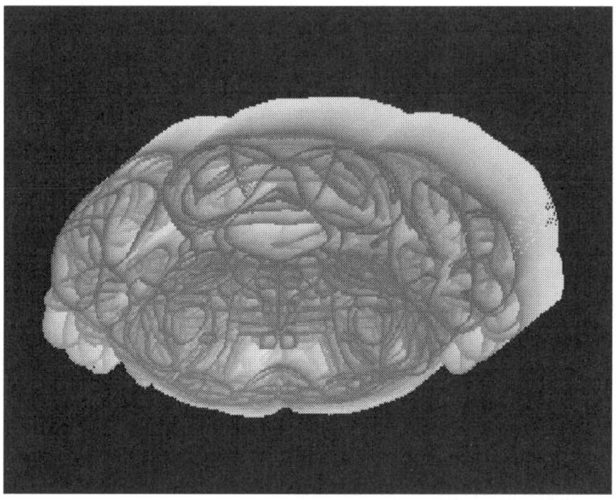

Fig. 3. Partial 3D modeling of mouse cerebellum and medulla, based on Paxinos and Franklin's 2D coronal diagrams [13].

3.3. Partitioning the brain solid model by anatomical structure

A 3D model of mouse brain cortical areas and nuclei has been reconstructed from the contours extracted from the 2D image scans and diagrams of Paxinos and Franklin [13], where the image scans oriented perpendicular to any of the three axes are provided. Each volume element is modeled from the compilation of 2D contours associated with an anatomical structure. Our brain solid model is partitioned by these volume elements, each based on a delineated anatomical structure within a common coordinate system. Fig. 3 shows a 3D coronal section of mouse brain that contains the cerebellum, pons and medulla. Fig. 4 models a 3D coronal section of mouse brain showing parts of the corpus callosum, hippocampus, thalamus, hypothalamus and cerebral cortex.

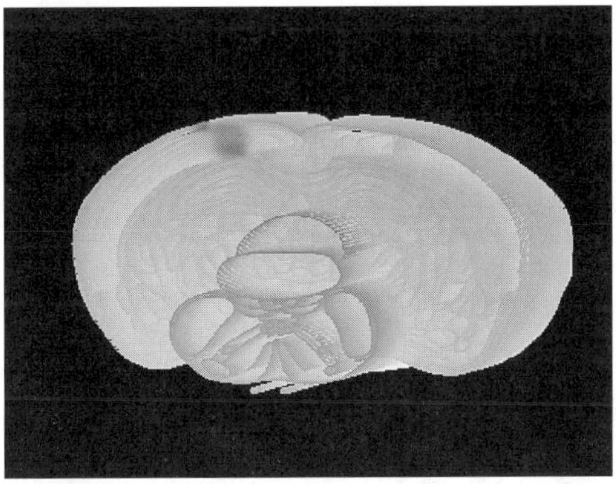

Fig. 4. Partial 3D modeling of mouse thalamus, hypothalamus, and hippocampus, based on Paxinos and Franklin's 2D coronal diagrams [13].

3.4. Registration of the brain solid model with the microstructure data

We spatially decompose the mouse brain, embedded in a 15 mm $\times$ 12 mm $\times$ 9 mm block, into a mesh of 7200 *big voxels*, where each big voxel is 0.625 mm$\times$0.625 mm$\times$ 0.625 mm in size. Microstructure data then are registered with our mesh of big voxels by assigning each neuron to the big voxel that houses its soma. The reconstructed microstructure data is then organized and indexed by the 3D brain atlas by superimposing our brain solid model over the mesh of big voxels. This registration allows global indexing and organization of microstructure data within a common coordinate framework at the gross anatomy level.

Our initial big voxels are equal sized and may intersect more than one gross anatomical structure when superimposed with the brain solid model. To accommodate the registration of finite elements that are smaller than our initial big voxel size, we use an octree-based refinement algorithm that is an extension of the grid-based algorithm [2,15]. The octree is a well-known tool for organizing spatial data with applications in solid modeling, mesh generation, and various other fields [18]. We use a modified octree concept [16,18], the "27-tree" method [16], where we start with a big voxel from the initial decomposition above. The big voxel is then split up into 27 sub-voxels (octants) that are refined recursively until the sizes of all octants fall below a predefined threshold that is set by the size of the smallest anatomical structure to be represented.

Acknowledgements

This paper has been supported in part by the Texas Higher Education Coordinating Board grant ATP-000512-0146-2001.

References

[1] S. Chow, Finite Element Decomposition of the Human Neocortex, M.S. Thesis, Department of Computer Science, Texas A&M University, 1998.

[2] J.D. Foley, A. van Dam, S.K. Feiner, J.F. Hughes, Computer Graphics: Principles and Practice in C, 2nd Edition, Addison-Wesley, Reading, MA, 1995.

[3] D.M. Jacobowitz, L.C. Abbott, Chemoarchitectonic Atlas of the Developing Mouse Brain, CRC Press, Boca Raton, FL, 1997.

[4] W. Koh, B.H. McCormick, Organization and visualization of brain tissue volume data, Neurocomputing 38–40 (2001) 1643–1650.

[5] W. Koh, B.H. McCormick, Brain microstructure database system: an exoskeleton to 3D reconstruction and modeling, Neurocomputing 44–46 (2002) 1099–1105.

[6] S. Koslow, S. Hyman, Human brain project: a program for the new millennium, Einstein Q. J. Biol. Med. 17 (2000) 7–15.

[7] B.H. McCormick, Development of the Brain Tissue Scanner, Technical Report, Department of Computer Science, Texas A&M University, College Station, TX, January 4, 2002.

[8] B.H. McCormick, R.W. DeVaul, W.R. Shankel, J.H. Fallon, Modeling neuron spatial distribution and morphology in the developing human cerebral cortex, Neurocomputing 32–33 (2000) 897–904.

[9] Mouse Brain Library, http://www.mbl.org.

[10] National
Institute of Mental Health, The Human Brain Project, http://www.nimh.nih.gov/neuroinformatics/index.cfm.

[11] NeuronDB, http://ycmi.med.yale.edu/senselab/neurondb/

[12] Olfactory Receptor DataBase (ORDB), http://ycmi.med.yale.edu/senselab/ordb/

[13] G. Paxinos, K.B.J. Franklin, The Mouse Brain in Stereotaxic Coordinates with CDROM, Academic Press, New York, 2001.

[14] G. Paxinos, C. Watson, The Rat Brain in Stereotaxic Coordinates with CDROM, Academic Press, New York, 1997.

[15] R. Schneiders, A grid-based algorithm for the generation of hexahedral element meshes, Eng. Comput. 12 (1995) 168–177.

[16] R. Schneiders, Octree-based hexahedral mesh generation, Int. J. Comput. Geometry Appl. 10 (4) (2000) 383–398.

[17] L. Swanson, Brain Maps: Structure of the Rat Brain with 2 CDROMs, 2nd Edition, Elsevier, Amsterdam, 1999.

[18] J.F. Thompson, B.K. Soni, N.P. Weatherill, Handbook of Grid Generation, CRC Press, Boca Raton, FL, 1999.

[19] XANAT, http://redwood.ucdavis.edu/bruno/xanat/xanat.html.

W. Koh is a Ph.D. candidate in Computer Science at Texas A&M University. She received her B.S. degree in Computer Science and Mathematics from the University of Texas at Austin, and M.S. degree in Computer Science at Texas A&M University. Her research interests include computer graphics, brain mapping, cortical networks, visualization, and computational neuroscience.

B.H. McCormick received his B.S. and Ph.D. degrees in Physics from MIT and Harvard University, respectively. He was Professor of Computer Science and Physics at the University of Illinois at Urbana-Champaign. At the University of Illinois at Chicago and Texas A&M University, he headed the Information Engineering and the Computer Science departments, respectively. Currently, he is a professor of Computer Science and director of the Brain Networks Laboratory, at Texas A&M University. His research areas include computer graphics and visualization, brain mapping, and neural networks. His attention is currently focused on the brain tissue scanner and its potential to generate data for biologically accurate 3D reconstruction of brain networks.

Available at
www.ComputerScienceWeb.com
POWERED BY SCIENCE DIRECT®

ELSEVIER

Neurocomputing 52–54 (2003) 313–320

NEUROCOMPUTING

www.elsevier.com/locate/neucom

Variability of postsynaptic responses depends non-linearly on the number of synaptic inputs

Jutta Kretzberg[a,*], Terrence Sejnowski[a], Anne-Kathrin Warzecha[b], Martin Egelhaaf[b]

[a] *Computational Neurobiology Laboratory, The Salk Institute, 10010 N. Torrey Pines Road, La Jolla, CA 92037 USA*

[b] *Lehrstuhl Neurobiologie, Universität Bielefeld, Postfach 100131, 33501 Bielefeld, Germany*

Abstract

A conductance-based model for synaptic transmission and postsynaptic integration reveals how postsynaptic responses and their variability depend on the number of synaptic inputs. With increasing number of balanced stochastic excitatory and inhibitory inputs, the postsynaptic responses and their variance first increase and then decrease again. This non-linearity can be attributed to an anti-correlation between the total excitatory and inhibitory currents. The anti-correlation, which occurs even though the conductances of the individual synapses vary independently of each other, is determined by the total synaptic conductance and grows with the number of inputs. As the number of inputs increases, the membrane potential comes increasingly closer to the resting level.

Keywords: Model; Graded synaptic transmission; Number of inputs; Synaptic conductances; Correlated synaptic currents

1. Introduction

A neuron that is synaptically contacted by N input neurons does not necessarily receive input from all those neurons simultaneously. The presynaptic neurons contributing to the response of an integrating neuron are mainly those that are stimulated. This is a plausible assumption, for instance, when a stimulus is moved across the receptive

* Corresponding author.
E-mail address: jutta@salk.edu (J. Kretzberg).

0925-2312/03/$ - see front matter
doi:10.1016/S0925-2312(02)00797-X

field of a visual interneuron. We analyze the effect of the number of active synaptic inputs on the variability of the postsynaptic responses with a conductance-based model that was fitted to experimentally determined data of fly motion-sensitive interneurons [3]. These cells receive an equal number of excitatory and inhibitory synaptic inputs, that, depending on the stimulus, can both be activated simultaneously. Here we model one special situation by assuming that both types of inputs have equal maximum conductance. We assume all presynaptic signals to be independent from each other. The number of pairs of excitatory and inhibitory inputs was varied in the range from 1 to 1000, corresponding to stimuli of different sizes, ranging from a tiny stimulus to one that covers the entire receptive field of the fly visual interneuron.

With this conductance-based, one-compartment model of synaptic transmission and postsynaptic integration we show that not only the average postsynaptic responses but also their variability depend non-linearly on the number of active synaptic inputs. This non-linear dependence is accompanied by an anti-correlation between total excitatory and inhibitory currents that occurs even though the individual presynaptic membrane potentials are uncorrelated.

2. The model

2.1. Model of synaptic transmission and postsynaptic integration

To analyze how postsynaptic responses depend on the number of synaptic inputs we use a conductance-based, one-compartment model for synaptic transmission and postsynaptic integration. Analogous to the graded information transfer between neurons in the retina or the olfactory bulb of vertebrates (review Ref. [4]) or between fly visual neurons, synaptic transmission is not restricted to presynaptic spikes. The graded de- and hyper-polarizations of the presynaptic membrane potential induce conductance changes of the postsynaptic membrane via a sigmoidal transfer function.

The resulting postsynaptic potentials contribute to the continuously varying signal $V(t)$ according to a one-compartment model of a passive membrane.

$$\frac{\mathrm{d}V(t)}{\mathrm{d}t} = \frac{1}{C} \cdot \left(g_m \cdot (V_{\mathrm{rest}} - V(t)) + \sum_{k}^{N_e} s_{e_k}(V_{\mathrm{pre}_k}) \cdot g_e \cdot (E_e - V(t)) + \sum_{l}^{N_i} s_{i_l}(V_{\mathrm{pre}_l}) \cdot g_i \cdot (E_i - V(t))\right),$$

with postsynaptic membrane potential $V(t)$, membrane capacitance C, resting potential V_{rest}, leak conductance g_m, maximal synaptic conductances g_e and g_i, number of synaptic inputs N_e and N_i, reversal potentials E_e and E_i.

The maximum conductance of each individual synapse g_e or g_i is scaled at every time step with $s_{e_k}(V_{\mathrm{pre}_k})$ or $s_{i_l}(V_{\mathrm{pre}_l})$, a scaling factor depending on the presynaptic membrane potential V_{pre}. In the case of excitatory synapses the scaling factor s_e is determined according to the equations

$$\frac{\mathrm{d}s_e(V_{\mathrm{pre}})}{\mathrm{d}t} = \frac{1}{\tau_e} \cdot (\bar{s}_e(V_{\mathrm{pre}}) - s_e),$$

$$\bar{s}_e(V_{\mathrm{pre}}) = \frac{1}{1 + \exp((V_{h_e} - V_{\mathrm{pre}})/\Delta_e)},$$

where τ_e $(=\tau_i)$ denotes the synaptic time constant with which the steady state scaling factor $\bar{s}_e$ $(\bar{s}_i)$ for a given value of V_{pre} is reached. The steady state scaling factor $\bar{s}_e$ $(\bar{s}_i)$ is a sigmoidal function of V_{pre}. Δ_e $(=\Delta_i)$ determines the slope of the transfer function. V_{h_e} $(=V_{h_i})$ is the value of V_{pre} for which the half maximum conductance is reached. Equivalent equations are used for inhibitory synapses.

Unless otherwise specified parameter values are $C = 4.2 \cdot 10^{-10}$ F, $V_{\mathrm{rest}} = -50$ mV, $g_m = 2 \cdot 10^{-7}$ S, $g_e = g_i = 2 \cdot 10^{-9}$ S, $N_e = N_i = 1 \ldots 1000$, $E_e = 0$ mV, $E_i = -100$ mV, $\tau_e = \tau_i = 0.1$ ms, $V_{h_e} = V_{h_i} = 1$ mV, $\Delta_e = \Delta_i = 0.5$ mV. The parameter values are chosen to be in the range experimentally determined for visual interneurons of the fly [1]. Identical model parameters for excitatory and inhibitory synapses are a good approximation for this system.

2.2. Model of spike generation

A simple model of spike generation that was used before to reproduce and explain experimental data of fly visual interneurons [3,5] transformed the postsynaptic membrane potential $V(t)$ into spikes. This model consists of a variable threshold $\theta(t)$ that is compared with the membrane potential $V(t)$. A spike is generated if $V(t) > \theta(t)$. The spike threshold is calculated for each time step according to the equation

$$\theta(t) = \begin{cases} \infty & \text{if } t - t^* \leqslant \gamma^{\mathrm{ref}}, \\ \theta_0 + \eta(t - t^*) + \rho(t) & \text{if } t - t^* > \gamma^{\mathrm{ref}} \end{cases}$$

with t^* being the time when the previous spike occurred and γ^{ref} the absolute refractory period. θ_0 is a constant basis threshold like in a standard integrate and fire neuron, $\eta(t - t^*) = \eta_0/(t - (t^* + \gamma^{\mathrm{ref}}))$ is the influence of the relative refractoriness with weight constant η_0. $\rho(t_i) = -\rho_0/T \cdot \sum_{j=1}^{T} 1/j \cdot (V(t_i) - V(t_{i-j}))$ is the influence of the membrane potential changes within the last T data points, with weight constant ρ_0 and actual membrane potential $V(t_i)$. Parameter values: $\theta_0 = 1$ mV, $\gamma^{\mathrm{ref}} = 2$ ms, $\eta_0 = 20$ ms $\cdot$ mV, $\rho_0 = 3.75$, $T = 1$ ms [3].

2.3. Generation of input signals

The membrane potentials of the presynaptic neurons fluctuate stochastically and independently from each other. This membrane potential noise is modeled as Gaussian white noise that is low-pass filtered and has a variance of 2.5 mV^2. It was adjusted to fit the variance and the power spectrum of intracellular recorded membrane potential noise in visual interneurons of the fly [3].

The duration of the input traces amounts to 1.32 s at a temporal resolution of 0.01 ms. This resolution was also used for all model simulations. All results shown here are averages over at least 100 simulations with different random inputs (see legends).

3. Results

3.1. The response of the integrating neuron with balanced excitation and inhibition depends non-linearly on the number of active synaptic inputs

First, the spike count increases steeply with increasing input number, then it reaches a maximum and decreases (Fig. 1A). Also the spike count variance (Fig. 1B) and the variance of the graded fluctuating postsynaptic potential that underly the spike responses

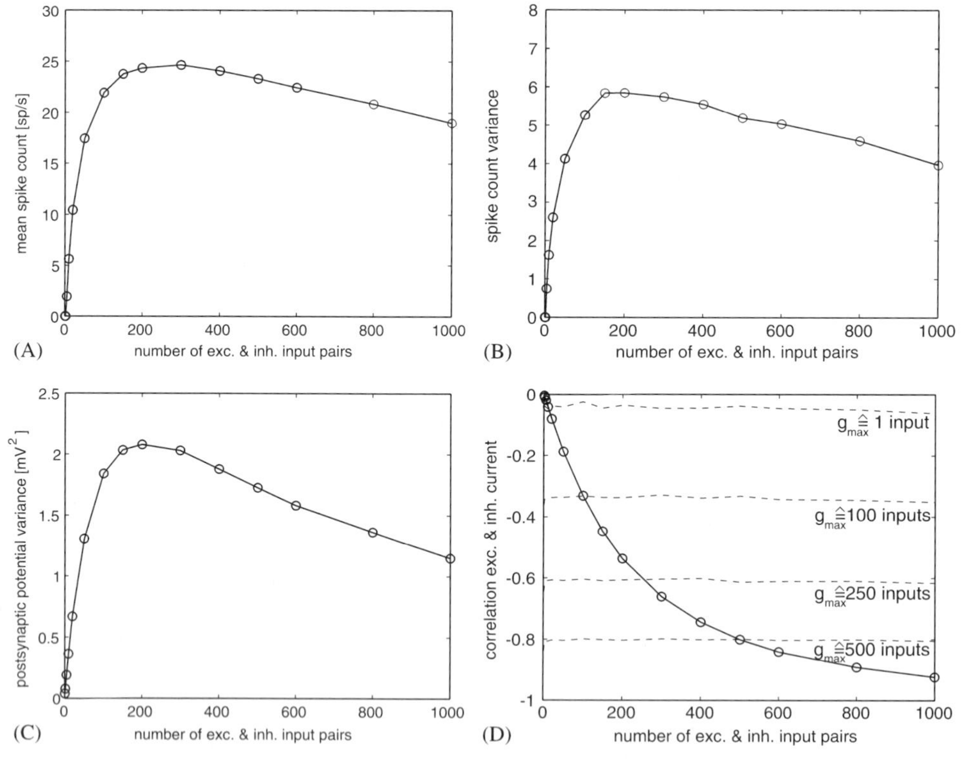

Fig. 1. The mean spike count (A), the spike count variance (B) and the variance of the postsynaptic potential (C) show a non-linear dependence on the number of active synaptic inputs. The anti-correlation between total excitatory and inhibitory currents assumes increasingly larger negative values with increasing number of inputs (D, solid line). When the total synaptic conductance is scaled by the number of inputs to keep it constant, the correlation between excitatory and inhibitory currents does not depend on the number of inputs (dashed lines). For the dashed lines 300 independent trials of 1.32 s duration were averaged, for the solid lines 3000 trials.

(Fig. 1C) depend in a non-linear way on the number of inputs. The similarity of Fig. 1C to Figs. 1A and B shows that the non-linear dependence is not due to the non-linearity of the spike threshold but is a property of the integration of synaptic inputs by a passive membrane.

3.2. Postsynaptic responses depend on the correlation of excitatory and inhibitory currents

The non-linear dependence of the postsynaptic responses on the number of inputs parallels the finding that the total excitatory and inhibitory currents are strongly anti-correlated for large numbers of synaptic inputs as is shown by the solid line in Fig. 1D. This anti-correlation occurs even though the membrane potentials of the individual presynaptic neurons are absolutely uncorrelated and the fluctuations of $V(t)$ are small compared to the distances to the reversal potentials. The reason for the anti-correlation between the total currents is the coupling of both excitatory and inhibitory currents to the postsynaptic membrane potential value. When the postsynaptic potential moves towards the excitatory reversal potential, the inhibitory driving force increases, leading to a large current opposing the excitatory current. This push-pull mechanism keeps the membrane potential close to the resting potential, resulting in fluctuations with a standard deviation as small as 1.5 mV.

The anti-correlation of the currents is determined by the maximum total synaptic conductance $g_{\max} = N_{\text{syn}} \cdot g_{\text{syn}}$ that can be reached when all N_{syn} synapses are maximally activated. The anti-correlation does not depend on the number of independent stochastic processes that are used as input signals. To obtain this result, the total maximum synaptic conductance $g_{\max}$ was kept constant by scaling $g_{\text{syn}} = g_e = g_i$ with the increase of the number of synaptic inputs N_{syn}. In this case the correlation between the total excitatory and inhibitory currents is constant for varying numbers of independent inputs and depends only on the value of $g_{\max}$ (dashed lines in Fig. 1D). The correlation function for the standard case (solid line in Fig. 1D) where $g_{\max}$ increases linearly with the number of inputs intersects the correlation function for a given $g_{\max}$ (dashed lines) exactly at the number of inputs that leads to the same maximum synaptic conductance.

3.3. Postsynaptic responses depend on the number of independent stochastic processes

Since the correlation between excitatory and inhibitory currents depends on the total maximum synaptic conductance it is not surprising that the postsynaptic potential variance depends quantitatively on the value chosen for the maximum conductance g_{syn} of one synapse. For a given number of synaptic inputs, the variance is larger for larger g_{syn}. Moreover, the maximum variance is reached at smaller numbers of inputs when g_{syn} is increased (Fig. 2A).

In contrast to the correlation of currents the membrane potential variance is not solely determined by the maximum total synaptic conductance $g_{\max} = N_{\text{syn}} \cdot g_{\text{syn}}$. The stars in Fig. 2A indicate two configurations with the same total synaptic conductance $g_{\max}$. The left star indicates a situation with half the number of independent synaptic

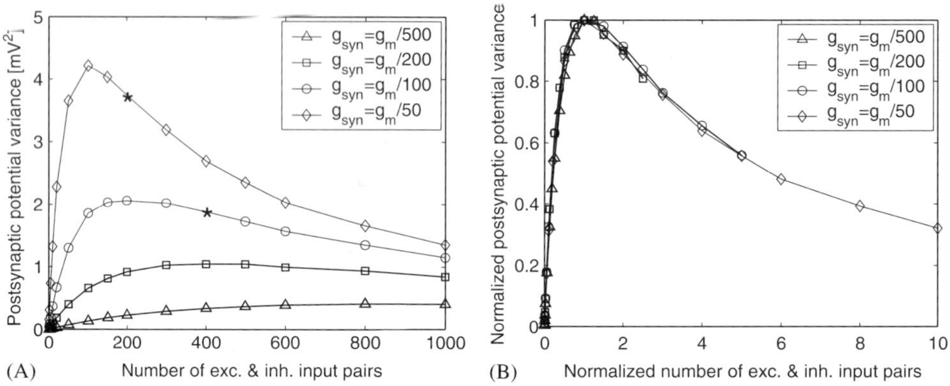

Fig. 2. (A) The variance of the postsynaptic potentials depends on the synaptic conductance. (B) When the curves shown in A are scaled in x- and y-direction by setting the maximum variance and the number of inputs at that maximum to 1, all curves have qualitatively the same shape. Between 100 and 300 independent trials of 1.32 s duration were averaged for each data point. Stars are explained in Section 3.3.

inputs N_{syn} and twice the conductance g_{syn} for each synapse as in the case indicated by the right star. The combination of stronger but less inputs leads to a dramatically higher postsynaptic variability than that of more but weaker synaptic inputs.

Despite the quantitative differences, qualitatively the variance curves do not differ for different values of g_{syn}. When the curves are scaled by setting both the maximum variance and the number of inputs at the maximum to one, all curves are similar (Fig. 2B).

The same analysis was done for the other model parameters. For all parameter combinations tested the postsynaptic response and its variance depends non-linearly on the number of inputs.

4. Discussion

We have shown that the strength and the variance of postsynaptic responses (membrane potential fluctuations as well as spike counts) depend in a non-linear way on the number of synaptic inputs. This non-linear dependence of postsynaptic spike responses on the number of synaptic inputs has recently been found experimentally in dynamic clamp stimulation of granule cells of the dentate gyrus in rats (S. Harney and M. Jones, "Temporally Precise Spiking in the Presence of Synaptic Noise", CNS abstracts, Elsevier Preprint server). Similar results were reported in rat somatosensory cortex pyramidal neurons for the amplitude of sub-threshold membrane potential fluctuations [2]. Hence, this non-linear dependence on the number of synaptic inputs is a basic property of neurons. It is not induced by the spike non-linearity but by the anti-correlation of the total excitatory and inhibitory synaptic currents, which depends on the total synaptic conductance.

Acknowledgements

We thank Matthias Bethge and Jörn Anemüller for helpful discussions, Jean-Marc Fellous for comments on the manuscript, and Jens Lindemann for computer support. This work was supported by the Sloan Foundation, the Swartz Foundation and the Deutsche Forschungsgemeinschaft.

References

[1] A. Borst, J. Haag, The intrinsic electrophysiological characteristics of fly lobula plate tangential cells: I. Passive membrane properties, J. Computat. Neurosci. 3 (1996) 313–336.

[2] F.S. Chance, L.F. Abbott, A.D. Reyes, Gain modulation from background synaptic input, Neuron 35 (2002) 773–782.

[3] J. Kretzberg, M. Egelhaaf, A.-K. Warzecha, Membrane potential fluctuations determine the precision of spike timing and synchronous activity: a model study, J. Computat. Neurosci. 10 (1) (2001) 79–97.

[4] A. Roberts, B.M.H. Bush, Neurones without Impulses, Cambridge University Press, Cambridge, London, New York, 1981.

[5] A.-K. Warzecha, J. Kretzberg, M. Egelhaaf, Reliability of a fly motion-sensitive neuron depends on stimulus parameters, J. Neurosci. 20 (23) (2000) 8886–8896.

Jutta Kretzberg studied 'applied computer science in the natural sciences' at the Department of Technology, University of Bielefeld, Germany, where she received a diploma in computer science in 1997. In her Ph.D. thesis she analyzed aspects of the reliability of spike responses, based on computer models for fly visual neurons. In 2001, she received a Ph.D. in biology from University of Bielefeld. J. Kretzberg currently is a Sloan-Swartz fellow at the Salk Institute for Biological Studies, La Jolla, CA, working in the groups of T. Sejnowski (Salk) and W. Kristan (UCSD). Her major research interests are the reliability of neuronal computations and the comparison of spiking and non-spiking neuronal information processing.

Terrence Sejnowski is an Investigator with Howard Hughes Medical Institute and a Professor at the Salk Institute for Biological Studies where he directs the Computational Neurobiology Laboratory. He is also Professor of Biology at the University of California, San Diego, where he is Director of the Institute for Neural Computation. Dr. Sejnowski received a B.S. in physics from the Case-Western Reserve University, a M.A. in physics from Princeton University, and a Ph.D. in physics from Princeton University in 1978. In 1988, Dr. Sejnowski founded Neural Computation, published by the MIT press. The long-range goal of Dr. Sejnowski's research is to build linking principles from brain to behavior using computational models.

Anne-Kathrin Warzecha received her diploma and doctoral degree in Biology from University of Tübingen, Germany in 1991 and 1994, respectively. During 1995 she held a postdoctoral position at the Centre for Visual Sciences, Australian National University, Canberra, Australia. Since then she has been working as a research fellow in the Dept. of Neurobiology at Bielefeld University, Germany. Major research interests of A.-K. Warzecha cover the reliability of neuronal computations, the mechanisms underlying motion vision and the visual control of behavior.

Martin Egelhaaf received his diploma and doctoral degree in Biology from Tübingen University (Germany) in 1980 and 1984, respectively. 1984-1993: research scientist at the Max-PlanckInstitute for Biological Cybernetics (Tübingen, Germany). 1994 - 1995: leader of a research group at the Centre of Visual Sciences (Australian National University, Canberra, Australia). Since 1995: Professor and head of the Department of Neurobiology at Bielefeld University (Bielefeld, Germany). Major research interest: visual information processing in biological systems.

Available at
www.ComputerScienceWeb.com
POWERED BY SCIENCE DIRECT®

ELSEVIER

Neurocomputing 52–54 (2003) 321–326

NEUROCOMPUTING

www.elsevier.com/locate/neucom

Synaptic plasticity determines the character of interaural-time-difference representation

Christian Leibold*, J. Leo van Hemmen

Physik Department, Technische Universität München, D-85747 Garching bei München, Germany

Abstract

Neuronal activity of cells in the first binaural brainstem nucleus, in mammals the MSO, depends on interaural time differences (ITDs). The neurons thereby resolve disparities of about 10 μs. Based on a computational model for the development of synaptic couplings, we give restrictions for a recently proposed coding hypothesis based on a monotonic dependence of the firing rate upon the ITD. We find that synaptic plasticity at MSO neurons with high best frequencies favors a rate-place code with a high variability of the tuning curves, as in the classical Jeffress model, whereas low best frequencies give rise to monotonic ITD dependence.

Keywords: Spike-timing-dependent synaptic plasticity; Jeffress model; Psychophysical modeling; Sound localization

1. Introduction

At a neuronal level, azimuthal sound localization operates through determining and representing (coding) interaural time differences (ITDs). Recently McAlpine et al. [10] have proposed an ITD representation mechanism for low-frequency cells. In contrast to the classical model of Jeffress [5], it is not the ITD at which a neuron fires at maximal rate and in this way represents stimulus direction but instead it is the *amount of firing activity in an assembly* that encodes the direction. Low rates thereby represent ipsilateral positions, high rates contralateral ones. The number of action potentials in a population of cells is thus a reliable measure of the actual stimulus azimuth φ. A prerequisite for this so-called *rate-gradient code* is that the tuning curves of all cells have a maximum at similar contralateral leading ITDs.

* Corresponding author.
E-mail address: cleibold@ph.tum.de (C. Leibold).

0925-2312/03/$ - see front matter
doi:10.1016/S0925-2312(02)00800-7

We have simulated a spike-timing-dependent learning algorithm [1] with $1/f$-periodic input for excitatory synapses, which suffices to explain ITD tuning curves at a sub-millisecond time scale, and find that the resulting distribution of best interaural phase differences (IPD) varies in dependence upon input frequency. Here IPD is defined as ITD times $2\pi f$ where f stands for frequency. Low-frequency inputs yield synaptic delay distributions that are in accordance with the hypothesis of McAlpine et al. [10] whereas high-frequency inputs favor a Jeffress-like interpretation [5]. As a result we provide a framework relating the coding concept to biological restrictions that are typical to the animal, such as head diameter and the evolutionary necessity of lateralization acuity.

2. Psychophysical model

McAlpine et al. [10] suggested that the amount of neuronal activity within one population of cells accounts for the sensation of a sound-source azimuth. Since a brain is symmetric, there are two inverse rate representations corresponding to the two hemispheres. The simplest assumption possible is that the difference $\Xi(\varphi)$ between the activities of right and left hemisphere is the key quantity for perception. We therefore approximate the rate difference Ξ by a linear function of the ITD τ so that $\Xi(\tau)=\gamma\tau$. This turns out to be a good estimate, at least in the range of naturally occurring ITDs. The respective ITDs depend on the sound speed c_S and the head diameter d through $\tau(\varphi)=[\varphi+\sin(\varphi)]\ d/(2c_S)$.

In order to test the above hypothesis regarding $\Xi(\varphi)$, we have modeled a simple psychoacoustic two-alternative forced-choice experiment where a subject has to discern two possible sound sources at angles φ and $\varphi+\delta\varphi$. Defining a discrimination criterion that is optimal in a sense of minimal decision error based on Gaussian statistics we obtain an analytic expression for the probability P_c of correct answers, viz.,

$$P_c(\varphi,\delta\varphi)=\frac{1}{2}\left[1+\mathrm{erf}\left(\frac{\gamma}{\sqrt{2}\Sigma}\,\frac{\tau(\varphi)-\tau(\varphi+\delta\varphi)}{2}\right)\right], \tag{1}$$

where $\mathrm{erf}(x)=\frac{2}{\sqrt{\pi}}\int_0^x \mathrm{d}y\exp(-y^2)$ is the error function. The above formula relates the psychophysical quantity P_c to the average tuning slope γ, a neuro-physiological model parameter, and Σ, representing the sum of two standard deviations, viz., of (i) the population rate Ξ and (ii) a reading out of Ξ by higher order brain stations. We define a system's time constant $t_S := \Sigma/\gamma$, which sets the psychophysical time resolution in (1) and serves as a fit parameter to psychoacoustical data. A comparison between theory and cat [3] as well as gerbil [4] data yields $t_S \approx 15$ μs.

3. Localization error and biological restrictions

For a given threshold criterion for P_c of getting $P_c > p$ for some given p (generic values are $p=0.75$ [3] or $p=0.85$), Eq. (1) yields a localization error $E(\varphi,p)$ that is implicitly defined as $p=P_c(\varphi,E)$, meaning that for angles $\delta\varphi > E$ the probability of

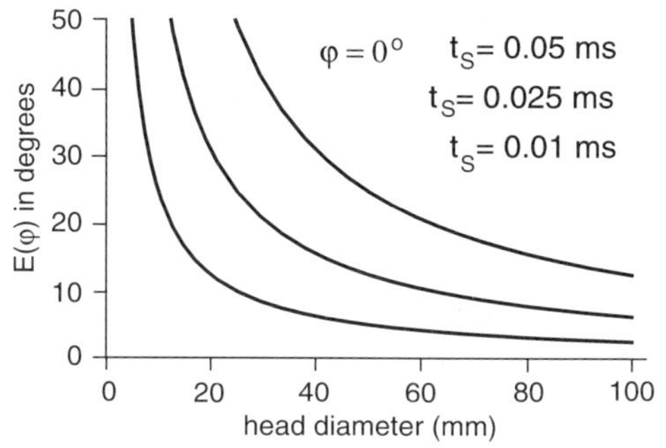

Fig. 1. Localization error $E(\varphi = 0, p = 0.75)$ from Eq. (2) as a function of head diameter d for various values of t_S (top graph $t_S = 50$ μs, middle $t_S = 25$ μs, bottom $t_S = 10$ μs).

correct answers exceeds p. We can find an analytic expression for E, if we approximate the error function by a Fermi-distribution with approximate inverse $\mathrm{erf}^{-1}(x) \approx 2.406^{-1}\ln[(1+x)/(1-x)]$. In this way we obtain

$$E(\varphi, p) = \frac{t_S}{d}\frac{2.35c_S}{1+\cos(\varphi)}\ln\left(\frac{p}{1-p}\right). \tag{2}$$

See Fig. 1 for a plot of (2). Given a time resolution t_S, the error E is inversely proportional to the head diameter d. If, on the other hand, both head diameter and an upper limit for E are given, the latter incorporating the evolutionary necessity of lateralization acuity, we have obtained a rigorous restriction of t_S based on the assumption of a rate-gradient code as suggested by McAlpine et al. [10]; see Introduction.

As an illustration of Eq. (2), we calculate the temporal resolution t_S for a few species, provided they implement a rate-gradient code. The great titmouse [8], a small bird belonging to the family *Paridae*, whose head diameter is $d \approx 2$ cm, has a temporal resolution of $t_S \approx 25$ μs. Although barn owls have a head diameter d that is about three times as large, their localization precision is so good ($E \approx 2^\circ$) [9] that temporal resolution should be as small as $t_S \approx 4$ μs. Also humans have a rather high lateralization acuity [3]. At least for angles $\varphi \approx 0$ Eq. (2) yields a time constant $t_S \approx 8$ μs.

The requirement of a high localization precision can thus be balanced by a short system time constant $t_S = \Sigma/\gamma$. Since the latter is defined to be the fraction of noise level Σ and the slope of the population tuning γ, a small t_S can be achieved by either reducing noise or increasing the steepness of tuning curves. If we roughly estimate the noise portion that is induced by the standard deviation of the population rate under the assumption of Poissonian spike generation we find that it is far too small in order to explain the psychophysically obtained system time constants t_S. The main portion of Σ is therefore realized by the discrimination reliability of the system, i.e., how reliable the "rest" of the brain "reads out" the rate information provided by the cell population under consideration.

Being a question beyond today's knowledge, we refrain from a discussion of how to improve the reliability of reading out a rate-gradient code and concentrate instead on how to achieve steeper slopes γ. To this end, we again refer to the paper of McAlpine et al. [10], who find that tuning slopes increase with higher best frequency of the cell. A strategy of reducing localization error is thus the recruitment of higher-frequency components for ITD representation. The next section, however, will show that, for frequencies exceeding an upper bound, this strategy is incompatible with a rate-gradient code, if we regard tuning curves as originating from spike-timing-induced synaptic plasticity.

4. Synaptic delay selection

In order to understand how ontogenetic development and selection of afferent delay distributions give rise to tuning curves that vary on a time scale of 10 μs, we propose that synaptic weights at binaural cells are subject to a spike-timing-dependent learning rule [1]. Synaptic strengths are increased (decreased), if presynaptic spikes arrive before (after) the postsynaptic neuron fires. This temporal constraint is implemented by means of a learning window; cf. Fig. 2. Its amplitude is measured in units of the learning parameter $\eta = 1/4000$, adjusting the speed of synaptic change. Furthermore, we use a weight increment of $w^{\rm in} = \eta/50$ following every presynaptic spike and a decrement $w^{\rm out} = -\eta/4$ of all synaptic strengths after each postsynaptic spike. There is a lower bound 0 and an upper bound 2 for each individual synaptic weight. We assume that the width of the learning window in the *medial superior olive* (MSO) is in agreement with the membrane time constants [2] so that it is significantly smaller than that observed in tectal [11] neurons. This spike-based Hebbian learning rule leads to a normalization of the mean output rate of a postsynaptic neuron and a selection of *synaptic delays* in a competitive self-organized process [6].

We have simulated the temporal evolution of solely excitatory synaptic efficacies using two populations (left, right) of input spike trains generated by periodically

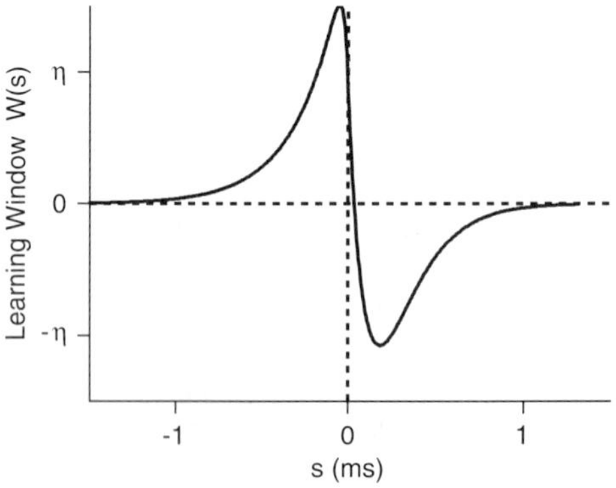

Fig. 2. Changes of excitatory synaptic strengths are governed by the time difference s between pre- and postsynaptic spike, $s < 0$ meaning that a presynaptic spike arrives before the postsynaptic neuron fires. The learning window W represents the idea of time-dependent synaptic plasticity [1] in dependence upon s; parameters have been taken from [7].

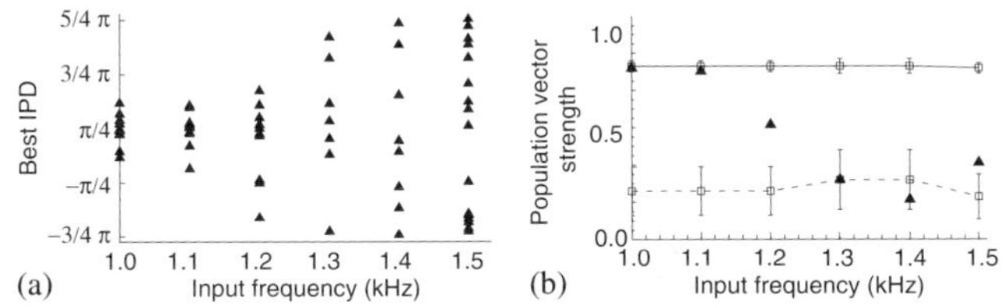

Fig. 3. (a) The distribution of best interaural phase differences (best IPDs) depends on the input frequency. (b) Population vector strength (solid triangles) due to Eq. (3). For frequencies below 1.2 kHz, the vector strength corresponds to the hypothesis of a best-IPD distribution with mean $\pi/4$ and standard deviation as for input frequency 1.0 kHz (solid line). For frequencies beyond 1.2 kHz, the hypothesis of randomly distributed IPDs (dashed line) is more suitable. Means (squares) and standard deviations are obtained through 20000 repetitions of the respective stochastic experiment.

(period length $1/f$) fluctuating Poisson processes until the synaptic configurations at the integrate-and-fire units had reached a stable state [1]. Firing rates are then obtained by averaging over 10 s. The maximum of a tuning curve defines the best phase difference. For each out of a given set of input frequencies $f \in \{1.0, 1.1, 1.2, 1.3, 1.4, 1.5\}$ kHz, we have performed several runs with changing random seeds and plotted the distribution of best interaural phase differences (IPDs); see Fig. 3a.

The initial distribution of synaptic weights was Gaussian with mean 1.0 and standard deviation 0.3. A transmission delay from ear to synapse is assigned to each weight. The distribution of transmission delays is also taken to be Gaussian with standard deviation 0.3 ms and mean 1 ms. The best IPD of $\pi/4$, as found by experiment [10] is achieved by an appropriate shift of the synaptic delays after learning; see Discussion. During learning, IPDs have been drawn randomly every 100 ms of formal time from an interval corresponding to ITDs in the range of ± 0.2 ms. In order to quantify the distribution of best ITDs Φ_k, $1 \leqslant k \leqslant K$, we define the population vector strength r to be

$$r = \left| \sum_{k=1}^{K} \exp(\mathrm{i}\Phi_k) \right| . \qquad (3)$$

Fig. 3b shows a phase transition in dependence upon the input frequency. Low frequencies yield narrow distributions with best IPDs in a small range whereas distributions stemming from simulations with higher input frequencies correspond to randomly drawn best IPDs. As a consequence, the rate-gradient code of Ref. [10] breaks down as high-frequency spectral components are used for the neuronal ITD representation, and a Jeffress representation [5] takes over.

5. Discussion

We have shown that a rate-gradient representation of interaural time differences is restricted to the low-frequency range $f < f_{\max} \approx 1.5$ kHz and that, as long as this representation is valid, the minimal localization error has a lower bound given by (2), with

$t_S \propto 1/f_{max}$. Synaptic development at higher input frequencies spawns a broad distribution of best IPDs and thus naturally favors a rate-place coding concept as proposed by Jeffress.

The asymmetry of tuning curves with preferred best IPDs at about $\pi/4$ contralateral leading phase is realized by shifting the means of the delay distributions originating from the contralateral ear to longer delays after learning. In MSO cells this tuning asymmetry can be induced by synaptic plasticity of *inhibitory* synapses.

References

[1] W. Gerstner, R. Kempter, J.L. van Hemmen, H. Wagner, A neuronal learning rule for sub-millisecond temporal coding, Nature 383 (1996) 76–78.
[2] B. Grothe, D.H. Sanes, Synaptic inhibition influences the temporal coding properties of medial superior olivary neurons: An in vitro Study, J. Neurosci. 14 (1994) 1701–1709.
[3] R.S. Heffner, H.E. Heffner, Sound localization acuity in the cat: Effect of azimuth, signal duration, and test procedure, Hearing Res. 36 (1988) 221–232.
[4] R.S. Heffner, H.E. Heffner, Sound localization and use of binaural cues by the Gerbil, Behavioral Neurosci. 102 (1988) 422–428.
[5] L.A. Jeffress, A place theory of sound localization, J. Comp. Physiol. Psychol. 41 (1948) 35–39.
[6] R. Kempter, W. Gerstner, J.L. van Hemmen, Intrinsic stabilization of output rates by spike-based Hebbian learning, Neural Comput. 13 (2001) 2709–2741.
[7] R. Kempter, C. Leibold, H. Wagner, J.L. van Hemmen, Formation of temporal-feature maps by axonal propagation of synaptic learning, Proc. Natl. Acad. Sci. USA 98 (2001) 4166–4171.
[8] G.M. Klump, Sound localization studies in non-specialized birds, in: G.M. Klump, R.J. Dooling, R.R. Fay, W.C. Stebbins (Eds.), Methods in Comparative Psychoacoustics, Birkhäuser, Basel/Switzerland, 1995, pp. 171–182.
[9] E.I. Knudsen, G.G. Blasdel, M. Konishi, Sound Localization by the Barn Owl (Tyto alba) measured with the search coil technique, J. Comp. Physiol. 133 (1979) 1–11.
[10] D. McAlpine, D. Jiang, A.R. Palmer, A neural code for low-frequency sound localization in mammals, Nat. Neurosci. 4 (2001) 396–401.
[11] L.I. Zhang, H.W. Tao, C.E. Holt, W.A. Harris, M.-M. Poo, A critical window for cooperation and competition among developing retinotectal synapses, Nature 395 (1998) 37–44.

Available at
www.ComputerScienceWeb.com
POWERED BY SCIENCE @ DIRECT®

ELSEVIER

Neurocomputing 52–54 (2003) 327–333

NEUROCOMPUTING

www.elsevier.com/locate/neucom

Cortico-thalamocortical operations of multi-target spatial working memory

Shuhei Miyashita, Yoshinori Tabuchi, Shoji Tanaka*

Department of Electrical and Electronics Engineering, Sophia University 7-1 Kioicho, Chiyoda-ku, Tokyo 102, Japan

Abstract

Representation and manipulation of multiple target locations would be important aspects of spatial working memory (Neurocomputing 44–46 (2002) 1001; Neural Networks 15 (2002) 573). It is, however, unknown how such processes are performed in the prefrontal cortical, corticocortical, and cortico-subcortical circuits. We here propose a model of the prefrontal cortical and cortico-thalamocortical circuits that represent and manipulate multiple targets. The simulation with this model shows that transient inputs to the thalamic relay cells increase the signal transmission of the thalamocortical afferents. This eliminates one or more of the targets represented in the prefrontal cortex. The number of the eliminated targets depends on the width of the directional tuning of the inputs to the relay cells.

Keywords: Mediodorsal nucleus; Multi-target; Selective; Spatial working memory; Thalamocortical

1. Introduction

There are intensive computational studies of the characteristics of the prefrontal cortical (PFC) circuits for working memory [1–3,7–12]. Tanaka [10,11] has suggested that the PFC circuit can perform fundamental cognitive operations of multiple targets. Although the PFC, especially the dorsolateral part, plays central roles, other cortical areas and subcortical structures are considered to be involved in working memory processing. The computer simulation by Tabuchi and Tanaka [6] shows that thalamocortical inputs interact with the cortical circuits to cause interesting phenomena, such

* Corresponding author. Tel.: +81-3-3238-3331; fax: +81-3-3238-3321.
E-mail address: tanaka-s@sophia.ac.jp (S. Tanaka).

0925-2312/03/$ - see front matter
doi:10.1016/S0925-2312(02)00809-3

as elimination of working memory representation by excitatory thalamocortical inputs. It would, therefore, be interesting to investigate how the cortico-thalamocortical system manipulates multi-target representation of working memory. This article shows computer simulation of a model cortico-thalamocortical circuit. The PFC in this model loads six targets simultaneously and the representation is maintained during the delay period. The representation is manipulated by transient inputs to the thalamic relay cells at the end of the delay period.

2. Model

2.1. Circuitry

The cortical model has three layers (the superficial, intermediate, and deep layers) and contains 1080 pyramidal cells and 240 interneurons [4,5,7–11,13]. The MD contains 36 relay cells and 36 inhibitory interneurons. The neurons are described by a leaky integrate-and-fire neuron model. The ion channels include: AMPA, NMDA, Nap, $GABA_A$, K(Ca), and leak. The excitatory conductance ratio is $g_{NMDA}/g_{AMPA} = 0.0575$. The intracortical inhibitory circuits are assumed to have two subtypes: the isodirectional and the cross-directional [9–11]. The conductance ratio of these circuits is $g_{GABAA,cross}/g_{GABAA,iso} = 0.31$. Fig. 1A shows the architecture of the model. The pyramidal cells in the deep layer send feedback projections to the MD relay cells and the interneurons. The interneurons have local inhibitory connections with the relay cells. The relay cells, in turn, send the thalamocortical projections to the pyramidal cells in the intermediate layer and the interneurons of the PFC with the same synaptic weights. The pyramidal cells in the intermediate layer of the PFC receive the cue-related inputs. The inputs cue the locations of the six targets to load and maintain during the delay period (Fig. 1B). The MD relay cells receive the transient excitatory inputs at the end of the delay period. These inputs are directionally tuned.

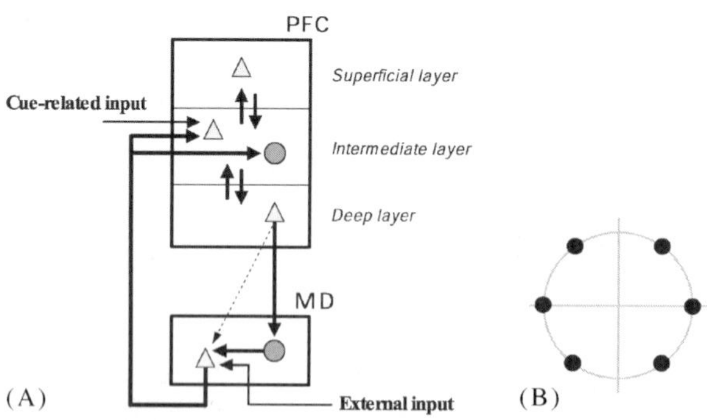

Fig. 1. (A) The circuit diagram of the model. Triangles in the PFC: the pyramidal cells, filled circle in the PFC: the inhibitory interneurons, triangle in the MD: the relay cells, filled circle in the MD: the inhibitory interneurons. (B) Six targets represented in the model PFC. The directions of these targets are equally spaced.

3. Results

The activity of the MD relay cells is transmitted to the cortical pyramidal cells in the intermediate layer. The cortical neurons respond to the transient thalamocortical inputs. The activity is processed intracortically in the PFC, then the pyramidal cells in the deep layer feedback the signal to the MD neurons. Although the pyramidal cells in the deep layer send feedback projections to the relay cells as well as the interneurons in the MD, the projections to the relay cells are assumed to be weak in this model. To clarify the roles of the thalamic interneurons, we show below the simulation under the condition that only the interneurons receive the feedback signals.

3.1. Simultaneous loading of multiple targets

Six targets were loaded successfully by receiving cue-related inputs. That is, six different populations of the cortical neurons were activated at the same time and the activity was sustained during the delay period. At the end of the delay period, the MD relay cells received transient excitatory inputs. Depending on the sharpness of the directionally tuned transient input to the MD relay cells, some of the represented targets were eliminated.

3.2. Selective elimination of target representation

The transient inputs to the MD with sharp directional tuning selectively eliminated one of the six targets represented in the PFC (Fig. 2). The target eliminated in this simulation is in the direction of 180° because the transient inputs had the peak at 180° in the directional space. The remaining five targets were continuously represented after the termination of the transient inputs.

3.3. Non-selective elimination of target representation

As the directional tuning of the transient inputs to the MD increases, more targets were eliminated by the inputs. When it was 90°, all of the six targets were eliminated as soon as the transient inputs were given (Fig. 3). During these inputs, the MD relay cells increased their firing rates transiently (Fig. 3F). Before that, the MD relay cells fired at low rates. The firing rates of the MD relay cells after the transient inputs became higher than those before. This is due to disinhibition of the MD relay cells (see the raster of the MD interneurons in Fig. 3G). The interneurons fired at higher rates in the delay period because they received stronger feedback from the PFC representing the six targets. The firing rates of the MD interneurons became lower after the elimination of the target representation due to weaker feedback inputs from the PFC to the MD interneurons.

4. Discussion

The PFC can represent multiple targets as working memory. However, how the representation is manipulated is unknown yet. This article studied the capability of

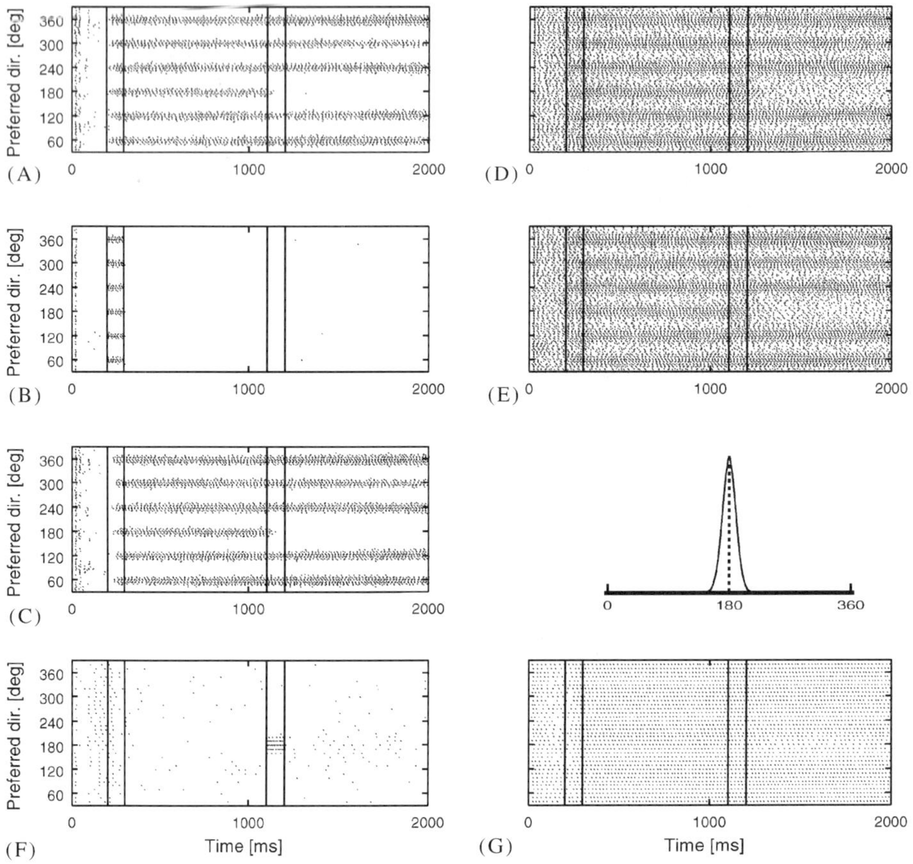

Fig. 2. Raster plots of the neuronal spikes in the model. (A) The pyramidal cells in the superficial layer of the PFC, (B) the pyramidal cells in the intermediate layer of the PFC, (C) the pyramidal cells in the deep layer of the PFC, (D) the inhibitory interneurons (isodirectional) of the PFC, (E) the inhibitory interneurons (cross-directional) of the PFC, (F) the MD relay cells, and (G) the MD interneurons. The cue-related inputs were given during 200–300 ms, followed by the delay period. The transient inputs to the MD were given during 1100–1200 ms. The inset shows the directional profile of the inputs, whose width was 10° (standard deviation). The figure shows transient activity at the beginning of the simulation ($0 < t < 50$ ms), which is irrelevant to the processes studied here.

the cortico-thalamocortical system to manipulate multi-target spatial working memory. The simulations showed that transient inputs to the MD relay cells eliminate some of the targets represented in the PFC. The selectivity depends on the sharpness of the directional tuning of the inputs. Sharply tuned inputs can eliminate only one of the six targets represented in the PFC. Broadly tuned inputs, on the contrary, eliminate all of the targets. These selective/non-selective processes of the spatial working memory would be generally used in cognitive operations.

This study suggests three important issues on the inhibitory actions in multi-target working memory processes. First, multiple target representation owes to the

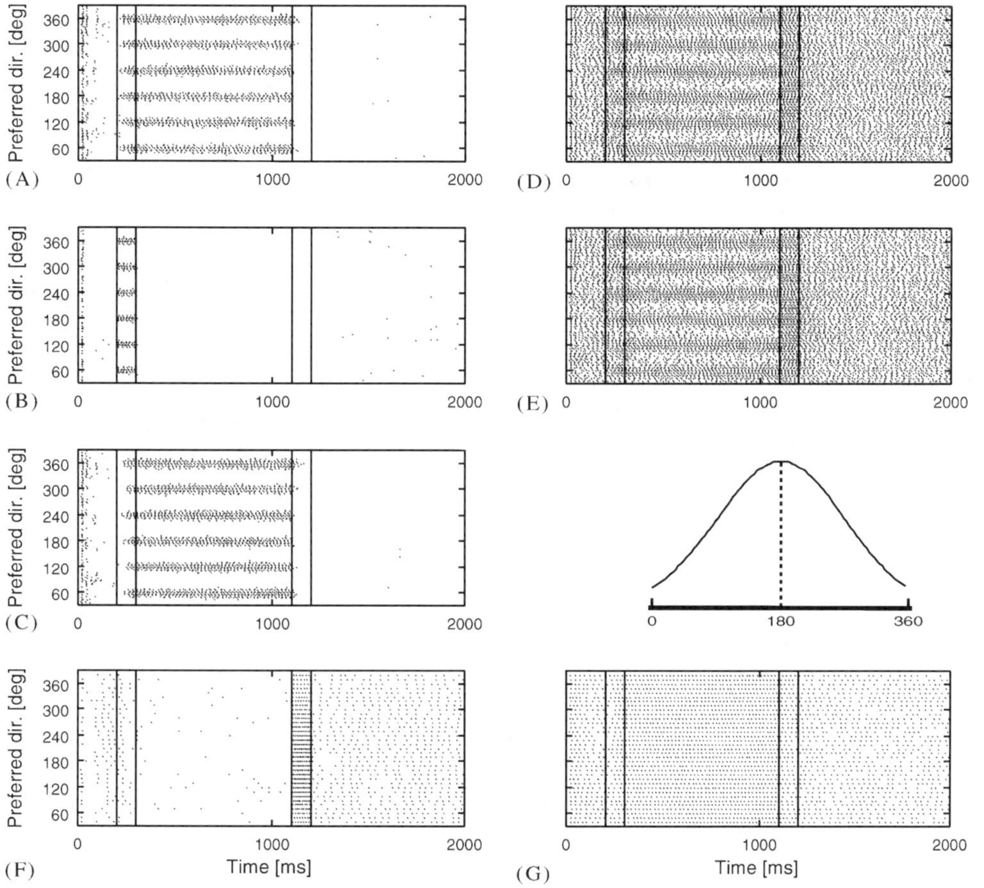

Fig. 3. Raster plots of the neuronal spikes in the model. (A) The pyramidal cells in the superficial layer of the PFC, (B) the pyramidal cells in the intermediate layer of the PFC, (C) the pyramidal cells in the deep layer of the PFC, (D) the inhibitory interneurons (isodirectional) of the PFC, (E) the inhibitory interneurons (cross-directional) of the PFC, (F) the MD relay cells, and (G) the MD interneurons. The cue-related inputs were given during 200–300 ms, followed by the delay period. The transient inputs to the MD were given during 1100–1200 ms. The inset shows the directional profile of the inputs, whose width was 90° (standard deviation). The figure shows transient activity at the beginning of the simulation ($0 < t < 50$ ms), which is irrelevant to the processes studied here.

regulation of the local inhibition because there is competition between targets [10,11]. The cross-directional inhibition mediates this competition in this model. The regulation of the inhibition becomes more and more critical as the number of representing targets increases.

Secondly, it is the intracortical inhibition that extinguishes one or more of the targets represented in the PFC. This is interesting because the thalamocortical afferents convey excitatory signals to the pyramidal cells and the interneurons in the PFC. For the selective elimination, the isodirectional inhibition works predominantly because the inputs are given to the population of neurons representing the target to be eliminated. For the non-selective elimination, on the contrary, both the iso- and the cross-directional inhibition work concurrently to eliminate all of the targets represented. The summated effects are so strong that the elimination of the representation is very quick as Fig. 3 shows.

Thirdly, the MD interneurons play important roles in this mode of the operation of the circuit. By receiving the feedback signals from the deep layer of the PFC, these interneurons tonically inhibit the MD relay cells in this model. As a result, the MD relay cells fire at low rates during the delay period. This is necessary because, otherwise, the thalamocortical inputs destroy the working memory representation in the PFC. After the elimination, the feedback to the MD interneurons becomes lower, then the interneurons become less active. This disinhibits the MD relay cells, then the thalamocortical afferents send stronger inputs (Fig. 3). The PFC neurons thus show noticeable background activity after the elimination of the working memory. At this stage, the PFC becomes ready to load another working memory into itself.

Acknowledgements

This work was supported by the Grants-in-Aid for Scientific Research on Priority Areas to S.T. (#13210123 and #14017083) from the Japanese Ministry of Education, Science, and Technology.

References

[1] N. Brunel, X.-J. Wang, Effects of neuromodulation in a cortical network model of object working memory dominated by recurrent inhibition, J. Comput. Neurosci. 11 (2001) 63–85.
[2] A. Compte, N. Brunel, P.S. Goldman-Rakic, X.-J. Wang, Synaptic mechanisms and network dynamics underlying spatial working memory in a cortical network model, Cereb. Cortex 10 (2000) 910–923.
[3] D. Durstewitz, J.K. Seamans, T.J. Sejnowski, Neurocomputational models of working memory, Nature Neurosci. 3 (2000) 1184–1191.
[4] M. Iida, S. Tanaka, Postsynaptic current analysis of a model prefrontal cortical circuit for multi-target spatial working memory, Neurocomputing 44–46 (2002) 855–861.
[5] K. Morooka, S. Tanaka, Correlation analysis of signal flow in a model prefrontal cortical circuit representing multiple target locations, Neurocomputing 44–46 (2002) 541–548.
[6] Y. Tabuchi, S. Tanaka, Intracortical vs. thalamocortical processing of spatial working memory. Neurocomputing, in this issue.
[7] S. Tanaka, Architecture and dynamics of the primate prefrontal cortical circuit for spatial working memory, Neural Networks 12 (1999) 1007–1020.
[8] S. Tanaka, Post-cue activity of prefrontal cortical neurons controlled by local inhibition, Neurocomputing 32–33 (2000) 563–572.
[9] S. Tanaka, Computational approaches to the architecture and operations of the prefrontal cortical circuit for working memory, [Review], Prog. Neuro-Psychopharm. Biol. Psychiat. 25 (2001) 259–281.
[10] S. Tanaka, Multi-directional representation of spatial working memory in a model prefrontal cortical circuit, Neurocomputing 44–46 (2002) 1001–1008.
[11] S. Tanaka, Dopamine controls fundamental cognitive operations of multi-target spatial working memory, Neural Networks 15 (2002) 573–582.
[12] X.-J. Wang, Synaptic reverberation underlying mnemonic persistent activity, Trends Neurosci. 24 (2001) 455–463.

[13] K. Yamashita, S. Tanaka, Circuit simulation of memory field modulation by dopamine D1 receptor activation, Neurocomputing 44–46 (2002) 1035–1042.

Shuhei Miyashita received B.E. from Sophia University, Tokyo, in 2002. He is a graduate student at Tokyo Institute of Technology. He is currently studying robotics.

Yoshinori Tabuchi received B.E. from Sophia University, Tokyo, in 2002. He is a graduate student at Sophia University. He is currently studying neuroscience and electrical engineering.

Shoji Tanaka received B.E., M.E., and Ph.D. degrees from Nagoya University, Japan. He is Professor at Department of Electrical and Electronics Engineering, Sophia University. During 1998–1999, he was a Visiting Science at the Section of Neurobiology, Yale University School of Medicine, USA.

Available at
www.ComputerScienceWeb.com
POWERED BY SCIENCE @ DIRECT®

ELSEVIER

Neurocomputing 52–54 (2003) 335–340

NEUROCOMPUTING

www.elsevier.com/locate/neucom

Internetplattform Neuroinformatik: a pilot study for the OECD neuroinformatics portal☆

R. Ritz*, R. Förster, A.V.M. Herz

Institute for Theoretical Biology, Humboldt-University Berlin, Invalidenstr. 43, D-10 115 Berlin, Germany

Abstract

Following the open source philosophy, more and more neuroscientists are willing to share their primary data as well as custom made software with the scientific community. To facilitate this interaction, we are establishing a website that links publicly available resources (such as experimental data, numerical tools, and computer models) and contains extensive annotation. In this paper, we explain step by step how interested scientists can actively contribute content to the site.

Keywords: Web publishing; Content management system; Workflow; Information repository; Metadata; Data sharing

1. Introduction

Novel experimental and computational techniques have led to major transitions in the neurosciences, all the way from the molecular to the system level. At the same time, more and more scientists are willing to share their data, analysis tools and computer models and have thus started a new research culture reflecting the open source philosophy. While there is still a debate within the community as to whether or not this paradigm shift should be made [1], it is obvious that there are many open questions ranging from technical to legal issues [2,5].

But even many of the data and computer programs already publicly available are not known to the general neuroscience community and other interested scientists.

☆ Supported by BMBF.
* Corresponding author.
E-mail address: ritz@itb.biologie.hu-berlin.de (R. Ritz).

0925-2312/03/$ - see front matter
doi:10.1016/S0925-2312(02)00824-X

To overcome this deficit, the *Working Group on Neuroinformatics* of the *OECD Global Science Forum* [3] has issued a proposal to create an internet-based knowledge management system for all data relating to nervous system structure and function through an *internet portal*. Here we explain how you as an interested individual can contribute content to this site. A more general introduction to the portal is available elsewhere [4].

2. The action box

One general feature of the neuroinformatics portal at `www.neuroinf.de` that you may notice immediately when surfing the site is the following: On the left-hand side of the screen you find a column entitled Action. In this so-called *action box* we first display who we currently think you are (i.e., your login name if you are logged in, or Guest otherwise) and then, more importantly, the possible actions that you are allowed to take. These actions are a filtered list depending on your personal privileges as well as on the page (or object in our parlance) you are currently looking at. As Guest you can View and Search the site. Things become more interesting when you Login. For this you need to have an account at the portal but that can easily be obtained by clicking on Join.

3. Joining the community

Once you have decided to join the portal community, we are asking you to provide a freely chosen login name (you will be warned if it is already in use) and a valid email address. This address will be used to send you a randomly generated password. Once you have logged in with this password, you may change it to anything you like. You are then taken to the Personal data page of your member preferences. Use this form to provide whatever research-related information you would like to make available about yourself via the portal. Otherwise just leave it blank. Note, however, that even if you enter informations and click on Save you still will be the only one who is able to see the page that you get by clicking on View (to get back to the edit form go to Personal data). To make this page publicly available and to be included in the People listing under Topics you also need to Submit your entry for publication. Look for the Submit option in the action box.

This demonstrates another important feature of the portal, namely that the general visibility of objects depends on a variable called State. Its value gets displayed in the action box and it is typically either private, pending, or published. The collection of possible states and the allowed transitions between them together with rules as to who is allowed to invoke which transition is commonly referred to as *Workflow* (details below). This is how we control what is made available to the general public through the portal.

With respect to publishing personal data we are required by law to obtain explicit permission from you. Therefore, when you submit personal information we go through a number of consistency checks to verify your identity, including an email back to you asking for confirmation (like most mailing lists do for subscriptions) before we publish your data. If, on the other hand, you happen to notice that you are already included in the PERSONS listing and you did not ask for this yourself, please notify us. You can reach us—for any portal related questions, critique, suggestions—via the FEEDBACK option in the top bar or the email link to `neuroinf-admin` at the bottom of each page.

Background: *a note on browser caching*: If you try to follow this step by step introduction and you get the impression that the options you are offered within the action box do not meet your expectations, one possible cause could be that your browser does not reload the page even though it might have changed. To reduce net traffic and page download times many browsers keep local copies of visited pages and only those copies get displayed when you return to one of those pages. Clicking on your browsers RELOAD button should circumvent the local cache and give you an updated view.

Triggered by the registration process you also obtain a private area within the portal, the so-called *member area*.

4. The member area

Now that you have an account at the site you also have your own member area. Conceptually, this is like a home directory in any other computer system. This is where you go if you want to contribute content to the portal site. To get to your *desktop* you need to be logged in first. Then you should click on MY STUFF.

At your first visit there should be just one object that has been created per default: a document called `index_html`, your personal and initially PRIVATE homepage within the portal. We recommend that you do not rename or delete it but you should of course edit it if you want to use it. Clicking on its name shows you what the homepage currently looks like. Via the EDIT action you can get to the editing interface. Here you can specify the format, and either upload a file that you want to import from your local system or you write (or cut and paste with your mouse) the page's source code into the text area field. Note, however, that even though `html` is provided as a default format some `html`-tags will not work. It is for security reasons that potentially harmful tags (like form elements) will simply be ignored.

The options that you have to organize your desktop (RENAME, CUT, COPY, DELETE,...) should be self-explaining but the possibilities offered after clicking on NEW... may deserve some clarification.

5. Content types

Selecting NEW... from your desktop area (MY STUFF) brings you to the *Create Content* page. Here you first need to select the type of content that you want to provide.

Background: *what content types to expect*: Currently there is only a limited number of different content types available for creation through regular members, like documents, news items, discussions, research groups, software, databases, events, etc. Expect this list to grow in the future as we extend the scope and coverage of the portal. In any case, however, we expect that the information you provide characterizes a resource that is already available **somewhere else** (preferentially online). We do not intend and do not have the resources to act as a primary data repository. The added benefit of providing information about resources that are available somewhere else anyway consists of better visibility within the community and the other features that we provide, like cross-linking, classification and discussion of entries (see below).

Next, you have to provide a name for the content object you are about to create. This name should be one word and as short and descriptive as possible. It will become **part of the URL** of the object.[1] Therefore, it should not contain any special characters, white-spaces, or whatever may cause trouble within URLs. After clicking on ADD you are taken to the appropriate entry form for the selected content type. Provide whatever information you see fit and click on SAVE. Now you see how your entry will be displayed through the portal. If you want to change any items, click on EDIT again. If you feel comfortable with your entry you should consider submitting it for publication. What happens then will be explained in the next section.

Background: *a note on publications*: The site is no preprint server and we also do not want to compete with dedicated literature data bases. You may, however, use the publication entry for pointing to documents that would otherwise be hard to find, like preprints, technical reports, master-, diploma-, or PhD-thesis. You can of course also point to regular publications but then we expect that they are available online and free of charge to anybody.

6. Workflow

To limit the number of irrelevant objects we implemented a web-based review process by defining the following workflow. As mentioned above, any content object can be in one of several states like PRIVATE, PENDING, or PUBLISHED and depending on the state, different groups have the right to see and manipulate the object and find it through the site's SEARCH facilities. While in the PRIVATE state, only the owner can view and edit the object, whereas PENDING objects can be viewed by the reviewers also. In addition, PENDING objects cannot be edited anymore by the owner. Only PUBLISHED objects are accessible to all users and they cannot be edited while in this state (but see below).

To change the state of an object you need to invoke a transition. Standard transitions that regular members are able to invoke are SUBMIT which takes objects from the PRIVATE to the PENDING state and RETRACT which moves objects back to the PRIVATE state. Use the latter if you want to edit or update an entry that you had published previously. When you are done with your changes, you need to SUBMIT it again in order to get the updated version reviewed and published. When selecting a transition

[1] Do not worry too much. With RENAME you can still change this name later.

from the action box you are taken to a form that first tells you what you are about to do and then allows you to add a comment which will be included in the STATUS HISTORY. Use this as you see fit. Only when you press SUBMIT ITEM or RETRACT THIS ITEM the transition will actually be invoked.

While this workflow is our main mechanism for quality control we use it only to check whether information offered is within the scope of the portal, whether the details given are sufficient to be meaningful, whether the links work, etc. We do not even try to judge the scientific quality of the resource described. To that end we offer the REPLY option to regular members. Any content item published within the portal can be made the starting point of a threaded discussion simply by selecting REPLY while viewing it. But you need to be logged in to have this option available. You can also REPLY to other comments within a particular thread. Your comments will be PUBLISHED immediately, they cannot be edited later and you will be identified as the author. We retain the right to remove comments if they are offending or otherwise not acceptable.

7. Advanced options

If you ever created a content object within our portal yourself you might have noticed already that there are further options that have not been described yet, such as CLASSIFY, METADATA or CROSSLINK. These we consider advanced and they are still being developed. Nevertheless, feel free to check them out. Use CLASSIFY to select keywords from our hierarchical topics list to further characterize your content object. With METADATA you can for instance set an effective or expiration date for the display of your resource within the portal or influence what we may pass on to third parties about your entry. CROSSLINK shall be used to establish **reciprocal** and visible links between different portal entries. For detailed explanation of these features you are referred to the online documentation.

8. Outlook

As the portal described here is currently a pilot study, expect it to be developed further. We may still introduce changes at various levels ranging from the appearance to the core functionality (without loosing data, of course). But the general design and the core features as presented here are most likely not to change. We invite you to browse our site and to join the portal community to actively contribute content. The more you do so, the more useful the portal will be to the whole community.

References

[1] S.H. Koslow, Should the neuroscience community make a paradigm shift to sharing primary data? Nature Neurosci. 3 (2000) 863–865.

[2] OECD Working Group on Neuroinformatics, Neuroscience data and tool sharing: a legal policy framework for neuroinformatics, Neuroinformatics, 2003, in press.

[3] Organization for Economic Co-operation and Development, Final Report of the OECD Global Science Forum: Working Group on Neuroinformatics, OECD, Paris, 2002.
[4] R. Ritz, R. Förster, A.V.M. Herz, Facilitating data and software sharing in the neurosciences—a neuroinformatics portal, in: R. Kötter (Ed.), Neuroscience Databases: A Practical Guide, Kluwer, Norwell, MA, 2002, pp. 293–306.
[5] A.W. Toga, Neuroimage databases: the good, the bad and the ugly, Nature Rev. Neurosci. 3 (2002) 302–309.

Available at
www.ComputerScienceWeb.com
POWERED BY SCIENCE @DIRECT®

ELSEVIER

Neurocomputing 52–54 (2003) 341 – 346

NEUROCOMPUTING

www.elsevier.com/locate/neucom

Sensitivity of AMPA receptor channel to calcium oscillations: a computational study

Gabriela Antunes[a], Fábio M. Simões-de-Souza[b], Antônio C. Roque[b,*]

[a] *Faculdade de Ciências, Universidade Estadual Paulista "Júlio de Mesquita Filho", Bauru, SP, Brazil*
[b] *Departamento de Física e Matemática, Faculdade de Filosofia, Ciências e Letras de Ribeirão Preto, Universidade de São Paulo, Avenida dos Bandeirantes 3900, Ribeirão, Preto, SP 14040-901, Brazil*

Abstract

We used a computational model of biochemical pathways that are involved in the phosphorylation/dephosphorylation of AMPA receptor to study the receptor responses to calcium oscillations. In the model, the biochemical pathways are assumed to be located immediately under the postsynaptic membrane and we included three states of AMPA receptor: dephosphorylated, and phosphorylated in one or in two sites. To characterize the effects of calcium oscillations on the AMPA receptor, we exposed the model to stimuli with three varying parameters, namely frequency, number of pulses and calcium spike duration. Our model showed sensitivity to all of these three parameters.

Keywords: AMPA receptor; Biochemical networks; Plasticity; Calcium oscillations

1. Introduction

Calcium spike trains are usual responses of cells to stimuli [9] and have been implicated with activity of some kinases such as CaMKII [3,5] and protein kinase C [8], which are sensitive to calcium oscillations. Protein kinases are involved in several intracellular processes, including phosphorylation of α-amino-3-hydroxy-5-methyl-4-isoxazolepropionic acid (AMPA) receptor channels [7], which are implicated in excitatory synaptic transmission in the central nervous system [12]. The AMPA receptor is a

* Corresponding author. Tel.: +55-16-6023768; fax: +55-16-6339949.
E-mail addresses: g_antunes@hotmail.com (G. Antunes), fabioms@neuron.ffclrp.usp.br (F.M. Simões-de-Souza), antonior@neuron.ffclrp.usp.br (A.C. Roque).

doi:10.1016/S0925-2312(02)00828-7

heteromer composed of four subunits (GluR1-4) [4] whose functions can be regulated by the phosphorylation and dephosphorylation of individual subunit proteins [2]. The AMPA receptor is phosphorylated by cAMP-dependent protein kinase (PKA), protein kinase C (PKC), calmodulin kinase II (CaMKII) and other unspecified kinases, and dephosphorylated by some phosphatases like protein phosphatase 1 (PP1), calcineurin and PP2A [4]. AMPA receptor channels mediate fast synaptic transmission in response to presynaptic glutamate release [11] and play an important role in synaptic plasticity [10].

Recently, Lee et al. [6] have proposed a model for bidirectional synaptic plasticity regulated by intracellular calcium concentration in which CaMKII, PKA, PP1 and PP2A act on AMPA receptor channels. In that model (Fig. 1A—top panel) Fig. 1 the phosphorylation and dephosphorylation of AMPA receptors occur in two sites (Serine 831- CaMKII site; and Serine 845 - PKA site).

In this work, we constructed a computational model incorporating the bidirectional model of Lee et al. [6] and the biochemical pathways of Bhalla and Iyengar [1] to study the responses of these receptors to calcium oscillations.

2. Methods

Bhalla and Iyengar's [1] model of biochemical pathways includes CaMKII, calmodulin (CaM), adenylate cyclase (AC), phosphodiesterase (PDE), protein phosphatase 1 (PP1), calcineriun, and cAMP-dependent protein kinase (PKA) (Fig. 1B—top panel). In our computational model, these pathways are assumed to be located immediately under the postsynaptic membrane.

Based on the work of Lee et al. [6] we also included three states of AMPA receptor channel (dephosphorylated, phosphorylated in one site, and phosphorylated in

Fig. 1. Top panel: (A) Bidirectional model of synaptic plasticity, modified from Lee et al. [6]. In this model, the AMPA receptor can be phosphorylated in two sites (Serine 831- CaM KII site; and Serine 845 - PKA site) and dephosphorylated by PP1 and PP2A. High frequency stimulation (HFS) and low frequency stimulation (LFS) shift the AMPA receptor state to the right or left, respectively. (B) Scheme of the biochemical pathways located immediately under the postsynaptic membrane. The interactions between these biochemical pathways suggest they can form a complex biochemical network, with feedback loops and crosstalks between the signaling pathways, and may have properties that are non-intuitive (modified from [1]). Middle panel: Responses of PKA, PP1 and CaMKII to stimuli with different numbers of pulses, frequencies and duration of calcium spikes. (A) PKA, PP1 and CaMKII responses to 6 pulses of 1s duration with 0.1 Hz (○), 0.5 Hz (△) and 0.8 Hz (□). (B) PKA, PP1 and CaMKII responses to 75 pulses of 80 ms duration with 2.5 Hz (○), 5 Hz (△) and 10 Hz (□). Responses for constant stimulation (∗) are also indicated for comparison. Bottom panel: Responses of AMPA receptors to stimuli with different numbers of pulses, frequencies and duration of calcium spikes. (A) Dephosphorylated AMPA (AMPA-0P), phosphorylated in one site AMPA (AMPA-1P) and phosphorylated in two sites AMPA (AMPA-2P) responses to 6 pulses of 1 s duration with 0.1 Hz (○), 0.5 Hz (△) and 0.8 Hz (□). (B) Dephosphorylated AMPA (AMPA-0P), phosphorylated in one site AMPA (AMPA-1P) and phosphorylated in two sites AMPA (AMPA-2P) responses to 75 pulses of 80 ms duration with 2.5 Hz (○), 5 Hz (△) and 10 Hz (□). Responses for constant stimulation (∗) are also indicated for comparison.

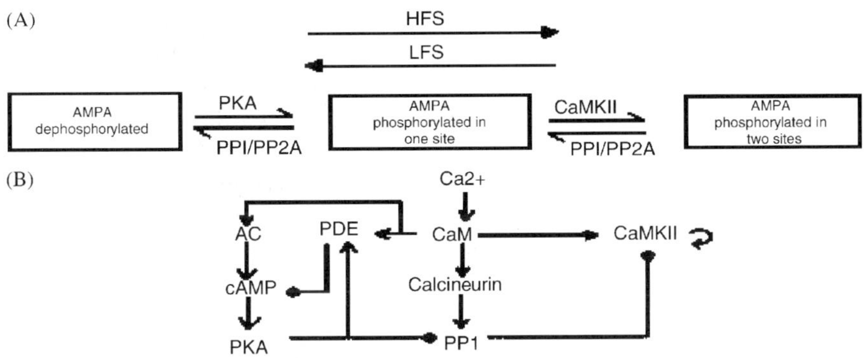
(A)
HFS
LFS
AMPA dephosphorylated
PKA
PPI/PP2A
AMPA phosphorylated in one site
CaMKII
PPI/PP2A
AMPA phosphorylated in two sites
(B)
Ca2+
AC
PDE
CaM
CaMKII
cAMP
Calcineurin
PKA
PP1

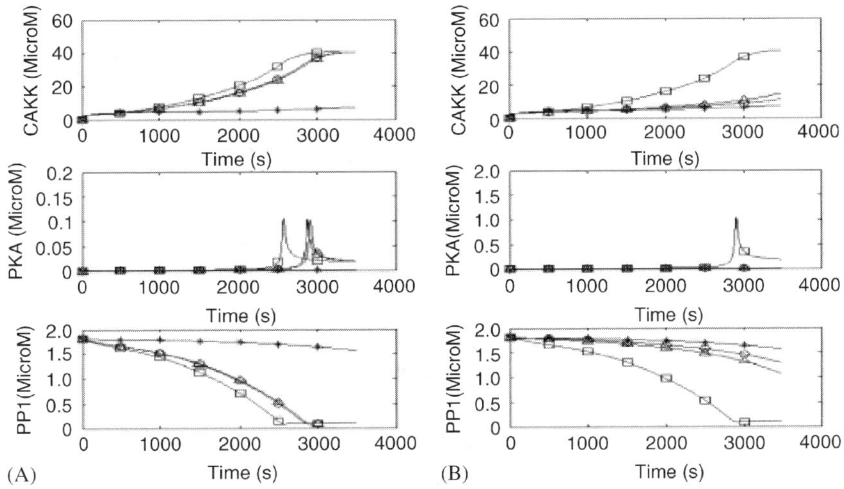
CAKK (MicroM)
PKA (MicroM)
PP1(MicroM)
Time (s)
(A)
(B)

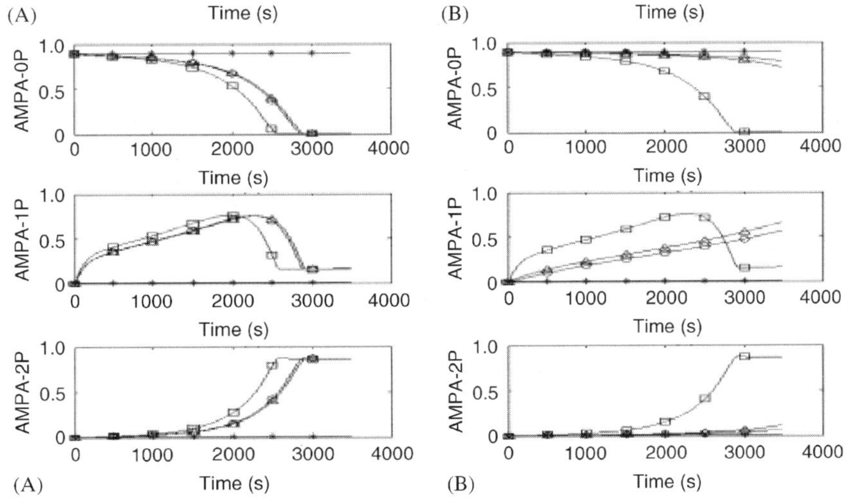
AMPA-0P
AMPA-1P
AMPA-2P
Time (s)
(A)
(B)

two sites) to study the behavior of the receptor channel exposed to different calcium stimuli.

To characterize the effects of calcium oscillations on the AMPA receptor channel we exposed our model to stimuli with different values of (i) frequency (measured by the inverse of the constant interspike interval), viz. 0.1, 0.5, 0.8, 2.5, 5.0, 10 Hz; (ii) number of pulses during the exposition time, viz. 6 or 75; and (iii) duration of calcium spike, viz. 80 ms or 1 s. The stimulation protocol was adapted from De Koninck and Schulman [3]. All stimuli have the same calcium concentration (1 μM) and total time of exposition to calcium (6 s). The simulation was done in the GENESIS 2.2 neural simulator.

3. Results

To observe the effects of calcium oscillation on the AMPA receptor channel, we measured the activity of CaMKII, PKA and PP1 (Fig. 1—middle panel) and the phosphorylation and dephosphorylation of AMPA receptor channel consequently to the effects of calcium oscillation on the biochemical pathways (Fig. 1—bottom panel). The results indicate that the phosphorylation and dephosphorylation of the AMPA receptor channel are affected by the oscillations.

In Fig. 1A—middle panel, the activities of PKA, CaMKII and PP1 show different responses to stimuli with the same number of pulses (6) and duration (1 s) but varying frequencies (0.1, 0.5 and 0.8 Hz). The increase in the stimulus frequency causes an increase in the activity of CaMKII, a decrease in the activity of PP1 and a reduction in the latency time of the first PKA spike. The increase in the PKA activity together with the decrease in the PP1 activity changes the AMPA receptor state (according to the model of Fig. 1A—top panel) from the dephosphorylated state (AMPA-0P) to the phosphorylated in one site state (AMPA-1P) (Fig. 1A—bottom panel). On the other hand, the increase in the CaMKII activity together with the decrease in the PP1 activity changes the AMPA receptor state (according to the model of Fig. 1A—top panel) from the phosphorylated in one site state (AMPA-1P) to the phosphorylated in two sites state (AMPA-2P) (Fig. 1A—bottom panel).

In Fig. 1B—middle panel, PKA, CaMKII and PP1 responded to another stimulus protocol in which stimuli with constant number of pulses (75) and duration (80 ms) were applied with different frequencies (2.5, 5 and 10 Hz). In this second protocol, the increase in the frequency of stimulus increases the activity of CaMKII, decreases the activity of PP1 and the PKA spike occurs only for the highest frequency stimulus (10.0 Hz). Consequently, the AMPA receptor channel does not change its state for lower frequencies (2.5, 5.0 Hz) but only for the highest frequency stimulus (10.0 Hz) (Fig. 1B—bottom panel).

4. Discussion and conclusion

All stimulation protocols had the same calcium concentration (1 μM) and total time of exposition to calcium (6 s), suggesting that the different responses observed in the model come from a sensitivity of the system to frequency, number of pulses and duration of calcium spikes.

Our results indicate that the AMPA receptor channels can be influenced by the calcium oscillations through the variation in the activity of some kinases and phosphatases involved in the phosphorylation/dephosphorylation of the receptor.

Acknowledgements

FMSS and ACR would like to thank FAPESP for funding this work.

References

[1] U.S. Bhalla, R. Iyengar, Emergent properties of networks of biological signaling pathways, Science 283 (1999) 339–381.

[2] G.C. Castellani, E.M. Quinlan, L.N. Cooper, H.Z. Shouval, A biophysical model of bidirectional synaptic plasticity: dependence on AMPA and NMDA receptors, Proc. Nat. Acad. Sci. USA 98 (2001) 12772–12777.

[3] P. De Koninck, H. Schulman, Sensitivity of CaM Kinase II to the Frequency of Ca^{2+} oscillations, Science 279 (1998) 227–230.

[4] R. Dingledine, K. Borges, D. Bowie, S.F. Traynelis, The glutamate receptor ion channels, Pharmacol. Rev. 51 (1999) 7–62.

[5] Y. Kubota, J.M. Bower, Decoding time-varying calcium signals by the postsynaptic biochemical network: computer simulation of molecular kinetcs, Neurocomputing 26–27 (1999) 29–38.

[6] H.K. Lee, M. Barbarosie, K. Kameyama, M.F. Bear, R.L. Huganir, Regulation of distinct AMPA receptor phosphorilation sites during bidirecional synaptic plasticity, Nature 405 (2000) 955–959.

[7] A.L. Mammen, K. Kameyama, K.W. Roche, R.L. Huganir, Phosphorylation of the α-amino-3-hydroxy-5-methylisoxazole-4-propionic acid receptor gluR1 subunit by calcium/calmodulin-dependent kinase II, J. Biol. Chem. 272 (1997) 32528–32533.

[8] E. Oancea, T. Meyer, Protein kinase C as a molecular machine for decoding calcium and diacylglycerol signals, Cell 95 (1998) 307–318.

[9] K. Prank, F. Gabbiani, G. Brabant, Coding efficiency and information rates in transmembrane signaling, BioSystems 55 (2001) 15–22.

[10] L. Scher, M. Frerking, Restless AMPA receptors: implications for synaptic transmission and plasticity, Trends Neurosci. 24 (2001) 665–670.

[11] S.H. Shi, AMPA receptor dynamics and synaptic plasticity, Science 298 (2001) 1851–1852.

[12] T. Soderling, V.A. Derkach, Postsynaptic protein phosphorylation and LTP, Trends Neurosci. 23 (2000) 75–80.

Gabriela Antunes received a Degree in Psychology from the State University of São Paulo (UNESP) at Bauru, SP, Brazil, in 2002. Her research interests are computational neuroscience applied in the study of the signaling pathways involved in the process of memory.

Fábio M. Simões-de-Souza received a Technical Degree in Industry informatics from the ETEP, São José dos Campos, SP, Brazil, in 1994. He received his B.Sc in Biology and M.Sc in Psychobiology from the University of São Paulo at Ribeirão Preto, SP, Brazil, in 1999 and 2002, respectively. His scientific interests are computational neuroscience applied in the study of the role of gap junctions in the nervous system and biologically detailed modeling of the olfactory system.

Antonio C. Roque was born in São Paulo, SP, Brazil. He received his B.Sc. degree in physics from the State University of Campinas, Brazil, in 1985, and his Ph.D. degree in cognitive science and artificial intelligence from the University of Sussex, Brighton, UK, in 1992. He joined the faculty of the Department of Physics and Mathematics of the University of São Paulo at Ribeirão Preto, Brazil, in June, 1993, where he founded and is the current coordinator of the Laboratory of Neural Systems. His research interests are computational neuroscience and neural networks applications in medicine.

Available at
www.ComputerScienceWeb.com
POWERED BY SCIENCE @DIRECT®

ELSEVIER

Neurocomputing 52–54 (2003) 347–353

NEUROCOMPUTING

www.elsevier.com/locate/neucom

The effect of spike redistribution in a reciprocally connected pair of neurons with spike timing-dependent plasticity

Gerardina Hernández[a,1], Jonathan Rubin[b,*,2], Paul Munro[c]

[a]*Intelligent Systems Program, University of Pittsburgh, Pittsburgh, PA 15260, USA*
[b]*Department of Mathematics, University of Pittsburgh, Pittsburgh, PA 15260, USA*
[c]*School of Information Science, University of Pittsburgh, Pittsburgh, PA 15260, USA*

Abstract

This work considers a system of two spike-response cells, mutually connected by excitatory synapses that modify according to a multiplicative form of spike timing-dependent plasticity (STDP). The units are driven by independent external inputs and one common input, representing a global cortical rhythm. Our simulations show that the synaptic weights converge to attractors that encode input and plasticity parameters, independent of initial conditions. In certain parameter regimes, the temporal correlation induced by coupling and the shared cortical signal lead to redistribution of spikes to favor doublets. When this occurs, the weight changes associated with these doublets dominate convergence of weights to their attractors. Thus, spike timing-dependent plasticity can allow subsets of spikes to play a predominant role in the evolution of synaptic weights.

Keywords: Spike timing-dependent plasticity; Synaptic weights; Spike redistribution

1. Introduction and model

Spike timing-dependent plasticity (STDP) is characterized by synaptic weight changes that depend on the precise timing of spikes fired by pre- and post-synaptic cells. Past

* Corresponding author. Tel.: +1-412-624-6157; fax: +1-412-624-8397.

E-mail addresses: gerardina.hernandez@highmark.com (G. Hernández), rubin@math.pitt.edu (J. Rubin), pmunro@mail.sis.pitt.edu (P. Munro).

[1] Current address: Highmark Blue Cross Blue Shield, Fifth Avenue Place, 120 Fifth Avenue, Ste. P7205, Pittsburgh, PA 15222-3099, USA.

[2] Partially supported by NSF grants DMS-9804447 and DMS-0108857; member of the Center for the Neural Basis of Cognition.

0925-2312/03/$ - see front matter © 2002 Elsevier Science B.V. All rights reserved.
doi:10.1016/S0925-2312(02)00830-5

computational studies have tallied changes in synaptic weights due to STDP by treating all timing differences within the timing window as relevant. Recently, the suggestion has been made that data on STDP may be better represented by a rule in which only certain pre- and post-synaptic spike pairs are counted [8,2].

We simulate two neurons, a and b, mutually connected by positive weights W_{ab} and W_{ba} (the first subscript denotes the presynaptic cell). Each neuron is governed by the spike response model [3], which is an instance of a "threshold-fire model". A spike train is characterized by the set of firing times $Z_x = \{t_x^1, \ldots, t_x^n\}$, with $t_x^i < t_x^j$ for $i < j$, where t_x^i is the ith firing time of neuron x. A neuron generates a spike when its membrane potential crosses a threshold θ. Here, membrane potential refers to the internal state $u_i(t)$ of a neuron i at time t, defined by

$$u_i(t) = \sum_{t_i^z \in Z_i} \eta_i(t - t_i^z) + \sum_j \sum_{t_j^z \in Z_j} W_{ji}\pi_{ji}(t - t_j^z) + w_i h_i(t) + w_{ctx} h_{ctx}(t) \quad (1)$$

for $t > \max\{\max(Z_i), \max(Z_j)\}$. In Eq. (1), a presynaptic spike at time t_j^z contributes an amount $W_{ji}\pi_{ji}(t - t_j^z)$ to the state u_i at time t. The weight W_{ji} is the strength of the corresponding connection, dynamically derived from a multiplicative spike timing-dependent learning process described below. The positively valued kernel π_{ji} describes the time course of an excitatory postsynaptic potential. The negatively valued kernel η_i in (1) describes the neuron's response to its own firing (reset of the membrane potential after each spike, together with neuronal refractoriness); see [3] for details. The term $h_i(t)$ represents a Poisson input signal, with fixed weight w_i; we will use f_i to denote the frequency of this signal. Finally, $h_{ctx}(t)$ is a periodic context signal, with weight w_{ctx}, applied to both cells in the network. This shared context signal represents a global cortical rhythm; such rhythms are associated with cognitive tasks (13 Hz sensory-motor rhythm: see [7,4]; 40 Hz gamma rhythm: reviewed in [1]).

Let τ denote the time at which a presynaptic spike occurs minus the time of a postsynaptic spike. To represent the temporally dependent potentiation (LTP) and depression (LTD) kernels for STDP, we define $K_s(\tau) = \beta_s|\tau|e^{-\alpha_s|\tau|}$ where $s = P$ for $\tau < 0$, $s = D$ for $\tau \geqslant 0$, and where $\alpha_P, \alpha_D, \beta_P > 0$ and $\beta_D < 0$. In our simulations, the transition between maximal potentiation and depression is 10 ms. The kernel parameters are $\alpha_P = 0.5$, $\alpha_D = 0.125$, $\beta_P = \alpha_P$ and $\beta_D = -0.18\alpha_D$.

We implement a multiplicative spike timing-dependent learning rule (mSTDP) which implicitly constrains weights between bounds $0 \leqslant \Omega_{\min} < \Omega_{\max}$. When the most recent presynaptic spike t_j^n occurs before the most recent postsynaptic spike t_i^m, the weight from cell j to cell i is potentiated according to

$$\Delta W_{ji}(t_i^m) = \lambda[\Omega_{\max} - W_{ji}(\max\{t_i^{m-1}, t_j^n\})] \left[\sum_{z=1}^{n} (K_P(t_j^z - t_i^m))\right], \quad (2)$$

where λ denotes a learning rate. When the most recent postsynaptic spike t_i^m fires before the most recent presynaptic spike t_j^n, the weight from j to i is depressed according to

$$\Delta W_{ji}(t_j^n) = \lambda[W_{ji}(\max\{t_i^m, t_j^{n-1}\}) - \Omega_{\min}]\left[\sum_{z=1}^{m}(K_D(t_j^n - t_i^z))\right]. \quad (3)$$

If $t_i^m = t_j^n$ computationally, then both Eqs. (2) and (3) are implemented, with max $\{t_i^{m-1}, t_j^{n-1}\}$ used as the argument of W_{ji} on the right-hand side of both. In our simulations, we take $\lambda = 0.1$, $\Omega_{\max} = 2$, $\Omega_{\min} = 0$, unless otherwise noted.

2. Results

2.1. Existence and properties of attractors

For fixed parameter values, the weights W_{ab}, W_{ba} of the excitatory synapses between the neurons converge under mSTDP to a fixed attractor in the (W_{ab}, W_{ba}) plane, irrespective of the initial weight values used. An example appears in Fig. 1, which shows trajectories of weights, both in the (W_{ab}, W_{ba}) plane and as functions of time.

The weight values in the attractor do depend on the intrinsic parameters and driving signals in the network. In particular, the input and context frequencies, the input and context weights, and the relative areas of the LTP and LTD kernels are encoded by the attractor mean, while the specific spike times in a particular spike train influence the details of the convergence to the attractor. These results hold with or without a context signal. An example illustrating the role of driving frequencies is shown in the upper left of Fig. 2. Additional examples appear in [6,5].

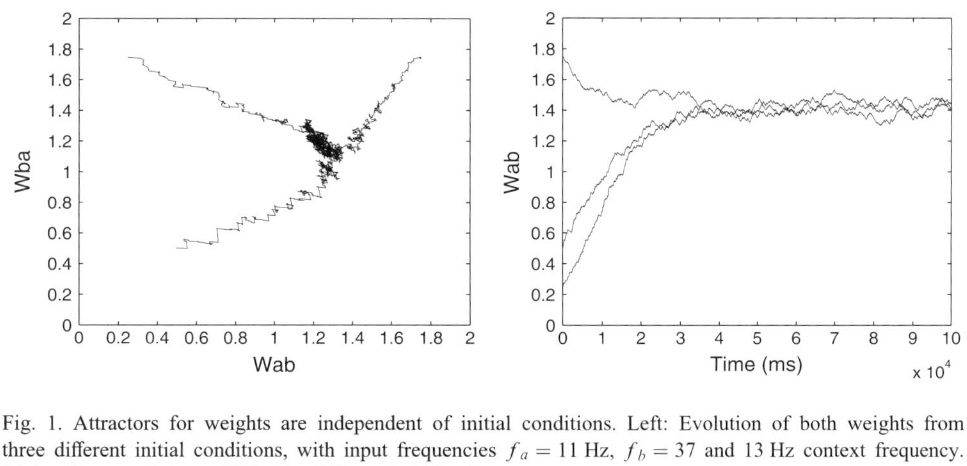

Fig. 1. Attractors for weights are independent of initial conditions. Left: Evolution of both weights from three different initial conditions, with input frequencies $f_a = 11$ Hz, $f_b = 37$ and 13 Hz context frequency. Right: Time evolution of W_{ab} for different initial values.

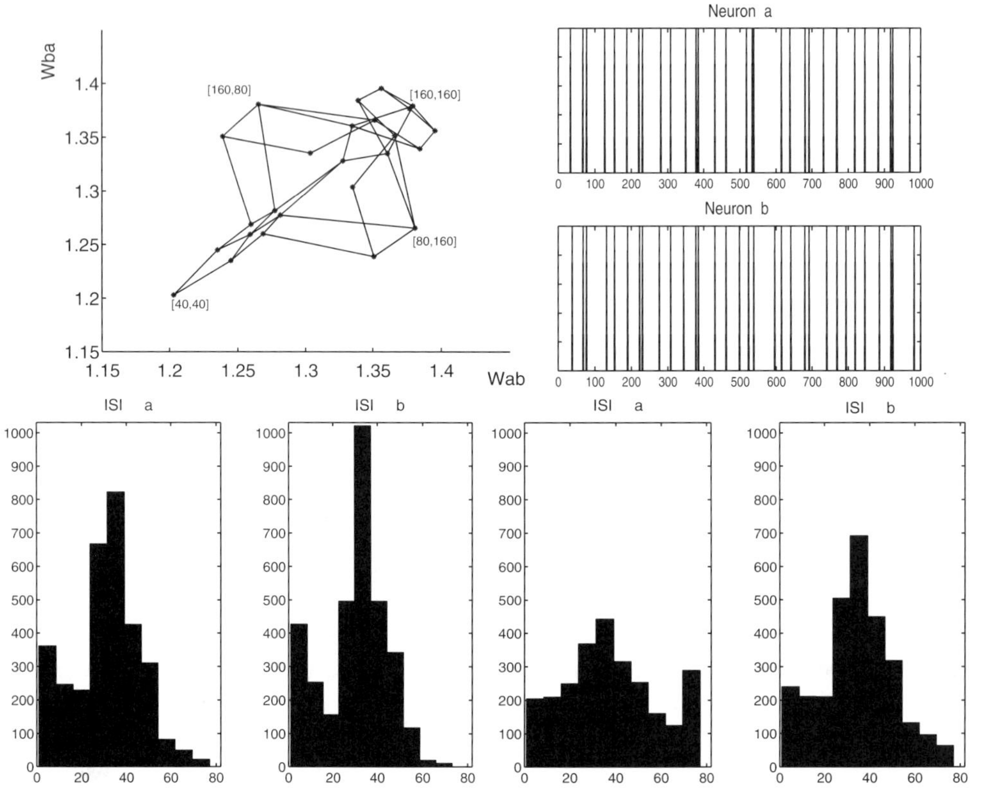

Fig. 2. Spike redistribution with 13 Hz context signal and strong w_{ctx} ($w_{ctx} = 15$). Top left: Mean attractor values. Different points correspond to different drive frequencies (f_a, f_b). Top right: Segments from spike trains (vertical lines denote spikes) of both cells in the large f_a, f_b regime, which favors doublets, often at similar times in both cells. Bottom row: Distributions of interspike intervals for neuron a (ISI_a) and neuron b (ISI_b). Left plots show a redistribution toward small ISIs for large f_a, f_b. Right plots show that this redistribution does not occur for small f_a, f_b.

2.2. Redistribution of spikes

The properties of the weight attractors, as well as the mechanisms underlying these results, will be further detailed elsewhere [6,5]. For our purposes here, it is important to note that the temporal correlation mechanisms of coupling and a shared context signal lead to redistribution of spikes, relative to baseline firing of uncoupled cells due to their independent Poisson inputs. The degree of redistribution depends on the parameter regime simulated. For example, when w_{ctx} is weak, we find little redistribution. When the context signal frequency is 13 Hz and w_{ctx} is large, the nature of the spike redistribution depends on the frequencies of the independent driving signals $h_a(t), h_b(t)$. The interspike interval (ISI) plots on the bottom row of Fig. 2 show that high input frequencies f_a, f_b to both cells cause a shift of ISIs toward lower values,

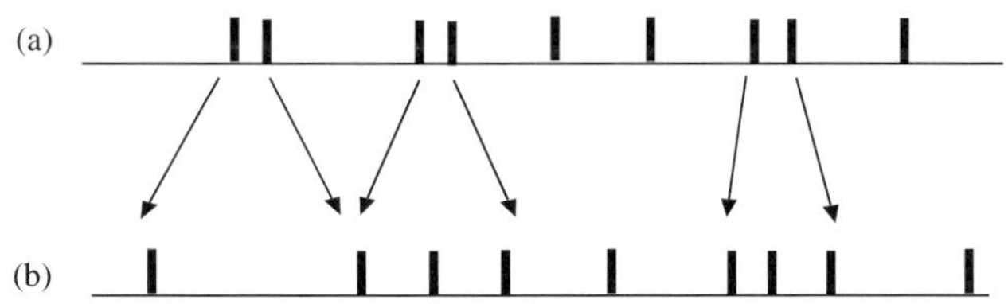

Fig. 3. Schematic representation of the selection of a subset of spikes. A subset of spikes from cell a are selected, corresponding to small ISIs, and then weight changes are computed from their interactions only with the spikes of cell b that are closest in time (between the arrows for each pair).

corresponding to a tendency for both cells in the network to fire spike doublets; this is not observed with low frequency drives to the cells or with context frequency of 40 Hz. When w_{ctx} is large, cells are entrained to fire together in response to most context signals, and the relatively low context frequency in the 13 Hz case allows them to recover sufficiently between signals to fire independent spikes, which may become doublets when context signals follow soon afterward. Further, the coupling between cells leads to a high probability that a doublet of one cell will elicit a doublet from the other, as illustrated in the spike trains in the upper right of Fig. 2.

2.3. Computations with a spike subset

To explore the role of doublets in the simulated recurrent network, we computed weight changes using a computational rule that only includes doublets. To do this, we examined the spike train of cell a, computed the ISIs between each pair of consecutive spikes, and kept only the spike pairs from the bottom 25% of the cell's ISI distribution. For each such pair, call them t_a^1, t_a^2, we formed a subset of postsynaptic spikes consisting of the last postsynaptic spike before t_a^1, the first postsynaptic spike after t_a^2, and all postsynaptic spikes between these. We then computed the changes in W_{ab} prescribed by our learning kernel for the interactions of each of t_a^1, t_a^2 with each spike in the postsynaptic subset; see Fig. 3. This gave an evolution of W_{ab} based on fewer than 25% of the spikes fired by the network. A symmetric procedure was used to compute the evolution of W_{ba}.

To evaluate how well the doublet procedure reproduces the weight changes seen with the full spike set, we also performed a control experiment. In the control, the same number of presynaptic spikes was used as in the doublet procedure, but these were selected randomly. The postsynaptic subset for a presynaptic spike consisted of the two postsynaptic spikes closest in time to the presynaptic one. The relative error between the original weights W_{ij} and the weights W_{ij}^p generated by procedure p was computed for both the doublet and the control procedures as $E_{ij}^p = 1/n \sum_{z=1}^{n} |W_{ij}(z) - W_{ij}^p(z)|/W_{ij}(z)$ for z ranging over all spike times (of both cells) in the full spike set.

We found that the doublet procedure always led to convergence of weights to the original attractor region (Fig. 4). Further, a key point, also illustrated in Fig. 4, is that the weight values in the doublet procedure tracked the original weight values much better than expected by chance: the relative errors in both weights with the doublet

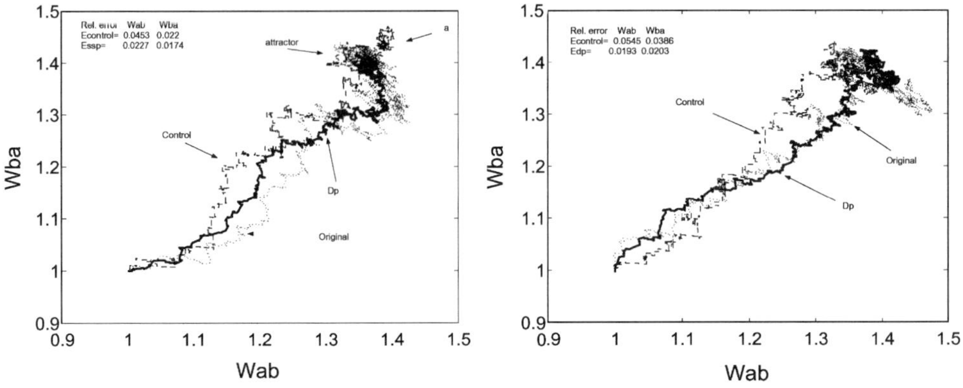

Fig. 4. Tracking of weight changes from a full simulation (Original—dotted curve) by the doublet procedure (Dp—solid curve) and the control procedure (Control—dashed curve). The input frequencies were 160 Hz to neuron *a* and 200 Hz to neuron *b*, with learning rate $\lambda = 0.02$. Left: The control procedure converges to region "*a*" rather than to the attractor region below it. Left and right: The doublet procedure tracks the weight changes from the full simulation better than the control procedure does; corresponding errors ($E_{\text{control}}, E_{dp}$) are displayed in the upper left of each plot.

procedure were approximately 2%, and these errors were found to be significantly smaller than those for the control procedure over 20 simulations with fixed parameters, as measured by a one-tailed t-test ($p < 0.0005$). The reliable reconstruction of original weight values by the doublet procedure implies that for learning events which occur over a short time period, when weights may be far from attractor values, evolution of weights is still dominated by effects of spike doublets.

3. Conclusions

The existence of weight attractors provides a synaptic encoding of incoming stimuli that combines rate information, encoded in attractor values, with temporal information, encoded in particular spike times. Our principal finding is that, in parameter regimes in which spikes are redistributed, by coupling and common context signals, to favor spike doublets, the dynamics of the weights as they converge to these attractors is dominated by the effects of these doublets. This arises simply from applying the learning rule (2) and (3). Thus, in addition to driving downstream units, bursts may induce longer-term effects on network behaviors.

Recent results indicate that the combined effect of multiple spikes is more complex than had previously been assumed [8,2], and experimental data may be better modeled by computational versions of STDP that do not count all spikes than by applications of STDP rules to complete spike sets [8]. Clearly, the explanation of this finding must involve the fact that different spiking patterns induce different responses in cellular machinery (calcium channels, NMDA receptors, and so on). Nonetheless, our results show that even a fixed STDP rule, which does not take into account the involvement of such factors, can allow a certain subset of spikes to play a predominant role in driving weight dynamics when sufficient spike redistribution occurs.

References

[1] A. Engel, P. Fries, W. Singer, Dynamic predictions: oscillations and synchrony in top-down processing, Nat. Neurosci. Rev. 2 (2001) 704–716.
[2] R.C. Froemke, Y. Dan, Spike timing dependent synaptic modification induced by natural spike trains, Nature 416 (2002) 433–438.
[3] W. Gerstner, Spiking neurons, in: W. Maass, C.M. Bisshop (Eds.), Pulsed Neural Networks, MIT Press, Cambridge, MA, 1999.
[4] D. Giannitrapani, The Electrophysiology of Intellectual Functions, Karger, Basel, 1985.
[5] G. Hernández, Spike-timing-dependent plasticity under temporal correlation mechanisms, Ph.D. Thesis, Intelligent Systems Program, University of Pittsburgh, 2001.
[6] G. Hernández, P. Munro, J. Rubin, in preparation.
[7] J.O. Lubar, J.F. Lubar, Electroencephalographic biofeedback of SMR and beta for treatment of attention deficit disorder in a clinical setting, Biofeedback Self-Reg. 9 (1984) 1–23.
[8] P. Sjöström, S. Nelson, G. Turrigiano, Rate, timing, and cooperativity jointly determine cortical synaptic plasticity, Neuron 32 (2002) 1149–1164.

ELSEVIER

Available at
www.ComputerScienceWeb.com
POWERED BY SCIENCE @DIRECT®

Neurocomputing 52–54 (2003) 355–362

NEUROCOMPUTING

www.elsevier.com/locate/neucom

The influence of neuron shape changes on the firing characteristics

Seiichi Sakatani, Akira Hirose

Department of Frontier Informatics, Graduate School of Frontier Sciences, The University of Tokyo, 7-3-1 Hongo, Bunkyo-ku, Tokyo 113-8656, Japan

Abstract

The neuron is highly full of variety in its shape and size. Each neuron has changed its shape to adapt to the region and function in the nervous system. The influence of neuron shape on the firing characteristics, however, has still been unclear. In this paper, we evaluate numerically the influence of the changes of the neuron shape on the firing characteristics. We construct neuronal micro-compartmental models of the pyramidal cell at the CA3 hippocampus to analyze quantitatively the shape dependence of the firing characteristics. We find that, the sharper the shape of the apical-dendrite hillock is, the higher the firing frequency becomes and, at the same time, the firing latency gets shorter for the apical-dendrite synaptic stimulus. These facts lead to a hypothesis that the neuron has changed its shape to respond more sensitively to the stimulus injected at the apical dendrite.

Keywords: Neuron shape; Pyramidal cell; Hippocampus; Firing rate; Firing latency

1. Introduction

The recent progress in physiological measurements has led to new insights into the neuronal behavior. In such efforts, we have recognized the usefulness of neuronal models and constructive methods for analysis by numerical calculation. The authors previously constructed a theory and reported spatiotemporal calculation results of a confluence of membrane potential and its dependence on neuron's configuration [2,4].

The neuron is highly full of variety in its shape and size. In the evolution process, each neuron has changed the shape to adapt to the region and function in the nervous system. Winslow et al. [7] reported that a large part of the variation of whole neuron input resistance is due to variation in the geometry. The authors [2] previously predicted

E-mail address: sakatani@eis.t.u-tokyo.ac.jp (S. Sakatani).

0925-2312/03/$ - see front matter
doi:10.1016/S0925-2312(02)00839-1

that the signal velocity v on a round soma has a dependence of $v \propto D^{0.77}$ on the conductive-layer thickness D. However, the influence of neuron shape on the firing characteristics has not been investigated in detail.

In this paper, we evaluate numerically the influence of the neuron-shape changes on the firing characteristics. We construct a new neuronal model of the pyramidal cell at the CA3 hippocampus by proposing micro-compartmental models to analyze the firing characteristics quantitatively.

Calculations reveal the following facts: (1) The firing frequency is higher when the shape of the apical-dendrite hillock is sharper. (2) The firing latency is shorter for the apical-dendrite synaptic stimulus when the shape of the apical-dendrite hillock is sharper. These facts lead to a hypothesis that the neuron has changed its shape to respond more sensitively to the stimulus injected at the apical dendrite.

In Section 2, we describe our cell models. Then we show calculation results of the firing rate and firing latency in Section 3. Section 4 gives a discussion. In Section 5, we make our conclusions.

2. Cell model

We construct our model of a guinea pig CA3 pyramidal cell by placing two types of dendrites, soma, and unmyelinated axon as shown in Fig. 1. The basal (left) and apical (right) dendrites in Fig. 1 are represented in bundles by equivalent cylinders. In the conventional compartmental model [6], the soma is expressed as one cylinder shown in Fig. 1A. The radii (R_k) are 2.42 μm for the basal dendrite, 4.23 μm for the soma, 2.89 μm [6] for the apical dendrite and 0.50 μm [5] for the axon. The lengths of each compartment (H_k) are 110 μm for the basal dendrite, 125 μm for the soma, 120 μm for the apical dendrite and 15 μm for the axon which is small enough to ensure a sufficient stability of calculation for the quick action-potential behavior. (Length constant λ calculated with the parameters described below is 160 μm.) The number of the axon compartments is 80, resulting in a total length of 1200 μm. The basal dendrite cylinder is represented by compartments nos. 1–8, the soma by no. 9 and the apical dendrite cylinder by compartments nos. 10–19, respectively. We call it *Model A* in this paper.

We present our micro-compartmental soma models shown in Figs. 1B(a)–(c) and call them *Model B(a)*, *Model B(b)* and *Model B(c)*, respectively. These models have almost the same surface areas. They are divided into cylinders whose heights are 2.5 μm and connected by resistance. In Model B(a), the soma is expressed as a frustum. Base radius is 12.5 μm, top radius is 2.9 μm and height is 60 μm. The base is divided into one circle and four rings. The frustum is divided into 24 cylinders. In Model B(b), the soma consists of two frustums. Base radii are 12.5 μm, top radii are 2.5, 3.1 μm and heights are 22.5, 47.5 μm, respectively. The smaller frustum is divided into nine cylinders and the bigger one is into 19. In Model B(c), the soma consists of two symmetrical frustums. Base radii are 12.5 μm, top radii are 2.9 μm and heights are 35 μm. Each frustum is divided into 14 cylinders. We index the somatic compartment nos. S1–S28 or S29. That is, compartment no. 9 is now expressed by nos. S1–S28 or

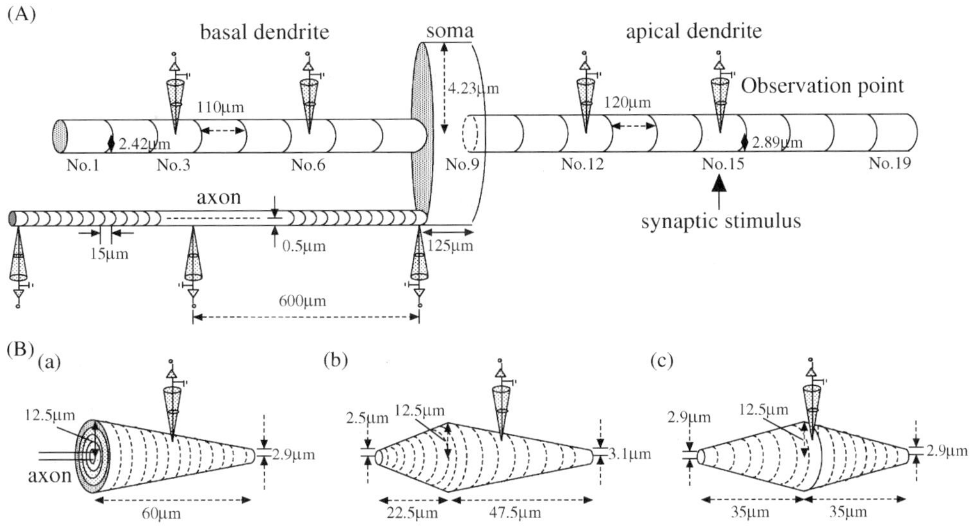

Fig. 1. A. Physical structure of the cell model. Bundled basal dendrites are laid on the left-hand side whereas apical one on the right-hand side. B(a)–(c). Our micro-compartmental soma models instead of the single cylinder. These models have almost the same surface areas. (a)Model B(a). The soma is expressed as a frustum. Radius is 12.5 μm, height is 60 μm. The base area is divided into one circle and four rings. The frustum is divided into 24 cylinders. (b)Model B(b). The soma consists of two frustums. Radii are 12.5 μm, heights are 22.5 and 47.5 μm, respectively. The smaller frustum is divided into nine cylinders and the bigger one is divided into 19. (c)Model B(c). The soma consists of two symmetrical frustums. Radii are 12.5 μm, heights are 35 μm. Each frustum is divided into 14 cylinders.

S29. Soma no. S1 is connected to the basal dendrite, while soma no. S28 or no. S29 to the apical one.

We determine the passive electric properties of the cell model, such as R_m (membrane resistivity), R_i (intracellular resistivity) and C_m (membrane capacitance), as $R_m = 1.0 \times 10^4\ \Omega\ \mathrm{cm}^2$ for the soma and dendrites, whereas $1.0 \times 10^3\ \Omega\ \mathrm{cm}^2$ for the axon, $R_i = 1.0 \times 10^2\ \Omega$ cm everywhere, and $C_m = 3\ \mu\mathrm{F/cm}^2$ everywhere again.

The dynamics of ionic channel and the distributions of the ionic channel conductance densities are the same as those in Traub et al. [6], and Sakatani and Hirose [4].

We assume excitatory N-methyl-D-aspartate (NMDA) and α-amino-3-hydroxy-5-methyl-4-isoxazole propionic acid (AMPA) synapses. We use a constant current as NMDA synaptic current for simplicity. The NMDA current is injected into the soma or the apical dendrite 0.66 mm distant from soma (compartment no.15). AMPA synaptic current is injected into the apical dendrite 0.66 mm distant from soma (compartment no. 15). The present current waveform equivalent to the AMPA synaptic stimulus is expressed in nA as

$$I_{\mathrm{synaptic}}(t) = G_{\mathrm{syn}} t\, \mathrm{e}^{-t/\tau}(V - V_{\mathrm{syn}}), \tag{1}$$

where G_{syn} has a unit of mS/s, V is the local membrane potential relative to the resting potential and $V_{\mathrm{syn}} = 60$ mV is the EPSP reversal potential relative to the rest. Time constant $\tau = 2$ ms is consistent with the voltage-clamp data on the mossy fiber EPSP. The full width half maximum (FWHM) pulse width is then about 5 ms.

The conductive-layer thickness is needed for calculation of the current flowing under the soma membrane because the central part of the soma including nucleus is not influenced by the membrane potential gradient. We estimate the effective thickness of the conductive layer D from the intracellular resistance on the condition that the intracellular resistivity is constant. In Model A, intracellular resistance ρ_k is calculated from intracellular resistivity $R_{\mathrm{i}} = 1.0 \times 10^2$ Ω cm, the cylinder-length $H = 1.25 \times 10^2$ μm and the radius $R = 4.23$ μm as

$$\rho_k = R_{\mathrm{i}}H/\pi R^2 = 2.22 \times 10^6[\Omega]. \tag{2}$$

The somatic intracellular resistance in each Model B(a)–(c) is fitted to 2.22 MΩ. The intracellular resistance in Model B(a) $\rho_{\mathrm{B(a)}}$ (in Ω) is expressed as the sum of the resistance at the bottom area and that of the surface area.

$$\rho_{\mathrm{B(a)}} = \sum_{j=S1}^{N_{\mathrm{soma}}} \rho_j$$
$$= \left\{ \frac{R_{\mathrm{i}}D}{\pi R_{S1}^2} + \sum_{j=S2}^{S5} \frac{R_{\mathrm{i}}}{2\pi D} \log\left(\frac{R_j}{R_{j-1}}\right) + \sum_{j=S6}^{S29} \frac{R_{\mathrm{i}}}{\pi D(2\Delta R_j - D)} \Delta H \right\} \times 10^4, \tag{3}$$

where R_{basal} is the radius of the basal dendrite, R_{soma} is that of the soma, j is the index of the somatic divided cylinder, ΔR_j is the radius of the divided cylinders, ΔH is the height of the divided cylinders, N_{soma} is the number of the soma micro-compartments. The intracellular resistances in Model B(b) and Model B(c) are expressed as

$$\rho = \sum_{j=S1}^{N_{\mathrm{soma}}} \rho_j = \sum_{j=S1}^{S28} \frac{R_{\mathrm{i}}}{\pi D(2\Delta R_j - D)} \Delta H \times 10^4. \tag{4}$$

As the fitting results with the values calculated by (3) and (4), the effective thicknesses of the conductive layer D in Model B(a)–(c) are 0.85 μm, 0.87 μm, 0.86 μm, respectively. In the following calculation of the firing characteristics, the time step is 10 ns.

3. Calculation results

In this section, we quantitatively analyze the average firing rate and the firing latency to elucidate the influence of neuron shape changes on the firing characteristics.

3.1. The average firing rate

We calculate the firing frequency when the soma or the distal apical dendrite is stimulated. The CA3 cells express three different modes of firing in response to steady

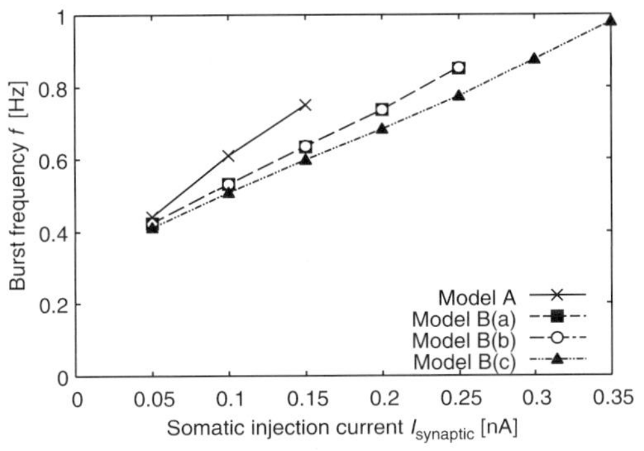

Fig. 2. Frequency-current (f–$I_{synaptic}$) curves for bursting mode induced by a small somatic current injection.

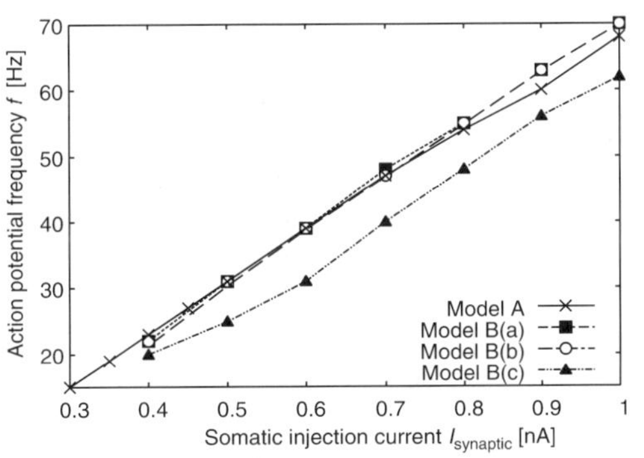

Fig. 3. Frequency-current (f–$I_{synaptic}$) curves for repetitive action potential mode induced by a large somatic current injection.

injection of current into the soma, depending on the magnitude of the injected current [8]. The soma generates rhythmic bursts at low frequency for small injected current 0–0.3 nA [1]. At somewhat higher currents 0.3–0.5 nA, rhythmic bursts with intercalated runs of fast spikes occur. At sufficiently large current more than 0.5 nA, rhythmic single action potentials are observed after an initial burst.

Figs. 2 and 3 show the relations between the firing frequency f and the somatic injection current $I_{synaptic}$ which mimics the NMDA mediated synaptic stimulus when the soma is stimulated by a constant current. Fig. 2 shows the frequency-current curves for the bursting mode induced by a small somatic current injection. In Model A, the burst frequency changes linearly to the injected currents in the range of 0–0.15 nA and the slope is about 3 Hz/nA. At a higher current (larger than 0.2 nA), the rhythmic bursts with intercalated runs of fast spikes occur. In Model B(a), the burst frequency is also proportional to the injected currents of 0–0.25 nA and the slope is about 2 Hz/nA. In Model B(b), the burst frequency also changes linearly to the injected currents of 0 –0.25 nA, while the slope is the same as that in Model B(a). In Model B(c), the burst

Table 1
Comparison of the firing latency when the apical dendrite is synaptically stimulated

Model	Time[ms] (small stimulus) $G_{syn} = 0.025$ mS/s	Time[ms] (large stimulus) $G_{syn} = 0.05$ mS/s
A	55.6	9.9
B(a)	23.3	9.8
B(b)	23.1	9.8
B(c)	29.9	10.3

frequency varies also linearly to the injected currents up to 0.3 nA and the slope is smaller than those in Model B(a) and Model B(b).

Fig. 3 shows the frequency-current curves for the repetitive action potential mode induced by the somatic larger current injection. In Model A, the repetitive action potential at 20–70 Hz occurs for the current range of 0.4–1.0 nA. In Model B(a), the repetitive action potential at 22–70 Hz is observed for the current 0.4–1.0 nA. The frequency is the same as that in Model A. The slope is also the same. In Model B(b), the frequency is 22–70 Hz for the current at 0.4–1.0 nA. The curve B(b) is almost the same as that of Model A again. In Model B(c), the frequency is 20–60 Hz for the current at 0.4–1.0 nA. The frequency is smaller than that in Model A.

3.2. The firing latency

In this section, we investigate the temporal evolutions of the membrane potentials at the observation points indicated in Fig. 1 when the apical dendrite is stimulated by AMPA synaptic current at 660 μm distant from soma. We report the result of the firing latency. We compare the firing latency in each model on two different conditions. One is for the small synaptic stimulus ($G_{syn} = 0.025$ mS/s) and the other is for the large ($G_{syn} = 0.05$ mS/s). Table 1 shows the comparison of the firing latency when the apical dendrite (660 μm distant from soma) is stimulated by the synaptic current expressed by (1). For the small stimulus ($G_{syn} = 0.025$ mS/s), the firing latency in Model A is larger than those in others. The firing latencies are identical with those in Model B(a) and Model B(b). That in Model B(c), however, is 6 ms larger when compared with those in Model B(a) and Model B(b). For the large stimulus ($G_{syn} = 0.05$ mS/s), the firing latencies are almost identical with those in all model. The difference of the firing latency is larger when the synaptic stimulus is smaller. In other words, the influence of neuron shape on the firing latency is large when the synaptic stimulus is small.

4. Discussion

The firing characteristics of Model B(a) and Model B(b) are almost the same as those of Model A except the case when the soma is stimulated by a small current injection. The frequencies in Model A is higher than those of Model B(a) and Model B(b). The result of Model B(a) and Model B(b) are good agreement with the experimentally obtained data [8,1] when the soma is stimulated by a small current injection.

We find two significant results. One is the fact that the firing rate in Model B(c) is lower than that of others when the soma is stimulated by the NMDA synaptic current. Pinsky and Rinzel [3] reported the number of the bursts increases in accordance with the level increase of maximal NMDA conductance so that the NMDA synapses provide a persistent excitation. The NMDA receptor at CA3 has a dominant roll of recalling process in the associative memory. If the pyramidal cell at CA3 change its shape like that in Model B(c), the recalling process may not work appropriately.

The other is that the firing latency in Model B(c) is longer when the apical dendrite is stimulated by the AMPA synaptic current. Pinsky and Rinzel [3] also reported that brief stimulation of a single cell in a resting network produces multiple synchronizing population bursts, with fast AMPA synapses providing the dominant synchronizing mechanism. On assumption that all the shapes expressed by Models B(a)–(c) existed at CA3, only those of Model B(c) would not synchronize with others because of their larger latency. For causing the synchronous population bursts, all the pyramidal cells at CA3 choose almost the same shape. Moreover, the actual CA3 cells are in a pyramidal shape, resulting a quick response to the apical stimulus. We can interpret this fact as such that the pyramidal cell at hippocampus CA3 selected its somatic pyramidal shape to respond to the AMPA mediated synaptic stimulus at the apical dendrite more sensitively. This accords with the widely recognized fact that the stimulus fed to the cell at the apical dendrite are the major input signals.

The difference in the frequencies and the latencies are caused by a large resistance gradient on the apical side membrane of Model B(c). If the resistance gradient of a soma is so large as that in Model B(c), the current diffuses from the apical-dendrite hillock to the central part of the soma membrane. For this reason, it takes more time to charge the membrane capacity. Therefore, it takes more time that the membrane potential exceeds its threshold in the nonlinear depolarization dynamics. Therefore, the firing frequency is lower and the firing latency is larger in Model B(c) than those of others.

5. Conclusion

We analyzed the influence of neuron shape on the firing rate and the firing latency. The following two facts have been revealed. First, the firing frequency is higher when the shape of the apical-dendrite hillock is sharper. Secondly, the firing latency is shorter for the apical-dendrite synaptic stimulus when the shape of the apical-dendrite hillock is sharper. These facts lead to a hypothesis that the neuron has changed its shape to respond more sensitively to the stimulus at the apical dendrite.

References

[1] J.J. Hablitz, D. Johnston, Endogenous nature of spontaneous bursting in hippocampal pyramidal neurons, Cell. Mol. Neurobiol. 1 (1981) 325–334.

[2] A. Hirose, S. Murakami, Spatiotemporal equations expressing microscopic two-dimensional membrane-potential dynamics, Neurocomputing 43 (2002) 185–196.
[3] P.F. Pinsky, J. Rinzel, Intrinsic and network rhythmogenesis in a reduced traub model for CA3 neurons, J. Comput. Neurosci. 1 (1994) 39–60.
[4] S. Sakatani, A. Hirose, A quantitative evaluation of the magnetic field generated by a CA3 pyramidal cell at EPSP and action potential, IEEE Trans. Biomed. Eng. 49 (2002) 310–319.
[5] R.D. Traub, J.G.R. Jefferys, R. Miles, M.A. Whittington, K. Toth, A branching dendrite model of a rodent CA3 pyramidal neurone, J. Physiol. 481 (1994) 79–95.
[6] R.D. Traub, R.K.S. Wong, R. Miles, H. Michelson, A model of a CA3 hippocampal pyramidal neuron incorporating voltage-clamp data on intrinsic conductances, J. Neurophysiol. 66 (1991) 635–650.
[7] J.L. Winslow, S.E. Jou, S. Wang, J.M. Wojtowicz, Signals in stochastically generated neurons, J. Comput. Neurosci. 6 (1999) 5–26.
[8] R.K.S. Wong, D.A. Prince, Afterpotential generation in hippocampal pyramidal cells, J. Neurophysiol. 45 (1981) 86–97.

Available at
www.ComputerScienceWeb.com
POWERED BY SCIENCE @DIRECT®

ELSEVIER

Neurocomputing 52–54 (2003) 363–370

NEUROCOMPUTING

www.elsevier.com/locate/neucom

Learning structured representations☆

Lokendra Shastri*, Carter Wendelken

International Computer Science Institute, 1947 Center Street, Suite 600, Berkeley, CA 94704, USA

Abstract

SHRUTI is a connectionist model that demonstrates how a network of neuron-like elements can encode a large body of semantic, episodic, and causal knowledge, and rapidly make decisions and perform explanatory and predictive reasoning. To further ground this model in the functioning of the brain it must be shown that components of the model can be learned in a neurally plausible manner. Previous work has already demonstrated the rapid learning of episodic facts via cortico-hippocampal interactions. Here we discuss how other SHRUTI representations such as causal rules, statistical and semantic knowledge, and categories might be learned.

Keywords: Reasoning; Category learning; Relational learning; Recruitment learning

1. Introduction

Although a great deal is known about neural representations in sensory, somatosensory, and motor cortices, the neural structures underlying higher-order cognitive processes are largely unknown. The SHRUTI model [7,11] provides a plausible set of proposals for what these representations might be, and how they could be employed in a wide range of cognitive processes such as semantic memory, reasoning and decision making. An important consideration for any neural representation scheme is its learnability. In this paper we discuss how several key representational elements of SHRUTI might be learned.

☆ This work was supported by NSF grants SBR-9720398 and ECS-9970890.
* Corresponding author. Tel.: +1-510-666-2910; fax: +1-510-643-7684.
E-mail address: shastri@icsi.berkeley.edu (L. Shastri).

0925-2312/03/$ - see front matter © 2003 Elsevier Science B.V. All rights reserved.
doi:10.1016/S0925-2312(02)00840-8

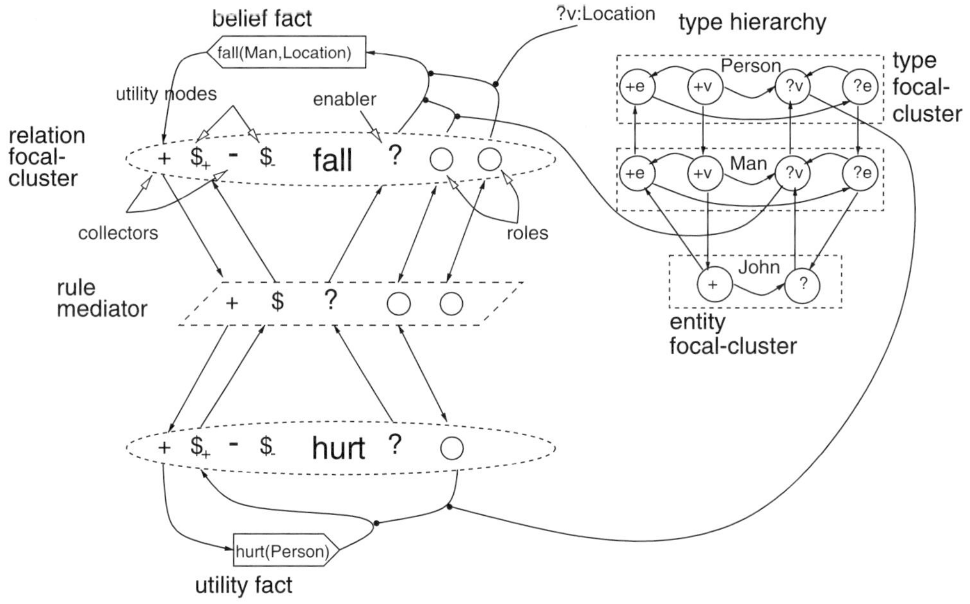

Fig. 1. Diagram showing major structures of the SHRUTI architecture.

1.1. The SHRUTI model

SHRUTI is a connectionist model that demonstrates how a network of neuron-like elements could encode a large body of structured knowledge and rapidly make decisions and perform explanatory and predictive reasoning. SHRUTI can encode different types of conceptual knowledge including relational schemas/frames for encoding action and event types (e.g., buying); causal rules between relational schemas (e.g., if you buy something you own it); types (categories); individual entities; and different types of facts such as episodic facts that record specific events (I saw John at the library today), taxon facts that record general statistical knowledge (Soccer moms own minivans), and reward and value facts that record associations between situations and rewards (or punishment).

SHRUTI suggests that the encoding of relational information (e.g., event schemas) is mediated by neural circuits composed of structured cell ensembles, termed *focal-clusters* (see Fig. 1). A relation focal-cluster (e.g., for the relation *fall*) consists of (i) role cells (e.g., for the role *fallee*) whose *synchronous* firing with entity cells (e.g., for the entity *John*) encodes role-entity *bindings* (e.g., the binding *fallee* = *John*) comprising the currently active relational *instance* (e.g., John fell in the hallway), (ii) + and − collector cells whose firing signifies belief and disbelief, respectively, in the currently active relational instance, (iii) enabler cells (?) whose firing signifies a search for support for this relational instance, and (iv) $\$_+$ and $\$_-$ utility cells that encode the desirability of this relational instance and its negation, respectively. Focal-clusters of entities and types are structured in a similar manner. The grouping of cells within a focal-cluster highlights their functional cohesiveness, but does not imply physical proximity.

The processing of relational information involves the transient propagation of *rhythmic* activity across relation and entity focal-clusters. Causal knowledge (e.g., a rule) is encoded by links that enable the propagation of rhythmic activity across focal-clusters, and persistent facts in long-term memory are realized as temporal pattern-matching circuits.

1.2. Learning in structured connectionist models

Two forms of learning, both related to hebbian learning, form the basis of learning structured representations in SHRUTI. These are: recruitment learning [2,6] and causal hebbian learning [13].

Recruitment learning (and also vicinal algorithms [12]) can be described informally as follows: learning occurs within a partially structured network containing a large number of richly interconnected nodes. *Recruited* nodes in the network are nodes that have acquired distinct functionality by virtue of their *strong* interconnections to other recruited and sensorimotor nodes. Unrecruited (*free*) nodes are connected via weak links to a large number of free, recruited, and sensorimotor nodes. Free nodes form a pool of nodes from which suitably connected nodes are recruited for representing new functional units. The recruitment process transforms a quasi-structured network into a collection of nodes and circuits with specific functionalities [8]. It has been shown [9] that recruitment learning can be grounded in long-term potentiation (LTP) [3].

Learning relatively complex structures of the SHRUTI model poses a significant challenge for the recruitment learning approach because learning such structures involves recruiting not just individual cells but also structured ensembles or functional circuits. The probability of extracting a particular structure from a set of random connections quickly approaches zero as the complexity of the structure increases. Consequently, some level of pre-existing organization is required. The sorts of pre-existing organization required for the SHRUTI model to be learnable involve certain recurring patterns of interconnections among nodes. Interestingly, computational modelling has shown that such organization could result from a genetically based developmental process [4]. It is well known that many different regions of the brain are organized differently prior to experience, and these differences are likely to be related to the functional needs of different cognitive processes. An excellent example of this may be found in the idiosyncratic architecture and local circuitry of the hippocampal formation which has been shown to be ideally suited for supporting the rapid encoding of episodic memories [8–10].

A second learning mechanism, causal hebbian learning (CHL), has been proposed as a partial solution for the learning of cause–effect relationships based on experience. In CHL, synaptic strength (weight) updates depend on the relative timing of pre- and post-synaptic firing (cf. differential hebbian learning [5] and spike-timing dependent plasticity [1]). Furthermore, different connection types may exhibit different dependences; some may update their weights only when pre-synaptic firing precedes post-synaptic firing, while others may do so only in the reverse scenario.

2. Learning SHRUTI structures

In the following we sketch how the learning mechanisms discussed above can be applied to learning various components of the SHRUTI model. While a detailed circuit-level design, analysis and implmentation has been carried out for the learning of episodic facts, only a preliminary account has been developed for the learning of other components focusing mainly on initial structures and network dynamics that would enable learning to occur.

2.1. Episodic facts

A key step in learning the model of an environment is remembering events as they happen. This is the goal of SMRITI, an anatomically and physiologially grounded and cell-level model of episodic memory formation in cortico-hippocampal circuits [8–10]. SMRITI demonstrates how the propagation of a rhythmic pattern of activity (corresponding to an activity-based encoding of an event) can lead to the recruitment of a persistent episodic memory trace in the hippocampal system as a result of LTP. SMRITI meshes fully with SHRUTI's representations and provides the encoding of episodic facts. The development of an anatomically and physiologically plausible account of episodic fact learning is important, since it illustrates the efficacy of mapping SHRUTI structures to realistic neural circuitry. Moreover, the presence of a storehouse of observations and events provides a strong foundation for other instance-based learning mechanisms.

2.2. Types

Learning of new types (categories) occurs on the basis of ongoing experience and previously memorized events or observations. As a starting point, we assume that the system has the ability to create representations for unique entities or individuals, and also that it has a repertoire of simple relational concepts. Type learning takes place most readily when the system is in a dedicated learning state during which random subsets of memorized events are simultaneously activated. A new type structure encompassing a set of entities is recruited when (i) these entities repeatedly fill the same roles in multiple events, and (ii) these events are instances of multiple event types. For example if the system were to observe *hasFeathers*(*Sam*), *hasFeathers*(*Tweety*), *hasBeak*(*Sam*), and *hasBeak*(*Tweety*), it might recruit a new type structure (*FeatheredBeakedThing*?) encompassing *Sam* and *Tweety* (see Fig. 2.). In general, a large number of entities and observations would be required in order to trigger recruitment. The occurrence of reinforcement signals—local reinforcement measures (represented by activity within a particular structure) as well as global indicators (communicated by neuromodulators)—can impact the learning of types and other SHRUTI structures. Type learning in SHRUTI can be characterized informally as follows:

1. Periodically, the system enters "type learning" mode, during which random sets of existing memories (episodic facts) are activated.

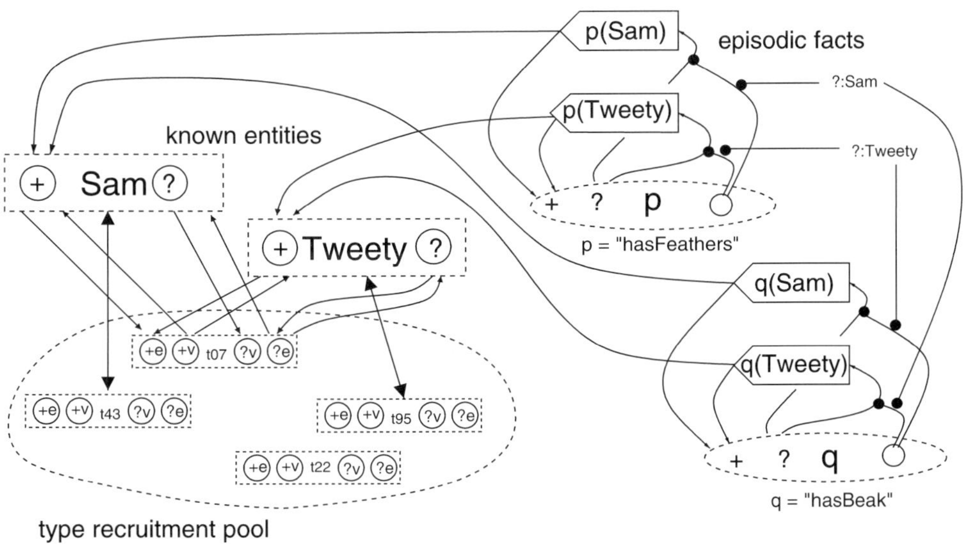

Fig. 2. Connectionist structure involved in the learning of a new type/category. Shown are two entity focal-clusters labeled Sam and Tweety, two relations *p* and *q* each with a pair of episodic facts, and a pool of free type focal-clusters connected randomly to existing entity representations. The focal-cluster labeled *t*07 is connected to both entities and can therefore be recruited as the representation of their common type. Bidirectional arrows represent multiple connections along both directions.

2. Existing types and entities receive input from active episodic facts with which they are associated, and become active upon receiving sufficient input from multiple facts.
3. Each type and entity is connected to numerous *free* type focal-clusters in the type recruitment pool. A free type focal-cluster receiving sufficient input during multiple phases gets recruited to serve as the parent type for the types and/or entities that caused it to become active.
4. A new type representation is strengthened with use, particularly when it is active in conjunction with reinforcement signals.
5. When not used, connections stengths may decay over time, thereby making structures become available for re-recruitment.

2.3. *Taxon facts and value facts*

Taxon facts represent statistical knowledge extracted from multiple occurrences of the same type of events [7]. A possible neural structure of taxon facts is described in [8,10]. Taxon facts are recruited when multiple episodic facts involving the same relational concepts and the same *types* of entities become active simultaneously. As with the learning of new types, this situation obtains most readily when the system is in a dedicated learning state, and stored instances are randomly activated. Building on the previous example, if the system were to observe *flys(Sam)* on one occasion and *flys(Tweety)* on another, where *Sam* and *Tweety* share a common supertype (let us

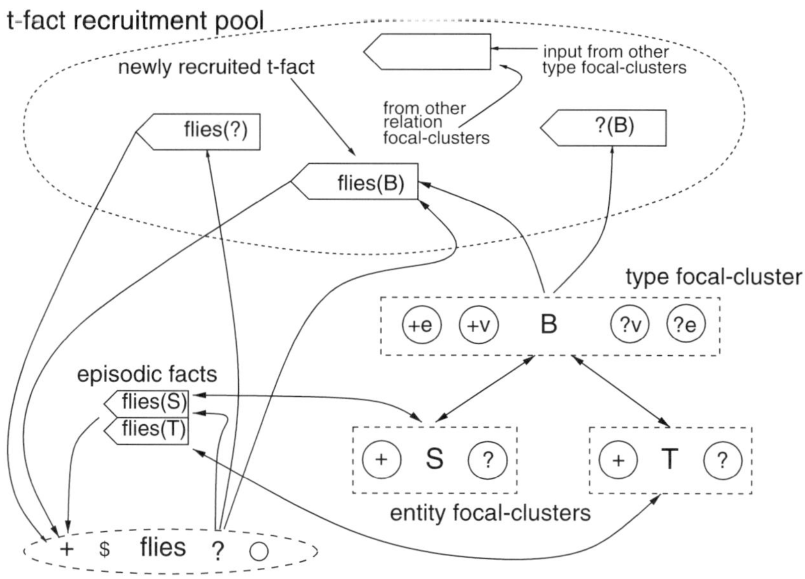

Fig. 3. A diagram of the major structures involved in the recruitment of taxon facts (*t*-facts). Shown are the relation focal-cluster *flies* with associated episodic facts, entities labeled *S* and *T* with learned supertype labeled *B*, and a pool of free taxon fact structures connected to these.

call it "Bird"), then it might create a new taxon fact *flies(Bird)* in response to these observations (see Fig. 3).

Value facts associate situations with utilities, and like taxon facts, are statistical in nature [14]. Where taxon fact learning depends on the accumulation of episodic facts, value fact learning depends on the accumulation of analogous reward facts, that is, memories of specific instances of reward or punishment.

In addition to their initial recruitment, an important aspect of taxon fact and value fact learning is the updating of their weights as new evidence arrives. Learning is triggered whenever an associated relational instance is active, with fact weight adjusting towards the current level of belief or utility. In this way, taxon (value) fact weight comes to represent prior probability (expected utility) of the relational instance.

2.4. *Causal knowledge (rules)*

The existence of a CHL mechanism that allows for different connection types with different temporal dependences is key to learning the bidirectional, asymmetric links of SHRUTI's causal rules. Specifically, connections between collector cells are enhanced whenever the source is active before the target (within a certain window of time), whereas connections between enabler cells are enhanced whenever the target is active before the source. Thus, when event A is observed preceding event B, both the forward and backward links for the rule $A \Rightarrow B$ are strengthened. The initial condition consists of a set of relation focal-clusters, with weak connections linking each cell to others of its type (collector cells linked to other collector cells, role cells linked to other role cells, etc.) The observation of a large number of events as they occur in time (or temporally compressed re-observation of sequences of events from memory) leads to the formation of rules reflecting the apparent causal structure of the environment.

3. Conclusion

A sketch of how major structures of the SHRUTI model may be learned has been presented. This is a preliminary account, and much work remains to be done. In particular, a cell-level implementation of SHRUTI is required in order to explore learning in detail. The recruitment of additional model structures such as relation focal-clusters and rule mediators also needs to be addressed. As a greater understanding of the neural substrate required for learning various structures is obtained, it will be possible to map these structures onto specific brain regions and projections. The success achieved in the development of SMRITI, an anatomically and physiologically grounded and cell-level model of episodic fact learning within cortico-hippocampal structures, provides strong guidance and well-founded hope that similarly detailed and neurally grounded accounts of learning various SHRUTI structures will emerge.

References

[1] G. Bi, M. Poo, Synaptic modification by correlated activity: Hebb's postulate revisited, Annu. Rev. Neurosci. 24 (2001) 139–166.
[2] J. Feldman, Dynamic connections in neural networks, Bio-Cybernet. 46 (1982) 27–39.
[3] R. Malenka, R. Nicoll, Long-term potentiation—a decade of progress? Nature 285 (1999) 1870–1874.
[4] G. Marcus, Plasticity and nativism: towards a resolution of an apparent paradox, in: S. Wermter, J. Austin, D. Wilshaw (Eds.), Emergent Neural Computational Architectures, Springer, Berlin, 2001, pp. 368–382.
[5] P. Roberts, Computational consequences of temporally asymmetric learning rules: I. Differential hebbian learning, J. Comput. Neurosci. 7 (1999) 235–246.
[6] L. Shastri, Semantic Networks: An Evidential Formalization and its Connectionist Realization, Morgan Kaufmann, Los Altos/Pitman Publishing Company, London, 1988.
[7] L. Shastri, Advances in SHRUTI—a neurally motivated model of relational knowledge representation and rapid inference using temporal synchrony, Appl. Intell. 11 (1999) 79–108.
[8] L. Shastri, From transient patterns to persistent structure: a model of episodic memory formation via cortico-hippocampal interactions, In revision, Behavioral and Brain Sciences, 2001. http://www.icsi.berkeley.edu/ shastri/psfiles/ShastriEMBBS.pdf.
[9] L. Shastri, A computationally efficient abstraction of long-term potentiation, Neurocomputing 44–46 (2002) 33–41.
[10] L. Shastri, Episodic memory and cortico-hippocampal interactions, Trends Cognitive Sci. 6 (2002) 162–168.
[11] L. Shastri, V. Ajjanagadde, From simple associations to systematic reasoning, Behav. Brain Sci. 16 (3) (1993) 417–494.
[12] Leslie Valiant, Circuits of the Mind, Oxford University Press, Oxford, 1994.
[13] C. Wendelken, L. Shastri, Probabilistic inference and learning in a connectionist causal network, in: Proceedings of the Second International Symposium on Neural Computation, Berlin, Germany, May 2000.
[14] C. Wendelken, L. Shastri, Combining belief and utility in a structured connectionist agent architecture, in: Proceedings of the 24th Annual Conference of the Cognitive Science Society, George Mason University, Fairfax, Virginia, 2002.

Lokendra Shastri is a Senior Research Scientist and Principal Investigator at the International Computer Science Institute (ICSI). Before moving to ICSI, he was on the faculty of the Computer and Information Science Department, University of Pennsylvania. Shastri received a Bachelor of Engineering degree in Electronics from the Birla Institute of Technology and Science, India in 1977, an M.S. in Computer Science from the Indian Institute of Technology, Madras in 1981 and a Ph.D. in Computer Science from the University of Rochester in 1986. He is interested in neurally motivated computational models of learning, memory, inference, and language understanding; the role of temporal synchrony in neural information processing; and spatio-temporal pattern recognition. Shastri is on the advisory board of *Neural Computing Surveys* and the editorial board of *Connection Science*. He has served as a consultant for the National Science Foundation and the United Nations Development Programme.

Carter Wendelken is a computer science graduate student at the University of California, Berkeley. He received a B.S. in Mathematical Sciences from the University of North Carolina at Chapel Hill. Wendelken is working toward his dissertation on neurally plausible models of inference and decision-making. He expects to complete his Ph.D. in 2003 and pursue a neuroscientific investigation of ideas emerging from this work.

Available at
www.ComputerScienceWeb.com
POWERED BY SCIENCE @DIRECT®

ELSEVIER

Neurocomputing 52–54 (2003) 371–379

NEUROCOMPUTING

www.elsevier.com/locate/neucom

A computational model of the interaction between external and internal cues for the control of hippocampal place cells

Simona Doboli[a,*], Ali A. Minai[b], Phillip J. Best[c]

[a] *Computer Science Department, Hofstra University, 210 Adams Hall, Hempstead, NY 11549, USA*
[b] *ECECS Department, University of Cincinnati, Cincinnati, OH 45221-0030, USA*
[c] *Department of Psychology, Miami University, Oxford, OH 45056, USA*

Abstract

Place cell activity in the hippocampus depends in a complex way on the information from two major sensory sources: External (e.g., visual) cues and internal cues (self-motion, vestibular cues). Dissociation of the two inputs in the presence of the animal modifies the place fields in ways that depend on the magnitude of the difference: If the mismatch is small, place fields typically follow the external cues, but if the mismatch is large, their activity is determined mainly by the internal cues. The dynamics underlying the interaction between the external and internal cues in the hippocampus is not completely known, but is an important issue in understanding the hippocampus' role in spatial learning and navigation. In this paper, we propose an attractor model of the hippocampus which captures the interaction between external and internal sensory inputs to the hippocampus. The proposed mechanism accounts for the difference seen in the experimental results in the case of a small versus a large discrepancy between the visual and idiothetic information.

Keywords: Remapping; Idiothetic cues; Visual cues; Path integration; Continuous attractor model

1. Introduction

Place cells in the rodent hippocampus [13,14] are thought to encode spatial cognitive maps, with their activity providing a distributed representation of the animal's position in the environment [14]. Place cell activity depends on multiple factors such as

* Corresponding author.
E-mail address: cscszd@hofstra.edu (S. Doboli).

0925-2312/03/$ - see front matter
doi:10.1016/S0925-2312(02)00843-3

sensory, behavioral, contextual and motivational cues [5,11,15,8,4,18,7,10,2]. Although the significance of single cue dependencies has mostly been explained through experiments and computational models, the mechanisms of interaction between cues and multiple sensory modalities for the control of the place cells are not well understood. In this paper, we propose a computational model of the rodent hippocampus to study the effect of the interaction between external (i.e., visual, olfactory, tactile) and internal (i.e., self-motion, vestibular and idiothetic) sensory cues on the activity of place cells. The main aim is to account for the experimental results that show a dependence of the behavior of place cells on the magnitude of mismatch between the external and internal cues when the mismatch takes place in the presence of the animal [18,7,6].

2. Review of experimental results

Many experimenters have investigated which of the two sensory inputs—external or internal—controls the firing of place cells when the information received by the animal from the two streams is not consistent [20,4,18,1,7,6]. In a stable and familiar environment, the external and internal inputs are concordant, with small errors in the idiothetic system corrected by the visual input [4,19,16]. A mismatch between the two inputs can appear in three situations: (1) the animal's internal cues are updated based on self-motion cues, without a change in the visual input [18,7], (2) the visual input changes (e.g., by rotation of walls) without activation of the animal's internal sense [6], or (3) both inputs change by different amounts [6].

If the mismatch occurs in the absence of the animal and the environment has previously been perceived as stable, the visual cues always orient the place cells. Type (1) mismatch was tested in the presence of the animal [18,7] by rotating either the only available cue [18], or the whole environment [7]. In both cases, a dependence of the control of place cells on the magnitude of the mismatch was observed [18,7]. When the discordance was *small*, place cells followed the visual input, and the place fields rotated with respect to the fixed reference frame [18,7]. After a *large* discordance, the visual input was no longer dominant and the place fields either remained in the same position with respect to the fixed reference frame—complete idiothetic control [18]—or remapped [7]. The model proposed here explicitly addresses type (1) mismatch while being consistent with the results of the other types of discordances.

2.1. Theoretical model

Our model, shown in Fig. 1, has three types of inputs to place cells: (a) the *external sensory input* which encodes distances to perceived landmarks [12] and allocentric bearings to them [9,22], (b) the *internal sensory* input from the path integration module, which, based on vestibular and self-motion cues, biases the place cells according to the animal's location in idiothetic coordinates [22,16], and (c) the *recurrent input* which embeds in the place cell layer a set of continuous attractors with centers uniformly distributed on the environment, [23,19,17]. Such a pattern of connectivity can arise through Hebbian learning in the synapses between co-active place cells [17,3].

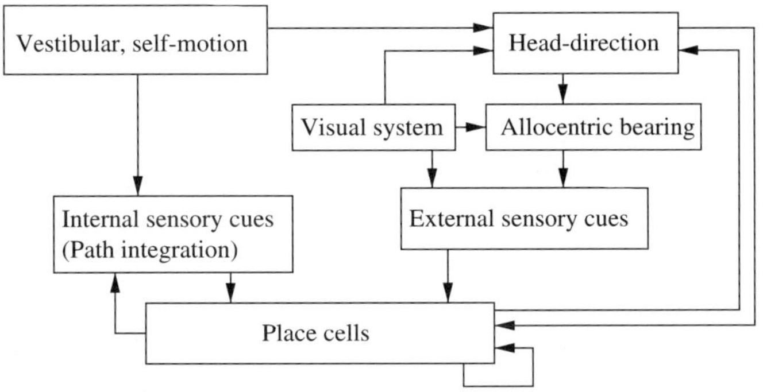

Fig. 1. The architecture of the hippocampal model.

The structure of the model is consistent with existing experimental evidence about the activity of place cells [5,21,15,4] and current models of the hippocampus [9,22,16,19,17]. The main assumptions of the model are: (a) the external sensory input encodes distances to perceived landmarks as well as allocentric bearings to them [9,22], and (b) distinct external and internal pathways to place cells in the hippocampus [16].

The model explains the dependence of the place representation on the magnitude of a type (1) mismatch between external and internal sensory cues as follows. When the simulated rat is introduced in a stable and familiar environment, the external visual input aligns the place code, and through it, the idiothetic input and, implicitly, the allocentric bearing or the rat's internal sense of direction (tuning curves of head direction cells are constant across trials in a familiar environment [21]). Experiments in which rotations of the environment take place in the presence of the animal activate its idiothetic system, but do not change the egocentric position of visual cues. If the environment transformations do not disorient the rat by altering its sense of direction, the *allocentric bearings* to landmarks are updated correctly by the internal cues, but they differ from the learned bearings in the initial, stable orientation of the environment. Thus, a type (1) mismatch can affect the location encoded by the internal sensory input, and the strength of the external input—which decreases with the magnitude of the difference between the learned allocentric bearing to landmarks and the perceived ones after the environment modification. The recurrent input has an inertial role by biasing the place cells active before the environment transformation.

In the model, a mismatch between external and internal sensory inputs is resolved by a competitive mechanism for the activity of place cells, since only a small number of place cells fire at any time. A *small mismatch* decreases the strength of the external input only slightly, so that, with the inertial tendency of the recurrent input, it wins the competition over the internal input and maintains the same set of active place cells as before the change in the environment and place fields rotate in the fixed frame of reference by the same angle as the rotation of the environment. A *large mismatch*, in contrast, decreases the magnitude of the external input significantly, so that the internal input takes control over the activity of place cells and place fields do not rotate with respect to the fixed frame of reference.

Experiments also show that a large mismatch can remap the place code if the idiothetic system cannot correctly follow the displacement of the animal (e.g., in the case of a very abrupt and fast rotation [7]). The model explains this by a random reset of the location encoded by the internal sensory input, though biased by the location indicated by external cues ([7] shows that in approximately half of the large, fast rotation experiments the place representation rotates by the same amount, while in the other half it remaps). If the new coordinates of the internal system are close to the location specified by the external input, then place fields will rotate with the magnitude of the rotation, exhibiting external control. If the internal input indicates a position far away from that of the external input, the active place cells will switch to that location. This results in a remapping of the place representation, equivalent to a translation of the cognitive map to the position of the idiothetic input. In this case, new place cells will start firing, some place cells will become silent, and others will change their place fields.

We consider this type of remapping distinct from the remapping seen in dissimilar environments where a new attractor map is selected. Based on experimental results, the overlap between the set of active place cells in distinct environments is much smaller [23] than the one seen after a large mismatch between internal and external inputs [7]. In the latter case, approximately 50% of the place cells have (different) place fields both before and after the rotation [7]. If the set of active place cells in different maps is chosen by chance, the percentage of common place cells should be lower than 50%.

2.2. Computational model

The place cell layer has N cells with centers distributed uniformly on a circular grid representing a cylindrical environment [18,7]. Some place cells have centers inside the cylinder and some outside. The dendritic input of a place cell is as follows: $h_i(t) = g_{\text{ext}} h_i^{\text{ext}}(t) + g_{\text{int}} h_i^{\text{int}}(t) + g_{\text{rec}} h_i^{\text{rec}}(t)$, where $h_i^{\text{ext}}(t)$ is the external sensory input, $h_i^{\text{int}}(t)$ is the internal sensory input, $h_i^{\text{rec}}(t)$ is the recurrent input and $g_{(.)}$ are the corresponding gains. At each time step, a competitive rule fires the K_N most active neurons ($K_N < 0.02N$).

The external sensory input is modeled as a vector encoding the distances and allocentric bearings to perceived landmarks [9,22,16]. A cell's external receptive field is formed by a product of Gaussian functions, each representing the response to a landmark:

$$h_i^{\text{ext}}(t) = \exp\left[-\sum_{k=1}^{n_i} w_k \left(\frac{[\mathrm{d}(r_{\text{ext}}(t), l_k) - \mathrm{d}(c_i, l_k)]^2}{2\sigma_{\text{ext}}^2(c_i)} + \frac{[\alpha_{rk}(t) - \alpha_k]^2}{2\sigma_{\text{dir}}^2(c_i)}\right)\right], \tag{2}$$

where n_i is the number of landmarks the cell responds to and w_k is the significance of the landmark. The distance between the rat's position ($r_{\text{ext}}(t)$) and landmark position (l_k) is denoted by $\mathrm{d}(r_{\text{ext}}(t), l_k)$ (in the environment reference frame). The distance between the cell center (c_i) and landmark l_k is $\mathrm{d}(c_i, l_k)$. The perceived allocentric bearing to landmark l_k is $\alpha_{rk}(t)$, and α_k is the learned one. These angles are expressed with respect to an arbitrarily chosen reference direction. All positions are expressed in Cartesian coordinates.

The internal sensory input encodes the position of the rat in path-integrator coordinates:

$$h_i^{\text{int}}(t) = \exp(-\mathrm{d}(r_{\text{int}}(t), c_i)^2 / 2\sigma_{\text{int}}^2), \tag{3}$$

where $\mathrm{d}(r_{\text{int}}(t), c_i)$ is the distance between the simulated rat's position in path-integrator coordinates and the place cell center. It is assumed that, initially, external and idiothetic frames are superimposed and identical with the position of the rat in absolute coordinates: $r_{\text{ext}}(t) = r_{\text{int}}(t) = r_{\text{abs}}(t)$.

The third input is formed by the feedback connections onto place cells. Their strength is set through associative learning, such that it decays exponentially with the distance between connecting place cell centers. This establishes a set of continuous attractors spread uniformly on a familiar environment [19,17]. The relation of the recurrent input is:

$$h_i^{\text{rec}}(t) = \sum_{j=1, j\neq i}^{N} w_{ij} x_j(t-1), \tag{4}$$

with $w_{ij} = \beta_{ij} \exp[-\mathrm{d}(c_i, c_j)^2 / (2\sigma_w^2)]$, the recurrent weights and $x_j(t) \in \{0, 1\}$ the state of the neuron j at time t. The value of β_{ij} expresses the strength of the recurrent connection.

3. Simulation results

We have used simulations to replicate the experimental results obtained by Rotenberg and Muller [18] and Knierim et al. [7]. The environment was simulated as a circular surface of radius R_{cyl}. In the model there are $N = 2912$ place cells with centers spread uniformly on a 2 R_{cyl} circular surface. Initially, the cue card is oriented to the east, symmetrically across the horizontal axis, and specified by three points representing the landmarks: two at each end of the cue card and one in the middle.

The rotation phase was simulated for the [18] experiment—denoted RM97—by slowly rotating counterclockwise the position of the cue card along the wall (increments of 5°) while the simulated rat followed the margin of the cue [18] such that the relative position of the rat with respect to the card remained constant during the rotation (as in the RM97 experiment).

Place fields were reconstructed from the activity of a subset of the simulated place cells as follows: With the card in the initial position, after a 45° counter-clockwise manipulation, and after a 180° counter-clockwise manipulation. Fig. 2(i) shows the place fields of four place cells in the three situations corresponding to the simulation of the RM97 experiment. The place fields after the 45° rotation (column (b)), rotate by the same amount, while those after 180° rotation remain unchanged (column (c)).

For the simulation of the experiment by Knierim et al. [7]—denoted K98—the rotation of the environment is modeled by a sudden jump in the path-integrator coordinates. The external input remains in the same relative position as before the rotation, but its

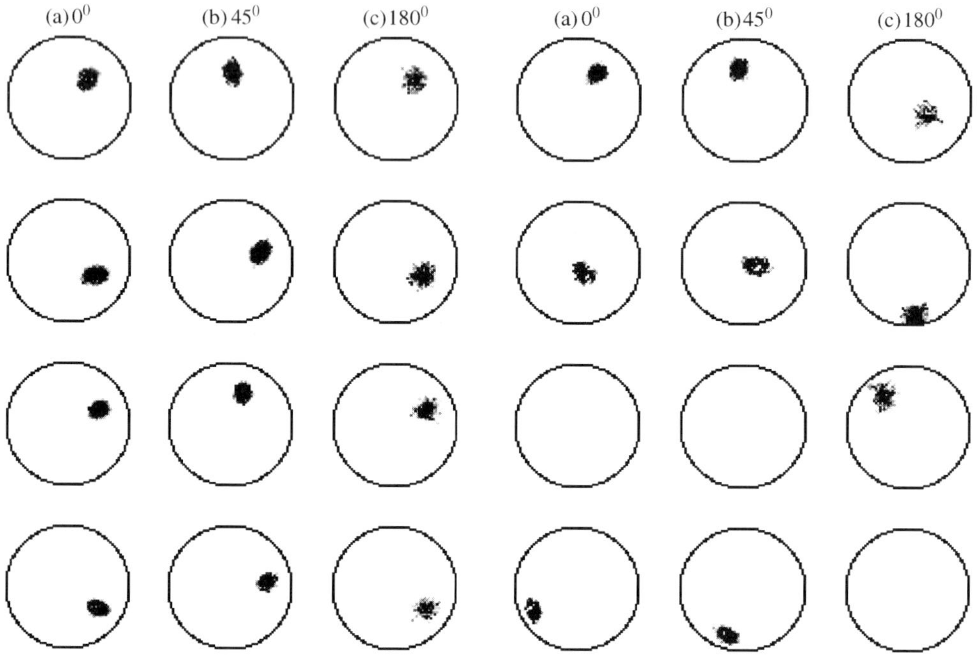

Fig. 2. Example of place fields for the RM97 simulation (i) and for the K98 simulation (ii). Each row shows a different place cell. (a) Place fields when the cue card is in the familiar position, (b) after the 45° rotation, (c) after the 180° rotation.

strength is modulated by the change in the sense of direction perceived by the animal. Depending on the initial coordinates of the internal sensory input after the rotation, the place representation can either appear rotated by 180°, or remapped. Fig. 2(ii) shows place fields obtained after a small and a large rotation, where a remapping of the place representation occurs in the latter one. After the 45° rotation (column (b)), the place fields rotate by the same angle. After the 180° rotation (column (c)) some place cells change their field, new place cells start firing, and place cells that lose their place field. The simulation initializes the internal sensory input at a large distance relative to the absolute position of the simulated rat immediately before the rotation and a perceived rotation angle of 90°.

In the simulation, the proportion of place cells that stop firing (26%), change their place fields (46%) and start firing (26%) is very similar to results of remapping given in [7].

4. Conclusions

We have presented an attractor model of the hippocampus that accounts for the interaction between the external and internal sensory cues for the control of place cells when the two inputs become discordant. In particular, the model accounts for the difference in the experimental results with the magnitude of the discordance between external and internal cues in the presence of the animal. The model also provides a simple explanation for the kind of remapping seen in these situations. This remapping is considered different from that seen between two distinct environments.

The following predictions can be made based on the model:

- After a remapping following a large mismatch between internal and external inputs, the model predicts that the position of the modified place fields can be obtained by the same linear translation from the original position. The model simulates remapping by a translation of the path-integrator coordinates such that the resulting attractor map does not overlap entirely with the original one. By this transformation place cells acquire new place fields, some cells stop firing and some cells start firing. What this implies is that the position of the modified place fields can be obtained by a linear translation from the initial location. Of course, in reality there are cumulative errors in the path integration, and the initial distance and angle of the translation could be different for each place cell. Whether the modified place fields can be reconstructed from the original ones by means of the same transformation could be easily checked by analyzing data coming from simultaneous recordings of place cells.
- The locus of the competition—hippocampus or HD cells—following similar environmental manipulations might be determined by inactivating the hippocampus during and after the rotation and recording from HD cells. If the HD cells still show similar responses as with the hippocampus active [7] (i.e., predominant idiothetic control to large rotations of the environment), then the HD system is capable of resolving the mismatch between the internal and external sensory cues.

Acknowledgements

This research was supported by Grants No. IBN-9634424, IBN-9808664, and IBN-9816612 from the National Science Foundation.

References

[1] J. Bures, A.A. Fenton, Y. Kaminsky, J. Rossier, B. Sacchetti, L. Zinyuk, Dissociation of exteroceptive and idiothetic orientation cues: effect on hippocampal place cells and place navigation, Philos. Trans. R. Soc. London 352 (1997) 1515–1524.

[2] S. Doboli, A.A. Minai, P.J. Best, Latent attractors: a model for context-dependence place representations in the hippocampus, Neural Comput. 12 (5) (2000) 1009–1043.

[3] S. Doboli, A.A. Minai, P.J. Best, A.M. White, An attractor model for hippocampal place cell hysteresis, in: Proceedings of Computational Neuroscience Meeting (CNS'2000), Bruges, Belgium, 2000.

[4] K.L. Gothard, W.E. Skaggs, B.L. McNaughton, Dynamics of mismatch correction in the hippocampal ensemble code for space: interaction between path integration and environmental cues, J. Neurosci. 16 (1996b) 8027–8040.

[5] A.J. Hill, P.J. Best, The effect of deafness and blindness on the spatial correlates of hippocampal unit activity in the rat, Exp. Neurol. 74 (1981) 204–207.
[6] K.J. Jeffery, J.M. O'Keefe, Learned interaction of visual and idiothetic cues in the control of place field orientation, Exp. Brain Res. 127 (1999) 151–161.
[7] J.J. Knierim, H.S. Kudrimoti, B.L. McNaughton, Interactions between idiothetic cues and external landmarks in the control of place cells and head-direction cells, J. Neurophysiol. 80 (1998) 425–446.
[8] E.J. Markus, Y.-L. Qin, B. Leonard, W.E. Skaggs, B.L. McNaughton, C.A. Barnes, Interactions between location and task affect the spatial and directional firing of hippocampal neurons, J. Neurosci. 15 (1995) 7079–7094.
[9] B.L. McNaughton, J.J. Knierim, M.A. Wilson, Vector encoding and the vestibular foundations of spatial cognition: neurophysiological and computational mechanisms, in: M.S. Gazzaniga (Ed.), The Cognitive Neurosciences, The MIT Press, Cambridge, MA, 1995, pp. 585–595.
[10] A.A. Minai, P.J. Best, Encoding spatial context: a hypothesis on the function of the dentate gyrus-hilus system, in: Proceedings of the International Joint Conference on Neural Networks, Anchorage, 1998, pp. 587–592.
[11] R.U. Muller, J.L. Kubie, The effects of changes in the environment on the spatial firing properties of hippocampal place cells, J. Neurosci. 7 (1987b) 1951–1968.
[12] J. O'Keefe, N. Burgess, Geometric determinants of the place fields of hippocampal neurons, Nature 381 (6581) (1996) 425–428.
[13] J. O'Keefe, J. Dostrovsky, The hippocampus as a spatial map: preliminary evidence from unit activity in the freely moving rat, Brain Res. 34 (1971) 171–175.
[14] J. O'Keefe, L. Nadel, The Hippocampus as a Cognitive Map, Clarendon Press, Oxford, UK, 1978.
[15] G.J. Quirk, R.U. Muller, J.L. Kubie, The firing of hippocampal place cells in the dark depends on the rat's recent experience, J. Neurosci. 10 (1990) 2008–2017.
[16] A.D. Redish, D.S. Touretzky, Cognitive maps beyond the hippocampus, Hippocampus 7 (1997) 15–35.
[17] A.D. Redish, D.S. Touretzky, The role of the hippocampus in solving the Morris water maze, Neural Comput. 10 (1998) 73–111.
[18] A. Rotenberg, R.U. Muller, Variable place-cell coupling to a continuously viewed stimulus: evidence that the hippocampus acts as a perceptual system, Philos. Trans. R. Soc. London B 352 (1997) 1505–1513.
[19] A. Samsonovich, B.L. McNaughton, Path integration and cognitive mapping in a continuous attractor neural network model, J. Neurosci. 17 (1997) 5900–5920.
[20] P.E. Sharp, T.B. Hugh, D. Etkin, D.B. Tzanetos, Influences of vestibular and visual motion information on the spatial firing patterns of hippocampal place cells, J. Neurosci. 15 (1) (1995) 173–189.
[21] J.S. Taube, R.U. Muller, J.B. Ranck Jr., Head-direction cells recorded from the postsubiculum in freely moving rats. II. Effects of environmental manipulations, J. Neurosci. 10 (1990) 436–447.
[22] D.S. Touretzky, A.D. Redish, Theory of rodent navigation based on interacting representations of space, Hippocampus 6 (1996) 247–270.
[23] M.A. Wilson, B.L. McNaughton, Dynamics of the hippocampal ensemble code for space, Science 261 (1993) 1055–1058.

Simona Doboli received her Ph.D. in Electrical Engineering at the University of Cincinnati in 2001. She is currently an Assistant Professor in the Computer Science Department at Hofstra University, Hempstead, NY. Her research interests include computational models of the hippocampal function, neural models of sequence processing, and modeling non-linear systems using neural networks.

Ali A. Minai received his Ph.D. in Electrical Engineering and post-doctoral training in Computational Neuroscience, both at the University of Virginia. He is currently an Associate Professor in the Electrical and Computer Engineering and Computer Science Department at the University of Cincinnati, and directs the Complex Adaptive Systems Laboratory. His research interests include neural computation, non-linear dynamics and complex systems. Dr. Minai is a member of Sigma Xi, Eta Kappa Nu, Tau Beta Pi, IEEE and INNS. He is an associate editor of the IEEE Transactions on Neural Networks.

Phillip J. Best is a Professor in the Psychology Department and the Center for Neuroscience at Miami University, Oxford, OH. He received his Ph.D. in Psychology at Princeton University, and has served on the faculties of the University of Michigan, the University of Virginia, the University of New Orleans, and Miami University. His research has focused on hippocampal cellular activity during spatial navigation, as well as the effects of lesions and ethanol on hippocampal activity and navigational behavior. Dr. Best has served on various review panels for NSF and NIH. He has been an associate editor of the Psychonomic Bulletin and Review since 1993, and was a consulting editor for Psychobiology from 1986 to 1997.

Available at
www.ComputerScienceWeb.com
POWERED BY SCIENCE @ DIRECT®

ELSEVIER

Neurocomputing 52–54 (2003) 381 – 387

NEUROCOMPUTING

www.elsevier.com/locate/neucom

NNET: linking small- and large-scale network models☆

Frances K. Skinner[a,b,*], Jiang Brandon Liu[a]

[a] *Toronto Western Research Institute, University Health Network, University of Toronto, 399 Bathurst St., MP 13-317, Toronto, Ont., Canada M5T 2S8*

[b] *Departments of Medicine (Neurology) and Physiology, Institute for Biomaterials and Biomedical Engineering, University of Toronto, Toronto, Ont., Canada*

Abstract

To examine how mechanistic principles derived in small networks are manifest in larger networks, we developed a versatile neuronal network simulator, NNET, in portable C++ code. NNET allows easy, fast and accurate characterization of any neuronal population and coupling pattern of electrical and chemical synapses. Using NNET, we simulated 30-cell networks of interneurons coupled with both gap junctions and GABAergic synapses, based on a mechanism derived in two-cell networks.

Keywords: Neuronal network simulator; Gap junction; Interneuron; Synchrony; Hippocampus

1. Introduction

The complexity of brain networks and the many details inherent at the many levels make it a monumental challenge to understand the many interacting dynamics that give rise to the different patterns that have been recorded in the brain. One major aspect of computational neuroscience is the use of well-constructed computer simulations as a mode of mathematical expression to understand the operation of the brain [15]. Although it is not clear how much model detail (and subsequent simulations) is

☆ This work was supported by NSERC and DCIEM of Canada. We would like to thank Jenni Chung for performing some of the heterogeneous network simulations.

* Corresponding author. Toronto Western Research Institute, University Health Network, 399 Bathurst St., MP 13-317, Toronto, Ont., Canada, M5T 2S8. Tel.: +1-416-603-5800x5107; fax: +1-416-603-5745.

E-mail address: fskinner@uhnres.utoronto.ca (F.K. Skinner).

0925-2312/03/$ - see front matter
doi:10.1016/S0925-2312(02)00844-5

appropriate and required to determine underlying principles in neuronal network operation, it is clear that small networks with some biophysical detail can provide insight and understanding in larger networks. For example, the mechanistic understanding of synchrony in inhibitory networks determined by Wang and Rinzel [20] has been invoked in large thalamic and hippocampal networks [6,19].

Networks of inhibitory interneurons or GABAergic cells are recognized as being essential units in cortical functioning, likely responsible for controlling the various EEG rhythmic frequencies [2,13]. However, it is also abundantly clear that interneurons form a diverse population of neurons in terms of morphology, biochemical characteristics, channel types and distribution and connectivity profiles [16]. In particular, it is known that interneurons form networks via both GABAergic synapses and gap junctions (g.j.) (e.g. [7–10]) and that g.j.s are critical for synchronous activity [4,11].

Experimental work indicates the involvement of bursting oscillations in interneuronal networks with both gap junctional and inhibitory coupling in setting a slow population rhythm [22]. This motivated us to develop a mechanism by which this could occur. We found that synchronized bursting could be produced robustly in two-cell networks connected by both GABAergic and g.j. synapses in which persistent sodium and slowly inactivating potassium currents ($I_{K_{s-i}}$) were present in the individual neurons [17]. With such intrinsic properties, synchronized bursting occurs when inhibition sufficiently removes the potassium inactivation. The g.j. coupling acts to stabilize the synchronous, intraburst spiking, so that synchronized bursting occurs. However, with only a two-cell network we could not examine how the connectivity architecture contributes to possible population bursting.

In this paper we describe a newly developed neuronal simulator software, N-cell NETwork (NNET), which can be used to build networks of arbitrary size and connectivity, and in which the individual neurons have a conductance-based formalism. We use NNET to simulate large networks of interneurons connected with g.j.s and GABAergic synapses based on the mechanism derived in [17].

2. NNET Software

Design goal and "intermediate level of detail" model networks: Most neuronal network simulation packages place emphasis either on the connective configuration of networks of simple nodes (i.e, artificial neural networks) or on the elaborate physiological and molecular details of the membrane current dynamics of a single neuron (e.g., NEURON or GENESIS). Neither genre suits our purpose of examining an intermediate level (of detail) neuronal network with sufficient details to achieve the phenomenological characterization of each individual neuron, but rid of the staggering details which hinder the study of a network of meaningful size and architecture. We designed and implemented a software package, NNET, to bridge this gap [14]. Furthermore, within this realm of models, we were not aware of any network simulators that included g.j.s as a feature.

Outline of design and implementation: NNET is portable, scalable, and re-usable. It solves a prototyped system of stiff ordinary differential equations (ODEs) (which

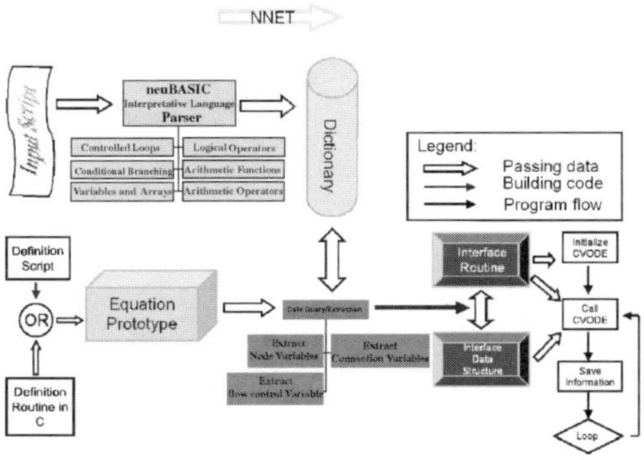

Fig. 1. Schematic of NNET design.

must be a subset of the equations in [17]) using CVODE [3]. The network can have both synaptic and g.j. couplings of arbitrary architectures. The default output is a binary file readable by MATLAB. Output includes dynamic variables, coupling currents and field potentials. As in [5], the extracellular field potential, V_{ext}, is calculated as $V_{\text{ext}} = R_{\text{e}}/4\pi \sum I_k/r_k$ where R_{e} is the extracellular resistivity, I_k is the total postsynaptic current on the kth cell and r_k is the distance between the location of the probe and the kth cell.

Fig. 1 illustrates NNET's design. The C++ NNET program provides a scripting environment (neuBASIC) in which the values for the coefficients of the equations and various parameters for the manipulation of CVODE are set for later retrieval. Essentially, NNET provides a versatile programming environment for constructing and numerically solving a prototyped system of stiff initial value set of ODEs. Given a set of equations that describes the neuron and one connection between any two neurons, called an Equation Prototype, NNET parses a script (neuBASIC) that is similar to the BASIC procedural programming language, in which the coefficients of the equation system are specified and then extracted to construct an abstract data structure and associated routines that are subsequently supplied to the CVODE numerical solver to generate the results. The support for loops, conditional branching and common arithmetic functions eases the effort and shortens the time in specifying a system. NNET is freely available (send email to FKS).

3. Application to inhibitory network simulations

The effect of gap-junctional coupling in the generation and maintenance of slow field potentials in inhibitory networks was examined. Using model cell and couplings as described in [17], we built networks of up to 30 cells using our NNET software.

Scalability, multistability and bursting: As noted in our earlier work, our two-cell model network can exhibit other stable solutions besides the synchronized bursting solution (for which $I_{\text{K}_{s-i}}$ must actively participate, i.e., be sufficiently deinactivated).

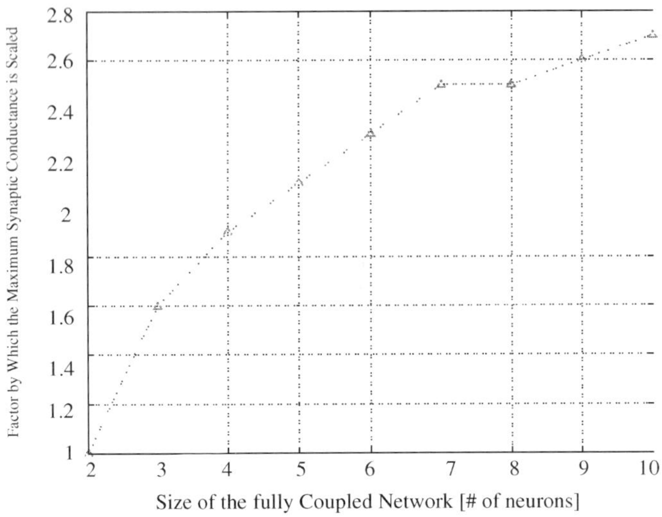

Fig. 2. Scaling factor necessary to maintain population bursts in a network of N fully coupled cells.

Other solutions of antiphase spiking and unsynchronized firing can be uncovered via perturbations of the g.j. coupling (not shown). Since we wanted to examine the synchronized bursting mechanism, we started with the same parameters as in the two-cell network of Skinner et al. [17]. We then performed a scalability study of 3–10 all-to-all coupled networks (in total, 46 networks were examined). We found that population bursting was maintained only if the maximal g.j. conductance, g_{gap}, was increased as the network size was increased (possibly due to accumulation of numerical errors), approaching an asymptotic factor of 3 (see Fig. 2). (Note that a normalization of synapses is done such that the total synaptic drive per cell on average remains the same as network size is changed, i.e., g_{gap} (or g_{syn}) is divided by the number of synapses (M) in the network. Thus, $M_{syn} = N$ for all-to-all coupling and $p = M_{syn}/N$ is the coupling probability (similarly for M_{gap}).) Thus, in subsequent network simulations, we increased g_{gap} by a factor of 3 so that we know that our networks can exhibit synchronized bursting with all-to-all inhibitory and g.j. coupling via the mechanism described in [17].

Large network bursting possible with minimal g.j. connectivity: Fig. 3 shows plots of field potentials in 30-cell networks for various g.j. coupling architectures (M_{syn} the same). Population bursts (PBs) occur with similar frequencies and durations as in the two-cell case. Bursting cells are perfectly synchronized, but the number of cells that are bursting during the PB changes (not shown). One can get at least one PB with as little as 10% g.j. connectivity (Fig. 3, $M_{gap} = 3$). We know from the two-cell work that the g.j.s are playing a stabilizing role, i.e., with a large enough g_{gap}, synchronized bursts can occur and be stably maintained in the face of perturbations (see Fig. 8 in [17] and Fig. 2 in [18]). In the 30-cell network simulations, we see that the amount of g.j. connectivity plays a similar (stabilizing) role. Note that in Fig. 3, g_{gap} per se is not

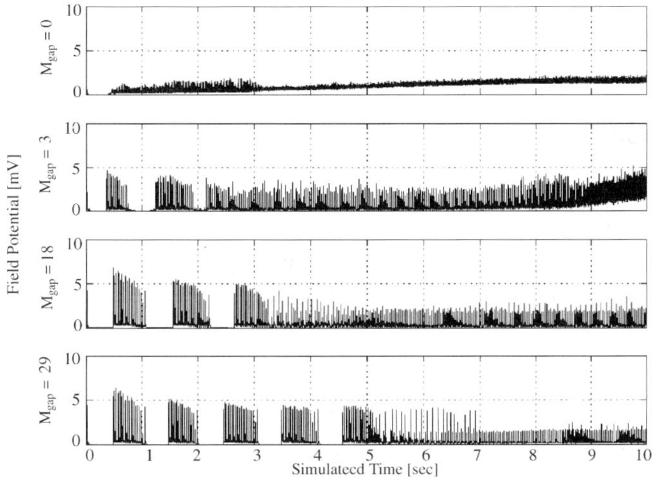

Fig. 3. PBs occur in larger networks and require g.j. connectivity. Field potentials of four 30-cell networks with (top to bottom) $M_{gap} = 0$, no g.j.s; $M_{gap} = 3$; $M_{gap} = 18$; and $M_{gap} = 29$, all-to-all coupling ($M_{syn} = 18$ in all cases).

being changed, just the degree of g.j. connectivity, or the number of cells connected by g.j.s. In the two-cell case, the presence of stable bursts relies on a large enough g_{gap}. In the large 30-cell networks, the presence of PBs also requires g.j.s, but the degree of g.j. connectivity dictates the length of time for which PBs occur (as shown in Fig. 3).

Mechanism of PB: In the two-cell situation, the bursting behaviour terminates because of the removal of the inactivation leading to an increase in the potassium current which eventually prevents the individual cell from spiking. This is also true in the large network, but because of the particular (not all-to-all) connectivity, all cells do not have g.j.s and so some are tonically firing and others are suppressed (as would be the case without g.j.s and other possible stable solutions). Unlike the two-cell network, the PBs decline in amplitude since less cells are participating as the burst progresses. In summary, the field potential decreases during a burst due to the gradual loss of network synchrony, which leads to a suppression of a subpopulation of neurons and ultimately the loss of PBs.

Heterogeneity: In the above simulations, all cells in the network are identical. However, theoretical work indicates that heterogeneity makes synchrony fragile [21]. Therefore, we might expect that for the same g.j. coupling, but with heterogeneous cells, PBs would not last as long. In 30-cell network simulations, we found that this was not necessarily the case. Indeed, it was possible for PBs to last longer with less g.j. connectivity. To understand this, we performed a preliminary study with smaller (6-and 12-cell) networks in which we examined exactly which cells were interconnected. We found that this phenomenon occurred when g.j. connections were present between cells that had more diverse firing patterns, presumably because then an appropriate balance between high and low firing cells could be achieved to extend the length of time for which PBs could occur.

4. Discussion and conclusion

In summary, a versatile network simulator, NNET, has been developed. There is no limit on network size except for memory and computing power, which is not a formidable issue nowadays. Besides the simulations described here, NNET has been modified for other specific uses [1,12]. Future plans include the incorporation of synaptic plasticity and parallel versions of CVODE. Since NNET includes gap junctions (g.j.s), networks of simple multi-compartment cells can also be simulated.

Although 30 cells is still a relatively small network, it is large enough to allow us to make some assessment of connectivity issues. It is intriguing to think about g.j.s as "switches" for controlling the frequency of network output, i.e., output can switch from slow rhythms (synchronized bursts) to faster rhythms (no bursts) by manipulating the g.j. connectivity. Given that gap junctions are not static pores, this suggests an interesting mode of control of network frequency (over orders of magnitude).

References

[1] P.M. Baker, P.S. Pennefather, B.A. Orser, F.K. Skinner, Disruption of coherent oscillations in inhibitory networks with anesthetics: the role of $GABA_A$ receptor desensitization. J. Neurophysiol. 88 (2002) 2821–2833.
[2] G. Buzsáki, J.J. Chrobak, Temporal structure in spatially organized neuronal ensembles: a role for interneuronal networks, Curr. Opin. Neurobiol. 5 (1995) 504–510.
[3] S.D. Cohen, A.C. Hindmarsh, CVODE, a stiff/nonstiff ODE solver in C, Comput. Phys. 10 (1996) 138–143.
[4] M.R. Deans, J.R. Gibson, C. Sellitto, B.W. Connors, D.L. Paul, Synchronous activity of inhibitory networks in neocortex requires electrical synapses containing connexin 36, Neuron 31 (2001) 477–485.
[5] A. Destexhe, Spike-and-wave oscillations based on the properties of $GABA_B$ receptors, J. Neurosci. 18 (1998) 9099–9111.
[6] A. Destexhe, D. Contreras, T.J. Sejnowski, M. Steriade, A model of spindle rhythmicity in the isolated thalamic reticular nucleus, J. Neurophysiol. 72 (1994) 803–818.
[7] T. Fukuda, T. Kosaka, Gap junctions linking the dendritic network of GABAergic interneurons in the hippocampus, J. Neurosci. 20 (2000) 1519–1528.
[8] M. Galarreta, S. Hestrin, A network of fast-spiking cells in the neocortex connected by electrical synapses, Nature 402 (1999) 72–75.
[9] J.R. Gibson, M. Beierlein, B.W. Connors, Two networks of electrically coupled inhibitory neurons in neocortex, Nature 402 (1999) 75–79.
[10] A.I. Gulyás, N. Hájos, T.F. Freund, Interneurons containing calretinin are specialized to control other interneurons in the rat hippocampus, J. Neurosci. 16 (1996) 3397–3411.
[11] S.G. Hormuzdi, I. Pais, F.E.N. LeBeau, S.K. Towers, A. Rozov, E.H. Buhl, M.A. Whittington, H. Monyer, Impaired electrical signaling disrupts gamma frequency oscillations in connexin 36-deficient mice, Neuron 31 (2001) 487–495.
[12] M. Jazayeri, L. Zhang, F.K. Skinner, A dynamical feedback mechanism for the generation of hippocampal field rhythms, Neurocomputing 38–40 (2001) 683–689.
[13] J.G.R. Jefferys, R.D. Traub, M.A. Whittington, Neuronal networks for induced 40 Hz rhythms, Trends Neurosci. 19 (1996) 202–208.

[14] J. Liu, Electrical coupling in populations of inhibitory interneuronal networks—a simulation design and study, B.A.Sc. Thesis, University of Toronto, 2001.
[15] M.D. Mauk, The potential effectiveness of simulations versus phenomenological models, Nature Neurosci. 3 (2000) 649–651.
[16] C.J. McBain, A. Fisahn, Interneurons unbound, Nature Rev. Neurosci. 2 (2001) 11–23.
[17] F.K. Skinner, L. Zhang, J.L. Perez Velazquez, P.L. Carlen, Bursting in inhibitory interneuronal networks: a role for gap-junctional coupling, J. Neurophysiol. 81 (1999) 1274–1283.
[18] F.K. Skinner, L. Zhang, J.L. Perez Velazquez, P.L. Carlen, Gap junctions are needed to stabilize slow bursting behaviour, Neurocomputing 26–27 (1999) 517–523.
[19] R.D. Traub, M.A. Whittington, S.B. Colling, G. Buzsáki, J.G.R. Jefferys, Analysis of gamma rhythms in the rat hippocampus in vitro and in vivo, J. Physiol. 493 (1996) 471–484.
[20] X.-J. Wang, J. Rinzel, Alternating and synchronous rhythms in reciprocally inhibitory model neurons, Neural Comput. 4 (1992) 84–97.
[21] J.A. White, C.C. Chow, J. Ritt, C. Soto-Treviño, N. Kopell, Synchronization and oscillatory dynamics in heterogeneous, mutually inhibited neurons, J. Comput. Neurosci. 5 (1998) 5–16.
[22] Y. Zhang, J.L. Perez Velazquez, G.F. Tian, C.-P. Wu, F.K. Skinner, P.L. Carlen, L. Zhang, Slow oscillations ($\leqslant$ 1 Hz) mediated by GABAergic interneuronal networks in rat hippocampus, J. Neurosci. 18 (1998) 9256–9268.

Dr. Frances Skinner is a Senior Scientist in the Toronto Western Research Institute, University Health Network, with appointments in Medicine (Neurology), Physiology and Biomedical Engineering at the University of Toronto. She graduated from the University of Waterloo (B.Math.) and Toronto (M.A.Sc., Ph.D.) and did 4 years of postdoctoral work in Boston and California. In general, she enjoys collaborative work and is interested in cellular-based mechanisms underlying neuronal network dynamics.

Jiang Brandon Liu obtained his B.A.Sc. degree in Engineering Science from the University of Toronto in 2000 and is currently working towards his M.A.Sc. degree in computer engineering at the same university. He worked as a research assistant at the Neural Computational Lab in the summers of 1999 and 2000.

Available at www.ComputerScienceWeb.com
POWERED BY SCIENCE @ DIRECT®

ELSEVIER

Neurocomputing 52–54 (2003) 389–395

NEUROCOMPUTING

www.elsevier.com/locate/neucom

Exploring the functional significance of dendritic inhibition in cortical pyramidal cells

M.W. Spratling*, M.H. Johnson

Centre for Brain and Cognitive Development, Birkbeck College, London, UK

Abstract

Inhibitory synapses contacting the soma and axon initial segment are commonly presumed to participate in shaping the response properties of cortical pyramidal cells. Such an inhibitory mechanism has been explored in numerous computational models. However, the majority of inhibitory synapses target the dendrites of pyramidal cells, and recent physiological data suggests that this dendritic inhibition affects tuning properties. We describe a model that can be used to investigate the role of dendritic inhibition in the competition between neurons. With this model we demonstrate that dendritic inhibition significantly enhances the computational and representational properties of neural networks.

Keywords: Dendrites; Inhibition; Competition; Receptive fields; Learning

1. Introduction

Lateral inhibition between cortical cells is known to play an important role in determining the receptive field properties of those cells [4,10]. Such lateral inhibition provides a mechanism through which neurons compete to respond to the current pattern of stimulation. The vast majority (93%) of inhibitory synapses targeting neocortical pyramidal cells terminate on the dendrites [3,5]. This dendritic inhibition is likely to have strong effects only on more distal inputs along the same dendritic branch [1,11,12,17,19,22], and will thus selectively inhibit specific groups of excitatory inputs, or dendritic compartments. In contrast, many neural network models assume that dendrites have passive conductance properties [16], and hence, that all inhibitory contacts are equally effective at suppressing responses to any excitatory inputs. For such

* Corresponding author.
E-mail address: m.spartling@bbk.ac.uk (M.W. Spratling).

0925-2312/03/$ - see front matter © 2003 Elsevier Science B.V. All rights reserved.
doi:10.1016/S0925-2312(02)00847-0

point-neuron models, all inhibitory contacts act as if they target the soma or axon initial segment.

While somatic inhibition non-selectively inhibits responses to all stimuli, dendritic inhibition selectively inhibits specific patterns of excitatory inputs. Recent physiological data suggests that the latter mechanism is important in determining the response properties of cortical pyramidal cells since the blockade of GABAergic synapses (in monkey area TE) results in specific dis-inhibition of responses to particular stimulus features rather than removal of non-specific inhibition [25]. We propose a simple neural network architecture which can be used to explore the role of dendritic inhibition as a mechanism of competition between pyramidal cells.

2. Method

Fig. 1(a) illustrates the architecture used by conventional models of lateral inhibition. In some implementations, competition is achieved (as shown) by using explicit lateral connections between the nodes [6,14,24], while in others, competition is implemented implicitly through a selection process which chooses the 'winning' node(s) [7,13,18]. In all of these algorithms, nodes attempt to 'block' other nodes from generating a response to the current stimulus. A node's success in this competition is dependent on the total strength of the stimulation it receives and nodes which compete unsuccessfully have their output activity suppressed.

Fig. 1(b) shows the architecture of our model of dendritic inhibition. As with conventional models, we simplify reality by assuming that the role of interneurons can be approximated by direct inhibitory connections between excitatory nodes. Furthermore, we group together all the synapses contributing to a dendritic compartment to form a single input to the node. Dendritic inhibition is then modeled as (linear) inhibition of this input. This mechanism thus enables nodes to selectively inhibit other nodes from responding to particular input features. Full details of this algorithm can be found in [21].

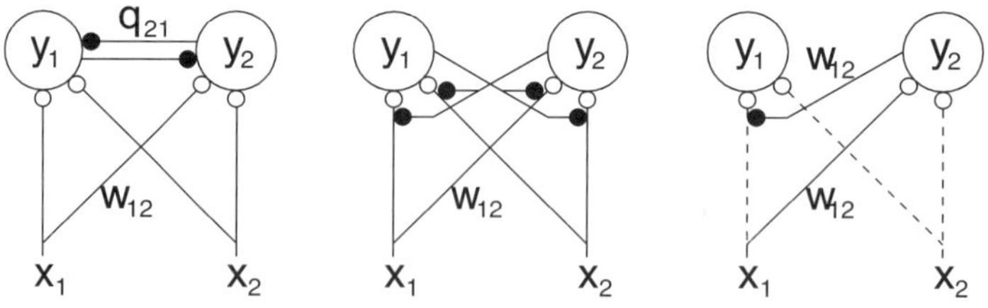

Fig. 1. Mechanisms of lateral inhibition. Simple, two-node, networks are presented to illustrate different mechanisms of lateral inhibition. Nodes are shown as large circles, excitatory synapses as small open circles, and inhibitory synapses as small filled circles. (a) The conventional model of lateral inhibition provides competition between node outputs. (b) Dendritic inhibition provides competition for inputs. (c) For the simulations presented here, each lateral weight (such as the one shown) has a strength that is equal to the corresponding afferent weights (shown as a solid line for this example).

3. Results

By making the lateral weights equal in strength to the corresponding afferent weights (as shown in Fig. 1(c)) each node can 'block' its preferred inputs from activating other nodes. With this arrangement of weights, if a node is strongly activated by the stimulus, and it has a strong synaptic weight to a certain feature of that stimulus, then it will inhibit other nodes from responding to that feature. On the other hand, if an active node receives a weak synapse from a certain input, then it will weakly inhibit other nodes from responding to that specific input. Hence, if individual nodes have learnt weights that are selective to certain stimuli then when multiple stimuli are simultaneously presented to the network each of the nodes representing one of these stimuli can be simultaneously active. A network using dendritic inhibition can thus respond appropriately to any combination of input patterns. Furthermore, such a network deals naturally with stimuli which share sub-features in common (i.e., patterns which overlap in the input space). This capacity is important if different sets of neurons are to respond to distinct sensory events which may share many features in common. Both these properties are illustrated in Fig. 2. In this example six nodes each represent one of six patterns defined across six input features (labelled 'a' to 'f'). Each node responds exclusively to its preferred input pattern despite the strong overlap between stimuli. Furthermore, multiple nodes responds when multiple stimuli are presented.

Competition not only makes responses more selective (in the short-term), but since learning is activity-dependent it also makes the receptive fields of individual nodes more distinct (in the long-term). Improved response properties result in more correct learning episodes [14], and hence, the advantageous coding properties that arise from using

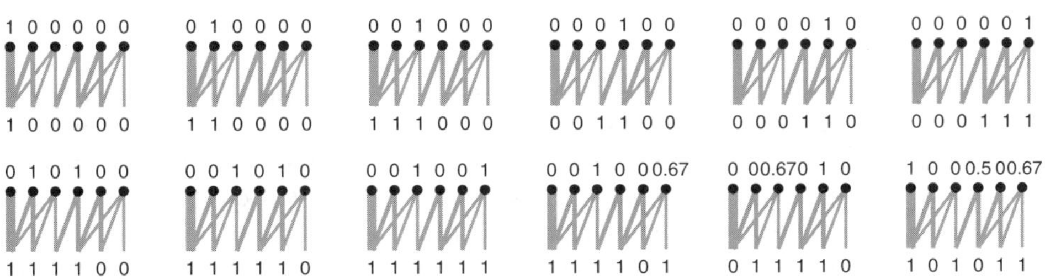

Fig. 2. Representing multiple, overlapping, input patterns. A network consisting of six nodes receiving six inputs ('a', 'b', 'c', 'd', 'e', and 'f') is wired up so that nodes are selective to the patterns 'a', 'ab', 'abc', 'cd', 'de', and 'def'. The response of the network to each of these input patterns is shown on the top row. Dendritic inhibition (lateral weights have been omitted from the figures) enables each node to respond exclusively to its preferred pattern. In addition, the response to multiple and partial patterns is shown on the bottom row. Pattern 'abcd' causes the nodes representing 'ab' and 'cd' to be active simultaneously, despite the fact that this pattern overlaps strongly with pattern 'abc'. Input 'abcde' is parsed as 'abc' together with 'de', and input 'abcdef' is parsed as 'abc' + 'def'. Input 'abcdf' is parsed as 'abc' + two-thirds of 'def', hence the addition of 'f' to the pattern 'abcd' radically changes the representation that is generated. Input 'bcde' is parsed as two-thirds of 'abc' plus pattern 'de'. Input 'acef' is parsed as 'a' + one half of 'cd' + two-thirds of pattern 'def'.

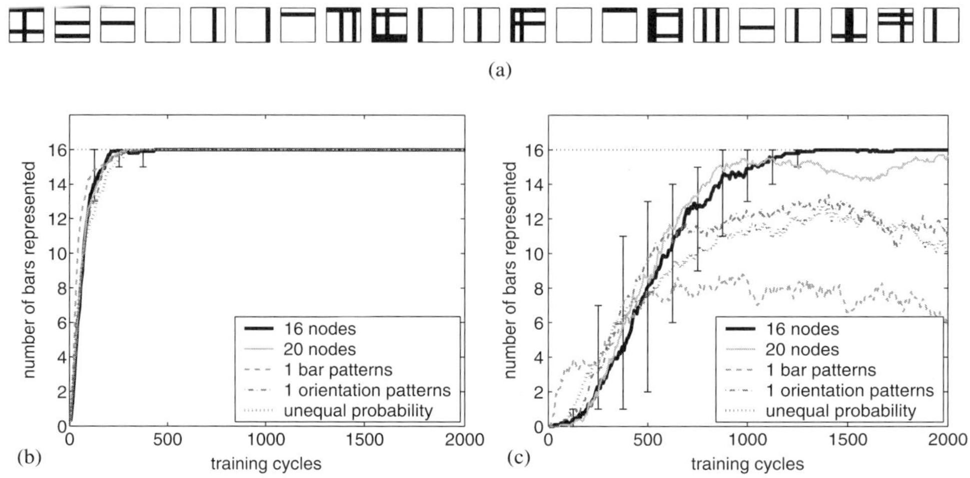

Fig. 3. The bars problem. (a) Examples of typical input patterns used in the bars problem. Bars in an 8×8 pixel image are active with probability $\frac{1}{8}$. Dark pixels indicate active inputs. The speed of learning is illustrated for (b) our algorithm, and (c) Földiák's algorithm [6]. Each plot shows the change, over the course of training, in the number of bars correctly represented by exactly one node in the neural network. Mean performance, over ten trials (using different randomly generated sequences of training patterns) is shown. Performance is plotted for networks containing 16 and 20 nodes trained with the standard bars data (like that illustrated in (a)), and for 16-node networks trained with images each containing a single bar, with images which contained only horizontal or only vertical bars, and with images in which horizontal bars occurred with half the probability of vertical bars. For each of these variations the dendritic inhibition model learns reliably and quickly, while the model using the conventional mechanism for lateral inhibition fails to learn any of the tasks except the standard bars problem for which it was specifically designed. The error bars show the best and worst performance over all trials for a 16-node network trained on the standard bars problem.

dendritic inhibition can result in efficient, unsupervised, learning [21]. For example, Fig. 3(b) shows the performance of our algorithm when applied to the bars problem [6]. Corresponding results for a conventional neural network architecture are shown in Fig. 3(c). The speed of learning when using dendritic inhibition compares favourably, not only to networks that use the conventional form of lateral inhibition, but with all neural network algorithms that have been applied to this task [21]. More significantly, a single algorithm can learn to represent the independent components of the training data (i.e., the individual bars) under a number of different conditions (such as the number of nodes in the network, the frequency with which bars appear and the number of bars making up each image). In contrast, a conventional network architecture requires a priori information about the task in order to define learning rules that will find appropriate weights for a particular task. For example, to solve the bars problem, Földiák [6] found it is necessary to use the value for the probability of a bar being active within the learning rules. Hence, this algorithm fails to learn even when the statistics of the input data are only slightly different from that expected (as illustrated in Fig. 3(c)). Furthermore, even with a priori knowledge of each task, it would not be possible to define appropriate learning rules for each case. Algorithms which use the conventional mechanism of lateral inhibition are thus severely restricted in the class of problems that they can successfully learn, and require a priori knowledge to succeed even on those tasks.

4. Conclusion

A network of neurons competing through dendritic lateral inhibition is capable of generating correct representations based on the 'knowledge' stored in the synaptic weights of the neurons. Specifically, it is capable of generating a local encoding of individual input patterns as well as responding simultaneously to multiple patterns, when they are present, in order to generate a distributed encoding. It can produce an appropriate representation even when patterns overlap and respond to partial patterns such that the response is proportional to how well that input matches the stored pattern. Not only can dendritic inhibition provide appropriate coding properties, it can also be used to efficiently learn such representations. Hence, competition via dendritic inhibition significantly enhances the computational and representational capacities of networks of neurons and may be exploited by cortical circuits. Furthermore, such a neural network has significant practical advantages over algorithms that employ the conventional mechanism of competition.

Our architecture assumes that all afferent synapses, carrying information about a feature of the input space, can be modelled as a single input to a node. This can be justified since it is presumed that related synapses cluster together within dendritic trees so that local operations are performed by multiple, functionally distinct, dendritic subunits before integration at the soma [8,9,12,15,16,19,20]. Dendritic inhibition could thus act to 'block' the output from individual functional compartments. It has long been recognized that a dendrite composed of multiple subunits would provide a significant enhancement to the computational powers of an individual neuron [15,16] and that dendritic inhibition could contribute to this enhancement [11,12,20]. However, the role of dendritic inhibition in competition between cells and its subsequent effect on neural coding and receptive field properties has not previously been investigated.

The idea embodied in our model is that pyramidal cells inhibit the activity of dendritic compartments in other pyramidal cells within the same cortical area. It has been shown that cortical pyramidal cells innervate inhibitory cell types which in turn form synapses on the dendrites of pyramidal cells [2,23]. Our model predicts that it should be possible to find pairs of pyramidal cells for which action potentials generated by one cell induce inhibitory post-synaptic potentials within the dendrites of the other. More complex models, which include a separate inhibitory cell population, and which use multi-compartmental models of dendritic processes could relate our proposal more directly with physiology. We hope that our demonstration of the computational and representational advantages that could arise from dendritic inhibition will serve to stimulate more detailed studies of this type.

Acknowledgements

This work was funded by MRC Research Fellowship number G81/512.

References

[1] L.T. Borg-Graham, C. Monier, Y. Fregnac, Visual input evokes transient and strong shunting inhibition in visual cortical neurons, Nature 393 (6683) (1998) 369–373.
[2] E.H. Buhl, G. Tamas, T. Szilagyi, C. Stricker, O. Paulsen, P. Somogyi, Effect, number and location of synapses made by single pyramidal cells onto aspiny interneurones of cat visual cortex, J. Physiol. 500 (3) (1997) 689–713.
[3] J. DeFelipe, I. Fariñas, The pyramidal neuron of the cerebral cortex: morphological and chemical characteristics of the synaptic inputs, Prog. Neurobiol. 39 (6) (1992) 563–607.
[4] U.T. Eysel, I.A. Shevelev, N.A. Lazareva, G.A. Sharev, Orientation tuning and receptive field structure in cat striate neurons during local blockade of intrcortical inhibition, Neuroscience 84 (1) (1998) 25–36.
[5] J. Fiala, K. Harris, Dendritic structure and spines, in: G. Stuart, N. Spruston, M. Häusser (Eds.), Dendrites, Oxford University Press, Oxford, UK, 1999, pp. 1–34 (Chapter 1).
[6] P. Földiák, Forming sparse representations by local anti-Hebbian learning, Bio. Cybernetics 64 (1990) 165–170.
[7] S. Grossberg, Competitive learning: from interactive activation to adaptive resonance, Cognitive Sci. 11 (1987) 23–63.
[8] M. Häusser, Synaptic function: dendritic democracy, Curr. Biol. 11 (1) (2001) R10–2.
[9] M. Häusser, N. Spruston, G.J. Stuart, Diversity and dynamics of dendritic signalling, Science 290 (5492) (2000) 739–744.
[10] B. Jagadeesh, Inhibition in inferotemporal cortex: generating selectivity for object features, Nat. Neurosci. 3 (8) (2000) 749–750.
[11] C. Koch, T. Poggio, V. Torre, Nonlinear interactions in a dendritic tree: localization, timing, and role in information processing, Proc. Nat. Acad. Sci. USA 80 (9) (1983) 2799–2802.
[12] C. Koch, I. Segev, The role of single neurons in information processing, Natl. Neurosci. 3 (Suppl.) (2000) 1171–1177.
[13] T. Kohonen, Self-Organizing Maps, Springer, Berlin, 1997.
[14] J.A. Marshall, Adaptive perceptual pattern recognition by self-organizing neural networks: context, uncertainty, multiplicity, and scale, Neural Networks 8 (3) (1995) 335–362.
[15] B.W. Mel, Information processing in dendritic trees, Neural Comput. 6 (1994) 1031–1085.
[16] B.W. Mel, Why have dendrites? A computational perspective, in: G. Stuart, N. Spruston, M. Häusser (Eds.), Dendrites, Oxford University Press, Oxford, UK, 1999, pp. 271–289 (Chapter 11).
[17] W. Rall, Theoretical significance of dendritic trees for neuronal input-output relations, in: R.F. Reiss (Ed.), Neural Theory and Modeling, Stanford University Press, Stanford, CA, 1964, pp. 73–97.
[18] D.E. Rumelhart, D. Zipser, Feature discovery by competitive learning, Cognitive Sci. 9 (1985) 75–112.
[19] I. Segev, Dendritic processing, in: M.A. Arbib (Ed.), The Handbook of Brain Theory and Neural Networks, MIT Press, Cambridge, MA, 1995, pp. 282–289.
[20] I. Segev, W. Rall, Excitable dendrites and spines: earlier theoretical insights elucidate recent direct observations, Trends Neurosci. 21 (11) (1998) 453–460.
[21] M.W. Spratling, M.H. Johnson, Pre-integration lateral inhibition enhances unsupervised learning, Neural Comput. 14 (9) (2002) 2157–2179.
[22] N. Spruston, G. Stuart, M. Häusser, Dendritic integration, in: G. Stuart, N. Spruston, M. Häusser (Eds.), Dendrites, Oxford University Press, Oxford, UK, 1999, pp. 231–271 (Chapter 10).
[23] G. Tamas, E.H. Buhl, P. Somogyi, Fast IPSPs elicited via multiple synaptic release sites by different types of GABAergic neurone in the cat visual cortex, J. Physiol. 500 (3) (1997) 715–738.
[24] C. von der Malsburg, Self-organisation of orientation sensitive cells in the striate cortex, Kybernetik 14 (1973) 85–100.
[25] Y. Wang, I. Fujita, Y. Murayama, Neuronal mechanisms of selectivity for object features revealed by blocking inhibition in inferotemporal cortex, Nat. Neurosci. 3 (8) (2000) 807–813.

Michael Spratling received a B.Eng. degree in Engineering Science from Loughborough University, and M.Sc. and Ph.D. degrees in Artificial Intelligence from the University of Edinburgh. He is currently a postdoctoral research fellow in the Center for Brain and Cognitive Development, Birkbeck College, University of London. His primary research interests are: neural information processing and coding (in the visual system), unsupervised learning, cortical feedback mechanisms, and attention.

Mark H. Johnson is Professor of Psychology and Director of the Centre for Brain and Cognitive Development at Birkbeck College, University of London, and a member of the UK Medical Research Council's External Scientific Staff. He received his B.Sc. in Biological Sciences at Edinburgh and his Ph.D. in Neurobiology from Cambridge. He was Research Scientist (1985–1990), and then Senior Research Scientist (1993–1998) at the MRC Cognitive Development Unit, London. From 1989 to 1990 he was a Research Fellow at the University of Oregon, Eugene, and from 1991 to 1995 Associate Professor of Psychology at Carnegie Mellon University, Pittsburgh. He has published over 100 scholarly articles and four books on brain and cognitive development in human infants and other species, including Developmental Cognitive Neuroscience: An Introduction (Blackwell, 1997). He also edits the international journal Developmental Science.

Available at
www.ComputerScienceWeb.com
POWERED BY SCIENCE @DIRECT®

ELSEVIER

Neurocomputing 52–54 (2003) 397–401

NEUROCOMPUTING

www.elsevier.com/locate/neucom

Quantal synaptic failures improve performance in a sequence learning model of hippocampal CA3

D.W. Sullivan[a], W. B Levy[a,b,*]

[a]Neurosurgery Department, University of Virginia Health System, P.O. Box 800420, Charlottesville, VA 22908-0420, USA

[b]Psychology Department, University of Virginia Health System, P.O. Box 800420, Charlottesville, VA 22908-0420, USA

Abstract

Quantal synaptic failures are random, independent events in which the arrival of an action potential fails to release transmitter. Although quantal failures destroy information, there exist conditions where such failures enhance learning by a neural network model of the hippocampus. In particular, we show how the appropriate failure rate can allow the model to learn the hippocampally dependent task of transverse patterning. Usefully, since lowered activity levels produce higher memory capacity, synaptic failures lead to robust performance at lowered activity levels. Thus, the synaptic failure mechanism is another example of a random fluctuation that improves neural network computations.

Keywords: Neural network; Fluctuations; Noise; Configural learning; Transverse patterning

1. Introduction

Examined in isolation, a noisy communications channel is less efficient than noise-free communications channels, as it requires more energy to produce equivalent capacity. Although noisy communication is bad for the users of a single communication channel, it can have a positive effect on a postsynaptic neuron with many inputs [2] or a network of neurons. In two studies [5,6] a hippocampal model learning the transverse patterning (TP) problem was improved by adding a random component in a recurrently connected, Hebbian learning, neural network model of hippocampal CA3.

* Corresponding author.
E-mail address: wbl@virginia.edu (W. B Levy).

0925-2312/03/$ - see front matter
doi:10.1016/S0925-2312(02)00848-2

This paper investigates the effects of introducing a biologically accurate, multiplicative form of noise, quantal synaptic failures, to the recurrent excitatory synapses of a hippocampal CA3 model. By adding the synaptic failure process, learning by the model is more robust, and TP can be learned at lower activity levels when a suitable synaptic failure rate is used.

2. Methods

Our model of hippocampal CA3 consists of a sparsely connected (10%) network of McCulloch-Pitts neurons. External inputs represent projections from the dentate gyrus and entorhinal cortex to CA3.

Previously a neuron's excitation was defined as the sum of the weights of its recurrent inputs that were active on time step $(t-1)$. However, with the addition of synaptic failures, a neuron's excitation is the sum of the weights of its active inputs that did not undergo synaptic failure. We model synaptic transmission at each synapse as independent random Bernoulli events. For $(y_j(t))$, the internal excitation of neuron j on time step t,

$$y_j(t) = \sum_{i=1}^{N} C_{ij} W_{ij}(t-1) Z_i(t-1) \Phi_{ij}(t),$$

where N is the total number of neurons. A synaptic connection from neuron i to neuron j is represented by C_{ij} (1 if a connection is present, 0 if a connection is not present). The weight of the synapse from neuron i to neuron j on time step t is represented by $W_{ij}(t)$. The variable $Z_i(t-1)$ represents the binary, $[Z_i(t-1) \in \{0,1\}]$, firing of neuron i on time step $(t-1)$. The binary variable $\Phi_{ij}(t)$ implements the random failure process at the synapse from neuron i to neuron j. All $\Phi_{ij}(t)$ are randomly and independently determined on each trial such that $\Phi_{ij}(t) = 1$ with probability $(1-f)$, where f is the probability of synaptic failure.

Once internal excitation has been calculated, a deterministic binary (0/1) output firing decision occurs, representing the absence or presence of an action potential respectively. Here this decision is determined according to a k-winners-take-all rule.

This model learns via locally controlled synaptic modification, using a minimally time-spanning associative synaptic modification rule [3,4]. On each time step of each training trial, the weight of every synapse in the network is updated according to the following equation:

$$W_{ij}(t) = W_{ij}(t-1) + \mu Z_j(t)(Z_i(t-1)\Phi_{ij}(t-1) - W_{ij}(t-1)),$$

where the synaptic modification rate, μ, is equal to 0.05. Synaptic modification occurs only if postsynaptic neuron j fires. At the beginning of a simulation, all weights have a value of 0.4.

Transverse patterning is a hippocampally dependent cognitive task that requires the subject to assign meaning to stimulus pairs that cannot be derived from the individual stimuli [1]. For a detailed description of how we simulate transverse patterning, see [5] or [7].

3. Results

Fig. 1 demonstrates the model's sensitivity, at a particular set of parameters, to the rate of synaptic failure. As rate of synaptic failure is increased from 0% to 10%, TP learning increases from 0% to 80% of the simulations. For a wider range of failure rates, 30–70%, five out of five simulations successfully learn TP. However, when synaptic failure rates exceed 70%, performance drops abruptly. Thus, very small differences in synaptic failure rate can lead to very large differences in TP performance rate.

Synaptic failure rate interacts with the other simulation parameters, such as number of neurons, activity level, learning rate, initial weight, and size of the input code. The range of synaptic failure rates that are beneficial changes according to the other parameters. The number of neurons in the network is especially important. Although synaptic failures do not benefit networks smaller than 2000 neurons, the effects of synaptic failures become more robust as network size increases. This effect is likely due to an insufficient number of active connections per neuron in the smaller networks.

The interaction of synaptic failure rate with lowered activity levels is, perhaps surprisingly, beneficial. Although one might expect that lowered activity together with synaptic failures would lead to too few active inputs per neuron (and this can happen when failure rate is too high and activity level is too low), it is not always so. Fig. 2 shows TP performance rates for simulations run at four activity levels and two synaptic failure rates. Without synaptic failures, simulations require activity levels of 10% in order to achieve good (> 80%) TP performance. But when synaptic failures are present (here, set to 70%), simulations can learn TP with 100% success at each of the four activity levels plotted.

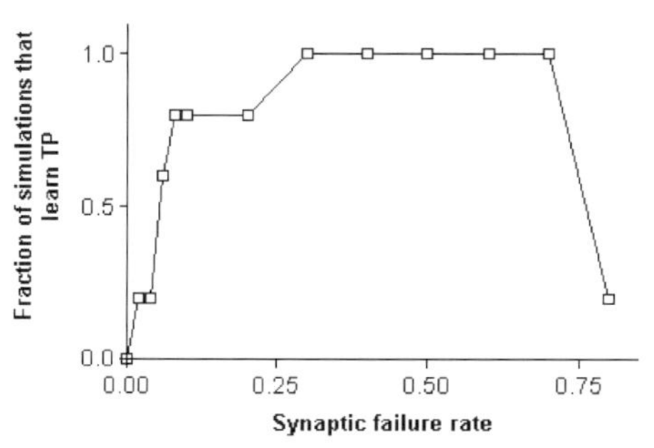

Fig. 1. Synaptic failure improves TP performance. Running at a relatively low activity level (7%), simulations are incapable of successfully completing the TP problem without synaptic failures. However, as the probability of synaptic failures is increased, TP performance rate rises quickly. While simulations display robust performance when synapses fail between 30% and 70% of the time, a synaptic failure rate that is too high (>80%) destroys performance. Simulation parameters: $N = 8192$, Activity $= 0.07$, number of external inputs $= 172$, and $\mu = 0.05$.

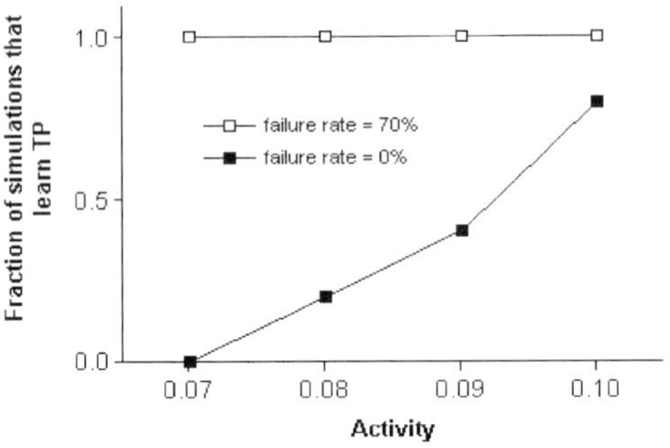

Fig. 2. Synaptic failures allow better TP performance, with the greatest improvement found at lower activity levels. When synaptic failure rate is set to zero, performance at 7%, 8%, and 9% activity is poor, with less than three out of five simulations successfully completing the TP task. However, when the probability of synaptic failure is set to 70%, five out of five simulations at each of the activity levels plotted learned TP. Each point is the average success rate of five simulations, each simulation using a different random connectivity matrix. For each simulated activity level, the external input comprises 30% of the active neurons. Simulation parameters: $N = 8192$, $\mu = 0.05$.

4. Discussion

Synaptic failures can benefit computation via greater memory capacity and more robust performance. Apparently, the rate of synaptic failure is a powerful parameter that can have profound effects on the model's ability to learn transverse patterning. Specifically, the simulations here, with low activity levels, can learn TP only if synaptic failures are present in the model. While information is lost—individually—at each failing synapse, the overall computation is more robust. This is in line with the conclusion of [2] that there is more than enough information going into a neuron even when connections are relatively few (low hundreds). Future studies will investigate the basis of this effect, but the idea developed and quantified in [5] concerning fluctuation driven code search during training seems to be a good candidate.

Acknowledgements

This work was supported by NIH MH48161 and MH63855 to WBL, and MH57358 to N. Goddard (with subcontract # 163194-54321 to WBL).

References

[1] M.C. Alvarado, J.W. Rudy, Rats with damage to the hippocampal-formation are impaired on the transverse patterning problem but not on elemental discriminations, Behav. Neurosci. 109 (1995) 204–211.

[2] W. B Levy, R.A. Baxter, Energy-efficient neuronal computation via quantal synaptic failures, J. Neurosci. 22 (2002) 4746–4755.

[3] W. B Levy, O. Steward, Synapses as associative memory elements in the hippocampal formation, Brain Res. 175 (1979) 233–245.
[4] W. B Levy, O. Steward, Temporal contiguity requirements for long-term associative potentiation/depression in the hippocampus, Neuroscience 8 (1983) 791–797.
[5] A.P. Shon, X.B. Wu, D.W. Sullivan, W. B Levy, Initial state randomness improves sequence learning in a model hippocampal network, Phys. Rev. E 65 (2002) 1–15.
[6] X.B. Wu, W. B Levy, Enhancing the performance of a hippocampal model by increasing variability early in learning, Neurocomputing 26–27 (1999) 601–607.
[7] X.B. Wu, J. Tyrcha, W. B Levy, A neural network solution to the transverse patterning problem depends on repetition of the input code, Biol. Cybern. 79 (1998) 203–213.

Available at
www.ComputerScienceWeb.com
POWERED BY SCIENCE @ DIRECT®

ELSEVIER

Neurocomputing 52–54 (2003) 403 – 409

NEUROCOMPUTING

www.elsevier.com/locate/neucom

Intracortical vs. thalamocortical processing of spatial working memory

Yoshinori Tabuchi, Shoji Tanaka*

Department of Electrical and Electronics Engineering, High-Tech Research Centre, Sophia University, 7-1 Kioicho, Chiyoda-ku, Tokyo 102-8554, Japan

Abstract

Working memory processes contain not only maintenance but manipulation of the working memory contents. The system of the mediodorsal (MD) nucleus and the prefrontal cortex is considered to contribute to some aspects of the working memory processes. This article studies computationally the characteristics of the integrated cortical and thalamocortical system. The model circuit is operated in the "thalamocortical mode", in which the thalamocortical inputs can control the cortical working memory representation. Our simulation shows that transient excitatory inputs to the MD relay cells can erase the working memory representation through the intracortical processes, especially the cross-directional inhibition.

Keywords: Corticothalamic; Delay-period activity; Intracortical; Spatial working memory; Thalamocortical

1. Introduction

The characteristics of the prefrontal cortical (PFC) circuits for working memory have been investigated intensively [1–3,8–12,14]. However, working memory would not be processed only in the PFC circuits; other cortical areas and subcortical structures are also considered to be involved in working memory processing. Among these, the cortico-thalamocortical system of the PFC and the mediodorsal nucleus (MD) would be important because these areas are reciprocally connected [4]. It is, therefore, interesting to investigate how this system contributes to working memory. To do so, we have constructed a model cortico-thalamocortical system. This model contains the

* Corresponding author. Tel.: +81-3-3238-3331; fax: +81-3-3238-3321.
E-mail address: tanaka-s@sophia.ac.jp (S. Tanaka).

0925-2312/03/$ - see front matter
doi:10.1016/S0925-2312(02)00849-4

intracortical circuits, the corticothalamic feedback connections, the intrathalamic local circuits, and the thalamocortical feedforward connections. In this article, the characteristics of these circuits are investigated by computer simulation of the model circuit. This article focuses on spatial working memory, so that neuronal activities in the model have selectivity in the directional space.

2. Model

The cortical model has three layers (the superficial, intermediate, and deep layers) and contains 1080 pyramidal cells (360 in each layer) and 240 interneurons [5,7–12,15]. The MD model contains 36 relay cells and 36 interneurons. The architecture of the model is shown in Fig. 1. The pyramidal cells in the deep layer send feedback projections to the MD relay cells and interneurons. The interneurons have local inhibitory connections with the relay cells. The relay cells, in turn, send the thalamocortical projections to the pyramidal cells in the intermediate layer of the PFC. The neurons are described by a leaky integrate-and-fire neuron model. The ion channels include: AMPA, NMDA, Nap, $GABA_A$, K(Ca), and leak. The ratio of the NMDA conductance to the AMPA conductance is $g_{NMDA}/g_{AMPA}=0.05$ and the ratio of the cross-directional inhibition to the isodirectional inhibition [10–12] is $g_{GABAA,cross}/g_{GABAA,iso}=0.36$.

3. Results

3.1. Omni-directional inputs to MD cells

We tested four types of transient, omni-directional inputs to the MD cells: (1) an excitatory input to the MD relay cells, (2) an inhibitory input to the MD relay cells, (3) an excitatory input to the MD interneurons, and (4) an inhibitory input to the MD interneurons. Because of the local connections in the MD, inputs to the MD interneurons influence the activity of the MD relay cells. For example, the inhibitory input to the MD interneurons disinhibits the MD relay cells. In general, both the excitatory and inhibitory inputs tend to terminate the sustained activity in the PFC. But the dynamics responding to the thalamocortical inputs are significantly different. We report, in the following, the case of (1) because it contains interesting but somewhat paradoxical results.

3.2. Cortical and thalamic activity

The MD relay cells respond to the transient input (Fig. 2). The activity of the MD relay cells is transmitted to the cortical pyramidal cells in the superficial layer. As a result, the working memory representation changes significantly by receiving the input. The excitatory input to the MD relay cells increases briefly the activity of all the cortical neurons that receive the inputs. That is, the intermediate layer exhibits omnidirectional

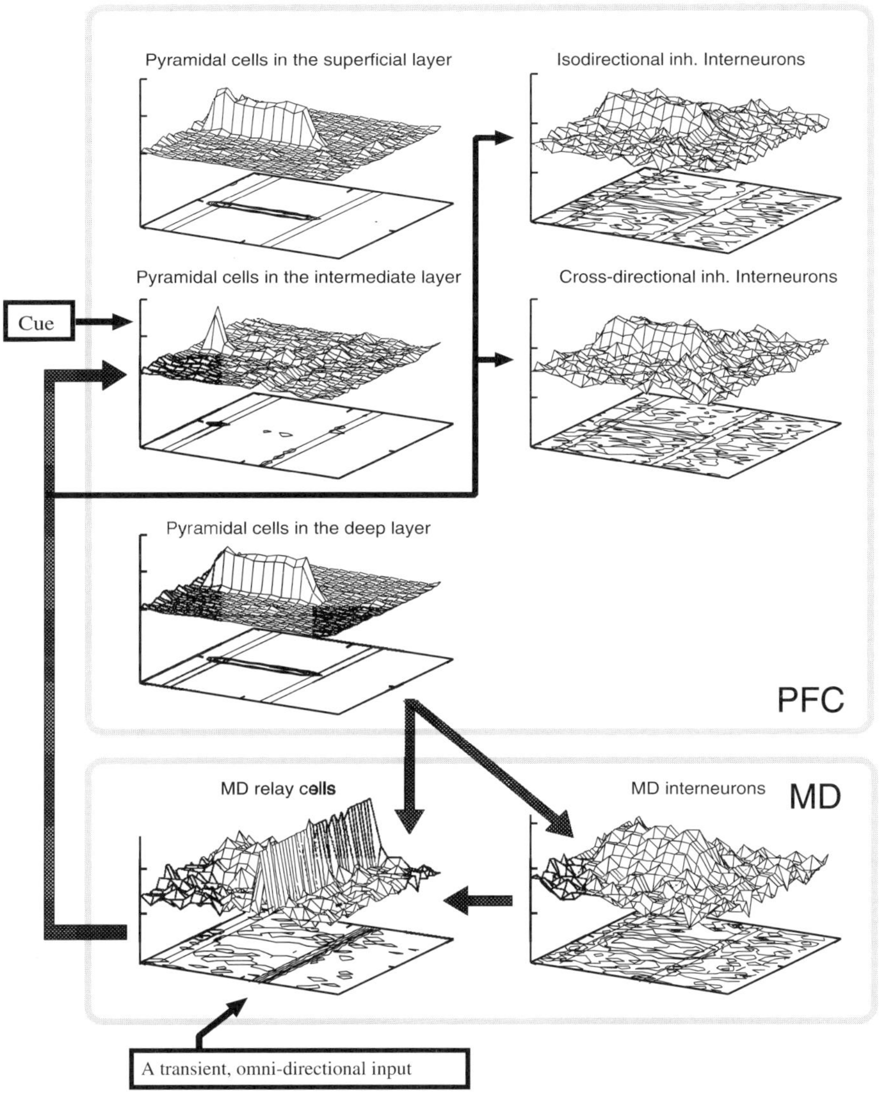

Fig. 1. Model architecture with the spatio-temporal profiles of the neuronal activities. The pyramidal cells in the deep layer of the PFC send feedback projections to the MD cells (both the relay cells and the interneurons). The MD relay cells send feedforward projections to the pyramidal cells in the intermediate layer and the inhibitory interneurons of the PFC. The thalamocortical projections to the interneurons is weaker than those to the pyramidal cells (the ratio is 0.21). The input, cueing the target location, is given to the pyramidal cells in the intermediate layer of the PFC. A transient, omnidirectional input is given to the MD relay cells.

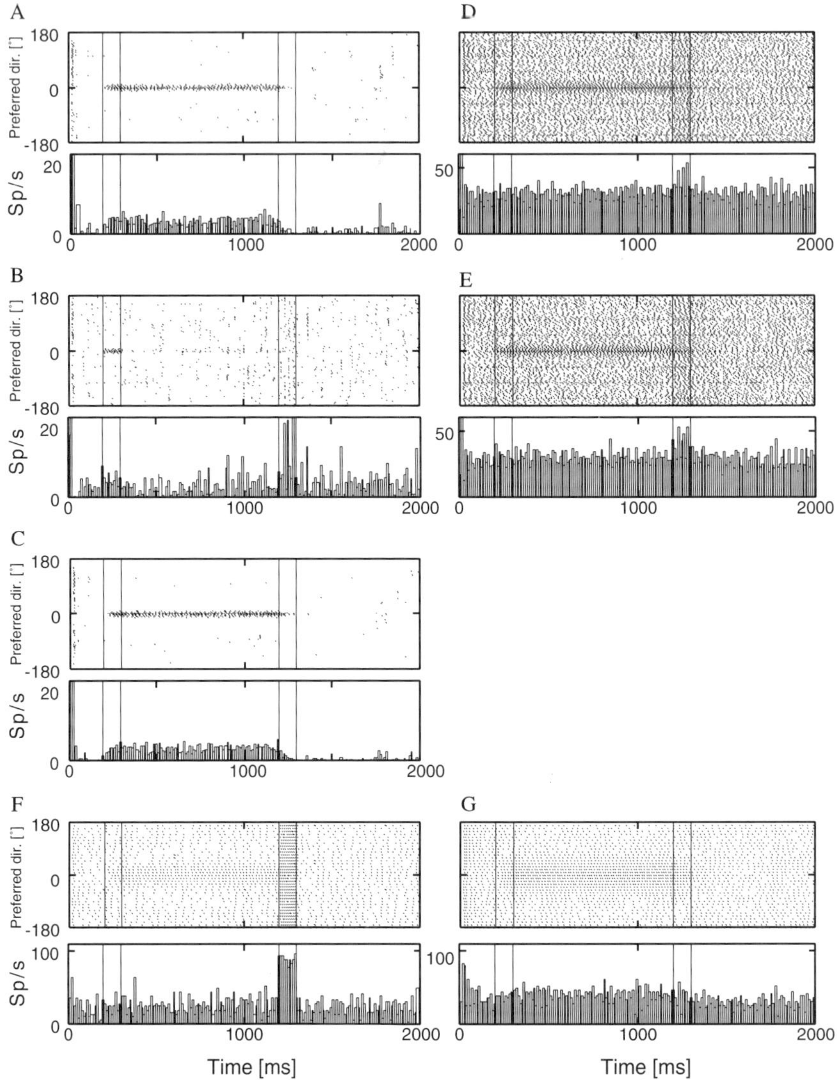

Fig. 2. Raster plots and the population activities of the neurons in the PFC and the MD: (A) the pyramidal cells in the superficial layer of the PFC, (B) the pyramidal cells in the intermediate layer of the PFC, (C) the pyramidal cells in the deep layer of the PFC, (D) the isodirectional inhibitory interneurons, (E) the cross-directional inhibitory interneurons, (F) the MD relay cells, (G) the MD inhibitory interneurons. The input, cueing the target location, is given to the pyramidal cells in the intermediate layer of the PFC during 200–300 ms. The MD relay cells receive a transient omnidirectional during 1200–1300 ms. The time bin for the population activity is 10 ms. The figure shows transient activity at the beginning of the simulation ($0 < t < 30$ ms), which is irrelevant to the processes studied here.

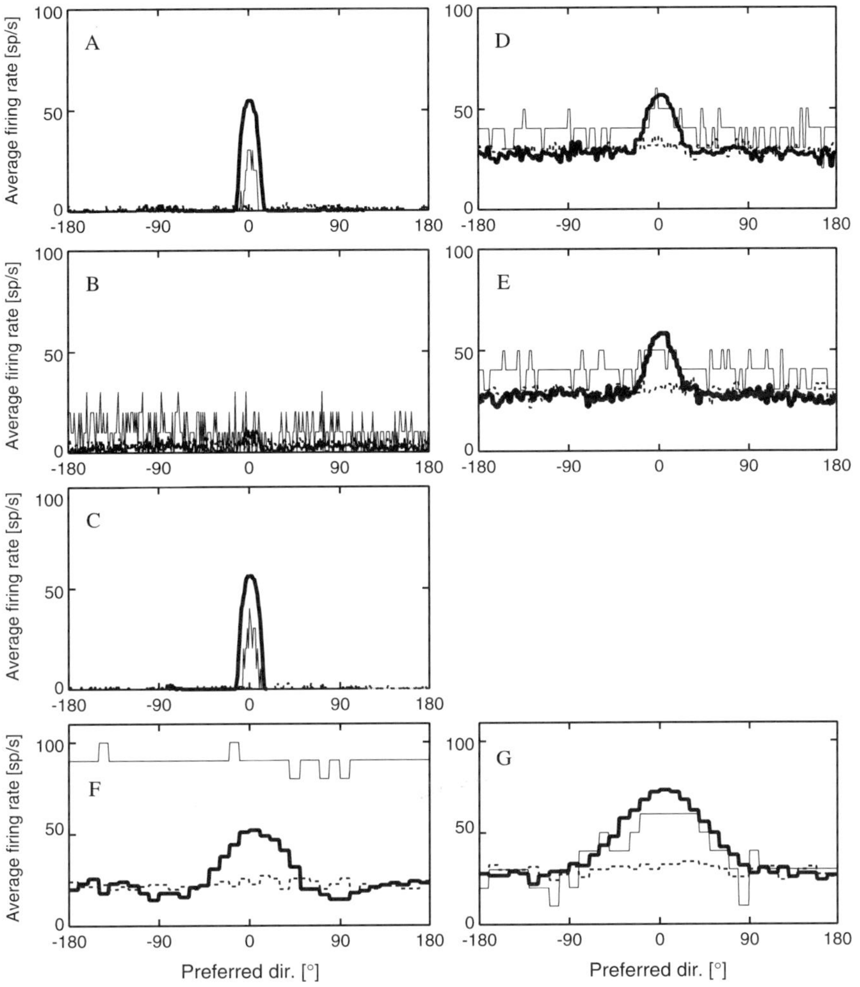

Fig. 3. Activity profiles of the neurons in the PFC and the MD: (A) the pyramidal cells in the superficial layer of the PFC, (B) the pyramidal cells in the intermediate layer of the PFC, (C) the pyramidal cells in the deep layer of the PFC, (D) the isodirectional inhibitory interneurons, (E) the cross-directional inhibitory interneurons, (F) the MD relay cells, (G) the MD inhibitory interneurons. Thick solid lines: the activity profiles averaged over the first delay period (300–1200 ms), thin solid lines: the activity profiles averaged over the MD-input period (1200–1300 ms), broken lines: the activity profiles averaged over the second delay period (1300–2000 ms).

activity by reflecting the thalamocortical input. This activity turns off the sustained activities in the superficial and the deep layers. This is due to the intracortical inhibition, because the cortical interneurons show transient increase in the firing rates during the thalamocortical-input period (1200–1300 ms). The activity profiles of the model neurons are shown in Fig. 3.

4. Discussion

4.1. Thalamocortical mode

This cortical model has strong excitatory projections from the intermediate layer to the superficial layer [8–13]. These projections amplify the activity, by which the transient activity in the intermediate layer is converted to the sustained activity in the superficial layer. This study suggests two interesting features in the processing of the cortical neuronal activity. First, the thalamocortical inputs assist the sustainment of the delay-period activity. We call this mode of operation the "thalamocortical mode". In this mode, the thalamic neurons, receiving sustained inputs from the cortical neurons in the deep layer, are activated tonically and forward the tonic inputs to the intermediate layer of the cortex. Although the firing rates of the neurons in the intermediate layer are fairly low during the first delay period, there is transmission of signals from the intermediate to the superficial layer. When blocked the thalamocortical inputs, the activity of the cortical neurons does not sustain any longer.

4.2. Intracortical inhibition

The second interesting feature is that the intracortical inhibition works effectively when the cortex receives omni-directional excitatory inputs. In this model, the thalamocortical projections terminate on the pyramidal cells in the intermediate layer and the interneurons. Then, the thalamocortical inputs activate both the excitatory and inhibitory circuits. Note, however, that the strength of the thalamocortical projections to the interneurons is 0.21 relative to that to the pyramidal cell. Nevertheless, the intracortical inhibition terminates the sustained activity. Slight change in the strength of the intracortical inhibition alters the cortical dynamics significantly (not shown). This suggests that the regulation of the intracortical inhibition is critical to the control of the cortical dynamics [6,10–13]. In the case presented here, the cross-directional inhibition plays more important roles in the termination of the sustained activity because many neurons activated during the MD-input period exert inhibitory influences through the cross-directional inhibitory circuits. The intracortical inhibition and the intrathalamic inhibition contributing to multi-target spatial working memory processes are argued by Miyashita et al. [6].

Acknowledgements

This work was supported by the Grants-in-Aid for Scientific Research on Priority Areas (#13210123 and #14017083) from the Japanese Ministry of Education, Science, and Technology. The authors acknowledge valuable discussions with S. Funahashi at Kyoto University and H. Barbas at Boston University.

References

[1] N. Brunel, X.-J. Wang, Effects of neuromodulation in a cortical network model of object working memory dominated by recurrent inhibition, J. Comput. Neurosci. 11 (2001) 63–85.

[2] A. Compte, N. Brunel, P.S. Goldman-Rakic, X.-J. Wang, Synaptic mechanisms and network dynamics underlying spatial working memory in a cortical network model, Cereb. Cortex 10 (2000) 910–923.
[3] D. Durstewitz, J.K. Seamans, T.J. Sejnowski, Neurocomputational models of working memory, Nature Neurosci. 3 (2000) 1184–1191.
[4] M. Giguere, P.S. Goldman-Rakic, Mediodorsal nucleus: areal, laminar, and tangential distribution of afferents and efferents in the frontal lobe of rhesus monkeys, J. Comp. Neurol. 277 (1988) 195–213.
[5] M. Iida, S. Tanaka, Postsynaptic current analysis of a model prefrontal cortical circuit for multi-target spatial working memory, Neurocomputing 44–46 (2002) 855–861.
[6] S. Miyashita, Y. Tabuchi, S. Tanaka, Cortico-thalamocortical operations of multi-target spatial working memory, Neurocomputing, this issue.
[7] K. Morooka, S. Tanaka, Correlation analysis of signal flow in a model prefrontal cortical circuit representing multiple target locations, Neurocomputing 44–46 (2002) 541–548.
[8] S. Tanaka, Architecture and dynamics of the primate prefrontal cortical circuit for spatial working memory, Neural Networks 12 (1999) 1007–1020.
[9] S. Tanaka, Post-cue activity of prefrontal cortical neurons controlled by local inhibition, Neurocomputing 32–33 (2000) 563–572.
[10] S. Tanaka, Computational approaches to the architecture and operations of the prefrontal cortical circuit for working memory, [review], Prog. Neuro-Psychopharm. Biol. Psychiat. 25 (2001) 259–281.
[11] S. Tanaka, Multi-directional representation of spatial working memory in a model prefrontal cortical circuit, Neurocomputing 44–46 (2002) 1001–1008.
[12] S. Tanaka, Dopamine controls fundamental cognitive operations of multi-target spatial working memory, Neural Networks 15 (2002) 573–582.
[13] S. Tanaka, A. Yoshida, Signal flow in a prefrontal cortical circuit model for working memory loading, Neurocomputing 38–40 (2001) 957–964.
[14] X.-J. Wang, Synaptic reverberation underlying mnemonic persistent activity, Trends Neurosci. 24 (2001) 455–463.
[15] K. Yamashita, S. Tanaka, Circuit simulation of memory field modulation by dopamine D1 receptor activation, Neurocomputing 44–46 (2002) 1035–1042.

Yoshinori Tabuchi received B.E. from Sophia University, Tokyo, in 2002. He is a graduate student at Sophia University. He is currently studying neuroscience and electrical engineering.

Shoji Tanaka received B.E., M.E., and Ph.D. degrees from Nagoya University, Japan. He is Professor at Department of Electrical and Electronics Engineering, Sophia University. During 1998–1999, he was a Visiting Science at the Section of Neurobiology, Yale University School of Medicine, USA.

Available at
www.ComputerScienceWeb.com
POWERED BY SCIENCE @ DIRECT®

ELSEVIER

Neurocomputing 52–54 (2003) 411 – 417

NEUROCOMPUTING

www.elsevier.com/locate/neucom

Novelty detection in a Kohonen-like network with a long-term depression learning rule

Dionyssios Theofilou*, Volker Steuber, Erik De Schutter

Laboratory of Theoretical Neurobiology, Born-Bunge Foundation, University of Antwerp-UIA, Universiteitsplein 1, Antwerp 2610, Belgium

Abstract

In the cerebellar cortex, long-term depression (LTD) of synapses between parallel fibers (PF) and Purkinje neurons can spread to neighboring ones, independently of their activation by PF input. This spread of non-specific LTD around the activated synapses resembles how units are affected in the neighborhood of the winner in a Kohonen Network (KN). However in a classic KN the weight vectors become more similar to the input vector with learning, while in the LTD case they should become more dissimilar. We devised a new LTD-KN where units, opposite to the classic KN, decrease their response (LTD-like) when a pattern is learned and we show that this LTD-KN functions as a novelty detector.

Keywords: Cerebellum; Long-term depression; Self-organizing map; Kohonen network; Novelty detector

1. Introduction

Cerebellar Purkinje cells receive input from 150 000 parallel fibers, the axons of granule cells, and a single climbing fiber, the axon of an inferior olivary neuron. When a climbing fiber signal reaches the Purkinje cell repeatedly in conjunction with a signal from a parallel fiber, the synapse that received parallel fiber input when the climbing fiber signal arrived becomes depressed. According to the Marr–Albus–Ito theory [1,6,9] this mechanism may support motor learning, with the climbing fiber signal representing an error signal, and LTD leading to the depression of synapses responsible for the inaccuracy or miscalculation of the motor command.

* Corresponding author.
E-mail address: dion@bbf.uia.ac.be (D. Theofilou).

0925-2312/03/$ - see front matter © 2002 Elsevier Science B.V. All rights reserved.
doi:10.1016/S0925-2312(02)00855-X

However this hypothesis has been challenged in the past [2] and seems to be in conflict with recent experimental results: First, parallel fiber stimulation alone can cause significant calcium influx and lead to LTD, without a climbing fiber signal present [4,3]. Second, the depression caused by the parallel fiber activity may not be specific to the synapses that receive the stimulus, but it may spread further to adjacent synapses that were not activated by the PF input [4,7,8].

According to the experimental findings, this non-specific LTD can be described by a Gaussian function with a half-width of about 50 μm around the activated synapses. It has been suggested that LTD may modify 600 times more inactive neighboring synapses than active synapses on a single Purkinje cell [8].

The way how LTD spreads to the neighboring synapses resembles the way neighboring units are influenced around the winner node in a Kohonen neural network [5]. During learning, the node with the strongest response when a pattern is presented to the Kohonen network is selected as the central winning output node. This node is altered as to maximize future response and increase the probability that it will be selected again when a similar input is presented. Moreover, it is not only the winner node which is affected during learning, but also all surrounding nodes, selected by a neighborhood function, usually a Gaussian. In contrast to the Kohonen network where the weight vectors of the output units become more similar to the input vector, cerebellar LTD should lead to a movement of weight vectors away from the input. In the following we will describe a modified version of a Kohonen network with an LTD learning rule.

2. The LTD-like learning rule

First, we normalize all input vectors before applying them to the network and the weight vectors at the initial phase and after every presentation of a pattern:

$$\sum_i (x_i)^2 = 1, \quad \sum_j (w'_{r'j})^2 = 1.$$

When a pattern is presented, the response of every node is given by the dot product of input vector and weight vector. The dot product is the measure of similarity between weight and input.

$$y_{r'} = \vec{x}\vec{w}_{r'}.$$

The node with the maximum response and therefore the smallest distance to the current input vector is the node r:

$$y_r = \max_{r'}(y_{r'}).$$

This node is chosen as the winner of the competition in the LTD-like network. Then the change of the weight vectors for winner and neighborhood-nodes is determined by an inverse and negative version of the classic Kohonen learning rule:

$$\Delta\vec{w}_{r'} = -ah_{rr'}g(\vec{x} - \vec{w}_{r'}),$$

where

$$g(\vec{x}-\vec{w}_{r'}) = \begin{cases} \dfrac{\vec{x}-\vec{w}_{r'}}{\|\vec{x}-\vec{w}_{r'}\|\delta_1} & \text{for } \|\vec{x}-\vec{w}_{r'}\| \leqslant \delta_1, \\ \dfrac{\vec{x}-\vec{w}_{r'}}{\|\vec{x}-\vec{w}_{r'}\|^2} & \text{for } \delta_1 < \|\vec{x}-\vec{w}_{r'}\| < \delta_2, \\ 0 & \text{for } \|\vec{x}-\vec{w}_{r'}\| \geqslant \delta_2, \end{cases}$$

δ_1, δ_2 are constants, points of maximum and minimum values of the g function, $h_{rr'}$ is a Gaussian neighborhood function centered at the winning unit r.

$$h_{rr'} = \mathrm{e}^{-((r-r')^2/2\sigma^2)},$$

where σ defines the radius of the neighborhood which decreases over time and a is a positive learning rate that decreases toward zero as the learning progresses. The evolution of the weight vectors during learning is shown in Figs. 1 and 2.

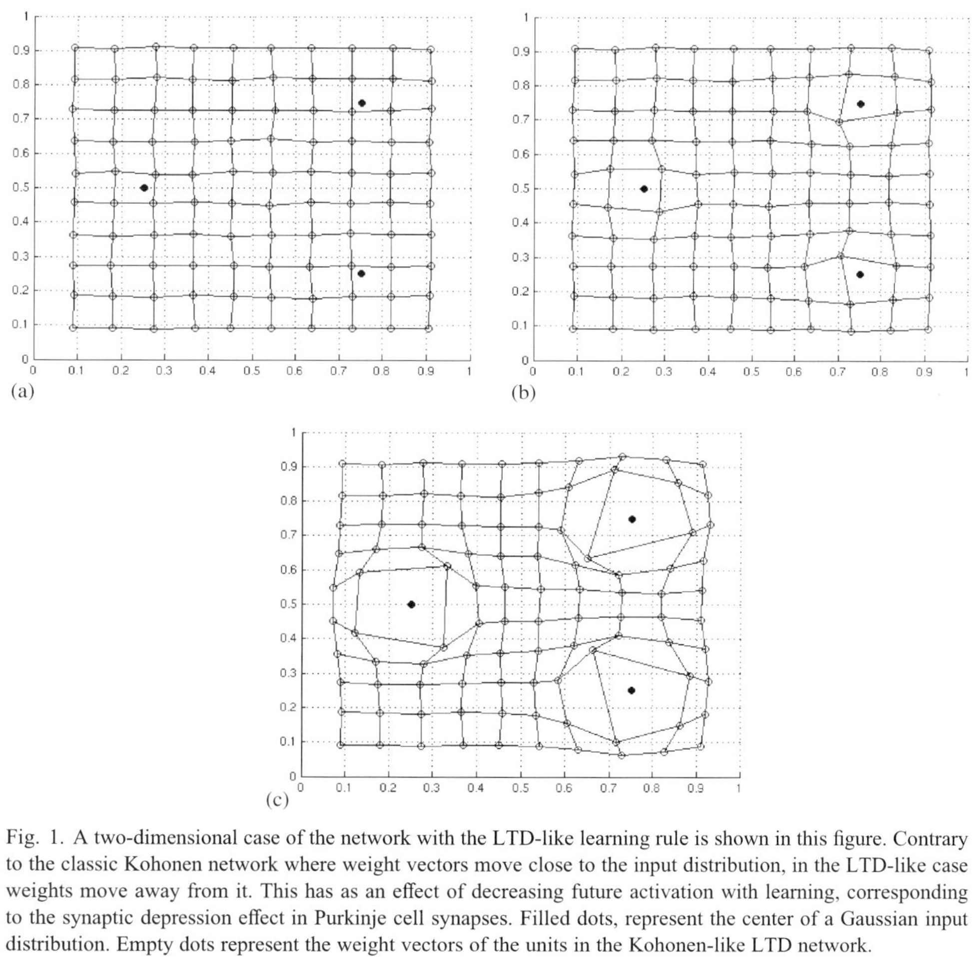

Fig. 1. A two-dimensional case of the network with the LTD-like learning rule is shown in this figure. Contrary to the classic Kohonen network where weight vectors move close to the input distribution, in the LTD-like case weights move away from it. This has as an effect of decreasing future activation with learning, corresponding to the synaptic depression effect in Purkinje cell synapses. Filled dots, represent the center of a Gaussian input distribution. Empty dots represent the weight vectors of the units in the Kohonen-like LTD network.

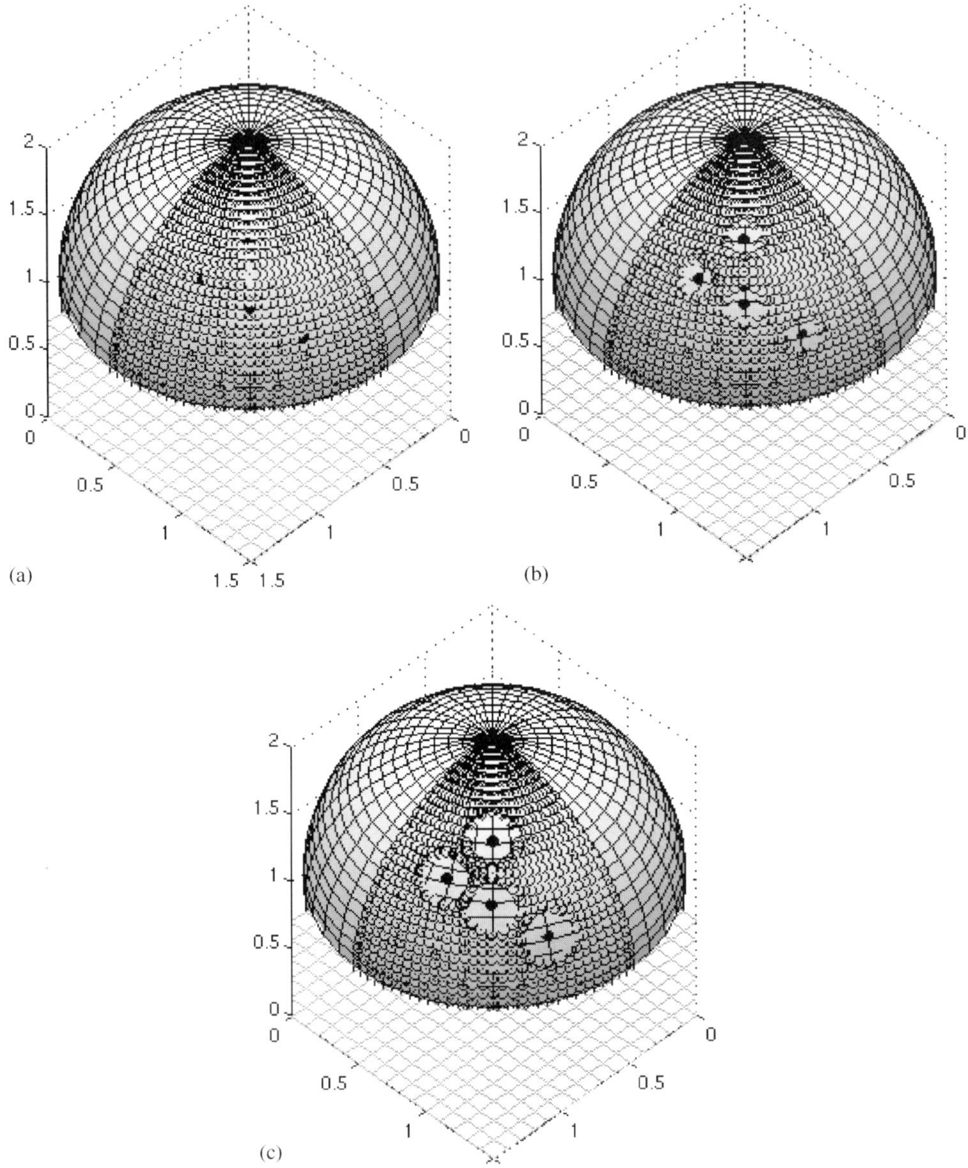

Fig. 2. The three-dimensional case of the network with the LTD-like learning rule is shown in this figure. The grid of empty dots represents the weight vectors, while the filled dots are the centers of input distributions. Additionally, all vectors are normalized (thus projected onto the surface of a unitary sphere) and comprise components in the range [0, 1]. On the left, the network before learning is shown, where all weights are randomly distributed. As learning progresses the weight vectors move away from the input vectors (center and right figure). With respect to the excitatory parallel fiber—Purkinje cell synapses, only positive weights are biologically plausible, thus only the first quadrant of the sphere contains weight units.

3. A novelty detector

As an effect of learning, all patterns used in the training set and all patterns similar to them give decreasing activation values as shown in Fig. 3. All patterns dissimilar to the training set result always in a stable high activation, both during and after learning. The differentiation of responses between patterns similar to training-patterns (known) and patterns not ever used in the training set (novel) comes eventually as a consequence of learning. Thus, an important property of the LTD Kohonen-like network is that it can function as a novelty detector. To demonstrate this we have made a map of all possible input patterns for the network (differing on small discrete step intervals) and show these results in Fig. 4. This activation map of possible inputs shows that the network can discriminate the known patterns used for learning and all patterns similar to them by providing a lower activation than for novel ones.

Novelty detection can be used to support both motor control and cognitive functions of the cerebellum. For example, the ability to make fast distinctions between familiar and novel patterns can support learning in many contexts. Moreover control systems can use novelty detection to distinguish between expected and unexpected patterns (even if the latter are not truly novel). Therefore this new interpretation of cerebellar learning could allow the cerebellar circuitry to perform many useful tasks fast.

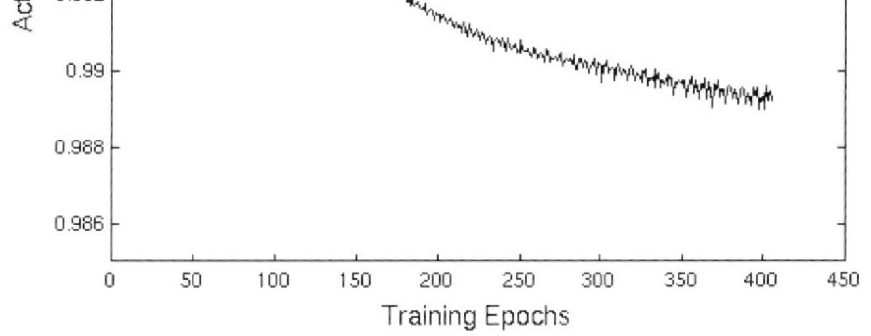

Fig. 3. The decrease of activation as learning progresses. The graph shows how the activation for a specific input similar to one of the learning patterns evolves during the first 400 training epochs.

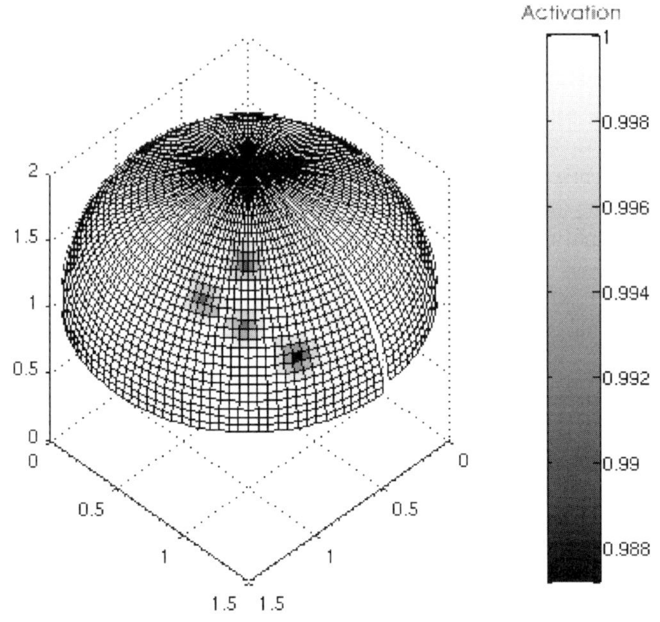

Fig. 4. In this figure all possible activation responses of the network for the input space after learning are shown. The input space contains darker areas which give lower activation and represent inputs similar to the ones used for training (learned patterns), while the areas with light color represent inputs not similar to the ones used in training (novel patterns). The network is a novelty detector for all possible patterns not similar to the ones used for training, by providing a higher activation response for them.

4. Conclusions

In the cerebellar cortex, long-term-depression of synapses between parallel fibers (PFs) and Purkinje cells is analogous to moving the synaptic weight vector away from the PF input vector. We have developed a Kohonen-like network with a LTD learning rule and shown that this network can function as a novelty detector. This novelty detection suggests new interpretations of cerebellar function in motor control and cognitive tasks

Acknowledgements

Thanks to Avrama Blackwell for her help and ideas when this project was initially considered at the EU computational neuroscience course in Trieste. This project was financially supported by the EC grant QLRT-2000-02256.

References

[1] J.S. Albus, A theory of cerebellar function, Math. Biosci. 10 (1971) 25–61.

[2] E. De Schutter, Cerebellar long-term depression might normalize excitation of Purkinje cells: a hypothesis, Trends Neurosci. 18 (1995) 291–295.
[3] J. Eilers, H. Takechi, E.A. Finch, G.J. Augustine, A. Konnerth, Local dendritic Ca2+ signaling induces cerebellar long-term depression, Learn. Mem. 4 (1) (1997) 159–168.
[4] N.A. Hartell, Strong activation of parallel fibers produces localized calcium transients and a form of LTD that spreads to distant synapses, Neuron 16 (1996) 601–610.
[5] T. Kohonen, Self-organised formation of topologically correct feature maps, Biol. Cybernet. 43 (1982) 59–69.
[6] D. Marr, A theory of cerebellar cortex, J. Physiol. 202 (1969) 437–470.
[7] T. Reynolds, N.A. Hartell, An evaluation of the synapse specificity of long-term depression induced in rat cerebellar slice, J. Physiol. 527 (3) (2000) 563–577.
[8] S.-H. Wang, L. Khiroug, G.J. Augustine, Quantification of spread of cerebellar long-term depression with chemical two-photon uncaging of glutamate, Proc. Natl. Acad. Sci. 97 (15) (2000) 8635–8640.
[9] M. Ito, The cerebellum and neural control. Raven Press, New York, 1984.

Available at
www.ComputerScienceWeb.com
POWERED BY SCIENCE DIRECT

ELSEVIER

Neurocomputing 52–54 (2003) 419–424

NEUROCOMPUTING

www.elsevier.com/locate/neucom

A model of visual working memory in PFC

Frank van der Velde*, Marc de Kamps

Cognitive Psychology Unit, Leiden University, Wassenaarseweg 52, Leiden 2333AK, The Netherlands

Abstract

We present a model of visual working memory in ventral prefrontal cortex. Activation in ventral PFC consists of reverberating activity. Representations in ventral PFC are conjunctions of location and (partial) identity representations. With many objects, representations in ventral PFC interfere, which results in loss of information. Therefore, the number of objects in working memory is limited. However, because ventral PFC is connected to the 'identity' levels in the visual cortex, the number of features for each object is unlimited. The 'blackboard' architecture of ventral PFC results in a unification (binding) of the feature representations of the objects maintained in memory.

Keywords: Blackboard architecture; Binding problem; Prefrontal cortex; Working memory

1. Introduction

Recent investigations [6] have shown that humans have the ability to maintain a number of visual objects in visual working memory. A remarkable characteristic of this finding is that the number of objects that can be maintained in working memory without interference (i.e., loss of information) is limited (to about four), but the number of object features (e.g., color, form, location) is unlimited for each of the objects. Functional neuroimaging studies [1] have shown that in particular the ventral prefrontal cortex (V-PFC) is involved in the maintenance of information in visual working memory. In contrast with animal studies [7], these neuroimaging studies did not reveal a distinction between spatial information and object (feature) information within PFC. However, a distinction between V-PFC and dorsal PFC (D-PFC) was found. Whereas V-PFC was mainly involved in maintenance of information, D-PFC appeared to be

* Corresponding author.
E-mail address: vdvelde@fsw.leidenuniv.nl (F. van der Velde).

0925-2312/03/$ - see front matter © 2003 Elsevier Science B.V. All rights reserved.
doi:10.1016/S0925-2312(02)00858-5

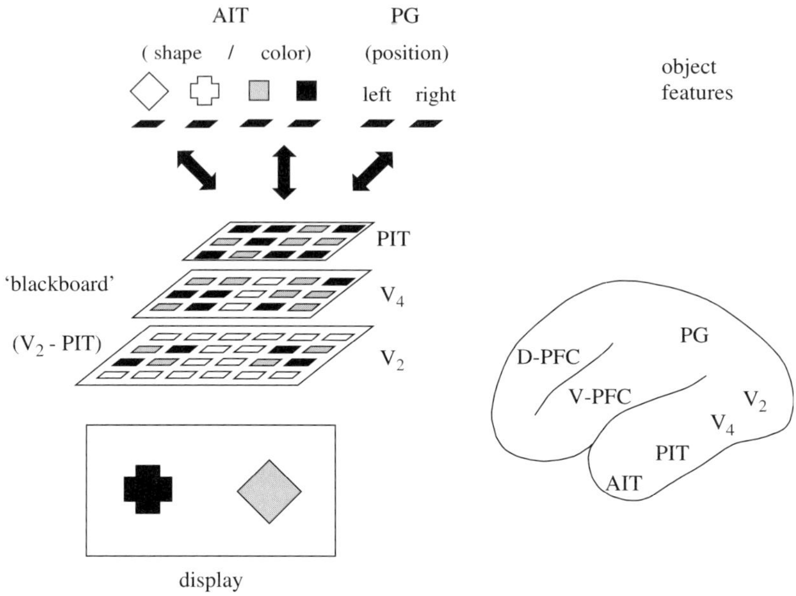

Fig. 1. Blackboard architecture of visual cortex. PIT = posterior infero-temporal cortex; AIT = anterior infero-temporal cortex; PG = parietal cortex; V-PFC = ventral prefrontal cortex; D-PFC = dorsal prefrontal cortex. For explanation see text.

involved in operations that manipulate the information maintained in visual working memory.

In line with these observations, we present a model of visual working memory in PFC. The model is based on a neural 'blackboard' architecture that we used in our simulation of object-based attention in the visual cortex [5], illustrated in Fig. 1. The visual cortex consists of the ventral pathway, which includes the areas V_2, V_4, the posterior infero-temporal cortex (PIT) and the anterior infero-temporal cortex (AIT), and the dorsal pathway, which leads into the parietal cortex (PG). The ventral pathway is involved in the processing and selection of 'object features' (e.g., shape and color) and the dorsal pathway is involved in the processing and selection of spatial information (e.g., positions of objects in a display) and spatial transformations (e.g., for making eye movements). Both pathways start from the primary visual cortex (V_1), but they are also interconnected on the levels of V_2, V_4 and PIT. Both pathways project to the prefrontal cortex. In the ventral pathway, objects are identified through a feedforward network of areas, going from the primary visual cortex (V_1) to the higher areas in the temporal cortex (AIT). In this network, the retinotopic representation in the lower areas is gradually transformed (in V_2 to PIT) into a location-invariant identity (e.g., shape, color) representation (in AIT).

In object-based attention, an object is selected as a target, and a feedback process is initiated in the networks that produce object identification. This feedback process carries information about the identity of the object to the lower areas in the visual cortex

(V_2 - PIT). The feedback process interacts in these areas with the feedforward process described above, by means of microcircuits that connect the feedforward and feedback networks [5]. This interaction results in the enhanced activation of target representations in these areas, which results in the selection of target-related location information [5]. The representations in these lower 'in-between' areas consist of conjunctions of (partial) identity information and location information. This is a direct consequence of the fact that these areas gradually transform the retinotopic information in the primary visual cortex into identity-based information. By means of the interaction process described above, information about the location of an identified object can be recovered, even though that information was lost at the level of object identification. In a converse manner, the selection of a location in the 'in-between' areas can be used to select the information about the identity of an object, which will result in the identification of the object on the given location. In computational terms, the 'in-between' areas provide a 'blackboard' architecture because they link different 'processors' to one another [4]. The 'processors' in this case are networks for feature identification (e.g., shape, color, location, motion, etc.). The blackboard serves to unify (bind) the information processed in each of the specialized processors.

2. Blackboard architecture of working memory in PFC

Here, we propose a similar 'blackboard' architecture for the maintenance of information in visual working memory (Fig. 2).

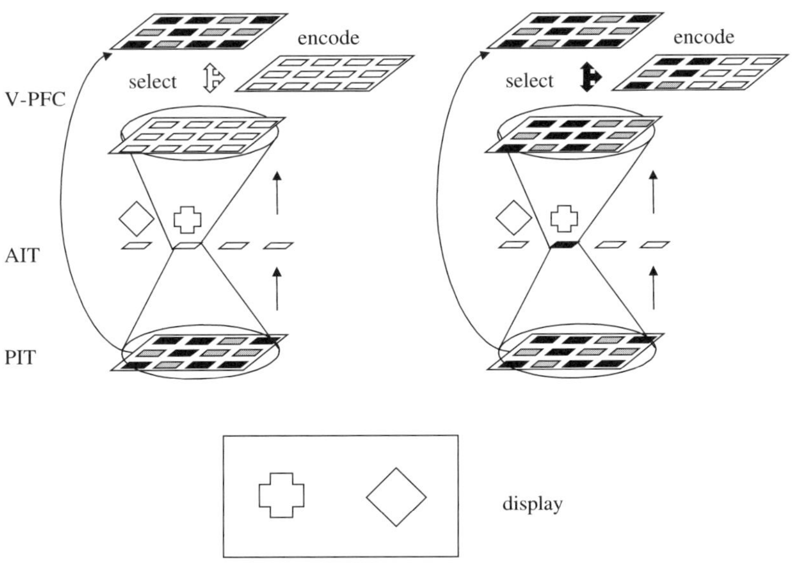

Fig. 2. A blackboard architecture in prefrontal cortex (PFC). PIT = posterior infero-temporal cortex; AIT = anterior infero-temporal cortex; V-PFC = ventral prefrontal cortex. For explanation see text.

In this model, V-PFC has a layered structure with representations similar to the representations in the visual (temporal) cortex (Fig. 1). First, PIT connects to one of the layers in V-PFC (for the purpose of illustration: the top layer in Fig. 2). As in PIT itself, the representations in this layer of V-PFC consist of conjunctions of location and (partial) identity (object-feature) representations. In turn, another layer of V-PFC (the bottom layer in Fig. 2) is connected to the higher-level areas in the visual cortex, in which (location-invariant) object identity and location information is processed and represented (in Fig. 2 illustrated for AIT). These connections are similar to the connections of the feedback network of the visual cortex [5]. Thus, they have the same 'fan-out' structure, in contrast with the 'fan-in' structure of the connections in the feedforward network of the visual cortex (as illustrated with the cones in Fig. 2). As a result, the representations in the bottom layer in V-PFC consist of distributed 'identity' representations, in line with the representations in the feedback network of the visual cortex [5]. The bottom and top layer in V-PFC interact in a manner similar to the interaction between the feedforward and feedback networks of the visual cortex, using similar microcircuits [5]. This interaction results in the activation of a third layer in V-PFC (the 'encode' layer in Fig. 2). The representations in this layer are similar to the representations in the top layer. In this model, the encode layer is the layer in which the sustained (reverberating) activity occurs, typical of working memory in the cortex [7].

Fig. 2 illustrates that a similar selection process occurs in the V-PFC model as in the blackboard model of the visual cortex. In Fig. 2 (left), two objects are processed in the visual cortex, and their PIT representations also activate the representations in the top layer in V-PFC. However, none of the objects activates its identity representation in AIT (e.g., for shape in Fig. 2). As a result, the interaction between the top and bottom layer in V-PFC does not produce activation in the encode layer. In Fig. 2 (right), one of the objects (the cross) does activate its identity representation in AIT, which activates the bottom layer in V-PFC. As a result, the interaction between the top and bottom layer activate the representations in the encode layer that are selective for the features (e.g., shape, color, position) of the cross. Thus, objects have to activate their identity representations in AIT (i.e., have to be 'attended') before they can be stored in working memory.

3. Feature binding in working memory

The nature of the representations in ventral PFC and the connections with the higher-level areas in the visual cortex produces the behavioral effects described above. When too many objects are present in a display, their representations in ventral PFC interfere, which results in loss of information. This interference results in a limitation of the number of objects that can be maintained in working memory. However, because the ventral PFC is connected to the 'identity' levels in the visual cortex, the number of features for each object is unlimited. The blackboard architecture of ventral PFC results in a unification (binding) of the feature representations of the objects maintained in memory, similar to the blackboard architecture in the visual cortex. In this

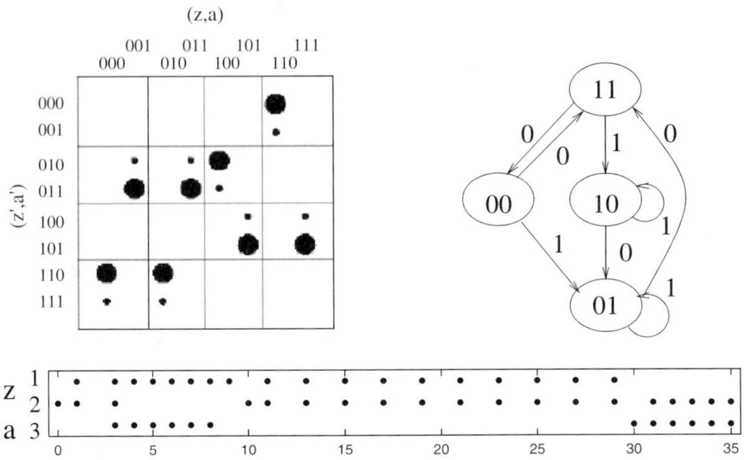

Fig. 1. Optimized example system with 2 internal units (1,2) and 1 input (3). Upper left: Markov kernel. The total kernel can be segregated into blocks defined by internal states (z, z') only. Right: corresponding automaton with internal states (z) as node and input (a) as edge symbols. Below: sample trajectory.

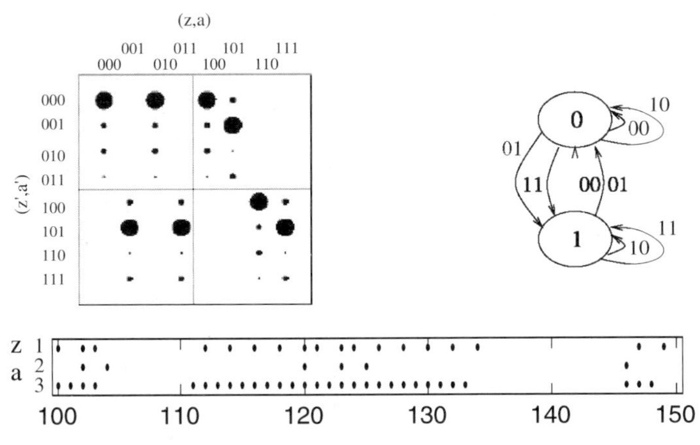

Fig. 2. Similar to Fig. 1 but with two independent input units (2,3). Firing of the internal third unit (1) depends on both inputs but strongly reflects unit 3 activity.

Both systems converge to comparable DFAs. Nonetheless, in Fig. 2, the externally induced temporal correlations are still expressed in the precise transition probabilities, e.g., three strong transitions to 000 because all units fire relatively seldom, or three strong transitions from states with unit 3 active to 101, because unit 3 fires long bursts but unit 2 is mostly silent. Similarly, spatial correlations are expressed in Fig. 3 by the dominating transitions 000 → 000 and 100 → 100, which reflect that input units 2 and 3 are simultaneously silent in most steps. Activation of one input unit, however,

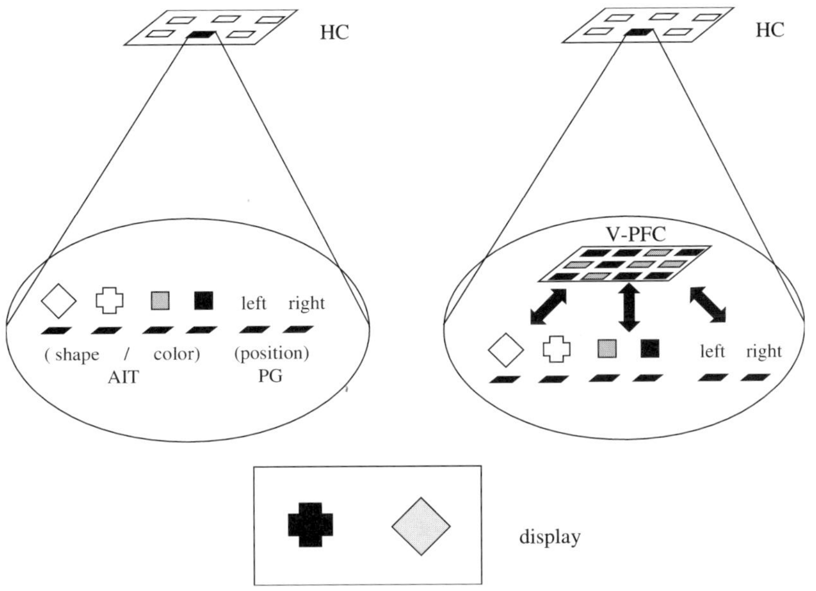

Fig. 3. Feature binding in working memory. AIT = anterior infero-temporal cortex; PG = parietal cortex; V-PFC = ventral prefrontal cortex; HC = hippocampus complex. For explanation see text.

way, all features of an object can be maintained in working memory as long as the representations of the object in ventral PFC do not interfere.

The binding of feature representations in V-PFC is illustrated in Fig. 3. In a display of two objects, a blue cross and a yellow diamond, the features are cross, diamond, blue, yellow, left and right. To maintain both objects in (working) memory, a correct binding between these features is needed (e.g., a blue cross instead of a yellow cross, a diamond on the right instead of a diamond on the left). The hippocampus and surrounding areas, referred to as the 'hippocampus complex' (HC) in Fig. 3, form an episodic memory of the feature units that are concurrently active in the visual cortex. However, representations in HC are conjunctive representations [3], as indicated with the 'cone' in Fig. 3. This entails that they can represent a collection of features, but not their relations (binding). For instance, on the basis of the representation illustrated in Fig. 3 (left), one cannot determine whether the display consisted of a blue cross and yellow diamond or of a yellow cross and blue diamond.

Binding (or relations) between features can be represented with a compositional form of representation, which will result from embedding the individual feature representations in a computational architecture such as a 'blackboard' architecture. The relations between the features can then be retrieved from the interactions within the 'blackboard' [4], as in case of the visual cortex illustrated in Fig. 1 [2,5]. In the model illustrated in Fig. 3, the encode layer in V-PFC subserves the role of a 'blackboard' in visual working memory, with representations comparable to those in PIT (or perhaps V_4). Relations between features in working memory are retrieved by the interactions between V-PFC and the individual feature representations, as illustrated in Fig. 3 (right). Fig. 3 (right) also illustrates that the HC can form a conjunctive ('cone') representation that preserves relations (binding) between features if it includes the V-PFC. This HC representation can be the basis for an episodic memory in which relations between objects are preserved.

4. Event coding in working memory

The (temporal) manipulation of information in working memory will depend on D-PFC in de model presented here. Briefly stated, V-PFC represents (maintains) the feature information in working memory (as discussed and illustrated above), whereas D-PFC is used to represent the temporal structure of information in working memory. HC now forms a conjunctive representation of V-PFC and D-PFC. In this way, two displays, for instance, can be stored in sequence in working memory. The order of the sequence is represented in D-PFC and the content of each display is represented in the blackboard representation of V-PFC (similar to Fig. 3, right). D-PFC will determine which of the two display representations is active in V-PFC. In this way, an object in one display can be used as the target in a visual search task, without an interference with the objects in the other display stored in working memory [8].

References

[1] M. D'Esposito, Functional neuroimaging of working memory, in: R. Cabeza, A. Kingstone (Eds.), Handbook of Functional Neuroimaging of Cognition, MIT Press, Cambridge, MA, 2001, pp. 293–327.

[2] M. de Kamps, F. van der Velde, Using a recurrent network to bind form, color and position into a unified percept, Neurocomputing 38–40 (2001) 523–528.

[3] R.C. O'Reilly, Y. Munakata, Computational Explorations in Cognitive Neuroscience: Understanding the Mind by Simulating the Brain, MIT Press, Cambridge, MA, 2000.

[4] F. van der Velde, On the use of computation in modelling behavior, Network: Comput. Neural Systems 8 (1997) 1–32.

[5] F. van der Velde, M. de Kamps, From knowing what to knowing where: modeling object-based attention with feedback disinhibition of activation, J. Cognitive Neurosci. 13 (4) (2001) 479–491.

[6] E.K. Vogel, G.F. Woodman, S.J. Luck, Storage of features, conjunctions, and objects in visual working memory, J. Exp. Psychol.: HPP 27 (2001) 92–114.

[7] F.A.W. Wilson, S.P.O. Scalaidhe, P.S. Goldman Rakic, Dissociation of object and spatial processing domains in primate prefrontal cortex, Science 260 (1993) 1955–1957.

[8] G.F. Woodman, E.K. Vogel, S.J. Luck, Visual search remains efficient when visual working memory is full, Psychol. Sci. 12 (2001) 219–224.

Available at
www.ComputerScienceWeb.com
POWERED BY SCIENCE @ DIRECT®

ELSEVIER

Neurocomputing 52–54 (2003) 425–430

NEUROCOMPUTING

www.elsevier.com/locate/neucom

Desynchronization of neural activity in a network model

Wim van Drongelen*, Kurt Hecox

Department of Pediatrics, The University of Chicago, 5841 South Maryland Av Mc, Chicago, IL 60637, USA

Abstract

Electric brain activity during seizures is often associated with hyperactive and synchronous bursts of activity. We investigated different protocols of electric stimulation in a model network as potential seizure control mechanisms. Our network consisted of a chain loop of cells with excitatory synapses, representing a neocortical microcircuit. Random fluctuations of the membrane potential in each neuron simulated activity originating from other areas. Overall depolarizing and hyperpolarizing currents were not effective in stopping synchronized activity in the network. An electric current proportional to the inverted field potential was an effective desynchronizing stimulus in smaller networks.

Keywords: Epilepsy; Electrostimulation; Network activity

1. Introduction

A major goal in epilepsy treatment is to control high levels of activity and synchronous oscillations of underlying neural networks in the brain. Several studies have investigated electric stimulation as a possible therapy for seizure control [2,4,5,8,10,14, 16,19,21]. With few exceptions, it can be concluded that direct stimulation to areas where epileptiform activity occurs is the most effective.

In a clinical setting, a large-scale parametric study on the relationship between electrostimulation and seizure activity is not always feasible or ethical. Electric stimulation in slices of brain tissue is studied under the assumption that the behavior of small networks can provide insight into the activity of the intact brain. Simulated activity of

* Corresponding author.
E-mail address: wvandron@peds.bsd.uchicago.edu (W. van Drongelen).

0925-2312/03/$ - see front matter © 2002 Elsevier Science B.V. All rights reserved.
doi:10.1016/S0925-2312(02)00860-3

neural networks provides an additional route to study aspects of synchronized network behavior and the relationship between seizure control and various paradigms of electric stimulation. Current simulation studies seek to explain how synchronous activity may be generated and sustained, by looking into intrinsic properties of individual neurons, local cortical networks, and thalamo-cortical loops [e.g. [3,6,11–13,20]. Recently, Kudela et al. [9] explored the relationship between open loop electrostimulation and recurrent bursting in a model of connected sub-networks.

In this study we simulate synchronized activity within a small network representing local epileptiform activity in the cortex microcircuitry before it propagates to a larger area. We examined different types of electrostimulation as a tool to stop population discharges by looking for desynchronization of the activities.

2. Methods

To study the effects of electrostimulation in a network, the following elements were included in our simulation: (1) model neurons connected within (2) a network that is capable of displaying synchronous activity, and (3) a method to stimulate electrically using different paradigms. The paradigms we examined are (a) depolarization, (b) hyperpolarization, and (c) field-potential-related currents. The model is implemented as a MatLab (The MathWorks Inc, Natick, MA) script file and C programs created from these scripts. The accuracy of simulation is 1/25 ms.

(1) *Neural elements*: The unit of our model network is a Hodgkin–Huxley-type membrane model [7] representing one neuron. Each element consists of membrane capacitance (C) and ion channels. The total ionic current (I) is the sum of the individual ionic currents: $I = i_{\mathrm{Na}} + i_{\mathrm{K}} + i_{\mathrm{L}}$, where i_{Na} is the sodium current, i_{K} the potassium current, and i_{L} the leakage current.

(2) *Network structure*: Isolation of tissue components in in vitro experiments established that the microcircuitry of the neocortex is capable of generating and sustaining epileptiform bursts [1]. This indicates that seizure activity may arise in small networks before it is propagated to involve a larger area. Our model network was a chain loop of a small number of units (< 10) coupled via excitatory synapses. Activity in excitatory synapses was simulated by a depolarizing current (i_s) for a duration of 5 ms. Because the distance between neighboring neurons was assumed to be small (< 100 μm), synaptic transmission was responsible for the delay between activities of connected units, there was no extra delay representing the conduction over the axon. Additional activity from other areas in the brain was simulated by independent random excitatory and inhibitory activity (i_{R}) on each neuron. The small network could be considered as a set of neocortical pyramidal cells in layers II/III and/or layers V/VI that interconnect in a local microcircuit [15,17].

(3) *Electric stimulation*: Current to mimic electric stimulation is added to the transmembrane currents.

(a) A *depolarization* is caused by an inward transmembrane current (i_{D}).
(b) A *hyperpolarization* is created by an outward transmembrane current (i_{H}).

(c) We calculate the extracellular *field potential* by assuming that the recording electrode is equidistant to all the neurons. Their contribution to the recorded field potential is proportional to the sum of their individual electric activities [18,20,22]. The summated potential was scaled and subjected to a bandfilter (0.5–70 Hz) to mimic the effects of the impedances of the extracellular medium and the filters that are typically applied to EEG signals. The applied transmembrane feedback current (i_{FB}) was inversely proportional to the processed field potential and applied to each of the network elements. The equation describing the membrane potential (V) of one network element in time (t) is

$$C(\mathrm{d}V/\mathrm{d}t) + I + i_s + i_R + i_D + i_H + i_{FB} = 0.$$

3. Results

The effect of electrostimulation was studied by starting synchronized activity with an initialization pulse in a simulation with a total duration of 700 ms. Different types of test current were applied during a 100 ms epoch starting at 500 ms after the start of the simulation. This allowed us to observe behavior of the network before, during and after the test current. A typical result of each stimulus paradigm on a synchronized network of five neurons is shown in Fig. 1A (depolarization), B (hyperpolarization), and C (EEG feedback). Each example shows the applied current in the upper panel. The following five panels show individual activity of each neuron in the network. The bottom trace in each column of Fig. 1 depicts the individual activities superimposed. This trace is to visually determine the degree of synchrony in the network. A depolarization initializes synchronous activity 1 ms after the start of the simulation. In the examples shown, the overall activity of the network is in synchronous mode at the start of the test current. In Figs. 1A and C several small intervals of 'spontaneous' desynchronizations, prior to the test current onset, can be observed.

The results of the three stimulus paradigms can be summarized as follows.

1. The depolarizing current (Fig. 1A) causes a slight increase of activity in the individual neurons, and causes some desynchronization during the stimulus. However, immediately after the stimulus offset the network is in synchronous state.
2. A hyperpolarization of the neurons (Fig. 1B) stops the generation of action potentials during the application of the stimulus. The stimulus offset triggers the network's elements and creates synchronous activity.
3. The initial effect of the feedback current (Fig. 1C) is inhibition of the elements, followed by an excitation. Interestingly, the activity (in action potentials/s) arising after the initial inhibition is slightly higher than before the stimulus onset. More importantly this activity is desynchronized, and the network remains to be desynchronized after the stimulus offset. This is a reproducible finding in the smaller networks that we used in our simulation study.

4. Discussion

Our study demonstrates that EEG-like feedback current desynchronizes networks whereas depolarization and hyperpolarization do not desynchronize the neural

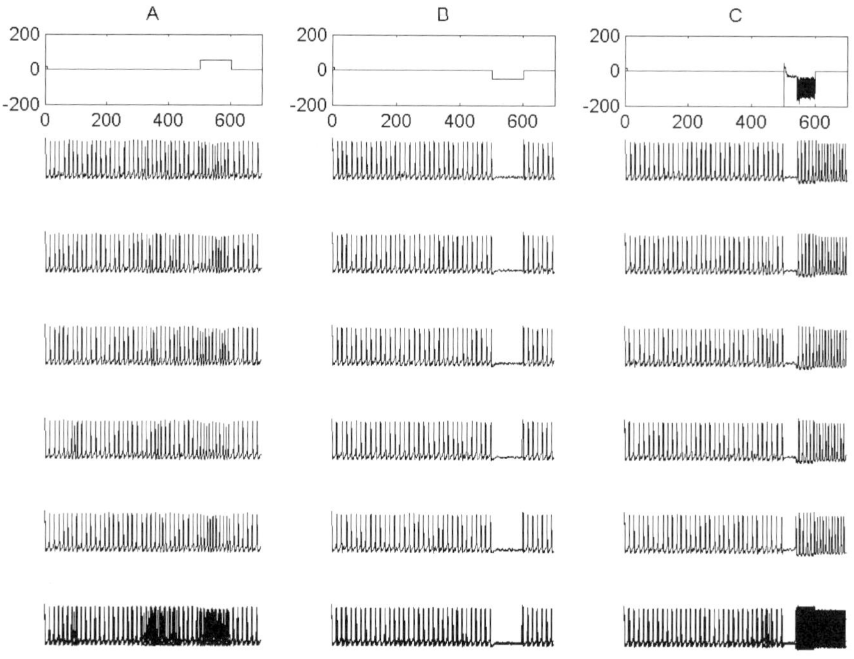

Fig. 1. The effects of three different electrostimulation paradigms ((A) depolarization, (B) hyperpolarization, and (C) EEG feedback). Stimulus current in μA/cm^2 (upper panel) and activities of five neurons in the network (panels 2–6). The bottom trace shows the superimposed cellular activities. In the case of EEG feedback (C), the network remains desynchronized after stimulus offset.

activity. Our data agree with results from in vitro studies where electrostimulation was used to modify bursting behavior in populations of neurons. Gluckman and colleagues [5] found that adaptive field control, similar to our feedback current paradigm (Fig. 1C), is an effective inhibitor of seizure activity. Although the details of their experiment and findings differ from our simulation, the overall conclusion is similar. Nakagawa and Durand [16] found that hyperpolarization inhibits seizure-like activity in hippocampal slices, but that population burst activity rebounds after stimulus offset (their Fig. 6). This finding is identical to the result following hyperpolarization (Fig. 1B). To our knowledge, a depolarizing current pulse has not been tested as a desynchronizing stimulus in smaller networks. Simulations by Kudela et al. [9] show propagation of epileptiform activity in a larger network. In this simulation, correctly timed depolarization can stop a propagating burst. In experimental models, the effects of a depolarizing current was used to evoke bursts in order to place network bursting activity in cortical slices [8,19].

One might hypothesize about the reason why EEG feedback is an effective desynchronizing stimulus. Clearly, it cannot be simply explained by inhibition caused by a hyperpolarization of the membrane (Fig. 1B). It seems that the inverted field potential creates the right 'dose' to stop synchronous firing of the neurons. In a population that is synchronously active, the field potential is proportional to each individual activity, and therefore the inverse of that activity may be an effective desynchronizing stimulus for the individual elements in the network. It is likely that high levels of synchrony between cells only arise in small populations, and as soon as larger populations are considered the cells spread over a larger distance and conduction delays destroy synchronous firing. Indeed in larger networks of sequentially coupled neurons, oscillations occur but the synchrony between neurons breaks down due to conduction and transmission delays. The activity pattern of larger networks was not visibly affected by an EEG feedback current.

Acknowledgements

This work was supported by a grant from the Falk Center for Advanced Study and care of Pediatric Epilepsy.

References

[1] B.W. Connors, Y. Amitai, Generation of epileptiform discharge by local circuits of neocortex, in: P.A. Schwartzkroin (Ed.), Epilepsy, Cambridge University Press, Cambridge, 1993, pp. 388–423.

[2] I. Cooper, I. Amin, M. Riklan, J.M. Waltz, T. Poon, Chronic cerebellar stimulation in epilepsy, Arch. Neurol. 33 (1976) 559–570.

[3] A. Destexhe, D. Pare, Impact of network activity on the integrative properties of neocortical pyramidal neurons in vivo, J. Neurophysiol. 81 (1999) 1531–1547.

[4] R. Fisher, S. Uematsu, G. Krauss, B. Cysyk, R. McPherson, R. Lesser, B. Gordon, P. Schwerdt, M. Rise, Placebo-controlled pilot study of centromedian thalamic stimulation in treatment of intractable seizures, Epilepsia 33 (1992) 841–851.

[5] B. Gluckman, H. Nguyen, S. Weinstein, S. Schiff, Electric field control of epileptic seizures, J. Neurosci. 21 (2001) 590–600.

[6] D. Golomb, Models of neuronal transient synchrony during propagation of activity through neocortical circuitry, J. Neurophysiol. 79 (1998) 1–12.

[7] A.L. Hodgkin, A.F. Huxley, A quantitative description of membrane current and its application to conduction and excitation in the nerve, J. Physiol. (London) 117 (1952) 500–544.

[8] K. Jerger, S. Schiff, Periodic pacing an In Vitro epileptic focus, J. Neurophysiol. 73 (1995) 876–879.

[9] P. Kudela, P.J. Franaszczuk, G.K. Bergey, External termination of recurrent bursting in a model of connected local neural sub-networks, Neurocomputing 44–46 (2002) 897–905.

[10] R.P. Lesser, S.H. Kim, L. Beyderman, D.L. Miglioretti, W.R.S. Webber, M. Bare, B. Cysyk, G. Krauss, B. Gordon, Brief bursts of pulse stimulation terminate afterdischarges caused by cortical stimulation, Neurology 53 (1999) 2073–2081.

[11] W. Lytton, D. Conteras, A. Destexhe, M. Steriade, Dynamic interactions determine partial thalamic quiescence in a computer network model of spike-and-wave seizures, J. Neurophysiol. 77 (1997) 1679–1696.

[12] W. Lytton, T. Sejnowski, Simulations of cortical pyramidal neurons synchronized by inhibitory interneurons, J. Neurophysiol. 66 (1991) 1059–1079.

[13] W. Lytton, E. Thomas, Modeling thalamocortical oscillations, in: P.S. Ulinski, E.G. Jones, A. Peters (Eds.), Cerebral cortex 13, Kluwer Academic/Plenum Publishers, New York, 1999, pp. 479–509.

[14] R. McLachlan, Vagus nerve stimulation for intractable epilepsy: a review, J. Clin. Neurophysiol. 14 (1997) 358–368.

[15] V.B. Mountcastle, The columnar organization of the neocortex, Brain 120 (1997) 701–722.

[16] M. Nakagawa, D. Durand, Suppression of spontaneous epileptiform activity with applied currents, Brain Res. 576 (1991) 241–247.

[17] R. Nieuwenhuys, The neocortex. An overview of its evolutionary development, structural organization and synaptology, Anat. Embryol. 190 (1994) 307–337.

[18] P.L. Nunez, Electric Fields of the Brain, Oxford University Press, New York, 1981.

[19] S.J. Schiff, K. Jerger, D.H. Duong, T. Chang, M.L. Spano, W.L. Ditto, Controlling chaos in the brain, Nature 370 (1994) 615–620.

[20] R.D. Traub, J.G.R. Jefferys, R. Miles, M.A. Whittington, K. Toth, A branching dendritic model of a rodent CA3 pyramidal neurone, J. Physiol. (London) 481 (1994) 79–95.

[21] M. Velasco, F. Velasco, A. Velasco, B. Boleaga, F. Jimenez, F. Brito, I. Marquez, Subacute electrical stimulation of the hippocampus blocks intractable temporal lobe seizures and paroxymal EEG activities, Epilepsia 41 (2000) 158–169.

[22] H.R. Wilson, J.M. Bower, Cortical oscillations and temporal interactions in a computer simulation of piriform cortex, J. Neurophysiol. 67 (1992) 981–995.

Available at
www.ComputerScienceWeb.com
POWERED BY SCIENCE @ DIRECT®

ELSEVIER

Neurocomputing 52–54 (2003) 431 – 436

NEUROCOMPUTING

www.elsevier.com/locate/neucom

Temporal Infomax on Markov chains with input leads to finite state automata

Thomas Wennekers*, Nihat Ay

Max-Planck-Institute for Mathematics in the Sciences, Insel Strasse 22-26, D-04103 Leipzig, Germany

Abstract

Information maximization between stationary input and output activity distributions of neural ensembles has been a guiding principle in the study of neural codes. We have recently extended the approach to the optimization of information measures that capture spatial *and* temporal signal properties. Unconstrained Markov chains that optimize these measures have been shown to be almost deterministic. In the present work we consider the optimization of stochastic interaction in constrained Markov chains where part of the units are clamped to prescribed processes. Temporal Infomax in that case leads to finite state automata.

Keywords: Markov model; Information maximization; Finite state automata

1. Introduction

One of the most basic questions in computational neuroscience is that for the nature of neural codes. Experiments suggest a considerable interaction of neurons already on the level of spikes, e.g., expressed by spatio-temporal correlations [1,4,6,7]. A well-known measure that quantifies relations of interacting units is the so-called *mutual information*: The Kullback–Leibler divergence

$$I(p) := D(p\|p_1 \otimes \cdots \otimes p_N) = \sum_{\nu=1}^{N} H(p_\nu) - H(p), \tag{1}$$

where $H(\cdot)$ denotes the Shannon entropy and p_ν the νth marginal of p measures the "distance" of p from the factorized distribution $p_1 \otimes \cdots \otimes p_N$. It is a natural measure

* Corresponding author. Tel.: +49-341-9959-50; fax: +49-341-9959-555.
E-mail address: wenneker@mis.mpg.de (T. Wennekers).

0925-2312/03/$ - see front matter
doi:10.1016/S0925-2312(02)00862-7

for "spatial" interdependence of N stochastic units and a starting point of recent approaches to neural complexity [5,6]. In order to capture intrinsically temporal aspects of dynamic interaction, I in (1) has been extended by Ay [2] to the dynamical setting of Markov chains, where it is referred to as *(stochastic) interaction*. The optimization of stochastic interaction in Markov chains has been shown to result in almost deterministic dynamical systems [3]. This work neglected external input into the considered systems. The present study, therefore, presents optimized Markov chains, where a part of the system is clamped to externally prescribed stochastic processes. Surprisingly, the optimized processes turn out to be finite state automata.

2. Temporal infomax on constrained Markov chains

Consider a set $V=\{1,\ldots,N\}$ of binary units with state sets $\Omega_v=\{0,1\}$, $v\in V$. For a subsystem $A\subset V$, $\Omega_A:=\{0,1\}^A$ denotes the set of all configurations restricted to A, and $\bar{P}(\Omega_A)$ is the set of probability distributions on Ω_A. Given two subsets A and B, where B is non-empty, $\bar{K}(\Omega_B|\Omega_A)$ is the set of all Markov transition kernels from Ω_A to Ω_B. If $A=B$ we use the abbreviation $\bar{K}(\Omega_A)$. For a probability distribution $p\in\bar{P}(\Omega_A)$ and a Markov kernel $K\in\bar{K}(\Omega_A)$, the *conditional entropy* of (p,K) is defined as

$$H(p,K)=-\sum_{\omega,\omega'\in\Omega_A}p(\omega)K(\omega'\mid\omega)\ln K(\omega'\mid\omega). \tag{2}$$

$H(p,K)$ has been used in [2] to generalize (1) to Markov transitions. For that purpose, marginal kernels $K_v\in K(\Omega_v)$ are defined for strictly positive p as

$$K_v(\omega'_v\mid\omega_v):=\frac{\sum_{\substack{\sigma,\sigma'\in\Omega_V\\ \sigma_v=\omega_v,\sigma'_v=\omega'_v}}p(\sigma)K(\sigma'\mid\sigma)}{\sum_{\substack{\sigma\in\Omega_V\\ \sigma_v=\omega_v}}p(\sigma)},\qquad \omega_v,\omega'_v\in\Omega_v,\ v\in V. \tag{3}$$

Then, the *stochastic interaction measure* of K with respect to p is given by

$$I(p,K):=\sum_{v\in V}H(p_v,K_v)-H(p,K), \tag{4}$$

continuously extended to $\bar{P}(\Omega_V)\times\bar{K}(\Omega_V)$. Eq. (4) provides the required generalization of (1) to Markov transitions. $I(p,K)$ is large if the marginal transitions have high entropy, but that of the full transition is low. Thus, supposed the current state $\omega\in\Omega_V$ is known, the next state is predictable with high probability, but, conversely, not much information is gained from knowledge about single units, ω_v.

In the following we consider Markov chains $X_n=(X_{v,n})_{v\in V}$, $n=0,1,2,\ldots$, given by an initial distribution $p\in\bar{P}(\Omega_V)$ and a kernel $K\in\bar{K}(\Omega_V)$. We restrict attention to *parallel* Markov chains. A Markov kernel $K\in\bar{K}(\Omega_V)$ is called *parallel* if there exist kernels $K^{(v)}\in\bar{K}(\Omega_v\mid\Omega_V)$, $v\in V$, such that

$$K(\omega'\mid\omega)=\prod_{v\in V}K^{(v)}(\omega'_v\mid\omega)\quad\text{for all }\omega,\omega'\in\Omega_V. \tag{5}$$

Parallel Markov chains are a more natural assumption in neural modeling than general Markov chains, because the action potential generation of a neuron depends only on its own input and internal dynamics. Therefore, each unit v in (5) is given an individual kernel $K^{(v)}$. Spatial correlations in firing patterns may nonetheless result from common input, i.e., the global state ω in (5).

In [3] we studied parallel Markov chains which maximized $I(p,K)$ under no further constraint regarding K. The optimized chains were shown to be almost deterministic and representable in phase space by more or less complex graphs consisting of transient trajectories and nested loops which correspond with repeating firing patterns. In the present work, in contrast, we show simulations of parallel Markov chains which maximize $I(p,K)$ under the *additional constraint* that the kernels $K^{(v)}$ of a subset of units in (5) are fixed during the optimization process. The simulations show that Markov chains with such clamped "input units" converge to finite state automata.

3. Simulations

The simulations displayed in the following implement the usual Markov dynamics on N binary units together with a random search scheme to optimize the stochastic interaction of the Markov chains: The interaction, I, is computed with respect to an induced stationary probability distribution of a parallel Markov kernel, and starting from initial random values the kernel is iteratively perturbed such that I increases (cf. [3]). In contrast to [3], however, the optimization here is not unconstrained, but $B < N$ kernels $K^{(v)}$ in (5) are fixed during the optimization. Especially, these kernels are chosen independent of the state of the internal units. In the simplest case the 'input units' are just B-independent Bernoulli processes with fixed rates. We define 'burst processes' by $a_v := K^{(v)}(1|0) = 1 - K^{(v)}(0|0)$ and $b_v := K^{(v)}(0|1) = 1 - K^{(v)}(1|1)$. If $a_v = 1 - b_v$, they implement Bernoulli processes with rate a_v. For other a_v, $b_v \in]0,1[$, they generate burst-like firing patterns with a_v related to the mean interburst interval and $1 - b_v$ to the mean burst duration. Spatial correlations between input units can be introduced by appropriate input kernels.

Fig. 1 shows an optimized system with three units, where unit 3 has been clamped to a burst process with $a_3 = 0.1$ and $b_3 = 0.1$. The unit has a firing probability of $\lambda = a_3/(a_3 + b_3) = 0.5$, and if it has fired the probability that it fires again in the next step is quite high, $K^{(v)}(1|1) = 1 - b_3 = 0.9$. The converged Markov transition kernel is displayed in the upper left of Fig. 1 where circle area represents transition probability. Note that the kernel can be divided into blocks according to the substates z, z' of the internal units, i.e., the first $N - B$ bits of the total binary state representation: Given an internal state z and an input a (the B least significant bits of the total state), the internal target state z' in the next step is completely determined, but only the next input a' is random. This means that the Markov chain can be represented by a deterministic finite state automaton (DFA) with nodes labeled by the internal states and edges labeled by the (random) input patterns.

Figs. 2 and 3 display examples with three units each and $B = 2$. In Fig. 2, unit 2 is a Bernoulli process with $\lambda = 0.1$ and unit 3 an independent but temporally correlated burst process with $a_3 = b_3 = 0.1$. In Fig. 3, the inputs (2,3) are spatially correlated near Bernoulli processes. The internal unit 1 combines activity from both input units.

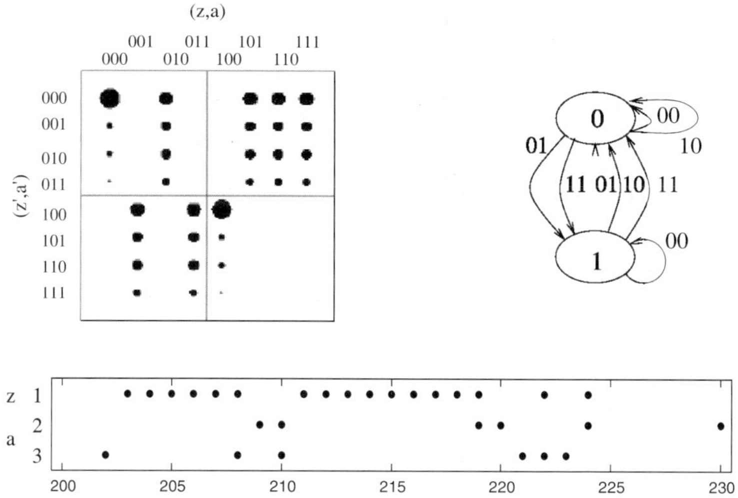

Fig. 3. Similar to Fig. 2 but with correlated inputs (2,3) favoring coactivation. The DFA is nonetheless similar to that in Fig. 2. Note also that the internal unit 1 acts as a latch (hysteresis, memory) activated from state 0 to 1 by an input of 1 from unit 3 and inactivated from 1 to 0 by any further additional input of 1.

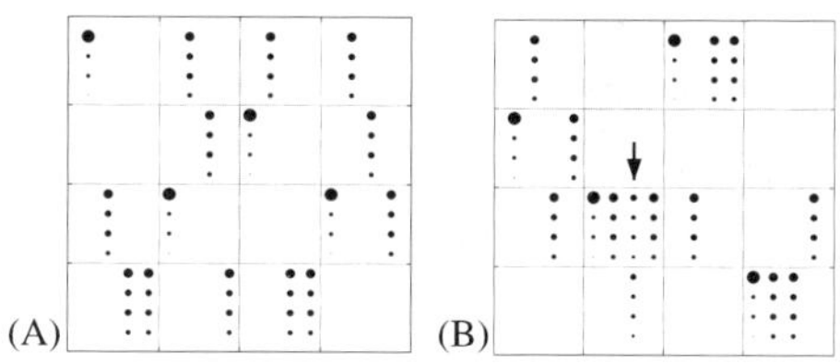

Fig. 4. Two optimized Markov matrices for $N = 4$, $B = 2$. The example in A is deterministic, but B reveals non-deterministic internal transitions (arrow).

often induces transitions to states where the same or other input unit fires, since both units excite themselves and each other.

Finally, if N grows, it turns out that the optimization does not always converge to *deterministic* automata, but internal states become increasingly likely that reveal transitions to several targets for the same input, cf. Fig. 4. Such systems correspond with non-deterministic finite state automatas (NFAs). However, it can be shown mathematically that these NFAs must always be "almost" deterministic revealing a number of targets per node that grows at most linearly in N as compared to the possible 2^N target states [8].

References

[1] M. Abeles, Corticonics: neural circuits of the cerebral cortex, Cambridge University Press, Cambridge, 1991.

[2] N. Ay, Information geometry on complexity and stochastic interaction, 2002, submitted for publication.
[3] N. Ay, T. Wennekers, Dynamical properties of strongly interacting Markov chains, 2002, submitted for publication.
[4] R. Eckhorn, Neural mechanisms of scene segmentation: recordings from the visual cortex suggest basic circuits for linking field models, IEEE Trans. Neural Networks 10 (1999) 464–479.
[5] R. Linsker, From basic network principles to neural architecture, Proc. Nat. Acad. Sci. USA 83 (1986) 7508–7512.
[6] F. Rieke, D. Warland, R.R. van Steveninck, W. Bialek, Spikes: Exploring the Neural Code, MIT Press, Cambridge, 1998.
[7] W. Singer, C.M. Gray, Visual feature integration and the temporal correlation hypothesis, Ann. Rev. Neurosci. 18 (1995) 555–586.
[8] T. Wennekers, N. Ay, Temporal infomax and determinism in Markovian neural systems, in preparation.

Thomas Wennekers studied physics at the University of Düsseldorf and computer science at the University of Ulm, where he received a Ph.D. in 1998. At present he is doing research at the Max-Planck-Institute for Mathematics in the Sciences in Leipzig in the fields of computational neuroscience and brain theory.

Nihat Ay studied Mathematics at the Ruhr-University Bochum and obtained a Ph.D. in Mathematics from the University of Leipzig in 2001. He currently works at the Max-Planck-Institute for Mathematics in the Sciences in Leipzig on information geometry and its applications in complex adaptive systems.

Available at
www.ComputerScienceWeb.com
POWERED BY SCIENCE @ DIRECT®

ELSEVIER

Neurocomputing 52–54 (2003) 437–440

NEUROCOMPUTING

www.elsevier.com/locate/neucom

Calcium as the associative signal for a model of Hebbian plasticity: application to multi-input environments

L.C. Yeung[a,b,c,*], B.S. Blais[d], L.N. Cooper[a,b], H.Z. Shouval[a,b]

[a] *Department of Physics, Brown University, 184 Hope Street, Providence, RI 02912, USA*
[b] *Institute for Brain and Neural Systems, Brown University, Providence, RI 02912, USA*
[c] *Brain Sciences Program, Brown University, Providence, RI 02912, USA*
[d] *Bryant College, Providence, RI 02917, USA*

Abstract

The sign and magnitude of bi-directional synaptic plasticity have been shown to depend on: the rate of presynaptic stimulation, the level of postsynaptic depolarization, and the precise relative timing between pre- and postsynaptic spikes. It has been proposed that these different induction paradigms can coexist, and be accounted for by a single learning rule that depends on the dynamics of intracellular calcium concentration. We extend this rule to a multi-synaptic environment, where collective properties such as cooperativity, competition and selectivity can be investigated.

Keywords: Synaptic plasticity; Hebbian learning; STDP; Calcium; NMDA

1. Introduction

The intracellular ionic calcium concentration ([Ca^{2+}]) is known to act as a mediator of various cascades of metabolic activities in many cellular systems. In particular, it plays an important role in activity-driven synaptic changes. For example, different magnitudes and patterns of postsynaptic [Ca^{2+}] can selectively induce long-term potentiation (LTP) or long-term depression (LTD) [2,8,11,19]. The functional dependence of plasticity on [Ca^{2+}] has been described as U-shaped [6,7]: low levels of [Ca^{2+}]

* Corresponding author. Department of Physics, Brown University, 184 Hope Street, Providence, RI 02912, USA.
E-mail address: yeung@cns.brown.edu (L.C. Yeung).

0925-2312/03/$ - see front matter © 2003 Elsevier Science B.V. All rights reserved.
doi:10.1016/S0925-2312(02)00866-4

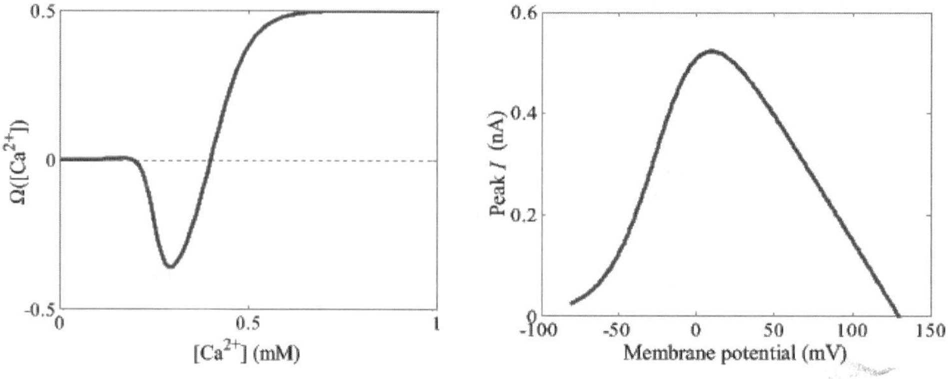

Fig. 1. Left: the shape of the learning rule. Right: the calcium current as a function of the postsynaptic potential.

induce no synaptic changes, while modest levels lead to LTD and higher ones, to LTP. The rise of $[Ca^{2+}]$ concentration depends on back-propagating action potentials (BPAPs) [9,13] in the postsynaptic cell; most interestingly, when these action potentials are paired with sub-threshold excitatory postsynaptic potentials (EPSPs), the dendritic $[Ca^{2+}]$ transient is substantially higher than the one produced by EPSPs or APs alone [12]. It is intuitive that this property should be related to the spike time-dependent type of plasticity (STDP) [3,14].

Experimental evidence, therefore, suggests a model of bi-directional synaptic plasticity that depends directly on the dynamics of $[Ca^{2+}]$. This more biophysical approach rids the model of ambiguities as to which aspects of the cellular activity (rate, depolarization or spike-timing) one should be consider in a learning rule. The model we have proposed, denoted *calcium control hypothesis* [15], describes this dependence as

$$\frac{dW_i}{dt} = \eta(Ca_i)(\Omega(Ca_i) - \lambda W_i), \tag{1}$$

where W_i is the weight of the synapse of index i, $Ca \equiv [Ca^{2+}]$, η is a calcium-dependent learning rate, Ω is a U-shaped difference of sigmoids (Fig. 1, left) and λ is the decay constant. Throughout the simulations below, $\lambda=0$, and stability is attained by the use on hard boundary conditions of W_i. These hard boundary conditions may not be necessary if metaplasticity (see Discussion) is included. We assume that the primary source of Ca influx into the spine is mediated by NMDA receptors, as described by the simple linear differential equation below

$$\frac{d(Ca_i)}{dt} = I_i(V,t) - \frac{Ca_i}{\tau_{Ca}}. \tag{2}$$

The temporal dependence of the NMDA current I describes the kinetics of the NMDA channels: it peaks at the arrival of a pre-spike and decays as a sum of two exponentials [5]. The voltage dependence is the standard function for the Mg^{2+}-block dynamics

(Fig. 1, right) [10]. Finally, V obeys the canonical non-leaking integrate-and-fire model. When a post-spike is elicited, a doubly exponential form of BPAP is added to V.

Eq. (1) has correctly reproduced the traditional experimental plasticity-inducing protocols in a one-dimensional input space [15]. Here, we extend this to a multidimensional environment. This allows us to examine receptive field formation and its significant biological implications.

2. Selection of correlated inputs

Selectivity is a general feature of many cortical neurons, and underlies the formation of receptive fields and topographic mappings. Therefore, a biologically plausible learning rule must be able to account for the formation of selective receptive fields. A simulated neuron is called *selective* to a specific input if it responds strongly to that input and weakly to others. We have chosen to induce selectivity in our simulated neuron by choosing different spatiotemporal structures for different groups of input neurons.

We use a neuron with 100 excitatory synapses, half of which receive Poisson spike-trains with correlated instantaneous rates with fixed mean $\bar{r} = 10$ Hz (group A), and the remaining receive uncorrelated input with the same $\bar{r}$ (group B). The correlation function had the same magnitude across synapses of group A, but decays exponentially in time [18]:

$$\langle r_i(t) r_j(t') \rangle = \bar{r}^2 + \bar{r}^2(\sigma^2 \delta_{ij} + (1 - \delta_{ij})c^2) \mathrm{e}^{-|t-t'|/\tau_c} \quad \text{if } i,j \in A, \tag{3}$$

where c is the correlation coefficient and δ_{ij} is the Kronecker delta. In addition, this neuron receives 20 inhibitory uncorrelated Poisson inputs with the same $\bar{r}$.

After 100 s of simulated time, we observe that, for 10 Hz inputs, the synapses of group A are potentiated, while group B is depressed (Fig. 2, center). This indicates that spikes from the correlated group cooperatively elevate the calcium level above

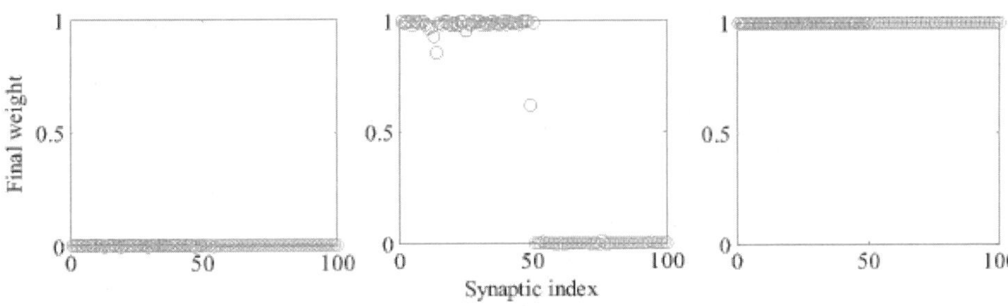

Fig. 2. Input selectivity at various stimulation frequencies: left, 8 Hz; middle, 10 Hz and right, 12 Hz. Group A (the righter half of each histogram) is the group with input correlation; group B (the lefter half of each histogram) is the group with no correlation.

the potentiating threshold, through spatial and temporal integration of local depolarizations. Furthermore, such depolarization increases the probability of causally eliciting a postsynaptic spike, which will associate with the presynaptic spike through the BPAP in a Hebbian manner. It was observed that, when segregation occurs, the postsynaptic spiking rate is stable, and has a high degree of variability (coefficient of variation ≈ 1, not shown).

However, consistent with the rate-based and the rate-dependent STDP protocols [17], for low and high enough $\bar{r}$, all the synapses are depressed and potentiated, respectively.

3. Discussion

Both simulations [15] and biophysical analysis [16] lead us to a learning model of the form of Eq. (1). Our previous work, however, assumed a simplified, one-dimensional input space. In this paper, we show that this learning rule can give rise to selective receptive fields in a more realistic multidimensional input space. However, this selectivity is highly sensitive to the model parameters, and for low input rates all synapses are driven to their lower saturation limit, whereas for high input rates they are driven to their upper limit. Metaplasticity, the activity-dependent modification of the functional form of synaptic plasticity, has successfully added robustness in other systems, as described by the BCM theory [1,4]. We are currently developing the appropriate form of metaplasticity for the system we described.

References

[1] W.C. Abraham, M.F. Bear, Trends Neurosci. 19 (1996) 126–130.
[2] M.F. Bear, L.N Cooper, F.F. Ebner, Science 237 (1987) 42–48.
[3] G. Bi, M. Poo, J. Neurosci. 18 (24) (1998) 10464–10472.
[4] E.L. Bienenstock, L.N. Cooper, P.W. Munro, J. Neurosci. 2 (1982) 32–48.
[5] G. Carmignoto, S. Vicini, Science 258 (1992) 1007–1011.
[6] K. Cho, J.P. Aggleton, M.W. Brown, Z.I. Bashir, J. Phys. 532.2 (2001) 459–466.
[7] R.J. Cormier, A.C. Greenwood, J.A. Connor, J. Neurophysiol. 85 (2001) 399–406.
[8] J.A. Cummings, R.M. Mulkey, R.A. Nicoll, R.C. Malenka, Neuron 16 (1996) 825–833.
[9] Y. Isomura, N. Kato, J. Neurophysiol. 82 (1999) 1993–1999.
[10] C.E. Jahr, C.F. Stevens, J. Neurosci. 10 (1990) 3178–3182.
[11] J. Lisman, Proc. Natl. Acad. Sci. USA 86 (1989) 9574–9578.
[12] J.C. Magee, D. Johnston, Science 275 (1997) 209–213.
[13] H. Markram, P.J. Helm, B. Sakmann, J. Phys. 485 (1) (1995) 1–20.
[14] H. Markram, J. Lübke, M. Frotscher, B. Sakmann, Science 275 (1997) 213–215.
[15] H.Z. Shouval, M.F. Bear, L.N. Cooper, PNAS 99 (16) (2002) 10831–10836.
[16] H.Z. Shouval, G.C. Castellani, B.S. Blais, L.C. Yeung, L.N. Cooper, Biol. Cybern. 87 (2002) 383–391.
[17] J. Sjöström, G.G. Turrigiano, S.B. Nelson, Neuron 32 (2001) 1149–1164.
[18] S. Song, K.D. Miller, L.F. Abbott, Nature Neurosci. 3 (2000) 919–926.
[19] S. Yang, Y. Tang, R.S. Zucker, J. Neurophysiol. 81 (1999) 781–787.

Available at
www.ComputerScienceWeb.com
POWERED BY SCIENCE @ DIRECT®

ELSEVIER

Neurocomputing 52–54 (2003) 441–445

NEUROCOMPUTING

www.elsevier.com/locate/neucom

Adaptation of the transfer function of the Hodgkin–Huxley (HH) neuronal model☆

Yuguo Yu[a,*], Tai Sing Lee[b]

[a] *Center for the Neural Basis of Cognition, Carnegie Mellon University, Pittsburgh, PA 15213, USA*
[b] *Computer Science Department, Carnegie Mellon University, Pittsburgh, PA 15213, USA*

Abstract

The transfer functions of sensory neurons are known to adapt to the statistics of the input signals. It is however unclear whether such adaptation arises from the nonlinear dynamics of the neuron or emerges from the collective interaction among neurons embedded in a network. We investigated the Hodgkin–Huxley (HH) neuronal model's response to Gaussian white noise signals of different variances and found that the recovered kernel adapt its preferred temporal frequency and its energy gains according to noise variance. This adaptation is likely a consequence of the cooperative interaction between the noises and the bifurcation dynamics of the neurons.

Keywords: Adaptation; Receptive field; Wiener kernel

1. Introduction

Recent neurophysiological research suggests that the stimulus-response functions of visual and auditory neurons adapt to the statistics of stimuli [1–6]. The adaptation of retinal neurons, for example, allow the visual system to operate effectively over an enormous dynamic range of lighting conditions. In blowfly, it was found that the input–output relationship of the H1 neuron adapt to the variance of the input noises [3,4]. Such adaptation has been suggested to play a role in maximizing information transmission by matching the sensitivity of the neurons to the spatial and temporal range of the stimuli and noises [7]. However, it is not clear how this adaptation comes

☆ This research was supported by NSF CAREER 9984706, NIH Vision Research core grant EY08098, and a NIH grant on the Pittsburgh Supercomputing Center.
* Corresponding author.
E-mail address: yug@cnbc.cmu.edu (Y. Yu).

0925-2312/03/$ - see front matter
doi:10.1016/S0925-2312(02)00867-6

about. Does it emerge from the interaction of a network of neurons, or does it emerge from the intrinsic dynamical mechanisms inside a neuron? Neurons communicate and compute with spikes, and the generation of spikes involves nonlinear mechanisms. Could the nonlinear dynamics of the neuron itself naturally produce a certain degree of adaptation and plasticity?

2. Methods

To investigate these issues, we apply an advanced system identification technique based on Laguerre expansion [8] to examine the transfer function of the Hodgkin–Huxley (HH) model [9] under the stimulation of Gaussian white noise (GWN) stimuli. The HH model provides an empirical description of real neuron that captures both of its spiking behavior and refractory properties.

Based on the input GWN, i.e., $x(t)$ and the output spike trains of the HH neuron, i.e., $y(t)$, we can characterize the cell's transfer function $h_\tau, h_{\tau_1,\tau_2}$ (the first and the second order kernel respectively) with memory length L as follows:

$$y(t) = h_0 + \sum_{\tau=1}^{L} h_\tau x(t-\tau) + \sum_{\tau_1=1}^{L} \sum_{\tau_2=1}^{L} h_{\tau_1,\tau_2} x(t-\tau_1)x(t-\tau_2), \tag{1}$$

where h_0 is the mean firing rate, h_τ is the first order kernel and h_{τ_1,τ_2} is the second order kernel. We restricted τ to be positive so we only consider causal filters. We used GWN signals with a fixed, short correlation time (cut frequency = 500 Hz) and zero mean, so the only free statistical parameter is the standard deviation σ.

3. Results

We varied σ systematically from 1 to 20 $\mu A/cm^2$, and recorded the spike trains generated by the HH model. Fig. 1(a) and (b) shows an example of the input signal (with $\sigma = 5$) and the output of the HH neuron in response to this input signal. Using 200 s of these input and output data, we recovered the first and second order kernels of the HH neuron using the Laguerre expansion technique, shown in Fig. 1(c) and (d). To verify the appropriateness of the method, we applied the method to recover a mathematically defined static kernel (e.g., the first kernel $K1 = \sin(\pi t/10)\exp(-t/10)$, the second kernel $K2 = K1 \otimes K1$) with noises of different σ. We found the static kernels recovered are invariant over a large range of σ. But for a HH neuron, the kernels recovered using different noise σ are found to change. Specifically, when we increase σ, the kernel is found to contract in time, i.e. the preferred temporal frequency of the filter increases. Fig. 2(a) shows the kernels recovered for three different σ's. The amplitude of the kernel first increases and then decreases with an increase in σ, reaching a global maximum at an intermediate σ. This phenomenon is related to the reported variance or contrast adaptation in electrophysiological experiments [1,4].

To evaluate this adaptation phenomenon systematically, we compute the power spectral density (PSD) of the first order kernel for noise with different σ as shown in

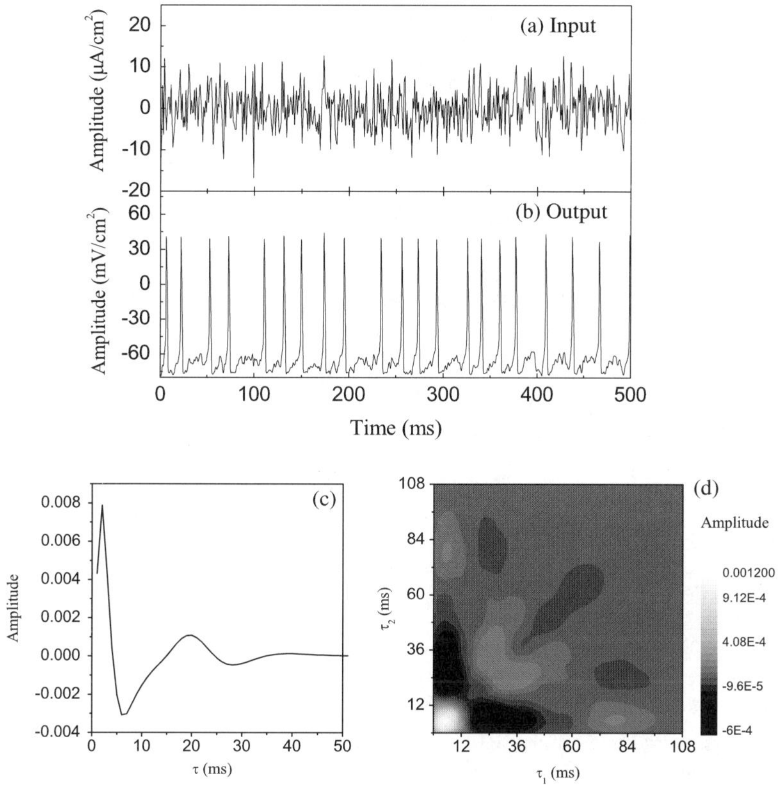

Fig. 1. (a) Input GWN with $\sigma = 5$. (b) The spike trains of the HH neuron in response to this stimulus. (c) The first order kernel of HH neuron recovered. (d) The second order kernel of the HH neuron.

Fig. 2(b). The preferred frequency of the kernels was computed as the peak frequency in the PSD. Fig. 2(c) shows that the preferred frequency of the kernel increases gradually with an increase in σ, changing from 47 to 66 Hz as σ changes from 1 to 20. The kernel frequency saturated with further increase in σ. These findings suggest that the frequency tuning of a neuron is adaptive to the variance of the noise input.

Fig. 2(b), also revealed that the peak of the PSD for noise with $\sigma = 3$ is much higher than that for $\sigma = 2$, and 10, suggesting that more spectral energy is allowed to pass through for input noise of $\sigma = 3$. That is, for some optimal stimulus statistics, the spike activities of the neuron become more coherent and resonant with the driven stimulus. The system absorbs the energy of the environment to maximize its response to stimulus, resulting in a largest gain in the transfer function for $\sigma = 3$ (see Fig. 2(d)). The total energy of the transfer function is given by,

$$E = \sum_{f=0}^{L-1} (\mathrm{PSD}(f))/L, \tag{2}$$

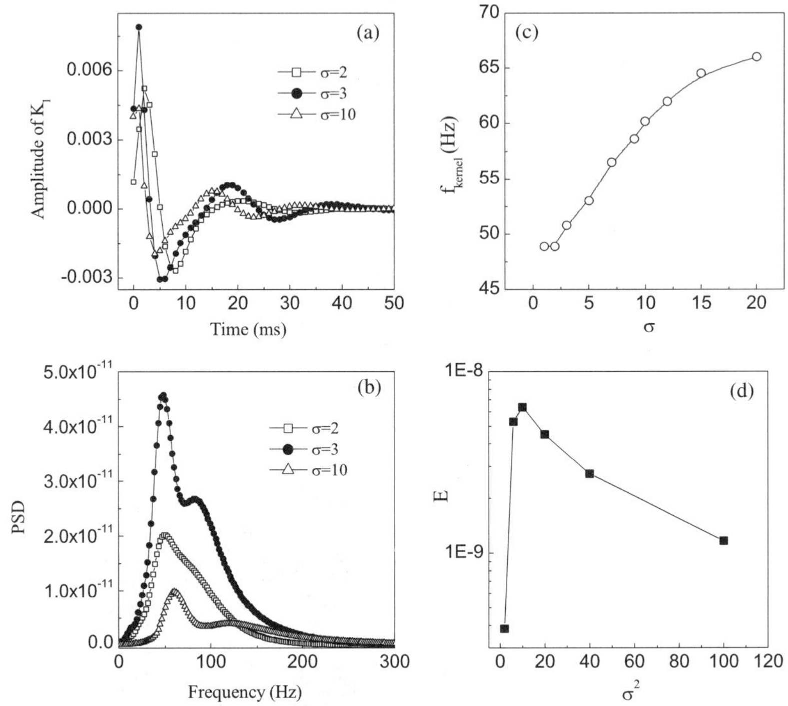

Fig. 2. (a) The first order kernels of the HH neuron for noises with $\mu = 0$; $\sigma = 2, 3, 10$, respectively. (b) The power spectra densities of the first order kernels for the three sets of noises. (c) The preferred frequency of the first order kernel as a function of σ. (d) The energy E of the first order kernel as a function of σ^2.

where f is the index of frequency and L is the length of the PSD. Fig. 2(d) shows the energy E of the first order kernel is maximal at $\sigma^2 = 9$. This phenomenon is called a coherence resonance (CR) [10].

When the noise level is low, the system is inert. Spikes will occasionally occur because the noise disturbance can drive the system through the spiking threshold. When the noise level is high, the neuron is mainly in the spiking state. The cooperation of the noise and the excitable dynamics of the system drives the system to fire spikes quasi-regularly. At too high a noise level, the quasi-regular spiking activity of the neuron is disrupted by the large stochastic nature of the noise, resulting in a decrease in the coherence of the system. At a tradeoff point, with an intermediate stimulus statistics, the neuronal dynamics is driven optimally by absorbing the energy of the environment maximally, with a maximum sensitivity to signals. In coherent resonance, the noise levels effectively put the system into different states, changing the neurons' frequency tuning as well as the gain of its transfer function.

4. Discussion

These observations suggest that the transfer function of HH neuron is not static but dynamic. The time scale of the transfer function varies with the statistics of the stimuli.

The variance of the stimulus puts the neuron into different coherent states, reflecting different degrees of synergistic interaction between the stimulus and the nonlinear dynamics of the system. The neuron's transfer function varies with the coherent state: the preferred frequency increases with σ and the gain of the transfer function reaches a maximum at $\sigma = 3$ for the standard HH model. It is expected a different set of model parameters will give a different range of frequency tuning variation as well as a different σ for the maximum cooperation with the nonlinear dynamics of the neuron.

The observed adaptation in the input–output relationship in various neurophysiological experiments [1–6] is therefore likely due, in part, to the adaptation of the transfer function of the single neuron to the different noise levels. The first and second order kernels recovered from the HH neuron, however, are significantly dilated in time scale than the adaptation found in monkey cortical neurons [11] or blowflies [3,4]. The difference in neuronal parameters may be a factor. Furthermore, given the neuron measured is part of the network with many layers of neurons lying between it and the photoreceptors on the retina, network delay and interaction would likely produce a low-pass effect, dilating the effective kernel of the measured neuron.

Our simulation suggests that synergistic cooperation between the statistics of the noises with the nonlinear dynamics of the neuron is an important and maybe crucial contributing factor to the adaptation of the neuronal transfer function to the stimulus statistics.

References

[1] H.M. Sakai, K.-I. Naka, J. General Physiol. 105 (1995) 815–835.

[2] S.M. Smirnakis, M.J. Berry, D.K. Warland, W. Bialek, M. Meister, Nature (London) 386 (1997) 69–73.

[3] N. Brenner, W. Bialek, R. de Ruyter van Steveninck, Neuron 26 (2000) 695–702.

[4] A.L. Fairhall, G.D. Lewen, W. Bialek, R. de Ruyter van Steveninck, Nature (London) 412 (2001) 787–792.

[5] K.J. Kim, F. Rieke, J. Neurosci. 21 (2001) 287–299.

[6] R.L. Jenison, J.W.H. Schnupp, R.A. Reale, J.F. Brugge, J. Neurosci. 21 (2001) 4408–4415.

[7] M.J. Wainwright, Vision Res. 39 (1999) 3960–3974.

[8] V.Z. Marmarelis, Identification of nonlinear biological systems using Laguerre expansions of kernels, Ann. Biomed. Eng. 21 (1993) 573–589.

[9] A.L. Hodgkin, A.F. Huxley, J. Physiol. 117 (1952) 500–540.

[10] A.S. Pikovsky, J. Kurths, Coherence resonance without external periodic force, Phys. Rev. Lett. 78 (1997) 775–778.

[11] R.D. Romero, Y. Yu, P. Afshar, T.S. Lee, Dynamic adaptation of temporal kernels of Macaque V1 neurons, Neurocomputing, in press.

Yuguo Yu is a Ph.D. in Physics from Nanjing University (2001), PRC and is currently a postdoctoral fellow in the Center for the Neural Basis of Cognition, Carnegie Mellon University.

Available at
www.ComputerScienceWeb.com
POWERED BY SCIENCE @ DIRECT®

ELSEVIER

Neurocomputing 52–54 (2003) 447–453

NEUROCOMPUTING

www.elsevier.com/locate/neucom

Learning control units for invariant recognition

Junmei Zhu*, Christoph von der Malsburg[1]

Computer Science Department, University of Southern California, Los Angeles, CA 90089-2520, USA

Abstract

A control unit is a group of synapses that are consistent with each other in terms of transformation parameters. We showed previously that these high-order interactions led to very fast dynamic link matching (DLM) for self-organization of mappings between two patterns, as needed in invariant recognition and many other visual tasks. We present here a model to learn from examples control units and the connection between them, by generalized Hebbian plasticity. Examples for learning are consistent mappings formed, for instance, by conventional DLM. Simulation results for 1D patterns showed that the learned results resemble those designed by hand, and thus were very encouraging.

Keywords: Dynamic link matching; Invariant recognition; Control unit; Hebbian learning

1. Introduction

We are interested in object recognition in the presence of variations such as transformations (translation, scale, orientation), illumination, and deformations. Although this task is solved fast and effortlessly by humans, it is computationally very difficult because of the enormous pictorial changes due to the variations. The most important process underlying invariant recognition, as well as other visual tasks such as motion estimation and stereo correspondence, we believe, is to create a mapping between two patterns, such that points projected from the same physical spot are connected.

This is the well-known correspondence problem in computer vision, where a lot of algorithms have been developed. That most of the algorithms do not work well

* Corresponding author.
E-mail addresses: junmeizh@organic.usc.edu (J. Zhu), malsburg@organic.usc.edu (C. von der Malsburg).
[1] Also at Institute for Neuroinfomatik, Ruhr-University Bochum, Germany

0925-2312/03/$ - see front matter © 2002 Elsevier Science B.V. All rights reserved.
doi:10.1016/S0925-2312(02)00869-X

outside of well-controlled laboratory settings, unfortunately, suggests the drawbacks of this algorithmic schema. In general, the two aspects of solving a problem, algorithm and implementation, are handled by humans and computers, respectively. Human creativity produces algorithms as well as the corresponding codes to be fed into the computer, whereas computers follow the codes blindly. In consequence, computers on their own cannot respond to situations that are not programmed explicitly. Human intervention is required to get a new code whenever a new problem comes up. To reduce this bottleneck, a new computing paradigm should be developed to shift the creativity aspect to the computer. The inspiration of this new paradigm comes from biological systems that provide ingenious solutions to many of the problems we are interested in, such as invariant recognition, and is called organic computing.

Our work is an attempt in the direction of organic computing, the aim of which is, by using only basic organization principles, to build a system that can learn and develop the equivalent of human created algorithms within itself, to perform whatever task it may encounter.

Dynamic link matching (DLM) is such an example system that creates efficient solutions based on only basic self-organizing rules. The basic principle it uses is rapid synaptic plasticity, in which a mapping between two patterns is controlled by signal correlations, which are shaped by neighboring connections and by the mapping itself. DLM is intrinsically invariant to translation and is robust against illumination and deformation, and was proven to be very successful. Although not especially designed for a particular task, its performance in face recognition outperformed all other human created algorithms [1].

However, a difficulty for DLM as a biological model, although based on biologically plausible principles, is its relative slowness in comparison to the split-second recognition time in adults. We previously presented an extension to DLM that has significantly shorter convergence time [3,4]. The extension is based on *synaptic control units*, each of which stands for a group of synapses that are consistent with each other in terms of transformation parameters. Control units and their connections represent stored knowledge of transformations that leads to the speedup of DLM convergence: significant local connectivity patterns can be retrieved by control units and be propagated rapidly through the interconnections between control units.

In the previous studies, control units and the connections between them were all designed manually, which was possible because the variations we considered then were scale, translation, and in-plane rotation. This kind of analytical solution, nonetheless, not only is impossible when deforming patterns become complicated, but also belongs to the algorithmic schema that should be replaced. Therefore, we present here a model to learn control units and the connections between them from examples.

This paper is organized as follows: Section 2 provides the necessary background in DLM and control units, making clear what are expected to be learned. Section 3 gives the learning rules, and an overview of the learning system. Section 4 provides some simulation results. Finally, Section 5 is the conclusion.

2. Background: control units in dynamic link matching

In this section, we provide some background on DLM and the control units, serving as a basis for what we are going to learn.

2.1. DLM

DLM self-organizes a mapping between two patterns, an image and a model [2]. The mapping is represented by a connection function $x(r,t)$, where r is the position in the image domain, and t in the model domain. This function is called a matrix if we consider discrete r and t, which are 2D vectors if the patterns are 2D. For each $x(r,t)$ is called a dynamic link, or a synapse, and they are the system variables that can have rapid changes. Synaptic growth is controlled by cooperation and competition. For each $x(r,t)$ the cooperation term is obtained by gathering support from all synapses that are its spatial neighbors in the image domain and the model domain.

2.2. Control units

Control units were introduced to speedup DLM convergence and to deal with transformations, especially scale and in-plane rotation, more efficiently [3,4]. Control units stand for groups of synapses, which represent stored significant patterns. They control synapses by regulating how synaptic cooperation is computed. The speedup of DLM is achieved by the specific long-range synaptic cooperative interactions carried out by the interaction between control units, where the specificity comes from the fact that any single synapse is ambiguous as to the transformation parameters, but a group of synapses, as represented by control units, are unambiguous.

There are two aspects of a control unit we need to learn: a control function, specifying what synapses are under its control, and its connection to other control units. Detailed description of control units can be found in Ref. [4]. In short, a control function has the same domain as the connection matrix, and is represented by a real function $K(r,t)$. It is specific to transformation parameters, and is localized in variables (r,t), with the center of its compact support called the control center. As an example, for 1D patterns, a control function may have the form of a 2D Gaussian, whose center is controlled by the translation parameter, and the axes by the scale parameter. The connections between control units are also intuitive. In theory, two control units have strong connections between them if their control centers are close, and they are neighbors in the transformation space, i.e., similar translation, scale, and orientation.

3. Learning control unit connections

This section presents the learning rules and the learning system. Both control units and their connections are learned by generalized Hebbian plasticity.

3.1. Learning examples

The idea of this learning system is as follows. It is reasonable for a baby to stare at an object for several minutes before it can recognize it. But once a connectivity pattern is formed, it should be stored into the control units to speed up the convergence of future map creation. Therefore, examples used for learning control units are consistent mappings formed, for instance, by slow conventional DLM. Each mapping is represented by a synaptic matrix $x(r,t)$. An advantage of learning mappings, instead of learning patterns as in most learning systems, is that the weight of a connection in a mapping has a physical meaning, while the intensity level of a pixel in a pattern does not.

3.2. System structure

In reality, the distribution of sizes or centers of control units may be random as a result of competitive learning, but here we consider a simplified and regular structure.

The domain of mappings is divided into small regions called control regions, indexed by s, with membership function $A_s(r,t)$. Different regions can overlap with one another. For each s there is a set of control units, with control functions $K_{si}(r,t)$. The response of unit $K_{si}(r,t)$ to input $x(r,t)$, y_{si}, is

$$y_{si} = \sum_{r,t} x(r,t)A_s(r,t)K_{si}(r,t) + \sum_{s',i'} c(s,i,s',i')y_{s'i'}, \quad (1)$$

where $c(s,i,s',i')$ is the connection strength between control units K_{si} and $K_{s'i'}$.

3.3. Learning rules

3.3.1. Control functions

For each input mapping $x(r,t)$, the learning rule for control functions is

$$\Delta K_{si}(r,t) = \alpha A_s(r,t)x(r,t)y^*_{si}, \quad (2)$$

where

$$y^*_{si} = \begin{cases} y_{si} & \text{if } \text{argmax}_j(y_{sj}) = i \\ 0 & \text{otherwise} \end{cases}$$

and α is the learning rate.

In order to avoid uncontrolled growth, this update is followed by a normalization:

$$\sum_{r,t} K_{si}(r,t) = \text{const.} \quad (3)$$

3.3.2. Connection between control units

The connection between control units K_{si} and $K_{s'i'}$, $c(s,i,s',i')$, is also learned by Hebbian rule:

$$\Delta c(s,i,s',i') = \beta y^*_{si} y^*_{s'i'}, \quad (4)$$

where β is the learning rate. This is also followed by normalization:

$$\sum_{s',i'} c(s,i,s',i') + \sum_{s,i} c(s,i,s',i') = \text{const.} \tag{5}$$

3.4. The learning system

The learning system is summarized as follows:

- Initialization
  - control function: random values within its control region, 0 otherwise,
  - control unit connections: 0.
- For each training example
  - compute control unit responses (Eq. (1)),
  - update control functions (Eqs. (2) and (3)),
  - update connections (Eqs. (4) and (5)).

4. Experiments

The development of control units and their connections is simulated in an example in which the goal is to learn translation invariance for 1D patterns.

4.1. Examples for learning

Each learning example is a mapping between a pair of relatively shifted image and model. Image and model are identical patterns up to a scale difference. Accordingly the synaptic matrix has value 1 along a straight line, whose slope depends on the scale difference, and 0 elsewhere. Some examples are shown in Fig. 1. Mappings in the upper row are from identical image and model, while those in the lower row have a size difference of 0.5 octave. The position of the line corresponds to the translation parameter, which is generated randomly and uniformly over all possible translations. Noise in the mapping is simulated by randomly setting elements next to the ideal line value 1.

4.2. Results

In this simulation, the synaptic matrix is of size 20×20, and is divided into 4×4 small regions, each of which has size 5×5. In each small region there are 9 control units, to cover all possible translations. In the following simulation, different scales are learned separately, which means in each run all examples have the same and fixed scale. Learning examples are presented sequentially, and the system learns according to the procedure in Section 3.4. We use normalization constant $= 1$ for control functions, and 0.2 for control unit connections. $\alpha = 0.1$, $\beta = 0.05$.

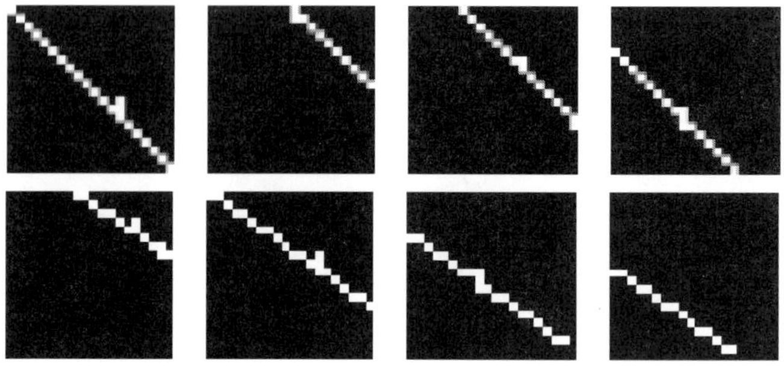

Fig. 1. Examples of synaptic matrices used for learning translation invariance: mappings with different translation parameter. Row 1: identical image and model; Row 2: image and model has scale difference of 0.5 octave.

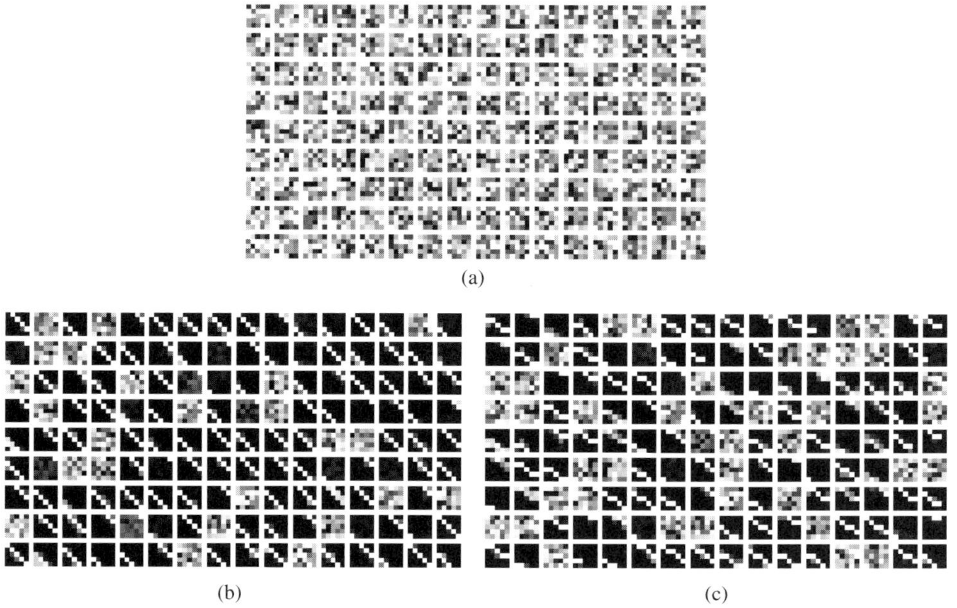

Fig. 2. Control functions. (a) initial values; (b) and (c) learned control functions after 1000 iterations; (b) identical image and model; (c) image and model has scale difference of 0.5 octave.

The initial random control functions are shown in Fig. 2(a). The learned control functions after 1000 learning examples are shown in Figs. 2(b) and (c). What is plotted is only the control functions within their respective control regions. Fig. 2(b) shows the functions learned with examples with an image and model of the same scale, while in Fig. 2(c) the scale difference is 0.5 octave. Each small block represents the control function of one control unit. In the horizontal direction runs s, the index of control center position in the synaptic matrix. In the vertical direction is i, the index of control units sharing the same control center. We can see that the learned functions have reasonable shapes, resembling those designed by hand previously [4]. Also it seems that control units which share the same control center learned different translation parameters, which often span the whole range of possible translations. But we also noticed that even after 1000 training examples, it is still possible to have a few uncommitted slots, i.e., control units with unstructured control functions.

The learned interconnections between units are evaluated in the following way. For any control unit indexed by si, compute the summation of the control functions of all control units, weighted by their connection strength to unit si: $\sum_{s',i'} c(s,i,s',i')A_{s'}(r,t)K_{s'i'}(r,t)$. In most instances, the reconstructed result resembles one of the training examples, i.e., a straight line with appropriate slope, meaning that all control units with the same translation are connected, while those with different translation are not.

We also did simulations on learning simultaneously shift and scale invariance for 1D patterns. The results are very encouraging, but still more work is needed.

5. Conclusion

This work demonstrates the feasibility to learn control unit functions and the connections between control units. Both can be learned by generalized Hebbian plasticity. The same learning rules apply to 2D patterns and to more complicated invariances.

Future work includes learning simultaneously multiple invariances, and eventually using these learned results to replace the hand crafted ones in the previous system.

Acknowledgements

The authors would like to thank the developers of the FLAVOR software environment, which served as the platform for this work. This work was supported by ARO-WASSP, contract DAAD19-00-1-0356.

References

[1] P.J. Phillips, H. Moon, S.A. Rizvi, P.J. Rauss, The FERET evaluation methodology for face-recognition algorithms, IEEE Trans. Pattern Anal. Mach. Intell. 22 (2000) 1090–1104.

[2] L. Wiskott, C. von der Malsburg, Face recognition by dynamic link matching, in: J. Sirosh, R. Miikkulainen, Y. Choe (Eds.), Lateral Interactions in the Cortex: Structure and Function, Electronic book, 1996.

[3] J. Zhu, C. von der Malsburg, Fast Dynamic Link Matching, ICCNS*01, Boston, 2001.

[4] J. Zhu, C. von der Malsburg, Synapto–synaptic interactions speed up dynamic link matching, Neurocomputing 44 (2002) 721–728.

Available at
www.ComputerScienceWeb.com
POWERED BY SCIENCE DIRECT®

ELSEVIER

Neurocomputing 52–54 (2003) 455–459

NEUROCOMPUTING

www.elsevier.com/locate/neucom

A computational model for the biological underpinnings of infant vision and face recognition

Laura A. Mueller[a,1], Liya Wang[b], Amir H. Assadi[a,*,2]

[a]*Department of Mathematics, University of Wisconsin, 480 Lincoln Dr., Madison, WI 53706, USA*
[b]*Department of Biomedical Engineering, University of Wisconsin, Madison, WI, 53706, USA*

Abstract

A scientists' challenge is to understand how an infant's brain learns to efficiently process visual information. An infant's range of focus is short for several months, and faces enter into this range more often than any other scene. Within a few months of birth, the brain can differentiate faces from other objects and an infant can recognize a known face from a stranger's. The repetition of face stimulus in conjunction with the relatively short time it takes for recognition to occur suggests there may be more regularity among facial stimuli, or perhaps a more sensitive information processing mechanism biased towards facial stimuli.

Keywords: Face recognition; Infant vision; Principal component analysis; Backpropagation neural network

Vision has been a popular subject of study, but most research has focused on the fully developed visual systems of adult humans. To further understand the brain's role in perceptual processes such as face recognition, one needs focus on the visual system's development from infancy to adulthood. Infants experience the world through all senses, some more developed than others. The visual learning of a mother's face or other caretakers is accomplished by feedback from a combination of touch, taste, smell, and sound. During early developmental stages, there are communication pathways between

* Corresponding author. Fax: +1-608-263-8891.
E-mail address: ahassadi@factstaff.wisc.edu (A.H. Assadi).
[1] Laura Mueller was partially supported through NSF-KDI-LIS and other grants to Amir Assadi, by the NSF-VIGRE to UW-Madison Mathematics Department, and by the Office of Provost and Vice Chancellor for Academic Affairs, UW-Madison for the Symmetry Project.
[2] Amir Assadi and Liya Wang are partially supported through an NSF-KDI-LIS grant.

0925-2312/03/$ - see front matter © 2002 Elsevier Science B.V. All rights reserved.
doi:10.1016/S0925-2312(02)00730-0

the visual and other sensory areas of the cortex, showing how the biological network is self-organizing. For theoretical study, a model with feedback, as if from other senses, provides comparable "behavioral" output.

In this project, we investigate the learning process of face recognition for an infant's brain, using simulation by an artificial neural network (ANN) as a model. The biological hypotheses for this model are based on plasticity and well-established behavioral research on the role of response to low frequencies in early stages [1]. Learning in the brain occurs through creation or elimination of synapse connections or changes in synapse connection strengths in a neural pathway. If a neuron repeatedly contributes to the firing of another neuron, the strength between those two neurons increases. The Hebbian learning rule reflects this biology in ANNs by strengthening the connection weights between connected neurons. After training the network on a set of "typical faces" and also on a few non-facial stimuli, the ANN with tested with a previously unseen face as well as new objects. Higher success rates in tests for recognizing faces compared to other objects by the ANN suggests regularity amongst low frequency components of faces is essential for the brain to learn more efficiently to recognize a face.

The model for infant vision consists of a multilayer perceptron, based on a Back-propagation Algorithm, used to classify images as a face or non-face (Fig. 3). Images are filtered prior to being fed to the neural network (NN). In accordance to behavioral research, early infant vision responds to low frequencies only. A low-pass filter is first used to remove high frequency components in the images (Figs. 1 and 2). Then Generalized Hebbian Learning is applied for principal components extraction, using the statistical technique principal component analysis (PCA). PCA consists of solving the correlation matrix of a matrix of images for eigenvalues and eigenvectors. The number of eigenvalues carrying at least 90% of the information indicates the number of eigenfaces to be used as inputs for the NN (Fig. 3). The back propagation algorithm involves the feedforward step, where each input is multiplied by a weight matrix and the hyperbolic tangent function is taken of the resulting matrix. Backpropagation involves finding the error between the expected output and actual output, altering the weight matrices, and feeding the network forward again. This cycle (termed an epoch) continues until the error is no greater than 10% (Fig. 4).

All the training and testing samples images represent 5% of the entire data set after passing the same low-pass filter. During the training process, the network performed with 100% accuracy for 30 training samples (Fig. 4), of which 16 were face images and 14 were non-face images. During testing, the network performed with 100% accuracy for the test samples, of which four were face images and four were non-face images.

Equipped with no previous knowledge of facial structure, the NN learned to distinguish faces from non-faces accurately in a similar manner that an infant's brain builds the neural connections after birth. Preprocessing images to remove high frequency eliminated unnecessary information from the images, helping the NN to efficiently learn the face versus non-face distinction. Successful results suggest this ANN serves as a biologically viable model to explain the learning process an infant brain undergoes for face recognition.

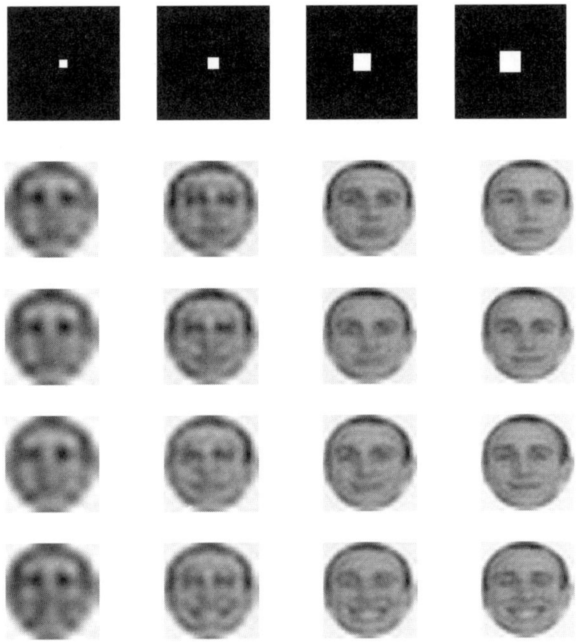

Fig. 1. Low-pass frequency filters displayed on the top row. Corresponding filtered faces displayed below the filters.

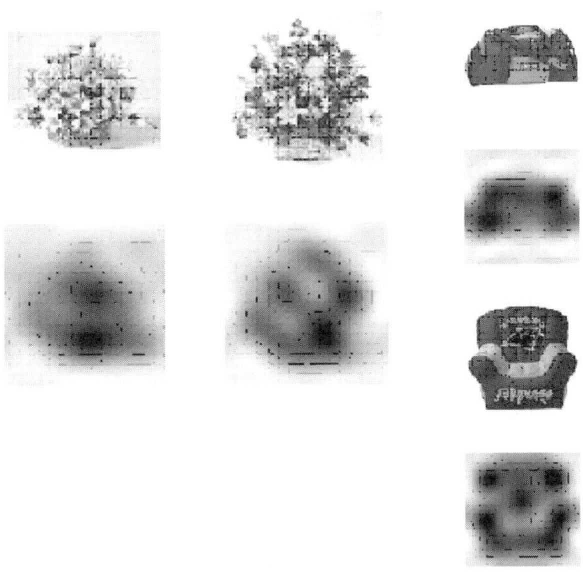

Fig. 2. Examples of filtered non-face images.

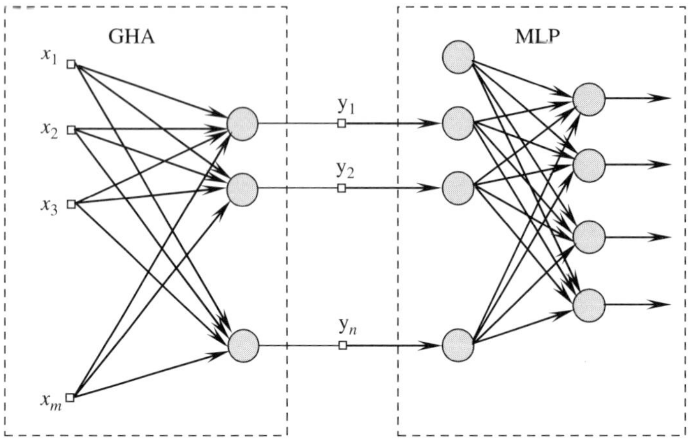

Fig. 3. Neural network design.

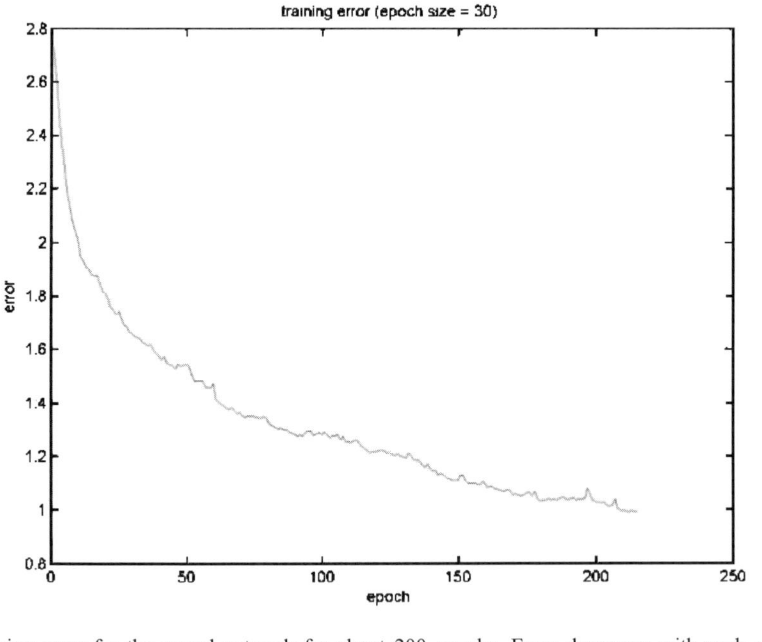

Fig. 4. Training error for the neural network for about 200 epochs. Error decreases with each epoch until all the training samples are correctly identified.

1. For further reading

The following references may also be of interest to the reader: [2–7].

Acknowledgements

The authors acknowledge helpful and inspiring discussion with Dr. Michael Struck of Department of Ophthalmology, UW-Madison.

References

[1] M. Banks, The role of low frequency in infant vision, PNAS (1981).
[2] H. Barlow, The exploitation of regularities in the environment by the brain, Brain Behav. Sci. May (2001).
[3] H. Barlow, S.P. Tripathy, Correspondence noise and signal pooling in the detection of coherent visual motion, J. Neurosci. 17 (20) (1997) 7954–7966.
[4] W. Crick, The Astonishing Hypothesis: The Scientific Search for the Soul, Touchstone, New York, 1994.
[5] Y. Dan, et al., Efficient coding of natural scenes in the lateral geniculate nucleus: experimental test of a computational theory, J. Neurosci. 16 (10) (1996) 3351–3362.
[6] S. Palmer, Vision Science: Photons to Phenomenology, MIT Press, Cambridge, MA, 1999.
[7] M. Smith, Neural Networks for Statistical Modeling, International Thomson Computer Press, Boston, MA, 1996.

Available at
www.ComputerScienceWeb.com
POWERED BY SCIENCE @ DIRECT®

ELSEVIER

Neurocomputing 52–54 (2003) 461–466

NEUROCOMPUTING

www.elsevier.com/locate/neucom

Temporal infomax leads to almost deterministic dynamical systems

Nihat Ay*, Thomas Wennekers

Max-Planck-Institute for Mathematics in the Sciences, Inselstr. 22–26, D-04103 Leipzig, Germany

Abstract

The well-known Kullback–Leibler divergence of a random field from its factorization quantifies spatial interdependences of the corresponding stochastic elements. We introduce a generalized measure called 'stochastic interaction' that captures also temporal interdependences. Maximization of stochastic interaction in the setting of Markov chains is shown analytically and by simulations to result in an almost deterministic global dynamics, but almost unpredictable single unit activity.

Keywords: Markov model; Stochastic interaction; Information maximization

1. Introduction

Information processing in the brain essentially depends on interrelations between neurons expressed by cooperation and competition in neural assemblies. The generation of spatio-temporal activation patterns, for instance, is at the base of statistical analyses as well as functional theories for neural synchronization and associative propagation of activity [1,6,9,11]. The hypothesis of strong interrelations is contained also in many conceptual approaches to an understanding of first principles for neural cooperation and learning. Information theory provides an appropriate framework for an analysis of such principles [4,7,8,10]. A well-known measure that quantifies relations of interacting units is the so-called *mutual information*: The Kullback–Leibler divergence

$$I(p) := D(p \,\|\, p_1 \otimes \cdots \otimes p_N) = \sum_{\nu=1}^{N} H(p_\nu) - H(p), \tag{1}$$

* Corresponding author.

E-mail addresses: nay@mis.mpg.de (N. Ay), wenneker@mis.mpg.de (T. Wennekers).

0925-2312/03/$ - see front matter
doi:10.1016/S0925-2312(02)00732-4

where $H(\cdot)$ denotes the Shannon entropy and p_v the vth marginal of p, measures the "distance" of p from the set of factorized random fields. It is a natural measure for "spatial" interdependence and a starting point of many approaches to neural complexity. In [2,3] it has been studied from the *information-geometric* point of view, where it is referred to as (*stochastic*) *interaction*. In order to capture also intrinsically temporal aspects of dynamic interaction, I in (1) has been extended by Ay (cf. [4]) to the dynamical setting of Markov chains. In the present paper we consider processes that optimize the temporal version of stochastic interaction. This leads to analytical results concerning the most fundamental feature appearing in strongly interacting stochastic systems, namely, the development of the system dynamics towards determinism. Thereby, a necessary first step is provided for an investigation of learning processes also in more detailed recurrent neural network models.

2. Theoretical framework and results

Consider the set $V = \{1,\dots,N\}$ of binary units with state sets $\Omega_v = \{0,1\}$, $v \in V$. For a subsystem $A \subset V$, $\Omega_A := \{0,1\}^A$ denotes the set of all configurations on A, and $\bar{P}(\Omega_A)$ is the set of probability distributions on Ω_A. Given two subsets A and B, where B is non-empty, $\bar{K}(\Omega_B \mid \Omega_A)$ is the set of all Markov transition kernels from Ω_A to Ω_B. If $A = B$, we use the abbreviation $\bar{K}(\Omega_A)$. For a probability distribution $p \in \bar{P}(\Omega_A)$ and a Markov kernel $K \in \bar{K}(\Omega_B \mid \Omega_A)$, the *conditional entropy* of (p, K) is defined as

$$H(p,K) = -\sum_{\omega \in \Omega_A, \omega' \in \Omega_B} p(\omega) K(\omega' \mid \omega) \ln K(\omega' \mid \omega). \tag{2}$$

We set $H(Y \mid X) := H(p,K)$ if X and Y are random variables with $\mathrm{Prob}\{X = \omega, Y = \omega'\} = p(\omega)K(\omega' \mid \omega)$ for all $\omega \in \Omega_A$ and $\omega' \in \Omega_B$. In the following we mainly consider the case $A = B = V$, where $H(p,K)$ measures the average uncertainty about the next state of the system given the present state. Assume p is strictly positive. Then H vanishes iff K is deterministic—that is, if each support set, $\mathrm{supp}\, K(\cdot \mid \omega)$, contains only a single possible target configuration, $|\mathrm{supp}\, K(\cdot \mid \omega)| = 1$, for all ω. H attains its maximal value $N \ln 2$ iff $K(\omega' \mid \omega) = 2^{-N}$ for all $\omega, \omega' \in \Omega_V$.

A Markov kernel $K \in \bar{K}(\Omega_V)$ is called *parallel* if there exist kernels $K^{(v)} \in \bar{K}(\Omega_v \mid \Omega_V)$, $v \in V$, such that

$$K(\omega' \mid \omega) = \prod_{v \in V} K^{(v)}(\omega'_v \mid \omega) \quad \text{for all } \omega, \omega' \in \Omega_V. \tag{3}$$

A kernel $K \in \bar{K}(\Omega_V)$ is called *split* if there exist $K^{(v)} \in \bar{K}(\Omega_v)$, $v \in V$, with

$$K(\omega' \mid \omega) = \prod_{v \in V} K^{(v)}(\omega'_v \mid \omega_v) \quad \text{for all } \omega, \omega' \in \Omega_V. \tag{4}$$

The split kernels are a proper subset of the parallel kernels and represent a dynamical version of the factorized probability distributions. Thus, in analogy to (1) we define the *stochastic interaction* of the units with respect to a distribution $p \in \bar{P}(\Omega_V)$,

p strictly positive, and a transition kernel $K \in \bar{K}(\Omega_V)$ as the p-divergence of K from being split: for this purpose, we define the marginal kernels $K_v \in K(\Omega_v)$, $v \in V$, of K by

$$K_v(\omega_v' \mid \omega_v) := \frac{\sum_{\substack{\sigma,\sigma' \in \Omega_V \\ \sigma_v = \omega_v, \sigma_v' = \omega_v'}} p(\sigma) K(\sigma' \mid \sigma)}{\sum_{\substack{\sigma \in \Omega_V \\ \sigma_v = \omega_v}} p(\sigma)}, \quad \omega_v, \omega_v' \in \Omega_v. \tag{5}$$

Then, the *stochastic interaction measure* I is defined as

$$I(p,K) := \sum_{v \in V} H(p_v, K_v) - H(p,K), \tag{6}$$

continuously extended to $\bar{P}(\Omega_V) \times \bar{K}(\Omega_V)$ (cf. [4]). Eq. (6) generalizes (1) to Markov transitions. $I(p,K)$ is large if the marginal transitions have high entropy, but that of the full transition is low. Supposed the current state $\omega \in \Omega_V$ is known, this corresponds with high predictability of the next global state, but, conversely, not much information is gained from knowledge about the states of single units, ω_v. The predictability and, hence, degree of determinism of systems that optimize I is further characterized by the following Theorem:

Theorem 1. *Consider a probability distribution* $p \in \bar{P}(\Omega_V)$ *and a transition kernel* $K \in \bar{K}(\Omega_V)$. *If* (p,K) *is a local maximizer of the interaction measure* I *then for all* $\omega \in \operatorname{supp} p$ *one has* $|\operatorname{supp} K(\cdot \mid \omega)| \leqslant 1 + N$.

Theorem 1 presents a restriction to binary units of a more general theorem proved in [5]. Note, that the expression $|\operatorname{supp} K(\cdot \mid \omega)|$ counts the number of transitions with non-vanishing probability from an arbitrary state $\omega \in \operatorname{supp} p$ to its target states. Since there are exponentially many possible target states, namely 2^N, the estimate in Theorem 1, which is only linear in N, provides a strong upper bound on the number of transitions in strongly interacting systems. This has a direct implication for the entropy in such systems:

Corollary 2. *In the situation of Theorem 1*: $H(p,K) \leqslant \ln(1+N)$.

3. Simulations

Consider a Markov chain $X_n = (X_{v,n})_{v \in V}$, $n = 0, 1, 2, \ldots$, given by an initial distribution $p \in \bar{P}(\Omega_V)$ and a kernel $K \in \bar{K}(\Omega_V)$. A probability distribution $p \in \bar{P}(\Omega_V)$ is called *stationary* with respect to $K \in \bar{K}(\Omega_V)$ iff $\sum_{\omega \in \Omega_V} p(\omega) K(\omega' \mid \omega) = p(\omega')$ for all $\omega' \in \Omega_V$. In the present section we consider only Markov chains that are induced by parallel kernels and corresponding stationary distributions. In that case (cf. [4]):

$$I(p,K) = \sum_{v \in V} (H(X_{v,n+1} \mid X_{v,n}) - H(X_{v,n+1} \mid X_n)). \tag{7}$$

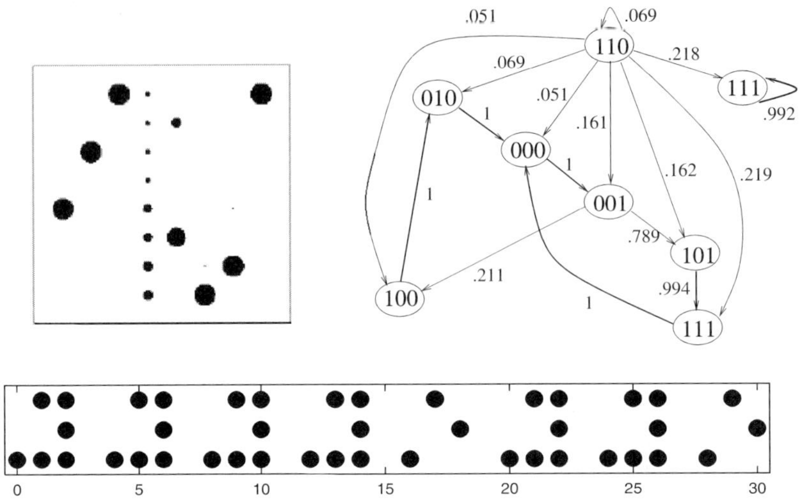

Fig. 1. Markov matrix (upper left), state transition graph (right), and sample activity (bottom) for a system of $N = 3$ strongly interacting units.

In (7) the term $H(X_{v,n+1}\,|\,X_{v,n}) - H(X_{v,n+1}\,|\,X_n)$ measures the reduction of uncertainty about unit v's next state by the additional knowledge of the current states of all other units. In the following we show representative simulations of small complex systems with strong interaction. The simulations implement the usual Markov dynamics on a set of N binary units together with a random search scheme to optimize the stochastic interaction: in every simulation step I is computed for a given parallel Markov kernel and induced stationary probability distribution, and starting from initial random values that kernel is iteratively perturbed such that I increases (cf. [5]).

Fig. 1 displays an optimized system for $N = 3$ that is typical also for larger systems. The approached entropies and interaction were $H(p,K) = 0.125$, $H(p_1,K_1) = 0.620$, $H(p_2,K_2) = 0.642$, $H(p_3,K_3) = 0.666$, $I(p,K) = 1.803$. The interaction, I, comes near to its theoretical upper bound of $N \ln 2 \approx 2.07$ because the marginal kernel entropies are all near their maximum of $\ln 2 \approx 0.69$, and the full kernel entropy is very small (cf. (6)). This indicates that given the current state ω, the next state can be quite reliably predicted, but in contrast, not much information is gained from knowledge about the state of any individual unit alone. The Markov matrix in Fig. 1 (circle area represents transition probability) and the derived state transition graph (node-labels indicate states, ω, edge-labels transition probabilities) exhibit the determinism even more clearly: most columns in the Markov matrices have only a single entry of probability 1 and, accordingly, most nodes in the transition graphs have just one outgoing edge.

Nonetheless, the transition graph reveals that the dynamics consists of an ergodic component comprising two cycles of deterministic transitions nested by a *branching state* (state 0 0 1). Also visible is a *transient state* (1 1 0), which once left is never occupied again—thus, $p(\omega) = 0$ for transient states. Theorem 1 bounds the maximum number of outgoing transitions to $\leqslant N + 1$ for nodes with $p(\omega) > 0$ (the ergodic

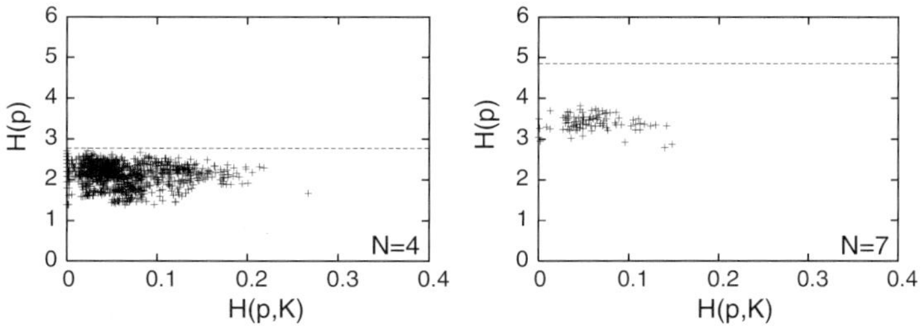

Fig. 2. Left: $H(p)$ and $H(p,K)$ for a series of simulations with $N=4$ and $N=7$. A theoretical upper bound for I and $H(p)$ is $N \ln 2 \approx 2.77$ and 4.85, respectively.

component). In a large number of simulations with N up to 8, we observed an increasing complexity in the transition graph structure, but seldomly ergodic branching nodes with more than 2 outgoing transitions, and never ergodic nodes with $> N+1$ transitions. Transient states, in contrast, can be deterministic as well as project to an arbitrary number of targets up to all possible ones (e.g., 1 1 0 in Fig. 1). Branching nodes and nested loops result in activity patterns that switch randomly between repetitive deterministic configuration sequences of various lengths (Fig. 1, lower plot).

Fig. 2 shows distributions of $H(p)$ and $H(p,K)$ for $N = 4$ and 7. $H(p,K)$ is always small compared to the theoretical upper bound $N \ln 2$. So, the systems are almost deterministic. The ergodic component as in Fig. 1 consists of nested loops, but the graph structure gets increasingly complex and cycles long with system size (not shown). Also, the size of the transient component increases. Therefore, $H(p)$ falls below its maximal value of $N \ln 2$ for larger N.

Overall, Theorem 1 and the simulation results show that the maximization of stochastic interaction leads to Markov chains with highly deterministic global dynamics, but randomness in single unit activity. Future work will address stochastic interaction in more realistic neural network models.

References

[1] M. Abeles, Corticonics: Neural Circuits of The Cerebral Cortex, Cambridge University Press, Cambridge, 1991.

[2] S.-I. Amari, Information geometry on hierarchy of probability distributions, IEEE Trans. Inform. Theory 47 (2001) 1701–1711.

[3] N. Ay, An information-geometric approach to a theory of pragmatic structuring, Ann. Probab. 30 (2002) 416–436.

[4] N. Ay, Information geometry on complexity and stochastic interaction, 2002, submitted for publication.

[5] N. Ay, T. Wennekers, Dynamical properties of strongly interacting Markov chains, 2002, submitted for publication.

[6] R. Eckhorn, Neural mechanisms of scene segmentation: recordings from the visual cortex suggest basic circuits for linking field models, IEEE Trans. Neural Networks 10 (1999) 464–479.

[7] R. Linsker, From basic network principles to neural architecture, Proc. Nat. Acad. Sci. USA 83 (1986) 7508–7512.

[8] F. Rieke, D. Warland, R. Ruyter van Steveninck, W. Bialek, Spikes: Exploring the Neural Code, MIT Press, Cambridge, 1998.

[9] W. Singer, C.M. Gray, Visual feature integration and the temporal correlation hypotheses, Annu. Rev. Neurosci. 18 (1995) 555–586.
[10] G. Tononi, O. Sporns, G.M. Edelman, A measure for brain complexity: relating functional segregation and integration in the nervous system, Proc. Nat. Acad. Sci. USA 91 (1994) 5033–5037.
[11] T. Wennekers, G. Palm, Cell Assemblies, Associative memory and temporal structure in brain signals, in: R. Miller (Ed.), Time and the Brain, Harwood Academic Publishers, New York, 2000, pp. 251–273.

Nihat Ay studied Mathematics at the Ruhr-University Bochum and obtained a Ph.D. in Mathematics from the University of Leipzig in 2001. He currently works at the Max-Planck-Institute for Mathematics in the Sciences in Leipzig on information geometry and its applications in complex adaptive systems.

Thomas Wennekers studied physics at the University of Düsseldorf and computer science at the University of Ulm, where he received a Ph.D. in 1998. He does research at the Max-Planck-Institute for Mathematics in the Sciences in Leipzig in the fields of computational neuroscience and brain theory.

Available at
www.ComputerScienceWeb.com
POWERED BY SCIENCE @ DIRECT®

ELSEVIER

Neurocomputing 52–54 (2003) 467–472

NEUROCOMPUTING

www.elsevier.com/locate/neucom

Second-order statistics of natural images

Hauke Bartsch*, Klaus Obermayer

Department of Electrical Engineering and Computer Science, Technical University Berlin, Germany

Abstract

Assuming adaptation of the visual cortex to its environment, we analyze the invariance found in natural images to explain the selective response of visual cortical neurons.
We argue that the invariant structure of images can be formally expressed by dot-products. Utilizing the specific structure of the proposed model we show how non-linear functions can be learned efficiently from natural images.
In addition to localized edge detectors we found model neurons that respond to changes in texture and cells that detect edge curvature. The analysis suggests new types of features neurons in primary visual cortex may be selective to.

Keywords: Symmetry; Invariance; Texture; Curvature

1. Introduction

Under the assumption that the visual cortex adapts to its environment, the invariances found in natural images offer a likely explanation for the emergence of cortical wiring patterns and the operations performed therein [1,4,2,5].

By referring to an object as invariant or symmetric we usually indicate that the object appears unchanged if observed from different perspectives. Symmetry of an object must therefore be defined with respect to an operation or transformation. A high similarity of an object with its transformed counterpart will indicate high symmetry of this object given the transformation.

* Corresponding author. Dept. of Computer Science, Technische University, Sekretariat FR 2-1, Fachbereich 13-Informatik, Franklinstr. 28/29, Berlin D-10587, Germany. Tel.: +49-30-31473310; fax: +49-30-31473121.

E-mail address: hauke@cs.tu-berlin.de (H. Bartsch).

doi:10.1016/S0925-2312(02)00734-8

2. One-dimensional symmetry detectors

Transformations will be expressed by a square symmetric matrix A applied onto a data vector $\mathbf{x}$ ($A\mathbf{x}=\mathbf{y}$). Whereas x represents an image, throughout this paper we will assume that $\mathbf{x}$ is a vector, appending successive columns. If the transformation A leaves the structure in $\mathbf{x}$ unchanged (with respect to its position, form and orientation), we will claim A as coding the symmetry found in $\mathbf{x}$. A well chosen A with respect to symmetry is indicated by $A\mathbf{x} \approx x$. Searching (non-trivial) solutions for A can therefore be done by learning a quadratic form $\mathbf{x}^{\mathrm{T}}A\mathbf{x}$.

Assuming an ensemble X of gray level images ($\mathbf{x} \in X$), what is the information about the image that is explored by A? Using an eigenvalue decomposition on A it is simple to show that calculating the expectation of the quadratic form we arrive at

$$\langle x^{\mathrm{T}}Ax\rangle_x = \left\langle \sum_i^N \lambda_i \mathbf{n}_i^{\mathrm{T}}\mathbf{x}\mathbf{x}_i^{\mathrm{T}}\mathbf{n} \right\rangle_x = \sum_i^N \lambda_i \mathbf{n}_i^{\mathrm{T}} \langle \mathbf{x}\mathbf{x}^{\mathrm{T}}\rangle_x \mathbf{n}_i = \sum_i^N \lambda_i \mathbf{n}_i^{\mathrm{T}} C\mathbf{n}_i. \quad (1)$$

Only the correlation matrix C of the data will be of importance for defining A which codes therefore statistics of second order.

2.1. Rotation, scaling and translation by a binary quadratic form

Before learning the quadratic form from natural images we now try to implement symmetries connected with the rigid transformations rotation, scaling and translation. In analyzing the resulting covariance structure of A we find sets of orthogonal basis functions.

In the following, the coordinates of a pixel i will be denoted by (π-periodic) polar coordinates (θ_i, r_i) in order to simplify the definition of binary matrices A. Because we are only interested in illustrating the representation power of the model we will incorporate the respective covariance structure by a crude approximation into the model. In particular this explains the indefiniteness of the matrix A.

Rotate. The goal is to let A^{rot} encode rotations of the image pixels around their mean position. We have to set the matrix A to the covariance matrix C obtained from a data set X that contains rotationally symmetric figures. In a first approximation we will assume that a_{ij}^{rot} $(i \neq j)$ will be set to one if $|r_i - r_j| < \varepsilon$ $(i \neq j)$ and otherwise to zero. ε was arbitrarily chosen to be the distance of two neighboring pixels on the outer perimeter of the circular receptive field.

Scale. In a similar way A^{sca} is constructed to encode scaling. Entries of matrix a_{ij} will be set to one only if the angular distance between i and j is sufficiently small, i.e. $|\theta_i - \theta_j| < \varepsilon$ $(i \neq j)$. We do not take into account that by this we also code for inversion symmetries: true scaling would not allow for pixels to cross the origin. It follows that the choice of a specific pixel pair subset does not uniquely define a single symmetry operation.

Shift. A^{shift} is constructed to encode the pixel pairs that occur for translations of pixels along one direction (here 90°). a_{ij} $(i \neq j)$ is set to one only if $(x_i = x_j)$ and otherwise to zero.

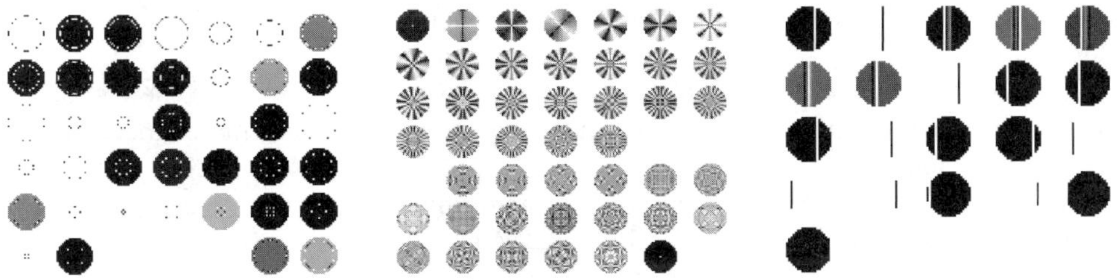

Fig. 1. Orthogonal basis functions for rotation (left), scaling (center) and shift (right) computed as eigenvectors of a quadratic form.

All orthogonal basis functions found by the analysis (see Fig. 1) show responses for pattern with the respective symmetries, circles for rotational symmetries, crosses and fan-like figures for scale invariance and lines for shift invariance.

3. Multi-dimensional symmetry detection

An object may be symmetric with respect to more than one transformation. Thus, we expect to find different instantiations of transformations to be mixed in an ensemble of example patches from natural images. Therefore, a more appropriate model is to find a number i of symmetries each one expressed by its own matrix A^i. Let us define Φ as a vector valued function mapping the data into the space of different symmetry operators:

$$\Phi(\mathbf{x}) = (x^T A^1 x, x^T A^2 x, \ldots, x^T A^N x)^T. \tag{2}$$

Additional assumptions can be made for the distribution of Φ incorporating for example sparseness, independence or uncorrelatedness. But before doing this, we first cast the problem into the notion of dot-products. This will give us the vantage of analyzing a linear model in which assumptions can be justified more easily.

Any quadratic form can be written as a linear weighting in the space of dot-products $\mathbf{x}^T A^i \mathbf{x} = (a^i_{11}, \ldots, a^i_{kk})(x_1 x_1, \ldots, x_n x_m)^T = \mathbf{w}^i \mathbf{s}$. $\mathbf{s}$ consists of monomials of constant order 2. In reducing the number of calculations needed we will restrict ourself to commutative monomials eliminating the respective entries from $\mathbf{w}$ ($\mathbf{w}$ is of the dimensionality of $\mathrm{vech}(\mathbf{x}^T \mathbf{x})$). Adding more dimensions to the model Φ, e.g. more quadratic forms $x^T A^i x$, we arrive at

$$\Phi = W\mathbf{s}. \tag{3}$$

Here, W is a square matrix. It can be viewed as a linear mixture expressed by the mixing matrix W of the signals $\mathbf{s}$ in the space of dot-products.

Linear methods for monomial spaces of constant order: By the model in Eq. (3) the input data $\mathbf{x}$ is projected into a manifold of the higher dimensional space of $\mathbf{s}$. Assuming linear sources in two dimensions source directions (u, v) are mapped into $(u^2, uv, v^2) = u^2(1, a, a^2)$, $a = v/u$ which is again a linear direction in the space of dot-products. Thus the detection of linear direction in the manifold of **s** selects linear directions in **x**.

Notably, because the dimension of **s** is larger than the dimension of **x** in solving (3) we solve an equivalent problem in **x** where the goal is to detect more linear sources than observations.

There is no a priori choice for a specific algorithm solving Eq. (3). The system is ill-posed (W and **s** are unknown), thus additional assumptions about the underlying sources are required. Assuming statistical independence of the transformations in space of dot-products lead us in the next section to methods of ICA.

3.1. Results

To perform an ICA we used the FastICA Matlab package described in [3] (symmetric approach with tanh as non-linearity). 100,000 example image patches $\tilde{y}$ of size 7×7 were extracted at random positions from images depicting natural scenes. We performed no pre-processing of the images except a shift to zero mean for each image separately. We assumed that $\tilde{y}_i\tilde{y}_j = \tilde{y}_j\tilde{y}_i$ and $\tilde{y}_i\tilde{y}_i = 0$ together with a circular rather than rectangular receptive field which reduces the number of valid pixel combinations to $7\pi(7\pi - 1)/2 \approx 700$. A dimension reduction was performed by doing an eigenvalue decomposition of Φ and keeping only the eigenvectors belonging to the 49 largest eigenvalues (this keeps 94% of the variance).

Doing ICA in the space of dot-products results in finding independent sources in dot-product space. To visualize each component found by ICA we plotted examples from the resulting orthogonal basis function set (see Fig. 2). The figure displays in each case only the absolute largest and second largest eigenvector of the 7×7 matrices A^i, which were found as lines in the mixing matrix W.

Among others, edges with different orientation, spatial frequency and position are prominent structures found by the model (see Fig. 2). A further analysis of the filters is given in the discussion.

4. Discussion

Whereas coding of edges is trivially implemented by a linear filter, our model relies solely on statistics of second order, e.g. products of pixel values. Edges are found to be implemented by a simple mechanism of 'be the same' and 'be different'. The pixels on each side of the edge are pooled together by assigning only positive weights between the pixels. Together they code for the same coherent region of space. Contrarily, the weights between the two groups are negative, thus forcing the two regions to code for different structure.

Coding of texture properties: Fig. 3 displays more eigenvectors of a selected component. Sub-structure is concentrated either on the upper or lower integration field of the filter. The filter distinguishes in its response non-textured regions (objects) from textured regions (background) or vice versa.

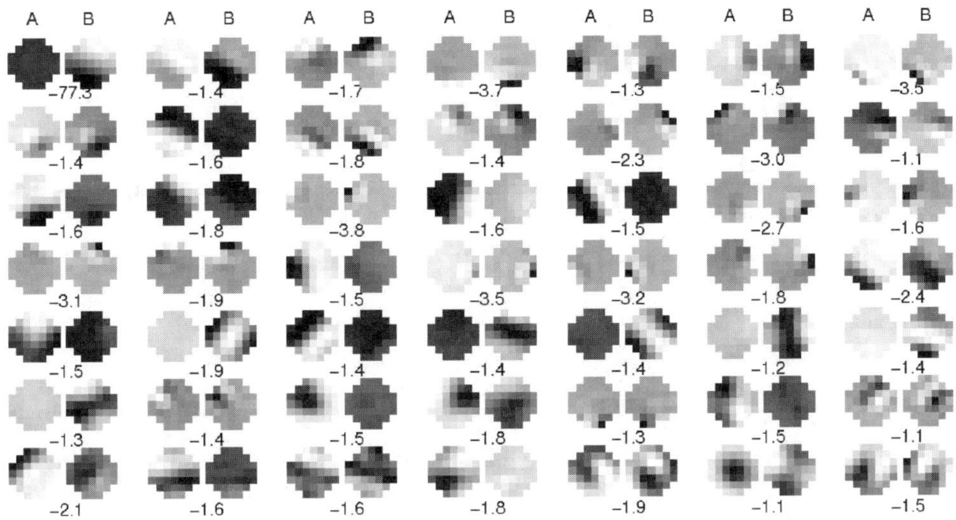

Fig. 2. Learned symmetry detectors by FastICA. For each of 49 found independent component two images (A, B) are shown. The number below each component codes for the ratio of the largest to the second largest eigenvalue.

Fig. 3. Texture dependent sub-fields. Eigenvectors with negative eigenvalues concentrate onto the lower right part of the receptive field, eigenvectors with positive eigenvalues onto the upper left part.

Detecting curvature. More difficult to interpret are filters which show a *yin-yang*-like pattern. They are stable features in our simulations. We offer an interpretation to their function in terms of forming curvature detection units. We consider the orbit that is spanned by the largest positive and negative eigenvector $\tilde{n}(\gamma) = \gamma \mathbf{n}_0 + (1-\gamma)\mathbf{n}_N$, $\gamma = [0 \dots 1]$. As shown in Fig. 4, (first row), maximum (left, $\gamma = 0$) and minimum (right, $\gamma = 1$) together constitute an edge detector (center, $\gamma \approx 0.5$). Because of orthogonal eigenvectors the unit's response to each image $\tilde{n}(\gamma)$ is the sum of weighted (by γ) individual responses. Increasing or decreasing γ 'bends' the line along the edge in directions of positive and negative curvature—larger curvature results in larger (absolute) output.

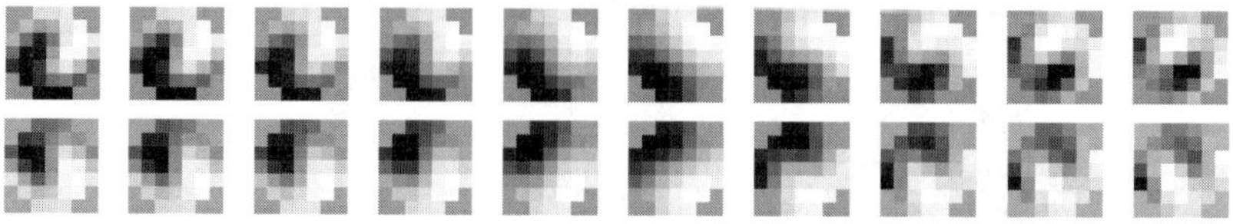

Fig. 4. Two dot-product filters that code for curvature of a line. The *top row* shows the orbit that is spanned by the largest and smallest eigenvector of the ICA-component no. 47 shown in Fig. 2. In the *bottom row* same is done for the ICA-component no. 49.

References

[1] P.C. Bressloff, J.D. Cowan, M. Golubitsky, P.J. Thomas, M.C. Wiener, Geometric visual hallucinations, euclidean symmetry, and the functional architecture of striate cortex, Phil. Trans. R. Soc. London B 356 (2001) 299–330.

[2] W. Einhäuser, C. Kayser, P. König, K.P. Körding, Extracting slow subspaces from natural videos leads to complex cells, in: G. Dorffner, H. Bischof, K. Hornik (Eds.), Artificial Neural Networks—ICANN 2001, Lecture Notes in Computer Science, Vol. 2130, Springer, Berlin, 2001.

[3] A. Hyvärinen, Fast and robust fixed-point algorithms for independent component analysis, IEEE Trans. Neural Networks 10 (3) (1999) 626–634.

[4] Rajesh P.N. Rao, Daniel L. Ruderman, Learning lie groups for invariant visual perception in: M.S. Kearns, S.A. Solla, D.A. Cohn (Eds.), Advances in Neural Info Processing Systems, Vol. 11, MIT Press, Cambridge, MA, 1999, pp. 810–816.

[5] L. Wiskott, T. Sejnowski, Slow feature analysis: unsupervised learning of invariances, Neural Comput. 14 (2002) 715–770.

Hauke Bartsch is a member of the Neural Information Processing Group at the Department of Computer Science of the Technical University of Berlin. He received his Masters degree in 1997 at the Technical University Berlin, Germany. His research interests are models for the structure of neuronal networks in the visual system.

Klaus Obermayer is Professor for Neural Information Processing at the Department of Computer Science at the Technical University of Berlin. He received his Masters degree in Physics in 1987 from the University of Stuttgart and his Ph.D. in 1992 from the Technical University in Munich, Germany. His research interests cover the areas of computational neuroscience, the theory of artificial neural networks, and applications in image and signal processing.

ELSEVIER

Available at
www.ComputerScienceWeb.com
POWERED BY SCIENCE @DIRECT®

Neurocomputing 52–54 (2003) 473–480

NEUROCOMPUTING

www.elsevier.com/locate/neucom

Self-organization of spatiotemporal receptive fields and laterally connected direction and orientation maps

James A. Bednar*, Risto Miikkulainen

Department of Computer Sciences, The University of Texas at Austin, Austin, TX 78712, USA

Abstract

Studies of orientation maps in primary visual cortex (V1) suggest that lateral connections mediate competition and cooperation between orientation-selective units, but their role in motion perception has not been established. Using a self-organizing model of V1 with moving oriented patterns, we show that (1) afferent weights of each neuron organize into Gabor-like spatiotemporal receptive fields with ON and OFF lobes, (2) these receptive fields form realistic joint direction and orientation maps, and (3) lateral connections develop between patches with similar orientation and direction preferences. These results suggest that a single self-organizing system may underlie the development of orientation selectivity, direction selectivity, and lateral connectivity.

Keywords: Direction maps; Orientation maps; Spatiotemporal receptive fields; Direction selectivity; Lagged cells

1. Introduction

Most simple cells in the primary visual cortex (V1) are selective for the direction and orientation of a moving stimulus. Recent measurement techniques have made it possible to plot the neurons' full spatiotemporal receptive fields, which include specific excitatory (ON) and inhibitory (OFF) subregions that vary over time [5]. The functional properties of these cells form a mosaic across V1, with patches of nearby neurons preferring similar directions and orientations [13].

* Corresponding author.
E-mail addresses: jbednar@cs.utexas.edu (J.A. Bednar), risto@cs.utexas.edu (R. Miikkulainen).

doi:10.1016/S0925-2312(02)00735-X

In addition to their afferent input from the LGN, the neurons in these maps are connected intracortically through specific long-range lateral connections [7]. The lateral connections have been found to link cells with similar orientation preferences [3,7], which can allow the connections to suppress redundancy in the input and improve the cells' ability to detect changes in a stimulus [2,8,12]. However, the role of these connections in the development and adult function of directional selectivity is not yet clear.

Several computational models have shown that directional selectivity and interleaved orientation and direction maps can develop through activity-dependent self-organization [6,10,14]. However, these simulations have only modeled afferent connection learning, and have treated lateral connections as fixed. In prior work with the LISSOM self-organizing model (Laterally Interconnected Synergetically Self-Organizing Map), we have shown how a Hebbian learning process can develop topographic maps, ocular dominance, orientation, and size preference columns, and patterned lateral connections between them [8,11,12]. We have also shown that these self-organized maps and lateral connections can function in adult visual perception to segment and bind coherent objects and reduce redundancy in the input, and that visual illusions and aftereffects arise through this process [2,4,12]. In this paper we extend the model to develop motion and direction preferences through the self-organization of spatiotemporal receptive fields. Together, these results show that activity-dependent self-organization can explain many of the anatomical and functional characteristics of the cortex.

2. HLISSOM model

These simulations are based on the HLISSOM model [1], which extends LISSOM to include the ON and OFF channels of the LGN. The main idea is that directional selectivity can develop from LGN cells whose output arrives at the cortex after a delay. In the LGN, many cells fire soon after a retinal stimulus. However, other *lagged* cells have recently been found in cat LGN that respond only after a fixed delay [9]. The delay time in these lagged cells varies over a continuous range up to hundreds of milliseconds [15]. We will show below that V1 neurons can use these timing differences to develop spatiotemporal receptive fields.

The architecture for the HLISSOM model is shown in Fig. 1, and will be briefly reviewed below. (For more details, see [2].) The model consists of a hierarchy of two-dimensional sheets of neural units modeling different areas of the visual system: a sheet of retinal photoreceptors, several paired sheets of ON-center and OFF-center LGN units (with a different lag for each pair), and a sheet of cortical units ("neurons") representing V1. Because the focus is on the two-dimensional organization of V1, each neuron corresponds to a vertical column of cells through the six anatomical layers of the cortex.

The model simulates behaviour at three different timescales, denoted by variables f, t, and s in the equations below. Variable f represents a single fixation out of a sequence of thousands presented to the model. At each fixation, motion in the image

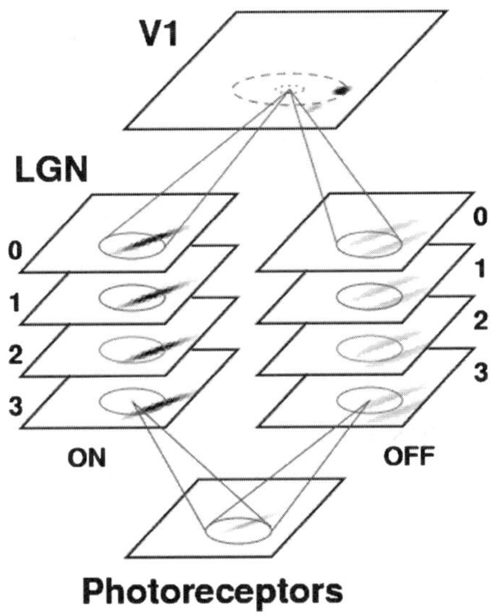

Fig. 1. HLISSOM model of orientation and direction selectivity. The model is a hierarchy of sheets of neural units, modeling the early visual pathway. In this implementation, there are eight sheets of LGN units. The LGN units receive input from their circular receptive fields (RFs) on the photoreceptors. Connections to a sample photoreceptor RF are plotted for two LGN units; all units at the same position in the other six LGN sheets also receive input from that same RF. Each LGN sheet contains a homogenous group of units, with either ON- or OFF-center organization and a lag of 0, 1, 2, or 3 timesteps. Similarly, neurons in V1 have circular RFs on all eight LGN sheets; connections to two of the RFs are shown. V1 neurons also have lateral excitatory (small dotted circle) and lateral inhibitory (large dashed circle) connections to their neighbors. Moving input patterns are drawn on the photoreceptor sheet in discrete timesteps, like frames of a movie. At the first timestep, the ON and OFF LGN cells with time lag 3 compute their activity. At each subsequent timestep, the input pattern is moved slightly and LGN cells with lags 2, 1, and 0 compute their activity in turn. Once all LGN cells have been activated, initial V1 activity is computed from the LGN responses, and the activity then spreads laterally within V1. Both afferent and lateral V1 weights are adapted when the activity settles.

is simulated at several discrete timesteps t. For a given time t, lateral interactions in the cortex are computed for several timesteps s.

The input to the model is a series of activity patterns on the sheet of photoreceptors, such as two-dimensional oriented Gaussians or grayscale photographic images. A sample input is drawn in the photoreceptor sheet of Fig. 1. For a given moving input, the photoreceptor activity is drawn at multiple discrete timesteps. At each such time step t, the activity levels of all LGN cells with lag t are calculated. Each LGN cell (i,j) with lag t computes its response η_{ij} as a scalar product of a fixed weight vector and its receptive fields on the photoreceptor sheet at time t:

$$\eta_{ij} = \sigma\left(\sum_{\rho ab} \gamma_\rho X_{\rho ab} w_{ij,\rho ab}\right), \tag{1}$$

where ρ identifies the input sheet (the photoreceptors, in this case), σ is a piecewise linear sigmoid activation function, γ_ρ is a constant scaling factor, $X_{\rho ab}$ is the activation of input unit (a,b) on sheet ρ at timestep t, and $w_{ij,\rho ab}$ is the corresponding weight value. Each V1 neuron computes its initial response like that of an LGN cell, except that ρ identifies one of the ON or OFF LGN sheets. After the initial response, the V1 activity settles through short-range excitatory and long-range inhibitory lateral interaction:

$$\eta_{ij}(s) = \sigma\left(\sum_{\rho ab} \gamma_\rho X_{\rho ab}(s-1) w_{ij,\rho ab}\right), \tag{2}$$

where ρ identifies either an LGN sheet or the lateral excitatory or inhibitory weights to V1, γ_ρ is a constant scaling factor for each ρ (negative for inhibitory lateral weights), and $X_{\rho ab}(s-1)$ is the activation of input unit (a,b) during the previous settling step. The V1 activity pattern starts out diffuse, but within a few iterations of Eq. (2), converges into a small number of stable focused patches of activity, or activity bubbles. After the activity has settled, the connection weights of each V1 neuron are modified. All V1 weights adapt according to the Hebb rule, normalized so that the sum of the weights from each sheet ρ is constant for each neuron (i,j):

$$w_{ij,\rho ab}(f+1) = \frac{w_{ij,\rho ab}(f) + \alpha_\rho \eta_{ij} X_{\rho ab}}{\sum_{ab}[w_{ij,\rho ab}(f) + \alpha_\rho \eta_{ij} X_{\rho ab}]}, \tag{3}$$

where η_{ij} stands for the activity of neuron (i,j) in the final activity bubble, $w_{ij,\rho ab}(f)$ is the connection weight from the previous fixation, α is the learning rate for each type of connection, and $X_{\rho ab}$ is the presynaptic activity. The larger the product of the pre- and post-synaptic activity $\eta_{ij} X_{\rho ab}$, the larger the weight change. At long distances, few neurons have correlated activity and therefore most long-range connections eventually become weak. The weakest connections are eliminated periodically, resulting in patchy lateral connectivity similar to that observed in the visual cortex.

For the experiments reported in this paper, four 36×36 ON-center and four 36×36 OFF-center cell sheets received input from a 54×54 photoreceptor sheet. Each ON/OFF cell had a fixed Difference of Gaussians receptive field (RF) within the photoreceptor array. Initially, the afferent weights of the 142×142 V1 neurons were random, and the lateral weights had a smooth circular Gaussian profile. The learning parameters were the same as in our earlier V1 orientation model [2], scaled for this cortex size using the model scaling methodology presented in [1].

3. Results

Fig. 2 shows the self-organized afferent weights for a representative neuron after 20,000 image fixations. Nearly all neurons developed spatiotemporal receptive fields selective for orientation, and most were also selective for direction. These RFs are similar to those found experimentally in the cortex [5]. As in V1 cells, the model RFs

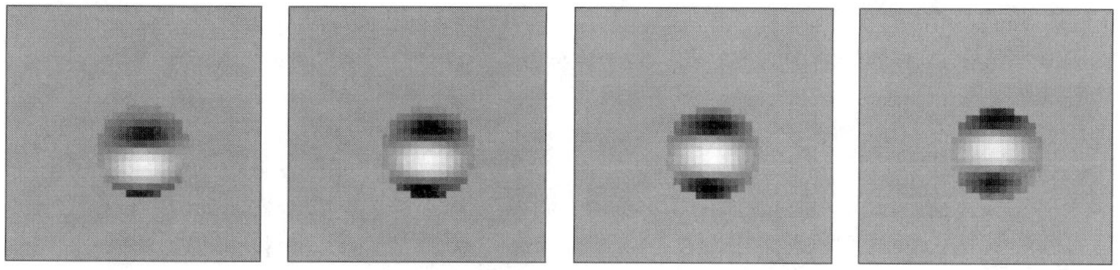

Fig. 2. Self-organized spatiotemporal receptive field. These images show the afferent weights for a typical neuron, plotted on the retina. As a visualization of the preferred stimulus of this neuron, the OFF weights were subtracted from the ON for each lag. In the resulting plots, white indicates a net ON subregion of the retina, i.e. an area that will excite the neuron if it is brighter than its surround. Black indicates OFF subregions. Together, the afferent weight plots show that the most effective stimulus for this neuron would be a horizontal bar or grating moving upwards. More specifically, this neuron will be highly active at time t if there was a bright bar aligned with the ON subregion in (a) at time $t-3$, a bright bar aligned with the ON subregion of (b) at time $t-2$, and so on until the present time t, where the bar will be aligned with the ON subregion of (d). Visual cortex neurons in animals have similar spatiotemporal properties [5].

can only rarely be expressed as a function of a fixed spatial RF and a time-dependent scaling factor. The RFs are thus *space-time inseparable* [5]. Also as in the cortex, the orientation and direction preferences of a neuron were generally perpendicular.

Fig. 3 shows that the self-organized global orientation and direction maps are also similar to those found in animals [13]. For instance, a patch of neurons highly selective for one orientation and direction of motion will often have an adjacent or contiguous patch selective for the same orientation, but the opposite direction. Long-range lateral connections within the map follow this global organization, connecting neurons similar in both orientation and direction preference. These connection patterns represent the correlations between units over the course of self-organization. As will be described below, the self-organized lateral weight patterns are likely to play a crucial role in adult visual perception as well as development.

4. Discussion and future work

The results demonstrate that a single Hebbian learning algorithm can explain how topographic maps, spatiotemporal receptive fields, and lateral connections synergetically self-organize from moving stimuli. The model predicts that long-range lateral connections in V1 will be found to connect neurons with similar direction preference as well as orientation preference. In future work we will validate the model against experimental results from particular species and study how the orientation, direction preference, and lateral connectivity interact with other stimulus dimensions, such as spatial frequency preference.

Other studies will investigate how the direction map and its connections operate in the adult. The self-organized connection patterns should allow moving objects to be bound together and segmented from other objects, as previously shown for static

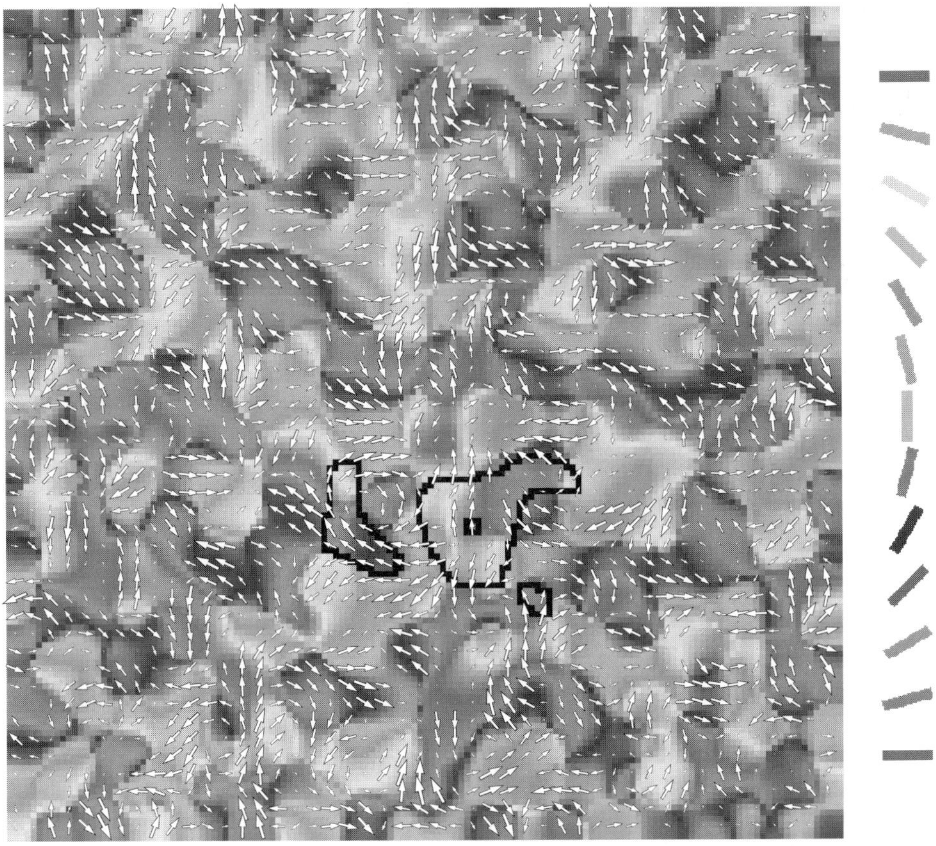

Fig. 3. Self-organized orientation and direction map. Through self-organization, a smoothly varying map of orientation and direction preferences emerged. Each of the 142 × 142 neurons is colored with its preferred orientation, according to the key at right. The orientation map is similar to those found experimentally in animals, containing features such as pinwheels, linear zones, and fractures [13]. The direction preferences for every third neuron are plotted as white arrows overlaid on the orientation map. The size of each arrow is proportional to the neuron's directional selectivity, and the arrow points in the neuron's preferred motion direction. The preferred orientation is generally perpendicular to the preferred direction. As an example, the neuron from Fig. 2 is marked with a black square in the lower center of the plot. The neurons in that region of the map prefer horizontal stimuli moving upwards. Neurons in the patch just below the marked neuron also prefer nearly horizontal lines, but moving in the opposite direction. Other paired patches can be seen elsewhere in the map, as found experimentally [13]. The black outline shows the extent of the self-organized lateral weights for the marked neuron. The short-range connections target neurons with a variety of preferences, but long-range connections (a) target similar orientations and direction preferences, (b) extend along the orientation axis (horizontal), and (c) avoid orthogonal orientations and opposite directions. Animations of this plot are available at www.cs.utexas.edu/~nn.

patterns [4]. Short-term direction-specific adaptation of the lateral connections should replicate the motion aftereffect, also called the waterfall illusion, as we found for the tilt aftereffect [2]. This work will help unify explanations of development and adult visual function into a coherent theory of the visual cortex.

5. Conclusion

Hebbian learning of moving, oriented patterns in a self-organizing model can explain how orientation selectivity, direction selectivity, and lateral connectivity develop synergetically in V1. The model provides specific predictions for the role of the lateral connections in the development and function of the direction map. This study and future work will help strengthen our understanding of the visual cortex as a continuously adaptive, self-organizing system.

Acknowledgements

Supported in part by the National Science Foundation under grants IRI-9309273 and IIS-9811478 and by the National Institutes of Mental Health under Human Brain Project grant 1R01-MH66991.

References

[1] J.A. Bednar, Learning to See: Genetic and Environmental Influences on Visual Development, Ph.D. Thesis, Department of Computer Sciences, The University of Texas at Austin, 2002.

[2] J.A. Bednar, R. Miikkulainen, Tilt aftereffects in a self-organizing model of the primary visual cortex, Neural Comput. 12 (7) (2000) 1721–1740.

[3] W.H. Bosking, Y. Zhang, B. Schofield, D. Fitzpatrick, Orientation selectivity and the arrangement of horizontal connections in tree shrew striate cortex, J. Neurosci. 17 (6) (1997) 2112–2127.

[4] Y. Choe, R. Miikkulainen, Self-organization and segmentation in a laterally connected orientation map of spiking neurons, Neurocomputing 21 (1998) 139–157.

[5] G.C. DeAngelis, I. Ohzawa, R.D. Freeman, Receptive-field dynamics in the central visual pathways, Trends Neurosci. 18 (1995) 451–458.

[6] I. Farkas, R. Miikkulainen, Modeling the self-organization of directional selectivity in the primary visual cortex, Proceedings of ICANN-99, Springer, Berlin, 1999, pp. 251–256.

[7] C.D. Gilbert, T.N. Wiesel, Columnar specificity of intrinsic horizontal and corticocortical connections in cat visual cortex, J. Neurosci. 9 (1989) 2432–2442.

[8] R. Miikkulainen, J.A. Bednar, Y. Choe, J. Sirosh, Self-organization, plasticity, and low-level visual phenomena in a laterally connected map model of the primary visual cortex, in: Goldstone, et al. (Eds.), Perceptual Learning, Vol. 36 of Psychology of Learning and Motivation, Academic Press, New York, 1997, pp. 257–308.

[9] A.B. Saul, A.L. Humphrey, Evidence of input from lagged cells in the lateral geniculate nucleus to simple cells in cortical area 17 of the cat, J. Neurophys. 68 (4) (1992) 1190–1208.

[10] H. Shouno, K. Kurata, Formation of a direction map by projection learning using Kohonen's self-organization map, Biol. Cybernetics 85 (4) (2001) 241–246.

[11] J. Sirosh, R. Miikkulainen, Cooperative self-organization of afferent and lateral connections in cortical maps, Biol. Cybernetics 71 (1994) 66–78.

[12] J. Sirosh, R. Miikkulainen, J.A. Bednar, Self-organization of orientation maps, lateral connections, and dynamic receptive fields in the primary visual cortex, in: Sirosh, et al. (Eds.), Lateral Interactions in the Cortex: Structure and Function, Austin, TX, The UTCS Neural Networks Research Group, ISBN 0-9647060-0-8, www.cs.utexas.edu/~nn/web-pubs/htmlbook96 (1996).

[13] M. Weliky, W.H. Bosking, D. Fitzpatrick, A systematic map of direction preference in primary visual cortex, Nature 379 (1996) 725–728.

[14] S. Wimbauer, O.G. Wenisch, J.L. van Hemmen, K.D. Miller, Development of spatiotemporal receptive fields of simple cells: II, Biol. Cybernetics 77 (6) (1997) 463–477.
[15] J. Wolfe, L.A. Palmer, Temporal diversity in the lateral geniculate nucleus of cat, Visual Neurosci. 15 (4) (1998) 653–675.

Jim Bednar is a postdoctoral researcher at the University of Texas at Austin. He completed his Ph.D. at UT in spring 2002, and also has an M.A. in Computer Science, a B.A. in Philosophy, and a B.S. in Electrical Engineering. His research focuses on computational modeling of cortical map development. Jim is the lead author of the forthcoming Topographica modeling software package, under development through a Human Brain Project grant from the National Institutes of Health.

Risto Miikkulainen is a Professor of Computer Sciences at the University of Texas at Austin, and is on the faculty of the Institute for Neuroscience and the Center for Perceptual Systems. He received his Ph.D. in Computer Science from UCLA in 1990, and his M.S. in Applied Mathematics from the Helsinki University of Technology in 1986. His research focuses on applications of neural networks and genetic algorithms to cognitive science, computational neuroscience, and neuroevolution.

Available at
www.ComputerScienceWeb.com
POWERED BY SCIENCE DIRECT®

ELSEVIER

Neurocomputing 52–54 (2003) 481–487

NEUROCOMPUTING

www.elsevier.com/locate/neucom

Temporal dynamics of attention-modulated neuronal synchronization in macaque V4

Hualou Liang[a,*], Steven L. Bressler[b], Mingzhou Ding[b], Robert Desimone[c], Pascal Fries[d]

[a] *School of Health Information Sciences, University of Texas Health Science Center at Houston, Houston, TX, 77030, USA*

[b] *Center for Complex Systems and Brain Sciences, Florida Atlantic University, Boca Raton, FL 33431, USA*

[c] *Laboratory of Neuropsychology, National Institute of Mental Health, National Institutes of Health, Bethesda, MD 20892, USA*

[d] *F.C. Donders Centre for Cognitive Neuroimaging, University of Nijmegen, 6525 EK Nijmegen, The Netherlands*

Abstract

It was recently observed that neurons in area V4 exhibited enhanced gamma band (35–90 Hz) synchronization when monkeys attended to a visual stimulus as compared to when they were not attending to the same stimulus (Science 291 (2001) 1560). Spike-triggered averaging of local field potentials (LFPs) was used to show attentional modulation in an early period from 50 to 150 ms after stimulus onset (Science 291 (2001) 1560). In this work we further studied the fine temporal structure in the same data by focusing only on the LFPs without reference to the concurrent spike trains. With the method of adaptive multivariate autoregressive (AMVAR) modeling, we discovered that attentional modulation of gamma power (~65 Hz) in V4 can be as brief as about 25 ms. Gamma coherence between two V4 recording sites revealed similar attention effects, as well as a second peak around 45 Hz. Directional influences between two V4 populations revealed that one can play a more dominant role than another. These results implicate gamma oscillation as a possible agent in carrying out attention-biased competition among visual stimuli in favor of those that are behaviorally relevant. The AMVAR method was instrumental in revealing the dynamics of gamma frequency synchronization with high temporal and frequency resolution.

Keywords: Selective visual attention; Multivariate autoregressive modeling; Gamma synchronization; Coherence; Bottom-up; Top-down

* Corresponding author. Tel.: 713-500-3914; fax: 713-500-3915.
E-mail addresses: hualou.liang@uth.tmc.edu (H. Liang).

0925-2312/03/$ - see front matter © 2003 Elsevier Science B.V. All rights reserved.
doi:10.1016/S0925-2312(02)00741-5

1. Introduction

Selective visual attention is the process whereby a particular visual stimulus is selected from multiple stimuli for further processing. In this process, attention biases competition among visual stimuli in favor of those that are most behaviorally relevant [1]. Recent work shows that gamma band (35–90 Hz) synchronization among neurons in cortical area V4 may play a role in this mechanism [3] through elevation of the level of population synchrony in response to the attended stimulus. By applying the technique of spike-triggered averaging to local field potentials (LFPs) it was determined that attentional modulation is already present from 50 to 150 ms poststimulus [3]. Since spike-triggered averaging treats every spike inside the 100 ms interval on an equal footing, the fine temporal dynamics of synchronization in that 100 ms interval is not clear, which the earlier analysis did not attempt to further resolve. Here, we explore the possible finer dynamics in the LFP data related to attention deployment by a parametric spectral analysis method referred to as adaptive multivariate autoregressive (AMVAR) modeling [2].

LFPs from four closely spaced electrodes in area V4 were simultaneously recorded from two macaque monkeys trained to perform a visual spatial attention task. As the monkey fixated centrally, two patches of moving high contrast square wave grating were presented at equal eccentricity, one inside and one outside the receptive field (RF) of the recorded neurons. On different trials, the monkey was cued to attend to one or the other stimulus location. The result was two attention conditions: attention inside the RF vs. attention outside the RF. The analysis described here used 300 correctly performed trials in each attention condition from one monkey.

Local field potential recordings were analyzed to determine power, coherence, and directed transfer function (DTF), a statistical measure of directional influence between two given recordings spectra, derived from AMVAR models. The particular focus was on the detailed time course of attention-modulated neuronal synchronization.

2. Methods

We begin by recapitulating the AMVAR modeling technique introduced in [2] and describe the spectral quantities used in this study.

In MVAR modeling, after an analysis window is chosen LFPs are assumed to be the output of linear filters fed by white noises processes. Thus the p LFP recordings, $\mathbf{X}_t = (x_{1t}, x_{2t}, \ldots, x_{pt})^{\mathrm{T}}$, can be modeled as AR processes of unknown coefficients, which are given by

$$\sum_{k=0}^{m} \mathbf{A}_k \mathbf{X}_{t-k} = \mathbf{E}_t,$$

where $\mathbf{E}_t$ is a temporally uncorrelated residual error with covariance matrix Σ, and $\mathbf{A}_k$ are $p \times p$ coefficient matrices which are obtained by solving the multivariate Yule–Walker Eq. (of size mp^2) using the Levinson, Wiggins and Robinson (LWR) algorithm [8]. The model order m is determined by the Akaike information criterion (AIC) [7].

Data from multiple trials are treated as realizations of the same underlying (approximately) stationary stochastic process and are combined to produce the estimation of the model coefficients. The term Adaptive means that the above procedure is repeated for each analysis window along the time course of the experiment.

Once the model coefficients $\mathbf{A}_k$ and Σ are estimated, the spectral matrix can be written as

$$S(f) = \langle X(f)X^*(f)\rangle = H(f)\Sigma H^*(f),$$

where the asterisk denotes matrix transposition and complex conjugation, and $H(f) = (\sum_{j=0}^{m} A(j)e^{-2\pi i j f})^{-1}$ is the transfer function of the system. A suite of tools for spectral analysis can be derived from the spectral matrix [4]. For the purpose of this article, we will only consider the power, coherence, and DTF spectra.

The power spectrum of channel l is given by $S_{ll}(f)$ which is the lth diagonal element of the spectral matrix $S(f)$. The coherence spectrum between channel l and channel k is given by

$$C_{lk}(f) = |S_{lk}(f)|/(S_{ll}(f)S_{kk}(f))^{1/2}.$$

The value of coherence can range from 1, indicating maximum interdependence between channel l and channel k at frequency f, down to 0, indicating no interdependence. The DTF [5] is defined as: $\gamma_{kl} = |H_{kl}(f)|^2$, which measures the statistical influence (information flow) from channel l to channel k. Applications of DTF on cortical field potential time series have shown promise for understanding information transfer between cortical areas [6].

AMVAR analysis was performed with a 50-ms-long window (10 time points) that was stepped point by point from 500 ms prestimulus to 500 ms poststimulus. In each window, data from all 300 trials and 4 sites were used to estimate the MVAR model which yielded all the spectral quantities. The result was a time-frequency representation of the neural dynamics. A bootstrap resampling procedure, with a resampling size of 100, was employed to assess the variability of all the spectral quantities.

3. Results

The power spectrum, as the simplest yet most used measure, reveals how the activity at a site is distributed in frequency with peaks, if any, indicating synchronous oscillatory activity. An example of time-frequency power spectrum from one V4 site (Fig. 1, left) for attention within the RF is used for illustration. At approximately 110 ms poststimulus, there was a brief, narrow-band oscillatory event in the power, peaking around 66 Hz. A similar power spectrum was observed for attention outside the RF condition, but with a lower level of gamma synchronization. This can be clearly seen by taking either the spectrum from the analysis window centered at 110 ms poststimulus (Fig. 1, middle) or power at 66 Hz as a function of time for the two conditions (Fig. 1, right). We see, with great temporal precision, the enhanced gamma synchronization when the monkey directed attention within the RF compared to attention outside the RF, a result consistent with that in [3].

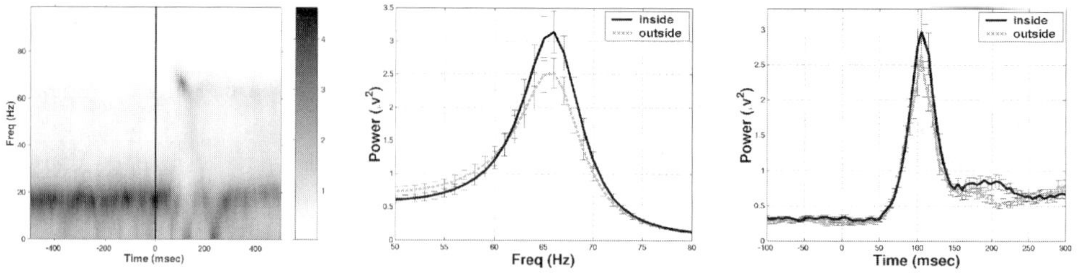

Fig. 1. Left: Time-frequency power plot for V4 recording site 1 for attention inside the RF using a 50-ms (10 point) sliding window. Time 0 indicates the stimulus onset. Middle: Power spectra of site 1 for the two conditions, attention inside the RF (solid lines) and attention outside the RF (dashed lines), from a 50-ms long window centered 110 ms after stimulus onset. Note that both conditions have peaks at 66 Hz, but the peak is larger for the "inside" condition. Right: Power at 66 Hz as a function of time for the two conditions. An initial episode of elevated gamma power related to attention occurred in an interval centered 110 ms poststimulus, lasting about 25 ms, and a second episode of attention modulated gamma power elevation, although at a relatively low overall power level, started around 150 ms poststimulus and lasted around 90 ms. Error bars were obtained by the bootstrap resampling method.

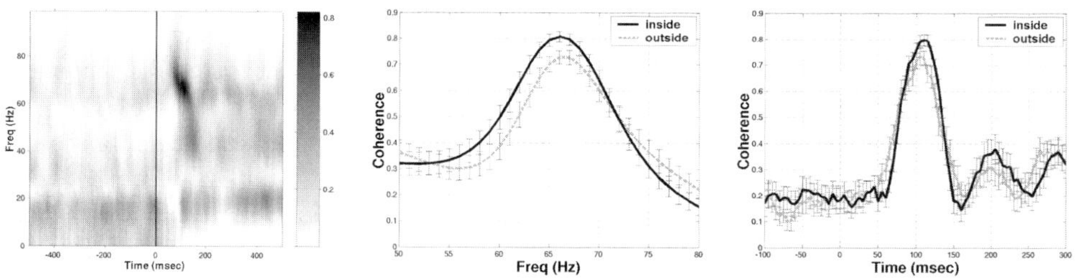

Fig. 2. Left: Time-frequency plot of spectral coherence between V4 recording sites 1 and 2 for attention inside the RF, using a 50-ms (10 point) sliding window. Time 0 indicates the stimulus onset. Middle: Coherence of two conditions, attention inside the RF (solid lines) and attention outside the RF (dashed lines), sliced at 110 ms poststimulus. Note that both conditions have peaks at 66 Hz, but that the peak is larger for the "inside" condition, the same effect as that seen in the power spectra in Fig. 1. Right: Time course of coherence at the peak frequency of 66 Hz for the two conditions. Error bars were obtained by the bootstrap resampling method.

Whereas the power spectrum reflects the activity at a single recording site, the coherence spectrum reflects the functional coupling of neuronal activity between two different sites. An example of time-frequency coherence between the LFPs from two sites for the condition of attention within the RF is shown in Fig. 2 (left). Note that the 66-Hz peak in coherence occurred at approximately the same time and frequency as that seen in the power spectra for site 1 (Fig. 1, left). In addition, a second peak, with a lower coherence, emerged also in the gamma range, near 45 Hz, at a later time. The coherence spectra taken from the analysis window at 110 ms poststimulus for both conditions are shown in Fig. 2 (middle) and the coherence at 66 Hz as a function of time is shown in Fig. 2 (right). At approximately 110 ms poststimulus, the 66-Hz coherence peak is larger for the "inside" condition, an effect similar to that of

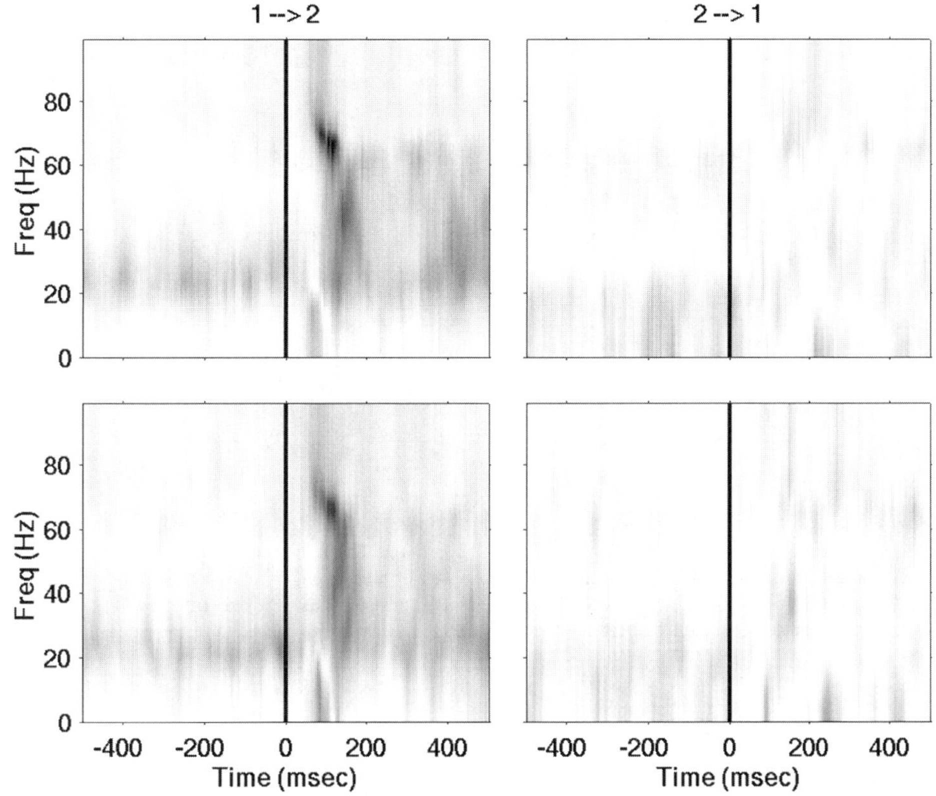

Fig. 3. Time-frequency plots of the DTF between sites 1 and 2 for the two conditions, attention inside the RF (top) and attention outside the RF (bottom), using a 50-ms (10 point) sliding window. DTF values are represented on a color scale from 0 (white) to the maximum value (black). The vertical solid line indicates the stimulus onset.

the power at this frequency (Fig. 1), indicating that attention modulation also affected the coupling between spatially extended V4 ensembles.

The DTF, unlike coherence, is a directional measure of the statistical influence between two processes. Fig. 3 shows two time-frequency plots for each condition: one showing influence from site 1 to 2 (left column), and the other showing the influence from 2 to 1 (right column). In the $1 \rightarrow 2$ plots, we see a pattern similar to the coherence plots for these sites (Fig. 2, left), with the 66- and 45-Hz poststimulus gamma peaks. However, both peaks are largely absent in the reverse ($2 \rightarrow 1$) direction. Therefore, it seems that the 66- and 45-Hz coherences could be attributed to the directional influence from site 1 to site 2.

The time course of DTF for 1 to 2 and from 2 to 1, at 66 Hz for the condition of attention inside the RF is shown in Fig. 4. This pattern suggests that location 1 may drive location 2 with a small, delayed feedback.

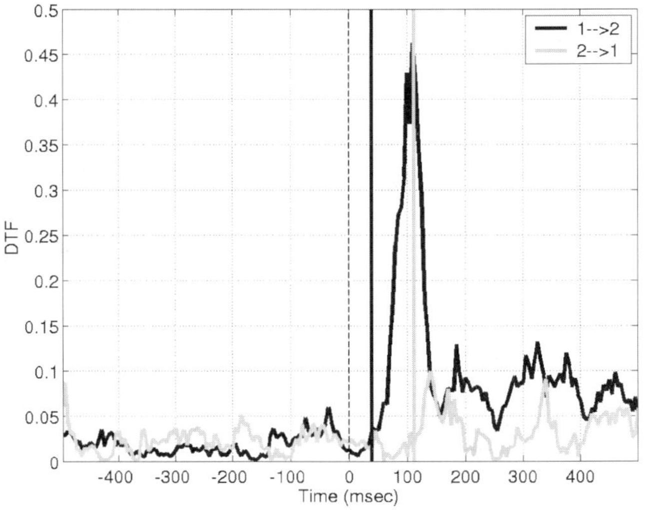

Fig. 4. Time course of DTF at 66 Hz for the "inside" condition. The influence from site 1 to site 2 (black) shows a large peak near 100 ms poststimulus. The onset of this event is marked by the vertical black solid line. The 2-to-1 DTF (gray) shows a much smaller poststimulus increase (marked by the vertical gray line), that is delayed with respect to the 1-to-2 DTF. The vertical dashed line indicates the stimulus onset.

4. Conclusions

In this brief report we examined the temporal structure of attention-modulated neuronal synchronization in macaque area V4 by AMVAR spectral analysis, and found that gamma power (~65 Hz) was elevated when attention was directed to a stimulus within the RF of the neurons at a recording site. We also found elevated LFP coherence in the gamma frequency range due to attention at the same gamma peak frequency, as well as at a second gamma peak (~45 Hz) frequency, among a subset of recording sites. Directional influences between two V4 subpopulations revealed that one direction was more dominant than the other for these two gamma-frequency events. Overall, these results, combined with the recent idea that heightened levels of synchrony among multiple pre-synaptic spike trains increase their post-synaptic efficacy on target neurons, suggest a possible mechanism concerning how attention biases competition among visual stimuli in favor of those that are behaviorally relevant [1].

This report, although preliminary, suggests that the AMVAR methodology, together with the accompanying set of spectral quantities such as power, coherence and DTF, provides a useful tool with high temporal and frequency resolution for characterizing functional relations among neural ensembles during visual selective attention tasks. Currently, the analysis is being expanded to systematically assess the precise timing of attention modulation in the gamma range and other quantitative information from all the recordings sites for all sessions from all monkeys.

Acknowledgements

S.L.B. and M.D. are supported by NSF grant IBN0090717, NIMH grants MH58190 and MH42900, and ONR grant N00014-99-1.

References

[1] R. Desimone, J. Duncan, Neural mechanisms of selective visual-attention, Annu. Rev. Neurosci. 18 (1995) 193–222.
[2] M. Ding, S.L. Bressler, W. Yang, H. Liang, Spectral analysis of cortical event-related potentials by adaptive multivariate autoregressive modeling: model order, stability and consistency, Biol. Cybernet. 83 (2000) 35–45.
[3] P. Fries, J.H. Reynolds, A.E. Rorie, R. Desimone, Modulation of oscillatory neuronal synchronization by selective visual attention, Science 291 (2001) 1560–1563.
[4] G.M. Jenkins, D.G. Watts, Spectral Analysis and its Applications, Holden Day, San Francisco, 1968.
[5] M.J. Kaminski, K.J. Blinowska, A new method of the description of the information flow in the brain structures, Biol. Cybernet. 65 (1991) 203–210.
[6] H. Liang, M. Ding, R. Nakamura, S.L. Bressler, Causal influences in primate cerebral cortex during visual pattern discrimination, NeuroReport 11 (2000) 2875–2880.
[7] S.L. Marple, Digital Spectral Analysis with Applications, Prentice-Hall, Englewood Cliffs, NJ, 1987.
[8] M. Morf, A. Vieira, D. Lee, T. Kailath, Recursive multichannel maximum entropy spectral estimation, IEEE Trans. Geosci. Electron. 16 (1978) 85–94.

Available at
www.ComputerScienceWeb.com
POWERED BY SCIENCE @DIRECT®

ELSEVIER

Neurocomputing 52–54 (2003) 489–495

NEUROCOMPUTING

www.elsevier.com/locate/neucom

A neural model of frontostriatal interactions for behavioural planning and action chunking

Nicola De Pisapia*, Nigel H. Goddard

Institute for Adaptive and Neural Computation, School of Informatics, University of Edinburgh, 5 Forrest Hill, Edinburgh EH1 2QL, UK

Abstract

Neurobiology and neuropsychology interpret the Prefrontal Cortex as a key region, in connection with others, for behavioural planning, i.e. for the generation and evaluation of goal oriented complex courses of actions, without actually executing them. We illustrate an abstract reward-based computational model of planning and its possible neural counterparts. We then focus on how the Dorsolateral PFC and the Striatum learn incrementally to chunk sequences of actions, thus allowing fast and hierarchically structured planning. We test the abstract model by simulating internal model formation and action chunking in motor and cognitive planning problems.

Keywords: Prefrontal cortex; Striatum; Dopamine; Planning; Chunking

0. Introduction

The prefrontal cortex (PFC) and the basal ganglia (BG) are known to play a fundamental role in cognitive control and in the organization of complex behaviour [6,2]. Within this view, we emphasize that the PFC is crucially involved in generating and selecting purposeful complex sequences of actions, i.e. in behavioural planning of how to reach goals in a changing environment. In this paper we propose an abstract computational model of how the PFC, together with the BG, might achieve a planning capacity, and show two simulations.

* Corresponding author.
E-mail address: pisapia@anc.ed.ac.uk (N.De Pisapia).

0925-2312/03/$ - see front matter
doi:10.1016/S0925-2312(02)00753-1

1. Temporal difference learning and temporal abstraction

The BG, and in particular the Striatum, which constitutes the input nucleus receiving connections from the PFC, are known to be involved in the acquisition of sensory motor habits and behavioural repertoires [2]. Experimental data show a high degree of modifiability in the corticostriatal connections, and the key role of the dopaminergic signal coming from the Substantia Nigra Pars Compacta (SNPC) into the Striatum, which acts as an error in the prediction of rewards. This signal has led researchers to construct computational models of the BG using reinforcement learning (RL) [14], and in particular actor/critic models using Temporal Difference (TD) learning, e.g., [9]. TD models in RL learn to associate states of the environment with a value function: $V^{\pi}(t)=E[r(t)+\gamma r(t+1)+\gamma^2 r(t+2)+\cdots]$, which gives an expectation at time t of the present and future rewards r, according to a discount factor $\gamma<1$, and according to a mapping of states into actions π, called a *policy*. The value function allows the system to select which actions lead to the maximal reward, using an error signal in the expectations of the rewards: $\delta(t)=r(t)+\gamma V^{\pi}(t+1)-V^{\pi}(t)$, which corresponds to the signal associated with the dopaminergic activity from the (SNPC), and which also allows the construction of sequences of actions.

To investigate planning in mammals, we make the further central hypothesis that a motor sequence, if repeatedly successful in achieving its goal, is chunked into a higher order action. Chunking has been studied in Artificial Intelligence in a variety of guises (e.g., SOAR [4]). Here we adopt TD learning together with Temporal Abstraction (TA), a recent theoretical development in RL (see [8] for more details).

TA is a mathematical description of how to construct hierarchies of chunks, called *options*. Options are defined by the set of states I in which the option applies, a decision rule μ which specifies what actions are executed by the option, and a completion function β which specifies the probability of completing the option on every time step. Note that also single actions can be treated as options, and that two options a and b can also be composed to produce a new option, ab.

To allow planning, RL agents must predict what happens when one or another option is executed over temporally abstract courses of actions. Expectations of reward and state at the end of the execution of an option o are given by:

$$r_s^o = E[r(t+1)+\gamma r(t+2)+\cdots+\gamma^{(k-1)} r(t+k)\mid \varepsilon(o,s,t)],$$

$$p_{ss'}^o = \sum_{k=1}^{\infty} p(s',k)\gamma^k, \quad \forall s' \in S,$$

where $t+k$ is the random time of termination o, $\varepsilon(o,s,t)$ is the event of o being started in state s at time t, S is the set of all the states, and $p(s',k)$ is the probability that o terminates in s' after k time steps. Note that $p_{ss'}^o$ is a combination of the likelihood that s' is the terminal state for o (e.g., information about *what* events to expect and *where* in the environment), together with a measure of how delayed that outcome will be, relative to γ (*when* to expect the end of an option to happen). These two predictions, one for the reward and the other for the final state, are called a *multi-time model* of an option.

The new extended definition of the value function becomes:

$$V^{\mu}(t) = E[r(t+1) + \cdots + \gamma^{k-1} r(t+k) + \gamma^{k} V^{\mu}(t+k) \mid \varepsilon(\mu, s, t)],$$

where k is the duration of the first option selected by μ. Methods used to learn value functions and selection of the optimal actions, such as TD learning, can now be generalized to learn value functions over models of options.

Given these definitions, one simple method for an agent to plan, i.e. to simulate and select options to be undertaken to achieve goals, is *Synchronous Value Iteration* [14,8]. This consists of a step by step evaluation process of the states of the environment, that starts from the goal state, until it reaches the current state of the agent. After this process, the best current policy can be easily chosen by the agent by taking options that in each time step lead to the higher evaluation.

The key insight here is that by allowing TA, a RL agent can evaluate not only what is the state and the reward after a primitive action is taken, but also which is the state and the reward after a sequence of actions, acting at varying time scales, is taken.

We connect neural responses to our abstract account of planning by modelling modules that build options, learn their internal models and plan as just described.

2. The general model of planning

The modules of our model are motivated by the RL computational requirements for planning, as described in the previous section. Our neuroanatomical hypotheses map each of these functional modules to a specific brain area. While these mappings are supported by experimental evidence, we are well aware of the possibility that any or all of these functions may be subserved by a *network* of brain regions. Nevertheless, we hypothesise:

(1) An Internal Model module, corresponding to Dorsolateral PFC, which updates (multi-time) internal models of the world and maintains them in working memory. These models give information, precisely as in the RL description seen in Section 3, about *where*, *what* and *when* to expect goal related events to happen. The *where* and *what* aspects have been largely studied in monkeys (see [1,9] for example). The *when* aspect is consistent with findings by Watanabe [16] (see next section for a simulation).

(2) An Anticipated Rewards module, corresponding to Orbitofrontal [15] and Anterior Cingulate Cortex [5], which actively maintains goal representations, and therefore evaluates task relevance of different chunks.

(3) An Action Chunks module, corresponding to Striatum in the BG, which incrementally chunks the actions, under the directives of the previous modules, and organizes these chunks following a hierarchical constructive development. The idea of chunking in the BG has been suggested in [2], and tested in interesting works on the dorsolateral Striatum of rats [3].

(4) External inputs to the model are given by high level information about space and object identity in the Perception States module, corresponding to Parietal and Temporal Cortex.

(5) A Prediction Error module which allows learning by sending an error in the prediction of reward occurrence and of task related stimuli. This mimics phasic dopaminergic effects respectively from the SNPC to the Striatum, and from the Ventral Tegmental Area to the PFC [13]. A reward signal is sent at the time of occurrence of a reward to the chunking module [10,11].

(6) Finally an Actions module, which stores the actual actions that define each chunk. This corresponds to Motor Cortex, and also the Cerebellum. This module consolidates internal model representations of the environment once the activation based models in the Internal Model module have been several times successful in its predictions. The shift of the neural activity from frontal areas of the brain to posterior areas has been experimentally verified in motor tasks (for example in [12]).

3. Two simulations

Here we report the results of two simulations, implemented to test parts of the illustrated abstract model. The actual implementations are simple connectionist networks as in [7]. The units are binary, and in different modules they can represent single stimuli, single actions or single options. Learning is based on TD and TA.

First we tested our model to see if the construction of multi-time internal models could match cellular recordings in Dorsolateral PFC. We simulated Watanabe's task [16] to look into temporal aspects of the construction of internal models. In his experiments, the animals (*Macaca fuscata*) had to learn a delayed response task, where they received a specific food if, after a short left or right light presentation, a fixed delay, and a go signal, they pressed a left or right button, depending on where the light was presented. The specific food kept fixed in each session of 50 trials, but then modified in subsequent sessions. Some of the neurons in Dorsolateral PFC displayed spatial specificity, namely they showed activity related to which side the light was presented. Some other neurons, as the one simulated here, didn't show this spatial specificity, but they were related to the specific kind of food that the animal was expecting. These Dorsolateral PFC neurons showed, after training, an anticipatory activity, which started at a low level at the beginning of the task, grew maximally at the expected time of specific food occurrence, and then decayed rapidly. We interpret this activity as the neural codification of multi time models, namely it is a combination of the likelihood that a specific food is the terminal state for an action, together with a measure of how delayed that outcome will be, as reported in Section 1.

In the simulation we divide the task into 11 discrete time steps. The three vertical lines in Fig. 1 indicate from the left respectively cue onset (light on the left, in this case) (step 1), go signal (step 7), and food delivery (step 9). In the figure we see activity of a unit, corresponding to the prediction relative to the probability of terminating *press-left* actions with the specific food. What is showed is actually the learning process of this prediction, trial after trial. The higher curves show the prediction in later trials, and therefore are those that must be compared with Watanabe's plots. As in the biological recordings, in this unit we see that the expectation of option termination grows steadily until the time of delivery.

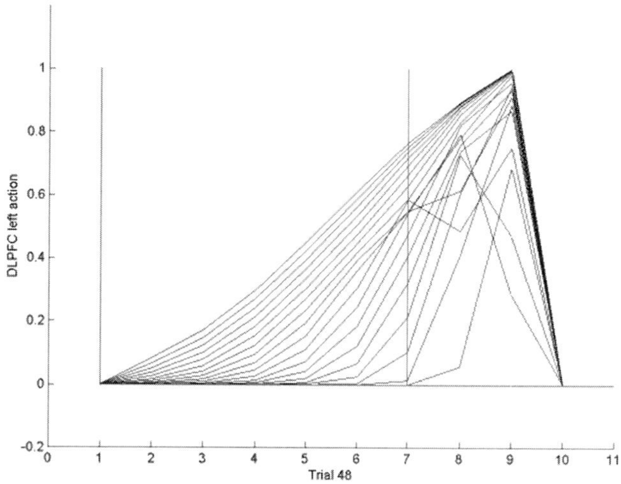

Fig. 1. The learning of the expectation of *when* specific food will be delivered in a simulation of [15].

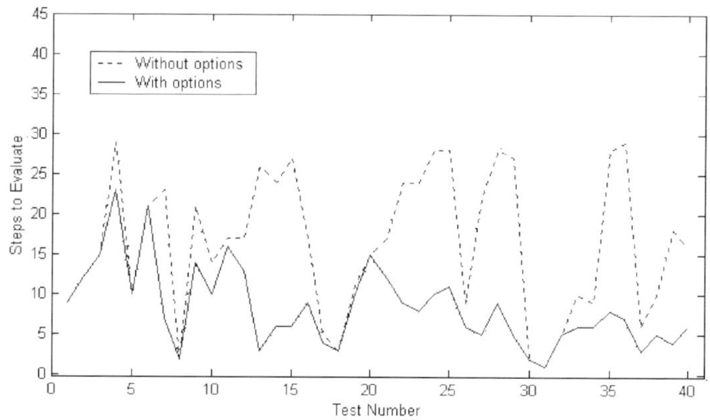

Fig. 2. The Tower of Hanoi simulation using planning and chunking.

In the second simulation we test the chunking capacity of our general model. An agent plays the Tower of Hanoi, a game used in cognitive psychology to study planning capacities in humans. There are five disks of different size and three pegs. Any disk can be moved to any empty peg or onto any other disk of bigger size. Given any starting configuration, the task is to reach a goal configuration in the shortest number of moves. In our simulation initially the agent is capable only of basic actions, and therefore of planning only about simple goals. But as learning takes place, the agent chunks sequences of actions, and therefore plans also about more and more complex (distant) goals.

In the plot in Fig. 2 we compare performance by first generating a sequence of 60 tests (in the form of start/end pairs); we then run the simulation on this specific sequence both with and without the chunking mechanism. What we report here is the number of steps it takes to the evaluative planning process, as in *Synchronous Value Iteration* (see Section 1), to choose the best policy.

The chunking process improves performance considerably during the test sequence, as the model learns chunks from earlier tests (lower numbers). These chunks are used in later tests to plan more efficiently.

4. Discussion

This general reward based model, composed an internal model construction module, an action chunking mechanism, and the capacity to shift action representations to other modules during skill acquisition, is the first attempt to account for actual cellular recordings in Dorsolateral PFC and also computational efficiency in planning.

Several modifications to the general model can be imported. For example, different ways to learn internal models using TA can be configured. The described methods require an option to be executed until termination, others, called *intra-option*, where learning that takes place *off-line*, learn about an option from small fragments of experience consistent with that option, even if the option itself is not executed. In these methods, options not executed and not relevant to the reward that the system is trying to maximize, can nonetheless be improved for a future use.

Another modification regards when the system should actually chunk a sequence of actions. This is currently a major area of research in RL [8]. In the simulation reported here the chunking mechanism comes into play indiscriminately, every time a sequence of actions is successful in achieving its goal. Even this indiscriminate mechanism improves computationally the performance, but more sophisticated procedures can be designed.

Current work is focussing on implementing these modifications, extending the simulations to other parts of the model, and on designing animal and human experiments to further test the hypotheses. Future work will concentrate on enriching the biophysical aspects of the model.

References

[1] P.S. Goldman-Rakic, Cellular basis of working memory, Neuron 14 (1995) 477–485.

[2] A.M. Graybiel, The basal ganglia and chunking of action repertoires, Neurobiol. Learn. Memory 70 (1998) 119–136.

[3] M.S. Jog, Y. Kubota, C.I. Connolly, V. Hillegaart, A.M. Graybiel, Building neural representations of habits, Science 286 (1999) 1745–1749.

[4] J.E. Laird, P.S. Rosenbloom, A. Newell, Chunking in Soar: the anatomy of a general learning mechanism, Machine Learn. 1 (1986) 11–46.

[5] A.W. MacDonald III, J.D. Cohen, V.A. Stenger, C.S. Carter, Dissociating the role of the dorsolateral prefrontal and anterior cingulate cortex in cognitive control, Science 288 (2000) 1835–1838.

[6] E.K. Miller, J.D. Cohen, An integrative theory of prefrontal cortex function, Ann. Rev. Neurosci. 24 (2001) 167–202.

[7] P.R. Montague, P. Dayan, T.J. Sejnowski, Framework for mesencephalic dopamine system based on predictive hebbian learning, J. Neurosci. 16 (1996) 1936–1947.

[8] D. Precup, Temporal abstraction in reinforcement learning, Ph.D. thesis, University of Massachusetts Amherst, 2000.
[9] S.C. Rao, G. Rainer, E.K. Miller, Integration of what and where in the primate prefrontal cortex, Science 276 (1997) 821–824.
[10] W. Schultz, Predicted reward signal of dopamine neurons, J. Neurophysiol. 80 (1998) 1–27.
[11] W. Schultz, Multiple reward signals in the brain, Nature 1 (2000) 199–207.
[12] R. Shadmehr, H. Holcomb, Neural correlates of motor memory consolidation, Science 277 (1997) 821–825.
[13] R. Suri, Anticipatory responses of dopamine neurons and cortical neurons reproduced by internal model, Experiment. Brain Res. 140 (2001) 234–240.
[14] R.S. Sutton, A.G. Barto, Reinforcement Learning: An Introduction, MIT Press, Boston, 1998.
[15] L. Tremblay, W. Schultz, Relative reward preference in primate orbitofrontal cortex, Nature 398 (1999) 704–706.
[16] M. Watanabe, Reward expectancy in primate prefrontal neurons, Nature 382 (1999) 629–632.

Nicola De Pisapia is a Ph.D. student in the Institute for Adaptive and Neural Computation, School of Informatics, at the University of Edinburgh. He received a Master in Logic and Computation from Carnegie Mellon University in 1999, and a Master in Philosophy and Artificial Intelligence from the University of Napoli in 1995. His research interests include neurocomputational models of behavioural planning.

Nigel Goddard is a Reader at the Institute for Adaptive and Neural Computation, School of Informatics, at the University of Edinburgh. He received his Ph.D. in Computer Science from the University of Rochester in 1992. His research interests include collaborative modelling and databasing software for the neurosciences, and neurally-plausible modelling of high-level cognitive function.

Available at
www.ComputerScienceWeb.com
POWERED BY SCIENCE @ DIRECT®

ELSEVIER

Neurocomputing 52–54 (2003) 497–503

NEUROCOMPUTING

www.elsevier.com/locate/neucom

Tuning properties of noisy cells with application to orientation selectivity in rat visual cortex

Axel Etzold*, Christian W. Eurich, Helmut Schwegler

Institut für Theoretische Physik, Universität Bremen, Postbox 330 440, D-28334 Bremen, Germany

Abstract

Common measures for the tuning of cells that are used in the neuroscience literature break down even in the case of moderately noisy neurons. For this reason, a considerable proportion of recorded neuronal data remains unconsidered. One reason for the unreliability of tuning measures is that least-squares fitting of a function for the tuning curve is likely to give too much influence to outliers. We present an algorithm using a rank-weighted norm to construct a tuning curve which weighs outlying data less strongly. As a model function for the tuning curve, we take a trigonometric polynomial, whose coefficients can be determined using a linear approximation. This approach avoids the occurrence of multiple local minima in the optimization process. A test criterion is given to answer the question whether a trigonometric polynomial of lower degree can account for the data. Throughout, we apply our findings to our own experimental data recorded from a population of neurons from area 17 of the rat.

Keywords: Orientation tuning; Stochastic neural responses; Visual cortex

1. Introduction

Since the discovery of orientation selectivity in visual cortical neurons by Hubel and Wiesel [4], experiments to determine orientation tuning curves have been made in many animals. In experimental studies, different measures are used to assess the tuning of cells. As a result, the number of cells from a particular area classified as selective to a particular stimulus can differ greatly from one study to another. We applied some

* Corresponding author.
E-mail address: aetzold@physik.uni-bremen.de (A. Etzold).

0925-2312/03/$ - see front matter
doi:10.1016/S0925-2312(02)00759-2

commonly used measures to our own electrophysiological recordings from area 17 of the rat and found that they pose problems when applied to the noisy responses we obtained. Here, we address two problems associated with these measures: their sensitivity to outliers in the neural responses and difficulties in finding minima in the optimization due to particular classes of tuning functions.

Sections 2–4 treat index methods for the assessment of tuning, problems related to least-squares fits and to fits using particular classes of functions, respectively. In Section 5, we give an algorithm to construct tuning curves avoiding the problems addressed before. Fits based on this algorithm are discussed in the last section.

2. Index methods for the assessment of orientation tuning

In experiments, it is only possible to measure the tuning of cells at a limited number of points. Some researchers use an index method to describe the orientation selectivity of cells [2]:

$$\mathrm{ORI} = \frac{\text{max resp} - \text{orth resp}}{\text{max resp}}, \tag{1}$$

where 'max resp' and 'orth resp' are responses to the optimal orientation and to the orientation orthogonal to the optimal one, respectively. Cells are classified as orientation selective if ORI exceeds a pre-defined threshold [2]. Yet, the cell's true preferred orientation might not be among the measured points and the minimal response need not lie orthogonally to the optimal response, e.g. in the case of additional directional tuning. If the firing of the neuron is noisy, values used to calculate ORI obtained from averaging over a limited number of trials still have considerable jitter. This means that often, the classification of cells may be determined because of noise present at the time of the recording rather than due to the actual stimulus. The index method does not indicate how secure the classification of a cell with respect to its sensitivity is. We suggest that this is a reason why many papers using index methods give different results in the proportion of cells from a population which they classify as sensitive to a particular stimulus.

3. Breakdown of least-squares fits

Generally, cells respond differently to repeated presentations of the same stimulus and typically, some of these answers differ greatly from the rest. Whereas an expert biologist would reject these outliers, least squares estimation gives them great importance in the fit of the model function. A single data point lying sufficiently far apart can suffice to significantly change an estimation based on the least squares method, even if the estimation is based on many data. This difficulty can be addressed by using a different norm, which we are going to present in Section 5.

4. Problems related to the choice of a model function for the tuning curve

Tuning curves are often bell-shaped and very roughly symmetrical. Although it may be useful in a special case, the choice of a particular function as a model for a tuning curve is somehow arbitrary and needs to be done a new in every case. For example, in contrast to our data obtained from the rat, the data obtained in [9] from the cat have very little noise and the variant of a von Mises function [7] used there [9] fits the data well. If the choice of a particular function is not appropriate, the fitted tuning curve can lie far from the measured data, and special features of the tuning like asymmetry or multiple maxima might not be reflected by the model function.

Due to the structure of the modified von Mises function, an estimation of its parameters by solving normal equations is not possible. Applying algorithms that use stochastic elements (e.g. simulated annealing) for the fit instead cannot guarantee that the minimum found by the algorithm is actually a global minimum. Further, the noisier the data are, the more likely they are to induce several fitting parameter sets giving different tuning curves. We address this problem in the following Section.

5. An algorithm for fitting tuning curves using trigonometric polynomials

In the first step, we specify a class of functions for the tuning curve whose parameters can be estimated by an approximation in a linear subspace. One such class of functions is a trigonometric polynomial.

For an experiment with c different orientations, it is given by

$$a_0 + \sum_{k=1}^{d} (a_k \cos k\theta + b_k \sin k\theta)$$

[5,6], where $d = c/2$ if c is even and $d = (c-1)/2$ if c is odd.

Let Y_{ij} be the response of a neuron in the ith trial $i \in \{1, \ldots, c\}$ upon the presentation of a stimulus with orientation $j \in \{1, \ldots, r\}$. Y_{ij} can be written in the following form:

$$Y_{ij} = a_0 + \sum_{k=1}^{c} (a_k \cos k\theta_i + b_k \sin k\theta_i) + \varepsilon_{ij}. \tag{2}$$

We write

$$\mathbf{Y} = (Y_{11} \quad \cdots \quad Y_{1r} \quad Y_{21} \quad \cdots \quad Y_{2r} \quad \cdots \quad \cdots \quad Y_{c1} \quad \cdots \quad Y_{cr})^{\mathrm{T}},$$

$$\boldsymbol{\beta} = (a_0 \quad a_1 \quad \cdots \quad a_d \quad b_1 \quad \cdots \quad b_d)^{\mathrm{T}} = (a_0 \quad \mathbf{a} \quad \mathbf{b})^{\mathrm{T}},$$

$$\boldsymbol{\varepsilon} = (\varepsilon_{11} \quad \cdots \quad \varepsilon_{1r} \quad \varepsilon_{21} \quad \cdots \quad \varepsilon_{2r} \quad \cdots \quad \cdots \quad \varepsilon_{c1} \quad \cdots \quad \varepsilon_{cr})^{\mathrm{T}}$$

and define a matrix $\mathbf{X}$ in such a way that the problem of fitting a tuning curve is equivalent to finding the parameter vector β in the equation $\mathbf{Y}=\mathbf{X}\beta+\varepsilon$, which minimizes $\|\mathbf{Y} - \mathbf{X}\beta\|$. Thus, the residual vector ε which accounts for the stochasticity of the data not explained by the parameter β is minimized in norm. To reduce the impact of single outliers, we construct a different norm in the following way. We rank the entities

$|\varepsilon_{ij}| := |Y_{ij} - \sum_{j=0}^{2d+1} X_{ij}\beta_j|$ in a descending order. If two or more of these numbers are equal, we assign the average of their ranks to all of them. This rank is denoted by $R(|\varepsilon_{ij}|)$. Now we can use

$$\|\boldsymbol{\varepsilon}\|_{w1} = \sum_{i=1,\dots,c\ j=1,\dots,r} R(|\varepsilon_{ij}|)|\varepsilon_{ij}|. \tag{3}$$

as a norm underweighing outliers. A proof that this function is a norm can be found in [3]. Based on this norm, the following algorithm for estimating β can be applied:

1. Set $k=0$. Obtain an initial estimate $\hat{\beta}^{(0)}$. This can be done by making a singular value decomposition of **X** [8]. From this, get an estimate of the variance $\hat{\sigma}^{(0)}$ by estimating the variance of $\varepsilon_{ij}^{(0)} := Y_{ij} - \sum_{j=0}^{2d+1} X_{ij}\hat{\beta}_j^{(0)}$.
   In a least-squares analysis, the algorithm would end here, but note that we use a different norm which weighs distances differently. The first step yields the residuals $\varepsilon = \hat{\varepsilon}^{(0)}$.
2. Based on the singular value decomposition of $\mathbf{X} = \mathrm{UWV}^{\mathbf{T}}$ performed in step 1, compute the matrix $\mathbf{H} = \mathbf{UW}$ which spans the column space of **X**.
3. Let $\mathbf{k} \to \mathbf{k}+1$. Calculate new residuals $\hat{\varepsilon}^{(\mathbf{k})}$ by setting
   $$\hat{\boldsymbol{\varepsilon}}^{(\mathbf{k})} = \hat{\boldsymbol{\varepsilon}}^{(k-1)} - \hat{\sigma}^{(0)} Hw(R(\hat{\boldsymbol{\varepsilon}}^{(k-1)})), \tag{4}$$
   where $w(R(\hat{\varepsilon}^{(k-1)}))$ denotes a vector-valued function which determines the influence of outliers according to the norm chosen. In our case, it is given by $w(i) = \phi(i/(n+1))$, $\phi(u) = \sqrt{12}(u - \frac{1}{2})$ [10]. If the new residuals display a lower dispersion around the fitted model as measured by the norm $\|\cdot\|_{w1}$ the step has been successful. Otherwise a linear search can be made along a direction to find a value which minimizes the dispersion. This is the new residual in step k.
4. By replacing the norm in the normal equations used for approximation in least squares-analysis, it can be shown that the dispersion is minimized by
   $$\beta = \sigma(X^{\mathrm{T}}X)^{-1}\|Y\|_{w1}. \tag{5}$$
   In this equation, we replace **Y** by the residuals $\varepsilon^{(\mathbf{k})}$ in order to obtain a correction term for β which is added to the last estimate of β to produce the new estimate for $\beta^{(\mathbf{k})}$. Based on the new value for $\beta^{(\mathbf{k})}$, we get a new estimate of the variance $\hat{\sigma}^{(\mathbf{k})}$ like in step 1.
5. If the relative drop of the dispersion from one step to the next falls below a predefined threshold, stop the algorithm and accept the last values $\beta^{(\mathbf{k})}$ and $\hat{\sigma}^{(\mathbf{k})}$ as final estimates $\hat{\beta}$, $\hat{\sigma}$. Otherwise, go to step 3.

6. Can a polynomial of lower degree account for the data?

While the algorithm described in the last section uses a very general class of functions from which diverse tuning curves can be fitted, it is desirable to have a simple function for the modeling of a tuning curve. Especially, we expect many of the coefficients of the higher terms in the trigonometric polynomial (3) to be small. To check whether the measured data allow for a simpler model, we use the algorithm described

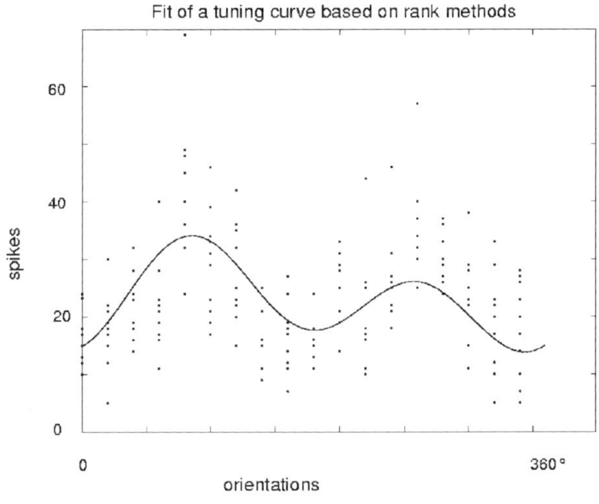

Fig. 1. Examples of recordings from a simple cell. Data were obtained from a total of 80 neurons in rat area 17 (ten animals, Brown Norway) [1]. Stimuli consisted of whole screen black and white gratings, moving with constant velocity (5–20 deg/s) and spatial frequency (0.08–0.6 cycle/deg). Background illumination was kept below 1 cd/m^2 and stimulus intensities ranged from 7 to 10 cd/m^2. Each experiment consisted of several blocks of trials in which 18 stimuli with particular moving direction were presented in a pseudo-random order. In the figure shown, each dot corresponds to a value Y_{ij}. Every orientation condition was repeated 10 times. Equal responses are not shown separately. Thus, the maximum on the right in the scatterplot looks stronger on this plot than it actually is, for the lowest point in the fifth and the second lowest point in the sixth condition from the right occurred three times and twice, respectively, whereas double answers laid more to the median of the scattering distribution in the other conditions. The curve shows a tuning curve obtained by fitting a tuning curve of degree 2 by reducing the degree of the fitting polynomial according to Section 7 using a critical value of $\alpha = 10\%$. The residual distribution (not shown) still presents some trend, which could be avoided by choosing a lower critical value in the test described in Section 7.

in the previous section to fit a trigonometric polynomial (3) of lower degree than d. This gives a distribution of residuals ε_{ij}^{red} for the reduced model. If the reduced model can account for the data, these residuals will have an approximately symmetrical distribution around zero. Based on the residual distribution, we calculate the sum of the signs of the residuals in all trials. If the fitted model function accounts well for the data, the distribution of the residuals is a binomial distribution of cr residual points around zero with a probability that the sign of a residual is positive or negative which is $\frac{1}{2}$. We then accept the best approximating polynomial of the lowest degree as fitted by the algorithm presented as a fitting tuning curve, whose residual distribution cannot be identified as non-symmetrical around zero by this test.

In our recordings, we found that 14 out of 80 cells were sharply tuned ($\alpha < 1\%$), 9 further cells moderately tuned ($1\% < \alpha < 5\%$) and the rest untuned (i.e. the tuning curve was constant for $\alpha > 5\%$). Figs. 1 and 2 show examples of a tuned and an untuned cell, respectively.

As another example, we show a fit of a cell classified *not tuned* by using the trigonometric polynomial of the highest degree. For this neuron, this test gave a constant fit for critical levels higher than 1%.

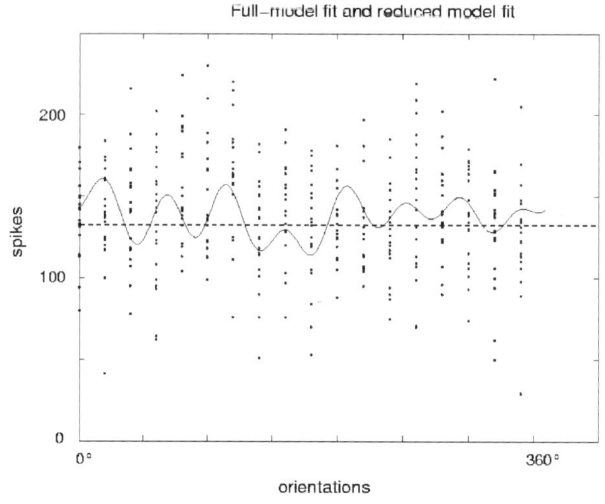

Fig. 2. Examples of recordings from a simple cell. Every orientation condition was presented 20 times, equal answers are not shown separately. The continuous curve shows the fitted tuning curve using a fit of a polynomial of maximal degree. *However, none of the higher terms was significant.* We accept the dashed constant function as the tuning curve.

7. Discussion

The algorithm presented will deterministically find an approximation of the tuning curve by a trigonometric polynomial. This avoids using stochastic optimization methods. It also gives a criterion for the reduction of parameters. The approximation polynomial of maximal degree allows to have asymmetries and several maxima in the tuning curve. If the scattering of the residuals around the fitted polynomial of maximal degree does not differ significantly from a symmetric distribution, a simpler model is chosen for the tuning curve.

However, a class of polynomials approximating the tuning function more closely with respect to the given norm might be found. This work is currently in progress.

Acknowledgements

Data were kindly provided by Winrich Freiwald, Heiko Stemmann and Aurel Wannig from the Department of Biology, University of Bremen, Germany.

References

[1] W.A. Freiwald, H. Stemmann, A. Wannig, A.K. Kreiter, U.G. Hofmann, M.D. Hills, G.T.A. Kovacs, D.T. Kewley, J.M. Bower, C.W. Eurich, A. Etzold, S.D. Wilke, Stimulus representation in rat primary visual cortex: multi-electrode recordings with micro-machined silicon probes and estimation theory, Neurocomputing 44–46 (2002) 407–416.

[2] S.V. Girman, Y. Sauvé, R.D. Lund, Receptive field properties of single neurons in rat primary visual cortex, J. Neurophysiol. 82 (1999) 301–311.
[3] G.H. Hardy, J.E. Littlewood, G. Pólya, Inequalities, 2nd Edition, Cambridge University Press, Cambridge, 1952.
[4] D.H. Hubel, T.N. Wiesel, Receptive fields of single neurons in the cat's primary visual cortex, J. Physiol. 148 (1959) 574–591.
[5] K. Itô (Ed.), Encyclopedic Dictionary of Mathematics, MIT Press, Cambridge, MA, and London, 1996.
[6] D. Jackson, The Theory of Approximation, American Mathematical Society Colloquium Publications, Providence, RI, 1930.
[7] K.V. Mardia, P.E. Jupp, Directional Statistics, Wiley, Chichester, 2000.
[8] W.H. Press, S.A. Teukolsky, W.T. Vetterling, B.P. Flannery, Numerical recipes in C, 2nd Edition, Cambridge University Press, Cambridge, 1992.
[9] N.V. Swindale, Orientation tuning curves: empirical description and estimation of parameters, Biol. Cybernetics 78 (1998) 45–56.
[10] H. Witting, G. Nölle, Angewandte Mathematische Statistik, Teubner, Stuttgart, 1970.

Axel Etzold, born in 1974, got his master's diploma in Mathematics in 2000 from the University of Giessen (Germany). He currently is a Ph.D. student in Prof. Dr. Schwegler's and Dr. Eurich's group at Bremen university. His research interests include signal processing and encoding in neural populations, dynamical systems modeling and statistical data analysis methods.

Christian W. Eurich, born in 1965, got his Ph.D. in Theoretical Physics in 1995 from the University of Bremen (Germany). As a postdoc, he worked in the Departments of Mathematics and Neurology at the University of Chicago, and he was guest researcher at the Max-Planck Institut für Strömungsforschung in Göttingen and at the RIKEN Brain Institute in Tokyo. In 2001, he held a professorship for Cognitive Neuroinformatics at the University of Osnabrück. Currently, Christian Eurich is Research Assistant at the Institute for Theoretical Neurophysics at the University of Bremen. His research interests include signal processing and encoding in neural populations, neural dynamics, and motor control problems such as balancing tasks and postural sway.

Helmut Schwegler, born in 1938, studied mathematics and physics at München and Darmstadt, he was Professor of Physics at the Technical University of Darmstadt, since 1971 he is Professor of Theoretical Physics and Theoretical Biophysics at the University of Bremen. Beside his activities in the Physics Department he was a participant in the interdisciplinary research project "Biosystems Research" of the university until 1990, afterwards in a project "Cognitive Autonomous Systems", and since 1995 he is a member of the interdisciplinary "Center of Cognitive Sciences". He is also a member of the "Center of the Philosophical Foundations of Science" of the university. His scientific work concerns different fields of Theoretical Physics and Theoretical Biology, Cognition Research, Systems Theory and Philosophy of Science.

Available at
www.ComputerScienceWeb.com
POWERED BY SCIENCE @ DIRECT®

ELSEVIER

Neurocomputing 52–54 (2003) 505–510

NEUROCOMPUTING

www.elsevier.com/locate/neucom

Power spectral analysis of very slow brain potential oscillations in primary visual cortex of freely moving rats during darkness and light

Igor V. Filippov*

Department of Physiology and Biophysics, Yaroslavl State Medical Academy, Revolutsionnaya st. 5, Yaroslavl City 150000, Russia

Abstract

Previous reports demonstrated the presence of very slow oscillations (frequencies less than 1 Hz) in primary visual cortex (V1) neurons. In this work the hypothesis that very slow brain potential (VSBP) oscillations in V1 accompany illumination changes was tested. In darkness (0 lux), the results revealed the presence of oscillations in the range of seconds, multisecond and minute fluctuations. Continuous light exposure (2000–2500 lux) induced statistically significant ($P < 0.001$) dramatic changes only in the frequency domain of seconds. These observations permit the conclusion that in V1 VSBP reproducible changes in the range of seconds occur in response to light and darkness.

Keywords: Slow oscillatory activity; Very slow brain potential oscillations; Primary visual cortex

1. Introduction

Very slow oscillatory phenomena (with frequencies less than 1 Hz) were described in cortical neurons in different current research articles [15]. Investigations of electroencephalographic (EEG) recordings also revealed existence of similar rhythms in this type of CNS activity [2,11,15]. The accumulating evidence demonstrates a wide range of frequencies of brain activity, including very slow spontaneous extracellular potential fluctuations (oscillations) ranging from several seconds to more than several minutes [4]. Formerly, such signals were referred to as "DC-shifts" or "slow"

* Tel.: +7-852-305-763; fax: +7-852-305-013.
E-mail address: filippov@yma.ac.ru (I.V. Filippov).

0925-2312/03/$ - see front matter
doi:10.1016/S0925-2312(02)00761-0

potentials, and there are many reports dedicated to this topic [1,8,10,12,14]. These processes, also known as very slow brain potential (VSBP) oscillations, are categorized into different frequency bands. These are "minute (60 s)" oscillations or fluctuations (less than 0.0167 Hz), "multisecond" oscillations or fluctuations (0.1–0.0167 Hz), and the frequency band (0.1–0.5 Hz) whose period is ca. 1 s, known as "second" oscillations or fluctuations [2,9].

Albeit several recent reports have shown the presence of very slow oscillatory activity at the different levels of the visual system [3,15,16], the extracellular dynamics of VSBP oscillations in visual system is mainly undocumented, and the functional significance of this activity remains poorly known. In this work, the hypothesis that VSBP oscillation changes in the primary visual cortex should accompany changes in illumination was tested.

2. Methods

The experimental part of this research was carried out in accordance with the "Principles of laboratory animal care" (NIH publication No. 86-23, revised 1985).

Animal surgery and general experimental design: Experiments were performed on 10 adult freely moving male albino rats. Under general anesthesia (sodium pentobarbital, 40 mg/kg i.p.) and according to the rat brain atlas [13], two electrodes were stereotaxically implanted across the primary visual cortex (V1) for long-term chronic extracellular recordings. Experimental studies were initiated after 14 days from the implantation. During experiments, animals were placed in a large Faraday cage and isolated in special individual compartments with a light source on the top and light-reflecting non-transparent walls. Freely moving within this cage the animal was then exposed to different background illumination levels: darkness (0 lux) or continuous ambient light (2000–2500 lux). Electrophysiological recordings of the VSBP from the primary visual cortex were performed in both darkness ($n = 70$) and light ($n = 70$) conditions in each experimental animal daily. After the end of an experimental series, the rats were devitalized using an injection of sodium pentobarbital (100 mg/kg i.p.). The brain was prepared in order with the methods described in detail in handbook [5], so that classical light microscopy could be used to determine the electrode tracks in the aforementioned brain area together with visual inspection of electrode positions using the topographical cytoarchitectural maps of the rat cortex [13].

Electrophysiological technique, special test procedures and measurements: Extracellular bipolar recordings were obtained with the implanted trans-cortical gold electrode pair (horizontal inter-electrode distance, 2 mm). Electrodes were manufactured from 98% gold wire and had the diameter 0.35 mm, with the active area length 2 mm. High input impedance (2×10^8 Ω), differential multi-channel low-noise (internal noise in 0–0.5 Hz bandpass, less than 3 μV), low-bias current (less than 10^{-12} A), and high-stable (zero line drift not more than 15 μV/h after setting the temperature mode) universal AC/DC-amplifier with DC-offset compensation (model UU-93, EPM IEM RAMN, St.-Petersburg, Russia) was used for the electrophysiological measurements. Amplified signals were continuously digitized on-line at a sampling frequency of 1 Hz by analog-to-digital converter board (model KPCI-3101, Keithley Instruments, Inc., Cleveland, Ohio, USA) and subjected to the personal computer for storing, preprocessing, and further off-line analysis.

Some authors [6,8,9] have shown previously that gold metal electrodes in conjunction with high input impedance amplifiers are adequate tools for chronic VSBP oscillation recordings (up to frequencies of 0.008 Hz). However, since this type of metal electrode was used for very slow potential recordings, special attention was paid to measures aimed to eliminate possible artifacts because, apparently, very slow potential fluctuations and shifts may have an artifactual origin. These special measures were: (1) all electrodes were absolutely identical. (2) Before the implantation procedure, careful individual selection of electrodes for the pairs was used: all electrodes were tested in physiological saline (recordings were made with the same electrophysiological research equipment settings and in the absolutely same conditions as were used for experimental animal studies) and only those electrodes that had the stable standing inter-electrode potential without any transient shifts, fluctuations, noise, etc. were used for the pairs. (3) Other tests ($n = 20$) with the same paradigm were aimed to exclude the possible artifact effect(s) of metal electrode illumination on the inter-electrode potentials in the very slow frequency domain. For these purposes the bath with physiological saline solution was placed into experimental cage (where the animals were tested) and recordings were performed in darkness and during illumination in the same conditions as during animal tests with illumination changes. As a result, no statistically significant differences between inter-electrode potentials in the very slow frequency domain in darkness and during illumination in these control experimental series has been found. (4) Only those time fragments of electrograms when animal moving activity was minimal or absent were analyzed.

Statistical analysis: For analysis, the following length of acquired data was selected. For the ca. 1 Hz oscillation band, 256 data points (=256 s) and for the multisecond band, 512 points (=512 s). Each segment was subjected to the fast Fourier transformation (FFT) and power spectral analysis. The obtained data were analyzed by two-way ANOVA. Statistical difference with $P < 0.05$ was considered significant.

3. Results

In general, all recordings (both in darkness and during light exposure) typically displayed the presence of spontaneous VSBP oscillations (Figs. 1A and B) in the primary visual cortex with the following properties: regular quasi-sinusoidal ca. 1 Hz oscillations with peak-to-peak amplitude up to 0.3 mV, relatively regular quasi-sinusoidal multisecond oscillations with peak-to-peak amplitude up to 1 mV, and seldom irregular spontaneous sinusoidal minute (60 s) fluctuations with peak-to-peak amplitude up to 3 mV.

In darkness, results of the power spectra analysis (Figs. 1C and D) demonstrated that: (1) The main ca. 1 Hz or second range oscillations were within 0.1–0.21 Hz frequency domain, with several peaks in this band and three significant peaks around 0.11, 0.12 and 0.15 Hz; the middle one was the most prominent. (2) In the multisecond oscillation

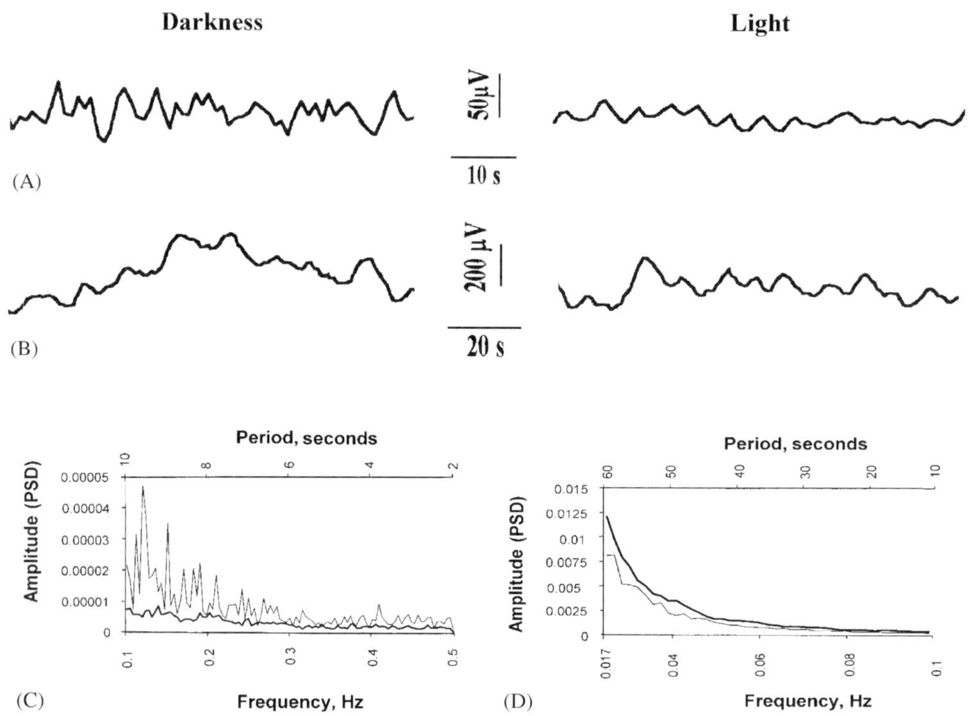

Fig. 1. Examples of typical very slow oscillation recordings from primary visual cortex of the rat in darkness and during illumination in two frequency domains: (A) second oscillations and (B) multisecond oscillations. In this figure, graphical results of power spectral analysis for the different frequency domains during darkness (thin line) and light exposure (thick line) are also presented: (C) second oscillations, averaged over $n = 70$ graphs and (D) multisecond oscillations, averaged over $n = 50$ graphs. Note: Abbreviation PSD on C and D means power spectral density; amplitude on these graphs is expressed using absolute scale units (mV/Hz). The resolution of the power spectra on C and D is not proportional in the three scales (e.g. number of points per octave).

range, waves in the band of 0.0167–0.03 Hz without particular peaks prevailed. But it is necessary to note that the relative power was higher in the slowest types of this activity.

During illumination, there were observed statistically significant ($P < 0.001$) dramatic changes only in the ca. 1 Hz or second frequency domain (Fig. 1C). It can be seen that, in contrast to the darkness, light exposure induced a remarkable decrease in the power of oscillations in the frequency domain of 0.1–0.21 Hz. During illumination, no statistically significant differences ($P = 0.3187$) in the multisecond frequency domain of the primary visual cortex in comparison to darkness was found (Fig. 1D).

4. Discussion

Aforementioned findings are consistent with the modern data regarding the very slow spontaneous oscillatory activity in the visual system. Albrecht and colleagues [3] have described the very slow oscillations in the dorsal and ventral lateral geniculate nucleus of the urethane anesthetized and awake freely moving rats during darkness and continuous illumination. It has been found that many lateral geniculate neurons display a strong very slow oscillatory behavior in the range of 0.025–0.01 Hz. It was also demonstrated that such activity could be significantly altered by the change from darkness to light in the direction that continuous illumination blocked these very slow oscillations. There are some data (also agreeing with this work results) regarding the presence of very slow fluctuations in the neocortical extracellular recordings during wakefulness, i.e. oscillations of 0.02–0.068 Hz were determined in human cortical EEG at rest and were described as very slow EEG modulations [11]. Very slow fluctuations of visual threshold in the frequency domain of 0.02–0.068 were also shown in psychophysiological tests [16].

The results on the band whose period is ca. 1 s or second oscillations (0.1–0.5 Hz) demonstrate correlation between this activity in primary visual cortex and the level of illumination. Accordingly, it is logical to propose that this very slow frequency domain may possibly reflect some brain mechanisms implicated in the process of vision. However, it is possible to assume that the very slow oscillations of ca. 1 Hz range could be associated with the mechanisms of visual attention sustaining and switching during visual environment changes. Indirect support for this hypothesis is the report of spontaneous attentional fluctuations with frequencies of 0.083–0.33 Hz [2].

Basing on the experimental data, it is possible to suggest that VSBP multisecond oscillations and minute (60 s) fluctuations are associated with spontaneous fluctuations of functional states in primary visual cortex neurons. These spontaneous fluctuations could reflect periodical influences from the brainstem nuclei (e.g. locus coeruleus, dorsal raphe nuclei, etc.), which form massive axonal projections within the primary visual cortex [13]. This is in agreement with the previously proposed hypothesis [11] that the very slow EEG modulations occur as a result of spontaneous fluctuations in the excitability of cortical neurons, controlled by brain stem structures. Previous studies made at our laboratory in which the presence of very slow oscillations in the locus coeruleus-frontal neocortex of corresponding frequencies was discovered [7] also provide strong support for the above proposal.

In conclusion, the obtained results show the presence of a variety of different forms of VSBP oscillations (ca. 1 Hz or second, multisecond oscillations, and minute fluctuations) in the extracellular recordings from V1 of rats. Different functional significance of the different frequency domains of this activity in response to sensory stimulation of the change to darkness or light was determined. These data allow author to hypothesize possible relations between oscillations in the band whose period is ca. 1 s and mechanisms of specific visual attention shifts, while the multisecond oscillations, minute (60 s) fluctuations, obviously, reflect processes of global excitability of cortical neuronal networks during different illumination levels.

Acknowledgements

I am very grateful to Professor Theodore H. Bullock (Department of Neurosciences, University of California, San-Diego, California, USA) for his helpful and valuable discussions on this manuscript prototype. I also appreciate technical help of Eric E. Bloam (Application Engineer at Keithley Instruments, Inc., Cleveland, Ohio, USA).

References

[1] W.R. Adey, Slow electrical phenomena in the central nervous system, Neurosci. Res. Program Bull. 7 (2) (1969) 75–180.
[2] N.A. Aladjalova, Psychophysiological Aspects of Brain Infra-slow Rhythmical Activity, Nauka, Moscow, 1979 (in Russian).
[3] D. Albrecht, G. Royl, Y. Kaneoke, Very slow oscillatory activities in lateral geniculate neurons of freely moving and anesthetized rats, Neurosci. Res. 32 (1998) 209–220.
[4] T.H. Bullock, Slow potentials in the brain: still little understood but gradually getting analytical attention, Brain Res. Bull. 50 (5–6) (1999) 315–316.
[5] J. Bures, O. Buresova, J.P. Huston, Techniques and Basic Experiments for the Study of Brain and Behavior, Elsevier Science Publishers B.V, Amsterdam, NY, 1983.
[6] R. Cooper, J.W. Osselton, J.C. Shaw, EEG Technology, Butterworth, London, 1969.
[7] I.V. Filippov, Infraslow brain physiological processes—a new approach to neurophysiology of locus coeruleus and cerebral cortex in stress, in: Materials of 21st International Summer School of Brain Research, Cognition, Emotion and Autonomic Responses, Amsterdam, August 23–27, 1999, p. 101.
[8] R.J. Gumnit, H. Matsumoto, C. Vaconetto, DC activity in the depth of an experimental epileptic focus, Electroenceph. Clin. Neurophysiol. 28 (1970) 333.
[9] V.A. Iliukhina, Slow Electrical Activity of the Human Brain, Nauka, Leningrad, 1977 (in Russian).
[10] M.E. Lickey, Localization and habituation of sensory evoked DC responses in cat cortex, Exp. Neurol. 15 (1966) 437–454.
[11] P. Novak, V. Lepicovska, Slow modulations of EEG, Neuroreport 3 (1992) 189–192.
[12] J.L. O'Leary, S. Goldring, DC-potentials of the brain, Physiol. Rev. 44 (1964) 91–125.
[13] S.N. Olenev, Construction of the Brain, Leningrad, Meditsina, 1987 (in Russian).
[14] V. Rowland, Slow potential phenomena of cortex, in: G.C. Quarton, T. Melnechuk, F.O. Schmitt (Eds.), The Neurosciences—A Study Program, Rockefeller University Press, New York, 1967, pp. 482–495.
[15] M. Steriade, A. Nunez, F. Amzica, A novel slow (< 1 Hz) oscillation of neocortical neurons in vivo: depolarizing and hyperpolarizing components, J. Neurosci. 13 (1993) 3252–3265.
[16] F. Thoss, B. Bartsch, J. Stebel, Analysis of oscillation of the visual sensitivity, Vision Res. 38 (1998) 139–142.

Available at
www.ComputerScienceWeb.com
POWERED BY SCIENCE @ DIRECT®

ELSEVIER

Neurocomputing 52–54 (2003) 511–518

NEUROCOMPUTING

www.elsevier.com/locate/neucom

The neural multiple access channel

Brian J. Fischer*, M. Brandon Westover

Washington University in St. Louis, St. Louis, USA

Abstract

In many neural systems, independently encoded information must at some point be transmitted over the spike train of one neuron. We introduce a method for quantitatively studying the effects of the signal encoding and transmission processes on the rates of transmission of multiple sources of information over one spike train, using the multiple access channel model from network information theory. To illustrate this method we study the effects of a small set of synaptic input patterns and input signal power spectra on the information capacity region of a simple three-neuron system.

Keywords: Neural systems; Signal encoding; Transmission; Multiple access channel

1. Introduction

The functional properties of sensory neurons are generally characterized by how they transmit and transform information. In the case of transformations, neurons are characterized by tuning curves or receptive field maps with such descriptions giving insight into how neurons might perform computations on the input signals. In the case of transmission, fundamental limits to the amount of information throughput are calculated without regard to what specific computations are performed on the input. Such information theoretic analyses are important in that they place limits on the types of information processing tasks that the neuron, or system, can perform.

One information processing task that has yet to be analyzed is the simultaneous transmission of multiple signals over the spike train of one neuron. In this paper we

* Corresponding author.

E-mail addresses: fischerb@thalamus.wustl.edu (B.J. Fischer), westovem@medicine.wustl.edu (M.B. Westover).

0925-2312/03/$ - see front matter
doi:10.1016/S0925-2312(02)00762-2

develop a method for quantifying the ability of a single neuron to transmit information about multiple source signals using information theoretic principles. Specifically, by considering a simple three neuron system we develop a mapping between this neural coding problem and a standard model of network information theory, the multiple access channel. Using our simple example we show how one can approach the questions: Supposing multiple "senders" of neural information must transmit their information through a common neuron, what strategies will allow (1) maximal transmission of information from only one of the senders, or (2) maximum simultaneous transmission of information from both senders?

1.1. Multiple access channel

The multiple access channel is characterized by the presence of two or more sources of information and one receiver; here we discuss the case of two sources. The multiple access channel communication system consists of two message sets, two encoding functions, a collection of probability transition functions relating the input to the output, and a decoding function (see Fig. 1). Each message set is a set of integer message indices, $\{1, 2, \ldots, 2^{nR_i}\}$. The encoding functions map the indices of the messages to signals appropriate for transmission by the channel. Communication of information over physical devices is inherently a stochastic process and thus the relationship between the input and output of the channel must be expressed in terms of probability transition functions. The deterministic decoding function produces an estimate of the input from the output. The message sets are included for completeness in the description of the multiple access channel and should be thought of as indices of the signals that will be transmitted over the channel.

The important well-known theoretical result regarding the multiple access channel is the fact that the capacity region is determined by the mutual information between the input and the output random variables. This is stated precisely in the following theorem taken from Cover and Thomas [1].

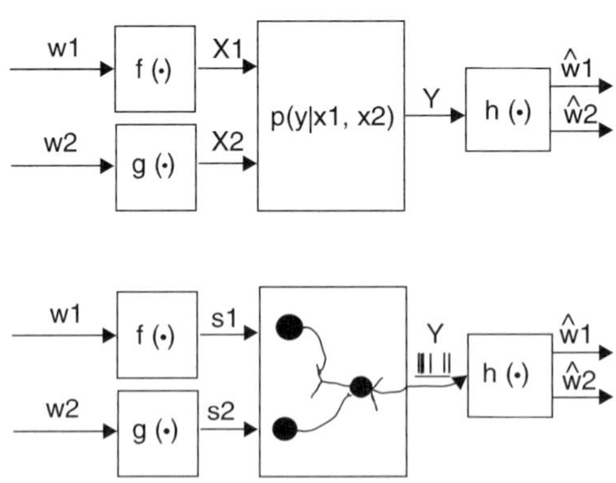

Fig. 1. Diagram for the multiple access channel (top) and neural equivalent (bottom).

Theorem 1. *The capacity region of the multiple access channel* $(X_1 \times X_2, p(y|x_1,x_2), Y)$ *is the closure of the convex hull of all* (R_1, R_2) *satisfying*

$$R_1 < \max_{p(X_1)p(X_2)} I(X_1; Y|X_2), \qquad R_2 < \max_{p(X_1)p(X_2)} I(X_2; Y|X_1)$$

$$R_1 + R_2 < \max_{p(X_1)p(X_2)} I(X_1, X_2; Y)$$

for some product distribution $p(x_1)p(x_2)$ *on* $X_1 \times X_2$.

In this paper we consider a set of possible encoding strategies used by a three neuron system limited to the variable arrangement of synaptic inputs from two presynaptic neurons onto a postsynaptic neuron. Each presynaptic neuron encodes one of two source signals and we consider the effects of the power spectrum of each signal on information transmission. This problem maps onto the above model in the following way. Each selection of a connection pattern between presynaptic and postsynaptic neurons defines a new channel. The message sets remain sets of integers representing the indices of the messages to be sent. The encoding functions map the elements of the message sets to stationary Gaussian random signals characterized by one of three power spectra defined below. The output is taken to be the spike train of the postsynaptic neuron. To quantify the performance of the system, we approximate the capacity region of the channel for each input location combination. We compute the maximum mutual informations in Theorem 1 by performing a complete search over pairs of signals from the three power spectra.

2. Methods

2.1. Model

Our neural multiple access channel consisted of two presynaptic neurons and one postsynaptic neuron. Signals, $s_i(t)$ $i=1,2$, of length 5 s were generated using MATLAB and were encoded into spike trains by two identical presynaptic leaky integrate and fire (LIF) neurons. Signals were Gaussian random signals with one of three power spectra: (1) low pass, (2) $1/f^2$, and (3) bandpass, peaked at 30 Hz. The two LIF neurons formed the presynaptic input to a more biophysically detailed model neuron, created using NEURON [2]. The postsynaptic neuron consisted of a soma, axon, two basal dendrites, an apical trunk, and two distal apical branches. The axon, soma, and apical trunk contained Hodgkin Huxley type Na^+ and K^+ channels while the remaining branches were passive.

Although synaptic dynamics and the reliability of synaptic transmission are factors in determining the amount of information transmission between real neurons [4,5], in this investigation we used only a simplified, deterministic, model of synaptic transmission. The input signals, $s_i(t)$, were encoded by the LIF neurons; the output of these neurons were taken as lists of spike times $\{t_1, \ldots, t_n\}$. At each synaptic contact made by a presynaptic LIF neuron onto the postsynaptic neuron, the onset of a conductance increase occurred at each time that the presynaptic neuron spiked. The synapses were modelled as alpha synapses with a time constant of $\tau = 3$ ms.

Input LIF spike trains carrying each signal $s_i(t)$ were transmitted to the postsynaptic neuron via 60 (except in the default case of no input) identical synaptic inputs arranged in one of the four following patterns: (1) no input, (2) uniformly distributed over the soma and all dendritic branches, (3) uniformly distributed over the distal apical branches, and (4) uniformly distributed over the soma. We emphasize that this setup is a simplification. However, the paradigm and methods developed here apply equally well to more realistic models.

2.2. Calculating the capacity region

To analyze the effects of both the signal power spectrum and the location of the presynaptic inputs on information transmission, we have defined separate multiple access channels for each of the 16 combinations of location patterns. To calculate the boundary of the capacity region for each of these channels we compute three quantities:

$$\max_{p(s_1)p(s_2)} I(s_1;Y|s_2), \qquad \max_{p(s_1)p(s_2)} I(s_2;Y|s_1), \qquad \max_{p(s_1)p(s_2)} I(s_1,s_2;Y),$$

where s_1, s_2 are the input signals and Y is the list of spike times $\{t_1,\ldots,t_n\}$ of the postsynaptic neuron in our model system. For each of these quantities the maximum over all distributions on the input signals will be taken with respect to the set of three possible distributions defined above.

Calculating the mutual information between Gaussian random signals and spike trains exactly is impractical because of the large amount of data required to estimate spike train probability densities [3]. Consequently, we calculate lower bounds on the mutual information using the reconstruction method advocated by Rieke et. al. [3]. In brief, this involves computing optimal linear reconstruction filters for the spike trains; the mutual information is then computed by comparing the signal with the spike filtered estimate and assuming a standard Gaussian channel model relationship between the signal and estimate. The novelty of our calculation here is that we compute the *conditional* mutual information. To our knowledge, this is the first presentation of such calculations. First we show that an estimator of the type used in our calculations results in a lower bound.

Let $\tilde{s}_1 = \tilde{s}_1(s_2, Y)$ be any estimator of s_1 using s_2, and Y. Then,

$$\begin{aligned} I(s_1;Y|s_2) &= I(s_1;Y,s_2) - I(s_1;|s_2) \geqslant_a I(s_1;\tilde{s}_1,s_2) - I(s_1;|s_2) = I(s_1;\tilde{s}_1|s_2) \\ &= H(s_1|s_2) - H(s_1|\tilde{s}_1,s_2) \geqslant_b H(s_1|s_2) - H(s_1|\tilde{s}_1) \\ &=_c H(s_1) - H(s_1|\tilde{s}_1) = I(s_1;\tilde{s}_1), \end{aligned}$$

where we have used: (a) the data processing inequality because $(\tilde{s}_1, s_2)$ is a function of (Y, s_2), (b) conditioning reduces entropy, and (c) s_1 and s_2 are independent.

Now, let $\tilde{s}_1 = s_1 + g(s_2, \eta)$ where $g(s_2, \eta)$ represents the noise in the approximation of s_1 due to the presence of s_2 and noise inherent in the approximation procedure, η. Let $\tilde{s}_1 = \tilde{s}_1(\omega)$, $s_1 = s_1(\omega)$, and $g = g(\omega)$. Then, the information as a function of frequency satisfies:

$$I(s_1;\tilde{s}_1) = H(\tilde{s}_1) - H(\tilde{s}_1|s_1) = H(\tilde{s}_1) - H(s_1 + g|s_1) = H(\tilde{s}_1) - H(g|s_1)$$
$$\geqslant_a H(\tilde{s}_1) - H(g) =_b \tfrac{1}{2}\, log_2(1 + SNR(\omega)),$$

where we have used: (a) conditioning reduces entropy and (b) assuming that g is Gaussian; $SNR(\omega)$ is the ratio of the power spectra of the output to the input. The total information is obtained by integrating over ω. To approximate $I(s_1;Y|s_2)$ we treat s_2 as noise and average over trials. Specifically, we fix s_1 and average the spectrum of the output spikes, $Y(\omega)$, over 10 independent realizations of s_2. We then estimate s_1 using the averaged spike train spectrum and compute the information as described above. Note that this estimate of the conditional mutual information is a lower bound, but is larger than what would be obtained without knowledge of s_2.

Similarly it can be shown that $I(s_1,s_2;Y) = I(s_2;Y) + I(s_1;Y|s_2) \geqslant I(s_2;\hat{s}_2) + I(s_1;\tilde{s}_1)$ where $\tilde{s}_1$ is the estimator obtained by averaging spike train spectra to reduce the noise and $\hat{s}_2$ is the estimator obtained in the standard way without averaging.

We perform these calculations for all s_1, s_2 pairs and search over these to find the distributions which achieve the maximum. These are the points that we use to construct the capacity regions for each channel.

3. Results and discussion

This experiment was performed to illustrate that formulating questions about the transmission of multiple signals over a neural system in terms of the multiple access channel model facilitates a quantitative interpretation of the performance of the system. Our experiment addressed the simple question of how the location of synaptic inputs onto a neuron and the content of the signals determines the maximum information rates at which the two signals can be sent over the spike train of one neuron.

As expected, the results show that it is possible to send more information about the signals with certain spatial configurations of synaptic inputs than in others. This is easily seen by the varying sizes of the capacity regions in Fig. 2. Points (i,j) on the figure indicate the boundary of the capacity region obtained when sender 1 uses location pattern i and sender 2 uses location pattern j. Surprisingly, the maximum rate that a single sender can use is increased by the presence of input from the other source. The points on the $R2$ axis, (1,2), (1,3), and (1,4), correspond to the maximum rate that a single sender can use when inputs are placed in patterns 2, 3, and 4, respectively. Contrasting these points on the $R2$ axis with those of the curves for location pairs (2,2), (4,4), (2,4), (3,3), and (2,3) shows that the maximum rate that a sender can use is increased when input from another source is present. Another unexpected result is that the maximum rate for a single source is achieved when input is located in the distal apical branches. This can be seen by looking at the curves for location pairs (2,3) and (3,3) on the $R2$ axis and pairs (3,3) and (3,4) on the $R1$ axis (Fig. 3). The similar maximum values attained in each of these cases suggests that input to the distal apical branches was affected in similar ways by input to the soma, the entire neuron, or the

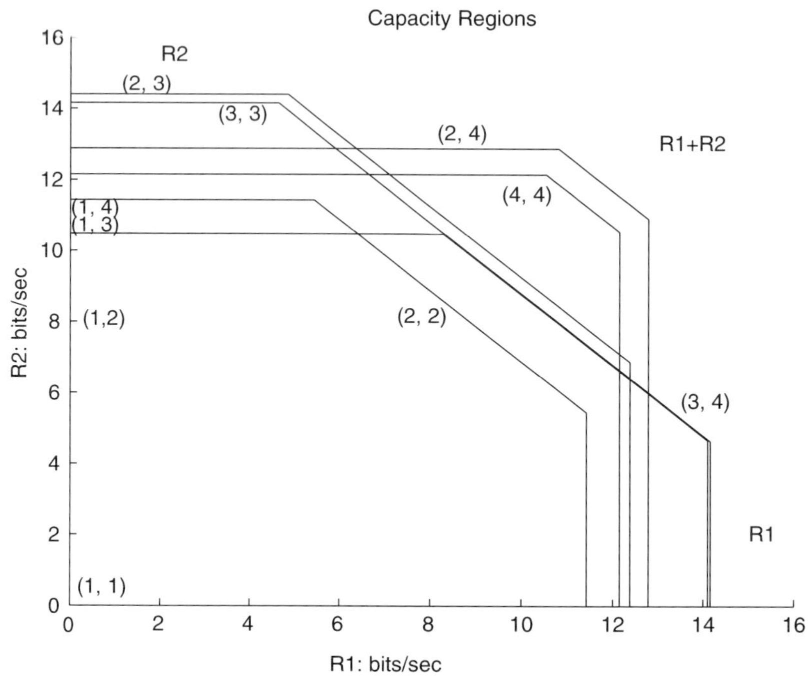

Fig. 2. Computed capacity regions for location pairs (n,m) where n (m) is the spatial pattern of synaptic inputs used by sender 1 (2).

distal apical branches from the second sender. For the case of the combined rates for the two senders, the maximum was achieved when both senders had either some or all of their input at the soma (Fig. 4). These results show quantitatively that different patterns of synaptic inputs provide the system with the capability to either transmit information from one source at a high rate or transmit information from multiple sources at high rates simultaneously.

In our experiment we have used input signals of varying power spectra. The results show that for a given location pair, the amount of information that can be sent reliably about the two input signals depends on their relative power spectra (Table 1). In Table 1 we present the power spectrum pairs that were used to achieve the maximum of the mutual informations at each location pair (taking into account symmetry). There are two interesting results. First, the power spectrum used to achieve the maximum rate for a single sender was the bandpass spectrum, #3, for each location pair. Second, the maximum combined rate for the two senders was achieved in all cases by the power spectrum pairs (3,3). This is contrary to what would be expected based on standard multiplexing wisdom, which dictates that the two power spectra should not overlap.

The methods developed here, and illustrated by our simple model system, are a starting point for analyzing the transmission of information from multiple sources over a common channel. We intend to apply this method to more detailed and realistic systems in future work.

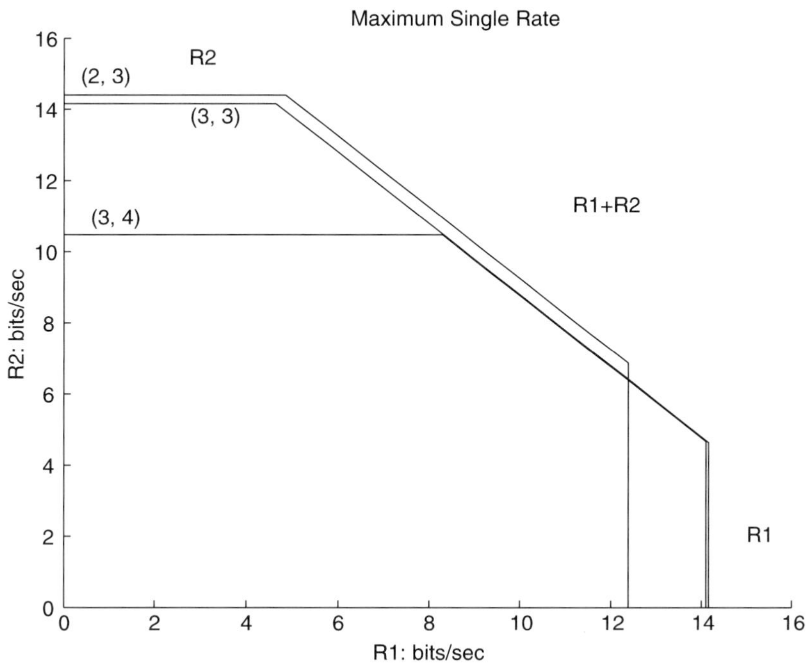

Fig. 3. Selected capacity regions from Fig. 2 showing the three largest single user rates achieved.

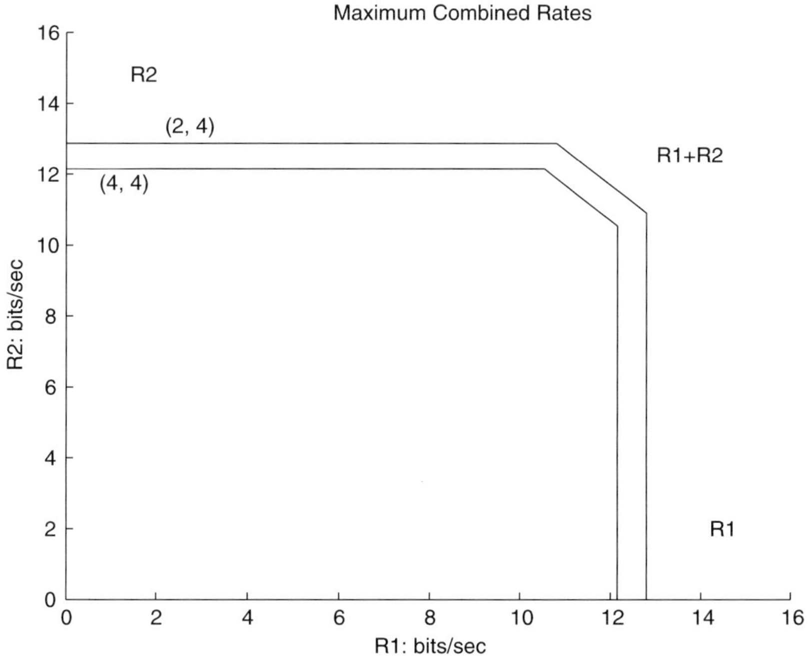

Fig. 4. Selected capacity regions from Fig. 2 showing the two largest combined rates achieved.

Table 1
Maximum-achieving power spectrum pairs

Location pairs	max $R1$	max $R2$	max $R1+R2$
(1,1)	NA	NA	NA
(1,2)	NA	(2,3)	NA
(1,3)	NA	(2,3)	NA
(1,4)	NA	(1,3)	NA
(2,2)	(3,2)	(2,3)	(3,3)
(2,3)	(3,1)	(1,3)	(3,3)
(2,4)	(3,3)	(3,3)	(3,3)
(3,3)	(3,1)	(1,3)	(3,3)
(3,4)	(3,1)	(2,3)	(3,3)
(4,4)	(3,3)	(3,3)	(3,3)

References

[1] T.M. Cover, J.A. Thomas, Elements of Information Theory, Wiley, New York, 1991.
[2] M.L. Hines, N.T. Carnevale, The NEURON simulation environment, Neural Comput. 9 (1997) 1179–1209.
[3] F. Rieke, D. Warland, R. de Ruyter van Steveninck, W. Bialek, Spikes. Exploring the Neural Code, MIT Press, Cambridge, MA, 1997.
[4] M.V. Tsodyks, H. Markram, The neural code between neocortical pyramidal neurons depends on neurotransmitter release probability, Proc. Nat. Acad. Sci. USA 94 (1997) 719–723.
[5] A. Zador, Impact of synaptic unreliability on the information transmitted by spiking neurons, J. Neurophysiol. 79 (1998) 1219–1229.

Available at www.ComputerScienceWeb.com
POWERED BY SCIENCE @ DIRECT®

ELSEVIER

Neurocomputing 52–54 (2003) 519–524

NEUROCOMPUTING

www.elsevier.com/locate/neucom

A model of surface detection and orientation tuning in primate visual cortex

Marc-Oliver Gewaltig*, Ursula Körner, Edgar Körner

Honda Research Institute Europe GmbH, Carl-Legien Str. 30, D-63073 Offenbach/Main, Germany

Abstract

In this contribution we propose that surface detectors, responding to areas of homogeneous gray-level may be used to support orientation selective cells in V1. By inhibiting orientation detectors within their input region, surface detectors act as local feedback system, emphasizing objects with large, continuous contours and enable the fast generation of an initial hypothesis about the input. The performance of this model is demonstrated, using a set of real-world images. We demonstrate that for images with a high degree of low-frequency clutter, the performance of orientation detection can be considerably improved.

Keywords: Visual system; Cortex; Koniocellular pathway; Segmentation

1. Introduction

In early visual processing, speed matters. If an animal is confronted with a visual scene, it is important to quickly arrive at an hypothesis about its main parts so that an appropriate reaction (e.g. escape) is possible. In order to do such a quick segmentation of the visual input, the information has to be reduced to its most salient parts. After a first object hypothesis has been established at higher processing levels, a top-down signal may then enable a refined analysis of object detail (see e.g. [4]).

This poses the question, how the visual system can perform this segmentation without knowledge of the types of objects which are in the scene. We propose that information on continuous image regions may be used to confine edge detection to borders between mid-sized objects.

* Corresponding author. Tel.: +49-69-890-11-739; fax: +49-69-890-11-749.
E-mail address: marc-oliver.gewaltig@honda-ri.de (M.-O. Gewaltig).

doi:10.1016/S0925-2312(02)00767-1

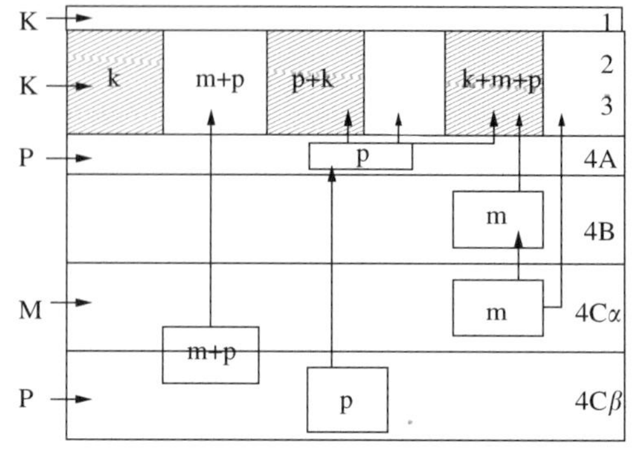

Fig. 1. Sketch of the major routes of input from the dorsal lateral geniculate nucleus (LGN) of the thalamus to the monkey primary visual cortex (V1). Capital letters indicate thalamic input. Small letters indicate further relay of these signals within cortex (adapted from [5,1]).

2. Biological background

In primates there are two major pathways through which information flows from the retina via the lateral geniculate nucleus (LGN) to the primary visual cortex: the parvocellular (P) and the magnocellular (M) pathway (Fig. 1). M-cells in the LGN have larger receptive fields and are highly contrast sensitive. P-cells have smaller receptive fields and their contrast sensitivity is comparatively low; however, many of them are sensitive to color (wavelength). M and P cells account for 90% of all LGN relay cells. The remaining 10% belong to a third, less well-investigated pathway, the so-called koniocellular (K) pathway, which is thought to be a phylogenetically old part of the visual system.

K-cells in LGN have very large receptive fields and their projections to V1 terminate in the cytochrome-oxidase (CO) blobs of the superficial layers [5,2]. In contrast, inputs from the M and P pathway terminate in layer IV. Moreover, while M and P cells exclusively project to V1, K cell inputs have been identified in higher areas of the ventral visual pathway [3].

Because of their large receptive fields, K cells are not able to resolve fine details, however, they may provide information about large homogeneous regions. Tuning of orientation selective cells in layer IV is relatively broad, while it is considerably sharper in layers II and III. We propose that surface information, provided by the koniocellular pathway, supports orientation selectivity in layer II/III and, thus, may contribute to the improved tuning.

3. The model

In this contribution, we present a computational model, which demonstrates how surface information can be used to support and enhance the response properties of orientation selective cells in V1. Starting from the retinal image, we model LGN

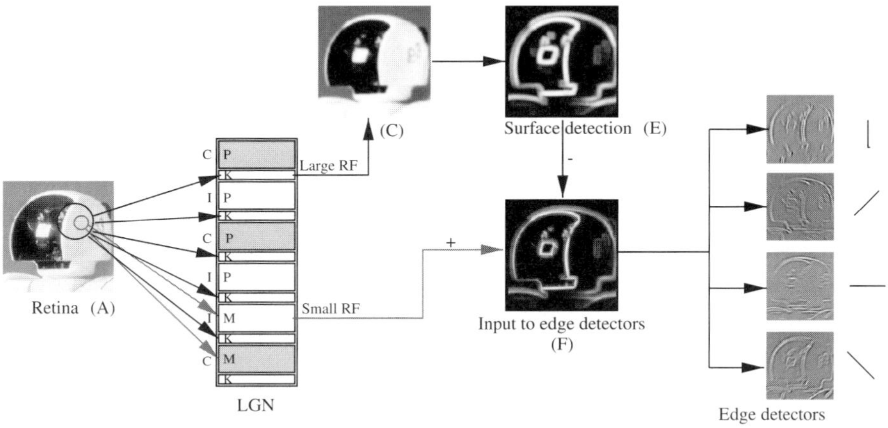

Fig. 2. Schema of the proposed model.

responses for two pathways: the koniocellular and the magnocellular pathway. The model system is derived from the schema shown in Fig. 2 and consists of four topographically organized layers:

(1) a retinal input layer A,
(2) a layer of koniocellular LGN cells C,
(3) a layer of K-driven surface detectors E, and
(4) a layer representing the input to orientation selective cells F.

Here, we restrict the discussion to the properties of the surface detectors and do not explicitly model orientation selective cells of the M-path. Rather, we show how the input F to these cells is modified by the surface detection system.

3.1. LGN activity

Receptive fields of K-cells in the LGN are very large and often have no antagonistic surround [7]. Thus, they will be fully activated, even if a stimulus is larger than their receptive field. We model the receptive field of LGN K-cells by a two-dimensional Gaussian $g(x, y)$ with standard deviation σ_1:

$$h(x, y) = \frac{1}{\sigma_1\sqrt{2\pi}} \exp\left(-\frac{x^2 + y^2}{4\sigma_1^2}\right). \tag{1}$$

The firing rate of an LGN K-cell at position (x, y) is given by $\mathsf{C}[x, y] = \Theta_1(\mathsf{B}[x, y])$ with the sigmoid activation function

$$\Theta_1(z) := \frac{1}{1 + \exp(-2b_1(z - \theta_1))} \tag{2}$$

and $\mathsf{B}[x, y] = (\mathsf{A} * h)(x, y)$ with threshold θ_1 and slope b_1. The threshold is determined by the mean activity in the image. Note that activities are normalized to unity.

3.2. Surface detectors

The K-cell layer C projects to a layer of surface detectors E. The receptive field of a surface detector at position (x, y) is modeled by a circular patch $\gamma(x, y)$ within C. Surface detectors respond best, if their receptive field is covered by an area of *homogeneous* color or gray-level. Surface detectors effectively evaluate the variance of color or gray-level in their circular receptive field:

$$\mathsf{D}[x, y] := \sqrt{\frac{1}{P} \sum_{(m,n)\in\gamma(x,y)} (\mathsf{C}[i-m, j-n] - \langle \mathsf{C} \rangle_{\gamma(i,j)})^2} \tag{3}$$

with

$$\langle \mathsf{C} \rangle_{\gamma(x,y)} := \frac{1}{P} \sum_{(m,n)\in\gamma(x,y)} \mathsf{C}[m-x, n-y]. \tag{4}$$

The activation of the surface detectors is given by $\mathsf{E}[x, y] = \Theta_2(\mathsf{D}[x, y])$, where $\Theta_2(\cdot)$ is a sigmoid activation function equivalent to (2). The threshold θ_2 is again chosen according to the mean activity of their input.

3.3. Edge detector input

The surface detectors project one-to-one to an interneuron layer. Each activated surface detector will thus trigger inhibition which will act on all edge detectors whose receptive field overlaps with that of the surface detector. Since the receptive fields of the surface detectors are considerably larger than those of the orientation selective cells, one surface detector will cause inhibition of a large number of orientation detectors. In the computational model this is achieved by masking the input image with the activity of the inhibition layer E, thus, arriving at the effective input to the edge detectors $\mathsf{F} := \mathsf{A} \cdot \mathsf{E}$.

Here, we used a multiplicative interaction to mask the input with the surface detector response. It is also possible to use an additive interaction. In this case, an additional normalization may be required, but the results are qualitatively identical.

4. Results

Fig. 3 shows how the model performs on natural images. The images are an arbitrary selection from a larger set of images on which the system was tested. Note that the images differ considerably in resolution and are scaled to fit to this display.

Fig. 3a shows an example from the COIL database [6]. The surface detectors emphasize the contours of the object. However, images from the COIL database are optimized

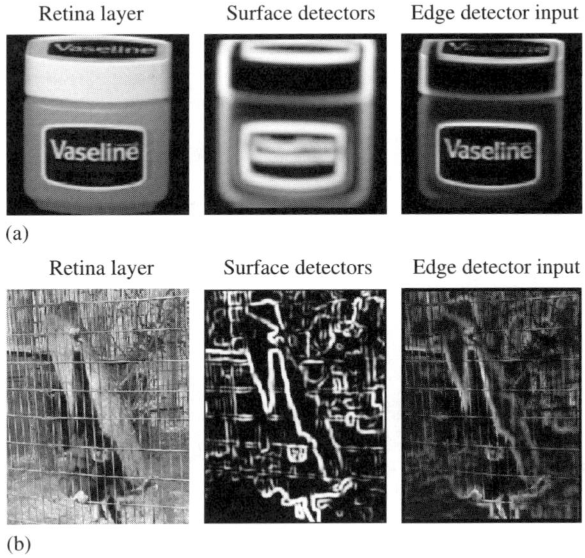

Fig. 3. Example images, demonstrating the operation of surface detectors. Original image (left), inverse surface detector response (middle), strongest response shown as black, and input to orientation selective cells (right). For all images, the same parameter configuration of the model was used. In all cases, the surface detectors emphasize the contours of the object. (a) Example from the COIL database [7]. (b) Object embedded in a cluttered environment. A distractor (the grid) overlaps with the main object in the scene (the monkey).

for edge-based object recognition. Here, almost any edge-based object recognition systems performs well, since the images are of high contrast and low noise.

The example in Fig. 3b is more realistic, since the main object (the monkey) is embedded in a cluttered environment. If an edge-detection system was directly applied to this image, the response would be dominated by the contributions from the rectangular grid. However, surface detectors respond best to homogeneous regions, contributed by the monkey and the tree-trunk. Thus, the image is reduced to the contours of these objects, suppressing much of the grid and the high-resolution clutter. This reduction in detail is an important guide for subsequent edge-based recognition systems.

5. Discussion

We have presented a model for surface detection which is based on local homogeneity. Surface detectors and edge detector cooperate according to the rule that surfaces and edges are mutually exclusive. In our model, this relation is hard-coded. In a natural system, however, it should be susceptible to learning. Surface detectors are activated by continuous regions of high local homogeneity. If a surface detector is activated, it inhibits all orientation detectors which have their receptive field inside the larger receptive field of the surface detector. At the borders of such regions, edge detection is possible. Since extended homogeneous regions can in most cases be attributed to mid-sized objects, the surface detection system emphasizes the *contours* of such objects. The spatial scale is defined by the receptive field size of the surface detectors.

Because surface and edge detection work as parallel feed-forward systems, the proposed mechanism does not require additional time. It may, thus, support the fast generation of an initial hypothesis about the input. Subsequent processing steps may trigger top-down feedback to suppress the surface-driven inhibition and enable access to fine detail of the input.

References

[1] J.D. Boyd, J. Mavity-Hudson, V.A. Casagrande, The connections of layer 4 subdivisions in the primary visual cortex (V1) of the owl monkey, Cerebral Cortex 10 (7) (2000) 644–662.

[2] S.H.C. Hendry, R.C. Reid, The koniocellular pathway in primate vision, in: W.M. Cowan, E.M. Shooter, C.F. Stevens, R.F. Thompson (Eds.), Annual Review of Neuroscience, Vol. 23, Annual Reviews, Palo Alto, CA, 2000, pp. 127–153.

[3] A. Hernandez-Gonzalez, C. Cavada, F. Reinoso-Suarez, The lateral geniculate nucleus projects to the inferior temporal cortex in the macaque monkey, Neuroreport 5 (18) (1994) 2693–2696.

[4] E. Körner, M.-O. Gewaltig, U. Körner, A. Richter, T. Rodemann, A model of computation in neocortical architecture, Neural Networks 12 (7–8) (1999) 989–1005.

[5] J.S Lund, T. Yoshioka, J.B. Levitt, Substrates for interlaminar connections in area v1 of macaque monkey cerebral cortex, in: A. Peters, K.S. Rockland (Eds.), Cerebral Cortex: Primary Visual Cortex in Primates, Plenum Press, New York, 1994, pp. 37–60.

[6] S.K. Nayar, S.A. Nene, H. Murase, Real-time 100 object recognition system, in: Proceedings of the ARPA Image Understanding Workshop, Palm Springs, 1996.

[7] R.W. Rodieck, The First Steps in Seeing, Sinauer Associates, Sunderland, MA, 1998.

Available at www.ComputerScienceWeb.com
POWERED BY SCIENCE DIRECT®

ELSEVIER

Neurocomputing 52–54 (2003) 525–530

NEUROCOMPUTING

www.elsevier.com/locate/neucom

Analyzing mechanosensory transduction by identifying invariant directions in stimulus space

T. Gollisch[a,*,1,2], A.V.M. Herz[a,2]

[a]*Institute for Theoretical Biology, Humboldt University, Invalidenstr 43, Berlin, 10115 Germany*

Abstract

Extended receptive fields are characteristic for the integrative properties of sensory neurons. How this integration is performed is an important aspect in understanding neural coding and can give insight into the computations and the biophysics involved. We present a method of systematically investigating the nature of spectral integration in auditory receptor neurons by analyzing how these neurons combine intensities contained in different parts of the stimulus spectrum. The method consists of identifying regions in parameter space that lead to the same neural response, and it is also applicable to other stimulus domains.

Keywords: Neural coding; Sensory transduction; Receptive field; Invariant directions; Stimulus space

1. Introduction

Auditory receptor neurons, such as hair cells and chordotonal sensilla, transduce vibrations of the basilar or tympanal membrane into deflections of the cell-membrane potential and, possibly, spikes. In doing so, the neuron performs an integration in the frequency domain as well as over a certain temporal window. Together, these two domains can be viewed as the spectro-temporal receptive field of the neuron. Auditory receptor neurons thus share with most sensory neurons that they respond to a multi-dimensional stimulus space.

There have been a number of approaches to describe and characterize the input–output relations of sensory neurons in order to understand the underlying coding

* Corresponding author.
E-mail address: t.gollisch@biologie.hu-berlin.de (T. Gollisch).
1 Supported by Boehringer Ingelheim Fonds.
2 Supported by the Deutsche Forschungsgemeinschaft.

0925-2312/03/$ - see front matter © 2002 Elsevier Science B.V. All rights reserved.
doi:10.1016/S0925-2312(02)00771-3

strategies [2,6]. The simplest phenomenological models describe both the input and output by single scalars, the amplitude A of a one-dimensional stimulus and the firing rate R, respectively: $R = g(A)$, where $g(x)$ denotes some, generally nonlinear function.

The simplest extension of those models is to include a multi-dimensional input space with stimulus components given by $A_1, A_2, \ldots, A_N$. The response R is now given by a (nonlinear) function of all stimulus components. This is often simplified by taking the effect of the receptive field explicitly into account and using a weighted sum of all stimulus components (cf. [1]):

$$R = g\left(\sum_{n=1}^{N} F_n \cdot A_n\right).$$

The model parameters $F_1, F_2, \ldots, F_N$ define the structure of the receptive field. This "ad hoc" integration over the receptive field is not the only way, of course, to combine the stimulus components. A better description may be found if one regards an appropriate nonlinear transformation $h(A_n)$ of the components:

$$R = g\left(\sum_{n=1}^{N} F_n \cdot h(A_n)\right).$$

Finding the right transformation $h(A_n)$ or some other description of how the stimulus components are combined may help to interpret the stimulus encoding in a mechanistic fashion. If, e.g., A denotes the amplitude of an acoustic stimulus and $h(A) = A^2$, the neuron can be interpreted as an energy detector.

In this work, we present an experimental approach to systematically discriminate different models of stimulus integration over the receptive field. This is applied to spectral integration in auditory receptors of locusts, i.e., the combining of intensities contained in different parts of the stimulus spectrum.

2. Stimulus-response curves and iso-activity regions

Sensory neurons are often characterized by stimulus-response curves where the response, such as the firing rate, is measured at different stimulus intensities. Fig. 1 shows the firing rate of a locust auditory receptor in response to pure-tone stimulation of $f_1 = 4$ kHz and $f_2 = 9.55$ kHz. For a generic model $R = g(\sum F_n \cdot h(A_n))$, where the A_n denote the amplitudes of the tones of frequency f_n, the shape of the response curves is determined by the combination of the functions g and h, while the parameters F_n set the sensitivity and therefore specify how far left or right on the x-axis a certain stimulus-response curve lies. (Note that in Fig. 1, the units on the x-axis, dB-SPL, represent a logarithmic measure of the sound-pressure amplitude.) The function h itself cannot be determined by pure-tone stimulation alone, but only by determining responses to combinations of tones.

To illustrate this, Fig. 2 shows the response of four simple input–output models to various stimulus combinations (A_1, A_2).

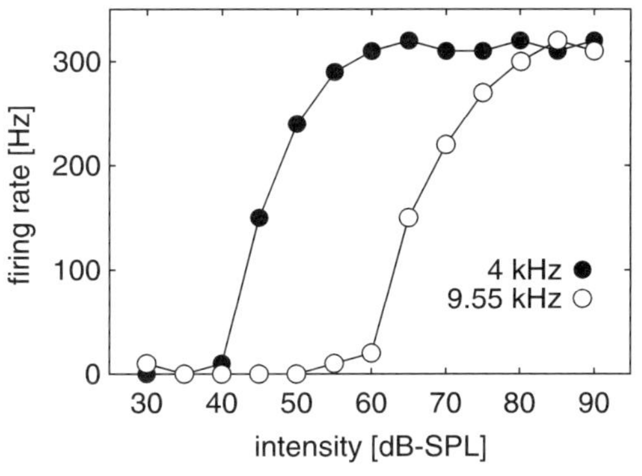

Fig. 1. Stimulus-response curves of a locust auditory receptor to 4-kHz and 9.55-kHz pure-tone stimululation.

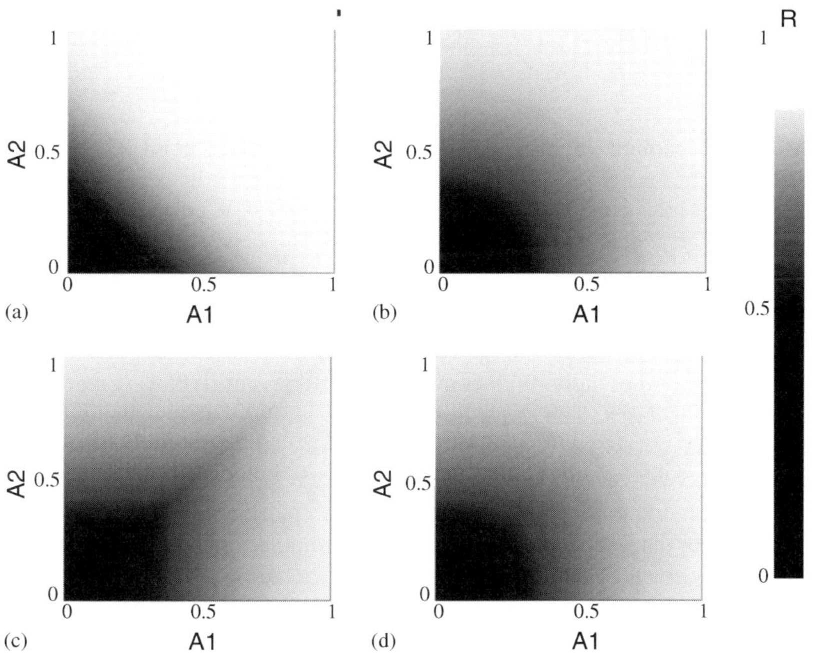

Fig. 2. Gray-scale-coded response strength R as a function of the stimulus combination (A_1, A_2) for four different models. In each case, a static nonlinearity $g(x) = \tanh(1.5 \cdot x)$ was applied, but the components were combined in different ways (see text). F_1 and F_2 were both set to 1.

In the four models of Fig. 2, the responses are determined according to

(a) $R = g(F_1 \cdot A_1 + F_2 \cdot A_2)$,

(b) $R = g(\sqrt{F_1 \cdot A_1^2 + F_2 \cdot A_2^2})$,

(c) $R = g(\max(F_1 \cdot A_1, F_2 \cdot A_2))$,

(d) $R = g(\frac{\pi}{2} \cdot \langle |F_1 \cdot A_1 \sin(2\pi f_1 t) + F_2 \cdot A_2 \sin(2\pi f_2 t)| \rangle_t)$,

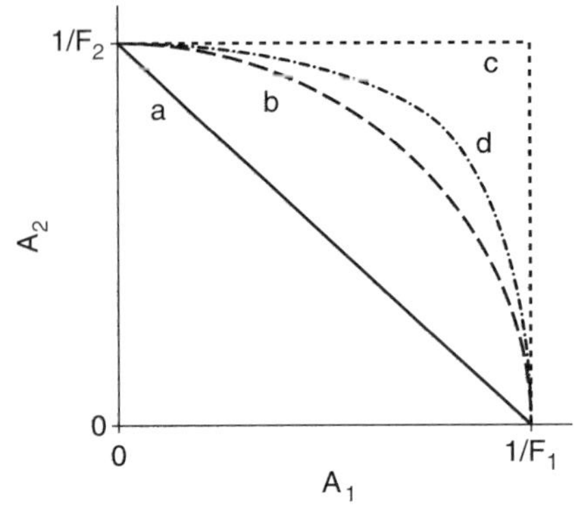

Fig. 3. Iso-activity regions for four different models. The four curves marked a–d correspond to lines of equal response (i.e., lines of equal shade of gray) in Fig. 2(a)–(d), respectively. The different shapes of the curves show that measuring these iso-activity regions can reveal how the integration over the receptive field is performed.

where $|x|$ is the absolute value of x, $\langle y(t)\rangle_t$ denotes the temporal average over $y(t)$, and $g(x) = \tanh(1.5 \cdot x)$. Models (a) and (b) are directly of the form

$$R = g\left(\sum_n F_n \cdot h(A_n)\right),$$

whereas (c) and (d) represent alternative ways of combining different stimulus components.

The responses to the individual components $(A_1, 0)$ and $(0, A_2)$, respectively, are identical in all four cases. This is reflected by the identical gray-scale gradients along the x-axis and the y-axis, respectively. Combinations of the stimulus components, on the other hand, reveal very different integrative behavior.

To capture this more clearly, we look at the shapes of the *iso-activity regions* (corresponding to the *invariant directions* in stimulus space). These are given by those stimulus-parameter combinations that leave the response invariant, and they are characteristic of the integration over stimulus components as illustrated in Fig. 3. These regions are independent of the static nonlinearity g, which is convenient as it allows accurate experiments and interpretation without having to fully describe the final non-linearity of spike generation.

3. Experiments

We performed intracellular recordings from the axons of receptor cells in the auditory nerve of *Locusta migratoria*. Stimuli were combinations of pure tones played over loudspeakers, $S(t) = A_1 \sin(2\pi f_1 t) + A_2 \sin(2\pi f_2 t)$, with sound frequencies $f_1 = 4$ kHz and $f_2 = 9.55$ kHz. Stimulus generation and spike detection were controlled by the custom-made *Online Electrophysiology Laboratory* (*OEL*) software. Stimulus length

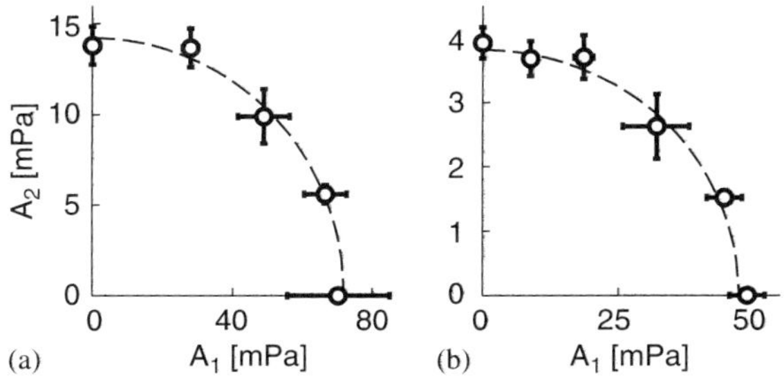

Fig. 4. Iso-activity regions in a two-dimensional stimulus space measured from two locust auditory receptors. The points display measured amplitude combinations (A_1, A_2) for two-tone stimuli that led to a firing rate of 150 Hz each. The dashed lines are fitted ellipses of the form $F_1 \cdot A_1^2 + F_2 \cdot A_2^2 = 1$. The parameters F_1 and F_2 were determined by a least-square fit for each data set.

was 100 ms, during which the receptor cells fire tonically with a small phasic part due to firing-rate adaptation. No offset response was observed.

For a given relative combination of A_1 and A_2 (constant ratio), the total stimulus intensity was tuned via an internal feedback loop until a desired output firing rate (usually 150 Hz) was reached.

4. Results

Examples of measured iso-activity regions in a two-dimensional stimulus space for locust auditory receptors can be seen in Fig. 4. Several amplitude combinations (A_1, A_2) were measured that all led to the same firing rate, and the data points are well fitted by ellipses. This corresponds to case (b) in Figs. 2 and 3.

The findings can be summarized in a phenomenological input–output relation for sound-intensity coding in the locust ear. In the generalized form, the firing-rate response R to a stationary stimulus consisting of N pure tones with amplitudes A_n is given by $R = g(\sum_{n=1}^{N} F_n \cdot A_n^2)$. The parameters F_n determine the tuning properties of the receptor neuron and are probably connected to the resonance properties of the tympanal membrane at the attachment site of the neuron (cf. [5]). The sum of squares of the amplitudes suggests that the receptors can be described as energy detectors. The function g reflects the nonlinear encoding into a firing rate (resulting in the stereotypical shape of the stimulus-response curves as in Fig. 1) and has to be separately determined experimentally. For a more detailed analysis and discussion of the experimental results, see [4].

5. Conclusion

Many sensory neurons integrate inputs over a large receptive field in producing their responses. Analyzing the nature of this integration can lead to valuable insights into the coding properties or the mechanisms of stimulus transduction. Determining iso-activity regions in stimulus space, i.e., regions that leave the response invariant, can be used for an analysis that is independent of static nonlinearities as typically occur when a stimulus is encoded into a firing rate. A systematic investigation of iso-activity regions can be achieved by predefining a desired response and tuning the stimulus until this response is reached. With increasing availability of computing power, this can be implemented efficiently by using online analysis of the responses and feedback loops in the experimental design.

Exploiting the invariance of the response along certain directions in stimulus space is somewhat reminiscent of the technique of *silent substitution* used in the visual system to single out responses of specific retinal receptors [3]. This is achieved by varying stimuli along invariant directions of some receptor types such that resulting changes in the measured signals can be interpreted as being due to the remaining receptors. In this approach, the invariant regions must be accurately known beforehand, though, and are not systematically explored as proposed here.

We have shown that the method of exploring invariant regions in stimulus space can be applied to the simple case of two-dimensional stationary stimuli in locust auditory receptors. The results support an energy-integration mechanism. Possible extensions of the approach include using higher-dimensional stimulus spaces and applying the same concepts to temporal integration. The latter can be done by systematically varying the intensity for different temporal parts of a (short) stimulus and analyzing the probability of eliciting a spike. In a further step, the method could be used to investigate down-stream and central neurons. A possible application in visual research is the analysis of how neurons combine stimulation in *on–off* receptive fields by subdiving stimulus space into the two dimensions of the *on* and the *off* part, respectively.

Acknowledgements

We thank J. Benda and C. Machens for making the framework of the *OEL* software available to us and for their assistance in maintenance and in the development of the algorithms used in this work, and H. Schütze for his support during the experiments.

References

[1] E.J. Chichilnisky, A simple white noise analysis of neuronal light responses, Network 12 (2001) 199–213.
[2] P. Dayan, L.F. Abbott, Theoretical Neuroscience: Computational and Mathematical Modeling of Neural Systems, MIT Press, Cambridge, MA, 2001.
[3] O. Estévez, H. Spekreijse, The 'silent substition' method in visual research, Vis. Res. 22 (1982) 681–691.
[4] T. Gollisch, H. Schütze, J. Benda, A.V.M. Herz, Energy integration describes sound-intensity coding in an insect auditory system, J. Neurosci. 22 (2002) 10434–10448.
[5] A. Michelsen, The physiology of the locust ear. II. Frequency discrimination based upon resonance in the tympanum, Z vergl Physiol. 71 (1971) 63–101.
[6] F. Rieke, D. Warland, R. de Ruyter van Steveninck, W. Bialek, Spikes: Exploring the Neural Code, MIT Press, Cambridge, MA, 1997.

Available at
www.ComputerScienceWeb.com
POWERED BY SCIENCE @ DIRECT®

ELSEVIER

Neurocomputing 52–54 (2003) 531–540

NEUROCOMPUTING

www.elsevier.com/locate/neucom

A dynamical model for the vertical vestibuloocular reflex and optokinetic response in primate

Yutaka Hirata[a,*], Ichiro Takeuchi[a], Stephen M. Highstein[b]

[a] *Department of Electronic Engineering, College of Engineering, Chubu University, 1200 Masumoto-cho, Kasugai, Aichi 487-8501, Japan*

[b] *Department of Otolaryngology, School of Medicine, Washington University, St. Louis, MO, USA*

Abstract

The vestibuloocular reflex (VOR) in concert with the optokinetic response (OKR) stabilizes vision during head motion. The VOR system characteristics are both compensatory and adaptively self-calibrated. A model was constructed to aid in the understanding of the roles of the cerebellum and other neuronal sites in the performance and adaptation of the vertical VOR. The model structure was based upon the known neuroanatomy, and model parameters were estimated using experimental data. The model can reproduce and predict eye movements and cerebellar Purkinje cell firing patterns during VOR, OKR, and various visual–vestibular mismatch paradigms.

Keywords: Eye movement; Monkey; Cerebellum; System identification; Motor learning

1. Introduction

The vestibuloocular reflex (VOR) in concert with the optokinetic response (OKR) stabilizes vision during head motion. The VOR produces compensatory eye movements under adaptive control that result in a VOR gain (eye velocity/head velocity) of 1 in the light; changes in VOR–OKR system characteristics are compensatory and self-calibrating, thus changes in VOR characteristics caused by aging, injury, or visual–vestibular mismatch are adjusted to achieve compensatory eye movements. The

* Corresponding author.
E-mail address: yutaka@isc.chubu.ac.jp (Y. Hirata).

0925-2312/03/$ - see front matter
doi:10.1016/S0925-2312(02)00777-4

cerebellar flocculus (FL) is necessary for VOR adaptation since its removal or inactivation precludes further changes in gain [5,10], yet the FL role in VOR adaptation has long been controversial [3,6]. FL neuronal circuitry is characterized by multimodal sensory–motor transformations involving multiple feedback loops making causality ambiguous. The complexity of this circuitry has been problematic in assigning a role to FL in the process of adaptation and performance of the VOR. The system identification approach combines electrophysiological experiments and mathematical modeling and might provide way to overcome the difficulty. Presently we built a model for the monkey vertical VOR/OKR based upon its well understood anatomy and physiology. The model is extended from our previous multiple linear regression model that was built to describe only a single frequency [2].

The present model consists of 6 subsystems each of which represents the signal processing of a sensory modality or the generation of motor and pre-motor commands in the cerebellum or in non-cerebellar structures. Characteristics of each subsystem were identified by fitting model output to experimental data with rich information in terms of frequency and amplitude, and the validity of the model was checked by predicting various experimental data that were not used for the model identification. It was shown that the model can reproduce or predict eye movements and FL Purkinje cell firing patterns during VOR, OKR, and visual–vestibular mismatch paradigms over a wide frequency range as well as those after lesion and inactivation of the FL.

2. Methods

2.1. Model structure

Fig. 1A illustrates the neuroanatomy of the vertical VOR/OKR in primate. In order to evaluate the possible roles of the cerebellum in VOR/OKR control and adaptation, we divided the circuit into pathways that include the cerebellum and ones that do not. Six subsystems form the model as illustrated in Fig. 1B: $G^{\text{vestib}}_{\text{preFL\&FL}}(s)$, $G^{\text{visual}}_{\text{preFL\&FL}}(s)$, and $G^{\text{ecopy}}_{\text{preFL\&FL}}(s)$ describe the characteristics of pre-FL and FL pathways that process vestibular, visual and efference copy signals, respectively and convert each signal into the output of the FL Purkinje cell activity. $G_{\text{postFL}}(s)$ describes the characteristics of the post-FL pathway that converts FL Purkinje cell activity into a component of the eye movement command. $G^{\text{vestib}}_{\text{nonFL}}(s)$ and $G^{\text{visual}}_{\text{nonFL}}(s)$ describe the characteristics of the non-FL pathways that process vestibular and visual signals, respectively, and convert these sensory signals into a component of the eye movement command. Neuronal circuitry that is included in $G^{\text{visual}}_{\text{nonFL}}(s)$ is not illustrated in Fig. 1A due to the lack of anatomical evidence, but it is apparent that such a pathway exists because lesion of the FL does not completely diminish eye movement in response to visual stimuli [10]. System equations of the model are described as follows:

$$F(s) = H(s)G^{\text{vestib}}_{\text{preFL\&FL}}(s) + R(s)G^{\text{visual}}_{\text{preFL\&FL}}(s) + X(s)G^{\text{ecopy}}_{\text{preFL\&FL}}(s),$$

$$X(s) = F(s)G_{\text{postFL}}(s) + H(s)G^{\text{vestib}}_{\text{nonFL}}(s) + R(s)G^{\text{visual}}_{\text{nonFL}}(s),$$

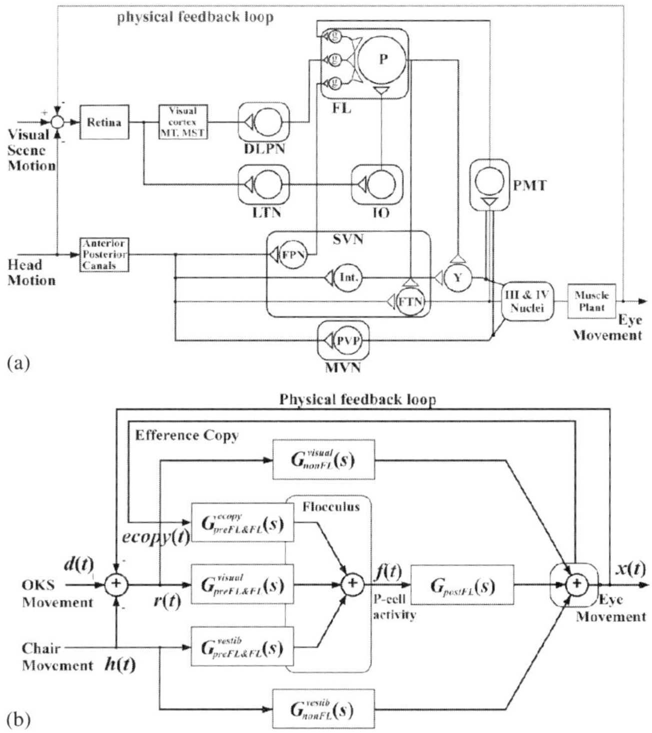

Fig. 1. Vertical VOR and OKR neuronal circuit (A) and model (B). In (A), FPN and FTN are the floccular (FL) projecting neurons and the FL target neurons in superior vestibular nuclei (SVN), respectively, Y is dorsal Y group. Int. is the inter-neurons in SVN projecting to Y group. P is a FL Purkinje cell and g is a granular cell. PVP is position vestibular pause neurons in medial vestibular nuclei (MVN). MT, MST and DLPN are the middle temporal visual area, the medial superior temporal area and the dorsolateral pontine nuclei, respectively. MN, extraocular motor neurons. LTN, lateral terminal nucleus of the accessory optic system. In (B), $G^{\mathrm{ecopy}}_{\mathrm{preFL\&FL}}(s)$, $G^{\mathrm{vestib}}_{\mathrm{preFL\&FL}}(s)$ and $G^{\mathrm{visual}}_{\mathrm{preFL\&FL}}(s)$ are pre-FL/FL subsystems each of which represents a transfer function of pre-FL/FL efference copy pathway, vestibular pathway and visual pathway, respectively. The three components are added in the FL and form the Purkinje cell SS output. $G_{\mathrm{postFL}}(s)$ represents a transfer function of post-FL pathway which transfers the Purkinje cell simple spike activity to a part of the motor command. $G^{\mathrm{visual}}_{\mathrm{nonFL}}(s)$ and $G^{\mathrm{vestib}}_{\mathrm{nonFL}}(s)$ represent transfer functions of non-FL visual and vestibular pathways, respectively. Corresponding neuronal circuit to $G^{\mathrm{visual}}_{\mathrm{nonFL}}(s)$ is not shown in (A). $h(t)$, $d(t)$, $r(t)$, $f(t)$, $ecopy(t)$, and $x(t)$ are head position, optokinetic stimulus position, retinal slip position, FL Purkinje cell simple spike firing pattern, efference copy signal, and eye position, respectively.

where $F(s)$, $H(s)$, $R(s)$, and $X(s)$ denote the Laplace transform of FL Purkinje cell firing rate, head position, retinal slip position, and eye position, respectively while s is the Laplace operator. Each subsystem is described as the following transfer function that was determined based on known physiological evidence:

$$G^{\mathrm{vestib}}_{\mathrm{preFL\&FL}}(s) = (\alpha_h s^2 + \beta_h s + \gamma_h)\mathrm{e}^{-p_h s}, \tag{1}$$

$$G^{\mathrm{visual}}_{\mathrm{preFL\&FL}}(s) = (\alpha_r s^2 + \beta_r s + \gamma_r)\mathrm{e}^{-p_r s}, \tag{2}$$

$$G^{\text{ecopy}}_{\text{preFL\&FL}}(s) = (\alpha_e s^2 + \beta_e s + \gamma_e)\text{e}^{-p_e s}, \quad (3)$$

$$G_{\text{postFL}}(s) = \text{e}^{-\tau_x s}/(a_x s^2 + b_x s + c_x), \quad (4)$$

$$G^{\text{vestib}}_{\text{nonFL}}(s) = (a_h s^2 + b_h s + c_h)\text{e}^{-\tau_h s}/(u_h s^2 + v_h s), \quad (5)$$

$$G^{\text{visual}}_{\text{nonFL}}(s) = (a_r s^2 + b_r s + c_r)\text{e}^{-\tau_r s}/(u_r s^2 + v_r s), \quad (6)$$

where α_h, β_h, and γ_h denote head acceleration, velocity and position sensitivity of the Purkinje cell, respectively while p_h denotes the latency of Purkinje cell firing in response to head motion. α_r, β_r, and γ_r denote retinal slip acceleration, velocity and position sensitivity of Purkinje cell, respectively while p_r denotes the latency of Purkinje cell firing in response to retinal slip. Higher-order terms relating to retinal slip were not included because it has been shown that retinal slip jerk does not contribute to Purkinje cell firing [4,9]. α_e, β_e, and γ_e denote eye acceleration, velocity and position sensitivity of the Purkinje cell while p_e denotes the latency of Purkinje cell firing in response to eye movement. Higher-order terms relating to eye movements were not included because it has been shown that eye jerk does not contribute to Purkinje cell firing [1,8]. Other parameters in the equations do not have specific physiological meaning.

Model parameters were determined by fitting the model output to experimental data (eye movement, and FL Purkinje cell firing rate) during a visual–vestibular mismatch paradigm and the validity of the model was checked by predicting experimental data during other paradigms described below.

2.2. Experiment

Two adult male squirrel monkeys were utilized for these experiments. For head fixation, a stainless-steel bolt was secured to the occiput. This bolt fit into a receptacle on the monkey chair to fix the head. The scleral search coil method was used for eye movement recording. A stainless-steel recording chamber aimed at the FL was implanted and secured to the skull. All these surgical procedures and experiments were approved by the Animal Welfare and Use Committee of Washington University.

Animals were seated in a primate chair with their heads-fixed. The chair was placed in the center of a white cylindrical screen 1 m in diameter (extending 36 cm above and 50 cm below the animal's head) on which black random dots were projected. This is the optokinetic stimulus (OKS). The right side of the animal was placed down so that rotation of the OKS about an earth vertical axis produced upward or downward nystagmus. Horizontal and vertical eye position, OKS velocity, and chair velocity were continuously digitized at a sampling frequency of 200 Hz with the use of a CED 1401 interface (Cambridge Electronic Design) for display and storage using the Spike-2 program. The output of a threshold detector, signaling the occurrence of action potentials, was sampled with a resolution of 0.01 ms and stored in the same way. Raw data were also stored on VCR tape (Neurodata PCM) for the offline isolation check.

FL and ventral parafloccular vertical zone Purkinje cells were identified by the occurrence of complex spikes and by their characteristic simple spike discharge patterns during vertical visual–vestibular interaction paradigms. Once a single unit was isolated, OKR, VOR and other visual–vestibular interaction stimuli were applied for 1 min each. The visual–vestibular interaction paradigms used were VOR suppression (VORs) and the VOR enhancement (VORe) in which OKS moves in-phase and 180° out of phase with the head rotation, respectively. The maximum amplitude of head rotation and OKS were limited to 40°/s to assure experimentation within the linear range of the system. We used sum-of-sines (SOS) stimuli that consisted of five frequencies (0.05, 0.27, 0.5, 1.27, 2.5 Hz) to identify the wide band responses of the VOR/OKR system. We also applied conventional single frequency stimuli (0.1, 0.5, 2.5 Hz).

2.3. Data handling

Purkinje cell firing rate was calculated from spike occurrence time by counting the number of spikes in a bin. The bin size 0.005 s (sampling frequency 200 Hz) corresponds to the sampling interval of other data. Saccades and post-saccadic drifts if any were eliminated from the eye movements data and corresponding Purkinje cell firing rate data. After the desaccading process, firing rate data from different cells measured during the same paradigm were averaged. Anatomical plausibility of the averaging is that multiple Purkinje cells converge to a target vestibular neuron called FL target neuron (FTN). The desaccaded eye movement data measured simultaneously with those Purkinje cell firings were also averaged. These averaged data were used for parameter estimation.

2.4. Parameter estimation and verification of the model

Model parameters were determined by fitting model output to experimentally measured eye movements and FL Purkinje cell firing pattern during the VORe with SOS paradigm, simultaneously. The VORe with SOS paradigm was chosen because all the subsystems in the model are activated in a wide frequency range, and the largest range of model output (eye movements) is induced during this paradigm. More than 20 samples of Purkinje cell firing rates were averaged to prepare the firing pattern for the parameter estimation. The parameter estimation was done by a non-linear optimization method (BFGS [7]) implemented on Matlab (Mathworks).

Once a best set of parameters was estimated, the validity of the model was tested by predicting eye movements and Purkinje cell firing patterns during other paradigms at other frequencies. In addition, VOR and OKR after inactivation of the FL were simulated to compare experimental results in monkey [10] for the same purpose.

3. Results

3.1. Parameter estimation

Fig. 2 illustrates results of the model fitting to experimental eye movement (upper panel) and Purkinje cell firing pattern (middle panel) during the VORe with the SOS

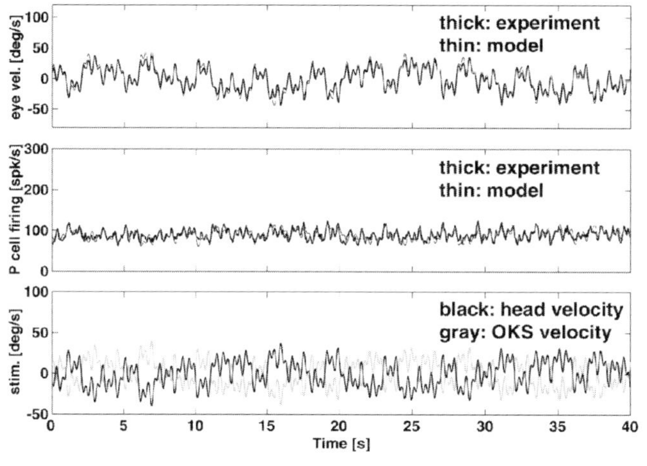

Fig. 2. Results of the model fitting to experimental eye movement (upper) and Purkinje cell firing pattern (middle) during the VORe with the SOS paradigm (lower). Thick lines are experimental data and thin lines are model output.

Table 1
Model parameters estimated by the model fitting shown in Fig. 2

Pre-FL pathway			Non-FL, post-FL pathway		
$G^{\text{visual}}_{\text{preFL\&FL}}(s)$	α_r	1.30E − 4	$G^{\text{visual}}_{\text{nonFL}}(s)$	a_r	−0.0470
	β_r	−0.0879		b_r	−0.0987
	γ_r	0.0904		c_r	−0.1821
	p_r	0.122		u_r	−0.1905
				v_r	−1.696
				τ_r	1.07E − 6
$G^{\text{vestib}}_{\text{preFL\&FL}}(s)$	α_h	−0.125	$G^{\text{vestib}}_{\text{nonFL}}(s)$	a_h	0.0395
	β_h	−0.609		b_h	1.507
	γ_h	0.162		c_h	−0.370
	p_h	0.0137		u_h	−0.0294
				v_h	−1.934
				τ_h	9.06E − 5
$G^{\text{vestib}}_{\text{preFL\&FL}}(s)$	α_e	−0.161	$G_{\text{postFL}}(s)$	a_x	−0.129
	β_e	−0.825		b_x	−1.044
	γ_e	−0.011		c_x	9.20E − 4
	p_e	0.027		τ_x	4.77E − 3

paradigm (lower panel). Thick lines indicate experimental data averaged over more than 20 samples and thin lines are model output. In both upper and middle panels, eye velocity and Purkinje cell firing patterns reproduced by the model fit the experimental data well. The estimated parameters are shown in Table 1.

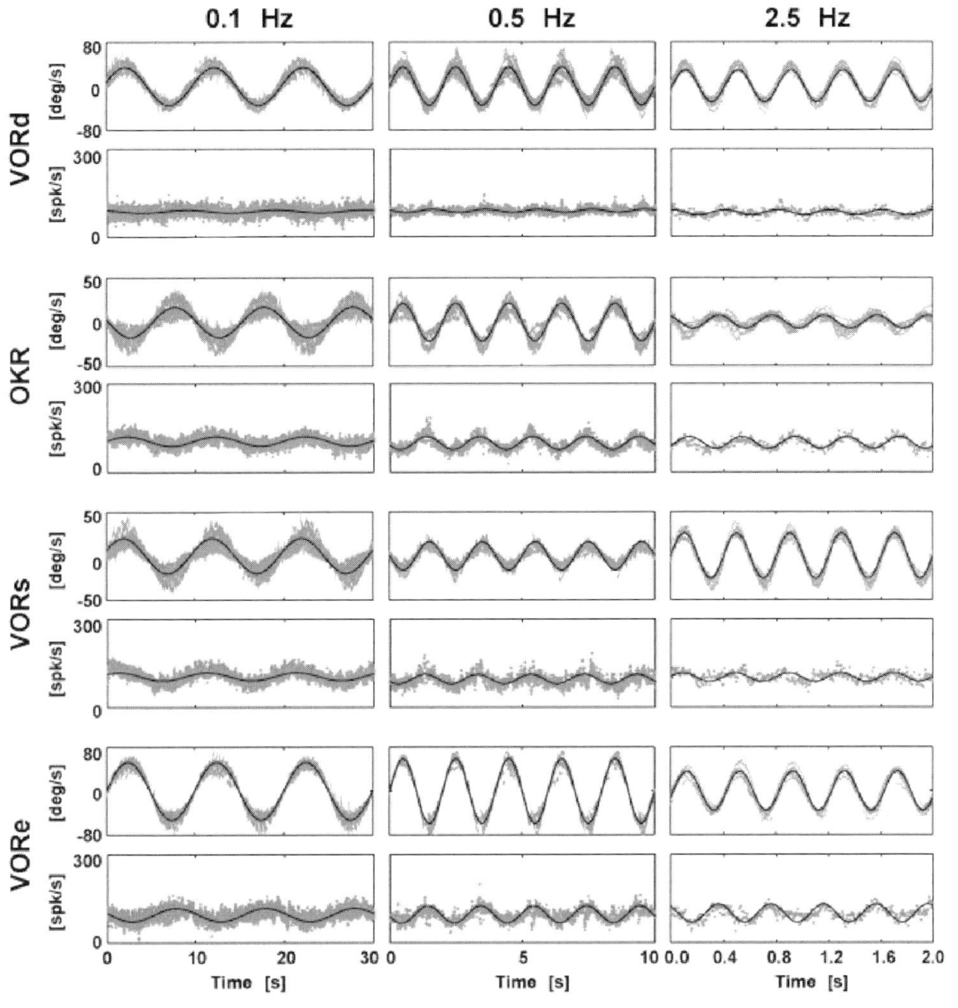

Fig. 3. Results of the prediction during VORd, OKR, VORs, VORe at 0.1, 0.5, and 2.5 Hz in the same format as in Fig. 2 except that all the experimental data are superimposed instead of showing their average and that the stimulus traces are omitted.

3.2. Verification of the model

Fig. 3 illustrates results of the prediction during VORd, OKR, VORs, VORe at 0.1, 0.5, and 2.5 Hz in the same format as in Fig. 2 except that all samples measured in each paradigm are superimposed to show the degree of variability in the experimental data instead of showing averaged traces and that the stimulus traces are omitted. In all the paradigms at all the frequencies examined, the model can predict experimental data well, demonstrating the generality of the model.

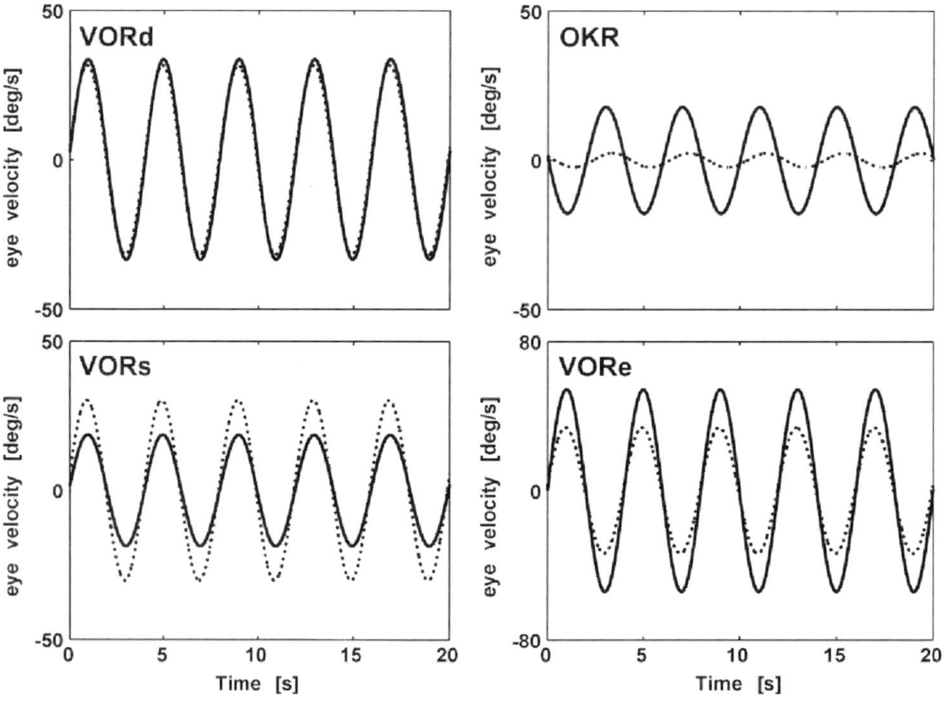

Fig. 4. Simulation of the flocculectomy. Eye velocities before (solid lines) and after the flocculectomy (dashed lines) are superimposed.

To further test the model validity, we simulated lesion experiments in monkeys. It has been demonstrated that flocculectomy did not change VOR gain while it reduced the OKR gain significantly. Also, the monkey's ability to make rapid modifications of the VOR such as enhancement or suppression was severely degraded [10]. In the model, flocculectomy can be simulated by disconnecting the output of FL to the post-FL subsystem.

Fig. 4 illustrates simulated eye velocities during VORd, OKR, VORs, and VORe before (solid line) and after flocculectomy (dashed line). As demonstrated in the behavioral experiment [10], the simulation results show small change in eye velocity during VORd, significant reduction during OKR and degraded enhancement and suppression during VORe and VORs, respectively.

4. Conclusion

VOR adaptation has been an attractive model for elucidating the mechanisms of neural motor control and learning because of its well documented anatomy and easiness of observation of input and output signals. However, the neuronal circuitry forms multiple feedbacks that recursively connect neural signal processing units and make it difficult to understand the signal flows and causality in the observed neuronal activity. A mathematical model whose structure reflects the anatomy is required to overcome these difficulties. There have been several models developed to simulate the VOR and its motor learning. However, only a few models [2] were quantitatively constructed based upon real experimental data and enable quantitative interpretation of their simulation results.

The model currently constructed is quantitative and anatomically plausible. It was described around the cerebellar flocculus (FL) that has been known to play a pivotal role in the VOR motor learning, that is, pathways that include FL (FL pathway) and those that do not (non-FL pathway) were separately formulated and the former was further divided into pre- and post-FL pathways. Also, in each FL and non-FL pathway, those processing each sensory modality were explicitly described. Thus the model may provide separate information on signal processing in each pathway and its role in the VOR and its motor learning. The model parameters were determined by using experimental FL Purkinje cell firing patterns and eye movements in response to visual–vestibular mismatch stimuli that cover most of the dynamic range of the VOR/OKR system in squirrel monkeys. We demonstrated the model validity and generality by predicting FL Purkinje cell firing patterns and eye movements during other various visual–vestibular interaction paradigms and reproducing those after FL lesion.

The model is linear in that all the subsystems are described as linear systems and combined linearly. The fact that the model could quantitatively reproduce eye movements and FL Purkinje cell firing patterns during various visual–vestibular interaction paradigms indicates that the monkey vertical VOR/OKR system behaves linearly within the stimulus range currently employed. The model can be used to elucidate multimodal signal transformation in the cerebellum and to pinpoint a site or sites responsible for VOR motor learning in future studies.

Acknowledgements

We thank Drs. Pablo Blazquez, Andrea Green, Gavin Perry and Hide Tsuneoka for their valuable discussion on the model. Pat Keller provided animal care and technical support. Supported by NIH EYE grant EY-05433, a Grant-in-Aid for Scientific Research on Priority Areas(C)-Advanced Brain Science Project-from Ministry of Education, Culture, Sports, Science and Technology, Japan, and the Hori information science promotion foundation.

References

[1] H. Gomi, M. Shidara, A. Takemura, Y. Inoue, K. Kawano, M. Kawato, Temporal firing patterns of Purkinje cells in the cerebellar ventral paraflocculus during ocular following responses in monkeys I. Simple spikes, J. Neurophysiol. 80 (1998) 818–831.

[2] Y. Hirata, S.M. Highstein, Acute adaptation of the vestibuloocular reflex: signal processing by floccular and ventral parafloccular Purkinje cells, J. Neurophysiol. 85 (2001) 2267–2288.

[3] M. Ito, Cerebellar flocculus hypothesis, Nature 363 (1993) 24–25.

[4] Y. Kobayashi, K. Kawano, A. Takemura, Y. Inoue, T. Kitama, H. Gomi, M. Kawato, Temporal firing patterns of Purkinje cells in the cerebellar ventral paraflocculus during ocular following responses in monkeys II. Complex spikes, J. Neurophysiol. 80 (1998) 832–848.

[5] S.G. Lisberger, F.A. Miles, D.S. Zee, Signals used to compute errors in monkey vestibulo-ocular reflex: possible role of flocculus, J. Neurophysiol. 52 (1984) 1140–1153.

[6] S.G. Lisberger, T. Sejnowski, Motor learning in a recurrent network model based on the vestibulo-ocular reflex, Nature 360 (1992) 159–161.

[7] E. Polak, Computational Methods in Optimization, Academic, New York, 1971.

[8] M. Shidara, K. Kawano, H. Gomi, M. Kawato, Inverse dynamics model eye movement control by Purkinje cells in the cerebellum, Nature 365 (1993) 50–52.

[9] A. Takemura, Y. Inoue, H. Gomi, M. Kawato, K. Kawano, Change in neuronal firing patterns in the process of motor command generation for the ocular following response, J. Neurophysiol. 86 (2001) 1750–1763.

[10] D.S. Zee, A. Yamazaki, P.H. Butler, G. Gucer, Effects of ablation of floccular and parafloccular on eye movements in primate, J. Neurophysiol. 46 (1981) 878–899.

ELSEVIER

Available at
www.ComputerScienceWeb.com
POWERED BY SCIENCE @DIRECT®

Neurocomputing 52–54 (2003) 541 – 546

NEUROCOMPUTING

www.elsevier.com/locate/neucom

Adaptation using local information for maximizing the global cost

Thomas Hoch*, Gregor Wenning, Klaus Obermayer

Department of Electrical Engineering and Computer Science, Technical University of Berlin, Franklinstr. 28/29, 10587 Berlin, Germany

Abstract

Recently the information transmission properties of noisy, parallel summing threshold arrays, have been investigated and interpreted in a neural coding context (see Stocks, Phys. Rev. Lett. 84 (2000) 2310; Phys. Rev. E 63 (2001) 1). The mutual information between certain stimuli and corresponding responses displays a maximum as a function of the noise level. This optimal noise level depends on the number N of neurons within the array, information that is not locally available for single neuron adaptation. We give an analytic expression for the optimal noise level, that only depends on locally available information. The result is based upon an approximation to the mutual information. In the large N limit both descriptions coincide.

Keywords: Stochastic resonance; Adaptation; Summing threshold array

1. Introduction

A noise induced maximum in the mutual information is a signature of stochastic resonance (see [3] for a review). Stochastic resonance has been extensively studied in the context of single neurons (see e.g. [4]). Recently the information transmission properties of noisy, parallel summing threshold arrays, have been investigated and interpreted in a neural coding context (see Stocks [5,7,6]). The mutual information between certain stimulus and response distributions is maximized as a function of the noise level. This optimal noise level depends on the number N of array elements. Adaptation of single neurons to the optimal noise level would require the knowledge of N. But this is information about the global architecture of the system and it is

* Corresponding author.
E-mail address: hoch@cs.tu-berlin.de (T. Hoch).

0925-2312/03/$ - see front matter © 2003 Elsevier Science B.V. All rights reserved.
doi:10.1016/S0925-2312(02)00779-8

not—in an obvious way—available locally at the single cell. In this study we give an analytic expression for the optimal noise level, which depends only on information which is—in principal—locally available at any single neuron. This local optimality criterium is based upon an approximation of the mutual information. In the large N limit both descriptions yield the same optimal noise level. The quality of the approximation is investigated numerically.

This paper is organized as follows. First we introduce the abstract model which describes the input–output relationship of a summing array of parallel threshold elements. In Section 3 the mutual information between a distribution of inputs and a distributions of outputs is introduced. Here we give an approximation to the standard mutual information, based on the work of Brunel et al. [1] and the model given in Section 2. Using this approximation we deduce an analytical expression for the optimal noise level which depends only on information, which is locally available at any single neuron. Section 4 contains numerical results in which we demonstrate the quality of our approximation. Section 5, finally, concludes with a brief discussion.

2. Model

The model [5] consists of a parallel summing array of N threshold devices (Fig. 1). The input to each threshold device is the sum of the signal X and the noise η_i, which is compared to a constant threshold Θ, to yield the following input–output relation:

$$y_i = sign(x + \eta_i - \Theta). \tag{1}$$

The signal X is the same for all threshold devices and is drawn from a gaussian distribution with variance σ_x. Each neuron is corrupted with gaussian noise η_i with variance σ_η, $i \in \{1,\ldots,N\}$, which is assumed to be mutually independent of the signal X and the other noise sources. Hence, Z represents the number of devices n that are

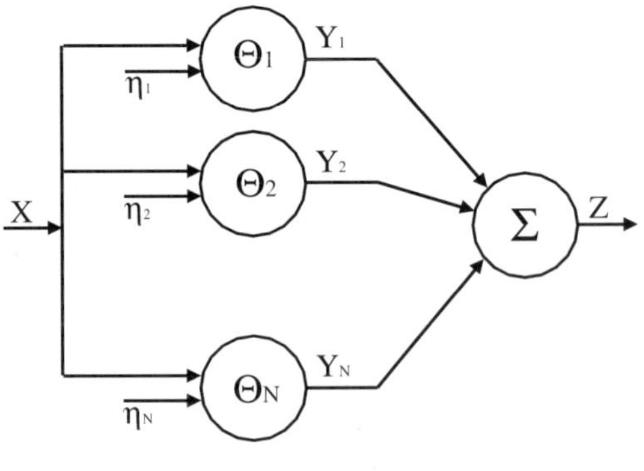

Fig. 1. The model consists of an array of threshold elements. Each neuron receives the same input (drawn from a gaussian distribution X) and is corrupted with additive gaussian noise η_i, $i \in \{1,\ldots,N\}$, which is assumed to be independent of the signal and the other noise sources.

set to one for a given realization of X. Then the conditional probability $P(Z=n|x)$ can be calculated as follows:

$$P(n|x)=\binom{N}{n}P_{1|x}^{n}(1-P_{1|x})^{N-n}, \tag{2}$$

where the conditional probability $P_{1|x}$, that the output of a neuron is set to one, is the cumulative distribution

$$P_{1|x}=\int_{\Theta-x}^{\infty}\frac{1}{\sqrt{2\pi}\sigma_\eta}\exp\left(-\frac{\eta^2}{2\sigma_\eta^2}\right)\,\mathrm{d}\eta. \tag{3}$$

3. Approximation of the mutual information

The average mutual information is an information theoretic measure which quantifies the amount of information the output contains about the input [2]. The mutual information I^{MI} between $P(X)$ and $P(Z)$ is

$$I^{\mathrm{MI}}=H(Z)-H(Z|X) \tag{4}$$

$$=-\sum_{n=0}^{N}P(n)\log_2 P(n)+\int_{-\infty}^{\infty}\mathrm{d}xP_X(x)\sum_{n=0}^{N}P(n|x)\log_2 P(n|x). \tag{5}$$

Stocks demonstrates [7], that maximum information transfer occurs at an optimal noise level ($\sigma_{\mathrm{opt}}^{\mathrm{MI}}$). To yield an analytical expression for the optimal noise level, we adapted an approximation introduced by Brunel and Nadal [1]. They have shown that, in the limit of large N, the mutual information between the signal and the output becomes equal to the mutual information between the signal and an efficient gaussian estimator. The maximum likelihood estimator from $P(n|x)$ exists and is asymptotically unbiased, efficient and gaussian distributed around its mean value with variance $1/F(x)$, and $F(x)$ is the Fisher information:

$$F(x)=E\left[-\frac{\partial^2\log_2 P_Z(n|x)}{\partial x^2}\right]_x=\left(\frac{\partial P_{1|x}}{\partial x}\right)^2\frac{N}{P_{1|x}(1-P_{1|x})}. \tag{6}$$

Then we can calculate the amount of information the maximum likelihood estimator contains about the stimulus:

$$I(X,\hat{X})=H(\hat{X})-\int_{-\infty}^{\infty}\mathrm{d}xP_X(x)H(\hat{X}|X). \tag{7}$$

In the limit of large N, $I(\hat{X}|X)\to I^F$, which is equal to

$$I^F=H(x)-\int_{-\infty}^{\infty}\mathrm{d}xP_X(x)\frac{1}{2}\log_2\left(\frac{2\pi e}{F(x)}\right). \tag{8}$$

Since processing cannot increase information, the mutual information between the signal and the output is at least equal to I^F [1].

$$I^{\mathrm{MI}}(X,Z)\geqslant I^F. \tag{9}$$

We will use Eq. (8) to obtain an analytical expression for the optimal noise level $\sigma^F_{\rm opt}$. From the condition

$$\frac{\partial I^F}{\partial \sigma_\eta} = \frac{\partial}{\partial \sigma_\eta}\left(H(x) - \int_{-\infty}^{\infty} \mathrm{d}x P_x(x)\frac{1}{2}\log_2\left(\frac{2\pi e}{F(x)}\right)\right) = 0 \tag{10}$$

we derived the optimal noise level by replacing the $\log_2(2\pi e/F(x))$ with its second order Taylor Series, which is expanded about $x_0 = 0$. This yields

$$\sigma^F_{\rm opt} \simeq \sqrt{\left(1 - \frac{2}{\pi}\right)(\mu_x^2 + \sigma_x^2)}. \tag{11}$$

Note that the optimal noise level depends on the first and second moment of the input distribution.

4. Results

In Fig. 2a the mutual information and I^F are plotted against $\sigma = \sigma_n/\sigma_x$ for various N and $\Theta = 0$. The curves indicate that for increasing N, I^F approximates the mutual information. Note that I^F can be negative, because the approximation is good only in the case $F(x) \gg 1$. That means, that the estimator is sharply peaked around its mean value. Furthermore, Fig. 2a displays, that the optimal noise level of the Fisher information $\sigma^F_{\rm opt}$ is independent from the number of threshold devices, where as the optimum of the mutual information shows a dependency on the number of threshold elements, especially for small N. According to our approximation, Eq. (8), the optimal noise level is overestimated compared to the optimum of the mutual information, as can be seen in Fig. 2b. But due to the asymmetry of the mutual information, the distance between the maxima of $I^{\rm MI}$ and I^F is relatively small and decreases with increasing N.

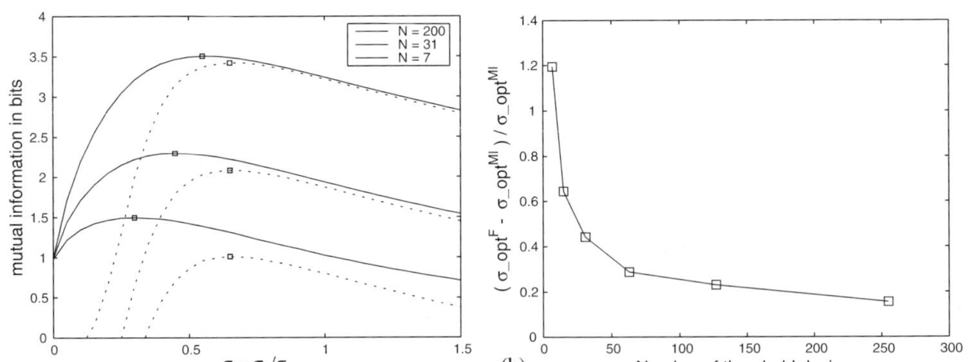

Fig. 2. (a) Displays the mutual information between the stimulus and the output (solid line) and the mutual information I^F according to Eq. (8) (dashed line) are plotted against $\sigma = \sigma_\eta/\sigma_x$ for various N and $\Theta = \mu_x$. (b) Displays the relative distance between the optimum noise level of the mutual information and $\sigma^F_{\rm opt}$ of I^F.

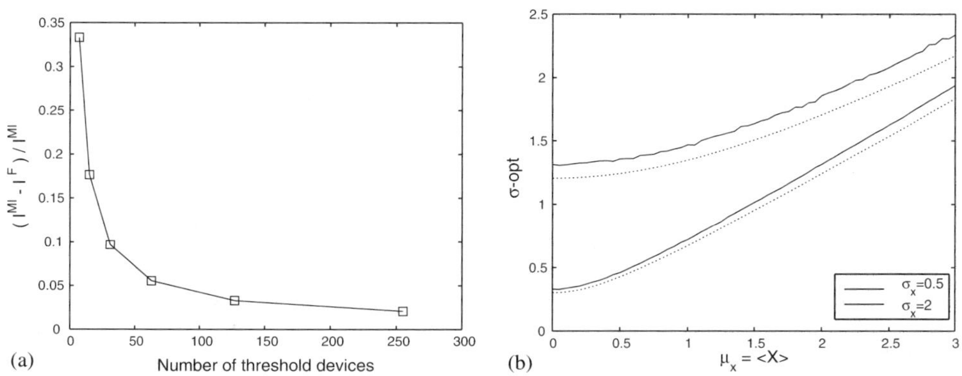

Fig. 3. (a) Displays the relative deviation of I^F from the maximum of the mutual information I^{MI}. (b) Presents the optimal noise level according to Eq. (8) (dotted line) and of I^F (solid line) for various μ_x and σ_x. Because of the Taylor expansion, σ_{opt} is slightly below the optimal noise level of I^F, obtained from numerical calculations.

In Fig. 3a the relative deviation from the maximum of the mutual information is presented for increasing N. In Fig. 3b, σ^F_{opt} calculated from Eq. (11), is compared to the optimal noise level of I^F, obtained from numerical calculations. Because of the Taylor expansion, σ^F_{opt} is slightly below the optimal noise level for I^F.

5. Discussion

In this model study we examined stochastic resonance in a summing parallel threshold array. We compared optimal noise levels of the mutual information I^{MI} between a distribution of inputs and a corresponding distribution of outputs, and I^F, what is an approximation to I^{MI}. The quality of this approximation increases with the number N of parallel nonlinearities. The optimal noise level $\sigma^{\mathrm{MI}}_{\mathrm{opt}}$ for I^{MI} depends on N, thus a neuron that should adapt to the optimal noise level needs to have some knowledge about the system size N. But this information is not locally available at a single neuron. The optimal noise level σ^F_{opt} in terms of I^F does not depend on N, but only on variables, which are in principle locally available at any single neuron.

For all N, σ^F_{opt} is larger than $\sigma^{\mathrm{MI}}_{\mathrm{opt}}$ and with increasing N this difference decreases. Due to the asymmetry of the stochastic resonance curves (information vs. noise) a moderate overestimation of the optimal noise level does not degrade the amount of information transmitted in a dramatic way. Thus adaptation to σ^F_{opt} instead of $\sigma^{\mathrm{MI}}_{\mathrm{opt}}$ is feasible and reasonable. Furthermore the analytic expression for the optimal noise level, in terms of I^F, does not only depend on the mean of the input distribution, as in [8], but also on the variance. More general, evaluating the integral in Eq. (8) using more terms in the Taylor expansion yields a weighted sum of the moments of the input distribution.

Acknowledgements

Supported by Wellcome Trust (061113/Z/00).

References

[1] N. Brunel, J. Nadal, Mutual information, Fisher information, and population coding, Neural Comput. 10 (1998) 1731–1757.

[2] T.M. Cover, J.A. Thomas, Elements of Information Theory, Wiley Series in Telecommunications, 2nd Edition, Wiley, New York, 1991.

[3] L. Gammaitoni, P. Hänggi, P. Jung, F. Marchesoni, Stochastic resonance, Rev. Modern Phys. 70 (1) (1998) 223–287.

[4] D.F. Russel, L.A. Wilkens, F. Moss, Use of behavioral stochastic resonance by paddle fish for feeding, Nature 402 (1999) 291–294.

[5] N.G. Stocks, Suprathreshold stochastic resonance in multilevel threshold systems, Phys. Rev. Lett. 84 (2000) 2310–2314.

[6] N.G. Stocks, Generic noise-enhanced coding in neuronal arrays, Phys. Rev. E 64 (2001) 1–4.

[7] N.G. Stocks, Information transmission in parallel threshold arrays: suprathreshold stochastic resonance, Phys. Rev. E 63 (2001) 1–9.

[8] G. Wenning, K. Obermayer, Adjusting stochastic resonance in a leaky integrate and fire neuron to subthreshold stimulus distributions, Neurocomputing 44–46 (2002) 225–231.

Available at
www.ComputerScienceWeb.com
POWERED BY SCIENCE @ DIRECT®

ELSEVIER

Neurocomputing 52–54 (2003) 547–552

NEUROCOMPUTING

www.elsevier.com/locate/neucom

Modeling receptive fields with non-negative sparse coding

Patrik O. Hoyer

Neural Networks Research Centre, Helsinki University of Technology, P.O. Box 9800, FIN-02015 HUT, Finland

Abstract

An important approach in visual neuroscience considers how the processing of the early visual system is dependent on the statistics of the natural environment. A particularly influential model in this respect has been sparse coding. In this paper we argue for a non-negative variant of the model. This is based partly on neurophysiological grounds and partly on the intuitive understanding of parts-based representations. We discuss the logic behind our reasoning and show experiments on natural images demonstrating the usefulness of the new model.

Keywords: Natural image statistics; Sparse coding; Non-negative representations

1. Introduction

A fundamental problem in visual neuroscience concerns understanding how the statistics of the environment is reflected in the processing of the early visual system. Research in this area originated almost half a century ago [1] and has recently attracted significant attention as the computational power and the theoretical tools needed to investigate these issues have become available. For a review of the field, see Ref. [14].

A large share of this research has concerned the mammalian primary visual cortex, also known as V1. This is largely due to the fact that, ever since the pioneering work of Hubel and Wiesel [6], much is known about the processing performed there. Neurons in V1 can be divided into simple- and complex-cells. Of these, the responses of simple-cells can be reasonably described as a linear filtering of the visual input. The spatial properties of these linear filters (called the receptive fields) can be characterized

E-mail address: patrik.hoyer@hut.fi (P.O. Hoyer).

0925-2312/03/$ - see front matter
doi:10.1016/S0925-2312(02)00782-8

as being localized, oriented, and bandpass [6,3]. The question of why this particular type of filtering is performed has undergone considerable debate during the years.

In a highly influential paper, Olshausen and Field [11] showed that linear sparse coding (SC) of natural images yielded features qualitatively very similar to receptive fields of simple-cells in V1. Subsequently, the very closely related model of independent component analysis (ICA) [7] was shown to give similar results [2]. This framework gave the observed receptive fields an *information-theoretic explanation* which had been lacking in previous work, demonstrating how sensory coding is related to the statistics of the environment.

The basic idea in sparse coding is relatively simple: The observed multidimensional data (random vector) $\mathbf{x}$ is modeled as a linear superposition of some features (called basis vectors) $\mathbf{a}_i$, as in $\mathbf{x} \approx \sum_{i=1}^{n} \mathbf{a}_i s_i$. In the basic image data case, each input vector $\mathbf{x}$ consists of the pixel intensities of a small image patch, and the $\mathbf{a}_i$ likewise contain the pixel intensities of basis patches. The latent variables s_i that give the mixing proportions are stochastic and differ for each observed input $\mathbf{x}$. The crucial assumption in the sparse coding framework is that these hidden variables exhibit sparseness. Sparseness is a property independent of scale (variance), and implies that the s_i have probability densities which are highly peaked at zero and have heavy tails. Essentially, the idea is that any single typical input vector $\mathbf{x}$ can be accurately described using only a few active (significantly non-zero) units s_i. However, all of the basis vectors $\mathbf{a}_i$ are needed to represent the data because the set of active units changes from input to input.

The model, as typically used, is completely symmetric around zero. The data is normalized to zero mean, and the sources s_i are assumed to have even-symmetric probability densities (i.e. $p(s_i) = f(|s_i|)$) with a high peak at zero and heavy tails. This also implies that the observed data is completely symmetric with respect to the origin. When this model is learned from natural image data, the learned basis patches have the principal properties of the spatial receptive fields of simple-cells in V1. The neural interpretation of the model is thus that simple-cells in V1 perform sparse coding on the visual input they receive, with the receptive fields closely related to the sparse coding basis vectors and the firing rates of the neurons representing the latent variables s_i.

Could the sparse coding model be used more generally to understand how cortical circuits represent their inputs? The fact that areas of the cerebral cortex analyzing quite different inputs nevertheless anatomically are relatively similar [13] suggests the existence of common computational strategies for cortical sensory coding. Could sparse coding perhaps explain further properties of the visual system, or possibly properties of other sensory modalities? We believe that to tackle such questions, the model must first be slightly modified, as explained in the next section.

2. Non-negative sparse coding

There are at least two obvious ways in which the standard sparse coding image model is unrealistic as a model of V1 simple-cell behavior. Perhaps the most glaring discrepancy is the fact that in the model each unit s_i can, in addition to being effectively silent (close to zero), be either positively or negatively active. This basically means that every feature contributes to representing stimuli of opposing polarity; for example, the same unit that codes for a dark bar on a bright background also codes for a bright bar on a dark background. This is in clear contrast to the behavior of simple-cells in V1: these neurons tend to have quite low background firing rates and, as firing rates cannot go negative, thus can only represent one-half of the output distribution of a signed feature s_i.

Another major difference between the model and neural reality is that the input data in the model is double-sided (signed), whereas V1 receives the visual data from the lateral geniculate nucleus (LGN) in the form of separated ON- and OFF-channels. Of course, as an abstract model of visual coding, the input data should indeed be (signed) image contrast. But if we on the other hand are interested in how V1 recodes its input signals, we must consider separate ON- and OFF-channel input.

Thus, if we would like to transform sparse coding from a relatively abstract model of image representation in V1 to a concrete model of simple-cell recoding of LGN inputs, the model must be changed. First, our input data should consist of hypothetical activities of ON- and OFF-channels in response to natural image patches. Second, all coefficients s_i should be restricted to non-negative values. As both the sources s_i and the data $\mathbf{x}$ thus have non-negative values only, it is logical to assume the same of the features $\mathbf{a}_i$. This is due to the fact that if the features contained negative values, this would inevitably be true of the observed data $\mathbf{x}$ as well.

Although we so far considered non-negativity constraints with the objective of making the model better fit known neurophysiology, there are also other equally important arguments for non-negativity. In particular, it has been argued that non-negativity can be important for learning parts-based representations [8]: In the standard sparse coding model, the data is described as a combination of elementary features involving both additive and subtractive interactions. The fact that features can 'cancel each other out' using subtraction is contrary to the intuitive notion of combining parts to form a whole. Non-negativity ensures that elementary object features combine additively.

Representations involving only additive combinations of object parts seem especially attractive for modeling higher-level features of images: In Ref. [5] it was shown how contour-coding could be learned from the statistics of complex-cell-like responses to natural images. This was possible by representing the model complex-cell responses by a non-negative, sparse, code.

Arranging all observed data vectors $\mathbf{x}$ into the columns of a matrix $\mathbf{X}$, the corresponding unknown sources into the matrix $\mathbf{S}$, and the unknown features $\mathbf{a}_i$ into the columns of $\mathbf{A}$, the linear model is given by $\mathbf{X} \approx \mathbf{AS}$. The non-negativity constraints require that all elements of $\mathbf{A}$, and $\mathbf{S}$ are zero or positive. The same is assumed of the observed data $\mathbf{X}$.

We suggest to use the following objective function for *non-negative sparse coding* (NNSC) [4,5]:

$$C(\mathbf{A},\mathbf{S}) = \frac{1}{2}\|\mathbf{X} - \mathbf{AS}\|^2 + \lambda \sum_{ij} S_{ij} \tag{1}$$

with λ equal to zero this objective reduces to the squared error version of non-negative matrix factorization (NMF) [12,8,9]. However, for positive λ this objective emphasizes decompositions where $\mathbf{S}$ is sparse.

Minimizing the objective with respect to both $\mathbf{A}$ and $\mathbf{S}$ without any constraints other than non-negativity yields the NMF solution. This is because scaling up $\mathbf{A}$ while correspondingly scaling down $\mathbf{S}$ eventually drives the effective value of λ to zero. Thus, some constraint is needed on the scales of $\mathbf{A}$ and $\mathbf{S}$. We suggest to fix the norm of each column of $\mathbf{A}$ to unity, i.e. $\|\mathbf{a}_i\| = 1$, $\forall i$. An efficient algorithm for minimizing the objective subject to the suggested constraints is given in Ref. [4].

3. Learning receptive fields from ON/OFF-channels

Here we show how features resembling simple-cell receptive fields can be learned from natural image data preprocessed into separate ON- and OFF-channels. We used a set of natural images kindly provided by Bruno Olshausen. These images can be found as part of the *sparsenet* software package [10]. The images had been preprocessed by a spatial filter, that approximately whitened the data, for details see Ref. [11]. We sampled 12×12 -pixel patches from the images, and then separated positive and negative values into separate channels. Each image patch was thus represented by a $2 \times 12 \times 12 = 288$ -dimensional vector, each element of which mimics the activity of an ON- or OFF-center neuron to the input patch. These vectors made up the columns of $\mathbf{X}$.

We then minimized the objective given in Eq. (1) under the stated constraints, with 72 sources, λ set to 5 and with each input channel scaled to have unit average squared activation. The complete code for performing the experiments can be found at `http://www.cis.hut.fi/phoyer/code/`. The resulting basis patterns (columns of $\mathbf{A}$) are shown in Fig. 1. Note how the basis vectors represent Gabor-like image input by elongated ON- and OFF-subregions. This is most clearly seen in the rightmost panel, where the weights from the OFF-channel have been subtracted from those from the ON-channel. These learned features are at least qualitatively very similar to the ones found with the standard sparse coding model applied to symmetric image data.

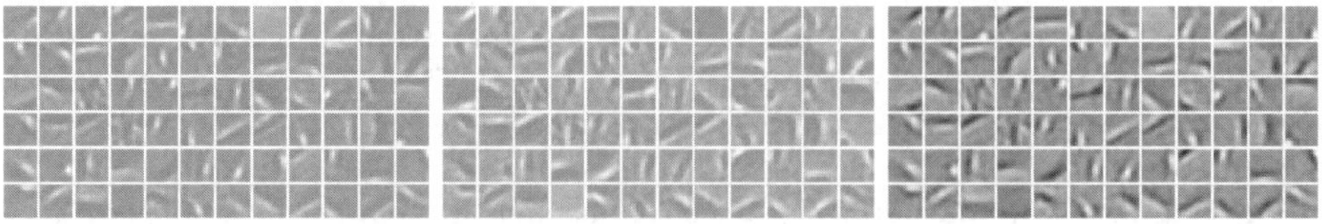

Fig. 1. Basis vectors from non-negative sparse coding of ON/OFF-channel image data. Left: Generative weights for the ON-channel. Each patch represents the part of one basis vector $\mathbf{a}_i$ corresponding to the ON-channel. Gray pixels denotes zero weight, brighter pixels are positive weights. Middle: Corresponding weights for the OFF-channel. Right: Weights for ON minus weights for OFF.

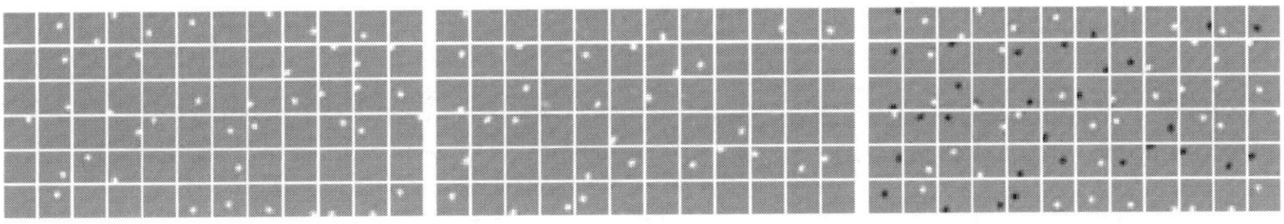

Fig. 2. Basis vectors learned by non-negative matrix factorization. Left: Weights for ON-channel. Middle: Weights for OFF-channel. Right: Weights for ON minus weights for OFF.

To gain some insight into the roles of the sparseness objective and the non-negativity constraints in these results we can compare the learned decomposition to those given by other methods. For example, in Fig. 2 we show the results of non-negative matrix factorization (NMF) on our data. This is actually a special case of our model, when $\lambda = 0$. Note that all features are unoriented; this held for all tested dimensionalities. Thus it seems that sparseness is important for learning oriented, localized features. The table below summarizes our findings on the behaviour of different decompositions on the data.

	NNSC	NMF	PCA	ICA/SC
Localized	**yes**	**yes**	**no**	**yes**
Oriented	**yes**	**no**	**yes**	**yes**
Bandpass	**yes**	**no**	**yes**	**yes**
Non-negative	**yes**	**yes**	**no**	**no**

4. Discussion and conclusions

Although we here showed only the learning of simple cell-like receptive fields, we believe that this modified sparse coding model can be used to explain higher-order visual receptive fields as well. For example, in Ref. [5] it was shown how contour coding and end-stopped receptive fields could be learned by non-negative sparse coding on the simulated responses of complex cells to natural images. It remains to be seen if the proposed model could also be useful in predicting or understanding properties of neurons analyzing other sensory modalities.

Finally, we would like to point out one common misconception about models with non-negativity constraints: A frequent objection to non-negativity constraints is the fact that inhibition is widespread in the cortex. However, it is important to note that the constraints are on the activities and the *generative* representation. In fact, inhibition is a vital part of our model: it is required during *inference* as an important part of the competition between units to represent the stimuli as sparsely as possible.

Acknowledgements

I wish to thank Aapo Hyvärinen and Jarmo Hurri for useful discussions and helpful comments on an earlier version of the manuscript.

References

[1] H.B. Barlow, Possible principles underlying the transformations of sensory messages, in: W.A. Rosenblith (Ed.), Sensory Communication, MIT Press, Cambridge, MA, 1961, pp. 217–234.
[2] A.J. Bell, T.J. Sejnowski, The 'independent components' of natural scenes are edge filters, Vision Res. 37 (1997) 3327–3338.
[3] R.L. DeValois, D.G. Albrecht, L.G. Thorell, Spatial frequency selectivity of cells in macaque visual cortex, Vision Res. 22 (1982) 545–559.
[4] P.O. Hoyer, Non-negative sparse coding, in: Neural Networks for Signal Processing, Vol. XII, Proceedings of the IEEE Workshop on Neural Networks for Signal Processing, Martigny, Switzerland, 2002, pp. 557–565.
[5] P.O. Hoyer, A. Hyvärinen, A multi-layer sparse coding network learns contour coding from natural images, Vision Res. 42 (12) (2002) 1593–1605.
[6] D.H. Hubel, T.N. Wiesel, Receptive fields and functional architecture of monkey striate cortex, J. Physiol. 195 (1968) 215–243.
[7] A. Hyvärinen, J. Karhunen, E. Oja, Independent Component Analysis, Wiley Interscience, New York, 2001.
[8] D.D. Lee, H.S. Seung, Learning the parts of objects by non-negative matrix factorization, Nature 401 (6755) (1999) 788–791.
[9] D.D. Lee, H.S. Seung, Algorithms for non-negative matrix factorization, in: Advances in Neural Information Processing, Vol. 13 (Proc. NIPS*2000), MIT Press, Cambridge, MA, 2001.
[10] B.A. Olshausen, The Sparsenet software package for MATLAB, Available at `http://redwood.ucdavis.edu/bruno/sparsenet.html`.
[11] B.A. Olshausen, D.J. Field, Emergence of simple-cell receptive field properties by learning a sparse code for natural images, Nature 381 (1996) 607–609.
[12] P. Paatero, U. Tapper, Positive matrix factorization: a non-negative factor model with optimal utilization of error estimates of data values, Environmetrics 5 (1994) 111–126.
[13] A.J. Rockel, R.W. Hiorns, T.P.S. Powell, The basic uniformity in structure of the neocortex, Brain 103 (1980) 221–244.
[14] E.P. Simoncelli, B.A. Olshausen, Natural image statistics and neural representation, Ann. Rev. Neurosci. 24 (2001) 1193–1215.

Available at
www.ComputerScienceWeb.com
POWERED BY SCIENCE @ DIRECT®

ELSEVIER

Neurocomputing 52–54 (2003) 553–559

NEUROCOMPUTING

www.elsevier.com/locate/neucom

A two-layer temporal generative model of natural video exhibits complex-cell-like pooling of simple cell outputs

Jarmo Hurri*, Aapo Hyvärinen

Neural Networks Research Centre, Helsinki University of Technology, P.O. Box 9800, 02015 HUT, Finland

Abstract

We present a two-layer dynamic generative model of the statistical structure of natural image sequences. The second layer of the model is a linear mapping from simple cell outputs to pixel values, as in most work on natural image statistics. The first layer models the dependencies of the activity levels (amplitudes or variances) of the simple cells, using a multivariate autoregressive model. The second layer shows emergence of basis vectors that are localized, oriented and have different scales, just like previous work. But our new model enables the first layer to learn connections between the simple cells that are similar to complex cell pooling: connections are strong among cells with similar location, frequency and orientation. In contrast to previous work in which one of the layers needed to be fixed in advance, the dynamic model enables us to estimate both of the layers simultaneously from natural data.

Keywords: Natural image sequences; Primary visual cortex; Generative models

1. Introduction

A central question in the study of sensory neural networks is how stimuli are represented or coded by neurons. One approach to studying the neural code is to examine how its properties are related to the statistics of natural stimuli. Within the past 10 years, computational principles relating the properties of cells in the primary visual cortex to the statistics of natural stimuli have been proposed. The most influential of

* Corresponding author.
E-mail addresses: jarmo.hurri@hut.fi (J. Hurri), aapo.hyvarinen@hut.fi (A. Hyvärinen).

0925-2312/03/$ - see front matter © 2002 Elsevier Science B.V. All rights reserved.
doi:10.1016/S0925-2312(02)00783-X

these theories have been sparse coding [2], independent component analysis [1,7], and temporal coherence [3]. The principle of temporal coherence is based on the idea that when processing temporal input, the representation changes as little as possible over time. In a recent paper [5] we have shown that a non-linear form of temporal coherence is related to the structure of simple cell receptive fields. According to the results presented in [5], simple cell receptive fields are optimally temporally coherent in the sense that the *activity levels* of simple cells are stable over short time intervals. By activity level we mean the amplitude or energy of the output of a linear filter that models a simple cell.

In the measure of temporal activity coherence introduced in [5] there was no possibility of an interaction between the activity levels of different cells. In this paper, we introduce a model which includes inter-cell activity dependencies. This is accomplished by a two-layer generative model in which the activity levels are generated in an autoregressive manner. We will show that estimation of the model from natural image sequence data yields simple-cell-like receptive fields, and a completely unsupervised complex-cell-like pooling between the outputs of simple cells.

2. Definition of the model

The generative model of natural image sequences introduced in this paper has two layers (see Fig. 1). The first layer is a multivariate autoregressive model of the activity levels (amplitudes) of simple cell responses at time t and time $t - \Delta t$. The signs of cell responses are generated by a latent signal between the first and second layer. The second layer is linear, and maps cell responses to image features.

We start the formal description of the model with the second, linear layer. We restrict ourselves to linear spatial models of simple cells. Let vector $\mathbf{x}(t)$ denote an image patch taken from natural image sequences at time t. (Vectorization of image patches can be done by scanning images columnwise into vectors.) Let the vector $\mathbf{y}(t) = [y_1(t) \cdots y_K(t)]^T$ represent the outputs of K simple cells. The linear generative

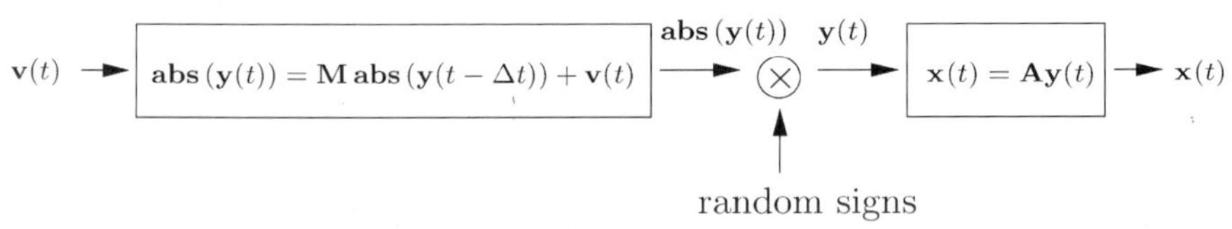

Fig. 1. The two layers of the generative model with spatiotemporal activity dependencies. Let $\mathbf{abs}(\mathbf{y}(t)) = [|y_1(t)| \cdots |y_K(t)|]^T$ denote the activity levels (amplitudes) of simple cell responses. In the first layer, the driving noise signal $\mathbf{v}(t)$ generates the activities of simple cells via a multivariate autoregressive model. Matrix $\mathbf{M}$ captures the spatiotemporal activity dependencies in the model. The signs of the responses are generated randomly between the first and second layer to yield signed responses $\mathbf{y}(t)$. In the second layer, natural image sequence $\mathbf{x}(t)$ is generated linearly from simple cell responses. In addition to the relations shown here, the generation of $\mathbf{v}(t)$ is affected by $\mathbf{M}\,\mathbf{abs}(\mathbf{y}(t - \Delta t))$ to ensure non-negativity of $\mathbf{abs}(\mathbf{y}(t))$. See text for details.

model for $\mathbf{x}(t)$ is similar to the one in [2]:

$$\mathbf{x}(t) = \mathbf{A}\mathbf{y}(t). \tag{1}$$

Here $\mathbf{A} = [\mathbf{a}_1 \cdots \mathbf{a}_K]$ denotes a matrix which relates the image patch $\mathbf{x}(t)$ to the outputs of simple cells, so that each column $\mathbf{a}_k$, $k = 1, \ldots, K$, gives the feature that is coded by the corresponding simple cell. When the parameters of the model are estimated, what we obtain first is the mapping from $\mathbf{x}(t)$ to $\mathbf{y}(t)$, denoted by

$$\mathbf{y}(t) = \mathbf{W}\mathbf{x}(t). \tag{2}$$

Conceptually, the set of filters (vectors) $\mathbf{w}_1, \ldots, \mathbf{w}_K$ corresponds here to the receptive fields of simple cells, and $\mathbf{W} = [\mathbf{w}_1 \cdots \mathbf{w}_K]^{\mathrm{T}}$ denotes a matrix with all the filters as rows. The dimension of $\mathbf{x}(t)$ is typically larger than the dimension of $\mathbf{y}(t)$, so that (2) is generally not invertible but an underdetermined set of linear equations. A one-to-one correspondence between $\mathbf{W}$ and $\mathbf{A}$ can be established by computing the pseudoinverse solution $\mathbf{A} = \mathbf{W}^{\mathrm{T}}(\mathbf{W}\mathbf{W}^{\mathrm{T}})^{-1}$.

In contrast to basic sparse coding [2] or independent component analysis [7] we do *not* assume that the components of $\mathbf{y}(t)$ are independent. Instead, we assume that the activity levels (amplitudes) of the components of $\mathbf{y}(t)$ are correlated. We model these dependencies with a multivariate autoregressive model in the first layer of our model. Let $\mathbf{abs}(\mathbf{y}(t)) = [|y_1(t)| \cdots |y_K(t)|]^{\mathrm{T}}$, and let $\mathbf{v}(t)$ denote a driving noise signal (the distribution of $\mathbf{v}(t)$ is constrained by the non-negativity of the process, and will be discussed in more detail below). Let $\mathbf{M}$ denote a $K \times K$ matrix, and let Δt denote a time lag. Our model for the activities is a *constrained multidimensional first-order autoregressive process*, defined by

$$\mathbf{abs}(\mathbf{y}(t)) = \mathbf{M}\,\mathbf{abs}(\mathbf{y}(t - \Delta t)) + \mathbf{v}(t) \tag{3}$$

and unit energy constraints $E_t\{y_k^2(t)\} = 1$ for $k = 1, \ldots, K$.

There are dependencies between the driving noise $\mathbf{v}(t)$ and $\mathbf{abs}(\mathbf{y}(t))$, caused by the non-negativity of $\mathbf{abs}(\mathbf{y}(t))$. Let $\mathbf{u}(t)$ denote a zero-mean random vector with components which are statistically independent of each other. We define $\mathbf{v}(t) = \mathbf{max}(-\mathbf{M}\,\mathbf{abs}(\mathbf{y}(t - \Delta t)), \mathbf{u}(t))$, where, for vectors $\mathbf{a}$ and $\mathbf{b}$, $\mathbf{max}(\mathbf{a}, \mathbf{b}) = [\max(a_1, b_1) \cdots \max(a_n, b_n)]^{\mathrm{T}}$. We assume that $\mathbf{u}(t)$ and $\mathbf{abs}(\mathbf{y}(t))$ are uncorrelated. To make the generative model complete, a mechanism for generating the signs of cell responses $\mathbf{y}(t)$ must be included. We specify that the signs are generated randomly with equal probability for plus or minus after the strengths of the responses have been generated.

Note that the unit energy constraints and the uncorrelatedness of the outputs can be represented by a single matrix equation

$$\mathbf{W}\mathbf{C}_{\mathbf{x}(t)}\mathbf{W}^{\mathrm{T}} = \mathbf{I}, \tag{4}$$

where $\mathbf{C}_{\mathbf{x}(t)} = E_t\{\mathbf{x}(t)\mathbf{x}(t)^{\mathrm{T}}\}$, and that they imply $E_t\{\|\mathbf{y}(t)\|^2\} = K$. Therefore, because the sign generation mechanism also implies that each $y_k(t)$ has zero mean, the variances of the outputs will also be constant.

In Eq. (3), a large positive matrix element $\mathbf{M}(i,j)$, or $\mathbf{M}(j,i)$, indicates that there is a strong dependency between the activities of cells i and j. Thinking in terms of grouping cells with large activity dependencies together, matrix $\mathbf{M}$ can be thought of as containing similarities (reciprocals of distances) between different cells. We will use this property in the experimental section to derive a spatial organization of the simple cells from **M**.

3. Estimation of the model

To estimate the model defined above we need to estimate both **M** and **W** (the pseudoinverse of **A**). In this section, we first show how to estimate **M**, given **W**. Then we describe an objective function which can be used to estimate **W**, given **M**. Each iteration of the estimation algorithm consists of two steps. During the first step **M** is updated, and **W** is kept constant; during the second step these roles are reversed.

First, regarding the estimation of **M**, consider a situation in which **W** has been fixed. It can be shown [6] that **M** can be estimated by using approximative method of moments, and that the estimate is given by

$$\mathbf{M} \approx \beta E_t\{(\mathbf{abs}(\mathbf{y}(t)) - E_t\{\mathbf{abs}(\mathbf{y}(t))\})(\mathbf{abs}(\mathbf{y}(t-\Delta t)) - E_t\{\mathbf{abs}(\mathbf{y}(t))\})^{\mathrm{T}}\}$$
$$\times E_t\{(\mathbf{abs}(\mathbf{y}(t)) - E_t\{\mathbf{abs}(\mathbf{y}(t))\})(\mathbf{abs}(\mathbf{y}(t)) - E_t\{\mathbf{abs}(\mathbf{y}(t))\})^{\mathrm{T}}\}^{-1}, \quad (5)$$

where $\beta > 1$. Because this scaling does not affect the optima of (6), and because we are more interested in the relative magnitudes of elements of **M** than their absolute values, we can discard β in the estimation process. See [6] for details.

The estimation of **W** is more complicated. A rigorous derivation of an objective function based on well-known estimation principles is very difficult because the statistics involved are non-Gaussian, and the processes have difficult interdependencies. Therefore, we derived an objective function heuristically, and verified its validity through simulations. The objective function is defined by

$$f(\mathbf{W},\mathbf{M}) = \sum_{i=1}^{K}\sum_{j=1}^{K} \mathbf{M}(i,j)\,\mathrm{cov}\{|y_i(t)|, |y_j(t-\Delta t)|\}. \quad (6)$$

In the actual estimation algorithm, **W** is updated by employing a gradient projection approach to the optimization of (6) under constraint (4). The initial value of **W** is selected randomly. The fact that the algorithm described above is able to estimate models following the two-layer model has been verified through extensive simulations (details are given in [6]).

4. Experiments with natural image sequences

The data and preprocessing used in the experiments were very similar to those in [5], so we will describe them only shortly here, and refer the reader to [5] for details. The natural image sequences used in data collection consisted of 129 image sequences, which were a subset of natural image sequences used in [4]. The sampling rate in these sequences was 25 Hz. Initially 200,000 image sequences with a duration of 440 ms, and spatial size 16×16 pixels, were sampled from these sequences. The fairly long

Fig. 2. Grouping similar to complex cell pooling of simple cell outputs emerges from spatiotemporal activity dependencies. Here we have plotted each of the basis vectors (columns of **A**) at a 2D position obtained by applying multidimensional scaling to the similarity values defined by **M**. As can be seen, nearby basis vectors seem to be mostly coding for similarly oriented features with similar frequencies at nearby spatial positions. In addition, some global topographic organization also emerges: those basis vectors which code for horizontal features are on the left in the figure, while those that code for vertical features are on the right. In this figure, some short distances have been extended in order to be able to show the basis vectors in a reasonable scale.

duration of these initial samples was necessary because of the temporal filtering used in preprocessing.

The preprocessing consisted of four steps: temporal decorrelation, subtraction of local mean, normalization, and dimensionality reduction. Temporal decorrelation, performed via temporal filtering, enhances temporal changes in the data, and differentiates our results from those obtained with static images [5]. It can also be motivated as a model of temporal processing at the lateral geniculate nucleus. After temporal decorrelation the spatial local mean (spatial DC component) was subtracted from each of the 400,000 image patches, and the patches were normalized to unit norm. This normalization can be considered as a form of contrast gain control. Finally, to reduce the effect of noise and aliasing artifacts, the dimensionality of the data was reduced to 160 using principal component analysis [7].

The estimation algorithm described in Section 3 was applied to the preprocessed natural image sequence data to obtain estimates for $\mathbf{M}$ and $\mathbf{A}$ (the pseudoinverse of $\mathbf{W}$). The extracted matrices $\mathbf{A}$ and $\mathbf{M}$ can be visualized simultaneously by using the interpretation of $\mathbf{M}$ as a similarity matrix (see Section 2), and by applying multidimensional scaling (MDS) to it (see [6] for details). The resulting spatial layout produced by the MDS procedure is shown in Fig. 2. First, regarding the properties of the resulting basis vectors, we can see that the basis vectors are localized, oriented, and have multiple scales. These are the most important defining criteria of simple cell receptive fields [8]. Second, regarding the similarity defined by the spatiotemporal dependency matrix $\mathbf{M}$, we can see that those basis vectors which are very close to each other seem to be mostly coding for similarly oriented features with the same frequencies at nearby spatial positions. This kind of grouping is characteristic of pooling of simple cell outputs at complex cell level, as well as of the topographic organization of the visual cortex [8]. In addition to the local topography described above, some global topography also emerges in the results: those basis vectors which code for horizontal features are on the left in Fig. 2, while those that code for vertical features are on the right.

5. Conclusion

We have described a two-layer dynamic generative model of image sequences, and an algorithm for estimating the model from sample data. Application of the algorithm to natural image sequences yields a set of linear filters, or basis vectors, which are similar to simple cell receptive fields, and connections between the simple cells that are similar to the way in which simple cell outputs presumably are pooled at the complex cell level. The basis vectors are learned in one layer of the model, and the pooling property in the other. Both layers are learned simultaneously and in a completely unsupervised manner.

References

[1] A. Bell, T.J. Sejnowski, The independent components of natural scenes are edge filters, Vision Res. 37 (23) (1997) 3327–3338.

[2] B.A. Olshausen, D. Field, Emergence of simple-cell receptive field properties by learning a sparse code for natural images, Nature 381 (6583) (1996) 607–609.

[3] P. Földiák, Learning invariance from transformation sequences, Neural Comput. 3 (2) (1991) 194–200.

[4] J. Hans van Hateren, D.L. Ruderman, Independent component analysis of natural image sequences yields spatio-temporal filters similar to simple cells in primary visual cortex, Proc. R. Soc. London B 265 (1412) (1998) 2315–2320.

[5] J. Hurri, A. Hyvärinen, Simple-cell-like receptive fields maximize temporal coherence in natural video, Neural Comput. 2003, in press.
[6] J. Hurri, A. Hyvärinen, A two-layer dynamic generative model of natural image sequences, submitted for publication.
[7] A. Hyvärinen, J. Karhunen, E. Oja, Independent Component Analysis, Wiley, New York, 2001.
[8] S.E. Palmer, Vision Science—Photons to Phenomenology, The MIT Press, Cambridge, MA, 1999.

Available at
www.ComputerScienceWeb.com
POWERED BY SCIENCE DIRECT

ELSEVIER

Neurocomputing 52–54 (2003) 561–566

NEUROCOMPUTING

www.elsevier.com/locate/neucom

A reconstruction method of projection image on worker honeybees' compound eye☆

Hidetoshi Ikeno*

School of Humanities for Environmental Policy and Technology, Himeji Institute of Technology, 1-1-12, Shinzaike-Honcho, Himeji, Hyogo 670-0092, Japan

Abstract

Visual perception by insect compound eyes is of interest in visual science for revealing neuronal principals of vision, and applying their small and simple structures to artificial vision systems. In order to investigate how animal behavior is related to sensory stimuli, it is important to take into account real input signals to the nervous system. In the case of an insect with compound eyes, sceneries projected onto compound eye are the basic signals for visual information processing. In this paper, a method to reconstruct the projected images on the honeybee compound eyes is presented. This method estimates the illuminance change of a single ommatidium surface. It may be applied to analyze the correspondence between input images and retinal responses.

Keywords: Compound eye; Ommatidium; Honeybee; Vision

1. Introduction

Flying insects like bees can not only control their posture, but also navigate to the target based on time-varying image signals during their flight. Insect visual mechanisms are of interest for revealing neuronal principals of image processing and applying their simple structure to artificial vision. It is well known that capabilities of scenery recognition differ across species, so insects with compound eyes are living in a different

☆ A part of this study was performed through Special Coordination Fund (Neuroinformatics Research in Vision, PI Prof. Shiro Usui) for the promotion of Ministry of Education, Culture, Sports, Science and Technology of the Japanese Government.
* Tel.: +81-792-92-1515; fax: +81-792-93-5710.
E-mail address: ikeno@hept.himeji-tech.ac.jp (H. Ikeno).

0925-2312/03/$ - see front matter © 2002 Published by Elsevier Science B.V.
doi:10.1016/S0925-2312(02)00785-3

visual world from mammals. However, only a few experiments take into consideration the difference in image perception by compound eyes [3]. It tends to be assumed that the same characteristics or information are used for recognition and behavioral control even in the insect brain. But if we want to reveal neural mechanisms for image processing in insect brains, it is essential to analyze and model based on real input signals into the photoreceptors.

In the case of honeybee flight, sceneries of her outside world are projected onto her compound eyes and received by photoreceptors in the ommatidium. Photoreceptor responses are processed through the visual signal pathway from retina to brain for detection of information and control of her behavior. Even though the projected images on her eyes are distorted by her position and posture, she can navigate easily with discrimination of patterns, recognition of shapes, and estimation of depth and distance. In my laboratory, we are building a mathematical model of image processing and flight control in the honeybee based on neural and behavioral properties. As a first step in the analysis of flight behavior to the visual stimulus, it is important to know real inputs to their eyes, that is, illuminance changes on the ommatidium surface. In this paper, a method to calculate the projection image on the honeybee compound eye is presented. A sequence of acquisition images through the flight path can be reconstructed by this method.

2. Model

A method to calculate the projection image on the compound eye of honeybees is presented and demonstrated by Giger [2]. In this method, a projection image of outside scenery was reconstructed by geometrical transformation, but I present a new method which is calculated based on illuminance changes on ommatidium lens surfaces.

2.1. Geometrical properties

Geometrical properties of compound eyes of worker honeybees were measured by Seidl and Kaiser [6]. In a frontal view of the honeybee head (Fig. 1(a)), a horizontal base plane (XY) was determined by the widest part of the head capsule. The boundaries of the visual receptive field based on the XY plane are 105.5° ventrally and 119.0° dorsally in vertical directions. The Z-axis is determined to pass through a point central to a virtual circle for approximating the horizontal arrangement of ommatidia (Fig. 1(b)). Boundaries of the horizontal receptive field are reported at 104.5° forward and 71.5° backward from the Y-axis.

In the illuminance change method, a projected image on the compound eye is reconstructed based on the calculation of illuminance changes on each ommatidium surface. For the calculation, it is necessary to determine the coordinates and the direction of view for each ommatidium. Ommatidia are arranged vertically from the ventral boundary at 45° to the dorsal one the 49° and horizontally from 9° to 43° on the virtual circle as shown in Fig. 1(b). Ommatidium are allocated along the circumference of radius 2 mm on YZ plane. Numbers of horizontally allocated ommatidia were

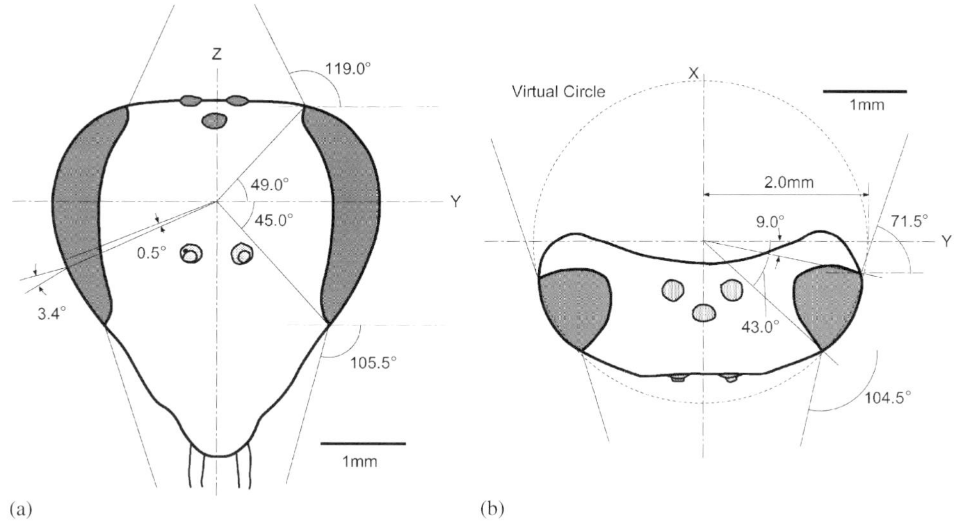

Fig. 1. Geometrical parameters of honeybees' compound eye, (a) frontal view, (b) dorsal view.

decreased from 38° both ventrally and dorsally. The number of ommatidia in each compound eye is known to be about 5000. Ommatidia were assigned at regular interval as 0.5°, within the boundaries. The assignment was staggered by half of interval, 0.25°, in horizontal arrangement.

2.2. Calculation of projection image

The direction of visual field vector (perpendicular line to the lens) for each ommatidium was determined by equally dividing the whole receptive field of the compound eye. The response sensitivity of each ommatidium was approximated by a Gaussian function with $mean = 0$, $standard\ deviation = 1.77°$.

The image presented to the honeybee was composed of black and white pixels as shown in Fig. 2. The illuminance change on each ommatidium lens due to each image pixel was calculated as follows:

$$\Delta I_i = L_i \frac{S_i}{S_0} \frac{1}{r_i^2} \cos \theta_i. \quad (1)$$

In this equation, ΔI_i is the illuminance produced change by one image pixel i. S_i/S_0 is the ratio of image pixel size within a receptive field to the receptive field size of the ommatidium. L_i is the brightness of the image pixel, and r_i is the distance between the pixel and the ommatidium. θ_i is the angle between the visual field vector and a line perpendicular to the image pixel. For each ommatidium, the total illuminance change caused by all image pixels is described by the sum of ΔI_i.

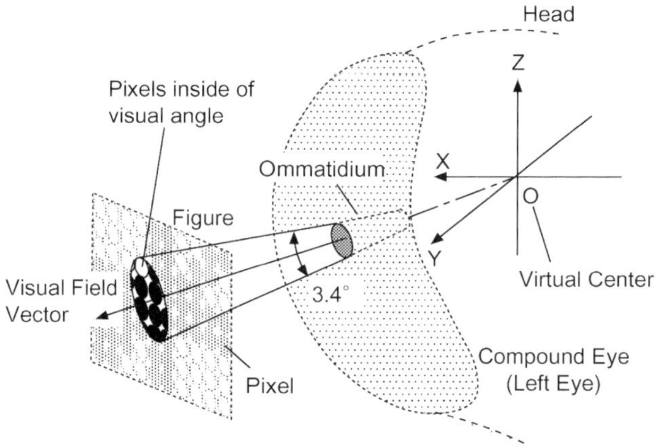

Fig. 2. Acquisition of light in the ommatidium.

3. Simulation results

Sceneries reconstructed by this illuminance change method were compared with the results of B-EYE [2]. It was confirmed that the illuminance change method also reconstructed a projection image on honeybee eyes. This method required a long time for calculations, because it estimated the illuminance for each ommatidia surface. However, it corresponded well to the photoreceptor model [1], since the presented method can produce real input stimuli for the photoreceptor.

An image compound of vertical, horizontal and diagonal lines with 100 mm length and 10 mm width was presented as a target in front of the honeybee flight path (Fig. 3(a)). Projection images were calculated on the compound eye in the case of approaching toward the target through five UVW coordinate points (a: $80, 0, 10$), (b: $70, -30, 30$), (c: $60, 0, 10$), (d: $50, 30, -30$), (e: $40, 0, 10$) (unit: mm) as shown in Fig. 3(b). In this simulation, posture was kept constant at 30° tilt to the vertical target plane. In the results of the simulation in Fig. 4, each picture is presenting the right side view of right eye, front views of both eyes and left side view of left eye. Numbers in the center show the UVW coordinates of bee positions.

At point (a) which is 80 mm from the target, a small image of the target was projected on both eyes, but the illuminance differed little from ommatidia receiving background light. When the bee flew in the upper right direction, to point (b), most of the image moved onto the left eye, but a small part of the target in the binocular visual field still remained on the right eye. At this point, the image of the line was curved and the target was asymmetric. At point (c), the image was more clear than at point (a), but vertical line was thicker in the lower part. The image became more distorted as the bee moved down to the left direction, point (d), and closer to the target, point (e).

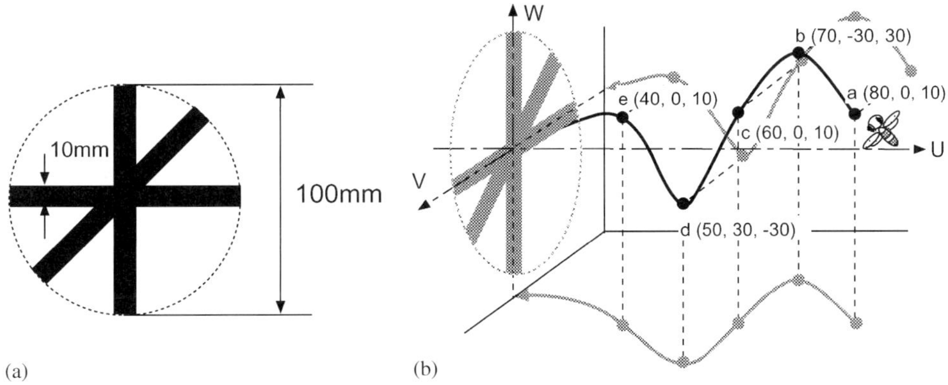

Fig. 3. Target pattern and flight route for simulation; (a) target pattern, (b) flight route.

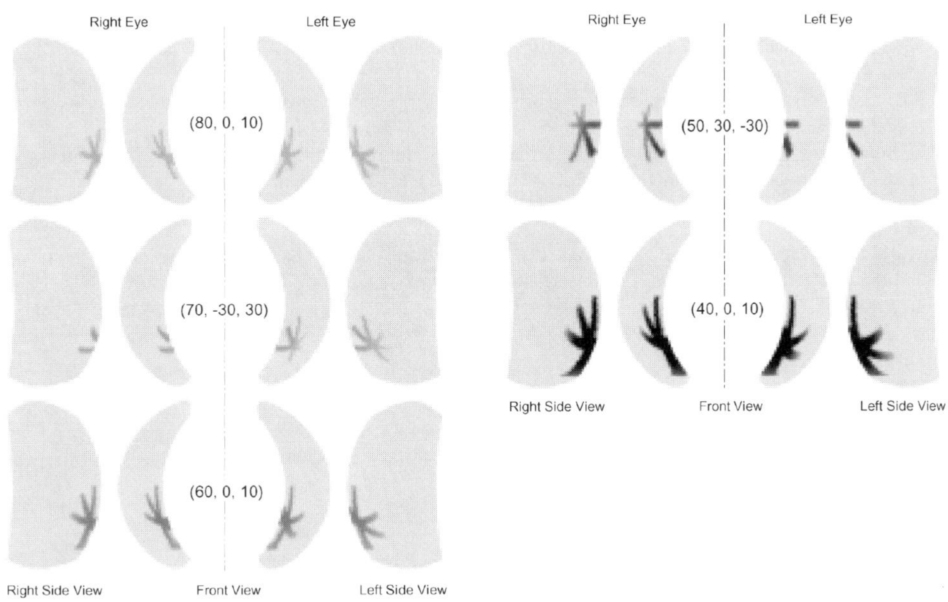

Fig. 4. Projected target pattern on compound eye.

4. Remarks

In this paper, a method was presented for reconstructing images projected onto the compound eye of worker honeybees. As shown in the simulation results, honeybees uses strongly distorted images whose distortion depends on the distance and direction from the target, but the honeybee shows skillful navigation based on this uncertain information. Presently, we are using a CCD camera to measure flight trajectories in the Y-shaped maze, which is used for conditioning honeybee [4]. Input images to the visual nervous system during the flight can be calculated by using this new method, which will be useful for analysis of behavioral control mechanisms based on visual stimuli and memory.

5. For further reading

The following reference may also be of interest to the reader [5].

References

[1] K. Becker, W. Backhaus, A physiological model of dark-adapted and light-adapted photoreceptors of the honeybee, Biol. Cybernet. 82 (2000) 153–160.

[2] B-EYE, http://cvs.anu.edu.au/andy/beye/beyehome.html.

[3] D. Efler, B. Ronacher, Evidence against a retinotopic-template matching in honeybees' pattern recognition, Vision Res. 40 (2000) 3391–3403.

[4] A. Horridge, Seven experiments on pattern vision of the honeybee, with a model, Vision Res. 40 (2000) 2589–2603.

[5] S.B. Laughlin, G.A. Horridge, Angular sensitivity of the retinula cells of dark adapted worker honey bee, Z. Vergl. Physiol. 74 (1971) 329–335.

[6] R. Seidl, W. Kaiser, Visual field size, binocular domain and the ommatidial array of the compound eyes in worker honey bee, J. Comput. Physiol. A 143 (1981) 17–26.

Hidetoshi Ikeno received the B.E., M.E. and D.E. degree from the Toyohashi University of Technology in 1983, 1985 and 1993, respectively. 1985–1989: Research Associate; 1989–1992: Assistant Professor; 1994–1998: Associate Professor at Maizuru National College of Technology. Assistant Professor at Toyohashi University of Technology, 1992–1994. Since April 1998, he has been an Associate Professor at Himeji Institute of Technology.

Available at
www.ComputerScienceWeb.com
POWERED BY SCIENCE @ DIRECT®

ELSEVIER

Neurocomputing 52–54 (2003) 567–572

NEUROCOMPUTING

www.elsevier.com/locate/neucom

Effects of baclofen on the directional selectivity of transient neurons in the catfish retina

Sung-Jong Lee*, Sun-Ho Bai

Physics and Biophysics Section, Department of Natural Sciences, College of Medicine, The Catholic University of Korea, Seoul 137-701, South Korea

Abstract

Effects of baclofen on the directional selectivity of the inner retinal neurons were studied using conventional intracellular recording techniques and computer simulations. Baclofen suppressed sustained components, but enhanced transient components of third-order neurons in the catfish retina. In addition, directional selectivity of transient neurons was enhanced by the application of baclofen. The simplified network model was tested to account for the details of directional selectivity observed in the inner retina. Results suggest that the directional selectivity in fish retina could be evoked by the asymmetric distribution of GABAergic third-order neurons.

Keywords: Baclofen; GABAb receptors; Directional selectivity; Third-order neurons

1. Introduction

Directional selectivity is a dynamic property of the receptive field, formed by retinal microcircuits. Directionally selective third-order neurons respond strongly to the motion of a bar stimulus moved across their receptive field in the preferred direction, which evokes little response to the same stimulus moved in the opposite direction. Several models have been proposed to explain the cellular origin and mechanism of the directional selectivity [1,2]. Starburst amacrine cells are well situated to account for pre-ganglionic models, since they have been shown to release both acetylcholine and GABA onto the bistratified dendrites of the direction selective ganglion cells [6,7].

* Corresponding author.
E-mail address: chris@catholic.ac.kr (S.-J. Lee).

0925-2312/03/$ - see front matter
doi:10.1016/S0925-2312(02)00798-1

On the other hand, it was reported that the directional selectivity was induced by the infusion of baclofen (GABAb receptor agonist) when 2-amino-4-phosphonobutyrate (APB) was used to isolate the OFF pathway in the amphibian retina [5]. GABAergic inhibitory system is known to play a key role in the mechanism of directional selectivity. Also, it has been reported that the blockade of GABAa receptors leads to the loss of directional selectivity [3]. But, it is unprecedented that the directional selectivity of third order neurons was affected by GABAb system. The GABAb system has been known to open potassium channels on third-order neurons, and close calcium channels on bipolar and ganglion cells [4]. Consequently, baclofen hyperpolarizes the membrane potentials and produces a selective suppression of portions of the light response while GABA and glycine suppress all components of the light response.

Retinal third-order neurons are dichotomized into transient and sustained cells according to their light responses. It has been proposed that transient cells carry information in the direction of motion, while sustained cells carry information on steady-state conditions, such as color or spatial acuity [4]. Interestingly, baclofen seems to make a transition from sustained to transient in propensity by the suppression of sustained components, and the enhancement of transient components of third-order neurons.

Therefore, we have explored the effects of baclofen on directional selectivity from the physiological recordings and computer simulation, to discern the functional roles of GABAb receptors on transient neurons in the inner retinal microcircuit.

2. Methods

The catfish (*Ictalurus punctatus*) retina was investigated by using the intracellular recording techniques. The stimuli were generated by a computer monitor, which was imaged through an optical system onto the retina. They consisted of moving bars of various sizes and directions. The same stimuli were repeated on a slave monitor during the experiment. The preferred-null axis was empirically determined as that plane that showed the greatest asymmetry in amplitude to back and forth movement of the bar. A gravity driven perfusion system was used to super fuse the solutions and a piece of absorbent tissue with a hole large enough to expose the retina was centered over the eyecup to serve as a wick to draw off the super-fusate. A model of the transient neuron was described in terms of the Hodgkin–Huxley kinetic model based on electrophysiological measurements. The network model was tested for its ability to account for the details of directional selectivity induced by GABAb receptor system in the inner retina. Simulations were carried out using NEURON package.

3. Results and discussion

Perfusion experiments were performed to show the effects of baclofen on neurons in the catfish inner retina. The action of baclofen (0.1 mM) on third-order neurons is

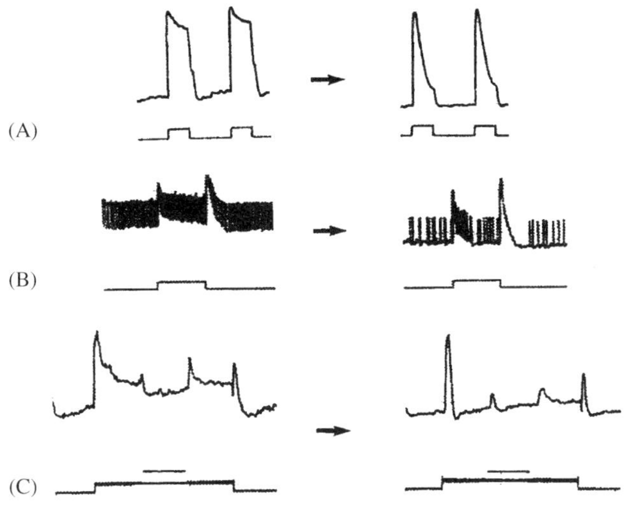

Fig. 1. Intracellular recordings from third-order neurons exposed to baclofen. Baclofen application was mainly associated with a suppression of sustained components and an enhancement of transient components.

illustrated in Fig. 1. When the retina was exposed to baclofen, sustained components of the ON-sustained cell were suppressed and spiking discharges of putative ganglion cell were reduced (A & B). The square pulses below the voltage trace indicate the timing of light stimuli. Baclofen hyperpolarized the dark membrane potential, suppressed sustained component and enhanced transient component of the ON-sustained cell with large transient component, but did not affect the surrounding antagonism of the cell (C). The application of baclofen was mainly associated with a suppression of sustained component and an enhancement of transient components. The responses were recovered after returning to the control solution.

To show the effects of baclofen for the transient components under moving stimulation, we applied a single light bar with appropriate width and speed (Fig. 2). In response to a single moving bar of light, transient neurons generated symmetrical patterns of EPSP that were dependent on the direction of the stimulus movement. The scan produced two groups of EPSP as the leading edge and trailing edges of the bar traversed the receptive field.

Under control conditions, the responses give vague indication of directional selectivity (A). However, after the retina was treated with baclofen, the cell showed a preference for motion of the slit in the 135° direction and a weak response to the opposite direction (B). The results suggest that the directional selectivity is mediated by GABAb receptors in the catfish inner retina. Baclofen revealed directional selectivity from the latent direction-selective neurons. Also, baclofen made a transition from sustained to transient by the suppression of sustained components and the enhancement of transient components in the third-order neurons. The transition plays an important role in the drug induced directional selectivity and modulates

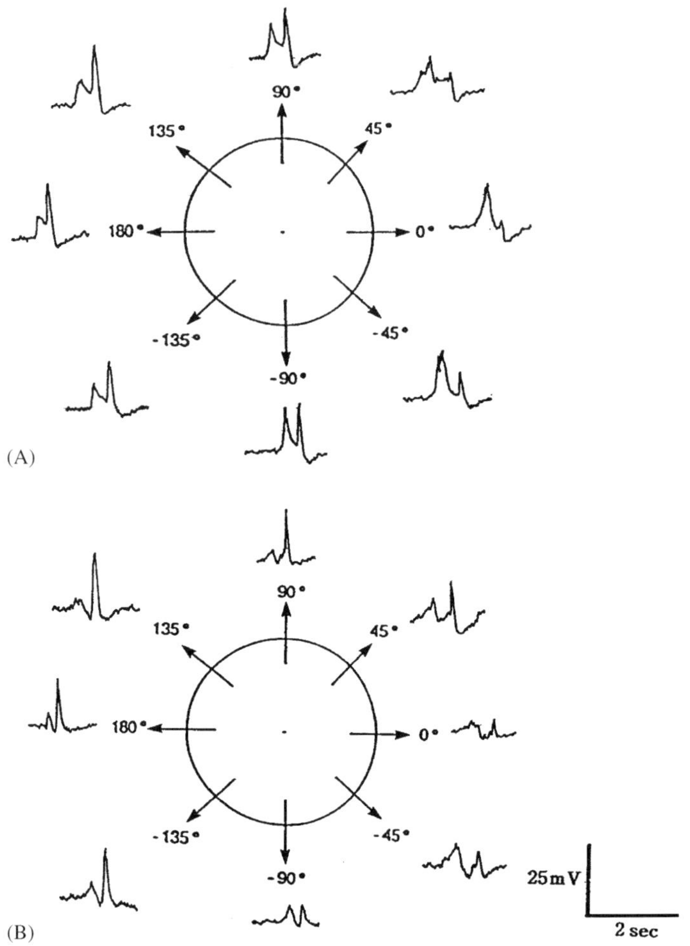

Fig. 2. Responses of an ON-OFF transient cell to a 350 μm-wide light slit moving across the retina in the directions indicated by the arrows. The long axis of the slit was orthogonal to the direction of motion.

the dynamic properties of inner retinal neurons. Accordingly, the network model was proposed to interpret the directional selectivity induced by baclofen (Fig. 3). We introduced GABAb sensitive third order neuron (X) in the modified preganglionic model.

The simulations show a consistency with the experimental data, though the direct physiological and anatomical bases for the baclofen induced directional selectivity are not presently known (Fig. 4).

Therefore, it is presumed that the asymmetric distribution of GABAb receptors and its nonlinear integration are essential parts of the mechanism of baclofen induced directional selectivity.

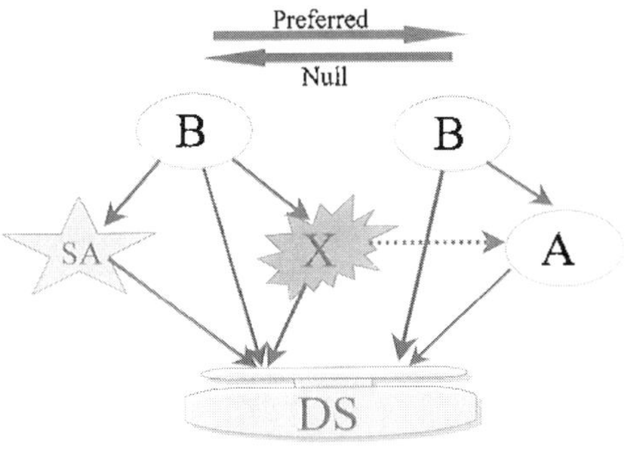

Fig. 3. Schematic diagram of a proposed neural circuit model for baclofen induced directional selectivity (B: bipolar cells; SA: starburst amacrine cell; A: amacrine cell; X: GABAb sensitive third-order neuron, DS: directionally selective ganglion cell, GABAb sensitive).

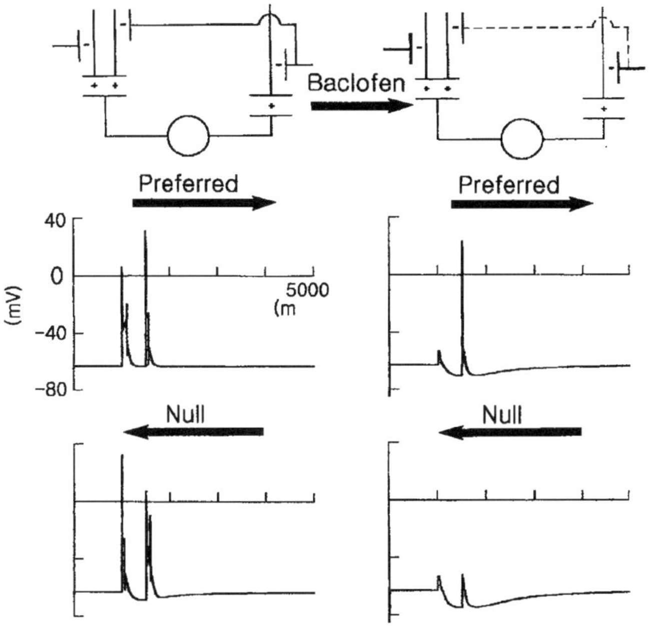

Fig. 4. The simulated responses of the transient neuron in the proposed neural circuit model for baclofen induced directional selectivity.

References

[1] N.M. Grzywacz, F.R. Amthor, D.K. Merwine, Necessity of acetylcholine for retinal directionally selective responses to drifting gratings in rabbit, J. Physiol. 512 (1998) 575–581.

[2] S. He, R.H. Masland, Retinal direction selectivity after targeted laser ablation of starburst amacrine cells, Nature 389 (1997) 378–382.

[3] C.A. Kittila, S.C. Massey, Pharmacology of directionally selective ganglion cells in the rabbit retina, J. Neurophysiol. 77 (1997) 675–689.
[4] Z.H. Pan, M.M. Slaughter, Control of retinal information coding by GABAB receptors, J. Neurosci. 11 (1991) 1810–1821.
[5] N. Tian, M.M. Slaughter, Pharmacology of the GABAB receptor in amphibian retina, Brain Res. 660 (1994) 267–274.
[6] G. Yang, R.H. Masland, Direct visualization of the dendritic and receptive fields of directionally selective retinal ganglion cells, Science 258 (1992) 1949–1952.
[7] K. Yoshida, D. Watanabe, H. Ishikane, M. Tachibana, I. Pastan, S. Nakanishi, A key role of starburst amacrine cells in originating retinal directional selectivity and optokinetic eye movement, Neuron 30 (2001) 771–780.

Sung-Jong Lee received his Ph.D. in Physiology from the Catholic University of Korea in 1997. Currently he is a researcher and teaching assistant at the Department of Natural Sciences, the Catholic University of Korea. His research interests include electrophysiology of retina, cellular and network properties to biochemical signaling pathways, large-scale simulations and artificial vision.

Sun-Ho Bai is Professor at the Physics and Biophysics Section, Department of Natural Sciences, College of Medicine, the Catholic University of Korea. He received his Ph.D. from the Catholic University of Korea in 1980. During 1986–1988, he was a visiting scientist at the Biophysics Department, SUNY/Buffalo, USA, where he did research on the electrophysiology of the amphibian retina. His present research in neuroscience focuses on the visual information processing in the retina.

Available at
www.ComputerScienceWeb.com
POWERED BY SCIENCE @DIRECT®

ELSEVIER

Neurocomputing 52–54 (2003) 573–582

NEUROCOMPUTING

www.elsevier.com/locate/neucom

Modeling the effects of nicotine on a continuous performance task

Daniel S. Levine[a,*], Nilendu Jani[b], David G. Gilbert[c]

[a]*Department of Psychology, University of Texas at Arlington, 501 S. Nedderman Drive, Arlington, TX 76019-0528, USA*

[b]*Iconoci Inc., Bedford, TX, USA*

[c]*Department of Psychology, Southern Illinois University, Carbondale, IL 62901, USA*

Abstract

We have developed a neurocomputational model of some attentional effects of nicotine. Our simulations reproduce results showing smoking abstinence to impair performance on a rapid visual information processing task involving detecting sequences of even or odd digits.

Our model treats presence of nicotine as enhancement of modulatory acetylcholine signals, which focus attention on task-relevant stimuli. Our network includes cholinergic signals from nucleus basalis; working memory representations in dorsolateral prefrontal cortex; inputs to nucleus basalis from entorhinal cortex; and recency signals. Digits are represented at entorhinal, nucleus basalis, and dorsolateral levels, along with dorsolateral representations of the concepts "even" and "odd."

Keywords: Nicotine; Acetylcholine; Attention; Working memory; Rapid visual information processing

1. Introduction

Nicotine, like other drugs of abuse such as cocaine and heroin, often leads to short-term increases in positive affect and decreases in negative affect [7]. However, it is different from these other drugs in its cognitive affects, particularly in its enhancement of attention (most notably, sustained attention) and working memory. The sustained attention task used most often in nicotine-related experiments is the Rapid Visual

* Corresponding author. Tel.: +1-817-272-3598; fax: +1-817-272-2364.
E-mail address: levine@uta.edu (D.S. Levine).

0925-2312/03/$ - see front matter
doi:10.1016/S0925-2312(02)00801-9

Information Processing task (RVIP), a type of continuous performance test. In the most common form of the RVIP, visual stimuli (individual digits from 1 to 9) are presented one at a time at a rate of 100 or more per minute at the center of a computer display monitor. Participants press a key whenever three odd or three even digits appear in sequence. Almost all well-designed RVIP studies have found nicotine (delivered by smoking, transdermal patch, or injection) to enhance target detection in habitual smokers [14,24].

Nicotine has effects on almost all major neurotransmitter systems (see [7] for review), particularly the cholinergic, dopaminergic, and noradrenergic systems. The close association of this substance with nicotinic cholinergic receptors, and the importance of acetylcholine in attention [14,19], however, make cholinergic modulation appear to be a particularly important substrate for the attention-enhancing effects of smoking. Hence, in the neural network model we have developed, for parsimony we treat the short-term deficit of recently quitting smokers on attentional tasks requiring continuous performance such as the RVIP as a deficit in cholinergic modulation of systems involved in both attention and working memory. Dopaminergic and noradrenergic systems take on more importance when we extend the model to other tasks, and affective states, that are also affected by nicotine.

2. Nicotine and cholinergic pathways

There are two major types of acetylcholine receptors: muscarinic (which are more prevalent) and nicotinic. Nicotine is, of course, a nicotinic receptor agonist. Two major sources of cholinergic innervation, involving both types of receptors, are the nucleus basalis of the midbrain, which modulates the neocortex, and the septum, which modulates the hippocampus.

Many investigators have posited that cholinergic modulation has some selective effects on some aspect of working memory. Everitt and Robbins [5] review evidence indicating that the nucleus basalis cholinergic system contributes to visual attention more than to memory per se. This attentional function seems to span both the nicotinic and muscarinic subsystems, as indicated by data showing that visual attentional dysfunction from nucleus basalis lesions is reversed both by nicotine and by the muscarinic agonist physostigmine [17].

There are also behavioral results that partly differentiate functions of the nicotinic versus muscarinic subsystems of the cholinergic system. Granon et al. [11] injected a nicotinic antagonist (bungarotoxin) or a muscarinic antagonist (scopolamine) into the prefrontal cortex of rats and studied their performance on various maze learning tasks. They found that scopolamine impaired working memory on all tasks but bungarotoxin only on tasks requiring responses different from prevailing ones. They concluded that "Nicotinic transmission appears to be important in delayed response tasks requiring effortful processing for response selection, while the muscarinic system is involved in general working memory processes" (p. 139). In humans, nicotine enhances recall of words that are semantically related to a prime but not of words that follow the prime automatically [22].

Since both prefrontal cortex (especially its dorsolateral part) and hippocampus are important areas in working memory, the effects of nicotine and other cholinergic agonists on working memory could operate through either the nucleus basalis-prefrontal or the septohippocampal cholinergic pathways (see [12] for evidence suggesting the latter). Newman and Grace [20] review evidence that prefrontal cortex is more directly involved with working memories for specific stimuli, rewards, and potential behaviors, whereas hippocampus is involved with memories for contexts. Hironaka et al. [13] studied acetylcholine efflux from both of those regions in rats. They concluded that prefrontal acetylcholine is related to working memory, which they defined in animals as "trial-unique memory of to-be-remembered items", whereas hippocampal acetylcholine is related to reference memory, defined as memory of consistent response rules. In general, nicotine has a stronger effect on working memory than on reference memory [16].

This combination of results from both rats and primates suggests to us that in the RVIP task, the hippocampus should store representations of contexts including the context of the experimental task itself. The dorsolateral prefrontal cortex should store representations of the specific stimuli relevant for the current task, in this case, odd and even digits and the previously learned concepts of "odd" and "even" themselves. Such a role for the prefrontal cortex is analogous to roles for the rat cortex reviewed by Sarter et al. [23], who found that lesions of cholinergic projections to the cortex lead to profound impairments in sustained attention. On a task such as the RVIP, sustained attention requires a focus on the presence or absence of stimuli relevant to the current task. We propose on the basis on the results we have discussed that this focus is achieved through selective enhancements within the set of dorsolateral prefrontal working memory representations of loci that represent task-relevant stimuli.

Our model, specific to the RVIP task, includes network analogs of the nucleus basalis, and we posit that cholinergic modulation from that region selectively enhances the task-relevant working memory loci in the dorsolateral prefrontal cortex. How the decision is made, on the basis of the experimenter's instructions, as to which loci are task-relevant (and therefore are enhanced) we leave to future extensions of our model. Based on the role of the hippocampus in setting contexts and episodes, we expect that this type of decision is mediated by the septohippocampal cholinergic pathways, including nicotinic pathways.

3. The neural network model

Fig. 1 shows the network we have designed to simulate the effects on RVIP of nicotine. It is assumed that nicotine has an agonist effect on nicotinic cholinergic activity. Our network combines some modules representing brain areas and others representing memory activations of particular target patterns (in this case, digits) or categories of these targets.

The digits 1–9 (0 is typically not included in the RVIP digit display) have representations in three different areas. The first area is the entorhinal cortex, gateway from sensory cortices to the hippocampus, where representations are labeled e_i, $i = 1$ to 9. Entorhinal is also one of the chief areas of afferent input to the cholinergic

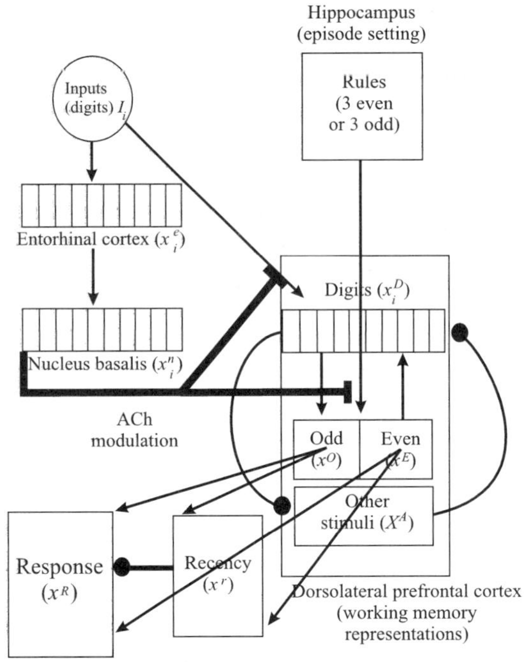

Fig. 1. Basic model of brain areas involved in RVIP performance. Arrows denote excitaiton; filled-in circles denote inhibition; bars denote modulation. Cholinergic modulation is assumed to selectively amplify working memory representations of task-relevant inputs (in this case, digits). "Odd" and "even" are abstract representations in positive feedback with nodes for odd and even digits, respectively.

nucleus basalis [17]. These same digits have representations at nucleus basalis, labeled n_i. Finally, these digits have working memory representations in dorsolateral prefrontal cortex, labeled d_i. These d_i compete via lateral inhibition with each other and with a generic dorsolateral representation D for random inputs other than the digits. Excitatory signals to d_i nodes from digit inputs are modulated by corresponding n_i, so that the amount of cholinergic activation determines attentional biases toward relevant (digit) inputs. Odd and even digit nodes are also in positive feedback with dorsolateral representations for the concepts of "odd" and "even". The dorsolateral representations for the abstract concepts of "odd" and "even" are each activated by dorsolateral nodes for the corresponding digits. Cholinergic signals modulate connections from digit nodes to odd and even nodes, mimicking nicotine's ability to enhance attention to the concepts (in this case, odd and even) relevant to the current task. We assume the hippocampal episode representation encodes the current task or rule which biases cholinergic modulation by determining which inputs are relevant, but do not incorporate this effect explicitly into the network equations.

Connections from odd and even nodes to a response node (perhaps in premotor cortex), labeled R, constitute a third locus of cholinergic modulation from the nucleus basalis nodes n_i. The "three in a row" rule is not explicitly incorporated; rather, a key press is interpreted as occurring when response node activity is above a given threshold. Based on the rule instructions (presumably relayed by hippocampus), connections from the odd (O) and even (E) representations to the response (R) representation are strengthened by cholinergic modulation. Each input of an even or odd digit increases R activation, and the threshold for an actual response (key press) is the amount of activation obtained by excitation from three digits. But every switch from even to odd or odd to even activates a "recency" representation r that inhibits the response node, reducing its activation back to a baseline level. The recency node is inspired by data hinting that there are cells in the rhinal cortex that discriminate familiarity, recency (having not been seen immediately before the current stimulus), and novelty (having never been seen) [2].

3.1. The RVIP data

Our model is based on a study, with female subjects, assessing the effects of quitting smoking on mood, EEG, and attention. All subjects had habitually smoked or more cigarettes per day for the preceding two years, and agreed to receive payment for their participation only if biochemically verified smoking abstinence was achieved across the 31-day abstinence period. Details of the study are provided by Gilbert et al. [9] and RVIP, EEG, mood, and individual difference findings are reported in Gilbert et al. [10].

Phase 1 was a 5-week duration baseline during which all participants smoked at their usual rate and attended weekly experimental sessions and biweekly monitoring sessions. After the completion of Phase 1, individuals were randomly assigned to the quit group (70% chance) or the smoke group (30% chance) that continued to smoke during the 31-day required abstinence period of the quit group. Phase 2 differed across the two groups only in terms of whether or not the participant smoked. Experimental sessions were held on days 3, 10, 17, and 31 of abstinence (or corresponding time for the smoke group).

The RVIP Task consists of successive vertical presentation of digits (not including 0) on a computer screen for 15 min. Digit presentation rate was 116/min. These digits included 160 target sequences consisting of three even or three odd digits in a row. The subject was instructed to press a particular key on the computer whenever he or she noticed such a target sequence.

Fig. 2 shows the results from Gilbert et al. [10] on this task. Note that smokers who continue to smoke detect around 120 of the 160 target sequences, whereas smokers who quit average 100–110.

4. Simulation results

The simulations are repeated for three different values of theta (5.0, 5.5 and 6.0). Lower theta values represent lesser nicotine ingestion, that is, the smoker has abstained a greater number of days from smoking. Consistent with the experimental data, as theta decreases, the network performance grows worse at detecting triplets of odds or evens.

For all three simulations, the following sequence of 200 digits was generated using a pseudo random number generator in MATLAB and input to the network. As shown

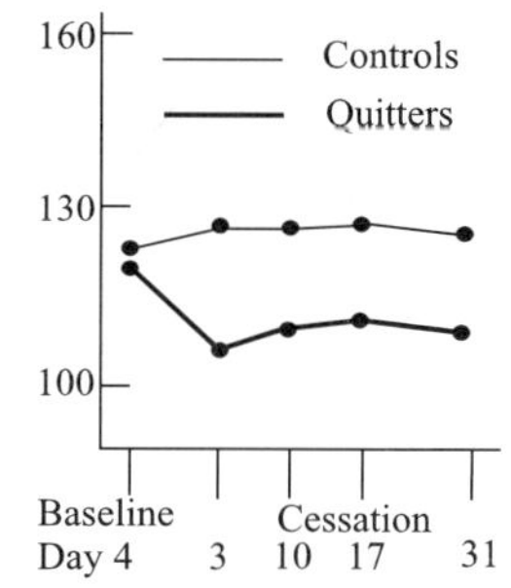

Fig. 2. Graph of the comparison between quitters and continuing smokers on the RVIP test.

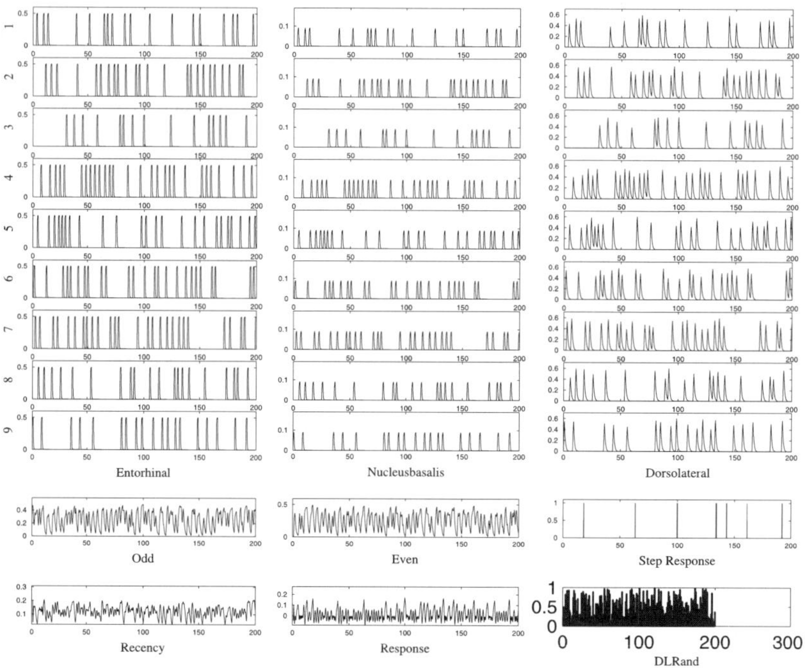

Fig. 3. Simulation results for $\theta = 6.0$. As the spikes in the graph of "Step Response" indicate, the network detects all 21 triplets of odd and even digits.

below in underline, there are a total of 21 triplets of odd or even digits (see Figs. 3 and 4).

7 5 6 1 4 6 5 8 9 3 7 4 2 7 5 2 8 1 9 8 4 1 5 7 2 5 8 3 7 1 2 5 3 4 9 5 6 2 3
6 4 2 3 9 4 8 5 2 4 6 1 9 2 6 5 3 1 6 2 8 7 2 6 4 7 8 5 9 7 8 6 9 8 4 7 1 3 4 9
7 6 4 2 9 5 7 8 6 4 5 6 9 7 4 8 1 7 2 5 6 3 9 5 4 2 7 9 2 8 6 3 8 1 6 4 9 2 1 3 9
8 3 7 2 5 7 2 4 1 7 4 2 1 6 4 1 5 6 2 4 7 3 2 4 9 7 8 4 1 6 7 4 6 7 3 4 9 5 6 2
1 7 3 8 9 6 5 7 6 3 5 4 6 5 8 4 5 7 2 4 5 7 8 2 4 5 1 6 2 8 1 9 4 7 8 5 1 7 6 9

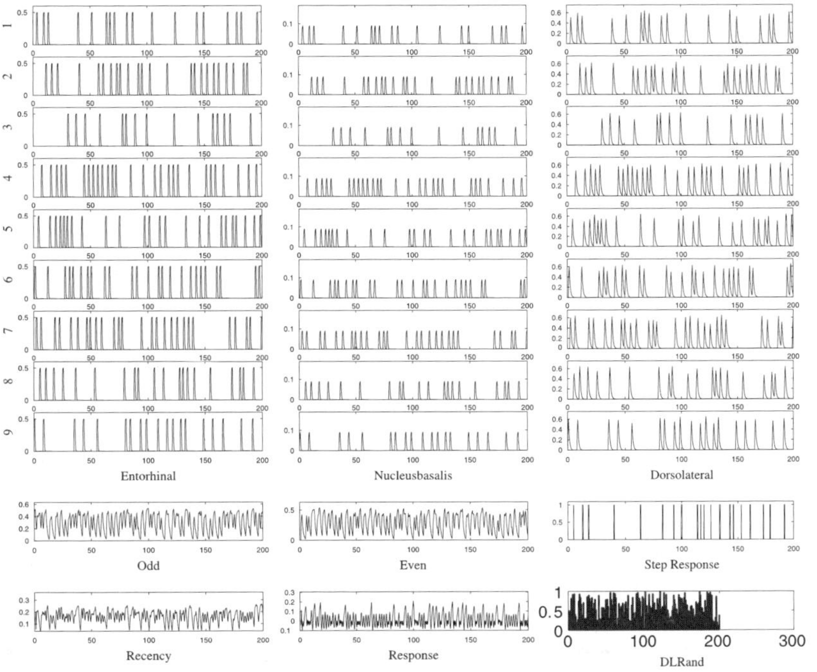

Fig. 4. Simulation results for $\theta = 5.0$. As the spikes indicate, with low theta values, the network detects a few of the triplets of odd and even digits. Specifically, in this case it detects 7 out of 21. For the intermediate value $\theta = 5.5$, the network detects 14 triplets out of 21 (graph not shown here).

The following diagrams show activations of network nodes as functions of time. The 3 columns with 9 panels each show activations in nodes representing digits 1–9. Starting from left, the columns represent nodes in entorhinal cortex, nucleus basalis, and dorsolateral prefrontal cortex. The rest of the panels (scanning first horizontally and then vertically) show activations in the odd and even nodes; step response (response above a certain threshold); activations in the recency and response nodes; and a random number generated every time unit (used for simulating distraction due to lack of nicotine). Each spike in the "step response" graph indicates that the network actually detects an even or odd triplet and responds with a key press.

5. Future extensions of the model

Coull et al. [4] found in a PET study that this task increased blood flow in a network of sites in both the prefrontal and parietal cortices involved in working memory and sustained attention. Lawrence, Ross, and Stein [15] studied this task using fMRI in chronic smokers under both nicotine and a placebo. They found that nicotine increased activation relative to the placebo in posterior parietal and occipital cortices but not in prefrontal cortex.

Based on these data we are now expanding our model to include posterior (parietal and occipital) as well as prefrontal representations of the digit stimuli, both influenced by nicotinic cholinergic inputs from nucleus basalis. Prefrontal representations are more subject to executive influences than posterior representations, as in other recent neurocognitive models (e.g. [21]). It seems to us a plausible conjecture that posterior cortex is primary engaged in the sustained spatial attention aspect of tasks such as the RVIP, whereas prefrontal (particularly dorsolateral prefrontal) is primarily engaged in the working memory aspects. Based on the observation that nicotine selectively enhances performance of effortful tasks [11,24]. Stein's results hint that the sustained attention aspects of the RVIP task are the most effortful, since they involve 15 min of concentration on a relatively unchanging visual scene. The working memory aspects of RVIP are less effortful, since most of us have easy access to the stored facts about which digits are even or odd. We predict that fMRI would show greater effect of nicotine on prefrontal activation for different digit-related cognitive tasks that tax working memory more; for example, if the subject is instructed to press a key when the third digit is equal to the sum of the first two.

Our model also needs to add how the brain decides which environmental stimuli are relevant or irrelevant to the current task. Several experimentalists and theorists have recently suggested such a role for parts of the ventral striatum, especially the nucleus accumbens [6]. The basal ganglia gating mechanism is thought to be influenced by inputs from the amygdala in the case of "hot cognition" wherein task relevance is influenced by strong affect. In the case of "cold cognition," such as following an experimenter's instructions in a task such as the RVIP, a more relevant input to the basal ganglia is from the hippocampus. Newman and Grace [20] suggested the hippocampus plays such a regulatory rule for working memory "throughput." This could be a mechanism for the operation of the simplified "hippocampal episode setting" shown in Fig. 1.

Finally, since nicotine affects the noradrenergic and dopaminergic systems as well as the cholinergic, future versions of the model will include effects on all three transmitter systems. Again we propose that while RVIP primarily engages nicotine's cholinergic effects, other cognitive tasks involve nicotine's effects on one of the other transmitter systems. As for norepinephrine, Kassel [14] reviewed data suggesting that acetylcholine is involved in narrowing attentional focus, whereas norepinephrine is involved in increasing processing capacity so that a wider range of stimuli can be selected from. Wider range is not likely to be advantageous in the RVIP. We conjecture that the nicotine-norepinephrine connection is more important for other tasks that involve either decisions among a wider range of stimuli (e.g., the RVIP with distractors [8]) or divided attention [3]. As for dopamine, there is evidence that nicotine enhances dopaminergic neuron activity in the ventral tegmentum and thereby enhances dopamine transmission in brain regions such as the nucleus accumbens and orbitofrontal cortex. Dopamine is likely to mediate nicotine's negative affect reduction and positive affect enhancement [7]. Yet again, dopamine does not appear to be a good candidate for mediating nicotine's effects on the RVIP. Dopamine has most effects on tasks requiring cognitive flexibility [1] or response to reinforcing or novel stimuli [18], neither of which are at a premium on the RVIP.

Thus our conjecture is that nicotine ingestion could operate by raising the release or circulating level of any of these three neurotransmitters, and the requirements of different cognitive tasks activate different combinations of transmitter systems. Also, different tasks tend to generate selective increases in blood flow to particular brain regions such as parietal or prefrontal cortex. Such task-triggered selective activations could involve feedback from prefrontal executive regions to other brain areas including midbrain modulatory transmitter nuclei [23].

References

[1] F.G. Ashby, A. Turken, A. Isen, A neuropsychological theory of positive affect and its influences on cognition, Psychol. Rev. 106 (1999) 526–550.

[2] R. Bogacz, M. Brown, C. Giraud-Carrier, Model of familiarity discrimination in the perirhinal cortex, J. Comput. Neurosci. 10 (2001) 5–23.

[3] J. Coull, Neural correlates of attention and arousal: insights from electrophysiology, functional neuroimaging, and psychopharmacology, Prog. Neurobiol. 55 (1998) 343–361.

[4] J. Coull, C. Frith, S. Frackowiak, P. Grasby, A fronto-parietal network for rapid visual information processing: a PET study of sustained attention and working memory, Neuropsychologia 34 (1996) 1085–1095.

[5] B. Everitt, T. Robbins, Central cholinergic systems and cognition, Ann. Rev. Psychol. 48 (1997) 649–684.

[6] M. Frank, B. Loughry, R. O'Reilly, Interactions between frontal cortex and basal ganglia in working memory: a computational model, Cogn. Affect. Behav. Neurosci. 1 (2001) 137–160.

[7] D. Gilbert, Smoking, Taylor & Francis, Washington, DC, 1995.

[8] D. Gilbert, J. Hammersley, A. Izetelny, R. Radtke, J. Huggenvik, T. Markus, S. Small, V. Stout, N.E. Rabinovich, A. Skerly, Effects of nicotine on attention and affect, Poster presented at Eighth Annual Scientific Sessions of the Society for Research on Nicotine, Savannah, GA, February, 2002.

[9] D. Gilbert, F.J. McClernon, N. Rabinovich, L. Plath, C. Masson, A. Anderson, K. Sly, Mood disturbance fails to resolve across 31 days of cigarette abstinence in women, J. Consult. Clin. Psychol. 70 (2002) 142–152.

[10] D. Gilbert, F.J. McClernon, N. Rabinovich, C. Sugai, L. Plath, G. Asgaard, J. Huggenvik, N. Botros, Y. Zuo, EEG, attention, and mood effects of quitting smoking last for more than 31 days and are more severe with stress, dependence, DRD2 A1 allele, and depressive traits, submitted.

[11] S. Granon, B. Poucet, C. Thinus-Blanc, J.-P. Changeux, C. Vidal, Nicotinic and muscarinic receptors in the rat prefrontal cortex, Psychopharmacol. 119 (1995) 139–144.

[12] R. Gray, A. Rajan, K. Radcliffe, M. Yakehiro, J. Dani, Hippocampal synaptic transmission enhanced by low concentrations of nicotine, Nature 383 (1996) 713–716.

[13] N. Hironaka, K. Tanaka, Y. Izaki, K. Hori, M. Nomura, Memory-related acetylcholine efflux from rat prefrontal cortex and hippocampus: a microdialysis study, Brain Res. 901 (2001) 143–150.

[14] J. Kassel, Smoking and attention, Clin. Psychol. Rev. 17 (1997) 451–478.

[15] N.S. Lawrence, T.J. Ross, E.A. Stein, Cognitive mechanisms of nicotine on visual attention, Neuron 36 (2002) 539–548.

[16] E. Levin, S. Kaplan, A. Boardman, Acute nicotine interactions with nicotinic and muscarinic antagonists: working and reference memory effects in the 16-arm radial maze, Behav. Pharmacol. 8 (1997) 236–242.

[17] M.-M. Mesulam, F. Mufson, Neural inputs into the nucleus basalis of the substantia innominata (Ch4) in the rhesus monkey, Brain 107 (1984) 253–274.

[18] J. Mirenowicz, W. Schultz, Importance of unpredictability for reward responses in primate dopamine neurons, J. Neurophysiol. 72 (1994) 1024–1027.

[19] J. Muir, B. Everitt, T. Robbins, Reversal of visual attentional dysfunction following lesions of the cholinergic basal forebrain by physostigmine and nicotine but not by the 5-HT_3 antagonist, ondansetron, Psychopharmacol. 118 (1995) 82–92.

[20] J. Newman, A. Grace, Binding across time: the selective gating of frontal and hippocampal systems modulating working memory and attentional states, Conscious. Cogn. 8 (1999) 196–212.
[21] R. O'Reilly, D. Noelle, T. Braver, J. Cohen, Prefrontal cortex and dynamic categorization tasks, Cereb. Cortex 12 (2002) 246–257.
[22] J. Rusted, L. Graupner, A. Tennant, D. Warburton, Effortful processing is a requirement for nicotine-induced improvements in memory, Psychopharmacol. 138 (1998) 362–368.
[23] M. Sarter, B. Givens, J. Bruno, The cognitive neuroscience of sustained attention: where top-down meets bottom-up, Brain Res. Rev. 35 (2001) 146–160.
[24] D. Warburton, Caffeine, nicotine, and attentional performance, in: J. Snel, M. Lorist (Eds.), Nicotine, Caffeine, and Social Drinking, Harwood, London, 1998, pp. 245–256.

Daniel S. Levine is Professor of Psychology at the University of Texas at Arlington. He has been on the Board of Governors of the International Neural Network Society (INNS) since 1995 and was President of INNS in 1998. He is the author of Introduction to Neural and Cognitive Modeling (2nd edition, Lawrence Erlbaum Associates, 2000) and of over 70 articles on neural networks and related areas.

Available at
www.ComputerScienceWeb.com
POWERED BY SCIENCE DIRECT

ELSEVIER

Neurocomputing 52–54 (2003) 583–590

NEUROCOMPUTING

www.elsevier.com/locate/neucom

Simulations of the cultured granule neuron excitability

Marja-Leena Linne[a,*], Tuula O. Jalonen[b,c]

[a]*Institute of Signal Processing, Tampere University of Technology, 33101 Tampere, Finland*
[b]*Institute of Medical Technology, FIN-33014 University of Tampere, Finland*
[c]*Department of Physics, FIN-40014 University of Jyväskylä, Finland*

Abstract

We have developed a biophysical model of a cultured rat cerebellar granule neuron and simulated its excitability under different experimental conditions. The basic excitability properties of such a small neuron; the specific action potential waveforms, the overall firing patterns induced by current stimulations, and the linear frequency-current relation, are the main model constraints. Simulations show that for a one-compartmental granule neuron model, the constraints are met using six voltage- and time-dependent ion channel types and calcium dynamics linked to BK_{Ca} ion channel function. This kind of model of a single neuron forms a solid basis for building the increasingly more complex network models of cultured neurons, and specifically for the cerebellar neuronal network.

Keywords: Biophysical model; Cell culture; Cerebellum; Excitability; Granule neuron

1. Introduction

Cerebellum is known to be important in controlling, fine-tuning and predicting movements, and may also be involved in higher cognitive functions. To study the properties of the interesting cerebellar circuitry, we created a biophysically-detailed model of one of the cerebellar neuronal types, the granule neuron (CGN), and simulated its excitability. The CGN model here presented, is capable of reproducing the major electroresponsive properties of its real in vitro counterpart when a minimum number of compartments/parameters is used (see also [11,12]). This is the first model developed

* Corresponding author.
E-mail address: marja-leena.linne@tut.fi (M.-L. Linne).

0925-2312/03/$ - see front matter
doi:10.1016/S0925-2312(02)00804-4

specifically for cultured CGN, in comparison to the earlier compartmental models for slice preparations [6,8,14]. The CGN model of cultured neurons can further be utilized for simulations of complex interactions between ion channels, intracellular signaling and neuronal excitability (as described by Bhalla [2]), for simulation of the neuronal excitability during differentiation [19], and for models of the behavior of complex networks of cultured neurons. The networks are at present increasingly studied by using multi-electrode arrays as culture platforms (see e.g. Fromherz [7]).

2. Methods

2.1. Model implementation

We used compartmental modeling techniques, Hodgkin–Huxley type reconstructions of ion channel/current behavior and a simple model for calcium dynamics, to mathematically describe the granule neuron excitability. The CGN model was implemented using the GENESIS script language and neuronal simulator [20]. The starting parameter values of the model equations were selected based on data from in vivo and in vitro experiments on cerebellar granule neurons. The basic excitability properties of the granule neuron were used as the main constraints of the model. Emphasis was put on data from neurons in culture.

2.2. Model assumptions and equations

Based on experimental evidence and electrotonic calculations the neuron was assumed as one-compartmental sphere containing six different voltage-dependent ion channel/current types ($BK_{Ca}, Ca_{HVA}, K_A, K_{Dr}, K_{ir}$ and Na_F). In summary, the total current through the neuronal membrane ($I_{tot}(t)$) was given as the sum of capacitive, passive and active ionic as well as externally applied currents:

$$I_{tot}(t) = d/dt(V_m(t)) \cdot C_M + (V_m(t) - E_M)/R_M + \sum_{i=1}^{n} g_i(V_m(t),t) \cdot (V_m(t) - E_i) + I_{app}(t), \tag{1}$$

where $V_m(t)$ is the transmembrane potential, C_M the cell membrane capacitance, R_M the linear membrane resistance and E_M its equilibrium potential, $g_i(V_m(t),t)$ the voltage- and time-dependent ion conductance and E_i its equilibrium potential, and $I_{app}(t)$ the externally applied current. $g_i(V_m(t),t) = (Y^i_m)^{p_i} \cdot (Y^i_h)^{q_i} \cdot G_i$ and $d/dt(Y^i_{m/h}) = \alpha^i_{m/h}(1 - Y^i_{m/h}) - (\beta^i_{m/h} \cdot Y^i_{m/h})$, where Y^i_m and Y^i_h are the gating particles for current activation (m) and inactivation (h), G_i is the maximum conductance, and α^i_m, α^i_h, β^i_m and β^i_h are the forward and backward rate constants for current activation (m) and inactivation (h). i is an index for specific ion channel type. The processing of calcium ions was assumed to take place only in a very narrow volume (shell) close to cell

Table 1
Constraints for the CGN model

Action potential waveform	Basic firing properties
1. $V_{th}(-0.045\ \text{V} \leqslant V_{th} \leqslant -0.035\ \text{V})$	1. Frequent, repetitive firing
2. $V_{peak}^{max}(0\ \text{V} \leqslant V_{peak}^{max} \leqslant +0.01\ \text{V})$	2. Linear conversion of f–I
3. $V_{peak}^{min}(V_{peak}^{min} < -0.05\ \text{V})$	3. Delay in firing at threshold stimuli (I_{app}^{th} between 11 and 12 pA)

V_{th} is the threshold membrane potential for firing, V_{peak}^{max} is the maximum (positive) peak value and V_{peak}^{min} the minimum (negative) peak value of the membrane potential, I_{app}^{th} is the applied current threshold for initiating firing.

membrane, according to equation:

$$\mathrm{d}/\mathrm{d}t([\mathrm{Ca}]) = (B \cdot I_{\mathrm{CaHVA}}/v_{\mathrm{shell}}) - (([\mathrm{Ca}] - [\mathrm{Ca}]_{\mathrm{rest}})/\tau_{\mathrm{Ca}}), \tag{2}$$

where B is a constant for calcium ion transfer into the cell, I_{CaHVA} the calcium current through voltage- and time-dependent calcium ion channels, v_{shell} the volume of the shell, $[\mathrm{Ca}]_{\mathrm{rest}}$ the intracellular calcium concentration at rest, and τ_{Ca} the time constant for the decay of intracellular free calcium.

2.3. Starting parameter and constraint selection

The selection of starting and fixed parameter values for passive and active membrane properties are presented in detail in the Ph.D. Thesis by Linne [11]. In short, the channel equation parameters are based on the data given in some earlier publications [1,4,15,17,18], but with slight modifications. The basic excitability properties of the granule neuron, used as the main constraints of the model, were gathered from several publications [3,5,10,16,19], as well as from our own recordings [9,13] (see Table 1).

2.4. Optimization and fine-tuning

The CGN model was optimized and fine-tuned based on a semi-automatic, iterative approach [11]. It utilizes the following sets of parameters (Θ; Table 2): (i) a set of *known parameters* (Θ_1) which are *fixed* at a physiologically reasonable value (reliable estimates based on large experimental data set are required), (ii) a first set of *unknown parameters* (Θ_2) which are chosen based on realistic physiological approximations (initial guesses) and *fixed*, and (iii) a second set of *unknown parameters* (Θ_3 and Θ_4) which are optimized during the *iterative fine-tuning.*

The final parameters used in the model simulations were: $d_{\mathrm{soma}} = 6 \times 10^{-6}$ m, $R_M = 5 \times 10^9\ \Omega$, $C_M = 3.5 \times 10^{-12}$ F, $E_M = -0.07$ V (passive model), $E_M = -0.025$ V (active model), $E_{\mathrm{Na}} = +0.07$ V, $E_{\mathrm{K}} = -0.075$ V, $E_{\mathrm{Ca}} = +0.14$ V, and $E_{\mathrm{BKCa}} = -0.085$ V [(i)]. Ion conductance parameters were given both fixed (p_i, q_i) parameter values

Table 2
Optimization and fine-tuning protocol for the CGN model

1. Define Θ.
2. Select the parameter subspace_1 ($\Omega_i \in \Theta_1$) with known, fixed parameters, and set each Ω_i its corresponding value.
3. Set the initial guesses for the parameter subspace_2, Θ_2, $G_i \in \Theta_2, i = \mathrm{Na_F}, \mathrm{K_{Dr}}, \mathrm{K_{ir}}$. Set G_{Kir} as constant initial guess.
4. Define the lower (V_{th}^1) and upper (V_{th}^2) bounds for V_{th}.
5. If $V_{\mathrm{th}}^1 \leqslant V_{\mathrm{th}} \leqslant V_{\mathrm{th}}^2$ goto 7.
6. Else keep adjusting Θ_2, but keep G_{Kir} constant, until 5 is met.
7. Define initial guesses for parameter subspace_3, Θ_3, $G_i \in \Theta_3$, $i = \mathrm{Ca_{HVA}}, \mathrm{BK_{Ca}}$, and adjust $0.5G_{\mathrm{KDr}}$
8. Define upper bounds for $V_{\mathrm{peak}}^{\mathrm{max}}(V_{\mathrm{peak}}^{\mathrm{max1}}$ and $V_{\mathrm{peak}}^{\mathrm{max2}})$.
9. If $V_{\mathrm{peak}}^{\mathrm{max1}} < V_{\mathrm{peak}}^{\mathrm{max}} < V_{\mathrm{peak}}^{\mathrm{max2}}$ and $V_{\mathrm{th}} > -0.045$ V goto 11.
10. Else keep adjusting Θ_3 until 9 is met.
11. Define initial guess for $G_i \in \Theta_4$, $i = \mathrm{K_A}$.
12. Adjust basic firing properties according to Table 1.

and initial guesses (G_i) according to the protocol in Table 2 (presented as 'channel name (p_i, q_i, G_i)'): $\mathrm{BK_{Ca}}(1, -, 30\ \mathrm{S/m^2})$, $\mathrm{Ca_{HVA}}(2, 1, 4.6\ \mathrm{S/m^2})$, $\mathrm{K_A}(3, 1, 10\ \mathrm{S/m^2})$, $\mathrm{K_{Dr}}(4, -, 120\ \mathrm{S/m^2})$, $\mathrm{K_{ir}}(1, -, 28\ \mathrm{S/m^2})$, and $\mathrm{Na_F}(3, 1, 400\ \mathrm{S/m^2})$ [(i), (ii) and (iii)]. Final calcium dynamics parameters were fixed in the beginning and given the following values: $B = 5.2 \times 10^{-6}$ mol/C, $[\mathrm{Ca^{2+}}]_{\mathrm{rest}} = 100 \times 10^{-6}\ \mathrm{mol/m^3}$, $\tau_{\mathrm{Ca}} = 1 \times 10^{-3}$ s, and $d_{\mathrm{shell}} = 1 \times 10^{-7}$ m.

3. Results

Simulations show, that for a one-compartmental neuron six voltage- and time-dependent ion channel types ($\mathrm{BK_{Ca}}$, $\mathrm{Ca_{HVA}}, \mathrm{K_A}, \mathrm{K_{Dr}}, \mathrm{K_{ir}}$, and $\mathrm{Na_F}$) and simple calcium dynamics reliably reproduce the experimentally recorded basic firing properties of the cultured granule neuron. This suggests that a one-compartmental representation is sufficient for reproducing the constraints (Table 1) for a small cultured neuron. In summary, the CGN model of this study is able to reproduce the following basic electroresponsive properties: (1) realistic action potential waveform, specific for a cultured granule neuron (Fig. 1), (2) delay in firing with small applied currents (I_{app}; Fig. 2), (3) fast frequent firing, up to 300 Hz (Fig. 2), and (4) linear frequency–current (f–I) relation, up to 200 Hz (Fig. 3). Furthermore, simulations by eliminating particular ion channel currents from the final fine-tuned model (i.e., resembling pharmacological blockade of ion channels) further validated the model robustness. For example, complete elimination of $\mathrm{BK_{Ca}}$ causes reduced hyperpolarization, thereby causing increase in firing frequency (not shown here). Similarly, complete elimination of $\mathrm{K_A}$ causes increase in firing frequency.

The current application threshold ($I_{\mathrm{app}}^{\mathrm{th}}$) for firing was set between 11 and 12 pA during the final step of the optimization process (Table 2). The CGN model of this study

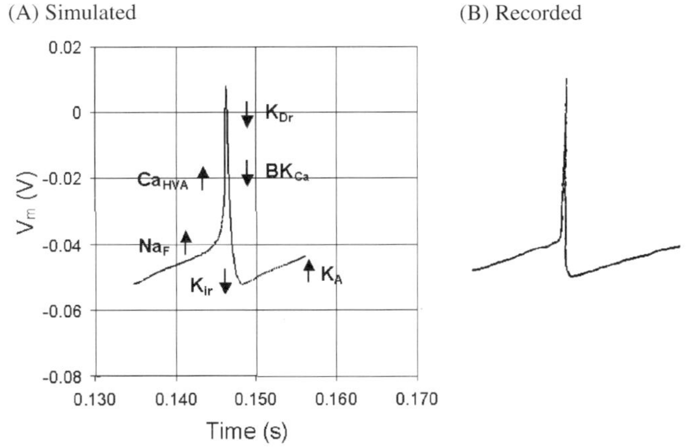

Fig. 1. Simulated (A) and recorded (B) action potential waveforms. The waveform of the recorded AP in (B) and its specific phases are satisfactorily reproduced: both the potential preceding the spike and after the spike are reproduced with the simulated model in (A). The threshold membrane potential for firing is approximately −40 mV. The positive peak of the AP is close to +10 mV and the negative peak to −55 mV. The experimentally recorded AP is taken from Lin and Moran [10].

was found to fire at 9 Hz frequency with 12 pA current stimulus. The so-called 'K_A effect', a delay in the firing due to the early, strong activation of the K_A type potassium channels with current stimulus pulses close to the threshold (I_{app}^{th}), is reproduced (Fig. 2a), similarly to the effect shown for granule neurons in slice preparations [1]. The constrained CGN model is frequently firing when current stimulus pulses of $I_{app} \geqslant 13$ pA are used. As can be seen from Fig. 2, the model neuron reproduced reasonably realistic, frequent firing, showing no adaptation. When the applied current increases, the model neuron is capable of reproducing frequent, repetitive firing up to 200 Hz without saturation ($I_{app} \leqslant 35$ pA). With relatively high current steps ($I_{app} \geqslant 40$ pA), the CGN model is still frequently firing, but shows mild saturation of the f–I curve. The highest firing rate the model neuron can attain is 300 Hz. The turtle granule neuron model was found to achieve 100 Hz without damped oscillations, when implemented as one-compartmental representation based on the original model by Gabbiani et al. [8] within the framework of this study using GENESIS (Fig. 3).

Fig. 3 shows the f–I relations for our model and other earlier CGN models. The present model reproduces the linear f–I relationship and fairly reliably follows the experimentally obtained f–I relations [3,6]. Therefore, CGN model of this study with linear input–output conversion of applied currents predicts reliable translation of inputs into action potential discharges, even with relatively high firing frequencies. The slopes of the f–I curves are as follows: 9 Hz/pA for the CGN model of this study, 8 Hz/pA for slice and a model granule neuron by D'Angelo et al. [5,6], 6 Hz/pA for slice granule neuron by Brickley et al. [3], and 4 Hz/pA for the turtle models when the model is reduced to one compartment and simulated with GENESIS neuronal simulator ([8,14]; linear part of the curve).

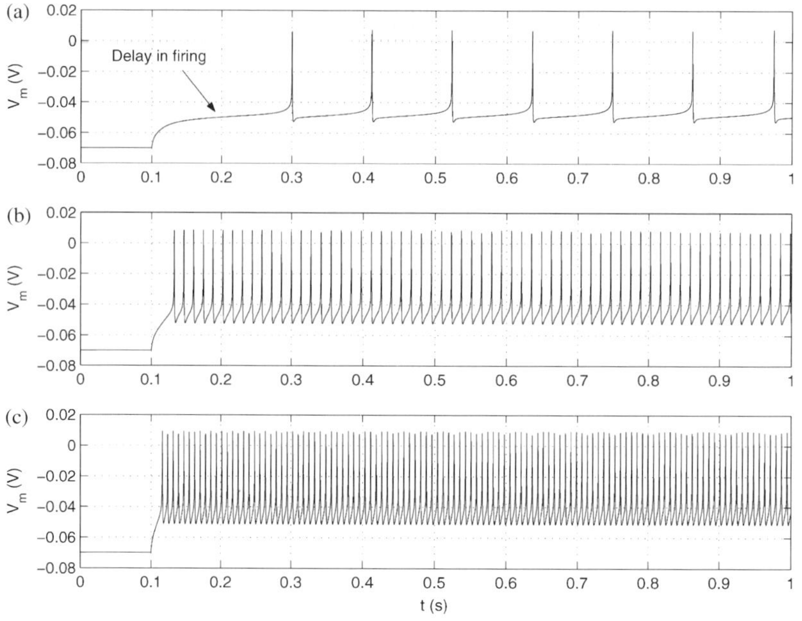

Fig. 2. Frequent firing of the CGN model showing no adaptation. I_{app} is 11.9 pA (a), 18 pA (b) and 24 pA (c). The delay in firing is successfully reproduced (a).

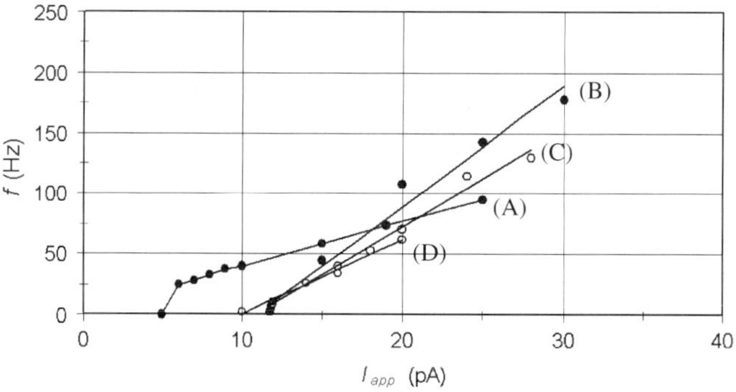

Fig. 3. Comparing *f–I* relations for existing model and in vitro granule neurons. (A) Turtle granule neuron model [8,14], (B) CGN model of this study, (C) in vitro granule neuron of slice preparation [5] and model neuron [6], and (D) in vitro granule neuron of slice preparation [3].

4. Discussion and conclusions

In conclusion, we have developed a biophysical model of a cultured rat cerebellar granule neuron, which is capable of reproducing the major electroresponsive properties of its real counterpart (see also [11,12]). The presented model has the following strengths: (1) it consists of a realistic set of ion channels, characterized for a rat cerebellar granule neuron, (2) it is able to reproduce the basic electroresponsive properties, specifically the linear f–I relation, under current stimulation conditions, (3) it is reliable and robust to changes in key parameters, being able to mimic the behavior of real neurons under similar experimental conditions, and (4) it is implemented and optimized on a GENESIS platform, which makes it more flexible to be used for further implementations at network and subcellular levels.

This model is a further addition to the earlier compartmental models for cerebellar granule neurons, which include models for a turtle [8] and for a rat [6,14]. Those three models, however, are mainly aimed at mimicking neuronal behavior in slice preparations. Furthermore, the turtle granule neurons differ from the rat granule neurons in that they have larger soma and dendritic volumes, thereby making the reptile turtle granule neurons electrotonically less compact compared to mammalian rat ones. In addition, turtle granule neurons express different types of ion channels, compared to rat CGNs. Single neuron models, such as the model here presented for a neuron in culture conditions, can be used for various kind of simulations of the complex interactions between ion channels, intracellular signaling and excitability. For the future subcellular modeling of e.g. cellular signaling pathways a lot of biochemical kinetic and molecular data, specifically obtained from the cultured granule neuron, is already available. Therefore, the present model would serve as an ideal template for simulating the excitability during neuronal differentiation in various culture preparations, such as the explant cultures used by Shibata et al. [19]. Single-cell culture models can be further utilized as basis for complex network models of cultured neurons, which have been recently increasingly studied by multi-electrode array systems.

Acknowledgements

Supported by Magnus Ehrnrooth Foundation and the grants 79854 and 80455 from the Academy of Finland for M.-L.L.

References

[1] R. Bardoni, O. Belluzzi, Kinetic study and numerical reconstruction of A-type current in granule cells of rat cerebellar slices, J. Neurophysiol. 69 (1993) 2222–2231.

[2] U.S. Bhalla, R. Iyengar, Emergent properties of networks of biological signaling pathways, Science 283 (1999) 381–387.

[3] S.G. Brickley, S.G. Cull-Candy, M. Farrant, Development of a tonic form of synaptic inhibition in rat cerebellar granule cells resulting from persistent activation of GABAA receptors, J. Physiol. 497 (1996) 753–759.

[4] S.G. Cull-Candy, C.G. Marshall, D. Ogden, Voltage-activated membrane currents in rat cerebellar granule neurones, J. Physiol. 414 (1989) 179–199.

[5] E. D'Angelo, G. De Filippi, P. Rossi, V. Taglietti, Ionic mechanism of electroresponsiveness in cerebellar granule cells implicates the action of a persistent sodium current, J. Neurophysiol. 80 (1998) 493–503.

[6] E. D'Angelo, T. Nieus, A. Maffei, S. Armano, P. Rossi, V. Taglietti, A. Fontana, G. Naldi, Theta-frequency bursting and resonance in cerebellar granule cells: experimental evidence and modeling of a slow K^+-dependent mechanism, J. Neurosci. 21 (2001) 759–770.
[7] P. Fromherz, Electrical interfacing of nerve cells and semiconductor chips, Chem. Phys. Chem. 3 (2002) 276–284.
[8] F. Gabbiani, J. Midtgaard, T. Knöpfel, Synaptic integration in a model of cerebellar granule cells, J. Neurophysiol. 72 (1994) 999–1009.
[9] T.O. Jalonen, S. Johansson, I. Holopainen, S.S. Oja, P. Århem, Single-channel and whole-cell currents in rat cerebellar granule cells, Brain. Res. 535 (1990) 33–38.
[10] F. Lin, O. Moran, Voltage dependent sodium currents in cultured rat cerebellar granule cells, Biosci. Rep. 10 (1990) 445–453.
[11] M.-L. Linne, Computational model for granule neuron excitability, Ph.D. Thesis, Department of Information Technology, Tampere University of Technology, Finland, Publications 342, 151 pp., 2001.
[12] M.-L. Linne, T.O. Jalonen, Mathematical model of a small neuron, in: P. Kuosmanen, M. Gabbouj (Eds.), *X* EUSIPCO 2000 Proceedings, Vol. I, Tampere University of Technology, Tampere, Finland, 2000, pp. 63–66.
[13] M.-L. Linne, S.S. Oja, T.O. Jalonen, Simultaneous detection of action potential current waveforms and single ion channel openings in rat cerebellar granule cells, Int. J. Neural Systems 7 (1996) 377–384.
[14] R. Maex, E. De Schutter, Synchronization of Golgi and granule cell firing in a detailed network model of the cerebellar granule cell layer, J. Neurophysiol. 80 (1998) 2521–2537.
[15] E. Moczydlowski, R. Latorre, Gating kinetics of Ca^{2+}-activated K^+ channels from rat muscle incorporated into planar lipid bilayers, J. Gen. Physiol. 82 (1982) 511–542.
[16] M. Robello, C. Carignani, C. Marchetti, A transient voltage-dependent outward current in cultured cerebellar granules, Biosci. Rep. 9 (1989) 451–457.
[17] P. Rossi, E. D'Angelo, J. Magistretti, M. Toselli, V. Taglietti, Age-dependent expression of high-voltage activated calcium currents during cerebellar granule cell development in situ, Pflügers Arch. 429 (1994) 107–116.
[18] P. Rossi, G. De Filippi, S. Armano, V. Taglietti, E. D'Angelo, The weaver mutation causes a loss of inward rectifier current regulation in premigratory granule cells of the mouse cerebellum, J. Neurosci. 18 (1998) 3537–3547.
[19] R. Shibata, K. Nakahira, K. Shibasaki, Y. Wakazono, K. Imoto, K. Ikenaka, A-type K^+ current mediated by the Kv4 channel regulates the generation of action potential in developing cerebellar granule cells, J. Neurosci. 20 (2000) 4145–4155.
[20] M.A. Wilson, U.S. Bhalla, J.D. Uhley, J.M. Bower, GENESIS: a system for simulating neural networks, in: D. Touretzky (Ed.), Advances in Neural Information Processing Systems, Morgan Kaufmann, San Mateo, CA, USA, 1989, pp. 485–492.

Available at
www.ComputerScienceWeb.com
POWERED BY SCIENCE @ DIRECT®

ELSEVIER

Neurocomputing 52–54 (2003) 591–597

NEUROCOMPUTING

www.elsevier.com/locate/neucom

Quantifying olfactory perception: mapping olfactory perception space by using multidimensional scaling and self-organizing maps

Amir Madany Mamlouk[a,*], Christine Chee-Ruiter[b], Ulrich G. Hofmann[c], James M. Bower[d,e]

[a]*Institute for Neuro- and Bioinformatics, University of Lübeck, Germany*
[b]*Infant Perception Laboratory, California Institute of Technology, Pasadena, CA, USA*
[c]*Institute for Signal Processing, University of Lübeck, Germany*
[d]*Cajal Neuroscience Research Center, University of Texas, San Antonio, TX, USA*
[e]*Research Imaging Center, University of Texas Health Science Center, San Antonio, TX, USA*

Abstract

In this paper we describe an effort to project an olfactory perception database onto the nearest high dimensional Euclidean space using multidimensional scaling. This yields an independent Euclidean interpretation of odor perception, whether this space is metric or not. Self-organizing maps were then applied to produce two-dimensional maps of the Euclidean approximation of olfactory perception space. These maps provide new knowledge about complexity and potentially the functionality of the sense of smell from the point of view of human odor perception. This report is based on a recent thesis by Madany Mamlouk, Quantifying olfactory perception, at the University of Lübeck, Germany.

Keywords: Olfactory perception; Multidimensional scaling; Self-organizing maps

1. Introduction

While considerable effort has been spent trying to relate the chemical structure of odorants to odor perception, very little is yet understood about the larger organization

* Corresponding author.
E-mail addresses: madany@inb.uni-luebeck.de, madany@gmx.net (A. Madany Mamlouk).

0925-2312/03/$ - see front matter © 2002 Elsevier Science B.V. All rights reserved.
doi:10.1016/S0925-2312(02)00805-6

of human odor perception. Instead, both the perfume industry and olfactory researchers rely on sets of data such as Dravnieks' Atlas of Odor Character Profiles [3] or other compilations which describe the odors elicited by a particular chemical, e.g. Hexyl Butyrate, using a set of odor descriptors, e.g. *fruity*, *sweet* and *pineapple*.

Over the last several years we have been examining the relationship between the odor descriptors found in these data sets in an attempt to determine whether they might reveal an underlying structure in human odor perception. Our initial efforts using an information theoretic approach suggested that, in fact, there might be an overall organization to the space of human olfactory perception [2].

In this paper we extend this effort using a combination of mapping techniques. The structures revealed could provide scientists a more rigorous way to select odorants for human psychophysical experiments as well as a more solid foundation for systems, cellular and molecular studies of the biology of olfaction.

2. Method

Our analysis of odor descriptors and olfactory perception space is based on the data set found in the Aldrich Flavor and Fragrances Catalog [1], including 851 stimuli using 278 odor descriptors. For estimating dissimilarities between different odors, intuitively most satisfying results have been obtained using the subdimensional distance d_s, defined as

$$d_s(O_x, O_y) = \frac{\sum_{i=1}^{n} (|o_i^x - o_i^y| \cdot o_i^x) + (\sum_{i=1}^{n} (|o_i^x - o_i^y| \cdot o_i^y))/\sum_{i=1}^{n} o_i^y}{\sum_{i=1}^{n} o_i^x + 1} \qquad (1)$$

where $O_j = (o_1^j, \ldots, o_n^j)$ is the n-dimensional binary profile or feature vector of odor j. O_x is assumed to be less prominent than O_y, which is expressing the subdimensional characteristic of this dissimilarity measure. Also, it can be interpreted as a weighted version of a cross-entropy measure [2].

We used multidimensional scaling (MDS) to extract information for a reasonable map from a (278×278) odor dissimilarity matrix. There is no a priori reason to expect that the odor space is metric [7], but MDS is projecting these probably non-metric dissimilarities to the nearest Euclidean space, which then is a metric space. MDS is a common method for dimensional reduction and graphical representation of multidimensional data, but it can be used to estimate the dimensionality of a data set as well [8], which turned out to be an interesting feature in analyzing this data set.

To avoid local minima, Monte-Carlo simulation techniques were used. Because of the rotational invariance of MDS the standard deviation of the resulting distance matrices has been used to control stability during simulation. Surprisingly, the high dimensional scaling of odor space was found to be statistically stable with an average of only 4% standard deviation using 95% confidence intervals. In Fig. 1, this stability can be seen for MDS stress values, which describe the tension between the given dissimilarities and the resulting Euclidean distances [5]. A dimension of about 32 appeared to be a good estimation in terms of the trade-off between a high and statistically stable stress relaxation and the highest sustainable dimensional reduction. It should be mentioned again that the database is deduced from psychophysical data, therefore while there is

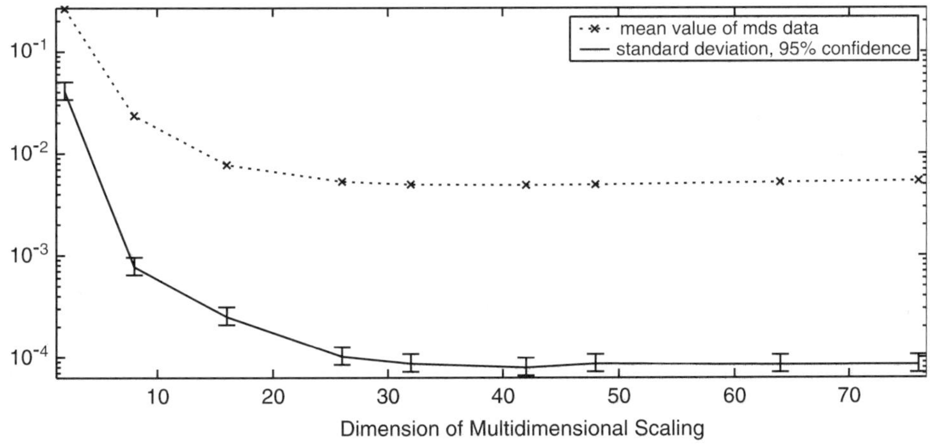

Fig. 1. Standard deviation and mean values of stress values for Monte-Carlo simulation runs of MDS on Aldrich database. Stress values become smaller the more the scaled distances and the given dissimilarities correspond.

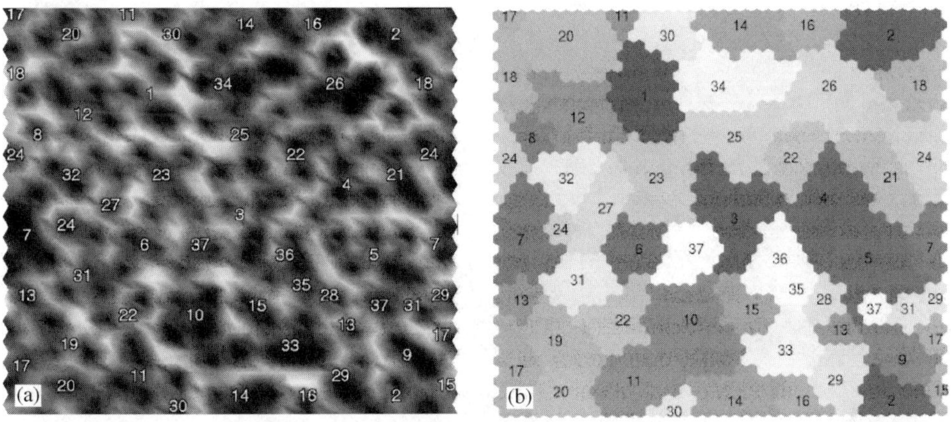

Fig. 2. Olfactory perception map for scaled data of the Aldrich database. (a) The distances between neighboring units of the SOM are shown. Large distances are bright. (b) SOM is reassigned using k-means clustering to visualize fragmentations, e.g. cluster 15 is located in the lower right corner and right below the center.

no noise, inaccuracies are introduced into the dataset because subjects profiling their olfactory sensations used discrete labels.

To visualize the high dimensional results of MDS on a low dimensional map, we used a two-dimensional self-organizing mapping (SOM, [6]), which preserves the general topology of the odor data, i.e. if two points are close-by on the map, they should be close-by in the high dimensional input space as well.

In Fig. 2, the grid elements are reassigned after learning by k-means clustering [4] to denote potential fragmentations on the final map. We used the "*SOM Toolbox for Matlab*" by Vesanto et al. [9].

3. Results

The use of MDS onto the given odor dissimilarity matrix resulted in an appropriate thirty-two-dimensional Euclidean approximation of olfactory perception space. The quality of the dimensional reduction was tested using stress-values as well as scatter plots. This analysis indicated that MDS on higher dimensions did not increase the quality of the projection, while lower dimensions increased the stress-values significantly. Fig. 2 shows the resulting toroid map of the olfactory perception space, i.e. its sides are continuous.

Because SOMs have been used to reduce the dimensionality of the olfactory perception space, some distortion of the map is inevitable. In particular, in some cases the lower dimensional map shows odor descriptors as neighbors, when they are not in the high dimensional space. This condition is indicated in Fig. 2a by the bright shading of common links.

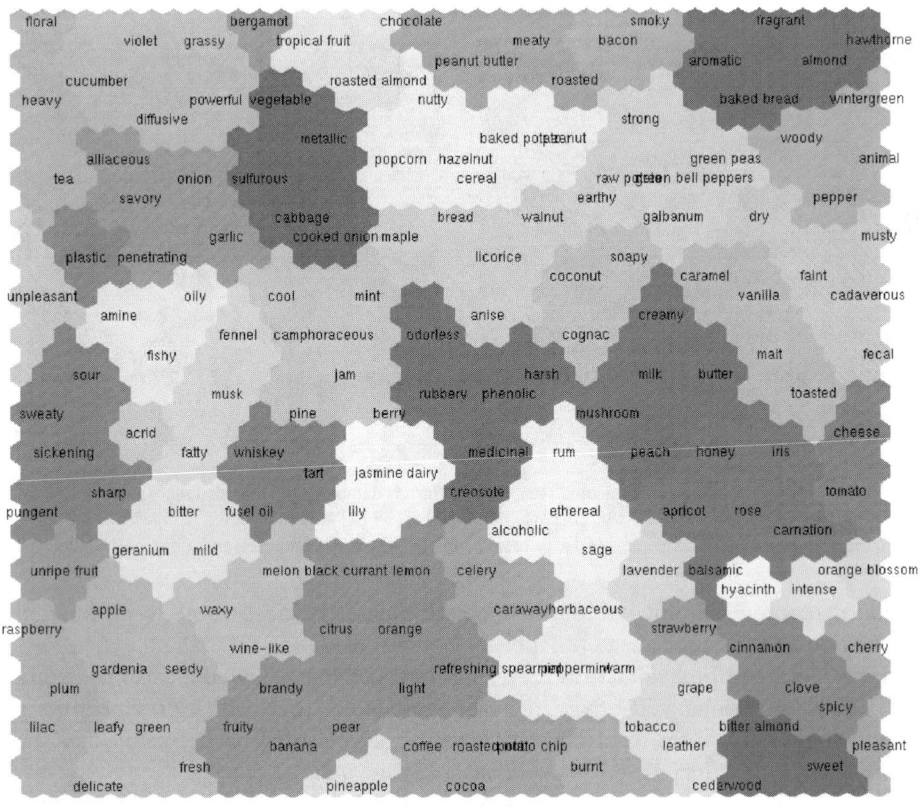

Fig. 3. Odor perception map labeled with odor descriptors, each descriptor has been plotted onto its nearest neighbor on the SOM. The shaded areas are corresponding to the clusters in Fig. 2.

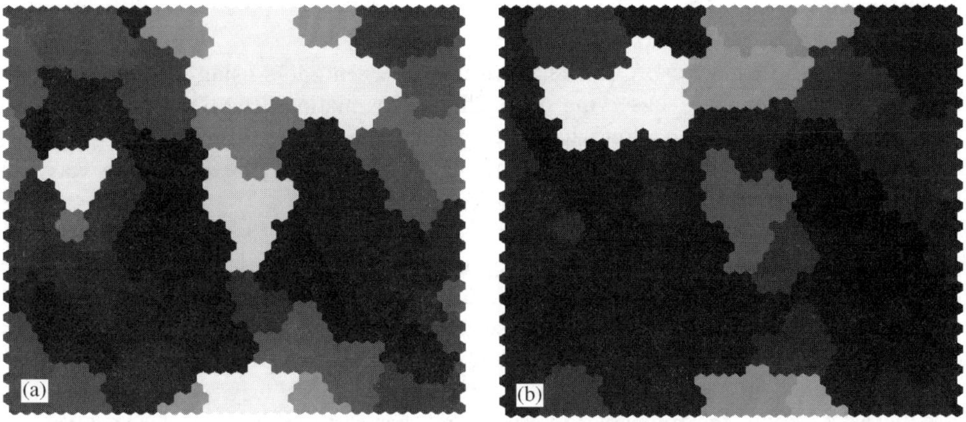

Fig. 4. Localization of all odors evoked by compounds carrying Nitrogen and Sulfur. The brighter a cluster is, the higher is the percentage of its odors that are evoked by odorants containing Nitrogen (a) and Sulfur (b). Compounds that contain both Nitrogen and Sulfur are included as well.

Conversely, in some cases neighbors in odor space might not be neighbors on the map. This is indicated by Fig. 2b, where the additional clustering of the SOM by k-mean clustering can be used to find such fragmented clusters.

Please note that fragmentations still persist in the labeled map as shown in Fig. 3, for example *celery*, *caraway* and *pleasant* are embedded in fragmented cluster 15, thus they are neighbors in odor space. Odor descriptors that carry only redundant global information are omitted on this map, for example *grapefruit*, which is only evoked by a single stimulus in our data set.

One interesting feature of these maps is that they suggest an ordering of relatedness in odor perception. For example, without such a map, it is not clear how the odor perception of *apple*, *banana* and *cherry* might be ordered. Using the label map in Fig. 3 and the cluster map in Fig. 2b, we find that *cherry* belongs to cluster 17, *apple* to cluster 19 and *banana* to cluster 11. Because of the toroid character of the map, clusters 17 and 19 are neighbors, and clusters 19 and 11 are next to each other. However, there is always at least one cluster between clusters 11 and 17.

Thus, these odor maps suggest that the odor percept *cherry* is more closely related in perceptual space to *apple* than to *banana*, while *apple* shares an association with both *cherry* and *banana*.

Finally, perhaps the most surprising result of our previous analysis of odor perception space [2] was the finding that the descriptors used to classify molecules containing nitrogen or sulfur were clearly segregated in the odor perception maps. Because these molecules are key atoms in different metabolic cycles, it was proposed that human olfactory perception reflected the organization of animal and plant metabolism. As shown in Fig. 4, this result is also apparent using these new mapping techniques.

4. Discussion

In this study, we have expressed relationships between odors using techniques which generate a solid topology conserving map. The approximation via MDS provides strong quantitative support for the long held belief that olfactory perception space is high dimensional. The resulting maps also allow us to order odor perception in a way not previously possible.

We believe that these maps can provide a new foundation for studies of the olfactory system that is related to the structure of olfactory perception rather than strictly based on the structure of chemical compounds. Many efforts to construct an understanding of olfactory perception based primarily on structural differences in chemical compounds have failed. Using the techniques described here, relationships between single odors can be quantified in a fundamentally new and more rigorous way. New questions can also be asked about the molecular, cellular and systems biology underlying olfactory perception.

To date, most efforts, focusing on the organization of receptors or neural responses, have constructed maps based on general properties of chemicals [7,10]. This approach however, has had only marginal success in relating the structure of chemical stimuli to perception. The hypothesis we continue to pursue, that human olfactory perception reflects the metabolic relationships between molecules in the natural environment has been supported by this further analysis and potentially represents a major step forward in understanding olfaction.

References

[1] Aldrich, Ed., Flavor and Fragrances Catalog, Sigma Aldrich Chemicals Company, Milwaukee, WI, 1996.
[2] C.W.J. Chee-Ruiter, The biological sense of smell: olfactory search behavior and a metabolic view for olfactory perception, Ph.D. Thesis, California Institute of Technology, Pasadena, CA, 2000.
[3] A. Dravnieks, Atlas of Odor Character Profiles, Data Series DS 61, ASTM, Philadelphia, PA, 1985.
[4] A.K. Jain, R.C. Dubes, Algorithms for clustering data, Prentice-Hall Advanced Reference Series: Computer Science, NJ, 1988.
[5] J.B. Kruskal, M. Wish, Multidimensional Scaling, Sage Publications, Beverly Hills, CA, 1978.
[6] H. Ritter, T. Martinetz, K. Schulten, Neural Computation and Self-Organizing Maps: An Introduction, Addison-Wesley, Reading, MA, 1992.
[7] S.S. Schiffman, Contributions to the physicochemical dimensions of odor: a psychophysical approach, Ann. N. Y. Acad. Sci. 237 (1974) 164–183.
[8] R.N. Shepard, A.K. Romney, S.B. Nerlove, Multidimensional Scaling: Theory and Applications in the Behavioural Sciences, Vol. I—Theory, Seminar Press, New York, 1972.
[9] J. Vesanto, J. Himberg, E. Alhoniemi, J. Parhankangas, SOM Toolbox for Matlab 5, Report A57, Helsinki University of Technology, Neural Networks Research Centre, Espoo, Finland, 2000.
[10] M.H. Woskow, Multidimensional scaling of odors, Ph.D. Thesis, University of California Los Angeles, Los Angeles, CA, 1964.

Amir Madany Mamlouk graduated in computer science with a minor in biomathematics and bioinformatics at the University of Lübeck, Germany. He performed research for his diploma thesis at the California Institute of Technology. Currently he is research associate with the Institute for Neuro- and Bioinformatics at the University of Lübeck. His main research interests are methods for dimensional reduction, artificial neural networks and self-organizing structures. He focusses on their appliance in biological systems, especially on the understanding of the sense of smell.

Christine Chee-Ruiter received a B.S. in General Engineering from Harvey Mudd College, and a Ph.D. in Computation and Neural Systems from the California Institute of Technology. She is currently completing a postdoctoral research project on the psychophysics of multimodal (vision and audition) sensory development in infants at Caltech. Besides her current work, her research interests include olfactory perception and search.

Ulrich G. Hofmann received his Diploma in Physics 1993 and his Ph.D. in Biophysics 1996 from the Technical University of Munich, Germany. He stayed at the Abo Akademi, Finland, as an EU-fellow (HCM) and as Feodor-Lynen-Fellow at the California Institute of Technology. Since 1999 he is research associate ("Habilitand") at the Institute for Signal Processing of the Medical University of Lübeck and project coordinator and initiator of VSAMUEL. His long term research aims to interface brains with computers.

James Bower has a dual-professorship within Radiology at the University of Texas Health Science Center in San Antonio, and Biology at the University of Texas, San Antonio. Research projects in his laboratory employ a range of techniques from detailed realistic biological modeling, through electrophysiology to human psychophysics. The laboratory studies both the olfactory system and the cerebellum.

Available at
www.ComputerScienceWeb.com
POWERED BY SCIENCE @ DIRECT®

ELSEVIER

Neurocomputing 52–54 (2003) 599–604

NEUROCOMPUTING

www.elsevier.com/locate/neucom

CINNIC, a new computational algorithm for the modeling of early visual contour integration in humans

T. Nathan Mundhenk*,[1], Laurent Itti[1]

Computer Science Department, University of Southern California, HNB 06 3641 Watt Way, Los Angles, CA 90020, USA

Abstract

In order to gain a better understanding of visual saliency, we have developed algorithm which simulates the phenomenon of contour integration for the purpose of visual saliency. The model developed consists of the classical butterfly pattern of connection between orientation selective neurons in the primary visual cortex. In addition, we also add a local group suppression gain control to eliminate extraneous noise and a fast plasticity term which helps to account for closure effect often observed in humans exposed to closed contour maps. Results from real world images suggest that our algorithm is effective at picking out reasonable contours from a scene. The results improved with the introduction of both the fast plasticity and group suppression. An addition of multi-scale analysis has also increased the effectiveness as well.

Keywords: Contour; Integration; Visual; Saliency; Model

1. Introduction

We have created an algorithm, which will integrate contours in real world images. The idea is to simulate the way the human brain integrates contours for visual salience in early visual pre-processing. To this end, our goal is to simulate saliency for a given contour. The important components of this model are not only its abilities to find contours in an image, but for it to find contours that a human finds salient as

* Corresponding author.
E-mail addresses: mundhenk@usc.edu (T.N. Mundhenk), itti@pollux.usc.edu (L. Itti).
[1] More information at http://iLab.usc.edu.

0925-2312/03/$ - see front matter © 2002 Elsevier Science B.V. All rights reserved.
doi:10.1016/S0925-2312(02)00853-6

well. Such things include contour continuity, length, closure and the uniqueness of the contour when compared to its background.

Our approach has been to start with a simple model of neural connections. We use here a standard butterfly shape for connections, which has been tried with success in the past (for instance, [2]). The elements of the butterfly pattern are connections that branch out like wings to collinear neurons, which creates excitation on the neurons it connects to. However, suppression is also added with a set of orthogonal wings that are used to suppress parallel contours. The end result should be that the butterfly model picks out salient contours in an image using co-linearity. However, it seems that the original butterfly shape of connections is difficult to manually tune for multiple images suggesting that the brain has adaptive centers to do this. In addition to this, the butterfly shape also does not necessarily account for contour closure or the observation that contour effects seem to extend beyond their receptive field.

To address these factors, we propose to use several devices that should aid in explaining several observed effects. We have added an adaptive layer using group suppression for when a group of neurons seems to get to excited. We have also added fast plasticity described by Braun [1], for neural weight to allow connections to dynamically adapt to local neural activity. In addition, we are using the model at multi-scales, which should help in preventing otherwise salient contours from being excluded.

2. Model

The current model which we have named Carefully Implemented Neural Network for Integrating Contours (CINNIC) starts with the basic butterfly local connections, but in addition to this, we have added the use of multi-scale analysis, a local group suppression gain control and fast plasticity for long-range effects. Fig. 1 shows a schematic description of our model, which is also described by Eqs. (1)–(5). The first step is analyzing the input image using Gabor filters tuned for 12 different angles. This produces 12 images that represent elements from the original image at increments of 15°. A noise factor of approx. 2% is introduced at this stage. These 12 images are then reduced into three different scales 64×64, 32×32 and 16×16 pixels in resolution, which are run separately and do not interact. A 4D convolution is run to simulate the interaction between the different orientation images. The convolution is done using a set of 144 kernels that represent all possible interactions between pairs of the 12 orientation images. These kernels specify the excitation and suppression that should occur between two elements in the images. The kernels take into account the co-linearity of two elements as well as their separating distance. For simplicity, interaction strength decreases as a ramp function as two elements are further separated. The kernels are statically specified at the beginning of a program run by input parameters and do not interact.

Each scale is run separately from the other. Each element is convolved against every other element within its range in such a way that co-linear elements tend to excite, while parallel elements tend to suppress each other (Eq. (1)). This is expressed as $x_{ij\alpha}$ being the source image pixel at location (i,j) and orientation α, and $x_{kl\beta}$ being the other image pixel at location (k,l) and orientation β then taking the product of these two by the kernel $k_{\alpha\beta(k-i)(l-j)}$. It should be noted that m and n equal the image

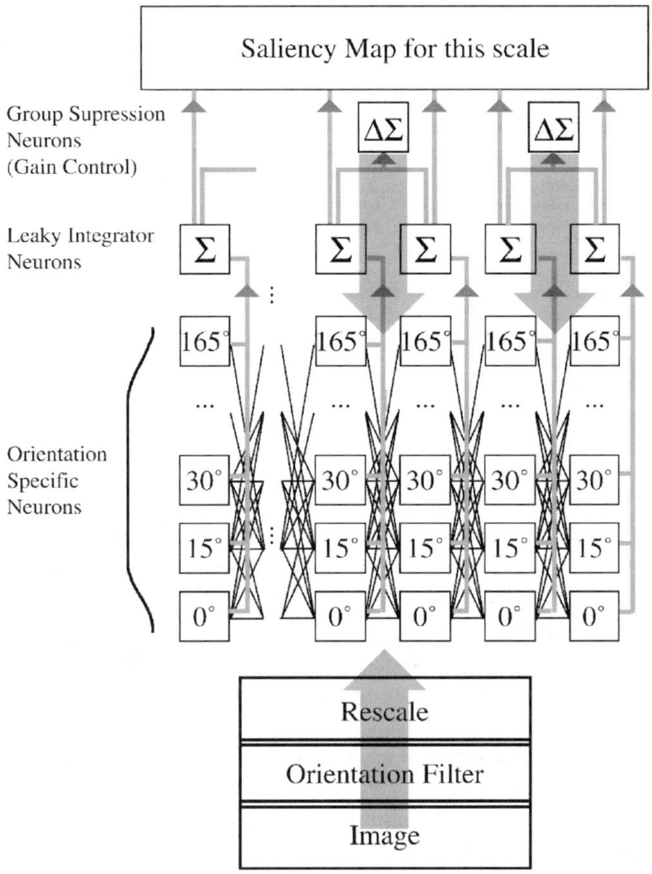

Fig. 1. This is a basic representation of the CINNIC algorithm. An input image is filtered, rescaled then interacted with images at other orientations including itself. The output goes to a saliency map of leaky integrator neurons. Group suppression is fed back from the change in group potential.

scale for instance 64, 32 or 16. Further, $(S_{kl})^+$ is a group suppression term for the current group (detailed below) with t being the current iteration. $(P_{kl\beta})^+$ is a plasticity term (also detailed below). The resulting potential from a single iteration is sent to a saliency map $(V_{ij})^{t+1}$. Each pixel in the saliency map represents a column of pixels from each of the 12 orientation images summed. The saliency map itself is made up of leaky integrator neurons, which lose

$$(v_{ij\alpha})^{t+1} = (x_{ij\alpha}) \sum_{\substack{k\in[[0,m]]\\ l\in[[0,n]]\\ \beta\in[[0,11]]}} (x_{kl\beta})(k_{\alpha\beta(k-i)(S_{kl})^+(l-j)})(P_{kl\beta})^+, \tag{1}$$

$$(x_{kl\beta}) \begin{cases} (x_{kl\beta}) & \text{iff } (k_{\alpha\beta\ldots}) < 0 \\ 1 & \text{otherwise.} \end{cases}$$

$$(V_{ij})^{t+1} = \sum_{\substack{k\in[[0,m]] \\ l\in[[0,n]] \\ \beta\in[[0,11]]}} (v_{kl\alpha})^{t+1} - L \tag{2}$$

$$(S_{ij})^t = \upsilon \left[\sum_{(k,l)\in N_i\times N_j} ((V_{kl})^t - (V_{kl})^{t-1} \right] - T \tag{3}$$

with

$$N_i = [[i - (m/8); i + (m/8)]]$$

$$N_j = [[j - (m/8); j + (m/8)]]$$

$$(P_{ij\alpha})^t = (v_{ij\alpha})^t(C) \tag{4}$$

$$I_{ij} = \mathrm{sig}((V_{ij})^t) \tag{5}$$

some constant potential L from one iteration to the next (Eq. (2)). To form a final saliency map for one of the three image resolutions, the potential from the leaky integrator neurons are fed through a sigmoidal function that simulates neural firing patterns (Eq. (5)) with I_{ij} being the final saliency map pixel for this scale.

Non-linearities are introduced in the form of the group suppression gain control (Eq. (3)) where T is the threshold constant and $(V_{kl})^t$ is the potential for a neuron in this group which are all summed for that group with υ as a constant multiplied by that sum. m and n represent the image size at that scale. The suppression is based upon the rate of the change of excitation. Fast plasticity (Eq. (4)) is introduced as $v_{ij\alpha}$ being the potential this neuron had multiplied by a constant C. The fast plasticity works by increasing all weights for a single simulated neuron, proportionally to the excitation it received in the previous iteration. That is, neurons that are stimulated more tend to stimulate co-linear neighbors more as well as to suppress parallel neighbors more. This function is introduced to recreate non-local interactions that are observed in human subjects in an attempt to account for observed contour closure effect. The fast plasticity used here is bounded to 5 times the original connection strength for any given neuron.

In our model the usage of fast plasticity was chosen for several reasons. The first was the suggestion by Braun [1] that other methods that attempted to explain contour closure either occurred too fast such as cumulative propagation [2], or were too slow such as temporal synchronization [3] as to explain the time it takes for closure effect to happen which Braun measured at about 250 ms. Second we wished to test the idea of fast plasticity and find if it was a viable mechanism for explaining closure effect.

Another non-linearity introduced is a simple local gain control using group suppression. Neurons are grouped into local neighborhoods of size $1/8 \times 1/8$ pixels of the image size at the current scale (e.g. 8×8 pixels for a 64×64 pixel scale). If the total change in potential from a group surpasses a threshold then the neurons increase their suppression of parallel neighbors proportionally to the increase past threshold. The group includes all neurons in all orientation maps for a given visual location, which report to the same image location. There is no current cap on how much additional suppression can be added using this method.

The algorithm runs for eight iterations, which was a number chosen based upon its observed optimality. After the final iteration, the three scales are brought back together and combined using a fixed weighted average. This average is the total saliency map of contours for the input image. The entire process takes approximately 2 min using an Athlon 1400 MHz based PC running Linux. The time is mostly due to the enormous amount of computation needed to compute interactions between neurons from all possible pairs of the 12 images using 144 2D kernels.

3. Results and conclusion

Initially, the model had been tested on a number of both real world images as well as contour simulation images. The results at face value seemed to yield positive results. Large amounts of noise are cleaned out with the adaptive group suppression. The fast plasticity enhances the contours of the test images and the multi scale model insures that fewer contours are missed.

Quantitative analysis has also been conducted. Twenty-four real world images representing a wide variety of contexts were used. A priori we used Photo Shop or GIMP to trace the outline of what seemed to us to be reasonable salient contours. The outline images were taken and then compared with the output images from CINNIC. It should be noted that, all images from CINNIC were run with the same values and no manual tuning was done between the time images were run.

Analysis was done between the test image and CINNIC output image for all 24 images using Euclidian distance as well as linear regression correlation under several different conditions. Without group suppression the mean Euclidian distance was poorer. Also, the variance was twice as much without group suppression for both correlation and Euclidian distance suggesting that group suppression increases the consistency of outcomes. Fast plasticity also seemed to help. Without fast plasticity the minimum correlation was negative among the 24 images. However, with fast plasticity, all correlational values were positive.

The data also supports the notion of improvement from the use of multi scales. Each of the scale images by themselves was compared with the outline images. In neither case did the mean correlational value or Euclidian distance surpass the values of the final max selected composite image.

Finally, Fig. 2 shows the output from the algorithm on three real world images used in the sample of 24 images. The five most salient points are circled by the algorithm and shown against the original image in the top row. The bottom row represents the raw saliency map output from which the five most salient points are selected using the five brightest regions.

Qualitative analysis suggests that CINNIC has an affinity for longer straighter contour segments with greater continuity, which seems to agree with literature on contour integration. CINNIC is also more effective at finding contours that are more unique given its region, as is also suggested by literature on contour integration. CINNIC's abilities with closure effect are currently being assessed, but the data regarding it as of yet is not conclusive.

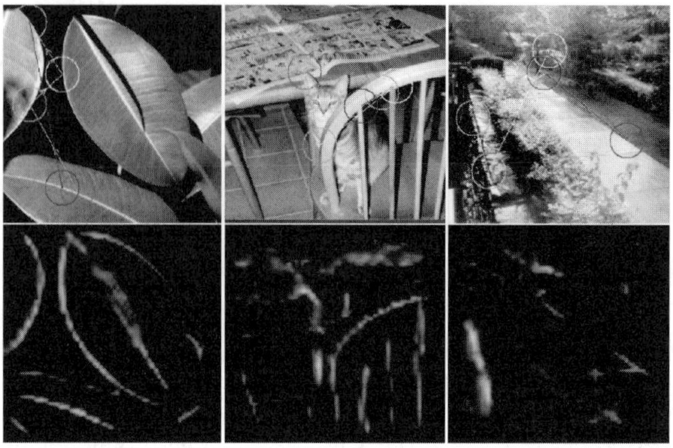

Fig. 2. The top row are real world images with the top five most salient points circled by the algorithm. The bottom row represents the raw saliency map for each image.

Acknowledgements

We would like to acknowledge Jochen Braun, Christof Koch and Mike Olson for their help and suggestions. This research is supported by the National Imagery and Mapping Agency, the National Science Foundation, the National Eye Institute, the Zumberge Faculty Innovation Research Fund and the Charles Lee Powell Foundation.

References

[1] J. Braun, On the detection of salient contours, Spatial Vision 12 (2) (1999) 211–225.
[2] S. Yen, L.H. Finkel, Extraction of perceptually salient contours by striate cortical networks, Vision Res. 38 (5) (1998) 719–741.
[3] Z. Li, A neural model of contour integration, Neural Comput. 10 (1998) 903–940.

T. Nathan Mundhenk received his BA in psychology from the University of Cincinnati in 2000 and was team leader of the University of Cincinnati Robotics Team. He is a current member of iLab at the University of Southern California (Los Angeles, CA) where he is a graduate student in computer science. He is currently involved in research into contour integration, vision in ego motion and is the team leader for the Beobot robotics project at iLab.

Laurent Itti received his MS in image processing from the Ecole Nationale Super-euredes Telecommunications (Paris) in 1994, and Ph.D. in computation and neural systems from Caltech (Pasadena, CA) in 2000. He is currently an assistant professor of computer science at the University of Southern California in Los Angeles. He is the Director of iLab, which pursues research in biologically motivated computer vision and neuromorphic robotics.

Available at
www.ComputerScienceWeb.com
POWERED BY SCIENCE DIRECT

ELSEVIER

Neurocomputing 52–54 (2003) 605–614

NEUROCOMPUTING

www.elsevier.com/locate/neucom

The role of cortico-basal-thalamic loops in cognition: a computational model and preliminary results

Srini Narayanan

ICSI, 1947 Center Street, Berkeley 94704, USA

Abstract

Clinical and experimental research over the last decade has implicated neuroanatomic loops connecting the frontal cortex to the basal ganglia and thalamus in various aspects of planning and memory. We report on computational model whose central aspects are: (1) a model of cortical-striatal-thalamic loops in planning and executive control, and (2) a fine-grained model of basal-ganglia function that exploits specific component connectivity and dynamics. The model is biologically plausible given current literature on the neurophysiology and disease pathology of the relevant brain regions. Specifically, our model has implications for subjects with diseases affecting the relevant brain regions (Parkinson's disease and Huntington's disease).

1. Introduction

There is by now robust evidence that the pre-frontal cortex plays a key role in various aspects of working memory and executive control [3]. As Middleton and Strick [5] points out, there is also clear evidence that the basal ganglia are closely involved with prefrontal cortex activity. From a functional viewpoint, while damage to the basal ganglia seems to produce cognitive deficits comparable to prefrontal cortex malfunction, teasing out the individual contributions has proven more problematic.

This report describes an ongoing effort to build a computational model fleshing out the role of cortical-basal-thalamic loops in planning and executive control. A distinguishing feature of the approach is a fine-grained model of basal-ganglia function that exploits specific component connectivity and dynamics. After introducing the relevant

E-mail address: snarayan@icsi.berkeley.edu (S. Narayanan).

0925-2312/03/$ - see front matter © 2003 Published by Elsevier Science B.V.
doi:10.1016/S0925-2312(02)00813-5

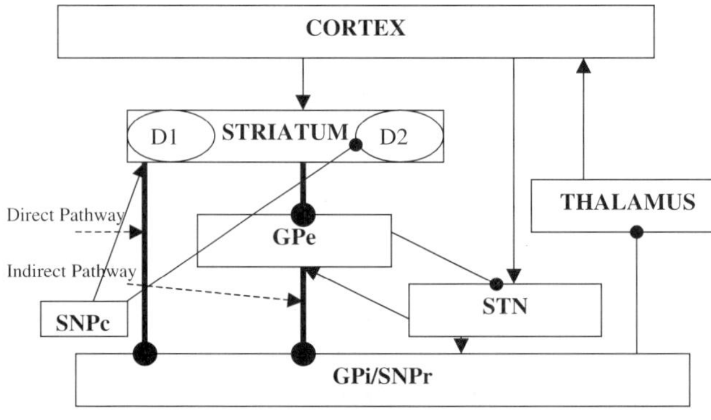

Fig. 1. This shows the block diagram view of the components and connectivity of the Cortico-basal loops. Inhibitory projections are shown with rounded tips while excitatory connections are shown with pointy tipped arrows. The direct and indirect pathways through the BG complex are highlighted and identified within the figure. The Striatum is shown with the two types of dopamine receptors D1 and D2. The Striatum, STN, GP (GPe and GPi), STN and SN (SNPr and SNPc) together comprise the primate BG complex.

biological facts, this report describes the model and preliminary results of applying the model to published behavioral data from Parkinson's (PD) and Huntington's (HD) subjects on a standard cognitive test (the Wisconsin card sorting task (WCST)).

1.1. The basal ganglia (BG) complex

The BG complex consists of several component structures; the *striatum*, the *subthalamic nucleus* (*STN*), the *globus pallidus internal* (*GPi*) *and globus pallidus external* (*GPe*) segments, and *the substantia nigra* (*par compacta* (*SNPc*) *and par reticula* (*SNpr*). Fig. 1 shows the various components and their interconnections. The description below is necessarily brief and the reader is referred to [1,4–6] for more details.

The striatum: receives excitatory input from many/all cortical regions (including ITP, PP, PFC, PMC, MC), from the midline and intralaminar nuclei of the thalamus, and from the limbic system (amygdala and hippocampus). Ninety percent of striatal cells are GABAergic medium spiny cells. The striatum provides phasic inhibitory output to other basal ganglia structures (GPe, GPi, SNr). Striatal cells have bistable states. In the tonic (steady) state, striatal cells do not respond to low levels of input (down state). In the presence of substantial, coordinated excitation, these cells can provide substantial output to low levels of additional input (up state). Dopaminergic transmission in the striatum is mediated by D1 (+) and D2 (−) receptors. While the striatum has in the order of 110 million cells, downstream structures including output structures of the BG complex have around 100,000 cells. Thus the *fan-in/fan-out ratio in the striatum is* > 1000, suggesting a fairly substantial data compression in the BG complex.

The subthalamic nucleus (*STN*): is the major source of excitation in the BG. Receives projections from cortical regions as well as from the external segment of the globus palladus (GPe).

The globus palladus (*GP*): consists of two segments; the external segment (GPe) is an internal segment (GPi) participating with a loop with the STN. The internal segment is one of the two principal output nuclei of the BG with GABAergic (−) projections to functionally related areas in the ventral thalamus.

The subtantianigra (*SN*): has two component sub-structures, the SubtantiaNigra-ParsCompacta (SNPc) which provides dopaminergic input to the Straitum (+ or − depending on receptor type) and the substantianigraparsreticula (SNPr), which are the other (along with GPi) major output nuclei of the BG.

1.2. Multiple segregated cortico-basal-thalamic loops

A significant neuroanatomic fact about the primate BG complex is their involvement in five parallel cortical-thalamic loops namely two motor loops (skeletomotor and occulomotor), and three non-motor loops. The non-motor loops include a *dorso-lateral PFC loop*, a *lateral orbito-frontal loop* and an *anterior cingulate loop*. The *dorso-lateral pre-frontal loop* (DLPFC (area 9 and 10)-BG (caudate nucleus-Gpi-SNPr)-thalamus (VA,MD)-DLPFC) is believed to be involved in executive functions, planning and working memory. The *lateral orbitofrontal loop* (LPFC-BG (VM caudate-GPi-SnPr)-thalamus-LPFC) mediates empathetic and socially appropriate responses. Damage to this loop produces irritability, lack of empathy and is implicated in obsessive compulsive disorder (OCD). The *anterior cingulate loop* (Ant.Cing.Gyrus-BG (ventral striatum-rostromedial palladium-SNPr)-thalamus (MD thalamic nucleus)-ACC) is believed to be involved in communicating reinforcement signals from VT and SNpc. Damage to this loop may result in akinetic mutism (profound defects in movement initiation).

1.3. Model

Many recent models have identified the role of the basal ganglia as one of selective disinhibition [1,2,4,7]. [1] These models assume that cortical afferents are integrated by the striatum (the main input nuclei of the basal ganglia). Cells from the striatum project topographically through inhibitory connections to the palladal output nuclei of the BG. When a striatal match is found (where all the appropriate input cortical areas are active), The tonic activity of the appropriate palladal cells is suppressed through increased inhibition. This suppression in turn disinhibits the thalamus and through it the appropriate prefrontal region (through segregated and topographic projections). Thus, the main role of the basal-ganglia in these models is one of integrating inputs from multiple cortical areas through conjunctive cortico-striatal connections and then selectively disinhibiting particular areas in the prefrontal cortex based on the result of the match.

Our model of the basal ganglia differs from previous models in that it incorporates the specific connectivity and dynamics of the various basal ganglia modules. Specifi-

[1] The connection between this model and the other primary role of the BG complex in reinforcement learning is outside the scope of this paper.

cally, the model incorporates multiple pathways within the basal ganglia including (a) the direct pathway (Sriatum-Sbstantia Nra pars Rticula (SNPr)-Gobus Palladus interna Gpi, (b) the indirect pathway (striatum-globus pallidus externa (GPe) and SNPr/GPi), (c) the internal loop between GPe and the STN, and (d) direct excitation of the STN. In the model, SNr/GPi nuclei are tonically active and inhibit the thalamus. The projections from the SNr/GPi to the thalamus are topographically organized. When active, the direct pathway results in inhibiting the basal ganglia output nuclei and thus disinhibiting the thalamus and the appropriate cortical column. The indirect pathway modifies this behavior by exciting the output nuclei and inhibiting the thalamus and cortical outputs. In our model, the direct pathway is involved in gating (through selective disinhibition) while the indirect pathway and other circuits regulate the selection process by controlling the amount of activation to scale to contextual factors controlled by the afferent cortical activity (similar to [4]). Our model is consistent with known disease pathologies (such as decreased amplitude movements or tremors) in certain PD patients. The model makes specific predictions about the role of different pathways and circuits within the basal-ganglia and potentially offers computational explanations for the mysterious success of certain interventions (such as deep brain stimulation of the STN [6] for PD).

The basic assumptions behind the model are:

(1) When excited the striatal input to the output nuclei results in promoting specific action structures through *disinhibition* while *maintaining inhibitory control* over others. The direct and indirect pathway work through a "*push-pull*" model of control. Specific disease pathologies may contribute to degradation of the push or pull aspects depending on whether they impact the direct of indirect pathways.

(2) Projections in the parallel cortico-basal-thalamic loops are topographically organized.

(3) The role of the BG complex in cognitive control is analogous to its role in motor control. The basic operation is *selectively disinhibit specific cognitive synergies* (*analogous to their role in selectively disinhibiting specific motor synergies*).

1.4. Model implementation

Our model of the BG complex is implemented on a simulation framework based on Stochastic Petri Nets. A Petri Net is a bipartite graph containing *places* (drawn as circles) and *transitions* (drawn as rectangles). Places hold *tokens* and represent predicates about the world state or internal state. Transitions are the active component. Transitions spread activation by firing (based on a stochastic firing function). The most relevant features of Petri Nets for our purposes are their ability to model events and states in a distributed system and to cleanly capture sequentiality, topographically structured connections, inhibition, concurrency and event-based asynchronous control. Fig. 2 shows the basic computational primitive (node in the network) of the BG model.

Our extensions to the basic Petri Net formalism include typed arcs, hierarchical control, durative transitions, parameterization, typed (individual) tokens. For this paper, the crucial fact about our representation is the ability of SPNs to capture structured connections, complex temporal dynamics. Using this basic model, we implement the various

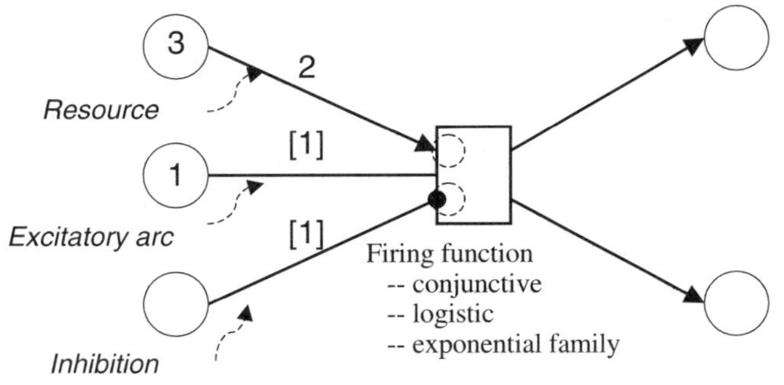

Fig. 2. The Basic Model of collections of BG cells. The input sites may be excitatory, inhibitory or weighted (the resource link). The firing of cells can be conjunctive (there is excitation in all the excitatory cites, no inhibition and sufficient activation in the resource arcs, then the cell fires). Weights on the output links (default 1) represent the *Gain* parameter. Different regions in the BG are modeled by different types of nodes in the model. The striatal cells in the model are conjunctive matching cells. Other excitatory nuclei (such as STN) use the logistic firing function used in many neural network models. Other inhibitory cells (such as the output nuclei) use other firing functions based on the exponential family of functions. Also, different deficits of specific regions in the BG are modeled changing the gain, firing function and weights of the relevant region in the model.

components and connections of the BG in our computational framework. Specific components use specific firing functions (striatal cells use the conjunctive firing function, STN cells use the logistic firing function and other nuclei use negative exponential firing functions (all the functions are from the exponential family of functions).

2. Results

We applied the model to data from normal, PD, HP and schizophrenic patients on the WCST to compare our model predictions to previous results [1]. Since we were mainly interested in the role played by the basal ganglia, out model of the prefrontal cortex is similar in many respects to [1].

In the WCST, subjects sort a deck of 128 cards according to specific criteria. Cards vary by color (4 values), form (4 values), Number (4 values) (ex. red triangle 2). The subject places each of the 128 stimulus cards in front of a target card and gets a correct/incorrect response from the experimenter. Once the subject has made 10 correct responses a category is achieved and the rule (sort according to one of the features) is arbitrarily changed. The test is completed when 6 categories are achieved or all 128 cards are used.

Many results show patients with frontal damage including schizophrenics perform poorly on the WCST. Patients with *Huntington's disease* (*HD*) which affects striatal cells show impaired performance. Patients with *Parkinson's disease* (*PD*) which affect the substantia nigra have been shown to exhibit impaired performance. Normals tend to do well, almost all subjects complete the six categories. Schizophrenic patients exhibit perseverative errors. PD patients without dementia show erratic/random errors.

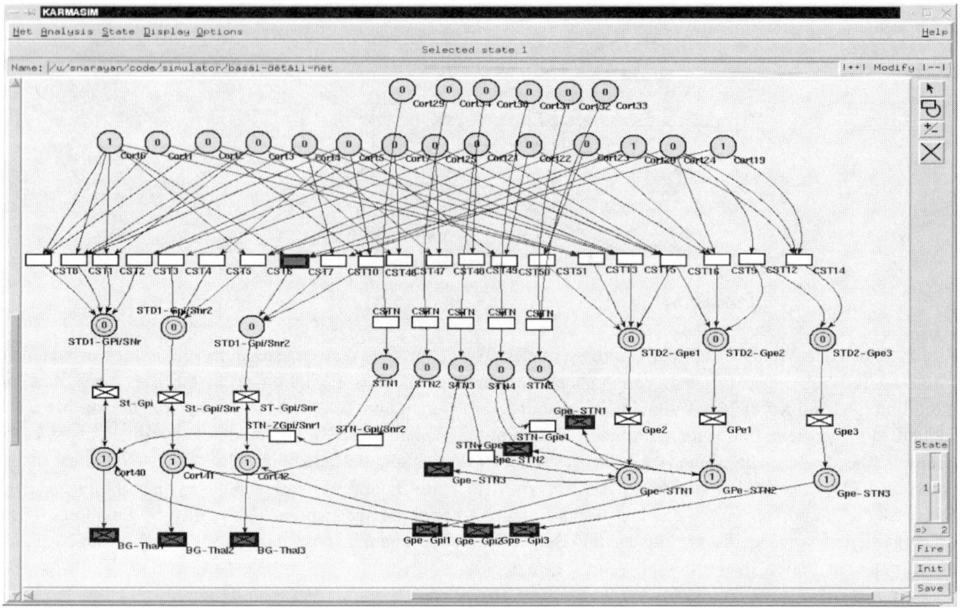

Fig. 3. Implementation and computational simulation of the BG complex. The figure shows part of the Basal Ganglia model. The striatal match cells are implemented as conjunctive transitions (shown as CorticalStriatal (CS) transitions in the Stochastic Petri Net based model). The striatal cells are bi-stable, and when in the up state can inhibit the output nuclei (the St-Gpi transitions) with input. This is the direct pathway which results in disinhibiting the thalamus (through the BG-thal transitions in the model. The inhibitory effect of the indirect pathway can be seen through the GPe cells. Other pathways include the direct stimulation of STN and the STN-GPe loops.

PD patients with dementia show both random and perseverative errors. HD patients show both perseverative and random errors.

2.1. Baseline model

Most computational models of the WCST are concerned with PFC details and deficits. Some models attempt to include the PFC-BG-Thalamus-PFC loop. Amos [1] surveys the state-of-the-art and introduces a model that explicitly deals with the published behavioral data on PD, HD and Schizophrenic patients on the WCST task. Amos describes and implements a model for the cortical–basal–thalamic interaction for the WCST test. The model of the BG complex in this work is relatively simple and considers *only the direct pathway*. Our initial results reported in this paper use the Amos model as a baseline. All parts of the model except for the BG are kept similar to the baseline. We replaced the BG part of the baseline with the more biologically realistic model implemented as shown in Fig. 3. We then simulated the various deficits (HD and PD) in our model and compared it both to the behavioral data and to previous modeling results.

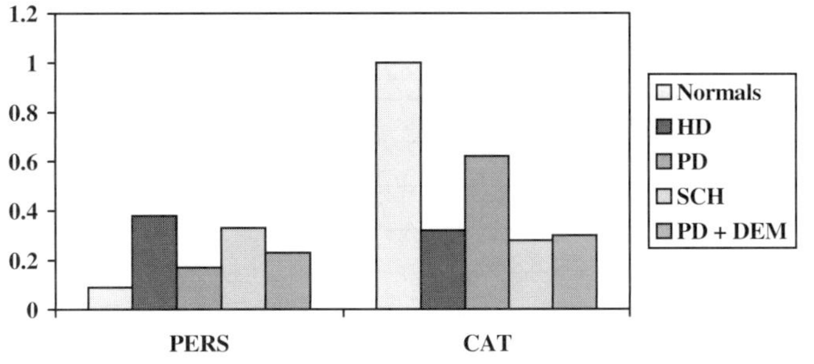

Fig. 4. The model's performance of the different patient data. The PERS label in the chart shows the relative performance of different populations in terms of the perseverative errors. Notice that the model predicts high percentage of perseverative errors for the HD and Schizophrenic patients compared to the normals. Notice also that the normals perform quite well in that the get all the categories right (percentage of categories right is shown by the value corresponding to the label CAT. Notice as well that the model performs badly for all other patient populations.

Fig. 4 shows the results made by our model based on specific deficits in the patient populations. PD was modeled as a decreased gain (see Fig. 2) of the indirect pathway and the SN regions, HD as noisy striatal matching coupled with decreased gain in the direct pathway, and Schizophrenia by decreasing the frontal bias as well as decreasing the output gains of the Substantia Nigra cells. Fig. 5a and b show the model predictions compared to the empirical data on the different patient populations (from data surveyed in [1]).

Our model showed results that suggested increased perseveration (with respect to card sorting strategy) for schizophrenic patients, erratic shifting of strategies for PD patients and a combination of errors for HD patients. While the overall results are compatible with previous efforts, our model predicts that the role of the basal ganglia is far more intricate than previously described. Our results showed that the complex dynamics both from the modifying influence of the indirect pathway and the projection back from the thalamus on the probability of shifting or maintaining the current strategy captured much finer grained detail than previous work. In particular, our model predicts greater erratic errors for PD patients due to a combination of inappropriate matching at the striatum (low dopamine levels decrease the gain and hence the efficacy of a clear winner match) as well as enhanced inhibition provided by the indirect pathway. HD patients, in contrast our model also makes empirical predictions that match experimental evidence on different stages of PD [6]. None of the previous models are able to make such detailed predictions. Interestingly, degradation of the Substantia Nigra regions in our model in combination with the frontal deficits produces a better match to the data of schizophrenic patient populations (compared to previous purely frontal lobe models). This is an intriguing finding of our work which we are subjecting to more detailed analysis.

Results from our model suggest that the basal ganglia may be involved in problem solving and planning that have complex dynamics and require prediction or simulation.

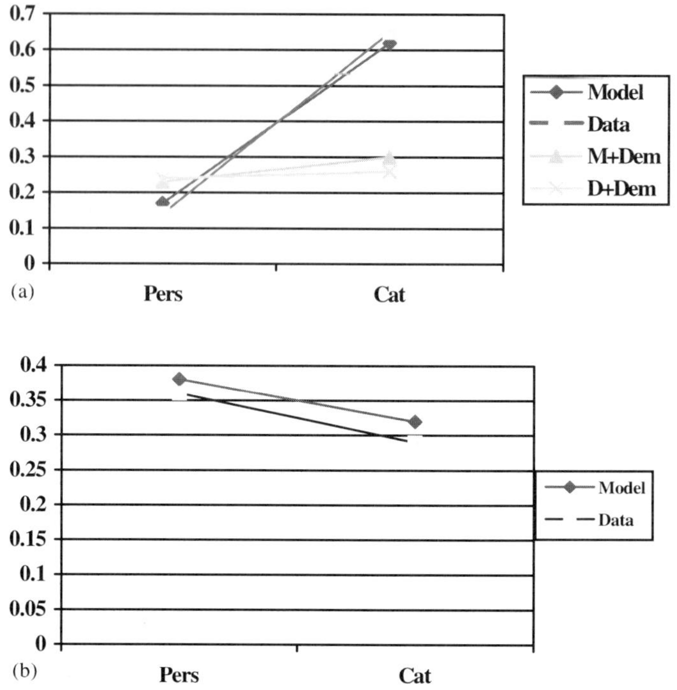

Fig. 5. (a) Performance of Model (M) compared to published data (from [1]) on the *PD* patient population. The chart shows the comparison of the model predictions with the data for patients with and without dementia. (b) Performance of Model (M) compared to published data (from [1]) on the *HD* patient population. The chart shows the comparison of the model predictions with the data for patients with and without dementia.

To better understand the implications and evaluate our model, we are continuing efforts in the following directions: (1) Design cognitive tests for which our models of planning, working memory and executive control are likely to predict non-obvious results. (2) Apply these tests on subjects with and without diseases affecting relevant brain regions (PD,HD) and evaluate the model with respect to the results.

2.2. Comparison to the baseline

The more detailed model described here is able to explain HD and PD data fitting the actual quantitative data better (where available). The model is able to explain differences in different stages of PD (something the baseline model is unable to do and to incorporate feedback from Thalamic input to the cortex, so makes finer-grained predictions about perseveration. Crucially, the model makes different predictions about the possible deficits in Schizophrenic patients, suggesting (unlike the baseline) that BG deficits may play a role in their cognitive performance. Some evidence that supports our model comes from [7]. After surveying the literature on the clinical symptoms of a variety of types of Schizophrenic patients (both negative and positive, and unmedicated/medicated) Middleton and Strick [5] conclude:

> We propose that dysfunction of SNPr can lead to all of these symptoms. At the very least, we believe that abnormal activity in the nigral output channels participates in the symptoms associated with Schizophrenia.

3. Discussion

Cortico–basal–thalamic loops are increasingly being implicated in a variety of seemingly unconnected cognitive (as opposed to motor) disorders including obsessive compulsive disorder, Tourette's syndrome, Autism, Schizophrenia, attention deficit disorder (ADD), and even certain forms of depression. Our model indicates that

(1) The Basal Ganglia are part of a execution/simulation loop, where the next motor/cognitive action is selected based on *striatal matching of context* and the push-pull interaction of the two main pathways. The result is a *selective disinhibition* of the thalamus and cortical regions (both motor and non-motor depending on the specific loop involved).

(2) *Thought selection and updating* uses the *same sub-cortical structure* (*different loops*) *and mechanism* as action selection and updating.

(3) Detailed computational modeling can explicate this hypothesis and make specific and testable predictions.

The work described in this paper, while preliminary, suggests the utility of detailed modeling. We are addressing several shortcomings in the model as well as trying to identify, acquire data and test our model on cognitive tasks that have a greater temporal aspect such as planning or language understanding. Modeling such tasks, we believe is the next step to understanding the role of Cortico-subcortical interactions in cognition.

References

[1] A. Amos, A computational model of information processing in the frontal cortex and basal ganglia, J. Cognitive Neurosci. 12–3 (2000) 505–519.

[2] M. Frank, B. Loughry, R. O'Reilly, Interactions between frontal cortex and basal ganglia in working memory: a computational model, Institute for Cognitive Science, TR 00-01, University of Colorado, Boulder, 2000.

[3] P.S. Goldman-Rakik, Circuitry of Primate Prefrontal Cortex and the Regulation of Behavior by Representational Memory, Handbook of Physiology—The nervous system, Vol. 5, American Physiological Society, Bethesda, MD, 1987, pp. 373–417.

[4] K. Gurney, T.J. Prescott, P. Redgrave, A computational model of action selection in the basal ganglia. I. A new functional anatomy, Biol. Cybern. 84 (2001) 410–410.

[5] F.A. Middleton, P.L. Strick, Basal ganglia output and cognition: evidence from anatomical, behavioral, and clinical studies, Brain Cognition 42 (2000) 183–200.

[6] A.M. Owen, J.L. Iddon, J.R. Hodges, T.W. Robbins, Spatial and non-spatial working memory at different stages of Parkinson's disease, Neuropsychologia 35 (1997) 519–532.

[7] T.G. Taylor, N.R. Taylor, Analysis of recurrent cortico-basal ganglia-thalamic loops for working memory, Biol. Cybern. 82 (2000) 415–432.

Srini Narayanan is a Senior Research Scientist at ICSI, Berkeley where he works on neural models of cognition. He obtained his Ph.D. from the University of California, Berkeley in 1998. As part of his dissertation, Dr. Narayanan implemented a neural model which made specific predictions on the role of motor control representations in abstract thought including reasoning about metaphoric discourse. As part of an interdisciplinary group at ICSI/UC Berkeley, Srini Narayanan (http://www.icsi.berkeley.edu/~snarayan) has continuing collaborations with Experimental and Clinical Neuroscientists as well as Computer Scientists and Linguists testing and refining the model and building new computational models exploring the Neural Theory of Language and Thought (NTL) (http://www.icsi.berkeley.edu/NTL).

Available at
www.ComputerScienceWeb.com
POWERED BY SCIENCE @ DIRECT®

ELSEVIER

Neurocomputing 52–54 (2003) 615–620

NEUROCOMPUTING

www.elsevier.com/locate/neucom

An engineering model of the masking for the noise-robust speech recognition

Ki-Young Park[a,b], Soo-Young Lee[a,b,c,*]

[a] *Brain Science Research Center, Korea Advanced Institute of Science and Technology, Taejeon 305-701, South Korea*

[b] *Department of Electrical Engineering, Korea Advanced Institute of Science and Technology, Taejeon 305-701, South Korea*

[c] *Department of Bio-Systems, Korea Advanced Institute of Science and Technology, Taejeon, South Korea*

Abstract

The masking effect of human hearing is modeled by lateral and unilateral inhibition, and tested for isolated word recognition tasks. Frequency masking suppresses unwanted signals close to the dominant signal of interest in frequency domain, and the weak signals following dominant ones in the time are suppressed by temporal forward masking. The masking effect filters out unimportant signals, which may improve the performance of speech recognition systems. With the parameters derived from the psychological observations, proposed model shows good analogy to psychoacoustic masking effects as well as superior recognition performance.

Keywords: Masking; Lateral inhibition; Automatic Speech recognition

1. Introduction

Although tremendous attempts have been made to make machines which recognizes the human speeches, it is still very difficult for real world environments with the background noises. The modeling of human auditory system is one of the successful approaches which have been proposed to solve the problem. Critical-band filterbank and mel-scale frequency sampling are the most widely used properties from the psychoacoustics.

* Corresponding author. Department of Bio-Systems, Korea Advanced Institute of Science and Technology, Taejeon 305-701, South Korea.

E-mail address: sylee@ee.kaist.ac.kr (S.-Y. Lee).

0925-2312/03/$ - see front matter © 2003 Elsevier Science B.V. All rights reserved.
doi:10.1016/S0925-2312(02)00791-9

One of the useful aspects which had been seldom utilized is the "masking" effect of human hearing system. Masking has been investigated for ages and used to quantify the frequency selectivity of the auditory system. However, there have been few approaches which utilize the masking for the recognition tasks. The nature of the masking, that the spectral components of high intensity level suppress the adjacent spectral components, helps the recognition performance. The low-intensity level signals are usually unimportant for speech recognition and may be noises to be suppressed. In this paper, the time–frequency masking is modeled with lateral inhibition and incorporated into the current auditory model, mel-frequency cepstral coefficients (MFCC) model, which is the most widely used speech features, and tested on the task of isolated word recognition. The proposed algorithm does not require extensive computation, and results in much better recognition performance, especially in noisy environments.

Masking has been defined as a process in which the audible threshold for one sound is raised by the presence of another (masking) sound. Frequency masking means that signals are masked by the masking sound occurring at the same time. With temporal masking, signals can also be masked by the sound preceding it, called forward masking, or even by the sound following it, called backward masking. Frequency masking helps to discriminate signals from the other by enhancing the spectral resolution. Also by suppressing the adjacent signal in spectral domain, it reinforces the signal of critical interest so that the unimportant signals are filtered out. Forward masking, a short-term adaptation process of the auditory system, helps in the discrimination capability between signals by emphasizing time-dependent variations.

2. Frequency masking with lateral inhibition model

The essence of the masking is to reinforce the dominant signal components and to suppress the adjacent components. To implement this concept into the current auditory system, a lateral inhibition is introduced with a simple Mexican-hat convolutional filter as shown in Fig. 1(a). The sharp peak at the center reinforces the very close stimuli, and the negative values at neighborhoods inhibit the stimuli in the range.

To apply the inhibition filtering in spectral domain, the blocked speech signals are first transformed into frequency information using discrete Fourier transforms, and the inhibition is applied between adjacent frequency components. It can be easily incorporated into the popular MFCC speech features as shown in Fig. 1(c).

The Mexican-hat filter is modeled by the difference of Gaussian:

$$w(n) = A\exp(-n^2/\sigma_1^2) - B\exp(-n^2/\sigma_2^2), \quad n = -L,\ldots,0,\ldots,L, \tag{1}$$

where $2L + 1$ is the length of the masking filter. The most crucial problem is to determine the shape of the filter, i.e., the parameters of A, B, σ_1 and σ_2. By applying two constraints, i.e., zero mean and the maximum value of one, one can shape the filter for the remaining two physical parameters of the excitatory range R and inhibition

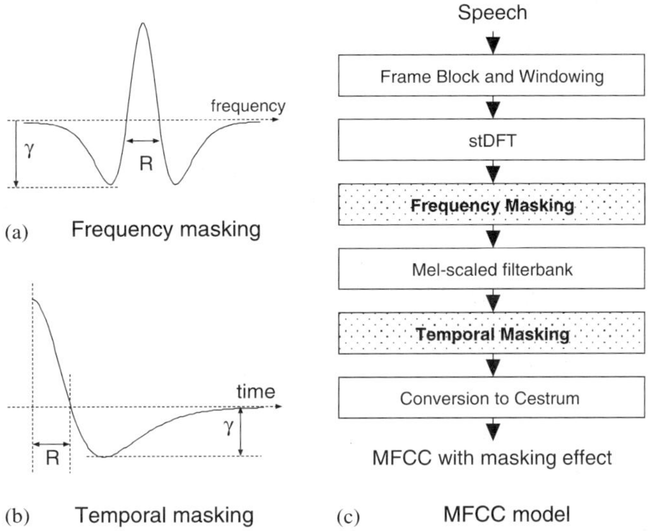

Fig. 1. The filters for the frequency (a) and temporal (b) masking and the MFCC model with the proposed masking mechanism (c).

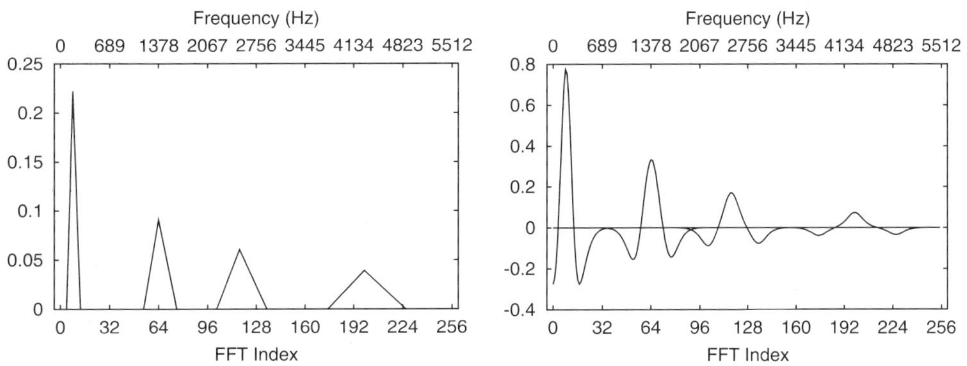

Fig. 2. Mel-scale filterbank with masking effect (right) compared to the original mel-scale triangular filterbank (left).

gain γ. The masked spectral magnitude is computed as

$$m(k) = \sum_{n=-L}^{L} w(k-n)S(k), \qquad k = 0, 1, \ldots, N-1, \tag{2}$$

where $S(k)$ is the magnitude of the FFT output of the blocked speech signal and N is the number of FFT points. The mel-scale filterbank in the MFCC is originated from the critical band analysis. Since the implementation of the filterbank can be regarded as the weighted sum of the frequency components, the two process can be combined into one as shown in Fig. 2. The figure shows that the Mexican-hat filter in frequency domain results in the modification of the mel-scaled filterbank. The output value of each critical band is inhibited by those of adjacent bands, and the inhibition range increases nonlinearly as the center frequency increases. These are consistent with psychoacoustic observations.

3. Temporal masking

The short-term adaptation and the temporal integration [4] are the possible mechanisms of the temporal masking. Many researches have modeled the temporal masking as the temporal integration of the response of the auditory nerve [1,5]. If we assume that the temporal masking is due to the temporal integration of the response of the auditory nerves, temporal masking can be modeled as

$$y(n) = x(n) + A\sum_{k=1}^{\infty} \alpha^{-k}x(n-k) - B\sum_{k=1}^{\infty} \beta^{-k}x(n-k), \tag{3}$$

where $x(n)$ is the output signal before temporal masking, $y(n)$ the output of the temporal masking, α and β are time constants of integration. The second term on the right side of Eq. (3) denotes the slow response of the neurons, in other words, the responses of the neurons could not catch up the changes of the stimuli, and the previous inputs are accumulated to affect the current output. The third term indicates the masking integration. The current response of the neurons are suppressed by the preceding signal of high intensity.

The time constants of two integration terms, α and β, represent the extent to which the previous neural signals affects the current signals. It is known that α is much greater than β. Typically, α is a few tens of milliseconds and β a few hundreds of milliseconds. The gain factor A and B imply the amount of accumulation and masking, respectively. Eq. (3) is represented in z domain as

$$H(z) = \frac{1 - [(1-A)\alpha + (1+B)\beta]z^{-1} + (1-A+B)\alpha\beta z^{-2}}{1 - (\alpha+\beta)z^{-1} + \alpha\beta z^{-2}}. \tag{4}$$

Again for the preliminary experiments α is set to 0.72 which corresponds to 30 ms, and β is set to 0.97 corresponding to 300 ms. A and B are set to comply with the constraints of zero-mean and the maximum value of one as in the case of frequency masking. Fig. 3(a) and (b) show the impulse response of the time integration filter of Eq. (4) with the parameters given above. From the frequency response of the model, it is shown that the proposed model with the parameters given actually do the role of bandpass filtering which cuts off the signal components which vary too fast or too slow. Then the model can be regarded as the generalized version of RASTA model proposed by Hermansky and Morgan, where they used the bandpass filtering method in feature domain to enhance the recognition performance [3]. They also showed that the temporal filtering of the sequence of feature vectors could give the masking effect [2]. The temporal masking filter is applied at the logarithm filterbank output. Fig. 1(c)

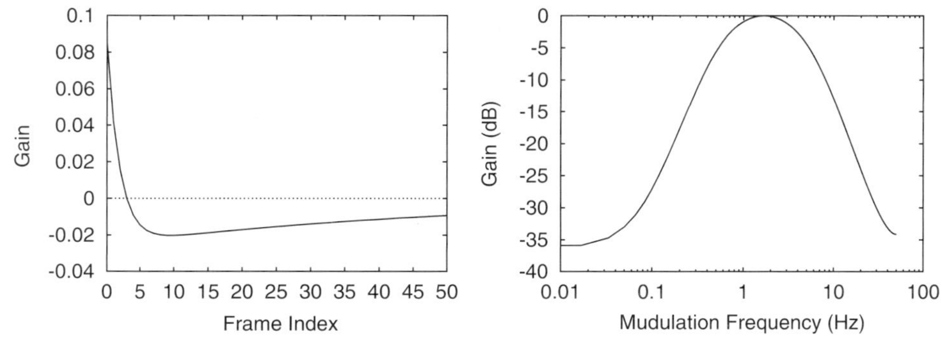

(a) Impulse response. (b) Frequency response.

Fig. 3. Characteristics of the filter used for the integration model of temporal masking.

shows the modified MFCC model including both the temporal masking stage (upper shaded box) and the frequency masking (lower shaded box).

4. Experimental results

The temporal masking using the integration model in the feature domain, as well as the frequency masking, is applied to the isolated word recognition task. For the simulation, 50 Korean words spoken by 13 men three times each were tested by nearest-neighbor classifier for simplicity. Fig. 4 compares the false recognition rates of the baseline without any masking, with the frequency masking, and with both the frequency and the temporal masking. Frequency masking reduces the misclassification rate greatly and the temporal masking reduces the error rate even further. We found the optimal shape for the recognition task with the exhaustive calculation by varying two control variables, R and γ. The simulation showed that the value of excitatory range R between 60 and 100 Hz and the value of inhibition gain γ between -0.4 and -0.2 gave much better performance than the baseline system for frequency masking. For temporal masking, the values of R between 40 and 100 ms and γ between -0.05 and -0.15 gave the consistent improvements on the recognition performance. We also found that parameters deviated from the optimal values gave much better performance than the baseline system. Therefore, broad ranges of parameter values could be used.

The values of parameters were found by the brute-force search, and after many trials of simulations, we found that parameters that deviated from the optimal values also gave much better performance than the baseline system which meant broad ranges of values for each parameters could be for practical use.

5. Conclusions and discussions

In this paper, the psychoacoustical phenomena of masking is modeled by simple convolutional filtering in both spectral and time domain. The concepts of lateral

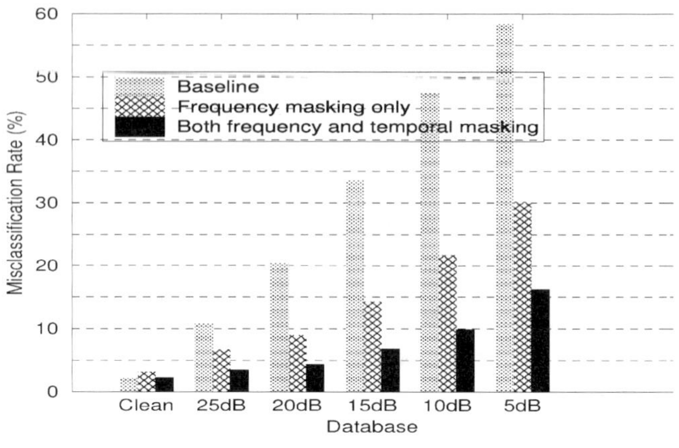

Fig. 4. False recognition rates of the baseline system without any masking, with the frequency masking only and both the temporal and frequency masking.

inhibition in spectral domain and of unilateral inhibition in time domain model the frequency masking and temporal masking, respectively. Proposed model results in efficient features for speech recognition and provides much better performance than popular MFCC features especially in noisy environments.

Acknowledgements

This research was supported by Korean Ministry of Science and Technology as Brain Neuroinformatics Research Program.

References

[1] T. Dau, D. Püschel, A quantitative model of the "effective" signal processing in the auditory system. I. Model structure, J. Acoust. Soc. Am. 99 (6) (1996) 3615–3631.

[2] H. Hermansky, Should recognizers have ears? Speech Commun. 25 (1) (1998) 3–27.

[3] H. Hermansky, N. Morgan, RASTA processing of speech, IEEE Trans. Speech Audio Process. 2 (4) (1994) 587–589.

[4] A.J. Oxenham, Forward masking: adaptation or integration? J. Acoust. Soc. Am. 109 (2) (2001) 732–741.

[5] B. Strope, A. Alwan, A model of dynamic auditory perception and its application to rebust word recognition, IEEE Trans. Speech Audio Process. 5 (5) (1997) 451–464.

Available at
www.ComputerScienceWeb.com
POWERED BY SCIENCE DIRECT®

ELSEVIER

Neurocomputing 52–54 (2003) 621 – 629

NEUROCOMPUTING

www.elsevier.com/locate/neucom

Modeling the spinal cord neural circuitry controlling cat hindlimb movement during locomotion

D.G. Ivashko[a], B.I. Prilutsky[b], S.N. Markin[a], J.K. Chapin[c], I.A. Rybak[a,*]

[a]*School of Biomedical Engineering, Science and Health Systems, Drexel University, Philadelphia, PA 19104, USA*

[b]*Center for Human Movement Studies, Georgia Institute of Technology, Atlanta, GA 30332, USA*

[c]*Department of Physiology and Pharmacology, SUNY Health Science Center, Brooklyn, NY 11203, USA*

Abstract

We have developed a computational model of the spinal cord neural circuitry that controls locomotor movements of simulated cat hindlimbs. The neural circuitry includes two central pattern generators integrated with reflex circuits. All neurons were modeled in the Hodgkin–Huxley style. The musculoskeletal system includes two three-joint hindlimbs and the trunk. Each hindlimb is actuated by nine one- and two-joint muscles (a Hill-type model). Our simulations allow us to suggest a specific network architecture in the spinal cord and a pattern of feedback connectivities (from Ia and Ib fibers and touch sensors) that provide stable locomotion and realistic patterns of muscle activation and kinematics of limb movements.

Keywords: Spinal cord; Neural control of locomotion; Locomotory CPG; Cat hindlimbs; Musculoskeletal model

1. Introduction

The central nervous system controls locomotion and other automatic movements in a hierarchical fashion. The lower-level controller in the spinal cord generates the

* Corresponding author.
E-mail address: rybak@cbis.ece.drexel.edu (I.A. Rybak).

0925-2312/03/$ - see front matter © 2002 Elsevier Science B.V. All rights reserved.
doi:10.1016/S0925-2312(02)00832-9

motor program for the neuromuscular apparatus. This low-level controller interacts with proprioceptive feedback and receives descending signals from the higher-level (supra-spinal) centers. The higher centers, in turn, may select and initiate the appropriate motor programs from the spinal cord's repertoire. The descending commands from supra-spinal centers to spinal interneurons are automatically integrated into the current state of proprioceptive and exteroceptive information [8].

The neuronal circuits in the mammalian spinal cord can generate rhythmic motor patterns that drive locomotor movements even in the absence of descending inputs from higher brain centers and sensory feedback [3,6]. This supports the concept of the *central pattern generator* (CPG), which presumably is located in the spinal cord and generates a basic locomotor rhythm (for review see [5]). According to the contemporary biological view, the CPG is a complex, distributed network of interneurons in the spinal cord integrated into the system of multiple reflex circuits [8]. The basic locomotor pattern, generated by the CPG, provides a coordinated activation of various muscles, which in turn control and coordinate joint movement within and between the limbs. Therefore, locomotion results from a complex interplay between the CPG, reflex circuits and multiple feedback and feedforward modulatory signals. The proprioceptive signals modulate the locomotor rhythm and pattern by providing necessary correction of the locomotor rhythm and pattern to maintain the walking animal in a proper relationship to the environment [14]. They regulate the timing of phase transitions and reinforce the generation of motoneuronal activity during ongoing phases of locomotion [11]. Previous modeling studies have demonstrated that stable and adaptive locomotion involves a global entrainment of the musculoskeletal system to rhythms generated by the CPGs [9,19]. The objective of this work was to develop and analyze a computational model of neural control of locomotion at the spinal cord level using realistic models of the network of neurons, muscles, and limb biomechanics.

2. Model

We have developed a systems model that includes realistic models of neurons and neuronal circuits in the spinal cord as well as a realistic model of the musculoskeletal system. Specifically, we focused on modeling the neural control of the movement of cat hindlimbs during locomotion. Each hindlimb was modeled as a system of three rigid segments interconnected by three frictionless joints: hip, knee and ankle (Fig. 1a). Two hindlimbs were connected to the common segment (pelvis) whose other end was connected to the trunk segment (Fig. 1a). The distal end of the trunk was held at the constant vertical distance from the ground to compensate for the lack of forelimbs. Each hindlimb was actuated by nine one- and two-joint muscles (Fig. 1b). The dynamics of muscle contraction was described by the Hill/Zajac model [20] that incorporated force–length–velocity properties of the muscles, muscle geometry, and the properties of the tendon. Geometrical parameters characterizing muscle origin and insertion sites were measured in a representative cat. Physiological muscle parameters (optimal fiber length, maximal isometric force, maximum shortening velocity, angle of pennation, resting tendon length, etc.) were estimated from the literature [2,16–18]. The dynamics of

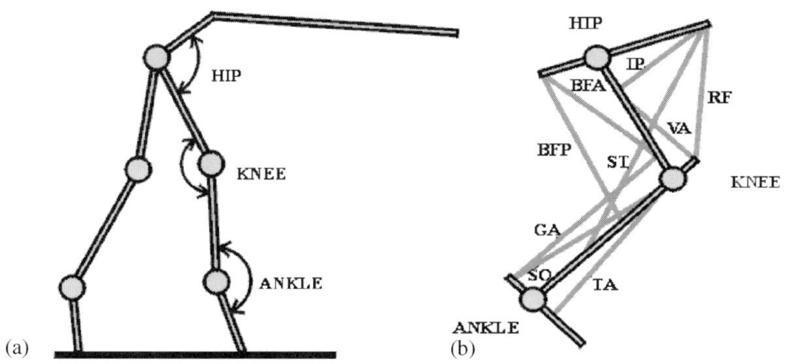

Fig. 1. Schematic of musculoskeletal system: (a) Two three-joint hindlimbs with pelvis and trunk and (b) nine muscles used in the model to control a single hindlimb: IP, iliopsaos; BFA, biceps femoris anterior; RF, rectus femoris; ST, sartorius medial; BFP, biceps femoris posterior; VA, vastii; GA, gastrocnemius; TA, tibialis anterior; and SO, soleus.

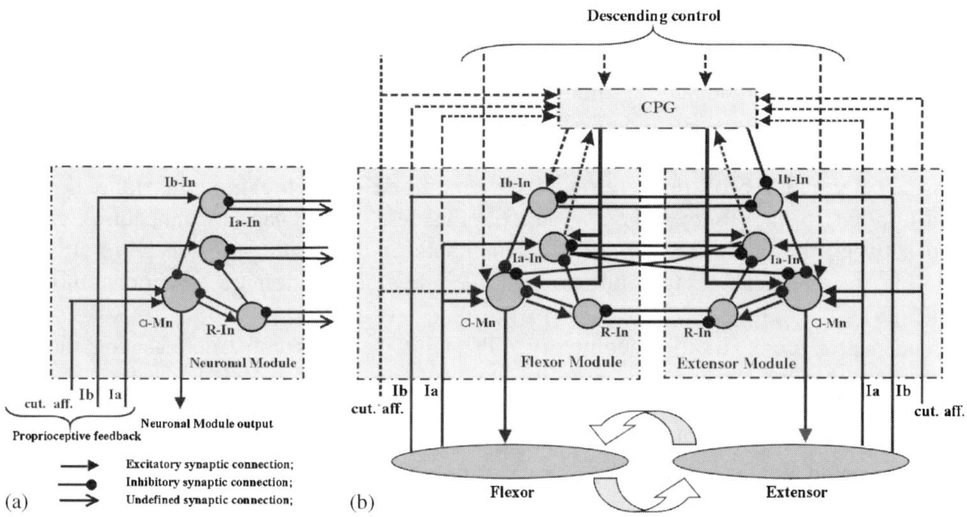

Fig. 2. Integration of CPG and reflex circuits: (a) The NM controlling one muscle and (b) an example of two interconnected NMs, controlling a pair of antagonistic flexor and extensor muscles actuating the same joint.

body (including two legs, pelvis, and trunk) was described by Lagrange equations. The mass-inertia parameters of body segments were calculated using empirical equations by Hoy and Zarnicke [7].

The model of the spinal cord neural circuitry has a modular structure. The model of a single neuronal module (NM) was constructed as a minimal neural network necessary for the formation of basic reflex circuits and their integration with the CPG (Figs. 2a and b). Each NM controls one muscle and contains an α-motoneuron

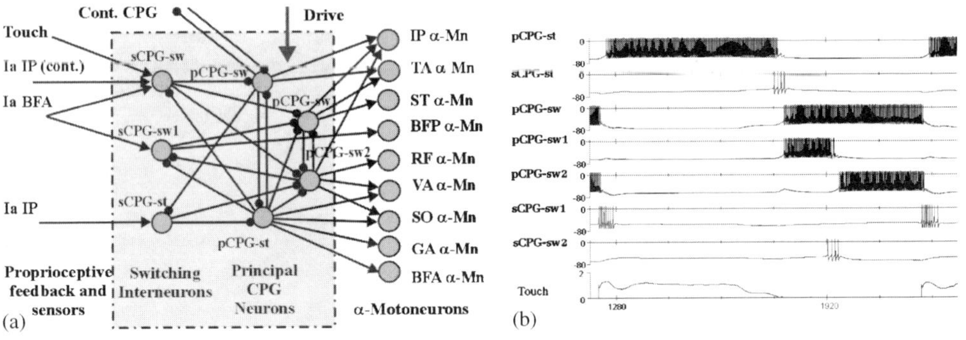

Fig. 3. CPG: (a) Schematic of neural connections and (b) activities of all neurons of the CPG (traces of membrane potentials) during the locomotor cycle. The activity of the touch sensor is shown at the bottom to identify the stance phase of locomotion.

(α-MN), actuating the controlled muscle, and several interneurons, including a Renshaw cell and Ia and Ib interneurons (Fig. 2a). Ia and Ib interneurons were included to mediate proprioceptive feedback from of Ia and Ib afferens, respectively (Fig. 2b). All neurons were modeled in the Hodgkin–Huxley style (single compartment models) and included the fast sodium, delayed-rectifier potassium, transient potassium-A, high-threshold calcium, calcium-dependent potassium, leakage and synaptic ionic channels. The schematic of reflex circuits was modified from the previous models [1,4] and applied to each antagonistic group of muscles. The synaptic connections within and between the NMs and the structure of inputs of Ia and Ib proprioceptive afferents provide the classical stretch reflexes and reciprocal activation of antagonist muscles (Fig. 2b).

The neural model of the locomotory CPG (Fig. 3a) was constructed using the hypothesis that each hindlimb is controlled by one complex CPG [10], which in turn is connected with the CPG controlling another hindlimb via a coordinating neural network. The CPG (for each hindlimb) was incorporated into the spinal cord neural circuitry and integrated with the circuits of spinal reflexes via direct synaptic interconnections and through multiple proprioceptive feedbacks.

Because of the paucity of available information on interneuronal connectivities and function in the mammalian spinal cord, we utilized information obtained from studies of the homologous, but much better understood respiratory CPG in the medulla. Specifically, we assumed that some general architectural principles and particular neural schematics discovered in studies of the respiratory CPG (e.g. those for phase transitions) might be useful and applicable for the construction of the locomotory CPG (see also [10,14]). With respect to the respiratory CPG, both experimental [13] and modeling [15] studies have demonstrated that, in addition to the "principal" CPG elements (whose activity explicitly defines each phase of the cycle), the CPG may contain special "switching" neural elements that fire during phase transitions and, in fact, produce these transitions via inhibition of the corresponding principal CPG elements. Moreover, we suggest that the switching interneurons operate (fire) under control of various proprioceptive and descending control signals and hence significantly contribute to the shaping of the locomotor pattern (timing of phase transitions, shaping motoneuronal firing busts, etc.).

Each CPG in our model contains four principal CPG neurons: pCPG-st, active during the stance phase of locomotion; pCPG-sw, firing during the entire swing phase; pCPG-sw1 and pCPG-sw2 active during the early and late swing phases, respectively (Fig. 3a). The locomotor movement is initiated by applying the "descending" drive to all principal CPG neurons. Activation of each principal neuron in the particular phase of the locomotor cycle initiates and supports the firing activity of the corresponding motoneurons within NMs, which in turn actuates the corresponding muscle group (Figs. 2b and 3a). Activation of the corresponding switching neuron terminates the currently active phase of locomotion and hence produces phase transition. Firing of the switching neurons (and hence the timing of phase transitions) is controlled by proprioceptive feedbacks from the same hindlimb, touch sensors and some contralateral proprioceptive signals (Fig. 3a). The latter, together with the direct neural connections between the CPGs, provide coordination of activities of two CPGs and, therefore, coordination of movements of the hindlimbs (Fig. 3a). The structure and weights of synaptic connections to each switching neurons was manually adjusted to provide stable locomotion. Fig. 3b shows activity of all neurons of one CPG during the locomotor cycle.

Overall, the CPGs in our model operate under regulation of multiple reflex feedbacks. At the same time, they provide control and modulation of reflexes during locomotion. Specifically, co-activation of the Ia interneuron and the α-motoneuron in the same NM by the CPG (Fig. 2b) suppresses the classical stretch reflexes for the antagonist muscles in the corresponding phase of locomotion. Moreover, during the extension phase of locomotion, the CPG inhibits the extensor Ib interneuron and hence breaks the "classical" negative feedback loop of Ib fibers to the extensor α-motoneurons (Fig. 2b). At the same time, Ib feedback provides excitation of the extensor α-motoneuron via the CPG during the extensor phase of locomotion. Therefore, during locomotion the Ib feedback loop to the extensor α-Mn changes from negative to positive, which is consistent with the experimental data [11,12].

3. Model performance: comparison with experimental data

In order to evaluate the model performance, an experiment was conducted on one adult cat (mass 3 kg) (the methods used were consistent with the American Physiological Society Animal Care Guidelines). The cat was trained to walk on a walkway with two embedded miniature force platforms. Two electronically synchronized high-speed video cameras (60 Hz, Peak Performance Technologies) were placed on both sides of the walkway with their optical axes aligned perpendicular to the plane of progression. The cameras recorded 2D displacements of 28 reflective markers on both sides of the cat's body. After the experiment, marker coordinates were digitized and the kinematics and kinetics of cat walking were calculated. A comparison between locomotor variables generated by the model and those recorded experimentally

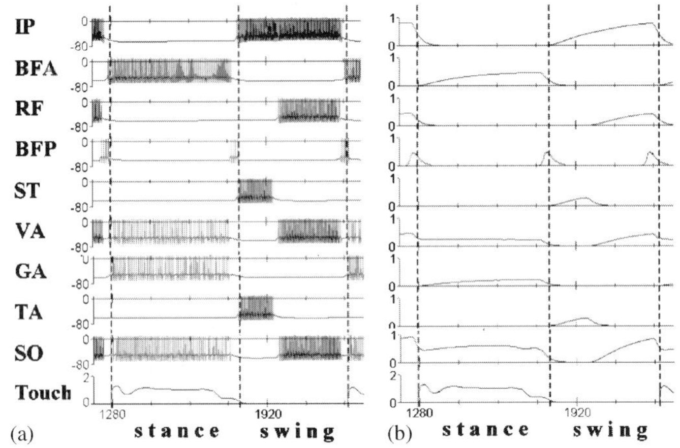

Fig. 4. Activation of α-motoneurons (a) and integrated EMG (b) of the corresponding muscles during a cycle of walking generated by the model. Touch is a feedback signal from the touch receptor which is proportional to the vertical component of the ground reaction force. (The rest of abbreviations are similar to Fig. 1.)

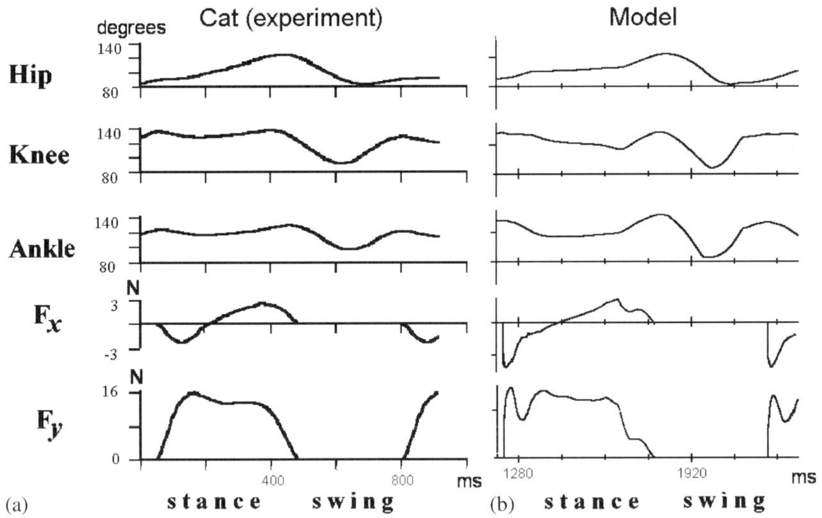

Fig. 5. Comparison between patterns of joint angles and ground reaction forces recorded experimentally (a) and generated by the model (b). F_x and F_y are the anterior–posterior and vertical components of the ground reaction force, respectively.

demonstrated a qualitative agreement. The developed model was able to provide the control of stable locomotor movement of the hindlimbs. Fig. 4 shows the dynamics of activity of α-motoneurons providing activation of muscles controlling one hindlimb and the corresponding electromyographic activity (EMGs). The latter was similar to

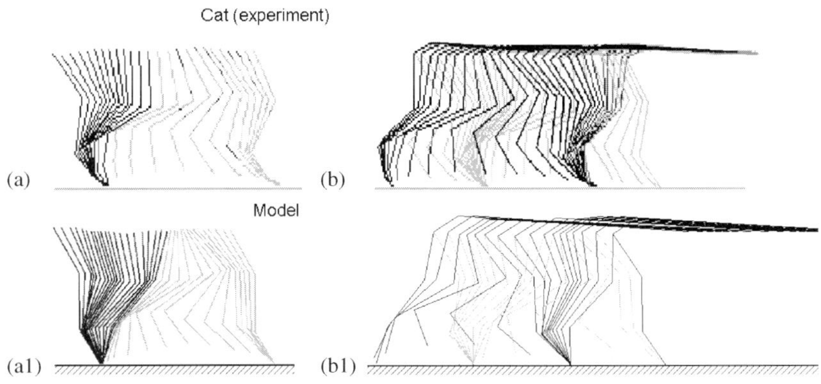

Fig. 6. Stick-diagrams of one leg (a and a1) and two legs and the trunk (b and b1) during walking cycle recorded experimentally in one cat (a and b) and generated by the model (a1 and b1).

the patterns described in the literature [10] with few exceptions (activation of IP and RF was shifted in the cycle and SO became active too early in swing, Fig. 4b). The anterior–posterior component of ground reaction force was negative in the first half and positive in the second half of the stance in both model and experiment (Figs. 5a and b). However, both the anterior–posterior and vertical force components generated by the model in early stance were much higher than the actual forces. The model also demonstrated kinematical patterns (joint angles and stick diagrams) similar to those obtained in the experiment (Figs. 5 and 6). In particular, the ankle and knee joints in both model and the cat underwent flexion and then extension in the stance phase. The maximum flexion of these joints occurred in the middle of the swing for both the model and experiment. Hip angles generated by the model and recorded experimentally increased monotonically throughout the stance phase and decreased to the minimum in the swing. The model demonstrated the flexibility necessary for the adaptive adjustment of locomotor movements to characteristics of the environment.

Acknowledgements

This work was supported by ONR (N000140210086) and NSF (0091942) Grants to I.A. Rybak, and ONR (N000149810679) and NINDS (NS40543, NS24707) Grants to J.K. Chapin. Boris I. Prilutsky was supported by the Center for Human Movement Studies, Georgia Tech, Atlanta.

References

[1] D.P. Bashor, A large-scale model of some spinal reflex circuits, Biol. Cybern. 78 (1998) 147–157.

[2] R.V. Baratta, M. Solomonow, R. Best, M. Zembo, R. D'Ambrosia, Architecture-based force–velocity models of load-moving skeletal muscles, Clin. Biomech. 10 (1995) 149–155.

[3] T.G. Brown, The intrinsic factors in the act of progression in the mammal, Proc. R. Soc. London Ser. B 84 (1911) 309–319.

[4] Ch.-P. Chou, B. Hannaford, Study of human forearm posture maintenance with a physiologically based robotic arm and spinal level neural controller, Biol. Cybern. 76 (1997) 285–298.

[5] J. Duysens, H.W. Van de Crommert, Neural control of locomotion: the central pattern generator from cats to humans, Gate Posture 7 (1998) 251–263.

[6] S. Grillner, P. Zangger, On the central generation of locomotion in the low spinal cat, Exp. Brain Res. 34 (1979) 241–261.

[7] M.G. Hoy, R.F. Zernicke, Modulation of limb dynamics in the swing phase of locomotion, J. Biomech. 18 (1985) 49–59.

[8] D.A. McCrea, Supraspinal and segmental interactions, Can. J. Physiol. Pharmacol. 74 (1996) 513–517.

[9] N. Ogihara, N. Yamazaki, Generation of human bipedal locomotion by a bio-mimetic neuro-musculo-skeletal model, Biol. Cybern. 84 (2001) 1–11.

[10] G.N. Orlovsky, T.G. Deliagina, S. Grillner, Neural Control of Locomotion. From Mollusk to Man, Oxford University Press, Oxford, 1999.

[11] K.G. Pearson, Common principles of motor control in vertebrates and invertebrates, Ann. Rev. Neurosci. 16 (1993) 265–297.

[12] A. Prochazka, D. Gillard, D.J. Bennett, Implications of positive feedback in the control of movement, J. Neurophysiol. 77 (1997) 3237–3251.

[13] D.W. Richter, Neural regulation of respiration: rhythmogenesis and afferent control, in: R. Gregor, U. Windhorst (Eds.), Comprehensive Human Physiology, Vol. II, Springer, Berlin, 1996, pp. 2079–2095.

[14] S. Rossignol, J.P. Lund, T. Drew, The role of sensory inputs in regulating patterns of rhythmical movements in higher vertebrates: a comparison among locomotion, respiration, and mastication, in: A.H. Cohen, S. Rossignol, S. Grillner (Eds.), Neural Control of Rhythmic Movements in Vertebrates, Wiley, New York, 1988, pp. 201–283.

[15] I.A. Rybak, J.F.R. Paton, J.S. Schwaber, Modeling neural mechanisms for genesis of respiratory rhythm and pattern: II. Network models of the central respiratory pattern generator, J. Neurophysiol. 77 (1997) 2007–2026.

[16] R.D. Sacks, R.R. Roy, Architecture of the hind limb muscles of cats: functional significance, J. Morphol. 173 (1982) 185–195.

[17] S.H. Scott, G.E. Loeb, Mechanical properties of aponeurosis and tendon of the cat soleus muscle during whole-muscle isometric contractions, J. Morphol. 224 (1995) 73–86.

[18] S.A. Spector, P.F. Gardiner, R.F. Zernicke, R.R. Roy, V.R. Edgerton, Muscle architecture and force–velocity characteristics of cat soleus and medial gastrocnemius: implications for motor control, J. Neurophysiol. 44 (1980) 951–960.

[19] G. Taga, A model of the neuro-musculo-skeletal system for anticipatory adjustment of human locomotion during obstacle avoidance, Biol. Cybern. 78 (1998) 9–17.

[20] F.E. Zajac, Muscle and tendon, properties, models, scaling and application to biomechanics and motor control, Crit. Rev. Biomed. Eng. 17 (1989) 359–411.

Dmitry G. Ivashko received his M.S. in Applied Mathematics and Ph.D. in Physical and Mathematical Sciences from Moscow Institute of Physics and Technology (Russia) in 1993 and 1999, respectively. In 1996–1997 he was a junior researcher at the Computing Center of the Russian Academy of Sciences (Moscow). From 1997 to 2001 he was a research assistant in the Department of Theoretical Mechanics at Moscow Institute of Physics and Technology. From 2001 he is a postdoctoral researcher in Ilya Rybak's group at the School of Biomedical Engineering, Science and Health Systems, Drexel University, Philadelphia, PA. His research interests include nonlinear control systems, mathematical modeling, biomechanics and control of locomotion.

Boris I. Prilutsky received his B.S. in Physical Education from Central Institute of Physical Culture, Moscow and in Applied Mathematics and Mechanics from Moscow Institute of Electronic Engineering in 1978 and 1987, respectively, and Ph.D. in Biology and Biomechanics from Research Institute of Traumatology and Orthopedics, Riga, in 1990. From 1978 to 1992 he was a research scientist and a lecturer at the Biomechanics Laboratory and the Department of Biomechanics of Central Institute of Physical Culture in Moscow. He was a postdoctoral fellow at the Department of Kinesiology of University of Calgary, Canada in 1992–1995 and the Department of Health and Performance Sciences of Georgia Institute of Technology, Atlanta, GA, USA in 1995–1998. He is currently a senior research scientist at the Center for Human Movement Studies at Georgia Institute of Technology, Atlanta, GA. Dr. Prilutsky is a member of the American Association for the Advancement of Science, the American Society of Biomechanics (Young Scientist Award, 1995), International Society of Biomechanics, and the Society for Neuroscience. His research interests include biomechanics and neuromuscular control of movement.

Sergey N. Markin received his M.S. in Radiophysics and Ph.D. in Computer Science from Rostov State University (Russia) in 1992 and 1997, respectively. From 1992 to 2000 he was a research scientist in AB Kogan Research Institute for Neurocybernetics at Rostov State University. In 1998–1999 he was a visiting scientist in the Center for Artificial Vision Research at Korea University in Seoul. From 2000 he is a postdoctoral researcher at Drexel University (Philadelphia, PA). He is currently conducting research in groups of Ilya Rybak and Simon Gizster. His research interests include modeling of neural network, neural control, visual perception and biologically inspired robotics.

John K. Chapin received his Ph.D. in Physiology from the University of Rochester in 1979. He has held faculty positions at the University of Texas Southwestern Health Science Center and the Medical College of Pennsylvania and is currently Professor of Physiology and Pharmacology at the SUNY Downstate Medical College in Brooklyn, New York. He is an expert in somatosensory and motor functions of the CNS. He has also developed techniques for large-scale simultaneous multi-neuron recording in the motor system and the use of these recorded motor command signals to control robotic devices.

Ilya A. Rybak received his Ph.D. in Biophysics from St. Petersburg University in 1988. In 1977–1991 he was a senior researcher and head of the lab for Neural Network Modeling in Vision Research in AB Kogan Research Institute for Neurocybernetics at Rostov State University (Russia). In 1992 he received "The Best Paper Award" from "Neurocomputing" for his work on neural network model of visual perception. In 1993–1998 he was a visiting scientist at the Central Research Department at the DuPont Company. He is currently a Research Professor at the School of Biomedical Engineering, Science and Health Systems, Drexel University, Philadelphia, PA. His scientific interests include computational modeling of biological neurons and neural networks with a research focus on neural control of respiration and locomotion, visual perception, biologically inspired robotics and brain–machine interfaces.

ELSEVIER

Available at
www.ComputerScienceWeb.com
POWERED BY SCIENCE @ DIRECT®

Neurocomputing 52–54 (2003) 631–636

NEUROCOMPUTING

www.elsevier.com/locate/neucom

A visual system for invariant recognition in animated image sequences

Luis F. Lago-Fernández[a,*], Manuel A. Sánchez-Montañés[a], Eduardo Sánchez[b]

[a] *Escuela Politécnica Superior, Universidad Autónoma de Madrid, Carretera de Colmenar Viejo Km 15, Madrid 28049, Spain*

[b] *Facultad de Ciencias Físicas, Universidade de Santiago de Compostela, Santiago de Compostela 15706, Spain*

Abstract

The problem of invariant recognition in simple animated image sequences is investigated using a biologically inspired model. It integrates a bottom-up attention module, a synchronization-based segmentation mechanism, a normalization network and a recognition module. The results show how the system provides the following properties: (1) change-based attention in animated sequences, (2) multiple object segmentation, and (3) translation and scale invariant recognition.

Keywords: Visual system; Attention; Synchronization; Invariant recognition

1. Introduction

There are two basic approaches to the problem of invariant recognition: normalization [7], and invariant feature extraction [1,8]; see [10] for a review. In this work we propose a normalization-based recognition mechanism that tries to preserve all the advantages of these models, while minimizing the complexity of the required network. Additionally, we have addressed how invariant recognition can be achieved when animated image sequences are presented. This situation requires from the system: (1) to determine the relevant information over time, and (2) to adapt the invariant mechanisms to work with dynamic inputs. In our approach, the relevant information is selected by

* Corresponding author.
E-mail address: luis.lago@ii.uam.es (L.F. Lago-Fernández).

0925-2312/03/$ - see front matter
doi:10.1016/S0925-2312(02)00845-7

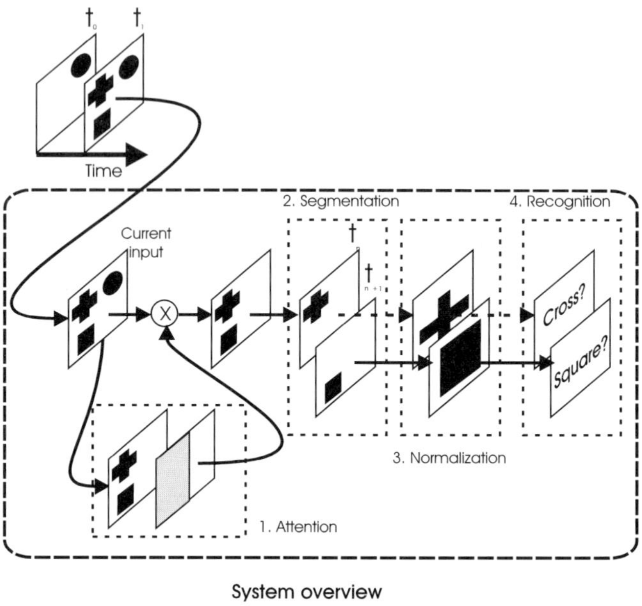

Fig. 1. System overview: Four main parts can be distinguished: (1) change-based attention module, which eliminates information about static regions; (2) segmentation module, which separates the different objects in the scene; (3) normalization module, which provides position, scale and orientation invariant object representations; and (4) recognition module.

a bottom-up attention module, which consists of a change-based saliency map and a Winner Takes All (WTA) mechanism inspired by Koch and Ullman [4]. The dynamic inputs provided by the attention module are processed by a synchronization-based segmentation network [5]. Segmented objects are normalized by a simple recurrent network to achieve position, scale and orientation invariant representations, which are then fed into the recognition module. Fig. 1 shows a general overview of the system. In the following section we discuss the four main modules.

2. The system

In our approach a change-based feature map is proposed to deal with dynamic images and detect changing regions. Fig. 2 shows how these regions drive a WTA process which facilitates relevant locations for further detailed processing. The attention module has been adapted from [3] to deal with changing images. The temporal derivative of the input image, $\dot{x}_{ij}$, is approximated by means of a set of interneurons that store the previous image and provide, at any time t, an inhibitory input to the neurons in the saliency map. The output d_{ij} of the change-based saliency map is then defined by $\dot{x}_{ij}$, being active ($d_{ij} = 1$) if this derivative is greater than a threshold Θ, and silent ($d_{ij} = 0$) otherwise. In the WTA map, each control neuron attends a predefined region

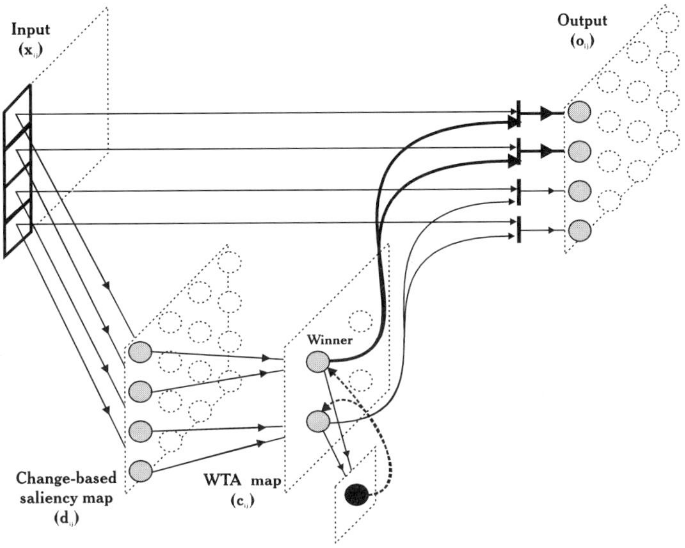

Fig. 2. Attention mechanism. Neurons in the change-based saliency map perform a temporal derivative of the input, and their output drives the activation of the control neurons in the WTA map. These neurons are globally inhibited by an interneuron in charge of selecting a winner, which signals the most prominent region/s of change. Finally, presynaptic facilitation triggered by the WTA map output c_{ij}, allows information about relevant regions to pass to further processing stages.

of the saliency map. The control neurons compete among themselves through a global feedback inhibition (see Fig. 2), and have recurrent connections (not shown in the figure) to recall the previous winner locations. These connections are disabled when afferent activity is detected. Finally, the connectivity between the input neurons and the output neurons is controlled by a presynaptic facilitation mechanism triggered by the WTA map. Thus the connection between x_{ij} and o_{ij} is only activated when the activity of the corresponding control neuron, c_{ij}, is greater than 0.

The segmentation module exploits the general idea that the brain could be using temporal correlations to bind the different features that compound single objects [6,2]. It consists of a network of integrate and fire neurons which are locally connected among themselves via excitatory synapses, and globally connected to a common inhibitory interneuron that sends back inhibition to the whole network. Local excitatory connections provide a fast synchronization of all the neurons responding to a connected region in the input scene, while global inhibition prevents disconnected regions from oscillating in synchrony. This kind of network is inspired by Wang and Terman [9], but here we consider a simplified integrate and fire neuron model [5].

The normalization module is composed of the translation-invariant network, and the scale/orientation-invariant network. The image given by the segmentation module is fed into the translation-invariant module, which is composed of a 2D network of excitatory neurons and four control neurons which modulate the intrinsic recurrent connectivity of the network (see Fig. 3). The “up” control neuron activates the synapses between

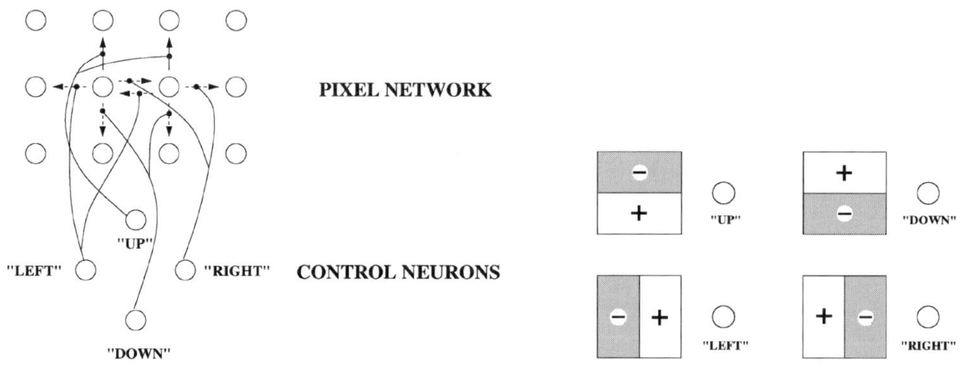

Fig. 3. Translation-invariant network: Left: each neuron represents a pixel of the image given by the segmentation module, and excitates its four nearest neighbours. Each one of the four control neurons activates a particular connectivity (up, down, left or right). Right: receptive fields of the control neurons.

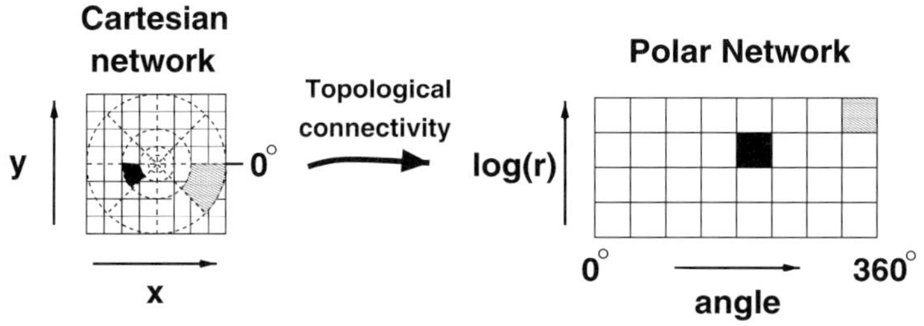

Fig. 4. Scale/orientation invariant network input. The information in the translation-invariant network is represented in Cartesian coordinates. The connectivity to the scale/orientation invariant network is adjusted so that the information is now represented in polar coordinates. This connectivity is local and topological.

each pixel neuron and its upper neighbour. The "down" control neuron activates the synapses between each pixel neuron and its bottom neighbour, and analogously for the "left" and "right" control neurons. These neurons compete among themselves through a WTA mechanism. Therefore, when the "up" control neuron wins, the network performs a translation of 1 pixel in the "up" direction. The receptive fields of these control neurons are shown in Fig. 3. For example, the "up" neuron receives excitation from the lower pixels, and inhibition from the others. Therefore, this neuron is activated as long as there are more pixels in the lower part of the image than in the upper one. The system is then moving the image until its "center of gravity" coincides with the origin. Then, no control neuron wins the competition and the computation is finished.

The output of this network is fed into the scale/orientation-invariant network. The connectivity between these two networks can be easily chosen so that the information is now represented in polar coordinates (see Fig. 4). The structure and dynamics of this network is exactly the same as in the translation-invariant network, including the control neurons. Since the dilation/contraction operations, as well as the rotation operations, can be described as translations in polar coordinates, this network is performing normalization with respect to both size and orientation.

Finally, the output from the normalization module is fed into the recognition network. This module is composed of two layers of neurons. The first one develops internal representations in order to recognize previously shown stimuli. The second module learns associations between these representations and the label of the stimulus ("circle", "square" or "triangle"), which is given by the teacher. The dynamics of this module is biologically realistic, for details see [5].

3. Results and discussion

To test the model, we stimulated the network with sequences of input images consisting of black objects on a white background. We used different size versions of each object, which were presented at various locations with simultaneous appearance of up to four objects. The successive processing steps achieved attention fixation on changing regions; segmentation of the different objects that appear in these regions; normalization to standard size, position and orientation; and finally recognition. An example of the output from the different modules for a typical trial is shown schematically in Fig. 5.

Since the attention module was inspired by the pioneering model of Koch and Ullman [4] and its extensions [3], our approach could be integrated in these models in order to enhance their functionality by allowing transients processing. The change-based saliency map, and its associated WTA map, could be naturally understood as a new feature map to be added to the color, orientation and intensity maps. These maps could be integrated in a unique saliency map in charge of encoding not only static patterns, but also animated image sequences.

The routing complexity problem can be stated as the incredible amount of control neurons needed in classical routing networks [7]. Contrary to these feedforward models, our approach introduces recurrent connections, which keep the system very simple, as

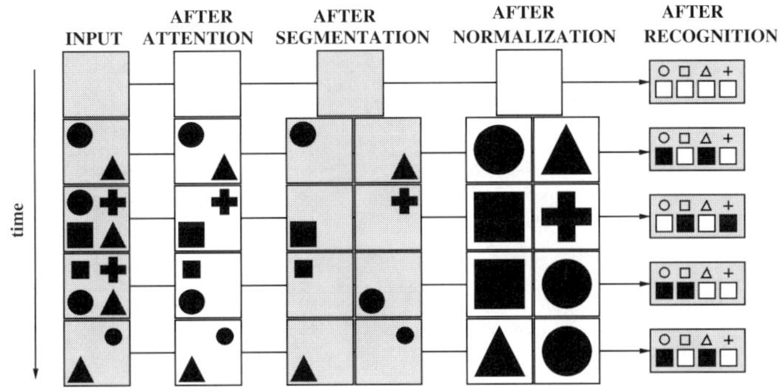

Fig. 5. Schematics of the network processing on a typical input sequence.

well as closer to biology. This solution drastically reduces the number of required control neurons down to 8.

Acknowledgements

We acknowledge financial support from MCT grant BFI2000-0157.

References

[1] K. Fukushima, Neocognitron: a self-organizing neural network model for a mechanism of pattern recognition unaffected by shift in position, Biol. Cybernet. 36 (1980) 193–202.
[2] C.M. Gray, Synchronous oscillations in neuronal systems: mechanisms and function, J. Comput. Neurosci. 1 (1994) 11–38.
[3] L. Itti, C. Koch, Computational modelling of visual attention, Nat. Rev. 2 (2001) 1–9.
[4] C. Koch, S. Ullman, Shifts in selective visual attention: towards the underlying neural circuitry, Hum. Neurobiol. 4 (1985) 210–227.
[5] L.F. Lago-Fernández, M.A. Sánchez-Montañés, F.J. Corbacho, A biologically inspired visual system for an autonomous robot, Neurocomputing 38–40 (2001) 1385–1391.
[6] C. von der Malsburg, The correlation theory of brain function, in: E. Domany, J.L. van Hemmen, K. Schulten (Eds.), Models of Neural Networks II, Springer, Berlin, 1994, pp. 95–119.
[7] B.A. Olshausen, C.H. Anderson, D.C. Van Essen, A neurobiological model of visual attention and invariant pattern recognition based on dynamic routing of information, J. Neurosci. 13 (11) (1993) 4700–4719.
[8] M. Riesenhuber, T. Poggio, Hierarchical models of object recognition in cortex, Nat. Neurosci. 2 (11) (1999) 1019–1025.
[9] D. Wang, D. Terman, Locally excitatory globally inhibitory oscillator networks, IEEE Trans. Neural Networks 6 (1) (1995) 283–286.
[10] L. Wiskott, How does our visual system achieve shift and size invariance? in: L. Van Hemmen, T.J. Sejnowski (Eds.), Problems in Systems Neuroscience, Oxford University Press, Oxford, to appear. Available at http://www.cnl.salk.edu/~wiskott/homepage.html

Available at
www.ComputerScienceWeb.com
POWERED BY SCIENCE @ DIRECT®

ELSEVIER

Neurocomputing 52–54 (2003) 637–641

NEUROCOMPUTING

www.elsevier.com/locate/neucom

Real-time estimation of predictive firing rate

David C. Tam*

Department of Biological Sciences, University of North Texas, P.O. Box 305220, Denton, TX 76203, USA

Abstract

Population vector analysis of multi-unit firing code in the motor cortex provides a real-time estimate of the desired voluntary movement direction and trajectory used in neuroprosthesis for assisting paralyzed patients. Accurate estimation of instantaneous firing rate in each epoch used in the population vector calculation is crucial. Yet real-time instantaneous firing rate is often unknown until the next spike has fired. We have developed statistical based estimator for predicting the waiting time for the next spike to fire. Instantaneous firing rate is computed based on the conditional probability density function (pdf), providing a better estimate of the intended movement trajectory.

Keywords: Spike train analysis; Probability density function; Firing rate

1. Introduction

Population vector analysis of multi-unit firing code in motor cortical neurons is a measure developed by Georgopoulos et al. [1–4] to estimate movement direction and trajectory based on a vector sum of the weighted-average firing rate with respect to the "preferred movement direction" of each individual neuron in the ensemble. This measure is a powerful means to predict the movement trajectory to the intended target. In other words, the spike-firing rate of the motor cortical neurons can be decoded to represent the intended voluntary arm movement using the population vector analysis.

Simultaneous recording of spike trains recorded from 64-channel electrodes implanted in the motor cortex provided a means for real-time prediction of arm movement trajectory [5]. This real-time prediction of arm movements based on population vector

* Tel.: +1-940-565-3261; fax: +1-940-565-4136.
E-mail address: david@david.tam.name (D.C. Tam).

0925-2312/03/$ - see front matter
doi:10.1016/S0925-2312(02)00850-0

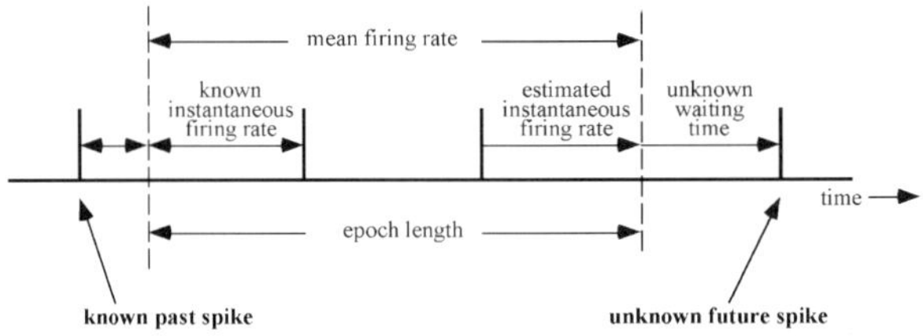

Fig. 1. Diagram showing the knowledge of boundary-delimiting spikes needed to compute the mean firing rate. The waiting-time for the future spike is unknown in real-time when the mean firing rate is computed for the current epoch.

analysis of multiple spike trains will be implemented as a neuroprosthesis so that a robotic arm can be moved by "mental efforts" of the patient implanted with 64-electrodes in the motor cortex.

One of the challenges in obtaining the firing rate of the spike train in real-time is due to the typically low firing rate in motor cortical neurons. Because of the low firing rate, or the "sparse code," accurate estimate of the true firing rate of each individual neuron is important so that the population vector sum can be computed accurately. This challenge is usually a non-issue when the population vector is computed off-line, based on a posteriori information of the spike firings, since the instantaneous firing rate can always be computed within a given spike train.

But when real-time estimate of the population vector is needed, the instantaneous firing rate for each neuron is required. Traditionally, the population vector is computed incrementally for each epoch of fix time-interval, such as every 30 ms. The mean firing rate for each neuron is then computed for each epoch, and the population vector based on the preferred direction of each neuron is computed using this mean firing rate for each epoch.

Using this method, estimation errors exist at the boundary of the epoch because spikes do not often fall on the epoch boundary. The true instantaneous firing rate can be computed if the two boundary-delimiting spikes exist (see Fig. 1). In off-line analysis, the boundary-delimiting spikes are known; thus, there would not be any ambiguous estimation of the mean firing rate for the epoch.

But in real-time analysis, the future delimiting spike is unknown for the current epoch calculation of mean firing rate (see Fig. 1). Thus, it necessitates the estimation of the waiting time for the next spike to occur. The accuracy of this estimation will influence on the estimate of the mean firing rate for the current epoch used in the population vector calculation. Since the overall firing rate is low, which is typically found in motor cortex, the error in estimating the waiting-time for the unknown future spike can be large.

There are many existing methods for resolving this ambiguity. One of them is to use a Gaussian filter to convert the spike firings into firing rate, which is effectively a smoothing function for the discrete spike rate. This method is highly dependent on the parameter used for the Gaussian filter, thus it is user-dependent.

Another method commonly used is to extrapolate the next firing interval based on the previous firing rate (or firing interval). This is often a good estimate since it is dependent on the neuron's own firing rate rather than an arbitrary parameter.

The current paper proposes an alternate predictor of the waiting-time for the unknown future spike so that the instantaneous firing rate of the neuron during the current epoch can be better estimated.

2. Theoretical methods—adaptive statistical estimation

One of the better predictors of future events is based on the history of past events. In other words, since we are recording on-going spikes from the neuron in real-time, we can collect the firing statistics of the neuron. The interspike-interval (ISI) statistic characterizes the probability of firing of a next spike based on the firing of a current (reference) spike. Normalizing the ISI by the number of spike gives the probability density function (pdf) for the neuron, which will be used as the predictive function for the waiting-time for the next spike firing.

Let the spike train be, $a(t)$, recorded from the neuron with a total of N spikes be represented by

$$a(t) = \sum_{n=1}^{n=N} \delta(t - t_n), \tag{1}$$

where t_n is the occurrence times of nth spikes in spike train $a(t)$, and $\delta(t)$ is a delta function denoting the occurrence of a spike at time t.

The ISI, τ_n, is defined as the time-interval between two adjacent spikes in the spike train:

$$\tau_n = t_n - t_{n-1}. \tag{2}$$

Then, the probability density function (pdf) of the first-order ISI is defined as:

$$p(\tau) = \frac{1}{N} \sum_{n=1}^{N} \delta(\tau_n - \tau) \tag{3}$$

and the probability distribution function, PDF, which is given by

$$P(\tau) = \int_0^{\tau} p(t)\,\mathrm{d}t. \tag{4}$$

Since the spike train is recorded in real-time, the total number of spikes, N, increases with time. Thus, the pdf function also changes with time, with a better approximation when the sampling size, N, increases with time.

3. Statistical estimation

Given that the firing pdf for the neuron is known, the probability of the next spike firing can be estimated for the future spikes. We also have additional information of the firing probability, since we are predicting the next spike firing at the boundary-delimiting conditions. Because the delimiting condition is that a spike did not fire at the end of the epoch-window period, the probability of next spike firing becomes the conditional probability given that a waiting-time, τ', has elapsed since the last spike and a spike fired at the next time increment, $\tau' + \Delta t$. In other words, the probability given that a spike had not fired for τ' is essentially $1 - P(\tau)$.

4. Comparison between current method and other traditional methods

There are many other alternate methods for computing the mean firing rate. One of the most common methods is simply compute the mean firing rate (number of spikes fired within the epoch-window divided by the window-length) truncating the extra time between spikes at both ends of the epoch-window. This always results in over-estimating the firing rate since the extra times at the boundary condition are ignored.

Another traditional method is to use the averaged firing rate during the epoch as the predictor to extrapolate the future firing rate for the next spike. In other words, it uses the mean firing rate as predicted firing rate for the next spike. This method is better than the truncated firing rate method above since it uses the past averaged firing rate as the predictor for the future spike rate.

In comparison, the present method provides a better estimate of the future spike rate since it takes advantage of the instantaneous firing rate for the prediction rather than the mean firing rate. The current method provides a means to increase the statistical significance as N (total number of spikes in the sample size) increases with real-time on-line recording session. Furthermore, the predicted estimated rate is operator-independent and self-adaptive. The future firing rate is estimated entirely based on how long a spike had not occurred at the end-boundary.

Given these advantages, real-time estimate of the firing rate at each epoch can be performed with higher accuracy. With the improved accuracy of estimated instantaneous firing rate for each individual neuron, the vector sum of the population ensemble can be more accurately represented. With a better representation of the population vector, the intended movement trajectory can be predicted more accurately for sparse-code encoding in the cortex.

References

[1] A.P. Georgopoulos, R.E. Kettner, A.B. Schwartz, Primate motor cortex and free arm movements to visual targets in three-dimensional space. II. Coding of the direction of movement by a neuronal population, J. Neurosci. 8 (1988) 2928–2937.

[2] A.P. Georgopoulos, A.B. Schwartz, R.E. Kettner, Neuronal population coding of movement direction, Science 233 (1986) 1416–1419.

[3] R.E. Kettner, A.B. Schwartz, A.P. Georgopoulos, Primate motor cortex and free arm movements to visual targets in three-dimensional space. III. Positional gradients and population coding of movement direction from various movement origins, J. Neurosci. 8 (1988) 2938–2947.

[4] A.B. Schwartz, R.E. Kettner, A.P. Georgopoulos, Primate motor cortex and free arm movements to visual targets in three-dimensional space. I. Relations between single cell discharge and direction of movement, J. Neurosci. 8 (1988) 2913–2927.
[5] A.B. Schwartz, D.M. Taylor, S.I. Tillery, Extraction algorithms for cortical control of arm prosthetics, Curr. Opin. Neurobiol. 11 (2001) 701–707.

Dr. David C. Tam is an associate Professor at the University of North Texas. He holds a Ph.D. in physiology, and three B.S. degrees in computer science, physics and astrophysics, all from the University of Minnesota. His current research interest is developing statistical multiple spike trains analysis techniques and applying these analytical methods to both simulated and experimental neurophysiological data in using emotional behavioral paradigms to understand the signal processing functions of the central nervous system (CNS) as well as extracting the encoding schemes used by the CNS.

Available at
www.ComputerScienceWeb.com
POWERED BY SCIENCE @ DIRECT®

ELSEVIER

Neurocomputing 52–54 (2003) 643–648

NEUROCOMPUTING

www.elsevier.com/locate/neucom

Semantic rewiring mechanism of neural cross-supramodal integration based on spatial and temporal properties of attention

T. Koshizen*, So Yamada, Hiroshi Tsujino

Wako Research Centre, Honda R& D Co. Ltd., 1-4-1 Chuo, Wako-Shi, Saitama-ken 02111-1208, Japan

Abstract

This paper proposes cross-supramodal integration system which is hypothetically dedicated by a neuronal network where prefrontal cortex is interacted with hippocampus and cortical areas in human brain. The substrate of the network is basically computed by rewiring mechanism consisting of a bidirectional processing network dedicated by PFC. The computational meaning of the rewiring allows the bidirectional network to learn cross-supramodal integration among different sensors. The cross-supramodality is represented by possible attention classes across space and time. We speculate that the rewiring mechanism may particularly be crucial as consolidating of spatial and temporal attention based on the bidirectional processing network where PFC integrates expected reward outcomes and accurate representations of spatial information together to attain accurate behavior. Conclusively, the leverage on the nature of human rewiring mechanism may essentially relate to cognitive issues such as semantic priming and language acquisition.

Keywords: Spatiotemporal attention; Rewiring mechanism; Cross-supramodal integration; PFC-based bidirectional network; Cognitive function

1. Introduction

Cross-modal integration is the one of the remarkable issues in brain computations because it allows us to contribute to endanger adaptive and flexible behavior in various forthcoming events of our daily life. It is, however, still not elucidated about neural mechanism of how underlying cross-supramodal integration can be taken into account

* Corresponding author.
E-mail address: takamasa_koshizen@n.w.rd.honda.co.jp (T. Koshizen).

0925-2312/03/$ - see front matter
doi:10.1016/S0925-2312(02)00794-4

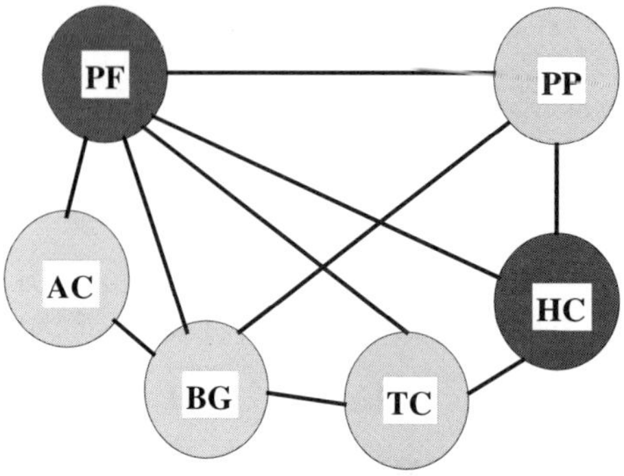

Fig. 1. Schematic connection of hypothetical network.

of attention mechanisms as well as behavior tasks. With respect to the neuronal network of attention mechanisms, Cohen et al. and Robertson et al. [2,6] showed neurobiologically the attention demanding and modulation process. In CNS*2001, we suggested the computational model of anterior cingulate (AC)–prefrontal (PF)–posterior parietal (PP) cortex in relation to attention demanding and modulation mechanism [4]. In human object recognition, PP is important for encoding object locations while temporal cortex (TC) is for object shape. The computational work also suggested that such attention mechanism could be processed by bottom-up and top-down mechanisms of attention in order to enhance visual responses and to guide accurate behavioral sequences at any given time. From neuronal evidence, reward information from behavior assessment and spatial information are integrated as working memory in the PFC so that spatial information becomes more accurate when reward outcome is expected. Fig. 1 shows presumably a neuronal network which is related to spatial and temporal characteristic of attention. We infer that the role of the network centered by PF might consolidate both spatial and temporal characteristic of attention by interacting with hippocampus (HC), where the appropriate relationship between behavior sequences and different modalities is learned to come out of indelible memory. Indeed, HC encodes both the spatiotemporal relationships by involving PP and PF cortex, for example, see [8]. We in this paper aim to provide the computational model of spatiotemporal attention in accordance with the cross-supramodal integration. Although Sharma et al. suggested neurophysiologically the relationship between the rewiring mechanism and cross-modal integration in ferret [7], we attempt in this paper, the learning scheme of the rewiring mechanism as well as its meaning presumably. In summary, we postulate that the rewiring mechanism may play the crucial role to train the nonlinearity of cross-supramodal integration by a bidirectional processing network with respect to spatiotemporal attention, which is attained by PFC-HC interaction.

2. Computational model across cross-supramodal integration

In this section, we present the computation model based on the cross-supramodal integration. Fig. 2 (right) illustrates the learning architecture of the cross-supramodal

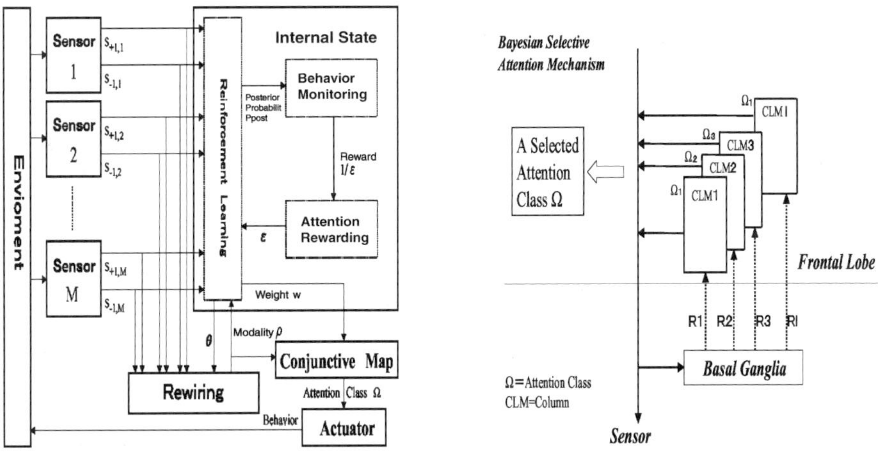

Fig. 2. Cross-supramodal rewiring system (left) and BSA model (right).

integration system. The system is mainly composed of the three parts, namely rewiring mechanism, attentive reinforcement learning and behavior control. Essentially, it contains bayesian selective attention (BSA) with sensorimotor system [3,5]. That is, the system as shown in Fig. 2 (right) is derived from the bidirectional processing network where PF presumably generates the top-down information processing for selecting the attention classes Ω, which are computed by the bottom-up information processing based on the reinforcement learning at basal ganglia (BG). To deal with the cross-supramodal integration, the rewiring mechanism is especially taken into account with the BSA's sensorimotor system. In the rewiring mechanism each sensory modality allows to be weighted by the (symmetric/asymmetric) parameter w through the bidirectional processing network (Fig. 2 (left)). In addition, the cross-supramodal integration system employs an attentive reinforcement learning scheme, which trains the parameter (θ) of each attention class across the visual shape and visual motion modalities.

Now, we assume that the attention class Ω_i^t of the ith supramodality consisting of visual shape ($i=-1$) and visual motion ($i=+1$) is mathematically modelled as follows:

$$\Omega_i^t \propto \rho_i(\mathbf{q}_i^t; \theta_i), \tag{1}$$

where sensory vector over i supramodality $\mathbf{q}_i$ belongs in $\mathscr{R}^M$. The probability density function $|\rho|$ quantifies the number of population of $\mathbf{q}_i$, which is parameterized by θ, which is comprised of the mean and standard deviation of density ρ. Let ρ be the Gaussian distribution function as follows:

$$\rho_i(\mathbf{q}_i^t; \theta_i) = y_i(\mathbf{q}_i^t) \sum_k^M \gamma_{i,k} \exp\left(-\frac{\|\mathbf{q}_{i,k} - \mu_{i,k}\|^2}{2\sigma_{i,k}} \right), \tag{2}$$

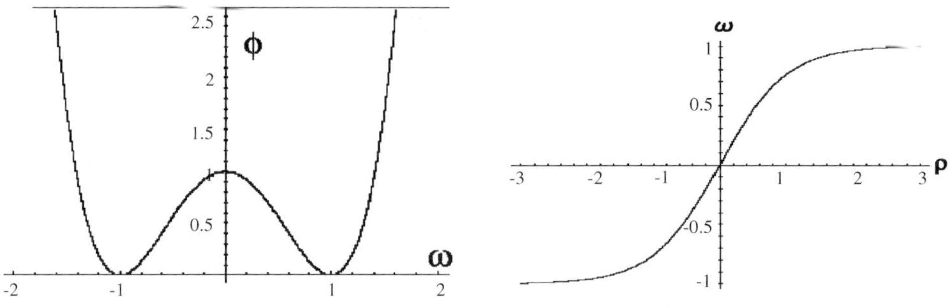

Fig. 3. Potential function (left) and weight value (right).

where the probabilistic parameter $\theta=(\gamma,\mu,\sigma)$ is computed by expectation maximization (EM) algorithm [3], γ denotes mixing coefficient.

$$\Omega_i^t = \int_{\mathscr{S}} w_i^t \rho_i(\mathbf{q}_i^t;\theta_i)\,\mathbf{dq}^t, \tag{3}$$

where the weight value w_i^t of the rewiring mechanism is optimized by Allen–Cahn (A–C) equation [1]. $\mathscr{S}$ denotes the whole sensory space where the velocity component V_x, V_y is obtained from the *first-order* differentiation of the position component x, y. Assume $\mathscr{S}=\mathscr{S}(x,y)\cup\mathscr{S}(V_x,V_y)$. The A–C equation is basically applied for phase transitions whose solution is provided by minimizing the following cost function:

$$\mathscr{E}(\mathbf{q}^t) = \int_S \phi(w_i^t)\,\mathbf{dq}^t + \varepsilon^2\int_S |\nabla w_i^t|^2\,\mathbf{dq}^t. \tag{4}$$

The quantity $\mathscr{E}$ can be derived from the double-well potential function, which is illustrated by Fig. 3 (left). Here, we suggest the kind of potential function that may be endowed by HC. Then, the updating equation of w_{ij}^t is can be resolved by the above equation: Notably the solution of A–C equation becomes the tangential sigmoidal-type function (Fig. 3 (right)):

$$w_i \equiv \tanh(\rho_i(\mathbf{q}_i^t;\theta_i)/\varepsilon), \tag{5}$$

where ε is the parameter minimized to *zero*. Eq. (5) shows the predictive weights $w_{i,j}^t$, which is trained throughout the perturbation across supramodal densities ρ_i.

Since the elicitation of attention demanding depends on the expectation posterior probability ε, which is predictive for expected reward outcomes based on the synchronization in the cross-supramodal rewiring system:

$$\varepsilon = \left|\log\left[\frac{1}{M}\sum_{k=1}^{M}\left\{\frac{\int_{S(x,y)} P_{-1,k}^{\text{post}}\,\mathrm{d}x\,\mathrm{d}y}{\int_{S(V_x,V_y)} P_{+1,k}^{\text{post}}\,\mathrm{d}V_x\,\mathrm{d}V_y}\right\}\right]\right|, \tag{6}$$

where P^{post} denotes the posterior probability in conditional to supramodality i and sensory modality k.

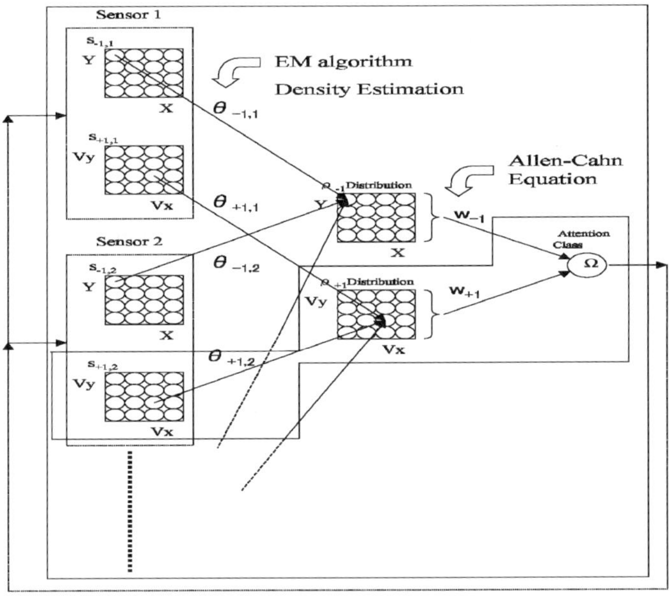

Fig. 4. Learning phase of cross-supramodal mapping.

We assume the expectation posterior errors are bounded by the following exponential function because of 'cumulative' aspects of the prior probability:

$$\varepsilon \approx \lambda \exp(-\lambda\tau), \varepsilon \geqslant 0, \tag{7}$$

where λ is positive constant and τ denotes the learning feedback iteration in cross-supramodal rewiring system. λ implicates amplitude of the expectation posterior error where the curvature of expectation posterior errors monotonically decrease according to Eq. (7).

Fig. 4 shows the learning architecture where all the sensor information can be organized into the supramodality. Some comments would be appropriate upon the training and memory processes of the cross-supramodal mapping shown in Fig. 4. It allows sensory information to be mapped by $N \times N$ matrices for both the supramodal space. In summary, the proposed scheme can effectively constrain the calculation by $O(2MN^2)$ as well as the memory by $O(N^2)^2$ no matter how the number of sensors are addictive. As a virtue, $O(N^2)^{2M}$ is required in the both training and memory processes unless the cross-supramodal mapping is incorporated with the learning scheme shown in Fig. 4.

3. Conclusion

In this paper, our computational model suggest how the spatial and temporal properties of attention can be taken into account of cross-supramodal integration system. That is, our neuronal network in relation to PFC-HC shows the two different modalities such as visual shape and visual motion, which are eventually bound by the spatiotemporal attention. Accordingly, the semantic role of rewiring mechanism may be to approximate the underlying nonlinearity between the cross-supramodality through the spatiotemporal attention mechanism that induces the bidirectional processing network mediated from PF. This also implicates such visual processing can be attained by other sensory modalities such as tactile and audition, based on the rewiring mechanism. In other words, this made it possible to mediate the tactile sensor cortex for visual shape but the auditory sensor cortex for visual motion.

References

[1] S.M. Allen, J.W. Cahn, A macroscopic theory for antiphase boundary motion and its application to antiphase domain coarsening, Acta. Metall. 27 (1979) 1085–1095.
[2] J.D. Cohen, M. Botvinick, C.S. Carter, Anterior cingulate and prefrontal cortex: who's in control, Nature Neurosci. 3 (5) (2000) 421–423.
[3] A.P. Dempster, N.M. Laird, D.B. Rubin, Maximum likelihood from incomplete data via the EM algorithm, J. R. Stat. Soc. Ser. B 39 (1977) 1–38.
[4] T. Koshizen, K. Akatsuka, H. Tsujino, A computational model of attentive visual system induced by cortical neural network, Neurocomputing, 44–46 (2002) 879–885.
[5] T. Koshizen, H. Tsujino, in: Klette, Pelag, Sommer (Eds.), Hypothetical modelled perceptual sensory modality of human visual selective attention scheme by PFC-based network, Lecture Notes in Computer Science, Vol. 1998, Springer, Berlin, 2001, pp. 261–269.
[6] E.M. Robertson, J.M. Tormos, F. Maeda, A. Pascual-Leone, The role of the dorsolateral prefrontal cortex during sequence learning is specific for spatial information, Cerebral Cortex 11 (2001) 628–635.
[7] M. Sur, S.L. Pallas, A.W. Roe, Cross-modal plasticity in cortical development: differentiation and specification of sensory neocortex, Trends Neurosci. 13 (1990) 227–233.
[8] N.M. White, M.-C. Ouellet, Roles of movement and temporal factors in spatial learning, Hippocampus 7 (1997) 501–510.

Available at
www.ComputerScienceWeb.com
POWERED BY SCIENCE DIRECT

ELSEVIER

Neurocomputing 52–54 (2003) 649–654

NEUROCOMPUTING

www.elsevier.com/locate/neucom

How does the information-geometric measure depend on underlying neural mechanisms?☆

Masami Tatsuno[a,b,*], Masato Okada[b]

[a] *ARL Division of Neural Systems, Memory and Aging, The University of Arizona, Life Sciences North Building, P.O. Box 245115, Tucson, AZ 85724-5115, USA*

[b] *Laboratory for Mathematical Neuroscience, Brain Science Institute, RIKEN, 2-1 Wako-shi, Saitama 351-0198, Japan*

Abstract

To analyze how information is represented among neuron groups, a recently presented information-geometric method (IGM) has attracted growing attention. However the detailed properties underlying the information-geometric measure have not yet been elucidated because of the ill-posed nature of the problem. Here the underlying neural mechanism of the information-geometric measure is investigated with an isolated pair of model neurons. For the symmetric network, the information-geometric measure is solely dependent on the underlying anatomical connections between the recorded neurons. For the asymmetric network, however, the information-geometric measure is dependent both on the intrinsic connections and on the external inputs to it. In other words, there are multiple neural mechanisms corresponding to the same information-geometric measure. In addition, the relation between IGM and conventional cross-correlation is also investigated.

Keywords: Information geometry; Correlation function; Spikes; Neural firing

1. Introduction

Development of an analytical method for neuronal population activity is critical to understand how information is represented in the brain [1,7]. To this end, the novel analytical methods based on information geometry were proposed by Amari [2,3], and

☆ This work was supported by Research Grant 626-52045 from RIKEN.

* Corresponding author. The University of Arizona, Life Sciences North Building, P.O. Box 245115, Tucson, AZ 85724-5115, USA. Tel.: +1-520-626-5916; fax: +1-520-626-2618.

E-mail address: masami@email.arizona.edu (M. Tatsuno).

0925-2312/03/$ - see front matter
doi:10.1016/S0925-2312(02)00852-4

were extended to be applicable to real neuron data by Nakahara et al. [5,6]. The advantages of the information-geometric measure can be summarized as follows: (1) The modulation in mean firing rate and in coincident firing can be decomposed orthogonally. (2) Information carried by modulation in the mean firing rate and in coincident firing can be calculated. (3) Biologically plausible null hypothesis can be used for statistical test. (4) Systematic extension to any number of neurons is possible. Therefore, this method could be a powerful analytical tool for multiple spike data.

However, the possible underlying neural mechanisms for generating the information-geometric correlation components have not yet been elucidated because there could be multiple neural mechanisms producing the same information-geometric measure. For example, in the system of two neurons, the information-geometric measure corresponds to the network with symmetric connections, and the network with asymmetric connections is beyond the information-geometric measure. However, if the equilibrium state exists, the asymmetric network can be mapped onto symmetric networks. Using this property, the possible neural mechanisms underlying the information-geometric measure were investigated with an isolated pair of model neurons.

2. Information-geometric measure

Let us briefly introduce the information-geometric measure. For details, see Refs. [3,2]. Let X_1 and X_2 are 2 binary variables and let $p = p(x_1, x_2)$, $x_i = 0, 1$, be its probability. Their joint probability is given by $p_{ij} = \mathrm{Prob}\{x_1 = i; x_2 = j\} \geqslant 0$, $i, j = 0, 1$, with a constraint $p_{00} + p_{01} + p_{10} + p_{11} = 1$. Let us now expand $\log p(x_1, x_2)$ in the polynomial of x_1 and x_2 as follows:

$$\log p(x_1, x_2) = \theta_1 x_1 + \theta_2 x_2 + \theta_3 x_1 x_2 - \psi. \qquad (1)$$

This is called a log-linear expansion and it is an exact expansion since x_i takes the binary values 0, 1. The coefficients θ's and ψ are given by

$$\theta_1 = \log \frac{p_{10}}{p_{00}}, \quad \theta_2 = \log \frac{p_{01}}{p_{00}}, \quad \theta_3 = \log \frac{p_{11} p_{00}}{p_{01} p_{10}}, \quad \psi = -\log p_{00}. \qquad (2)$$

θ_3 represents the information-geometric measure for 2 neuron interactions.

3. Model and correlation function

The isolated pair of model neurons was used in this study, because this simple setting allows us to investigate the system with mathematical clarity.

Following Ginzburg and Sompolinsky [4], the state of each neuron at time t takes one of two states, denoted by $S_i(t) = 0, 1$, where $i = 1, 2$. The neurons are assumed to have a stochastic nature, and the transition rates for the ith neuron takes the form

$$w(S_i \to (1 - S_i)) = \frac{1}{2\tau} \{1 - (2S_i - 1)[2g(u_i) - 1]\}, \qquad (3)$$

where $g(u)$ is a monotonically increasing, differentiable function such as $g(u) = (1 + \tanh(u))/2$. The local field acting upon the ith neuron at time t is

$$u_i(t) = J_{ij}S_j(t) + h_i, \tag{4}$$

where J_{ij} denotes the connection from the jth presynaptic neuron to the ith postsynaptic one, and h_i represents the local input. In the equilibrium limit, the average activities are given by

$$\langle S_i \rangle = \langle S_j \rangle \Delta g_i + g(h_i), \tag{5}$$

where $\Delta g_i = g(J_{ij} + h_i) - g(h_i)$.

4. Results

The relationship between the information-geometric measure and the underlying neural mechanisms was elucidated in the equilibrium limit. The information-geometric measures,

$$\theta_1 = \log \frac{p_{10}}{p_{00}}, \quad \theta_2 = \log \frac{p_{01}}{p_{00}}, \quad \theta_3 = \log \frac{p_{11}p_{00}}{p_{01}p_{10}}, \tag{6}$$

were expressed by the underlying neural mechanisms

$$p_{00} = 1 - \langle S_1 \rangle - \langle S_2 \rangle + \langle S_1 S_2 \rangle, \tag{7}$$

$$p_{01} = \langle S_2 \rangle - \langle S_1 S_2 \rangle, \tag{8}$$

$$p_{10} = \langle S_1 \rangle - \langle S_1 S_2 \rangle, \tag{9}$$

$$p_{11} = \langle S_1 S_2 \rangle, \tag{10}$$

where

$$\langle S_1 \rangle = \frac{g(h_1) + \Delta g_1 g(h_2)}{1 - \Delta g_1 \Delta g_2}, \tag{11}$$

$$\langle S_2 \rangle = \frac{g(h_2) + \Delta g_2 g(h_1)}{1 - \Delta g_1 \Delta g_2}, \tag{12}$$

$$\langle S_1 S_2 \rangle = \tfrac{1}{2}\{\langle S_1 \rangle g(J_{21} + h_2) + \langle S_2 \rangle g(J_{12} + h_1)\}. \tag{13}$$

For the symmetric network ($J_{21} = J_{12} = J$), the above equation reduced to $\theta_1 = 2h_1$, $\theta_2 = 2h_2$ and $\theta_3 = 2J$, and therefore the modulation in θ_3 reflects the change in the anatomical connections. On the other hand, for the asymmetric network, the underlying neural mechanisms cannot be determined uniquely, because $(\theta_1, \theta_2, \theta_3)$ are mapped from $(J_{21}, J_{12}, h_1, h_2)$. This ill-posed nature was depicted in Fig. 1.

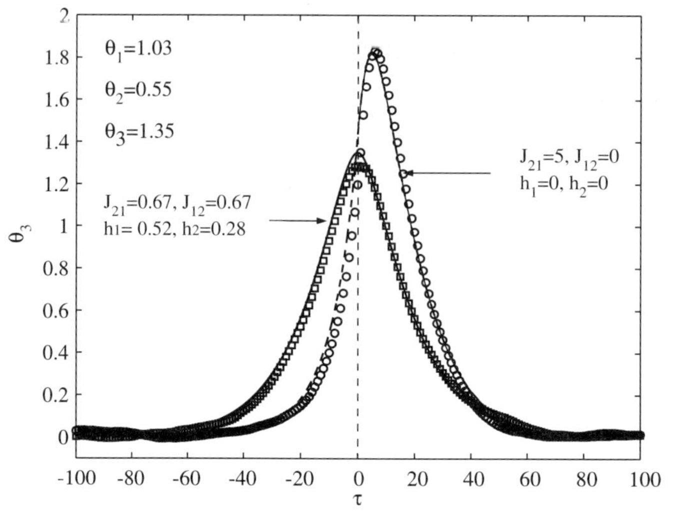

Fig. 1. Multiple neural mechanisms generate the same information-geometric measure parameters. The set of information-geometric measure parameters $(\theta_1, \theta_2, \theta_3) = (-1.03, 0.55, 1.35)$ was generated from the two neural mechanisms parameters $(J_{21}, J_{12}, h_1, h_2) = (5, 0, 0, 0)$ and $(J_{21}, J_{12}, h_1, h_2) = (0.67, 0.67, -0.52, 0.28)$. The dotted and solid curves were obtained by theoretical calculation, and the circle and square by computer simulation.

The relationship between the information-geometric measure and the correlation function was also obtained. It was given by

$$\begin{aligned}\theta_3(\tau) = {} & \log(C_{12}(\tau) + \langle S_1(t)\rangle\langle S_2(t+\tau)\rangle) \\ & + \log(C_{12}(\tau) + 1 - \langle S_1(t)\rangle - \langle S_2(t+\tau)\rangle + \langle S_1(t)\rangle\langle S_2(t+\tau)\rangle) \\ & - \log(-C_{12}(\tau) + \langle S_1(t)\rangle - \langle S_1(t)\rangle\langle S_2(t+\tau)\rangle) \\ & - \log(-C_{12}(\tau) + \langle S_2(t+\tau)\rangle - \langle S_1(t)\rangle\langle S_2(t+\tau)\rangle),\end{aligned} \tag{14}$$

where,

$$C_{12}(\tau) = \langle (S_1(t) - \langle S_1(t)\rangle)(S_2(t+\tau) - \langle S_2(t+\tau)\rangle)\rangle, \quad \tau \geqslant 0, \tag{15}$$

It is easy to show that these two measures have a nonlinear but monotonic relationship. As was shown in Fig. 2, both measures are almost identical and therefore depict a similar feature of neural firing. However, the information geometric measure has an advantage because it is easily extended to more than three neuron interactions, it provides a biologically plausible statistical test, and it generates orthogonal decomposition correlation components.

5. Summary

The relationship between the information-geometric measure parameters and the possible neural mechanisms was elucidated using a pair of stochastic neuron model. The

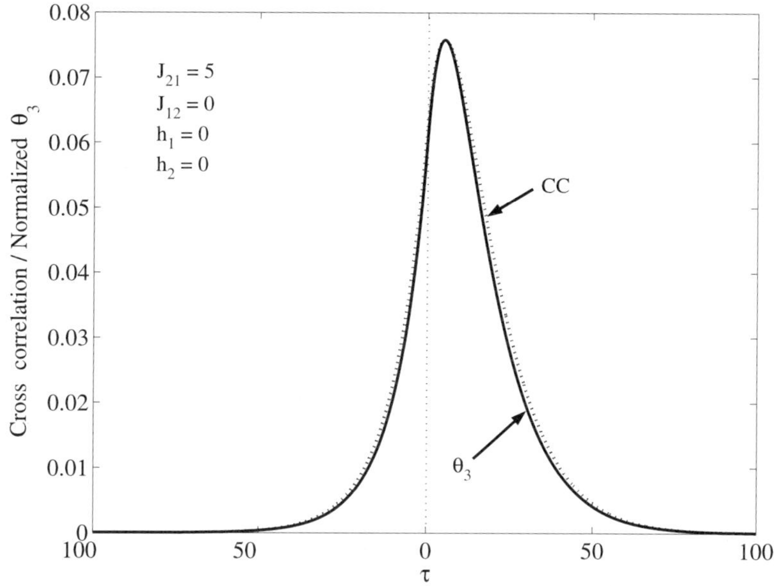

Fig. 2. Relationship between the information-geometric measure and the correlation function. Both the information-geometric measure and the correlation function were calculated for the neural mechanisms parameters $(J_{21}, J_{12}, h_1, h_2) = (5, 0, 0, 0)$. The information-geometric parameter was normalized to have the same peak value with the correlation function.

results are summarized as follows; For symmetric connections, θ_3 is only dependent on the anatomical connections ($J_{21} = J_{12}$), and θ_1 and θ_2 are only dependent on the local inputs h_1 and h_2, respectively. Therefore, the change in information-geometric parameter θ_3 reflects the change in the anatomical connections. For asymmetric connections, θ_3 is dependent both on the anatomical connections and on the local inputs. The multiple neural mechanisms $(J_{21}, J_{12}, h_1, h_2)$ can generate the same information-geometric measure parameters $(\theta_1, \theta_2, \theta_3)$, and therefore neural mechanisms cannot be determined uniquely from the information-geometric measure parameters. However, the shape of the time-delayed information-geometric measure parameters is informative to determine the direction of the connection. The relationship between the information-geometric measure parameters $\theta_3(\tau)$ and the correlation function $C_{ij}(\tau)$ was also investigated, and the nonlinear but monotonic relationship was elucidated.

References

[1] A. Aertsen, G. Gerstein, M. Habib, G. Palm, Dynamics of neuronal firing correlation: modulation of effective connectivity, J. Neurophysiol. 61 (5) (1989) 900–917.
[2] S. Amari, H. Nagaoka, Methods of Information Geometry, AMS and Oxford University Press, 2000.
[3] S. Amari, Information geometry on hierarchy of probability distributions, IEEE Trans. Inf. Theory 47 (5) (2001) 1701–1711.

[4] I. Ginzburg, H. Sompolinsky, Theory of correlations in stochastic neural networks, J. Phys. Rev. E 50 (4) (1994) 3171–3191.
[5] H. Nakahara, S. Amari, Information-geometric measure for neural spikes, Neural Comput. 14 (10) (2002) 2269–2316.
[6] H. Nakahara, S. Amari, M. Tatsuno, et al., Information geometric measures for spike firing, Soc. Neurosci. Abstr. (2001) 2178.
[7] F. Rieke, D. Warland, R.D.R.V. Steveninck, W. Bialek, Spikes, MIT Press, Cambridge, MA, 1997.

Masami Tatsuno received his Ph.D. in Physics from Waseda University, Japan, in 1997. From 1996 to 1999, he was a Research Associate of the Department of Applied Physics, Waseda University, and from 1999 to 2002, he was a postdoctoral researcher at RIKEN Brain Science Institute, Japan. Since 2002, he has been a postdoctoral researcher at the University of Arizona, USA. He is currently working on the possible neural mechanisms underlying the information-geometric measure, and the information representation of the rat cortex.

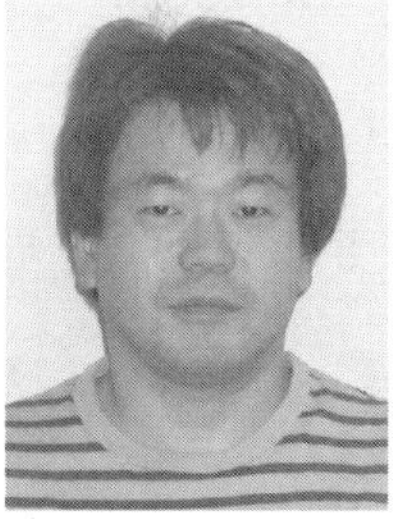

Masato Okada received his Ph.D. in Science from Osaka University, Japan, in 1997. From 1991 to 1996, he was a Research Associate of School of Engineering Science, Osaka University, Japan, and from 1996 to 2001, he was a Research Fellow of Japan Science and Technology Corporation. Since 2001, he has been the deputy lab head of Laboratory for Mathematical Neuroscience, RIKEN Brain Science Institute, Japan. His research field is physics and neuroscience including computational neuroscience, statistical physics, information theory, image processing, and learning theory.

Available at
www.ComputerScienceWeb.com
POWERED BY SCIENCE @DIRECT®

ELSEVIER

Neurocomputing 52–54 (2003) 655–660

NEUROCOMPUTING

www.elsevier.com/locate/neucom

A computational approach to arm movement on the sagittal plane performed by parietal lobe damaged patients: an attempt to examine a computational model for handwriting for its neurobiological plausibility from a neuropsychological symptom

Satoko Tsunoda[a,*], Akira Matsui[b], Yasuhiro Wada[b], Takeyuki Aiba[c], Chiyoko Nagai[d], Yumiko Uchiyama[d], Makoto Iwata[d], Shigeki Tanaka[e], Michiyo Kozawa[f], Masae Kamiyama[g], Kazuyoshi Fukuzawa[a]

[a] *Waseda University, 1-24-1 Toyama, Shinjyuku-ku, Tokyo 162-8644, Japan*
[b] *Nagaoka University of Technology, Niigata 940-2188, Japan*
[c] *Fujitsu Hyper Software Technologies, Kanagawa 222-0033, Japan*
[d] *Tokyo Women's Medical University, Tokyo 162-8666, Japan*
[e] *Jin-ai University, Fukui 915-8586, Japan*
[f] *Yukioka Hospital, Osaka 530-0021, Japan*
[g] *Kyorin University Hospital, Tokyo 181-8611, Japan*

Abstract

We measured arm movement on the sagittal plane of parietal lobe damaged patients and control subjects. In neurological patients, (1) the hand height did not become lower as the motion duration became shorter; (2) the hand jerk of measured trajectory increased when motion duration lengthened. These results suggested that the parietal lobe is involved with realizing smooth arm movement. The performance of patients with parietal lobe lesions may reflect their difficulties in planning a trajectory with smooth movement.

Keywords: Parietal lobe damaged patients; Sagittal plane movement; Hand height

* Corresponding author.
E-mail address: satoko@ruri.waseda.jp (S. Tsunoda).

0925-2312/03/$ - see front matter © 2002 Elsevier Science B.V. All rights reserved.
doi:10.1016/S0925-2312(02)00857-3

1. Introduction

Wada and Kawato [5] proposed a computational models on handwriting and reaching movement based on the minimum commanded torque change model where we plan an arm trajectory minimizing the time integral of the square of the rate of the commanded torque change.

The object function of the minimum commanded torque change model is expressed as follows:

$$C_{\mathrm{CTC}} = \frac{1}{2}\int_0^{t_f} \sum_{i=1}^{N} \left(\frac{\mathrm{d}\tau_i}{\mathrm{d}t}\right)^2 \mathrm{d}t. \tag{1}$$

Here, t_f indicates the motion duration. N is the number of joints, and τ_i is the commanded torque generated in joint i. The following two facts are mathematically deduced from the model: (1) when movement distance is constant, the time integral of the square of the rate of the commanded torque change becomes larger as the motion duration becomes shorter; (2) when movement distance is constant, the time integral of the square of the rate of the commanded torque change becomes smaller as the hand height becomes lower.

The minimum commanded torque change model reproduces smooth arm movement during reaching movement well [2–4]. Wada et al. [3,4] measured hand trajectory of the arm movement in control subjects on the sagittal plane and confirmed that hand height varies as motion duration varies. They also demonstrated that the minimum commanded torque change model explains the relationship between hand height and motion duration.

In the present study, we first measured and quantitatively compared arm movements on the sagittal plane in the parietal lobe damaged patients with agraphia and the control subject. We also attempted to explain the differences in arm movement of control subjects and the parietal lobe damaged patients using the minimum commanded torque change model.

2. Purpose

The purposes of the present study are as follows: (1) To quantitatively analyze and compare arm movements on the sagittal plane in the neurological patients with agraphia and the control subjects. (2) To investigate the functional role of the parietal lobe in arm movement.

3. Methods

3.1. Subjects

Seven neurological patients: They had the left parietal lobe lesions accompanying agraphia. Table 1 shows lesions of the patients. Control subjects: Seven normal males,

Table 1
Lesions of the patients

Patients	Age	Lesions
K	77	Left angular gyrus
H	74	Right parietal lobe lesion extending into the superior portion of the parietal lobe, left lateral of occipital lobe lesion extending into the angular gyrus prietooccipital lesion extending into left lateral horn of the ventricle
Y	60	Superior parietal lobule, postcentral gyrus, precentral gyrus area
S	41	Left angular gyrus, left temporal lobe
M	26	Subcortical area of left parietal lobe
T1	57	Extensive parietal lobe lesion
T2	53	Parietal lobe

five of which also served as subjects in Wada et al.'s studies [3,4] served as control subjects. All subjects were right-handed.

3.2. Procedure and apparatus

The subjects were required to move their hands back and forth between the starting point and the final point marked on a desk along a board placed at a right angle to a horizontal plane. They were asked to raise their hands in the air during the movement except that they were asked to touch the two points. Each trial consisted of seven back and forth movements. The subjects repeated about sixty trials with three different distances (0.200, 0.175 and 0.150 m) and three different speeds (control, fast and slow). The position of the subject's shoulder and hand were magnetically measured using FASTRAK (Polhemus Inc.) with a sampling frequency of about 60 Hz. The apparatus used in this study was described in detail in Wada et al. [3].

3.3. Data analysis

We analyzed the obtained data of hand position using following three ways.

3.3.1. Position data

We extracted the data of hand position in the Z direction (= hand height data) from the obtained position data. We represented hand height data as the exponential function of motion duration, then, we calculated the correlation coefficient between the exponential function and the hand height data of measured trajectory.

3.3.2. Hand jerk

We calculated hand jerk [1] (an index of movement smoothness) of a measured trajectory. Then, we calculated standardized hand jerk which is the hand jerk of the measured trajectory divided by the hand jerk of the minimum hand jerk trajectory. The equation of the standardized hand jerk is given by

$$\frac{C_{\mathrm{data}}}{C_{\mathrm{mj}}} = \frac{\frac{1}{2}\int_0^{t_f}\left\{\left(\frac{\mathrm{d}^3 y_{\mathrm{data}}}{\mathrm{d}t^3}\right)^2 + \left(\frac{\mathrm{d}^3 z_{\mathrm{data}}}{\mathrm{d}t^3}\right)^2\right\}\mathrm{d}t}{\frac{1}{2}\int_0^{t_f}\left\{\left(\frac{\mathrm{d}^3 y_{\mathrm{mj}}}{\mathrm{d}t^3}\right)^2 + \left(\frac{\mathrm{d}^3 z_{\mathrm{mj}}}{\mathrm{d}t^3}\right)^2\right\}\mathrm{d}t}, \tag{2}$$

where data is measured trajectory, mj is minimum hand jerk trajectory, t_f is movement time and x, y, z shows hand position.

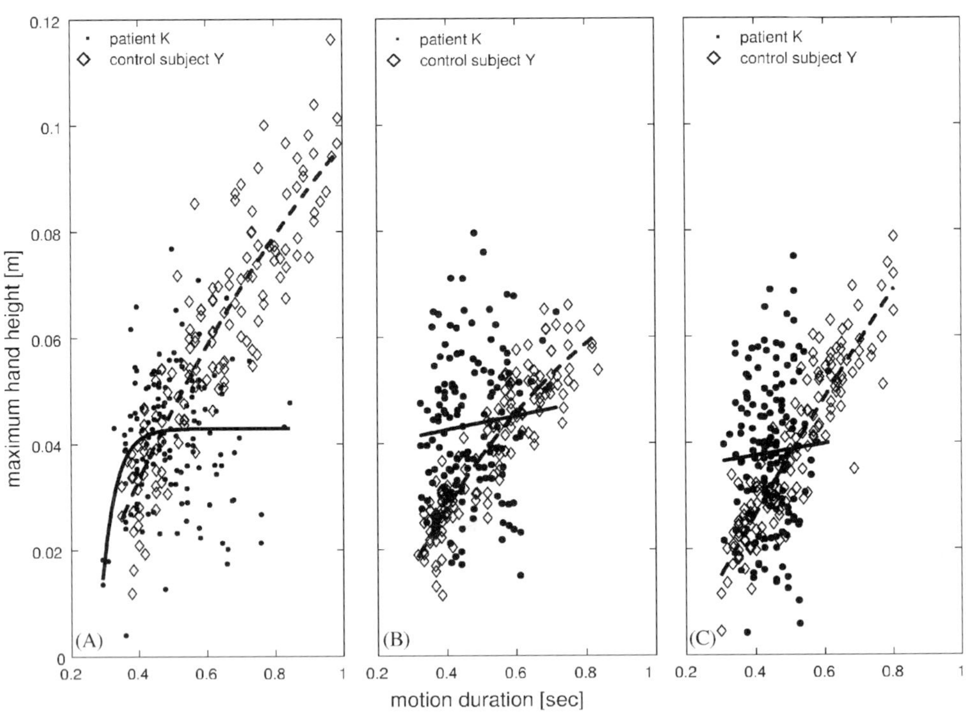

Fig. 1. Maximum hand height data and approximated exponential curve in (A) distance = 0.200 m, (B) distance = 0.175 m and (C) distance = 0.150 m. In control Y (diamonds and dashed lines), hand height tended to become lower as the motion duration became shorter in all movement distances. However, in patient K (dots and lines), hand height did not become lower as the function of the motion duration.

[1] We employed hand jerk model [1] to analyze the data obtained in the present study for mathematically simplicity.

4. Result and discussion

4.1. Prediction of C_{CTC} and the relationship between hand height and motion duration

In the measured trajectory of control subjects, hand height tended to become lower as the motion duration became shorter in all movement distances (Fig. 1). In neurological patients, however, hand height did not become lower as the function of the motion duration. In control subjects the correlation coefficient between the exponential function and the measured proved to be high while that of the neurological patients proved to be low. The low correlation coefficient in the neurological patients suggests that they could not minimize the commanded torque which is produced as motion duration became shorter.

4.2. Movement smoothness

We analyzed the relationship between standardized hand jerk and motion duration to examine movement smoothness. The standardized hand jerk (an index of movement smoothness) in the neurological patients became larger when motion duration became longer (Fig. 2). On the other hand, the standardized hand jerk of the normal subjects was smaller than that of the neurological patients. Furthermore, the hand jerk of measured trajectory of normal subjects varied only slightly when motion duration became longer.

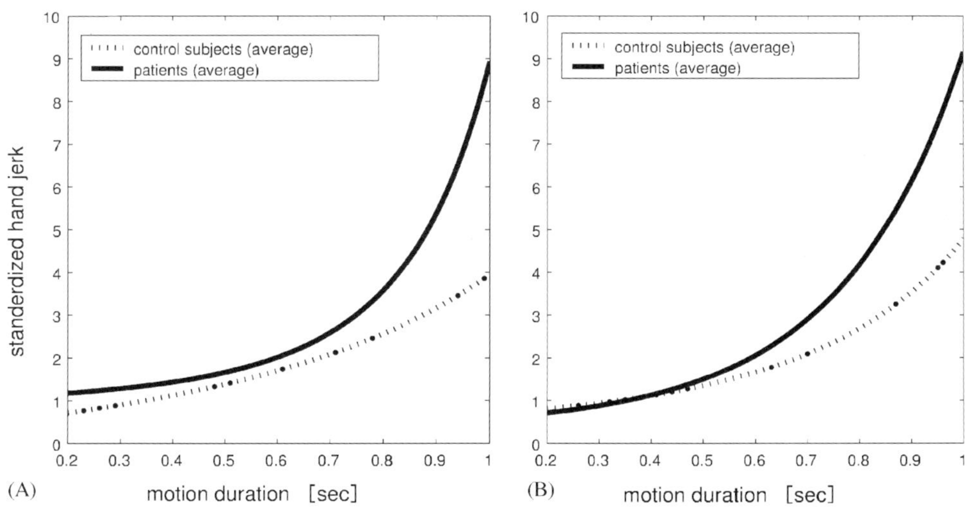

Fig. 2. Comparison between the standardized hand jerk of the control subjects and that of the patients in (A) movement from starting point to end point, (B) movement from end point to starting point. The standardized hand jerk in the neurological patients increased when motion duration became longer. On the other hand, the standardized hand jerk of the normal subjects was rather small in longer motion duration.

The results suggest that the movement of the neurological patients was not as smooth as in control subjects when the motion duration became longer assuming that the value of standardized hand jerk is an index of movement smoothness.

5. Conclusion

The result of the present study suggests that the control subjects can realize the smooth arm movement on the sagittal plane by lowering hand height as motion duration became longer. On the other hand, it is suggested that the neurological subjects cannot move their hands like the control subjects. It is further suggested that the parietal lobe is involved in realizing smooth arm movement.

References

[1] T. Flash, N. Hogan, The coordination of arm movements: an experimentally confirmed mathematical model, J. Neurosci. 5 (1985) 1688–1703.

[2] E. Nakano, H. Imamizu, R. Osu, Y. Uno, H. Gomi, M. Kawato, Quantitative examinations of internal representations for arm trajectory planning: minimum commanded torque change model, J. Neurophysiol. 81 (5) (1999) 2140–2155.

[3] Y. Wada, T. Aiba, K. Fukuzawa, An examination of a relation between hand height position and motion duration on the human sagittal plane movement based on the minimization principle, Neurocomputing 38–40 (2001a) 1581–1587.

[4] Y. Wada, T. Aiba, K. Fukuzawa, A computational trajectory model for the human sagittal plane movement at various motion durations, Neurocomputing 38–40 (2001b) 1589–1594.

[5] Y. Wada, M. Kawato, Theory for cursive handwriting based on the minimization principle, Biol. Cybernet. 73 (1995) 3–13.

Satoko Tsunoda

Available at
www.ComputerScienceWeb.com
POWERED BY SCIENCE @DIRECT®

ELSEVIER

Neurocomputing 52–54 (2003) 661–669

NEUROCOMPUTING

www.elsevier.com/locate/neucom

Behaviors from an electrically stimulated spinal cord neuronal network cultured on microelectrode arrays

A. Novellino[a,*], M. Chiappalone[a], A. Vato[a], M. Bove[b], M.B. Tedesco[a], S. Martionia[a]

[a]*Neuroengineering and Bio-nanoTechnology, NBT, Department of Biophysical and Electronic Engineering, DIBE, University of Genova, Genova, Italy*

[b]*Department of Experimental Medicine-DIMES, Section of Human Physiology, University of Genova, Genova, Italy*

Abstract

The spontaneous electrophysiological activity of neural networks seems to play an important role in the Central Nervous System (CNS) developing, subsequent maturation and learning. Learning a new behavior is an exploration process that involves the modulation and the formation of association set between stimuli and responses. Here, we analyze how the electrophysiological activity of cultured spinal cord neurons (14 DIV) from the chick embryo is affected by electrical stimulation. Active neurons show a typical high frequency activity pattern called burst. Induced changes in the patterns of electrophysiological activity are described.

Keywords: Microelectrode array; Spinal cord network; Burst activity; Electrical stimulation; Network dynamics

1. Introduction

It is well known that neuronal networks in the developing spinal cord are spontaneously active. Evidence is also accumulating that such activity plays an important role in the maturation of the networks [4]. Cultured spinal neurons from the chick embryo are chosen as a neurobiological system quite appropriate for long-term recording.

* Corresponding author. University of Genova, Via all' Opera Pia 11a, Genova, Italy. Tel.: +39-010-353-2765; fax: +39-010-353-2133.

E-mail address: antonio.novellino@dibe.unige.it (A. Novellino).

0925-2312/03/$ - see front matter © 2002 Elsevier Science B.V. All rights reserved.
doi:10.1016/S0925-2312(02)00861-5

In this paper, we analyze the signals generated by neurons dissociated from the embryo spinal cord in terms of electrophysiological activity, by taking advantages of arrays of planar microelectrodes [5].

Two main features make microelectrode arrays (MEAs) a valuable tool for electrophysiology [1], namely: (a) they are non-invasive and therefore, under appropriate conditions, they can register the electrophysiological activity of neurons for a long period of time (i.e., from several minutes up to several hours) and (b) allow a multi-site recording [7]. Electrical activity from the network is simultaneously recorded by 8 electrodes. Thus, it is possible to analyze changes induced by the electrical stimulation and correlate the activity between different channels (i.e., different area in the neural network), in order to define the dynamics of the network.

Electrophysiological rhythmic activity mostly ranges from asynchronous spiking to organized patterns in the form of "bursting". A burst is characterized by a rapid sequence of several spikes separated from each other by a few ms (Inter Spike interval, ISI) and the interval among bursts is in the scale of seconds or even minutes. In other words, from the point of view of the signal patterns, we deal with time sequences where episodes in the range of seconds (bursts) are separated by periods of silence in the tens of seconds range (Inter Burst interval, IBI).

It should be underlined that, because of a number of sources of biological variability, the signals resulting from one specific experiment show a similar pattern but not a one to one reproduction of other experiments. The main aim of this work is to show that is possible to reversibly modify the neurons spatio-temporal dynamics by electrically stimulating the neural network.

2. Methodology

Dissociated neurons were obtained from the spinal cord of chick embryos after 7–8 days of incubation, according to a modified version of the protocol described in [3]. Neurons were seeded on microelectrode arrays covered with adhesion promoting molecules (Polylisyne, laminin).

Electrophysiological signals were recorded after 20 days in vitro (DIV), which need to allow the formation of synaptic contacts among the cells.

2.1. Experimental protocol

In order to investigate the role of the activity-dependent modification of synaptic strength in the developing Central Nervous System, an electrical stimulation protocol was used [8]. The experimental protocol consisted of the following steps:

(1) Control condition, corresponding to the spontaneous activity in culture medium (Neuro Basal medium, Sigma).
(2) Electrical stimulation test, single pulse stimulus (amplitude: 0.3–0.1 V, lasting 100 μs) delivered for 2 min at 0.5 Hz.

(3) Post electrical stimulation control condition.
(4) High K^+ Condition, corresponding to the activity under the effects of K^+, added at the final concentration of 9 mM.
(5) Resting, the network is left 2 h in the incubator to verify healthy condition of neurons.
(6) Post incubator control condition.
(7) Tetanic stimulation, delivered by trains of stimuli (each ones of 0.3–0.1 V, lasting 100 μs, at 20 Hz) each train at 0.2 Hz.
(8) Post-tetanic stimulation control condition.

2.2. Microelectrode array and measurement system

The array used for our experiments is made of glass with 60 gold microelectrodes. An experimental set up, based on the microelectrode array and constituted by the following functional elements, was developed:

(1) Microelectrode array, which is itself an interface between the biological and the electrical environment;
(2) Faraday Cage, to avoid electromagnetic interference;
(3) 8 channels amplifier and filtering stage (gain = 1000);
(4) Long-term acquisition instrumentation: Digital Tape Recorder (BioLogic DTR-1802) with a maximum of 8 recording channels at the sampling frequency of 12 kHz and GPIB Interface. Each channel is connected to the MEA via the amplification and filtering system. For an easier identification, we named them according to the labels on the DTR display: L1, L2, L3, L4, R1, R2, R3, R4.
(5) Oscilloscope for real time monitoring of signals;
(6) System for network electrical stimulation (MUX, Stimulating Interface, and Isolator);
(7) PC for data management, equipped with National Instruments$^{(TM)}$ AT-MIO device, used for generating electrical stimuli.

2.3. Signal processing

Signals collected from a microelectrode array have typical amplitudes in the range of 0.1–0.4 mV and are embedded in biological and thermal noise ranging from 10 V up to 40 V peak-to-peak. To extract spike features we processed data by an ad hoc algorithm of peak-detection, which recognizes a potential of action when the difference between the maximum and the minimum, of the considered window (bins of 5 ms.) of the signal, is higher than peak-to-peak set threshold. Moreover, to investigate burst patterns [6], we developed an algorithm for their automatic detection, utilizing pre-processed data by the peak-detection algorithm. A window is shifted along the processed signal and when a spike cluster is detected a graph is shown to the user. This graph represents the time of occurrence of spike clusters and their amplitudes in arbitrary units, which represent the sum of spikes amplitudes that belong to the same cluster.

While in spiking-processing we extracted the number of spikes vs. bins, the extracted features for each burst detected are the following:

- Time of occurrence (ms).
- Burst duration (ms).
- InterBurst interval–IBI (ms), defined as the time length between the end of a burst and the beginning of the next one.
- Burst amplitude (arbitrary units).

3. Results

Chicken spinal cord dissociated neurons were cultured on the substrates and the experiment was done after 14 DIV (days in vitro) cultures.

Firstly we processed the spontaneous activity, control condition, by means of a bursting (Figs. 1–3) and ISI (Figs. 4, 8, 9) analyses. This is a useful step to define a reference state for further analyses.

It should be observe that the electrophysiological pattern in control condition is about the same for all the channels: bursts, lasting about 5.6 s, characterized by about 26.6 s of IBI. Synchronization of network may be appreciated observing the ISI behaviors (Fig. 4), which are about the same for each channel and show that two spike are mostly time separated by less than 10 ms.

After the electrical stimulation phase, the spinal neurons respond with significantly different signals, in which bursting seems to be suppressed (Figs. 5, 8, 9), as shown by Jimbo et al. [9].

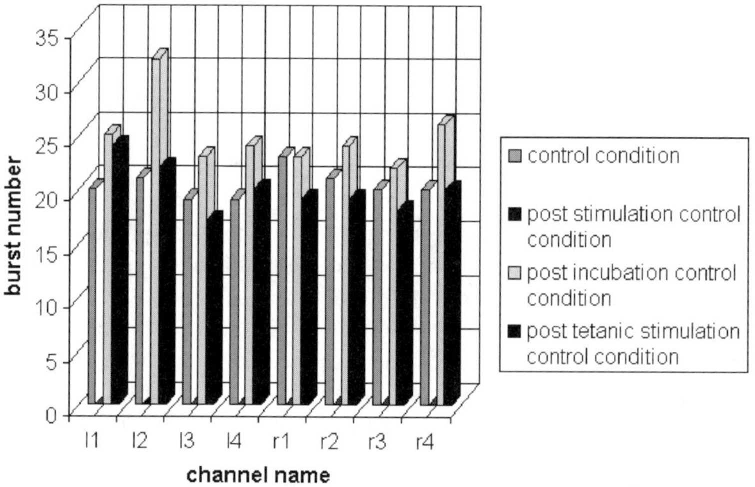

Fig. 1. Thanks to electrical stimulation is possible to change the network activity in term of number of bursts.

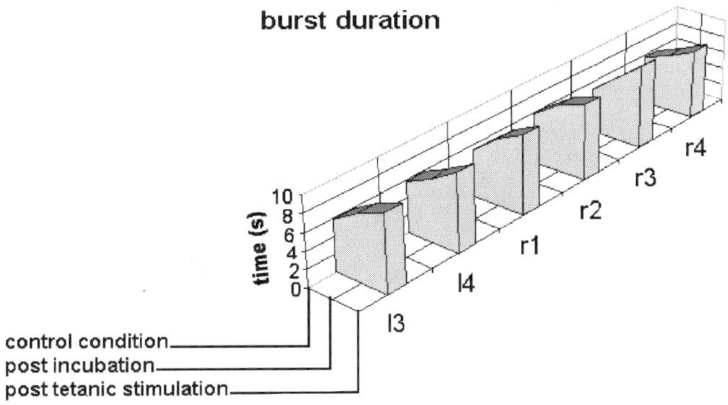

Fig. 2. During the three phases observed the burst duration increases.

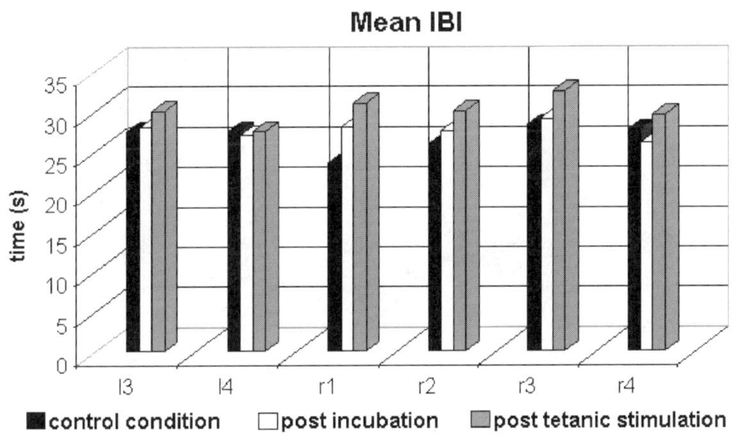

Fig. 3. The mean inter burst interval is a parameter to indicate the intensity of network activity.

A high concentration of K^+ is added to culture medium. This is done to block intra cellular K^+ natural flux to the extra cellular sites and to increase the excitability of network depressed by electrical stimulation. Initially the network continues post stimulus behavior, while after 2 h of incubation it is observable that the spinal neurons deliver a spontaneous activity (Figs. 1–3) characterized by IBI of about 27.6 s, higher (Fig. 3) than control condition one, with burst duration of about 7.4 s (Fig. 2). In this experimental phase, the spontaneous activity is higher than the previous one: neurons seem to recover their synchronized spontaneous activity, in term of burst duration

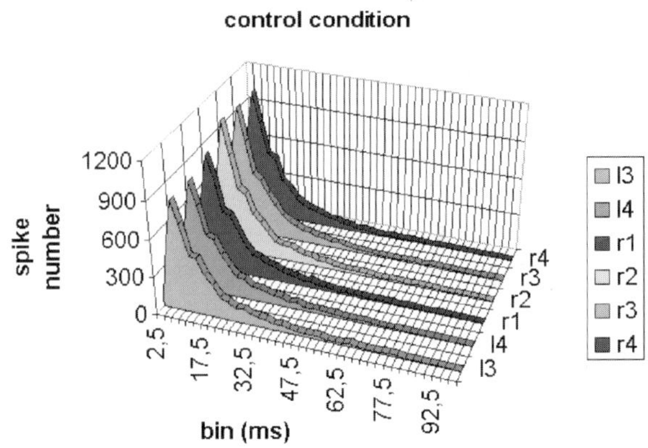

Fig. 4. The inter spike interval histogram is useful to define the level of the synchronization in the network behavior.

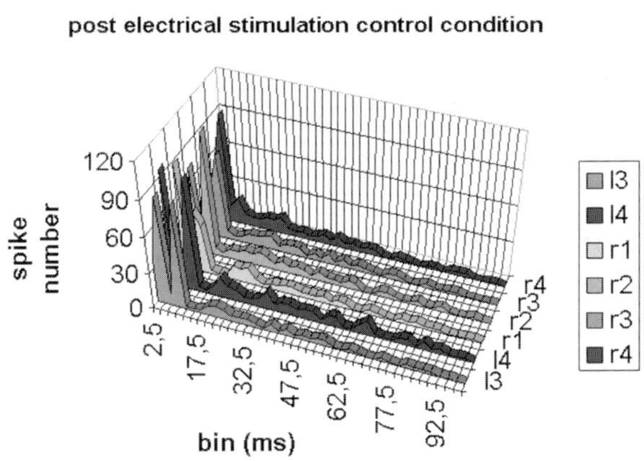

Fig. 5. After the electrical stimulation the activity decreases and the ISI histogram becomes very different from the control condition one.

(Fig. 2) and ISI (Fig. 6), but some changes inside burst can be observed: ISI is now mostly ranged between 5 and 7.5 ms, as shown in Figs. 8 and 9. Moreover, it may observable that the new state is characterized by more and longer burst (Figs. 1 and 2) with lower IBI (Fig. 3) than control condition.

The tetanic phase shows that the electrical stimulation induces a new depression of activity, both in terms of bursting (Figs. 1–3) and spiking (Figs. 7–9), which is less drastic than previous, post electrical stimulation, depression.

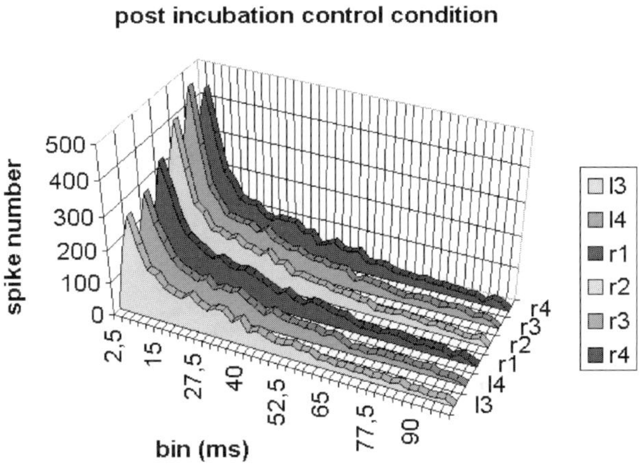

Fig. 6. After 2 h of incubation the spinal neurons deliver a spontaneous activity with a different level of synchronization observable by a different shape of the ISI histogram.

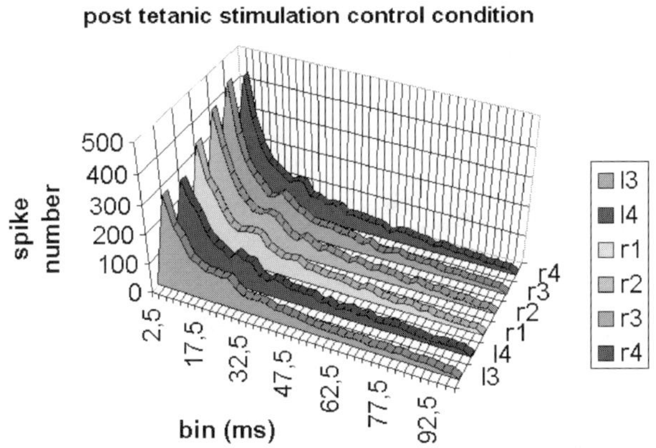

Fig. 7. This figure shows that after the titanic stimulation the shape of the ISI histogram is very similar to the post incubator one.

4. Discussion and prospects

In this study, we have examined how a large number of synaptic pathways are affected by electrical and tetanic stimulation, with the aim to characterize how a complex spinal cord network respond to an external input and whether learning will occur. Our result gives some insight into the changes that happen inside in the network, in particular we find that electrical stimulation produces long lasting changes in the neurons, which are measured as changes in the number of extra cellular spikes recorded from the system.

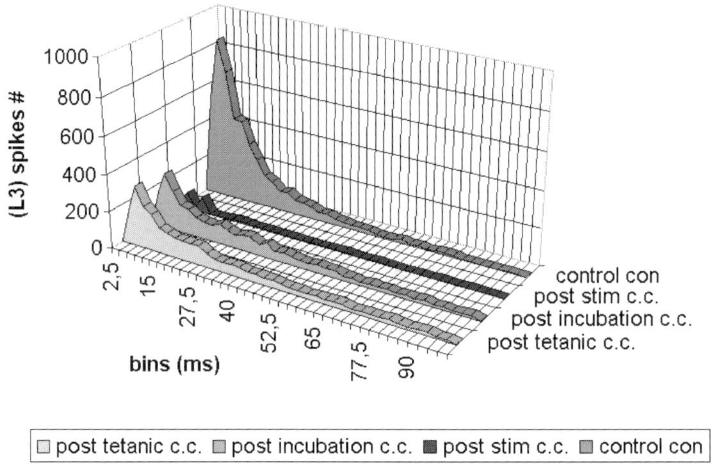

Fig. 8. During the experiment the shape of the ISI histogram changes in all channels.

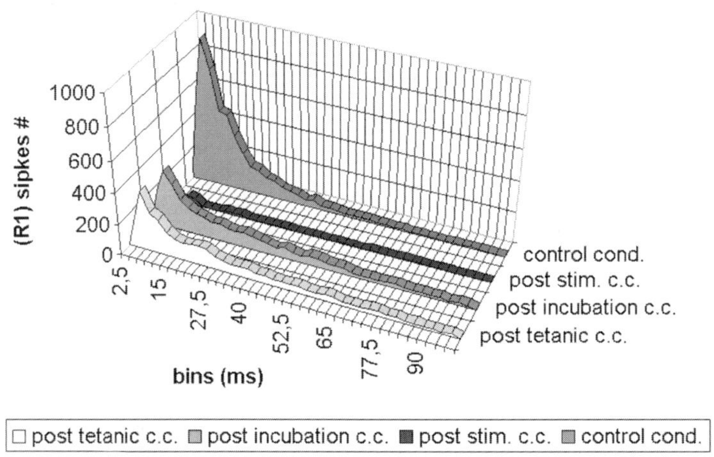

Fig. 9. The ISI histogram trend in the channel R1 is very similar to the channel L3 one.

First of all we observe that the spontaneous activity of network shows a high level of synchronization (Fig. 4).

After the electrical stimulation, the network changes behavior, in which bursting seems to be depressed (Figs. 1 and 5).

This can be explained, in the activity-dependent modification terms as a long-term-depression (LTD) [8]. Another possibility might be that electrical stimulation produces totally release of synaptic neurotransmitters. After the incubation, network increases its activity (Figs. 1–2, 3,6). This fact can validate the hypothesis of high release of neurotransmitters and confirm that electrical stimulation produce a long lasting but reversible effect on neural network dynamics. The tetanic stimulation induces depression of activity, but network is still synchronized in terms of events (Fig. 7). This depression is less drastic than previous one, so it is possible to note that network is delivering stimulation frequencies sensitive; in other words it seems to mean that it may be possible to induce different kind of activity-dependent changes according the exposed frequencies of electrical stimulation. The difference among these responses and the frequencies of these patterns (the IBI between patterns and ISI, between events, inside patterns) represent the current state of the network, so these behaviors may underline whether learning is occurring and how the network is encoding new features.

Moreover, we observe that in all the experimental phases the network shows time synchronized activity among all considered sites. It seems to mean that induced, by the electrical stimulation, modifications are not a local effect, but refer to a complex spatio-temporal network dynamics.

As very recently suggested in literature [11,2,10], this kind of approach will open new opportunity for study in learning in-vitro and for using bio-artificial neural networks as new possible information processing tool [10].

References

[1] M. Bove, M. Grattarola, S. Martinioa, G. Verreschi, Interfacing cultured neurons to planar substrate microelectrodes: characterization of the neuron-to-microelectrode junction, Bioelectrochem. Bioeng. 38 (1995) 255–265.

[2] T.B. DeMarse, D.A. Wagenaar, A.W. Blau, S.M. Potter, The neurally controlled animat: biological brains acting with simulated bodies, Autonomous Robots 11 (2001) 305–310.

[3] M. Dichter, G. Fischbach, The action potential of chick dorsal root ganglion neurons maintained in cell culture, J. Physiol. 276 (1977) 281–298.

[4] R.D. Fields, P.G. Nelson, Activity-dependent development of the vertebrate nervous system, Int. Rev. Neurobiol. 34 (1992) 133–214.

[5] G.W. Gross, Simultaneous single unit recording in vitro with a photo etched laser deinsulated gold multielectrode surface, IEEE Trans. Biomed. Eng. 26 (1979) 273–279.

[6] G.W. Gross, Internal dynamics of randomized mammalian networks in culture, in: D.A. Stenger, T.M. McKenna (Eds.), Enabling Technologies for Cultured Neural Networks, Academic Press, New York, 1994 (Chapter 13).

[7] Y. Jimbo, A. Kawana, Electrical stimulation and recording from cultured neurons using a planar array, Bioelectrochem. Bioeng. 40 (1992) 193–204.

[8] Y. Jimbo, H.P.C. Robinson, A. Kawana, Strengthening of synchronized activity by tetanic stimulation in cortical culture: application of planar electrode arrays, IEEE Trans. Biomed. Eng. 45 (1998) 1297–1304.

[9] Y. Jimbo, T. Tateno, H.P.C. Robinson, Simultaneous induction of pathway-specific potentiation and depression in networks of cortical neurons, Biophys. J. 76 (1999) 670–678.

[10] S.M. Potter, Distributed processing in cultured neuronal networks, Prog. Brain Res. 130 (2001) 49–62.

[11] G. Shahaf, S. Marom, Learning in networks of cortical neurons, J. Neurosc. 21 (2001) 8782–8788.

Available at
www.ComputerScienceWeb.com
POWERED BY SCIENCE @ DIRECT®

ELSEVIER

Neurocomputing 52–54 (2003) 671–676

NEUROCOMPUTING

www.elsevier.com/locate/neucom

Layer 4C in monkey V1 may linearize the output of the LGN

M. Brandon Westover*, Charles H. Anderson

Department of Anatomy and Neurobiology, School of Medicine, Washington University, 4496 Maryland Avenue 1-8, St Louis, MO 63110, USA

Abstract

In primates, most LGN fibers terminate in cortical layer 4C, an anatomically prominent structure of unexplained function. We hypothesize that the enormous number of cells in layer 4C of monkey primate visual cortex functions as a neural network "hidden layer" that inverts distortions introduced by transmitting visual signals through the LGN. This hypothesis helps explain how simple cells respond (quasi-) linearly to visual inputs in spite of nonlinearities present in LGN responses. Linearization averts prematurely discarding visual information, in keeping with the role of primary visual cortex as *the* source of raw visual information to the rest of the brain.

Keywords: Layer 4; Visual cortex; Nonlinearity; Distortion

The transfer of sensory information to the visual cortex occurs primarily at the interface between afferents from the LGN and a large population of tiny, densely packed "spiny stellate" cells in cortical layer 4C. There is an enormous mismatch between the number of incoming LGN fibers and the number of recipient 4C neurons, leading to the estimate that each LGN fiber branches to connect with 30–100 4C neurons [2,5]. Layer 4C also serves as the primary source of visual input to the simple cell population in cortical layer 2/3, each simple cell receiving inputs from numerous 4C cells.

It would appear that layer 4C is strategically situated to perform some kind of conditioning of visual signals before "turning them over" for further analysis by the other cortical layers. However, the function of layer 4C has been hard to decipher. For example, at other stages in the visual hierarchy, new computations are reflected in a

* Corresponding author.
E-mail address: westovem@medicine.wustl.edu (M.B. Westover).

0925-2312/03/$ - see front matter
doi:10.1016/S0925-2312(02)00863-9

progressively more complex selectivity of cellular responses for features of the visual input. On the other hand, layer 4C cells appear to simply inherit the receptive fields of their LGN drivers.

There have been two main previous hypotheses. First, based on the large fanout of incoming LGN fibers, Horace Barlow suggested that layer 4C might serve as an "interpolating grid" [2]. Image samples conveyed by LGN afferents would be interpolated by the more numerous, densely packed 4C cells, and this higher resolution neural image would serve as the readout underlying visual hyperacuity. In another hypothesis, Anderson and Van Essen proposed that layer 4C serves as a grid for rescaling and rerouting visual information, as part of the operation of a "shifter circuit" [1]. Subsequent experimental work has failed to validate key predictions of either proposal (c.f. [3]).

In this paper we outline a new proposal for the function of layer 4C. To motivate our proposal, we first discuss in detail an often overlooked problem encountered in early visual processing, namely, that visual signals undergo *nonlinear distortion* during transmission from retina to cortex. This fact at first appears inconsistent with the well-known observation that simple cells show an essentially *linear* dependence on visual inputs. By what mechanism is linearity achieved, and why? *We propose that layer 4C functions to reverse the distortion, and by doing so, prevents the loss of visual information.* We provide illustrations from computer simulations to demonstrate the principles of our proposal, and highlight ways in which our hypothesis may help to explain the experimentally observed linearity of simple cell physiology.

The early stages of visual processing admit a straightforward computational description: In response to images projected on the retina, the firing rates of ganglion-, lateral geniculate-, spiny stellate-, and simple cells are well predicted using models which compute the inner product (spatial integral of the product) of a linear filter with the incoming image, and pass the result through a static nonlinear function. We use the notation $\mathscr{I}, \mathscr{F}$, and g for the image, linear filter, and static nonlinearity, respectively. Then we express a neural firing rate in terms of the input image as

$$r(\mathscr{I}) = g[< \mathscr{I}, \mathscr{F} > - \theta]$$

where θ is a threshold, and $\langle , \rangle$ denotes the inner product. Here we have neglected the effects of noise, but will discuss these later. For simple cells, the underlying linear filter is usually described as a Gaussian-weighted sine wave, i.e. a Gabor function. Current evidence suggests that these receptive fields arise basically in a feedforward fashion from alternating stripes of ON and OFF centers of LGN receptive fields, as originally proposed by Hubel and Wiesel [4]. However, completely explaining the linear behavior of simple cells with the Hubel and Wiesel model is complicated by the presence of the static nonlinearity in the retinal ganglion and LGN cells, g. The essential features of the LGN nonlinearity are saturation at high contrasts, and nonnegativity (firing rates are positive quantities). We can conveniently model this nonlinearity using the sigmoid function $g(x) = kx^2/(\sigma + x^2)$, where k and σ are constants determining the maximum firing rate and half saturation level.

The essential problem introduced by saturation is illustrated in Fig. 1. Consider a time-varying sinusoidal signal transmitted down the retinogeniculate pathway by pairs

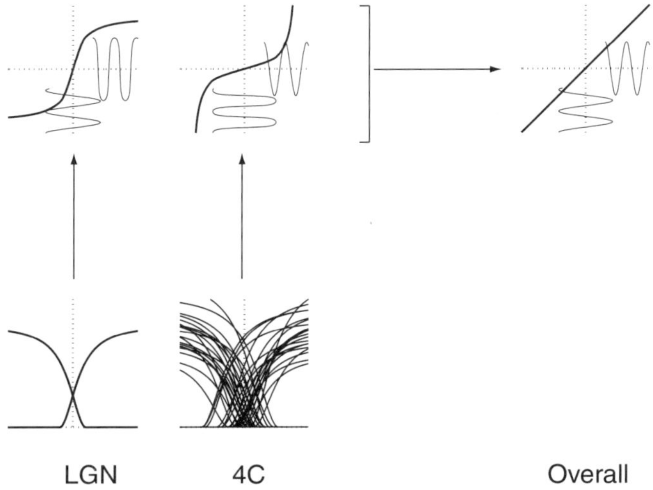

Fig. 1. Bottom: Response curves for an ON/OFF pair of LGN relay neurons (left) and their postsynaptic layer 4C targets. Top: Amplitude transfer function for visual signals using responses of the ON/OFF pair (left), and using layer 4C as a distortion inverter (middle). Combining these in tandem yields an overall linear transfer (right).

of ON and OFF RGCs and LGN RCs. As shown on the left in Fig. 1, the ON and OFF responses can be combined (by taking the difference) to capture both the positive and negative components of the visual signal. Using this combined representation highlights the effects of saturation. Saturation at high signal amplitudes produces a compressed sinewave as the output of RGCs, and the effect is cascaded through the LGN to produce more pronounced distortion in the output of the ON/OFF pair of LGN cells. In the following paragraphs we first explain how layer 4C can "linearize" the overall transfer of signals from the retina to simple cells by reversing the effects of saturation. We postpone addressing the *benefits* of linearization to allow us to frame the discussion in the context of the hypothetical question, "what if layer 4C did not exist?"

The implementation of our hypothesis is simple. As stated above, each LGN cell arborizes to contact 30–100 layer 4C cells. Due to this direct feedforward connection, each spiny stellate cell will inherit the basic ON or OFF, difference-of-Gaussians-shaped receptive field of its presynaptic LGN afferent. Suppose, however, that there is moderate diversity amongst either the form of the static nonlinearities g for the cells, e.g. different maximum firing rates, or some scatter among the thresholds θ_i. Then by taking an appropriate synaptically weighted sum of ON and OFF spiny stellate cell inputs to a simple cell, the overall amplitude transfer function characterizing the relay LGN to the simple cell input can be made to approximate the inverse of the preceding distortion function, achieving an overall linear transfer of the visual signal to the target simple cell. In other words, we propose that layer 4C serves as a "hidden layer", in which the response curves of its neurons serve as basis functions with which the inverse of the RGC/LGN amplitude transfer function is computed. For our example, we have used

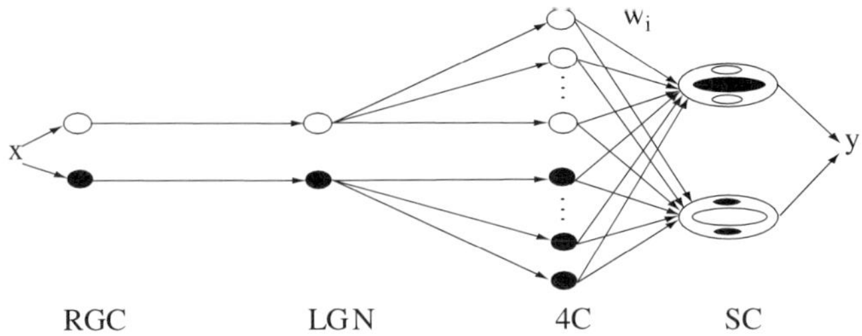

Fig. 2. Schematic of the feedforward neuronal implementation of our proposal. Pairs of neurons are used to represent both positive and negative signal components.

30 ON and 30 OFF neurons, and have computed weights w_i by the method of least squares, i.e. by minimizing

$$E(\{w_i\})\left\langle\left\|g^{-1}(x)-\sum_i w_i a_i(x)\right\|^2\right\rangle_x,$$

where a_i are the response functions $(''g's'')$ for layer 4 cells, $x=\langle\mathscr{I},\mathscr{F}\rangle$, and $\langle,\rangle_x$ indicates an average over the ensemble of possible scalar inputs x to the cells, which for simplicity we assumed as uniform. In this case, the simple cell response will not show evidence of the distortion, but will have a linear response, consistent with experimental observations.

We now discuss the benefits of removing distortions, which we refer to as "linearization". The most immediate benefit is that linearization reformats LGN signals in such a way that the responses of simple cells saturate less. The consequence is that simple cell responses provide more information about the initial retinal input than if they were directly fed the LGN output. To qualitatively illustrate this point, we performed the following numerical experiments: Suppose that the visual input to the static nonlinearity g of a single ON/OFF pair of RGC is a scalar valued, zero-mean random variable x, with variance σ^2; i.e., $x\sim\mathscr{N}(0,\sigma^2)$. The variance is chosen so that the RGC and LGN responses occasionally saturate. Also, assume that at each synapse, the signal is corrupted by additive Gaussian noise, where the noise sources at each stage are independent. The firing rate of an "antiphase pair" of simple cells (obtained by taking the difference of the responses of the simple cells in Fig. 2) is then a (non-Gaussian) random variable y (Fig. 3).

We produced 1,000,000 random sample input/output pairs (x,y) and used these to form Monte-Carlo histogram estimates of the probability densities $p(y)$ and $p(y|x)$. We used these histograms and the assumed pdf for x, $p(x)$, to compute the Shannon information (mutual information) between the visual signals and the firing rates of the model simple cell pair, given by

$$I(y;x)=\sum_{x,y}p(x)p(y|x)\log\frac{p(y|x)}{p(y)}.$$

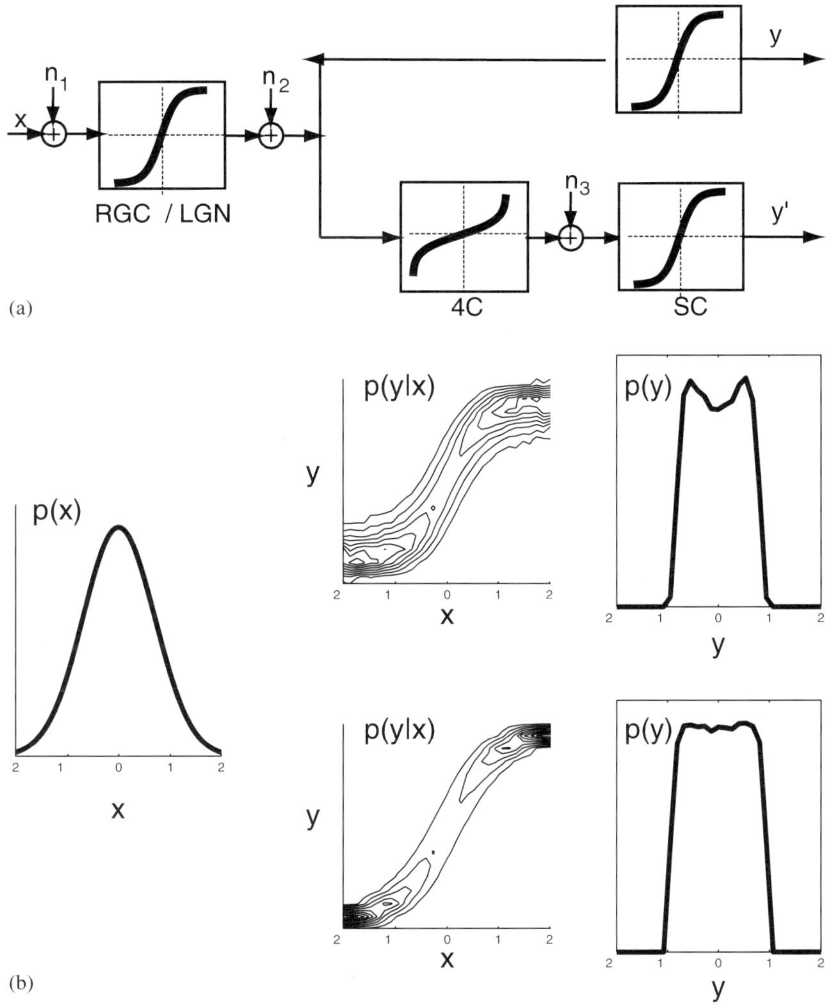

Fig. 3. Top: Block diagram for the numerical experiments described in the text. Top pathway, without layer 4C; bottom pathway, layer 4C acting as distortion inverter. Bottom: Plots and contour plots for estimated pdf's (see text).

We repeated these experiments for two cases: with and without the layer 4C "inverter", for several different noise levels. We find generally that so long as the noise involved in the inversion process is small compared to the noise already present in the incoming signals, linearization by layer 4C produces substantial improvements in the overall information transfer $I(y;x)$. Notice that the benefits will be greatest for simple cells responding to low spatial frequency information. This is because these simple cells suppress the noise.

More speculatively, we suppose that distortion inversion for low spatial frequency information is important for visual perception. Most of the information in natural images resides in the low spatial frequency bands; residual distortions would be noticeable if not inverted.

We have outlined a novel computational hypothesis for the function of layer 4C in primate visual cortex. A testable prediction of our hypothesis is that there exists a diversity of neural response curves in layer 4C. To our knowledge, the data to confirm of refute this prediction does not yet exist.

A final interesting related question for further study is whether a similar operation takes place in cats, which lack the densely packed layer 4C found in primates [3]. Instead, a similar though less dramatic fanout ($1 \rightarrow 4$–10 in cats vs. $1 \rightarrow 30$–100 in primates) occurs in cat LGN. We hypothesize that this fanout in cat LGN may serve the same purpose as we have hypothesized for layer 4C in primates.

References

[1] C.H. Anderson, D.C. Van Essen, Shifter circuits: a computational strategy for dynamic aspects of visual processing, PNAS 84 (1987) 6297–6301.

[2] H.B. Barlow, Critical limiting factors in the design of the eye and visual cortex. The Ferrier Lecture, 1980, Proc. R. Soc. London Ser. B 212 (1981) 1–34.

[3] S. Edelman, Y. Weiss, Vision: Hyperacuity, in: M. Arbib (Ed.), The Handbook of Brain Theory and Neural Networks, MIT Press, Cambridge, MA, 1998.

[4] D. Ferster, K.D. Miller, Neural mechanisms of orientation selectivity in the visual cortex, Ann. Rev. Neurosci. 23 (2000) 441–471.

[5] J.S. Lund, Local circuit neurons of Macaque monkey striate cortex: I. Neurons of Laminae 4C and 5A, J. Comp. Neuol. 257 (1987) 60–92.

Available at
www.ComputerScienceWeb.com
POWERED BY SCIENCE @ DIRECT®

ELSEVIER

Neurocomputing 52–54 (2003) 677–682

NEUROCOMPUTING

www.elsevier.com/locate/neucom

Comparing neurophysiological measurements of simulated and real brains

Peter Andras*

Department of Psychology, School of Computing Science, Neural Systems Group, University of Newcastle upon Tyne, Claremont Tower, Newcastle upon Tyne, Ridley Building, Newcastle NE1 7RU, UK

Abstract

The paper presents the comparison of simulated brains with real brains by means of statistical properties of neurophysiological measurements. Two neural models are selected to build a simulated column of the primary visual cortex. The first simulation uses classical neuron models, the second uses more complex neural assembly models that perform computations by using emerging spatio-temporal firing patterns. The comparison indicates that the second, more complex model produces results that are more similar to recordings from real brains.

Keywords: Chaos; Cortex model; Neural networks; Neurophysiological comparison

1. Introduction

There are several neuron and neural assembly models [3]. An important issue is to investigate to what extent these models produce similar results to real brains in simulated natural conditions. Such evaluations can provide the basis for selection of models for more detailed investigations.

Here we analyse two simulated cortical columns. The first simulated cortex uses classical model neural networks (e.g. [5]). The neurons have simple tuning curves, enabling them to select local features of the stimulus, and they interact to select the global properties of the incoming information (e.g., edges in the visual input). The second model is built by neural networks that use spatio-temporal activity patterns

* Tel.: +44-191-222-5790; fax: +44-191-222-5622.
E-mail address: peter.andras@ncl.ac.uk (P. Andras).

0925-2312/03/$ - see front matter © 2002 Elsevier Science B.V. All rights reserved.
doi:10.1016/S0925-2312(02)00729-4

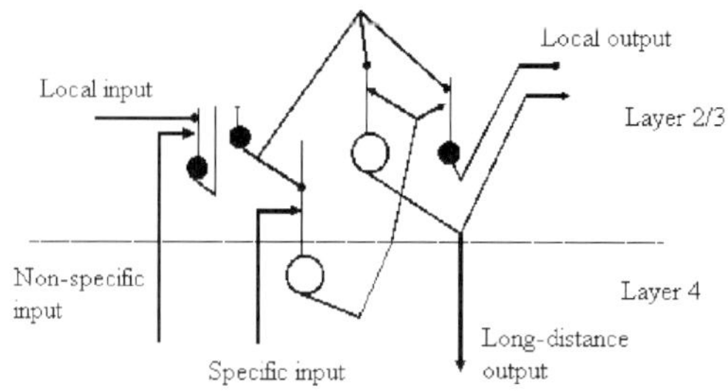

Fig. 1. The classic neural network model of a cortical column structure.

to process information [1]. We made simulated single electrode recordings and surface EEG recordings from both model columns. The results were compared to similar measurements from real brains.

The rest of the paper is structured as follows. We present the neural network models in Section 2. The data measurements are discussed in Section 3. The comparison of the simulation data with the real data is presented in Section 4. The paper is closed by the conclusion section.

2. Simulation methods

Here we describe the two neural network models. In both cases we built a simulated cortical column of the visual cortex.

2.1. Classical neural network model

Classical neural network models of the cortex build on idealized neuron models characterized by a tuning curve that represents the activity (e.g., firing rate) of the neuron in response to various input stimuli that are comparable by some appropriate metric (e.g., the orientation of short bars).

The model cortical column is made up of many excitatory neurons with similar tuning curves [6]. The model also contains a set of inhibitory neurons that mediate lateral inhibition, lateral excitation being performed by the excitatory neurons. The layered structure of the model column is presented in Fig. 1.

2.2. Sierpinski network model

Experimental evidence suggests that neural activity patterns play an important role in neural information processing [2]. Consequently, in order to reproduce more faithfully the operations of biological neural networks we consider models that simulate such behavior, i.e., computation by activity patterns of neurons. One such model is the Sierpisnki neural network [1] that contains a mixture of excitatory and inhibitory

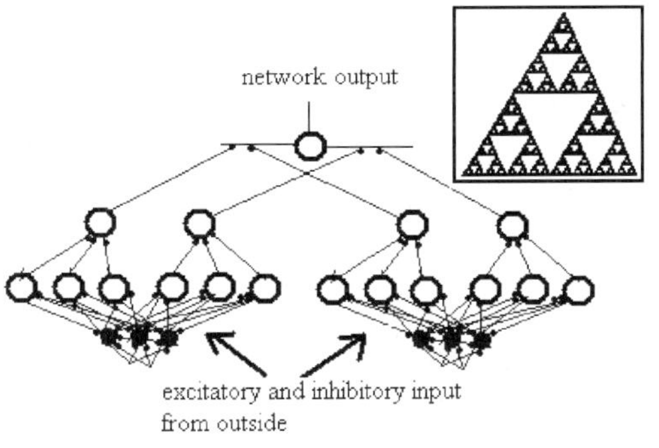

Fig. 2. The Sierpinski neural network. Inset shows a Sierpinski triangle.

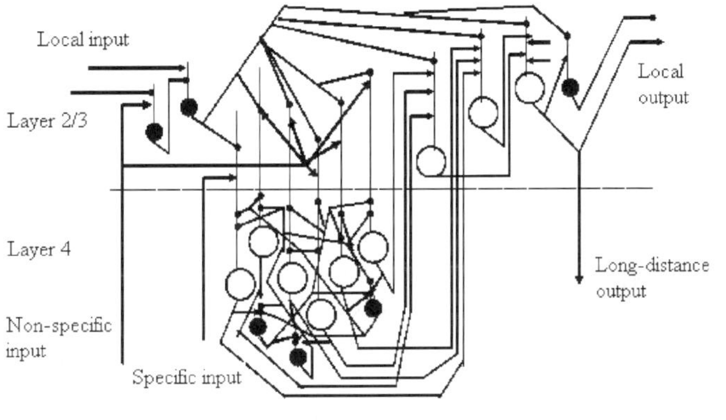

Fig. 3. The Sierpinski model of the cortical column structure.

neurons organized in a particular network architecture. The structure of the Sierpinski network is presented in Fig. 2.

The bottom layer of the sub-networks is formed by three pairs of excitatory neurons and three inhibitory neurons. This excitatory–inhibitory complex is organized such that at any time only one pair of excitatory neurons is spiking together with one of the inhibitory neurons. The pair of the active excitatory neurons is selected randomly. The two neurons in the upper layer of the sub-networks integrate the output of the excitatory–inhibitory complex, and send excitatory output to the top neuron of the network. The top neuron detects the coincident firing of the sub-network output neurons. The temporal output pattern of the sub-networks can be represented as a Sierpinski triangle (see inset in Fig. 2) [1], and the output of the top neuron of the network represents the number of intersections between the triangles of the sub-networks. The behavior of the top neuron can be characterized by a Sierpinski tuning curve [1].

The second cortical column model is built by Sierpinski neural networks together with additional inhibitory neurons that mediate the lateral inhibition. The layered structure of the Sierpinski cortical column is presented in Fig. 3.

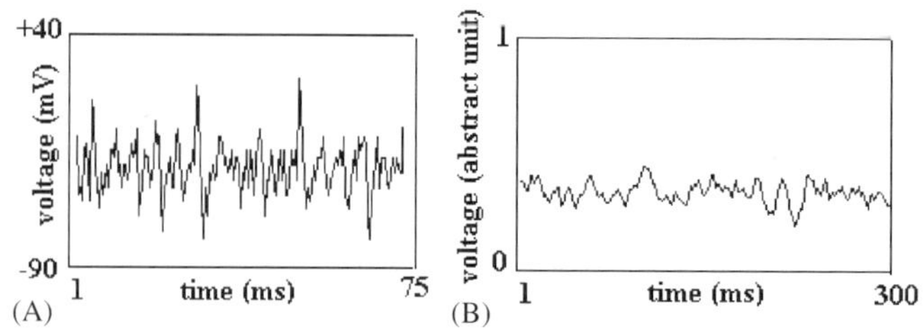

Fig. 4. Brain recordings: (A) single electrode recording from macaque visual cortex; and (B) surface EEG recording from cat visual cortex.

3. Data measurement

In this section we present the data that we used to do the comparison of the models with the real brain. We considered single electrode recordings and surface EEG measurements.

3.1. Real data

The single electrode recordings represent the activity of multi-units in macaque primary visual cortex [4]. The EEG recordings represent single channel data of 64-electrode surface EEG measured on the visual cortex of cats [7]. Twenty recordings were considered in both cases. Fig. 4 shows examples of the real data.

3.2. Simulation data

We simulate the electrophysiological measurements for the model columns. To calibrate the time of the simulations we matched the average spike time spans to the average time spike times of the real data. In all cases we made 20 recordings with different randomized initial conditions.

For the single electrode recordings we selected a point in layer 4 of the model column and measured the variation of the axonic signal of all neurons situated in a relatively close neighbourhood of the simulated electrode tip. We summed the axonic signals considering a distance dependent weighting of the individual signals. For the EEG recordings we considered the dendritic signals in the whole simulated column. We summed the signals with distance dependent weighting. Fig. 5 shows examples of the simulated data.

4. Comparison

We compared the time and frequency structure of the real and simulated data, by comparing the autocorrelation and Fourier coefficients of the data series. The means

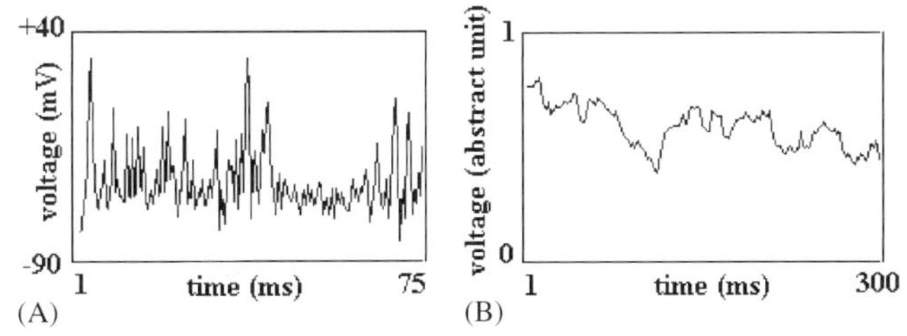

Fig. 5. Simulated brain recordings: (A) single electrode recording from layer 4 of the classical model; and (B) surface EEG recording from the Sierpinski model.

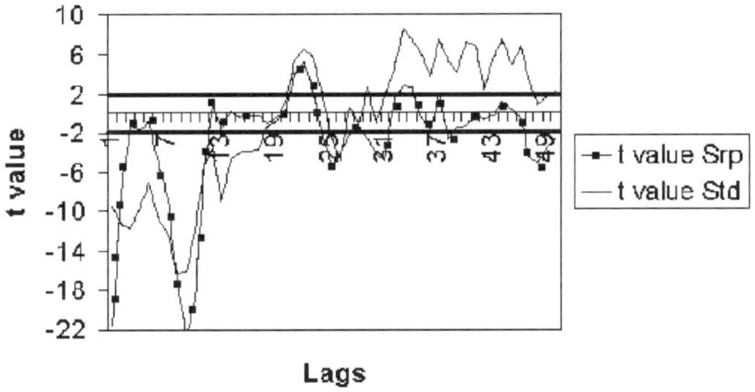

Fig. 6. The comparison of the single electrode recordings: Srp stands for the Sierpinski model, Std stands for the classical model. The graphs show how significantly the autocorrelation coefficients of the simulated data series differ from those of the real data series. The horizontal thick lines show the 5% significance boundaries.

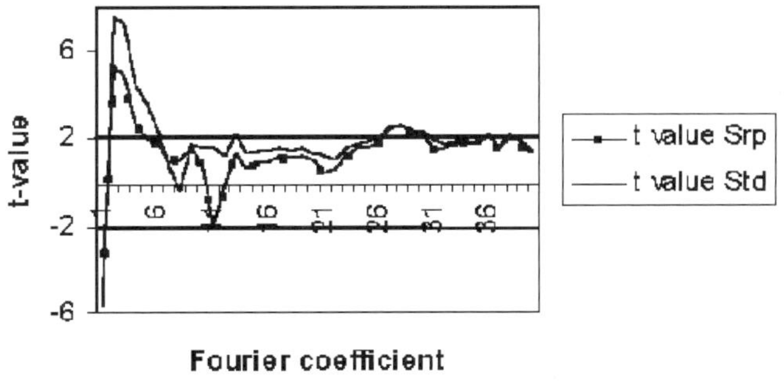

Fig. 7. The comparison of the surface EEG recordings: Srp stands for the Sierpinski model, Std stands for the classical model. The graphs show how significantly the Fourier coefficients of the simulated data series differ from those of the real data series. The horizontal thick lines show the 5% significance boundaries.

and standard deviations of the coefficients were calculated for each data type. The mean values were compared with the t-test.

The results of the comparison of autocorrelation coefficients for single electrode recordings are shown in Fig. 6. In this case the majority of the Fourier coefficients were situated in the 5% confidence interval of the real values for both models. The results show that the Sierpinski model data are more similar to the real data in terms of the autocorrelation structure than the data from the classical model.

The results of the Fourier coefficient comparison for surface EEG recordings are presented in Fig. 7. These results indicate that in these terms there is not very much difference between the two models. In terms of autocorrelation coefficients both models show significant difference from the real data for almost all coefficients.

5. Conclusions

We analyzed two neural network models to see to what extent can they produce realistic simulated electrophysiological measurements. The comparison shows that the Sierpinski model performs slightly better than the classical model. The results suggest that more realistic activity pattern-based models may lead to even better fits with the real data.

References

[1] P. Andras, Computation with chaotic patterns, Neurocomputing 44–46 (2002) 263–268.
[2] W.J. Freeman, Role of chaotic dynamics in neural plasticity, Prog. Brain Res. 102 (1994) 319–333.
[3] G.L. Gerstein, K.L. Kirkland, Neural assemblies: technical issues, analysis, and modelling, Neural Networks 14 (2001) 589–598.
[4] K. Guo, R.G. Robertson, A. Thiele, S. Panzeri, S. Mahmoodi, M.P. Young, V1 neurons use spatio-temporal prior probabilities of stimuli in dynamic scenes and compute by Bayesian inference. Abstracts of the 32nd Annual Meeting of the Society for Neurosciences, abstract no. 557.2, 2002.
[5] D.J. Heeger, E.P. Simonicelli, A. Movshon, Computational models of cortical visual processing, Proc. Nat. Acad. Sci. 93 (1996) 623–627.
[6] A. Nevado, M.P. Young, S. Panzeri, Functional imaging and neuronal information processing, Neurocomputing 44–46 (2002) 1127–1131.
[7] F.W. Ohl, H. Scheich, W.J. Freeman, Change in pattern of ongoing cortical activity with auditory category learning, Nature 412 (2001) 733–736.

Available at
www.ComputerScienceWeb.com
POWERED BY SCIENCE @ DIRECT®

ELSEVIER

Neurocomputing 52–54 (2003) 683–690

NEUROCOMPUTING

www.elsevier.com/locate/neucom

Extraction of nonlinear features in MEG and fMRI data of human brain

Liya Wang[a], Boris Baryshnikov[b], Hamid Eghbalnia[c], Amir H. Assadi[d,*]

[a] *Department of Biomedical Engineering, University of Wisconsin, Madison, WI 53706, USA*
[b] *Department of Medical Physics, University of Wisconsin, Madison, WI 53706, USA*
[c] *Department of Biochemistry, University of Wisconsin, Madison, WI 53706, USA*
[d] *Department of Mathematics, University of Wisconsin, Madison, WI 53706, USA*

Abstract

We have determined a new method for quantifying the nonlinearities of brain imaging data sets to allow comparison of measures of nonlinearity across comparable experiments. We have an efficient algorithm for finding nonlinear features of high dimensional data from variations in two- dimensional projections and slices. By computing the second-order statistics of these features, we are able to construct an analogue of the Riemann Curvature Tensor, thereby describing the nonlinearity of the data set. We demonstrate the usefulness of this technique on real MEG and fMRI data.

Keywords: Principal components analysis; Multi-channel; Data analysis; Biomagnetism; Auditory perception

1. Introduction

Working with massive "real world" data is inherently difficult because we often do not know important factors that are significant in any pattern recognition and feature extraction tasks. These factors range from the signal-to-noise ratio in the data, lack of a statistical distribution for the uncontaminated data, to situations such as not having a robust statistical model for the noise. A simplifying assumption is often helpful, but we must be certain to assess the impact of the assumptions that are made along the way on the final outcome of the analysis. Bayesian methods have the advantage that

* Corresponding author. Fax: +1-608-263-8891.
E-mail address: ahassadi@facstaff.wisc.edu (A.H. Assadi).

0925-2312/03/$ - see front matter
doi:10.1016/S0925-2312(02)00731-2

they make such assumptions explicit in the model. Nonetheless, the Bayesian models will reach their limits in the absence of better estimates on the distribution of data or at least, partial knowledge of such distributions. We propose a method for obtaining relevant probabilistic distributions by extracting information about nonlinearity in those representations of data that lend themselves to certain operations that we call for short "local-to-global integration" and "local linearization". We apply this technique to a variety of data-types that could fit under the general category of "multi-channel data". By multi-channel data registration we mean any methodology that extracts the time-series signals from the brain or any other biological source of electromagnetic emission, including EEG, MEG, MCG, fMRI, and multi-electrode recording.

In the application of dynamic PCA to MEG data, we propose a new method for quantifying the nonlinearities of the human brain response during auditory perception and with focus on auditory attention. The results permit a rigorous quantitative comparison of the brain's processing of auditory information across subjects, and potentially, it provides further insight into general principles underlying processing mechanisms of the human brain. We apply our method to fMRI data and show that for a set of 8 randomly selected voxels, a small temporal window of eigenvectors corresponding to the first three eigenvales can "describe" the data. Our preliminary analysis here illustrates that dynamics PCA can be used as a powerful detection and classification algorithm (see [2,3] for related applications).

2. Method

The theoretical underpinning of our algorithms is traced back to viewing data as a geometric entity. A data set is represented by the vectors that register observable features of data points, and forms a geometric subset of the Euclidean space of possibly high dimension. Other representations of the data set could be obtained from geometric properties of this subset, provided that we account for key geometric entities such as metric and topological properties required from the physical properties, as well as biological/behavioral constraints. The intuition behind our method for capturing nonlinear features in data could be briefly outlined as follows. From differential geometry, we know that the Riemannian curvature tensor (RCT) gives a comprehensive description of the nonlinearity in the underlying spaces endowed with certain abstractions of the infinitesimal calculus. A possible approach to description of nonlinearity in data is to look for a concept similar to this marvelous discovery of Riemann. In particular, we confine our study to data sets that can be represented as a collection of points in the Euclidean space. In this context, we define a new geometric structure that lends itself to associate to the data at various points something that is analogous to the Riemann curvature tensor that we refer to as "data curvature tensor" (dCT). If the data is sampled sufficiently finely from the points on a Riemannian manifold in the standard sense, then the Riemann curvature tensor and the dCT are within the approximation bounds defining the sampling of data from the manifold. Thus, the Riemann curvature tensor, and hence nonlinearities of the manifold can be recovered by computing the dCT from available data points. We prove a result that under reasonable circumstances, data a

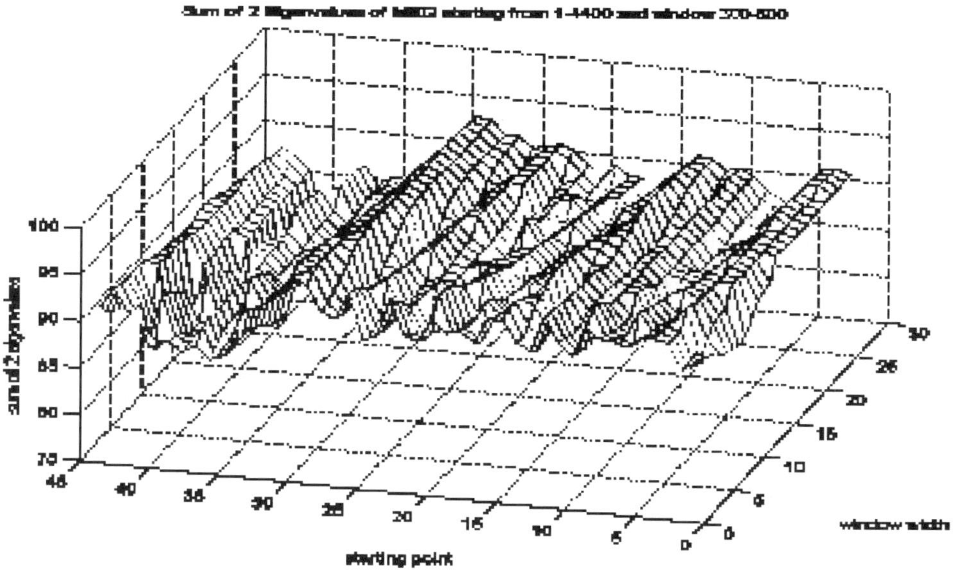

Fig. 1. The surface representing the variation of the sum of the largest two eigen values in terms of changes in initial point and the length of the window.

collection of two-dimensional subspaces that linearly approximate local statistics of data suffices for estimation of dCT. Therefore, in the dCT approach, we must determine the nonlinearity of features in the prescribed family of two-dimensional planes in order to have the basic estimates from which we calculate estimates for other nonlinear features of the data. We consider, therefore, estimation of nonlinearities of data in the plane.

In this paper, we propose an information-theoretic estimate for describing the nonlinear dependencies of data in the family of two-dimensional linearization. With this in mind, we must remark that extracting the feature from a general data set is an ill-posed problem in the mathematical sense, and we make certain assumptions to transform it to a well-defined estimate. Thus, it is important to recover such assumptions when we have implicit knowledge from the system that gives rise to the data, here, the MRI instrumentation and basic facts about neurophysiology of the brain underlying the acquired signals. The technical aspect outside of mathematics is to justify the realistic nature of such estimates from fMRI data. We propose to a large-scale data analysis experiment that will either validate our hypotheses, or provide alternative estimates that could be used for analysis of nonlinearity in fMRI in the dCT framework.

We are given time series data from N sources (such as MEG channels) arranged spatially. Each x is the time series from one source and t is sampled to generate L data points. They can be written

$$X(t) = \{x_1(t), x_2(t), \ldots, x_N(t)\},$$
$$t = (t_1, t_2, \ldots, t_L),$$

as

$$x_{ij} = x_i(t_j).$$

Each row of this matrix, where x_{ij} is the element in both ith row and jth column. So each row of this matrix is the time course of one recorded source and each column represents the data recorded for all sources at that instant of time.

Consider this matrix as a large window. We choose a small window that has the same height but smaller width. Next, we move this small window across the matrix and also change the window width every time, while at the same time, we do Principal component analysis (PCA) to get the sum of the largest two eigenvalues. This process yields a surface as shown in Fig. 1.

The original matrix for this picture is constructed from fMCG (fetal MCG) signals. The surface depicts that the information captured by two eigenvalues ranges from 79% to 96%. One reason for improvement of data analysis results by dynamic PCA is that, data represented by X is always noisy and often non-stationary. However, we may assume that X is at least quasi-stationary, that is, for each signal from a single source, the local covariance structure varies sufficiently slowly.

3. Results

In our paper we demonstrate these techniques when applied to real data from fMRI, MEG of human brain and MCG of fetal heart signals. The outcome describes the nonlinearity of the temporal relationship between regions of the human brain during a motor activation task, and in turn, we show what the results reveal about nonlinear dynamics in fMRI.

A 37-channel SQUID magnetometer (Magnes II, Biomagnetic Technologies) in a magnetically shielded room was used for MEG recordings of the brain activity. Five young adults participated in the study. In our protocol subjects were presented with two types of tones. In the active phase of the experiment base tones (1000 Hz, 200 ms) and target tones (1050 Hz, 200 ms) were randomly intermixed in each run. Subjects were asked to count target tones and report the number at the end of the experiment. The brain activity was recorded over auditory cortex area for the corresponding hemisphere. At least 150 epochs were collected per single run with average inter-stimulus interval of 1.5 s; sampled at 520.8 Hz.

During this experiment evoked (time-locked to the stimulus onset) and induced (not time-locked) oscillatory gamma activity was observed in all subjects. Preliminary analysis showed differences in topography of gamma oscillations and slow auditory evoked response.

Based on the concept of dynamic PCA outlined above, we parameterize spatially and temporally the search space intended for feature extraction from the MEG data. Spatial parameterization determines two independent parameters that specify the subset of channels to be considered, namely, spatial radius and channel location on the sensor. In order to avoid very large search spaces, we use several fixed space configurations for the data analysis below: (1) all channels (maximal radius, central position); (2) the subset of channels that exhibit maximum power in a particular frequency range (slow

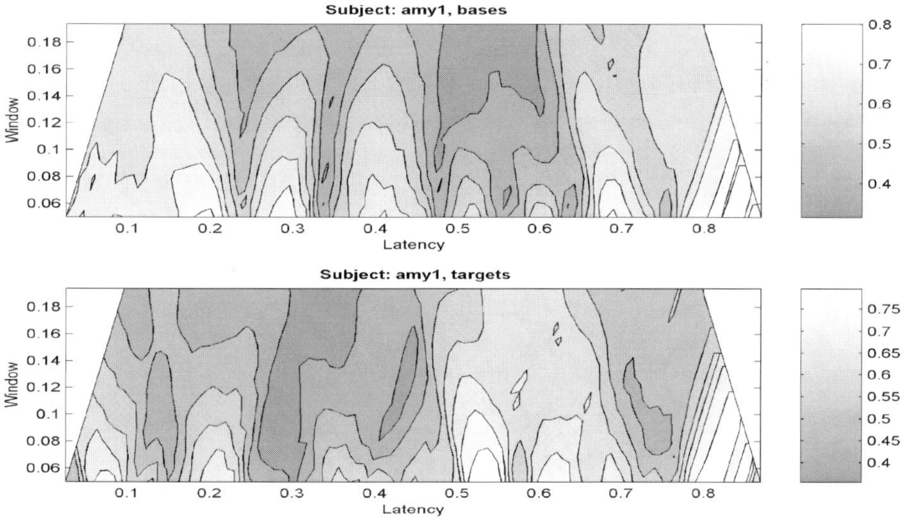

Fig. 2. Characteristic surfaces of an averaged gamma band response. In the active listening phase (targets) there is a clear region of complex brain activity (likely to be a high-dimensional space) around 0.3 s. Both latency and window are in seconds. Stimulus onset is at 0.1 s on these graphs.

auditory evoked response or gamma-band response). The latter decision is made based on a preliminary time-frequency analysis of the data. Time parameterization includes two parameters: the size of the window and its origin relative to the beginning of the epoch.

It is believed that gamma oscillations (high frequency oscillatory EEG or MEG activity in the range of approximately 25–60 Hz) represent an important mechanism by which the brain binds or integrates spatially distributed processing related to the same percept. Throughout the 1990s in studies in humans and in animals it was shown that oscillatory synchronization in the gamma range occurs between neurons responding to the same percept. This synchronization of neural activity, which was mediated by gamma-band oscillations, was shown to be associated with various stimulus properties such as continuity, vicinity, and common motion [1,4,5]. Fig. 2 demonstrate the difference in characteristic surfaces for gamma band response in base and target tones presentations. There is a clearly identifiable difference in base vs. target surface.

For analysis of fMRI data we use a public data set available via www.fmridc.dartmouth.edu. The data provided in this data set is not highly resolved ($64 \times 64 \times 19 \times 91$) but is sufficient to illustrate the application of our method. We selected the time series of 8 voxels and used the standard Bartlett test for significance of the number of eigenvectors for the entire time series to describe the data to within $p < 0.05$. The significance results indicated that 4 eigenvectors were needed. We then used our method and tested for significance of one to four eigenvectors. These results indicated that three eigenvectors with the window size of approximately 18 were sufficient to achieve the same significance results. The results are illustrated in the Fig. 3 below.

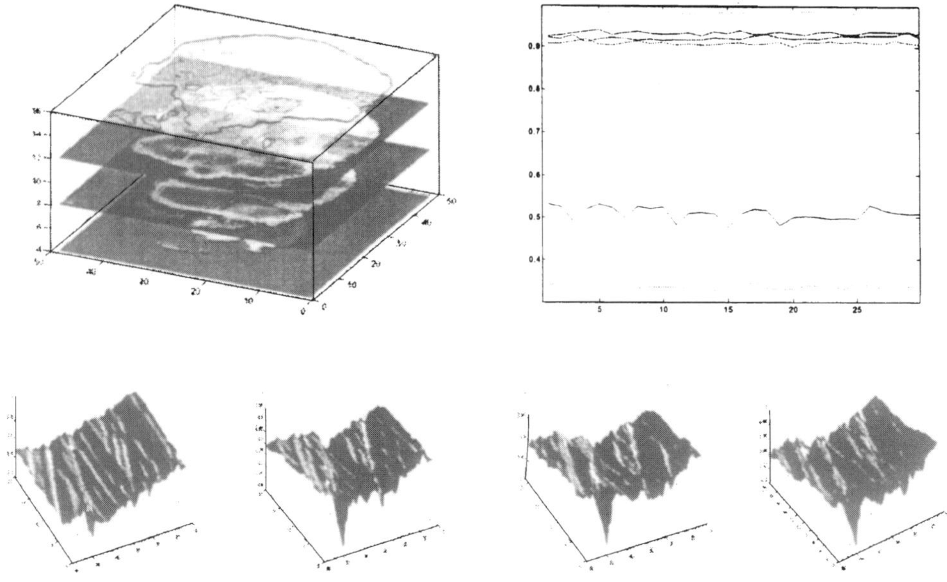

Fig. 3. The top row shows a sample of coronal slices from the dataset at time index 12 on the left and a sample of 6 Voxel time series chosen for analysis on the right for an initial segment of 30 points (out of a total of 91). Voxel values are normalized to one and shown on the y-axis. The bottom row illustrates the variation of total information measured by the percentage of total variance as a function of window size and window location for 8 Voxles (partially shown above). The leftmost window is the first eigenvalue followed by the sum of first two eigenvalues, sum of three, etc. The height in the figures shows the percentage of total variance. The axis labeled with 0–60 indicates the position of the window and the axis labeled with 0 –20 indicates the increase in size of the window beyond a nominal size of 10.

The following figure (Fig. 4) show more examples of such surfaces from real data. Our analyses visualize the surfaces such that we can distinguish two key properties: first, the information content exhibits sensitivity to the choice of the initial temporal point; (b) the possible durations could be selected from a range of intervals that proves robustness of the results with respect to the duration of the signal. Depending on applications, other features could be also distinguished.

4. Conclusion

We have proposed a novel approach to identification of the high-dimensional nonlinearity of a data set that embraces ideas from both differential geometry, probability (information geometry). Our algorithm recovers high-dimensional nonlinearity from information-theoretic features in a family of two-dimensional surfaces. This general method is bound to have many applications in multi-channel data analysis. Nonlinearity of information processing in brain could be estimated in terms of the dynamics of information content (e.g. by generalizing Shannon's mathematical theory of communication) from a multi-channel communication system. The encoding part of the

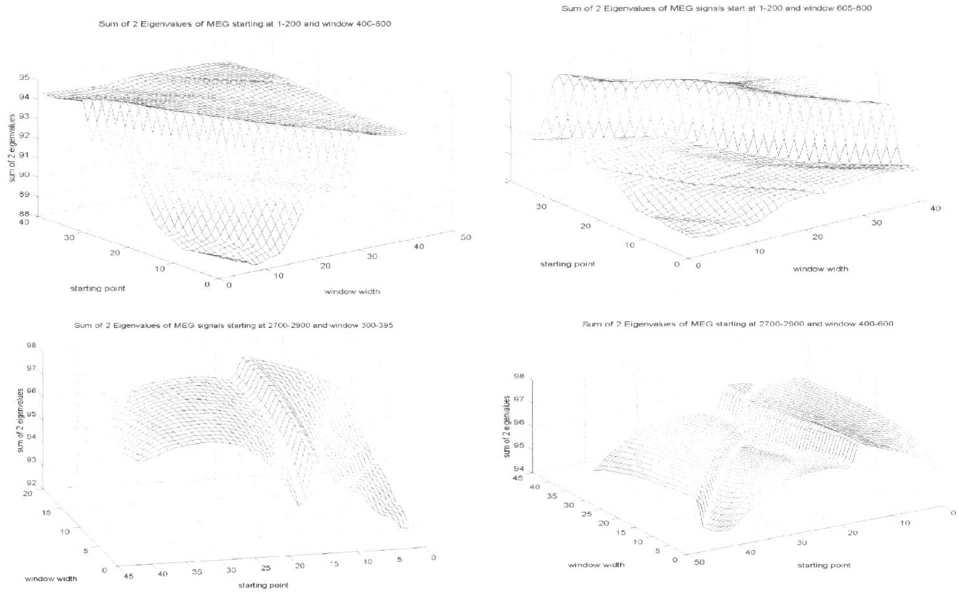

Fig. 4. Examples of information surfaces that visualize sensitivity and stability with respect to changes in initial point and length of windows.

multi-channel is in terms of spiking neurons and their secondary measurements as in the BOLD effect in fMRI, electromagnetic fields in MEG or EEG. The decoding side (receiver) guesses the coarse patterns of messages by optimizing extraction of information content in a sparse set of virtual channels (i.e. the algebraically weighted superposition of the output of the physical channels). The paper has given a summary of the progress by our research group to extract biologically significant features from multi-channel data using the above-mentioned framework.

Acknowledgements

Li Ya Wang, Hamid Eghbalnia and Amir Assadi were partially supported through NSF-KDI-LIS and other grants to Amir Assadi. Boris Baryshnikov is partially supported through grants to Professor Ron Wakai. The authors are grateful to Professor Ron Wakai of Dept of Medical Physics (UW-Madison) for providing the MEG facilities and scientific supervision of experimental protocols.

References

[1] A.K. Engel, P. Fries, W. Singer, Dynamic predictions: oscillations and synchrony in top-down processing, Nature Reviews Neuroscience 2 (10) (2001) 704–716.

[2] D. Fallahati, H. Eghbalnia, M. Backonja, A. Assadi, Dynamic PCA for network feature extraction in multi-electrode recording of neurophysiologic data in cortical substrates of pain, J. Neurocomputing, 2002, to appear.
[3] S. Pasha, H. Eghbalnia, A. Assadi, Dynamical models for nonlinear feature extraction and applications to face recognition, J. Neurocomputing, 2002, to appear.
[4] W. Singer, Neuronal synchrony: a versatile code for the definition of relation?, Neuron 24 (1999) 49–65.
[5] C. von der Malsburg, The what and why of binding: the modeler's perspective, Neuron 24 (1999) 95–104.

Available at
www.ComputerScienceWeb.com
POWERED BY SCIENCE DIRECT®

ELSEVIER

Neurocomputing 52–54 (2003) 691–698

NEUROCOMPUTING

www.elsevier.com/locate/neucom

A bifurcation of a synchronous oscillations into a torus in a system of two mutually inhibitory aVLSI neurons: experimental observation

Vladimir E. Bondarenko[a,*], Gennady S. Cymbalyuk[b,1], Girish Patel[c], Stephen P. DeWeerth[d], Ronald L. Calabrese[b]

[a] *Department of Physiology and Biophysics, School of Medicine and Biomedical Sciences, State University of New York at Buffalo, 124 Sherman Hall, Buffalo, NY 14214-3078, USA*
[b] *Biology Department, Emory University, Atlanta, GA 30322, USA*
[c] *Microtune Inc., 2201 10th Street, Plano, TX 75074, USA*
[d] *School of Electrical and Computer Engineering, Georgia Institute of Technology, Atlanta, GA 30332-0250, USA*

Abstract

We studied a system of two 'identical' oscillatory aVLSI neurons with mutually inhibitory connections. The system demonstrates different oscillatory behaviors depending on the strength of the inhibitory connections: anti-phasic, synchronous, phase-shifted, and quasiperiodic oscillations. We experimentally observed a bifurcation of synchronous oscillations into quasiperiodic oscillations with two independent frequencies. This bifurcation was confirmed by the analysis of the phase between neuronal outputs, the cross-correlation function, the amplitude spectrum, and the correlation dimension. The observation of this bifurcation in a physical system suggests that this scenario might also occur in living half-center oscillators, such as those found in central pattern generators.

Keywords: Quasiperiodic oscillations; Half-center oscillator; Central pattern generator; Silicon neuron

* Corresponding author. Tel.: +1-716-829-3640; fax: +1-716-829-2344.
E-mail addresses: vyb2@buffalo.edu, bond@sai.msu.su (V.E. Bondarenko).
[1] On leave from Institute of Mathematical Problems in Biology RAS, Pushchino, Moscow Region, Russian Federation, 142290.

0925-2312/03/$ - see front matter © 2003 Elsevier Science B.V. All rights reserved.
doi:10.1016/S0925-2312(02)00739-7

1. Introduction

Neuronal systems with mutually inhibitory connections between two units are common building blocks that play an important role in invertebrate and vertebrate central pattern generators (CPGs) [10,9]. Models of such systems can produce oscillatory output with different form and phase relationships between the two units [15,14], depending on the strength of inhibitory coupling. Similar phenomena have been observed in experiments. For example, the segmental swim CPGs of the lamprey comprise two reciprocally inhibitory units that are oscillatory neural networks located on opposite sides of the spinal cord. Under normal conditions, the two units oscillate in anti-phase. However, in the presence of strychnine, which blocks glycinergic inhibitory coupling, they oscillate synchronously [4]. Thorough understanding of the dynamics of such a system must encompass all behaviors of the system including those under experimentally modified conditions such as that with strychnine application.

Previously, we developed and studied a two-neuron system implemented via aVLSI technology [12,5]. Experiments with the aVLSI two-neuron system showed synchronous oscillations for weak synaptic coupling and anti-phasic oscillations for strong synaptic coupling. In addition, phase-shifted oscillations and quasiperiodic oscillations were observed for moderate strength of coupling [5]. We suggested that the synchronous oscillations in lamprey swim CPGs observed in strychnine could be accounted for by weak mutually inhibitory interactions between the two units. While stationary oscillations are well acknowledged in neuroscience and commonly represented by limit cycles in models (in-phase or anti-phase oscillations), quasiperiodic oscillations are still treated as rather obscure dynamic regimes which belong strictly in the realm of modeling. Recent theoretical and experimental studies of neuronal networks demonstrated a potential role for quasiperiodic oscillations in the proper functioning and control of networks [2,11,13,3,8,1,6].

Here we study a mechanism by which quasiperiodic oscillations can be generated by a two-unit neural network with the ubiquitous mutually inhibitory connections discussed above. In our previous work [5], we proved that in a mathematical model of our system of two identical mutually inhibitory aVLSI neurons, the synchronous limit cycle gives rise to a stable two-dimensional torus (quasiperiodic oscillations) through a sub-critical Neimark–Sacker torus bifurcation. Here, we prove experimentally that the aVLSI system undergoes the same transition from synchronous to quasiperiodic oscillations.

2. Methods

Description of our silicon aVLSI neurons and methods of measurement is provided in detail in [5].

Pearson's correlation coefficient r_{Pearson} is defined by the equation

$$r_{\mathrm{Pearson}} = \frac{S_{ij}}{(S_{ii}S_{jj})^{1/2}}, \tag{1}$$

where $S_{ii}=\sum_{k=1}^{N}(u_{ki}-\bar{u}_i)^2$, $S_{jj}=\sum_{k=1}^{N}(u_{kj}-\bar{u}_j)^2$, and $S_{ij}=\sum_{k=1}^{N}(u_{ki}-\bar{u}_i)(u_{kj}-\bar{u}_j)$ for the ith and jth neuronal outputs, $\bar{u}_i$ and $\bar{u}_j$ are their averaged values. Pearson's correlation coefficient is used for the evaluation of the degree of synchronization between the two neuronal outputs.

The Shannon entropy is calculated from the equation

$$S_{\mathrm{Sh}} = -\sum_{k=1}^{K} p_k \ln p_k, \tag{2}$$

where p_k are the probabilities of finding the trajectory in the kth subinterval of the interval of the amplitude variation. Here, $K=64$.

The frequency spectra are calculated using the digital fast Fourier transform from the time series of $N=2048$–16384 points. The correlation dimension ν [7] is calculated using $N=2048$–16384 points as well.

3. Results

First, we studied the synchronization properties of two aVLSI neurons with mutually inhibitory connections using Pearson's correlation coefficient, when the strength of the synaptic connections, as determined by the aVLSI neuron parameter I_{Bsyn}, is varied. The data are shown in Fig. 1. For I_{Bsyn} from 3.2 pA to 2.7 nA, we observe synchronous

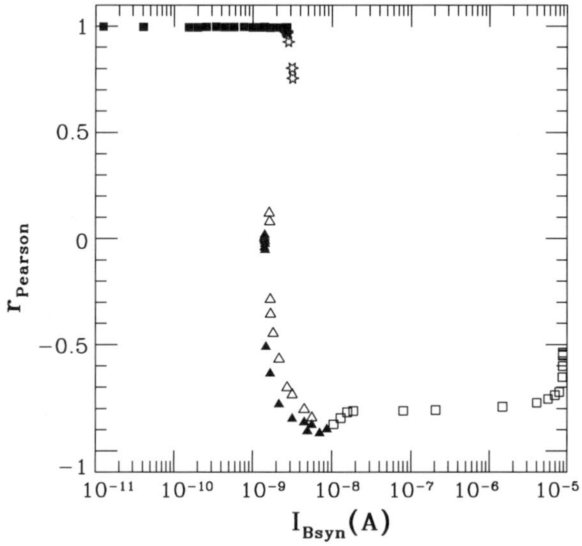

Fig. 1. Pearson's correlation coefficients between the first and second neuronal outputs versus the synaptic strength parameter, I_{Bsyn}. Here, closed squares, stars, triangles, and open squares show synchronous, quasiperiodic, phase-shifted, and anti-phasic oscillations, respectively. Open and closed triangles represent two branches of phase-shifted oscillations (see Fig. 6 from [5]).

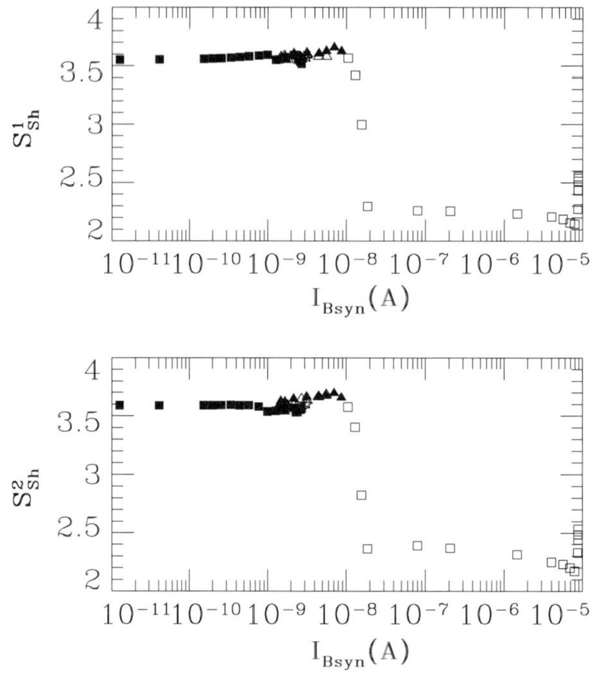

Fig. 2. Shannon entropy for the two neuron outputs as functions of I_{Bsyn}. S^1_{Sh} and S^2_{Sh} stand for the first and second neuron, respectively. Symbols are the same as in Fig. 1.

oscillations with $r_{Pearson}$ close to 1 (closed squares in Fig. 1). In the interval of I_{Bsyn} from 1.4 to 8.7 nA, two branches of phase-shifted oscillations appear (open and closed triangles in Fig. 1). They co-exist with synchronous and quasiperiodic oscillations. The latter develop in the narrow interval from 2.7 to 3.15 nA (stars in Fig. 1). When I_{Bsyn} exceeds 10 nA, only anti-phasic oscillations exist in our two-neuron system (squares in Fig. 1).

To estimate the degree of order of the observed neural outputs, the Shannon entropy S_{Sh} was calculated as a function of synaptic strength (Fig. 2). We found an increase in Shannon entropy when I_{Bsyn} is decreased from 10 μA to about 10 nA. S_{Sh} increases steeply for both neural outputs at the transition from anti-phasic to phase-shifted oscillations. Shannon entropy remains almost unchanged for further decrease of I_{Bsyn} up to 3 pA, when oscillations become synchronous.

The principal result of the paper is the observation of a bifurcation of synchronous oscillations into quasiperiodic oscillations with two independent frequencies in the behavior of the two-cell aVLSI oscillator. The transition occurs in a relatively narrow interval of I_{Bsyn} from 2.7 to 3.15 nA. Several characteristics of the neural outputs are calculated to locate the bifurcation: maps of V_1 versus V_2 (Figs. 3a–d), phase between the outputs of the two neurons (Figs. 3e–h), amplitude spectra (Figs. 3i–l), and correlation dimensions ν.

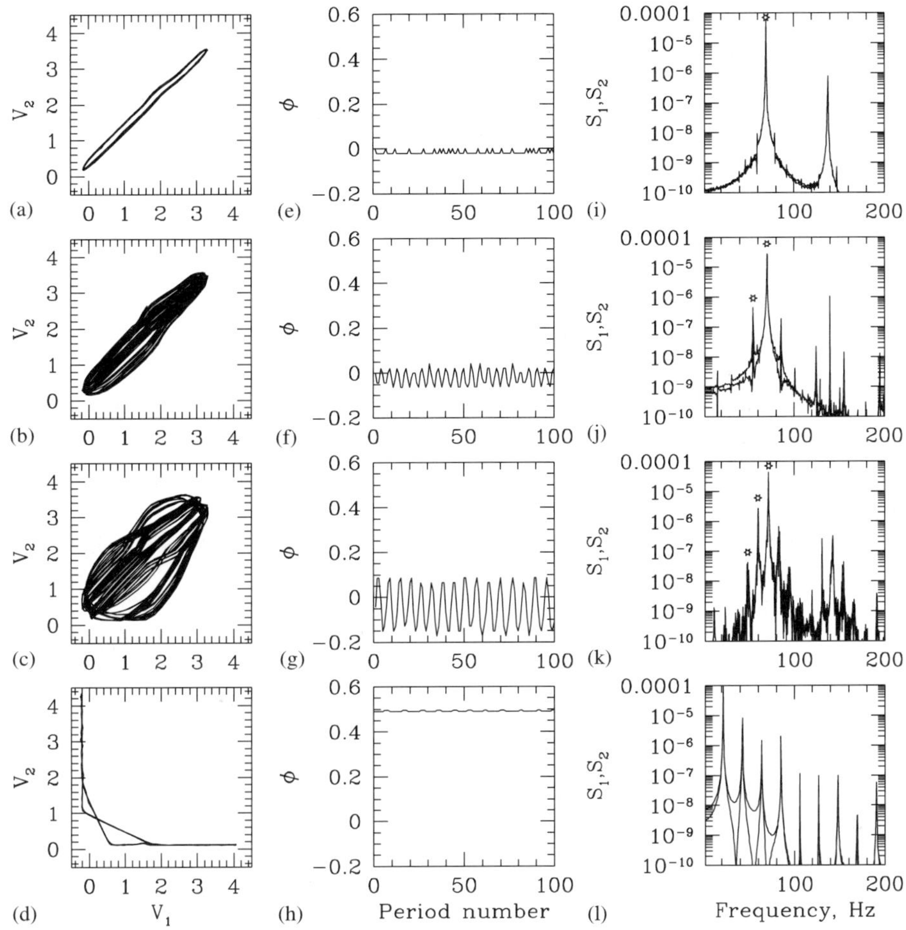

Fig. 3. Maps of neuron potentials V_1 versus V_2 (a–d), the phase between the outputs of the two neurons (e–h), and amplitude spectra (i–l) for the aVLSI oscillator system. I_{Bsyn}: 2.6 nA (a, e, i); 2.7 nA (b, f, j); 3.15 nA (c, g, k); and 79 nA (d, h, l).

Figs. 3a, e and i show characteristics of synchronous oscillations at $I_{Bsyn} = 2.6$ nA. In this case, the amplitude spectrum (Fig. 3i) shows only one component at 69.2 Hz with a second harmonic at 138.0 Hz, and correlation dimension $\nu = 1$. With an increase of I_{Bsyn} to 2.7 nA, we observe quasiperiodic oscillations (torus) with frequencies 54.8 and 70.2 Hz and their harmonics. The map of quasiperiodic activity covers an area of the plane (V_1, V_2) (Fig. 3b). Calculations of correlation dimension ν shows the appearance of a two-dimensional process, and the fluctuations in phase of the neuronal outputs are close to periodic (Fig. 3f). More developed quasiperiodicity is observed at $I_{Bsyn} = 3.15$ nA (Figs. 3c, g and k). The correlation dimension ν becomes equal to 2. The spectrum of oscillations shows multiple peaks from harmonics of two main frequencies 60.0 and 71.6 Hz. The fluctuations in the phase of the neural output signals become larger (ranging from −0.17 to 0.09) but are still close to periodic. At relatively strong coupling, $I_{\mathrm{Bsyn}} = 79$ nA, the oscillations become anti-phasic and square wave in form. This deviation from a sine-like waveform is clear from the amplitude spectrum which contains multiple equidistant peaks (Fig. 3l). Correlation dimension v for this process is equal to 1 and the phase of the neuronal outputs slightly fluctuates around a value of 0.5 (Fig. 3h).

4. Conclusions

Our investigations show that a physical system of two 'identical' aVLSI neurons with mutually inhibitory connections can produce different oscillatory behaviors depending on the strength of the inhibitory connections: anti-phasic, synchronous, phase-shifted, and quasiperiodic oscillations. We experimentally observed a bifurcation of periodic neuronal outputs into quasiperiodic activity with two independent frequencies. This type of activity is confirmed by the analysis of phase between neuronal outputs, maps of V_1 versus V_2, amplitude spectra, and correlation dimensions. The bifurcation of periodic activity into torus found in our aVLSI two-neuron system suggests that this scenario may also occur in living mutually inhibitory networks, like those found in central pattern generators. In future work, this system of aVLSI neurons may provide insight into the potential role of this transition mechanism in the function and control of the nervous system.

Acknowledgements

This work was supported by NIH Grant NS43098 and NSF Grant DBI 9873173.

References

[1] V.E. Bondarenko, Control and 'anticontrol' of chaos in an analog neural network with time delay, Chaos Solitons Fractals 13 (2002) 139–154.
[2] G.N. Borisyuk, R.M. Borisyuk, A.I. Khibnik, D. Roose, Dynamics and bifurcations of two coupled neural oscillators with different connection types, Bull. Math. Biol. 57 (1995) 809–840.
[3] R.J. Butera Jr., J. Rinzel, J.C. Smith, Models of respiratory rhythm generation in the pre-Botzinger complex. II. Populations of coupled pacemaker neurons, J. Neurophysiol. 82 (1999) 398–415.
[4] A. Cohen, R.M. Harris-Warrick, Strychnine eliminates alternating motor output during fictive locomotion in the lamprey, Brain Res. 293 (1984) 164–167.
[5] G.S. Cymbalyuk, G.N. Patel, R.L. Calabrese, S.P. DeWeerth, A.H. Cohen, Modeling alternation to synchrony with inhibitory coupling: a neuromorphic VLSI approach, Neural Comput. 12 (2000) 2259–2278.
[6] C.A. Del Negro, C.G. Wilson, R.J. Butera, H. Rigatto, J.C. Smith, Periodicity, mixed-mode oscillations, and quasiperiodicity in a rhythm-generating neural network, Biophys. J. 82 (2002) 206–214.
[7] P. Grassberger, I. Procaccia, Characterization of strange attractors, Phys. Rev. Lett. 50 (1983) 346–349.

[8] E.M. Izhikevich, Weakly connected quasi-periodic oscillators, FM interactions, and multiplexing in the brain, SIAM J. Appl. Math. 59 (1999) 2193–2223.
[9] E. Marder, Motor pattern generation, Curr. Opin. Neurobiol. 10 (2000) 691–698.
[10] E. Marder, R.L. Calabrese, Principles of rhythmic motor pattern generation, Physiol. Rev. 76 (1996) 687–717.
[11] M. Matsugu, J. Duffin, C.S. Poon, Entrainment, instability, quasi-periodicity, and chaos in a compound neural oscillator, J. Comput. Neurosci. 5 (1998) 35–51.
[12] G.N. Patel, G.S. Cymbalyuk, R.L. Calabrese, S.P. DeWeerth, Bifurcation analysis of a silicon neuron, in: S. Solla, T. Lenn, K.-R. Muller (Eds.), NIPS'99 12, MIT Press, Cambridge, MA, 2000, pp. 731–738.
[13] W. Senn, T. Wannier, J. Kleinle, H.R. Luscher, L. Muller, J. Streit, K. Wyler, Pattern generation by two coupled time-discrete neural networks with synaptic depression, Neural Comput. 10 (1998) 1251–1275.
[14] F.K. Skinner, G.G. Turrigiano, E. Marder, Frequency and burst duration in oscillating neurons and two-cell networks, Biol. Cybern. 69 (1993) 375–383.
[15] X.-J. Wang, J. Rinzel, Alternating and synchronous rhythms in reciprocally inhibitory model neurons, Neural Comput. 4 (1992) 84–97.

Vladimir E. Bondarenko received a PhD degree from the Physics Department at Moscow State University. Currently, he is a Research Assistant Professor at the Department of Physiology and Biophysics, SUNY at Buffalo. His research interests include neural networks, plasma physics, polymer science, and cardiovascular physiology.

Gennady S. Cymbalyuk received a PhD from the Physics Department at Moscow State University and the Institute of Mathematical Problems in Biology in Pushchino, Russia. Presently, he is a Research Associate at Emory University in the Biology Department. He is mainly interested in living oscillatory dynamical systems.

Girish N. Patel received a PhD from the School of Electric Engineering at Georgia Institute of Technology in 1999. He is currently working at Microtune, Inc. designing high speed radio frequency circuits. His primary area of research is in the design of analog integrated circuits.

Ronald L. Calabrese is a Professor of Biology at Emory University. His research interests include membrane biophysics, neuronal networks, motor pattern generators, neuromodulation and computational approaches to understanding neuronal and network function.

Available at
www.ComputerScienceWeb.com
POWERED BY SCIENCE DIRECT®

ELSEVIER

Neurocomputing 52–54 (2003) 699–705

NEUROCOMPUTING

www.elsevier.com/locate/neucom

On duration and dopamine modulation of sustained activity in prefrontal cortex using conductance-based network models

Marcelo Camperi[a,*], María Virginia Manías[b,1]

[a] Physics Department, University of San Francisco, 2130 Fulton St., San Francisco, CA 94117, USA
[b] Physics Department, UNLP and IFLP, Conicet, Argentina

Abstract

We investigate the cellular mechanisms underlying the generation, duration, and dopaminergic modulation of working-memory sustained activity in prefrontal cortex by means of computational modeling. We use a conductance-based network model endowed with all relevant prefrontal cortex conductances found in the experimental literature, considering two-population-networks with AMPA-, NMDA-, and GABA-like synaptic connections. We study dopamine modulation by means of a mechanism in which delay-period activity is modulated locally by dopaminergic levels. These levels in turn are modulated by local neuronal activity and afferent and external inputs.

Keywords: Working memory; Delay-period activity; Dopamine modulation

1. Introduction

Sustained, memory-related activity is observed in many brain areas, such as parietal cortex, hippocampus, inferotemporal cortex and motor areas [7,13]. However, there is overwhelming evidence locating it predominantly in the dorsolateral prefrontal cortex (PFC), associated with short-term memory processes, where it has been shown to be more robust and to persist even in the presence of interfering stimuli [13]. Moreover,

* Corresponding author. Tel.: +1-415-422-2057; fax: +1-415-422-5155.
E-mail address: camperi@usfca.edu (M. Camperi).
[1] Present address. Theoretical Physics Group, LBNL (USA).

0925-2312/03/$ - see front matter
doi:10.1016/S0925-2312(02)00743-9

some PFC neurons fire maximally in tune with, for example, particular spatial memory cue locations ("memory fields" [4,7,8]).

Working memory processes—short-term memory processes involved in organizing behavior, language, and thinking—are sensitive to dopaminergic modulation [4–6,11,13]. Neuronal activity is controlled by the neurotransmitter and experiments have revealed that iontophoresis of D_1 dopamine receptor agonists in PFC produces an increase in persistent activity, while D_1 antagonists decrease it [10]. In addition, dopamine (DA) may also stabilize neural functions in PFC, increasing the robustness of sustained delay activity with respect to distracting input and noise [5]. There appears to be an optimal level of DA modulation and, under differential D_1 activation of synaptic transmission on pyramidal cell and interneurons, working memory displays a characteristic inverted-*U* curve [9].

2. The model

We present a conductance-based network model exhibiting delayed-period sustained activity in the form of network bistability brought forth by graded recurrent excitation and inhibition. The network consists of N_E pyramidal neurons and N_I interneurons, in a setup akin to that in the tasks in [7,8]. Spatially localized transient inputs bring the network to an excited state, whereby synaptic recurrent excitation sustains the network activity after the input has been turned off.

The model equations (see Appendix A) incorporates all relevant PFC conductances. In modeling DA modulation, we concentrate on D_1 receptors, since they are much more abundant in PFC and, more importantly, both working memory performance and delay-period activity recorded in vivo are susceptible mainly or exclusively to D_1 receptor agonists and antagonists [1,9,10]. We simulate natural and forced changes in local DA levels via the introduction of both local and global dynamically modulated changes in maximal conductances, thresholds, and synaptic strengths.

All dopaminergic effects are contained in a dynamical equation that depends on external sources and internal afferent factors, as well as on local cortical activity:

$$\tau_{\text{dopa}}^{\text{E,I}} \frac{d\eta_i^{\text{E,I}}}{dt} = \alpha_{\text{dopa}}^{\text{E,I}}(1 - \eta_i^{\text{E,I}})D_i^{\text{E,I}} - f_{\text{step}}(V_i^{\text{E,I}})\eta_i^{\text{E,I}},$$

where the η_i represent the local DA concentration (per neuron), $f(V)$ is a "step" spiking threshold function, and D_i is a function representing local and global DA and DA receptor effect, external and afferent. The variables η are then dynamically linked to the various equations needed to reproduce the effects reported for PFC neurons, which include enhancing the persistent Na current (shifting its activation threshold toward more hyperpolarized potentials), reducing the slowly inactivating K+ current in pyramidal cells, reducing the high-voltage-activated (HVA) Ca-current, and enhancing NMDA-like synaptic currents while reducing AMPA-like ones (the overall effect of combined DA-induced changes in AMPA and NMDA currents reduces the EPSP amplitude but prolongs the duration). DA also enhances GABA-like synaptic currents, increasing the spontaneous activity of interneurons and increasing IPSP amplitude. DA-affected maximal conductances, synaptic strengths, and thresholds are thus modulated via equations of the form

$$g^i = g\{1 \pm \sigma f_{\text{sat}}^{\text{E,I}}(\eta_i^{\text{E,I}})\},$$

where σ is a maximally observed DA effect and f is a saturating function. Here, the g_X^i represent the different "constants" that are affected by DA, which becoming local parameters for each neuron.

These complex interdependences and the flexibility we have in choosing the function $D^{\text{E,I}}$ provide a dynamical loop that models enhancement and duration effects during delay-period activity, also providing a foundation for a plausible mechanism that explains the observed "inverted-U" behavior (see below).

3. Simulation results

For testing, we consider a small network of 40 excitatory and 20 inhibitory neurons.

Fig. 1 shows excitatory and inhibitory neuronal behaviors in the fully connected network model. In the presence of an additional stimulation, a state of high and sustained activity appears. This activity has a frequency of approximately 40 Hz and lasts for a few seconds after the input is removed. Recurrent synapses are mediated primarily by NMDA channels [2,12]. However, we find that stability requires also small amounts of AMPA. Moreover, we find that if the model is to reproduce observations from physiological studies, overall inhibition should dominate [2,12] (see Fig. 2).

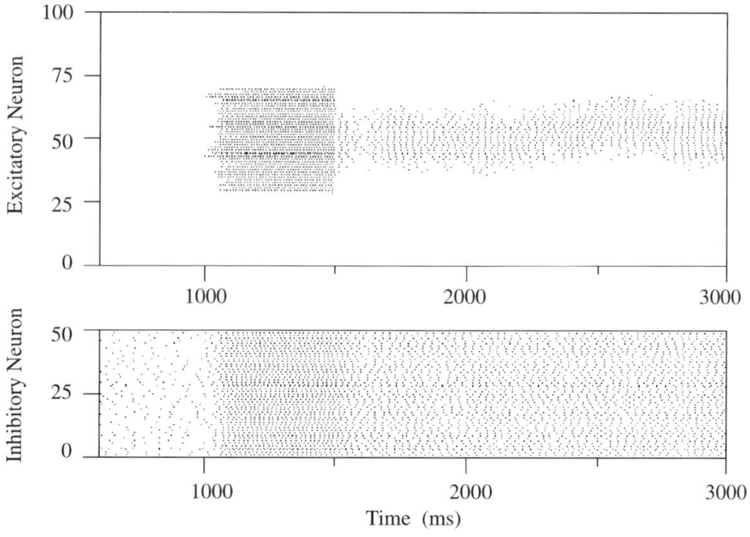

Fig. 1. Rastergrams showing basic results (without DA). A transient input lasting 500 ms is applied uniformly to a group of E-neurons. The network keeps the activity going after the input stops. The exact details of the transient input are not important.

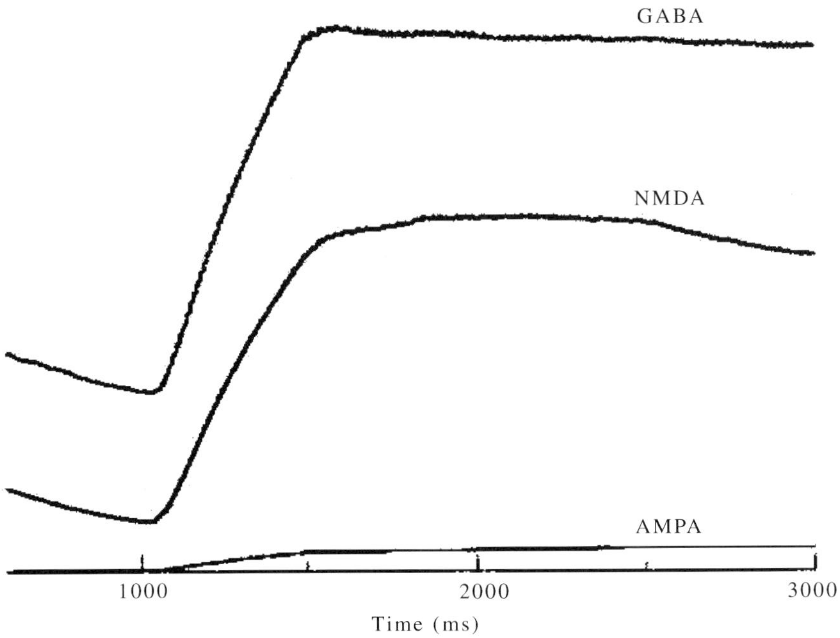

Fig. 2. Evolution of the time and network averages for the three types of synaptic inputs present. Overall inhibition is dominant.

We also study the DA modulation in our model.

The three panels in Fig. 3 demonstrate timing effects and the existence of the inverted-U effect: On the top panel, the overall DA level is not enough to kick-start the delay-period effect. On the bottom panel, there is too much global DA to start with and thus too much inhibition. The delay-period activity is more erratic and stops relatively quickly. On the middle panel, the overall DA level is just right, and the delay-period activity starts and is sustained. No "off signal" of any kind was used to stop delay-period activity. Its duration is determined by the interplay between the different DA-related dynamical effects present (see below).

4. Conclusions

The network model is dominated by inhibition, in the sense that, for the working parameter regime, overall recurrent excitatory–inhibitory synaptic interactions are balanced toward inhibition. DA increases the firing of interneurons innervating both interneurons and pyramidal cells and therefore enhances action potential-dependent GABAergic inhibitory synaptic transmission. Both AMPA and NMDA excitatory synapses are necessary, although the model produces more stable results when NMDA is dominant. AMPA still needs to be present, albeit in small amounts. We demonstrate the existence of an "inverted-U" optimal dopamine level (or DA-receptor activation) for

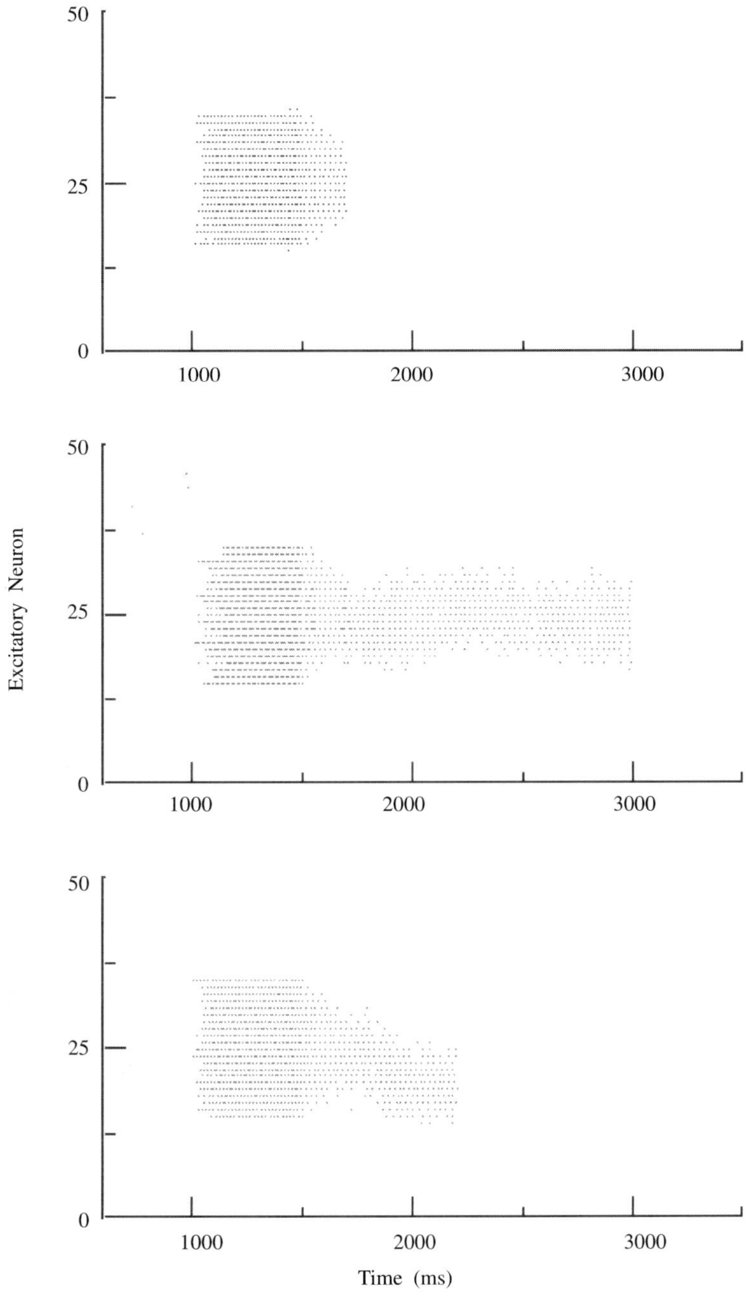

Fig. 3. The functions $D^{E,I}$ (see text) consisted of a global constant (the same for all E and I neurons) indicating overall DA levels in PFC, plus a transient, local (graded) factor into E neurons only. This factor represents afferent inputs (for example from the LGN) which communicate to the PFC the external, transient, localized visual input. The network is brought to high activity just by action of this localized extra DA levels.

working memory processes. D_1 receptor modulation must affect differentially pyramidal cells and interneurons. It enhances excitatory inputs to both, but this enhancement is more effective on pyramidal cells than in interneurons. In this way, increasing levels of DA stimulation of D_1 receptors will result in enhanced of working memory performance. However, at higher activation levels, modulation of NMDA conductances on pyramidal cells saturates, and further increases in modulatory dopamine levels will result in an enhancement of interneuron activity, since the glutamatergic inputs are enhanced for both pyramidal cells and interneurons, leading to a reduction in pyramidal cell activity by feed-forward inhibition. The complex dynamical interplay between afferent DA modulation and local spike-dependent effects may be responsible in setting the timing of delay-period activity in PFC.

Appendix A. Model equations

Model equations are of the Hudgkin–Huxley type, where for each neuron j we have

$$\frac{dV_j^X}{dt} = -I_{\text{ionic},j}^X - I_{\text{syn},j}^X + I_{\text{random},j}^X,$$

where $X = E$ or I. For E-neurons, we consider fast, spike-generating sodium, delayed-rectifier potassium, persistent sodium, HVA calcium, slowly inactivating potassium, and fast Ca^{2+}-and voltage-dependent potassium currents. For I-neurons, we include just the spiking fast sodium and a delayed-rectifier currents. The "random" currents are independent poisson processes that simulate general neuronal inputs. Finally, the synaptic currents include AMPA, NMDA, and GABA components:

$$I_{\text{syn},i}^{\text{E}} = \sum_{j=1}^{N_{\text{E}}} W_{ij}^{\text{EE}} \left(g_{\text{AMPA},ij}^{\text{EE}} s_{\text{AMPA},j} + \frac{g_{\text{NMDA},ij}^{\text{EE}} s_{\text{NMDA},j}}{1 + [Mg] \exp(-\sigma V_i^{\text{E}})} \right)$$

$$\frac{(V_i^{\text{E}} - V^{\text{EE}})}{N_{\text{E}}} + \sum_{j=1}^{N_{\text{I}}} g_{\text{GABA},ij}^{\text{IE}} s_{\text{GABA},j} \frac{(V_i^{\text{E}} - V^{\text{IE}})}{N_{\text{I}}},$$

$$I_{\text{syn},i}^{\text{I}} = \sum_{j=1}^{N_{\text{I}}} g_{\text{GABA},ij}^{\text{II}} s_{\text{GABA},j} \frac{(V_i^{\text{I}} - V^{\text{II}})}{N_{\text{I}}}$$

$$+ \sum_{j=1}^{N_{\text{E}}} \left(g_{\text{AMPA},ij}^{\text{EI}} s_{\text{AMPA},j} + \frac{g_{\text{NMDA},ij}^{\text{EI}} s_{\text{NMDA},j}}{1 + [\text{Mg}] \exp(-\sigma V_i^{\text{I}})} \right) \frac{(V_i^{\text{I}} - V^{\text{EI}})}{N_{\text{E}}}.$$

The variables s are standard dynamical synaptic variables [2,5,6,12]. The grading function W_{ij}^{EE} in the synaptic equations recreates the experimental conditions found in classic oculomotor tasks: E-cells have effective memory fields on a circle [3]. This is just a simplifying assumption that is not fundamental to our arguments.

References

[1] A.F. Arnsten, J.X. Cai, B.L. Murphy, P.S. Goldman-Rakic, Dopamine D_1 receptors mechanisms in the cognitive performance of young and aged monkeys, Psychopharmacology 116 (1994) 143–151.

[2] N. Brunel, X.J. Wang, Effects of neuromodulation in a cortical network model of object working memory dominated by recurrent inhibition, J. Comput. Neurosci. 11 (2001) 63–85.

[3] M. Camperi, X.J. Wang, A model of visuospatial working memory in prefrontal cortex: recurrent network and cellular bistability, J. Comput. Neurosci. 5 (1998) 383–405.

[4] R. Desimone, Nature 376 (1995) 549–550.

[5] D. Durstewitz, M. Kelc, O. Gunturkun, A neurocomputational theory of the dopaminergic modulation of working memory functions, J. Neurosci. 19 (1999) 2807–2822.

[6] D. Durstewitz, J.K. Seamans, T.J. Sejnowski, Dopamine-mediated stabilization of delay-period activity in a network model of prefrontal cortex, J. Neurophysiol. 83 (2000) 1733–1750;
D. Durstewitz, J.K. Seamans, T.J. Sejnowski, Neurocomputational models of working memory, Nature Neurosci. 3 (2000) 1184.

[7] S. Funahashi, C.J. Bruce, P.S. Goldman-Rakic, Mnemonic coding of visual space in the monkey's dorsolateral prefrontal cortex, J. Neurophysiol. 61 (1989) 331–349.

[8] P.S. Goldman-Rakic, Cellular and circuit basis of working memory in prefrontal cortex of nonhuman primates, Prog. Brain Res. 85 (1990) 325–326;
P.S. Goldman-Rakic, Cellular basis of working memory, Neuron 14 (1995) 477–485.

[9] P.S. Goldman-Rakic, E.C. Muly, G.V. Williams, D_1 receptors in prefrontal cells and circuits, Brain Res. Rev. 31 (2000).

[10] T. Sawaguchi, The effects of dopamine and its antagonists on directional delay-period activity of prefrontal neurons in monkeys during an oculomotor delayed-response task, Neurosci. Res. 41 (2001) 115–128.

[11] T. Sawaguchi, P. Goldman-Rakic, Sci. 251, 947–950.

[12] X.J. Wang, Synaptic basis of cortical persistent activity: the importance of NMDA receptors to working memory, J. Neurosci. 19 (1999) 9587–9603;
X.J. Wang, Synaptic reverberation underlying mnemonic persistent activity, Trends Neurosci. 24 (2001) 455–463.

[13] G. Williams, P. Goldman-Rakic, Nature 376, 572–575.

Marcelo Camperi is associate professor of physics at the University of San Francisco. He received his Ph.D. in Physics from Boston University for work in Mathematical Physics. His current interests include simulating very large networks on parallel computers, dynamical mechanisms for working memory in the prefrontal cortex, and uses of information theory in spike train analysis.

María V. Manías obtained her M.Sc. and Ph.D. degrees from de University of La Plata, Argentina. She worked on Quantum Field Theories, until she moved to Berkeley on September 2000, where she focused her interests in Theoretical Neuroscience. Currently, she is interested in modeling dynamical mechanisms for working memory in prefrontal cortex.

Available at www.ComputerScienceWeb.com
POWERED BY SCIENCE DIRECT®

ELSEVIER

Neurocomputing 52–54 (2003) 707–712

NEUROCOMPUTING

www.elsevier.com/locate/neucom

The role of postsynaptic potential decay rate in neural synchrony☆

Yoonsuck Choe[a,*], Risto Miikkulainen[b]

[a] *Department of Computer Science, Texas A&M University, 3112 Tamu, College Station, TX 77843-3112, USA*

[b] *Department of Computer Sciences, The University of Texas at Austin, Austin, TX 78712, USA*

Abstract

Integrate-and-fire neurons have been used widely to model large-scale networks with temporal dynamics. Previous work has focused on temporal *delays* in modulating synchronization behavior; not much attention has been given to the postsynaptic potential (PSP) *decay rate*. In this paper, we show that varying the PSP decay rate has the same effect as adjusting the axonal conduction delay. The decay rate can be adjusted independently at different locations in the neuron, allowing precise fine tuning of synchronization behavior. Also, because the adjustments can be local and small, this process can be more efficient than adjusting the larger-scale axonal delays.

Keywords: Integrate-and-fire neurons; Decay adaptation; Synchronization

1. Introduction

Experimental evidence suggests that temporally correlated activity contributes to binding and segmenting features in sensory input [4,10]. Thus, temporal information may be very important for neural function. Integrate-and-fire neurons have been studied extensively in this role because their simplicity makes it possible to study synchronization behavior in large-scale networks [2,8,11,13].

Some of this work assumes that spikes are transmitted instantaneously [2], but others take into account axonal conduction delays [11,13]. It turns out that the delay

☆ This research was supported in part by National Science Foundation under grant #IRI-9811478.

* Corresponding author.

E-mail address: choe@tamu.edu (Y. Choe).

doi:10.1016/S0925-2312(02)00747-6

characteristic of different types of connections (e.g. excitatory and inhibitory) determine the synchronization behavior in a highly interconnected network of neurons. The main results are: (1) excitatory connections with no delay cause synchrony [2,9,11], (2) excitatory connections with delay cause desynchrony [13], (3) inhibitory connections without delay cause desynchrony [11,13], and (4) inhibitory connections with delay cause synchrony [11–13,17]. Previous models of spiking neurons have either adapted or selected the axonal delays to regulate synchronization behavior [7,16]. Although delay can be adapted by changing the axonal morphology (length, thickness, and myelination; [6]), the fine degree of delay tuning needed in the above models may not be easy to achieve in a macrostructure as an axon in biological neurons.

An alternative to delay adaptation is to change the decay rate of the postsynaptic potential. Decay may be easier to alter in biological neurons since ion channels can be added or removed to tune the leakage of currents through the cell membrane. The number and distribution of ion channels can change through various mechanisms including activity-dependent gene expression and activity-dependent modulation of assembled ion channels (see [1] for a comprehensive review). Nowak and Bullier [14] studied various mechanisms of decay (or integration time), and further investigations of these mechanisms may well reveal how decay rate can be controlled. Some models already utilize synaptic decay [5,15], but the influence of different levels of decay on synchronization has not been fully tested.

Can decay rate affect synchronization behavior just as delay does? This paper shows that different decay rates can indeed cause greatly different synchronization behavior even for the same type of connections (i.e. excitatory or inhibitory). The results suggest that PSP decay rates should be studied in more detail in order to understand how synchronization is modulated in biological neural networks.

2. Model

The neuron model used in this paper is a generalization of the popular integrate-and-fire neuron. Instead of a single integration mechanism per neuron, each individual postsynaptic membrane has a separate integration term with independent PSP decay rates (Fig. 1).

Each connection between neurons is a leaky integrator that continuously calculates an exponentially decayed sum of incoming spikes: [1]

$$s(t) = x(t) + s(t-1)e^{-\lambda}, \tag{1}$$

where $s(t)$ and $s(t-1)$ are the current and the previous decayed sum (i.e. at time step t and $t-1$), $x(t)$ is the input spike (either 0 or 1), and λ is the decay rate. The excitatory connections and inhibitory connections have separate decay rates λ_e and λ_i. Such a leaky integrator models the post-synaptic potential (PSP) that decays exponentially over time in biological neurons. The formulation is a finite difference approximation of the leaky integrate-and-fire neuron when λ is the same for all synapses.

[1] This equation has the same form as the one derived from convolution equations in [5].

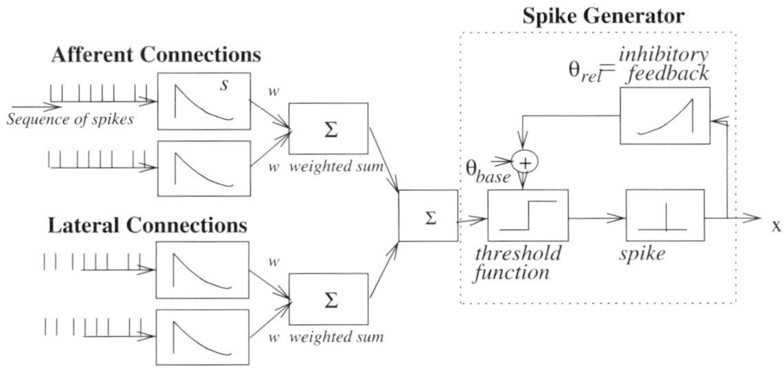

Fig. 1. *The Neuron Model*: Leaky integrators at each synapse perform decayed summation of incoming spikes, and the outgoing spikes are generated by comparing the sum of weighted sums to the dynamic spiking threshold. The dynamic threshold consists of the base threshold θ_{base} and the θ_{rel} that represents the refractory period.

Each cortical neuron in the model is connected to one input neuron and to all other cortical neurons through excitatory or inhibitory lateral connections. The weighted sum of the afferent and lateral leaky synapses are calculated and passed through a squashing function: 0 if sum $< \delta, 1$ if sum $> \beta$, and sum $- \delta/\beta - \delta$ otherwise. The resulting activation $\sigma(t)$ is compared to the dynamic threshold:

$$\theta(t) = \theta_{\text{base}} + \tau\theta_{\text{rel}}(t), \tag{2}$$

where θ_{base} is the base threshold, $\theta_{\text{rel}}(t)$ represents the refractory period, and τ is a scaling constant. The term $\theta_{\text{rel}}(t)$ is also a leaky integrator as in Eq. (1), but with a different decay rate λ_{rel}. If $\sigma(t)$ is greater than $\theta(t)$, an output spike is generated.

3. Experiments and results

Thirty fully connected neurons were simulated for 500 iterations. Each neuron received a constant input of value 1.0 throughout the simulation, and the connection weights were uniform with a value of 1/30. Afferent input was scaled by 0.8 and lateral input by 0.01 prior to passing through the squashing function to make the total afferent vs. lateral contributions to have a similar range. The squashing function parameters were $\delta = 0.0$ and $\beta = 3.0$. The threshold parameters were $\theta_{\text{base}} = 0.1$, $\tau = 0.65$, and $\lambda_{\text{rel}} = 0.05$. Membrane voltage of each neuron was randomly initialized.

Four separate experiments were conducted to test the effect of altered decay rate on synchronization behavior: (1) excitatory lateral connections alone with slow decay ($\lambda_e = 0.1$), (2) inhibitory lateral connections alone with slow decay ($\lambda_i = 0.1$), (3) excitatory lateral connections alone with fast decay ($\lambda_e = 1.0$), and (4) inhibitory lateral connections alone with fast decay ($\lambda_i = 1.0$). The results are summarized in Fig. 2a–d, respectively.

Two conditions, excitatory connections with fast decay and inhibitory connections with slow decay, resulted in synchrony. In contrast, excitatory connections with slow

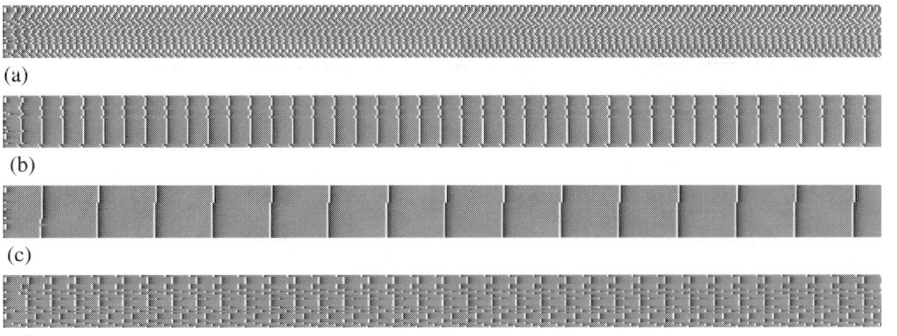

(a) (b) (c) (d)

Fig. 2. *Effect of connection type and decay rate on synchronization*: Thirty neurons with full lateral connections were simulated for 500 iterations. Four separate experiments were conducted to determine how decay in inhibitory and excitatory lateral connections affects synchronization. The x-axis represents the simulation iteration, and y-axis the neuron index. The membrane voltage is plotted in grayscale, from low to high (black to white). (a) Excitatory connections with slow decay result in desynchronized activity, (b) excitatory connections with fast decay result in synchronized activity, (c) inhibitory connections with slow decay result in synchronized activity and (d) inhibitory connections with fast decay result in desynchronized activity. The results show that synchronization behavior can vary greatly even for the same connection type if the decay rate differs.

decay and inhibitory connections with fast decay resulted in desynchrony. These results are similar to the results with conduction delays as reported in [11,13]. Since decay rate may be easier for neurons to adjust than conduction delay, neurons may indeed employ this process to fine-tune their synchronization behavior.

4. Discussion

Besides PSP decay, several other factors affect synchronization behavior in a network of integrate-and-fire neurons. For example, (1) noisy initialization of membrane voltage, or on-going addition of noise helps desynchronization between different populations [3,11,18], (2) higher levels of excitation can overcome moderate levels of noise to synchronize activity, and (3) longer absolute refractory period can help overcome noise and help synchronize neurons. All of these factors were systematically tested with the same model presented in this paper, but because of limited space the results are not presented here (see [3] for the full data).

An interesting future direction is to see whether adjusting the decay rate can counter the effects of conduction delays. A combined model with both adjustable PSP decay rate and adjustable conduction delay can be developed and the interaction of the two processes can be studied. It is possible that adjusting the decay rate can help overcome various side-effects such as unintended desynchronization introduced by conduction delay by resynchronizing the neurons.

Biological experiments can be done to verify whether the dendritic membrane potential can decay at different rates at different locations, and also whether there is such a difference between different types of synapses (e.g. glutaminergic vs. GABA-ergic synapses). If there is a difference, that data can be compared to the results presented in this paper, and it can help us understand the role of these different kinds of connections in modulating synchrony.

5. Conclusion

In this paper, we have shown that synchronization behavior can be modulated by adjusting the PSP decay rate for different types of synapses. Adjusting the decay rate at a local scale can allow a finer degree of synchronization behavior tuning, and it may be a more efficient process than adjusting axonal conduction delays which may require a macro-level change. The computational results presented in this paper calls for further investigations into the role of PSP decay rate in modulating neural behavior.

References

[1] L.F. Abbott, Eve Marder, Activity-dependent regulation of neuronal conductances, in: M.A. Arbib (Ed.), The Handbook of Brain Theory and Neural Networks, MIT Press, Cambridge, MA, 1995, pp. 63–65.

[2] S.R. Campbell, D.L. Wang, C. Jayaprakash, Synchrony and desynchrony in integrate-and-fire oscillators, Neural Comput. 11 (1999) 1595–1619.

[3] Y. Choe, Perceptual grouping in a self-organizing map of spiking neurons, Ph.D. Thesis, Department of Computer Sciences, The University of Texas at Austin, Austin, TX, Technical Report AI01-292, 2001.

[4] R. Eckhorn, R. Bauer, W. Jordan, M. Kruse, W. Munk, H.J. Reitboeck, Coherent oscillations: a mechanism of feature linking in the visual cortex?, Biol. Cybernet. 60 (1998) 121–130.

[5] R. Eckhorn, H.J. Reitboeck, M. Arndt, P. Dicke, Feature linking via synchronization among distributed assemblies: Simulations of results from cat visual cortex, Neural Comput. 2 (1990) 293–307.

[6] C.W. Eurich, K. Pawelzik, U. Ernst, J.D. Cowan, J.G. Milton, Dynamics of self-organized delay adaptation, Phys. Rev. Lett. 82 (1999) 1594–1597.

[7] W. Gerstner, Hebbian learning of pulse timing in the barn owl auditory system, in: W. Maass, C.M. Bishop (Eds.), Pulsed Neural Networks, MIT Press, Cambridge, MA, 1998, pp. 353–377 (Chapter 14).

[8] W. Gerstner, Spiking neurons, in: W. Maass, C.M. Bishop (Eds.), Pulsed Neural Networks, MIT Press, Cambridge, MA, 1998, pp. 3–54 (Chapter 1).

[9] W. Gerstner, J.L. van Hemmen, Associative memory in a network of spiking neurons, Network 3 (1992) 139–164.

[10] C.M. Gray, P. Konig, A. Engel, W. Singer, Oscillatory responses in cat visual cortex exhibit inter-columnar synchronization which reflects global stimulus properties, Nature 338 (1989) 334–337.

[11] D. Horn, I. Opher, Collective excitation phenomenon and their applications, in: W. Maass, C.M. Bishop (Eds.), Pulsed Neural Networks, MIT Press, Cambridge, MA, 1998, pp. 297–320 (Chapter 11).

[12] A.B. Kirillov, D.J. Woodward, Synchronization of spiking neurons: transmission delays, noise and nmda receptors, in: Proceedings of the World Congress on Neural Networks, Portland, OR, 1993, pp. 594 –597.

[13] A. Nischwitz, H. Glünder, Local lateral inhibition: a key to spike synchronization?, Biol. Cybernet. 73 (1995) 389–400.

[14] L.G. Nowak, J. Bullier, The timing of information transfer in the visual system, Cereb. Cortex 12 (1997) 205–239.

[15] H. Reitboeck, M. Stoecker, C. Hahn, Object separation in dynamic neural networks, in: Proceedings of the IEEE International Conference on Neural Networks, Vol. 2, San Francisco, CA, IEEE, Piscataway, NJ, 1993, pp. 638–641.

[16] T. Tversky, R. Miikkulainen, Modeling directional selectivity using self-organizing delay-adaptation maps, Neurocomputing 44–46 (2002) 679–684.
[17] C. van Vreeswijk, L.F. Abbott, When inhibition not excitation synchronizes neural firing, J. Comput. Neurosc. 1 (1994) 313–321.
[18] D.L. Wang, Emergent synchrony in locally coupled neural oscillators, IEEE Trans. Neural Networks 6 (1995) 941–948.

Yoonsuck Choe is an Assistant Professor of Computer Science at Texas A&M University (since the Fall of 2001). He received his B.S. degree in Computer Science from Yonsei University (Korea) in 1993, and his M.S. and Ph.D. degrees in Computer Sciences from the University of Texas at Austin in 1995 and 2001, respectively. His current research interests are self-organizing models of the visual cortex, perceptual grouping, and neural basis of analogy.

Risto Miikkulainen is a Professor of Computer Sciences at the University of Texas at Austin. He received an M.S. in Applied Mathematics from the Helsinki University of Technology, Finland, in 1986, and a Ph.D. in Computer Science from UCLA in 1990. His current research includes models of sentence processing, self-organization of the visual cortex, episodic memory, and evolving neural networks with genetic algorithms.

Available at
www.ComputerScienceWeb.com
POWERED BY SCIENCE DIRECT®

ELSEVIER

Neurocomputing 52–54 (2003) 713–719

NEUROCOMPUTING

www.elsevier.com/locate/neucom

Analogical cascade: a theory on the role of the thalamo-cortical loop in brain function☆

Yoonsuck Choe

Department of Computer Science, Texas A&M University, 3112 Tamu, College Station, TX 77843-3112, USA

Abstract

The cortico-thalamic connections and the nucleus reticularis thalami (nRt) have gained much attention lately because of their integrative and modulatory functions. This particular architecture has been thought to perform analysis and synthesis of memories, work as active blackboards or a global workspace, or gate attention and give rise to consciousness. In this paper, I show that this circuitry can be implementing a general *analogical functionality*. The concrete connection between analogical function and its exact neural basis established in this paper can help us better understand the brain function.

Keywords: Active neurons; Analogy; Thalamus; Nucleus reticularis thalami

1. Introduction

Current computational neuroscience research is focused on discovering what kind of information is carried by neurons. In other words, we are interested in what the neurons represent. However, this viewpoint makes the neurons *passive*. In such a passive view, neurons require further interpretation (of what they represent) by others and that can cause the problem of infinite regress. To overcome this problem, I propose that we assign a more *active* role to the neurons. Active, in the sense that neurons receive input (temporal or spatial) and immediately *invoke* other neurons if the input was preferable. With this slight change of viewpoint, a surprising functionality can be derived; that

☆ This research was supported in part by the Texas Higher Education Coordinating Board ARP/ATP program under Grant #000512-0217-2001.

E-mail address: choe@tamu.edu (Y. Choe).

0925-2312/03/$ - see front matter © 2003 Elsevier Science B.V. All rights reserved.
doi:10.1016/S0925-2312(02)00748-8

of *analogy*. In this paper, I will describe how as a collection such active neurons can perform analogical function, and show that the inhibitory mechanism in the nucleus reticularis thalami (nRt) can carry out activity gating necessary for proper analogical function.

2. Active neurons: a primitive for analogical computing

Instead of focusing on understanding what kind of information the neurons encode and process, we can ask what *action* is taken when they sense a certain feature in the incoming input, be that temporal or spatial. The action performed by neurons is basically *invoking* activity in other neurons. In this way, neurons represent a certain input feature, and take immediate action by invoking other neurons once the feature is detected. An important observation here is that such invocation establishes a relational context among neurons, because connections between neurons tend to strengthen when they are causality related [9].

The question then is what kind of general principle can such active neurons implement? Such a unit alone cannot achieve much, neither can a serial chain of such units. The true power of this simple unit is revealed when it is used in a massively parallel way. It turns out that the collective effort of these simple units can embody a simple yet powerful functional principle of analogy. We have to simplify matters to see how such neurons can process analogy.

Let us assume there are six neurons in an imaginary creature's brain inhabiting the world of fruits (Fig. 1). After the fruit brain experiences the world of fruits, it will learn the co-occurrences between features and establish relational arrows as shown in the figure (arcs with arrows). Also suppose that the brain is partitioned into several specialized map areas (or partitions), as in cortical maps. Now, suppose ⟨apple⟩, ⟨orange⟩, and ⟨word-red⟩ were presented to the creature simultaneously. If we track the activation, we can see that these detectors will turn on: apple detector, orange detector, color-red detector, color-orange detector, and finally, word-red detector. These activations are *input-driven*. Because the neurons are active, as soon as they detect what they are familiar with, they send out signals through the relational arrows horizontally across the cortex. As a result of this second-order activation, the word-orange detector turns on, even without input. Now, here is the crucial moment. We can ask this question: *which neuron's firing was purely cortically-driven?* Note that this question can be viewed as a filtering (or a gating) process. The result of the filtering is then ⟨word-orange⟩. The significance of this observation is that this process is very similar to solving analogical problems. The input presented to the creature is basically an analogical query: ⟨apple⟩ : ⟨orange⟩ = ⟨word-red⟩ : ⟨?⟩. The filtered cortical response ⟨word-orange⟩ can then be the *answer* to this query. [1] Thus, active neurons can perform a rudimentary analogical function when the responses are filtered properly.

[1] There is an issue of how the presence of ⟨word-red⟩ can affect the outcome at all. This problem will be discussed later in the discussion section.

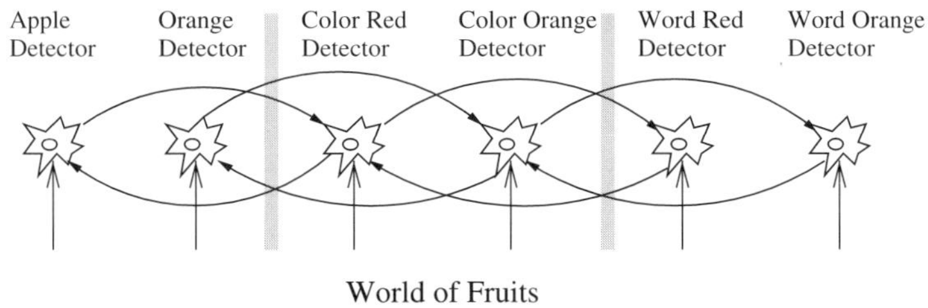

Fig. 1. World of fruits: A brain with object, color, and word detector neurons is shown. The six neurons each respond to these input features as labeled above. At the bottom is the fruit world, and the thick vertical arrows represent afferent input. The horizontal arcs are the relational arrows that point to their most frequently co-occurring counterparts that have been learned through experience. The gray vertical bars represent the partitioning of the brain into separate map areas (from the left to right, object map, color map, and word map). Note that for simplicity, the word-orange detector connects only to the color-orange detector, but not the orange detector, i.e. it is a word-color-orange detector, not a word-object-orange detector.

However, things can get complicated when combinations of objects are used as a query. Let us extend the creature's feature detectors to include concepts of small and big (not shown in the figure). Then we can allow the creature to learn the relations again. We can then present an analogical query like this: ⟨big⟩⟨apple⟩ : ⟨small⟩⟨apple⟩ = ⟨big⟩⟨orange⟩ : ⟨?⟩. In this case, if we follow the same steps as above, we come across a problem. Because the answer we expect (i.e. ⟨small⟩⟨orange⟩) already appeared in the query, if we look for purely cortically driven activations, the answer will be ⟨word-red⟩⟨word-orange⟩. However, we can overcome this problem if we ask: *what are the most cortically driven activities in each partition of the brain*? Because ⟨big⟩ and ⟨apple⟩ appeared in the input twice but ⟨small⟩ and ⟨orange⟩ appeared only once, the latter two can be selected, as well as the purely cortically driven activities listed above. Thus, even for derived activities that are input-driven, those that are less input-driven can survive and the correct analogical response can still be found among such activities that are more cortically driven within each partition (or area). Note that ⟨color-orange⟩ also survives the filtering, but what is more important here is that a simple filtering process as described above can generate a *small subset of potential answers* to analogical queries. Although the simple analogical query presented above has a straight forward answer, in more complex analogical problems, there can be multiple answers depending on the interpretation [6].

In this section, I have shown that active neurons that encode input features and relational contexts can collectively perform rudimentary analogical functions.[2] But does the brain function in such a way? In fact, an exact circuit that may be implementing such a function exists in the brain.

[2] Analogical tasks can become much more complex than the ones shown here. The example in this paper is decidedly simple to clearly illustrate the basic mechanism.

3. Neural basis of analogical completion and filtering

Two basic mechanisms are needed to account for the proposed analogical function: completion and filtering. Below, I will discuss how the cortico-cortical connections and the thalamo-cortical loop can implement these two mechanisms.

Completions may be accomplished by the long-range cortico-cortical connections. Synapses are strengthened when the presynaptic activity precede postsynaptic activity [9], thus the connections can implement causal relations. Also, specific patterns of connections observed in animals (e.g. in the primary visual cortex of monkeys [1]) show how such patterns can implement specific completion functions. Computational models also showed how such connectivity patterns can encode feature co-occurrence and how they can dictate the performance of the model [3,4].

For filtering, a separate mechanism is necessary. It turns out that the feedforward and feedback connections from the thalamus to the cortex together with the nRt inhibitions can filter feedbacks from the cortex to promote the most cortically driven feedback,

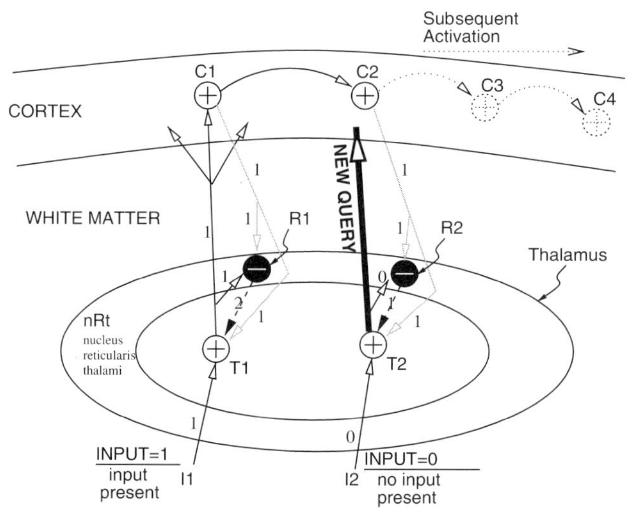

Fig. 2. *Analogical filtering in the thalamus*: The diagram shows a simplified thalamo-cortical loop that can perform analogical completion and selection, and propagate the selection back to the cortex. All connections shown are based on known anatomy of the thalamus and the cortex [7]. I1 and I2 are input fibers, T1 and T2 are thalamic relay cells, R1 and R2 are inhibitory nRt cells. C1, C2, C3, and C4 are cortical neurons (each is a set of neurons ranging multiple layers in a single cortical column). The neurons are either excitatory (+) or inhibitory (−), and the arrows are axons (pointing in the direction of action potential propagation). The numbered labels on each arc show the activity being carried. Black solid arrows are ascending fibers to the cortex and the cortico-cortical connections (relational arrows), and gray solid arrows are cortico-thalamic feedbacks. Black dashed arrows are inhibitory connections. The diagram shows a scenario when input was presented to C1, which excites C2, and in turn generates the feedback from C2 to T2, which is then retransmitted to the cortex as a new query (ascending thick black arrow). The selection decision for further propagation to the cortex depends on the relative excitation and inhibition T1(T2) receive from C1(C2) and R1(R2). On the right of C2 (...) in the cortex is the subsequent cascade of analogical completions. Note that to avoid clutter, reciprocal connections in the cortex as well as disinhibiting connections within the nRt layer are not shown.

i.e. the analogical answers. Let us first see how the purely cortically driven activities are selected (Fig. 2). In the thalamus, ascending fibers (T1 to C1) branch out and excite the inhibitory nRt neuron R1 (T1 to R1). When the feedback from C1 to T1 comes back, it branches and stimulates R1. As a result, if the descending feedback had a matching ascending signal, the inhibition T1 receives is twice as high as other neurons in the thalamus that are activated by purely cortically driven feedback (e.g. T2). If the synaptic weights are appropriate (i.e. $w_{TC} = 2$ and $w_{TR} = 1$),[3] at T1 the feedback will cancel out, but at T2 the feedback will survive the inhibition and will be retransmitted to the cortex (the *new query* arrow). Such a surviving cortical feedback together with the input stimulus at the next moment form a new analogical query to the cortex, and the same process is repeated. That is, C2 elicits activities in C3, and in turn C4 through the thalamo-cortical loop (note that they can be quite far away). For the selection of the *most cortically driven* feedback, the mutual inhibitions in the nRt layer (e.g. between R1 and R2) may disinhibit (inhibiting an inhibitory neuron results in less net inhibition) each other and allow the more cortically driven feedback to go back to the cortex, even when all current cortical activities are input driven.

4. Discussion

The neural mechanisms described in this paper can only account for simple kinds of analogies, and in some case it can even seem as simple pattern completion. For example, ⟨orange⟩=? will result in the same answer ⟨word-orange⟩ as in Section 2. How can the term ⟨word-red⟩ in the original query affect the outcome at all? For this, I believe that among many possible completions, the general map area (i.e. the partitions in Fig. 1) that are activated by input gets higher preference. In this example, the object-map, word-map and color-map will turn on, thus purely cortical activations in other general maps (say odor-map, etc.) will not be as salient as that of ⟨word-orange⟩. Thus, in this way, the presence of ⟨word-red⟩ can indeed affect the outcome of the analogical query. A more precise neural mechanism for this kind of selection of areas (or maps) needs to be investigated further.

Researchers regard the analogical capability as the crux of high-level cognition (see [5] for a collection of current work on analogy). However, analogy does not need to be limited to high-level cognition [2]. With this new view, we can start to understand perception, cognition, and motor functions under the unifying framework of analogy instead of trying to understand those as embodying separate functional principles. How can such a diverse functionality be integrated under the general principle of analogical processing? Massive connections exist within and across different functional areas in the brain, and the sensory/motor maps are topologically organized [8]. When the sensory, cognitive, and motor maps are connected in an orderly way preserving their local topology, analogies within and (more importantly) *across* different domains can be drawn [5].

[3] Here, w_{YX} is the synaptic connection strength from neuron X to neuron Y.

Within this huge number of maps specializing in different tasks, a cascade of multiple analogical completions can be going on in parallel, synchronized at each moment by the 40 Hz rhythm to hold an instantaneously coherent state [7]. Such state can then pose as another analogical query, and that process can repeat. When that cascade reaches a motor area, behavior will be generated. Memory content can also enter the analogical cascade, and this quasi-static contribution can prevent the continuously changing input stream from causing random cascades, thereby maintaining a more goal-directed and stable behavior. Specific mechanisms of how the memory content enters the thalamo-cortical loop and how completed analogies are archived in memory through the interactions with subcortical centers such as the hippocampus should be studied further.

5. Conclusion

In this paper, I assigned an active invocational role to neurons, and it turned out that collectively they can perform an *analogical function*. The thalamic and thalamo-cortical circuits were found to be ideal for implementing such a function, and especially, the gating inhibition in the nRt was found to be crucial for analogical processing. With analogy, different domains can be mapped and related, thus we can start to understand perception, cognition, and motor function under a unifying framework. This new framework will enable us to take a more focused approach in studying the brain function.

References

[1] G.G. Blasdel, Orientation selectivity, preference, and continuity in monkey striate cortex, J. Neurosci. 12 (1992) 3139–3161.

[2] D.J. Chalmers, R.M. French, D.R. Hofstadter, High-level perception, representation, and analogy, J. Exp. Theor. Artif. Intell. 4 (1992) 185–211.

[3] Y. Choe, Perceptual grouping in a self-organizing map of spiking neurons, Ph.D. Thesis, Technical Report AI01-292, Department of Computer Sciences, The University of Texas at Austin, Austin, TX, 2001.

[4] W.S. Geisler, J.S. Perry, B.J. Super, D.P. Gallogly, Edge co-occurrence in natural images predicts contour grouping performance, Vision Res. 41 (2001) 711–724.

[5] D. Gentner, K.J. Holyoak, B.N. Kokinov (Eds.), The Analogial Mind: Perspectives from Cognitive Science, MIT Press, Cambridge, MA, 2001.

[6] D.G. Hofstadter, Fluid Concepts and Creative Analogies: Computer Models of the Fundamental Mechanisms of Thought, Basic Books (a division of Harper Colins), New York, 1995.

[7] D. Mumford, Thalamus, in: M.A. Arbib (Ed.), The Handbook of Brain Theory and Neural Networks, MIT Press, Cambridge, MA, 1995, pp. 153–157.

[8] J. Sirosh, R. Miikkulainen, Y. Choe (Eds.), Lateral Interactions in the Cortex: Structure and Function, The UTCS Neural Networks Research Group, Austin, TX, 1996, Electronic book, ISBN 0-9647060-0-8, http://www.cs.utexas.edu/users/nn/web-pubs/htmlbook96.

[9] S. Song, K.D. Miller, L.F. Abbott, Competitive hebbian learning through spike-timing-dependent synaptic plasticity, Nature Neurosci. 3 (9) (2000) 19–926.

Yoonsuck Choe is an Assistant Professor of Computer Science at Texas A&M University (since the Fall of 2001). He received his B.S. degree in Computer Science from Yonsei University (Korea) in 1993, and his M.S. and Ph.D. degrees in Computer Sciences from the University of Texas at Austin in 1995 and 2001, respectively. His current research interests are self-organizing models of the visual cortex, perceptual grouping, and neural basis of analogy.

Available at
www.ComputerScienceWeb.com
POWERED BY SCIENCE DIRECT

ELSEVIER

Neurocomputing 52–54 (2003) 721 – 726

NEUROCOMPUTING

www.elsevier.com/locate/neucom

Estimating the afferent and efferent temporal interval entropy of neuronal discharge for single spike trains☆

Allan D. Coop*, George N. Reeke

Laboratory of Biological Modelling, The Rockefeller University, 1230 York Avenue, New York, NY 10021, USA

Abstract

We define a biological communication system at the level of a single neuron and quantify the temporal variability of afferent and efferent impulse patterns by means of an interval entropy measure. Two signal transmission conditions which bound a physiologically plausible range of transmission possibilities are explored. The number of efferent synapses is predicted by matching estimates of the mean efferent entropy to the total afferent entropy.

Keywords: Cerebellum; Purkinje cell; Model; Timing; Entropy

1. Introduction

The application of information theory to the analysis of neural communications systems is predicated on the mapping of theoretical components to the biological structures subserving neural function. Recently, several different definitions of signal source and destination have been employed, e.g. [6,8]. Here we present a communication system defined at the level of the single neuron. It extends from a source provided by active afferent fibers to a destination composed of the efferent synaptic terminals of a given cell. The input/output relations of this model system are explored by means of the temporal analytic-distribution interval method [9] which provides estimates of the entropy of afferent and efferent impulse activity.

☆ This work was supported by Neurosciences Research Foundation.
* Corresponding author. Tel.: +1-212-327-7646; fax: +1-212-327-7469.
E-mail address: adc@lobimo.rockefeller.edu (A.D. Coop).

0925-2312/03/$ - see front matter
doi:10.1016/S0925-2312(02)00749-X

The interval methodology is based on the idea that by fitting observed interval data to a suitable continuous analytic distribution, extrapolations to long intervals and long data records are both effectively carried out in compliance with the observed impulse activity. The procedure provides the so-called "interval entropy" measure which can be used to quantify the temporal variability of afferent impulse activity and evoked cell discharge under the assumption that intervals in a sequence are uncorrelated. The approach is demonstrated by application to a composite cerebellar Purkinje cell model [4].

2. Methods

The relation between a theoretical communication system comprising five elements including a source, transmitter, channel, receiver, and destination [12] and the analyzed biological communication system is given in Fig. 1. The messages in the biological system consist of the discrete temporal intervals which exist between successive impulses on individual afferent fibers.

Calculation of the interval entropy, H_I, proceeds by assuming (i) the distribution of intervals in an arbitrarily long sample can be estimated by fitting a sum of one (Q, Eq. (1)) or more suitable continuous analytic probability distribution functions to an observed discrete cumulative interval probability distribution (O) by minimization of the RMS error (E, Eq. (2)) for a total of M intervals. For the case of a single incomplete gamma function,

$$Q(x;a,s,\tau) = \gamma\left(\frac{x-s}{\tau}, a\right) \equiv \frac{1}{\Gamma(a)} \int_0^{\left(\frac{x-s}{\tau}\right)} t^{a-1} \mathrm{e}^{-t}\,\mathrm{d}t \quad (a,s>0), \tag{1}$$

where a, s, and τ are the order, time axis shift, and time scaling parameters, respectively, and

$$E = \left\{ \frac{1}{M} \sum_{j=1}^{M} [O(x_j) - Q(x_j;a,s,\tau)]^2 \right\}^{1/2} \times 100\%. \tag{2}$$

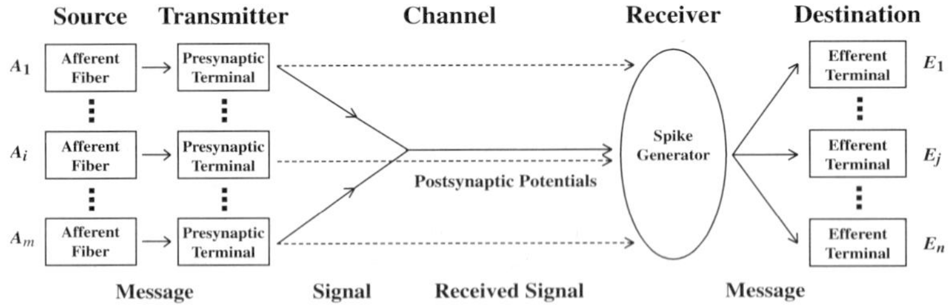

Fig. 1. Relation between the theoretical and the modeled biological communication system. Dashed lines indicate preservation of interval duration for impulse sequences occurring at individual afferent synapses as might occur when multiple dendrites attach independently to the soma (non-interleaved case). Solid lines indicate the interleaving of impulse times as might occur when dendritic branches converge to form a single dendritic trunk (interleaved case).

The probabilities (p_i) that interspike intervals lie in the intervals $x = [i\,\Delta t, (i+1)\,\Delta t]$, $i = 0, \ldots, N$, can be calculated from the analytic distribution for an appropriate choice of temporal resolution (Δt) where

$$p_i = [Q((i+1)\,\Delta t; a, s, \tau) - Q(i\Delta t; a, s, \tau)]. \tag{3}$$

The interval entropy is obtained from the discrete formulation given by Shannon [12]:

$$H_I = -\sum_{i=0}^{N} p_i \log_2 p_i. \tag{4}$$

The sum is truncated when $1 - \sum_i p_i < \varepsilon$, where ε is the required tolerance.

The mean afferent interval entropy, H_A, may be calculated by assuming the received signal is a temporally interleaved composite of the transmitted signal (Fig. 1, converged unbroken lines). This treatment is justified when impulses received at presynaptic terminals have equivalent effects at the soma, as for the one-compartment case studied here. Other treatments may be required when synapses are not equivalent. An alternative non-interleaved case is illustrated in Fig. 1 (dashed lines) and described in the figure caption. If it is assumed that neuronal activity is reliable [3,7,1,13,11] and that each transmitted signal always gives the same received signal with no two transmitted signals producing the same received signal, any distortion introduced by signal mismatch may be corrected at the receiver and the channel is noise free [12].

Simulated spike trains were obtained from a model cerebellar Purkinje neuron (PC) [4] in response to two extensively investigated activation paradigms [5]. In the stochastic or 'Poissonian' case, the impulse activity across afferent fibers is temporally uncorrelated. Alternatively, in the 'pulsed' case the stimulus is composed of pulse 'packets' delivered with a fixed interpulse interval. The impulses within each packet are distributed as a truncated Gaussian in time across the afferent fibers and their temporal correlation is determined by the coefficient of variation (CV = standard deviation/mean) of the Gaussian. Each afferent fiber contributes a single impulse to a packet. In both paradigms the number of active fibers controls cell discharge rate. Neither afferent transmission failure nor quantal variation is included [6], but the amplitude of each postsynaptic potential is modulated by paired pulse facilitation according to the model of Atluri [2]. The impulse rate is identical for all fibers, held constant for the duration of a simulation, and given as the mean rate/s per fiber.

3. Results

The total afferent interval entropy per second is given by

$$H_{\mathrm{A(tot)}} = H_{\mathrm{A}} \times N_{\mathrm{aff}} \times \text{impulse rate},$$

where N_{aff} is the afferent fiber number. For the interleaved case under Poissonian stimulation, $H_{\mathrm{A(tot)}}$ increases as the afferent stimulation rate is raised. However, doubling the stimulus rate from 64 to 128 impulses/s has little effect (Fig. 2A). Alternatively, under pulsed stimulation $H_{\mathrm{A(tot)}}$ is effectively independent of the stimulus rate (Fig. 2B).

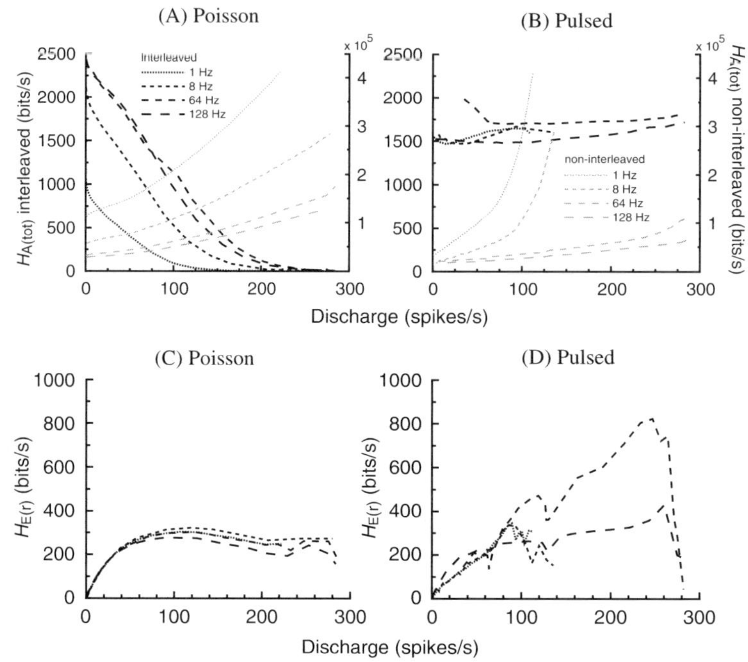

Fig. 2. Total afferent interval entropy, $H_{\mathrm{A(tot)}}$, for the interleaved case under: (A) Poissonian and (B) pulsed stimulation (magnitude given by left axes). The equivalent non-interleaved cases (see Fig. 1) are also given (gray, magnitude given by right axes). Mean efferent entropy rate per second, $H_{\mathrm{E(r)}}$, for (C) Poissonian and (D) pulsed stimulation. Legends give stimulation rate. The legend in A applies to all panels. Legend in B applies to panels A and B.

In the absence of knowledge about efferent synapse number, the mean efferent interval entropy rate per second is given by

$$H_{\mathrm{E(r)}} = H_{\mathrm{E}} \times \text{discharge rate}.$$

Under Poissonian stimulation $H_{\mathrm{E(r)}}$ is little influenced by the stimulation rate and is maximal at a discharge rate of about 100–120 spikes/s (Fig. 2C). Alternatively, changing the rate of pulsed stimulation has little effect on $H_{\mathrm{E(r)}}$ for discharge rates below about 100 spikes/s (Fig. 2D). However, as the discharge rate increases, the magnitude of $H_{\mathrm{E(r)}}$ becomes increasingly sensitive to stimulation rates in the range of 64–128 impulses/s.

When the total number of efferent synapses made by a single neuron is unknown, by assuming that entropy equivalence is maintained throughout the system, it can be estimated as

$$E_{\mathrm{syn}} = H_{A(\mathrm{tot})}/H_{\mathrm{E(r)}},$$

where E_{syn} gives the number of efferent synapses required to match $H_{\mathrm{E(r)}}$ to $H_{\mathrm{A(tot)}}$. In the Possionan interleaved case, there is a range of discharge rates for which less than one efferent synapse is sufficient to maintain entropy equivalence (Fig. 3A). This

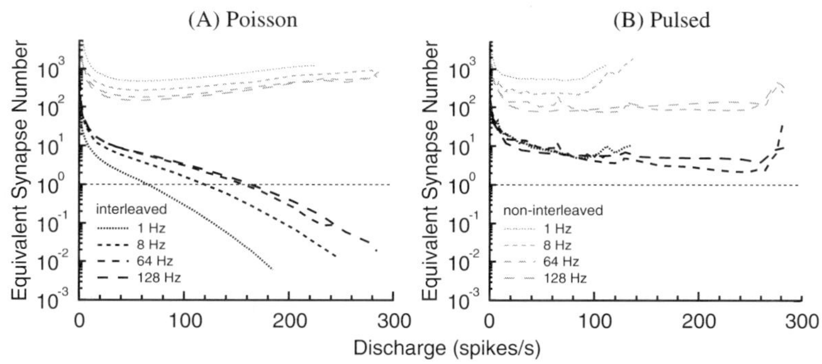

Fig. 3. Equivalent synapse number, E_{syn}, gives the number of efferent synapses required to match the mean efferent entropy rate per second, $H_{E(r)}$, to the total afferent entropy, $H_{A(tot)}$, for the interleaved case (see Fig. 1): (A) Poissonian and (B) pulsed stimulation. Horizontal line indicates one efferent synapse. Also given are the equivalent non-interleaved cases (gray). Legends give stimulation rates, each applies to both panels.

operating regime may be associated with the somewhat puzzling observation that transmitter release probabilities in many neurons are significantly less than the naively expected 1.0 [10]. Alternatively, in the interleaved case under pulsed stimulation, approximately 10 synapses are sufficient for entropy equivalence at a discharge rate of 40 spikes/s. At higher stimulation (64–128 Hz) and discharge rates (125–265 spikes/s), the number falls to 2.5–5.5 synapses (Fig. 3B).

4. Conclusions

The entropy of signal transmission within a neuronal communication system may be determined by synaptic connectivity. Conditions are identified for which, under Poissonian stimulation, the total afferent ($H_{A(tot)}$) and efferent ($H_{E(tot)}$) entropies are matched with less than one efferent synapse. It is unlikely that this would occur under conditions of near-synchronous pulsed activation.

References

[1] A. Arieli, A. Sterkin, A. Grinvald, A. Aertsen, Dynamics of ongoing activity: explanation of the large variability in evoked cortical responses, Science 273 (1996) 1868–1871.

[2] G. Atluri, Determinants of the time course of facilitation at the granule cell to Purkinje cell synapse, J. Neurosci. 16 (1996) 5661–5671.

[3] T.H. Bullock, The reliability of neurons, J. Gen. Physiol. 55 (1970) 565–584.

[4] A.D. Coop, G.N. Reeke, The composite neuron: a realistic one-compartment Purkinje cell model suitable for large scale neuronal network simulations, J. Comput. Neurosci. 10 (2001) 173–186.

[5] A.D. Coop, G.N. Reeke, Deciphering the neural code: neuronal discharge variability is preferentially controlled by the temporal distribution of afferent impulses, Neurocomputing 38 (2001) 153–157.

[6] W.B. Levy, R.A. Baxter, Energy-efficient neuronal computation via quantal synaptic failures, J. Neurosci. 22 (2002) 4746–4755.

[7] Z.F. Mainen, T.J. Sejnowski, Reliability of spike timing in neocortical neurons, Science 268 (1995) 1503–1506.

[8] A. Manwani, C. Koch, Detecting and estimating signals in noisy cable structures, II: Information theoretical analysis, Neural Comput. 11 (1999) 1831–1873.

[9] G.N. Reeke, A.D. Coop, Estimating the temporal interval entropy of neuronal discharge, Neural Computation, submitted.

[10] W.G. Regehr, C.F. Stevens, Physiology of synaptic transmission and short-term plasticity, in: G. Stuart, N. Sprouston, M. Häusser (Eds.), Dendrites, OUP, New York, 1999, pp. 135–175.

[11] P. Reinagel, R.C. Reid, Precise firing events are conserved across neurons, J. Neurosci. 22 (2002) 6837–6841.

[12] C.E. Shannon, W. Weaver, The Mathematical Theory of Communication, University of Illinois Press, Urbana, IL, 1949.

[13] C.F. Stevens, A.M. Zador, Input synchrony and the irregular firing of cortical neurons, Nature Neurosci. 1 (1998) 210–217.

Available at
www.ComputerScienceWeb.com
POWERED BY SCIENCE DIRECT®

ELSEVIER

Neurocomputing 52–54 (2003) 727–731

NEUROCOMPUTING

www.elsevier.com/locate/neucom

Dynamic aspects of delay activity

Marc de Kamps[a,*], Frank van der Velde[b]

[a]*Realtime Systems and Robotics, Institut für Informatik, Technische Universität München, Boltzmannstr. 3, D-85748, Garching bei München, Germany*

[b]*Faculty of Social Sciences, Department of Psychology, Section Cognitive Psychology, Leiden University, Wassenaarseweg 52, 2333 AK Leiden, The Netherlands*

Abstract

The population density equation (PDE) has generated considerable interest, recently. The PDE allows efficient simulation of populations with a large number of neurons. We apply the PDE to an existing model of delay activity, which produces stationary firing rates for delay assemblies. In order to observe transient phenomena, we apply the PDE to this model. Here we present the preliminary results. We find that a novel algorithm to solve the PDE is instrumental for this particular application. Our results agree with stationary rates that can be found using a diffusion approximation.

Keywords: Delay activity; Diffusion approximation; Population density; Partial differential equation; Response curve

1. Motivation

To allow the modeling of large-scale cortical networks, it is necessary to simulate the behavior of large groups of neurons efficiently. One interesting technique to describe the activity of large groups of neurons is the population density approach. In this approach a large population of neurons is described by a single density function $\rho(v,t)$. The neurons that we consider in this paper are leaky-integrate-and-fire (LIF) neurons, which are point model neurons described by a single dynamical variable v, the (rescaled) membrane potential. The probability for a single neuron in the population to have its membrane potential v in $[v, v+\mathrm{d}v]$ is given by $\rho(v,t)\,\mathrm{d}v$. In the mean-field

* Corresponding author.
E-mail addresses: swerner@xs4all.nl (M. de Kamps), vdvelde@fsw.leidenuniv.nl (F. van der Velde).

0925-2312/03/$ - see front matter
doi:10.1016/S0925-2312(02)00751-8

approximation a partial differential equation (PDE) [1] can be found (e.g., [5]) that describes the evolution in time of ρ in terms of the population's input. ρ may be used, for instance, to infer the population's firing rate, in other words, the population's response to input may be calculated. This technique has been applied successfully to a model of orientation tuning [4].

We intend to apply this technique to a model of delay activity, described in [1]. Delay activity is a stimulus-induced population response above baseline activity, which persists after the stimulus is removed. It is considered to be an important candidate for the neural substrate of short-term memory. The main reason to consider this model is that we want to investigate the usefulness of the population density technique to model large-scale cortical networks. The firing rates, that were found in [1], are stationary rates. In order to describe transient phenomena, quite extensive simulations are necessary. These simulations were performed, and therefore it is interesting to see if these results also can be obtained, at lower computational cost, with the population density technique.

Our results, so far, have resulted in a novel algorithm to solve the PDE, which seems to be more stable than, and as least as efficient as the finite difference methods that are currently in use. Moreover, it seems that for large-scale cortical networks, it is advantageous to use a single input population, that emulates a Gaussian white noise, rather than to include all input contributions of a population separately.

2. A model for delay activity

In the model of delay activity, formulated by Amit and Brunel [1], a cortical column is modeled with an excitatory population and an inhibitory population. The excitatory population has a substructure: a small part of this population responds with an increased firing rate, with respect to baseline, to specific external stimuli. In [1] it is shown that such a stimulus-specific population can retain its increased firing rate after the stimulus that induced it has vanished. Importantly, this firing rate is above baseline, but much lower than the maximal firing rate of the neurons in that population. Amit and Brunel consider LIF neurons in the diffusion approximation, i.e. as an Ornstein–Uhlenbeck process. In this approximation, it is possible to find an output firing rate of a neuron, in terms of the mean and variance of the input current. This approximation is particularly useful if input rates are high and synaptic efficacies are small, so that the input approximates a Gaussian white noise, which is the case for the parameter range considered by Amit and Brunel. For LIF neurons a response function may be derived, in terms of the first exit time of an Ornstein–Uhlenbeck process. The firing rate v is given in terms of the mean μ and variance σ of the input: $v = \phi(\mu, \sigma)$,

$$\phi = \left(\tau_{\text{ref}} + \tau\sqrt{\pi} \int_{(V_r-\mu)/\sigma}^{(\theta-\mu)/\sigma} \exp(u^2) erfc(u)\, \mathrm{d}u \right)^{-1}. \tag{1}$$

[1] We use the abbreviation PDE both for partial differential equation and population density equation. This is unlikely to cause confusion.

Here θ is the threshold potential in mV, and V_r the reset potential in mV. τ is the neuron's time constant in s and τ_{ref} its absolute refractive period. μ and σ are the mean and variance, respectively, of the input current during time τ. For a single neuron with one input with efficacy J, which receives a Poisson distributed input spike train with rate $r(t)$, one has

$$\mu = \tau J r(t),$$
$$\sigma^2 = \tau J^2 r(t). \tag{2}$$

In the mean-field approximation, where every neuron in a population is assumed to receive similar input, one can replace the entire population by a single neuron, representing the entire population. Amit and Brunel consider a network of four populations. As each population's output contributes to the other population's input, it is possible to find a set of closed equations which describe the stationary rates in the network. It turns out that it is possible to find network parameters, such that both spontaneous and delay activity are stable in the network. As this result depends mainly on the form of the response curve (Eq. (1)), our first goal is to see if the stationary solutions of the PDE are on this curve.

3. Population density equation

LIF neurons are described by a single dynamical equation:

$$\frac{\mathrm{d}v}{\mathrm{d}t} = -\gamma v. \tag{3}$$

The rescaled membrane potential is defined by $v = (V - V_{\text{rev}})/(V_{\text{th}} - V_{\text{rev}})$. Here V_{th} is the threshold potential and V_{rev} is the reversal potential, so that for the rescaled potential $-\infty \leqslant v < 1$ holds. If a neuron has its membrane potential driven past V_{th}, a spike is generated and its potential is reset to a (rescaled) potential value v_{reset}.

In its simplest form, the population density equation for LIF neurons is given by

$$\frac{\partial \rho(v,t)}{\partial t} - \gamma \frac{\partial}{\partial v}(\rho(v,t)v) = -\sigma(t)[\rho(v,t) - \rho(v - \bar{h}, t)]. \tag{4}$$

A necessary boundary condition is: $\rho(1,t) = 0$, as no decay from $v > 1$ into the system is possible. In this version of the equation we assume a single input synapse, with efficacy $\bar{h}$, which receives a spike train which is Poisson distributed with rate $\sigma(t)$. It is easy to generalize this case to an arbitrary number of inputs, with efficacies distributed according to a given probability density function (PDF), but Eq. (4) is adequate for this paper.

In this form the flux across the threshold potential, due to input, is equal to the (population) firing rate of the population: $r(t) = \sigma(t) \int_{1-\bar{h}}^{1} \rho(v,t)\,\mathrm{d}v$.

As neurons which are driven past threshold spike and then are reset to potential v_r, we add an extra term $r(t)\delta(v - v_r)$ to the right-hand side of Eq. (4), which reflects that probability density which is lost, due to flux across threshold, is balanced by probability density which appears at $v = v_r$, at the same time. There are various methods available to solve this equation (see e.g., [4,5]), but for the application of the PDE in this paper, we came up with a novel algorithm for reasons explained below.

4. A novel numerical method to solve the population density equation

The Gaussian white noise approximation has advantages for the approach presented here. Instead of considering four input populations, we only have to consider one, whose mean and variance are determined by the firing rates of the four input populations and the network parameters. As we consider a Gaussian white noise as an input current for neurons, we can pick the input rate $\sigma(t)$ and the input parameter $\bar{h}$, according to Eq. (3), which reproduces Gaussian white noise to a good approximation in our parameter range. For the μ and σ considered in this paper (see Fig. 1), this means one finds rather small values of $\bar{h}$ (typically in the order of $\bar{h} = 0.01$). This may produce a jagged density profile around $v = 0$ and a finite difference solution of the PDE may be sensitive to the large density gradients that occur there.

In order to avoid the stability problems caused by this, we consider a coordinate transformation:

$$\begin{cases} v \to v' = v\mathrm{e}^{-\gamma t}, \\ \rho(v,t) \to \rho'(v',t) = \mathrm{e}^{-\gamma t}\rho(v\mathrm{e}^{-\gamma t},t). \end{cases} \tag{5}$$

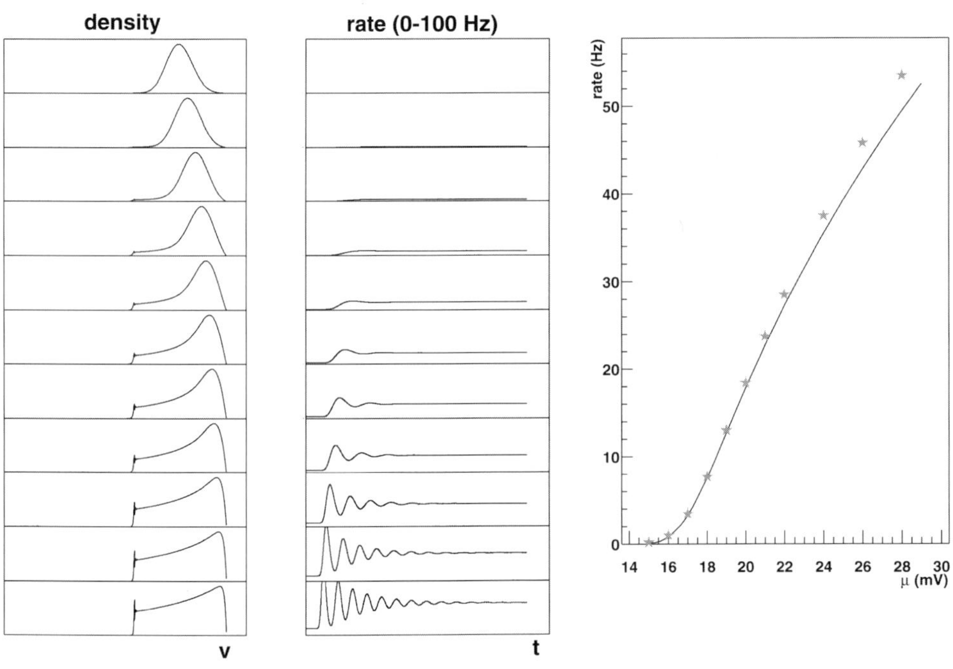

Fig. 1. Comparison between the solutions of the PDE and the diffusion approximation.

Eq. (14) changes under this transformation equation into

$$\frac{\mathrm{d}\rho'(v',t)}{\mathrm{d}t} = \sigma(t)\{\rho'(v' - \bar{h}e^{\gamma t}) - \rho(v')\}. \tag{6}$$

For a time period $\delta t \ll \gamma^{-1}$, so that $\bar{h}e^{\gamma t}$ may be considered constant, this corresponds to a set of ordinary differential equations, rather than a PDE, which is insensitive to density gradients. It possible to base an efficient numerical solution on this transformation which is described in detail in [3]. Another option would have been to solve the PDE in the Fokker–Planck approximation, which avoids the jagged profile at $v=0$. In the Fokker–Planck approximation, however, the PDE still remains sensitive to the density gradient, and stability problems may still occur [2].

5. First results

In Fig. 1, we consider a Gaussian white noise input with variable mean μ and a fixed variance $\sigma=2$ mV. We consider an excitatory population with membrane time constant $\tau=10$ ms, absolute refractive period $\tau_{\mathrm{ref}}=4$ ms. The figure shows the equilibrium density distribution in the left column, with μ increasing from top to bottom. In the middle column we plot the corresponding response rate as a function of time. The equilibrium values of the response rate are plotted in the right-hand figure using asterisks. The solid curve is the rate given by Eq. (1). There are several interesting observations: first of all, there is excellent agreement between the diffusion approximation prediction and the PDE solutions at low rates. At higher rates there is a slight disagreement, which can be explained by the somewhat simplistic treatment of the refractive period in Eq. (1). Secondly, at higher values of μ, the equilibrium rates are reached after substantial transient variation of the signal.

In conclusion, we have used to a novel algorithm to solve the PDE. Apart from the replication of earlier results, the good agreement between equilibrium rates given by PDE solutions and diffusion approximation in the description of response curves provides validation of the new algorithm. Since these response curves are the most important mathematical foundation of Amit and Brunel's model of delay activity, we are confident that we will be able to provide a dynamical description of their model in terms of PDE solutions in the future.

References

[1] D.J. Amit, N. Brunel, Model of global spontaneous activity and local structured activity during delay periods in the cerebral cortex, Cerebral Cortex 7 (1997) 237–252.

[2] A.R.R. Casti, A. Omurtag, A. Sornborger, E. Kaplan, B. Knight, J. Victor, L. Sirovich, A population study of integrate-and-fire-or-burst neurons, Neural Comput. 14 (2002) 957–986.

[3] M. de Kamps, A simple and stable numerical solution for the population density equation, 2002, submitted for publication.

[4] D.Q. Nykamp, D. Tranchina, A population density approach that facilitates large-scale modeling of neural networks: analysis and an application to orientation tuning, J. Comput. Neurosci. 8 (2000) 19–50.

[5] A. Omurtag, B.W. Knight, L. Sirovich, On the simulation of large populations of neurons, J. Comput. Neurosci. 8 (2000) 51–63.

Available at
www.ComputerScienceWeb.com
POWERED BY SCIENCE @ DIRECT®

ELSEVIER

Neurocomputing 52–54 (2003) 733–739

NEUROCOMPUTING

www.elsevier.com/locate/neucom

Analysis of dynamics and object recognition performance in coupled map networks

D. DeMaris

IBM Microelectronics, 11400 Burnet Rd, Austin, TX 78758, USA

Abstract

Coupled logistic map lattices can perform object recognition by mapping class members to a point in a space defined by densities (partition cell occupancies) measured at a readout time t for the entire lattice. Each unit represents average spiking dynamics in mixed excitatory–inhibitory neuronal populations. Non-stationary parameters improve peak recognition rate and recognition time, suggesting one role for interaction of multiple time scales in neural oscillations. A hypothesis that recognition functions by modulating the approach to an invariant measure was rejected by examining the response to noise. Moderate correlation ($r = 0.3$) between a configuration entropy measure and recognition rate is found.

Keywords: Density coding; Population codes; Nonlinear dynamics; Object recognition

1. Introduction

This report presents new results in ongoing study of oscillations in coupled map lattices (CML) used for the recognition of three-dimensional objects [1–3]. Such networks are interpreted as in the tradition of oscillatory neuronal ensemble models [8], where each units stands for the collective behavior of a mixed pool of excitatory and inhibitory neurons, and the entire network as a homogeneous spatial ensemble of locally connected column-like units. Previously it was shown that such systems could be trained via a genetic algorithm to map various viewpoints of paperclip objects to approximately the same distribution of at some readout time. Recognition rates of 85% were obtained with complete training on a set of 20 objects using a recognizer CML trained to recognize each object, and choosing as the match object the recognizer producing the nearest matching distribution to the characteristic distribution for the object.

E-mail address: demaris@well.com (D. DeMaris).

0925-2312/03/$ - see front matter © 2002 Elsevier Science B.V. All rights reserved.
doi:10.1016/S0925-2312(02)00752-X

Important features of the network and population coding method include rapid processing (only a few iterations), ease of learning, translation invariance, and generalization between outlines and filled objects.

The short range or local mean field coupled logistic map was used. In the locally coupled system, each unit is connected to nearest neighbors. The computation is divided into two steps; first a diffusive coupling step is applied:

$$S_{\mathrm{d}}(x,y) = (1-c)S_t(x,y) + \frac{c}{4}[S_t(x,y+1) + S_t(x,y-1) + S_t(x+1,y) + S_t(x-1,y)] \quad (1)$$

where S_{d} is a temporary array holding the state of the diffusion computation, t is the current time step, x, y are the spatial indices of the pixel array S at the center of the diffusion neighborhood, S is the continuous state variable at each pixel of the array restricted to the range [−1.0–1.0], and c is the coupling constant restricted to the range [0.0–1.0]. The factor $1-c$ is a squashing function, preventing the sum of the surround from exceeding the stable range of the map.

The second computational unit applied in each time step is the logistic map:

$$S_{t+1}(x,y) = 1 - bS_{\mathrm{d}}(x,y)^2 \quad (2)$$

where S, t, x, and y are as above and where b is the bifurcation parameter, restricted to the interval $[0.0 < b < 2.0]$. Initial and intermediate states S are restricted to the interval $[-1.0 < S < 1.0]$.

The basic iterations described above are applied at every lattice site, and are imbedded in an "outer loop" computation with the parameters b and c changing as a step function at time $t1$ and $t2$. This is interpreted in biological terms as a periodic modulation of the local and short range synchronization dynamics in a cortical column and its neighborhood, induced by bursting behavior from long range excitatory connections from memory areas which control the computation. Compared with constant parameters, these modulations have been demonstrated to provide small gains in both recognition rate and average recognition time. Each recognition parameter epoch consists of a triplet $\{b, c, t\}$, where t is number of iterations in the epoch. Two such epochs are applied in turn, and after the last iteration the occupancy statistics (density) of all units are read out. During evolutionary search for the parameter sets, each epoch is constrained to seven iterations. The total iterations required to create the representation space across all classifiers ranges from 6 to 14 iterations, with a mean of 12.7 across 50 learning trials.

Two follow-up studies are presented here. A key feature of the network is intrinsic non-stationarity (i.e. the two dynamical parameter epochs) uniformly applied to each cell in a nearest neighbor connected lattice. This non-stationarity was motivated by a hypothesis on the operational principle of such networks, and consideration of a possible functional role for interactions between slow (i.e. delta-alpha) and fast (gamma) oscillations in biological systems. The original hypothesis was that the coverage of state space would be uniformly expanding in the first parameter epoch and contracting in the second, and that in biological systems the slow oscillations repeatedly controlled computations in a "sample and iterate" scheme by varying the parameters. This conception proved to be an oversimplification, as revealed by fluctuating state space coverage in the time series of the network evolution, and by observing that stationary bifurcation and coupling parameters performed almost as well as non-stationary. Early results indicated that a small advantage in both average and peak recognition rates, as well as the number of iterations (i.e. recognition time) were obtained with non-stationarity. The following table summarizes a more extensive series of 50 learning trials with each condition; these advantages were confirmed.

Comparison of CML recognition: stationary and two parameter epochs

Scenario	Average rec. rate	Maximum rec. rate	Average time (iterations)	Time variance	Learning trials
Stationary	75.2	82.9	13.4	1.39	50
Two epochs	78.8	87.9	12.7	2.7	51

A second series of computational studies was undertaken based on the resulting large set of network parameters, with a goal of improved understanding of underlying network operating principles. While the network is understood from a computer science standpoint as a generalization of the dynamical recognizer framework [6], a better *physical* understanding might give insight into the variations in performance, and might lead to performance improvements within the genetic algorithm learning framework or suggest more direct adaptive learning methods.

2. Distance from noise response

During recognition, a particular recognizer network characterized by a parameter 6-tuple $\{b1, c1, s1, b2, c2, s2)$ must achieve two goals—to map diverse views of an object to a point in the space whose axes are defined by state space intervals, and to map other objects to different points in space. Differential rates in the approach to equilibrium of the target object and other stimuli might serve the latter.

It is known the state occupancy statistics of a large set of uncoupled maps at fully developed chaos may reach a stationary distribution in only a few iterations; with coupling or other nonlinearity conditions, the systems may not reach a stationary distribution so rapidly, but rather reach a state of statistical periodicity, in which a cycle of low period is observed in the population statistics. The ensemble evolution will depend on the stimulus.

The first set of measurements was formulated to test a hypothesis that networks ability to obtain a unique distribution for a particular object is predicated on an anomalous time course in an approach to statistical equilibrium.

I examined the Euclidean distance between the characteristic distribution for each object and the distribution created by that object's recognizer parameters applied to a noise initial configuration (a 300 × 300 matrix of states uniformly distributed in range—$1 < s < 1$). This distance was computed and averaged separately for correct and error object presentations in a given learning trial, and the ratio between distances of matching and non-matching trials was evaluated against the error rate. No significant correlations were found ($r < 0.1$).

This negative result suggests that the *density coding* hypothesis by Milton and Mackey [4] may not be applicable to highly structured initial conditions, where a line extraction subsystem is presumed to present initial conditions along object boundaries constrained to small regions of state space (i.e. either highly active or at background firing rate). However, their calling attention to the density evolution phenomena and the rapid approach to equilibrium is welcome and closely related to the present work. Instead, the deeper underlying mechanism for separation of classes may be similar to the Markov relaxation and information bottleneck framework recently described [7]. In Markov relaxation, clustering of stimulus configurations into classes is supported by the differential rates of mutual information loss, with stimuli in the same class reaching temporary plateaus in loss at a characteristic time. The present process proceeds orders of magnitude more rapidly, presumably due to the rapid state space coverage of chaotic systems.

3. Sum of configuration entropies

A well-known measure of dynamical systems is the information dimension, closely related to Shannon entropy and the capacity dimension of a dynamical system [5]. Normally, this measure is *defined* for infinite times, and practically *measured* for a long time series for a one-dimensional system. Given the transient nature of the processing in the present system, a different but related measure is required.

The measure employed here is the sum of the Shannon information for each configuration (lattice state at iteration i, computed over partition cells in the usual manner for information dimension: H_s, the Shannon entropy of the current signature (occupancy distribution at the readout time of a recognizer) with k bins is

$$H_s = \sum_{i=1}^{k} S_i \log_2 S_i \tag{3}$$

and the measure I designate as sum of configuration entropies is

$$H_c = \sum_{i=1}^{t} H_s \tag{4}$$

where t is the total number of iterates, in this case over two parameter epochs.

It is conjectured that the reason for the slight performance edge of the two epoch dynamics is attributable to a higher entropy obtainable by increasing the reachable states, increasing the chance of finding dynamical trajectories leading to the common subspace from different views, and of diverting undesired configurations and partial trajectories (of distractor memories) away from the subspace corresponding to the best match.

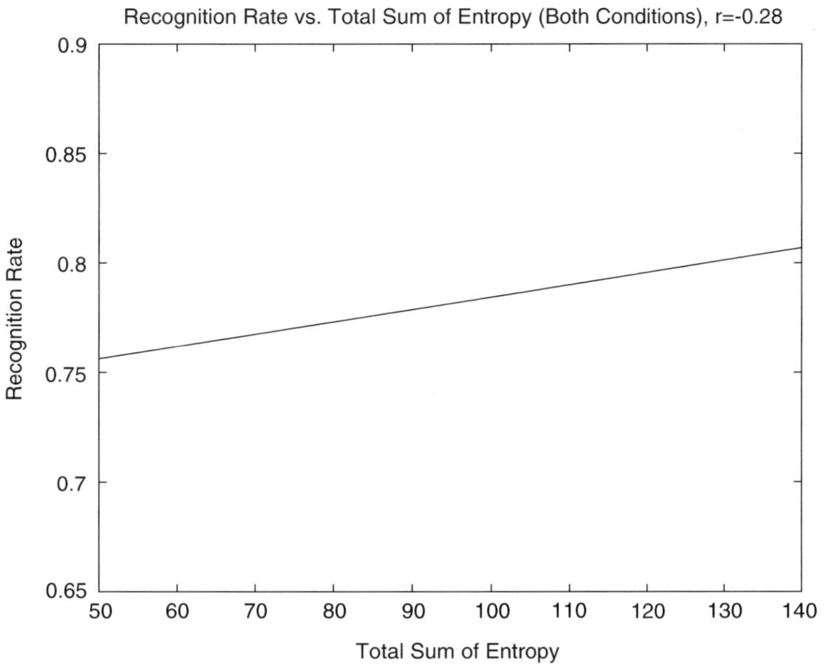

Fig. 1. Total sum of point wise entropy over all iterations vs. recognition rate for 16 trials of the CML recognition system applied to 20 paperclip objects.

In Fig. 1 below, this measure is plotted against recognition rate for a set of 16 two epoch trials, exhibiting a low-moderate positive correlation ($r = 0.28$); individual data points are shown along with a least squares linear fit. In Fig. 2, the ratio of this measure for correct identification trials to error trials is plotted. In this case, the correlation is negative, again with a low-moderate correlation ($r = 0.33$). This inverted relationship—with matching views having slightly lower entropy than error views—suggests that one error mechanism occurs when misidentified views spread out in state space during recognition dynamics; successful recognition is enhanced by limited entropy for distractor objects. It should be plausible to test whether this proposition holds in biological subsystems involved in coding object representations.

4. Conclusions

The lack of correlation between performance of matching and non-matching distributions suggests that the recent proposal of density coding, while close in spirit, does not reflect the underlying dynamics of this system. After examining the recognition dynamics, the nature of the processing in the homogeneous CML seems best conceived of as a parallel, distributed stochastic switching network in the state space of the recurrent spatio-temporal dynamics. The evidence of the point wise entropy measures

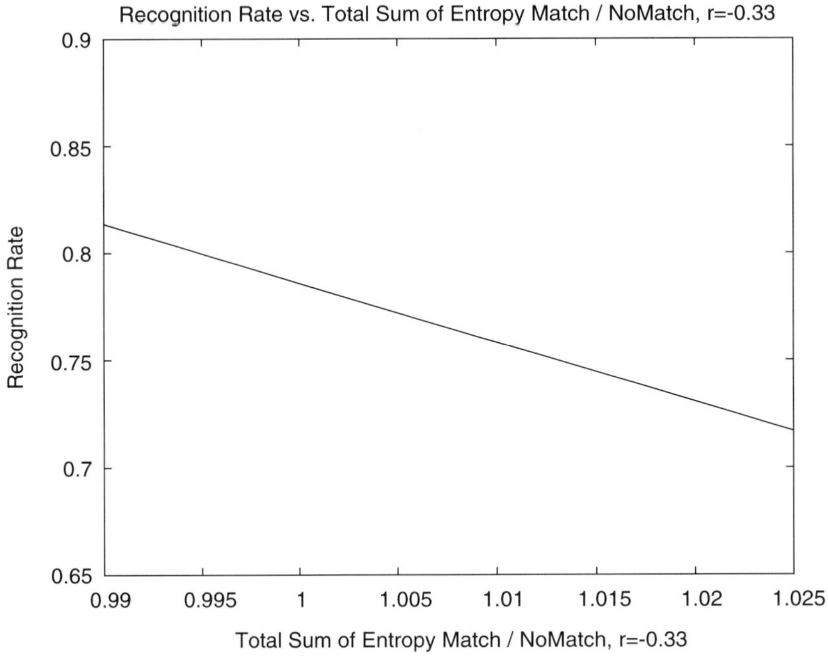

Fig. 2. Ratio of match to non-matching sum of point wise entropy over all iterations vs. recognition rate for 16 trials of the CML recognition system applied to 20 paperclip objects.

in particular (Figs. 1 and 2) suggests that specific pathways through slightly lower entropy configurations are visited in successful matches, while the overall high point wise entropy is correlated with higher performance.

Other questions of interest to computational neuroscience include what sorts of read-out mechanisms would be suitable to read such a code, and what constraints on the resolution would apply. Single neurons with dendritic structures and synaptic parameters which are sensitive to a particular distribution of inter-spike intervals received in a short time window are one plausible mechanism. One would like to know something about the discrimination between different spike intervals (corresponding to the state space interval or bin width in the distribution), and the discrimination for a particular subspace in the space of arrival time distributions.

References

[1] D. DeMaris, Neurocomputing 38–40 (2001) 547–554.

[2] D. DeMaris, Synchronization opponent networks: dynamics, computation and coding for similarity and object recognition, Ph.D. Thesis, Department of Electrical and Computer Engineering, University of Texas, 2001.

[3] D. DeMaris, B.F. Womack, Recognition of objects rotated in depth using partial synchronization of chaotic units, International Joint Conference on Neural Networks, Washington, DC, IEEE/INNS, 2001.

[4] J.G. Milton, M.C. Mackey, J. Physiol. (Paris) 94 (2000) 489–503.
[5] G. Nicolis, in: H. Atmanshpacher, H. Scheingraber (Eds.), Information Dynamics, Plenum Press, New York, 1991.
[6] J.B. Pollack, Mach. Learning 7 (1991) 227–252.
[7] N. Tishby, N. Slonim, Data clustering by Markovian relaxation and the information bottleneck method, in: T.K. Leen, T.G. Dietterich, V. Tresp (Eds.), Advances in Neural Information Processing, Vol. 13, MIT Press, Cambridge, MA, 2000, pp. 640–646.
[8] H.R. Wilson, J.D. Cowan, Biophys. J. 12 (1972) 1–24.

David DeMaris received the Ph.D. in Electrical and Computer Engineering from University of Texas Austin. His research interests include neural computation, nonlinear dynamics, cognitive neuropsychology, computer design and human computer interaction. He is presently employed with IBM Microelectronics in Austin TX performing applied research in vision systems and similarity.

Available at
www.ComputerScienceWeb.com
POWERED BY SCIENCE DIRECT

NEUROCOMPUTING

Neurocomputing 52–54 (2003) 741–745

www.elsevier.com/locate/neucom

Spike sorting the other way

Alexander G. Dimitrov*

Center for Computational Biology, Montana State University, 1 Lewis Hall, Bozeman MT 59717-3505, USA

Abstract

The problem of identifying neural activity from extracellular recordings is still with us. In recent years there have been a lot of improvements in clustering and identification over the first simple techniques. However, the manner in which data samples are selected as putative spike candidates has remained almost unchanged. This part of data processing still remains a largely heuristic procedure. In this work we offer a more quantitative approach to identifying putative spike shapes, which then serve as inputs to clustering methods.

Keywords: Spike sorting; Noise model; Likelihood distance

The usual approach of identifying putative spike events is to invent certain heuristics, which in many cases are sufficient to trigger on most of the events. These include simple amplitude triggers, triggers in alternate spaces (e.g., total Fourier power), coarse description of the shape of the spike (first go down, then go up, then wiggle just so) and others [5]. We decided to approach this problem by first identifying segments of data that do not contain spikes. The assumption behind this approach is that while regions containing spikes may have very complicated characteristics, spike-free regions are relatively simple, and can be described in a simple manner. We model these examples of "no spike" events (null hypothesis) with a maximum entropy [2], multivariate Gaussian model. We then search for events in the data which are "very different" from the null event model, using the negative log likelihood from the model as a distance function (Mahalanobis distance in this case). Peaks of the distance correlate well with positions of spike events in the data, and events extracted in this manner form a relatively clean dataset for subsequent clustering.

* Tel.: +1-406-994-6494; fax: +1-406-994-7438.
E-mail address: alex@cns.montana.edu (A.G. Dimitrov).

0925-2312/03/$ - see front matter
doi:10.1016/S0925-2312(02)00755-5

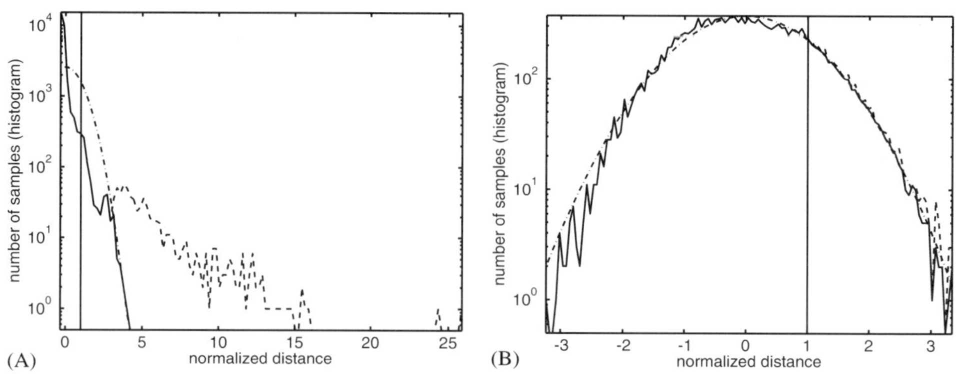

Fig. 1. Histograms of the initial distribution of d (A) and the distribution of d after resampling (B). (- -) The histogram of d. (- .) The expected histogram under the model. Samples on the right side of the plot are far from the model. The vertical line indicates a distance, to the right of which we resample the data. (–) The histogram of d from the resampled data. In (B) the same objects can be seen at point near the end of the resampling procedure. The data (–) now fits the expected distribution (- .) much better than before.

The following figures the important steps in this process. First, samples are selected at random from the dataset. The samples consist of data in temporal windows of length L (free parameter) across N electrodes. We model the samples as a multivariate Gaussian of dimension $N \cdot L$. To decrease the data size requirement, we perform probabilistic PCA, which preserves the eigenstructure for the top K principle components, and models the orthogonal space as a spherical Gaussian [1].

If the samples were truly Gaussian, the Mahalanobis distance d should be distributed according to the $\chi_{2N \cdot L}$ distribution [6]. Asymptotically in $N \cdot L$, $\hat{d} = (d - \text{mean}(d))/\text{std}(d)$ is a normal distribution with zero mean and variance of one ($\mathcal{N}(0,1)$). We assume that this is approximately correct for large window sizes L as well, and take this normal distribution as the expected one for $\hat{d}$ under the normality assumption. Under this assumption, we observe two types of outliers (Fig. 1). First, there are events that are extremely improbable under this model (all dashed line events with $\hat{d} > 5$ here fall in this category). Those events are discarded from the training dataset. Another set of events (events with $3 < \hat{d} < 5$ here) are likely to occur, but there is an excessive amount of them compared to the model. We resample these events at random, with probability $p(\text{sample}) = p(\text{expected})/p(\text{observed})$ (the discrepancy between expected and observed number of events). We then re-estimate the model using the resampled dataset. This procedure is repeated several times, until expected and observed distributions are "similar". At that point, we assume that we have built a satisfactory noise model. Here as a measure of similarity we use the Kullback–Leibler divergence, $KL(p(\text{observed}) \| p(\text{expected}))$ [4].

Once we have a noise model, we pass the full dataset through it, using a running window to collect samples and calculate the distance $\hat{d}$ of these samples to the noise model. The outcome of one such application is shown in Fig. 2. Peaks of the distance to the noise model (circles) seem to correlate well with spike events. This effect can be

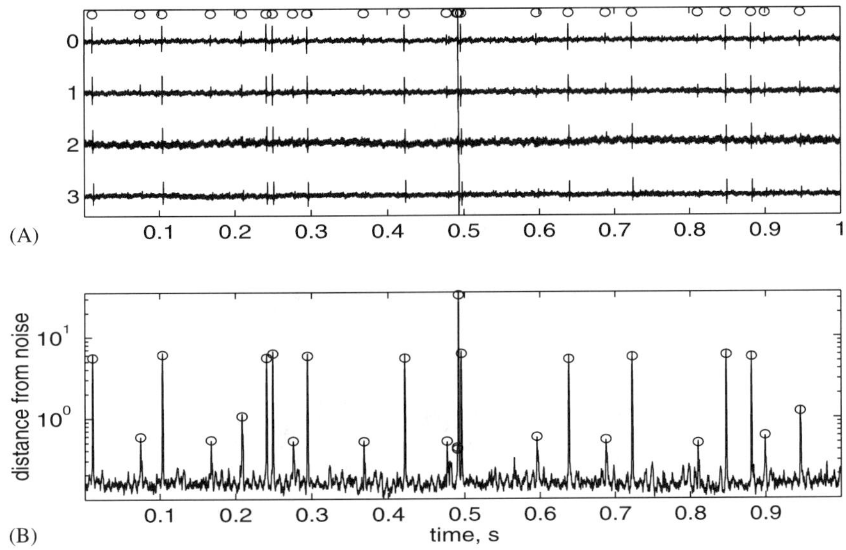

Fig. 2. Raw data traces (A) and their distance to the noise model (B). The distance function d was used as a filter on the data to obtain the trace in (B). Note that (B) is on a logarithmic scale. Peaks of $\hat{d}$ are identified and marked with circles on both panel (A) and panel (B).

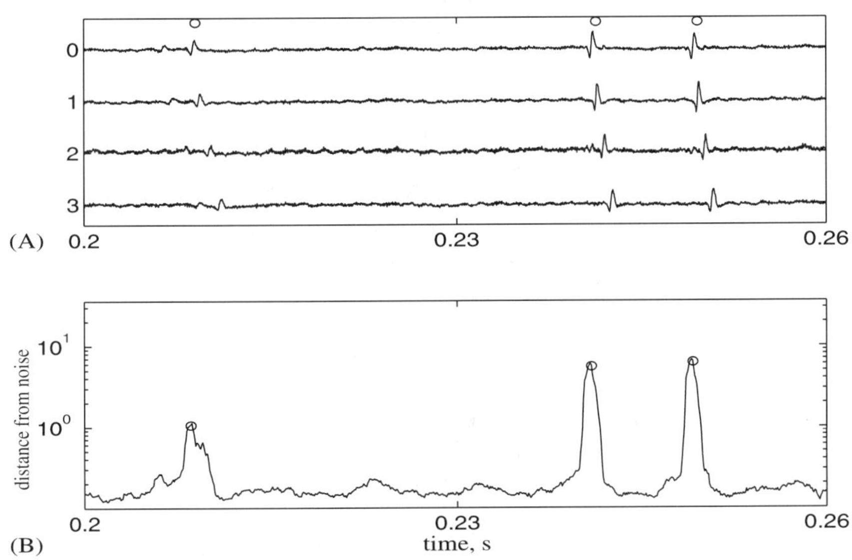

Fig. 3. An expanded portion of 60 ms of Fig. 2 shows a detailed view of the correspondence between raw data (A) and its distance to the noise model (B).

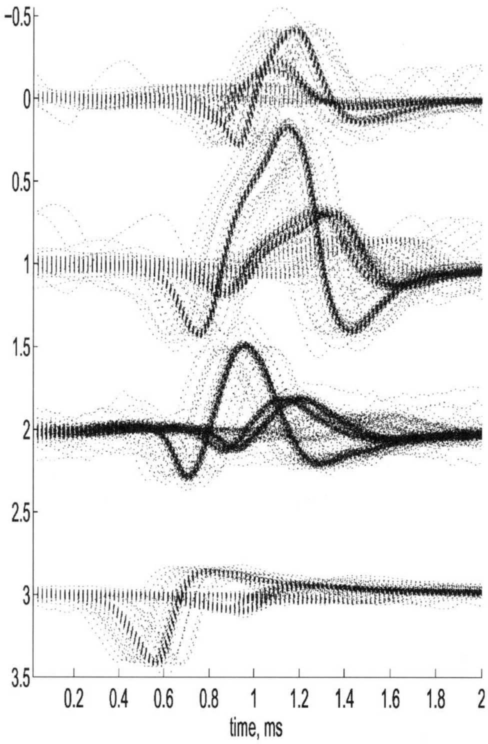

Fig. 4. Some of the events, marked as most unlikely under the noise model.

seen in detail in Fig. 3. The threshold above which we look for peaks can in principle be set automatically. Since we assume that $\hat{d} \propto \mathcal{N}(0,1)$, one can pick a threshold above a desired noise level (e.g., 3, which here indicates 3σ away from $\hat{d}=0$). The evidence in Fig. 3 suggests that such a threshold is usually too low. We interpret this to mean that the noise model is too broad and accepts certain spike events as noise (e.g., the first even near 0.2 s). The current algorithm relies on manual input of this threshold.

Given the results in the above figures, we assume that peaks of $\hat{d}$ indicate the presence of spikes. The next step collects all events that are marked by peaks in the distance to the noise model (locally most unlikely according to the model). Some of the events are shown in Fig. 4. As can be seen, they are mostly well aligned, and even visually one can discriminate several distinct event classes. Some of the alignment problems can actually be traced to the code we use to register peaks, so the true alignment is even better. The few apparently strongly misaligned events on both ends are actually second spikes occurring close to the main spike in the center.

The set of events shown in Fig. 4 can be used as a sample set to train a model with which to discriminate these events, by any of the methods reviewed in [5], or by other clustering approaches [3,6].

Acknowledgements

Research supported in part by NSF grants EIA0129895 and MRI9871191. The author thanks Maneesh Sahani for pointing out that the resampling algorithm can be expressed as a mixture model which explicitly includes the second hypothesis as a uniform density component.

References

[1] C.M. Bishop, Neural Networks for Pattern Recognition, Oxford University Press, New York, 1998.
[2] E.T. Jaynes, On the rationale of maximum-entropy methods, Proc. IEEE 70 (1982) 939–952.
[3] D.H. Johnson, D.E. Dudgeon, Array Signal Processing, Prentice-Hall, Englewood Cliffs, NJ, 1993.
[4] S. Kullback, Information Theory and Statistics, Wiley, New York, 1959.
[5] M. Lewicki, A review of methods for spike sorting: the detection and classification of neural action potentials, Network: Comput. Neural Systems 9 (4) (1998) R53–R78.
[6] L.L. Scharf, Statistical Signal Processing, Addison-Wesley, Reading, MA, 1991.

Available at www.ComputerScienceWeb.com
POWERED BY SCIENCE DIRECT
ELSEVIER
Neurocomputing 52–54 (2003) 747–753
NEUROCOMPUTING
www.elsevier.com/locate/neucom

Dynamics of neuronal populations modeled by a Wilson–Cowan system account for the transient visibility of masked stimuli

Udo Ernst[a,*], Axel Etzold[a], Michael H. Herzog[b], Christian W. Eurich[a]

[a] *Institute for Theoretical Physics, University of Bremen, Otto-Hahn-Allee 1, Bremen 28334, Germany*
[b] *Institute for Human Neurobiology, University of Bremen, Argonnenstr. 3, Bremen 28211, Germany*

Abstract

We study the transient dynamics occurring in a Wilson–Cowan type model of neuronal populations to explain psychophysical masking effects, in which a briefly presented vernier is followed by a grating of a variable geometrical configuration. A change in the spatial layout of the stimuli yields dramatic differences in the performance of the observers. In our model, these differences are explained by variations in the strength and duration of transient neural activity.

Keywords: Population dynamics; Visual masking; Transient activity; Contextual modulation

1. Introduction

To survive in a complex and ever-changing environment, an organism has to cope with sensory stimuli often varying on a short time scale. Information processing must therefore be dynamical and fast. In numerous situations it is not feasible to wait until the neural activation pattern of the brain settles into a steady state before an appropriate reaction occurs.

In this contribution we employ a model of the well-known Wilson–Cowan type [7,8] which has already been used to account for phenomena connected with the dynamics of neural populations in visual cortex [1,6]. Here, we show that this model class is well suited to describe transient phenomena emerging from a dynamical stimulus pattern

* Corresponding author. Fax: +49-421-218-9104.
E-mail address: uernst@physik.uni-bremen.de (U. Ernst).

0925-2312/03/$ - see front matter
doi:10.1016/S0925-2312(02)00758-0

that in reality pushes the visual system to its spatio-temporal limits. The magnitude of the transient activity of model neuronal populations predicts the visibility of target elements reported by observers during psychophysical masking experiments. Due to the structural simplicity of the underlying model, one can analyze and identify possible mechanisms leading to the observed behavior.

2. The shine-through effect

To investigate transient dynamics psychophysically, a paradigm is needed that brings the visual system to the brink of its temporal limits. The recently discovered shine-through effect serves this need very well since performance can change dramatically even with a temporal parameter change of only about 5 ms [6,4].

In the psychophysical experiments a vernier, i.e. two abutting lines, precedes gratings with varying spatial layout (see Figs. 1A and 2). If the grating comprises less than seven elements the vernier is completely masked by the grating.

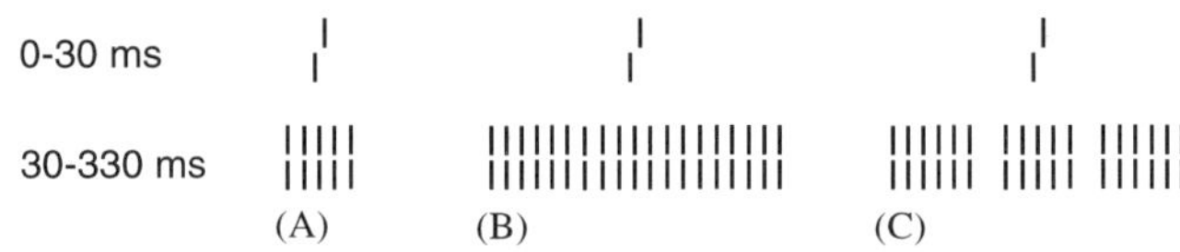

Fig. 1. A vernier precedes for a short time grating of a variable spatial configuration. (A) For gratings with less than seven elements the vernier is invisible. (B) For gratings with more than seven elements shine-through occurs. The vernier appears superimposed on the grating looking wider and brighter. (C) Shine-through diminishes strongly if an extended grating contains gaps. For a description of the methods, please refer to [3,5]. *Performance*: If the vernier is presented for the same duration in all three conditions, the performance of the observers is on average 91 ± 5%, 57 ± 3%, and 61 ± 3% correct responses for A, B, and C, respectively.

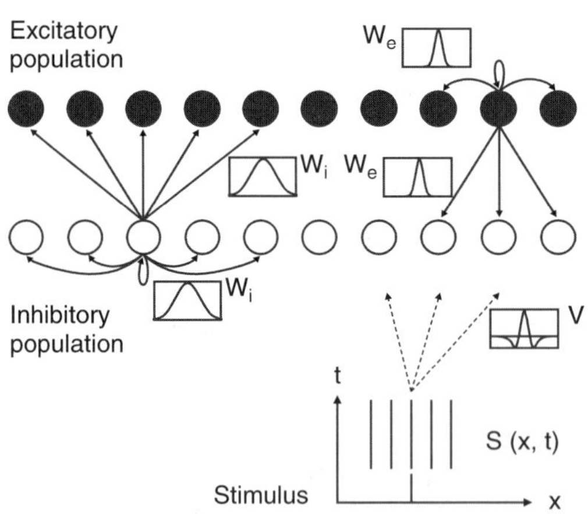

Fig. 2. Structure of model employed in the simulations. A spatio-temporal stimulus $S(x,t)$ is filtered by a difference of Gaussians and projected onto two populations in a one-dimensional neuronal layer. The two populations, an excitatory and an inhibitory one, are mutually coupled with synaptic weight functions described by the Gaussian kernels W_e and W_i, respectively.

If the grating comprises more than seven elements, the vernier becomes visible as a shine-through element, that appears to be superimposed on the grating looking wider and brighter than the vernier really is (shine-through; Fig. 1B). Performance with the 25 element shine-through grating is better than with a five element grating (Fig. 1). Shine-through depends crucially on the spatio-temporal homogeneity of the grating. Even subtle deviations from this homogeneity diminish or even abolish the shine-through effect. For example, leaving out two elements and adding them at the ends of the grating degrades performance strongly while the overall energy of the grating remains constant (Fig. 1C). Therefore, explanations of these effects cannot be attributed to low level stimulus cues such as mask energy. Surprisingly, the underlying neural mechanisms can, in spite of their apparent complexity, be described by very simple models—as the following sections show.

3. Model

In the model, we focus on the visibility of the vernier as a testbed for spatio-temporal interactions of neural populations. Since in the shine-through conditions the vernier appears as a bright flash superimposed on the grating, the processing of the vernier signal is expected to occur as a *transient* in the neural dynamics and not as a steady state. Our model employs the azimuthal axis only in order to simplify an analysis of the mechanisms underlying its dynamics.

The network model consists of excitatory and inhibitory populations of cortical neurons (described by subscripts e and i, respectively) arranged along a one-dimensional axis parametrised by the variable $x \in \mathbb{R}$. The dynamics of the system are given by a set of Wilson–Cowan type equations [8],

$$\tau_e \frac{\partial A_e(x,t)}{\partial t} = -A_e(x,t) + h_e(w_{ee}[A_e * W_{ee}](x,t) - w_{ie}[A_i * W_{ie}](x,t) + I(x,t)) \quad (1)$$

$$\tau_i \frac{\partial A_i(x,t)}{\partial t} = -A_i(x,t) + h_i(w_{ei}[A_e * W_{ei}](x,t) - w_{ii}[A_i * W_{ii}](x,t) + I(x,t)). \quad (2)$$

A_e and A_i denote the firing rates of the excitatory and inhibitory populations, respectively, τ_e and τ_i are the associated time constants, h_e and h_i are neural transfer functions, in this case choosen to be piecewise linear, w_{kl} indicate synaptic weights of population k acting on population l, $k,l \in \{e,i\}$, and W_{kl} are translation-invariant interaction kernels of population k targeting population j assumed to depend only on their distance $|x-x'|$. The symbol $*$ denotes a convolution, i.e. $[A_i * W_{ii}] := \int A_i(x',t)W_{ii}(x-x')\,dx'$. The interaction kernels are modeled as normalized Gaussians with standard deviations σ_e and σ_i. For simplicity, we assume the network to be highly symmetrical ($w_{ee} = w_{ei}$; $w_{ie} = w_{ii}$; $W_e := W_{ee} = W_{ei}$; $W_i := W_{ie} = W_{ii}$). Both neural populations receive the same spatio-temporal input $I(x,t)$ which is computed as the spatio-temporal stimulus $S(x,t)$

convolved with a Mexican-hat type of kernel function $V(x,t)$ whose integral vanishes,

$$V(|x-x'|) = \frac{1}{\sqrt{2\pi\sigma_E^2}} e^{-(x-x')^2/2\sigma_E^2} - \frac{1}{\sqrt{2\pi\sigma_I^2}} e^{-(x-x')^2/2\sigma_I^2}. \tag{3}$$

The width of the excitatory part of $V(t)$, σ_E, is chosen to be smaller than the width of the inhibitory part, σ_I, to take into account on-off receptive field properties of LGN neurons. $S(x,t)$ models the spatio-temporal stimulus intensity along the azimuthal component which is taken to be 1 whenever the vernier or a bar of the grating is presented, and 0 else.

Vernier visibility is assessed through the length of the time interval for which the neurons representing the vernier are above some threshold (i.e. determined by the background noise). This measure is motivated by the argument that the longer an activation associated with the vernier persists, the more information our visual system collects about the vernier, and this in turn increases visibility. The number of model parameters was reduced by considering symmetries, and the range of parameter values restricted by qualitative neurophysiological consideration. Parameters were then adjusted using a specific subset of stimulus conditions.

4. Results

Numerical results for the 5-element, shine-through, and gap grating conditions are given in Fig. 3. The grayscale-coded activities A_e of the excitatory populations show peaks at the position of the vernier and at the edges of the gratings, whereas almost no activity emerges for the inner grating elements. The cross-section (Fig. 3D) of the activity patterns in Figs. 3A–C taken at the center population reveals that the central peak in the 5-element condition decays faster than in the shine-through condition. This behavior is explained by the strong inhibition radiating from the active neurons representing the nearby edges of the grating (see arrows in Fig. 3A).

However, if the extended grating comprises 25 elements, the edges are too remote to exert a substantial inhibitory influence on the center (Fig. 3B). Thus, the activity elicited by the vernier is sustained by feedback excitation, and decays much slower than in the 5 element condition. Removing elements from the grating of 25 elements (see stimulus in Fig. 1C) re-introduces edges leading to an enhanced activation, whose inhibitory surround again suppresses the vernier activity as fast as in the feature inheritance condition (see arrows in Fig. 3C).

Perceptually, the fast suppression of the vernier activity by the small central grating shown in Figs. 3A and C leads to a complete masking of the vernier element. On the other hand, conditions which allow a longer persistence of the vernier activity like the one in Fig. 3B result in a conscious perception of the vernier and its displacement. Thus, the occurrence of shine-through is explained with the transient dynamics of a Wilson-and-Cowan type model.

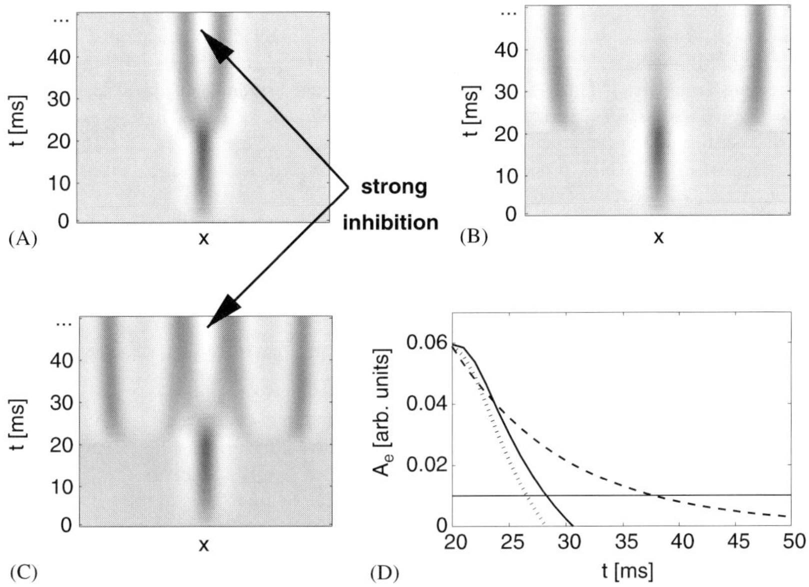

Fig. 3. Spatio-temporal activation pattern emerging from the Wilson–Cowan model for (A) the 5 element grating, (B) the shine-through, and (C) the gap condition. The activation levels $A_e(x,t)$ are coded in shades of grey (dark for high activation). The x-axis corresponds to the location of the neuronal population x, and time t in milliseconds is shown on the ordinates. While in (A) and (C), the central peaks are rapidly suppressed by the inhibition spreading from the two side peaks (located at the edges of the populations stimulated by the grating comprised of 5 elements), in (B) its activity persists while the two side peaks appear on the more distant edges of the large grating comprised of 25 double bars. The activation of the center population $A_e(0,t)$ is shown in (D), where the rapidly decaying solid and dotted curves correspond to the 5 element grating and the gap grating conditions, respectively. The dashed curve shows the slower decay in the shine-through condition.

5. Discussion

Our results demonstrate that even a structurally simple model based on only two partial differential equations is sufficient to explain psychophysical phenomena of object visibility and emergence. In particular, transient activation of neuronal population instead of fixed points of the dynamics determines the visibility of the target element.

Using quantitative stimulus conditions, it is possible to calibrate our model, and to make quantitative predictions [2,3]. Furthermore, our model explains experiments with irregular stimuli [5].

References

[1] R. Ben-Yishai, R. Bar-Or, H. Sompolinsky, Theory of orientation tuning in visual cortex, Proc. Nat. Acad. Sci. 92 (1995) 3844–3848.

[2] M.H. Herzog, U.A. Ernst, A. Etzold, C. Eurich, Local interactions in neural networks explain global effects in the masking of visual stimuli, Neural Comput., submitted for publication.
[3] M.H. Herzog, M. Fahle, C. Koch, Spatial aspects of object formation revealed by a new illusion, shine-through, Vision Res. 41 (2001) 2325–2335 (Please read Erratum: Vision Res. 42 (2001) 271).
[4] M.H. Herzog, M. Harms, U.A. Ernst, C.W. Eurich, S.H. Mahmud, M. Fahle, Extending the shine-through effect to classical masking paradigms, Vision Res. in revision.
[5] M.H. Herzog, C. Koch, Seeing properties of an invisible element: feature inheritance and shine-through, Proc. Nat. Acad. Sci. USA 98 (2001) 4271–4275.
[6] Z. Li, Computational design and nonlinear dynamics of a recurrent network model of the primary visual cortex, Neural Comput. 13 (2001) 1749–1780.
[7] U.A. Ernst, C.W. Eurich, Cortical population dynamics and psychophysics, in: The Handbook of Brain Theory and Neural Networks, ed. Michael A. Arbib, 2nd edition, MIT Press, Cambridge, MA, pp. 294–300 (2002).
[8] H.R. Wilson, J.D. Cowan, A mathematical theory of the functional dynamics of cortical and thalamic nervous tissue, Kybernetik 13 (1973) 55–80.

Udo Ernst got his Ph.D. in Theoretical Physics in 1999 from the University in Frankfurt, and is currently working as a postdoc in the group of Prof. Dr. K. Pawelzik at the Institute for Theoretical Neurophysics at the University of Bremen. He spent also two years in the Nonlinear Dynamics group of Prof. Dr. T. Geisel at the Max-Planck-Institute for Fluid Dynamics in Göttingen. His research interests include collective phenomena in neural networks, dynamical and structural properties of the primary visual cortex, data analysis of multisensory recordings from primates, and theoretical analysis and modeling of gestalt perception and feature binding in the visual system.

Axel Etzold got his diploma in Mathematics in 2000 from the University of Giessen (Germany). He currently is a Ph.D. student in Prof. Dr. Schwegler's and Dr. Eurich's group at Bremen university. His research interests include signal processing and encoding in neural populations, dynamical systems modeling and statistical data analysis methods.

Michael H. Herzog studied Mathematics, Philosophy, and Biology at the Universities of Erlangen (Germany), Tuebingen (Germany), MIT (USA), and Caltech (USA). He is a senior researcher at the University of Bremen and Osnabrueck (Germany). His research interests include perceptual learning, feature binding, visual masking, and schizophrenia.

Christian W. Eurich got his Ph.D. in Theoretical Physics in 1995 from the University of Bremen (Germany). As a postdoc, he worked in the Departments of Mathematics and Neurology at the University of Chicago, and he was guest researcher at the Max-Planck Institut für Strömungsforschung in Göttingen and at the RIKEN Brain Institute in Tokyo. In 2001, he held a professorship for Cognitive Neuroinformatics at the University of Osnabrück. Currently, Christian Eurich is Research Assistant at the Institute for Theoretical Neurophysics at the University of Bremen. His research interests include signal processing and encoding in neural populations, neural dynamics, and motor control problems such as balancing tasks and postural sway.

Available at
www.ComputerScienceWeb.com
POWERED BY SCIENCE @DIRECT®

ELSEVIER

Neurocomputing 52–54 (2003) 755–761

NEUROCOMPUTING

www.elsevier.com/locate/neucom

Power-law scaling in human EEG: relation to Fourier power spectrum

Thomas C. Ferree[a,*], Rudolph C. Hwa[b]

[a] *Department of Radiology, Dynamic Neuroimaging Laboratory, University of California San Francisco, 521 Parnassus Ave. Room C324B, San Francisco, CA 94143, USA*

[b] *Institute of Theoretical Science and Department of Physics, University of Oregon, Eugene, OR 97403, USA*

Abstract

We discuss a method of analyzing spontaneous human EEG time series, which emphasizes scale-independent behavior. We use detrended fluctuation analysis to quantify the temporal fluctuations as a function of window width, and show how power-law scaling behavior is frequently manifest over two distinct temporal ranges. These ranges encompass time scales associated with meaningful aspects of cortical physiology. This paper shows a simple way of quantifying the existence of such scaling behavior, and determining the characteristic time scale which separates the two regions. By making a qualitative connection with the discrete Fourier transform, we show how the violation of scaling between the two regions is associated with the normal human alpha rhythm, but that the existence scale-independent behavior on either side of the alpha rhythm enables a succinct description of the complex dynamics not accessible in the Fourier power spectrum.

Keywords: Electroencephalography; Fractal; Complexity; Alpha rhythm

1. Introduction

The human brain is obviously a complex system, and exhibits rich spatiotemporal dynamics. Among the noninvasive techniques for probing human brain dynamics, scalp electroencephalography (EEG) provides a direct measure of cortical activity with millisecond temporal resolution [9]. Because the spatial resolution at the scalp is low, EEG time series reflect the integrated activity of large neural populations. But such

* Corresponding author. Tel.: +415-502-3726.
E-mail address: tom.ferree@radiology.ucsf.edu (T.C. Ferree).

0925-2312/03/$ - see front matter © 2002 Elsevier Science B.V. All rights reserved.
doi:10.1016/S0925-2312(02)00760-9

large-scale integration is not necessarily a shortcoming of EEG, because it is well established that neural activity at this scale is correlated with brain function and disease. Thus, a practical goal is to effectively characterize the complex EEG wave forms with a minimum of assumptions, then study inferences from these dynamical measures.

We restrict our attention to the ongoing spontaneous EEG, rather than the event-related response functions. The most commonly used technique for analyzing spontaneous EEG is the Fourier power spectrum, based upon decomposing the signal into a linear sum of sinusoidal waves each with constant phase [12]. Clearly this has limitations for describing nonlinear and nonstationary systems. Recognizing the nonlinear aspect, chaos theory has also been applied [7]. While in certain cases the EEG does appear to exhibit low-dimensional chaotic dynamics, e.g., epileptic seizure [1], EEG dynamics are generally so high dimensional that establishing the existence of stable attractors is challenged by system nonstationarity, i.e., slowly changing variables or parameters which may change attractor properties [10,3].

We describe here an alternative approach, which avoids the assumptions of linearity and low-dimensional attractors, and emphasizes instead scale-independent behavior in a certain measure of the temporal fluctuations. Power-law scaling behavior is definitive of fractals, but is also nearly ubiquitous in complex systems, arising in a variety of different quantifications of the data [4]. From the viewpoint of theoretical physics, this fact is of great interest, because scaling phenomena are known to occur in thermodynamic systems near a critical state, with interesting implications for how perturbations may propagate through the system [2]. We make no claim here that our analysis shows the brain is in a critical state, although this theoretical backdrop did inspire our initial inquiry. Instead we emphasize here the practical determination of scaling behavior in real data, and make qualitative comparisons with the Fourier power spectrum. Other papers demonstrate the successful application of scaling analysis to the clinical problem of acute stroke detection [5,6].

2. Scaling analysis of EEG time series

The fluctuations F in a system, computed as a function of some scale k, are said to be scale independent if

$$F(k) \propto k^{\alpha} \tag{1}$$

because the dependence on k is invariant under a scale transformation $k \rightarrow ck$. If (1) holds, then the exponent α provides a succinct measure of the dynamics across a range of k. Eq. (1) implies $\ln F \propto \alpha \ln k$, thus linear regions in plots of $\ln F$ versus $\ln k$ indicate power-law behavior, with the slope of the line given by the exponent α. For ideal mathematical fractals, such behavior persists without limit in k. For physical systems, its range is always finite, and may in addition be interrupted by dynamical mechanisms which introduce characteristic time scales into the data.

To quantify the fluctuations in an EEG time series, we use a variant of detrended fluctuation analysis (DFA) [11], adapted to continuous time series [5,6]. The detrending renders the method relatively insensitive to nonstationarities in the data. To define $F(k)$, let an EEG time series be denoted by $V(t)$, where t is discrete time ranging from 1 to T. Divide the entire range of t to be investigated into B equal windows, discarding any remainder, so that each window has $k = \text{floor}(T/B)$ time points. Within each window, labeled b ($b = 1,\dots,B$), perform a least-square fit of $V(t)$ by a straight line, $\bar{V}_b(t)$, i.e., $\bar{V}_b(t) = \text{Linear-fit}[V(t)]$ for $(b-1)k < t \leqslant bk$. That is the semi-local trend for the bth window. The squared fluctuation from the semi-local trend in the bth window is defined

$$F_b^2(k) = \frac{1}{k} \sum_{t=1+(b-1)k}^{k+(b-1)k} [V(t) - \bar{V}_b(t)]^2. \tag{2}$$

This quantity is averaged over B windows to yield

$$F(k) = \sqrt{\frac{1}{B} \sum_{b=1}^{B} F_b^2(k)}. \tag{3}$$

The algorithm gives a measure of the fluctuations on scale k with a minimum of assumptions about the nature of the signal. The measure $F(k)$ has a useful theoretical limit: an uncorrelated random walk gives $\alpha = \frac{1}{2}$ analytically, and other values of α reveal temporal correlations [8].

Fig. 1a shows the first 0.5 s of three 10-s EEG times series for a normal subject, resting with eyes closed. The data were hardware filtered between 0.1 and 100 Hz, and digitized at 250 Hz. The semi-local trends $\bar{V}_b(t)$ are shown as dashed lines. Fig. 1b shows plots of $\ln F$ versus $\ln k$ for the three channels in Fig. 1a, for values of k from 5 to 250, in approximately equal steps of $\ln k$. For each channel shown, it is evident

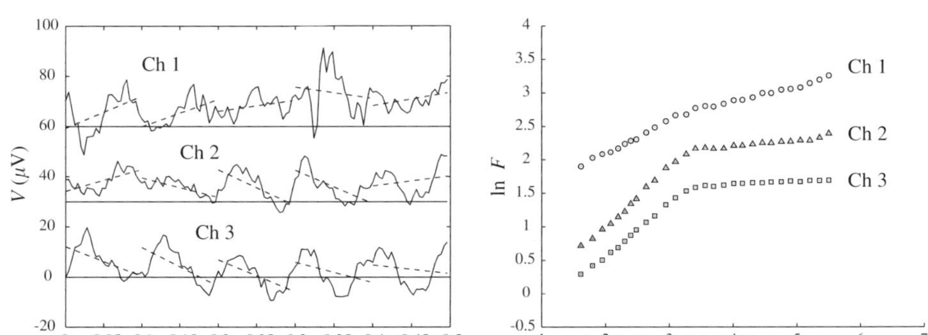

Fig. 1. (a) Examples of EEG time series in three channels for a single subject. The vertical scales of Chs. 1 and 2 are shifted upward by 60 and 30 μV, respectively. Dashed lines indicate semi-local trends for $k\Delta t = 0.1$ s. (b) $\ln F$ versus $\ln k$ for the three channels in (a). The vertical scales of Chs. 1 and 2 are shifted upward by 1.0 and 0.5 units, respectively. Errors in $\ln F$ are shown with vertical lines (mostly appearing as dots) inside plot symbols (see Section 3).

that power-law behavior exists over two temporal regions. We call the slopes of the two lines α_1 and α_2, corresponding to short and long time scales, respectively. We have analyzed 128 EEG time series for a total of 28 normal subjects [6], and found similar behavior in nearly all cases. There are certainly exceptions to the rule, the most common being that it may be necessary to allow for a transitional region in which the bend is not so sharp. Still this is strikingly simple behavior for such complex data sets. Even the rare violations of scaling might themselves prove informative.

3. Fitting scaling exponents

To determine the values of α_1 and α_2, we must decide how to fit lines to the curves in Fig. 1b. One way is to define two regions which are well separated from the bend so as to provide reasonable fits for all available data. This was the approach taken in [5,6] where it was found that $1 < \ln k < 2.5$ (Region 1) and $3.5 < \ln k < 5.75$ (Region 2) do a reasonable job for all 28 subjects studied so far. It is obviously of interest to automate this process, to obtain more objective channel-specific definitions of the two regions. When the bend is sharp as in Fig. 1b, this may be done simply by allowing a single match point between two contiguous regions to be a free parameter, which is varied to minimize the overall fitting error.

Optimally fitting lines to the curves in Fig. 1b requires estimates of the errors in $\ln F$. There are three main sources of error in the original voltage data $V(t)$. First, the EEG amplifiers introduce to each time point a Gaussian random variable with RMS amplitude 0.6 μV. Second, there is often a slow drift associated with changing quality of scalp-electrode contact, but fortuitously this tends to be removed by the detrending in DFA. Third, approximation of potentials relative to infinity via the average reference [9], used here, or any technique for deconvolving scalp potentials to the brain surface, introduces further errors in the signal to be analyzed. To illustrate the procedure here, we consider only errors due to amplifier noise. In accordance with data, we assume amplifier noise is stationary and normally distributed, with RMS amplitude denoted by $\delta V = 0.6$ μV.

In each window b, the local linear trend $\bar{V}_b(t) = a_b + m_b t$ is obtained using standard linear regression. Along with the maximum likelihood parameters of the line (a_b, m_b), errors in the estimates of these parameters $(\delta a_b, \delta m_b)$ may also be obtained. It is straightforward to account for each of these contributions when computing $F(k)$. We first define an array containing the complete list of quantities on which F depends: $\{r_j\} = \{V_b(t), \{a_b\}, \{m_b\}\}$, where $j = 1, \ldots, kB + 2B$, and consider $F = F(k; \{r_j\})$. Assuming each r_j has a Gaussian random error δr_j, and makes an independent contribution to $F(k; \{r_j\})$, these contributions combine in quadrature according to

$$\delta F(k) = \left[\sum_{j=1}^{(k+2)B} \left(\frac{\partial F}{\partial r_j} \right)^2 \delta r_j^2 \right]^{1/2} . \tag{4}$$

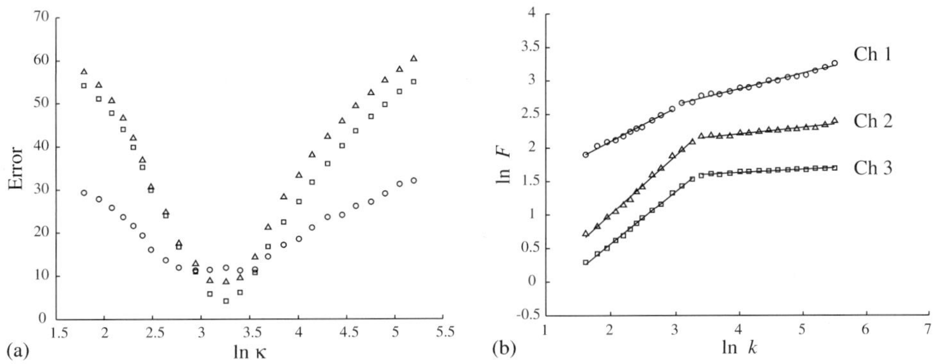

Fig. 2. (a) Fit error E versus bend location $\ln\kappa$ for Chs. 1–3. (b) Optimal linear fits for Chs. 1–3, with slopes equaling α_1 and α_2. Plot symbols are the same as in Fig. 1b.

Taking the derivates $\partial F/\partial r_j$ leads to

$$\delta F(k) = \left[\frac{1}{kB}\,\delta V^2 + \frac{1}{F^2}\,\frac{1}{B^2}\sum_{b=1}^{B}\left(\langle V-\bar{V}\rangle_b^2\,\delta a_b^2 + \langle (V-\bar{V})\tau\rangle_b^2\,\delta m_b^2\right)\right]^{1/2}, \tag{5}$$

where $\langle\,\rangle_b$ denotes an average over the bth window, and $\tau = (1,\ldots,k)$ is a vector containing the discrete time points in the local coordinates of window b. We allowed the trend errors δa_b^2 and δm_b^2 to depend upon b, since that was found numerically to be the case. To compute the errors in $\ln F$, we trivially compute

$$\delta \ln F(k) = \frac{1}{F(k)}\,\delta F(k), \tag{6}$$

which has the effect of increasing the error $\delta\ln F$ at small values of F, i.e., small values of k. This gives the error bars shown in Fig. 1b.

To fit the linear behavior in Fig. 1b, we assume there exist two linear regions, which meet at a single point $\ln k = \ln\kappa$. For each value of $\ln\kappa$, we use standard linear regression to compute best-fit lines for the two regions. In each region the best-fit line is defined as the one which minimizes χ^2, and the best overall fit is determined by the minimum of the total error $E = \chi_1^2 + \chi_2^2$. Fig. 2a shows the error E as a function of $\ln\kappa$, with a clear minimum for each channel. Figure 2b shows the optimal fits for each channel, the slopes of the lines giving the scaling exponents α_1 and α_2. When this procedure works, the value of κ reveals a characteristic time scale in the data, and the minimum error $E_{\min}$ gives a measure of the overall quality of the scaling behavior.

4. Relation to Fourier power spectrum

The most common method of analyzing spontaneous EEG time series is the Fourier transform. Because of how this analysis tool relates to linear systems, the emphasis is usually placed on individual peaks in the power spectrum, which are implicitly

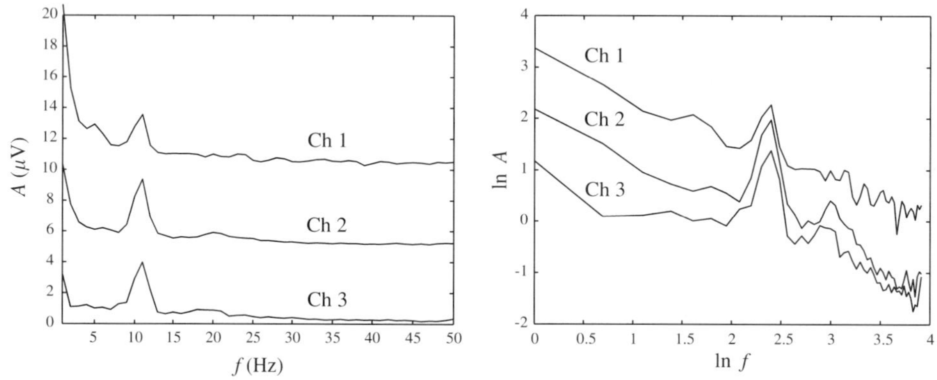

Fig. 3. Power spectrum amplitude of the three time series in Fig. 1a. (a) Linear axes: The vertical scales of Chs. 1 and 2 are shifted upward by 10 and 5 μV, respectively. (b) Log–Log axes: The vertical scales of Chs. 1 and 2 are shifted upward by 1 and 0.5 units, respectively.

interpreted as resonant frequencies. In contrast, scaling analysis considers the entire continuum of time scales when present, and emphasizes the often simple relationships *across* time scales [2,4]. Within Fourier analysis a close analogy is asking whether the power spectrum behaves as $P(f) \propto 1/f^{\beta}$. This is a form of scaling analysis, but still relies upon decomposing the signal into a linear superposition of sinusoidal waves.

We computed the Fourier power spectrum for the three time series in Fig. 1a. Each 10-s data segment was divided into ten 1-s segments. Each 1-s segment was Hann windowed, then Fourier transformed using a standard FFT algorithm. This gives 1 Hz frequency resolution. The Fourier amplitude $A(f) \equiv \sqrt{P(f)}$ was computed by taking the complex modulus and averaging over the ten 1-s segments. Fig. 3a shows A versus f on linear–linear axes. A clear peak is evident near 10 Hz. Fig. 3b shows A versus f on log–log axes. The peak near 10 Hz is still evident. At frequencies below the peak, $\ln A$ falls with $\ln f$, although not monotonically. At frequencies above the peak, the behavior is completely erratic. Clearly the DFA behavior in Fig. 1b is much more amenable to description with a few parameters, especially at shorter time scales and higher frequencies.

We have compared the results of DFA and Fourier analyses for the 28 subjects in Ref. [6]. The following observations are made: (1) The general tendency for $\ln F$ to increase with $\ln k$ is associated with the tendency for $\ln A$ to decrease with $\ln f$, because of the inverse relationship between time and frequency on the horizontal axis. (2) The characteristic time scale $\kappa\Delta t$ is associated with the dominant frequency $\ln f_{\max}$, which in these examples clearly represents the normal human alpha rhythm. (3) The behavior on either side of the characteristic time scale or frequency is always much smoother for $\ln F$ than $\ln A$, especially at shorter time scales (higher frequencies). This makes the results of the DFA approach more amenable to concise summary with a small number of parameters. Concise description is a stated objective when considering 128 channels across many subjects.

5. Discussion

We have described how a variant of detrended fluctuation analysis [5,6] can be used to reveal scale-independent behaviors in spontaneous EEG time series. Although any possible relation to self-organized criticality [2] is not yet established, the scaling exponents provide a concise summary of the complex dynamics across physiologically meaningful time scales.

For most subjects and channels, strikely good power-law behavior is seen over two temporal regions, spanning an interesting breadth of physiological time scales. The two regions are separated by a characteristic time scale, which appears associated with the normal human alpha rhythm. This paper describes an objective way to define the two regions, and obtain channel-specific estimates of the characteristic time scale and overall quality of scaling behavior.

Acknowledgements

We thank Dr. Phan Luu and Prof. Don Tucker for supplying the EEG data for our analysis. This work was supported, in part, by the National Institutes of Health under Grant Nos. NS-38829 and NS-27900, and by the US Department of Energy under Grant No. DE-FG03-96ER40972.

References

[1] A. Babloyantz, A. Destexhe, Low dimensional chaos in an instance of epilepsy, Proceedings of the National Academy of Sciences, USA, Vol. 83, 1986, pp. 3513–3517.

[2] P. Bak, How Nature Works: The Science of Self-Organized Criticality, Copernicus, New York, USA, 1996.

[3] W.J. Freeman, A proposed name for aperiodic brain activity: stochastic chaos, Neural Networks 13 (2000) 11–13.

[4] T. Gisiger, Scale invariance in biology: coincidence or footprint of a universal mechanism?, Biol. Rev. 76 (2001) 161–209.

[5] R.C. Hwa, T.C. Ferree, Fluctuation analysis of human electroencephalogram, Nonlinear Phenom. Complex Systems 5 (3) (2002) 302–308.

[6] R.C. Hwa, T.C. Ferree, Scaling properties of fluctuations in human electroencephalogram, Phys. Rev. E 66 (2002) 021901.

[7] K. Lehnertz, J. Arnhold, P. Grassberger, C.E. Elger (Eds.), Chaos in Brain?, World Scientific, Singapore, 2000.

[8] E.W. Montroll, M.F. Schlesinger, in: J.L. Lebowitz, E.W. Montroll (Eds.), Nonequilibrium Phenomena II: From Stochastics to Hydrodynamics, North-Holland, Amsterdam, 1984.

[9] P.L. Nunez, Electric Fields of the Brain, Oxford University Press, Oxford, 1981.

[10] M. Palus, Nonlinearity in normal human EEG: cycles, temporal asymmetry, nonstationarity and randomness, not chaos, Biol. Cybern. 75 (1996) 389–396.

[11] C.-K. Peng, J. Mietus, J.M. Hausdorff, S. Havlin, H.E. Stanley, L. Goldberger, Long-range anticorrelation and non-Gaussian behavior of the heartbeat, Phys. Rev. Lett. 70 (9) (1993) 1343–1346.

[12] D.O. Walter, Spectral analysis for electroencephalograms: mathematical determination of neurophysiological relationships from records of limited duration, Exp. Neurol. 8 (1963) 155–181.

Available at
www.ComputerScienceWeb.com
POWERED BY SCIENCE DIRECT®

ELSEVIER

Neurocomputing 52–54 (2003) 763–769

NEUROCOMPUTING

www.elsevier.com/locate/neucom

Coexistence of synchronized "oscillatory" and desynchronized "rate" activity in cortical networks

Erik Fransén

Department of Numerical Analysis and Computer Science, Royal Institute of Technology, SE-100 44 Stockholm, Sweden

Abstract

The basis of MRI and PET experiments is the finding that neuronal cell firing levels are modulated in a task dependent manner. Results from EEG and MEG experiments on the other hand point to the importance of synchrony, e.g. the peak frequency may depend on the difficulty of the task. In most models only one of these activity modes of firing is desirable or possible to produce. In this work we show how a cortical microcircuit can produce either synchronized or desynchronized firing, and how this solves problems of present day rate and synchronization models.

Keywords: Synchronization; Desynchronized rate activity; Cortical microcircuit

1. Introduction

The basis of MRI and PET experiments is the finding that neuronal cell firing levels are modulated in a task dependent manner. Results from EEG and MEG experiments on the other hand point to the importance of synchrony, e.g. the peak frequency may depend on the difficulty of the task. A coexistence of synchronized and desynchronized activity in the same region is also consistent with the observation in local field potentials, that the synchronization is limited in time and is preceded and succeeded by epochs of desynchronized activity.

Seen on a cellular scale, pyramidal cells in a local region of neocortex may thus fire more or less synchronized. In most models only one of these activity modes of firing,

E-mail address: erikf@nada.kth.se (E. Fransén).

0925-2312/03/$ - see front matter © 2002 Elsevier Science B.V. All rights reserved.
doi:10.1016/S0925-2312(02)00764-6

synchronized or desynchronized, is generated. In some cases only one mode can be generated, in other cases only one mode is desirable, the other is viewed as destructive or at least non-productive. In this work we describe a model that can produce either of the two modes depending on the drive to different local interneurons. We also show advantages from having the two modes coexist in a network.

2. Features and problems of present day models

Synchronization models (studied by Abeles, Bienenstock, Hertz, Singer, etc.): An advantage of synchronized (e.g. synfire chain) activity is its larger impact on a target cell, a desynchronized set of EPSPs may fail to elicit an action potential. Modulating the degree of synchrony may thus change the impact at the postsynaptic target without a need for change in rate. In synchronization models, one of the problems is how to keep a self-sustained activity for extended times (more than one or a few spikes). The proposed model offers the possibility to drive the synchronized population by desynchronized portions that have a stable activity.

Rate models (studied by Amit, Brunel, Fransén, Fuster, Lansner, etc.): An advantage of a rate (e.g. attractor) model is that the persistent firing may provide a basis for recurrent feed-back to settle the activity according to synaptic efficacies tuned by learning. In rate models, one of the problems is how to end the activity, and for example to go to a new stable state. The proposed model offers the solution of termination by synchronization.

Bump models: (studied by Abbott, Chow, Compte, Brunel, Tegnér, Wang, etc.): These models are used to model prefrontal cortex working memory, as well as M1 population vector activity [4], V1 orientation selectivity [5] and head direction cells [13]. One of the main problems in bump models is that a perturbation in the rate on one side of the bump, or including neuronal spike frequency adaptation or synaptic short-term plasticity, will send the bump traveling (Fransén, and Lansner unpublished observations) and [6,13]. In the proposed model, the bump may be stabilized by synchronized regions on the sides of the bump.

3. Cell and network model

Biophysical compartmental models of pyramidal cells and interneurons in cortex layer II/III were constructed. The cell models use Hodgkin–Huxley representations of intrinsic currents and 1-D diffusion models of the intracellular calcium. The cell morphology is reduced to six compartments. Simulations were done using the simulation package GENESIS.

The basic network is a set of pyramidal cells contacted by two different types of interneurons. It has 24 pyramidal cells and four interneurons of two types. Pyramidal cells are connected to the nearest two neighbors as well as to interneurons. The two types of interneurons connect to two different pools of 12 pyramidal cells. In the figures, the three upper interneurons connect to the upper pyramidal pool, the lower interneuron to the lower pool. The interneurons differ in their synchronizing effect on the pyramidal cells as described in the next section. A pyramidal cell contacts the two types of interneurons as well as other pyramidal cells. Anatomically, this is consistent with the synaptic triade as described by White [11]. Depending on the individual balance of inputs to the interneurons, a group of pyramidal cells may either synchronize or desynchronize when activated by afferent excitatory input or local recurrent activity.

4. Mechanisms of synchronization and desynchronization

Interneuron types show specificity in their localization on target pyramidal cells. Further, there are interneurons with only a $GABA_A$ component [7]. If dendritic localization, IPSP decay time constant ($GABA_A$ or $GABA_B$), synaptic efficacy or release probability varies among the contacts on pyramidal cells, the IPSP of the interneuron may be able to delay the upcoming spike of the pyramidal cell to a variable extent, and thereby desynchronize or synchronize the pyramidal cells to variable extents.

Based on this type of data, a number of mechanisms for synchronization have been proposed, including fast $GABA_A$ synapses [9], IPSP doublets [12] and short-term synaptic plasticity [8]. Likewise, a number of mechanisms for desynchronization have been proposed, including distal synaptic placement [3], modulator reduced spike frequency adaptation [2], and slow NMDA-like EPSPs [2,10].

We will here show that a local circuit of pyramidal cells, which are contacted by different types of interneurons, may either synchronize or desynchronize. Which activity mode that dominates is controlled by differences in drive onto the interneuron types. This drive may be external afferent input or local activity in the network. Thus, the same circuit can produce either type of activity, and different groups of pyramidal cells will subsequently interact on the basis of their firing mode.

In the present model, using moderate sized IPSPs, the interneuron giving somatically located $GABA_A$-type synapses produces a desynchronized pyramidal cell population, and the interneuron with dendritically located mixed $GABA_{A/B}$ synapses produces a synchronized population. We suggest that a slow time constant, here decay of $GABA_B$, produces a slow repolarization upon which fast depolarizations, here recurrent pyramidal inputs, produce a synchronizing effect of the type described by Ekeberg [1].

Using these mechanisms for synchronization and desynchronization, we have studied if synchronized and desynchronized activity can coexist but stay separated. In Fig. 1, we show an example of how synchronized and desynchronized populations can coexist in the same network without interfering or enforcing its own activity pattern on the other population. Different types of inhibitory input produces synchronization and desynchronization respectively.

We have also studied how a population of desynchronized cells can drive activity in a synchronized population, as well as how synchronizing a desynchronized population can end the activity. In Fig. 2, we show an example of how persistent rate activity is ended by synchronized activity. At time 1.4 s the inhibitory input is changed to the synchronizing type. We also demonstrate that persistent rate activity can drive the

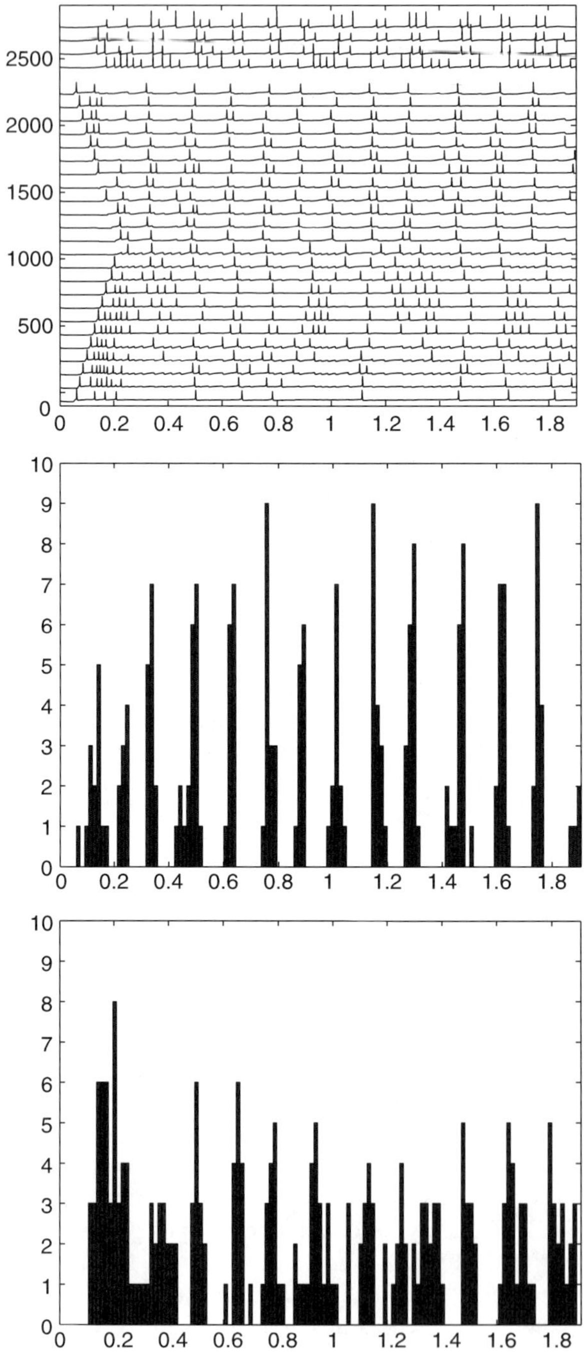
2500
2000
1500
1000
500
0
0 0.2 0.4 0.6 0.8 1 1.2 1.4 1.6 1.8
10 9 8 7 6 5 4 3 2 1 0
0 0.2 0.4 0.6 0.8 1 1.2 1.4 1.6 1.8
10 9 8 7 6 5 4 3 2 1 0
0 0.2 0.4 0.6 0.8 1 1.2 1.4 1.6 1.8

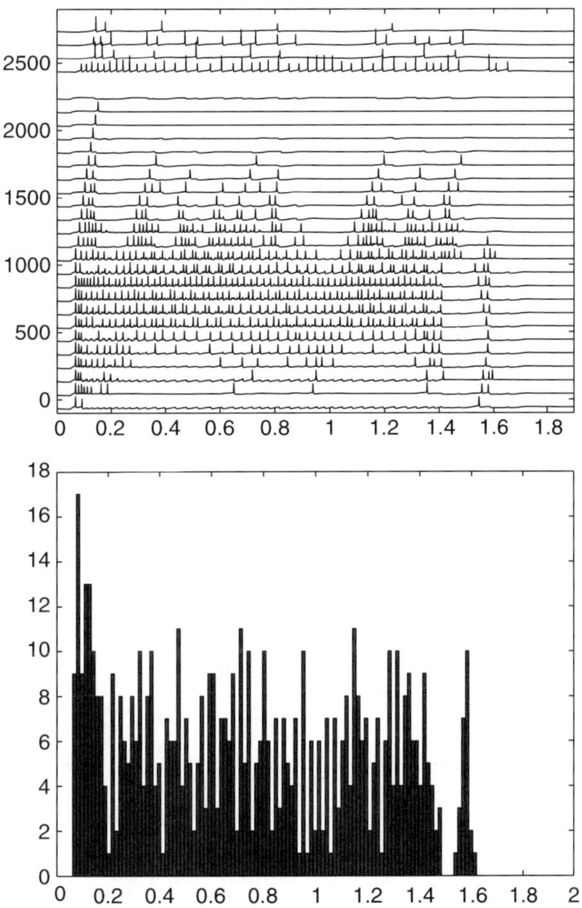

Fig. 2. (Top) To facilitate persistent rate activity, the network has, relative to Fig. 1, stronger intrinsic connections between pyramidal cells. Input is weaker relative Fig. 1, and is subthreshold to the synchronizing cells to show how these cells can be driven by the desynchronized cells. Soma membrane potential vs. time (s). (Bottom) Spike histogram of pyramidal cells. Note that input onset is simultaneous to all cells, this does however not lead to synchronized activity in the desynchronized population (lower 12 cells). The rate varies from 5 to 40 Hz depending on the location, cells near the borders (top and bottom) have fewer inputs from neighbors. Spikes per bin vs. time (s). Bin width is 15 ms.

Fig. 1. (Top) Bands of desynchronized activity do not invade the synchronized population. The synchronized activity does not lead to synchronicity in the desynchronized population. Note that there is no apparent difference in average frequency between the synchronized and the desynchronized population. Activity starts desynchronized by differential timing of onset of the external input. Top four cells are interneurons. Soma membrane potential vs. time (s). (Middle) Spike histogram of upper 12 pyramidal cells. These cells receive input from the top three interneurons. The input is medially at the dendrite and is of a mixed $GABA_A$ and $GABA_B$ type. Using three interneurons could have enabled a larger variation in spike timing, but synchronization dominates. Spikes per bin vs. time (s). Bin width is 15 ms. (Bottom) Spike histogram of bottom 12 pyramidal cells. These cells receive input from the lowest interneuron. The input is at the soma and is of the $GABA_A$ type. Using only one interneuron could have enabled synchronization, but desynchronization dominates. Spikes per bin vs. time (s). Bin width is 15 ms.

neighboring synchronized cells to spiking, but that the synchronizing cells themselves cannot sustain any activity. Thus, a synchronized population can prevent rate activity to spread, and could therefore stabilize the position of a bump.

5. Discussion

We have shown how a cortical microcircuit can generate both synchronized and desynchronized activity depending on the excitability or synaptic drive onto the interneurons. In this model, fast IPSPs produce desynchronized and slow IPSPs produce synchronized activity. The model displays the following features: (1) *Coexistence*: One and the same network can generate synchronized as well as desynchronized activity. (2) *Stabilization of bump*: Synchronizing regions surrounding a rate "bump" may prevent the bump to drift. (3) *Input drive*: Desynchronized regions may provide the continuous input to maintain synchronized activity. (4) *End of rate activity*: A shift toward synchronization can end the persistent activity in a rate model. (5) *Stop of propagation*: A shift toward desynchronized activity can prevent cells (in a synfire chain) from producing spiking at their target cells.

References

[1] Ö. Ekeberg, Response properties of a population of neurons, Int. J. Neural Syst. 4 (1) (1993) 1–13.

[2] E. Fransén, A. Lansner, Low spiking rates in a population of mutually exciting pyramidal cells, Network 6 (2) (1995) 271–288.

[3] A. Lansner, E. Fransén, Modeling Hebbian cell assemblies comprised of cortical neurons, Network 3 (2) (1992) 105–119.

[4] A.V. Lukashin, A.P. Georgopoulos, A dynamical neural network model for motor cortical activity during movement: population coding of movement trajectories, Biol. Cybernet. 69 (1993) 517–524.

[5] H. Sompolinsky, Mechanisms underlying orientation selectivity in vision, Soc. Neurosci. Sat. Symp. Dynam. Neurosci. IX (2001) 6.

[6] J. Tegnér, X.J. Wang, Role of dendritic disinhibition by calretinin interneurons in a conductance-based cortical network model for spatial working memory, Soc. Neurosci. Abstr. 26 (2000) 596.10.

[7] A.M. Thomson, A. Destexhe, Dual intracellular recordings and computational models of slow inhibitory postsynaptic potentials in rat neocortical and hippocampal slices, Neuroscience 92 (1999) 1193–1215.

[8] M. Tsodyks, A. Uziel, H. Markram, Synchrony generation in recurrent networks with frequency-dependent synapses, J. Neurosci. 20 (2000) RC50.

[9] C. Van Vreeswijk, L.F. Abbott, G.B. Ermentrout, When inhibition not excitation synchronizes neural firing, J. Comput. Neurosci. 1 (1994) 313–321.

[10] X.J. Wang, Synaptic basis of cortical persistent activity: the importance of NMDA receptors to working memory, J. Neurosci. 19 (1999) 9587–9603.

[11] E.L. White, Cortical Circuits, Birkhäuser, Boston, 1989.

[12] M.A. Whittington, R.D. Traub, N. Kopell, B. Ermentrout, E.H. Buhl, Inhibition-based rhythms: experimental and mathematical observations on network dynamics, Int. J. Psychophysiol. 38 (2000) 315–336.

[13] K. Zhang, Models for the head direction system as an angular integrator, Soc. Neurosci. Sat. Symp. Dynam. Neurosci. IX (2001) 12.

Erik Fransén is an Associate Professor at the Department of Numerical Analysis and Computer Science at the Royal Institute of Technology, Stockholm, Sweden. He was a post doctoral fellow at the Department of Psychology at Harvard University during 1997–1998. He received his Ph.D. in Computer Science from the Royal Institute of Technology, Stockholm, Sweden in 1996. He received his B.Sc. in Physics from Uppsala University, Sweden in 1987. His research interests include mathematical modeling and biophysical simulation of neurons and networks of neurons. He studies the ionic basis of cellular and network properties and the functional role of neuromodulators.

Available at
www.ComputerScienceWeb.com
POWERED BY SCIENCE DIRECT
ELSEVIER
Neurocomputing 52–54 (2003) 771–777
NEUROCOMPUTING
www.elsevier.com/locate/neucom

Analysis of higher-order correlations in multiple parallel processes

Gaby Schneider, Sonja Grün*
Department of Neurophysiology, Max-Planck-Institute for Brain Research, Deutschordenstr 46 Frankfurt/Main 60528, Germany

Abstract

The 'unitary event' method analyzes multiple spike trains to identify neuronal groups whose coherent activity does not conform with full independence. Here we distinguish 'genuine' coincidences from those due to subgroup correlations. The introduced model describes a neuron's firing as a superposition of independent and stationary Bernoulli processes, each representing the synchronous activity of one of all possible subsets of neurons. Using maximum likelihood and normal approximation, genuine correlations are identified under a null-hypothesis that respects lower-order correlations. Evaluation of the approach includes test power and adequateness of significance levels.

Keywords: Spike synchronization; Higher-order correlation; Significance; Test power

1. Introduction

The hypothesis that temporally structured coherent firing is an indication of assembly activity motivates the analysis of (near-)coincident firing activity observed in simultaneously recorded spike trains. The 'unitary event' method [2,3] analyzes the probability of coincident spiking of groups of neurons based on the null-hypothesis of full independence of the processes. Detection of unitary events implies existence of correlations between the processes. However, deviation from expectation due to a correlation caused by a subgroup of neurons is not identified.

* Corresponding author. Present address: Institute for Biology, Neurobiology, Freie Universität Berlin, Königin_Luise Str. 28/30, 14195 Berlin, Germany.
E-mail address: gruen@mpih-frankfurt.mpg.de (S. Grün).

doi:10.1016/S0925-2312(02)00772-5

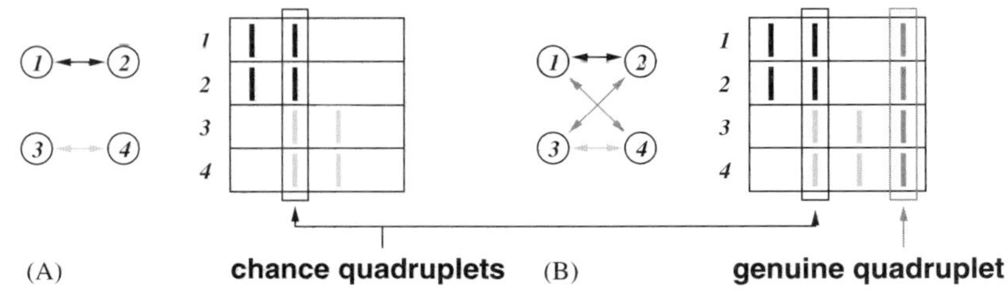

Fig. 1. Illustration of coincidence quadruplets originating in genuine quadruplet correlations and chance quadruplets. In (A) a system is assumed that contains only pairwise correlations between neurons 1–2 and 3–4, which may lead to chance co-activation of both subgroups of neurons. These by chance quadruplets have to be distinguished from genuine quadruplets that originate from four neurons being involved in a common process (B). Coexistence of involvements in processes of different neuron compositions is possible.

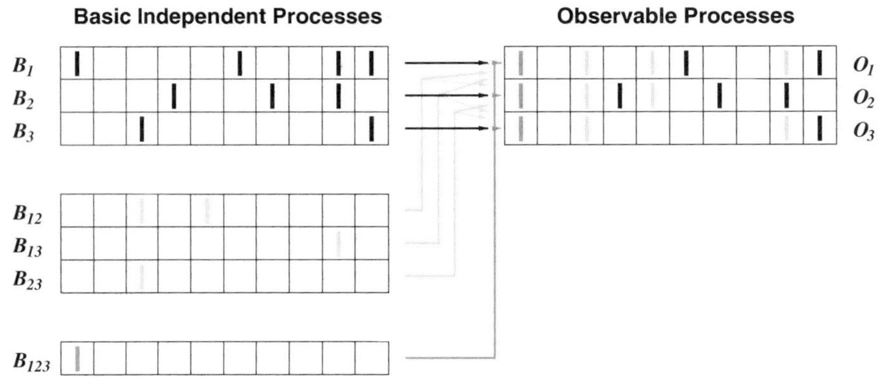

Fig. 2. MIIP illustrated for three neurons. Observed spiking (O_1, O_2, O_3) results from several 'basic' independent processes: the neurons' background activity (B_1, B_2, B_3), correlation processes between pairs B_{ij} and of all three neurons B_{123}. Thus, a certain spike-pattern at a time can be due to several different 'basic events', i.e. firing patterns of the basic processes.

The work presented here focuses on the identification of 'genuine' higher-order correlations defined here as the coincidences that cannot be explained by a chance co-activation of lower-order correlations (Fig. 1). The order of a genuine correlation is the number of neurons involved in a common process.

2. The model

As the basis for our study of higher-order correlations we define a model of independent interaction processes (MIIP) that regards firing activity of each neuron to be describable as a superposition of independent and stationary Bernoulli processes (Fig. 2). Observable processes (O_1, O_2, etc.) are distinguished from underlying basic processes of which there are independent 'background' processes (B_1, B_2, etc., with firing probabilities λ_1, λ_2, etc.) and correlation processes ($B_{12}, B_{13}, B_{23}, B_{123}$, etc., with coincidence rates $\lambda_{12}, \lambda_{13}, \lambda_{23}, \lambda_{123}$, etc.) producing coincidences among all possible subgroups of neurons.

3. Analysis method

The first goal is to estimate the parameters of the underlying Bernoulli processes based only on the observable processes. We do this using the maximum-likelihood (ML) principle. The estimates are chosen such that they maximize the likelihood of the given data piece. Coding the occurrence of a spike in a bin as a one and a non-spike as a zero, the sufficient statistics for a piece of data are the numbers of all 2^n possible observable binary vectors of length n across the n neurons observed in the T time steps (time resolution here assumed to be 1 ms).

The ML-estimates of the basic processes' firing probabilities for any number of neurons can be derived with the following formula: Let $N := \{1, \ldots, n\}$ be the set of neurons, $M, M_0 \subseteq N$, and $S_M := \sum_{t=1}^{T} I(\{O_i(t) = 0 \ \forall i \in N - M\})$ the pattern counts of all possible subsets $M \subseteq N$ then

$$1 - \hat{\lambda}_{M_0} = \frac{\prod_{M \subseteq M_0, |M| \neq |M_0| \bmod 2} S_M}{\prod_{M \subseteq M_0, |M| = |M_0| \bmod 2} S_M} \tag{1}$$

(see also e.g. [10]). Thus, we can estimate the probability of coincident firing of every specified subset of the observed neurons. Eq. (1) together with the δ-Method (e.g. [1]) first imply asymptotic normality of all ML-estimates of the basic processes' parameters due to asymptotic normality of the counts S_M. One can secondly use the δ-Method to get the asymptotic variance. We need to assign an index $i \in \{1, 2, \ldots, 2^{|M_0|}\}$ to each of the sets $M \subseteq M_0$, such that $S_i := S_M$, where $i \neq j$ for $M \neq M'$. Let the relative frequencies of the events $\hat{p}_i := S_i/T$ be the estimates for the corresponding probabilities. Writing the right-hand side of formula (1) as a function $f(p_i)_{i=1,2,\ldots,2^{|M_0|}}$ of the probabilities p_i and deriving the asymptotic covariance matrix $\Sigma = (\sigma_{ij}) = (1/T)\mathrm{Cov}(S_i, S_j)$, the asymptotic variance of $\hat{\lambda}_{M_0}$ can be written as

$$\sigma^2(\hat{\lambda}_{M_0}) = \frac{1}{T} \sum_{k,k'=1}^{2^n-1} \sigma_{kk'} \left(\frac{\partial f}{\partial p_k}\right) \left(\frac{\partial f}{\partial p_{k'}}\right). \tag{2}$$

We can thus Z-transform the ML-estimate to get a standard normally distributed random variable Z. Choosing a significance level for the test (e.g. 2.5%), we can compute the asymptotic test power, which is the probability to correctly reject a false H_0 (i.e. absence of the one inspected genuine higher-order correlation) in favor of the true alternative hypothesis, i.e. to detect higher-order coincidences.

$$\text{test power} = P\left(\frac{\hat{\lambda}}{\hat{\sigma}(\hat{\lambda})} > 1.96\right) \doteq P\left(Z > 1.96 - \frac{\lambda}{\sigma(\hat{\lambda})}\right). \tag{3}$$

By rearrangement of (3) and insertion of (2), the minimal number of time steps required to reach a requested level of the test power can be calculated.

3.1. Illustration for $n = 2$

We now want to illustrate the proposed method for two parallel processes. In the 2-neurons-case, we simply need the counts $S_{00} := S_{\{\}}$, $S_{0+} := S_{\{2\}}$ and $S_{+0} := S_{\{1\}}$ of the corresponding patterns. We have

$$p_{00} := P(O_1 = O_2 = 0) = (1 - \lambda_1)(1 - \lambda_2)(1 - \lambda_{12}), \tag{4}$$

$$p_{0+} := P(O_1 = 0) = (1 - \lambda_1)(1 - \lambda_{12}), \tag{5}$$

$$p_{+0} := P(O_2 = 0) = (1 - \lambda_2)(1 - \lambda_{12}). \tag{6}$$

This gives us the following formulas for the underlying parameters:

$$1 - \lambda_1 = \frac{p_{00}}{p_{+0}}, \quad 1 - \lambda_2 = \frac{p_{00}}{p_{0+}}, \quad 1 - \lambda_{12} = \frac{p_{0+}\,p_{+0}}{p_{00}}. \tag{7}$$

The counts S_{ij} of all possible patterns follow a multinomial distribution, which implies that the ML-estimates of the probabilities used in Eq. (7) are just the relative frequencies of the corresponding events, leading to Eq. (1).

To decide whether all observed coincidences are chance coincidences, we derive the distribution of the estimates under the null-hypothesis that only lower-order correlations exist. For $n = 2$, H_0 is equivalent to independence of the processes, and the interesting parameter is λ_{12}. According to Eq. (2), the asymptotic variance of $\hat{\lambda}_{12}$ is

$$\sigma^2(\hat{\lambda}_{12}) = \frac{(1 - \lambda_{12})(\lambda_{12}(1 - \lambda_1)(1 - \lambda_2) + \lambda_1 \lambda_2)}{T(1 - \lambda_1)(1 - \lambda_2)}. \tag{8}$$

We can thus make a test by Z-transforming the estimate $\hat{\lambda}_{12}$. For more than two neurons, one can proceed analogously by applying Eqs. (1) and (2). This allows to parallely and independently investigate all subgroup correlations of interest, irrespective of their order.

4. Comparison with simulations

Using simulated spike trains, we tested the empirical significance and test power and compared the results to the asymptotic levels (Fig. 3). Simulation experiments consisting of parallel spike trains, that were generated according to the model (Fig. 2) by 'injecting' coincidences [2,4] of given rates, were analyzed for coincident events of the order under investigation. The percentage of the number of significant experiments in relation to the total number of performed experiments yields the probability to reject H_0. In case of no injection of coincidences of the evaluated order, this yields the empirical significance (Fig. 3A and B), whereas for injected coincidences it yields the test power (Fig. 3C and D). For relatively short data segments (e.g. $T = 10^4$, black lines in Fig. 3A and B) and small background rates, the empirical significance (solid lines in Fig. 3A and B) turns out to be smaller than the applied significance level (dashed lines in Fig. 3A and B). For larger T (grey lines in Fig. 3A and B), the given significance level is approached from below for increasing rates, i.e. the test is conservative as long as the asymptotics are not reached.

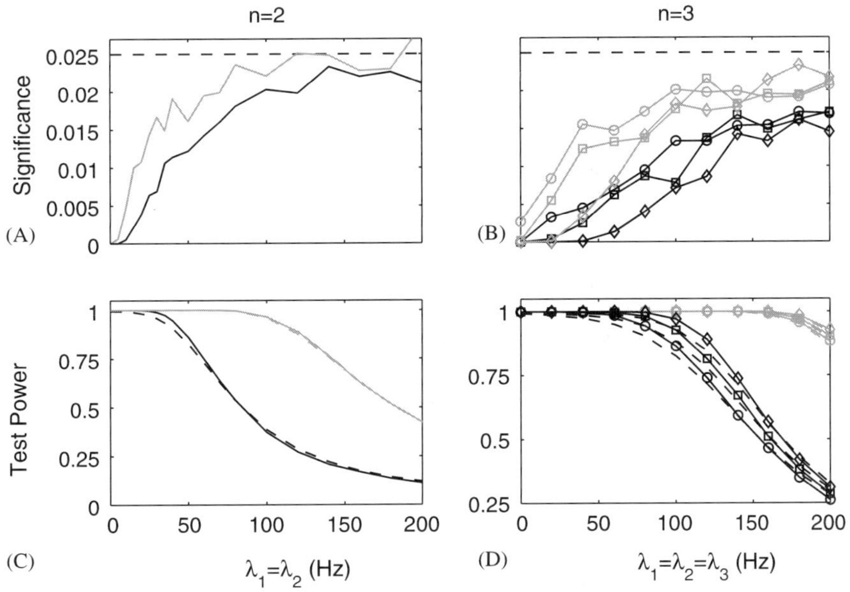

Fig. 3. Comparison of empirical significance level (A and B) and test power (C and D) to asymptotic levels as a function of background rate level. Empirical values are extracted as the relative number of significant experiments out of 10^4. The number of time steps considered are indicated by the color of the graphs (black: $T=10^4$, grey: $T=5\times10^4$). (A) For $n=2$ the empirical significance level ($\lambda_{12}=0$ Hz) is illustrated as a function of background rates ($\lambda_1=\lambda_2$). (B) For $n=3$, the empirical significance level ($\lambda_{123}=0$ Hz) is in addition shown for different pairwise coincidence rate levels ($\lambda_{12}=\lambda_{13}=\lambda_{23}$ chosen as 0 Hz ($\Diamond$), 2 Hz ($\Box$) and 4 Hz ($\circ$)). In both cases, the significance level is 2.5% (dashed lines). Empirical (solid) and asymptotic (dashed) test power (C) for $n=2$ with $\lambda_{12}=2$ Hz, and (D) for $n=3$ with $\lambda_{123}=2$ Hz, other parameters and coding as in (A and B).

The test power decreases with increasing rates of the background, and in case of $n>2$, also with increasing coincidence rates of lower order than evaluated. The asymptotic test power (dashed curves in Fig. 3C and D) describes the empirical test power (solid curves in Fig. 3C and D) quite well. For $n=2$ and $n=3$, the test power decreases with increasing background rates. For $n=3$, the larger the coincidence rate of order 2, the faster the decrease.

5. Conclusions and outlook

We presented an approach for the identification of genuine higher-order correlations as expressed by the existence of higher-order coincidences in multiple parallel spike trains. For the investigation of correlation within a set of parallel processes we formulated a model that is composed of a set of independent processes each of which is representing a process for coincidences of different order and composition. Using maximum-likelihood estimates and their asymptotic variance we developed a method to identify 'genuine' higher-order correlations. It proved to be conservative as long as the asymptotics do not hold and allows to parallely and independently investigate all subgroup correlations of interest.

The approach is related to methods that use log-linear models (cf. [5–8]) due to similarity of estimates. Still, in the proposed information-geometric measures it is not possible to distinguish between different subgroups of a given number of elements. Those approaches are extremely useful if one is interested in the existence of correlations of a given order. However, if we would like to identify the group(s) of neurons that are involved in the computation of a specific task ('assembly hypothesis'), we need to explicitly identify the special subgroups that express correlations.

All approaches discussed are restricted to the analysis of exact coincidences. As experimental data indicate that synchronous spiking activity may occur with a small temporal jitter up to a few ms (e.g. [4,9]), we currently work at an extension of the model for near-coincidences, i.e. coincidences with a small temporal jitter. The basic idea is to introduce additional independent processes per time offset, subgroup of neurons and order of spikes [11]. The approach can be extended to systems of more than two parallel processes.

As the analysis of higher-order correlations in itself implies an exponential increase in the number of parameters if one wants to separately consider all different subgroups of neurons, one needs to find means to reduce the parameter space. Furthermore, for an application onto experimental data, problems of nonstationarity and jitter size need to be considered carefully and are addressed in current projects.

Acknowledgements

We thank Gordon Pipa for stimulating discussions, and Brooks Ferebee, Wolf Singer and Anton Wakolbinger for supporting this project. This work has been partially supported by the Hermann–Willkomm Foundation and by the Volkswagen Foundation.

References

[1] Y. Bishop, S.E. Fienberg, P.W. Holland, Discrete Multivariate Analysis. Theory and Practice, MIT Press, Cambridge, MA, London, 1991.

[2] S. Grün, M. Diesmann, A. Aertsen, Unitary events in multiple single-neuron activity. I. Detection and significance, Neural Comput. 14 (2002) 43–80.

[3] S. Grün, M. Diesmann, A. Aertsen, Unitary events in multiple single-neuron activity. II. Non-Stationary data, Neural Comput. 14 (2002) 81–119.

[4] S. Grün, M. Diesmann, F. Grammont, A. Riehle, A. Aertsen, Detecting unitary events without discretization of time, J. Neurosci. Meth. 94 (1999) 67–79.

[5] R. Gütig, A. Aertsen, S. Rotter, Analysis of higher-order neuronal interactions based on conditional inference, submitted for publication.

[6] L. Martignon, G. Deco, K. Laskey, M. Diamond, W. Freiwald, E. Vaadia, Neural coding: higher-order temporal patterns in the neurostatistics of cell assemblies, Neural Comput. 12 (2000) 2621–2653.

[7] L. Martignon, H. von Hasseln, S. Grün, A. Aertsen, G. Palm, Detecting higher-order interactions among the spiking events in a of group of neurons, Biol. Cybernet. 73 (1995) 69–81.

Available at
www.ComputerScienceWeb.com
POWERED BY SCIENCE DIRECT®

ELSEVIER

Neurocomputing 52–54 (2003) 779–786

NEUROCOMPUTING

www.elsevier.com/locate/neucom

Linear and non-linear measures of the human neonatal EEG

K. Hecox*, S. Nayak, K. Gin, A. McGee, W. van Drongelen

Department of Pediatrics, University of Chicago, 5839 S. Maryland, Chicago 60637, USA

Abstract

Non-linear measures (Kolmogorov entropy, correlation dimension and a surrogate based "*z*" score) of the underlying structure of the human neonatal EEG were made and compared to linear measures (eigenvalues) and to EEGs obtained from older children (greater than 3 years of age). Strong age-dependent differences were noted for Kolmogorov entropy, the surrogate-based measure of non-linearity and eigenvalues. The eigenvalues decrease with age and the Kolmogorov entropy increases with age. The measure of global non-linearity decreased with age so that it was positive in only a small fraction of the greater than 3-year-old subjects.

Keywords: Kolmogorov entropy; Non-linear measures; Linear measures; Human neonatal EEG

1. Introduction

One of the most challenging areas of clinical EEG has been the interpretation of the EEG in human neonates. The range of the EEG waveforms and patterns considered normal is broad and the identification of abnormal waveforms quite conservative. Waveforms considered potentially epileptogenic in older infants and adults are interpreted as signs of underlying tissue abnormalities but not necessarily potentially epileptogenic (e.g. spikes or sharp waves) in neonates. The frequency spectrum of normal neonatal EEG activity encompasses frequencies considered abnormal in the older patient. The diversity of neonatal EEG waveforms has resulted in a broadening of the boundaries of what is accepted as normal, and a commensurate decline in sensitivity to pathological states. Whether alternative EEG measures, such as dynamical systems measures, might improve this circumstance is a matter of active inquiry.

* Corresponding author.
E-mail address: khecox@peds.bsd.uchicago.edu (K. Hecox).

0925-2312/03/$ - see front matter
doi:10.1016/S0925-2312(02)00774-9

Recently, there has been an acceleration of research in the area of EEG analysis, fueled by emerging technologies and methodologies. Development of the MEG, the introduction of new mathematical methods for the location of intracranial sources and the application of non-linear time series analysis to the EEG have all fueled an increase in EEG research. Of particular interest is the use of deterministic chaos to predict the time of occurrence of seizures.

Early reports of the use of deterministic chaos measures to detect seizures were quite optimistic. As more investigators became interested in the application of deterministic chaos to EEG analysis our understanding of the limitations of these metrics also increased. There followed, multiple demonstrations of circumstances in which spurious chaos values were generated, suggesting that the interpretation of these metrics should be cautious. It has now been shown, in multiple laboratories, that ictal events usually have dynamical properties unlike that seen in the interictal period. In addition, there are now several reports that deterministic chaos measures can predict seizures. These dynamical system measures seem particularly well suited as an indirect measure of the behavior of large numbers of neurons, perhaps acting as a network. The question of the relative contribution of single unit versus large networks to the unstable patterns of electrical behavior in epilepsy is a core issue in epilepsy research. Despite the high level of interest in these deterministic chaos measures we have found only one article on their developmental dependency and one article specifically directed to their application to epilepsy in children. This is a small proportion of the more than one hundred articles published in this field.

The pediatric epilepsies differ from those seen in adults in a number of ways, including the higher prevalence of a neocortical origin to the events. It is not known whether the developmental changes in the EEG signal, including its underlying structure, will have substantial impact on the utility of these measures in infants and children. Therefore, we have undertaken a program to determine whether these metrics can be reliably preformed in the pediatric aged patients, whether developmental differences preclude the application of the measurements and whether there are other age-specific considerations in their clinical use.

We report on the correlation dimension, eigenvalues and Kolmogorov entropy (KE). The correlation dimension metric was chosen since it allows us to compare our results with those obtained at other laboratories. The Kolmogorov 2 entropy was selected since preliminary studies in our laboratory have suggested that it may be particularly useful in seizure prediction, and is conceptually distinct from the correlation dimension. The eigenvalue provides a measure of the strength of linear models to explain the EEG.

2. Methods

Patient selection: All patients were obtained from the University of Chicago neonatal intensive care unit, and were born from 29 to 38 week gestation. Patients were selected among those who had undergone an EEG as part of a clinical evaluation. None of the patients were receiving CNS active medications, none had seizures, none had intra-ventricular hemorrhage and none (retrospectively) had significant neurological problems. All of the EEGs were interpreted as normal. The majority of infants were being screened as part of an apnea evaluation. Similar criteria were applied to older infants (greater than 3 years of age) who were obtained from the records of the routine ambulatory EEG laboratory.

Segment Selection: EEG segments were selected by scrolling through the record and selecting the first 30 s epoch in which the patient's state was clear and free of artifacts. Three replications of wakefulness were selected. The segments were each 30 s in duration, sampled at 400 Hz, filtered from 1 to 70. The electrodes chosen for analysis were from the neonatal modified 10–20 system—Fp1, T3, T4 O1.

EEG Analysis: Measures included correlation dimension, Kolmogorov entropy, eigenstructure and a normalized measure of non-linearity (z). Time delays and embedding dimension were data adaptive and the entropy measures are noise corrected.

Algorithms: The estimation of the non-linear variables was based on attractor reconstruction. The state of the system (i.e. the brain) that generates the EEG or ECoG can be represented by a projection of all variables in a multi-dimensional state space. The EEG signal consists of sampled data; in the state space, two points of the sampled time series with an inter-distance of k sampled points are plotted against each other, thus showing the relationship between the values of samples i and $i+k$. In general, if there is no relationship between the points in a time series (e.g. a random signal), one expects the points to be distributed evenly over the space. Takens [9] proved that the dynamic state of a system could be reconstructed from a time series generated by the system, by using time delay coordinates.

The technique can be described in a general fashion as embedding a time series $x_1, x_2, x_3, \ldots, x_N$ into a set of vectors X_i with m elements:

$$X_i = (x_i, x_{i+k}, x_{i+2k}, \ldots, x_{i+(m-1)k})^{\mathrm{T}}, \tag{1}$$

where k is the delay in number of samples and T is the epoch of the time series embedded in the vector. The value of T should not be too large so that the first and last values in the epoch are practically uncorrelated. On the other hand, T should be large enough to cover the dominant frequency in the time series. We determined the embedding parameters following the procedures described by Schouten et al. [7]. We use a delay k equal to unity, therefore the number of dimensions (m) in the time window T is determined by the sample frequency of the EEG (400 samples/s). The number of vectors (i) determines the total length of the time series required for the attractor reconstruction procedure. Our time series of N points were normalized by subtracting the mean ($x_m = [1/N]\sum x_i$) and dividing by the average absolute deviation dx:

$$dx = [1/N]\sum |x_i - x_m|. \tag{2}$$

The number of crossings of the mean divided by two was used as an estimate for the number of cycles (N_{cycles}). The value of m was chosen as the rounded value of N/N_{cycles}. Using these criteria, our values for m varied between 45 and 200, close to the recommended values (50–200) by Schouten et al. [7]. To obtain sufficient vectors X_i as the basis for estimation of the non-linear metrics, the total number of points per reconstruction was set to 12,000 (30 s of EEG).

One of the variables that can characterize an attractor is entropy [2]. Entropy estimates the rate at which information about the state of the system is lost. Order-2 Kolmogorov Entropy can be estimated by examination of two initially close orbits on an attractor, and measuring the time (t) required for the orbits to diverge beyond a set distance ε. One could present it as an index of chaos in the signal. For example, entropy of a static system is zero, noise has infinite entropy, and chaotic systems have entropy between the two extremes. The distribution of the time intervals $f(t)$ satisfies:

$$f(t) \sim \mathrm{e}^{-\mathrm{KE}\,t}, \tag{3}$$

where KE is the Kolmogorov entropy. Using the property in (3) and based on the maximum norm, Schouten et al. [6] developed a computationally efficient maximum-likelihood estimator for KE, which we have incorporated in our studies.

A commonly used characterization is the correlation dimension. Calculation of this metric is based on the correlation integral [1]:

$$C(s) = \{1/(N(N-1))\} \sum \Theta(s - |X_i - X_j|). \tag{4}$$

With Θ = Heaviside step function, and N = the number of points. The term $|X_i - X_j|$ denotes the distance between the points in state space. The distance between points in space may be evaluated with the Euclidian norm or the maximum norm; in terms of dimension, the results obtained with the two techniques are equivalent. The value of $C(s)$ is a measure of the number of pairs of points (X_i,X_j) on the reconstructed attractor whose distance is smaller than s. Because the correlation integral counts the pairs of closely related points in the plot, the distribution of the points over the space is characterized for the attractor. The values of s are data dependent in the sense that they are scaled by the average absolute deviation (Eq. (2)) between $dx/400$ and $dx/100$. If N approaches infinity and s approaches zero in Eq. (4), $C(s)$ scales according to a power law $C(s) \sim s^{\mathrm{D}}$, where D is the correlation dimension of the attractor.

In addition to the non-linear metrics, a linear one was extracted. The largest eigenvalue (Ei) of the time series was determined with a principal component analysis (PCA). The largest eigenvalue indicates the amount of variance explained by the first principal component. By monitoring Ei, we evaluate existence and contribution of a dominant component around the occurrence of a seizure. The eigenvalue gives us one relative measure of the system's behavior that is explained by linear metrics [5].

To test the non-linearity of the data, the recorded time series were compared with surrogate data sets. The test is based on creating surrogate data series with the same power spectrum as the original signal but with no phase correlation [3]. Surrogate data sets were introduced as a precautionary measure against the over-zealous search for low-dimensional chaos. The surrogate time series, which yields the same power spectrum and autocorrelation as the original time series, was obtained by the inverse transform of the phase randomized Fourier spectrum. The growth of inter-point distances was used as the statistic to test the null hypothesis whether the original could be described by linearly correlated noise [10]. From comparisons between the original and surrogate series, a quantity z was calculated from the Mann–Whitney U rank-sum

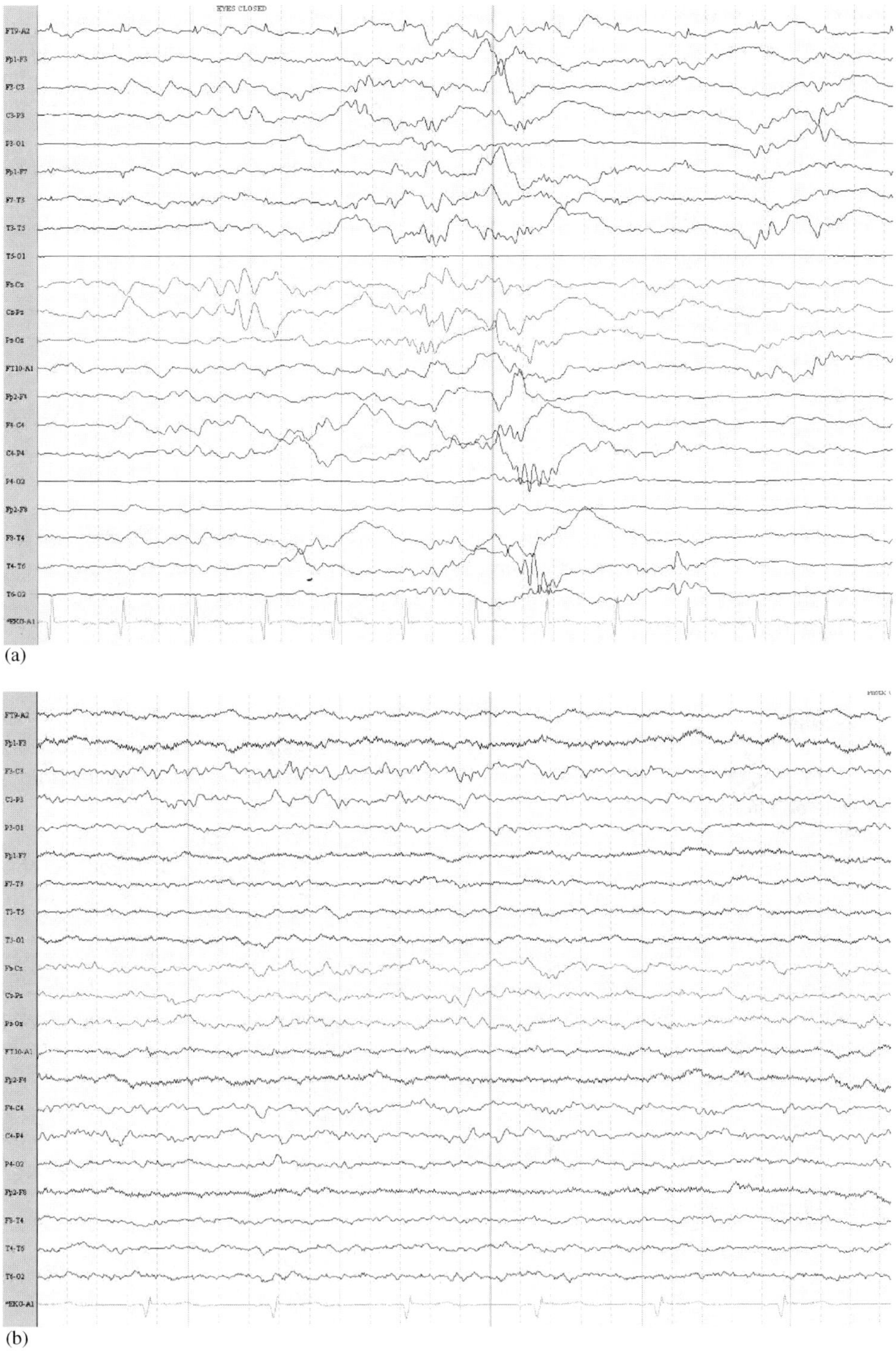
EYES CLOSED
FT9-A2
Fp1-F3
F3-C3
C3-P3
P3-O1
Fp1-F7
F7-T3
T3-T5
T5-O1
Fz-Cz
Cz-Pz
Pz-Oz
FT10-A1
Fp2-F4
F4-C4
C4-P4
P4-O2
Fp2-F8
F8-T4
T4-T6
T6-O2
*EKG-A1
(a)
(b)

Fig. 1.

statistic. By comparing the recorded data with 50 surrogate sets, a mean value of z can be obtained. The z-value is normally distributed with zero mean and unit variance. While it is not our intent to determine whether the EEG is a low-dimensional chaotic system, if the non-linearity index is low we are especially cautious in our interpretation of more specific non-linear tests.

The algorithms for data-conversion, and the wave editing process were implemented in MatLab (The Math Works Inc., Natick, MA). The largest eigenvalue and the variables based on non-linear dynamics were calculated with a software package developed for the analysis of time series that include noise [8].

3. Results

Examples of raw waveforms are shown in Fig. 1. A generalized linear analysis of variance fixed effects model was used (electrode x age) for data analysis. The significance level was fixed for all of the comparisons, at $p < 0.01$. The means are shown in Table 1.

There were no significant differences for the correlation dimension for age or electrode position. For the Kolmogorov entropy, the effect of age and electrode were significant. The neonates had a lower KE than the older children (KE of 37 versus 21 for the children and infants, respectively). The KE also showed a significant electrode effect, with the temporal leads showing the highest values for KE for both age groups.

The eigenvalues also showed significant overall age dependency (0.54 versus 0.74 for the children versus newborns, respectively). There was no significant effect of electrode position for eigenvalues, although the mean for the frontal electrode was highest among the four positions.

Our surrogate (z), used to determine if the data set is consistent with various linear systems, also showed a large age-dependant change (−0.91 versus −2.90 for the children versus neonates, respectively). Finally, there was an electrode effect, due to the fact that the frontal electrode had a higher z than other electrodes.

The relationships amongst the measures is consistent. As the eigenvalue decreases, with increasing age, the Kolmogorov entropy increases. There is a similar inverse relationship between KE and the eigen value for electrode position, also. The relationship between the values of z and KE and eigen value are more complex. As the eigen value decreases with age, z decreases and KE increases.

Table 1
Means of dynamical systems measures

	CD	KE	z	Eigen value
Neonates	4.0	21	−2.90	0.74
Children	4.3	37	−0.91	0.54

4. Discussion

While knowledge of the development of single unit behavior and cell–cell interactions has advanced at a steady pace, relatively little is known regarding the development of population or large neural network responses. In part, this is due to a weakness in the conceptual framework, but also hampered by the paucity of analytical tools available for characterizing large network behavior. Given the large number of variables that influence brain development, that these factors interact with one another in rather complex ways and that the cellular activity underlying the population behavior is known to be non-linear, it is not surprising that analytical tools derived from dynamical systems analysis are under evaluation. The overall outcome of this study provides encouragement for the continued application of dynamical systems analysis tools to characterizing the development of the human EEG. Reliable responses were obtained from even the youngest population-neonates. Additionally, KE, eigenvalues and surrogate-based measures of non-linearity (z) all showed large and consistent age dependant changes. These measures were also sensitive to regional differences in brain development.

There is only one prior article to which we can compare our results, and then only for the correlation dimension. Meyer-Lindenberg [4] described the evolution of the CD, Lyapunov exponent and non-linearity as measured using surrogate data analysis. The CDs for frontal, temporal and occipital leads were 4.0, 4.0 and 3.8 in the Meyer-Lindenberg study [4] versus 3.6, 4.3, and 4.0 for our neonatal group. Large differences were seen in the older age group 6.1, 7.7, and 7.7 versus our results of 4.2, 4.5 and 4.2. The methods for calculating the CD were different in the two studies (e.g. embedding, dimension, tau, number of samples, etc.) and the data selection and editing process was clearly different. We selected samples without eye movement contamination and specified patient state, while the earlier study used an autoregressive technique for removing the artifact and state was not an explicit variable.

The results of our study in the older age group are consistent with the frequent observation that normal EEG segments taken from neurologically normal adults have only modest evidence of nonlinearity. This is clearly not the case in the newborn where z was frequently greater than -3.0. Inspection of raw waveforms clearly demonstrates the dramatic differences in the appearance of the waveforms as a function of age. The discontinuities seen in the premature and term infant recordings are compelling, and never seen in normal older patients. The presence of discontinuities shows non-stationarity. Hence, we use entropy and correlation dimension only in an informal sense. The eigenvalues showed a clear age dependency. The older population had lower eigenvalues.

Finally, the age dependency for KE was quite large. Clearly KE increases with age, nearly doubling over the age range in this study. The biological significance of the increase in entropy with age is not yet clear. As entropy increases, the number of possible states of the system increases and the "information" loss is higher. Whether the increasing state possibilities are directly related to the maturation of anatomic connectivity, as a function of age, requires direct verification. Nevertheless, we are encouraged to continue to use KE as an index of complexity of the EEG signal and are continuing to investigate whether the KE may prove more sensitive than CD in the pediatric population. This is directly relevant to the application of seizure detection and anticipation algorithms.

Acknowledgements

This work was supported by The Falk Medical Trust Foundation.

References

[1] P. Grassberger, I. Procaccia, Characterization of strange attractors, Phys. Rev. Lett. 50 (1983) 346–349.
[2] P. Grassberger, I. Procaccia, Estimation of the Kolmogorov entropy from a chaotic signal, Phys. Rev. A 28 (1983) 2591–2593.
[3] D. Kaplan, L. Glass, Understanding Nonlinear Dynamics, Springer, New York, 1995.
[4] A. Meyer-Lindenberg, The evolution of complexity in human brain development: an EEG study, Electroenceph. Clin. Neurophysiol. 99 (1996) 405–411.
[5] D.F. Morrison, Multivariable Statistical Methods, McGraw-Hill, New York, 1967.
[6] J.C. Schouten, F. Takens, C.M. van den Bleek, Maximum-likelihood estimation of the entropy of an attractor, Phys. Rev. E 49 (1994) 126–129.
[7] J.C. Schouten, F. Takens, C.M. van den Bleek, Estimation of the dimension of a noisy attractor, Phys. Rev. E 50 (1994) 1851–1861.
[8] J.C. Schouten, C.M. van den Bleek, RRCHAOS: a Menu-driven Software Package for Chaotic Time Series Analysis, The Netherlands, 1993.
[9] F. Takens, Detecting strange attractors in turbulence, in: D.A. Rand, L.S. Young (Eds.), Lecture Notes in Mathematics, Vol. 898, Springer, Berlin, 1981, pp. 366–381.
[10] J.R. Van Ommen, Monitoring fluidized bed hydrodynamics, Thesis, Delft University, The Netherlands, 2001.

Available at
www.ComputerScienceWeb.com
POWERED BY SCIENCE DIRECT®

ELSEVIER

Neurocomputing 52–54 (2003) 787–792

NEUROCOMPUTING

www.elsevier.com/locate/neucom

Anomalous response variability in a balanced cortical network model

John Hertz[a,*], Barry Richmond[b], Kristian Nilsen[a]

[a]*Nordita, Blegdamsvej 17, 2100 Copenhagen, Denmark*
[b]*Laboratory of Neuropsychology, NIMH, Bethesda, MD 20892 USA*

Abstract

We use mean field theory to study the response properties of a simple randomly connected model cortical network of leaky integrate-and-fire neurons with balanced excitation and inhibition. The formulation permits arbitrary temporal variation of the input to the network and takes exact account of temporal firing correlations. We find that neuronal firing statistics depend sensitively on the firing threshold. In particular, spike count variances can be either significantly greater than or significantly less than those for a Poisson process. These findings may help in understanding the variability observed experimentally in cortical neuronal responses.
Published by Elsevier Science B.V.

PACS: 87.18.Sn; 87.19.La; 87.80.Vt

Keywords: Response variability; Cortical dynamics; Asynchronous firing

1. Introduction

Cortical neuronal responses often exhibit a puzzling variability (see, e.g. [4]): spike count distributions obtained for repeated presentations of a stimulus are frequently broader than Poisson distributions with the same mean counts. In this paper we study the statistics of neuronal responses in a simple model network with balanced excitation and inhibition and find that they also show large variability.

To study the responses in our model, we utilize mean field theory, which is exact in the limit of a large network with homogeneous connection probabilities [7].

* Corresponding author.
E-mail addresses: hertz@nordita.dk (J. Hertz), bjr@ln.nimh.nih.gov (B. Richmond), kristnil@nordita.dk (K. Nilsen).

0925-2312/03/$ - see front matter Published by Elsevier Science B.V.
doi:10.1016/S0925-2312(02)00775-0

The mean field theory of balanced cortical networks has been studied by a number of authors [1,2,8,6], but generally in or near steady firing states and assuming that the self-consistent current input is uncorrelated. Here, like Ref. [7], we consider time-dependent external drive (as in an experimental trial) and color the noise correctly.

2. The model

For these exploratory investigations, we use a network of N mutually inhibitory neurons with a 10% connection probability. This is probably the simplest model for which one can achieve balanced asynchronous activity. The nonzero connections are of equal strength, and the synaptic currents are assumed to act instantaneously, i.e., each presynaptic spike depresses the postsynaptic potential discontinously by a fixed amount. We do not include transmission delays.

We find it convenient to scale variables in the way used by van Vreeswijk and Sompolinsky [8]: When the mean number of neurons presynaptic to a given one is $K \gg 1$, each synaptic strength is given a value $J/\sqrt{K}$ and the average synaptic current scales like $\sqrt{K}$. It is inhibitory and is counterbalanced by an external excitatory current which is also proportional to $\sqrt{K}$. The fluctuations in the synaptic current are smaller than the mean by a factor of order $1/\sqrt{K}$, i.e., of order 1. In steady state, the two currents nearly cancel, leaving a net current of order 1. If the steady-firing state is stable, disturbances of the rates relax very rapidly, in a time of order $1/\sqrt{K}$.

The (subthreshold) membrane potentials of leaky integrate-and-fire neurons are described by

$$\dot{u}_i(t) = -\frac{1}{\tau} u_i + \sqrt{K} I_i^0(t) - \frac{J}{\sqrt{K}} \sum_{k,s} c_{ij} \delta(t - t_j^s), \qquad (1)$$

where τ is their (common) membrane time constant, $I_i^0(t)$ is the excitatory external input current felt by neuron i, c_{ij} is 1 or 0 according to whether there is a connection from neuron j to neuron i, and t_j^s is the sth spike time of neuron j. The external input $I_i^0(t)$ represents input from elsewhere in the brain (e.g., the preceding stage in a sensory pathway), and as such is noisy itself. However, as the recurrent connections in the randomly diluted network generate dynamical noise on their own, this extrinsic noise does not have a big qualitative effect, so we take $I_i^0(t)$ to be constant here. We also take it to be uniform across the population.

3. Mean field theory

In mean field theory, one studies a single neuron for which the recurrent synaptic current (the last term in (1)) is replaced by a self-consistent Gaussian current with self-consistent mean and variance. It is Gaussian because of the central limit theorem: it is the sum of a large number of (what can be proved to be) independent contributions from the other neurons. Explicitly, the subthreshold membrane potential of this single neuron obeys

$$\dot{u} = -\frac{1}{\tau}u_i + \sqrt{K}[I^0 + I^1(t)]. \tag{2}$$

Whenever u reaches a threshold θ, it fires a spike and u is reset to zero. The effective recurrent current $I^1(t)$ has a mean $\langle I^1(t)\rangle = -Jr(t)$ proportional to the instantaneous firing rate $r(t)$ of the neuron and a covariance

$$\langle \delta I^1(t)\delta I^1(t')\rangle = \frac{J^2(1-K/N)}{K}C(t,t'), \tag{3}$$

where $C(t,t')=\langle[S(t)-r(t)][S(t')-r(t')]\rangle$ is the autocorrelation function of the neuronal firing $S(t)=\sum_s \delta(t-t^s)$. The K in the denominator in (3) comes from the averaging over K independent inputs, and the $1-K/N$ in the numerator is a correction for finite connection concentration K/N. It can be derived using the methods of Ref. [5].

This model cannot be solved exactly analytically, but it is simple to solve numerically, using the method first introduced for spin glasses by Eisfeller and Opper [3]. We start with a guess at the form of the mean and covariance function of the random current $I^1(t)$ and run a series of "trials", in each of which we integrate (2) for an independent realization of $I^1(t)$. By averaging the output of the neuron over the trials, we get an estimate of $r(t)$ and $C(t,t')$, which is used to generate new examples of $I^1(t)$ for another set of trials. This loop is then iterated until the statistics converge to self-consistency. In our calculations here, we used 10 000 trials per iteration and up to 30 iterations.

Each trial was 100 integration steps (which we call "milliseconds") long. We chose parameters $K=500$, $N=5000$, and $\tau=10$ ms. The external excitatory input $I^0(t)$ was constant at a low value (evoking a background firing rate of 3 Hz) during the first and last 10 ms. In the middle 80 ms, an additional "stimulus" input, which peaked about 15 ms after onset, was added. It typically evoked about 4–5 spikes, with peak rates around 100 Hz. The spike count distributions, PSTHs and covariance functions were computed for a number of values of the firing threshold θ.

4. Results

For all cases studied, the relaxation of the network to its state of balanced excitation and inhibition was very rapid, as expected, so the response tracked the time course of the excitatory external input closely. The overall response strengths vary only weakly with threshold: a factor of 16 difference between the smallest and largest threshold values produced only an 18% difference in mean response. This is because the increased firing that would be produced by lowering the threshold is largely compensated by the comcomitant increase in inhibition.

However, varying the threshold had a strong effect on the irregularity of the firing. Fig. 1 shows the spike count distributions for three threshold values with ratio 1:4:16.

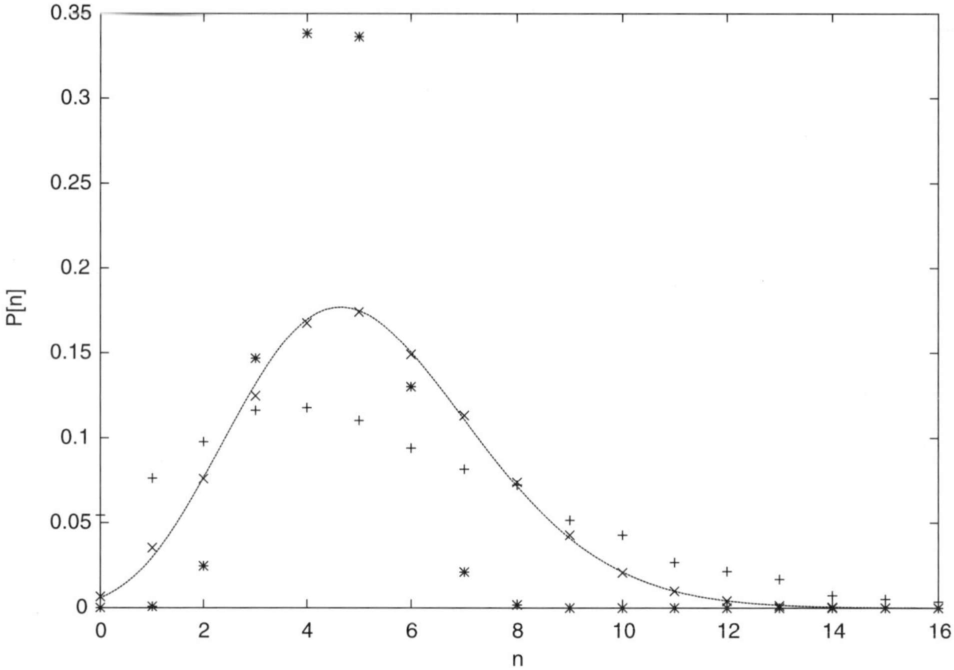

Fig. 1. Count statistics for threshold 0.125(+), 0.5(×) and 2.0(∗). The solid line shows the good fit to a Poisson distribution for the mean spike count (5.1574) obtained for threshold 0.5.

While the intermediate value fits a Poisson distribution well, the low threshold leads to an anomalously broad distribution (Fano factor (variance/mean ratio) $F = 2.3$) and the high one to an anomalously narrow one ($F = 0.25$).

These differences are also evident in the autocorrelation function $C(t, t')$. Fig. 2 shows $C(t', t)$ as a function of t for a fixed value of t' for the same three threshold values as in the preceding figure. For the lowest threshold, there is a "hill" centered around $t = t'$, while for the highest there is a valley. These are indicative of spike "bunching" and "antibunching", respectively, leading naturally to the higher and lower spike count fluctuations seen in Fig. 1. The intermediate threshold value shows very little correlation (apart from the delta-function peak at $t = t'$), consistent with the nearly Poisson count distribution found in this case.

5. Discussion

Measured Fano factors in visual and IT cortex [4] vary over a range at least as large as the one-order-of-magnitude difference between those for the smallest and largest

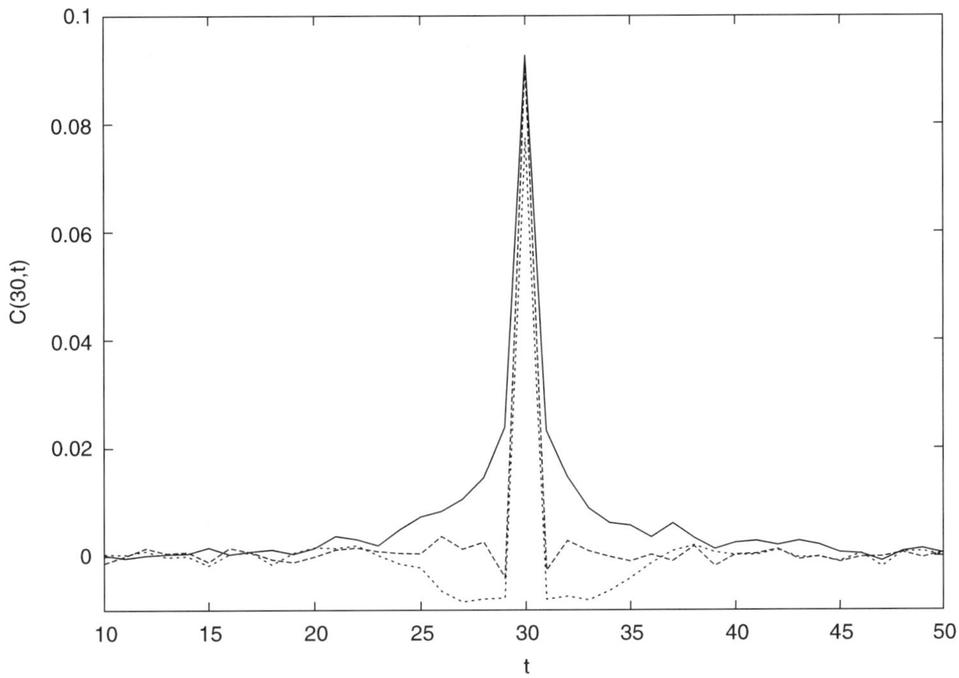

Fig. 2. Autocorrelation functions $C(30,t)$ for thresholds 0.125 (solid), 0.5 (longer dashes) and 2.0 (shorter dashes).

thresholds described above. Of course, threshold differences are not the only possible source of such response variability. We have also explored the effects of varying synaptic strengths, with similar results, and it seems likely that differences in a wide variety of single-neuron properties can have the same kind of effect.

Neurons in a local cortical network cannot all be expected to have the same threshold, and furthermore, their thresholds (or other parameters) may fluctuate (uncontrollably) from trial to trial. We have also found large variations in the Fano factor in a model where these fluctuations are assumed independent for different neurons and in different trials.

All these results are only suggestive, and more systematic work, both experiments and modeling, is called for. However, they do point to the possibility that the observed response variability of cortical neurons may be accounted for in terms of natural variations in properties from neuron to neuron and trial to trial.

References

[1] D.J. Amit, N. Brunel, Cereb. Cortex 7 (1997) 237–252.
[2] N. Brunel, J. Comput. Neurosci. 8 (2000) 183–208.

[3] H. Eisfeller, M. Opper, Phys. Rev. Lett. 68 (1992) 2094 2097.
[4] E.D. Gershon, M.C. Wiener, P.E. Latham, B.J. Richmond, J. Neurophysiol. 79 (1998) 1135–1144.
[5] R. Kree, A. Zippelius, Phys. Rev. A 36 (1987) 4421–4427.
[6] P.E. Latham, B.J. Richmond, P.G. Nelson, S. Nirenberg, J. Neurophysiol. 83 (2000) 808–827.
[7] C.F. Mari, Phys. Rev. Lett. 85 (2000) 210–213.
[8] C. van Vreeswijk, H. Sompolinsky, Science 274 (1996) 1724–1726;
C. van Vreeswijk, H. Sompolinsky, Neural Comput. 10 (1998) 1321–1371.

Available at
www.ComputerScienceWeb.com
POWERED BY SCIENCE DIRECT

ELSEVIER

Neurocomputing 52–54 (2003) 793–798

NEUROCOMPUTING

www.elsevier.com/locate/neucom

Bilateral competitive processing of visual spatial attention in the human brain

Claus C. Hilgetag[a,*], Rolf Kötter[b], Hugo Théoret[c], Joseph Claßen[d], Alexander Wolters[e], Alvaro Pascual-Leone[c]

[a]*School of Engineering and Science, International University Bremen, Campus Ring 6, Room 116, 28759 Bremen, Germany*

[b]*Voigt Brain Research Institute and Institute of Anatomy II, Heinrich Heine University, Universitätsstr. 1, 40225 Düsseldorf, Germany*

[c]*Beth Israel Deaconess Medical Center and Harvard Medical School, Department of Neurology, 330 Brookline Avenue, Boston, MA 02115, USA*

[d]*Department of Neurology, Bavarian Julius Maximilians University, Josef-Schneider-Str. 11, 97080 Würzburg, Germany*

[e]*Department of Neurology, Rostock University, Gehlsheimer Str. 20, 18147 Rostock, Germany*

Abstract

Visual spatial attention is an essential brain function that is produced through the interactions of several cortical and subcortical regions. Using the reversible deactivation technique of transcranial magnetic stimulation (TMS) in combination with a visual stimulus detection task, we demonstrated that parietal as well as occipito-parietal cortices in the human brain contribute to spatial attentional behavior. The functional role of left- and right-hemispheric regions appeared to be mirror-symmetric, although the strength of contributions from left and right cortices might differ. Reaction times for baseline and experimental conditions suggested that TMS interfered with an early stage of attentional processing. The experiments also demonstrated an ipsilateral enhancement of spatial attention after unilateral TMS. This observation supports a theoretical model for inter-hemispheric competition in the attentional network.

Keywords: Transcranial magnetic stimulation (TMS); Hemineglect; Extinction; Parietal cortex; Occipital cortex; Inter-hemispheric competition

* Corresponding author.
E-mail address: c.hilgetag@iu-bremen.de (C.C. Hilgetag).

0925-2312/03/$ - see front matter © 2002 Elsevier Science B.V. All rights reserved.
doi:10.1016/S0925-2312(02)00776-2

1. Introduction

Mechanisms of spatial attention underlie visual perception as well as many visual behaviors, such as search, orienting or avoidance reactions. Several cortical and subcortical regions have been implicated in attentional processing (reviewed in [21]). Knowledge about the extent and function of this attentional network in the human brain stems mainly from patients with localized brain damage who show deficits in redirecting spatial attention, mostly towards the contralesional space. Depending on whether the deficits predominantly appear for unilateral stimuli presented in contralateral space or for bilateral stimuli, such syndromes have been classified as neglect or extinction, respectively [22].

Using the "virtual lesion" technique of TMS, e.g., [15,18], we aimed to study contributions and interactions of different cortical stages in the attentional network more systematically. Reversible impairment of cortical processing by TMS can be created in an event-related way, or by prestimulation of cortical sites with 1 Hz repetitive TMS (rTMS) [16]. While the mechanisms underlying TMS deactivation are not yet fully understood, reversible deactivation approaches avoid a number of conceptual difficulties posed by traditional studies of patients with permanent lesions, such as an absent baseline before lesion onset, plastic reorganization of the brain after damage and the frequent non-systematic and wide-spread behavioral impact of permanent lesions.

We used TMS deactivation approaches to (i) identify cerebral cortical regions contributing to spatial attention and (ii) explore the specific contribution of different cortical regions in the attentional network. Our experiments were guided by the knowledge about the locations of lesions in neglect patients, earlier TMS studies [14], as well as the predictions from a theoretical model. This model [4–6], which was derived originally from structural information and behavioral data of studies in the cat, e.g., [19,10], suggests that spatial attention is based on a topographic neural representation of visual space and bilateral, inter-hemispheric competition. A simple prediction resulting from this concept is the disinhibition of structures in the unimpaired hemisphere through deactivation of structures in the impaired hemisphere. Such a functional release should also manifest itself in a measurable behavioral enhancement.

2. Methods

Subjects: Experiments were in line with NIH guidelines for human studies and were approved by the local Ethics Review Boards at Beth Israel Deaconess Medical Center, Boston and the Department of Neurology, Rostock University. Subjects belonged to two sets of healthy volunteers, consisting of 7 and 10 subjects, respectively. All subjects but one were male; the majority of them were right-handed.

Stimulus detection test: In a simple behavioral paradigm, subjects had to detect small rectangular stimuli presented briefly (for 40 ms) on a cathode ray tube (CRT) monitor either unilaterally in the left or right visual periphery (lateral eccentricity approximately 25°), or bilaterally in both fields. Subjects' performance in stimulus detection was assumed to reflect the efficiency of their attentional mechanisms. Stimuli consisted of small black rectangles and stimulus presentation was interspersed with catch trials in which no stimuli were present. Subjects used their dominant hand to press one of the three keys of a response box to indicate if they had detected a left, right, or bilateral stimulus. Stimulus sizes were adapted to the subjects' visual acuity and attentional capability to be at peri-threshold stimulus strength (with a resulting median level of correct stimulus detection around 20% for small stimuli and 70% for slightly larger ones), so that correct detection during the experimental trials could vary up or down. For further details see [7].

TMS protocols: We employed two different TMS approaches. The first used an event-related design (that is, magnetic stimulation was synchronized with stimulus onset), whereas the second employed a 'chronic' repetitive TMS approach (stimulating the target site with 90% motor threshold at 1 Hz for 10 min before experimental testing). This repetitive paradigm has been shown to result in behavioral effects consistent with transient suppression of cortical excitability, e.g., [1,11,9,20]. Consequently, we tested spatial detection performance in a block design, at baseline and after rTMS, during the window of cortical dysfunction.

TMS locations: We explored the contributions of two cortical regions to visual spatial attention, (i) occipito-parietal cortex and (ii) parietal cortex. Locations for TM stimulation were determined within the 10–20 relative EEG coordinate system. Stimulation sites in parietal cortex were verified by post-experimental structural MRI. In the first, chronic approach, seven subjects were tested after right parietal (P4) rTMS, six subjects after left parietal (P3) rTMS. In the event-related TMS approach, ten subjects were tested on bilateral sites in occipito-parietal cortex. Here the location of stimulation was determined as the maximally effective point for inducing contra-lesional hemi-extinction (at 90% phosphene threshold). As a control, we also performed sham experiments and stimulated right primary motor cortex (M1).

3. Results

No significant changes in the subjects' detection performance were observed for sham stimulation or stimulation of M1. For both sets of deactivated cortical regions, however, we found a trend to diminished detection of unilateral stimuli opposite to the cortical TMS site. This trend was significant in the acute TMS approach. The contralesional deficit was even more pronounced for the contralateral stimulus in bilateral stimulus pairs, and led to a significant decrease in correct responses for bilateral stimuli after functional deactivation of occipito-parietal as well as parietal cortical regions. Taken together, these behavioral effects present a picture similar to the clinical syndromes of contralesional hemi-neglect and hemi-extinction.

We also found enhanced detection of unilateral stimuli on the side of the TM stimulation, both in terms of error rates as well as reaction times (Fig. 1). The amplification trend was seen across both sets of studies and TMS locations; however, the effects only reached statistical significance for the chronic rTMS approach. The general response trends for left and right stimulation indicated symmetrical effects, although not all

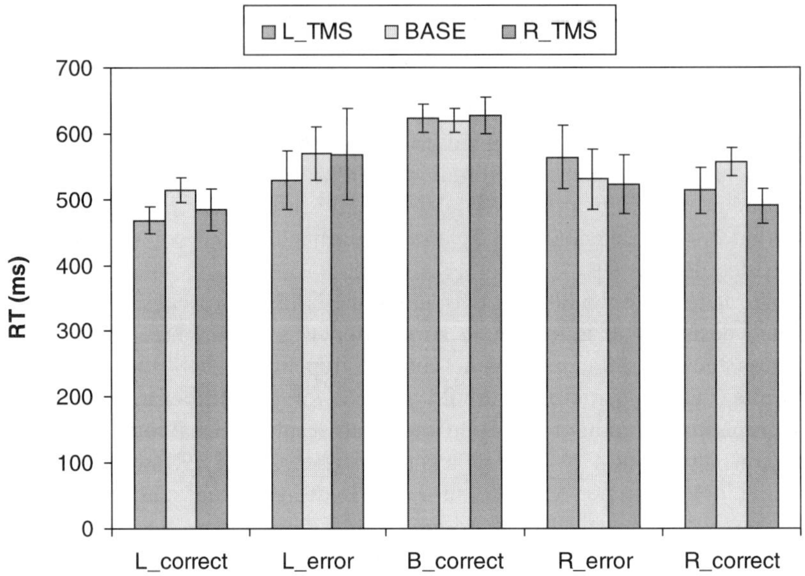

Fig. 1. Average reaction times and standard error for correct bilateral, as well as for correct and incorrect unilateral responses, at baseline and after rTMS of parietal regions P3 ("L_TMS") and P4 ("R_TMS") in the chronic rTMS approach. "L_correct", "B_correct" and "R_correct" denote reaction times for the respective correct detection responses. "L_error" and "R_error" indicate reaction times of incorrect left and right detection responses for presented bilateral stimuli (involving extinction of the contralateral stimulus). Correct unilateral responses were significantly faster than bilateral responses ($p < 0.05$). After rTMS, incorrect responses for only the ipsilateral stimulus in bilateral stimulus pairs were also significantly faster than correct bilateral responses. Left- and right-hemispheric TMS generally introduced mirror-symmetric reaction time effects, and speeded up detection of ipsilateral stimuli.

observations during the two TMS approaches reached statistical significance, and effects were generally more pronounced for right-hemispheric stimulation. Different stimulus conditions and the impact of TMS deactivation could also be distinguished in terms of reaction times (presented in Fig.1 for the rTMS approach).

Results obtained during the event-related TMS approach were more difficult to interpret due to errors caused by (i) interference of the TMS discharge with the CRT stimulus display as well as (ii) TMS-induced transient visual field effects (such as phosphenes and scotomas) experienced by the subjects.

4. Conclusions

Our results provide support for the concept of a distributed, bilateral, competitive network mediating visual spatial attention in humans. Different visual cortical stages, such as occipito-parietal and parietal cortex, contribute to the functioning of this network, each probably within the context of their individual functional roles. It appears likely that even more stations are involved, e.g., [13]. Largely mirror-symmetric effects for the deactivation of left- and right-hemispheric sites indicated the principal similarity of attentional processing in both hemispheres, although processing may be partly lateralized to the right hemisphere.

The results also suggested that the behavioral consequences resulting from the local impact of event-related TMS and 1 Hz rTMS are comparable, even though at least the latter approach also results in global changes of neural activation patterns [12].

The obtained reaction times allowed us to address another question: Are incorrect unilateral responses for bilateral stimuli more similar to correct unilateral detection or to correct bilateral detection? In other words, are both stimuli of a bilateral pair first processed by the brain after which the response to one of them is omitted, or is the initial processing of one stimulus suppressed altogether? The present results lend support to the latter hypothesis, as incorrect unilateral responses to bilateral stimuli were found to be significantly faster than correct bilateral responses. The finding suggests an interference of parietal cortical TMS with an early stage of attentional processing, and is in line with simple feedforward models of evoked attention, such as [17].

Future experiments should aim to replicate the results obtained for event-related TMS of occipito-parietal cortex, due to the technical difficulties we experienced. Additional experiments could use either a chronic rTMS paradigm or a modified event-related approach that minimizes interference of the TMS field with the stimulus display through the use of a projection system. We are, however, encouraged in our interpretation of the current results by independent findings [3,2] which also suggest involvement of occipito-parietal regions in attentional processing.

Enhanced ipsilateral stimulus detection was seen in error rates as well as reaction times and was found across studies and locations, hinting on the general nature of competitive inter-hemispheric interactions in spatial attention. It will be an interesting challenge for the future to determine the exact structural basis of the competitive inter-hemispheric interaction. Is it being mediated directly between the homotopic parietal and occipital sites? Alternatively, competition might be created via a number of intermediate stages (even subcortical ones, as in the cat, [5,19]), or just between some of the cortical pairs. These effects could then be transmitted ipsilaterally to other cortical regions in the network [13].

Functional competition between different brain structures, as seen here, might be a general principle of brain function [23] and may help to explain the 'paradoxical' behavioral enhancement or recovery observed after various brain lesions [8]. Future studies will have to describe systematically the specific contributions that different neural components make in the widely distributed network for spatial attention, and should explore their functional interactions.

Acknowledgements

Supported in part by the Wellcome Trust, DFG, and NIH.

References

[1] R. Chen, J. Classen, C. Gerloff, P. Celnik, E.M. Wassermann, M. Hallett, L.G. Cohen, Depression of motor cortex excitability by low-frequency transcranial magnetic stimulation, Neurology 48 (1997) 1398–1403.

[2] S. Ferber, H.O. Karnath, Size perception in hemianopia and neglect, Brain 124 (2001) 527–536.

[3] G.R. Fink, J. Driver, C. Rorden, T. Baldeweg, R.J. Dolan, Neural consequences of competing stimuli in both visual hemifields: a physiological basis for visual extinction, Ann. Neurol. 47 (2000) 440–446.

[4] C.C. Hilgetag, Spatial neglect and paradoxical lesion effects in the cat—a model based on midbrain connectivity, NeuroComputing 32–33 (2000) 793–799.

[5] C.C. Hilgetag, R. Kötter, M.P. Young, Inter-hemispheric competition of sub-cortical structures is a crucial mechanism in paradoxical lesion effects and spatial neglect, Prog. Brain Res. 121 (1999) 121–141.

[6] C.C. Hilgetag, S.G. Lomber, B.R. Payne, Neural mechanisms of spatial attention in the cat, NeuroComputing 38–40 (2001) 1281–1287.

[7] C.C. Hilgetag, H. Theoret, A. Pascual-Leone, Enhanced visual spatial attention ipsilateral to rTMS-induced 'virtual lesions' of human parietal cortex, Nat. Neurosci. 4 (2001) 953–957.

[8] N. Kapur, Paradoxical functional facilitation in brain-behaviour research. A critical review, Brain 119 (1996) 1775–1790.

[9] S.M. Kosslyn, A. Pascual-Leone, O. Felician, S. Camposano, J.P. Keenan, W.L. Thompson, G. Ganis, K.E. Sukel, N.M. Alpert, The role of area 17 in visual imagery: convergent evidence from PET and rTMS, Science 284 (1999) 167–170.

[10] S.G. Lomber, B.R. Payne, Removal of 2 halves restores the whole—reversal of visual hemineglect during bilateral cortical or collicular inactivation in the cat, Visual Neurosci. 13 (1996) 1143–1156.

[11] F. Maeda, J.P. Keenan, J.M. Tormos, H. Topka, A. Pascual-Leone, Interindividual variability of the modulatory effects of repetitive transcranial magnetic stimulation on cortical excitability, Exp. Brain Res. 133 (2000) 425–430.

[12] F.M. Mottaghy, G. Schlaug, M. Gangitano, B.J. Krause, A. Pascual-Leone, Functional connectivity of the mid-dorsolateral prefrontal cortex (MDLPFC) and temporal dynamics of rTMS induced effects as revealed by fMRI, Society of Neuroscience Abstracts 525.16, San Diego, 2001.

[13] M. Oliveri, P.M. Rossini, R. Traversa, P. Cicinelli, M.M. Filippi, P. Pasqualetti, F. Tomaiuolo, C. Caltagirone, Left frontal transcranial magnetic stimulation reduces contralesional extinction in patients with unilateral right brain damage, Brain 122 (1999) 1731–1739.

[14] A. Pascual-Leone, E. Gomez-Tortosa, J. Grafman, D. Alway, P. Nichelli, M. Hallett, Induction of visual extinction by rapid-rate transcranial magnetic stimulation of parietal lobe, Neurology 44 (1994) 494–498.

[15] A. Pascual-Leone, V. Walsh, J. Rothwell, Transcranial magnetic stimulation in cognitive neuroscience—virtual lesion, chronometry, and functional connectivity, Curr. Opin. Neurobiol. 10 (2000) 232–237.

[16] A. Pascual-Leone, E. Wasserman, N. Davey, Handbook of Transcranial Magnetic Stimulation, Oxford University Press, Oxford, 2002.

[17] A. Pouget, T.J. Sejnowski, A new view of hemineglect based on the response properties of parietal neurones, Philos. Trans. R. Soc. London B Biol. Sci. 352 (1997) 1449–1459.

[18] R. Rafal, Virtual neurology, Nat. Neurosci. 4 (2001) 862–864.

[19] J.M. Sprague, Interaction of cortex and superior colliculus in mediation of visually guided behavior in the cat, Science 153 (1966) 1544–1547.

[20] H. Theoret, J. Haque, A. Pascual-Leone, Increased variability of paced finger tapping accuracy following repetitive magnetic stimulation of the cerebellum in humans, Neurosci. Lett. 306 (2001) 29–32.

[21] A.W. Toga, F.S.J. Frackowiak, J.C. Mazziotta, Action and visio-spatial attention: neurobiological bases and disorders (Special issue) Neuroimage 14 (Issue 1, Part 2), (2001) 1–146.

[22] G. Vallar, Spatial hemineglect in humans, Trends Cognitive Sci. 2 (1998) 87–97.

[23] V. Walsh, A. Ellison, L. Battelli, A. Cowey, Task-specific impairments and enhancements induced by magnetic stimulation of human visual area V5, Proc. R. Soc. London B Biol. Sci. 265 (1998) 537–543.

Available at
www.ComputerScienceWeb.com
POWERED BY SCIENCE DIRECT®

ELSEVIER

Neurocomputing 52–54 (2003) 799–804

NEUROCOMPUTING

www.elsevier.com/locate/neucom

Modeling neural spatiotemporal behavior

David Horn[a,*], Brigitte Quenet[b], Gideon Dror[c], Orit Kliper[d]

[a] *School of Physics and Astronomy, Tel Aviv University, Tel Aviv 69978, Israel*
[b] *Lab. d'Electronique, Ecole Superieure de Physique et Chimie Industrielles, Paris 75005, France*
[c] *Department of Computer Science, The Academic College of Tel-Aviv-Yaffo, Tel Aviv 64044, Israel*
[d] *School of Mathematical Sciences, Tel Aviv University, Tel Aviv 69978, Israel*

Abstract

We study some aspects of the dynamic neural filter (DNF), a recurrent network that produces spatiotemporal sequences in reaction to sets of constant inputs. The biological motivation for this study came from the observation of spatiotemporal patterns in the locust antennal lobe. Some of the aspects of these results can be reformulated and characterized by the DNF. Studying deterministic dynamics we find differences between low and high numbers of neurons. For low numbers there exists clear correlation between distances in input space and edit distances of spatiotemporal sequences. For large numbers of neurons we observe divergence between close-by spatiotemporal sequences. Nonetheless neuronal correlations survive for small changes in input space.

Keywords: Temporal coding; Dynamic neural filter; Recurrent network; Olfaction

1. Introduction

An interesting issue in neuroscience is the question of spatiotemporal coding. Its existence has been demonstrated [7] in the locust olfactory system, where the spatiotemporal behavior of projection neurons encodes the odor presented to its receptor neurons. This transformation from odor-input to spatiotemporal activity occurs in the antennal lobe, that is the first module of the olfactory system. This system may therefore be regarded as a dynamic neural filter that turns spatial information distributed over its many glomeruli, that are fed by the receptor neurons, to specific spatiotemporal outputs. It is interesting to point out that, although this is a complicated biological

* Corresponding author.
E-mail address: horn@post.tau.ac.il (D. Horn).

0925-2312/03/$ - see front matter
doi:10.1016/S0925-2312(02)00780-4

system, it has an important simplifying feature that allows it to be represented by a simplified model of mathematical neurons, i.e. the fact that the activity of the projection neurons is limited to temporal bins defined by an oscillatory local field potential. Hence a model with binary neurons obeying Hopfield–Little dynamics [4] can provide a valid first-order approximation of the spatiotemporal behavior of the system. We have presented this model in a previous work [6] and demonstrated how it can be used to generate the spatiotemporal behavior of projection neurons observed by Wehr and Laurent [7]. An extensive study of the model was presented in [5]. Here we wish to expand on several aspects of the model that can be of particular interest to computational neuroscience.

2. The dynamic neural filter

The dynamics of the model has the following structure:

$$n_i(t+1) = H(h_i(t+1)) = H\left(\sum_j w_{ij} n_j(t) + R_i - \theta_i\right), \tag{1}$$

where w_{ij} is the synaptic coupling matrix, R_i is the external constant input (specifying odor activation) and θ_i is the threshold. H is the Heaviside step function taking the values 0 for negative arguments and 1 for positive ones. For simplicity we choose w_{ij} and R_i as integers, fixing $\theta_i = \frac{1}{2}$.

Consider the problem presented by Table 1, where we observe $K = 6$ series, representing the results of six inputs (e.g. odors), over $T = 4$ time steps in a problem of $N = 5$ neurons. The first two neurons correspond to the observed ones of Wehr and Laurent.

The modeler is challenged to find a matrix **w** and six input vectors R that solve this problem. We have presented a solution to this problem in [5] based on the perceptron algorithm. The principle is to generalize the N-dimensional neural space to an $N+K$ dimensional one, such that each one of the K series is presented by one 1 and $K-1$ null entries in these new axes. In this new space we define for each neuron i, a vector of perceptron weights $\vec{w}^i$

$$(\vec{w}^i)_j = w_{ij} \quad \text{for } j = 1, \ldots, N, \quad (\vec{w}^i)_{N+k} = R_i^k - \theta_i \quad \text{for } k = 1, \ldots, K. \tag{2}$$

Table 1
Six spatiotemporal sequences defined for four time-steps in a system with five neurons

Time/code	1	2	3	4	5	6
1	1 1 0 0 0	1 0 0 0 0	1 1 1 0 0	1 0 0 0 0	1 0 1 1 0	1 0 0 0 0
2	1 1 0 0 1	1 1 0 0 1	1 1 1 1 0	1 0 1 0 0	1 0 0 0 0	1 1 1 0 0
3	1 1 0 1 1	1 1 0 1 1	0 1 1 1 0	0 1 1 0 0	1 1 1 0 0	0 1 1 1 1
4	0 0 0 1 0	0 0 0 1 0	0 0 1 1 0	0 1 1 1 0	1 1 1 1 1	0 0 0 1 0

We assume that the K series start from an initial null state, and we use the perceptron algorithm to construct these vectors $\vec{w}^i$, whose $(N+k)$th component includes the information about R_i^k.

Thus we obtain a synaptic weight-matrix, that serves to define a dynamic neural filter (DNF), as well as K inputs that lead to the sequences of Table 1. But this can work only if each column in Table 1 does not include repetitive states, and if no xor-like contradictions of perceptron dynamics occur. If such situations occur we have to add hidden neurons. In fact this is the case with the Wehr–Laurent series, described by the first two neurons in Table 1. Since these sequences cannot be generated by an $N=2$ DNF, we added three hidden neurons in the example of Table 1, leading to a system that can be generated by a DNF. In general, for any problem of K binary sequences with T time steps one can find hidden neurons that allow for a DNF representation. A solution is guaranteed for high values of T and K if the total number N of neurons (observed + hidden) obeys

$$N \geqslant \tfrac{1}{2}K(T-2), \tag{3}$$

which can be derived from the Cover limit [1].

3. Spatiotemporal distances

The DNF can be viewed as a generative tool for production of spatiotemporal sequences. Every such sequence can also be used to reconstruct a DNF that can serve as its generator. This relationship can be exploited to serve as a measure in the generalized space of all spatiotemporal sequences.

In [5] we have numerically studied an $N=5$ model and observed that, for a fixed $\mathbf{w}$, distances in R correlate well with edit distances in the spatiotemporal sequence. This is easily demonstrated in Table 2, using a binary representation for every state of the sequence in the form $1+\sum_{i=1}^{N} n_i 2^{N-i}$. These states are then represented by numbers ranging from 1 (the null state) to 2^N when all neurons are active.

These sequences are generated by changing one R value. The initial state was always chosen to be the null state. Note that this set of sequences contains several 5-cycles. This is to be expected from a synaptic matrix that is asymmetric, i.e. far from being

Table 2
A list of six neighboring sequences in a system with five neurons, displayed over seven time steps

$t=0$	1	2	3	4	5	6	7
1	17	22	6	8	3	17	22
1	17	22	14	8	3	17	22
1	17	22	14	16	3	17	22
1	17	30	16	3	17	30	16
1	25	30	16	3	17	30	16
1	25	30	16	11	3	17	30

either symmetric or antisymmetric. Asymmetry may be defined by the parameter $\alpha = tr(\mathbf{w} \cdot \mathbf{w})/tr(\mathbf{w} \cdot \mathbf{w}^{\mathrm{T}})$ whose extreme values 1 and -1 correspond to full symmetry or antisymmetry, respectively. The matrix that has generated Table 2 and, incidentally, provides a solution to the problem of Table 1, has $\alpha = -0.41$. Large cycles are to be expected when $\alpha \approx 0$ [3].

So far we have discussed examples with small numbers of neurons. It is of interest to ask what happens if the number of neurons in our system is higher, e.g. $N = 50$. Investigating such a system with $\alpha \approx 0$ we find that close-by R values can generate divergent spatiotemporal sequences. In one such example, we varied R_1 and produced 27 different sequences that were tested over 1000 time steps. None of the sequences has shown any periodicity within this large range. Moreover, two sequences generated by neighbors in R space, coincided for several tens or hundreds of time steps until different states appeared. From this point onwards the states in the two sequences formed completely disjoint sets, i.e. complete divergence of the sequences occurred.

Obviously, there seems to be a large difference between small and large N values. This is due to the fact that the space of all available states increases exponentially as 2^N. Hence, for large N, once a change occurs in some state in the sequence, it becomes almost impossible to return to the original sequence.

4. Neuronal correlations

It is interesting to note that divergence of spatiotemporal representations was observed experimentally by Friedrich and Laurent [2] in the olfactory bulb of zebrafish. The authors have studied the response of this system to similar odors, characterized by small changes in molecular structures of the relevant chemicals. One of their interesting results is that the correlation between temporal patterns of similar odors over the mitral cells reduces with time (between the first and second 500 ms after odor presentation). This leads us to wonder whether in our large N system we can find transitions in time from large to medium correlations, rather than abrupt transitions from equality of sequences to maximal inequality.

The answer is affirmative. To understand it we have to realize that the relevant volume of R space is limited by constraints [5]

$$-\sum_j w_{ij} H(w_{ij}) \leqslant R_i \leqslant -\sum_j w_{ij} H(-w_{ij}) + 1. \tag{4}$$

Outside this range the dynamics become trivial because the input determines directly the neuronal values. The center of R space is the region where the system leads to large cycles (provided $\mathbf{w}$ is asymmetric) and where the chaotic-like phenomena of total divergence occur. As we move out of the center of this space, small changes in R may lead to a divergence, but finite correlations survive. An example of this behavior is shown in Fig. 1 which describes the Hamming distance between two neighboring sequences, that were generated by R values off the center, in an $N = 40$ problem. Once divergence occurred, none of the states in one system appeared in the other, nonetheless

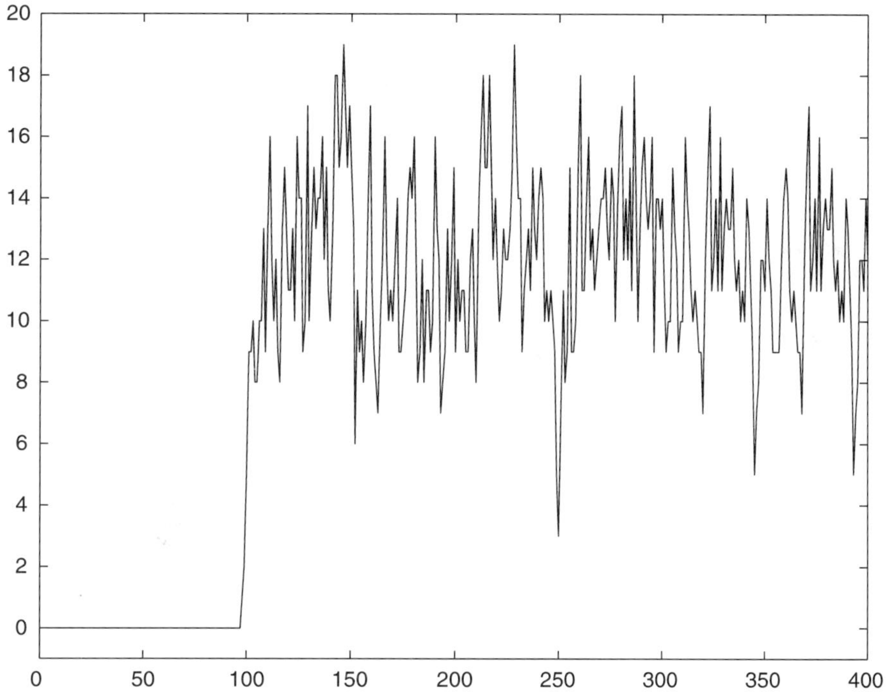

Fig. 1. Hamming distance between two neighboring sequences, in an $N=40$ problem, plotted vs. the temporal evolution of the sequences.

a finite correlation survived as indicated by the fact that the mean Hamming distance is around 12 and not 20, the number expected if no correlation had survived. In the olfaction problem, small changes in R may be thought of as small changes in the type or concentration of odor. We may then expect the latter to be traceable through neuronal correlations even if their spatiotemporal patterns look quite different.

5. Summary

We saw that the DNF can serve several purposes. For low N values it serves as a simple model for generating a set of binary spatiotemporal sequences. Any given set can be accommodated by a DNF. The latter can be explicitly constructed using a generalized perceptron rule, but may necessitate invoking hidden neurons. For large N, the DNF can be used as a simple testbed for discussing experimental observations of neuronal correlations in an almost chaotic regime. Both cases were exemplified by results of spatiotemporal behavior observed in olfaction.

References

[1] T.M. Cover, Geometrical and statistical properties of systems of linear inequalities with applications in pattern recognition, IEEE Trans. Electron Comput. 14 (1965) 326–334.

[2] R.W. Friedrich, G. Laurent, Dynamic optimization of odor representations by slow temporal patterning of mitral cell activity, Science 291 (2001) 889–894.
[3] H. Gutfreund, D.J. Reger, P.A. Young, The nature of attractors in an asymmetric spin glass with deterministic dynamics, J. Phys. A 21 (1988) 2775–2797.
[4] P. Peretto, An Introduction to the Modeling of Neural Networks, Cambridge University Press, Cambridge, MA, 1992.
[5] B. Quenet, D. Horn, The dynamic neural filter: a binary model of spatiotemporal coding, Neural Comput. 15 (2003), in press.
[6] B. Quenet, D. Horn, G. Dreyfus, R. Dubois, Temporal coding in an olfactory oscillatory model, Neurocomputing 38–40 (2001) 831–836.
[7] M. Wehr, G. Laurent, Odour encoding by temporal sequences of firing in oscillating neural assemblies, Nature 384 (1996) 162–166.

Available at
www.ComputerScienceWeb.com
POWERED BY SCIENCE DIRECT®

ELSEVIER

Neurocomputing 52–54 (2003) 805–811

NEUROCOMPUTING

www.elsevier.com/locate/neucom

Multimodality in a dynamic feedback network with stochastic input☆

M. Jazayeri[a,*], F.K. Skinner[b,c,d]

[a] *Center for Neural Science, New York University, 4 Washington Place Rm 809, New York 10003, NY, USA*

[b] *Toronto Western Research Institute, University Health Network, Toronto, Ont., Canada*

[c] *Departments of Medicine (Neurology) and Physiology, University of Toronto, Toronto, Ont., Canada*

[d] *Institute of Biomaterials and Biomedical Engineering, University of Toronto, Toronto, Ont., Canada*

Abstract

Different rhythmic activities in CA1 characterize the neuronal correlate of several behavioral states. Recently, in an in vitro preparation of whole hippocampus, generation of spontaneous slow rhythms, similar to in vivo hippocampal EEGs, has been reported. We previously proposed a dynamical feedback mechanism where populations of synchronized interneurons entrain spatially dynamic subpopulations of pyramidal cells. Here, using a stochastic phenomenological model, we show that the rhythm is inherently multimodal and that the number of excitatory cells as well as the excitability of the interneurons allow the network to differentially amplify/suppress the modal structure of the emergent rhythm.

Keywords: Feedback; Network; Stochastic; Multimodal; Rhythm

1. Introduction

Different population rhythms, as reflected by spontaneous field potentials, are present in several brain structures. These rhythms correlate with certain behavioral states and as such, their mechanisms have been the subject of intense investigations. It is not clear if these rhythms are active players in neural processing or concurrent epiphenomena emerging from the neuronal substrate while the underlying processing is taking place.

☆ This work was supported by NSERC and DCIEM of Canada.

* Corresponding author.
E-mail address: mjaz@cns.nyu.edu (M. Jazayeri).

0925-2312/03/$ - see front matter © 2002 Elsevier Science B.V. All rights reserved.
doi:10.1016/S0925-2312(02)00787-7

Neuronal networks in the hippocampus exhibit various oscillatory patterns including sharp waves, delta (< 4 Hz), theta (5–10 Hz), gamma (20–80 Hz) and ultrafast (140 –200 Hz) rhythms [2,3,5,6,8,10–12] which are likely the result of synchronous activity across populations of interacting excitatory and inhibitory neurons.

Recently in an in vitro preparation of the whole hippocampus, spontaneous field rhythms in the slow frequency band ($0.5 < 4$ Hz) have been recorded in the CA1 region. These frequencies closely resemble specific EEG recordings in behaving rats and might share similar neural mechanisms [10].

Based on the phase relationships of intra- and extracellular recordings, the synchronous firing of the recorded interneurons, the hyperpolarizations of pyramidal cells and the importance of the $AMPA$ and $GABA_A$ currents in the mediation of this rhythm, a Dynamical Feedback Mechanism was proposed [7]. Using the space-clamped version of the nonlinear Hodgkin–Huxley differential equations along with additive stochastic noise (a standard four-dimensional temporally homogenous Markov process), the postulated mechanism underlying the observed rhythmic activity was investigated [7].

In the current study, after characterizing the network mechanism using numerical simulations, a stochastic phenomenological model will be used to investigate the multimodal characteristics of the network in a stochastic environment. Moreover, the importance of the multimodality coupled with the spatial dynamicity inherent in the network behavior will be discussed. We will argue that the rhythm, by providing a specific temporal context, can affect the spatial content of the neuronal ensemble (population coding paradigm) and therefore can have important implications in neural information coding.

2. Model

The network is composed of a single inhibitory cell (I-cell) and multiple excitatory cells (E-cells). From the experimental data [10], it is known that the interneurons display effectively synchronous behavior so that our single I-cell represents a population of synchronized interneurons or GABAergic cells.

The excitatory synaptic connections between the pyramidal cells of the CA1, except for the nearby interconnections, are known to be sparse [9]. Due to their proximity, the nearby pyramidal cells are approximately sensing similar inputs, therefore it is further assumed that during the rhythmic activity, each group of these nearby pyramidal cells can be represented as a single E-cell with no interconnection.

Contrasted to pyramidal-pyramidal connection, a disproportionate number of CA1 pyramidal cells reciprocally synapse on to nearby interneurons [9,4]. In the model, global reciprocal connection is assumed between all E-cells and the I-cell.

All cells are modeled as single compartment Hodgkin–Huxley (HH) units with sodium, potassium, and leakage currents.

$$C\,\frac{\mathrm{d}V}{\mathrm{d}t} = i_{\mathrm{bias}} - g_{Na}\,m^3\,h\,(V - E_{Na}) - g_K\,n_K^4\,(V - E_K) - g_{\mathrm{leak}}\,(V - E_{\mathrm{leak}}) - i_{\mathrm{syn}},$$

where i_{syn} and i_{bias} represent the synaptic currents. The former is mediated by the synaptic connections in the network and the latter, by nonspecific background activity provided by distant brain regions and is modeled with a stochastic Gaussian distributed fluctuating current, reflecting the nonspecific nature of this component; i.e. $i_{bias} \sim Gaussian(\mu_{bias}, \sigma_{bias})$.

For the local synaptic connections, the EPSPs are modeled as fast $AMPA$ receptors. With regard to the IPSPs, the slow recurring hyperpolarizations of the pyramidal cells suggests a biphasic IPSP, which in our approach is modeled by $GABA_{A_{fast}}$ and $GABA_{A_{slow}}$. These two GABAergic currents have been reported to coexist in the CA1 region [1] and may be important in mediating the feedback mechanism in CA1.

All local synaptic connections are modeled as two-state kinetic markov processes. The total synaptic input is the linear summation of individual synaptic currents; i.e. $i_{syn} = \Sigma_{i,j} g_{syn}\, s\,(V - E_{syn})$. The kinetics of the dynamical variables as well as their associated parameters are all reported in previous work [7].

2.1. Dynamic feedback mechanism

In the simulations, due to their nonspecific background current (i_{bias}), multiple E-cells fire to excite the I-cell. Subsequently, because of the biphasic inhibitory feedback signal, all E-cells get hyperpolarized and the I-cell stops firing. After the effect of the inhibition wears off, E-cells, because of their basal activity start depolarizing up to the point where they start firing again. Another cycle starts anew when the number of firing E-cells reaches the point where they can drive the I-cell to fire. By changing the maximum excitatory synaptic conductance g_{AMPA} in a controlled way, the minimum number of firing E-cells that are needed to make the I-cell fire, n, can be controlled. Since n signifies the minimum number of firing E-cells needed to drive the I-cell above threshold, it can be interpreted as the level of excitability in the I-cell.

An epoch of the network behavior is illustrated in Fig. 1, where a total of 5 E-cells are reciprocally connected to the I-cell ($N = 5$), yet in each cycle the firing of only one E-cell is needed to make the I-cell fire ($n = 1$).

3. Stochastic phenomenological model

The aforementioned mechanism is essentially dependent on two factors. One is the stochastic firing of individual E-cells after inhibition which is achieved through the nonspecific background synaptic activity and the other, is the excitation of the I-cell due to firing of multiple E-cells. It was further discussed that the excitability of the I-cell can be measured by the minimum number of firing E-cells needed to actuate the I-cell; i.e. n. We take this implicit parameter, n, in conjunction with the total number of E-cells, N, to build a stochastic phenomenological model which captures the essence of the dynamical feedback mechanism.

In this model, to satisfy the second condition, i.e. excitability of the I-cell, it is necessary and sufficient to have an n that is less than or equal to the total number of E-cell in the network, N ($n = 1, 2, 3, \ldots \leqslant N$). The first condition, i.e. the stochastic

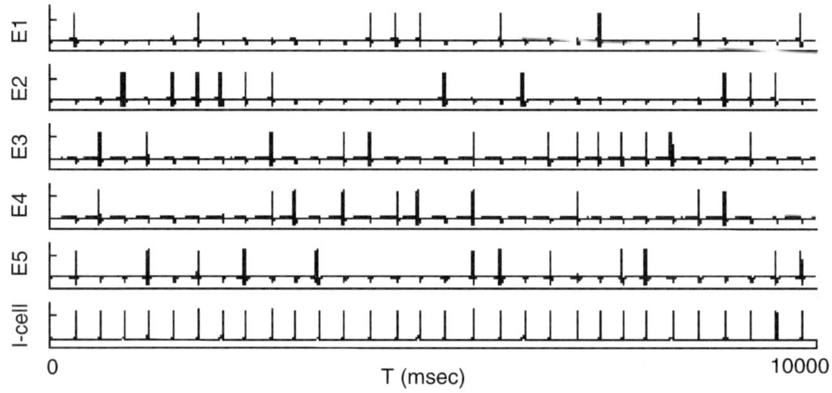

Fig. 1. An epoch of the network behavior with five E-cells and one I-cell. g_{AMPA} is chosen so that any of the E-cells can drive the I-cell to fire; i.e. $n = 1$ ($g_{\mathrm{AMPA}} = 0.02\ \mathrm{mS\,cm^{-2}}$). The input current is stochastic with a Gaussian distribution ($\mu = 4.3$, $\sigma = 0.2\ \mu\mathrm{A\,cm^{-2}}$). The rhythm has a stochastic period (mean/STD = 326.84/20.82 ms).

firing of the E-cell after inhibition, calls for a probability distribution function (pdf) that can describe the stochastic firing of each E-cell in the network. The network scheme that has been used throughout assumes no connection between E-cells, therefore, it can be concluded that as far as the activity of individual E-cells is concerned, between successive inhibitions, the activity of different E-cells is essentially independent. In other words, the network of N E-cells and one I-cell acts as N disjoint networks of each, one E-cell and one I-cell and the firing times of each E-cell in the network is stochastically similar to the firing time of the E-cell in the two-cell network; i.e. $N = 1$ and $n = 1$.

In the two-cell network, the firing time of the E-cell can be accurately approximated with the pdf of the network period ($P_{N=1,n=1}(T)$). To characterize the latter, a statistical approach has been adopted. The interspike intervals of the I-cell firings are taken as samples of the network period. A large sample of per-cycle ISIs is obtained by running the simulation of a two-cell network (one E-cell and one I-cell) for a total of 500 s. The histogram of these samples is further smoothed by cubic (spline) interpolation and normalized. The result corresponds to the pdf of the network period, for the two-cell configuration; i.e. $P_{1,1}(T)$ (Figs. 2(a1) and (b1)).

Now we have all the necessary elements for the stochastic phenomenological model: a total of N E-cells from which n are necessary and sufficient to drive the I-cell. Furthermore, after firing of the I-cell, the next round firing of each E-cell is characterized by $P_{1,1}(T)$. We would like to find $P_{N,n}(T)$, the distribution of the network period (which is the same as the pdf of the firing time of the nth E-cell).

In each cycle, E-cells are enumerated according to the order which they fire, from E_1 to E_N. Any of the N E-cells can be the nth E-cell to fire, therefore, there are N different possible choices for E_n. From the remaining $N - 1$ E-cells, $n - 1$ of them should fire before E_n and the remaining $N - n$ E-cells should all fire after E_n. Taking

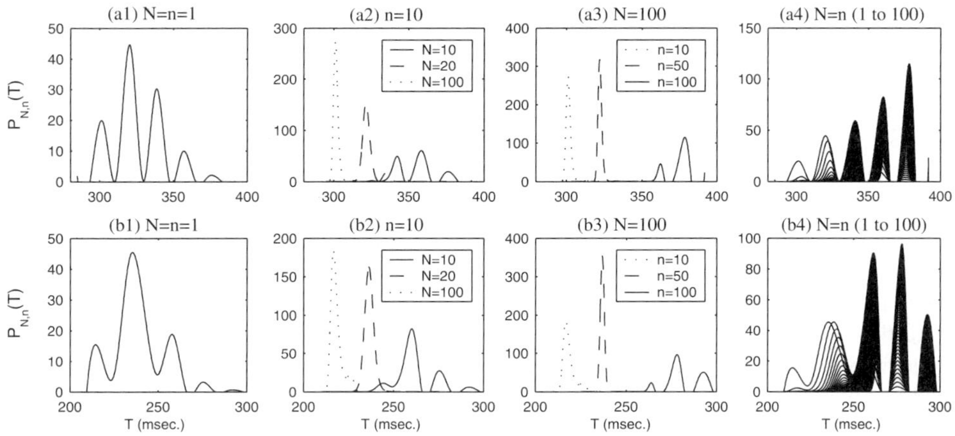

Fig. 2. The distribution of the network period for several different values of N and n and different Gaussian inputs. The left column (a1) and (b1) shows the original distribution of a two-cell network, $P_{1,1}(T)$. The second column (a2) and (b2) shows $P_{N,n}(T)$ for $N=10$ (solid), $N=20$ (dashed) and $N=100$ (dotted) with $n=10$. The third column (a3) and (b3) shows $P_{N,n}(T)$ for $n=10$ (dotted), $n=50$ (dashed) and $n=100$ (solid) with $N=100$. In (a4) and (b4), $P_{N,n}(T)$ for all $N=n=1,2,3,\ldots,100$ are superimposed. Increasing N (a2) and (b2) amplifies the lower peaks whereas, increasing n (a3) and (b3) suppresses lower peaks and amplifies higher peaks. Irrespective of the network size, the distribution embodies the peaks of the original pdf $P_{1,1}(T)$ (a4) and (b4). First row and second row correspond to input Gaussian distributions of mean/STD $=4.3,\ 0.19$ and mean/STD $=4.515, 0.2\ \mu\text{A cm}^{-2}$, respectively ((a1) and (b1) insets).

all these combinations into account, $P_{N,n}(T)$ will be

$$P_{N,n}(T)=P_{1,1}(T)P_{1,1}(\tau\leqslant T)^{(n-1)}P_{1,1}(\tau\geqslant T)^{(N-n)}\frac{N!}{(N-n)!(n-1)!}, \qquad (1)$$

where

$P_{1,1}(T)$,	prob. of one of the E-cells firing at T,
$P_{1,1}(\tau\leqslant T)^{(n-1)}=[\int_0^T P_{1,1}(\tau)\,d\tau]^{(n-1)}$,	prob. of $(n-1)$ E-cells firing before T,
$P_{1,1}(\tau\geqslant T)^{(N-n)}=[\int_T^\infty P_{1,1}(\tau)\,d\tau]^{(N-n)}$,	prob. of $(N-n)$E-cells firing after T,
$\frac{N!}{(N-n)!(n-1)!}$,	number of permutations.

3.1. *Stochastic phenomenological model: results*

The results show that $P_{1,1}(T)$ is highly dependent on the stochastic nature of the input bias current such that slight changes in the input current distribution can result in unimodal, bimodal, trimodal or in general multimodal characteristics in the firing time of individual E-cells, $P_{1,1}(T)$ (two examples shown in Figs. 2(a1) and (b1)).

On the other hand, the network, regardless of the modal structure of $P_{1,1}(T)$, depending on both N and n, differentially amplifies $P_{1,1}(T)$ to extract one or several of the modal components while suppressing the others.

Figs. 2(a1) and (b1) shows two different pdfs for the firing time of the E-cells that correspond to two different input Gaussian distributions (Figs. 2(a1) and (b1) insets). For a given I-cell excitability (n constant), as the number of E-cells in the network increases, the pdf of the network period, $P_{N,n}(T)$, moves towards lower periods and amplifies the higher frequency modes (Figs. 2(a2) and (b2)). Conversely, when in a network of N (constant) E-cells, the I-cell excitability increases, $P_{N,n}(T)$ suppresses lower periods and amplifies the higher ones (Figs. 2(a3) and (b3)). In Figs. 2(a4) and (b4), the results of $P_{N,n}(T)$ for all values of $1 \leqslant N = n \leqslant 100$ are superimposed. It is evident that the modes of rhythmicity in the network follow the multimodal structure of the original $P_{1,1}(T)$.

4. Discussion

The underlying mechanism described here hypothesizes that synchronized inhibition of multiple pyramidal cells in hippocampus might be the natural result of a dynamic feedback imposed by a population of synchronized interneurons. Our results demonstrate that depending on both the network size and the excitability of the interneuronal population, recurrence of this feedback can result in emergence of a robust rhythmic behavior from an underlying stochastic and modally nonspecific basal activity.

This rhythmicity originally depends on the interaction of the synaptic input and the intrinsic properties of individual cells, however, the dynamics of the network is able to extract and amplify certain preferred frequencies while suppressing the others.

Due to the spatial dynamicity of the E-cells, at each cycle a different subpopulation of E-cells may become activated. From an information coding perspective, the formation of such subpopulations might contain specific neural codes. Although existence of such population codes is highly speculative, it would be interesting to confront the results of the current phenomenological model with other models of population coding to investigate the dynamical feedback mechanism as a vehicle to code information. In the current model, the relationship between the rhythmic temporal context and the spatial dynamicity provides a potential substrate for spatiotemporal coding.

References

[1] M.I. Banks, T.B. Li, R.A. Pearce, The synaptic basis of GABA-A, slow, J. Neurosci. 18 (1998) 1305–1317.

[2] B.H. Bland, The physiology and pharmacology of hippocampal formation theta rhythms, Prog. Neurobiol. 26 (1986) 1–54.

[3] A. Bragin, G. Jandó, Z. Nádasdy, J. Hetke, K. Wise, G. Buzsáki, Gamma [40–100 Hz] oscillation in the hippocampus of the behaving rat, J. Neurosci. 15 (1995) 47–60.

[4] E.H. Buhl, K. Halasy, P. Somogyi, Diverse sources of hippocampal unitary inhibitory postsynaptic potentials and the number of synaptic release sites, Nature 368 (1994) 823–828.

[5] G. Buzsáki, Z. Horváth, R. Urioste, J. Hetke, K. Wise, High-frequency network oscillation in the hippocampus, Science 256 (1992) 1025–1027.
[6] G. Buzsáki, L. Leung, C.H. Vanderwolf, Cellular bases of hippocampal EEG in the behaving rat, Brain. Res. Rev. 6 (1983) 139–171.
[7] M.J. Jazayeri, L. Zhang, F.K. Skinner, A dynamic feedback mechanism for the generation of hippocampal field rhythms, Neurocomputing 38–40 (2001) 683–689.
[8] J.G.R. Jefferys, R.D. Traub, M.A. Whittington, Neuronal networks for induced '40 Hz' rhythms, Trends Neurosci. 19 (1996) 202–208.
[9] R.D. Traub, M.A. Whittington, Fast Oscillations in Cortical Circuits, Cambridge University Press, Cambridge, MA, 1999.
[10] C. Wu, H. Shen, W.P. Luk, L. Zhang, A fundamental oscillatory state of isolated rodent hippocampus, J. Physiol. 540 (2002) 509–527.
[11] A. Ylinen, A. Bragin, Z. Nádasdy, G. Jandó, I. Szabó, A. Sik, G. Buzsáki, Sharp wave-associated high-frequency oscillation [200 Hz] in the intact hippocampus: network and intracellular mechanisms, J. Neurosci. 15 (1995) 30–46.
[12] Y. Zhang, J.L. Perez Velazquez, G.F. Tian, C.-P. Wu, F.K. Skinner, P.L. Carlen, L. Zhang, Slow oscillations ($\leqslant$ 1 Hz) mediated by GABAergic interneuronal networks in rat hippocampus, J. Neurosci. 18 (1998) 9256–9268.

Mehrdad Jazayeri is a graduate student in the Center for Neural Science at the New York University. He was previously part of the Toronto Western Research Institute (TWRI), University Health Network working on phenomenological models of cellular mechanisms underlying rhythmicity in neuronal networks. Currently he is working on the dynamics of cortical neural networks.

Dr. Frances Skinner is a Senior Scientist in the Toronto Western Research Institute (TWRI), University Health Network with appointments in Medicine (Neurology), Physiology and Biomedical Engineering at the University of Toronto. She graduated from the University of Waterloo (B.Math.) and Toronto (M.A.Sc., Ph.D.) and did 4 years of postdoctoral work in Boston and California. In general, she enjoys collaborative work and is interested in cellular based mechanisms underlying neuronal network dynamics.

Available at
www.ComputerScienceWeb.com
POWERED BY SCIENCE DIRECT

ELSEVIER

Neurocomputing 52–54 (2003) 813–818

NEUROCOMPUTING

www.elsevier.com/locate/neucom

Nonlinear determinism of spiking activity recorded from rat *suprachiasmatic nucleus* neurons in vitro

Jaeseung Jeong, Yongho Kwak, Kyoung J. Lee*

Department of Physics, Center for Neurodynamics, Sungbuk-Gu-Anham-Dong 5-1, Korea University, Seoul 136-701, South Korea

Abstract

The possible existence of nonlinear determinism in complex interspike interval (ISI) patterns of rat suprachiasmatic nucleus (SCN) neurons is investigated in hypothalamic slice preparations. ISI sequences are recorded from 173 neurons using a cell-attached patch recording technique, and their correlation dimensions (D_2) are estimated. These values are then compared with those of the randomly shuffled surrogate data. Among 173 neurons, 16 neurons are found to exhibit deterministic ISI patterns of spikes. We show, using clustering analysis, that SCN neurons is divided into two subgroups of neurons each having distinct values of skewness (SK) and coefficient of variation (CV): a group of irregularly spiking SCN neurons having large values of SK ($2.0 < \mathrm{SK} < 10.0$) and CV ($0.4 < \mathrm{CV} < 0.8$) and the other group of regular SCN neurons with smaller SK ($-1.0 < \mathrm{SK} < 2.0$) and CV ($0.1 < \mathrm{CV} < 0.4$) values. Interestingly, most deterministic SCN neurons ($\frac{14}{16}$) belong to the group of irregularly spiking neurons.

Keywords: Suprachiasmatic nucleus; Nonlinear determinism; Heterogeneity

1. Introduction

The superchiasmatic nucleus (SCN) contains pacemaker neurons imposing circadian rhythmicity in mammals [7,8]. The circadian rhythm of SCN neurons is expressed by the slow sinusoidal modulation of their mean firing rates. This modulation is closely associated with rhythmic changes in endogenous and behavioral activities of the

* Corresponding author. Tel.: +82-2-3290-4288; fax: +82-2-3290-3534.
E-mail address: kyoung@nld.korea.ac.kr (K.J. Lee).

0925-2312/03/$ - see front matter
doi:10.1016/S0925-2312(02)00788-9

mammalian brain and body. Unlikely the slowly varying mean firing rates, interspike intervals (ISIs) of SCN neurons are highly complex, often accompanied by intermittent bursts of spikes [9]. So far, the relationship between the complex sequence of action potentials operating on a fast time scale and the mean firing rates varying over a much slower time scale is unknown.

To gain some insights into this question, it is quite necessary to characterize the complex spiking sequences of SCN neurons. In this study, we used nonlinear time series methods to determine whether the complex ISI patterns of SCN neurons are truly random or deterministic, a critical issue for understanding its underlying cellular mechanism. Nonlinear dynamical theory has shown that systems of deterministic chaos can exhibit a very irregular behavior as complex as stochastic systems do. Indeed, a number of recent studies have demonstrated that irregular spiking activities recorded from various brain regions emerge from a deterministic chaos [1,6]. To our knowledge, this is the first study investigating the spiking patterns of SCN neurons based on nonlinear dynamical measures.

2. Methods

Male Sprague-Dawley rats ($n=52$; 40–100 g) were housed in a temperature-controlled room (22–24°C) under a 12/12-h light/dark cycle (light on 07:00–19:00) for at least 2 weeks prior to use. The rats were anesthetized with Nembutal (6 mg/100 g body weight) in the daytime of subjects, and then the brains were quickly removed and submerged in ice-cold artificial cerebrospinal fluid. Using a vibrating tissue slicer (Vibratome 1000, Technical Products International, USA), a block of hypothalamic tissue was cut into slices coronally at the thickness of 120–150 μm. The experimental procedures described above were in accordance with the guideline set by the Korea University College of Medicine Animal Research Policies Committee.

After 1-h incubation in the recording chamber, extracellular recordings were commenced at room temperature (25–27°C). The recording electrodes made of borosilicate tubings (Sutter Inst. Co. USA) had a tip diameter of 2–4 μm with a resistance of 3–5 MΩ. Cell-attached patch (CAP) configuration without membrane rupture was achieved for extracellular, single-unit recording. In a CAP mode, a single action potential caused a transient capacitive current over the patch of membrane sucked into the pipette tip. This was recorded under voltage clamp conditions with a pipette potential of 0 mV. The recordings were performed using Axo-patch 200B amplifier (Axon Instruments, USA) in track mode from 173 SCN neurons for 20–40 min, and the ISI data were stored using pClamp software. The mean number of data points was 7356 ± 2961 (range: 1422–19 213).

The D_2 reflects the number of independent variables that are essential for describing the dynamics of the concerned system. In general, the larger the value of D_2, the more complex the behavior of the system. The D_2 values of ISI sequences were estimated using the Grassberger–Procaccia algorithm [2]. The D_2 of a deterministic system is preserved as the embedding dimension increases, while that of a stochastic system increases without any saturation.

The surrogate data test was used to confirm the presence of deterministic nature in the ISI sequences of SCN neurons. Surrogate data are a randomized sequence of the original data with all nonlinear determinism destroyed [10]. A statistically significant difference in the D_2 values between the original data and the surrogate data would indicate the presence of nonlinear determinism in the original data. Pairwise t-tests of the difference between the D_2 values of the original ISI sequences and the mean D_2 values of their 19 surrogates were applied to test the null hypothesis of a stochastic behavior.

3. Results

All observed neurons ($n = 173$) produced a highly irregular ISI sequence, whose distribution was unimodal with a mean of 0.24 ± 0.09 s. D_2 values were estimated for all data sets as the embedding dimension (d_e) increased from 2 to 15. Among 173 neurons, 16 neurons exhibited saturated D_2 values as the d_e increased, indicating that the behavior of this neuron was quite deterministic with finite degrees of freedom, despite the high irregularity of the ISI sequence. The mean D_2 values of the 16 deterministic neurons were 8.15 ± 1.36. The D_2 differences between the original data and their surrogate data increased as the d_e increased (Fig. 1(a)). The other neurons (157/173) exhibited truly random ISI patterns. Fig. 1(b) presents that D_2 values for the original ISI data and their surrogate data showed no significant differences as the d_e increased.

Based on the histogram and regularity analysis of ISIs, we test a possibility that there are two subgroups within the populations of SCN neurons. The regularity in spontaneous firing patterns was quantified by the coefficient of variation (CV) in spike intervals, which is defined as the standard deviation divided by the mean ISI. Higher CV values of ISI sequences indicate more irregular firing patterns. Fig. 2 clearly demonstrates that SCN neurons are divided into two clusters in the parametric space of CV and SK: the group with skewed ISI histograms towards more long ISIs ($2.0 < \mathrm{SK} < 10.0$) and high CV values ($0.4 < \mathrm{CV} < 0.8$), which was denoted as ‘cluster I’, and the neurons with normally distributed ISI histograms ($-1.0 < \mathrm{SK} < 2.0$) and low CV values ($0.1 < \mathrm{CV} < 0.4$), which was labeled ‘cluster II’. We quantitatively examined the presence of two subgroups using a hierarchical clustering analysis. The significant differences in CV and SK values between two clusters was confirmed, in Table 1 ($P < 0.001$; Student t-test). Surprisingly, the majority of deterministic SCN neurons (14/16) belonged to cluster I. These findings indicate that SCN neurons are heterogeneous in properties of firing patterns.

4. Discussion

A deterministic (or predictable) temporal structure of ISI sequences has been repeatedly reported in various neuronal preparations [3,4]. In fact, our present study demonstrated that some SCN neurons also generate deterministic ISI sequences. In other words, interspike intervals and overall patterns of neuronal spike trains are

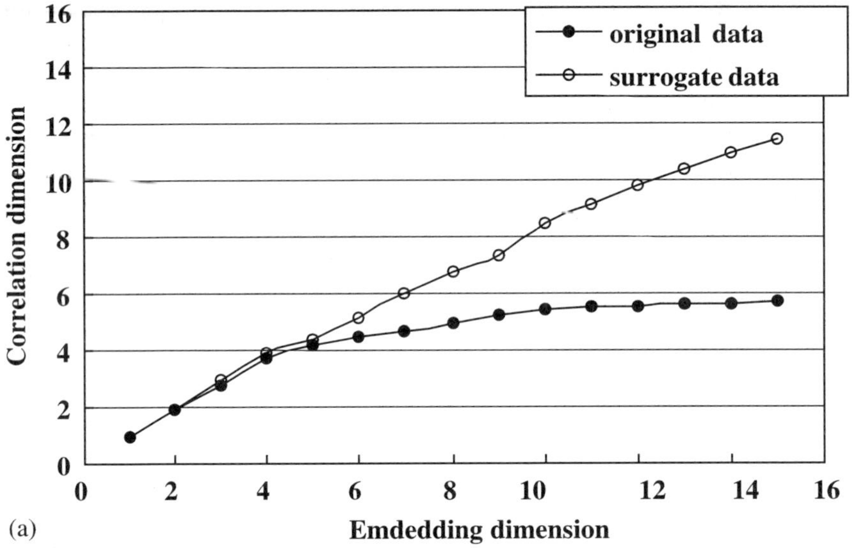

(a)

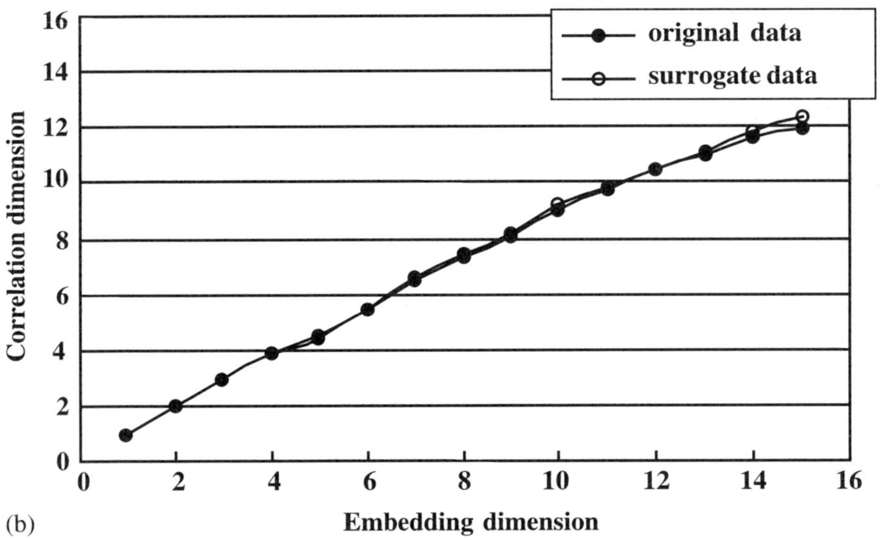

(b)

Fig. 1. The average D_2 values of the original ISI data recorded from (a) a typical deterministic SCN neuron and (b) a stochastic SCN neuron and their surrogate data as a function of the embedding dimension.

often generated and/or exquisitely regulated by deterministic rules, regardless of their highly irregular and complex behavior. This can be quite significant for SCN neurons that play a critical role for imposing circadian rhythms of physiological functions and behavior by providing modulated output signals to target brain regions through neural efferents or in a diffusible manner [5]. For the periodic modulation in the spiking

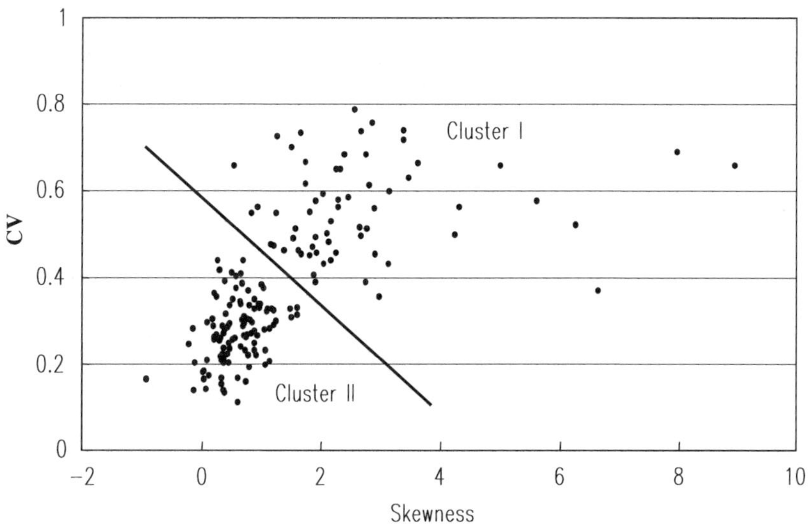

Fig. 2. Plot of CV of interspike intervals versus SK for 173 SCN neurons. Two clusters can be discerned having small CV and SK values, and large CV and SK values, respectively. The values on the axes are not standardized.

Table 1
Spike train properties (mean ± standard deviations) per cluster of SCN neurons ($n = 173$)

Parameters	Cluster I	Cluster II	t-Value
Number of cells	60	113	—
Mean ISI (s)	0.33 ± 0.14	0.21 ± 0.06	6.14*
Skewness	2.64 ± 1.76	0.67 ± 0.86	7.18*
CV	0.55 ± 0.12	0.26 ± 0.08	13.92*
Mean D_2	12.11 ± 3.54	14.56 ± 1.56	−4.59*
Probability of deterministic neurons	14/60	2/113	—

Student t-test; CV: coefficient of variation. *: $P < 0.001$.

rates, individual SCN neurons should be able to monitor spike trains at every moment and control overall patterns of spikes on a short time scale. Thus, we may speculate that SCN neurons must employ a deterministic rule to regulate interspike intervals to impose circadian rhythmicity on their overall firing patterns.

The other interesting finding is that SCN neurons are heterogeneous. This result is quite consistent with the earlier findings of Pennartz et al. [9]. They have examined electrophysiological and morphological properties of SCN neurons in hypothalamic slices to find that SCN neurons can be partitioned into three clusters. Although we cannot compare all the electrophysiological properties of the clusters in our data with those of Pennartz group, the corresponding clusters have similar ranges of mean firing rates and CV values. The heterogeneity of nonlinear determinism in SCN neurons suggests that the SCN is composed of distinct neuronal subgroups each of which makes a unique functional contribution to the circadian timing. For further investigation, the biochemical, electrophysiological, and morphological properties of deterministic SCN neurons will be examined to improve our understanding of the functional roles of distinct groups of SCN neurons.

Acknowledgements

This work was supported by Creative Research Initiatives of the Korean Ministry of Science and Technology.

References

[1] P. Faure, H. Korn, Is there chaos in the brain? Concepts of nonlinear dynamics and methods of investigation, C. R. Acad. Sci. III 324 (2001) 773–793.
[2] P. Grassberger, I. Procaccia, Measuring the strangeness of strange attractors, Physica D9 (1983) 189–208.
[3] R.E. Hoffman, W.-X. Shi, B.S. Bunney, Nonlinear sequence-dependent structure of nignal dopamine neuron interspike interval firing patterns, Biophys. J. 69 (1995) 128–137.
[4] J. Jeong, B.S. Peterson, J.C. Gore, R.E. Hoffman, W.-X. Shi, B.S. Bunny, Bursting as a source of nonlinear deterministic structure in interspike interval firing patterns of nigral dopamine neurons, Soc. Neurosci. Abstr. (2000) 26.
[5] J. Lesauter, R. Silver, Output signals of the SCN, Chronobiol. Int. 15 (1998) 535–550.
[6] T.M. Mckenna, T.A. Mcmullen, M.F. Shlesinger, The brain as a dynamic physical system, Neuroscience 60 (1994) 587–605.
[7] J.H. Meijer, W.J. Rietveld, Neurophysiology of the suprachiasmatic circadian pacemaker in rodents, Physiol. Rev. 69 (1989) 671–707.
[8] L.P. Morin, The circadian visual system, Brain Res. Rev. 19 (1994) 102–127.
[9] C.M. Pennartz, M.T. de Jeu, A.M. Geurtsen, A.A. Sluiter, M.L. Hermes, Electrophysiological and morphological heterogeneity of neurons in slices of rat suprachiasmatic nucleus, J. Physiol. 506 (1998) 775–793.
[10] T. Schreiber, A. Schmitz, Surrogate data methods, Physica D 142 (2000) 346–382.

Jaeseung Jeong (born 1972) received a Ph.D. in Physics for his work on "Computational modeling of the cerebral cortex damaged by Alzheimer's disease" at the Korea Advanced Institute of Science and Technology in 1999. Until 2001, he has worked on computational modeling of dopamine neurons in the *Subtantia nigra* as a postdoctoral associate at the Yale University, School of Medicine. He is currently as a research professor at the Department of Physics, Korea University, with a research focus on underlying mechanisms of the circadian rhythm generation in the Suprachiasmatic nucleus.

Yongho Kwak (born 1965) received a Ph.D. in Physics at the Korea University for his work on "Firing activity of Suprachiasmatic nucleus neurons". Currently he is working as a postdoctoral fellow at the University of Virginia to continue his research activity.

Kyoung J. Lee received a Ph.D. in Physics for his work on nonequilibrium pattern formation at the University of Texas at Austin in 1994. As a research associate at Princeton University, he worked on the developmental process of *Dictyostelium discodium* amoebae population. He became a faculty member of Korea University in 1996, and is now the director for the Center for Neurodynamics at Korea University. His current research is focused at dynamical aspects of various neural systems.

ELSEVIER

Available at
www.ComputerScienceWeb.com
POWERED BY SCIENCE @DIRECT®

Neurocomputing 52–54 (2003) 819–826

NEUROCOMPUTING

www.elsevier.com/locate/neucom

The KIV model—nonlinear spatio-temporal dynamics of the primordial vertebrate forebrain

Robert Kozma[a,*], Walter J. Freeman[b], Peter Erdi[c,d]

[a]*Department of Mathematical Sciences, Institute for Intelligent Systems, University of Memphis, Memphis TN 38152, USA*

[b]*Division of Neurobiology, University of California at Berkeley, LSA 142, Berkeley, CA 94720-3200, USA*

[c]*Center for Complex System Studies, Kalamazoo College, 1200 Academy Street, Kalamazoo, MI 49006-3295, USA*

[d]*Department of Biophysics, KFKI Research Institute for Particle & Nuclear Physics of the Hungarian Academy of Sciences, Budapest, Hungary*

Abstract

EEG measurements indicate the presence of common-mode, coherent oscillations in various cortical areas. In previous studies the KIII model has been introduced, which interprets the experimental observation as nonlinear, spatially distributed dynamical oscillations of coupled neural populations. In this paper we combine multiple KIII sets into the KIV model, which approximates the operation of the basic vertebrate forebrain together with the basal ganglia and motor systems. This paper outlines a summary description of the essential components of the KIV model, as the basis for future modeling of their cooperative dynamics guided by analysis of multichannel EEG in animals and humans.

Keywords: Neurodynamics; Chaos; Cortex; Hippocampus; Spatio-temporal EEG

1. Introduction

The discovery that brain dynamics exhibits chaotic features has profound implications for the study of higher brain function [15,14]. The KIII model is a working example of the implementation of these chaotic principles in a computer environment. KIII exhibits a number of experimentally observed behaviors of brains, like robust pattern recognition

* Corresponding author. Tel.: +1-901-678-2497; fax: +1-901-678-2480.
E-mail address: rkozma@memphis.edu (R. Kozma).

0925-2312/03/$ - see front matter © 2002 Elsevier Science B.V. All rights reserved.
doi:10.1016/S0925-2312(02)00796-8

and classification of input stimuli, and fast transitions between brain states [4,6,12,13]. KIII shows very good performance in the several kinds of learning needed to categorize input data, and it can generalize efficiently in various classification problems.

The operation of the KIII model can be described as follows. In the absence of stimuli the system is in a high dimensional state of spatially coherent basal activity, which is described by an aperiodic (chaotic) global attractor. In response to external stimuli, the system can be kicked-off the basal state into a local memory wing. The system resides in the localized wing for some time, then it returns to the basal state. This is a temporal burst process of the duration of up to 200 ms; see [16] for the olfactory bulb, and [2] for neocortex.

The next highest level of the K-sets is the KIV model. As in the case of all other K-sets (Freeman, 1975), the architecture and functionality of KIV is strongly biologically motivated. The data that are required for modeling neurodynamics at this level are derived by analysis of EEG recorded simultaneously from multiple electrodes in sensory and limbic structures [11,7–9]. KIV provides the neuro-architecture that is needed to model the interactions of key neural populations in the primordial vertebrate forebrain. Among these parts are the sensory cortices, the motor cortices and nuclei, and the hippocampal formation, which is essential for cognitive processes such as orientation, learning and memory.

In this paper, the architecture of the KIV model is outlined, starting with a description of the internal organization of the hippocampus. There follows the functional description of the interaction of hippocampus with the sensory cortices and basal nuclei comprising the primitive forerunner of the limbic system at the level of the salamander [10]. Several types of essential learning processes in the basic vertebrate forebrain are described using illustrative examples.

2. Outline of the KIV model

Three types of sensory signals are considered in KIV: exteroceptors, interoceptors (including proprioception), and orientation signals; e.g., gravity, visual flow, magnetic fields. Each of these sensory signals provide stimuli towards the brain, namely the sensory cortices, midline forebrain (MF) unit, and the hippocampal formation (HF), respectively; see Fig. 1. The present model is motivated by the architecture and putative function at the level of the amphibian brain. It is not intended to mimic all the biological details; rather it is used to incorporate the main elements required for operation of brains at the KIV-level of functionality.

The hippocampus is the main focus of cognitive mapping that supports spatial navigation and temporal orientation (short term memory). There is a rich literature of hippocampal-based navigation models [3,1]. In our model, the following parts of the hippocampus are modeled: Dentate Gyrus (DG), CA3, CA1, and CA2. We need CA2 in our model to generate the hippocampal KIII dynamical system, serving as its chaotic controller. CA1, CA2 and CA3 are modeled as KII units, while DG will be a KI unit. KII units are shown to generate point attractors, limit cycle attractors, and even chaotic attractors (though lacking robustness) in the gamma band.

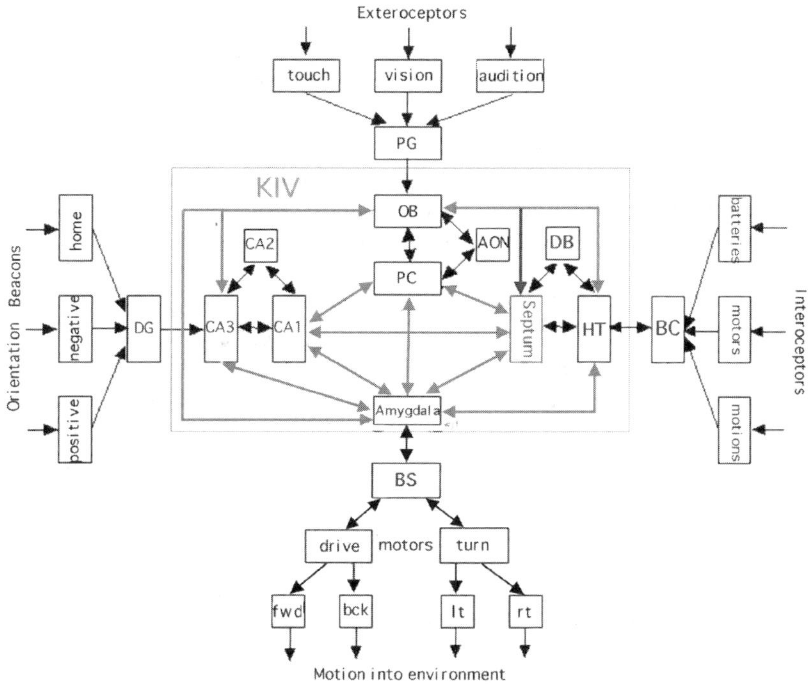

Fig. 1. Structure of the KIV model. Abbreviations: DG, dentate gyrus; CA1–CA3, Cornu Ammonis (hippocampal sections); PG, periglomerular; OB, olfactory bulb; AON, anterior olfactory nucleus; PC, prepyriform cortex; Spt, septum; DB, diagonal band; HT, hypothalamus; BC, basal ganglia, corpus striatum including thalamus; BS, brain stem. The sparse long connections that comprise the KIV set are shown as bi-directional, but they are not reciprocal. The entorhinal cortex is omitted, because that is a neocortical structure found only in mammals.

We model both the sensory cortical (SC), midline forebrain (MF), and the hippocampal formation (HF) systems as KIII sets. Each KIII set has three KII units as components and exhibits robust aperiodic oscillations in the gamma range. Each shows spatial coherence in the form of a shared, spatially distributed, aperiodic wave form, with amplitude modulation patterns occurring in sequential frames over time, indicating the existence of landscapes of chaotic attractors corresponding to categories of sensory stimuli that have been learned. These AM patterns are manifestations of self-organizing dynamics that creates coherent activity in the form of "wave packets" [5] as vectors of information in perception. The gating of bursts of KIII activity is governed by a limit cycle attractor in the KII set modeling the septum, that is fixed at frequency in the theta band, here 5 Hz (analogous to a sniff or saccade).

The cortical KIII system initiates the function of pattern recognition by the agency of sensory input-induced destabilization of high-dimensional dynamics. This actualizes an attractor landscape formed by previous experience in the OB/PC, which in our model is the common sensorium for all distance receptors, as it is in the salamander [10]. The hippocampal KIII system, thereafter, uses the categorization embodied in the outputs of the OB and PC as its content-laden input, to which the DG contributes the temporal and spatial location of the environmental events.

Another KIII component of the integrated KIV system, the Midline Forebrain formation, receives the interoceptor signals through the basal ganglia, and processes them in the hypothalamus and the septum. MF provides the value system of the KIV, using information on the internal goals and conditions in the animal. It provides the "Why?" stream to the amygdala, which combines this with the "What?" and "Where?" information coming from the cortex and the hippocampus to make a decision about the next step/action to be taken.

The motor part of the model limbic system is driven by the simulated amygdala. The direction of motion that it determines is based on the combined information from the three sensory systems, which collectively form the architecture of the global KIV. From EEG studies we infer that a cooperative state emerges from the collective interaction among the CA1, PC, Septum, and Amygdala, by which various behavioral patterns are formed and executed. The model given in Fig. 1 is designed to provide the platform with which to study by simulation this behavior formation and action selection mechanism. In a very simple approach, however, we define initially only three basic behaviors: wall following, object avoidance, and backup. Backup behavior is invoked if the robot is stuck or cannot execute a chosen action. A wide range of problems of intentional action can be solved with these three simple behaviors.

3. Hippocampal model of navigation

A continuous inflow of information at a high sampling rate is stored in a short-term memory in the form of stack data storage/register containing up to 10,000 temporal frames. In this memory, always the oldest items are deleted if the memory capacity is exceeded. This is continuously updated and maintained in the HF without inducing phase transitions or reinforcements learning. DG plays a key role in preprocessing for the conversion of orientation information and environmental cues into location information that is contributed by in the CA1 and CA3 units of the hippocampus.

To illustrate the short-term learning mechanism, consider an environment with 3 reference points/landmarks provided by orientation beacons. These could be three point odor sources; three radio frequencies; three colors: red, green, blue; or three sound transmitters. One of these reference points is the base (home) location, the starting point for exploratory behavior. The others are learned environmental support cues. Based on the distance and direction information for the landmarks, the present location of the robot is estimated. The estimation is performed by considering the DG as a feedforward NN, which learns the position by supervised learning, for example, backpropagation.

A practical example is for the robot to learn to identify its location using 3 landmarks. In the encoding use not only the instantaneous readings of the 6 inputs (3 distances and 3 angles) but the sequence of the most recent 8 vectors. This means that the input vector will be $48 = 6 \times 8$ dimensional. Let the robot randomly walk in the environment and record the 6 sensory readings continuously. Apply these inputs continuously to the DG MLP for several thousand steps and perform continuous learning with BP, using

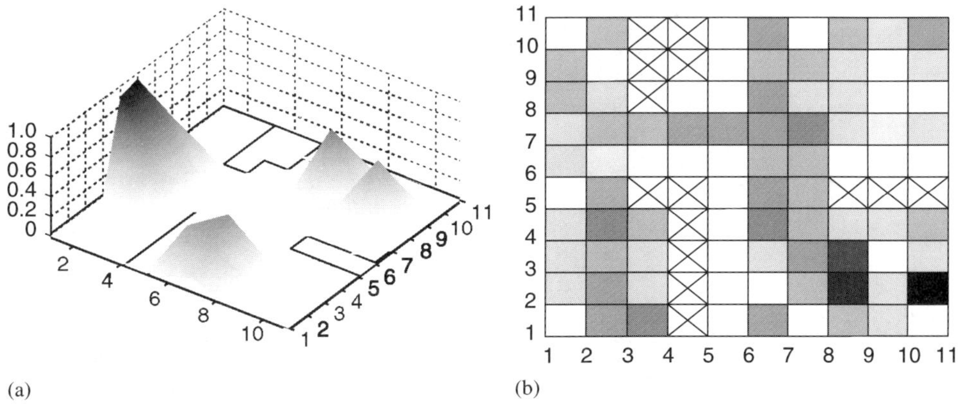

Fig. 2. (a) Example of place fields generated in numerical simulations. (b) Average localization error. White color indicates high accuracy (within 1 lattice point); black denotes the largest error, i.e., $d=4$ displacements in lattice geometry: $d = |\Delta x| + |\Delta y|$.

the known location coordinates $[x, y]$. During the testing phase, we turn off the learning and check if the output of the MLP is, indeed, indicates the correct actual location of the robot. Examples of the generated place fields are given in Fig. 2a. The average localization error is illustrated in Fig. 2b.

We apply reinforcement learning combined with non-Hebbian habituation for category formation. This learning is episodic, not continuous, long-term, and irreversible. It occurs when the device is moving into its environment and encounters important or unexpected changes in its sensory inflow. The KIII reinforcement learning [12] takes place in CA3/CA1 during the active periods provided by the theta rhythm, if reinforcement signal is present. In the numerical experiments, we use the same 2D environment, as in the previous section. The robot has 8 simulated near-field sensors. We use the most recent 8 time steps for learning. Reinforcement learning takes place during the 100 ms window defined by the theta rhythm.

At first, the only landmark the animal is given is the ‘home’ beacon, which is set by the human controller. In an explorative mode, the home acts as a repeller with a monotonic gradient field centered at the home, that drives the animal away from home. Constrained by the obstacle, it continues its path along the steepest possible gradient. Soon or later it will not be able to move further, it stuck. That is a conflict, which generates a reinforcement signal for learning.

The above learning mechanisms are complemented with the following algorithm to form additional landmarks based on the experience during exploration. When the animal is stuck, the controller is notified about this event and its location. As a result, a new landmark is generated and its position is added to the existing ones. From now on, the animal gets orientation signals from all the beacons, including this new one. At each episode of being stuck, as behavioral response, we use is the ‘back up’ motion. At the same time, reinforcement learning has been initiated. An example of such exploration is shown in Fig. 3a. It took the system about 500 steps to get from ‘Home’ [0,0] to the ‘Goal’ [80,60].

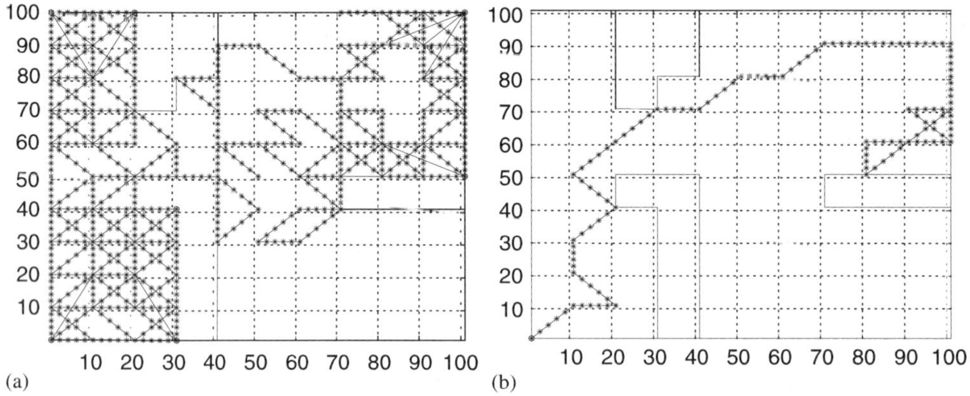

Fig. 3. (a) Trajectory during exploration of the environment; starting from 'Home' at [0,0], and the 'Goal' state is at [80,60]. (b) Trajectory from 'Home' to 'Goal' after completing a learning and exploration session.

In order to test the robot after training for about 500 steps, we re-start it from 'Home' and give a 'Goal' location. If the robot properly learned the environment, it will navigate efficiently and find a reasonably optimal path to the goal based on the internally formed cognitive map. This is illustrated in Fig. 3b. After learning, the length of the trajectory from 'Home' to 'Goal' is reduced to about 30 steps.

4. Discussion and future perspectives

The KIV-set is intended to model the cerebral hemisphere in the vertebrate brain at the evolutionary level of the salamander, which is the locus of goal-directed behavior. The exteroceptors give situation reports, the beacons give space-time orientation, and the septum organizes the valence from interoceptors. Generic goals are established by the human controller of the device, which must determine the specific location to which it should move by using the beacons to locate itself in relation to specific exteroceptive signals. The device is to be familiarized with a significant signal by training to recognize them by categorization and association with particular beacons (tone sound, light color, etc.) as signifying their location, for example, that of a fuel depot or hazard. The device will approach its intended goal flexibly, choosing its path among known hazards and avoiding new hazards by learning about them.

The most critical unsolved problem that we face in development of the KIV model is the dynamics of association between classified exteroceptive signals in the context of value laden interoceptive signals and the cognitive space established with the help of the orientation beacons. The synthesis of these signals must lead to a decision that is to be conveyed to the motor systems for guidance of a sequence of steps. The synthesis implies the formation of a global state variable, that consists of a sequence of temporal frames, each of which is spatially patterned to express the necessary information on which to base a decision. To construct a model that supports the formation of the proposed KIV state variable, we need to study the global dynamics of vertebrate brains that are engaged the guidance of goal-directed behavior. The best available source of this kind of information is multi-channel EEG from high-density arrays of electrodes in multiple sensory and limbic structures [11,8,9]. At the present stage of building the KIV model, we have successfully tested the operation of the hippocampal components in a simplified implementation. We will concentrate our future efforts on defining, describing and modeling the macroscopic KIV state variable.

Acknowledgements

The assistance of Prashant Ankaraju in conducting numerical simulations is appreciated. This research is supported by NASA grant NCC-2-1244, by NSF EIA-0130352, and by Hungarian Scientific Research Fund OTKA grant T-038140.

References

[1] A. Arleo, W. Gerstner, Spatial cognition and neuro-mimetic navigation: a model of hippocampal place cell activity, Biol. Cybernet. 83 (2000) 287–299.

[2] J.M. Barrie, W.J. Freeman, M.D. Lenhart, Spatiotemporal analysis of prepyriform, visual, auditory, and somesthetic surface EEGs in trained rabbits, J. Neurophysiol. 76 (1996) 520–539.

[3] N. Burgess, M. Recce, J. O'Keefe, A model of hippocampal function, Neural Networks 7 (6/7) (1994) 1065–1081.

[4] H.J. Chang, W.J. Freeman, Parameter optimization in models of the olfactory system, Neural Networks 9 (1996) 1–14.

[5] W.J. Freeman, Mass Action in the Nervous System, Academic Press, New York, NY, 1975.

[6] W.J. Freeman, Neurodynamics. An Exploration of Mesoscopic Brain Dynamics, Springer, London, UK, 2000.

[7] W.J. Freeman, B.C. Burke, M.D. Holmes, Application of Hilbert transform to scalp EEG with EMG, Human Brain Mapping, submitted for publication.

[8] W.J. Freeman, G. Gaál, R. Jörsten, A neurobiological theory of meaning in perception. Part 2. Multiple cortical areas synchronize without loss of local autonomy, Int. J. Bifurc. Chaos (2003), to appear.

[9] W.J. Freeman, L.J. Rogers, A neurobiological theory of meaning in perception. Part 4. Multicortical patterns of phase modulation in gamma EEG, Int. J. Bifurc. Chaos (2003), to appear.

[10] C.J. Herrick, The Brain of the Tiger Salamander, University of Chicago Press, Chicago, IL, 1948.

[11] L.M. Kay, W.J. Freeman, Bidirectional processing in the olfactory-limbic axis during olfactory behavior, Behav. Neurosci. 112 (1998) 541–553.

[12] R. Kozma, W.J. Freeman, Chaotic resonance: methods and applications for robust classification of noisy and variable patterns, Int. J. Bifurc. Chaos 11 (6) (2001) 2307–2322.

[13] R. Kozma, et al., Emergence of un-correlated common-mode oscillations in the sensory cortex, Neurocomputing 38–40 (2001) 747–755.

[14] S.J. Schiff, Controlling chaos in the brain, Nature 370 (1994) 615–620.

[15] C.A. Skarda, W.J. Freeman, How brains make chaos in order to make sense of the world, Behav. Brain Sci. 10 (1987) 161–195.

[16] G. Viana Di Prisco, W.J. Freeman, Odor-related bulbar EEG spatial pattern analysis during appetitive conditioning in rabbits, Behav. Neurosci. 99 (1985) 962–978.

Robert Kozma holds a Ph.D. in applied physics from Delft University of Technology (1992). Presently he is Associate Professor at the Department of Mathematical Sciences, Institute for Intelligent Systems, University of Memphis. Previously, he has been on the faculty of Tohoku University, Sendai, Japan (1993–1996); Otago University, Dunedin, New Zealand (1996–1998); he held a joint appointment at the Division of Neuroscience and Department of EECS at UC Berkeley (1998–2000). His research focus is spatio-temporal neurodynamics and emergence of intelligent behavior in biological and computational systems.

Walter J. Freeman studied physics and mathematics at M.I.T., philosophy at the University of Chicago, medicine at Yale University (M.D. *cum laude* 1954), internal medicine at Johns Hopkins, and neurophysiology at UCLA. He has taught brain science in the University of California at Berkeley since 1959, where he is Professor of the Graduate School. He received the Pioneer Award from the Neural Networks Council of the IEEE, and he is IEEE Fellow. He is the author of more than 350 articles and four books: "Mass Action in the Nervous System" (1975), "Societies of Brains" (1995), "How Brains Make Up Their Minds" (1999), "Neurodynamics: An Exploration of Mesoscopic Brain Dynamics" (2000).

Péter Érdi is the Henry R. Luce Professor in Kalamazoo College. He also the head of Department of Biophysics, KFKI Research Institute for Particle and Nuclear Physics, Hungarian Academy of Sciences, and Széchenyi Professor at the Eötvös University. His main scientific interest is the computational approach to the functional organization of the hippocamal formation. He is the co-author with Arbib and Szentagothai of the popular volume "Neural Organization—Structure, Function, and Dynamics" (1997).

ELSEVIER

Available at
www.ComputerScienceWeb.com
POWERED BY SCIENCE @ DIRECT®

Neurocomputing 52–54 (2003) 827–832

NEUROCOMPUTING

www.elsevier.com/locate/neucom

Searching for an IT model with columnar organization

Jeffrey L. McKinstry

Department of Mathematics and Computer Science, Point Loma Nazarene University, San Diego, CA 92106, USA

Abstract

The inferotemporal visual cortex (IT) is at the top of the 'what' stream. Cells in IT are known to respond to complex features and have a columnar organization; nearby columns prefer stimuli with similar features (Neurosci. Res. 32 (1998) 33). This research demonstrates that a hill-climbing search strategy can discover a model of IT that exhibits columnar organization, assuming that the patterns do not overlap significantly. This is a first step towards a more complete model of IT. Following others who have used search techniques to adjust neurocomputational model parameters successfully (Prog. Neurobiol. 61 (2000) 557), this work further demonstrates the benefit of search methods in computational neuroscience.

Keywords: Hill-climbing; Inferotemporal cortex; Pattern recognition; Winner-take-all network

1. Introduction

Hill-climbing [3], a search strategy which always chooses to go up-hill with respect to some objective function, was used to search for a set of parameters that would develop models of IT with a columnar organization [5].

2. The network

The network is similar to the biologically plausible self-organizing map used by Sirosh and Miikkulainen [4] to model column formation in primary visual cortex. A mean-firing rate neural model was chosen. A small (1 by 10 excitatory and 1 by 10

E-mail address: jeffmckinstry@ptloma.edu (J.L. McKinstry).

0925-2312/03/$ - see front matter © 2002 Elsevier Science B.V. All rights reserved.
doi:10.1016/S0925-2312(02)00807-X

inhibitory neurons) IT network was used to allow the search algorithm to complete in several hours time on a 1.8 GHz Pentium 4 processor. Each excitatory neuron received input from a common 20 by 20 visual area. The excitatory and inhibitory IT neurons were reciprocally connected. The firing rate of each neuron is calculated as follows:

$$\mathbf{f}(t+1) = \tanh(p\mathbf{f}(t) + (1-p) \times (c_{\text{input}}\mathbf{w}_{\text{input}} \cdot \mathbf{x} + c_{\text{inhib}}\mathbf{w}_{\text{inhib}} \cdot \mathbf{f}(t) + c_{\text{recurrent}}\mathbf{w}_{\text{recurrent}} \cdot \mathbf{f}(t)) - \theta),$$

where $c_{\text{conn\_type}}$ is the connection influence for connection type *conn_type*, θ is the threshold, and $w_{\text{conn\_type}}$ is the weight vector for connection type *conn_type*. Weights were updated after each cycle with the following formula:

$$\mathbf{w}(t+1) = \mathbf{w}(t) + \alpha\mathbf{x}(t),$$

where $\mathbf{x}(t)$ represents either the visual input, inhibitory input, or recurrent input depending on the connection type. The learning rate, α, was set independently for each connection type. Each connection type was normalized independently such that the weight vectors had unit length.

3. Training

In order to evaluate the quality of a given network, three training runs were made with the same parameter settings, but different random number seeds and training sequences. Two training stimuli is the minimum number required to demonstrate columnar organization. Diamonds and squares were used in the experiments. Each stimulus was presented to the network for 4 consecutive cycles before randomly switching to a new stimulus. A blank image was presented to the network for two cycles between each stimulus. Each stimulus was presented to the network an equal number of times. Networks were trained for 100 total cycles during each training run. The network quality was evaluated based on the response of the network during the 100 training cycles as described next.

4. Searching

The search algorithm set the following network parameters [2]. Search ranges are shown in brackets.

Neuron parameters: threshold $(\theta)[0, 1]$ andpersistence $(p)[0, 1]$
Connection parameters: connection radius [0, 6], learningrate $(\alpha)[0.01, 0.1]$, and connection influence $(c)[0.1, 1]$.

In addition, an annulus region was created (radius of 2) for connections from excitatory neurons to inhibitory neurons.

Since the same parameters applied to all neurons of the same type, the number of parameters is independent of the number of neurons in IT. Therefore this method should apply to larger, two-dimensional, networks.

Automated search requires a method of generating new candidate networks and a function to evaluate candidates. New parameters were generated each iteration by randomly modifying the best network found thus far. A Cauchy random number generator was used in order to concentrate the search around the existing solution while allowing for occasional large jumps. In addition, the standard deviation of the distribution was inversely proportional to the search cycle (0.25 at the beginning, and 0.05 at the end).

Once a new candidate network was generated, a quality metric was applied. Only improvements in quality were accepted. The quality measure, q, for a single training run was defined as

$$q = \sum_{i \in S} m_i + \prod_{i \in S} m_i,$$

where S is the set of all stimulus patterns (2 in this case), and m_i is the winning margin for stimulus i:

$$m_i = g\left(\sum_{t \in I} f(t) - \sum_{t \notin I} f(t)\right),$$

where I is the set of all time steps with input stimulus i, and $g(x)$ is half-wave rectification. The function q encourages networks with cells which respond strongly to only a few stimuli, but yet respond to all of the stimuli. The second term is maximized when the winning margins for each category are equal, rewarding networks which divide up the neurons in proportion to the input stimulus distribution.

5. Results

The search algorithm was allowed to run until 2000 networks had been evaluated. The best set of parameters was saved and used to train a new network shown below. This network exhibits a columnar organization, as shown in the Fig. 1.

Note that neurons that prefer the same stimulus cluster together (see the first row showing connections to the visual input area), and that after training, each neuron responds to a particular input pattern. In addition, the network divides up the neurons such that some respond to each pattern rather than all responding to the same stimulus.

To assess the robustness of the parameters obtained, the search algorithm was run five times. For each run, the resulting parameters were used to train 10 different networks starting from different initial weights. The networks converged such that both objects were represented by different groups of neurons $\frac{8}{10}$, $\frac{9}{10}$, $\frac{9}{10}$, $\frac{0}{10}$ and $\frac{8}{10}$ times with the five different parameter settings. The parameter settings obtained from all five runs are given in Tables 1 and 2. Parameters which are similar across most of the runs are given in italics.

The parameters obtained by searching yielded significantly better results than 5 random sets of parameters ($p < 0.05$, Wilcoxan Rank Sum). Each of these random parameter settings converged zero out of 10 times with different initial weights. This demonstrates that parameter tuning was necessary for this problem.

Fig. 1. Weights for all input connections from V4 (first row), lateral connections from IT (second row), and inhibitory connections from IT_i (last row) for each of the 10 excitatory neurons in IT, where white is maximum and black is the minimum value.

Table 1
Neuron parameters

IT_e threshold	IT_e persistence	IT_i threshold	IT_i persistence
0.94	*0.0*	0.89	0.83
1	*0.0007*	0	0
0.91	*0.25*	0.64	1
1	*0*	0.02	0.68
1	*0.02*	0.17	0.77

Table 2
Connection parameters

V4 to IT_e		IT_i to IT_e			IT_e to IT_e			IT_e to IT_i			IT_i to IT_i		
α	c	rad.	α	c	rad.	α	c	rad.	α	c	rad.	α	c
0.08	*0.64*	*5.97*	0.1	*−0.28*	*1.69*	0.05	*0.1*	*6*	0.01	*0.95*	2.00	*0.02*	*−0.55*
0.07	*0.70*	*5.56*	0.01	*−0.48*	*0.74*	0.01	*0.97*	*5.27*	0.04	*0.30*	0.01	*0.01*	*−0.75*
0.10	*0.63*	*3.76*	0.01	*−1*	*0.71*	0.08	*0.12*	*0.80*	0.01	*0.99*	5.79	*0.03*	*−0.98*
0.08	*0.66*	*5.79*	0.10	*−1*	*1.01*	0.10	*0.80*	*4.30*	0.06	*0.30*	0.78	*0.01*	*−0.90*
0.10	*0.69*	*4.27*	0.06	*−0.83*	*6.00*	0.01	*0.11*	*4.88*	0.05	*1.0*	6	*0.03*	*−0.97*

Several patterns can be found in the parameters from the five different runs. The learning rate and connection influence from the input layer (V4) to the excitatory IT neurons are remarkably consistent. It also appears that the networks are basically winner-take-all networks, with large inhibitory radius values between the excitatory and inhibitory neurons in IT, causing strong competition between the neurons in the network. Finally, the lateral connections from the excitatory IT neurons tend to be local (small radius), which would help to make nearby neurons respond together. There is a reciprocal relationship between the connection influences of the lateral connections and the excitatory to inhibitory connections; strong local excitation is balanced by a reduction in the connection strength to inhibitory neurons thereby maintaining a similar overall level of inhibition. Hill-climbing was successful in discovering effective parameters in this moderately complex, recurrent network.

The two patterns used in the first experiment were nearly orthogonal, overlapping in only 4 pixels. Can this class of network solve a more difficult pattern discrimination problem in which the patterns are more similar? The experiments above were repeated with two patterns that shared approximately 50% of their features in common: the square from the first experiment and an hour-glass shape which shared the top and bottom lines with the square. All of the five sets of parameters found by the search procedure produced the same result: zero out of ten networks converged to a correct solution. All of the neurons in the networks ended up responding to either the same object (e.g. all were hour-glass detectors) or to both objects, instead of dividing the two patterns among the neurons. These results were significantly worse than the results of the first experiment with highly dissimilar patterns ($p < 0.05$, Wilcoxan Rank Sum). This does not prove that no parameters exist that will allow this network to discriminate similar objects, but it does suggest that it is difficult to find them. If one assumes that biological neural networks should be robust to changes in parameters (and that realistic input patterns will not necessarily be orthogonal), this result suggests that a different, perhaps more complex network architecture is required.

6. Conclusions

It was shown that a simple search strategy can find the parameter settings for a modestly complex network which discriminates between two nearly-orthogonal patterns. The constraint that nearby neurons should respond to similar patterns was enforced by the objective function optimized by the search procedure.

No parameters were found that could separate the patterns with a high degree of overlap. It should be noted that such patterns can be discriminated with networks which assume non-local signaling, such as the self-organizing feature map [1]. It would therefore be of interest to expand the complexity of the network, searching for solutions to this problem in the expanded search space.

The most serious drawback of the method is the computational complexity, which limited the complexity of the network. Larger networks and harder problems can be solved with the use of parallel processing and more efficient coding; this method will become more attractive as computing power continues to grow exponentially.

References

[1] T. Kohonen, Self-Organization and Associative Memory, Springer, New York, NY, 1988.
[2] E.T. Rolls, S.M. Stringer, On the design of neural networks in the brain by genetic evolution, Prog. Neurobiol. 61 (2000) 557–579.
[3] S. Russell, P. Norvig, Artificial Intelligence: A Modern Approach, Prentice-Hall, Upper Saddle River, New Jersey, 1995.
[4] J. Sirosh, R. Miikkulainen, Self-organization and functional role of lateral connections and multisize receptive fields in the primary visual cortex, Neural Process. Lett. 3 (1996) 39–48.
[5] G. Wang, et al., Functional architecture in monkey inferotemporal cortex revealed by in vivo optical imaging, Neurosci. Res. 32 (1998) 33–46.

Jeffrey L. McKinstry received his B.A. in Computer Science from Point Loma Nazarene University (1987). He received an M.S. in Computer Science from USC in 1991. In 1999 he received a Ph.D. in Computer Engineering from the University of California, San Diego. His dissertation was entitled, "A model of Primary Visual Cortex: From Single Cells to Feature Maps". He is currently working as a postdoc in theoretical neurobiology at The Neurosciences Institute in La Jolla, California and teaching computer science at Point Loma Nazarene University in San Diego.

Available at
www.ComputerScienceWeb.com
POWERED BY SCIENCE DIRECT
ELSEVIER
Neurocomputing 52–54 (2003) 833–836
NEUROCOMPUTING
www.elsevier.com/locate/neucom

Neural mechanisms to improve timing

C.C. Mitchell

Duke University, Box 90320, Durham, NC 27708, USA

Abstract

We use probabilistic techniques to study neural systems which improve the timing accuracy of action potential firing. A striking example of such accuracy was discovered in the columnar region of the VNLL by Covey and Casseday (J. Neurosci. 11 (1991) 3456). In response to sound stimulation, these VNLL cells fire an action potential whose latency under repeated trials has a standard deviation of approximately 100 μs. This is remarkable because the typical standard deviation in the auditory nerve is 1 ms. In simple systems, we show how the improvement in timing depends on the number of incoming fibers and the number of hits required in a small time window.

Keywords: Coincidence detection; Octopus cells; Probability

1. Introduction

A fundamental problem in neurobiology is to understand how the central nervous system (CNS) performs accurate calculations with components whose properties vary in time and from cell to cell and which give variable responses under repeated trials. Local noise would compromise the information processing capabilities of successive levels of the CNS unless strong compensatory mechanisms were in place. Therefore, situations in which experimental evidence shows that information is sharpened as it progresses inward from the periphery to the CNS are particularly interesting for they may provide clues about such compensatory mechanisms.

A particularly striking example of increasing accuracy as one proceeds into the CNS from the periphery in the auditory system was discovered by Covey and Casseday [2]. Neurons in the auditory nerve (AN) synapse on cells in the cochlear nucleus (CN) and

E-mail address: ccm4@duke.edu (C.C. Mitchell).

0925-2312/03/$ - see front matter
doi:10.1016/S0925-2312(02)00808-1

certain cells in the CN, notably octopus cells, send projections to cells in the columnar region of the ventral nucleus of the lateral lemniscus (VNLLc). Covey and Casseday discovered that in bats these VNLLc cells fire a single action potential in response to sound stimulation with a precise latency; the standard deviation of this latency under repeated trials is typically about 100 μs and in some cases as low as 30 μs. This is remarkable because in the AN of mammals typical standard deviations for latency under repeated trials in a single fiber are approximately 1 ms. Oertel and co-workers [3,7] have proposed that much of this improvement in accuracy is created by coincidence detection by octopus cells of incoming signals from converging AN fibers.

For many years, synchronous firing of groups of neurons has been proposed as the mechanism by which sensory patterns are represented in the cortex. One mechanism by which stable patterns of synchronous firing could be created is the synfire chain model introduced by Abeles [1] and further developed by Aertsen, Diesmann, and others. These investigators have shown that if the number of firing neurons is large enough and the initial standard deviation of firing times is small enough, then stable propagation is possible. Our research explores the improvement of the standard deviation of spike timing at a single target neuron which receives convergent input. Thus we are continuing the line of investigation by Marsalek et al. [4]. Detailed understanding of the improvement of the standard deviation for firing times at single neurons will be of value for understanding the properties of synfire chains using less restrictive hypotheses.

Our report describes the mathematical properties of coincidence detection in a highly simplified neural system [8]. Consider the network shown in Fig. 1. There are n identical input neurons stimulated at the same time. Each responds by sending a single

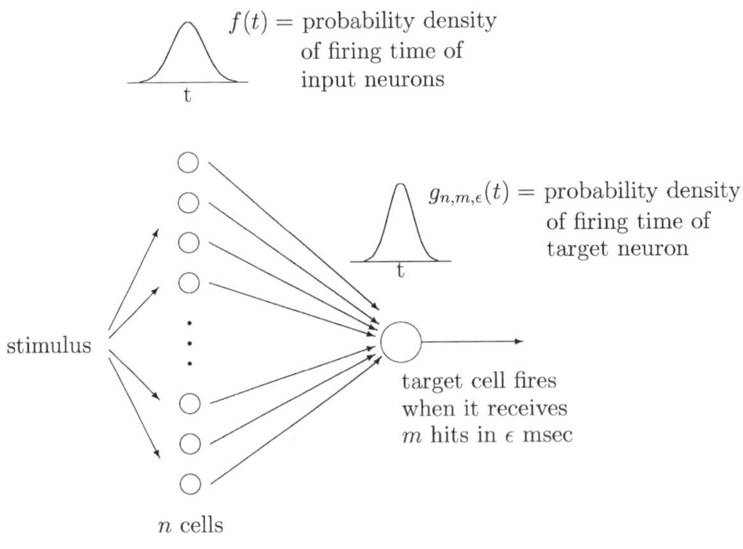

Fig. 1. The connectional pattern. n input neurons, whose stochastic times of firing are selected from the probability density, f, project to a target cell. The probability density for the firing time of the target cell is $g_{n,m,\varepsilon}(t)$. (Taken with permission from [8].)

spike to the target neuron at a random time t selected from a distribution f with standard deviation σ. We assume that the target neuron fires one spike (if it fires at all) the first time it receives m spikes in the preceeding ε milliseconds. Let $\sigma_{n,m,\varepsilon}$ denote the standard deviation of the time of firing of the target cell. If $\sigma_{n,m,\varepsilon}$ is smaller than σ, then the accuracy of the time of firing of the target cell has improved compared to the accuracy of its inputs. We report in this study the behavior of $\sigma_{n,m,\varepsilon}$ as a function of n, m, and ε.

Unfortunately, even in this highly simplified biological situation, one cannot compute explicit formulas for $\sigma_{n,m,\varepsilon}$ in terms of n, m, ε, and f, except in special cases.

One millisecond which is reasonable since the standard deviation under repeated trials in auditory nerve fibers is of this magnitude. One can show that for any value of σ, the parameters can be rescaled so that the probability density of the output is simply a rescaling of the output for the $\sigma=1$ case. Therefore, we fix $\sigma=1$ ms, which is the typical standard deviation under repeated trials in auditory nerve fibers. We begin by examining the behavior of $\sigma_{n,m,\varepsilon}$ for two special cases, $\varepsilon = 0$ and ∞. Then we investigate the general behavior of $\sigma_{n,m,\varepsilon}$ as a function of ε and as a function of m.

2. The case $\varepsilon \longrightarrow 0$

Under appropriate hypotheses on the properties of f, one can prove mathematically that

$$g_{n,m,\varepsilon}(t) \rightarrow \frac{f(t)^m}{\int_{-\infty}^{\infty} f(t)^m \, dt}, \tag{1}$$

as $\varepsilon \rightarrow 0$ [5]. Intuitively, taking the mth power of a continuous density f and renormalizing will emphasize the region near the peak and de-emphasize the regions away from the peak. One can show that for m large the standard deviation, $\sigma_{n,m,0}$, will be of order $1/m$ for input densities, f, which have non-zero first derivative at their maximum (e.g. the exponential) and order $1/\sqrt{m}$ or larger for input densities which have a first derivative of zero at their maximum (e.g. the normal) [6]. Clearly, the improvement in timing will depend strongly on the specific input density chosen.

3. The case $\varepsilon \longrightarrow \infty$

As ε gets large the time window restriction has less effect. Thus, in the limit when $\varepsilon \rightarrow \infty$, the target neuron will fire when the mth action potential arrives (this is equivalent to the simplest integrate and fire model considered by Marsalek et al. [4]). The mth largest of n selections from a density f is called an order statistic and this random variable has probability density:

$$g_{n,m,\infty}(t) = \frac{n!}{(m-1)!(n-m)!} F(t)^{m-1}(1 - F(t))^{n-m} f(t), \tag{2}$$

where $F(t)$ is the cumulative distribution function of $f(t)$, that is $F(t) = \int_{-\infty}^{t} f(s) \, ds$.

4. Dependence on ε and *m*

We now consider the effect of the size of the time window, ε, and the number of hits required, m. In this case the bulk of our investigation was done by Monte Carlo simulation. One would expect that as ε increases, the timing would become less accurate, i.e. $\sigma_{n,m,\varepsilon}$ would increase. In some cases, this is exactly what happens (e.g. $n=10$, $m=2$, f is exponential). On the other hand, there are also cases where exactly the opposite occurs (e.g. $n=10$, $m=8$, f is exponential). There are even cases where $\sigma_{n,m,\varepsilon}$ has a peak at some intermediate value of ε (e.g. $n=10$, $m=5$, f is exponential)! Similarly, one would expect that as m increases, more convergence of inputs would cause a decrease in $\sigma_{n,m,\varepsilon}$. In fact, our Monte Carlo simulations showed that in most cases $\sigma_{n,m,\varepsilon}$ is a non-monotone function of m! From these examples, we see that the dependence of $\sigma_{n,m,\varepsilon}$ on ε and m is complex and often counter-intuitive.

In conclusion, the complexity of the behavior that we have found suggests that determining the relationship between structure and function in the nervous system may be very difficult. In some circumstances, one could get improvement in timing either by increasing or by decreasing m. Similarly, one could get improvement either by increasing or by decreasing ε. The degree to which timing is improved depends in a complicated and interesting way on the input density f, the number of input neurons, n, the number of hits required, m and the duration of the time window, ε.

Acknowledgements

This research was done in collaboration with Michael C. Reed (Duke University, Department of Mathematics) and Jacob J. Blum (Duke University, Department of Cell Biology) and was supported by NSF grants #DMX-9709608 and #DMS-9805761.

References

[1] M. Abeles, Corticonics: Neural Circuits of the Cerebral Cortex, Cambridge University Press, Cambridge, 1991.

[2] E. Covey, J. Casseday, The monaural nuclei of the lateral lemniscus of an echolocating bat: parallel for analyzing temporal features of sound, J. Neurosci. 11 (1991) 3456–3470.

[3] N. Golding, D. Robertson, D. Oertel, Recordings from slices indicate that the octopus cells of the cochlear nucleus detect coincident firing of auditory nerve fibers with temporal precision, J. Neurosci. 15 (1995) 3138–3153.

[4] P. Marsalek, C. Koch, J. Maunsell, On the relationship between synaptic input and spike output jitter in individual neurons, PNAS 94 (1997) 735–740.

[5] C. Mitchell, Timing in Neural Systems: The Small ε limit, 2002, submitted for publication.

[6] C. Mitchell, Mathematical properties of time windowing in neural systems, Duke University Thesis, 2003.

[7] D. Oertel, R. Bal, S. Gardner, P. Smith, P. Joris, Detection of synchrony in the activity of auditory nerve fibers by octopus cells of the mammalian cochlear nucleus, Proc. Nat. Acad. Sci. 97 (2000) 11773–11779.

[8] M. Reed, J. Blum, C. Mitchell, Precision of Neural Timing: Effects of Convergence and Time-Windowing, J. Comput. Neurosci. 13 (2002) 35–47.

Available at
www.ComputerScienceWeb.com
POWERED BY SCIENCE @DIRECT®

ELSEVIER

Neurocomputing 52–54 (2003) 837–842

NEUROCOMPUTING

www.elsevier.com/locate/neucom

A network model of inhibitory effects induced by transcranial magnetic stimulation

Yoichi Miyawaki*, Masato Okada

RIKEN Brain Science Institute, 2-1, Hirosawa, Wako, Saitama 351-0198, Japan

Abstract

We have studied the inhibitory effects of transcranial magnetic stimulation (TMS) on a neural population. Because TMS can affect the electromagnetic activities inside our brain with high temporal resolution and noninvasively, it is widely used as a powerful tool both in the field of cognitive neuroscience and for clinical treatment. However, the neural mechanisms underlying these effects remain unclear, especially from a theoretical perspective. In our study, we employed a simple neural population model and computationally analyzed the responses to a TMS-like perturbation. When the perturbation was applied, the mean activity of the network transiently increased, and then decreased for a relatively long period followed by the loss of a localized activity pattern. When the afferent input had a strong transient component and a weak sustained component, there was a critical latency period during which the perturbation completely suppressed the network activity. These results suggest that the inhibitory effects typically observed in TMS studies can be yielded through dynamical interaction in a neural population.

Keywords: TMS; Inhibitory effect; Network model; Perturbation

1. Introduction

Transcranial magnetic stimulation (TMS) can noninvasively stimulate our neurons with high temporal resolution through a brief magnetic pulse generated by a coil placed on the scalp. Because of these outstanding characteristics, TMS has been widely used in experimental research regarding human functions, such as motor, memory, and perception. Typically, TMS has inhibitory effects on those functions; for example, TMS applied to a motor area suppresses EMG [5], and that applied to the visual cortex

* Corresponding author.
E-mail address: yoichi_miyawaki@brain.riken.go.jp (Y. Miyawaki).

0925-2312/03/$ - see front matter
doi:10.1016/S0925-2312(02)00810-X

creates a visual field deficit [2,7]. However, the neural substrates underlying these inhibitory effects are totally unknown.

Several biophysical studies have demonstrated the spike generation resulting from magnetic stimulation on a long straight axon model [10]. Recently, biophysical approaches were also used to examine cortical stimulation. Nagarajan described the spike generation in a finite cable model [9], and Kamitani demonstrated the long inhibitory period after the initial burst firing in realistic compartmental models of neocortical neurons [6]. All of these studies were based on the magnetic field effects on a single cellular level. In a real situation, however, the spatial extent of the magnetic field induced by a TMS coil is so large compared to the dimensions of a single neuron that a large number of neurons would be simultaneously stimulated under the coil. Therefore, we need to take into account the effect on the neural population and analyze the time course of the neural activity through their dynamical interactions.

A key issue to be considered at the population level is how to generate the optimal TMS timing to cause a suppressive effect. Most single pulse TMS studies have demonstrated that there is a certain period during which TMS has a suppressive effect. It is quite conceivable that the temporal properties of TMS interference are determined by the dynamical state of the neural population after the afferent input. In this work, we have analyzed the response of a simple analog neural network model to a TMS-like perturbation. We focused especially on the optimal timing and the effective range of the perturbation in suppressing network activity.

2. Methods

The simple analog neural network model we used here is represented by Eq. (1)–(4), which has been well analyzed as a model for a sensory feature detector system [4].

$$\tau_m \frac{\mathrm{d}}{\mathrm{d}t} m(\theta,t) = -m(\theta,t) + g[h(\theta,t)], \tag{1}$$

$$h(\theta,t) = \frac{1}{\pi}\int_{-\pi/2}^{\pi/2} J(\theta-\theta')m(\theta',t)\,\mathrm{d}\theta' + h_{\mathrm{ext}}(\theta,t), \tag{2}$$

$$J(\theta-\theta') = -J_0 + J_2\cos 2(\theta-\theta'), \tag{3}$$

$$h_{\mathrm{ext}}(\theta,t) = c(t)[1-\varepsilon+\varepsilon\cos 2\theta]. \tag{4}$$

Here, $m(\theta,t)$ is the output of neuron θ and $g[h]$ is the quasi-linear output function that is inactive ($g=0$) if $h<T$ (T: threshold), saturated ($g=1$) if $h>T+1/\beta$, and linearly operative as $\beta(h-T)$ between these points. $h(\theta,t)$ denotes the input to the neuron θ. Generally, the integral range is defined by the extent of the connections of neuron θ. Here, for simplicity, we assume that $m(\theta,t)$ has a periodic boundary condition ($-\pi/2 \leqslant \theta \leqslant \pi/2$), so that connections among each of the neurons are limited in this periodic range. Afferent input, $h_{\mathrm{ext}}(\theta,t)$, has a maximal amplitude $c(t)$ at $\theta=0$. ε measures the degree of modulation of afferent input to the target neural population ($0 \leqslant \varepsilon \leqslant 1/2$).

Synaptic weight and afferent input have only the zero- and second-order Fourier components, so that the dynamics of $m(\theta, t)$ can be fully described by the zero- and second-order Fourier coefficients of $m(\theta, t)$ ($m_0(t)$ and $m_{2c}(t)$, respectively). $m_0(t)$ represents the mean activity of the entire network and $m_{2c}(t)$ represents the degree of localization of an activity profile. These are the order parameters of this network model.

The spatial extent of the neural population that we deal with is supposed to be so small that the induced electric field will be almost uniform in such a limited area. Therefore, we assumed that TMS perturbation would be constant for all neurons and simply modified the input function as $\hat{h}(\theta, t) = h(\theta, t) + I_{\mathrm{TMS}}(t)$. Hence, the order parameter equations can be also modified by replacing h into $\hat{h}$. Here, we employ a rectangular input (amplitude: I_{TMS}, duration: W_{TMS}) as a TMS-like perturbation.

3. Results

To evaluate whether the TMS-like perturbation can suppress the network activity, we examined the equilibrium state of the network ($t \to \infty$). However, if the intensity of the afferent input is constant ($c(t) = \mathrm{const}$), there is only one attractor in this simple neural network. If the network has only one attractor, the final state of the network will be uniquely defined by the intensity of the afferent input even if any strong perturbations are delivered.

Thus, to achieve dependence on the past input, we adopted a large β and allowed the network to exhibit non-linearity. If an appropriate afferent input is given under this regime, the network will show bistability and converge either of the active state that the network holds a local excitation or the silent state that the whole network activity is zero. As a typical example of afferent input showing bistability, we can use the simplest input which consists of a suprathreshold transient component (amplitude: $A_{\mathrm{t}} > T$, duration: W_{t}) and a subthreshold sustained component (amplitude: $A_{\mathrm{s}} < T$). Here, we took a relatively small β but still in the non-linear range, so that the minimum value of A_{s} showing bistability was bounded to a non-zero value (Fig. 1). The time course of this afferent input model is shown at the bottom of Fig. 2a. This kind of time course of neural activity is commonly observed in the LGN, visual cortex, and other brain areas.

The network dynamics after the afferent input are illustrated in Figs. 2a and b. In the early period after the afferent input, the network goes to the attractor made by the transient component and then turns back to the relatively weak but stable attractor made by the sustained component, which is the final state of the network without the perturbation. When we applied the perturbation, the trajectory deviated from the control condition and converged either of the active or silent states depending on the timing of the perturbation. If the network was perturbed during the early period of the transient phase (SOA=1), it could not reach the active state, and finally dropped to the silent state. However, if the network was perturbed during the late period (SOA=2, 15), the trajectory temporarily turned away but could still maintain the activity and finally converged to the active state.

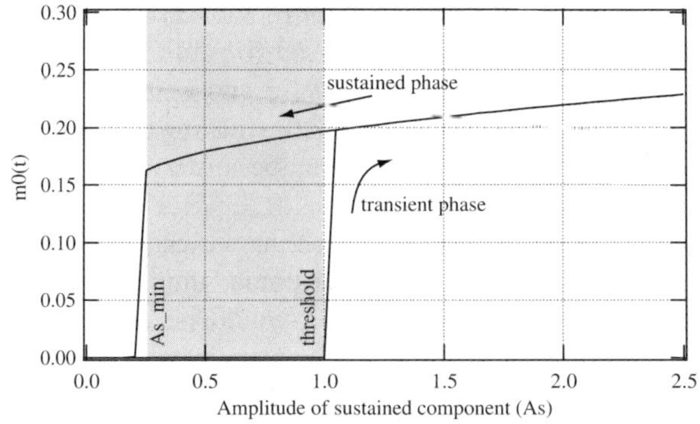

Fig. 1. Bistability of the network. If the transient component is sufficiently large, the final state simply depends on the sustained component. For A_s in the gray region, the network hold two attractors of the active and silent states. The network parameters used here were $\varepsilon = 0.1$, $\beta = 0.25$, $J_0 = 73$, $J_2 = 110$, $T = 1$ (see also Reference [4]).

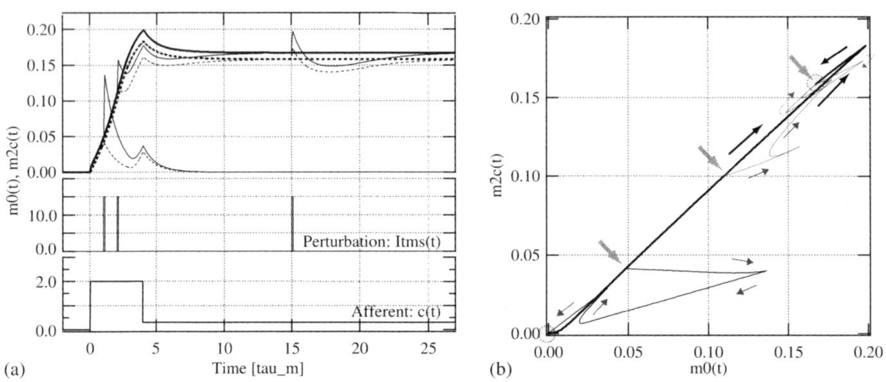

Fig. 2. (a) Time course of the order parameters with the perturbation. The results for SOA (stimulus onset asynchrony) = 1, 2, and 15 (thin lines) and the control condition without the perturbation (thick line) are shown. m_0 (solid line) transiently increased and then decreased for a relatively long period followed by a reduction of m_{2c} (dotted line). (b) The network state in the m_0-m_{2c} plane. Thin arrows indicate the direction of the network state. Thick arrows indicate the onset of the perturbation. The gray disks indicate the two attractors of the network. Depending on the timing of the perturbation, the network converged on either of the attractors. Parameters used here were $A_t = 2$, $W_t = 4\tau_m$, $I_{TMS} = 15$, $W_{TMS} = 0.1\tau_m$. Other parameters were the same as for Fig. 1.

These results indicate that there must be a critical SOA range where the perturbation can completely suppress the network activity. In addition, we can easily imagine that the susceptibility to the perturbation was not uniform in the critical period and there might be an optimal timing that maximizes the suppression effect of the perturbation. These characteristics can be measured through the minimum intensity of the perturbation needed to suppress the network activity for each SOA (Fig. 3). The minimum perturbation needed to suppress the network activity depended on the power of the

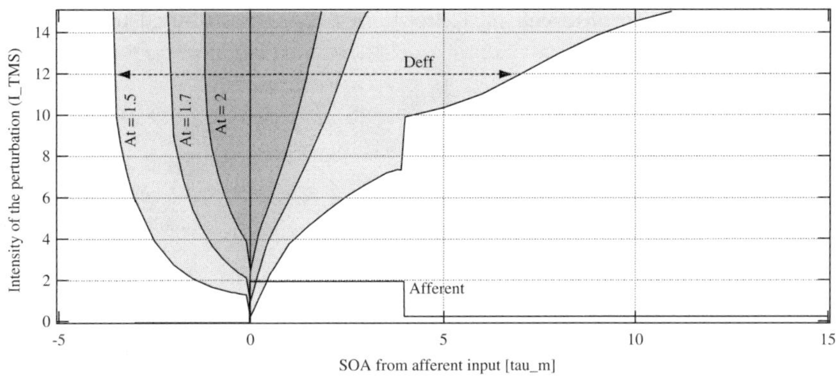

Fig. 3. Minimum perturbation intensity needed to suppress network activity. The width of each curve represents the SOA range (D_{eff}) within which network activity is suppressed. For example, when $A_t = 1.5$ and $I_{\mathrm{TMS}} = 12$, the perturbation delivered within an SOA range from -3.5 to 7 (i.e., $D_{\mathrm{eff}} = 10.5\tau_m$) can suppress network activity. The parameters were the same as for Fig. 2.

afferent input, so that the $I_{\mathrm{TMS}_{\min}}$-SOA curves fully shifted upward when a strong afferent input was given. Even so, the profiles of these curves consistently had basin shapes with a bottom at SOA=0 regardless of the intensity of the transient components. Inside these basin curves, the perturbation suppresses the activity, so the horizontal width of each basin represents the length of a critical SOA range. This length was over 10 times the time constant τ_m when the intensity of the transient component was relatively low.

4. Discussion

By observing the suppressive effects induced by a TMS-like perturbation in an simple neural population model, we have found that there is a critical period during which the perturbation can completely suppress network activity.

If we assume that the time constant τ_m is 10 ms, the duration of the perturbation would be 1 ms. This is consistent with the typical specification of commercially available TMS systems. The length of the critical SOA range would be over 100 ms at most. This is also consistent with typical experimental data of occipital TMS concerning the SOA range that can cause inhibitory effects on visual perception [2,3,7].

Parametric studies of occipital TMS [8] have shown that the degree of visual suppression depends on the TMS intensity and the brightness of the stimulus. In our computational study, we also found it difficult to suppress the network activity when the afferent input was strong. Amassian also reported that the suppression cannot be elicited if the characters are read too easily [1]. Therefore, the afferent input to the neural population (in this case, neurons in the visual cortex) might be slightly above the threshold. With regard to our simulation, it means that A_t should be limited to slightly above T, which was the condition under which the critical SOA range reached a large value (over 100 ms) comparable to the experimental data. These results suggest that the inhibitory effects observed in TMS studies can be attributed to interactions at the neural population level.

References

[1] V.E. Amassian, R.Q. Cracco, P.J. Maccabee, J.B. Cracco, in: A. Pascaul-Leone, N.J. Davey, J. Rothwell, E.M. Wasserman, B.K. Puri (Eds.), Handbook of Transcranial Magnetic Stimulation, Arnold Publisher, Paris, 2002, pp. 323–334.
[2] V.E. Amassian, R.Q. Cracco, P.J. Maccabee, J.B. Cracco, A. Rudell, L. Eberle, Electroencephalogr. Clin. Neurophysiol. 74 (1989) 458–462.
[3] G. Beckers, S. Zeki, Brain 118 (1995) 49–60.
[4] R. Ben-Yishai, R.L. Bar-Or, H. Sompolinsky, Proc. Natl. Acad. Sci. USA 92 (1995) 3844–3848.
[5] M. Inghilleri, A. Berardelli, G. Cruccu, M. Manfredi, J. Physiol. 466 (1993) 521–534.
[6] Y. Kamitani, V.M. Bhalodi, Y. Kubota, S. Shimojo, Neurocomputing 38–40 (2001) 697–703.
[7] Y. Kamitani, S. Shimojo, Nat. Neurosci. 2 (1999) 767–771.
[8] T. Kammer, H.G. Nusseck, Neuropsychologia 36 (1998) 1161–1166.
[9] S.S. Nagarajan, D.M. Durand, E.N. Warman, IEEE Trans. Biomed. Eng. 40 (1993) 1175–1188.
[10] B.J. Roth, P.J. Basser, IEEE Trans. Biomed. Eng. 37 (1990) 588–597.

Yoichi Miyawaki received his Ph.D. in Engineering from The University of Tokyo, Japan, in 2001. Since 2001, he has been a special postdoctoral researcher of Laboratory for Advanced Brain Signal Processing, RIKEN Brain Science Institute, Japan. His main focus of research is on human visual perception and noninvasive measurement of neural activity of the brain. He is currently working on the computational modeling of TMS-induced neural interference.

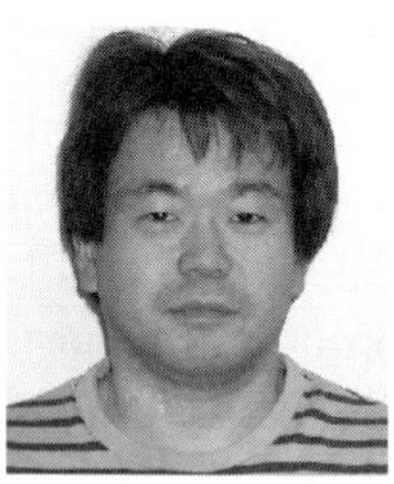

Masato Okada received his Ph.D. in Science from Osaka University, Japan, in 1997. From 1991 to 1996, he was a Research Associate of School of Engineering Science, Osaka University, Japan, and from 1996 to 2001, he was a Research Fellow of Japan Science and Technology Corporation. Since 2001, he has been the deputy lab head of Laboratory for Mathematical Neuroscience, RIKEN Brain Science Institute, Japan. His research field is physics and neuroscience including computational neuroscience, statistical physics, information theory, image processing, and learning theory.

Available at
www.ComputerScienceWeb.com
POWERED BY SCIENCE DIRECT

ELSEVIER

Neurocomputing 52–54 (2003) 843–848

NEUROCOMPUTING

www.elsevier.com/locate/neucom

Oscillatory modes in a neuronal network model with transmission latency

H. Nagashino[a,*], Y. Kinouchi[b]

[a]*School of Health Sciences, The University of Tokushima, Minami-Josanjima 2-1, Tokushima 770-8506, Japan*

[b]*Department of Electrical and Electronic Engineering, The University of Tokushima, Minami-Josanjima 2-1, Tokushima 770-8506, Japan*

Abstract

We analyzed the oscillatory activities in a neuronal network model as the basis of synchrony of the activities in the brain. The model consists of two groups of neurons that are interconnected. One group is composed of an excitatory and an inhibitory neuron which are expressed by Hodgkin–Huxley equations. The network shows different phase-locked oscillations and other activities depending on the structure and intensity of interconnection between groups or coupling of neurons in the group, or the value of transmission latency. In the present paper we show the oscillatory modes in symmetrical and asymmetrical networks with the mutual coupling from the inhibitory neuron in each group to the excitatory neuron in the other group.

Keywords: Neuronal network model; Oscillation; Synchrony; Hodgkin–Huxley equations; Latency

1. Introduction

The synchronization and phase-locked activities in different regions in the brain is one of the important features of information processing by temporal coding in the central nervous system [2]. To clarify the fundamental mechanism of the temporal coding [8], it is very significant to elucidate the phase-locked activities of the neuronal networks with different coupling configurations [4]. We analyzed the oscillatory activities in a neuronal network model which consists of two groups of neurons to investigate the basic characteristics of synchronization. The neurons are coupled with synaptic delay.

* Corresponding author. Fax: +81-886-56-7476.

E-mail address: nagasino@ee.tokushima-u.ac.jp (H. Nagashino).

0925-2312/03/$ - see front matter
doi:10.1016/S0925-2312(02)00812-3

Each group consists of excitatory and inhibitory neurons that are interconnected. The neurons that are synchronized in activities can be represented by a single neuron in the theoretical model. In the present simulation, therefore, each group in the model network is composed of an excitatory neuron and an inhibitory neuron for simplicity of analysis. The neurons are formulated by Hodgkin–Huxley equations [3].

We have demonstrated that in the network with mutual coupling of excitatory neurons in both groups, oscillatory activities of not only the in-phase mode which seems to be trivial but also anti-phase mode and the periodic solutions in which the phase difference of the oscillation in two groups of the neurons continuously changes with the coupling strength and latency [5].

In the present paper we show the results of the analysis of the oscillatory modes in symmetrical and asymmetrical networks with the mutual coupling from the inhibitory neuron in each group to the excitatory neuron in the other group. We demonstrate the regions of some parameters in the network in which different oscillatory solutions exist stably and show a variety of bistability in activities.

2. A neuronal network model

The dynamics of the membrane potentials of the neurons is formulated as

$$C_M \frac{\mathrm{d}\mathbf{V}_k}{\mathrm{d}t} = \mathbf{I}_{\mathrm{ext},k} + \mathbf{I}_{\mathrm{ion},k} + \sum_{j=1}^{2} \mathbf{W}_{kj}\mathbf{I}_{\mathrm{syn},j} \quad (k = 1, 2), \tag{1}$$

where

$$\mathbf{V}_k = \begin{pmatrix} V_{\mathrm{e}k} \\ V_{\mathrm{i}k} \end{pmatrix}, \qquad \mathbf{I}_{\mathrm{ext},k} = \begin{pmatrix} I_{\mathrm{ext,e}k} \\ I_{\mathrm{ext,i}k} \end{pmatrix},$$

$$\mathbf{I}_{\mathrm{ion},k} = \begin{pmatrix} I_{\mathrm{ion,e}k} \\ I_{\mathrm{i}k} \end{pmatrix}, \qquad \mathbf{I}_{\mathrm{syn},k} = \begin{pmatrix} I_{\mathrm{syn,e}k} \\ I_{\mathrm{syn,i}k} \end{pmatrix}.$$

In the above, $I_{\mathrm{ext},rk}$, $I_{\mathrm{ion},rk}$ and $I_{\mathrm{syn},rk}$ $(r = \mathrm{e,i})$ denote the external input current, the ion currents of the neuron and the synaptic current from the corresponding neuron in the network, respectively. The subscripts e and i denote excitatory and inhibitory neurons, respectively. The ion currents $I_{\mathrm{ion},rk}$ are formulated conventionally. Some formulations have been proposed for synaptic coupling [1,6,7]. The synaptic current $I_{\mathrm{syn},rk}$ is expressed here as

$$I_{\mathrm{syn}}(V) = \sum_J A_{\mathrm{syn}}(J) w \left\{ \exp\left(\frac{-t + t_{\mathrm{s}}(J) + \tau_{\mathrm{d}}}{\tau_+} \right) - \exp\left(\frac{-t + t_{\mathrm{s}}(J) + \tau_{\mathrm{d}}}{\tau_-} \right) \right\}, \tag{2}$$

where w is the coupling coefficient between the corresponding neurons and an element of the coupling matrix

$$\mathbf{W}_{kj} = \begin{pmatrix} a_{kj} & -c_{kj} \\ b_{kj} & -d_{kj} \end{pmatrix}.$$

The Jth instant when the membrane potential increases and exceeds -30 mV is denoted by $t_s(J)$ and the synaptic delay is denoted by τ_d. If the current of a previous firing is larger than 0.01, it is added to the present current and otherwise neglected. The factor $A_{syn}(J)$ is expressed as

$$A_{syn}(J) = \begin{cases} 1 & (\text{if } t \geqslant t_s(J) + \tau_d), \\ 0 & (\text{else}). \end{cases} \tag{3}$$

In the present simulations the values of the external input current $I_{ext,rk}$ and parameters C_M, τ_+ and τ_- are set as

$$I_{ext,rk} = 10\ \mu\text{A} \quad (r = \text{e}, \text{i};\ k = 1, 2),$$

$$C_M = 1\ \mu\text{F/cm}^2, \quad \tau_+ = 3\ \text{ms}, \quad \tau_- = 1\ \text{ms}.$$

3. Results

We analyzed the symmetrical networks of two identical groups. We made the coupling coefficients between the neurons in the groups denoted as $a_{11} = a_{22} = a_0$, $b_{11} = b_{22} = b_0$, $c_{11} = c_{22} = c_0$, and $d_{11} = d_{22} = d_0$. The bidirectional coupling coefficients between groups are equal and denoted as $a_{12} = a_{21} = a_1$, $b_{12} = b_{21} = b_1$, $c_{12} = c_{21} = c_1$, and $d_{12} = d_{21} = d_1$. The coupling coefficients inside the group are set as

$$a_0 = 10, \quad b_0 = 9, \quad c_0 = 10, \quad d_0 = 0,$$

when they are fixed. By the coupling given by a_0, it is assumed that a number of excitatory neurons are interconnected to act synchronously.

We investigated different configurations of mutual coupling between the groups. The network shows a variety of phase-locked oscillations and other oscillatory activities depending on the structure and intensity of interconnection between groups or coupling of neurons in the group, or the value of transmission latency. As in a number of oscillatory systems composed of two components, the present model generates in-phase mode, in which the phase difference of oscillation in two components is 0°, and anti-phase mode, in which the phase difference is 180°. We showed in a previous paper [5] the solution in which the two groups oscillate in continuously changing phase difference with the parameters of coupling and latency. This mode is referred to as intermediate-phase mode. In some networks chaotic oscillations and quasiperiodic oscillations are also observed. In the latter the phase difference of the oscillation gradually changes with time. This mode is referred to as varying phase mode. Fig. 1 shows the phase portrait and time series of an example of the varying phase mode.

In the present paper we show the results of the analysis of a network with one typical and symmetrical type of coupling. Both excitatory units are coupled bidirectionally, $c_1 \neq 0$, $a_1 = b_1 = d_1 = 0$. It is referred to as EI coupling. We analyzed the regions of oscillatory solutions in symmetrical networks with different values of coupling strength c_0 and c_1, and synaptic delay τ_d as well as asymmetrical networks in the sense that $c_{11} \neq c_{22}$. The regions of oscillatory solutions that are found in the model network

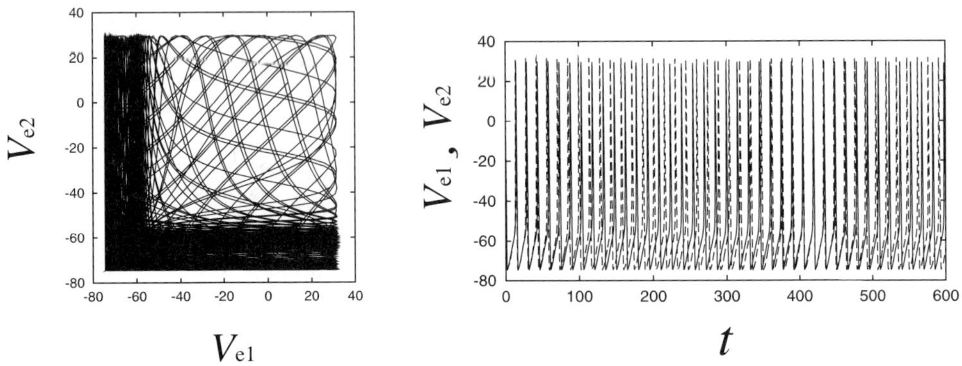

Fig. 1. An example of phase portrait and time series of varying phase mode.

are illustrated in Fig. 2. Each illustration consists of two rows. The upper row shows the ranges of the oscillatory solutions that emerge with the initial conditions where the in-phase mode would be likely to occur. The lower shows those with the initial conditions where the anti-phase mode would be likely to occur. The parameter value shown in the right end shows the one at which the analysis is started.

In the symmetrical networks in-phase mode and anti-phase mode coexist in large range of parameters c_0, c_1 and τ_d. There exists intermediate-phase mode in which the phase difference of the oscillation in the two groups varies from $0°$ to about $80°$ with the change of the parameters c_0, c_1 and τ_d. It coexists with in-phase mode or anti-phase mode. With the synaptic delay being larger than a critical value only in-phase mode exits. Chaotic activities are also observed with small values of c_0. It coexists with anti-phase mode. In the asymmetrical network in which c_{11} and c_{22} are different, the mode coexistence is similar to the case of the symmetrical network. In Fig. 2 (d) the center of the rows corresponds to the symmetrical network. The region of in-phase mode becomes smaller and that of intermediate-phase mode becomes larger. Varying phase mode emerges with the small values of b_{22} or large values of b_{22}, in other words, when the asymmetricity is large.

4. Conclusions

We analyzed the oscillatory activities in a neuronal network model as the basis of synchrony of the activities in the brain. The model consists of two groups of neurons that are interconnected. One group is composed of an excitatory and an inhibitory neuron which are expressed by Hodgkin–Huxley equations. The network shows different phase-locked oscillations and other oscillatory activities depending on the structure and intensity of interconnection between groups or coupling of neurons in the group, or the value of synaptic latency.

In this paper we showed the regions of oscillatory modes in the coupling strength and synaptic delay in symmetrical and asymmetrical networks with the mutual coupling

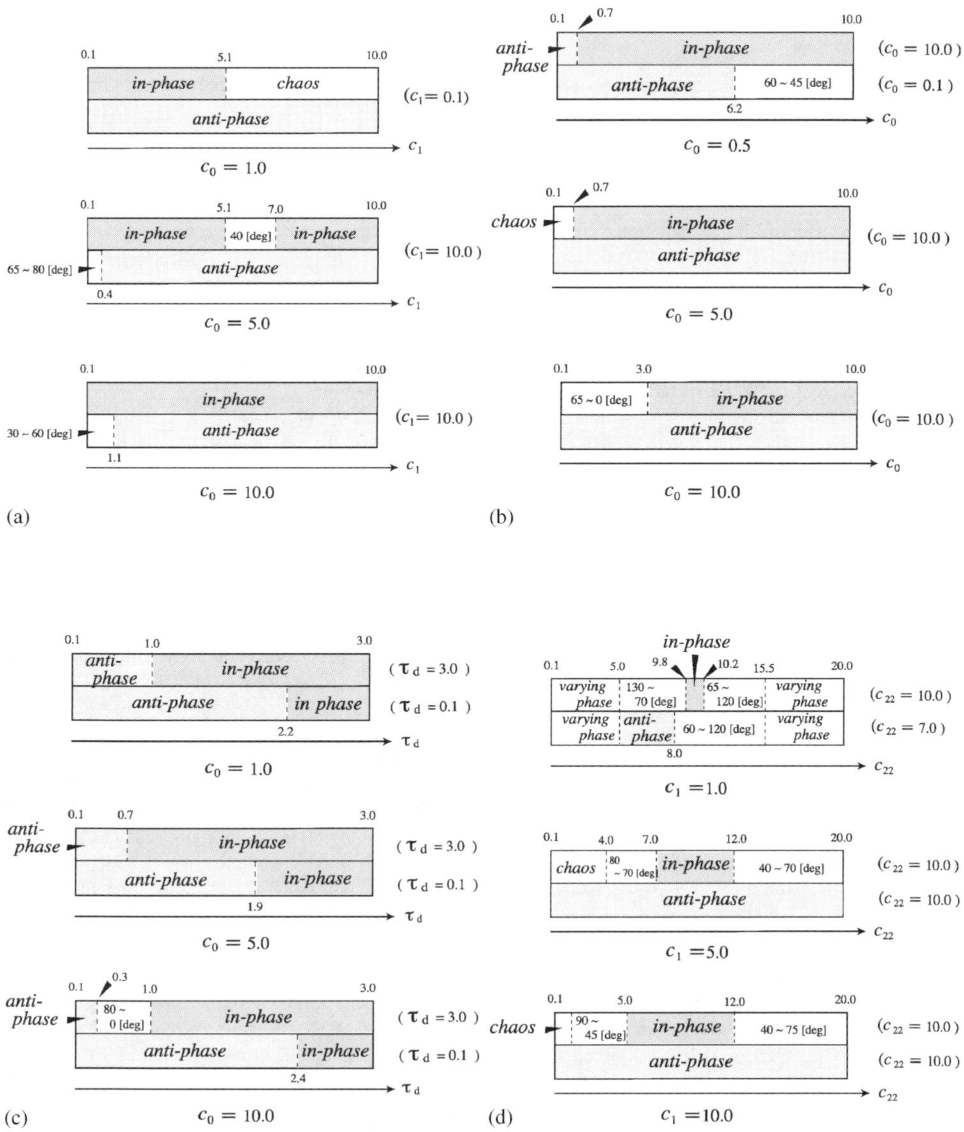

Fig. 2. The regions of different oscillatory modes in EI coupling: (a), (b) and (c) symmetrical networks; (a) and (b) $\tau_d = 1$; (c) $c_1 = 5$; (d) an asymmetrical network, $\tau_d = 1$, $c_{11} = 10$.

from the inhibitory neuron in each group to the excitatory neuron in the other group. The model possesses a variety of bistability in its activities. The rich variety of the results suggests that it is necessary to make further analysis of the networks with different coupling configurations and different asymmetricities.

Acknowledgements

The authors are most grateful to K. Kawamoto for his help with computer simulation.

References

[1] B. Cartling, Control of computational dynamics of coupled integrate-and fire neurons, Biol. Cybernet. 76 (1997) 383–395.

[2] C.M. Gray, P. Konig, A.K. Engel, W. Singer, Oscillatory responses in cat visual cortex exhibit intercolumnar synchronization which reflects global stimulus properties, Nature (London) 338 (1989) 334–337.

[3] A.L. Hodgkin, A.F. Huxley, A quantitative description of membrane current and its application to conduction and excitation in nerve, J. Physiol. 117 (1952) 500–544.

[4] K. Kawamoto, H. Nagashino, Y. Kinouchi, Oscillatory modes in coupled nonidentical neural oscillators, Proceedings of the 2000 International Conference on Nonlinear Theory and its Applications, Vol. 1, 2000, pp. 277–280.

[5] H. Nagashino, K. Kawamoto, Y. Kinouchi, Phase-locked oscillations in a neuronal network model, Neurocomputing 44–46 (2002) 555–560.

[6] I. Segev, J.W. Fleshan, R.E. Burke, Compartmental models of complex neurons, in: C. Koch, I. Segev (Eds.), Methods in Neuronal Modeling, Vol. 63, MIT Press, Cambridge, 1989.

[7] M. Stiber, R. Ieong, J.P. Segundo, Responses to transients in living and simulated neurons, IEEE Trans. Neural Networks 8 (6) (1997) 1379–1385.

[8] C. von der Malsburg, W. Schneider, A neural cocktail-party processor, Biol. Cybernet. 54 (1986) 29–40.

Hirofumi Nagashino received the B.E. and M.E. degrees in Electrical Engineering from The University of Tokushima in 1972 and 1974, respectively. He received the Doctor of Engineering degree from Osaka University, Japan, in 1982. In 1974 he joined Department of Electrical Engineering, Faculty of Engineering, The University of Tokushima. Currently he is a professor in Major in Radiologic Science, School of Health Sciences, The University of Tokushima, Japan. His research interest includes neural network models for synchronous activities and phase-locked oscillations in the nervous system, and application of artificial neural networks to biomedical engineering such as source localization from EEG, system identification from biological time series data and magnetic measurement system of movements.

Yohsuke Kinouchi received the B.E. and M.E. degrees in Electrical Engineering from The University of Tokushima in 1966 and 1968, respectively. He received the Doctor of Engineering degree from Kyoto University, Japan, in 1975. In 1968 he joined Department of Electrical Engineering, Faculty of Engineering, The University of Tokushima. Currently he is a professor in Department of Electrical and Electronic Engineering, The University of Tokushima, Japan. His research interest includes neural network models for oscillatory behaviors, influence of the magnetic fields on cells, dental application of magnet, measurement of bioimpedance, ultrasonic measurement system of arterial blood flow and application of artificial neural networks to biomedical engineering.

Available at
www.ComputerScienceWeb.com
POWERED BY SCIENCE @DIRECT®

ELSEVIER

Neurocomputing 52–54 (2003) 849–855

NEUROCOMPUTING

www.elsevier.com/locate/neucom

Change-point detection in neuronal spike train activity

Rama Ratnam[a], Jozien B.M. Goense[a,c], Mark E. Nelson[a,b,c,*]

[a]*Beckman Institute for Advanced Science and Technology, University of Illinois, 405 N. Mathews, Urbana, IL 61801, USA*

[b]*Department of Molecular and Integrative Physiology, University of Illinois, Urbana, IL 61801, USA*

[c]*Center for Biophysics & Computational Biology, University of Illinois, Urbana-Champaign, Urbana, IL 61801, USA*

Abstract

Animals respond to changes in their environment based on the information encoded in neuronal spike activity. One key issue is to determine how quickly and reliably the system can detect that a behaviorally relevant change has taken place. What are the neural mechanisms and computational principles that allow fast, reliable detection of changes in spike activity? Here we present an optimal statistical signal-processing algorithm for change-point detection, known as the cumulative sum (CUSUM) algorithm. We then show that the performance of a simple neuron model with leaky-integrate-and-fire dynamics can approach theoretically optimal performance limits under certain conditions.

Keywords: Spike train analysis; Signal detection; Change-point detection; CUSUM; Neural coding

1. Introduction

An important aspect of behavior is to evaluate choices and make decisions. At the neural level, the information to support these processes is carried in large part by the time-varying activity of spiking neurons. For an animal to respond quickly and reliably to changes in the environment, it must draw inferences in real time from neuronal spike train data. It is therefore useful to ask what sorts of neural algorithms

* Corresponding author. Beckman Institute for Advanced Science and Technology, University of Illinois, 405 N. Mathews, Urbana, IL 61801, USA.

E-mail address: m-nelson@uiuc.edu (M.E. Nelson).

doi:10.1016/S0925-2312(02)00815-9

might be employed for detecting changes in spike activity in real time, and how does the performance of these neural algorithms compare with known theoretical results on optimal signal detection?

Here we present an optimal statistical signal-processing algorithm for detecting abrupt changes in spike activity, based on an established technique known as the cumulative sum (CUSUM) method [1]. We then compare the detection performance of a simple neuron model with the optimal signal-processing algorithm. We show that the performance of a neuron model with leaky-integrate-and-fire dynamics can approach the theoretically optimal performance under appropriate conditions. Our interest in this topic is motivated, in part, by questions about how weakly electric fish detect and localize prey based on subtle changes in electrosensory afferent spike activity [6,8].

2. Defining the problem

Consider a spike train as a sequence of random interspike intervals (ISIs). Initially, the intervals $I_1, I_2, \ldots$ are distributed according to some probability density function (PDF) $f_0(I)$, which can be empirically estimated from the ISI distribution. At some unknown point in the ISI sequence, an abrupt shift occurs such that interval m and all subsequent intervals $I_m, I_{m+1}, \ldots$ are distributed according to a new PDF $f_1(I)$. The change in the underlying PDF will be reflected as a change in the observed ISI distribution following the change-point m. Before and after the change-point, the distributions are assumed to be stationary and the ISIs are assumed to be independent. Detection performance is measured in terms of the mean detection delay as a function of the mean time between false alarms. False alarms occur when the algorithm signals that a change has occurred in the absence of any real change in the underlying PDF.

3. The CUSUM algorithm

Consider sampling a single ISI I_k from a spike train without knowing whether it occurred before or after the change-point m. We wish to distinguish between two hypotheses: that the ISI occurred before the change (H_0), or that the ISI occurred after the change (H_1). The log-likelihood of H_1 relative to H_0 based on this single observation is

$$s(I_k) = \ln \frac{f_1(I_k)}{f_0(I_k)}. \tag{1}$$

The CUSUM algorithm is an iterative algorithm for processing consecutive samples based on a CUSUM of log-likelihood ratios. When the CUSUM crosses an upper threshold level h it indicates that the accumulated evidence in favor of H_1 is sufficient to signal that a change has occurred. Crossing a lower threshold level at zero indicates that the accumulated evidence favors hypothesis H_0 and that the CUSUM should be reset to zero. Starting with the first ISI in the spike train, I_1, and proceeding sequentially, the update rule for the CUSUM algorithm is

$$g_k = \begin{cases} g_{k-1} + s(I_k) & \text{if } g_{k-1} + s(I_k) > 0, \\ 0 & \text{otherwise,} \end{cases} \tag{2}$$

with the initial condition $g_0 = 0$. The detection of a change is signaled following observation of the nth ISI, if $g_n = h$. The detection delay d, measured in intervals, is defined by $d = n - m + 1$, where m is the index of the first ISI after the actual change-point. The CUSUM algorithm, which was first introduced by [7], has been shown to be asymptotically optimal for solving change detection problems involving independent and identically distributed samples drawn from two known PDFs [3]. Others have extended these results to establish non-asymptotic optimality under certain conditions [4,9] and optimality for processes that are not necessarily independent and identically distributed [5].

4. Applying CUSUM to spike train data

An application of the CUSUM algorithm to simulated spike train data is illustrated in Fig. 1. ISI values were randomly drawn from a gamma distribution of order n and mean $\bar{I}$. The PDF for a gamma distribution of order n and mean $\bar{I}$ is given by

$$f(I) = \frac{n^n}{\bar{I}^n (n-1)!} I^{n-1} \exp[-nI/\bar{I}]. \tag{3}$$

The parameters for the initial PDF f_0 were $n = 8$ and $\bar{I}_0 = 0.020$ s (corresponding to a mean rate $\bar{R}_0 = 50$ Hz); the resulting ISI distribution is illustrated in Fig. 1A. After the change-point, the new PDF f_1 had a mean $\bar{I}_1 = 0.015$ s (corresponding to a mean rate $\bar{R}_1 = 66.7$ Hz) and n remained unchanged; the resulting ISI distribution is shown in Fig. 1B. Fig. 1C shows a sample spike train before and after the change-point.

Following Eq. (1), the log-likelihood ratio for an ISI of duration I in this example is

$$s(I) = \ln \frac{f_1(I)}{f_0(I)} = \ln \frac{[n^n/\bar{I}_1^n(n-1)!]I^{n-1} \exp[-nI/\bar{I}_1]}{[n^n/\bar{I}_0^n(n-1)!]I^{n-1} \exp[-nI/\bar{I}_0]} = n\left(\ln \frac{\bar{R}_1}{\bar{R}_0} - \Delta R I\right), \tag{4}$$

where $\bar{R}_0 = 1/\bar{I}_0$, $\bar{R}_1 = 1/\bar{I}_1$, and $\Delta R = \bar{R}_1 - \bar{R}_0$. Substituting numerical values from this particular example yields $s(I) \cong 2.3 - 133I$. Using this expression for $s(I)$, the CUSUM update rule (Eq. (2)) is evaluated after the observation of each ISI, as illustrated by the solid points in Fig. 1D. When the value of g_k (solid points) crosses above the threshold level h (horizontal dashed line), a detection event is signaled and the update process is terminated.

Although the CUSUM output g_k is defined only at the end of each ISI, the form of the log-likelihood function $s(I)$ suggests an interpolation that will be useful when

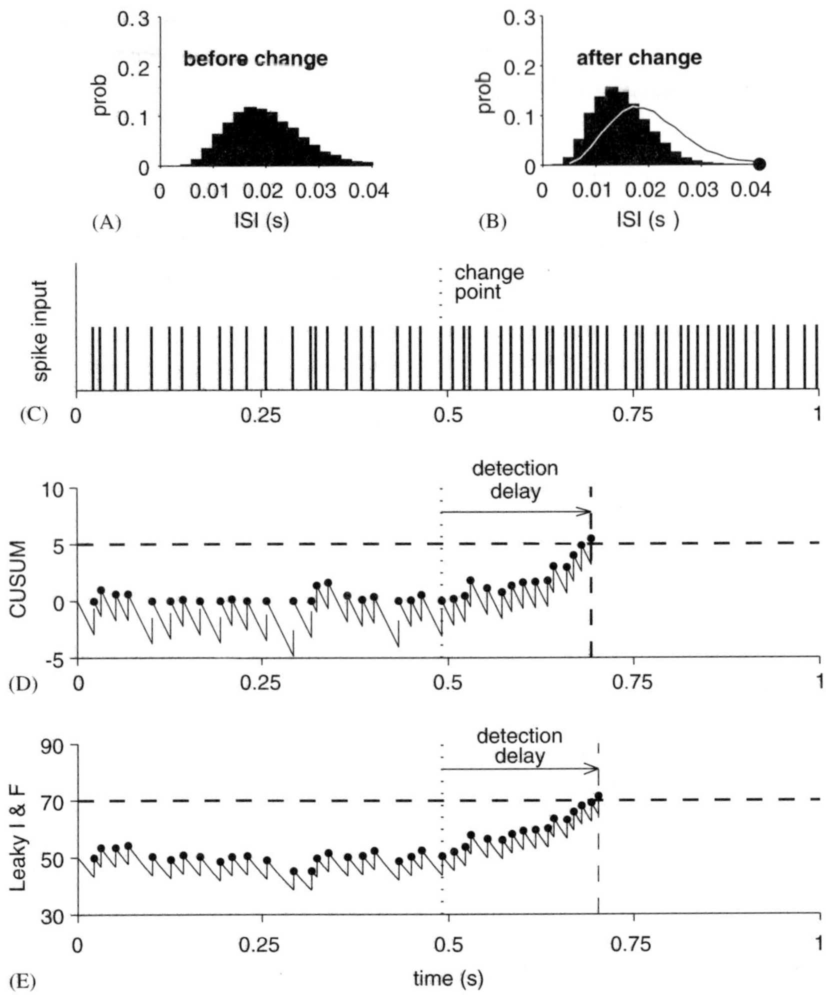

Fig. 1. Change-point detection in simulated spike train data: (A) ISI distribution before the change-point. (B) ISI distribution after the change-point. For reference the original ISI distribution is shown as a solid line. (C) Spike train data before and after the change-point, which is indicated by the vertical dotted line. (D) The CUSUM algorithm applied to the spike train data. Filled points represent the discrete values of g_k evaluated at the end of each ISI according to Eq. (2). The saw tooth pattern is the continuous interpolation of g described in the text. Horizontal dashed line is the threshold level h; vertical dashed line indicates the detection time. (E) Response of a leaky-integrate-and-fire neuron model ($\tau = 0.150$ s).

drawing analogies with neural implementations. Note that $s(I)$ consists of a constant term, $n\ln(\bar{R}_1/\bar{R}_0)$, and a term that scales linearly with the duration I of the ISI, $-n\,\Delta R I$. We can construct an interpolated version of g that decays linearly with a slope of $-n\,\Delta R$ between spikes and is boosted by a constant amount $n\ln(\bar{R}_1/\bar{R}_0)$ upon the arrival of a spike at the end of each ISI, as shown in Fig. 1D.

5. CUSUM performance and statement of optimality

Once the PDFs, f_0 and f_1, have been defined, the CUSUM algorithm has only one free parameter, the threshold level h. As h is increased, both the mean time between false alarms $\bar{T}$ and the mean detection delay $\bar{d}$ will increase. This tradeoff is illustrated in Fig. 2 for the change-detection problem formulated in Fig. 1. Empirically, the mean number of intervals between false alarms scales exponentially with threshold level, $\bar{T} \propto e^h$, while the mean detection delay (in intervals) scales linearly with threshold level, $\bar{d} \propto h$. Thus, a semi-logarithmic plot of $\bar{d}$ versus $\bar{T}$ for the CUSUM algorithm results in a nearly straight line, as shown in Fig. 2A.

We now consider a new quantity known as the *worst mean delay*, $\bar{d}_{wc}$. Just prior to the change-point, the state variable g_k (Eq. (2)) can potentially take on any value between 0 and h due to fluctuations in the baseline activity. The worst case, in terms

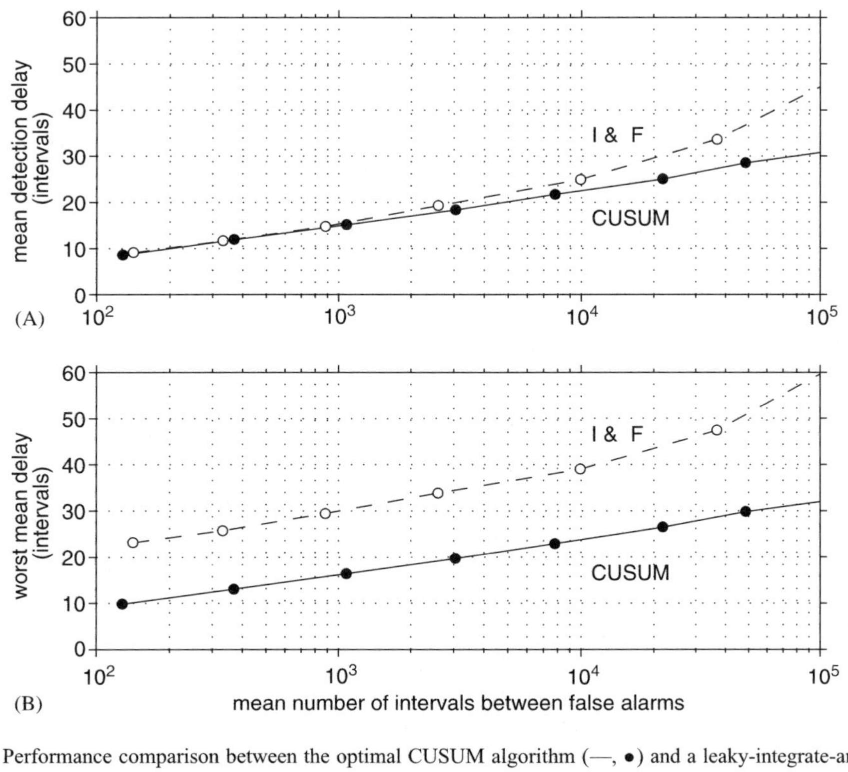

Fig. 2. Performance comparison between the optimal CUSUM algorithm (—, •) and a leaky-integrate-and-fire neuron model (- - -, ∘) for different threshold levels h. (A) Mean detection delay versus the mean time between false alarms. (B) Worst mean detection delay versus mean time between false alarms. The mean time between false alarms was computed from a random sequence of 10^6 ISIs drawn from the baseline distribution f_0. The mean detection delay was evaluated based on 10^3 random trials involving a change in the underlying PDF from f_0 to f_1. For evaluating worst-case delays, the relevant state variable (g or v) was set to zero just prior to the change-point.

of detection delay, will occur when g_k happens to start off at its lowest possible value, namely 0. The worst mean delay is always longer than the actual mean delay, and hence provides an upper bound. Asymptotically, as $\bar{T} \to \infty$, the CUSUM algorithm is known to produce the minimum worst mean delay $\bar{d}_{\mathrm{wc}}$ for a fixed mean time between false alarms $\bar{T}$ [1,3].

6. Leaky-integrate-and-fire as a biologically plausible implementation

How might a neuron in the brain implement a CUSUM-like detection algorithm? The saw tooth pattern of boosts and decays seen in the interpolated version of the CUSUM algorithm in Fig. 1D suggests a possible neural implementation. The time course is strikingly similar to the changes in membrane potential that are associated with integration of spike events by a postsynaptic neuron. This is illustrated using a leaky-integrate-and-fire neuron model (Fig. 1E) in which the membrane potential obeys

$$\frac{\mathrm{d}v}{\mathrm{d}t} = -\frac{v}{\tau} + \frac{x(t)}{\tau}, \tag{5}$$

where $x(t)$ is the input spike train, represented as a series of delta functions. For convenience, we have scaled the input give unity gain at DC. The membrane voltage is integrated according to Eq. (5) until it reaches a threshold level h, at which point an output spike is generated, signaling that a change has been detected. To parallel the structure of the CUSUM algorithm, we can recast Eq. (5) in terms of a discrete update rule that is evaluated at the end of each ISI I_k:

$$v_k = v_{k-1} \exp(-I_k/\tau) + 1/\tau. \tag{6}$$

Thus the membrane voltage decays exponentially between spikes and is boosted by a constant amount $1/\tau$ upon the arrival of a spike at the end of each ISI.

We can characterize the performance of the neuron model by plotting the mean detection delay versus the mean time between false alarms for different values of h. However, we now have an additional free parameter—the integration time constant τ. If τ is very small, the voltage will decay to zero before the next spike, and spike events will not summate; if τ is very large, summation will occur, but the system will be slow to respond. Intuitively, there should be an intermediate value of τ that will yield the best detection performance. A full parametric study of this issue is beyond the scope of this paper. For illustration purposes we have chosen $\tau = 0.150$ s, which was empirically determined to yield the best detection performance in this particular example. The dashed lines in Fig. 2 show the detection performance of the leaky-integrate-and-fire model, in comparison to the theoretically optimal CUSUM algorithm. The mean detection delay of the neuron model closely approaches the CUSUM model over much of the range (Fig. 2A), but the worst mean delay of the neuron model is considerably worse (Fig. 2B). Unlike the CUSUM algorithm, the neuron model does not have a reset mechanism when the accumulated evidence has swung in favor of hypothesis H_0. Thus random fluctuations in baseline activity just prior to the change-point can allow the membrane voltage to decay significantly below its baseline level, leading to longer detection delays. Despite poorer performance on worst-case scenarios, the typical performance of the leaky-integrate-and-fire model is near optimal.

7. Discussion

Based on the analytical formulation of the CUSUM update rule and log-likelihood ratio (Eqs. (2) and (4)), it might not be readily apparent how the nervous system could implement such an algorithm. Observing the time course of the CUSUM state variable (Fig. 1D), however, suggests that leaky integration of spike activity could provide a biologically plausible approximation to the CUSUM algorithm. Indeed, we have shown that a neuron model with leaky-integrate-and-fire model dynamics can approach the performance of the theoretically optimal algorithm in one particular example (Fig. 2A). Recently, we have used the leaky-integrate-and-fire neuron model to detect changes in empirical spike train data recorded from primary electrosensory afferents [2]. The results of this empirical study reveal an impressive sensitivity of the detection algorithm to small changes in electrosensory afferent spike activity.

Acknowledgements

This work was supported by a grant from the National Science Foundation (IBN 0078206).

References

[1] M. Basseville, I.V. Nikiforov, Detection of Abrupt Changes: Theory and Application, Prentice-Hall, Englewood Cliffs, NJ, 1993 (This out-of-print book can be downloaded at http://www.irisa.fr/sigma2/kniga/).

[2] J.B.M. Goense, R. Ratnam, M.E. Nelson, Burst firing improves the detection of weak signals in spike trains, Neurocomput., this issue.

[3] G. Lorden, Procedures for reacting to a change in distribution, Ann. Math. Stat. 42 (1976) 1897–1908.

[4] G.V. Moustakides, Optimal stopping times for detecting changes in distributions, Ann. Stat. 14 (1986) 1379–1387.

[5] G.V. Moustakides, Quickest detection of abrupt changes for a class of random processes, IEEE Trans. Inf. Theory 44 (1998) 1965–1968.

[6] M.E. Nelson, M. MacIver, Prey capture in the weakly electric fish *Apteronotus albifrons*: sensory acquisition strategies and electrosensory consequences, J. Exp. Biol. 202 (1999) 1195–1203.

[7] E.S. Page, Continuous inspection schemes, Biometrika 14 (1954) 100–115.

[8] R. Ratnam, M.E. Nelson, Nonrenewal statistics of electrosensory afferent spike trains: implications for the detection of weak sensory signal, J. Neurosci. 20 (2000) 6672–6683.

[9] Y. Ritov, Decision theoretic optimality of the CUSUM procedure, Ann. Stat. 18 (1990) 1464–1469.

Available at www.ComputerScienceWeb.com
POWERED BY SCIENCE @ DIRECT®

ELSEVIER

Neurocomputing 52–54 (2003) 857 – 862

NEUROCOMPUTING

www.elsevier.com/locate/neucom

Dynamics of human sleep EEG

E. Olbrich[a,*], P. Achermann[b], P.F. Meier[a]

[a] *Physics Institute, University of Zürich, CH-8057 Zürich, Switzerland*
[b] *Institute of Pharmacology and Toxicology, University of Zürich, CH-8057 Zürich, Switzerland*

Abstract

Several investigators of EEG time series reported a rejection of the null hypothesis of linear stochastic dynamics for epochs longer than 10 s. We examine whether this rejection is related to nonlinearity or to nonstationarity. Our approach is a combination of autoregressive (AR-) modeling and surrogate data testing. It is shown that the fraction of subsegments, for which the null hypothesis has to be rejected, increases with the length of the subsegments and can be related to fluctuations of the AR-coefficients on time scales in the range from 2 to 30 s.

Keywords: Sleep EEG; Nonlinearity; Nonstationarity; AR-model; Surrogate data

1. Introduction

During the last decades, many attempts were made to apply methods from nonlinear dynamics to analyze EEG time series. It turned out; however, that the assumption that the EEG can be described by a deterministic chaotic system does not hold in almost all cases. Today most researchers describe the EEG as a stochastic process. It is, however, still an open question to which extent one has to consider nonlinear effects.

If the EEG is given as a time series $x(t), x(t+\Delta t), \ldots, x(t+N\,\Delta t)$, a very general ansatz for a discrete time model of order m is given by

$$x(t+\Delta t) = f(x(t), x(t-\Delta t), \ldots, x(t-m\,\Delta t), \vec{p}, \vec{\xi}), \tag{1}$$

where $\vec{p}$ is a parameter vector and $\vec{\xi}$ represents a normal-distributed white noise term. The system is considered as nonstationary, if the parameters depend on time, and as

* Corresponding author. Tel.: +41-1-6354585; fax: +41-1-6355704.
E-mail address: olbrich@physik.unizh.ch (E. Olbrich).

0925-2312/03/$ - see front matter
doi:10.1016/S0925-2312(02)00816-0

nonlinear, if the function f is nonlinear. Note that this distinction is not unique. In certain cases it might be possible to transform a linear model with time-dependent parameters into a stationary nonlinear model by increasing the model order m.

For time series of human sleep EEG, the surrogate data analysis revealed evidences for a rejection of the null hypothesis of a linear stochastic process (see e.g. [2,4,6]). These analyses were performed on segments with a length larger than 10 s (16 s in [4] and 20.5, 41, 82 and 164 s in [2]). Because the null hypothesis includes also stationarity, its rejection can be both due to nonlinearity and nonstationarity and the test cannot distinguish between these two possibilities.

We study the relation between nonlinearity and nonstationarity by modeling the dynamics on short time scales by linear autoregressive (AR-) models. It was shown in [3] that AR-modeling of short segments of a duration of 1 s can catch essential features of human sleep EEG.

The possible influence of nonstationarity is investigated by studying how the rejection rate of the null hypothesis depends on the length of the subsegments.

2. Data and methods

The sleep EEG (C_3-A_2-derivation) of four healthy male subjects was recorded during 8 h in the sleep laboratory. Data were stored on a PC with a resolution of 128 Hz. For details of the recording and signal conditioning see [6]. Sleep stages were scored visually according to the criteria of Rechtschaffen and Kales [5]. 361 segments with a duration of 1 min (7680 points) were selected after visual inspection for lack of obvious nonstationarity due to artefacts and arousals. Only 1-min segments containing a definite sleep stage (REM sleep, stage 2 or slow-wave sleep (SWS, stages 3 and 4)) were considered. The segments from sleep stage 2 were divided into two classes: Class 2A consists of segments with sharp waves crossing the -100 μV threshold, which are mostly K-complexes, while the second class, 2B, comprises segments with small or no sharp waves.

For further analysis, the segments were divided into subsegments with length ranging from $T = 1$ to 30 s and AR-models were estimated using Burgs algorithm:

$$x(t + \Delta t) = \sum_{i=1}^{m} a_i x(t - (i-1)\Delta t) + \sigma_r \xi. \qquad (2)$$

The parameter vector $\vec{p}$ is spanned by the AR-coefficients a_i and the standard deviation of the residuals σ_r. The fitted AR-model represents the null hypothesis for the corresponding subsegment. To avoid over-fitting on the short segments the model order was estimated using Akaike's information criterion (AIC) [1]. However, to compare the fluctuations of the AR-coefficients (Fig. 3), the order of the model was fixed to six.

For every subsegment 100 surrogate data segments were produced by iterating the fitted model with randomly shuffled residuals. The model was tested using the correlation entropy based on the correlation sum

$$C(r) = \frac{2}{N(N-1)} \sum_{i=1}^{N} \sum_{j=i+1}^{N} \Theta(r - \|\vec{X}_i - \vec{X}_j\|).$$

$\vec{X}_i = (x(i), x(i-\tau), \ldots, x(i-(m-1)\tau))$ are the state vectors and $\| \cdots \|$ denotes the maximum norm. The delay time τ was chosen as $\tau = 3\Delta t$ with Δt being the sampling time. The embedding dimension was fixed to $m = 3$. The length scale r was set to $r = 0.5\sigma_x$, whereas σ_x is the standard deviation of the samples of the corresponding subsegment. There is a tradeoff between the embedding dimension and the length scale due to the finite segment length.

The correlation entropy is defined by

$$H = -\ln C(r).$$

It characterizes the marginal distribution of the $\vec{X}_i$.

The correlation entropy of the original data set was compared to the entropy of the surrogate data. If the entropy of the original data set was larger or smaller than the entropy of all of the 100 surrogate data sets, the null hypothesis had to be rejected with a probability $p < 0.02$ for false rejection.

3. Results

Firstly, we tested subsegments of length $T = 1$ s. The fraction of original segments for which the null hypothesis was rejected is smaller than 2% for all stages (Table 1) and therefore smaller than the false rejection rate due to statistical fluctuations. Therefore, we found no significant nonlinearity on this time scale. Interestingly, neither sharp waves nor sleep spindles or slow waves cause a significant rejection of the null hypothesis. However, their occurrence can be detected in the coefficients of the AR- models of the corresponding subsegments.

If the length of the subsegments is enlarged, the fraction of rejected segments increases. The first reason for this increase is that the distribution of the values of H_{surr} becomes more narrow. Its standard deviation σ_{surr} decreases approximately proportional to $T^{-1/2}$. This phenomenon can be observed also for surrogate data tests in stationary

Table 1
Fraction of subsegments for which the null hypothesis is rejected ($p < 0.01$)

Segment length T (s)	REMS (%)	Stage 2A (%)	Stage 2B (%)	SWS (%)
1	0.35	0.52	0.43	0.22
2	0.58	2.79	0.80	0.41
4	0.82	10.08	2.78	1.15
10	3.8	16.12	13.37	4.37
30	8.77	88.95	49.48	18.85
60	19.3	86.05	67.71	44.26

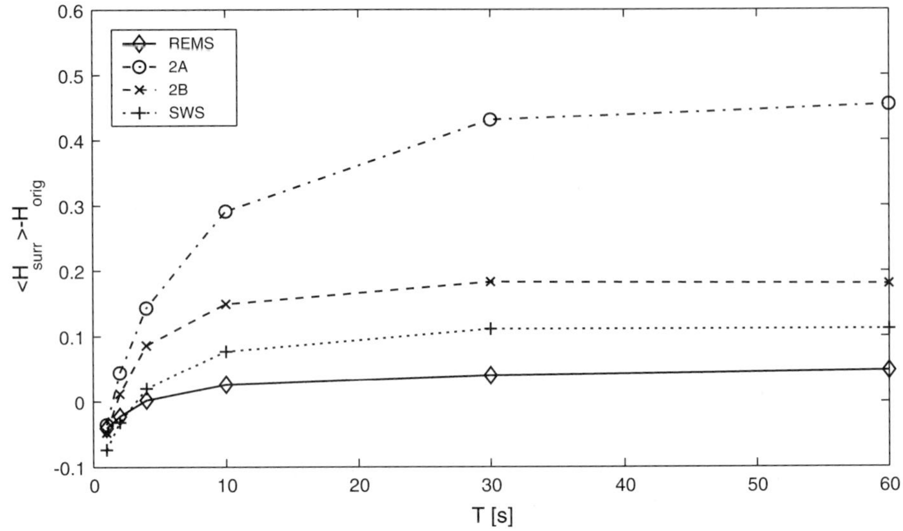

Fig. 1. Expectation values of $\langle H_{\mathrm{surr}}\rangle - H_{\mathrm{orig}}$ for different segment lengths.

nonlinear systems. A second possible reason for this increase is that the expectation value of the difference $\Delta H = \langle H_{\mathrm{surr}}\rangle - H_{\mathrm{orig}}$ itself becomes larger. This can be interpreted as an effect of nonstationarity, because for a sufficiently large sample from a stationary system, this difference is independent of the length of the segment. Fig. 1 shows the averaged behavior of these differences for the four groups depending on the length of the subsegments.

It is clearly seen that the largest differences ΔH can be observed for the two classes from sleep stage 2, which might be related to the nonstationarity introduced by the occurrence of K-complexes and sleep spindles. However, also in SWS and REM sleep significant nonstationarity is detected.

Fig. 2 shows the time course of ΔH over the night for one of the four subjects estimated from nonoverlapping 60-s segments. REM sleep and SWS are characterized by values of ΔH which lie almost inside the interval spanned by the surrogates.

Because we could not reject the null hypothesis for 1-s subsegments, AR-models (2) are a sufficiently good description for these subsegments. The nonstationarity concluded from the surrogate data analysis should be reflected in the fluctuations of its parameters. Fig. 3 shows the ratio between the significant part of this fluctuations and the statistical fluctuations for the coefficients a_1, a_2, a_3 and σ_r. It was estimated by averaging for every sleep stage the difference between the standard deviation of the parameters from the 60 1-s subsegments of every 1-min segment and the standard deviation for the estimator of the respective parameter normalized to the latter. The largest fluctuations are found in sleep stage 2A corresponding to the results of the surrogate data analysis.

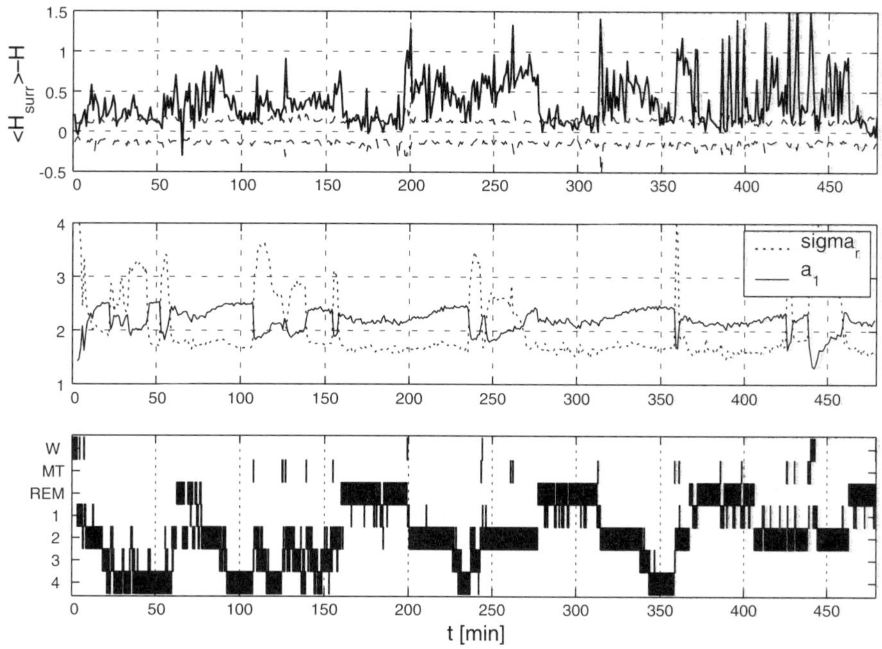

Fig. 2. Upper panel: ΔH of original data (—) and maximal and minimal ΔH of 100 surrogates (···) for nonoverlapping 60-s segments. Middle panel: σ_r and a_1 from 1-s subsegments averaged over 60-s segments. The large jumps are artefacts due to varying levels of tonic muscle activity. Lower panel: hypnogram; W=waking, MT=movement time, REM=REM sleep, 1–4: non-REM sleep stages.

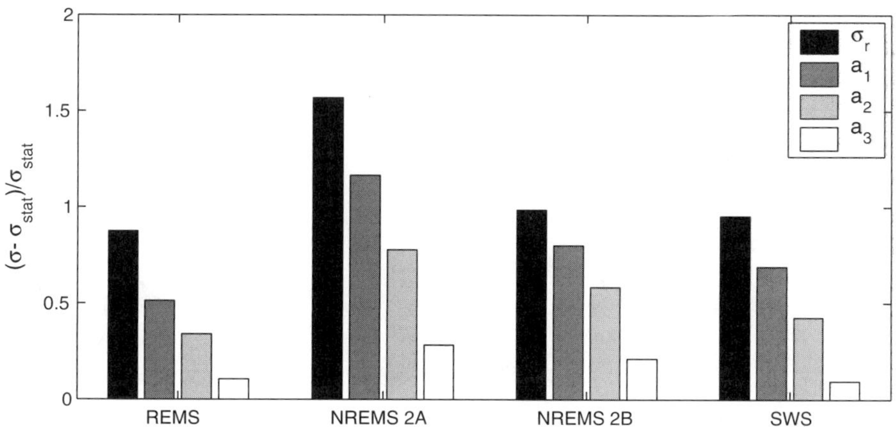

Fig. 3. Difference between the standard deviation of the distribution of four 1-s AR-parameters in the 1-min segments (the standard deviation of the residuals σ_r and the three first AR-coefficients) and the standard deviation of its estimates normalized to the standard deviation of the estimates.

4. Conclusions

Our results demonstrate that the nonlinear signatures found in segments of human sleep EEG with durations longer than 10 s should be explained rather by fluctuations of the AR-coefficients of the 1-s subsegments than by stationary nonlinearities. These parameter fluctuations render the EEG segments nonstationary on these time scales. However, this does not exclude the possibility that the fluctuations themselves might be described by at least approximately stationary processes on time windows of a duration up to several minutes within a definite sleep stage.

This leads to the picture of a hierarchy of three time scales: A first time scale up to 1 s contains the typical EEG patterns like spindles, K-complexes or alpha waves and can be modeled by AR-models. A second time scale up to several minutes, sometimes called microstructure of sleep [7], is characterized by the fluctuations of the AR-parameters. A third time scale from minutes to hours (macrostructure) can be related to the sleep stages and is reflected by the slow dynamics of the mean AR-parameters as it is shown in the middle panel of Fig. 2.

References

[1] H. Akaike, A new look at the statistical model identification, IEEE Trans. Automat. Control 19 (1974) 716–723.

[2] J. Fell, J. Röschke, K. Mann, C. Schäfer, Surrogate data analysis of sleep electroencephalograms reveals evidence for nonlinearity, Biol. Cybern. 75 (1) (1996) 85–92.

[3] J. Pardey, S. Roberts, L. Tarassenko, J. Stradling, A new approach to the analysis of the human sleep/wakefulness continuum, J. Sleep Res. 5 (1996) 201–210.

[4] E. Pereda, A. Gamundi, R. Rial, J. Gonzalez, Non-linear behaviour of human EEG: fractal exponent versus correlation dimension in awake and sleep stages, Neurosci. Lett. 250 (2) (1998) 91–94.

[5] A. Rechtschaffen, A. Kales, A manual of standardized terminology, technics and scoring system for sleep stages of human subjects, Public Health Services, NIH Public, 204.

[6] Y. Shen, E. Olbrich, P. Achermann, P.F. Meier, Dimensional complexity and spectral properties of the human sleep EEG, Clin. Neurophysiol. 114 (2) (2003) 199–209.

[7] M.G. Terzano, L. Parrino, Origin and significance of the cyclic alternating pattern (CAP), Sleep Med. Rev. 4 (1) (2000) 101–123.

Available at
www.ComputerScienceWeb.com
POWERED BY SCIENCE DIRECT®

ELSEVIER

Neurocomputing 52–54 (2003) 863–868

NEUROCOMPUTING

www.elsevier.com/locate/neucom

Speed-up methods for simulations of traveling waves in integrate-and-fire neural networks

Remus Osan*, Bard Ermentrout

Department of Mathematics, University of Pittsburgh, Pittsburgh, PA 15260, USA

Abstract

We consider single and multiple spike traveling waves solutions in an integrate-and-fire neural network. Due to the special properties of this system it is possible to deduce the dynamical state of each neuron in the full network from a small subset, using recurrence relationships. Since the dimension of the subset is on the order of the number of the traveling waves, this greatly reduces the computational time required for numerical simulations. These ideas are specific to the class of integrate-and-fire neural network models presented here.

Keywords: Integrate-and-fire; Traveling waves; Numerical methods; Optimization

1. Introduction

Synaptically generated traveling waves are one of the most studied large-scale dynamics of cortical tissue as they provide insight into its organization and functionality. Bressloff [1,2], Ermentrout [3], Golomb and Ermentrout [4,5], and Osan and Ermentrout [7], developed methods for studying the existence and evolution of such traveling waves of activity, under the assumption that each cell only fires once. The multiple traveling waves problem is much more complicated both from theoretical and computational points of view; however, some analytical and computational results were obtained for integrate-and-fire neural networks with finite support [8] and exponential [9] support connectivity functions. Computational models of neural tissue usually use reduced neurons coupled together via symmetric synapses whose strength decrease with distance. For these types of models large-scale simulations are needed to test the

* Corresponding author. Princeton University, Princeton, NJ 8544, USA.
E-mail address: rosan@princeton.edu (R. Osan).

0925-2312/03/$ - see front matter
doi:10.1016/S0925-2312(02)00817-2

existence and stability of such traveling waves. Conducting large-scale simulations in integrate-and-fire neural networks is often very time-consuming. It is thus important to devise methods that would decrease the overall computation time [6,10].

For the particular class of neural networks considered here, the firing times of the neurons can be computed 'exactly', that is, with arbitrary accuracy, by solving transcendental equations. A description of the integrate-and-fire neural network is presented in the next section. In the remaining sections we derive recurrence formulas that allow the computation of the dynamical state of one neuron as a function of the states of its neighbors.

2. Neural network description

The neural network we consider uses the integrate-and-fire model for a spiking neuron. Each 'neuron' is a linear integrator. Upon reaching a predefined threshold, V_T, the neuron emits a 'spike' and the voltage is reset to a lower voltage $V_R < V_T$. The effect of a spike on other neurons is to turn on a current whose time dependence is often a simple exponential function and whose magnitude is a function of the distance between the two connected neurons. Thus, the system evolves according to

$$\tau_1 \frac{\partial V(x,t)}{\partial t} = -V(x,t) + g_{\text{syn}} \int_D J(|y-x|) \sum_k \alpha(t - t_k(y))\,\mathrm{d}y, \tag{1}$$

along with the reset condition such that if $V(x,t)$ crosses a threshold V_T, it is reset to V_R. $t_k(y)$ represents the discrete set of times when neuron at position y fires the kth time. Here $\alpha(t)$ is the time-dependent current that arises from an impulse. The function J describes the space dependence of the coupling strength of neuron at positions x and y. The parameter g_{syn} sets the scale of the coupling strength. In the simplest models, the definitions of the functions used in Eq. (1) are

$$J(x) = \frac{\mathrm{e}^{-|x|/\sigma}}{2\sigma}, \tag{2}$$

$$\alpha(t) = \mathrm{e}^{-t/\tau_2} H(t). \tag{3}$$

H is the Heaviside step function $H(t) = 0$, if $t < 0$ and $H(t) = 1$, if $t \geqslant 0$.

If the specific form of the α function is given by Eq. (3), then by integrating the differential equation (1) we obtain the integral formulation

$$V(x,t) = \sum_{n=1}^{N(x,t)} \eta(t - t_n(x)) + g_{\text{syn}} \int_D J(|y-x|) \sum_{m=1}^{N(y,t)} A(t - t_k(y))\,\mathrm{d}y, \tag{4}$$

$$\eta(t) = (V_{\text{reset}} - V_T)\mathrm{e}^{-t/\tau_1} H(t), \tag{5}$$

$$A(t) = \frac{1}{1 - \tau_1/\tau_2}(\mathrm{e}^{-t/\tau_2} - \mathrm{e}^{-t/\tau_1})H(t) = \frac{\alpha(t) - \beta(t)}{1 - \tau_1/\tau_2}, \tag{6}$$

where $\beta(t) = e^{-t/\tau_1}H(t)$, and $N(y,t)$ represents the number of spikes the neuron at position y fired at time t. We use the following terminology when refering to the terms of Eq. (4): 'decaying reset' for the first one and 'synaptic integral' for the second one. For numerical simulations the integral form (4) is more convenient than the differential form (1).

In order to simulate Eq. (4), with the choice of functions from Eqs. (2) and (3), we need to use a discretized one-dimensional domain. One way to do that is to use a uniform grid with distance δ between nodes. By denoting spatial position of each neuron by $x = i\delta$, where $i \in \mathbf{N}$ represents the numbering of each neuron, we can rewrite Eq. (4) as

$$V(i,t) = \sum_{n=1}^{N(i,t)} \eta(t - t_n(i)) + g_{\text{syn}}\delta \sum_{j=1}^{\infty} J(|i-j|\delta) \sum_{m=1}^{N(j,t)} A(t - t_m(j)). \tag{7}$$

The computational time can be decreased by taking advantage of the symmetries that arise naturally in these types of networks for solitary and multiple travelling waves. We showed in [7] that for the type of network considered here the single spiking activity pattern arising after a domain is brought to firing is monotone, that is, neurons closer to the initial region fire before neurons further away. In this case we can completely determine the dynamics of the neural network by computing the synaptic integral from Eq. (7) only for the neighbor of the neuron that spiked last. In the next sections, we show how this idea can be implemented for the solitary traveling wave and extended for multiple traveling waves.

3. Single-spike traveling wave solutions

The single-spike assumption can be biologically justified by supposing strong synaptic depression. Using the convention $t(i\delta) = t_i$ and $V(i\delta, t) = V(i,t)$, and omitting the decaying reset term, Eq. (7) takes the following form:

$$V(i,t) = g_{\text{syn}}\delta \sum_{j=1}^{\infty} J(|i-j|\delta)A(t - t_j). \tag{8}$$

Let us consider a traveling wave moving towards the positive direction of the one-dimensional domain. We assume that at time t_{i_0} the traveling front reaches position i_0, that is, the neuron located there fires. We use Eq. (8) and $A(0) = 0$ to compute the potential at position $j_0 > i_0$ at time t_{i_0}:

$$V(j_0, t_{i_0}) = g_{\text{syn}}\delta \sum_{j=1}^{i_0} J(|j_0 - j|\delta)A(t_{i_0} - t_j) = V_T e^{-(j_0 - i_0)\delta/\sigma}. \tag{9}$$

Eq. (9) indicates that regardless of the details of the firing map, the potential on the right of the traveling front decays exponentially from V_T to zero as j_0 increases. Since the neuron one position ahead of the traveling front has the highest potential and receives the highest synaptic current, it follows that it will fire next. This is the intuitive explanation for the firing times monotonicity proof we used in [7].

We can obtain a recurrence relationship between the synaptic integrals from different spatial locations. At times $t_{i_0-1} < t < t_{i_0}$, the synaptic integrals at positions i_0 and $j_0 > i_0$ are

$$V(i_0,t) = \frac{g_{syn}\delta}{1-\tau_1/\tau_2}\sum_{j=1}^{i_0-1} J(|i_0-j|\delta)\sum_{\gamma\in\{\alpha,\beta\}} c_\gamma\gamma(t_{i_0}-t_j)\gamma(t-t_{i_0}), \tag{10}$$

$$V(j_0,t) = \left(\sum_{\gamma\in\{\alpha,\beta\}} S_\gamma(i_0,t_{i_0-1})\gamma(t-t_{i_0-1})\right) e^{-(j_0-i_0)\delta/\sigma}, \tag{11}$$

where $c_\alpha = 1,\ c_\beta = -1$ and

$$S_\gamma(i_0,t_{i_0-1}) = \frac{g_{syn}\delta}{1-\tau_1/\tau_2}\sum_{j=1}^{i_0-1} J((i_0-j)\delta)\gamma(t_{i_0-1}-t_j).$$

We use Eqs. (10) and (11) to compute the potential at position $y=(i_0+1)\delta$ for $t\in(t_{i_0},t_{i_0+1}]$, that is, after neuron at $x=i_0\delta$ fires:

$$V(i_0+1,t) = S_\alpha(i_0+1,t_{i_0})\alpha(t-t_{i_0}) - S_\beta(i_0+1,t_{i_0})\beta(t-t_{i_0}), \tag{12}$$

where $S_\alpha(i_0+1,t_{i_0})$ and $S_\beta(i_0+1,t_{i_0})$ are given by the recurrence formulas

$$S_\gamma(i_0+1,t_{i_0}) = S_\gamma(i_0,t_{i_0-1})e^{-\delta/\sigma}\gamma(t_{i_0}-t_{i_0-1}) + \frac{g_{syn}}{1-\tau_1/\tau_2}\delta J\left(\frac{\delta}{\sigma}\right). \tag{13}$$

We obtained a relationship (13) which allows us to compute the coefficients $S_\alpha(i_0+1,t_{i_0})$ of the synaptic integral in Eq. (12) at location $x=(i_0+1)\delta$, as a function of previous synaptic components $S_\gamma(i_0,t_{i_0-1})$ and the last interspike interval $\Delta t=t_{i_0}-t_{i_0-1}$. The intuitive meaning of this relationship is that the α and β components of the synaptic integral at the new location are obtained by right 'shifting' the components at the previous location and adding the effects of the last spike. Note that shifting decreases the synaptic components by a factor of $e^{\delta/\sigma}$. The time-decaying terms $\gamma(t_{i_0}-t_{i_0-1})$ also decrease the synaptic components, while the effects of the last spike lead to an increase in the overall synaptic integral.

4. Multiple traveling waves

The simple scheme outlined for the single-spike traveling waves solution does not extend directly for multiple waves. While the potential outside the region where neurons spiked at least once still decays exponentially with the spatial width of the cortical function J, the voltage in the region between successive traveling waves does not have a simple formula. However, for the traveling waves case, it is possible to devise a numerical scheme that extends the same basic ideas and allows the computation of firing times based on recurrence formulas. More precisely we require that the firing map $t_k(y)$ is monotone in y, that is, $t_k(y) > t_k(x)$, $y > x$. Under this assumption we monitor the set of neurons that will fire next, situated one position ahead of the traveling wave fronts.

Let us assume that the neuron located at position $x = i_0\delta$ is the last one to fire, at time t_{i_0}. The decaying reset term of a neuron located at position $j_0\delta$ depends only on its local spiking history and it can be computed as needed:

$$\eta(i_0, t) = \sum_{n=1}^{N(i_0,t)} \eta(t - t_n(i)). \tag{14}$$

The synaptic integral $S(j_0, t) = g_{\mathrm{syn}}\delta \sum_{k=1}^{\infty} J(|j_0 - k|\delta) \sum_{m=1}^{N(k,t)} A(t - t_m(k))$ can be expressed as a sum of left, central and right spatial contributions evaluated at the time of the last spike t_{i_0}:

$$S(j_0, t) = \sum_{\gamma \in \{\alpha,\beta\}} \left(\sum_{O=\mathrm{L,C,R}} S_{\gamma O}(j_0, t_{i_0}) \right) c_\gamma \gamma(t - t_{i_0}), \tag{15}$$

where

$$S_{\gamma \mathrm{L}}(j_0, t_{i_0}) = g_{\mathrm{syn}}\delta \sum_{k=1}^{j_0-1} J(|j_0 - k|\delta) \sum_{m=1}^{N(k,t_{i_0})} \gamma(t_{i_0} - t_m(k)),$$

$$S_{\gamma \mathrm{C}}(j_0, t_{i_0}) = g_{\mathrm{syn}}\delta J(0) \sum_{m=1}^{N(j_0,t_{i_0})} \gamma(t_{i_0} - t_m(j_0)), \tag{16}$$

$$S_{\gamma \mathrm{R}}(j_0, t_{i_0}) = g_{\mathrm{syn}}\delta \sum_{k=j_0+1}^{\infty} J(|j_0 - k|\delta) \sum_{m=1}^{N(k,t_{i_0})} \gamma(t_{i_0} - t_m(k)).$$

After the last spike, in the set of monitored neurons, the neuron at node i_0 is replaced by the neuron at node $i_0 + 1$. The synaptic integral at $i_0 + 1$ becomes

$$S(i_0 + 1, t) = \sum_{\gamma \in \{\alpha,\beta\}} c_\gamma \gamma(t - t_{i_0}) \left(\mathrm{e}^{-\delta/\sigma} \sum_{O=\mathrm{L,C}} S_{\gamma O}(i_0, t_{i_0}) + \mathrm{e}^{\delta/\sigma} S_{\gamma \mathrm{R}}(i_0, t_{i_0}) \right).$$

By computing the central self-contribution at the node $i_0 + 1$ at time $t = t_{i_0}$ and taking into account the effect of the last spike, we find the recurrence formulas that advance formula (15) from i_0 to $i_0 + 1$:

$$S_{\gamma L}(i_0 + 1, t_{i_0}) = \sum_{O=\mathrm{L,C}} S_{\gamma O}(i_0, t_{i_0-1}) \mathrm{e}^{-\delta/\sigma} \gamma(t_{i_0} - t_{i_0-1}) + \frac{g_{\mathrm{syn}}}{1 - \tau_1/\tau_2} \delta J \left(\frac{\delta}{\sigma} \right),$$
$$S_{\gamma \mathrm{R}}(i_0 + 1, t_{i_0}) = S_{\gamma \mathrm{R}}(i_0, t_{i_0-1}) \mathrm{e}^{\delta/\sigma} \gamma(t_{i_0} - t_{i_0-1}) - S_{\gamma \mathrm{C}}(i_0 + 1, t_{i_0}), \tag{17}$$

where t_k is the time of the last spike before t_{i_0}. The updated equations for the other neurons in the monitored set (located on at $j_0 > i_0$) are:

$$\eta(j_0, t_{i_0}) = \eta(j_0, t_k) \beta(t_{i_0} - t_k),$$

$$S_{\gamma \mathrm{L}}(j_0, t_{i_0}) = S_{\gamma \mathrm{L}}(j_0, t_k) \gamma(t_{i_0} - t_k) + \frac{g_{\mathrm{syn}}}{1 - \tau_1/\tau_2} \delta J \left(-\frac{(j_0 - i_0)\delta}{\sigma} \right), \tag{18}$$

$$S_{\gamma S}(j_0, t_{i_0}) = S_{\gamma S}(j_0, t_k) \gamma(t_{i_0} - t_k), \quad S = \mathrm{C, R}.$$

Only the dynamical state of a small subset of the network, situated one position ahead of the traveling wave fronts, is monitored. As a new neuron spikes the focus shifts to its right neighbor for which the synaptic integrals can be computed using Eqs. (14) and (17). For all the other neurons from the monitored set the synaptic integrals are updated using Eq. (18). In addition to a 'left shift', a 'right' shift is needed to compute the synaptic integrals.

5. Conclusions

We showed how the special spatial structure of traveling waves can be used to decrease the simulation time for an integrate-and-fire network, for both single- and multiple spike waves. Effectively this allows simulations where the computation for the state of thousands or more neurons is reduced down to computations on the order of the number of traveling waves. As an example, this procedure was used in numerical simulations with 50,000 neurons and 50 traveling waves, giving a speedup of around 1000.

The results of this paper apply to the widely used specific forms of the synaptic α function (3) and cortical connectivity function (2). However, our findings do not extend directly to more complicated functions. For the one-dimensional domain the use of a different connectivity function, such as gaussian, voids the shifting property. Furthermore, extensions to the two-dimensional domain are possible only if the 'Manhattan distance' ($d(\mathbf{a},\mathbf{b}) = |a_x - b_x| + |a_y - b_y|$) is used as a measure of the separation between two neurons.

References

[1] P.C. Bressloff, Phys. Rev. Lett. 82 (1999) 2979–2982.
[2] P.C. Bressloff, J. Math. Biol. 40 (2000) 169–198.
[3] G.B. Ermentrout, J. Comput. Neuro. Sci. 5 (1998) 191–208.
[4] D. Golomb, G.B. Ermentrout, Proc. Natl. Acad. Sci. USA 96 (23) (1999) 13480–13485.
[5] D. Golomb, G.B. Ermentrout, Network 11 (2000) 221–246.
[6] D. Hansel, G. Mato, C. Meunier, L. Neltner, Neural Comput. 10 (1998) 467–483.
[7] R. Osan, G.B. Ermentrout, Physica D 163 (2002) 217–235.
[8] R. Osan, J. Rubin, R. Curtu, B. Ermentrout, Neurocomputing, this issue.
[9] R. Osan, R. Curtu, J. Rubin, B. Ermentrout, J. Math. Biology, submitted.
[10] M.J. Shelley, L. Tao, J. Comput. Neurosci. 11 (2001) 111–119.

Available at
www.ComputerScienceWeb.com
POWERED BY SCIENCE DIRECT

ELSEVIER

Neurocomputing 52–54 (2003) 869–875

NEUROCOMPUTING

www.elsevier.com/locate/neucom

Traveling waves in a one-dimensional integrate-and-fire neural network with finite support connectivity

Remus Osan, Jonathan Rubin, Rodica Curtu, Bard Ermentrout*

Department of Mathematics, University of Pittsburgh, Pittsburgh, PA 15260, USA

Abstract

We study the existence of traveling waves in a one-dimensional network of integrate-and-fire neurons with finite support coupling. We show that when the reset after spiking is sufficiently low, the interspike intervals (ISIs) for a traveling wave with an explicit first wave front can be computed. Further, we analytically derive a self-consistency equation that generates an explicit dispersion relation between the velocity of periodic traveling waves and their corresponding ISIs. Finally, we prove that the ISIs of non-periodic traveling waves converge to the ISI of the periodic waves with the same speed.

Keywords: Traveling waves; Integrate-and-fire neurons; Synaptic coupling

1. Introduction

The neurons in the synaptically coupled integrate-and-fire network that we consider, on a one-dimensional domain D, obey the equation

$$\tau_1 \frac{\partial V(x,t)}{\partial t} = -V(x,t) + g_{\text{syn}} \int_D J(|y-x|) \sum_k \alpha(t - t_k(y))\,\mathrm{d}y. \tag{1}$$

In Eq. (1), the integral term represents synaptic coupling, with the summation taken over all input spikes from the cell(s) at position y to the cell(s) at position x, which occur at times $\{t_k(y)\}$. We denote this term by $I_{\text{syn}}(x,t)$. The function $J(x)$ encodes the spatial dependence of this coupling, while $\alpha(t)$ encodes the time course of the synaptic

* Corresponding author. Tel.: +1-412-624-8324; fax: +1-412-624-8397.
E-mail address: bard@math.pitt.edu (B. Ermentrout).

0925-2312/03/$ - see front matter © 2002 Elsevier Science B.V. All rights reserved.
doi:10.1016/S0925-2312(02)00818-4

current due to each spike. Each time that a cell's voltage reaches V_T, it is reset to V_R; that is, if $V(x,\tau^-) = V_T$, then we set $V(x,\tau^+) = V_R$. Traveling wave solutions to various forms of Eq. (1) have been studied by a variety of authors [1–6].

The standard choice of α function is

$$\alpha(t) = \mathrm{e}^{-t/\tau_2} H(t) = \begin{cases} 0, & t < 0, \\ \mathrm{e}^{-t/\tau_2}, & t \geqslant 0, \text{ and } \tau_2 > \tau_1 \end{cases} \tag{2}$$

where $H(t)$ is the Heaviside step function. The definition of the normalized coupling function for the finite support case is

$$J(x) = \begin{cases} 0, & |x| > \sigma, \\ \frac{1}{2\sigma}, & |x| \leqslant \sigma. \end{cases} \tag{3}$$

We conducted numerical simulations to explore the initiation of multi-spike traveling waves and to study how well such solutions approximate periodic waves. In these simulations, an initial region is 'shocked', or brought to firing by an application of current, generating a traveling wave of spiking activity. Due to the synaptic current received from all the neurons that fire, the neurons in the shocked region spike again, generating a second wave. The wave generation occurs faster and faster (as the accumulated synaptic current becomes greater and greater), but if V_R is sufficiently far from V_T, then the interspike intervals (ISIs) for each cell eventually converge toward a fixed value as illustrated in Fig. 1. The intuition behind this convergence is that

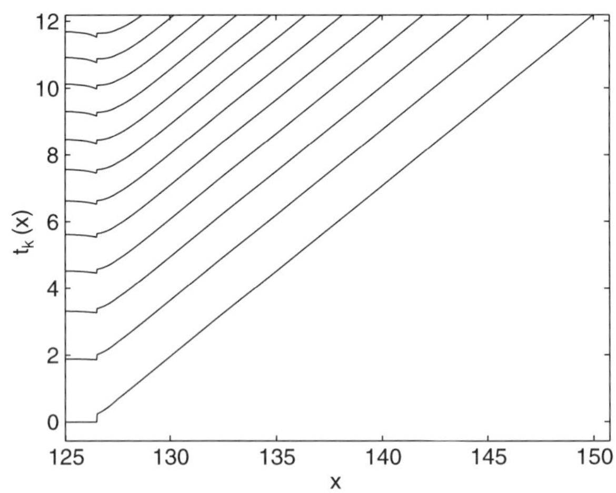

Fig. 1. Multi-spike traveling waves from simulation of Eq. (1). Space is represented on the horizontal axis and firing times on the vertical axis. The region near the spatial coordinate 125 (center of the shocked region) was given a transient excitatory input at time 0. All subsequent waves were generated by the dynamics of Eq. (1). The times between successive spikes at each spatial point decrease, approaching a limiting value away from the shocked region. The parameters used here are: $g_{\text{syn}} = 10$, $\tau_1 = 1$, $\tau_2 = 2$, $\sigma = 1$, $V_T = 1$, $V_R = -25$ and $L_0 = 3$ (the length of the shocked region).

synaptic current accumulates as firing rates increase, but the reset procedure provides an effective time delay that prevents firing rates from going to infinity.

2. Waves with finite numbers of spikes

Here we analytically investigate the case in which after firing the cells reset at low voltages $V_R < 0$, far below threshold. Numerical simulations reveal that in this particular case the distance between the fronts of the traveling waves is always larger than σ (Fig. 1). In other words, the propagation of a wave front through cells at position x is influenced only by the waves that have already passed through x and by the wave that is currently approaching, but *not by other waves that will reach x at some time in the future.*

It follows that the dynamics of the first wave front is essentially the same as the dynamics of the single-spike solution. Furthermore, the dynamics of the second wave front is due only to synaptic contribution from the first and second waves.

For the kth wave front, the firing time of cells at position x is given by $t_k(x) = x/c + T_k$. Let us assume that the cell to fire first at $t = 0$ is located at $x = 0$; that is, $T_0 = 0$ and $t_0(x) = x/c$. Consequently, we have

$$I_{\mathrm{syn}}(0,t) = \begin{cases} 0, & t < -t_0, \\ g_0(1 - \mathrm{e}^{-(t+t_0)/\tau_2}), & -t_0 < t < 0 \end{cases} \tag{4}$$

and

$$V(0,t) = \begin{cases} 0, & t < -t_0, \\ g_0(1 - \mathrm{e}^{-(t+t_0)/\tau_1} - \delta(\mathrm{e}^{-(t+t_0)/\tau_2} - \mathrm{e}^{-(t+t_0)/\tau_1})), & -t_0 < t < 0, \end{cases} \tag{5}$$

where we used the notation $g_0 = g_{\mathrm{syn}} c\tau_2/(2\sigma)$, $\delta = (1-\tau_1/\tau_2)^{-1}$ and $t_0 = \sigma/c$. The speed of the single-spike solution can be computed from Eq. (5), obtaining $V(0,0) = V_T$. For the numerical parameters listed in Fig. 1 we obtain two solutions: a fast wave (numerically stable) with $c \approx 1.944$ and a slow wave (numerically unstable) with $c \approx 0.102$.

We can now compute the ISI for a two-spike solution. Let us assume that the neuron at position $x = 0$ fires its second spike at time T_1. The synaptic contributions are due to the first and second wave only (Fig. 2).

The impact of the second wave is felt at $x = 0$ only after the second wave reaches within a distance of σ from $x = 0$, at time $T_1 - t_0$. Thus, the synaptic current received at $x = 0$, defined as $I_{\mathrm{syn}}(0,t) = g_{\mathrm{syn}} \int_{-\sigma}^{ct} \mathrm{d}y(1/2\sigma)\mathrm{e}^{-(t-y/c)/\tau_2}$, evaluates to

$$I_{\mathrm{syn}}(0,t) = \begin{cases} g_0(1 - \mathrm{e}^{-(t+t_0)/\tau_2}), & 0 < t < t_0, \\ g_0(1 - \mathrm{e}^{-2\sigma/c\tau_2})\mathrm{e}^{-t/\tau_2}, & t_0 < t < T_1 - t_0, \\ g_0\{(1 - \mathrm{e}^{-2\sigma/c\tau_2})\mathrm{e}^{-t/\tau_2} & \\ \quad +(1 - \mathrm{e}^{-(t+t_0-T_1)/\tau_2})\}, & T_1 - t_0 < t < T_1. \end{cases} \tag{6}$$

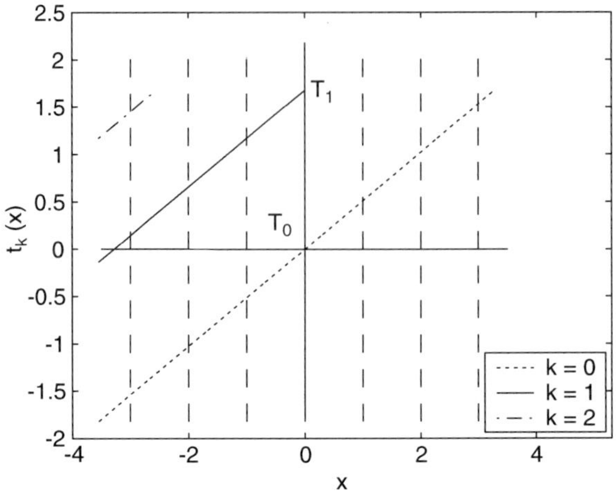

Fig. 2. Plot of the first, second and third traveling waves.

Integrating the currents yields

$$V(0,t) = \begin{cases} V_R e^{-t/\tau_1} + g_0((1 - e^{-t/\tau_1}) - \delta e^{-t_0/\tau_2}(e^{-t/\tau_2} - e^{-t/\tau_1})), \\ \quad 0 < t < t_0, \\ V_R e^{-t/\tau_1} + g_0 e^{-(t-t_0)/\tau_1}((1 - e^{-\sigma/c\tau_1}) - \delta e^{-\sigma/c\tau_2}(e^{-\sigma/c\tau_2} - e^{-\sigma/c\tau_1})) \\ \quad + g_0\delta(1 - e^{-2\sigma/c\tau_2})(e^{-(t-\sigma/c)/\tau_2} - e^{-(t-\sigma/c)/\tau_1}), \\ \quad t_0 < t < T_1 - t_0, \\ V_R e^{-t/\tau_1} + g_0 e^{-(t-t_0)/\tau_1}((1 - e^{-\sigma/c\tau_1}) - \delta e^{-\sigma/c\tau_2}(e^{-\sigma/c\tau_2} - e^{-\sigma/c\tau_1})) \\ \quad + g_0(1 - e^{-(t-T_1+t_0)/\tau_2} - \delta(e^{-(t-T_1+t_0)/\tau_2} - e^{-(t-T_1+t_0)/\tau_1})) \\ \quad + g_0\delta(1 - e^{-2\sigma/c\tau_2})(e^{-(t-\sigma/c)/\tau_2} - e^{-(t-\sigma/c)/\tau_1}), \\ \quad T_1 - t_0 < t < T_1. \end{cases} \tag{7}$$

The condition $V(0, T_1) = V_T$ allows us to compute the first ISI, $T_1 \approx 1.6828$, for fast wave speed obtained for the parameters in Fig. 1.

For computation of subsequent ISIs, let us define

$$I_0(t, n-1, n) = \begin{cases} g_0(1 - e^{-(t-T_{n-1}+t_0)/\tau_2}), & T_{n-1} < t < T_{n-1} + t_0, \\ g_0(1 - e^{-2\sigma/c\tau_2})e^{-(t-T_{n-1})/\tau_2}, & T_{n-1} + t_0 < t < T_n - t_0, \\ g_0((1 - e^{-2\sigma/c\tau_2})e^{-(t-T_{n-1})/\tau_2} & \\ \quad +(1 - e^{-(t+t_0-T_n)/\tau_2})), & T_n - t_0 < t < T_n \end{cases}$$

and

$$V_0(t,n-1,n)=\begin{cases} g_0(1-e^{-(t-T_{n-1})/\tau_1}-\delta e^{-t_0/\tau_2}(e^{-(t-T_{n-1})/\tau_2}-e^{-(t-T_{n-1})/\tau_1})), \\ \quad T_{n-1}<t<T_{n-1}+t_0, \\ g_0 e^{-(t-T_{n-1}-t_0)/\tau_1}(1-e^{-\sigma/c\tau_1}-\delta e^{-\sigma/c\tau_2}(e^{-\sigma/c\tau_2}-e^{-\sigma/c\tau_1})) \\ +g_0\delta(1-e^{-2\sigma/c\tau_2})(e^{-(t-T_{n-1}-\sigma/c)/\tau_2}-e^{-(t-T_{n-1}-\sigma/c)/\tau_1}), \\ \quad T_{n-1}+t_0<t<T_n-t_0, \\ g_0 e^{-(t-T_{n-1}-t_0)/\tau_1}((1-e^{-\sigma/c\tau_1})-e^{-\sigma/c\tau_2}\delta(e^{-\sigma/c\tau_2}-e^{-\sigma/c\tau_1})) \\ \quad +g_0(1-e^{-(t-T_n+t_0)/\tau_2}-\delta(e^{-(t-T_n+t_0)/\tau_2}-e^{-(t-T_n+t_0)/\tau_1})) \\ \quad +g_0\delta(1-e^{-2\sigma/c\tau_2})(e^{-(t-T_{n-1}-\sigma/c)/\tau_2}-e^{-(t-T_{n-1}-\sigma/c)/\tau_1}), \\ \quad T_n-t_0<t<T_n. \end{cases} \tag{8}$$

The current and voltage for the neuron at $x=0$ now take the form

$$I_{\text{syn}}(0,t)=I_0(t,n-1,n)+g_0(1-e^{-2\sigma/c\tau_2})e^{-(t-T_{n-2}+t_0)/\tau_2}\sum_{k=0}^{n-2}e^{-(T_{n-1}-T_k)/\tau_2},$$

$$V(0,t)=V_R e^{-(t-T_{n-1})/\tau_1}+V_0(t,n-1,n)$$
$$+g_0\delta(e^{\sigma/c\tau_2}-e^{-\sigma/c\tau_2})(e^{-(t-T_{n-1})/\tau_2}-e^{-(t-T_{n-1})/\tau_1})\sum_{k=0}^{n-2}e^{-(T_{n-1}-T_k)/\tau_2}, \tag{9}$$

where $k=0$ denotes the first traveling wave. It is easy to see how the ISIs can be computed using equation $V(0,T_n)=V_T$ recursively.

3. Periodic traveling waves and convergence of ISIs

Let us assume a periodic traveling wave with period T exists and the condition $T>t_0$ holds. Then a generalization of Eq. (9) can be used to compute the period. We index the waves with an index k such that larger k waves passed by a longer time ago. This yields $V(0,T)=V_T$, where

$$V(0,T)=V_R e^{-T/\tau_1}+g_0\{\delta(1-e^{-\sigma/c\tau_2})(e^{-T/\tau_2}-e^{-T/\tau_1})\sum_{k=1}^{\infty}e^{-kT/\tau_2}$$
$$+e^{-(T-\sigma/c)/\tau_1}\cdot[1-e^{-\sigma/c\tau_1}-\delta e^{-\sigma/c\tau_2}(e^{-\sigma/c\tau_2}-e^{-\sigma/c\tau_1})]+1-e^{-\sigma/c\tau_2}$$
$$-\delta(e^{-\sigma/c\tau_2}-e^{-\sigma/c\tau_1})+V_T\}. \tag{10}$$

Solving this equation for $c\approx 1.944$, corresponding to the fast wave speed for the single-spike wave for the parameter values in Fig. 1, we obtain $T_d\approx 0.553$. The sequence T_n, obtained from the recurrence formula $V(0,T_n)=V_T$ with $V(0,T_n)$ given by Eq. (9) for all n, converges nicely to this value. A graph of the sequence appears in Fig. 3.

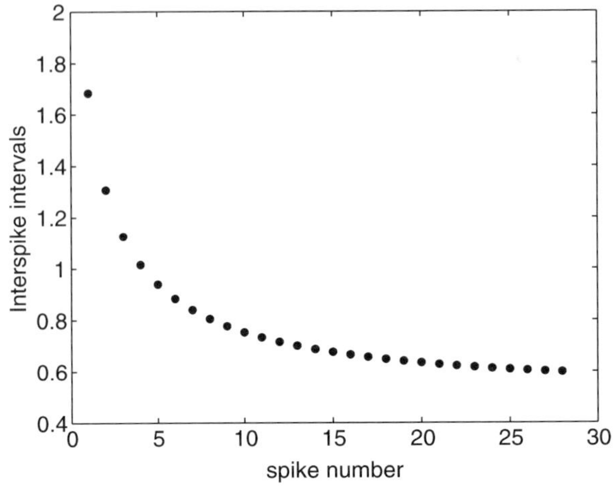

Fig. 3. ISIs converge to T_d. The first four ISIs are 1.682, 1.306, 1.126 and 1.015.

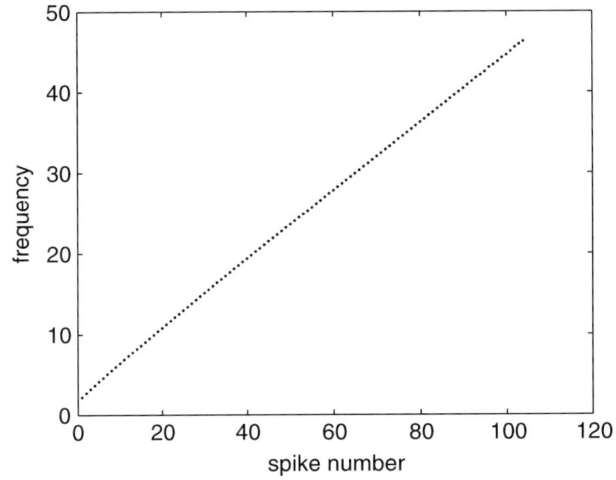

Fig. 4. Numerical simulations for $V_R = -10$ indicate that the inverse of the ISIs, the frequency, increases linearly with the spike number, towards infinity.

Under the assumption that $T > t_0$, the convergence of ISIs to the corresponding period T can be rigorously established. We sketch the proof here. The first ISI, T_1, solves $V(0, T_1) = V_T$, where $V(0, T_1)$ comes from Eq. (7). But Eq. (7) evaluated at $t = T_1$ is identical to Eq. (10) evaluated at $T = T_1$, taken without the infinite sum term. Denote the quantity from (10) without the infinite sum by $V_0(0, T)$, such that $V_0(0, T_1) = V_T$. Since $V_R + g_0 e^{\sigma/c\tau_1}[1 - e^{-\sigma/c\tau_1} - \delta e^{-\sigma/c\tau_2}(e^{-\sigma/c\tau_2} - e^{-\sigma/c\tau_1})] < 0$, the quantity $V_0(0, T)$ is monotone increasing in T, yielding $T_1 > T$. Now, the second ISI, T_2, solves $V(0, T_2) = V_T$, with $V(0, T_2)$ given by $V_0(0, T_2)$ plus the $k = 1$ term from the infinite sum in Eq. (10), with $T = T_2$. Adding this positive term to V_0 decreases the solution such that $T < T_2 < T_1$. Similarly, each subsequent ISI is smaller than the preceding one, but all are greater than T, since T solves the equation $V(0,T) = V_T$ with the full infinite sum from Eq. (10) included in $V(0,T)$. The monotone decreasing, bounded sequence $\{T_n\}$ of ISIs converges, say to $T' \geqslant T$. Since the terms in the sum in Eq. (10) decrease exponentially, one can show that $V(0,T_n)$ converges to V_T. By continuity, $V(0,T_n)$ converges to $V(0,T')$, so T' itself satisfies Eq. (10) (which has unique solution), and thus $T' = T$.

On the other hand, numerical simulations suggest that when the reset is not low enough, $T < t_0$; thus, the system loses stability and the firing rates diverge to infinity. Using Eq. (10) we obtain the critical value of the reset potential $V_c \approx -24.25$ for our usual parameter values. An example of this regime is illustrated in Fig. 4.

4. Conclusions

The special dynamics of the integrate-and-fire neural networks with finite support allows us to compute the ISIs for successive traveling waves and compare them with results from numerical simulations. We prove that the ISIs converge to a fixed value, that is, the asymptotically stable case is the one with equal separation between consecutive wavefronts. It is much more difficult to prove these analytical results for more complex connectivity functions, such as gaussian and exponential (although see (6) for computation of ISIs and existence of finite spiking solutions).

References

[1] P.C. Bressloff, Phys. Rev. Lett. 82 (1999) 2979–2982.
[2] P.C. Bressloff, J. Math. Biol. 40 (2000) 169–198.
[3] G.B. Ermentrout, J. Comput. Neurosci. 5 (1998) 191–208.
[4] D. Golomb, G.B. Ermentrout, Proc. Natl. Acad. Sci. USA 96 (1999) 13480–13485.
[5] R. Osan, R. Curtu, J. Rubin, B. Ermentrout (2002) preprint.
[6] R. Osan, G.B. Ermentrout, Physica D 163 (2002) 217–235.

ELSEVIER

Available at
www.ComputerScienceWeb.com
POWERED BY SCIENCE @ DIRECT®

Neurocomputing 52–54 (2003) 877–883

NEUROCOMPUTING

www.elsevier.com/locate/neucom

Noise-driven adaptation: in vitro and mathematical analysis☆

Liam Paninski*,[1], Brian Lau, Alex Reyes

Center for Neural Science, New York University, 4 Washington Place, New York, NY 10003, USA

Abstract

Variance adaptation processes have recently been examined in cells of the fly visual system and various vertebrate preparations. To better understand the contributions of somatic mechanisms to this kind of adaptation, we recorded intracellularly in vitro from neurons of rat sensorimotor cortex. The cells were stimulated with a noise current whose standard deviation was varied parametrically. We observed systematic variance-dependent adaptation (defined as a scaling of a nonlinear transfer function) similar in many respects to the effects observed in vivo. The fact that similar adaptive phenomena are seen in such different preparations led us to investigate a simple model of stochastic stimulus-driven neural activity. The simplest such model, the leaky integrate-and-fire (LIF) cell driven by noise current, permits us to analytically compute many quantities relevant to our observations on adaptation. We show that the LIF model displays "adaptive" behavior which is quite similar to the effects observed in vivo and in vitro.

Keywords: Adaptation; Noise; Integrate-and-fire; Fokker–Planck

It is widely understood that sensory neurons adapt to the prevailing statistics of their inputs [10]. Fairhall et al. [5] recently reported one such adaptation process in the fly visual system; they described a motion-sensitive neuron that appears to scale its input–output function to adapt its firing rate to the variance of the observed motion signal. However, the mechanisms underlying this type of contrast-dependent adaptation are unknown; specifically, it is unclear whether the observed phenomena arise from network

☆ This work was supported by NSF Grant IBN-0079619. LP and BL are supported by HHMI and NDSEG predoctoral fellowships, respectively. We thank E. Simoncelli for many interesting discussions.

* Corresponding author.

E-mail address: liam@cns.nyu.edu (L. Paninski).

[1] Contact: liam@cns.nyu.edu; http://www.cns.nyu.edu/~liam

0925-2312/03/$ - see front matter © 2003 Elsevier Science B.V. All rights reserved.
doi:10.1016/S0925-2312(02)00819-6

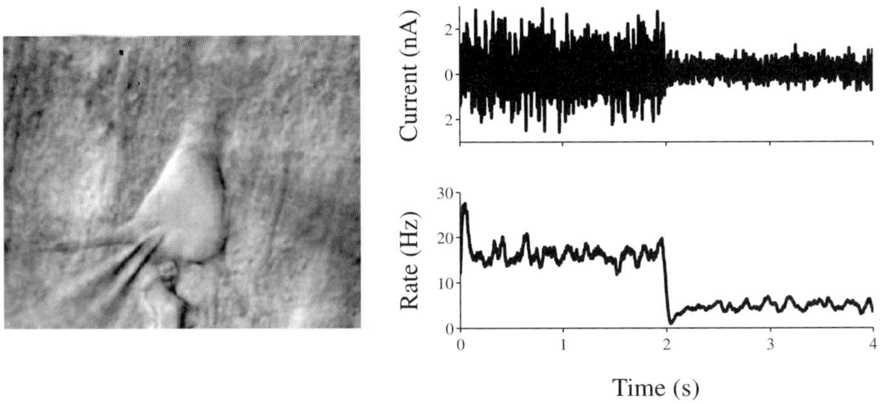

Fig. 1. Experimental details. Sagittal slices were prepared from adolescent and adult rats (P14-P24) as described in [8]. Briefly, slices were maintained at 30°C in artificial cerebrospinal fluid consisting of (in mM): 125 NaCl, 2.5 KCl, 25 glucose, 25 $NaHCO_3$, 1.25 NaH_2PO_4, 2 $CaCl_2$, and 1 $MgCl_2$. Cells were visualized using infrared differential interference contrast microscopy with a 40× water immersion objective. Dual-electrode whole-cell recordings were made using pipettes with 5–15 M resistance when filled with (in mM): 100 K-gluconate, 20 KCl, 4 ATP-Mg, 10 phosphocreatine, 0.3 GTP, and 10 HEPES, pH 7.3 (310 mOsm). Recordings were performed in current clamp using Axoclamp 2B amplifiers (Axon Instruments, Foster City, CA), and stimulus presentation and data acquisition was managed using IGOR (Wavemetrics, Lake Oswego, OR). Gaussian white noise current stimuli were delivered through one electrode, while voltage was recorded through the other electrode and processed on- and off-line. Left panel shows a photograph of a cell with the recording and stimulating electrodes partially visible; right panel shows a sample trace of the current input (including a jump between two values of noise variance), and the corresponding peri-event time histogram (note that the noise current was not "frozen," that is, a new noise current was drawn i.i.d. for each trial).

dynamics or from dendritic or somatic mechanisms in individual neurons. We hypothesized that (1) somatic mechanisms could account for at least part of the observed adaptation phenomena, and that (2) these somatic effects are general in the sense that they depend only weakly on the biophysical parameters governing a given neuron's behavior. To test hypothesis (1), we recorded intracellularly from layer V pyramidal neurons in sensorimotor cortex in vitro (see Fig. 1 for details), while stimulating with a noise current whose standard deviation (or "contrast") was varied parametrically. Hypothesis (2) will be addressed mathematically below.

For ease of comparison, we analyzed our data using the basic framework utilized in [5] (see also, e.g., [4]). For each neuron, we estimated a separate spike-triggered average (STA) at each current standard deviation [3], using data acquired after the neuron had reached a steady-state firing rate. We then projected our stimulus onto the normalized STA function (this operation is equivalent to a time-reversed convolution, or filtering), and estimated, via a nonparametric histogram approach, the conditional probability of a spike given each observed value of the projected current. These conditional firing rate functions (termed N-functions, for nonlinearity, in keeping with convention) will be the main object of our analysis (see Fig. 2 for an example).

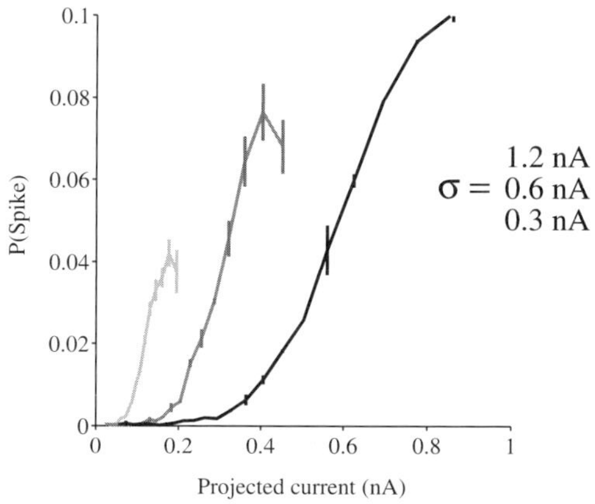

Fig. 2. Examples of N-functions for a single pyramidal neuron. Each curve represents data for a particular standard deviation of input current.

Our main observations are as follows: first, the STA (which is often thought of as a linear prefilter for the cell, the stimulus dimension to which the neuron is most sensitive) changed with the standard deviation of the injected current. As σ was increased, we observed a systematic reduction in the time-to-peak as well as the half-width of the STA, consistent with results seen in vitro [3] and in vivo [1]. We also observed changes in the N-functions (Fig. 2). If we define "gain" as the slope of the N-function, we have that the gain of the observed cortical cells was consistently inversely proportional to the standard deviation of the injected current; this result is strikingly similar to those of Fairhall et al. [5].

What could explain the gain changes described above? One common model for gain changes in cortical cells requires the presence of some channel whose conductance is dependent on the firing rate of the cell; it has been shown, for example, that calcium-dependent potassium channels can lead to changes in the input-dependent firing rate of the cell (see e.g. [11]). It seems plausible that such macroscopic changes in the firing rate could manifest themselves in changes at the more detailed level of the N-function (although, to our knowledge, these effects have not been studied in detail). However, we believe that an even simpler phenomenon is (at least partially) responsible here. Our hypothesis is that much of the "adaptation" phenomenon described above can be explained by the basic spiking dynamics of the cell, even in the absence of nonlinear ion channels. (See also [9], where a similar idea was proposed using different methods.)

To explain our results, we introduce some tools from the theory of stochastic dynamical systems. We start with perhaps the simplest widely-used neural model, the leaky integrate-and-fire (LIF) cell, described by the following equation:

$$\frac{\mathrm{d}V}{\mathrm{d}t} = \frac{1}{\tau_m}(V_L - V + R_m I) - (V_{th} - V_{reset})\delta(V - V_{th}),$$

where V denotes voltage, τ_m the membrane time constant, R_m the membrane resistance, V_L the leak reversal potential, V_{th} and V_{reset} the threshold and reset potential, respectively, and I is an input current which in our case is given by a standard white noise process of stationary mean μ and standard deviation σ.

A noise-driven dynamical system can, in general, be described either in terms of the pathwise (single-trial) behavior of the system or in a distributional (ensemble) sense. The pathwise behavior of the noise-driven LIF cell is easily understood: below threshold, the cell is a one-dimensional Ornstein–Uhlenbeck process [7]; at threshold, the voltage is reset instantaneously to V_{reset}. This system is clearly one-dimensional and strong Markov; that is, its behavior in the future depends on its past only through V at the present time. The distributional description of the noise-driven LIF model—that is, the equations governing the behavior of the probability distribution on voltage, $P(V)$, as a function of time—turns out to be most useful. The following Fokker–Planck equation completely characterizes the distributional behavior of the system, given initial conditions [7]

$$\frac{\partial P}{\partial t} = \frac{\sigma_0^2}{2}\frac{\partial P^2}{\partial^2 V} + \frac{1}{\tau_m}\frac{\partial[(V - V_0)P]}{\partial V} + R(t)(\delta(V - V_{reset}) - \delta(V - V_{th}))$$

with $R(t)$ the time-dependent mean firing rate of the cell, $\sigma_0 \equiv R_m\sigma/\tau_m$, and

$$V_0 = R_m\mu + V_L$$

the steady-state rest potential. The PDE above is a perturbed diffusion equation: the first term corresponds to diffusion (the effect of the injected noise on the voltage at time t), the second is a drift term (corresponding to the V-dependent steady-state driving force back towards the rest potential V_0), and the third corresponds to the voltage probability flux resulting from spiking activity, which is subtracted from $P(V)$ at V_{th} and added at V_{reset}. This equation has been introduced by several authors [2,6] as an approximation to the behavior of the LIF cell under a barrage of random synaptic currents. Note that since we are injecting Gaussian white noise and not a simulated superposition of PSPs, the above equation is *exact* within the LIF framework.

Surprisingly, many of the quantities of interest turn out to be analytically computable for this model. Most of what we need can be read directly from the following steady-state solution to the PDE:

$$P(V) = \frac{2R}{\sigma_0^2}\int_{\max(V,V_{reset})}^{V_{th}} dV' e^{[(V'-V_0)^2-(V-V_0)^2]/\tau_m\sigma_0^2},$$

where $P(V)$ denotes the invariant density and R is the equilibrium firing rate; note that $P(V)$ is in a sense a perturbed Gaussian, as expected of the solution to a perturbed diffusion equation with linear drift. We plot $P(V)$ for a few different values of noise in Fig. 3: when the injected noise is small (σ near zero), $P(V)$ looks like a Gaussian centered at V_0. However, as σ grows, $P(V)$ develops a kink at V_{reset}, dives nearly linearly to zero at V_{th}, and develops a large negative tail. We will see below that the

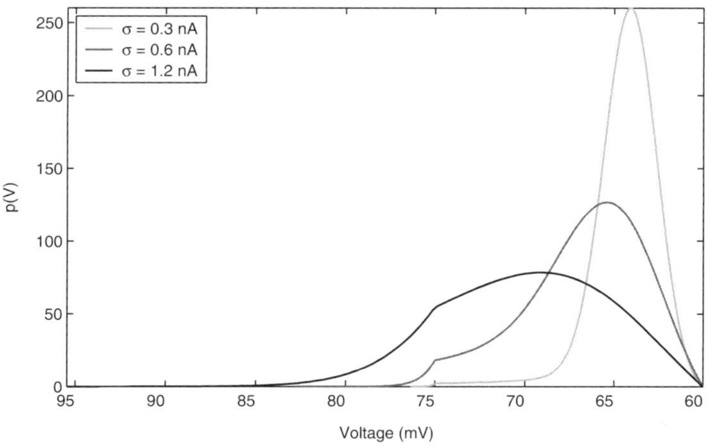

Fig. 3. Three examples of invariant densities, $P(V)$, for three different values of noise variance σ^2. V_{th} is -60 mV and V_{reset} is -75 mV here.

size of this negative tail—the probability that the Ornstein–Uhlenbeck process will have wandered into a highly hyperpolarized state—has a critical effect on the "gain" of the neuron, according to several reasonable definitions of gain.

Recall the definition of the N-function introduced above: this quantity is a conditional firing rate, given some filtered version of the recent stimulus. It is difficult to approach this gain function analytically when this filter is chosen by cross-correlation methods, as in the preceding section, because the STA of the LIF cell turns out to be a rather poorly behaved mathematical object (for example, it is not hard to show that this function depends rather strongly on the time step $\mathrm{d}t$ of the numerical simulation, with no well-defined limit in the continuous limit as $\mathrm{d}t \to 0$). Even when the filter is chosen to be a step function, detailed analysis of this conditional firing rate function seems to require a rather complicated analysis of the conditional pathwise behavior of the Ornstein–Uhlenbeck process, which has in our hands not yet led to any interesting conclusions. However, we can derive useful information about certain limits of the gain function, as the support of the step filter shrinks to zero or goes to infinity.

For example, we have an exact expression for the "transient gain function"

$$F_0(x,\sigma) \equiv \lim_{T\to 0} P\left(\text{spike} \in (-T, 0] \,\middle|\, \int_{-T}^{0} I(t)\,\mathrm{d}t = x\right) = \int_{V_{\text{th}} - \frac{xR_{\text{m}}}{\tau_{\text{m}}}}^{V_{\text{th}}} P(V)\,\mathrm{d}V$$

and the counterpart "long-time" gain function

$$F_\infty(x,\sigma) \equiv \lim_{T\to\infty} P\left(s \in (-\mathrm{d}t, 0] \,\middle|\, \int_{-T}^{0} I(t)\,\mathrm{d}t = xT\right)$$

$$= -\frac{\sigma_0^2}{2} \left.\frac{\partial P_{\mu+x}(V)}{\partial V}\right|_{V=V_{\text{th}}} \mathrm{d}t + o(\mathrm{d}t),$$

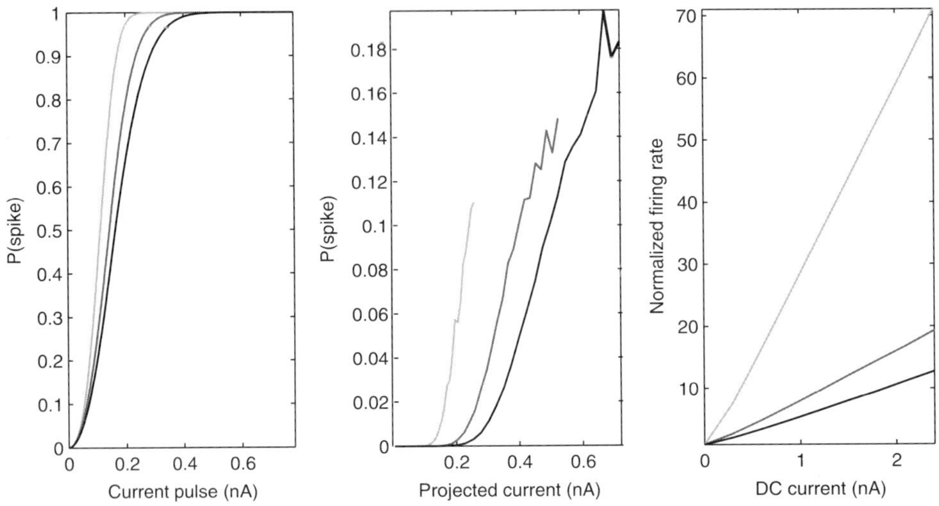

Fig. 4. Three "gain functions" for integrate-and-fire cell: middle panel shows N-function, computed by Monte Carlo; left panel is "transient" function $F_0(x,\sigma)$, and right is "long-time" function $F_\infty(x,\sigma)/F_\infty(0,\sigma)$, both computed analytically. (Note that $F_\infty(x,\sigma)/F_\infty(0,\sigma)$ is normalized so that the y-axis is a dimensionless ratio.)

where the subscript in the last expression indicates the dependence of the invariant density on the DC input current. The variable x in the two expressions above corresponds to the projected current (the x-axis in Fig. 2). The σ-dependence of these functions (computed analytically) is shown in Fig. 4; also shown are some sample N-functions, computed via Monte-Carlo. These gain functions are all σ-dependent, indicating strong adaptive phenomena in the standard LIF cell.

We have two related main conclusions. First, a very simple preparation displays "adaptation" to noise current input over a time scale of hundreds of milliseconds; this adaptation phenomenon must be independent of any "upstream" (e.g., synaptic) processes, since currents were injected and voltages measured directly at the same soma. Second, perhaps more surprisingly, a very simple model, devoid of any interesting dynamics save the (instantaneous) spiking process itself, adapts strongly to changing stimulus distributions; we can describe this behavior exactly, and it turns out to match the in vitro data qualitatively. The fact that a model as generic as the LIF cell displays adaptation so similar to that observed in vitro and in vivo seems to indicate that variance-dependent adaptation is, in fact, a general feature of spiking cells in the nervous system.

References

[1] W. Bair, J. Cavanaugh, J. Movshon, NIPS 9 (1997) 34–40.
[2] N. Brunel, V. Hakim, Neural Comput. 11 (1999) 1621–1671.
[3] H.L. Bryant, J.P. Segundo, J. Physiol. 260 (1976) 279–314.

[4] E. Chichilnisky, Network 12 (2001) 199–213.
[5] A. Fairhall, G. Lewen, W. Bialek, R. de Ruyter, Nature 412 (2001) 787–792.
[6] E. Haskell, D. Nykamp, D. Tranchina, Network 12 (2001) 141–174.
[7] S. Karlin, H. Taylor, A Second Course in Stochastic Processes, Academic Press, New York, 1981.
[8] A. Reyes, B. Sakmann, J. Neurosci. 19 (1999) 3827–3835.
[9] M. Rudd, L. Brown, Neural Comput. 9 (1997) 1047–1069.
[10] R. Shapley, Current Biol. 7 (1997) 421–423.
[11] M. Stemmler, C. Koch, Nature Neurosci. 2 (1999) 521–527.

Available at
www.ComputerScienceWeb.com
POWERED BY SCIENCE @ DIRECT®

ELSEVIER

Neurocomputing 52–54 (2003) 885–891

NEUROCOMPUTING

www.elsevier.com/locate/neucom

Using derivatives to compare cortical waves across preparations

Kay A. Robbins[a,*], David M. Senseman[b]

[a]*Department of Computer Science and Cajal Neuroscience Research Center, The University of Texas at San Antonio, 6900 North Loop 1604 West, San Antonio, TX 78249, USA*

[b]*Department of Biology and Cajal Neuroscience Research Center, The University of Texas at San Antonio, 6900 North Loop 1604 West, San Antonio, TX 78249, USA*

Abstract

Visual stimulation evokes waves of depolarization that propagate within and between cytoarchitectually distinct cortical areas in the turtle visual cortex. These waves appear to reflect underlying low-dimensional characteristics that can be captured using Karhunen–Loéve (KL) decomposition. This paper compares structure across in vivo, in vitro and simulated preparations of turtle visual cortex using derivative–response relationships. The derivatives of the projections on the low-order KL modes exhibit distinguishing characteristics based on the stimulus. The derived modes corresponding to these derivatives provide a low-order approximation to the initial response with structure similar to that of a traveling wave.

Keywords: Cortical wave; KL decomposition; Traveling wave

1. Introduction

In turtles, visual stimulation evokes a complex wave of depolarization within the cerebral cortex that originates in the rostral pole of the dorsal cortical area where the cortex receives direct afferent projections from the lateral geniculate complex of the thalamus. Waves evoked by visual stimulation propagate at a relatively uniform velocity from the rostral pole into the caudal pole of dorsal cortex. Waves with the same qualitative features have been observed for in vitro preparations [7], in vivo preparations [5] and for model simulations [3,4] of the turtle visual cortex. The current study

* Corresponding author. Tel.: +1-210-458-5543; fax: +1-210-458-4437.
E-mail addresses: krobbins@cs.utsa.edu (K.A. Robbins), senseman@utsa.edu (D.M. Senseman).

0925-2312/03/$ - see front matter
doi:10.1016/S0925-2312(02)00825-1

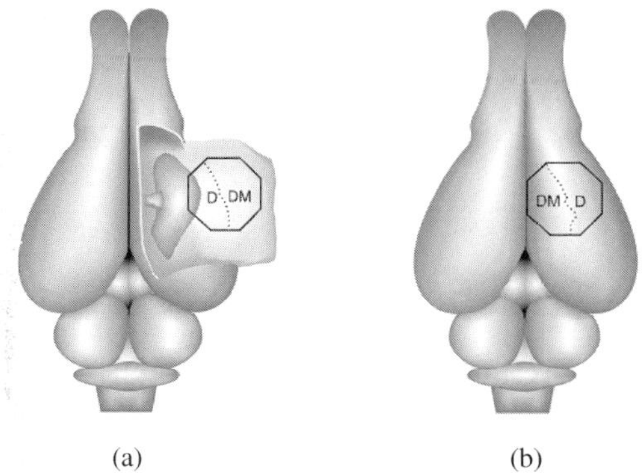

Fig. 1. Schematic of the recording area for the (a) in vitro and (b) in vivo preparations.

compares structural features of the waves across preparations in these three different realizations.

Fig. 1a shows a schematic of the in vitro experiments [6] in which the brain and the attached eyes were surgically removed. To facilitate optical recording, the cortical sheet was freed at the midline and folded flat against the transparent bottom of the recording chamber, taking care not to disturb the normal input from the eye. To monitor cortical activity, the cortical sheet was stained with an absorption voltage-sensitive dye (VSD) and trans-illuminated for optical recording. Fig. 1b shows a schematic of the in vivo experiments [5] in which the surface of the brain was exposed by a craniotomy and partially deafferented by sectioning the spinal cord. The cortical tissue was stained with a fluorescent VSD and excited using an epi-illumination system. Both experiments recorded with a 464-element silicon photodiode array that served as a fast video camera ($\leqslant 1$ ms/frame) with moderate spatial resolution.

Fig. 2 shows a schematic of the physiologically-based numerical model of the turtle visual cortex developed by Nenadic et al. [4]. The NGU simulation uses compartmental models of four neuronal types: lateral pyramidal cells (triangles), medial pyramidal cells (stars), stellate cells (circles) and horizontal cells (squares). The distribution, characteristics and connectivities of the cells were based on experimental measurements. The cortex model was driven by a linear array of integrate-and-fire lateral geniculate cells (LGN). In addition to the neuronal distributions, Fig. 2 shows 5 s time traces of four neurons from a trial in which a constant stimulus was applied to all of the LGN neurons for 3 s. Dotted vertical gray lines on each trace mark the starting and stopping points of the stimulus. Notice that the pyramidal neurons closest to the LGN (e.g., A and C) exhibit more vigorous firing patterns than neurons of the same type located more caudally (e.g., B and D). Furthermore, the lateral cells (e.g., A and B) exhibit more vigorous firing patterns than their medial counterparts (e.g., C and D). The initiation of activity at lateral neuron A occurs immediately after stimulus onset, while the activity at lateral neuron B shows a delay that is consistent with an initial wave of activation sweeping the cortex. These general features appear in the response to other types of stimuli as well. Notice that neurons B, C and D show secondary peaks

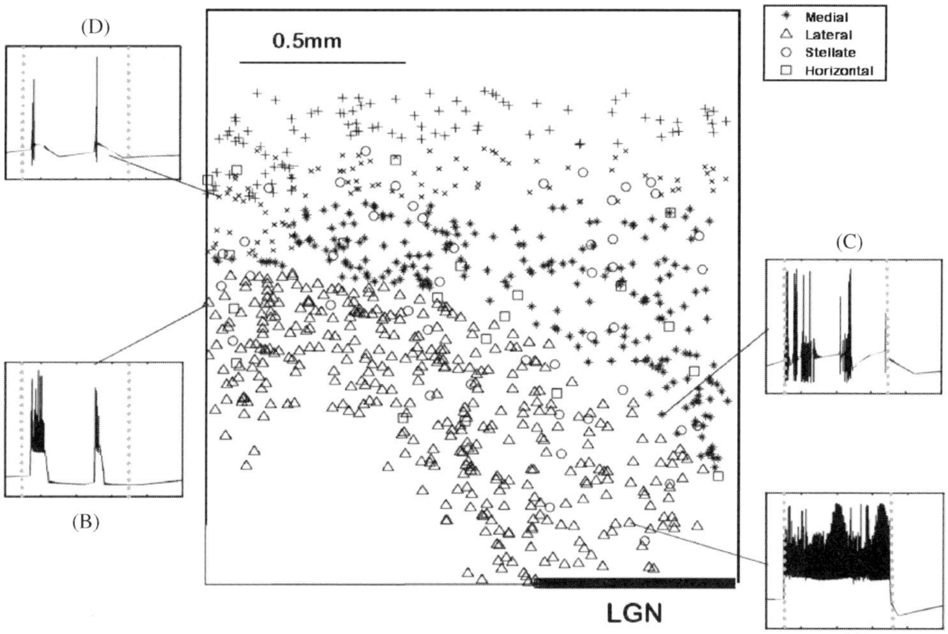

Fig. 2. Schematic of neurons in the NGU numerical model.

of response roughly 2 s after the initial stimulus, while A is continuously activated. These secondary peaks correspond to reexcitation or secondary waves of excitation that sweep through the cortex.

Although the three experimental setups and stimulus presentations are very different, the basic features of the initial response wave are similar in the three configurations. Table 1 summarizes the trials that are used in the analysis for this paper. The next section reviews Karhunen–Loéve (KL) decomposition and introduces "derived modes." The final section compares the results for the three experimental preparations.

2. Derivative relationships in KL decomposition

Our analysis of the wave motion first reduces the size of the response space by performing KL decomposition and then examines derivative relationships in the reduced space. For signals that are sampled at discrete points in space and time, the measured response $\mathbf{U}=(u_{x,t})$ can be collected in an $L\times N$ matrix where $x=1,\ldots,L$ and $t=1,\ldots,N$ represent the sample numbers in space and time, respectively. The discrete proper orthogonal decomposition of $\mathbf{U}$ is

$$\mathbf{U} = \sum_{k=1}^{r} \alpha_k \varphi^k (\psi^k)^{\mathrm{T}}, \tag{1}$$

Table 1
A summary of the trials used for the different preparations

Experiment	Sampling rate	Trials	Stimuli	Samples/Trial
In vitro	353 Hz	9	Diffuse flashes (100 ms)	576
		17	Spots on retinal midline (100 ms)	576
In vivo	1000 Hz	5	Looming ball	5000
		4	LED 3 s flash	5000
		2	LED train at 5.8 Hz	5000
		2	LED train at 11 Hz	5000
Model	1000 Hz	10	LGN patches (20 adjacent neurons)	1500
		1	Diffuse flashes	1500
		1	LED 3 s flash	5000
		1	LED train at 5.8 Hz	5000
		1	LED train at 11 Hz	5000

where r is the rank of $\mathbf{U}$, the φ^k's are $L \times 1$ orthonormal vectors representing spatial basis functions, and the ψ^k's are $N \times 1$ orthonormal vectors representing temporal basis functions. The φ^k's satisfy $\mathbf{R}\varphi^k = \lambda_k \varphi^k$ where $\mathbf{R} = \mathbf{U}\mathbf{U}^{\mathrm{T}}$ and $\lambda_k = \alpha_k^2$. The φ^k's are also the modes or spatial basis functions computed by KL decomposition [2].

The temporal basis function ψ^j can be obtained by left multiplying Eq. (1) by the transpose of φ^j:

$$(\varphi^j)^{\mathrm{T}}\mathbf{U} = \sum_{k=1}^{r} \alpha_k (\varphi^j)^{\mathrm{T}} \varphi^k (\psi^k)^{\mathrm{T}} = \sum_{k=1}^{r} \alpha_k \delta_{j,k} (\psi^k)^{\mathrm{T}} = \alpha_j (\psi^j)^{\mathrm{T}}.$$

Taking the transpose of this equation and dividing by α_j gives

$$\psi^j = \alpha_j^{-1} \mathbf{U}^{\mathrm{T}} \varphi^j. \tag{2}$$

Similarly the φ^j's can be recovered by multiplying Eq. (1) by ψ^j:

$$\mathbf{U}\psi^j = \sum_{k=1}^{r} \alpha_k \varphi^k (\psi^k)^{\mathrm{T}} \psi^j = \sum_{k=1}^{r} \alpha_k \varphi^k \delta_{k,j} = \alpha_j (\varphi^j)$$

or

$$\varphi^j = \alpha_j^{-1} \mathbf{U} \psi^j. \tag{3}$$

Eqs. (2) and (3) show that φ^j and ψ^j are duals of each other in the sense that one can be obtained from the other by projection on the original data set. If there are fewer time snapshots (N) than spatial points (L), it is more efficient to find the ψ^j's first [2].

We exploit the duality relationship in a slightly different context. We have observed that the time derivatives $\mathrm{d}\psi^p/\mathrm{d}t$ are correlated with ψ^q for certain p and q. We obtain *derived spatial basis functions* by assuming that the $\mathrm{d}\psi^p/\mathrm{d}t$ are temporal modes and use relationship (3) to compute corresponding spatial modes. We then examine how these *derived spatial modes* are related to the spatial modes obtained from the original KL decomposition. The spatial basis function derived by projecting the time derivative of ψ^j on the data set are $(\varphi^j)' = \alpha_j^{-1}\mathbf{U}\mathrm{d}\psi^j/\mathrm{d}t$. This paper focuses on the behavior of $\mathrm{d}\psi^1/\mathrm{d}t$ and its relationship with modes ψ^q for $q > 1$.

3. Results

As described in Ref. [8], we concatenate the responses for a collection of stimuli in a preparation and perform a KL decomposition on the composite, to obtain a "global basis". The "global" decomposition is not equivalent to averaging trials before performing KL decomposition. Rather, it is a way to create a common basis so that different trials from the same preparation can be directly compared. An overbar indicates basis functions computed from a dataset of concatenated trials for the preparation. A prime indicates differentiation with respect to time.

While the details differ quantitatively, the global KL decomposition in all three realizations gives a low-dimensional modal representation dominated by a large single-signed spatial basis function, $\bar{\varphi}^1 = \bar{M}^{1,1}$. We use the notation $\bar{M}^{i,j}$ to denote a spatial basis function with i extrema aligned with the major axes of $\bar{M}^{1,1}$ and j relative extrema aligned with the minor axes of $\bar{M}^{1,1}$ [8]. The modal notation is used because, while $\bar{M}^{1,1}$ always corresponds to the largest KL mode, $\bar{\varphi}^1$, the $\bar{M}^{1,2}$ mode corresponds to $\bar{\varphi}^2$ for the model, to mode $\bar{\varphi}^3$ for the in vitro data, and to mode $\bar{\varphi}^4$ for the in vivo data. We use $\bar{P}^{i,j}$ to denote the projection of the data set on $\bar{M}^{i,j}$, and $(\bar{P}^{i,j})'$ to denote the time derivative of $\bar{P}^{i,j}$. We represent the mode derived from $(\bar{P}^{i,j})'$ by $(\bar{M}^{i,j})'$.

The leftmost column of Fig. 3 shows $\bar{M}^{1,1}$ for the three experimental preparations. The model appears to have a bimodal $\bar{M}^{1,1}$ due to the gray scale color scheme, but in fact, the two dark areas are of the same sign. The second column shows $\bar{M}^{1,2}$ for the different preparations. In each case, $\bar{M}^{1,2}$ has positive and negative extrema separated by a zero response region. The extrema are aligned roughly along the principal axis of $\bar{M}^{1,1}$. We also observed in our data analysis that $\bar{P}^{1,2}$ for individual trials appears to be highly correlated with $(\bar{P}^{1,1})'$, the time derivative of the projection of the data set on $\bar{M}^{1,1}$. Additionally, $(\bar{P}^{1,1})'$ appears to clearly distinguish response characteristics. For example, trials in both the in vivo and model preparations that consisted of LED trains at a specified frequency had a $(\bar{P}^{1,1})'$ with power spectrum dominated by that frequency.

The middle two columns of Fig. 3 compare $\bar{M}^{1,2}$ with $(\bar{M}^{1,1})'$, the basis function derived from $(\bar{P}^{1,1})'$. The rightmost column of Fig. 3 shows the component of $(\bar{M}^{1,1})'$ that is orthogonal to $\bar{M}^{1,1}$. A comparison of columns 3 and 4 of Fig. 3 show that $(\bar{M}^{1,1})'$ is closely aligned with $(\bar{M}^{1,1})'^{\perp}$ indicating that $(\bar{M}^{1,1})'$ is nearly orthogonal to $\bar{M}^{1,1}$. The pair $\bar{M}^{1,1}$ and $(\bar{M}^{1,1})'^{\perp}$ span a two-dimensional space that appears to capture the main features of the initial wave motion.

A translationally invariant system (e.g., traveling waves in a homogeneous unbounded or periodic domain) has sinusoidal KL modes [2]. Sinusoidal basis

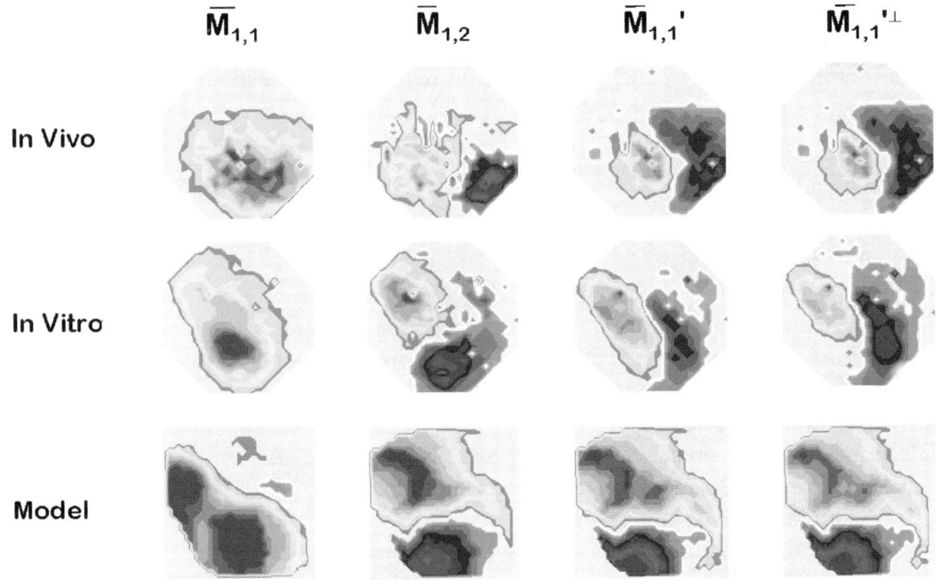

Fig. 3. A comparison of the structure of the KL modes and first-order derived modes.

functions satisfy simple derivative relationships at all orders, capturing the fact that basis functions come in pairs and are phase offsets of each other. The turtle cortex is not translationally invariant, nor do its basis images represent sinusoidal functions. Yet, the derivative relationships suggest that the global structure of the response wave can be analyzed in terms of a few low-order KL modes that are dominated by a traveling wave structure. Ermentrout [1] has analyzed several one-dimensional neural network models with traveling wave solutions and related the velocity of the wave to the strength of the synaptic coupling. The derived modes provide a mechanism for similar studies of traveling waves in experimental data and nonhomogeneous cortical models.

Acknowledgements

We thank L. Cohen, D. Kleinfeld and P. Mitra for permission to use their in vivo data sets. We are also grateful to P. Ulinski, B. Ghosh and Z. Nenadic for allowing us to use their model of the turtle cortex. This work was supported by the NIH (G12 RR13646), the NSF (ACI-9721348), the ONR (N00014-97-0029) and the UTSA Office of the Provost.

References

[1] B. Ermentrout, The analysis of synaptically generated traveling waves, J. Comput. Neurosci. 5 (1998) 191–208.

[2] M. Kirby, Geometric Data Analysis, Wiley, New York, 2001.
[3] Z. Nenadic, B. Ghosh, P. Ulinski, Modeling and estimation problems in the turtle visual cortex, IEEE Trans. Biomed. Eng. 49 (2002) 753–762.
[4] Z. Nenadic, B. Ghosh, P. Ulinski, Propagating waves in visual cortex: a large-scale model of turtle visual cortex, J. Comput. Neurosci. 14 (2) (2003) 161–184.
[5] J. Prechtl, L. Cohen, B. Pesaran, P. Mitra, D. Kleinfeld, Visual stimuli induce waves of electrical activity in turtle cortex, Proc. Nat. Acad. Sci. 94 (1997) 7621–7626.
[6] D. Senseman, Spatiotemporal structure of depolarization spread in cortical pyramidal cell populations evoked by diffuse retinal light flashes, Visual Neurosci. 16 (1999) 65–79.
[7] D. Senseman, K. Robbins, Modal behavior of cortical neural networks during visual processing, J. Neurosci. 19 (1999) RC3 1–7.
[8] D. Senseman, K. Robbins, High-speed VSD imaging of visually evoked cortical waves: decomposition into intra- and intercortical wave motion, J. Neurophysiol. 87 (3) (2002) 1499–1514.

Kay Robbins is a Professor at the University of Texas at San Antonio in the Department of Computer Science. Her research focuses on visualization and analysis of multimedia data sets. She collaborates with experimental scientists in the life and physical sciences to use visualization tools on real data. She holds a BS and PhD in Mathematics from the Massachusetts Institute of Technology.

David Senseman received his BS degree in Zoology from Kent State University and his Masters and Doctoral degrees in Biology from Princeton University. During his Post-doctoral fellowship in Biophysics at the University of Pennsylvania, he worked in Brian Salzberg's lab where he learned the "art" of voltage-sensitive dye (VSD) imaging. Since moving to the University of Texas at San Antonio in 1983, he has used VSD imaging to study sensory processing in cortical structures from a variety of mammalian and non-mammalian vertebrates.

Available at www.ComputerScienceWeb.com
POWERED BY SCIENCE DIRECT
ELSEVIER
Neurocomputing 52–54 (2003) 893–899
NEUROCOMPUTING
www.elsevier.com/locate/neucom

Effects of noise on recurrence in networks of spiking neurons

Patrick D. Roberts

Neurological Sciences Institute, Oregon Health Sciences University, 505 NW 185th Ave, Portland, OR 97006, USA

Abstract

We use mathematical analysis and numerical simulations to study the effects of recurrent synaptic connections in neural networks. These recurrent effects can be analyzed by expanding a spike probability function about the membrane potential in the absence of connections. This expansion can be written as a series of loops representing the recurrent signal transmission. Conditions for convergence of the series reveal the number of loops that are significant for the system's dynamics. This method is applied to a pair of mutually coupled spiking neurons to show how the membrane potential and cross-correlation function is changed by recurrence.

Keywords: Noise; Recurrence; Spiking neurons

1. Introduction

Recurrent connections are found in many biological neural systems. One type of recurrence arises from *lateral* connections between neurons within the same layer of processing [7]. The neurons within a layer are often interconnected, and the modulation of neighbors within the layer is an integral part of processing the sensory information. However, if these connections are too strong, then instabilities will develop that can be disastrous for sensory processing.

Another example of recurrence is due to *feedback* from layers at a higher level of processing leading to modulation of input layers [5]. The feedback gain itself must be modulated, because nonlinearities inherent in feedback systems can lead to

E-mail address: robertpa@ohsu.edu (P.D. Roberts).
URL: http://www.ohsu.edu/nsi/faculty/robertpa

0925-2312/03/$ - see front matter
doi:10.1016/S0925-2312(02)00826-3

uncontrollable consequences. Feedback control may improve feature identification by modulating sensory input, not by washing out the sensory information with feedback dynamics. We show in the following that the addition of noise can reduce the effects of recurrent nonlinearties by reducing the number of traversals that a signal can survive against noise.

Many studies of recurrence have concentrated on nonlinear effects such as bistability [3,4]. When two neurons are mutually coupled with strong synaptic connections, the output of the neurons has two stable states; one neuron is firing while the other is silent for inhibitory connections. Under certain conditions, the two neurons can switch between the two states. This type of neural dynamics is common in central pattern generators of invertebrate systems. By flipping between stable states, the system can generate a rhythmic output [9].

However, such bistable dynamics can be detrimental to sensory systems where lateral recurrence is used to sharpen contrasts. Nonlinear, bistable dynamics would blind the system with strong internal dynamics. In the mammalian cerebral cortex, it is likely that the nonlinear regime is a pathological state [1].

Methods of mathematical analysis are needed to determine how system characteristics, such as noise and synaptic strength, affect the dynamics of recurrence. Numerical methods have proven useful to find parameter sets where bistability and other nonlinear effects appear in specific models. Bifurcation analysis has been useful to determine the states of dynamical systems and analyze state changes [6]. However, many biological neural systems appear to have an equilibrium level of activity that is enforced by activity dependent mechanisms [8]. The presence of equilibrium fixed points in neural activity suggests that we should develop methods to compute corrections to neural activity arising from interconnectivity within neural networks.

In the following, we construct a model of synaptically coupled spiking neurons. Each neuron receives a non-adaptive input that maintains a baseline membrane potential resulting in a constant spike output rate. Gaussian noise is added to the membrane potential of the model neurons to randomly vary the instantaneous spike rate and simulate the presence of unpatterned synaptic inputs. We then expand the spike probability of one of the neurons about the baseline pike probability in the absence of the recurrent connections, and compute the correction due to the presence of the second neuron. The expansion is ordered in terms of recurrent loops so that the influence of information travelling through multiple recurrent loops becomes is quantified.

2. Model of coupled neurons

To represent the neurons, we use a spike-response model [2] with no relative refractory period. Time is discretized so that $t_n = n(\Delta t)$, with n an integer. The probability of a spike during the time interval t_n is a threshold (sigmoid) function of the membrane potential, $V(t_n)$. For neuron 1 in Fig. 1, with threshold θ, and noise parameter μ, the spike probability is given by the expression

$$f_1(t_n) = \frac{1}{1 + \exp[-\mu(V_1(t_n) - \theta)]}, \tag{1}$$

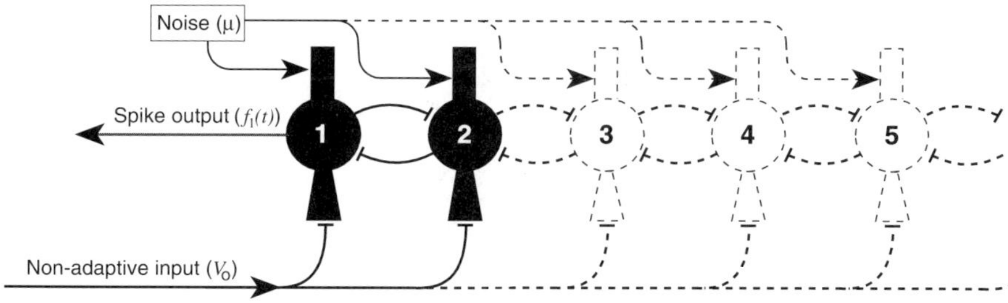

Fig. 1. Recurrent synaptic connectivity between model neurons. The model spiking neurons generate spikes that are the postsynaptic events coupling pairs of neurons. Each synaptic connection contributes a psp to the membrane potential represented by an psp waveform weighted by s. The text focuses on the first two neurons in the network.

representing the variability due to the Gaussian noise. The instantaneous spike frequency is obtained by multiplying the spike probability by the maximum spike frequency.

The membrane potential is the sum of all external inputs for each neuron. The baseline membrane potential maintained by non-adaptive inputs is represented by the term V_o. Synaptic connectivity is represented by weighted postsynaptic potential (psp) waveforms, $\varepsilon(t_n)$. The contribution from neuron 2 is represented by the spike probability convolved with the psp waveform, and weighted with the synaptic strength s. For an excitatory synapse $s > 0$, and for an inhibitory synapse, $s < 0$. The membrane potential of neuron 1 is

$$V_1(t_n) = +V_o + s\sum_m f_2(t_m)\varepsilon(t_m - t_n), \tag{2}$$

where the last term is the contribution from the activity of neuron 2. This formalism easily generalizes to any number of neurons in the network.

3. Results

The starting point of our analysis is to expand the spike probability about the spike probability that would result in the absence of recurrent synaptic connections, $\hat{f} = f_1(U_1(t_n))$,

$$f_1(t_n) = \hat{f}_1 + \mu(\hat{f}_1 - \hat{f}_1^2)(V_1(t_n) - U_1(t_n)) + \cdots . \tag{3}$$

The solution for the membrane potential of neuron 1 without the input of neuron 2 is the baseline membrane potential

$$\langle U_1(t_n)\rangle = V_o. \tag{4}$$

To compute the effects of recurrence, we include neuron 2 in our calculations and expand the membrane potential of 1 about the membrane potential in the absence or

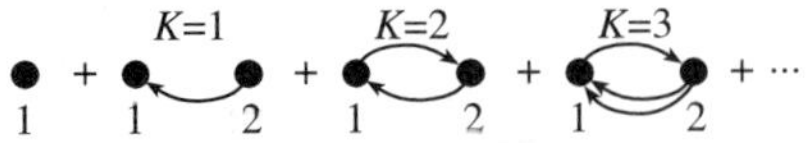

Fig. 2. Graphical representation of the loop expansion.

recurrence,

$$\langle V_1(t_n)\rangle = \langle U_1(t_n)\rangle + s\hat{f} + s\mu(\hat{f} - \hat{f}^2)\sum_m (\langle V_2(t_m)\rangle - V_o)\varepsilon(t_n - t_m). \tag{5}$$

Next, we expand $\langle V_2(t_m)\rangle$ about the baseline potential and collect terms to arrive at the recurrent loop expansion:

$$\langle V_1(t_n)\rangle = \langle U_1(t_n)\rangle + s\hat{f}\sum_{K=1}^{\infty}(s\hat{f})^{K-1}\mu^{K-1}(1-\hat{f})^{K-1}$$
$$\times\left[1+\mu(1-\hat{f})\sum_{i_1}\cdots\sum_{i_K}(U_a(t_n)-V_o)\varepsilon(t_{i_{K-1}}-t_{i_K})\cdots\varepsilon(t_n-t_{i_1})\right], \tag{6}$$

where

$$a = \left\{\begin{matrix} 1 & \text{if } K \text{ is odd} \\ 2 & \text{if } K \text{ is even} \end{matrix}\right\}. \tag{7}$$

Each term in this loop expansion represents the contribution to the membrane potential of 1 by going around the loop Fig. 2.

The first term has no recurrent influence, the second term ($K = 1$) includes only the contribution of only neuron 2 on neuron 1 as if neuron 2 had no other synaptic contact with neuron 1. The third term ($K = 2$) is the contribution of neuron 2 on 1 caused by neuron 1 on 2, etc.

The terms of expansion are weighted by the factor $[s\mu\hat{f}(1-\hat{f})]^K$. Thus, the series converges if the system has low synaptic strength or high noise. Low synaptic strength represents the weak coupling limit of the network. The convergence due to high noise is because the transmission of information is lost in the noise while traversing the recurrent loop. The series also converges faster if the fixed point is far from threshold because neurons are more sensitive to modulation near threshold.

Comparing the loop expansion with simulations: We may use this expansion to compute corrections to the neural dynamics due to coupling with other neurons in the network. Using reasonable values for the parameters of the model (see Fig. 3), the corrected membrane potential is

$$\langle V_1(t_n)\rangle = V_o + s\hat{f} + s^2\hat{f}^2\mu(1-\hat{f}) + s^3\hat{f}^3\mu^2(1-\hat{f})^2 + \cdots\,. \tag{8}$$

The membrane potential is increased for excitatory recurrence, and decreased for inhibitory recurrence. The correction to the baseline membrane potential is linear for recurrent synaptic weight values that are weak, or for noisy networks.

For stronger synaptic weights, the simulated results match the loop series closely up to $s \approx \pm 20$ (arbitrary units). When the synapses are excitatory, the higher order

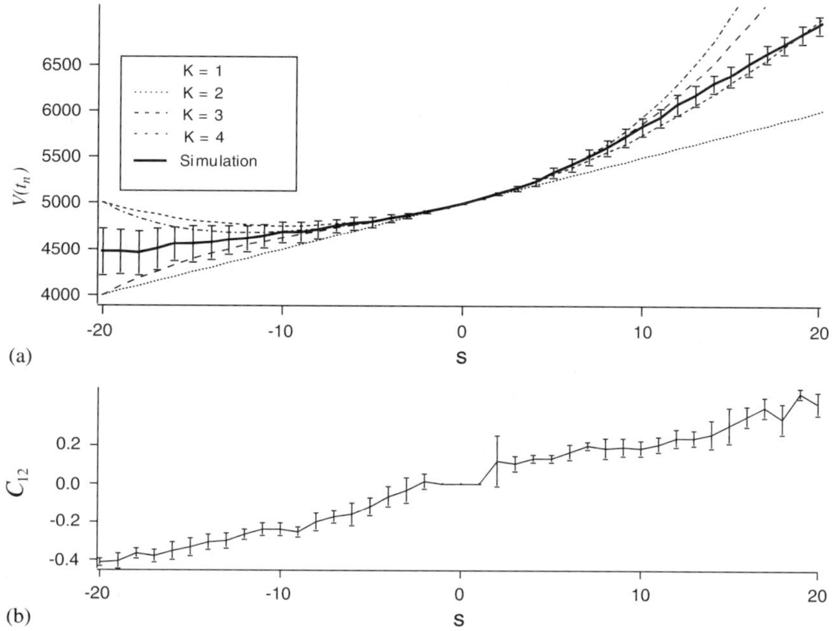

Fig. 3. Numerical simulation of the recurrent model: (A) The traces showing the membrane potential, V (arbitrary units), as a function of the recurrent synaptic weight, s. The solid is the result of a simulation and the broken lines are the series for the first two, three, four, and five terms and (B) the average of the cross-correlation function for five simulation runs. Parameter values: $\mu = 0.5$, $\hat{f} = 0.5$. The error bars represent one standard deviation.

corrections deviate more quickly than inhibitory synapses from the simulation result. This deviation is due to the saturation of the spike probability that was linearized in the first step of the analysis.

In the case of inhibitory synapses, the higher order terms seem to extend the accuracy of the prediction to stronger synaptic weights. However, at strong inhibition the membrane potential becomes bistable, and averaging over both states would yield a large variance in the simulation. The problem here is the distribution of the membrane potential is no longer Gaussian and higher order moments limit the reliability of the analysis.

How strong is 'weak'? In the noisy, weak limit, the first few terms of the loop expansion yield a good quantitative predictor of the membrane potential of neurons in a network with recurrent loops. To investigate the effect of one neuron on the firing rate of the other, we calculated the cross-correlation coefficient, C_{12}, for the two neurons in the simulation

$$C_{12} = \frac{\langle (V_1 - \langle V_1 \rangle)(V_2 - \langle V_2 \rangle) \rangle}{\sigma_1 \sigma_2}, \tag{9}$$

where, $\sigma_i^2 = \langle (V_i - \langle V_i \rangle)^2 \rangle$.

The averaged results of five simulations is shown in Fig. 3B. For the range of synaptic strengths where there is close agreement with the first few terms of the loop expansion, $-20 < s < 20$, the correlation coefficient value is, $-0.2 < C_{12} < 0.2$. This corresponds to a modulation of the spike probability of over 60%. Such a level of modulation could contribute significantly to the range of fluctuations observed in cortical neurons.

Another result of this analysis is that under these "weak and noisy" conditions, few neurons contribute to modulate the activity of a given neuron in a network. The match between the loop expansion and simulations tells us that neural modulations traverse only a few synapses in lateral recurrence, so that distant neurons have no effect. We have compared the graphs similar to Fig. 3A for networks of 4 and 50 neurons, and found the curves (V vs. s) to be within a standard deviation. This also implies that feedback connections will modulate those neurons across few.

4. Conclusion

The analysis and simulations presented here show that dynamical changes caused by recurrence can be calculated using an expansion in terms of signal transmission through recurrent loops. The loop expansion reveals what system parameters are responsible for convergence allowing one to truncate the series. This method can be used to estimate an upper bound on lateral synaptic connections in biological networks that do not exhibit nonlinear effects of recurrence.

The conclusion that the loop expansion series converges quickly for networks with high noise has consequences for sensory processing systems. Feedback recurrence in biological systems, such as cortical sensory systems, serve a first order role in modulating sensory input patterns so that only the first loop is important in the processing of sensory information.

Acknowledgements

The author would like to thank Gin McCollum, Gerhard Magnus, and Curtis Bell for insightful discussions. This research was supported in part by a National Institute of Health grant (R01 MH60364).

References

[1] B. Ermentrout, J. Cowan, Biol. Cybernet. 34 (1979) 137.
[2] W. Gerstner, J.L. van Hemmen, Network 3 (1992) 139.
[3] P. Getting, Annu. Rev. Neurosci. 12 (1989) 185.
[4] E. Marder, A.I. Selverston, in: R.M. Harris-Warrick, E. Marder, A.I. Selverston, M. Moulins (Eds.), Dynamic Biological Networks, The MIT Press, Cambridge, MA, 1992, pp. 161–196.
[5] R.P.N. Rao, T.J. Sejnowski, in: S.A. Solla, T.K. Leen, K.-R. Muller (Eds.), Advances in Neural Information Processing Systems, Vol. 12, MIT Press, Cambridge, MA, 2000, pp. 164–170.

[6] J. Rinzel, in: E. Teramoto, M. Yamaguti (Eds.), Mathematical Topics in Population Biology, Morphogenesis and Neurosciences, Springer, New York, 1987, pp. 261–281.
[7] J. Sirosh, R. Miikkulainen, Biol. Cybernet. 71 (1994) 66.
[8] G.G. Turrigiano, Trends Neurosci. 22 (1999) 22 1227.
[9] X.-J. Wang, J. Rinzel, Neural Comput. 4 (1992) 84.

Available at
www.ComputerScienceWeb.com
POWERED BY SCIENCE @ DIRECT®

ELSEVIER

Neurocomputing 52–54 (2003) 901–906

NEUROCOMPUTING

www.elsevier.com/locate/neucom

Information processing with spiking neurons in a cortical architecture framework under the control of an oscillatory signal

Tobias Rodemann*, Edgar Körner

Honda Research Institute Europe, Carl-Legien-Strasse 30, 63073 Offenbach (Main), Germany

Abstract

In this work we analyze information processing in spiking neurons in the cortex by investigating how short (10 ms) pulse packets of spikes are processed in cortical neurons. These signals are compliant with constraints of rapid feedforward recognition in the visual cortex. We add an oscillatory membrane potential modulation that simulates the effect of 40 Hz (evoked) gamma oscillations. We find that depending on the oscillation signal different input features are decoded, processed and encoded in a single output spike. We demonstrate that a latency-coding of information is possible.

Keywords: Latency-coding; Spiking neurons; Gamma oscillation

1. Introduction

Spiking neurons in a realistic cortical environment show a remarkably complex input–output mapping. A number of different functions like integrator (rate count), coincidence detector or, recently, latency decoder have been proposed. In this work we analyze the input–output relation for a neuron roughly matching principal neurons in the monkey visual cortex. Experiments by Thorpe et al. on ultra-fast image recognition in monkeys and humans have brought up serious temporal constraints [8]. The extremely fast recognition of complex stimuli in about 200 ms leaves an upper limit of just 10–20 ms for information transfer between two neurons. Taking into account that

* Corresponding author.
E-mail addresses: tobias.rodemann@Honda-RI.de (T. Rodemann).

0925-2312/03/$ - see front matter © 2003 Elsevier Science B.V. All rights reserved.
doi:10.1016/S0925-2312(02)00827-5

cortical neurons fire with rates $\ll 100$ Hz a maximum of one spike per neuron must suffice to transmit information. This prohibits the classic single neuron rate-coding approach. It was therefore proposed [5,8] that the first, fast categorization of inputs is purely feedforward, based on one spike per neuron only and makes use of a temporal coding scheme. Essential for the coding scheme proposed in [5] are the initial (evoked) gamma oscillations which are triggered even by simple stimuli [6,7]. It was proposed that this oscillation signal is used as a reference and control signal for a latency-coding scheme that is able to transmit information with single spikes within the time limit of 10 ms. In this work we will analyze how information is decoded and encoded in a cortical neuron subjected to an external oscillatory signal. We investigate how the number of spikes a neuron receives (roughly corresponding to an ensemble firing rate), their temporal dispersion/variance (a measure of coincidence) and their latencies influence the neuron's output. The analysis is done via computer simulations at the behaviorally relevant time-scale of 10 ms.

2. Network set-up

As the basic element of our circuit (Fig. 1) we employ a spiking neuron, simulated within the spike response model (SRM) [4] framework, which is closely related to the Integrate & Fire model. The neuron's status variable is its membrane potential. Due to the one spike per neuron limit we focus on the generation of the first output spike only and neglect all refractory effects from previous spikes. Incoming spikes trigger excitatory post-synaptic potentials of the form (α function):

$$\eta_i(t) = w\frac{(t-t_i)e}{\tau}\exp\left(-\frac{t-t_i}{\tau}\right) \quad (t > t_i), \tag{1}$$

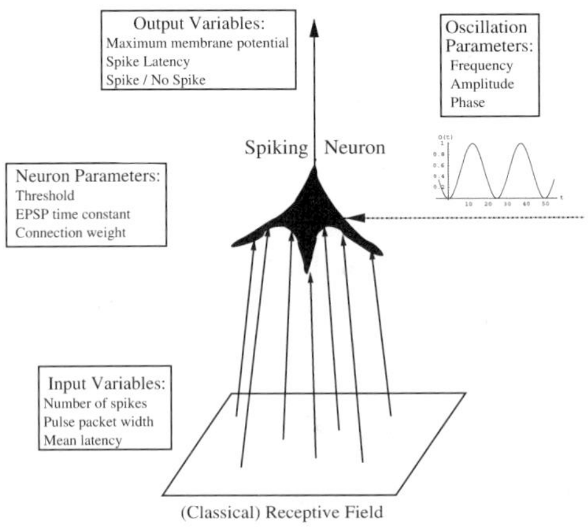

Fig. 1. The network element under investigation. A spiking neuron receives spikes from neurons within its receptive field and is furthermore subjected to a membrane potential oscillation.

with τ as the EPSPs rise time, w as a scale factor (connection weight) and t_i being the arrival time of spike i. Contributions from different synapses are summed up linearly. In addition to these spikes, corresponding to inputs from the previous stage in the processing hierarchy, neurons are also subjected to an external oscillatory signal (see also [2]). This oscillation signal is modeled as:

$$O(t) = \frac{A}{2}\left(1 + \sin\left(\omega t - \frac{\pi}{2} - \phi\right)\right), \tag{2}$$

with $\omega = 2\pi f$ and f being the frequency of the oscillation. The oscillation can be described by amplitude A, frequency f, and phase ϕ. This signal is added to the membrane potential. If the sum of EPSPs and oscillation exceeds the neuron's threshold θ, a spike is generated. The simulation is stopped at this point and the time (= latency) of the spike is recorded.

We describe input spike trains by a parameterized spike density function. We use the model of Gaussian pulse packets [1] for which spike arrival times follow a normal-distribution:

$$G(t) = N\frac{1}{\sqrt{2\pi}\sigma}\exp\left(-\frac{(t - t_m)^2}{2\sigma^2}\right). \tag{3}$$

The three density parameters correspond to the total number of spikes N to arrive at the neuron within the 10 ms time frame, the average arrival time t_m (mean input latency), and the width of the distribution σ (standard deviation of input spike latencies with σ^{-1} as a measure of spike coincidence). We note that the neuron takes three input variables (N, t_m, σ) and maps these to a single output spike. The neural output consists of two parts—whether a spike was generated or not and if yes, the latency of this spike. Neuronal parameters are taken as fixed and not modified. We changed oscillation parameters to see the effect on the mapping between input variables and output characteristics. We used the following parameter settings: $\tau = 2$ ms, $\Theta = 10$ mV, $w = 0.1$ mV, and $f = 40$ Hz.

3. Results

Without the oscillation signal the decision whether the neuron generates a spike or not is independent of the mean input latency t_m. The output latency can be approximated by t_m plus a fixed offset. Both N and σ have only a small influence on the output latency, but determine whether the threshold is exceeded or not. For our choice of parameters the number of spikes has a greater influence on the maximum value of the membrane potential (maximum over time) than the width of the pulse packet. The number of spikes and the variance of input spike latencies are decoded in the neuron and a two-dimensional "threshold" operation (Fig. 2) determines whether a spike is emitted or not. If a spike is emitted its latency will encode the average latency of incoming spikes.

When adding the oscillation signal the neuron's maximum membrane potential value becomes very sensitive to t_m. Therefore the spike generation process is latency dependent. The relative weighting of t_m compared to N and σ can be scaled up and down by

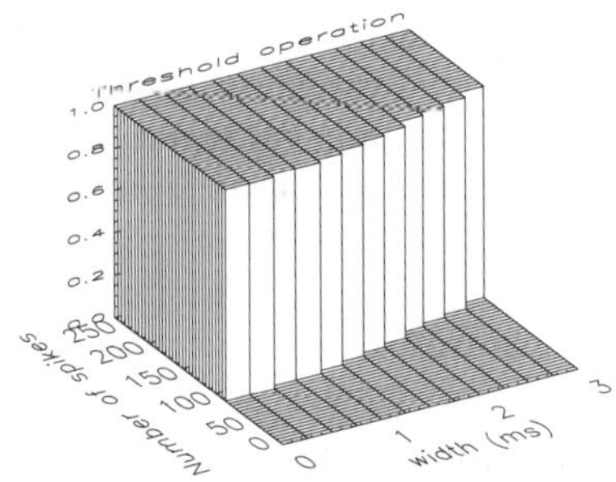

Fig. 2. Spike generation as a function of pulse packet width and number of spikes. The oscillation signal is disabled. The neuron spikes (output 1) if the number of spikes is high and the width of the pulse packet low enough.

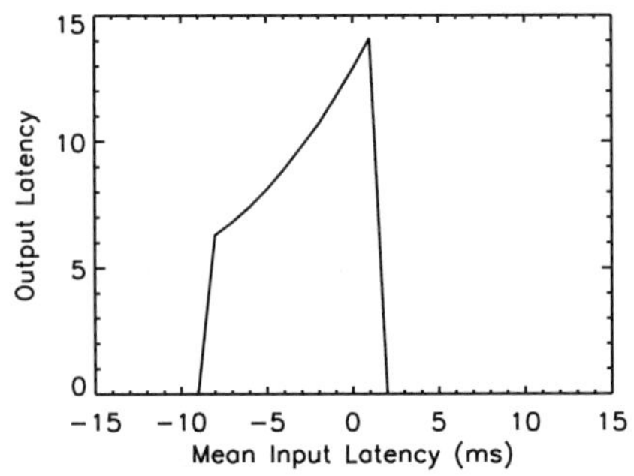

Fig. 3. Output latency as a function of the mean input latency relative to the peak of the oscillation signal. Parameters: $N = 200$, $\sigma = 1$ ms, $A = -18$ mV. Values of zero indicate that no spike was generated.

increasing or decreasing the absolute value of the oscillation amplitude. For a strong oscillation signal (excitatory or inhibitory) both the generation of an output spike and the latency of this spike depend strongly on input latencies. The neuron can thus decode input latencies. The output latency is still largely controlled by t_m but in a non-linear manner (Fig. 3). The maximum membrane potential is determined by all three input variables (see Fig. 4). A very interesting finding is the very strong dependence of the latency decoding mechanism on the phase relation between the pulse packet and the oscillation. If the pulse packet arrives in the rising phase of the oscillation, higher input latencies will lead to a higher maximum membrane potential (read: the later the better). If spikes arrive in the decreasing phase of the oscillation, earlier arriving spike will generate a higher membrane potential.

4. Summary and conclusion

We have used short pulse packets of spikes as inputs for biologically feasible spiking neurons in the visual cortex. These pulse packets are short enough to fit temporal

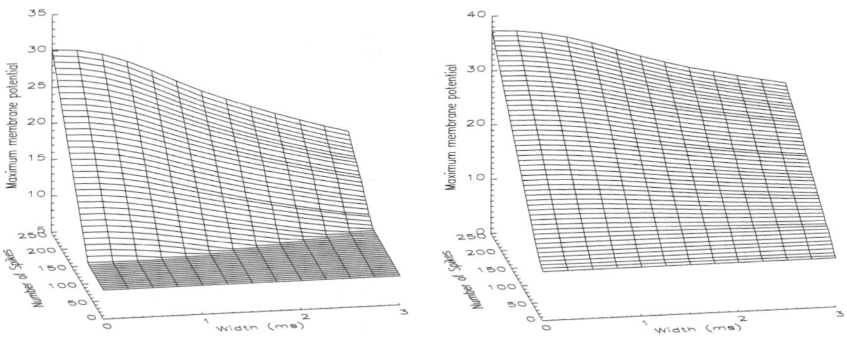

Fig. 4. Maximum membrane potential value as a function of pulse packet width σ and spike number N for mean input latencies of -5 ms (*left*) and $+5$ ms (*right*) relative to the minimum of the oscillation. Results are for a positive oscillation with 9 mV amplitude.

constraints put forward by Keysers et al. [3] and Thorpe et al. [8] for the fast feed-forward recognition in mammals. We also added an oscillatory membrane potential modulation that mimics (evoked) gamma oscillations. Under these conditions the neurons input–output relation does not conform to standard models of neural operation. Instead we find that the neurons' input–output is highly sensitive to all three input parameters that characterize the pulse packet. Information can be encoded and decoded in all three of these parameters. Probably the most efficient variable to encode information in is mean input spike latency, at least when the oscillation is strong. The neuron can very efficiently encode the mean input latency in its output latency and also make a latency dependent threshold operation. Spike latencies can play a role analogous to firing rates in classical models. However, latencies are far more efficient in terms of processing time, information capacity per spike and energy consumption.

References

[1] M. Diesmann, M.-O. Gewaltig, A. Aertsen, Characterization of synfire activity by propagating "pulse packets", in: J. Bowers (Ed.), Computational Neuroscience: Trends in Research, San Diego, CA, Academic Press, New York, 1996, pp. 59–64.
[2] J.J. Hopfield, Pattern recognition computation using action potential timing for stimulus representation, Nature 376 (1995) 33–36.
[3] C. Keysers, D.-K. Xiao, P. Földiak, D.I. Perret, The speed of sight, J. Cognitive Neurosci. 13 (1) (2001) 90–101.
[4] W. Kistler, W. Gerstner, J. Leo van Hemmen, Reduction of the Hodgkin–Huxley equations to a single-variable threshold model, Neural Comput. 9 (5) (1997) 1015–1045.
[5] E. Körner, M.-O. Gewaltig, U. Körner, A. Richter, T. Rodemann, A model of computation in neocortical architecture, Neural Networks 12 (7–8) (1999) 989–1005.
[6] M. Schürmann, C. Basar-Eroglu, E. Basar, Gamma responses in the EEG: elementary signals with multiple functional correlates, Neuroreport 8 (1997) 1793–1796.
[7] C. Tallon-Baudry, O. Bertrand, Oscillatory gamma activity in humans and its role in object representation, Trends Cognitive Sci. 3 (4) (1999) 151–162.
[8] S. Thorpe, A. Delorme, R.V. Rullen, Spike-based strategies for rapid processing, Neural Networks 14 (6–7) (2001) 715–725.

Tobias Rodemann received his masters degree in physics in 1998 from the Institut für Neuroinformatik at the University of Bochum. Since 1998 his is a graduate student at the Institut für Neuroinformatik and at the same time a member of HONDA's Future Technology Research division in Europe (HONDA R&D Europe (Germany)). His current research focus is information processing in systems of spiking neurons and the functional architecture of the neocortex.

Dr. Koerner received his Dr. of Engng. in 1977, and Dr. Sci. in 1984 from Technical University Ilmenau, Germany. From 1976 to 1984 he was a senior researcher at Technical University of Ilmenau, 1984–1987 a member of the Bioholonics Project of JRDC (Tokyo), and 1988–1992 professor and head of the department of neurocomputing and cognitive systems, Technical University of Ilmenau, Germany. 1992–1997 he worked as a Chief Scientist at Honda R&D Co., Wako Research Center, and since 1997 he is Head of the Future Technology Research Division at Honda R&D Europe (Germany). His research focus are brain-like artificial neural systems, smooth transition between signal-symbol processing, and selforganization of knowledge representation.

Available at
www.ComputerScienceWeb.com
POWERED BY SCIENCE @ DIRECT®

ELSEVIER

Neurocomputing 52–54 (2003) 907–912

NEUROCOMPUTING

www.elsevier.com/locate/neucom

Gain modulation and frequency locking under conductance noise

Michael Rudolph*, Alain Destexhe

Unité de Neurosciences Intégratives et Computationnelles, CNRS, UPR-2191, Bat. 33, Avenue de la Terrasse 1, 91198 Gif-sur-Yvette, France

Abstract

In the intact cortical network, neurons process information while receiving thousands of synaptic inputs. We investigated how this noise modulates the ability to resolve constant and oscillatory stimulation by using simplified neuron models. The cellular response was precise and reliable for a broad range of model parameters, while the output rate depended on the stimulus amplitude. The results suggest that both temporal and rate coding schemes are not exclusive but rather encode different aspects of the input at the same time. Conductance noise was found to play a modulating role, suggesting that high-conductance noise confers advantageous computational properties to neurons.

Keywords: Cerebral cortex; Precision; Reliability; Hodgkin–Huxley

1. Introduction

Cortical neurons in vivo are subject to a tremendous synaptic "noise" which was shown to markedly impact on the electrophysiological properties and the cellular response [8,5,9]. Recently, in the context of the general neural coding debate, neural responses to various more complex stimuli were investigated (for a review see [1]). Here, theoretical models and experiments suggest that cortical neurons are capable of precise and reliable spike responses even in the presence of noise (see e.g. [7]). However, these results were mainly obtained in low-conductance states, and it is not known if precise and reliable responses can be obtained in high-conductance states resembling in vivo conditions.

* Corresponding author.
E-mail address: michael.rudolph@iaf.cnrs-gif.fr (M. Rudolph).

0925-2312/03/$ - see front matter © 2003 Elsevier Science B.V. All rights reserved.
doi:10.1016/S0925-2312(02)00831-7

Here we address this question by using simplified Hodgkin–Huxley type models of cortical neurons subject to conductance noise and additional synaptic inputs. We show that the neuronal response shows differences depending on the type of noise, and we conclude that high-conductance noise confers advantageous computational properties to neurons, including the possibility of modulating their response between temporal and rate coding.

2. Methods

Simulations were performed using a simplified point-conductance model of cortical neurons [2], described by the membrane equation

$$C_m \dot{V}(t) = -g_{leak}(V(t) - E_{leak}) - \sum_{int} I_{int}(t) - I_{syn}(t) + I_{stim}(t),$$

where $V(t)$ denotes the membrane voltage ($C_m = 1\ \mu F/cm^2$ specific membrane capacity, $g_{leak} = 4.52 \times 10^{-5}\ s/cm^2$ leak conductance, $E_{leak} = -80$ mV reversal potential), and $\sum_{int} I_{int}(t)$ denotes the sum over intrinsic voltage-dependent currents. A fast Na^+, a delayed-rectifier K^+ current and a V_m-dependent K^+ current were included and described by Hodgkin–Huxley models.

$I_{syn}(t) = g_e(t)(V(t) - E_e) + g_i(t)(V(t) - E_i)$ denotes the membrane current due to synaptic noise, decomposed into excitatory $g_e(t)$ and inhibitory $g_i(t)$ time-dependent conductances (reversal potentials E_e=0 mV and E_i=−75 mV). Both were described by one-variable stochastic processes similar to the Ornstein–Uhlenbeck process: $\frac{dg_{i,e}(t)}{dt} = -\frac{1}{\tau_{i,e}}(g_{i,e}(t) - g_{i,e0}) + \sqrt{2/\tau_{i,e}}\,\sigma_{i,e} W_{i,e}$, where $g_{i,e0}$ are mean conductances, $\tau_{i,e}$ are time constants ($\tau_e = 10.49$ ms, $\tau_i = 2.728$ ms), $\sigma_{i,e}$ denote the standard deviation (SD) of $g_{i,e}(t)$. $W_{i,e}$ denotes Gaussian white noise of unit standard deviation and zero mean for inhibition and excitation, respectively. Background parameters were altered between 10% and 200% around standard values of g_{e0}=0.0121 μs, g_{i0}=0.057 μs, σ_e=0.012 μs, $\sigma_i = 0.0264$ μs. Stimulations were either an individual excitatory synaptic event with amplitude g_{amp} or (sinusoidal) excitatory conductance waveforms $I_{stim}(t) = -g_{amp}[1 + \sin(2\pi\nu_{stim}t)](V(t) - E_e)$, $0\ \text{ns} \leqslant g_{amp} \leqslant 30$ ns.

The responses for a fixed number of repetitions of a stimulus were accumulated in *peri-stimulus time histograms* (PSTHs) and the background subtracted, yielding reduced PSTHs (Fig. 1B). For single conductance pulses, the integrated reduced PSTH yields the probability for evoking spikes explicitly by the stimulus. For sinusoid stimuli the reduced PSTH contained statistically significant peaks (Fig. 1B), and the average SD of these events defined the precision of the response [6]. The reliability was defined as the ratio between the number of spikes in all events obtained from the reduced PSTHs and the total number of spikes during the stimulus.

3. Results

We first investigated the impact of various background properties on the response of the cell to an excitatory synaptic stimulus. The probability obtained by integration

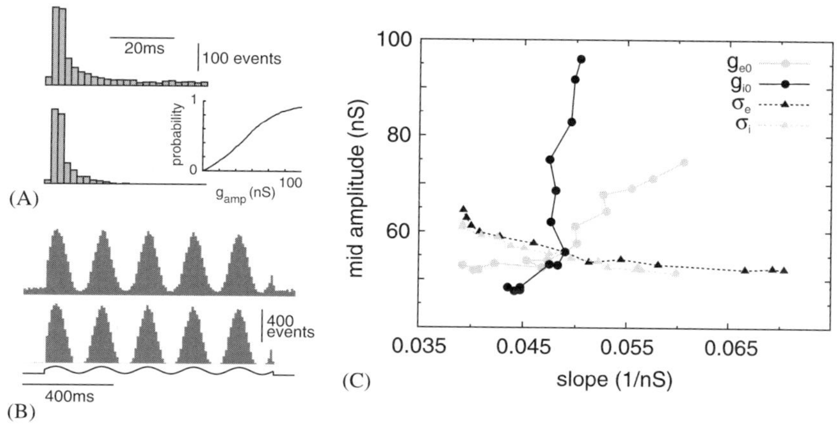

Fig. 1. Cellular response and gain modulation. (A) PSTH (upper plot) and reduced PSTH (lower plot) obtained from excitatory stimuli ($g_{amp} = 60$ ns, 1000 repetitions). Integration of the reduced PSTH yields the probability for spikes specifically evoked by the stimulus, which shows a sigmoidal behavior as a function of g_{amp} (inset). (B) PSTH (upper plot) and reduced PSTH (lower plot) obtained for sinusoidal excitatory conductance stimuli ($g_{amp} = 12$ ns, $v_{stim} = 5$ Hz, 1000 repetitions). (C) Gain modulation by background noise. The mid-amplitude and slope were obtained from sigmoidal fits to probability-g_{amp} plots (inset in A). Parameters of the background were changed between 10% and 200% of the standard values (see Methods).

of the reduced PSTHs shows a sigmoidal behavior as a function of the stimulation amplitude (Fig. 1A). From this, the amplitude g_{amp}^{mid} at which the probability reaches 0.5 as well as the slope at g_{amp}^{mid} can be defined. The phase plot (Fig. 1C) reveals that g_{i0} is the most effective parameter to change g_{amp}^{mid} and, thus, the *working point* of the cell. On the other hand, σ_e and σ_i impact mainly on the slope, this way modulating the sensitivity or *gain* of the cell. The mean excitatory conductance g_{e0} shows an effect on both measures.

Recently we have shown that simplified model neurons (see Methods) subject to oscillating current injection with fixed frequency were capable to respond with a high reliability and precision in the presence of both current as well as conductance noise in a broad range of model parameters. In accordance with previous experimental studies [3], a phase locking behavior was observed for a broad range of driving frequencies and noise levels. Two different modes of frequency locking can be distinguished: *spike-rate locking*, preferably observed for low noisy current, where cell phase locked to the external stimulus by modulating its firing rate, and *spike-time locking*, preferably in models with conductance noise, where the cell phase locked to the stimulus by an adjustment of individual spike times without change in its firing rate. Interestingly, changing the amplitude of the noise allowed to continously switch between both modes.

We next investigated the reliability and precision of high-frequency input signals. We calculated the response to sinusoid conductance waveforms with low stimulus amplitude g_{amp} in the presence of conductance noise. For increasing g_{amp}, the output frequency as well as the reliability increased monotonically (Fig. 2A). The mean excitatory and

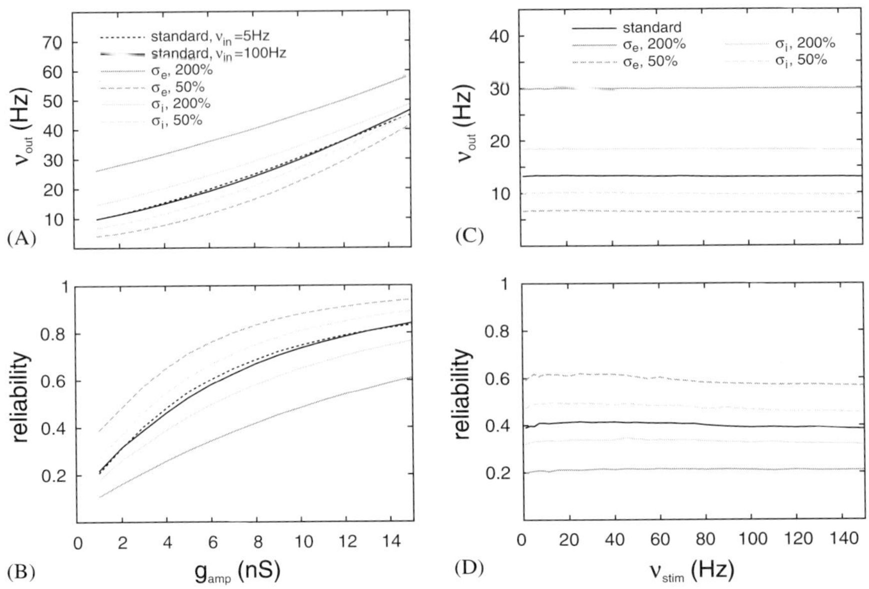

Fig. 2. Cellular response for different background parameters (relative to standard values, see Section 2). (A) Response frequency v_{out} as a function of stimulation amplitude g_{amp} for fixed input frequencies. (B) Reliability as a function of g_{amp} (legend as in A). (C) and (D) v_{out} (C) and reliability (D, for legend see C) as a function of v_{in} for fixed stimulation amplitude ($g_{amp} = 3$ nS).

inhibitory conductances had opposite effects on v_{out} and the reliability, with an increase in v_{out} and decrease in reliability with increasing g_{e0}. In the investigated parameter range, excitatory conductance was more efficient to evoke comparable changes in v_{out} or the reliability. This was also true for the effect of the noise SD, but here an increase in both σ_e and σ_i led to an increase in v_{out} (Fig. 2A) and overall decrease in the reliability (Fig. 2B).

The temporal resolution of the input, quantified by the mean SD of the output events, was low (< 2 ms) and decreased only slightly with increasing stimulation amplitude. Suprisingly, even large changes in the background parameters only minimally impacted on the ability to resolve weak periodic stimuli (Fig. 3A). Moreover, only minimal differences in the reliability and v_{out} were found when the input frequency was altered (see Fig. 2A and B, dotted lines).

For fixed stimulation amplitude, the cell locked to the sinusoid stimulus even for very high frequencies. In contrast, for cells subject to current noise, locking was lost after reaching a critical frequency (usually below 100 Hz for comparable stimulus and noise amplitudes). No dependence of the output frequency on v_{in} was found (spike-time locking, Fig. 2C). Indications for spike-rate locking could only be evidenced for very strong stimuli ($g_{amp} > 80$ nS). Noise showed a modulating impact similar to that found in simulations with variable g_{amp}, by shifting v_{out} and reliability (Fig. 2C and D). As expected, the SD of the output events decreased with increasing stimulus frequency,

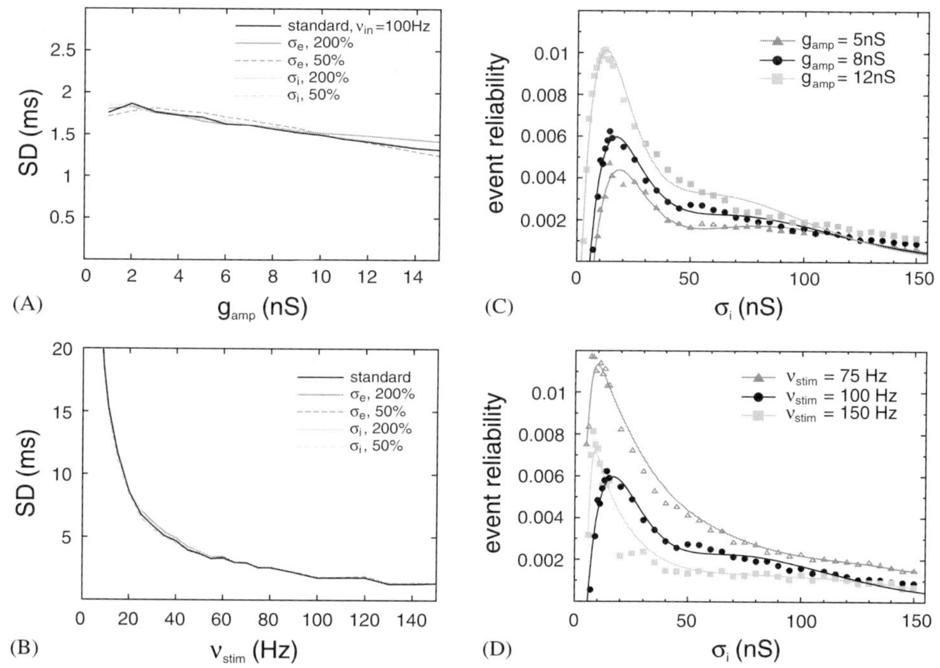

Fig. 3. (A) and (B) Average SD of the response as function of stimulation amplitude for fixed input frequencies (A), or stimulation frequency ν_{in} ($g_{amp} = 3$ ns, B) for different background parameters (given relative to standard values, see Section 2). (C) and (D) Reliability for the first event in the reduced PSTHs for $\nu_{in} = 100$ Hz and various amplitudes (C) as a function of the noise amplitude, and for fixed stimulation amplitudes (8 ns) but different ν_{in} (D). In all cases $\sigma_e = 0.4\sigma_i$.

following the stimulus time course, but was nearly unaffected by the background parameters (Fig. 3B).

To investigate further the impact of the background noise on the ability to resolve high stimulation frequencies, we continuously changed σ_e and σ_i while keeping the ratio between both parameters as well as other model parameters fixed. Whereas the precision of the response was very high and nearly unaffected by the noise and stimulation properties (not shown), the reliability of the first event was optimal only for a narrow range of noise amplitudes for both various stimulation amplitudes (Fig. 3C) and frequencies (Fig. 3D). This behavior is similar to stochastic resonance.

4. Conclusions

We investigated the impact of synaptic noise on the cellular response using a simplified models of cortical neurons. We found marked differences between the modulating effect of noise modeled as current or conductance. In the presence of conductance noise, the cell was able to resolve stimulating frequencies in the whole tested range (up to 150 Hz and beyond) with high precision and reliability, while keeping the average firing frequency nearly constant. This corresponds to a coding scheme built on the precise timing of spikes.

On the other hand, the strength of the input translated into a change of the firing rate, with the input frequency showing only minimal impact. This suggests that both coding dimensions are not exclusive but simultaneously present to encode different aspects of the input, a view also supported by recent experimental results [4]. Conductance noise showed a modulating effect on the firing rate and reliability, while keeping the precision nearly unaffected. How these effects impact on information processing on the network level remains to be answered (research supported by CNRS and NIH).

References

[1] R.C. deCharms, A. Zador, Neural representation and the cortical code, Annu. Rev. Neurosci. 23 (2000) 613–647.

[2] A. Destexhe, M. Rudolph, J.-M. Fellous, T.J. Sejnowski, Fluctuating synaptic conductances recreate in vivo-like activity in neocortical neurons, Neuroscience 107 (2001) 13–24.

[3] J.-M. Fellous, A.R. Houweling, R.H. Modi, P.H.E. Tiesinga, T.J. Sejnowski, The frequency dependence of spike timing reliability in cortical pyramidal cells and interneurons, J. Neurophysiol. 85 (2001) 1782–1787.

[4] A.A. Ghazanfar, C.R. Stambaugh, M.A. Nicolelis, Encoding of tactile stimulus location by somatosensory thalamocortical ensembles, J. Neurosci. 20 (2000) 3761–3775.

[5] N. Hô, A. Destexhe, Synaptic background activity enhances the responsiveness of neocortical pyramidal neurons, J. Neurophysiol. 84 (2000) 1488–1496.

[6] Z.F. Mainen, T.J. Sejnowski, Reliability of spike timing in neocortical neurons, Science 268 (1995) 1503–1506.

[7] L.G. Nowak, M.V. Sanchez-Vives, D.A. McCormick, Influence of low and high frequency inputs on spike timing in visual cortical neurons, Cereb. Cortex 7 (1997) 487–501.

[8] D. Paré D, E. Shink, H. Gaudreau, A. Destexhe, E.J. Lang, Impact of spontaneous synaptic activity on the resting properties of cat neocortical neurons in vivo, J. Neurophysiol. 79 (1998) 1450–1460.

[9] M. Rudolph, A. Destexhe, Do neocortical pyramidal neurons display stochastic resonance? J. Comput. Neurosci. 11 (2001) 19–42.

Available at
www.ComputerScienceWeb.com
POWERED BY SCIENCE DIRECT

ELSEVIER

Neurocomputing 52–54 (2003) 913–918

NEUROCOMPUTING

www.elsevier.com/locate/neucom

Self-sustained activity in networks of gain-modulated neurons

Emilio Salinas*,[1]

Department of Neurobiology and Anatomy, Wake Forest University School of Medicine, Medical Center Boulevard, Winston-Salem, NC 27157, USA

Abstract

Simulation studies have shown that recurrently connected neurons are capable of sustaining non-uniform profiles of activity in the absence of tuned input. These attractor networks are the basis for models of working memory and other processes where information about transient stimuli is stored temporarily. In addition, there is strong evidence that neurons often interact by affecting each other's gain. Here I study a minimal recurrent network that takes gain interactions into account. I show analytically that, in agreement with results of computer simulations, a center-surround organization gives rise to two types of stable solutions: a uniform state in which all neurons fire at the same rate, and a self-sustained profile of activity that may be centered at any point in the network. This theoretical framework based on nonlinear neuronal interactions is, in general, a powerful tool for investigating recurrent network dynamics.

Keywords: Gain fields; Divisive inhibition; Working memory; Neuron models

1. Introduction

Models of mnemonic activity are typically based on recurrent networks with so-called line attractors or bump attractors [10,7,9]. These are networks in which many states or firing rate distributions are possible. These states are stable, so they can be maintained for relatively long times. Which state is observed depends on initial conditions or on additional inputs that steer the network from one state to another. Such models have been successful at describing working memory processes in prefrontal cortex [2],

* Tel.: +1-336-713-5176; fax: +1-336-716-4534.
E-mail address: esalinas@wfubmc.edu (E. Salinas).
[1] Supported by startup funds from WFUSM.

0925-2312/03/$ - see front matter
doi:10.1016/S0925-2312(02)00834-2

responses of the head direction system [10] and responses of hippocampal place cells [6], among others [9]. However, analytical treatment of such attractors—particularly those that generate unimodal profiles of activity—has been rather limited [1,10,4]. Here I provide an analysis in which the attractor solution can be calculated explicitly in a straightforward way. The results show: (1) that the attractor solution may coexist with a state of uniform, low firing, (2) that the background input level has a strong influence on the steady-state solutions, and (3) that the mechanisms for generating persistent responses and traveling waves of activity are very similar.

2. Nonlinear network equations

Consider a network of N neurons where the response of unit i is given by

$$\tau \frac{\mathrm{d}r_i}{\mathrm{d}t} = -r_i + \frac{AH(\sum_j w_{ij}r_j - \theta) + h}{s + v\sum_j r_j^2}, \tag{1}$$

where H is the Heaviside step function ($H(x) = 1$ if $x > 0$ and is 0 otherwise), τ is a time constant, A and s are constants, and h represents an excitatory external input or background activity that comes from outside the network.

Neurons in the above equations have a hard activation threshold θ, as in Hopfield-type networks [3]. Crucially, however, they also incorporate gain interactions, for which there is abundant experimental evidence [5]. This is implemented through a mechanism similar to divisive normalization, which has served as an accurate empirical description of various experiments [8]. Thus, a neighboring neuron j can either directly drive neuron i to fire, through synaptic connection w_{ij}, or decrease its gain, through a synaptic connection of strength v. For simplicity, all gain interactions have equal strength, and no distinction is made between excitatory and inhibitory neurons.

The dynamics of this system depends on the network architecture, which is determined by the connectivity matrix $\mathbf{w}$. Here I will study a connectivity profile equal to a square window; that is, $w_{ij} = w_{\max}$ whenever $|i-j| \leqslant N_w$ and $w_{ij} = 0$ otherwise. Thus, each neuron excites equally $2N_w$ of its nearest neighbors. Periodic boundary conditions are used, so the array of neurons may be thought of as a ring [1,10,2]. Although all entries w_{ij} are positive or zero, the organization is center-surround because the divisive mechanism effectively makes long-range interactions inhibitory.

Next, the firing rates of this system at steady state will be determined.

3. Uniform-rate solution

The first solution that can be obtained is one in which all units fire at the same rate, denoted as R. To see this, set $r_i = R$ in Eq. (1)

$$\tau \frac{\mathrm{d}R}{\mathrm{d}t} = -R + \frac{AH(w_{\mathrm{tot}}R - \theta) + h}{s + vNR^2}. \tag{2}$$

Here w_{tot} is the total synaptic strength upon a postsynaptic neuron; that is, $w_{\mathrm{tot}} = \sum_j w_{ij} = 2w_{\max}N_w$. At steady state all derivatives are equal to zero. For the above

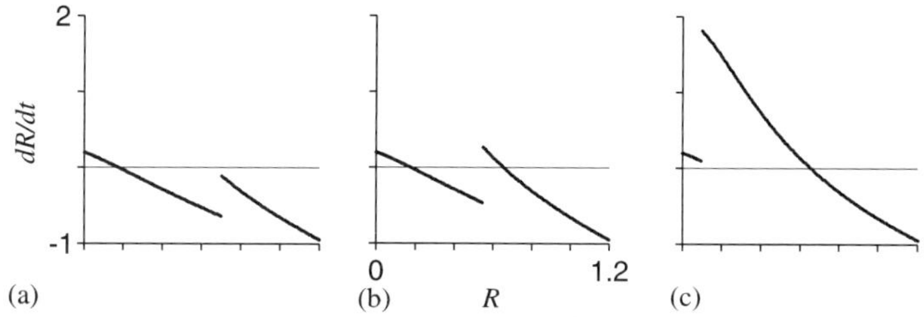

Fig. 1. Uniform-rate solution. Plots show time derivative of R vs. R. Points at which the curves intersect the horizontal line are stable steady-state values of R: (a) with $\theta = 2.10$ there is a single, low-rate steady state; (b) with $\theta = 1.65$ there are two stable solutions; and (c) with $\theta = 0.3$ there is a single, high-rate steady state. Curves were obtained from Eq. (2) with: $h = 0.11$, $A = 1$, $w_{tot} = 3$, $N = 100$, $v = 0.027$, $s = 0.56$, and $\tau = 1$.

expression there are two possible steady-state values, a low and a high one. This can be seen graphically in Fig. 1. The low firing level occurs when $Rw_{tot} < \theta$ and must satisfy $R = h/(s + vNR^2)$; the high firing level occurs when $Rw_{tot} > \theta$ and must satisfy $R = (A + h)/(s + vNR^2)$.

Note that the uniform-rate solution applies regardless of the actual connectivity pattern, because it only depends on the total synaptic strength being the same for all units. This is a normalization condition on the matrix **w**. As long as this condition is satisfied and the rest of the parameters are also equal across neurons, this solution will exist.

4. Self-sustained activity

The second solution to Eq. (1) occurs when the firing rates in the network form a self-sustained profile of activity. The shape of this profile is the same as for the connections: a square window (Fig. 2a, right). The reason is this. Suppose that initially the profile of activity is a square window centered at some point. The operation $\sum_j w_{ij} r_j$ then corresponds to a convolution of two square functions, which results in a function that looks like a trapezoid. The step function H takes this trapezoid minus the threshold θ as its argument, which results in another square function centered at the same point as the original. The divisive normalization mechanism determines the width, height and baseline of the activity profile, but not its shape or location. The width, height and baseline can be calculated analytically as follows.

Consider a stationary profile of activity that has a width $2N_r$. In this state, $2N_r$ neurons fire at a high rate r_{max} and $N - 2N_r$ neurons fire at a baseline rate r_{min}. From Eq. (1), at steady state these rates must satisfy

$$r_{min} = \frac{h}{s + v2N_r r_{max}^2 + v(N - 2N_r)r_{min}^2}, \tag{3}$$

$$r_{max} = \frac{A + h}{s + v2N_r r_{max}^2 + v(N - 2N_r)r_{min}^2}. \tag{4}$$

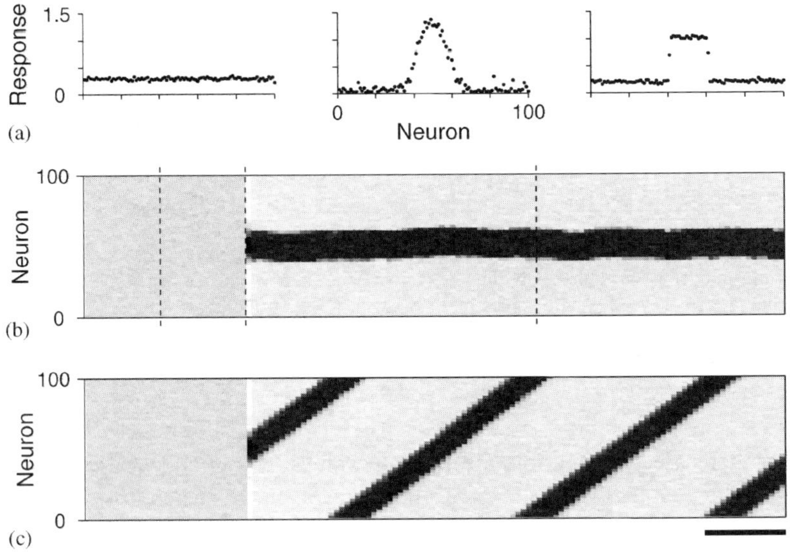

Fig. 2. Self-sustained activity: (a) Responses of all neurons at the three points in time indicated by dashed lines in the panel below. (b) Activity of all neurons as a function of time. Darkness level indicates response intensity. Initially all neurons fire at a low, uniform rate. A transient input (middle dashed line) switches the network to a square profile of activity. (c) As in (b) but with a connectivity pattern shifted by 1 unit. Scalebar equals 30 time units. Here $N_w = 15$, $w_{\max} = 0.1$, $h = 0.25$, $s = 0.63$, $\theta = 1.8$, $\Delta t = 0.1$; other parameters as in Fig. 1.

Three main results follow from this:

$$r_{\min} = r_{\max} \frac{h}{A+h}, \tag{5}$$

$$N_r = \frac{A + h - s r_{\max} - v N r_{\max} r_{\min}^2}{2 v r_{\max}(r_{\max}^2 - r_{\min}^2)}, \tag{6}$$

$$r_{\max} = \frac{\theta}{N_w w_{\max}} \frac{A+h}{A+2h}. \tag{7}$$

The first result is obtained by dividing Eqs. (3) and (4) by each other. The second result is the same as Eq. (4) but having solved for N_r. The third expression results from this observation: the difference between Eqs. (3) and (4) is whether the argument of the step function in Eq. (1) is larger or smaller than zero, so for a neuron at the edge of the square window the argument must be equal to zero. Hence, for such edge unit $\sum_j w_{ij} r_j = N_w(r_{\min} + r_{\max}) = \theta$; the second equality and Eq. (5) lead to Eq. (7). The first equality requires that N_w be smaller than $2N_r$, but by analyzing the output of H applied to the trapezoid function it can be seen that $N_r > N_w/2$, so the condition is always satisfied. In this way, $r_{\max}$, $r_{\min}$ and N_r are obtained in terms of the network parameters.

Figs. 2a and b show an example of the behavior of the network generated by simulating Eq. (1) with $N = 100$ neurons. To test for robustness, Gaussian noise ($\sigma = 0.02$) was added to all responses in each time step. Initially all neurons fire at a low level $R = 0.3$; then the responses around neuron 50 are increased transiently, after which a square pattern of activity develops. Its parameters are: $r_{\min} = 0.2$, $r_{\max} = 1$ and $N_r = 10$, which match the analytic calculations (for network parameters, see caption). Note that Fig. 2 shows only one example, but the self-sustained activity profile can indeed be centered anywhere within the array.

5. Discussion

In spite of its simplicity, the network model presented here displays interesting dynamic behaviors that can be studied analytically and that, qualitatively, match the activity of much more complex models with persistent activity [2,4]. The model also provides valuable insight into how the various network parameters influence output activity. For instance, the strength of inhibitory interactions v affects only the width of the square profile. The width also depends strongly on the background input h, although h influences $r_{\min}$, $r_{\max}$, and the uniform rates as well. Because h corresponds to input from external sources, it may change dynamically, and may hence be used for turning the network on or off. For instance, lowering h to 0.16 in Fig. 2 completely abolishes the self-sustained activity. Therefore, the uniform background may act as a control signal on the network.

Finally, an interesting extension of the results is the case of a square pattern of connections that is slightly shifted. That is, $w_{ij} = w_{\max}$ if $|i - j - \Delta| \leqslant N_w$ and $w_{ij} = 0$ otherwise. By applying H to the convolution between an initial square profile of activity and the shifted connections, it can be seen that the shift Δ does not change the shape of the profile; however, it moves it to a location Δ units away. Thus, the asymmetry generates a traveling wave. Fig. 2b shows an example. It is not difficult to show that the speed of the wave should be proportional to Δ, which is indeed what is found in simulations. At any instant, the activity profile is approximately square, with the same parameters as the stationary profile. This suggests that the problem of generating a stable, self-sustained pattern of activity is very similar to the problem of generating a stable, propagating wave of activity.

In conclusion, the present framework may be a powerful tool for analyzing the dynamics of recurrent neural networks.

References

[1] R. Ben-Yishai, R.L. Bar-Or, H. Sompolinsky, Theory of orientation tuning in visual cortex, Proceedings of the National Academy of Science, USA 92, 1995, pp. 3844–3848.

[2] A. Compte, N. Brunel, P.S. Goldman-Rakic, X.-J. Wang, Synaptic mechanisms and network dynamics underlying spatial working memory in a cortical network model, Cereb. Cortex 10 (2000) 910–923.

[3] J. Hertz, A. Krogh, R.G. Palmer, Introduction to the Theory of Neural Computation, Addison-Wesley, New York, 1991.

[4] C.R. Laing, C.C. Chow, Stationary bumps in networks of spiking neurons, Neural Comput. 13 (2001) 1473–1494.
[5] E. Salinas, P. Thier, Gain modulation: a major computational principle of the central nervous system, Neuron 27 (2000) 15–21.
[6] A. Samsonovich, B.L. McNaughton, Path integration and cognitive mapping in a continuous attractor neural network model, J. Neurosci. 17 (1997) 5900–5920.
[7] H.S. Seung, How the brain keeps the eyes still, Proceedings of the National Academy of Science, USA 93, 1996, pp. 13339–13344.
[8] O. Schwartz, E.P. Simoncelli, Natural signal statistics and sensory gain control, Nat. Neurosci. 4 (2001) 819–825.
[9] X.-J. Wang, Synaptic reverberation underlying mnemonic persistent activity, Trends Neurosci. 24 (2001) 455–463.
[10] K. Zhang, Representation of spatial orientation by the intrinsic dynamics of the head-direction cell ensemble: a theory, J. Neurosci. 16 (1996) 2112–2126.

Available at
www.ComputerScienceWeb.com
POWERED BY SCIENCE @DIRECT•

ELSEVIER

Neurocomputing 52–54 (2003) 919–924

NEUROCOMPUTING

www.elsevier.com/locate/neucom

The impact of thalamo-cortical projections on activity spread in cortex

Volker Schmitt[a,*], Rolf Kötter[b], Friedrich T. Sommer[a]

[a] *Department of Neural Information Processing, University of Ulm, Oberer Eselsberg, D-89069 Ulm, Germany*

[b] *Center for Anatomy and Brain Research, University of Düsseldorf, Universitätsstraße 1, D-40001 Düsseldorf, Germany*

Abstract

To what extent does thalamo-cortical connectivity influence global patterns of macroscopical activity spread in cat cortex? We address this question by simulation of activity flow in a network model representing cortical areas and thalamic nuclei as nodes with all or none activity. The connectivity schemes employed are based on data from anatomical tracer studies. For two networks, with and without thalamic nuclei, we investigate how well physiological experiments from strychnine neuronography can be reproduced. Our results indicate that thalamic interaction does influence cortical activity spread. However, while the relation between physiological activity spread and anatomy of the cortico–cortical connectivity is significant, the relation between activity spread and the anatomy of cortico–thalamic connections is not.

Keywords: Activity spread; Cortex; Thalamus

1. Introduction

The systematic collation of data from in vivo tracer techniques [3] reveals networks between cortical areas as well as between cortical areas and thalamic nuclei of the cat cerebral hemisphere. In an earlier study we used information about the cortico–cortical networks in a simulation study to reveal mechanisms of cortical activity spread [1]. Here we extend our earlier approach by including thalamic nuclei. Our question is how the addition of the thalamo–cortical loops influences the cortical spread of

* Corresponding author.

E-mail address: volker.schmitt@medizin.uni-ulm.de (V. Schmitt).

0925-2312/03/$ - see front matter
doi:10.1016/S0925-2312(02)00837-8

activity. To this end we examined the spread of activity by simulation experiments in two different networks. One network contained cortical areas and thalamic nuclei, the other contained just the cortical areas. We compared simulation outcomes to electrophysiologically ascertain topographic activation patterns that were induced by stimulation of the corresponding cortical regions.

2. Empirical data

2.1. The connectivity data

We used a collation of 823 connections between 53 cortical areas (called C) and 651 connections between the cortical areas and 42 thalamic nuclei (called C–T) [3]. The connection strengths were graded. Connections reported as dense or strong were given a weighting of 3. Connections of intermediate strength or for which no strength information was available were weighted as 2. Connections reported as weak or sparse were weighted as 1. Connections to thalamic nuclei were assumed to be reciprocal and of symmetric weight.

2.2. The electrophysiological data

To assess effects of connectivity on the global activity spread we compared the simulation results with relevant physiological findings. We used data from strychnine sulfate neuronographic analysis in cat [2]. Saturated strychnine sulfate was applied to small patches of cortex and induced stable and reproducible patterns of cortical activity. These patterns were recorded with bipolar electrodes in a total of 18 anesthetized cats. The activity resulting from stimulation of a single cortical area was mapped topographically marking active and silent areas. The remainder of the areas had not been explored or showed variable activity. In [2] 15 different areas have been strychninized. Due to difficulties with transfering areas in the parcellation of [3] we used 11 experiments out of 15.

3. The simulation model

We devised a minimal model for describing activation of cortical areas and thalamic nuclei. Each model unit describes the activity of one area or nucleus as "active" or "silent" by a binary variable. Thus, activity was represented by a binary pattern $x \in \{0,1\}^n$ with the number of areas and nuclei $n = 53$ or 95 depending on which connectivity model was used. A connection from area i to area j is denoted s_{ij}. The model prescribed that an area or nucleus was active in the next time step $(t+1)$ if the sum of its activating inputs exceeded a certain threshold Θ.

$$x(t+1)_j = H\left[\sum_{i=1}^{n} w_{ij}x(t)_i - \Theta\right], \tag{1}$$

where Θ is the activation threshold of a unit and $H[x]$ is the Heaviside function, w_{ij} denotes the coupling strength from unit i to j and $x(t)_i$ is the activity of the respective afferent unit. Simulation started with an initial activity pattern $x(0)$ containing the strychninized area of the corresponding experiment described in [2] as a single active unit. The coupling strength was aligned to the anatomical connection strength:

$$w_{ij} = (s_{ij})^{\gamma}, \tag{2}$$

γ and the values $w_{ii} = w_D$ for excitatory self-feedback were adjusted for the best results on average over all experiments. The threshold was controlled in a way that activity in the network spread as fast as possible from the single stimulated area to a set of a activated areas/nuclei, where a is chosen according to the individual experiment, i.e.

$$\Theta(t+1) = \max\{\Theta:\ \min\{|x(t+1) - a|\}\}. \tag{3}$$

After activity a is reached the iteration is continued until a fixed point or cycle is reached. Possible error types are "miss" errors if the simulation results in silent areas that have been reported to be active, and "add" errors if activity is found in areas reported to be silent. Areas with unknown or variable activity were ignored in the error statistics.

To judge the influence of a systematic connectivity structure like traced connectivity on the experimental outcome, we need to compare simulations based on the systematic connectivity with those based on random connectivity with same density and grading distributions. Therefore, we studied random connectivity models, for each comparison we run simulations with 20 random matrices. In the following the random model corresponding to the cortex is denoted rnd(C), and the random model corresponding to the cortico–thalamic network rnd(C–T). In a third random model, called C-rnd(T), we studied the influence of a randomization of thalamic connections only. To this end 20 matrices where generated with experimentally observed cortico–cortical connections but random connections to thalamic nuclei.

4. Results

The parameters γ and w_D in our model were adjusted for smallest errors on average over the 11 simulated experiments corresponding to different stimulation sites. Table 1 shows the optimized values and the resulting average errors.

The error percentages achieved by the two connectivity models for each stimulated area is shown in Fig. 1(a).

Table 1
Average errors of models with experimentally traced connections and optimized parameters

Model	ω_{ii}	γ	add %	miss %	error %
C	25	2.9	10.59	14.49	12.54
C–T	23	2.9	10.71	8.71	9.71

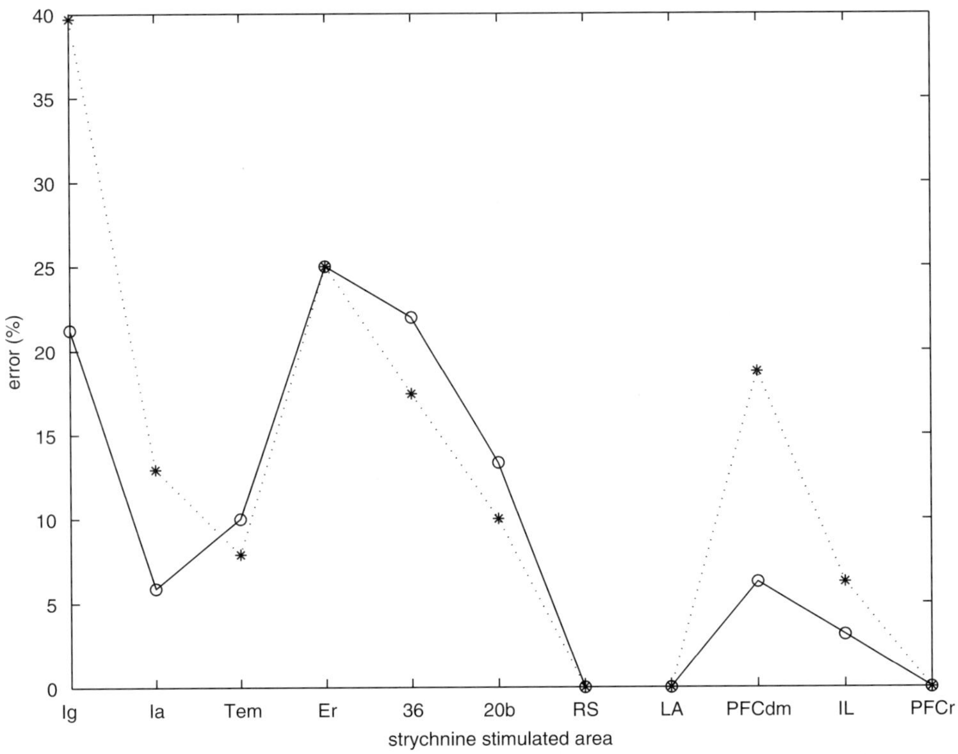

Fig. 1. (a) -○- C–T; · · ∗ · · C, (b) -○- C–T; -∗- C-rnd(T); · · + · · s.d. Errors obtained in simulations of activity spread using different connectivity models (s.d. = ±1 standard deviation).

In four experiments the error percentages were lower with thalamic nuclei and in three without connections to the thalamus. However, on average across all experiments the model including cortico–thalamic connections performed better. We controlled simulations for chance results. The random models rnd(C) and rnd(C–T) produced larger mean errors in replicating the electrophysical experiments than the "real" topographies, but had smaller standard deviations. Some "real" model simulation results were within the standard deviation range of "random" model results. Simple counting shows that the model without connections from and to thalamus has three such stimulated areas whereas the model with cortico–thalamic connections has only one result in standard deviation range (see Fig. 2a and b).

The C-rnd(T) "real/random" model produced mean errors of an order comparable to the "real" models (see Fig. 1b). We assessed mean errors obtained with "real" and "random" topographies by t-tests for equal or unequal variances depending on the outcome of an f-test. Each corresponding pair of "real" and "random" topographies showed significantly different mean errors, both "real" models compared to the mixed "real/random" C-rnd(T) model did not show these differences (Tables 2 and 3).

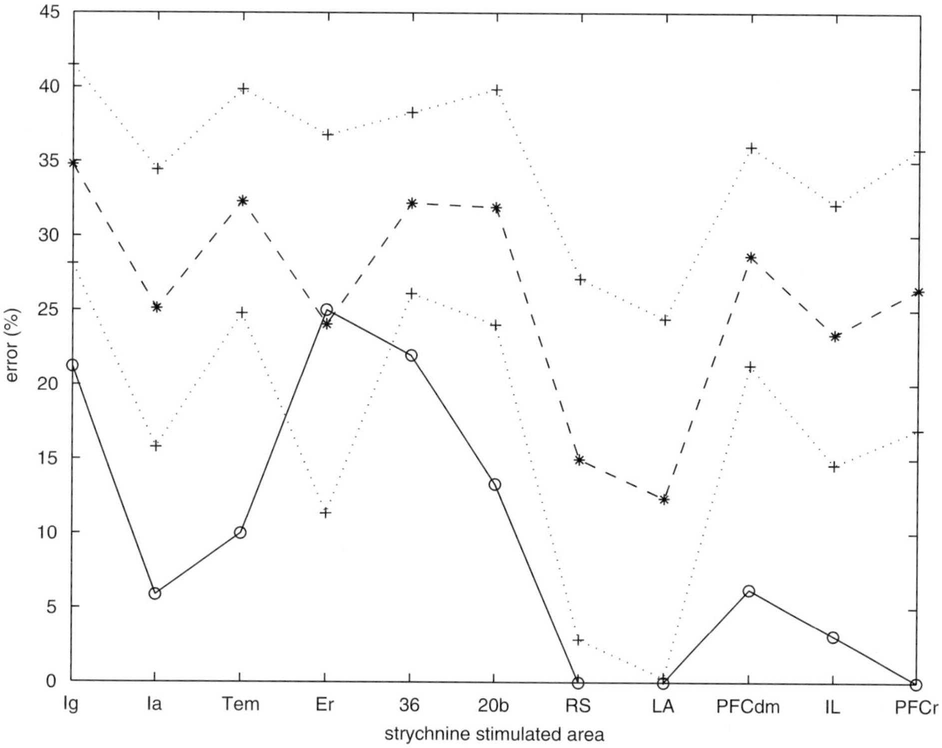

Fig. 2. (a) -○- C; -∗- rnd(C); · · + · · s. d. (b) -○- C–T; -∗- rnd(C–T); · · + · · s.d. Errors obtained in simulations of activity spread using the experimentally traced connectivity models and 20 models with random connectivities of the same density and grading (s.d. = ±1 standard deviation).

Table 2
Mean percentages and standard deviation of errors averaged across 11 experiments using different connectivity topologies

Model	Mean	s.d.
C	12.54	12.23
C–T	9.71	9.39
rnd(C)	28.08	9.32
rnd(C–T)	26.02	7.19
C-rnd(T)	12.44	10.41

Table 3
Significance level for different mean errors between the "real" and "random" topographies. In column 2 the first row was tested against rnd(C), the second against rnd(C–T)

significance level p	rnd(C)/rnd(C–T)	C-rnd(T)
C	0.003	0.984
C–T	0.000	0.526

5. Summary

We presented a simulation study to investigate the influence of thalamo–cortical projections on global pattern of activity spread in cortex. The simulation model was based on cortico–cortical and cortico–thalamic connectivity data from neuranatomical studies. We compared two connectivity models, a cortex network and a cortico–thalamic network. Both network models reproduced neuronographic experiments with good overall mean errors, the isolated cortex network with 12.5%, the cortico–thalamic network even better, with 9.7%. Systematic cortical connections always performed significantly better than random topologies with real densities, leading to errors of roughly 27%. The thalamo–cortical connections, however, do not act nearly as specifically: Networks with "real" cortico–cortical connections but random projections to thalamus perform almost as well as with the cortico–thalamic projections taken from the neuronanatomical database. Thus, our simulation experiments suggest that thalamic interactions only mildly influence the results of neuronography. The experimentally observed activity patterns are mostly determined by the cortico–cortical networks. The specific anatomy of thalamo–cortical connections has only a weak impact on activity propagation. In agreement with Scannell's analysis thalamo–cortical connections do not seem to modify the cortico–cortical organisation when tested in a simulated activity propagation paradigm.

References

[1] R. Kötter, F.T. Sommer, Global relationship between structural connectivity propagation in the cerebral cortex, Philos. Trans. Roy. Soc. London B 355 (2000) 127–134.

[2] P.D. MacLean, K.H. Pribram, Neuronographic analysis of medial and basal cerebral cortex I. cat, J. Neurophysiol. 16 (1953) 312–323.

[3] J.W. Scannell, G.A.P.C. Burns, C.C. Hilgetag, M.A. O'Neill, M.P. Young, The connectional organization of the cortico–thalamic system of the cat, Cerebral Cortex 9 (1999) 277–299.

Available at
www.ComputerScienceWeb.com
POWERED BY SCIENCE DIRECT®

ELSEVIER

Neurocomputing 52–54 (2003) 925–931

NEUROCOMPUTING

www.elsevier.com/locate/neucom

A new correlation-based measure of spike timing reliability

S. Schreiber[a,*], J.M. Fellous[b], D. Whitmer[c], P. Tiesinga[a,d], T.J. Sejnowski[b,c]

[a] *Sloan-Swartz Center for Theoretical Neurobiology and Computational Neuroscience Laboratory, The Salk Institute, 10010 N. Torrey Pines Rd., La Jolla, CA 92037, USA*
[b] *Computational Neurobiology Laboratory, Howard Hughes Medical Institute, The Salk Institute for Biological Studies, La Jolla, CA 92037, USA*
[c] *Department of Biology, University of California, San Diego, La Jolla, CA 92093, USA*
[d] *Department of Physics and Astronomy, University of North Carolina, Chapel Hill, NC 27599, USA*

Abstract

We introduce a new correlation-based measure of spike timing reliability. Unlike other measures, it does not require the definition of a posteriori "events". It relies on only one parameter, which relates to the timescale of spike timing precision. We test the measure on surrogate data sets with varying amounts of spike time jitter, and missing or additional spikes, and compare it with a widely used histogram-based measure. The measure is efficient and faithful in characterizing spike timing reliability and produces smaller errors in the reliability estimate than the histogram-based measure based on the same number of trials.

Keywords: Spike timing reliability; Correlation-based measure; Precision; Neural coding

1. Introduction

Spike timing unreliability limits the amount of information a neuron can potentially transmit about a stimulus. Intrinsic noise and extrinsic influences, such as synaptic noise, cause a neuron to respond with variable spike timing to the same stimulus. This variability is expressed as jitter in the timing of individual spikes, missing spikes,

* Corresponding author.
E-mail address: susanne@salk.edu (S. Schreiber).

0925-2312/03/$ - see front matter
doi:10.1016/S0925-2312(02)00838-X

and additional (noisy) spikes. Our goal is a reliability measure that captures all three phenomena.

Many measures have been used to characterize reliability of spike timing on the basis of the spike trains a neuron produces in response to repeated presentation of the same stimulus, see for instance [1–6]. Most measures are based on the post-stimulus time histogram (PSTH), which is insensitive to the structure of individual spike trains [5]. Here, we introduce a new measure of spike timing reliability, based on a measure of the similarity between pairs of individual spike trains. We test this measure with surrogate data sets designed to assess the performance of the measure with respect to additional spikes, missing spikes, and spike jitter.

2. Reliability measures

2.1. The histogram-based measure

Many measures of spike timing reliability rely on the construction of a PSTH, constructed from multiple trials. The PSTH depends on a particular choice of bin width, bin_{h}. "Events" are then defined as peaks in the PSTH. In order to reliably extract peaks, the PSTH is typically smoothed with a time window, τ_{h}. In practice, care must be taken to set the threshold, θ_{h}, to detect peaks, and exclude noise, which depends on bin_{h}. Once the events are detected, spikes are labeled as belonging to an event. This labeling depends on the choice of an allowable time window, w_{h}, around the time of peak outside of which spikes cannot be considered as part of the event. This parameter sets a time scale that depends on the phenomena of interest. Reliability, R_{hist}, is defined as the average number of spikes within events, divided by the total number of spikes, n_{total}, present in the histogram: $R_{\mathrm{hist}} = (\sum_{e \in \mathrm{events}} n_e)/n_{\mathrm{total}}$. This procedure depends on four parameters (bin_{h}, τ_{h}, θ_{h}, and w_{h}), and a fair amount of discretion from the part of the experimenter. Hence, it is difficult to compare reliability measures between studies, unless exactly the same procedure has been used.

For comparison with the correlation-based measure, we smooth the PSTH with a Gaussian filter (standard deviation τ_{h}), the threshold, θ_{h}, is fixed, and only bins with values exceeding θ_{h} are considered event bins. The allowable window is defined as twice the width of the peak, at mid-height.

2.2. The correlation-based measure

In contrast to histogram measures, which ignore information about the structure of individual spike trains by summing over all trials, the correlation measure presented here relies on the structure of individual trials and does not depend on identifying events a posteriori. It is based on a similarity measure between individual pairs of spike trains. The spike trains obtained from N repeated presentations of the same stimulus are convolved with a Gaussian filter of a given width, σ_{c}. After convolving all trials, the inner product is taken between all pairs of trials and each inner product is then divided by the norms of the two trials of the respective pair. Reliability, R_{corr}, is the average of all similarity values. Therefore, if the filtered spike trains $\vec{s_i}(i=1,\ldots,N)$ are represented as individual vectors, the correlation measure is

$$R_{\rm corr} = \frac{2}{N(N-1)} \sum_{i=1}^{N} \sum_{j=i+1}^{N} \frac{\vec{s_i} \cdot \vec{s_j}}{|\vec{s_i}|\,|\vec{s_j}|}.$$

It takes values between zero and one.

The measure depends only on the width of the Gaussian filter, σ_c, which is based on the phenomenological time scale of interest and directly tunes the measure with respect to the influence of spike jitter versus missing and additional spikes [6]. For a narrow filter, spike timing can yield high reliability values only if the spike jitter is less than or equal to the filter width. For a broad filter width, the influence of individual spike jitter (on a timescale below the filter width) is decreased and the value of reliability is dominated by the occurrence of additional and missing spikes. Therefore, the filter width is directly related to the precision of spike timing.

3. The surrogate test data sets

To test the new correlation measure we used a surrogate data set, consisting of seven firing events within one second, see Fig. 1. Each rastergram was comprised of $N=35$ spike trains. Without jitter, extra spikes, and missing spikes all trials are identical, so that reliability should always be 1 (upper rastergram in Fig. 1A and upper left most rastergram in Fig. 1B). We used this rastergram to derive additional surrogate sets of spike trains, by systematically varying the amount of jitter and the amount of missing and additional spikes. For the sets with variation in jitter, each spike of every trial in the reliable set was independently shifted in time, according to a Gaussian distribution with a standard deviation of σ_j, which corresponded to the amount of jitter introduced. For the sets with missing and additional spikes, additional spikes were introduced at random times and randomly selected spikes of the reliable set were removed.

4. Results

4.1. Reliability with jitter

First, the reliability of the test data sets was characterized as a function of the amount of jitter in the data (Fig. 2). For the correlation measure, reliability degraded smoothly with jitter. For higher values of the filter width, σ_c, the reliability tended to be higher (Fig. 2A).

For the histogram measure, the degradation of reliability with jitter was less smooth (Fig. 2B). The solid set of curves was obtained for a histogram measure with high threshold. The dashed curves were obtained with the same parameters of the histogram measure, but a threshold value lowered by 50%. Measured with a higher threshold, reliability as a function of jitter depended on the smoothing filter width of the histogram measure, τ_h. The jaggedness of the curves of the histogram-based measure arose

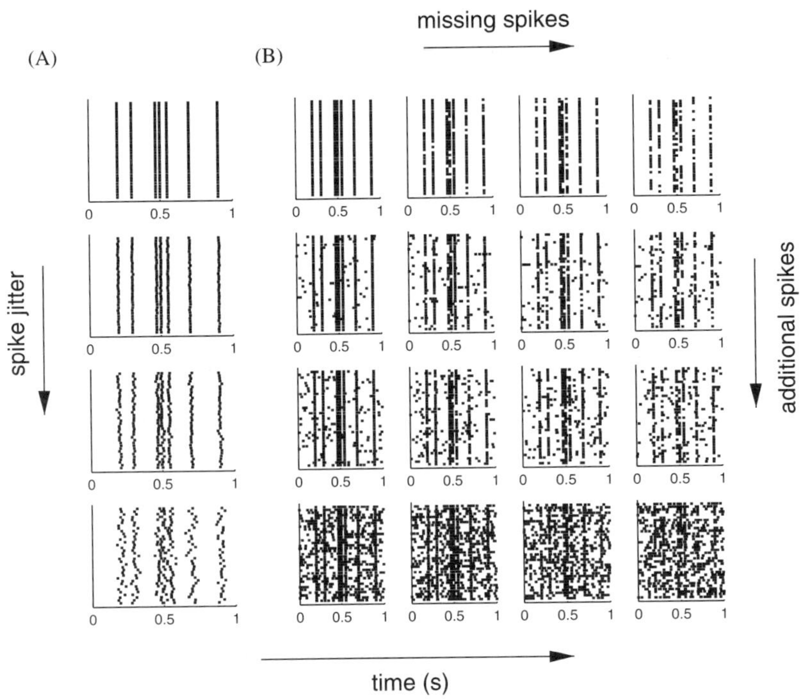

Fig. 1. (A) The basis surrogate data set (events at 200, 300, 470, 500, 550, 700, and 900 ms) and three examples of data sets with different jitter, σ_j (2, 6, and 16 ms); (B) Examples of data sets of different amounts of missing and extra spikes (0%, 2%, 4%, and 16% extra spikes, and 0%, 10%, 20%, 30% missing event-spikes).

because, from one jitter value to the next, some events fell below threshold and hence changed the reliability estimate discontinuously. For a lower threshold (dashed curves), the reliability curves were highly similar and all reliability values were high. Although this measure was stable over a broad range of filter width parameters, the reliability values did not reflect the true jitter. Similar observations could be made for reliability as a function of jitter when varying the bin size of the histogram measure (data not shown). Altogether, the histogram-based measure proved sensitive to the choice of threshold.

4.2. *Reliability with extra and missing spikes*

In Fig. 3 we compared how the reliability measures performed on the data sets that included extra and missing spikes. Fig. 3A shows the correlation-based reliability as a function of the percentage of missing event-spikes. Each curve was based on data sets with a fixed number of extra spikes per trial. Reliability values degraded smoothly as the number of missing spikes increased. The performance of the histogram-based measure on these data, shown in Fig. 3B, was similar. However, the degradation

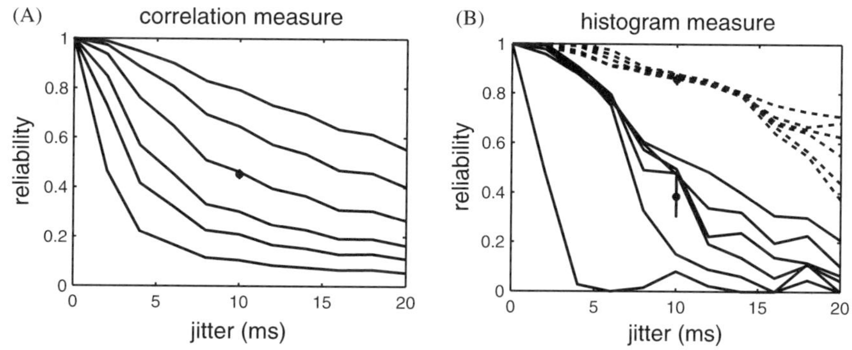

Fig. 2. (A) Correlation-based reliability as a function of spike jitter. Different curves correspond to different filter width, σ_c (1, 2, 3, 5, 8, and 12 ms, bottom to top). In a control calculation, 25 different rastergrams with 10 ms jitter were generated and the reliability was computed for a filter width of 5 ms (circle, and standard deviation). (B) Histogram-based reliability as a function of jitter ($bin_h = 2$ ms). Different curves correspond to different smoothing filter widths, τ_h (1, 2, 3, 5, 8, and 12 ms, bottom to top). The solid curves were calculated with a threshold of $\theta_h = 1200$ Hz. For the dashed curves the threshold was 600 Hz. Error bar based on 25 control data sets ($\sigma_j = 10$ ms, $\tau_h = 5$ ms).

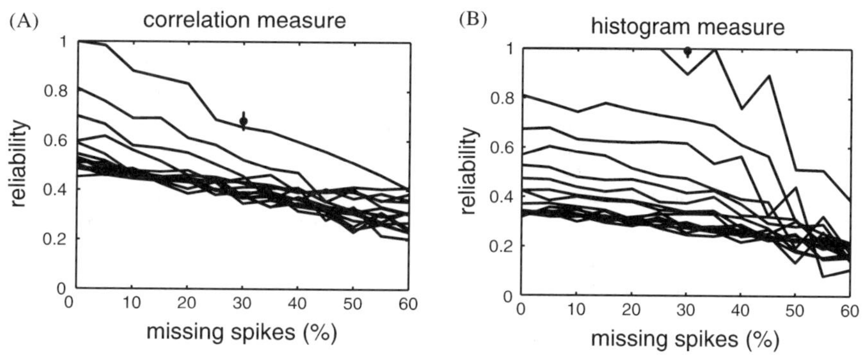

Fig. 3. (A) Correlation-based reliability as a function of missing spikes ($\sigma_c = 5$ ms). Different curves correspond to different levels of extra spikes (increasing from 0% to 30% from top to bottom). Error bar based on 25 different rastergrams (30% missing event-spikes, no extra spikes). (B) Histogram-based reliability as a function of missing spikes ($\tau_h = 5$ ms, $bin_h = 2$ ms, $\theta_h = 1500$ Hz). Different curves correspond to different levels of extra spikes (0% to 30% from top to bottom).

was less smooth and suffered from the same threshold choice sensitivity mentioned above.

4.3. Dependence on the number of trials

The estimate of spike timing reliability depended on the number of trials, N. The higher the number of trials, the smaller the (average) error in the estimate of reliability. Based on 208 surrogate data sets with extra and missing spikes (0–30% extra spikes,

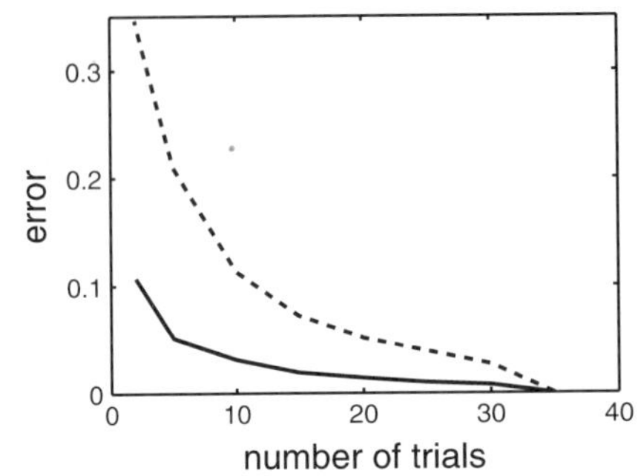

Fig. 4. The average error in reliability for the correlation-based measure (solid line, $\sigma_c = 5$ ms) and the histogram-based measure (dashed line, $\theta_h = 1500/35 \times N$) as a function of the number of trials, N. The error in the correlation-based estimate was always more than a factor of two lower than the error in the histogram-based estimate.

0–60% missing spikes), we compared the errors in the reliability estimate between the histogram measure and the correlation measure as a function of the number of trials per data set, N. An error free reliability estimate would be based on an infinite number of trials; however, for practical reasons, we assume the estimate based on 35 trials to be the reference value for a given data set. The error for estimates based on fewer than 35 trials ($N < 35$) was the standard deviation of the reliability values based on N trials chosen randomly out of the full 35 trials. Fig. 4 depicts the average error as a function of the number of trials.

The error rapidly decreased with increasing N for both reliability measures. However, the error in the correlation-based reliability estimate was less than half of the error in the histogram-based estimate. Hence, fewer trials need to be recorded in an experiment in order to obtain a good reliability estimate.

5. Conclusions

We introduced a new measure of spike timing reliability based on the correlation between pairs of individual responses of a neuron to repeated presentation of a stimulus. We showed that the new measure was robust in quantifying the reliability of data sets where the trial-to-trial variability was due to jitter in the timing of spikes or to the occurrence of additional or missing spikes. The correlation-based measure performed as well as traditional histogram-based measures [3] but needed 50% or fewer trials than the histogram-based measure for the same level of significance.

In contrast to traditional histogram-based measures it relies on only one parameter, σ_c, which is related to the precision of timing. σ_c can be varied to explore the data at different timescales. For a given data set, correlation-based reliability can be analyzed as a function of the filter width, σ_c. Regions of high and low slope indicate intrinsic times scales.

Histogram-based measures are not sensitive to slow variations in firing rate across cell trials, which might be characteristic of certain cells (e.g. stochastic switches between bursts and single spikes). A correlation measure, because it relies on individual trials rather than their average, is sensitive to such variations.

Acknowledgements

This study was supported by the Daimler-Benz Foundation, the Sloan-Swartz Center for Theoretical Neurobiology, and the Howard Hughes Medical Institute.

References

[1] M.J. Berry, D.K. Warland, M.T. Meister, The structure and precision of retinal spike trains, Proc. Natl. Acad. Sci. USA 94 (1997) 5411–5416.
[2] J.D. Hunter, J.G. Milton, P.J. Thomas, J.D.E. Cowan, Resonance effect for neural spike time reliability, J. Neurophysiol. 80 (1998) 1427–1438.
[3] Z.F. Mainen, T.J. Sejnowski, Reliability of spike timing in neocortical neurons, Science 268 (1995) 1503–1506.
[4] L.G. Nowak, M.V. SanchezVives, D.A. McCormick, Influence of low and high frequency inputs on spike timing in visual cortical neurons, Cereb. Cortex 7 (1997) 487–501.
[5] P.H.E. Tiesinga, J.M. Fellous, T.J. Sejnowski, Attractor reliability reveals deterministic structure in neuronal spike trains, Neural. Comput. 14 (2002) 1629–1650.
[6] J.D. Victor, K.P. Purpura, Nature and precision of temporal coding in visual cortex: a metric-space analysis, J. Neurophysiol. 76 (1996) 1310–1326.

Available at
www.ComputerScienceWeb.com
POWERED BY SCIENCE DIRECT®

ELSEVIER

Neurocomputing 52–54 (2003) 933–942

NEUROCOMPUTING

www.elsevier.com/locate/neucom

Computational modeling of bursting pacemaker neurons in the pre-Bötzinger complex

N.A. Shevtsova[a,*], K. Ptak[b], D.R. McCrimmon[b], I.A. Rybak[a]

[a]*School of Biomedical Engineering, Science and Health Systems, Drexel University, 3141 Chestnut Street, Philadelphia, PA 19104, USA*

[b]*Department of Physiology and Institute for Neuroscience, Feinberg School of Medicine, Northwestern University, Chicago, IL 60611-3008, USA*

Abstract

Bursting pacemaker neurons in the pre-Bötzinger complex (pBC) were modeled in the Hodgkin–Huxley style. The single neuron model included rapidly inactivating sodium, persistent sodium, and delayed-rectifier potassium currents. The kinetics of the rapidly inactivating and persistent sodium channels was modeled using experimental data obtained from whole-cell patch clamp recordings from pBC neurons in vitro. Our computational study focused on the conditions that could provide the generation of endogenous bursting activity in single pacemaker neurons and neural populations and on the specific roles of voltage-gated potassium and persistent sodium currents in triggering or suppression of endogenous population oscillations in the pBC.

Keywords: Computational modeling; Pre-Bötzinger complex; Endogenous oscillations; Potassium channels; Respiratory rhythm

1. Introduction

The pre-Bötzinger complex (pBC) is a region in the rostroventrolateral medulla that is considered an important part of the respiratory neural network [1–4,11,12,20,25,27]. As shown in vitro, this region, under certain conditions, can generate an intrinsic rhythmic bursting activity [3,11,12,20,25,27] that is resistant to blockade of synaptic inhibition [23]. It has been suggested, that this in vitro activity is driven by a

* Corresponding author.
E-mail address: natalia@cbis.ece.drexel.edu (N.A. Shevtsova).

0925-2312/03/$ - see front matter
doi:10.1016/S0925-2312(02)00841-X

sub-population of pacemaker neurons located in the pBC [1–3,11,20]. The theoretical analysis of possible intrinsic cellular mechanisms led to the suggestion that this endogenous rhythm is generated with a necessary contribution of the persistent sodium current [1,2] (for an opposite view see [4]). This current was recently described in pBC neurons [10,16,19,24] and its voltage-dependent properties were characterized [19,24].

Our computational study focused on the investigation of the possible conditions that may define the generation of pacemaker-driven oscillations in the pBC and on studying the possible roles of persistent sodium and voltage-gated potassium currents in triggering these endogenous oscillations.

2. Model description

A model of a single pBC pacemaker neuron was developed using the Hodgkin–Huxley formalism. The model was based on previous models [1–3]. However, in contrast to these models based on generic descriptions of sodium channels, we incorporated voltage-gated and kinetic parameters for sodium currents drawn from our in vitro studies of isolated pBC neurons [19,24].

The following ionic currents (and the corresponding channel conductances) were incorporated into the model: rapidly inactivating (fast) sodium (I_{Naf} with maximal conductance $\bar{g}_{\mathrm{Naf}}$); persistent sodium (I_{NaP} with maximal conductance $\bar{g}_{\mathrm{NaP}}$); delayed-rectifier potassium (I_{K} with maximal conductance $\bar{g}_{\mathrm{K}}$); leakage (I_{leak} with constant conductance g_{leak}), and synaptic excitatory (I_{synE} with conductance g_{synE}) and inhibitory (I_{synI} with conductance g_{synI}) currents, which together defined the dynamics of the neuron membrane potential. The voltage-gated and kinetic parameters for I_{Naf} and I_{NaP} were drawn from our in vitro studies [19,24]. The mean values of the maximal sodium conductances used in the model ($\bar{g}_{\mathrm{Naf}} = 150$ nS and $\bar{g}_{\mathrm{NaP}} = 4$ nS) were also set based our experimental measurements ($\bar{g}_{\mathrm{Naf}} = 34$–$170$ nS and $\bar{g}_{\mathrm{NaP}} = 0.5$–$5$ nS; see [19,24]). Similar to Model 1 by Butera et al. [1], the persistent sodium channel in our model had slow inactivation. Because the delayed-rectifier potassium channels have not been characterized in the pBC and other respiration related areas, we used the formal descriptions of this channel taken and adapted from the model of thalamocortical neurons by McCormick and Huguenard [15]. The mean value of $\bar{g}_{\mathrm{K}}$ was set to 50 nS. The reversal potentials for sodium and potassium currents were calculated using the Nernst equation, and the leakage reversal potential was obtained using the Goldman equation. The value of the whole-cell capacitance, $C = 36.2$ pF, was set from experimental data obtained on neurons dissociated from the rostral ventrolateral medulla of rat [9]. The mean value of the leakage conductance, $g_{\mathrm{leak}} = 2$ nS, was set to fit the experimentally measured input resistance $R_{\mathrm{in}} = 500$ MΩ [14].

To investigate firing behavior of pacemaker neuron populations we modeled a population of 50 neurons with all-to-all excitatory synaptic connections. Heterogeneity within the modeled neural population was set by random distribution of the maximal channel conductances ($\bar{g}_{\mathrm{NaP}}$, $\bar{g}_{\mathrm{K}}$ and g_{leak}), values of the external excitatory drive (g_{Edr}) and weights of the synaptic interconnections within the population (w_{ij}).

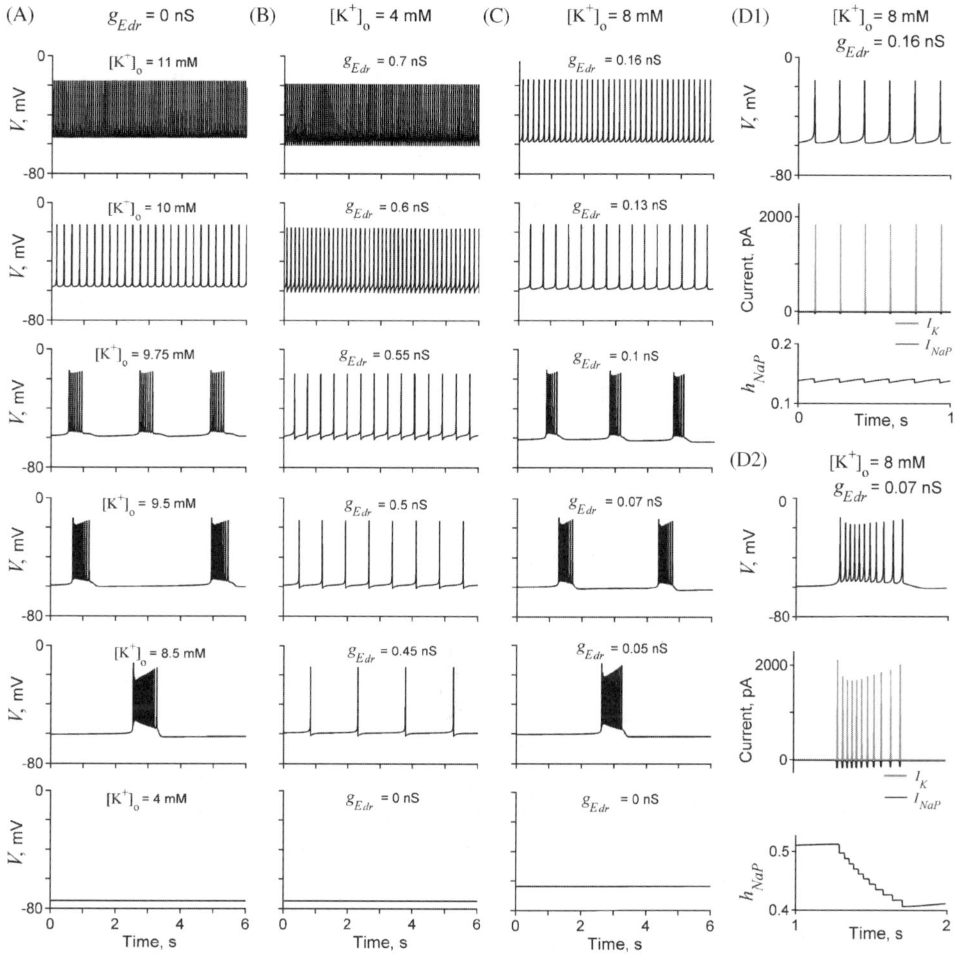

Fig. 1. Firing activity of a model of pBC pacemaker neuron under different conditions. (A) Firing behavior of the model without excitatory drive ($g_{\rm Edr} = 0$) at different levels of extracellular potassium concentration ($[K^+]_0$). $[K^+]_0$ increases bottom-up. An increase of $[K^+]_0$ triggers the rhythmic bursting activity at some threshold of $[K^+]_0$. Further increase in $[K^+]_0$ increases burst frequency and decreases burst duration. At a high level of $[K^+]_0$, bursting switches to tonic firing. (B) Firing behavior of the model at $[K^+]_0 = 4$ mM. $g_{\rm Edr}$ increases bottom-up. An increase in $g_{\rm Edr}$ causes an increase in the tonic firing frequency. Note the absence of bursting activity at any value of $g_{\rm Edr}$. (C) Firing behavior of the model at $[K^+]_0 = 8$ mM. $g_{\rm Edr}$ increases bottom-up. Busting activity is triggered when $g_{\rm Edr}$ exceeds some threshold. The frequency of bursts increases with an increase of $g_{\rm Edr}$ until the neuron switches to tonic firing at a higher level of $g_{\rm Edr}$. (D1) and (D2) Dynamics of the neuronal membrane potential (V, top trace), the persistent sodium ($I_{\rm NaP}$, middle trace, black) and delayed-rectifier ($I_{\rm K}$, middle trace, gray) currents, and the persistent sodium inactivation variable $h_{\rm NaP}$ (bottom trace) during tonic activity (D1) and bursting (D2).

3. Results

The model of a single pBC pacemaker neuron demonstrated the ability to generate endogenous bursting activity under certain conditions (within a particular area in the parameter space). Fig. 1A shows that an increase in extracellular potassium concentration ($[K^+]_0$) triggered a rhythmic bursting activity at some threshold of $[K^+]_0$. A further increase in $[K^+]_0$ increased burst frequency and decreased burst duration, and then, at a higher $[K^+]_0$, bursting switched to tonic firing (Fig. 1A). An increase in excitatory tonic drive to the neuron (g_{Edr}) at the basic level of $[K^+]_0$ (4 mM) did not produce bursting, but increased the tonic firing frequency (Fig. 1B). At a higher level of $[K^+]_0$ (e.g. at $[K^+]_0 = 8$ mM, Fig. 1C), busting activity was triggered when g_{Edr} exceeded some threshold. The frequency of bursts increased with an increase of g_{Edr} until the neuron switched to tonic firing at a high level of g_{Edr} (Fig. 1C). As described in detail by Butera et al. [1], the burst generation mechanism in this type of bursting pacemaker neuron model is explicitly dependent on the slow voltage-dependent inactivation of the persistent sodium current (h_{NaP}), which increases slowly during interburst periods and decreases during bust periods (Fig. 1D2). Comparison of the amplitudes of I_{NaP} and I_K during spikes showed that both the I_{NaP} amplitude and the ratio of amplitudes I_{NaP}/I_K is much higher in the bursting mode than in the mode of tonic activity (Figs. 1D1, 1D2).

To investigate firing behavior of a population of pacemaker neurons we modeled a population of 50 neurons with all-to-all excitatory synaptic connections. Our simulation showed that both an increase in the weights of excitatory synaptic interconnections and a randomization of neuronal parameters within the population increased the area of the parameter space that produced population busting compared to that for a single pacemaker neuron (see also [2,3]). Fig. 2 shows an example of our simulations at the population level. An increase of $[K^+]_0$ above some threshold (about 7.5 mM) triggered synchronized bursting activity in the population. Further elevation of $[K^+]_0$ produced increased burst frequency and decreased burst amplitude. At a higher level of $[K^+]_0$, the population bursting activity switched to high-frequency asynchronous firing (Fig. 2A). An increase of the g_{Edr} at the basic level of $[K^+]_0$ (4 mM) did not produce population bursting, but only increased the level of asynchronous activity (Fig. 2B). At higher values of $[K^+]_0$ (e.g. at $[K^+]_0 = 6$ mM, see Fig. 2C), population busting was triggered when g_{Edr} exceeded some threshold. The frequency of bursts increased with

→

Fig. 2. Firing activity of a model of population of pacemaker neurons under different conditions. The result of each simulation is represented by two diagrams: the top diagram is a raster plot for spike times in all 50 cells, sorted on the ordinate axis by cell index number; the bottom diagram is a corresponding integrated histogram of population activity (bin size = 10 ms). (A) An example of firing behavior of the model without excitatory drive ($g_{Edr} = 0$) at different $[K^+]_0$. $[K^+]_0$ increases bottom-up. An increase of $[K^+]_0$ triggers rhythmic bursting activity in the population at some threshold of $[K^+]_0$. A further increase in $[K^+]_0$ increases burst frequency and decreases burst amplitude. At a higher level of $[K^+]_0$, firing activity switches to asynchronous firing. (B) An example of firing behavior of the model at $[K^+]_0 = 4$ mM. Mean value of g_{Edr} in the population increases bottom-up. Note the absence of bursting at any value of g_{Edr}. (C) An example of firing behavior of the model at $[K^+]_0 = 6$ mM. Busting activity is triggered when g_{Edr} exceeds some threshold. At a higher level of g_{Edr} population activity switches to asynchronous firing.

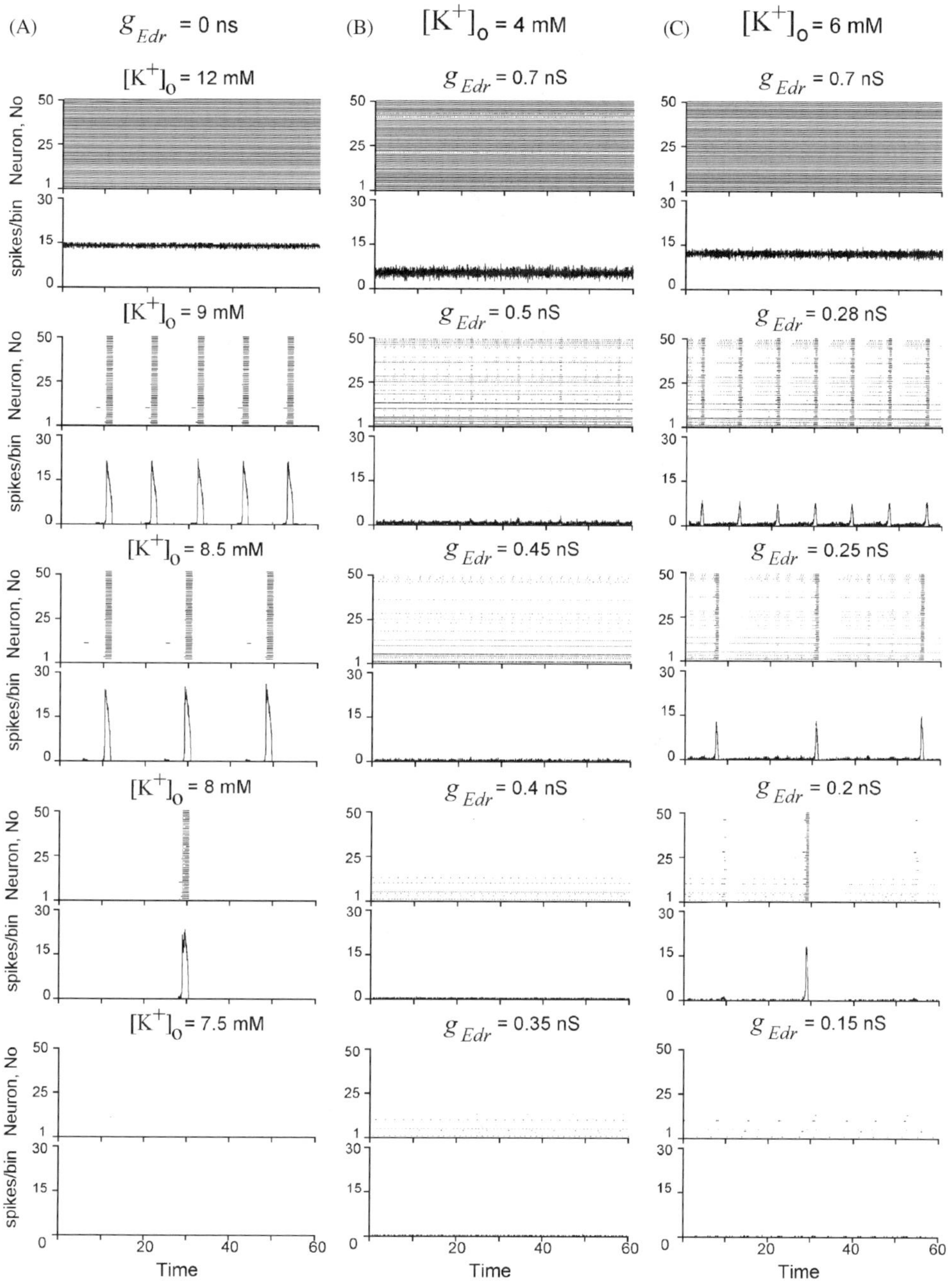
(A) g_{Edr} = 0 ns
(B) $[K^+]_o$ = 4 mM
(C) $[K^+]_o$ = 6 mM
$[K^+]_o$ = 12 mM
g_{Edr} = 0.7 nS
g_{Edr} = 0.7 nS
$[K^+]_o$ = 9 mM
g_{Edr} = 0.5 nS
g_{Edr} = 0.28 nS
$[K^+]_o$ = 8.5 mM
g_{Edr} = 0.45 nS
g_{Edr} = 0.25 nS
$[K^+]_o$ = 8 mM
g_{Edr} = 0.4 nS
g_{Edr} = 0.2 nS
$[K^+]_o$ = 7.5 mM
g_{Edr} = 0.35 nS
g_{Edr} = 0.15 nS
Neuron, No
spikes/bin
Time

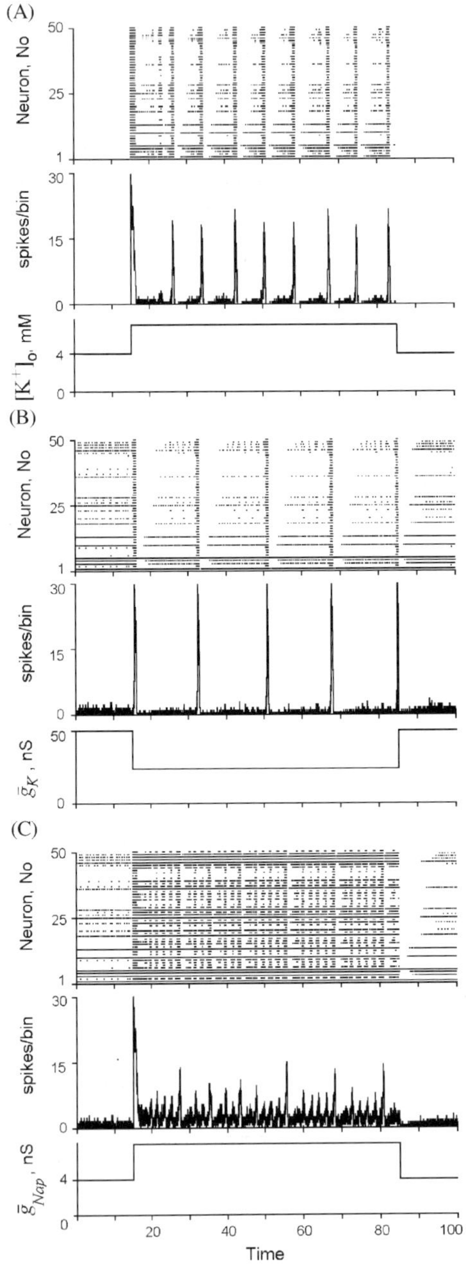

Fig. 3. Triggering endogenous bursting activity in the model of a population of pacemaker neurons. (A) An example of triggering endogenous bursting activity by elevation of $[K^+]_0$ (bottom traces). $[K^+]_0$ was increased from 4 to 8 mM. (B) An example of triggering endogenous bursting activity by reduction of the mean value of $\bar{g}_K$ in the population. The mean value of $\bar{g}_K$ was reduced from 50 to 25 nS to. (C) An example of triggering endogenous bursting activity by augmentation of the mean value of $\bar{g}_{NaP}$. The mean value of $\bar{g}_{NaP}$ was increased from 4 to 7 nS.

an increase in g_{Edr}. Finally, at some level of g_{Edr}, firing behavior of the population switched to high-frequency asynchronous firing (Fig. 2C).

The important conclusion from our modeling studies is that an increase of excitatory drive to the neuron and an increase of $[K^+]_0$ produce different effects on neuronal and population firing behavior. Specifically, an increase of excitatory drive at a normal physiological level of $[K^+]_0$ causes cellular depolarization without changing a balance between I_{NaP} and I_K. In this case, the normally expressed I_K provides sufficient membrane repolarization after each generated spike to reduce the membrane potential below the level of I_{NaP} activation and does not permit endogenous oscillations. In contrast, an increase of $[K^+]_0$ produces two simultaneous effects: one is the cellular depolarization (via shifting the leakage reversal potential to more positive values of voltage); the other is the reduction of I_K by shifting the potassium reversal potential to more positive values of voltage. Therefore, at higher levels of $[K^+]_0$, the reduced I_K does not produce a sufficient postspike repolarization and hence cannot restrain endogenous I_{NaP}-dependent bursting activity.

In this study, we focused on the investigation of the specific roles of the I_K and I_{NaP} in inducing or suppressing endogenous population bursting activity. Fig. 3 shows examples of our simulations. Elevation of $[K^+]_0$ from the basic level of 4–8 mM triggered bursting activity in the population. This bursting activity stopped when $[K^+]_0$ returned to the basic level (Fig. 3A). At the same time, the synchronized population bursting could also be triggered at $[K^+]_0 = 4$ mM by a reduction of $\bar{g}_K$ (Fig. 3B) or an augmentation of $\bar{g}_{\mathrm{NaP}}$ (Fig. 3C).

4. Conclusion and discussion

In summary, our modeling studies demonstrate that rhythmic bursting activity in a population of pacemaker neurons may be initiated by (1) an increase of the extracellular potassium concentration, or (2) a suppression of the voltage-gated potassium currents, or (3) an augmentation of the persistent sodium currents.

This conclusion is consistent with the majority of in vitro studies of endogenous rhythmic activity in the pBC, in which the researchers elevated $[K^+]_0$ to 7–9 mM in order to trigger and maintain a robust activity, e.g. see [11,12,27]. Our modeling predictions was recently confirmed by Pierrefiche et al. [18], who demonstrated that rhythmic bursting in the medullary slice could be triggered either by elevation of $[K^+]_0$ to 7 mM, or by application of different potassium current blockers (TEA or 4-AP) at $[K^+]_0 = 3$ mM.

Our results are also consistent with the suggestion that gasping during hypoxia is produced by a pacemaker-driven mechanism in the pBC [21,22,26]. Specifically, hypoxia is accompanied by elevation of $[K^+]_0$ [17], suppression of voltage-gated potassium currents [5,8,13], and augmentation of I_{NaP} [6,7,9], which are the same factors that produced the population oscillations in our modeling studies. The results of a recent study of the neurogenesis of gasping in situ [26] are also consistent with this idea. Specifically, application of 4-AP and strychnine in combination with an increase of $[K^+]_0$ converted the eupneic pattern of phrenic nerve discharge to a decrementing discharge similar to that recorded both in vitro and during ischemia-induced gasping in situ [26].

Acknowledgements

This work was supported by NSF (0091942) and ONR (N000140010719) grants to I.A. Rybak and NIH grants HL60097 and HL60969 to D.R. McCrimmon.

References

[1] R.J. Butera, J.R. Rinzel, J.C. Smith, Models of respiratory rhythm generation in the pre-Bötzinger complex: I. Bursting pacemaker neurons, J. Neurophysiol. 82 (1999) 382–397.

[2] R.J. Butera, J.R. Rinzel, J.C. Smith, Models of respiratory rhythm generation in the pre-Bötzinger complex: II. Populations of coupled pacemaker neurons, J. Neurophysiol. 82 (1999) 398–415.

[3] C.A. Del Negro, S.M. Johnson, R.J. Butera, J.C. Smith, Models of respiratory rhythm generation in the pre-Bötzinger complex: III. Experimental tests of model predictions, J. Neurophysiol. 86 (2001) 59–74.

[4] C.A. Del Negro, C. Morgado-Valle, J.L. Feldman, Respiratory rhythm: an emergent network property? Neuron 34 (2002) 821–830.

[5] C. Gebhardt, U. Heinemann, Anoxic decrease in potassium outward currents of hippocampal cultured neurons in absence and presence of dithionite, Brain Res. 837 (1999) 270–276.

[6] A.K. Hammarström, P.W. Gage, Inhibition of oxidative metabolism increases persistent sodium current in rat CA1 hippocampal neurons, J. Physiol. London 510 (1998) 735–741.

[7] E.M. Horn, T.G. Waldrop, Hypoxic augmentation of fast-inactivating and persistent sodium currents in rat caudal hypothalamic neurons, J. Neurophysiol. 84 (2000) 2572–2581.

[8] C. Jiang, G.G. Haddad, A direct mechanism for sensing low oxygen levels by central neurons, Proc. Natl. Acad. Sci. USA 91 (1994) 7198–7201.

[9] Y. Kawai, J. Qi, A.H. Comer, H. Gibbons, J. Win, J. Lipski, Effects of cyanide and hypoxia on membrane currents in neurons acutely dissociated from the rostral ventrolateral medulla of the rat, Brain Res. 830 (1999) 246–257.

[10] N. Koshiya, C. Del Negro, R.J. Butera, J.C. Smith, Persistent sodium current (INaP) in pre-pre-Bötzinger complex (pre-BötC) inspiratory neurons, Soc. Neurosci. Abs. 27 (2001) 243.3.

[11] N. Koshiya, J.C. Smith, Neuronal pacemaker for breathing visualized in vitro, Nature 400 (1999) 360–363.

[12] S.P. Lieske, M. Thoby-Brisson, P. Telgkamp, J.M. Ramirez, Reconfiguration of the neural network controlling multiple breathing patterns: eupnea, sighs and gasps, Nature Neurosci. 3 (2000) 600–607.

[13] J. Lopez-Barneo, R. Pardal, P. Ortega-Saenz, Cellular mechanisms of oxygen sensing, Annu. Rev. Physiol. 63 (2001) 259–287.

[14] E. Mazza Jr., N.H. Edelman, J.A. Neubauer, Hypoxic excitation in neurons cultured from the rostral ventrolateral medulla of the neonatal rat, J. Appl. Physiol. 88 (2000) 2319–2329.

[15] D.A. McCormick, J.R. Huguenard, A model of the electrophysiological properties of thalamocortical relay neurons, J. Neurophysiol. 68 (1992) 1384–1400.

[16] D.R. McCrimmon, A. Monnier, K. Ptak, G. Zummo, Z. Zhang, G.F. Alheid, Respiratory rhythm generation: pre-Bötzinger neuron discharge patterns and persistent sodium current, Adv. Exp. Med. Biol. 499 (2001) 147–152.

[17] J.E. Melton, S.C. Kadia, Q.P. Yu, J.A. Neubauer, N.H. Edelman, Respiratory and sympathetic activity during recovery from hypoxic depression and gasping in cats, J. Appl. Physiol. 80 (1996) 1940–1948.

[18] O. Pierrefiche, I.A. Rybak, W.M. St.-John, J.F.R. Paton, Neurogenesis of respiratory rhythm: effects of suppression of potassium currents and hypoxia, Soc. Neurosci. Abs. 28 (2002) 173.11.

[19] K. Ptak, G. Zummo, G.F. Alheid, J.D. Surmeier, D.R. McCrimmon, Biophysical and molecular properties of persistent sodium current in pre-Bötzinger complex respiratory neurons, Soc. Neurosci. Abs. 28 (2002) 221.8.

[20] J.C. Rekling, J.L. Feldman, Pre-Bötzinger complex and pacemaker neurons: hypothesized site and kernel for respiratory rhythm generation, Annu. Rev. Physiol. 60 (1998) 385–405.

[21] I.A. Rybak, J.F.R. Paton, R.F. Rogers, W.M. St.-John, Generation of the respiratory rhythm: state-dependency and switching, Neurocomputing 44–46 (2002) 605–614.

[22] I.A. Rybak, W.M. St.-John, J.F.R. Paton, Models of neuronal bursting behavior: implications for in vivo versus in vitro respiratory rhythmogenesis, Adv. Exp. Med. Biol. 499 (2001) 159–164.

[23] X.M. Shao, J.L. Feldman, Respiratory rhythm generation and synaptic inhibition of expiratory neurons in pre-Bötzinger complex: differential roles of glycinergic and gabaergic neural transmission, J. Neurophysiol. 77 (1997) 1853–1860.

[24] N.A. Shevtsova, K. Ptak, D.R. McCrimmon, I.A. Rybak, Study on sodium current in neurons of pre-Bötzinger complex (pBC), Soc. Neurosci. Abs. 28 (2002) 173.5.

[25] J.C. Smith, H.H. Ellenberger, K. Ballanyi, D.W. Richter, J.L. Feldman, Pre-Bötzinger complex: a brainstem region that may generate respiratory rhythm in mammals, Science 254 (1991) 726–729.

[26] W.M. St.-John, I.A. Rybak, J.F.R. Paton, Potential switch from eupnea to fictive gasping following blockade of glycine transmission and potassium channels, Am. J. Physiol. (Integr. Comp. Physiol.) 283 (2002) R721–R731.

[27] M. Thoby-Brisson, J.M. Ramirez, Identification of two types of inspiratory pacemaker neurons in the isolated respiratory network of mice, J. Neurophysiol. 86 (2001) 104–112.

Natalia A. Shevtsova started her scientific carrier in 1974 in AB Kogan Research Institute for Neurocybernetics at Rostov State University (Russia). She was a senior researcher in Ilya Rybak's group in 1972–1991. In 1992 she received "The Best Paper Award" from "Neurocomputing" for her work on neural network model of visual perception. She received a Ph.D. in Computer Science from Rostov State University in 1996. In 1996–1998 she was a research associate in James Reggia's group at the University of Maryland (College Park, MD). From 2001 she is a research associate in Ilya Rybak's group at the School of Biomedical Engineering, Science and Health Systems, Drexel University, Philadelphia, PA. Her research interests include computational modeling of biological neurons and neural networks, neural control of respiration and visual perception.

Krzysztof Ptak graduated in 1996 from Jagiellonian University in Cracow. After graduating, he joined Roger Monteau's group at the Aix-Marseille University. In 1999 he received a Ph.D. for "Substance P and modulation of central respiratory activity: in vitro study on rodents in the perinatal period" from both Aix-Marseille and Jagiellonian Universities. This work was awarded by the Prime Minister of the Polish government as the best doctoral thesis in 1999. He is currently a research associate in Donald McCrimmon's group at the Feinberg School of Medicine at Northwestern University in Chicago, IL. Presently he is studying the role of sodium conductance in generation of pacemaker activity in medullary pre-Bötzinger Complex neurons using patch-clamp recording and single-cell RT-PCR techniques.

Donald R. McCrimmon studied exercise physiology at the University of Calgary, Canada. After graduating in 1974, he joined David Cunningham's group at the University of Western Ontario and received an M.A. in 1976 for comparing the effects of intermittent versus continuous exercise training regimens on the cardiopulmonary systems. He then began work on the central neural control of breathing at the University of Wisconsin in Madison under the tutelage of Jerome Dempsey and Peter Lalley. After receiving a Ph.D. in Physiology in 1983 he traveled to Northwestern University in Chicago where he began postdoctoral work with Jack Feldman and in 1986 established a laboratory to study brainstem pathways controlling respiratory rhythm and pattern. He is currently an Associate Professor and Associate Chair of the Department of Physiology at Feinberg School of Medicine, Northwestern University (Chicago, IL).

Ilya A. Rybak received a Ph.D. in Biophysics from St. Petersburg University in 1988. In 1977–1991 he was a senior researcher and head of the lab for Neural Network Modeling in Vision Research in AB Kogan Research Institute for Neurocybernetics at Rostov State University (Russia). In 1992 he received "The Best Paper Award" from "Neurocomputing" for his work on neural network model of visual perception. In 1993–1998 he was a visiting scientist at the Central Research Department at the DuPont Company. He is currently a Research Professor at the School of Biomedical Engineering, Science and Health Systems, Drexel University, Philadelphia, PA. His scientific interests include computational modeling of biological neurons and neural networks with a research focus on neural control of respiration and locomotion, visual perception, biologically inspired robotics and brain–machine interfaces.

Available at
www.ComputerScienceWeb.com
POWERED BY SCIENCE @ DIRECT®

ELSEVIER

Neurocomputing 52–54 (2003) 943–948

NEUROCOMPUTING

www.elsevier.com/locate/neucom

A prefronto-parietal network model with feedforward and feedback connections

Shoji Tanaka*

Department of Electrical and Electronics Engineering, Sophia University, High-Tech Research Center, 7-1 Kioicho, Chiyoda-ku, Tokyo 102-8554, Japan

Abstract

The cortical network of the prefrontal cortex (PFC) and the posterior parietal cortex (PPC) has been suggested to play central roles in visuospatial working memory processing and attention. However, how these areas contribute to such processes remains obscure. To investigate this, I here propose a layered network model of the PFC and the PPC with feedforward/feedback-type corticocortical connections. Computer simulation with this model shows that the feedback and feedforward connections critically control the time course of the neuronal activity of the PFC and the PPC.

Keywords: Corticocortical; Feedback; Posterior parietal cortex; Prefrontal cortex; Visuospatial working memory

1. Introduction

The prefrontal cortex (PFC), with the assistance of neuromodulators, is considered to play central roles in cognition and working memory. The memory fields of spatial working memory in the PFC are modulated biphasically by dopamine over a wide range of the receptor activity [7,11]. This has been analyzed further by computer simulation [12]. However, how slight deviation of the dopamine level in the normal range alters the PFC circuit dynamics is unknown yet. Recently, Tanaka [10] addressed this issue via computer simulation of a model PFC circuit with dopamine effects. The results suggest the possibility that the PFC circuit performs several different modes of cognitive

* Tel.: +81-3-3238; fax: +81-3-3238-3321.
E-mail address: tanaka-s@sophia.ac.jp (S. Tanaka).

0925-2312/03/$ - see front matter © 2002 Elsevier Science B.V. All rights reserved.
doi:10.1016/S0925-2312(02)00851-2

operations of multi-target spatial working memory, such as "replacement", "addition", "rejection", and "elimination". The study further suggests that these modes are switched from one to another by slightly changing the dopamine level.

For visuospatial working memory processing in general, both the PFC and the posterior parietal cortex (PPC) are activated. However, how these two areas contribute to the processing remains obscure. Chafee and Goldman-Rakic [3,4] observed very similar neuronal activities of these areas of macaque monkeys during performing a visuospatial working memory task. It is, therefore, interesting to know if the two areas contribute equally to visuospatial working memory processing or if each area has distinct roles in the processing. To answer these questions, the circuit properties of the PFC–PPC network have to be studied. This article proposes a network model of the PFC and the PPC that are connected with each other. Computer simulation with this model investigates how the feedforward and feedback connections between the PFC and the PPC contribute to spatial working memory loading and representation.

2. Model

The PFC and the PPC are connected intimately with each other by the corticocortical fibers [6]. Earlier anatomical studies in the sensory areas have proposed at least two typical types of the corticocortical connectivity; i.e., the feedforward/feedback type and the lateral type [5]. Recently, this kind of study has extended to examine the connectivity between association areas, and has proposed a general rule of the laminar specificity of the corticocortical connections [2]. In this model, with some simplifications, the feedforward projections originate from the superficial layers of the PPC and terminate in the intermediate layer of the PFC and the feedback projections originate from the deep layers of the PFC and terminate in the superficial and deep layers of the PPC. Each of the model cortical areas has three layers (the superficial, the intermediate, and the deep) and contains pyramidal cells and interneurons. The circuitry in each area is based on the earlier model developed by Tanaka [9,10].

The model contains 2160 pyramidal cells and 720 interneurons (2880 in all). The neurons are described by a single compartment, leaky integrate-and-fire neuron model, which contains the following types of conductance: AMPA, NMDA, Nap, $GABA_A$, K(Ca), and leak. The model neuron is identical to the previously developed model [9,10] but with several different parameter values: the transmission and synaptic delay is chosen randomly from 2.0–10.0 ms for the intracortical connections and 10.0–20.0 ms for the corticocortical connections. The ratio of the NMDA conductance to the AMPA conductance [9,10,12] is $g_{\mathrm{NMDA,max}}/g_{\mathrm{AMPA,max}} = 0.05$. The ratio of the cross-directional inhibition to the isodirectional inhibition [8–10] is $g_{\mathrm{GABA_A(cross),max}}/g_{\mathrm{GABA_A(iso),max}} = 0.1$.

3. Results

The visual input conveying the target location is given to the neurons in the intermediate layer of the PPC. The input has a Gaussian directional profile with the standard

depend on the subareas more specifically, or they may differ across species to some extent. Because the model proposed here specifies neither the subareas (in the PFC and the PPC) nor the species, further arguments on the corticocortical connectivity is beyond the scope of this model.

An interesting notion suggested by this study is the capability of the cortical network to switch the activity pattern of the neurons between tonic and phasic by changing the strengths of the corticocortical connections. In its simplest view, the tonic activity with directional selectivity of the PFC neurons is a substrate for the spatial working memory. Then keeping the activity of the PFC neurons tonic is necessary for loading spatial working memory in the PFC. In contrast, the switching of the activity from tonic to phasic can make the network not load spatial working memory. This is mediated by the feedforward connections in the network. Weak feedforward connections make the PFC off-line from the sensory system. The feedback connections can have different roles. By boosting the activity of the PPC, both the PFC and the PPC exhibit concurrent working memory activity. Because the actual PFC is connected not only with the PPC but with downstream areas, this concurrent mode would be necessary for the PFC to exert top–down control of downstream areas. Taken together, changing the strengths of the feedforward and feedback connections alters the activity patterns or cortical dynamics modes of the PFC and the PPC qualitatively. It can further change the overall magnitudes of the activities of these areas quantitatively. The feedforward and feedback connections would thus change the signal flow in the cortical network to achieve cognitive control.

Acknowledgements

This work was supported by the Grants-in-Aid for Scientific Research on Priority Areas (#13210123 and #14017083) from the Japanese Ministry of Education, Science, and Technology. The author acknowledges valuable discussions with H. Barbas at Boston University.

References

[1] R.A. Andersen, G.K. Essick, R.M. Siegel, Encoding of spatial location by posterior parietal neurons, Science 230 (1985) 456–458.
[2] H. Barbas, N. Rempel-Clower, Cortical structure predicts the pattern of corticocortical connections, Cereb. Cortex 7 (1997) 635–646.
[3] M.V. Chafee, P.S. Goldman-Rakic, Matching patterns of activity in primate prefrontal area 8a and parietal area 7ip during a spatial working memory task, J. Neurophysiol. 79 (1998) 2919–2940.
[4] M.V. Chafee, P.S. Goldman-Rakic, Inactivation of parietal and prefrontal cortex reveals interdependence of neural activity during memory-guided saccades, J. Neurophysiol. 83 (2000) 1550–1566.
[5] D.J. Felleman, D.C. Van Essen, Distributed hierarchical processing in the primate cerebral cortex, Cereb. Cortex 1 (1991) 1–47.
[6] P.S. Goldman-Rakic, Topography of cognition: parallel distributed networks in primate association cortex, Annu. Rev. Neurosci. 11 (1988) 137–156.
[7] P.S. Goldman-Rakic, E.C. Muly III, G.V. Williams, D1 receptors in prefrontal cells and circuits, Brain Res. Rev. 31 (2000) 295–301.

[8] M.L. Schwartz, P.S. Goldman-Rakic, Callosal and intrahemispheric connectivity of the prefrontal association cortex in rhesus monkey: relation between intraparietal and principal sulcal cortex, J. Comp. Neurol. 226 (1984) 403–420.
[9] S. Tanaka, Multi-directional representation of spatial working memory in a model prefrontal cortical circuit, Neurocomputing 44–46 (2002) 1001–1008.
[10] S. Tanaka, Dopamine controls fundamental cognitive operations of multi-target spatial working memory, Neural Networks 15 (2002) 573–582.
[11] G.V. Williams, P.S. Goldman-Rakic, Modulation of memory fields by dopamine D1 receptors in prefrontal cortex, Nature 376 (1995) 572–575.
[12] K. Yamashita, S. Tanaka, Circuit simulation of memory field modulation by dopamine D1 receptor activation, Neurocomputing 44–46 (2002) 1035–1042.

Shoji Tanaka received his B.E., M.E., and Ph.D. degrees from Nagoya University, Japan. He is Professor at Department of Electrical and Electronics Engineering, Sophia University. During 1998–1999, he was a Visiting Scientist at the Section of Neurobiology, Yale University School of Medicine, USA.

Available at
www.ComputerScienceWeb.com
POWERED BY SCIENCE @ DIRECT®

ELSEVIER

Neurocomputing 52–54 (2003) 949–954

NEUROCOMPUTING

www.elsevier.com/locate/neucom

The spread of rate and correlation in stationary cortical networks

Tom Tetzlaff*, Michael Buschermöhle, Theo Geisel, Markus Diesmann

Department of Nonlinear Dynamics, Max-Planck-Institut für Strömungsforschung, Göttingen, Germany

Abstract

The analysis of the spatial and temporal structure of spike cross-correlation in experimental data is an important tool in the exploration of cortical processing. Recent theoretical studies investigated the impact of correlation between afferents on the spike rate of single neurons and the effect of input correlation on the output correlation of pairs of neurons. Here, this knowledge is combined to a model simultaneously describing the spatial propagation of rate and correlation, allowing for an interpretation of its constituents in terms of network activity. The application to an embedded feed-forward network provides insight into the mechanisms stabilizing its asynchronous mode.

Keywords: Rate model; Cross-correlation; Integrate-and-fire; Synfire chain

1. Introduction

Equilibrated activity states in models of cortical networks are often described in terms of firing rate or related measures. It has been shown (e.g. Refs. [5,7]) that the transmission of firing rates by a single neuron is modulated by temporal relationships between its afferents. Correlations between converging inputs generally affect the magnitude of post-synaptic current fluctuations. Since spiking is driven by both mean and variance of the membrane potential the output rate is not only determined by the firing rate of the inputs but also by the degree of correlation. Depending on the network architecture the sets of afferents of two different neurons exhibit a

* Corresponding author.
E-mail address: tom@chaos.gwdg.de (Tom Tetzlaff).

0925-2312/03/$ - see front matter
doi:10.1016/S0925-2312(02)00854-8

certain overlap. These common input channels induce correlations between the total inputs of the two neurons depending on the size of the overlap, the afferent firing rates, and on correlations inside each set of afferents. As shown e.g. in Ref. [7] such common input correlations are to some degree transmitted by the neuron pair. In summary, there is a complex interplay between firing rates and correlations. Therefore, it appears questionable whether pure rate models can reveal a realistic picture of network dynamics. In the present study we attempt to construct a model describing both variables simultaneously. While rate transmission is treated analytically, the dependence of output correlation on input rate and input correlation is obtained by two-neuron simulations. As an example application we investigate to what extent the rate propagation in a feed-forward network is distorted if correlations are taken into account.

2. Rate transmission

Both analytics and numerics are based on a leaky-integrate-and-fire neuron model with α-function shaped synaptic currents [3]. Since amplitudes of post-synaptic potentials $h(t)$ (PSPs) are chosen to be small compared to the distance ϑ between reset potential and spike threshold, membrane potential dynamics can be considered as a diffusion process. In this framework the output rate λ_{out} is approximately determined by the mean μ and variance σ^2 of the stationary membrane potential distribution (e.g. Ref. [4]). Because of the linearity of the subthreshold dynamics, μ and σ can be calculated by linear filter theory, using $h(t)$ as the systems impulse response. This way the statistical properties of the superposition $X = \sum_{i \in \mathscr{A}} x_i$ of all processes x_i arriving from the set $\mathscr{A}$ of afferents can be easily mapped onto those of the membrane potential. Here, we assume each process x_i to be stationary Poisson with rate λ_i. Hence, the mean weighted number of spikes $\langle X \rangle$ in a time interval Δt is given by $\sum_{i \in \mathscr{A}} g_i \lambda_i \Delta t$. For simplicity, the synaptic weights g_i are considered to be completely determined by the pre-synaptic neuron. Denoting the coefficient of correlation between processes x_i and x_j by r_{ij}, the variance of X is given by

$$\mathrm{Var}[X] = \sum_i g_i^2 \lambda_i \Delta t + \sum_{i \neq j} r_{ij} g_i g_j \sqrt{\lambda_i \lambda_j} \Delta t. \tag{1}$$

Thus, the output rate λ_{out} of a neuron depends on both the rates λ_i of all its afferents *and* the correlations r_{ij} between all afferent pairs. The dependence of the output rate λ_{out} on the rate Λ_{E} of the total excitatory input is shown in Fig. 1A for a fixed Poisson inhibitory input. The curves refer to different variance-to-mean ratios F (Fano factors) of the excitatory input process. The solid line represents the Poisson case ($F = 1$). An increase in variance, e.g. due to correlations between the input channels, raises the probability of reaching spike threshold at lower mean membrane potentials. Hence, the output rate starts rising at lower Λ_{E} (dashed curve: $F = 2$, dotted curve: $F = 3$).

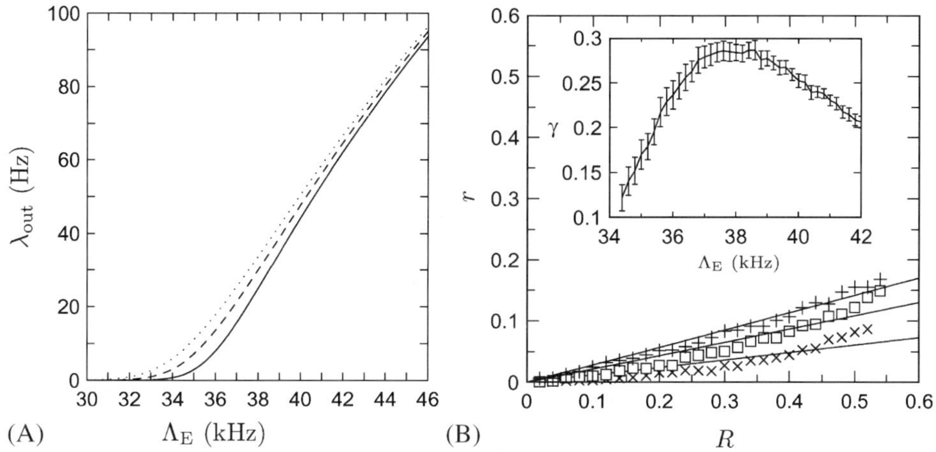

Fig. 1. (A) Response rate λ_{out} as a function of excitatory input rate Λ_{E} for different Fano factors $F = 1$ (Poisson, solid curve), $F = 2$ (dashed) and $F = 3$ (dotted). Inhibitory input is a Poisson process with a fixed rate $\Lambda_{\text{I}} = 30.24$ kHz. (B) Measured output correlation r (10,000 spikes/neuron) depending on input correlation R for three different excitatory Poissonian inputs with rates $\Lambda_{\text{E}} = 34.4$ kHz ($\times$), 36 kHz ($\square$) and 39 kHz (+). The inset summarizes the result of the linear regression $r = \gamma \cdot R$ (error bars indicate 95% confidence interval). $\Lambda_{\text{I}} = 30.24$ kHz, $\vartheta = 15$ mV, $\tau_{\text{m}} = 10$ ms, $\tau_{\text{ref}} = 2$ ms, membrane capacity: 250 pF, PSP amplitude: 0.14 mV, PSP rise time: 1.7 ms, correlation width $\Delta\tau = 5.1$ ms.

3. Correlation transmission

Given the sets $\mathscr{A}_k$, $\mathscr{A}_l$ of afferent channels of two different neurons k and l, we define the set $\mathscr{C}_{kl}$ of "common inputs" as the intersection $\mathscr{A}_k \cap \mathscr{A}_l$. We call the complements $\mathscr{B}_{kl} := \mathscr{A}_k \setminus \mathscr{C}_{kl}$ the sets of "background" processes. The total input processes $X_{k/l}$ can now be subdivided into a common and a background fraction: $X_{k/l} = C_{kl} + B_{kl/lk}$ (with $C_{kl} := \sum_{i \in \mathscr{C}_{kl}} x_i$ and $B_{kl/lk} := \sum_{i \in \mathscr{B}_{kl/lk}} x_i$). The coefficient of correlation R_{kl} between X_k and X_l is given by

$$R_{kl} = \frac{\text{Cov}[X_k, X_l]}{\sqrt{\text{Var}[X_k]\text{Var}[X_l]}} = \frac{\text{Var}[C_{kl}]}{\text{Var}[C_{kl}] + \text{Var}[B_{kl}]}. \tag{2}$$

For reasons of clarity, here we assumed that the three processes C_{kl}, B_{kl} and B_{lk} are pairwise independent and that the backgrounds are statistically equivalent, i.e. $\text{Var}[B_{kl}] = \text{Var}[B_{lk}]$. The degree of correlation induced by common input is thus determined by the ratio of the variances of the total common and background process. As shown in Eq. (1) these variances in turn depend on the rates and correlations inside the afferent populations.

Since the dynamics of our model neuron is completely deterministic, the input correlations $R_{kl} = 0$ and 1 are mapped 1:1 to the output correlation r_{kl}, independently of the statistics of the individual inputs. It is tempting to assume that this is true for all values of R_{kl}. However, several studies revealed that the gain of correlation is in general rather low and is modulated by the mean and variance of the input processes [6,7]. With the help of two-neuron simulations we measured the correlation transmission $R_{kl} \mapsto r_{kl}$ and its dependence on afferent firing rates. Here, both neurons k and l receive stationary Poisson inputs from a common excitatory channel C of rate Λ^{C} and individual background channels $B_{k/l}$. The latter are formed by superpositions of excitatory and inhibitory processes $B^{\mathrm{E}}_{k/l}$ and $B^{\mathrm{I}}_{k/l}$. Their spike rates $\Lambda^{\mathrm{B}}_{\mathrm{E}}$ and $\Lambda^{\mathrm{B}}_{\mathrm{I}}$ are chosen to be identical for both neurons. Excitatory and inhibitory synaptic weights differ only in sign. Under these conditions Eq. (2) reads $R = \Lambda^{\mathrm{C}}/(\Lambda^{\mathrm{C}} + \Lambda^{\mathrm{B}}_{\mathrm{E}} + \Lambda^{\mathrm{B}}_{\mathrm{I}})$. Therefore, the degree of input correlation can be tuned by adjusting the rates. Focussing on the dependence of the correlation transmission on the total excitatory input rate $\Lambda_{\mathrm{E}} = \Lambda^{\mathrm{C}} + \Lambda^{\mathrm{B}}_{\mathrm{E}}$ the inhibitory input rate $\Lambda_{\mathrm{I}} = \Lambda^{\mathrm{B}}_{\mathrm{I}}$ is kept constant. Simulations are performed on a temporal grid of 0.1ms resolution. Fig. 1B shows the measured output correlations r as a function of the input correlation R. At low input correlations R the output r is well approximated by the linear relation $r = \gamma \cdot R$. The inset in Fig. 1B shows the dependence of the fitted slope γ on the excitatory input rate Λ_{E}. Results are consistent with those of former studies [6,7]. It has to be emphasized that the maximum correlation gain $\gamma_{\max} < 0.3$ is small in the observed range. The findings in Ref. [7] revealed that correlation transmission is enhanced by increasing the magnitude of fluctuations of the stimuli. Our analysis is restricted to Poisson inputs with δ-type cross-correlations.

4. Application to feed-forward networks

In the following example we show how the simultaneous transmission of rate and correlation yields a model describing the spatial propagation of stationary spiking activity in a synfire chain [1]. The common input C_i of two neurons a and b in the $(i+1)$th layer of a complete synfire chain consists of the w channels $c_{i,j}$ $(j \in \{1, 2, \ldots, w\})$ from the preceding group i (see Fig. 2A). In addition, a and b receive independent

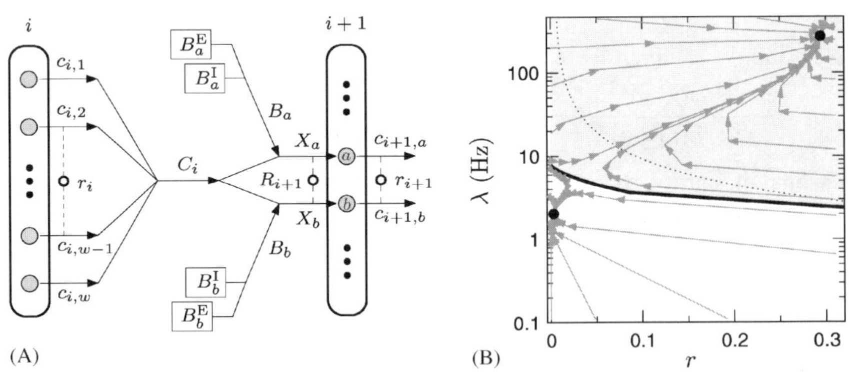

Fig. 2. (A) Synfire link, (B) (λ, r)-phase-space of a synfire chain of group size $w = 200$ (log-scaled ordinate). Gray trajectories indicate the spatial evolution from several initial conditions λ_0, r_0. Black dots mark the positions of the two attractors. Their basins meet at the separatrix (thick black curve). States below the dotted curve correspond to excitatory inputs with Fano factors $F < 2$ (parameters as in Fig. 1).

background inputs $B^{\rm E}_{a/b}$ and $B^{\rm I}_{a/b}$. Keeping the total number of excitatory synapses fixed at $K_{\rm E}$ and the number of inhibitory synapses at $K_{\rm I}$ mean and variance of the resulting input processes $X_{a/b}$ are given by

$$\langle X_{a/b}\rangle = [(K_{\rm E}-w)\lambda^*_{\rm E} + w\lambda_i - K_{\rm I}\lambda^*_{\rm I}]\Delta t,$$

$$\mathrm{Var}[X_{a/b}] = [(K_{\rm E}-w)\lambda^*_{\rm E} + (1+r_i[w-1])w\lambda_i + K_{\rm I}\lambda^*_{\rm I}]\Delta t, \quad (3)$$

Here, $\lambda^*_{\rm E}$ and $\lambda^*_{\rm I}$ denote the firing rates of the background neurons and λ_i those of the neurons in the ith layer. r_i is the coefficient of correlation between each pair of neurons in that group. Intra-chain connections are purely excitatory. Again, each individual process is assumed to be Poissonian. The resulting input correlation is

$$R_{i+1}(\lambda_i, r_i) = \left[1 + \frac{(K_{\rm E}-w)\lambda^*_{\rm E} + K_{\rm I}\lambda^*_{\rm I}}{w\lambda_i(1+r_i[w-1])}\right]^{-1}. \quad (4)$$

Using the approach presented in Section 2 and the simulation results for the correlation transmission an iterative map $(\lambda_i, r_i)_w \mapsto (\lambda_{i+1}, r_{i+1})_w$ can be constructed which simultaneously describes the propagation of firing rates and correlations along the chain. A typical result is shown in Fig. 2B. The phase portrait illustrates the spatial evolution of λ and r for several initial conditions λ_0, r_0 (gray trajectories). The map reveals two attractors (black dots): one at a low rate of about 2 Hz and a correlation of almost zero (we call this the "ground state" [8]) and one at large rates and higher correlations (gray area marks its basin of attraction). The thick black line shows the position of the separatrix. Its near horizontal shape is typical for the investigated system and arises as a result of the low correlation gain. This finding explains the results of a former study [8] which revealed that the transition from the asynchronous into the synchronous regime is well determined by a pure rate model ($r=0$). In our network model the stability of the ground state is barely affected by correlation perturbations and maintained for a reasonable range of group sizes w (however, cf. [2]). Due to the restrictions on Poisson processes our model is valid only at low rates and correlations. As a rough estimate for deviations from the Poisson assumption we considered the Fano factors $F(\lambda, r)$ of the total excitatory input process in the different regions of the (λ, r)-space (Fig. 2B).

5. Discussion

In the present study we analyzed the interplay between spike rates and correlations. Transmission characteristics serve as ingredients for a general framework to explore stationary states in cortical networks. Due to the requirement of Poissonian spike count statistics the scope of this framework is restricted to low rates and correlations. Since cross-correlations exhibit finite temporal widths the reduction to a single integrated quantity—the coefficient of correlation—constitutes an important constraint of the present work. In addition, higher-order correlations, e.g. those enforced by pairwise correlations, may play an important role in network dynamics. The present study exclusively incorporates pairwise correlations. As suggested by the example, the designed theory is nevertheless applicable to certain network structures and produces reasonable results.

Acknowledgements

We acknowledge stimulating discussions with M. Abeles, Y. Aviel, M. Herrmann, A. Kuhn, and C. Mehring.

References

[1] M. Abeles, Corticonics: Neural Circuits of the Cerebral Cortex, Cambridge University Press, Cambridge, 1991.

[2] Y. Aviel, E. Pavlov, M. Abeles, D. Horn, Synfire chains in a balanced network, Neurocomputing 44–46 (2002) 285–292.

[3] M. Diesmann, M.-O. Gewaltig, S. Rotter, A. Aertsen, State space analysis of synchronous spiking in cortical neural networks, Neurocomputing 38–40 (2001) 565–571.

[4] L.M. Ricciardi, L. Sacerdote, Diffusion Processes and Related Topics in Biology, Springer, Berlin, 1977.

[5] E. Salinas, T.J. Sejnowski, Impact of correlated synaptic input on output firing rate and variability in simple neuronal models, J. Neurosci. 20 (2000) 6193–6209.

[6] M.N. Shadlen, W.T. Newsome, The variable discharge of cortical neurons: implications for connectivity, computation, and information coding, J. Neurosci. 18 (1998) 3870–3896.

[7] S. Stroeve, S. Gielen, Correlation between uncoupled conductance-based integrate-and-fire neurons due to common and synchronous presynaptic firing, Neural. Comput. 13 (2001) 2005–2029.

[8] T. Tetzlaff, T. Geisel, M. Diesmann, The ground state of cortical feed-forward networks, Neurocomputing 44–46 (2002) 673–678.

Available at www.ComputerScienceWeb.com
POWERED BY SCIENCE DIRECT®

ELSEVIER

Neurocomputing 52–54 (2003) 955–961

NEUROCOMPUTING

www.elsevier.com/locate/neucom

Reliability and bifurcation in neurons driven by multiple sinusoids☆

Peter J. Thomas[a,b,*], Paul H.E. Tiesinga[a,b,c], Jean-Marc Fellous[a,b], Terrence J. Sejnowski[a,b]

[a] *Computational Neurobiology Lab, Howard Hughes Medical Institute, Sloan-Swartz Center for Theoretical Neurobiology, La Jolla, CA, USA*

[b] *The Salk Institute, 10010 N. Torrey Pines Rd, La Jolla, CA 92037, USA*

[c] *Department of Physics and Astronomy, University of North Carolina at Chapel Hill, 184 Phillips Hall, Chapel Hill, NC 27599, USA*

Abstract

The brain produces dynamical rhythms at many frequencies that shift in amplitude and phase. To understand the functional consequences of mixtures of oscillations at the single cell level, we recorded the spike trains from single rat cortical neurons in vitro in response to two mixed sine wave currents. The reliability of spike timing was measured as a function of the relative power, phase and frequencies of the sine wave mixture. Peaks in the reliability were observed at a preferred phase difference, frequency and relative power. These results have a natural interpretation in terms of spike train attractors and bifurcations.

Keywords: Neural code; Phase locking; Precision

1. Introduction

The responses of individual neurons depend on the ongoing background activity that varies with the behavioral state of an animal. The electroencephalogram (EEG) contains multiple rhythms that reflect spatially and temporally coherent neural activity

☆ This work was supported by the Sloan-Swartz Center for Theoretical Neurobiology at the Salk Institute and the Howard Hughes Medical Institute.

* Corresponding author. Salk Institute for Biological Studies, P.O. Box 85800, San Diego, CA 92186-5800, USA. Tel.: +1-858-453-1124; fax: +1-858-587-0417.

E-mail address: pjthomas@salk.edu (P.J. Thomas).

0925-2312/03/$ - see front matter
doi:10.1016/S0925-2312(02)00856-1

[1]. Slow oscillations with strong delta rhythms (0.5–2 Hz) are found in slow wave sleep, oscillations dominated by theta oscillations (4–12 Hz) are related to cognitive processing and cortico-hippocampal interactions, and gamma oscillations (30–80 Hz) are found in the cortex and other brain structures under attentive behavioral conditions [1,3]. Power in the gamma and theta bands is modulated by attention [3]. For example, the responses recorded from a macaque V4 neuron to the same stimulus were compared with two conditions, with attention inside or outside the neuron's receptive field. As attention shifted to the receptive fields of a cluster of neurons, these became more synchronized at high frequencies (30–70 Hz) and less so at low frequencies (0–10 Hz) [3]. In a human visual selective attention task, stimulus presentation induced phase resetting of different frequency components in the EEG [5].

The reason why the brain exhibits these dynamical rhythms is unknown, but recent experiments suggest that rapidly changing correlations may reflect internal events and regulate the flow of neural information, rather than its meaning [7]. The reliability of cortical neurons depends in part on the frequency content of their input [4,9] and pyramidal cells and interneurons are reliable in different frequency ranges when injected with pure sinusoidal currents of varying frequency [2]. The goal is to measure spike time reliability when neurons are stimulated by synaptic inputs containing the types of correlated and synchronous inputs that are observed in vivo. Here we report that reliability is greatly affected by the choice of the relative power, phase and frequencies of two mixed pure sine waves injected into pyramidal cells in vitro.

2. Methods

Experimental protocols were approved by the Salk Institute Animal Care and Use Committee and they conform to USDA regulations and NIH guidelines for humane care and use of laboratory animals. Regularly spiking layer five pyramidal neurons in 350 μm-thick coronal slices of rat pre-limbic and/or infra-limbic cortex were injected with 20 different stimulus waveforms, over multiple trials. Each stimulus waveform was a sum of two sinusoids:

$$I(t) = I_0 + I_1 \sin(2\pi f_1 t + 2\pi\Delta\phi) + I_2 \sin(2\pi f_2 t). \quad (1)$$

The relative phase ($\Delta\phi$), the relative frequency (f_2) and the relative power (I_1 and I_2) were varied (see Fig. 1). Whole-cell patch-clamp recordings using glass electrodes (4–10 MΩ) were performed under visual control at room temperature. Data were acquired in current clamp mode using an Axoclamp 2A amplifier (Axon Instruments, Foster City, CA). Data acquisition and current injection used standard computer protocols [2]. Programs were written in Labview 6.1 (National Instrument, Austin, TX), and data were acquired with a PCI-16-E1 data acquisition board (National Instrument, Austin, TX). Data acquisition rate was 10 kHz. Data were analyzed offline using MATLAB (The Mathworks, Natick, MA).

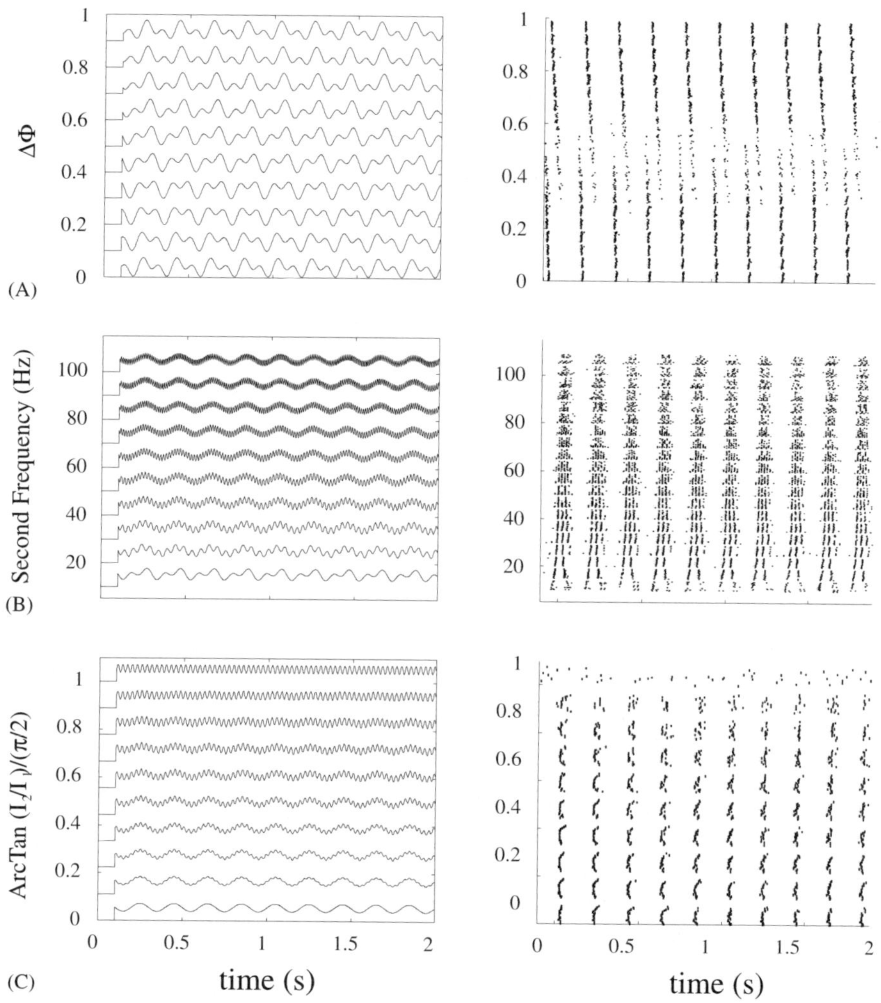

Fig. 1. Sample current stimuli and corresponding cell responses. (A) Variation in relative phase of two sine waves at 5 and 10 Hz. (B) Variation in the frequency of the second sine wave with a 5 Hz frequency for the first sine wave. (C) Variation in relative power between two sine waves at θ (5 Hz) and γ (35 Hz). Larger values correspond to more power in the γ sine wave. Left panels show 10 examples of the 20 stimulus waveforms used, the right panels show corresponding sample rastergrams obtained in response to the injection. Each response (A–C) was obtained from a different layer 5 pyramidal cell.

3. Results

3.1. Experiment 1

Cells were injected with a constant (100 pA) depolarizing current superimposed on the sum of two sinusoids of equal amplitude (50 pA each) with 20 different phase

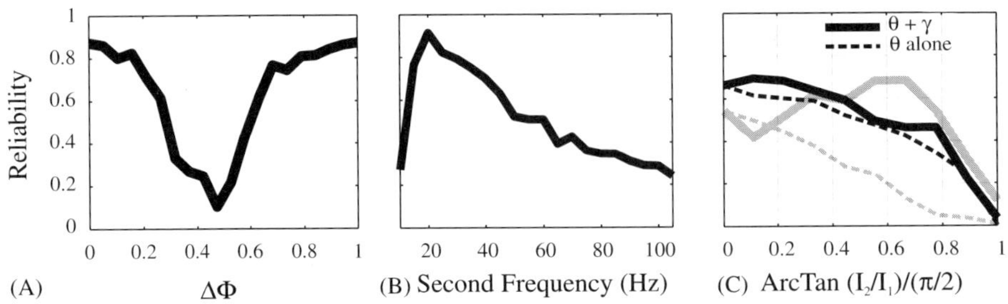

Fig. 2. Effects of parameter changes on reliability: (A) variation in relative phase of two sine waves at 5 and 10 Hz; (B) variation in the frequency of the second sine wave (the frequency of the first sine wave was 5 Hz); (C) variation in relative power between two sine waves at θ (5 Hz) and γ (35 Hz). Larger values correspond to more power in the γ sine wave. Solid curves represent reliability for θ and γ mixture, dashed curves represent the reliability for θ alone, with the same power as for the mixture. Two cells are shown (dark and grey curves).

offsets (ranging from $\Delta\phi = 0.0$ to 0.95) and fixed frequencies of 5 and 10 Hz, respectively. Fig. 1A shows a few of the injected waveforms and a typical cell's response. For $\Delta\phi = 0$, the cell exhibited a 1:1 firing pattern locked to the periodic stimulus, firing regularly at the fundamental frequency of the stimulus (5 Hz). As the phase offset $\Delta\phi$ increased, the cell fired earlier and earlier in the stimulus cycle. For $\Delta\phi = 0.30$, the cell exhibited a new firing pattern with times interspersed between the existing one. As $\Delta\phi$ varied from 0.30 to 0.55 the two sets of firing times shifted gradually while the fraction of trials with spikes lying in one versus the other shifted smoothly as well. At $\Delta\phi = 0.55$, the new pattern dominated, and for larger phase offsets the first pattern no longer occurred. As the phase offset continued to increase the spike times continued to shift until $\Delta\phi = 0.95$ where they nearly coincided with the spike times in the original pattern at $\Delta\phi = 0.0$. The reliability of the cell's response, calculated on a 5 ms time scale using a correlation-based reliability measure [8], showed a pronounced dip at $\Delta\phi = 0.45$ (Fig. 2A).

In an integrate-and-fire model neuron driven by the same stimulus, the spike times also varied smoothly until a critical value of the phase offset at which a *spike time bifurcation* occurred; that is, at the critical value of the phase offset, a small change in this parameter caused a discontinuous change in the resultant pattern of spike times (data not shown). The reliability is expected to show a minimum in the neighborhood of such a bifurcation point, where two alternative spike response patterns coexist. Because of the intrinsic noise of the recorded neuron from stochastic ion channel fluctuations and other noise sources transitions between the two coexisting patterns occur over a range of parameters near the critical parameter value.

3.2. Experiment 2

Cells were injected with a constant (100 pA) depolarizing current superimposed on the sum of two sinusoids of equal amplitudes (50 pA each) and fixed phase offset

($\Delta\phi = 0.5$), at two different frequencies: $f_1 = 5$ Hz and f_2 varied from 10 to 105 Hz in steps of 5 Hz. Fig. 1B shows some of the injected waveforms and a typical cell response. For $f_2 = 10, 15, 20$ Hz most cells fired two spikes on each f_1-cycle. For $25 \leqslant f_2 \leqslant 40$ Hz the cell reliably fired triplet bursts, and for higher values of f_2 it fired irregularly, with clusters of spikes occurring at the frequency f_1. The reliability had a maximum at $f_2 = 20$ Hz (Fig. 2B).

3.3. Experiment 3

Cells were injected with a constant (100 pA) depolarizing current superimposed on the sum of two sinusoids of frequencies $f_1 = 5$ Hz (θ) and $f_2 = 35$ Hz (γ). The relative amplitude of the two components was varied to keep the total stimulus constant, $I_1 = I\cos(\eta)$ and $I_2 = I\sin(\eta)$, with $I = 100$ pA. $\eta = \mathrm{ArcTan}(I_2/I_1)$ took 20 evenly spaced values from 0 to $\pi/2$. Fig. 1C shows 10 of the injected waveforms and the response from one cell. The response to this stimulus varied from cell to cell. For the cell shown here, a single spike occurred in each θ-cycle. Once the γ power I_2 exceeded the θ power I_1, the cell fired in one of two alternative γ cycles on each θ cycle, decreasing reliability.

Fig. 2C shows the reliability (on the 5 ms time scale) of the cell in Fig. 1C (black line) as well as a second cell (gray line). For comparison, the cells were also injected with θ frequency alone at varying values of I_1, with $I_2 = 0$ (dotted lines). Both cells steadily decreased in reliability as I_1 decreased, for $I_2 = 0$. The first cell (black line) had peak reliability for mixtures of θ and γ, with θ predominating. The second cell (gray line) *decreased* in reliability upon shift of power from θ to γ, and then showed a peak in reliability for a mixture of θ and γ with γ predominating.

4. Discussion

Cortical activity often contains multiple distinct frequency bands, whose relative phase [5] and power [3] change dynamically with behavioral state. To understand the effects of varying phase- and power-relationships between different frequency components on neuronal response, we studied the behavior of in vitro neurons driven by periodic stimuli containing multiple superposed frequency components. Our results show that when multiple input frequencies are present, the reliability and precision of the neuronal response is sensitive not only to the frequencies (Fig. 2B) but also to their relative phases (Fig. 2A) and power (Fig. 2C). In an earlier study, the addition of a γ-range frequency to a θ-range rhythm increased spike time precision [6].

Reliability is important for understanding neural coding. The precision of a neural response governs the amount of information about the stimulus that can be communicated and influences the flow of information in the network containing the neuron. Just as reliable timing is necessary for neural codes using the detailed temporal structure of spike-trains, unreliable timing erases information about the stimulus. The ability to modulate reliability by moving in and out of resonance with a given frequency component may be an important means of regulating correlation and gating information flow in cortical networks [7].

Depending on the relative phase relationships of different frequency components, a pyramidal cell can support either a precise spike-time code or a rate code in which spike-times show great variability but spike rate is conserved. This transition in behavior may be understood as a bifurcation between dynamical attractors that occur as the phase-offset varies. The precision and reliability of neural spike-timing depends on the interplay of synaptic input and the internal dynamics of a given neuron. When two dynamical attractors or spike patterns are close enough in phase space that intrinsic neuronal noise can induce transitions between them, then the timing of individual spikes becomes unpredictable. Thus, bifurcations are a general mechanism for controlling neural reliability and information flow.

References

[1] E. Basar, C. Basar-Eroglu, S. Karakas, M. Schurmann, Are cognitive processes manifested in event-related gamma, alpha, theta and delta oscillations in the EEG? Neurosci. Lett. 259 (1999) 165–168.

[2] J.-M. Fellous, A. Houweling, R. Modi, R. Rao, P. Tiesinga, T. Sejnowski, The frequency dependence of spike timing reliability in cortical pyramidal cells and interneurons, J. Neurophys. 85 (2001) 1782–1787.

[3] P. Fries, J.H. Reynolds, A.E. Rorie, R. Desimone, Modulation of oscillatory neuronal synchronization by selective visual attention, Science 291 (2001) 1560–1563.

[4] Z.F. Mainen, T.J. Sejnowski, Reliability of spike timing in neocortical neurons, Science 268 (1995) 1503–1506.

[5] S. Makeig, M. Westerfield, T.P. Jung, S. Enghoff, J. Townsend, E. Courchesne, T. Sejnowski, Dynamic brain sources of visual evoked responses, Science 295 (2002) 690–694.

[6] L.G. Nowak, M.V. Sanchez-Vives, D.A. McCormick, Influence of low and high frequency inputs on spike timing in visual cortical neurons, Cereb. Cortex 7 (1997) 487–501.

[7] E. Salinas, T.J. Sejnowski, Correlated neuronal activity and the flow of neural information, Nat. Rev. Neurosci. 2 (2001) 539–550.

[8] S. Schreiber, J.M. Fellous, D. Whitmer, P. Tiesinga, T.J. Sejnowski, A new correlation-based measure of spike timing reliability, Neurocomputing, this issue.

[9] J.D. Hunter, J.G. Milton, P.J. Thomas, J.D. Cowan, Resonance effect for neural spike time reliability, J. Neurophys. 80 (1998) 1427–1438.

Peter J. Thomas received his Ph.D. in Mathematics from the University of Chicago in 2000, under Jack Cowan. He is currently a Research Associate in the Computational Neurobiology Laboratory at the Salk Institute. His research includes single-neuron dynamics, signal transduction in biochemical networks, pattern formation in cortical maps, and mathematical biology.

Paul Tiesinga studied Theoretical Physics at Utrecht University in the Netherlands. His PhD thesis there was on the dynamics of Josephson-junction arrays. As a postdoctoral fellow at Northeastern University he worked on biophysically realistic modeling of thalamus and hippocampus. While at the Salk Institute in La Jolla as a postdoctoral Sloan fellow he studied the role of synchronized oscillations in cortical information processing. Presently he is an assistant professor in the Department of Physics & Astronomy at the University of North Carolina at Chapel Hill.

Jean-Marc Fellous did his undergraduate studies in Marseille, France. Graduate studies included a Master's degree in artificial intelligence (Paris VI University, France), and a PhD in Computer Science with an emphasis on brain theory and neural networks at the University of Southern California, Los Angeles. Postdoctoral studies included slice neurophysiology and neural modeling in the laboratories of John Lisman (Brandeis University) and Terrence Sejnowski (Salk Institute). Jean-Marc Fellous is currently a Howard Hughes Institute Research Associate in Prof. Sejnowski's laboratory. His interests include the computational roles of neuromodulation and rhythmogenesis.

Terrence Sejnowski is an Investigator with Howard Hughes Medical Institute and a Professor at the Salk Institute for Biological Studies where the directs the Computational Neurobiology Laboratory. He is also Professor of Biology at the University of California, San Diego, where he is Director of the Institute for Neural Computation. Dr. Sejnowski received a B.S. in physics from the Case-Western Reserve University, a M.A. in physics from Princeton University, and a Ph.D. in physics from Princeton University in 1978. In 1988, Dr. Sejnowski founded Neural Computation, published by the MIT press. The long-range goal of Dr. Sejnowski's research is to build linking principles from brain to behavior using computational models.

Available at
www.ComputerScienceWeb.com
POWERED BY SCIENCE DIRECT

ELSEVIER

Neurocomputing 52–54 (2003) 963–968

NEUROCOMPUTING

www.elsevier.com/locate/neucom

Is burst activity in cortical slices a representative model for epilepsy?

Wim van Drongelen[a,*], Henner Koch[b], Charles Marcuccilli[a], Kurt Hecox[a], Jan-Marino Ramirez[b]

[a] *Department of Pediatrics, The University of Chicago, 5841 South Maryland Avenue MC 3055, Chicago, IL 60637, USA*

[b] *Department of Organismal Biology and Anatomy, The University of Chicago, 1027 E-57th Street, Chicago, IL 60637, USA*

Abstract

Neocortical slices including sensory cortex of mouse were used to study cellular and network activity during bicuculline evoked seizure-like activity. The relationship between the activities of single cells and the network was quantified using entropy measures of spike trains. The network shows a large increase in synchronous burst activity during the seizure-like phase. Surprisingly, the individual cells do not seem to follow this pattern of increased synchrony. It is hypothesized that the recruitment of silent units during the seizure onset may explain this paradoxical finding. Our data agrees with recent findings in experimental seizures and intra-operative recordings in humans.

1. Introduction

Epilepsy is the most common serious pathological condition in neurology [2]. From the standpoint of electrophysiology, seizures are frequently associated with synchronized hyperactivity in neural networks and high levels of cellular depolarization. Synchronization in the nervous system is not always pathological. The normal sleep state is characterized by a slow and highly coordinated activation of millions of neurons [6], while the awake state is characterized by faster and less coordinated activity of neurons.

* Corresponding author.
E-mail address: wvandron@peds.bsd.uchicago.edu (W. van Drongelen).

0925-2312/03/$ - see front matter © 2003 Elsevier Science B.V. All rights reserved.
doi:10.1016/S0925-2312(02)00859-7

Since synchronization, high levels of activity and membrane depolarization are found in both epileptic and normal bursts, quantification of various aspects of burst activity might help to detect different mechanisms between normal and pathologic activity. To study burst activity at the tissue level, we performed simultaneous extra- and intracellular electrophysiological recordings on neocortical slices. This enabled us to quantify to what extent the cells are synchronized with a network. We evoked seizure-like bursts by application of bicuculline, an antagonist of the $GABA_A$ receptor [1].

2. Methods

Neonatal (P8-13) male or female CD-1 mice were deeply anesthetized with ether, decapitated at the C3/C4 spinal level and the cortex isolated in ice-cold artificial cerebrospinal fluid (ACSF). Neocortical slices (500 μm thick), encompassing the somatosensory cortex, were sectioned and transferred into a recording chamber and submerged under a stream of ACSF (temperature, 30°C; flow rate, 10 ml/min) containing (mM/l): 118 NaCl, 3 KCl, 1.5 $CaCl_2$, 1 $MgSO_4$, 25 $NaHCO_3$, 1 NaH_2PO_4 and 30 D-glucose equilibrated with carbogen (95% O_2 and 5% CO_2). After 30 min the potassium concentration was raised from 3 to 5 mM/l to obtain spontaneous rhythmic activity.

An overview of the experimental setup is shown in Fig. 1. Population activity was recorded extracellularly with suction electrodes positioned onto the surface of the cortical layers II/III. The signals were amplified 2000 times, filtered between 0.25 and 1.5 kHz, rectified and integrated using an electronic filter (time constant of 30 –50 ms). With these settings, the filter output was a measure for extracellular spike activity. Intracellular patch-clamp recordings were obtained from cortical neurons with the blind-patch technique. Patch electrodes were filled with a solution containing (mM/l): 140 K-gluconic acid, 1 $CaCl_2*6H_2O$, 10 EGTA, 2$MgCl_2*6H_2O$, 4Na_2ATP, 10 HEPES. In several recording sessions biocytin (4.5 mg/ml) was added to the solution. The recordings were low-pass filtered (0–2 kHz, Bessel 4-pole filter, −3 dB). An example of simultaneous intracellular recording and population activity is depicted in Fig. 1. The cells described in this study had membrane potentials ranging from −57 to −68 mV and action potentials with an overshoot.

At concentrations between 3 and 5 mM/l K^+ in the ACSF, the slice preparation generated rhythmic synchronized population activity. This activity was used as the baseline (BLN). During the BLN phase, the intrinsic firing properties of the neuron were determined by injecting depolarizing currents. After the blockade of GABA-ergic synaptic transmission with bicuculline (10 μM/l), seizure-like activity (SLA) bursts of population activity were generated, followed by a steady state (STD) of ongoing activity. An example of an extracellular recording including the three phases: BLN, SLA, and STD can be seen in Fig. 1B. In one of six recordings, the network did not transition to STD. This recording was not included in the quantitative analysis.

Signal processing was applied to describe the entropy and information content in the signals during different bursting patterns. Bursts were detected as they exceeded a visually determined threshold (Fig. 1C). The population activity and cellular signals

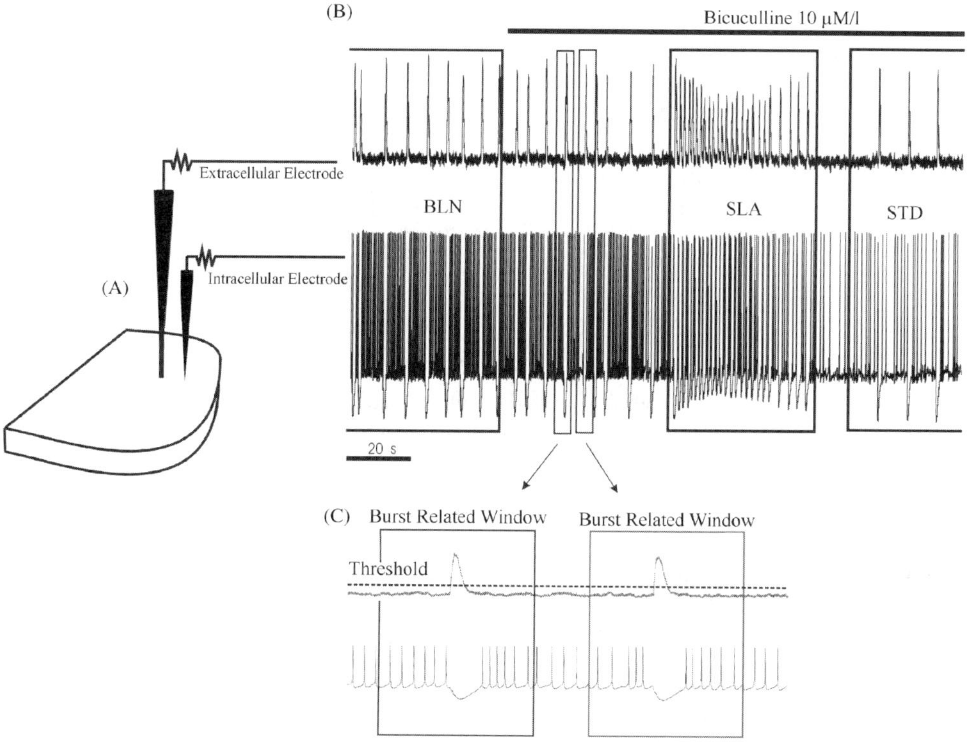

Fig. 1. Overview of the experimental setup and recording. (A) Intra- and extracellular electrodes were placed in a neocortical slice. (B) This setup allowed us to record an integrated measure of extracellular bursting (upper trace) and cellular activity (lower trace). After adding bicuculline, the activity transitioned from BLN to a SLA. In five out of six cases, SLA spontaneously transitioned into a STD of activity and (C) Details from (B) show the population burst and the cellular activity in analysis windows around the network burst.

were averaged and aligned on the basis of the activity bursts in the population (burst related window, Fig. 1C). Spike activity in each of the trials in the average was determined. An example of a series of burst-aligned trials is shown in the raster plot in Fig. 2. To study information content in the cellular activity, entropy was calculated using Shannon's formula for entropy:

$$H = -\sum_{i=0}^{\infty} p(i)\log_2(p(i)),$$

where H is the entropy in bits and $p(i)$ the probability of observing the ith activity pattern. The method we applied is similar to the one described by Reich and coworkers [5]. In short, the spike trains in all trials were aligned according to the population burst and subdivided into time bins that varied in size from 5 to 200 ms . The probabilities to observe i spikes was determined from the spike counts in the bins. The total entropy was determined by taking into account all bins, and represents the variability across time (H_t, Fig. 2). The noise entropy was determined by taking into account all bins

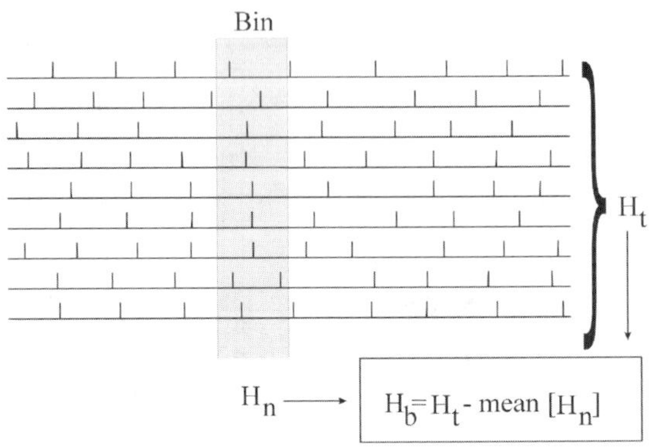

Fig. 2. Raster plot of subsequent spike trains aligned with the network burst. A value for the total entropy (H_t) is determined across time. A noise entropy value (H_n) is estimated across trials for each bin. Burst related entropy (H_b) is determined by subtracting the mean noise entropy from the total entropy.

across trials at a fixed time (H_n, Fig. 2). We used the mean value of all noise estimates as the noise estimate for further analysis. The bin size that generates the highest noise entropy rate was selected. The values were calculated in bits and were normalized by dividing by the bin size (bits/s). Burst-related information was calculated as the difference between the total and noise entropy:

$$H_b = H_t - \text{mean}[H_n].$$

Each of the values (H_t, H_n, H_b) was determined for each phase (BLN, SLA, STD) and for each cell. The significance of the changes of the measures between the phases was determined with a Wilcoxon test.

3. Results

The average network burst frequency during the BLN and STD is around 0.1 Hz and increases to and average of 0.4 Hz during the SLA. These differences in burst rate were significant ($p < 0.05$). The amplitude of the individual network bursts did not change significantly between states.

Five of the neurons were (according to their intrinsic properties) regular spiking neurons, one was an intrinsically bursting neuron. During the 'spontaneous' activity, the intrinsic burster could not be distinguished from the other neurons. Two regular spiking neurons we patched were labeled with biocytin, and found to be layer II/III pyramidal neurons. The relationship between the cellular activity and the population was recorded in each of the phases. A typical example of cellular BLN activity at the population burst (Fig. 3A) shows an excitation followed by inhibition. This is reflected in the time-course of the average membrane potential and the spike activity (firing rate and superimposed spike trains). During the SLA (Fig. 3B) the inhibition is earlier in the burst, and the excitation pattern becomes irregular. The STD (Fig. 3C) shows a strong synchrony between cell and population.

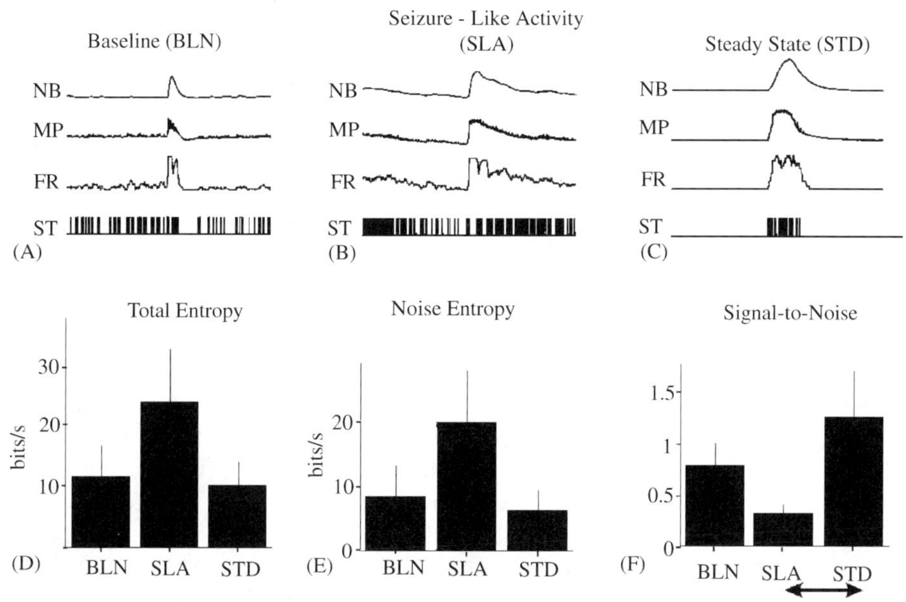

Fig. 3. Average results during the baseline (BLN, A), the seizure-like activity (SLA, B) and steady state (STD, C). Each graph shows the average network burst waveform (NB), the average membrane potential (MP), and the average firing rate (FR). The lower trace in each panel are the superimposed spike trains (ST). The total entropy (D), the noise related entropy (E), and the signal-to-noise ratio (F). The arrow indicates a significant difference ($p < 0.05$).

The total entropy per second (Fig. 3D) is highly correlated with the spike rate ($r = 0.99$, $p < 0.001$). The largest part of the total entropy is a noise component, i.e. not-burst-related (Fig. 3E). The difference between total entropy and noise entropy is the burst-related component, which is a measure for the similarity between the action potential activities of the cell and the network. The ratio between signal and noise (S/N) is minimal during the SLA and maximal during the STD (Fig. 3F). The difference between these maximum and minimum values is significant ($p < 0.05$).

4. Discussion

We demonstrated that the relationship between cellular and network activity can be quantified by entropy values of the spike trains, and that the S/N parameter changes from the SLA to the STD (Fig. 3F). It is remarkable that there is no significant increase in the burst-related entropy or S/N parameter during the SLA as compared to the BLN (signal-to-noise, S/N$= H_b/H_n$). This indicates that the activity of single cells do not become massively synchronized with the network at the onset of the SLA. Both visual inspection of the average activity (Fig. 3A–C) and the significant change in S/N (Fig. 3F) suggest that increased synchrony between unit and network occurs later during the STD.

The network burst rate is significantly increased during the SLA ($p < 0.05$). The amount of synchronous behaviour of the network can be assumed proportional to the product of burst rate and average amplitude of the burst. Considering the extracellular activity of the network during the SLA, one can therefore conclude that the synchronous network activity per unit time increases significantly during SLA ($p < 0.05$). Since this phenomenon cannot be explained from the activity of the individual units, there must be another factor playing a role. We think that the most likely explanation is an increased number of active units during seizure onset. Indeed an increase of evoked population activity after administration of bicuculline was shown in an intrinsic optical imaging study by Kohn et al. [3]. Due to recruitment of previously silent neurons, the network shows increased bursting during the SLA. Under this assumption, the SLA is characterized by a large number of active neurons, and each may show a low synchrony with the network bursting activity. Low synchrony between individual neurons during seizure like discharges was described by Netoff and Schiff [4]. Although in a general sense, our findings are similar to the results of this study, a detailed comparison is difficult because these authors used a different definition of SLA, and focused on the relationship between the membrane potential of cells in which the action potential generation mechanism was disabled.

Our results are in agreement with intra-operatively recorded seizure-related neuronal activity in humans, where low synchrony was found to be present at seizure onset [7]. It is compelling that recordings in slices demonstrated an important similarity between SLA and seizures in the intact brain. This similarity is an indication that (for some aspects) SLA may be a representative model for epilepsy, although at this stage it would be premature to answer the question in our title.

Acknowledgements

This work was supported in part by a grant from the Falk Center for Advanced Study and Care of Pediatric Epilepsy.

References

[1] B.W. Connors, Y. Amitai, Generation of epileptiform discharge by local circuits of Neocortex, in: P.A. Schwartzkroin (Ed.), Epilepsy, Cambridge University Press, New York, 1993, pp. 388–423.

[2] W.A. Hauser, J.F. Annegers, L.T. Urland, Prevalence of epilepsy in Rochester, Minnesota 1940–1980, Epilepsia 32 (1991) 429–445.

[3] A. Kohn, C. Metz, M. Quibrera, M.A. Tommerdahl, B.L. Whitsel, Functional neocortical microcircuitry demonstrated with intrinsic signal optical imaging in vitro, Neuroscience 95 (2000) 51–62.

[4] T.I. Netoff, S.J. Schiff, Decreased neural synchronization during experimental seizures, J. Neurosci. 22 (2002) 7297–7307.

[5] D.S. Reich DS, F. Mechler, J.D. Victor, Formal and attribute-specific information in primary visual cortex, J. Neurophysiol. 85 (2001) 305–318.

[6] M.V. Sanchez-Vives, D.A. McCormick, Cellular and network mechanisms of rhythmic recurrent activity in neocortex, Nature Neurosci. 3 (2000) 1027–1034.

[7] A.R. Wyler, G.A. Ojemann, A.A. Ward, Neurons in human epileptic cortex: correlation between unit and EEG activity, Ann. Neurol. 11 (1982) 301–308.

Available at
www.ComputerScienceWeb.com
POWERED BY SCIENCE DIRECT

ELSEVIER

Neurocomputing 52–54 (2003) 969–975

NEUROCOMPUTING

www.elsevier.com/locate/neucom

Circuit properties of the cortico-mesocortical system

Koki Yamashita, Shoji Tanaka*

Department of Electrical and Electronics Engineering, High-Tech Research center, Sophia University, 7-1 Kioicho, Chiyoda-ku, Tokyo 102-8554, Japan

Abstract

It has been shown that dopamine (DA) modulates memory fields of dorsolateral prefrontal cortex (PFC) neurons. The DAergic neurons which project to the PFC are localized in the midbrain. We here developed a computational model network which includes the PFC circuit and the mesencephalic DA unit to analyze the circuit property of the cortico-mesocortical system. In our computer simulation, the cortico-mesocortical system can regulate the DA level in the PFC and the sustained activity of the PFC neurons during the delay period. The simulation suggests that the stabilization of the DA level requires weak cortical feedback.

Keywords: Cortico-mesocortical; Dopamine; Midbrain; Prefrontal cortex; Working memory

1. Introduction

Recent studies have suggested that the dopaminergic (DAergic) modulation via D1 receptor activation in the prefrontal cortex (PFC) plays critical roles not only in the regulation of memory fields [1,11] but also fundamental cognitive operations [9,10]. Since the DAergic innervation of the frontal cortex comes from the midbrain [4,12], the investigation of the cortico-mesocortical system is useful to understand the mechanisms of the DAergic modulation of the cortical dynamics. Although interests in the DAergic modulation of the PFC have been growing, the circuit properties of the cortico-mesocortical system are still controversial. These circuit properties need to be examined in detail. In particular, the control of fundamental cognitive operations by DA has been proposed recently by Tanaka [10]. The question we ask here is how the

* Corresponding author: Tel.: +81-3-3238-3331; fax: +81-3-3238-3321.
E-mail address: tanaka-s@sophia.ac.jp (S. Tanaka).

0925-2312/03/$ - see front matter
doi:10.1016/S0925-2312(02)00865-2

DA release is controlled by the cortico-mesocortical system. To address this question, we construct a network model of the system. The part of model that includes the DAergic effects in the PFC is based on our previous study [13].

2. Model

2.1. Prefrontal cortex

Our PFC model contains 1080 pyramidal neurons and 240 inhibitory interneurons. All of the neurons are described by a single compartment, leaky integrate-and-fire neuron model [3,5,8–10,13]. We here simulate the oculomotor delayed-response (ODR) task (fixation period: 0–200 ms; cue period: 200–300 ms; delay period: 300–3000 ms). The DAergic modulation on each conductance is based on the experimental studies [2,6,7] and our previous research [9,10,13]. In the cases of g_{AMPA}, g_{NMDA}, and $g_{\mathrm{K(Ca)}}$, the DAergic modulation is expressed by: $g_{\mathrm{replace}} = r_{\mathrm{conductance}} g_{\mathrm{max,original}}$. Here $r_{\mathrm{conductance}}$ is the DA concentration-dependent coefficient which obeys the following equation: $r_{\mathrm{conductance}} = AZ_{\mathrm{DA}}^4 + BZ_{\mathrm{DA}}^3 + CZ_{\mathrm{DA}}^2 + DZ_{\mathrm{DA}} + E$ (Fig. 1A, Table 1). The effect on the

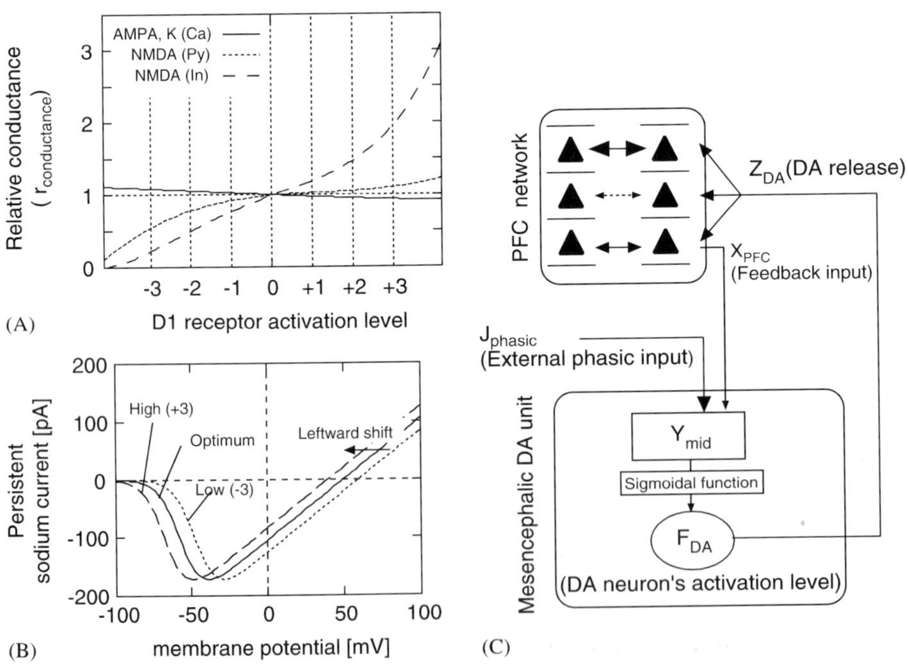

Fig. 1. (A) DAergic effects on the conductance values, (B) the DA-induces leftward shift of the persistent sodium V-I curve and (C) the architecture of the model network. The PFC network consists of 3 layers, and contains 360 pyramidal neurons in each layer and 240 inhibitory interneurons. The mesencephalic DA unit receives the feedback input from the deep layer pyramidal neurons (X_{PFC}), and the DA release level (Z_{DA}) is calculated by Eqs. (1)–(3).

Table 1
Coefficients of the polynomials (Fig. 1A)

Conductance	A	B	C	D	E
AMPA	0.000	0.015	−0.033	−0.092	1.110
NMDA(Py)	0.000	0.351	−1.375	1.902	0.112
NMDA(In)	0.742	−2.278	2.190	0.333	0.000
K(Ca)	0.000	0.015	−0.033	−0.092	1.110

persistent sodium current is characterized by the leftward shift of the voltage–current curve (Fig. 1B). The DAergic effects on the PFC neurons are due to the D1 receptor activation level.

2.2. Mesencephalic DA unit

Our model describes the interaction between the PFC and the midbrain by including the direct projection from the PFC to the mesencephalic DA unit (Fig. 1C). The model mesencephalic DA unit is initially driven by the external phasic input (J_{phasic}), and then send output to the PFC neurons to release DA. The DA release level in the PFC, $Z_{\text{DA}}(t)$, is given by

$$\tau_{\text{DA}} \frac{\mathrm{d}Z_{\text{DA}}(t)}{\mathrm{d}t} + Z_{\text{DA}}(t) - F_{\text{DA}}(Y_{\text{mid}}(t)) = 0, \tag{1}$$

$$\tau_{\text{P-M}} \frac{\mathrm{d}Y_{\text{mid}}(t)}{\mathrm{d}t} + Y_{\text{mid}}(t) - B_{\text{gain}} X_{\text{PFC}}(t) - J_{\text{phasic}} = 0, \tag{2}$$

$$F_{\text{DA}}(Y_{\text{mid}}(t)) = \frac{K}{1 + \exp(-(Y_{\text{mid}}(t) - sigm))}, \tag{3}$$

where the parameters are defined in Table 2.

Table 2
Model parameters

Symbol	Description
$\tau_{\text{P-M}}$	Time constant of the DA unit activation: 10 (ms)
B_{gain}	Feedback gain (from the PFC to the mesencephalic DA unit): 1.1 (a.u.)
J_{phasic}	Amplitude of the external phasic input: 100 (a.u.)
K	Maximum value of the sigmoidal function (DA neuron's activity): 87 (a.u.)
Sigm	Intermediate point of the sigmoidal function (DA neuron's activity): 43.5 (a.u.)
τ_{DA}	Time constant of the DA release: 800 (ms)

3. Results

3.1. *Modulation of PFC neurons' activity via D1 receptor activation*

While the D1 receptor activation level was in the optimum range, the model PFC neurons showed the delay-period activity owing to the recurrent excitation (Fig. 2B). However, the results in the cases of lower and higher receptor activation indicated suppressed firing of the PFC neurons (Fig. 2A,C). Moreover, the PFC neurons did not show sustained activity outside the sustainment range (between −3 and +3). These results show clearly how D1 receptor activation modulates the memory fields. This modulatory effect is characterized by the biphasic modulation (Fig. 2D), which is consistent with experimental findings [1,11] and our previous study [13].

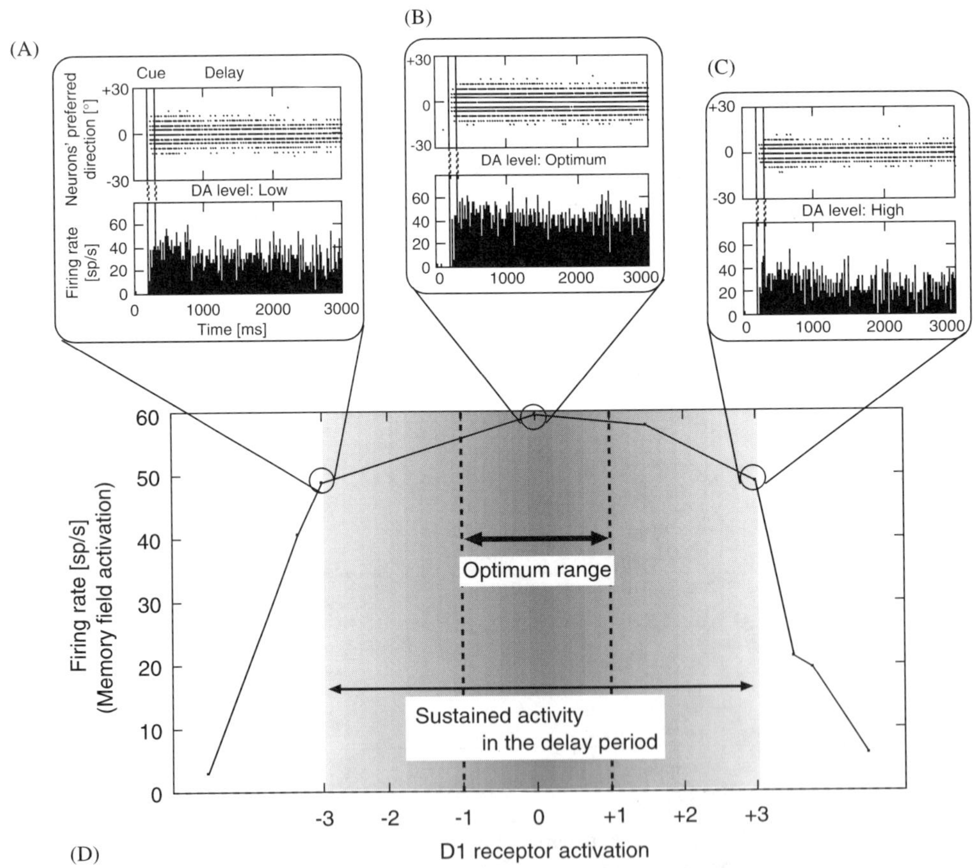

Fig. 2. Activity of the deep layer pyramidal neurons at the low (A), optimum (B), and high (C) levels of D1 receptor activation. The time bin width is 10 ms and (D) the biphasic DAergic modulation of the PFC neurons obtained in this simulation.

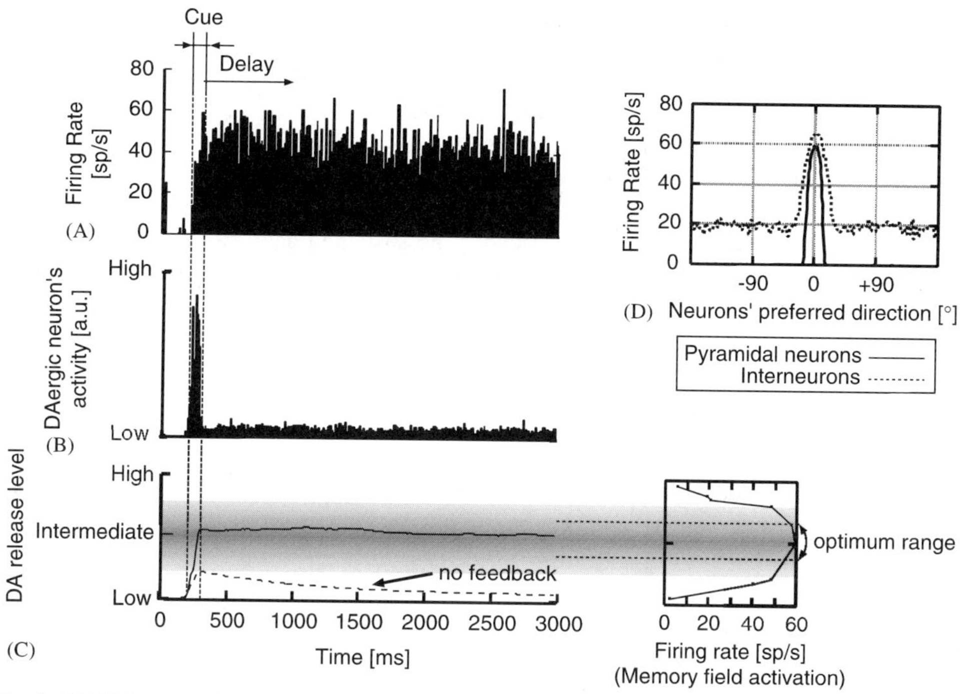

Fig. 3. (A) Firing rate of the pyramidal neurons in the deep layer of the PFC, (B) the activity of the DAergic neuron. The DAergic neuron shows the sustained activity at very low frequency after the burst activity. The time bin width is 10 ms, (C) the DA release level. (D) the activity profiles of the pyramidal neurons and the interneurons. The DAergic neuron shows the sustained activity at very low frequency following the burst activity. The DA release was stabilized at the optimum level during the delay period. The time bin width is 10 ms.

3.2. Regulation of DA release

As mentioned above, the model PFC has the inverted-U shaped property (Fig. 2D). With this issue in mind, we here investigate the mesencephalic DA unit and the DA release. In our simulation, the DAergic neuron shows two kinds of activity modes, which are the burst activity and the sustained activity at very low frequency (Fig. 3B). The burst activity occurred during the cue period in which the phasic input activated the mesencephalic DA unit. During the delay period, the DAergic neuron's activity was sustained but at a much lower firing rate than the initial bursting component. Nevertheless, the DA release level was stabilized in the optimum range (Fig. 3C). The DA release level was rapidly increased as the DAergic neuron exhibited the burst activity, and then this level was stabilized although the DAergic neuron's activity was fairly low during the delay period. As for the PFC neurons' activity, Fig. 3A and D show their well-tuned delay-period activity in this simulation.

4. Discussion

This study investigated the circuit properties of the cortico-mesocortical system. Our simulation showed that the DAergic neuron exhibits two kinds of activity modes

(Fig. 3B). The cause of the burst activity of the DAergic neuron is the external phasic input. On the other hand, the sustained activity at low frequency resulted from the cortical feedback input. Because the feedback gain was fairly low in our simulation, the activity of the DAergic neuron was much lower during the delay period than the initial bursting component (Fig. 3B). The DA level was stabilized in the optimum range of the inverted-U shaped curve only when the regulated cortical feedback input was present in the mesencephalic DA unit (Fig. 3C). That is, the feedback input would be necessary to control the DA release level. The sustained activity in the PFC is thus regulated such optimized DA release during the delay period. In conclusion, we suggest that the cortico-mesocortical closed-loop circuit regulates the DA release level in the PFC, and this system would stabilize the sustained activity of PFC neurons.

Acknowledgements

This work was supported by Grants-in-Aid for Scientific Research on Priority Areas to S. T. (#13210123 and #14017083) from the Ministry of Education, Science, and Technology, Japan.

References

[1] P.S. Goldman-Rakic, E.C. Muly, G.V. Williams, D1 receptors in prefrontal cells and circuits, Brain Res. Rev. 31 (2000) 295–301.

[2] N.A. Gorelova, C.R. Yang, Dopamine D1/D5 receptor activation modulates a persistent sodium current in rat prefrontal cortical neurons in vitro, J. Neurophysiol. 84 (2000) 75–87.

[3] M. Iida, S. Tanaka, Postsynaptic current analysis of a model prefrontal cortical circuit for multi-target spatial working memory, Neurocomputing 44–46 (2002) 855–861.

[4] D.A. Lewis, S.R. Sesack, Handbook of Chemical Neuroanatomy, Vol. 13, The Primate Nervous System, Part I, Elsevier Science, Amsterdam, 1997, pp. 263–375.

[5] K. Morooka, S. Tanaka, Correlation analysis of signal flow in a model prefrontal cortical circuit representing multiple target locations, Neurocomputing 44–46 (2002) 541–548.

[6] P. Pedarzani, J.F. Storm, Dopamine modulates the slow Ca^{2+}-activated K^{+} current I_{AHP} via cyclic AMP-dependent protein kinase in hippocampal neurons, J. Neurophysiol. 74 (1995) 2749–2753.

[7] J.K. Seamans, D. Durstewitz, B.R. Christie, C.F. Stevens, T.J. Sejnowski, Dopamine D1/D5 receptor modulation of excitatory synaptic inputs to layer V prefrontal cortex neurons, Proc. Nat. Acad. Sci. USA 98 (2001) 301–306.

[8] S. Tanaka, Computational approaches to the architecture and operations of the prefrontal cortical circuit for working memory (Review), Prog. Neuro-Psychopharm. Biol. Psychiat. 25 (2001) 259–281.

[9] S. Tanaka, Multi-directional representation of spatial working memory in a model prefrontal cortical circuit, Neurocomputing 44–46 (2002) 1001–1008.

[10] S. Tanaka, Dopamine controls fundamental cognitive operations of multi-target spatial working memory, Neural Networks 15 (2002) 573–582.

[11] G.V. Williams, P.S. Goldman-Rakic, Modulation of memory fields by dopamine D1 receptors in prefrontal cortex, Nature 376 (1995) 572–575.

[12] S.M. Williams, P.S. Goldman-Rakic, Widespread origin of the primate mesofrontal dopamine system, Cereb. Cortex 8 (1998) 321–345.

[13] K. Yamashita, S. Tanaka, Circuit simulation of memory field modulation by dopamine D1 receptor activation, Neurocomputing 44–46 (2002) 1035–1042.

Koki Yamashita received B.E. from Sophia University, Tokyo, in 2001. He is a graduate student at the Program of Electrical and Electronics Engineering, Sophia University. He is currently studying computational neuroscience and electrical engineering.

Shoji Tanaka received B.E., M.E., and Ph.D., degrees from Nagoya University, Japan. He is Professor at the Department of Electrical and Electronics Engineering at Sophia University. During 1998–1999, he was a Visiting Scientist at the Section of Neurobiology, Yale University School of Medicine, USA.

Available at
www.ComputerScienceWeb.com
POWERED BY SCIENCE DIRECT

ELSEVIER

Neurocomputing 52–54 (2003) 977–982

NEUROCOMPUTING

www.elsevier.com/locate/neucom

Localization of epileptic foci by means of cortical imaging using a spherical head model

X. Zhang[a], W. van Drongelen[b], K. Hecox[b], V.L. Towle[c,d], D.M. Frim[d], A. McGee[b], J. Lian[a], B. He[a,*]

[a] *Department of Bioengineering, University of Illinois at Chicago, MC-063, 851 S. Morgan Street, Chicago, IL 60607, USA*

[b] *Department of Pediatrics, The University of Chicago, Chicago, USA*

[c] *Department of Neurology, The University of Chicago, Chicago, USA*

[d] *Department of Surgery, The University of Chicago, Chicago, USA*

Abstract

We have applied a cortical imaging technique (CIT) with a three-sphere head model to estimate cortical potentials from scalp electroencephalogram (EEG) recordings during interictal spikes in a pediatric epilepsy patient. The CIT analysis was performed during the ascending limb of interictal spikes and localized areas of activity were observed, overlying the epileptogenic zone, as being confirmed by the electrocorticography recordings and neurosurgical resections of the patient. The present study suggests that CIT may become a useful alternative for noninvasive localization of intracranial sources generating epileptiform activity from pre-operative scalp EEG recordings.

Keywords: Cortical imaging technique; Source localization; Epilepsy; EEG

1. Introduction

Long-term electroencephalogram (EEG) and video monitoring of epileptic patients is a well-established procedure in the diagnosis and management of epilepsy. Due to the volume conduction effect of the skull, the scalp EEG is severely smeared, thus not being able to provide spatial details on the underlying brain electrical activity [3,11]. It is important to accurately localize the sites of seizure onset to improve the surgical

* Corresponding author. Tel.: +312-413-0030; fax: +1-312-996-5921.
E-mail address: bhe@uic.edu (B. He).

0925-2312/03/$ - see front matter
doi:10.1016/S0925-2312(02)00868-8

outcome and avoid neurologic deficits from surgical resections. Thus, electrocorticography (ECoG) has been employed for the direct inspection of the epileptogenic zone on the cerebral cortex and has become a kind of 'golden standard' for defining epileptogenic brain regions. However, invasiveness, risk of morbidity, limited availability, and the high cost of this technique limit its use in the clinical routine. Therefore, it is desirable to significantly enhance the spatial resolution of the scalp EEG, thus enabling a better prediction of the location of epileptic foci from noninvasive scalp EEG monitoring.

A recently developed cortical imaging technique (CIT) [6] can link the cortical potentials with scalp potentials by a lead field matrix. By solving the EEG inverse problem, the cortical potentials can be estimated from the scalp potentials with enhanced spatial resolution, by reducing the blurring effect of the head volume conductor. Compared with the 3-D inverse solutions [5,7,10,12,13,16], the CIT approach has the unique feature that it can be compared directly with the ECoG recordings. In the present study, we examine the feasibility of applying the previously developed CIT for noninvasive localization of epileptic foci from pre-operative EEG measurements.

2. Methods

2.1. Data acquisition

A 16-year-old male patient with medically refractory seizures was included in the present study according to a protocol approved by the IRB of The University of Illinois at Chicago and the University of Chicago. The long-term video EEG monitoring was conducted in the Pediatric Epilepsy Center at the University of Chicago Children's Hospital. Pre-operative scalp EEG (24 channels) was sampled at 400 Hz and band-pass filtered from 1–100 Hz. The scalp electrodes were uniformly distributed over the scalp according to the 10–10 system. The intracranial electrode pads were placed directly on the surface of brain, and post-operative ECoG was recorded for intracranial monitoring [15]. The artifact-free interictal spikes from pre-operative EEG data were visually detected. CIT analysis was applied to time points from 20 ms before to 5 ms after the largest negative peak.

2.2. Data analysis

The previously developed CIT algorithm was detailed in Ref. [6]. Briefly, the head volume conductor was approximated using an inhomogeneous three-concentric-sphere model, in which the radii of the brain, the skull, and the scalp spheres were taken as 0.87, 0.92, and 1.0, respectively (Fig. 1). The normalized conductivity of the scalp and the brain was taken as 1.0, and that of the skull as $\frac{1}{80}$. Each surface of the three spheres was discretized uniformly into 1280 triangular elements. The cortical potential $\boldsymbol{\Phi}$ and measured scalp potential $\boldsymbol{U}$ can be directly connected by lead field matrix $\boldsymbol{A}$ as $\boldsymbol{U} = \boldsymbol{A}\boldsymbol{\Phi}$. By solving the inverse problem using zero-order Tikhonov regularization

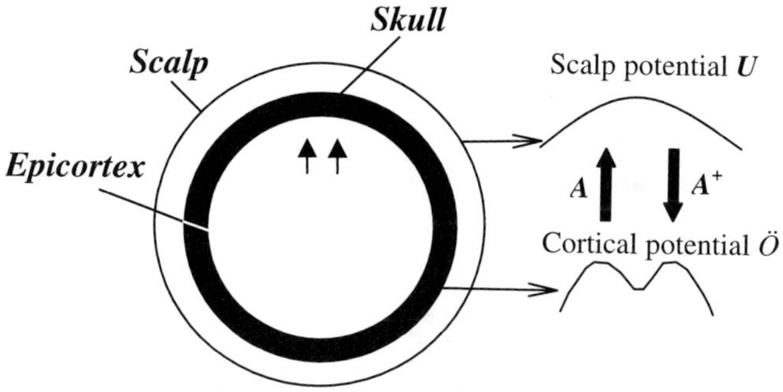

Fig. 1. Schematic diagram of the CIT.

[14], the cortical potentials can be estimated from the scalp potentials with enhanced spatial resolution as

$$\boldsymbol{\Phi} = \boldsymbol{A}^{+}\boldsymbol{U} = (\boldsymbol{A}^{\mathrm{T}}\boldsymbol{A} + \lambda\boldsymbol{I})^{-1}\boldsymbol{A}^{\mathrm{T}}\boldsymbol{U},$$

where $^{+}$ represents pseudoinverse, ^{T} represents matrix transpose, and $\boldsymbol{I}$ represents identity matrix.

The regularization parameter λ which is used to suppress the effect of noise, is determined by the L-curve approach [2].

3. Results

The intracranial monitoring of the patient suggested that the epileptogenic zone was in the left frontal lobe. Neurosurgical resection of a large portion of his left frontal lobe reduced this patient's seizure from many times per day to occasional once per week or per two weeks. One typical waveform segment of interictal spike and the scalp potential distributions at three time points are shown in Fig. 2. The largest negative peak was recorded at Fz, and was marked by black dot in Fig. 2(a). Notably, the scalp EEG maps showed a widely distributed pattern, and did not reveal the spatial details on the underlying epileptogenic zone. The CIT analysis results are shown in Fig. 3. At the correspondent time points with those in Fig. 2, the estimated cortical potentials were obtained with enhanced spatial resolution. It also clearly revealed that the activity was originated from the left superior frontal lobe, propagated toward midsaggital line, and spread there. The CIT results are consistent with the intracranial ECoG recordings, and are confirmed by the surgical resection.

4. Discussion

In this pilot study, we have applied the CIT [6] to estimate the cortical potentials during interictal spikes from pre-operative EEG recordings in a pediatric epilepsy

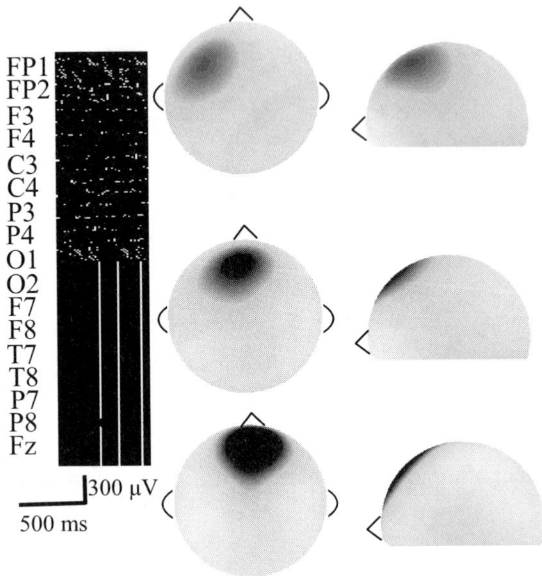

Fig. 2. (a) Waveforms of an interictal spike in an epilepsy patient. The largest negative peak is marked by the black dot and set to be 0 ms. (b) Top view of the scalp potential distributions at different time points (from top to bottom, −10, −5 and 0 ms). (c) Left-side view of the scalp potential distributions at corresponding time points.

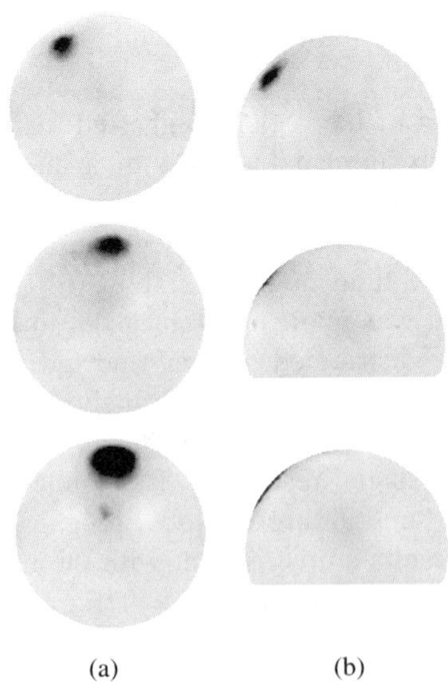

Fig. 3. Cortical potential imaging results: (a) Top view of the estimated cortical potential distributions by CIT at different time points (from top to bottom, −10, −5, 0 ms); (b) Left-side view of the estimated cortical potentials at the correspondent time points. The negative activity overlies with epileptogenic zone on the left frontal lobe.

patient. Enhanced spatial resolution was achieved in the estimated cortical potential maps as compared to the blurred scalp potential maps, thus facilitating noninvasive localization of the intracranial sources of epileptiform activity. Promising results have been obtained in the present study, in that the estimated cortical potential maps revealed localized area of activity overlapping with the cortical region displaying epileptiform activity, as being confirmed by the surgical resection of the patient. In comparison with the cortical current imaging [1,9], the present CIT analysis allows estimation of cortical potentials [3,4,6,8], thus providing a means of comparison with the ECoG [8]. More accurate inverse estimation would be expected to be achieved by employing more scalp recording electrodes, and by using the realistic geometry head model constructed from the patient's magnetic resonance images [8]. The present promising results suggest that the CIT approach merits further investigation for its application to noninvasive localization of epileptic foci from pre-operative EEG monitoring.

Acknowledgements

This work was supported in part by NIH EB00178, NSF CAREER Award BES-9875344 and a Falk grant.

References

[1] A.M. Dale, M.I. Sereno, Improved localization of cortical activity by combining EEG and MEG with MRI cortical surface reconstruction: a linear approach, J. Cog. Neurosci. 5 (1993) 162–176.

[2] P.C. Hansen, Analysis of discrete ill-posed problems by means of the L-curve, SIAM Rev. 34 (1992) 561–580.

[3] B. He, Brain electric source imaging—Scalp Laplacian mapping and cortical imaging, Crit. Rev. Biomed. Eng. 27 (1999) 149–188.

[4] B. He, J. Lian, K.M. Spencer, J. Dien, E. Donchin, A cortical potential imaging analysis of the P300 and novelty P3 components, Hum. Brain Mapping 12 (2001) 120–130.

[5] B. He, T. Musha, Y. Okamoto, S. Homma, Y. Nakajima, T. Sato, Electrical dipole tracing in the brain by means of the boundary element method and its accuracy, IEEE Trans. Biomed. Eng. 34 (1987) 406–414.

[6] B. He, Y. Wang, D. Wu, Estimation cortical potential from scalp EEG's in a realistically shaped inhomogeneous head model by means of the boundary element method, IEEE Trans. Biomed. Eng. 46 (1999) 1264–1268.

[7] B. He, D. Yao, J. Lian, D. Wu, An equivalent current source model and Laplacian weighted minimum norm current estimates of brain electrical activity, IEEE Trans. Biomed. Eng. 49 (2002) 277–288.

[8] B. He, X. Zhang, J. Lian, H. Sasaki, D. Wu, V.L. Towle, Boundary element method based cortical potential imaging of somatosensory evoked potentials using subjects' magnetic resonance images, NeuroImage 16 (2002) 564–576.

[9] H.J. Huppertz, S. Hoegg, C. Sick, C.H. Lücking, J. Zentner, A. Schulze-Bonhage, R. Kristeva-Feige, Cortical current density reconstruction of interictal epileptiform activity in temporal lobe epilepsy, Clin. Neurophysiol. 112 (2001) 1761–1772.

[10] J.C. Mosher, P.S. Lewis, R.M. Leahy, Multiple dipole modeling and localization from spatio-temporal MEG data, IEEE Trans. Biomed. Eng. 39 (1992) 541–557.

[11] P. Nunez, Electric Field of the Brain, Oxford University Press, London, 1981.

[12] R.D. Pascual-Marqui, C.M. Michel, D. Lehman, Low resolution electromagnetic tomography: a new method for localizating electrical activity in the brain, Int. J. Psychophysiol. 18 (1994) 49–65.

[13] M. Scherg, D. Von Cramon, Two bilateral sources of the AEP as identified by aspatio-temporal dipole model, Electroencephalogr. Clin. Neurophysiol. 62 (1985) 32–44.

[14] A.N. Tikhonov, V.Y. Arsenin, Solutions of Ill-posed Problems, Wiley, New York, 1977.

[15] V.L. Towle, S. Cohen, N. Alperin, K. Hoffmann, P. Cogen, J. Milton, R. Grzeszczuk, C. Pelizzari, I. Syed, J.P. Spire, Displaying electrocorticographic findings on gyral anatomy, Electroencephalogr. Clin. Neurophysiol. 94 (1995) 221–228.

[16] W. Van Drongelen, M. Yuchtman, B.D. Van Veen, B.C. Van Huffelen, A spatial filtering technique to detect and localize multiple sources in the brain, Brain Topogr. 9 (1996) 39–49.

Available at
www.ComputerScienceWeb.com
POWERED BY SCIENCE @ DIRECT•

ELSEVIER

Neurocomputing 52–54 (2003) 983–992

NEUROCOMPUTING

www.elsevier.com/locate/neucom

Author index to volumes 52–54 (2003)

0925-2312/03/$ - see front matter
PII: S0925-2312(03)00403-X

Available at www.ComputerScienceWeb.com
POWERED BY SCIENCE DIRECT

ELSEVIER

Neurocomputing 52–54 (2003) 993–997

NEUROCOMPUTING

www.elsevier.com/locate/neucom

Subject index to volumes 52–54 (2003)

0925-2312/03/$ - see front matter
PII: S0925-2312(03)00404-1

Available at
www.ComputerScienceWeb.com
POWERED BY SCIENCE DIRECT•

ELSEVIER

Neurocomputing 52–54 (2003)

NEUROCOMPUTING

www.elsevier.com/locate/neucom

NEUROCOMPUTING EDITORIAL BOARD MEMBERS

0925-2312/03/$ - see front matter
PII: S0925-2312(03)00405-3

NEUROCOMPUTING

Aims and scope

Neurocomputing publishes articles describing recent fundamental contributions in the field of neurocomputing. Neurocomputing theory, practice and applications are the essential topics being covered.
Neurocomputing welcomes theoretical contributions aimed at winning further understanding of neural networks and learning systems, including, but not restricted to, analysis of network dynamics, architectures, biological neural network modelling, brain research, computational learning theory, computational neuroscience, learning methods, self-organization, sensorimotor transformations, theories of learning and interdisciplinary topics with artificial intelligence, artificial life, bioinformatics, cognitive science, computational biology, fuzzy logic, genetic algo rithms, information theory, machine learning, neurobiology, pattern recognition, and statistics.
Neurocomputing covers practical aspects with contributions on advances in hardware and software development environments for neurocomputing, including, but not restricted to, simulation software environments, emulation hardware architectures, models of concurrent computation, neurocomputers, and neurochips (digital, analog, optical, and biodevices).
Neurocomputing reports on applications in different fields, including, but not restricted to, signal processing, speech processing, image processing, computer vision, control, robotics, optimization, scheduling, resource allocation and financial forecasting.
Neurocomputing publishes reviews of literature about neurocomputing and affine fields.
Neurocomputing reports on meetings, including, but not restricted to, conferences, workshops and seminars.
Neurocomputing reports on functionalityavailability of software, on comparative assessments, and on discussions of neurocomputing, software issues.
Neurocomputing includes the *Neurocomputing Letters* for the rapid publication of special short communications.
Neurocomputing serves scientists, researchers, engineers, technologists, and graduate students in broader areas of artificial and computational intelligence, biology, biophysics, brain studies, computer and cognitive science, electrical and electronic engineering, mathematics, neuroscien ces, philosophy, psychology, and statistics.

Authors' benefits

1. 25 reprints per contribution free of charge (additional reprints can be ordered).
2. 30% discount on all Elsevier Science books.

Orders, claims, and product enquiries: please contact the Customer Support Department at the Regional Sales Office nearest you:

New York: Elsevier Science, PO Box 945, New York, NY 10159-0945, USA; phone: (+1) (212) 633 3730 [toll free number for North American customers: 1-888-4ES-INFO (437-4636)]; fax: (+1) (212) 633 3680; e-mail: usinfo-f@elsevier.com

Amsterdam: Elsevier Science, PO Box 211, 1000 AE Amsterdam, The Netherlands; phone: (+31) 20 4853757; fax: (+31) 20 4853432; e-mail: nlinfo-f@elsevier.com

Tokyo: Elsevier Science, 9-15 Higashi-Azabu 1-chome, Minato-ku, Tokyo 106-0044, Japan; phone: (+81) (3) 5561 5033; fax: (+81) (3) 5561 5047; e-mail: info@elsevier.co.jp

Singapore: Elsevier Science, 3 Killiney Road, #08-01 Winsland House I, Singapore 239519; phone: (+65) 6349 0200; fax: (+65) 6733 1510; e-mail: asiainfo@elsevier.com.sg

Rio de Janeiro: Elsevier Science, Rua Sete de Setembro 111/16 Andar, 20050-002 Centro, Rio de Janeiro - RJ, Brazil; phone: (+55) (21) 509 5340; fax: (+55) (21) 507 1991; e-mail: elsevier@campus.-com.br [Note (Latin America): for orders, claims and help desk information, please contact the Regional Sales Office in New York as listed above]

⊗The paper used in this publication meets the requirements of ANSI/NISO Z39.48-1992 (Permanence of Paper). For more information re 'Permanence of Paper' and an example of the symbol used to show compliance with this standard you are referred to http://www.niso.org/standards/resources/Z39-48.pdf.

Published bimonthly 0925-2312/03 Printed in The Netherlands